Applied Optics

WILEY SERIES IN PURE AND APPLIED OPTICS

Founded by Stanley S. Ballard, University of Florida

ADVISORY EDITOR: Joseph W. Goodman, Stanford University

Applied Optics

A Guide to Optical System Design/Volume 2

LEO LEVI

Jerusalem College of Technology
Jerusalem, Israel

John Wiley & Sons, New York | Chichester | Brisbane | Toronto

Copyright © 1980 by John Wiley & Sons, Inc.

All rights reserved. Published simultaneously in Canada.

Reproduction or translation of any part of this work
beyond that permitted by Sections 107 or 108 of the
1976 United States Copyright Act without the permission
of the copyright owner is unlawful. Requests for
permission or further information should be addressed to
the Permissions Department, John Wiley & Sons, Inc.

Library of Congress Cataloging in Publication Data:

Levi, Leo, 1926-
  Applied Optics.

  (Wiley series in Pure and Applied Optics)
  Includes bibliographies.
  1. Optics. 2. Optical instruments. I. Title.
QC371.L48   535'.33   67-29942
ISBN 0-471-05054-7 (v. 2)

Printed in the United States of America

10 9 8 7 6 5 4 3 2 1

Preface

The overall goals of *Applied Optics* have already been stated in the first volume. Here I merely summarize these briefly, adding a few remarks specific to this volume.

Applied Optics is meant to serve a dual purpose: to permit self-study by the graduate engineer and to serve as a handbook to the experienced practitioner. Thus step-by-step derivations of all important results are included whenever consistent with space restrictions and, on the other hand, a concerted effort was made to include, in condensed form, a large amount of physical data, in the many graphs and 125 tables.

Here, even more than in the first volume, I have tried to summarize in individual chapters the results of many disciplines, such as photoelectric detection and photographic techniques usually treated in separate texts. Even peripheral topics, such as calculating the position of sun and stars, have been included in the chapter on atmospheric optics since this information is important in practice and often not readily available to the engineer.

Many of the topics treated here are in a state of rapid development, even more than most of those discussed in the first volume—the term "integrated optics" had not even been coined when I wrote the first volume, and a major part of a chapter had to be devoted to it in the second! This implies that the information is here scattered throughout the periodical literature even more—and this accounts for the delay in the completion of this volume despite my diligent efforts.

This volume, too, has benefited from discussions with many colleagues, impossible for me to enumerate. I feel, however, compelled to thank those who have reviewed complete chapters of the finished manuscript: Dr. M. Menat (Israel Defense Department), Chapters 12 and 16; Dr. J. Bodenheimer (Jerusalem College of Technology), Chapter 15; Dr. H. Arbel (National Physics Laboratory of Israel), Chapter 16; Dr. M. Goldman (Berke-Pathe-Humphries, Israel), Chapter 17; and Prof. A. Friesem (Weizmann Institute), Chapter 19. I also wish to thank Prof. H. Mandelbaum (Jerusalem), for reading Section 12.3.4, and Drs. J. J. McCann and W. E. Kock for supplying me with original prints of Figs. 15.20*a* and 19.14, respectively.

Again, I happily conclude with the acknowledgment of my wife's cheerful moral support through the many years when most of my spare time was taken up by the burden of writing this volume whose size exceeded our anticipations, by far.

LEO LEVI

Jerusalem, Israel

Contents

DETECTORS OF LIGHT

SYSTEMS

TABLES

Applied Optics

10

Optical Media

10.1 OPTICAL GLASSES

Optical glasses are the most popular medium in optical systems. They are relatively easy to manufacture and to work; they can be made with highly desirable optical and mechanical characteristics and many of them are very stable under ordinary conditions. They can be manufactured in sizes up to a meter in diameter (in some cases up to almost four meters) for lenses [1] and considerably larger for mirrors.

These glasses may be manufactured in a continuous process, yielding fair refractive index uniformity at economical prices. However, the classical method of producing the glass in crucibles, a batch at a time, is still necessary for the highest index uniformity ($\pm10^{-6}$) [2].

10.1.1 Structure

Glass is defined as an inorganic product of fusion cooled to a rigid condition without crystallizing [3]. It is obtained from a melt that is cooled in such a way that the molecular structure is irregular over long ranges, although it may be quite regular over a number of interatomic spaces. For instance, when crystalline quartz is melted and cooled slowly, a silica glass, "fused silica," is obtained. Its chemical formula is SiO_2 and its molecular structure is tetrahedral (a silicon atom with four oxygen atoms forming a regular tetrahedron around it). These tetrahedra are joined at their corners, so that each oxygen atom is shared by two tetrahedra; these, in turn, form a larger interconnected network. Whereas in a crystal the relative angular orientation of neighbouring tetrahedra (θ in the figure) is fixed, in a glass it varies from joint to joint (see Figure 10.1).

(a)

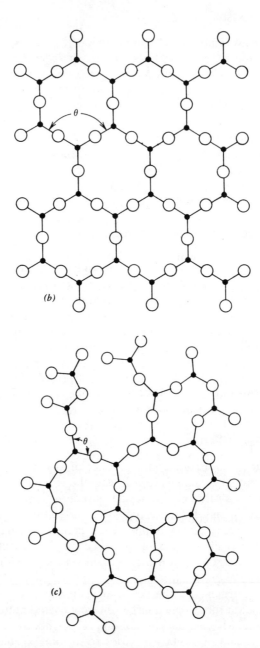

(b)

(c)

Figure 10.1 Illustrating the molecular structure of glass. Each tetrahedral cell is represented two-dimensionally by a unit as shown in (a). (b) The equivalent crystal structure. (c) The glass structure. After D. G. Holloway, *The Physical Properties of Glass*, Wykeham, London, 1973.

2

When significant amounts of other oxides, such as sodium oxide (Na_2O), are melted together with the silica, both the sodium atoms and the additional oxygen atoms will modify the large-scale network formation: the sodium ions, due to the associated electric fields, will contribute to the random distortion of the network and the extra oxygen atoms will tend to enter the network itself, causing breaks, since they will not find a second tetrahedron with which to connect. In this system, SiO_2 is called a *network-forming component* and Na_2O, a *network-modifying component*; it lowers the viscosity of the molten glass and, of course, affects the refractive index. When lime (CaO), too, is added the common soda-lime-silicate glass is obtained. In addition to nontechnical uses, this serves as plate glass (e.g., in instrument windows) and as optical crown glass.

Other substances used as network formers in optical glasses are:

1. Boric oxide (B_2O_3), which yields borosilicate glass with good thermal stability.

2. Phosphoric oxide (P_2O_5), which yields phosphate glass.

3. Lead oxide (PbO), which, when added in substantial amounts (>10%), results in flint glass with its high dispersive power.

4. Barium oxide (BaO), which produces barium crown glass, with its relatively high refractive index, and barium flint glass, when added in conjunction with PbO.

Detailed composition data of a number of glasses are given in Table 71. See Refs. 4–6 for additional examples.

Pure silica glass not only has a very high melting point, but also remains highly viscous up to temperatures at which it decomposes (cf. Figure 10.6). It is therefore difficult to manufacture, especially for optical components, where freedom from bubbles is required. "Vycor" is the trademark of a glass (manufactured by the Corning Glass Co.) that is 96% silica and approximately 3% boric oxide. The glass is first melted with a significant alkali content, making it relatively easy to work. After cooling, the alkali content is leached out by soaking the formed product in acid; subsequent reheating, with the attendant shrinking, yields the finished product. "Pyrex" is also a trademark of Corning, referring to a low-expansion borosilicate glass, having a silica content of over 80%. It has characteristics intermediate between fused silica and the usual soda-lime-silicate glasses, quite difficult to work, but distinguished by high mechanical, thermal, and chemical strength.

The rate at which the glass is cooled during solidification influences the final structure, so that the cooling (or annealing) rate is a significant determinant in the manufacturing process. This is discussed further in Section 10.1.3.

10.1.2 Optical Properties

10.1.2.1 Refractive Index—Theoretical Aspects. The phenomenon of refraction can be explained in terms of the atomic structure of matter as follows: To a first approximation, the electromagnetic wave passes undisturbed through the medium, which is mostly empty space containing widely spaced nuclei with clouds of electrons surrounding them. Only a small fraction of the wave's energy is abstracted by the atoms in which it sets up vibrations at the frequency of the wave itself. The vibrating electron clouds radiate their energy in the form of similar electromagnetic waves, whose phase, however, lags behind that of the incident wave. These secondary waves interfere with the incident wave and cause its phase to be retarded; the net phase velocity is lowered. This is refraction.

The size of the refractive index is determined by the size of the vibrations set up in the electron cloud and this, in turn, is proportional to the polarizability (α) of the electron clouds. The polarizability is here defined by the size (p) of the electric dipole set up in the molecule by a given electric field (E)

$$\alpha = \frac{p}{E},\tag{10.1}$$

where the dipole is equivalent to a pair of opposite charges ($\pm q$) separated in space by a small distance (d),

$$p = qd.\tag{10.2}$$

Specifically, the Lorentz-Lorenz formula fixes the relationship between the polarizability and the refractive index (n) as follows:

$$n^2 = \frac{1+2C}{1-C} = 1 + \frac{3C}{1-C},\tag{10.3}$$

where

$$C = \frac{4\pi N_A \rho \alpha}{3M},\tag{10.4}$$

and $N_A = 6.02217 \times 10^{23}$ is Avogadro's number,
ρ = the mass density of the medium, and
M = its molecular weight.

The frequency dependence of the oscillator amplitudes, and hence of the refractive index, is shaped by the resonance characteristics of the electron cloud. This characteristic has a Lorentz shape [Section 4.3.2 and (7.33)]:

$$\alpha(\omega) = \frac{1}{Q(\omega_0^2 - \omega^2)}, \qquad (10.5)$$

where, classically, the sharpness factor is given by

$$Q = \frac{\sqrt{m_e}}{q_e}, \qquad (10.6)$$

where m_e and q_e are mass and charge of an electron, respectively. On substituting

$$\omega = \frac{2\pi c}{\lambda}, \qquad (10.7)$$

(cf. 2.32), we obtain an equation that, far from the resonance, has the form

$$C \sim \alpha \sim \frac{\lambda \lambda_0}{\lambda^2 - \lambda_0^2}. \qquad (10.8)$$

On substituting this into (10.3), we obtain the dependence of the refractive index on wavelength [cf. Sellmeier's approximation to the refractive index (10.10d)].

10.1.2.2 Refractive Index—Practical Aspects. As already noted, the structure of a glass, and hence its refractive index and dispersion, are determined by its composition and cooling rate. Glasses with low dispersion (reciprocal dispersion index, $V > 55$) and high dispersion are called *crown* and *flint*, respectively [cf. (9.34) for the definition of V]. As a rule, flint glasses contain significant amounts of PbO (see Table 71). Within each group, glasses are subdivided according to their detailed characteristics and composition (see Figure 10.2 for the usual ranges in these classes). Table 72 [7] lists common abbreviations for these class names and also gives the refractive (n_d) and dispersion (V) indices and two P values for a prototype of each class. The P values are significant in achromatization when the secondary spectrum is to be reduced (cf. Section 9.2.3.1). The values given in Table 72 are defined by

$$P_{UV} = \frac{n_F - n_{UV}}{n_F - n_C}, \qquad P_{IR} = \frac{n_F - n_{IR}}{n_F - n_c}, \qquad (10.9)$$

where the n subscripts are defined in Table 63.

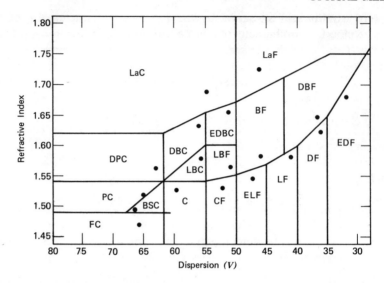

Figure 10.2 Types of optical glass. The prototypes of Table 72 are indicated by dots.

In any optical material, the spectral variation of the refractive index is determined experimentally. However, often it is convenient to have an analytical approximation to work with. A number of such approximations have been used. The best-known among these are

$$n = \sum_0 \frac{B_j}{\lambda^{2j}} \qquad \text{Cauchy [8]} \tag{10.10a}$$

$$n = n_0 + \frac{c}{(\lambda_0 - \lambda)^{1.2}} \approx n'_0 + \frac{c'}{\lambda'_0 - \lambda} \qquad \text{Hartmann [9]} \tag{10.10b}$$

$$n = B_0 + PL + RL^2 + A_1\lambda^2 + A_2\lambda^4, \quad L = (\lambda^2 - 0.028)^{-1} \quad \text{Herzberger [10]} \tag{10.10c}$$

$$n^2 = 1 + \sum \frac{Q_j\lambda^2}{\lambda^2 - P'_j} = B_0 + \sum \frac{P_j}{\lambda^2 - P'_j} \qquad \text{Sellmeier [8]} \tag{10.10d}$$

where $B_0 = 1 + \sum Q_j$,
$\quad P_j = P'_j Q_j$.

Here n is the refractive index for light at wavelength λ, and the other symbols are constants determined experimentally. Table 73 lists the appropriate constants for silica and arsenic trisulfide glasses and for a number of crystalline substances used as optical media. Also the spectral range of validity and an approximate upper limit on the inaccuracy is given. Refractive indices calculated on the basis of these formulae are listed in Table 74. The above approximations may yield poor fits when applied over wide spectral ranges including absorption bands. For such materials, a cubic-spline curve-fitting technique may give an improved approximation [11].

Uniformity of refractive index is a major factor in determining the performance of imaging systems. Microscopic nonuniformities cause light scattering, whereas small-area, high-gradient variations (striae) have a deleterious effect on imaging. Striae are relatively easily detected by visual inspection of a well-polished sample against a dark background near a bright background region.

Often the much less obvious slow index variations extending over large areas are even more serious since they affect major portions of the wavefront. These can be detected by interferometric techniques. They are caused by surface effects at the glass-crucible and the glass-air interfaces and by nonuniform cooling of the large glass mass. Special mixing and annealing techniques must be used to minimize all of these [12].

In certain glasses and with special precautions, inhomogeneities can be maintained below 10^{-6}, even in large samples [12].

10.1.2.3   *Transmittance.* The transmittance of an optical component, such as a lens or a prism, is less than unity due to reflection at the surfaces and absorption in the medium. The reflection is given by Fresnel's formulae (2.101–2.106). For normal incidence on a surface-dividing media of indices n and n', respectively, the reflectance is

$$\rho = \left(\frac{n - n'}{n + n'}\right)^2. \tag{8.3}$$

Considering the reflectance at the two surfaces, and neglecting the doubly reflected flux, the corresponding transmittance is

$$\tau_r = (1 - \rho)^2. \tag{10.11}$$

At any wavelength, the absorption increases with the distance, d, traversed according to (4.59):

$$\tau_i = \frac{\Phi_2}{\Phi_1} = e^{-\alpha d} \tag{10.12}$$

where $\Phi_{1,2}$ are the flux values at the beginning and end, respectively, of layer thickness, d, and α is the absorption coefficient. Thus the ratio of the flux (Φ_0) leaving the component to the flux (Φ_i) reaching it is, approximately[1]

$$\frac{\Phi_0}{\Phi_i} = \tau_r \tau_i = (1-\rho)^2 e^{-\alpha d} = \frac{16n^2}{(n+1)^4} e^{-\alpha d}, \qquad (10.15)$$

where the last step follows from (8.3).

Note that the attenuation does not follow (10.12) for nonmonochromatic flux extending over a spectral range over which α varies.

For a ray traversing a plane parallel plate, thickness d_0, at an angle of incidence θ, the distance traversed inside the plate is readily seen to be

$$d = \frac{d_0}{\cos \theta'} = \frac{nd_0}{\sqrt{n^2 - \sin^2 \theta}}, \qquad (10.16)$$

where θ' is the angle of refraction

$$\sin \theta = n \sin \theta'$$

and n is the refractive index of the medium. Values of the absorption

---

[1] This formula ignores the portion of the flux reflected at the second surface and again at the first surface, part of which will eventually leave the component after one or more such round trips. When the reflectance is high, it may be desirable to include this portion of the flux in the analysis. Its value may be found by considering the attenuation factor per round trip:

$$\rho^2 e^{-2\alpha d},$$

where $2d$ is the length of the round trip. This factor acts on the flux reaching the exit surfaces, to yield the flux reaching it after an additional round trip. Hence the total transmittance is

$$\tau = (1-\rho)^2 e^{-\alpha d}(1 + \rho^2 e^{-2\alpha d} + \rho^4 e^{-4\alpha d} + \cdots)$$

$$= \frac{(1-\rho)^2 e^{-\alpha d}}{1 - \rho^2 e^{-2\alpha d}}$$

$$= \frac{16n^2 e^{-\alpha d}}{(n+1)^4 - (n-1)^4 e^{-2\alpha d}}. \qquad (10.13)$$

With α vanishing, this simplifies to

$$\tau(0) = \frac{2n}{n^2 + 1}. \qquad (10.13a)$$

More generally, solving (10.13) for $e^{-\alpha d}$, we find

$$e^{-\alpha d} = \frac{\sqrt{64n^4 + \tau^2(n^2-1)^4} - 8n^2}{\tau(n-1)^4}. \qquad (10.14)$$

This may be used to find a more accurate value of α from the measured value of τ, especially when n is large.

coefficients at various wavelengths are listed in Table 80a for a number of optical glasses. These were calculated from published transmittance values [13]. Additional, commercially published data are listed in Table 80b.

The relatively high absorption in the ultraviolet is due to the electronic absorption edge of the glass, which corresponds to the electromagnetic frequency at which the electron clouds in the medium resonate. At that frequency the energy is rapidly absorbed. (The proximity of the absorption edge also affects the dispersion; specifically, it accounts for the accelerated rise of the refractive index at the short-wavelength end of the visible spectrum and the near ultraviolet.) A small change in the location of the absorption edge may then produce a large change in the spectral absorption, so that this part of the absorption spectrum is highly sensitive to small changes in glass composition and annealing.

Other absorption coefficient values can be found in Figure 10.3. For

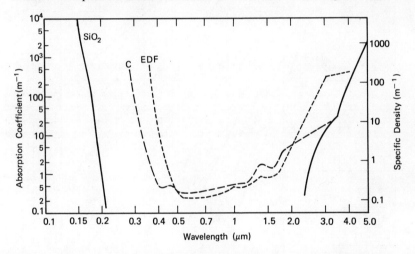

Figure 10.3 Absorption coefficients of representative glasses: silica (solid curve); ordinary crown (broken curve); extra dense flint glass (dotted curve). The curves are based on data for Suprasil-W (Heraeus), C 511/604 (Sovirel), and EDF 785/259 (Sovirel), respectively.

conversion of absorption coefficients to transmittance values, τ_i, as a function of thickness, see Figure 10.4.

10.1.2.4 Stress Birefringence. Stress induces birefringence in the glass (cf. Section 14.1). The difference in refractive index parallel and normal to the stress direction is readily calculated by the formula

$$\Delta n = SB = 10^{-12} S^* B^* \qquad (10.17)$$

where S is the stress (S^* when in units of N/m^2) and B is the stress-optic

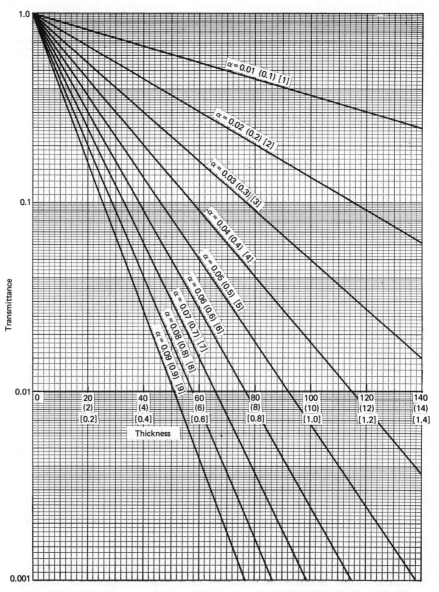

Figure 10.4 Transmittance versus thickness for various absorption coefficients, α. The α values in parentheses and brackets relate, respectively, to the thickness values marked similarly on the axis of the abscissae.

10

coefficient (sometimes called Brewster's constant) of the material (B^* when in units of brewsters, i.e., reciprocal pN/m^2). See Section 14.1.1.3 for details. Values of stress-optic coefficients for a number of glasses are included in Table 71. In general, the index in the direction of compression is less than that normal to it. This condition is reversed, however, in the densest flint glasses; a flint glass with refractive index of approximately 1.9 should have a vanishing birefringence.

10.1.2.5 Thermal Effects on Refraction. The refractive index of a glass changes with temperature due to two opposing effects:

1. Lowering of the refractive index due to the expansion of the glass
2. Displacement of the absorption edge from the ultraviolet toward the visible portion of the spectrum, raising the index.

The former effect predominates at lower temperatures and in crown glasses, where the absorption edge is far in the ultraviolet. Effect 2 dominates the index changes at higher temperatures and in flint glasses where the absorption edge lies close to the visible part of the spectrum. Extensive results for a number of optical glasses are shown in Table 81 [13] and representative curves in Figure 10.5. Note that in very dense flint glass the index change may be as high as $10^{-4}/°C$, which is far from insignificant. Data for the thermal coefficient of the refractive indices of a number of glasses are tabulated in Table 82 [14].

In addition to the reversible, thermal effects on refraction, permanent changes occur during the solidification process. These are discussed in conjunction with the annealing process in the next section.

10.1.3 Mechanical and Thermal Properties. Annealing

10.1.3.1 Mechanical Properties. The specific properties of some glasses are listed in Table 71.

Young's modulus for plate glass and pure silica glass is 69 GN/m^2 ($\approx 10^7$ psi ≈ 700 k bar) [6]. For 96% silica and low-expansion borosilicate glasses, it is 2–3% smaller, and for flint glass it may be as low as 55 GN/m^2, [5].

The *tensile strength* of glass may be as high as 7000 MN/m^2 under ideal conditions, but its average value under ordinary conditions is considerably lower than this due to microscopic surface scratches, surface damage due to atmospheric effects, and so on. It is approximately 35 MN/m^2 and 70 MN/m^2 for flint and borosilicate crown glasses, respectively. Tempering may raise these values by a factor of four. Silica and low-expansion borosilicate glasses have a tensile strength of approximately 100 MN/m^2. It should be noted that thin fibers show considerably greater tensile

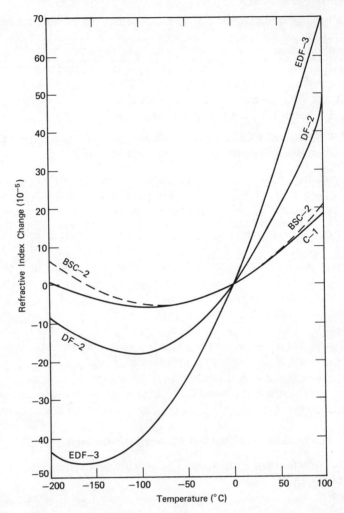

Figure 10.5 Refractive index (n_D) changes (relative to 0°) with temperature for glasses listed in Tables 80a and 81. Molby [13].

strengths; a 1.5-μm-diam fiber may exhibit a tensile strength in excess of 7000 MN/m^2 [6].

10.1.3.2 Thermal Expansion. The thermal coefficient of expansion of an optical glass may become a design consideration whenever mounted glass components are to be subjected to significant temperature changes. For instance, condenser lenses, which are often mounted close to a light source operating at a very high temperature may crack in use if their mount is not properly designed to accommodate the expansion and contraction resulting from the large temperature fluctuations to which this component is subjected. Together with mechanical strength, the thermal coefficient of expansion is also a major factor in determining the strength of a glass under thermal shock. For typical values, see Table 71.

For certain applications, such as very large optical components (e.g., telescope reflectors), extremely low thermal expansion is desirable. For such purposes, special *glass ceramics* (glasses converted, at least partially, into polycrystalline form) have been developed. One of these [15], containing titanium and zirconium oxides in a SiO_2—$Li_2Al_2O_4$ matrix, has a thermal coefficient averaging only 0.03×10^{-6} in the temperature range 5–50°C.

10.1.3.3 *Annealing and Viscosity.*

In contrast to crystalline substances, glasses exhibit a gradual transition from the liquid to the solid state, extending over a considerable temperature range; as it solidifies, the glass passes through a wide range of viscosities. If the cooling process is slow, the molecules rearrange themselves into a dense network with higher specific gravity and refractive index; if it is more rapid, the looser structure may be "frozen" into the glass. Also a slow transition reduces the formation of internal stresses, which are inevitable due to internal temperature nonuniformities. For near-optimum annealing, the cooling from 600°C to room temperature must be extended over weeks!

Figure 10.6 shows curves of viscosity as a function of temperature for a number of glasses. On this figure certain viscosity levels, important for engineering considerations, are indicated.

1. The *flow point* corresponds to the viscosity at which the glass can readily be poured (10^5 poise) [6].

2. At the *softening point* the glass will deform under its own weight [6] ($\sim 10^{7.6}$ poise) [3].

3. At the *annealing point*, the glass will relieve internal stresses within approximately 15 min [6] ($\sim 10^{13}$ poise) [3].

4. At the *strain point*, it will do so in several hours [3] ($\sim 10^{14.5}$ poise) [6].

The viscosities of glycerol and castor oil at room temperature are indicated in Figure 10.6 for comparison. For a number of glasses the temperatures at which they attain the above levels of viscosity are tabulated in Table 83. Pure silica glass has been found to retain a viscosity of 10^6 poise, up to a temperature of 2040°C [16]. Beyond this temperature there is a tendency for volatilization to set in.

When a softened glass is permitted to remain at a certain temperature for a time sufficient to permit stabilization of the molecular network and

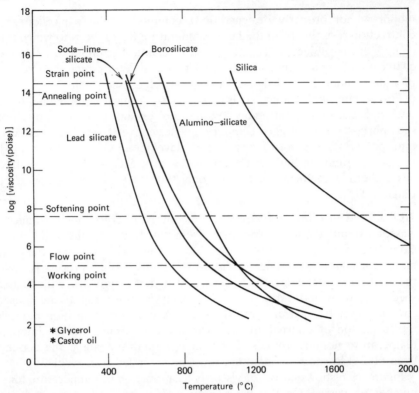

Figure 10.6 Viscosity versus temperature curves for some glasses. Curves from Ref. 5 except that for silica glass, which is from Ref. 16.

is then rapidly cooled and solidified, the state corresponding to the stabilization temperature tends to be "frozen" into the glass, and the glass will maintain it indefinitely, unless it is resoftened; see Table 84 [17] for an illustration. Here the room temperature refractive index obtained is tabulated as a function of the stabilization temperature from which the glass was cooled rapidly. The third column gives the length of time that the glass must be maintained at the stabilization temperature in order that it approach the equilibrium index within 10^{-5}, when rapidly cooled to room temperature. The sample was borosilicate crown glass (517/645)— the third glass in Tables 63 and 71.

Thermal effects on the absorption characteristics of glasses are discussed in Section 11.2.1.

10.1.4   Chemical and Radiation Effects [7], [18]

Silicate glasses are, generally, quite resistant to chemical action. Nevertheless they are susceptible to certain prevalent factors attacking their surfaces. Below are listed the most common causes of such damage.

1. Alkaline solutions and dilute hydrofluoric acid break the Si—O bonds and dissolve the glass slowly.

2. Acidic solutions tend to substitute hydrogen for the alkalis in the glass surface. This may lead to a film of lower refractive index there. Despite its esthetically offensive appearance, such a stain may actually be advantageous, acting as a low reflection coating.

3. High humidity, especially when alternating with periods of dryness, may cause the CO_2 in the air to interact with the glass to cause the growth of carbonate crystals on the glass surface. This is called *dimming*.

4. Water droplets on the glass surface begin by leaching as in 2. As the concentration of alkali ions in the water rises, the resulting hydroxides begin to dissolve the surface layer of the glass as in 1.

5. In tropical climates the glass surface may be attacked by fungus growth.

Optical glasses are often classified according to their resistance to the above effects. The *staining class* signifies the time required to form a low-index layer of approximately 0.1 μm thickness when the glass is immersed in dilute nitric acid at 25°C. Classes 1–5 correspond to times of >100 hr, 10–100, 1–10, 0.1–1 hr, <0.1 hr, respectively. Other manufacturers express the resistance of the glass to acid attack by means of the equation

$$t = ad + bd^2 \qquad (10.18)$$

where t is the time required to produce a low-index layer of thickness d when the sample is immersed in half-normal nitric acid; a and b are then the constants describing the resistance to chemical attack.

The *dimming classes* are established by comparing with standard ground glass samples the effects of a high-humidity environment on the light-scattering characteristics of a polished glass surface. Here Class 1 signifies no noticeable dimming and Class 5, a degree of dimming that interferes with clear vision through the sample. The other classes signify intermediate degrees of dimming, all these under specified conditions of exposure to humidity.

High-energy electromagnetic radiation, such as X and γ rays, often cause the formation of color centers in the glass. Silicate glasses tend to turn brown under irradiation of this type. The addition of cerium to the glass tends to reduce the effect significantly [19]. On this basis, a number of glasses have been developed specifically to eliminate the darkening; these are referred to as "protected." Their refractive indices are identical to those of the corresponding unprotected glass, but they do absorb more strongly at the shorter wavelengths. After heavy exposure to high-energy radiation, however, their transmittance is far superior to that of the unprotected glass. The spectral transmittance of protected and unprotected glasses are compared both before and after irradiation in Figure 10.7*a* and *b* [20], relating to representative crown and flint glasses, respectively.

Similar darkening effects are observed due to uv radiation, as in sunlight. These are called *solarization* and tend to produce increased uv absorption or purple coloration. The most common chemical reactions responsible for these are, respectively [21],

$$4FeO + As_2O_5 = 2Fe_2O_3 + As_2O_3$$

and

$$4MnO + As_2O_5 = 2Mn_2O_3 + As_2O_3.$$

Solarization effects may be reversed by heating. In some cases 200°C suffices; but occasionally up to 500°C are required.

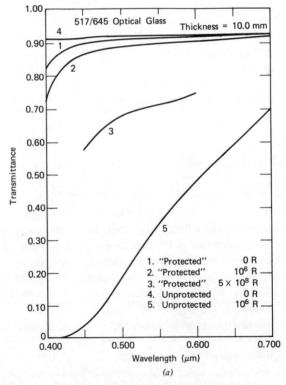

(a)

Figure 10.7 Effects of irradiation on the transmittance of "protected" and unprotected glasses. (*a*) Borosilicate crown. (*b*) Dense flint. The amount of irradiation [in roentgen (R)] is listed in the legend. Bausch & Lomb [20].

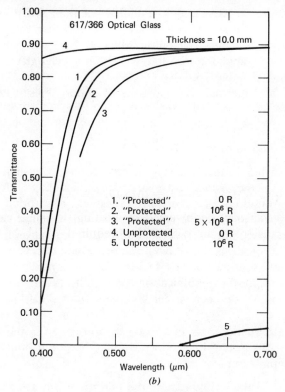

Figure 10.7 (*Continued*)

10.1.5   Performance of Glasses in the Ultraviolet (uv) and Infrared (ir)

The usual optical glasses, with the exception of very dense flint glasses, can be used in the uv down to 0.33–0.36 μm. The very dense flint glasses absorb significantly to 0.38 μm. In the ir, these glasses can be used readily up to 2.5 μm (borosilicate crown up to 2.7 μm). In silica glass there tends to be a strong absorption band at about 2.7 μm, which is assumed to be caused by a Si—OH bond, due to water introduced into the structure at the time of its preparation. Pure silica glasses transmit well throughout the range from 0.17 to 4 μm. See Figure 10.3 for its absorption coefficients.

Outside this range, crystals or special glasses must be used. For instance, arsenic trisulfide glass transmits well from the red portion of the visible spectrum to approximately 11 μm [22]. Data covering its refractive index are included in Tables 73 and 74 and its absorption spectrum is included in Figure 10.10d. Arsenic-modified selenium glass transmits well from approximately 1 μm to above 18 μm. Unless carefully purified, it exhibits an absorption band extending over the 12–13 μm region [22]. Its refractive index is tabulated in Table 75. Other special glasses for use in the ir are listed in Table 76.

Mechanical and thermal characteristics of materials used in the ir are listed in Table 87.

10.1.6 High Transmittance Glasses [23]

As noted earlier, optical glasses (except for the dense flint glasses) absorb slowly in the visible and near-ir spectral regions. However, for pathlengths of hundreds of meters, even these low absorption coefficients may be excessive. Such pathlengths occur in long-distance optical communication systems, where the radiation must be transmitted through "light pipes" (optical fibers) to avoid the disturbing effects of the atmosphere and physical obstacles. For a material to be considered suitable for a long-distance optical communication channel, its attenuation should not exceed substantially one octave per kilometer ($\alpha^* = 3$ dB/km or $\alpha_T = 7 \times 10^{-7}$ mm^{-1}).[2]

Radiation attenuation may be caused by absorption in the medium or by scattering. Absorption losses, in turn, may be due to impurities or to structural nonuniformities.

Attenuation. In the visible region, the *intrinsic* losses are due primarily to the skirts of the uv absorption edge that extend into the visible and beyond. The ir absorption bands are too narrow to be significant. From this point of view, "pure" silica glasses are optimum; their uv absorption is far into the uv and their absorption coefficients are known to be 1 dB/km or less at 0.633 μm.

Impurity losses may be due to contamination. Contaminants, such as the transition metals listed in Section 10.2.2.2, may cause significant attenuation (20 dB/km) at concentrations as low as 10^{-8}. In Table 85 [23] we present the concentration values causing such absorption, both at 0.8 μm and at the wavelength of peak absorption.

Structural nonuniformities cause carrier traps and may be due to mechanical disturbances or excessive ionization of one of the component species.

Scattering. Scattering losses may be classified as intrinsic or structural in nature. The intrinsic scattering is due to refractive index nonuniformities caused by thermal fluctuations, which affect the density and hence

---

[2] Note that 1 dB/km $= (10^{-7} \log 10)$ mm$^{-1} = (10^{-4} \log 10)$ m$^{-1} \approx 2.303 \times 10^{-4}$ m^{-1}

the refractive index of the material. These fluctuations grow linearly with absolute temperature. In a pure liquid the dependence is of the form

$$\alpha_s = \frac{8\pi^3}{3\lambda^4}(n^2 - 1)k\beta T. \tag{10.19}$$

where $k = 1.38 \times 10^{-23} \, J/K$ is Boltzmann's constant and β is the compressibility (fractional volume change per unit hydrostatic pressure).

In glasses, the effective temperature to be considered is the solidification temperature of the glass, usually approximately 1000 K (see Fig. 10.6). At 1 μm, this corresponds to 0.2–0.3 dB/km.

Structural nonuniformities may be due to imperfect mixing, insufficient annealing, or incipient crystallization of some of the components. In optical fibers, surface contamination and nonuniformities, too, are important scattering causes.

With extreme precautions glasses with attenuation coefficients of 0.2 dB/km ($\alpha = 46 \times 10^{-9} \, mm^{-1}$) have been attained at 1.55 μm [24]. The attenuation coefficients of two silica glasses, Ultrasil and Suprasil W1 [25] have been measured at 0.633 and 0.85 μm. They were found to be 12–13 dB/km in the red and 19 dB/km and 15 dB/km in the ir, for the two glasses, respectively [26]. A third silica glass, Corning No. 7940, has attenuation coefficients of 4.8, 1.9, and 0.6 dB/km at 0.633, 0.8 and 1.06 μm, respectively [23].

10.2 CRYSTALLINE OPTICAL MEDIA

As optical media, crystals are much less popular than glasses for the obvious reason that they are far more difficult to manufacture in large sizes. Their use is invariably motivated by the need for characteristics not readily obtainable in glass. Among these characteristics are the following:

1. Spectral transmittance characteristics extending beyond those obtainable with the usual glasses.
2. Special shape of the dispersion curve.
3. Birefringence, electro-, and magnetooptical characteristics.
4. Mechanical and thermal strength.

Figure 10.8 [27b] shows the ranges of usefulness for the crystalline media most frequently used in optics and Figure 10.9 [27a] shows their refractive indices.

The earliest isotropic crystalline material used in optical design is probably *fluorite* (calcium fluoride, CaF_2), which is outstanding both in spectral range (0.13–12 μm) and in its dispersion characteristics, which, in combination with optical glasses, permit elimination of the secondary

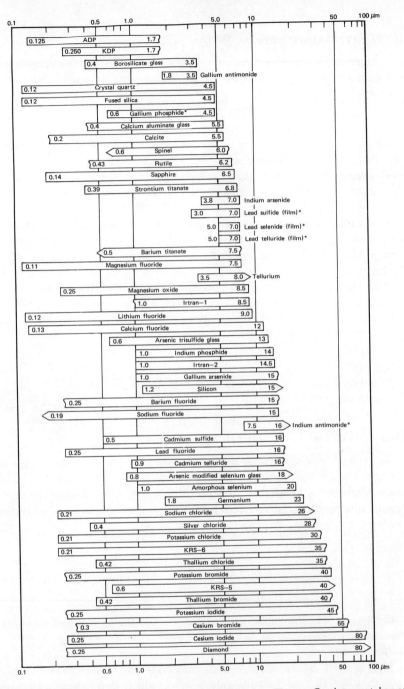

Figure 10.8 Spectral range of usefulness of optical materials. The cutoff points are taken at the wavelength where the transmittance of a 2-mm-thick sample drops to 10%. Those marked by an asterisk have a transmittance everywhere less than 10%. Ballard, Browder, and Ebersole [27b].

20

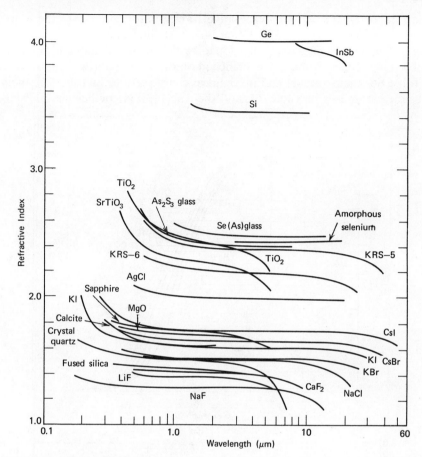

Figure 10.9 Refractive index versus wavelength for several optical materials. Ballard, Browder, and Ebersole [27a].

spectrum and apochromatization. *Quartz*, too, is a relatively popular material, combining a broad transmittance spectrum with great mechanical strength. It is both birefringent and optically active. *Calcite*, or Iceland spar (CaCO$_3$), is often used in polarization work because of its strong birefringence (see Section 8.3.5). Table 77 lists the refractive indices of the latter materials.

For work in ir spectroscopy, *rock salt* (NaCl) is often used. It is isotropic and transparent from 0.21 to 26 μm. It is, however, highly hygroscopic, and special precautions must be taken to maintain its surface quality. Values of its refractive index are listed in Table 74.

In modern ir systems, *germanium, silicon,* and clear synthetic *sapphire* are important media. Representative transmittance curves for these are shown in Figure 10.10 [28]. Often polycrystalline materials are used. A series of such materials was developed by the Eastman Kodak firm under the trade name "*Irtran.*" The composition of these materials are listed in Table 86, their thermal and mechanical characteristics in Table 87, their refractive indices in Table 74, and their absorption coefficients in Table

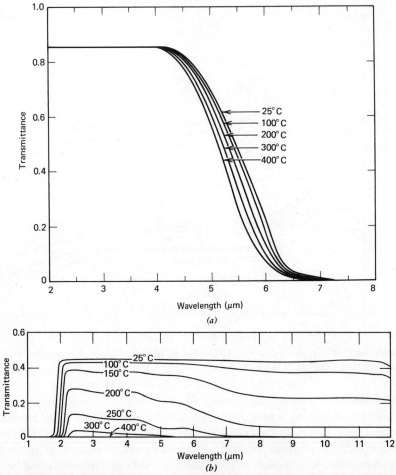

Figure 10.10 Transmittance spectra of crystalline materials and their dependence on temperature. (*a*) Sapphire; (*b*) germanium, and (*c*) silicon. The curves refer to samples 2.8 mm thick. The Ge was *p*-type, $\rho = 30 \, \Omega/\text{cm}$. The Si was *n*-type, $\rho = 5 \, \Omega/\text{cm}$. Gillespie, Olsen, and Nichols [28]. (*d*) Transmittance spectrum of (*A*) sapphire (1.2 mm thick); (*B*) Periklase (0.67 mm thick); (*C*) Arsenic trisulfide glass (5 mm thick). Smith, Jones, and Chasmar, *The Detection and Measurement of Infra-Red Radiation*, Clarendon, Oxford, 1968.

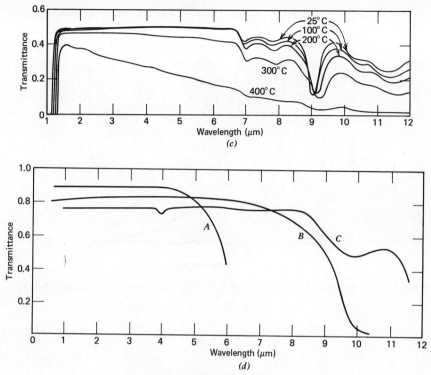

Figure 10.10 (Continued)

88. The absorption coefficients were calculated from manufacturer's data on the basis of (10.14). Other glasses and polycrystalline materials were developed by the Schott firm. They are included in Table 86, and their refractive indices are given in Table 76; the spectral range covered by that table indicates the approximate range of useful transmittance. KRS-5 (thallium bromide-iodide) and KRS-6 (thallium bromide-chloride) are also included in Table 87; the former is also listed in Tables 73 and 74.

The uv and ir absorption edges of several other crystals are shown in Figures 10.11 and 10.12 [29]. Absorption coefficients of several semiconductor materials are shown in Figure 10.13 [29]. Data on the reflection and transmission spectra of many crystalline materials have been published in Ref. 30[3]; refractive index data are included in Tables 73 and 74.

---

[3] Part I covers: NaCl, KBr, CsBr, CsI, SiO$_2$, LiF, CaF$_2$, Sapphire, As$_2$S$_3$, AgCl, KRS-5, Irtran 1 and 2, Si, Ge. Part III: CaCO$_3$, BaF$_2$, NaF, KCl, Ruby, Al$_2$O$_3$, TlCl, KRS-6, TlBr, Irtran 3–5, CuCl, CdSe. Part V: BN, CdS, Irtran 6, GaSb, CaAs, GaP, InSb, InP, RbBr, RbCl, RbI, SrTiO$_3$, Te. Parts II, IV, and VI are bibliographies of the material in the preceding parts. Data cover the region 2–50 μm.

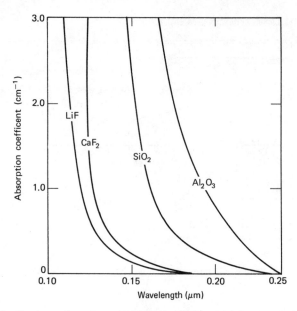

Figure 10.11 Uv absorption edges of LiF, CaF$_2$, SiO$_2$, and Al$_2$O$_3$ crystals, Smakula [29].

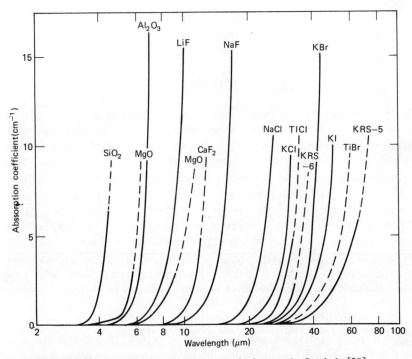

Figure 10.12 Ir adsorption edges of some ionic crystals. Smakula [29].

24

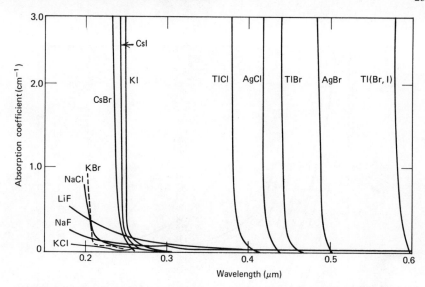

Figure 10.13 Absorption edges of some semiconductor materials. Smakula [29].

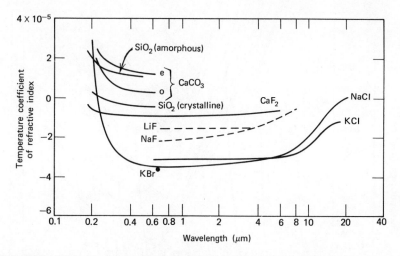

Figure 10.14 Temperature coefficient of refractive index for some crystals. Smakula [29].

The temperature dependence of the refractive index for a number of these is shown in Figure 10.14 [29] and their dispersion characteristics $(dn/d\lambda)$ in Figure 10.15 [29]. Table 87 lists mechanical and thermal characteristics of many materials used in ir optics. Achromatic doublets for use in the uv can be made of quartz and fluorite (or lithium fluoride or magnesium oxide). For use in the ir region, they are made of silicon and germanium or of arsenic trisulfide glass and lithium fluoride [31].

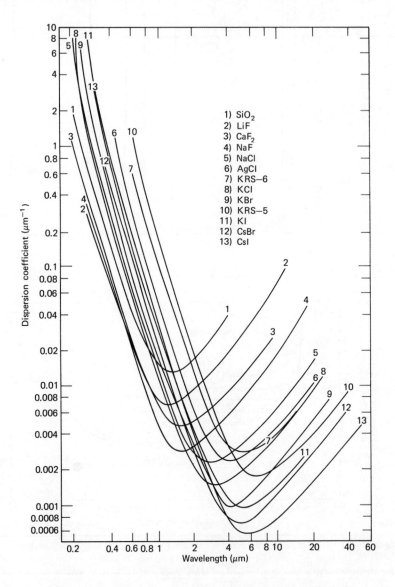

Figure 10.15 Dispersion curves for some crystals. Smakula [29].

10.3 OPTICAL PLASTICS

As the manufacturing technology of plastic materials advanced sufficiently to permit good stability and casting precision, these materials
became increasingly significant as economical optical media. Their major
advantages are as follows:

1. *Manufacturing Economy.* Precision casting can replace grinding and
polishing. Mounting devices can be manufactured integrally with the
optical component.

2. *Optical Advantages.* High relative dispersion (low V number) is
feasible even at relatively low refractive indices; many plastics are
superior to silicate glasses in uv and ir transmission.

3. *Mechanical Advantages.* The greater impact strength and lighter
weight of plastics make them superior to glass in some applications; the
specific gravity of plastics is generally about half that of glass.

On the other hand, there are several important disadvantages:

1. *Lower Precision.* The economical forming method, casting, limits the
precision as discussed below. Uniformity of refractive index is normally
no better than a part in 10^3 or 10^4—over a decade larger than in optical
glasses.

2. *Mechanical Limitations.* Plastics are far less resistant to scratching
and other forms of surface damage.

3. *Thermal Effects.* Thermal effects in plastics are far more pronounced than they are in silicate glasses, both in relative expansion (ca.
$10^{-4}/°C$) and in refractive index variation (ca. $1.5 \times 10^{-4}/°C$).

4. *Limited range of refractive indices.* Optical plastics with refractive
index above 1.6 do not seem to be practical.

In Table 89 [7a], [32] we present some optical, mechanical, and
thermal characteristics of popular optical plastics materials and in Table
78 [33] the variation of the refractive index with wavelength and temperature, for the most widely used plastics. (Their temperature coefficient of
index variation is approximately $1.4 \times 10^{-4}/°C$ as compared to about
$1.5 \times 10^{-6}/°C$ for crown glass.) Transmittance data for these are shown in
Figure 10.16 [32]. Refractive index data on a large number of these
materials are presented in Table 79 [33]. Note that polymethyl methacrylate is hygroscopic and that its refractive index may increase by as much
as 0.002 on prolonged soaking in water [33].

Manufacturing quality is claimed to approach that of precision optical
components [34] with refractive index tolerances held to ±0.001. For
example, for molded optics diameter and thickness tolerances are kept to

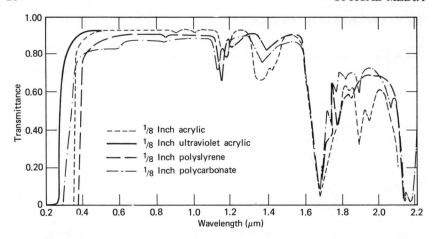

Figure 10.16   Transmittance spectra of some optical plastics [32].

$\pm 25\ \mu\text{m}$; scratch and dig width can be held to $30\ \mu\text{m}$ and $400\ \mu\text{m}$, respectively; and shrinkage varies from 0.1–0.6%. For plastic optical components finished by grinding and polishing, typical specifications met in practice are focal length, $\pm 2\%$; linear dimensions, ± 0.4 mm; haze, below 2% (based on ASTM method D1003–61); foreign particle inclusions and bubbles are held to below $50\ \mu\text{m}$ diam and are "rare." Curvature nonuniformities are held to within 0.5% of the mean value and in plano-convex (or concave) lenses, the plano surface is parallel to the tangent at the thickest (or thinnest) point to within $0.1°$ [32].

Thermosetting lacquers have been used to protect the soft plastic surface and its metallizing coatings, when possible; but these lacquers seem to be limited in applicability. Because of their tendency to accumulate static electric charges, plastic optical components are often provided with antistatic coatings [32].

REFERENCES

[1] J. G. Baker, "Planetary telescopes," *Applied. Opt.*, **2**, 111–129 (1963).

[2] H. Meyer, "Optical glass," in *Advanced Optical Techniques*, A. C. S. van Heel, ed., North-Holland, Amsterdam, 1967, pp. 493–502.

[3] *ASTM Standard*, Part 17, No. C162–71; ASTM, Philadelphia (1971).

[4] C. L. Babcock, *Silicate Glass Technology Methods*, Wiley, New York, 1977.

[5] D. G. Holloway, *The Physical Properties of Glass*, Wykeham, London, 1973.

[6] E. B. Shand, *Glass Engineering Handbook*, 2nd Ed., McGraw-Hill, New York, 1958.

[7] N. J. Kreidl and J. L. Rood, "Optical materials," in *Applied Optics and Optical Engineering*, Vol. 1, R. Kingslake, ed., Academic, New York, 1965, Chapter 5.

[8] M. Born and E. Wolf, *Principles of Optics*, 5th Ed., Pergamon, Oxford, 1975, Section 2.3.4.

[9] A. C. Hardy and F. H. Perrin, *Principles of Optics*, McGraw-Hill, New York, 1932. See *Opt. Spectra* **13³**, 83–85 (1979).

[10] M. Herzberger and C. D. Salzberg, "Refractive indices of infrared materials and color correction of infrared lenses," *J. Opt. Soc. Am.*, **52**, 420–427 (1962).

[11] B. W. Morrissey and C. J. Powell, "Interpolation of refractive index data," *Appl. Opt.*, **12**, 1588–1591 (1973).

[12] F. Reitmayer and E. Schuster, "Homogeneity of optical glasses," *Appl. Opt.*, **11**, 1107–1111 (1972).

[13] F. A. Molby, "Index of refraction and coefficients of expansion of optical glasses at low temperatures," *J. Opt. Soc. Am.*, **39**, 600–611 (1949).

[14] J. H. Wray and J. T. Neu, "Refractive index of several glasses as a function of wavelength and temperature," *J. Opt. Soc. Am.*, **59**, 774–776 (1969).

[15] D. A. Duke and G. A. Chase, "Glass-ceramics for high-precision reflective-optics applications," *Appl. Opt.*, **7**, 813–817 (1968).

[16] A. Dietzel, "Porzellan als Mittelpunkt der oxydischen Werkstoffe," *Ber. Dtsch. Keram. Ges.*, **36**, 301–304 (1959).

[17] N. M. Brandt, "Annealing of 517.645 borosilicate glass: I. Refractive index," *J. Am. Ceram. Soc.*, **34**, 332–338 (1951).

[18] K. Kinosita, "Surface deterioration of optical glasses," in *Progress in Optics*, Vol. 4, E. Wolf, ed., North-Holland, Amsterdam, 1965, pp. 85–143.

[19] J. S. Stroud, "Color Centers in a Cerium-Containing Silicate Glass," *J. Chem. Phys.*, **37**, 836–841 (1962).

[20] *Optical Glass*, Bausch & Lomb Inc., Rochester, NY.

[21] W. A. Weyl, *Coloured Glasses*, Dawson's, London, 1959.

[22] J. A. Savage and S. Nielson, "Chalcogenide glasses transmitting in the infrared between 1 and 20 μ—A state of the art review," *Infrared Phys.*, **5**, 195–204 (1965).

[23] R. D. Maurer, "Glass Fibers for Optical Communications," *Proc. IEEE*, **61**, 452–462 (1973).

[24] Anon., *Electronics*, 29 March 1979, p. 48.

[25] Trademarks of Heraeus GMBH, Hanau.

[26] W. Heitmann, "Attenuation measurement in low-loss optical glass by polarized radiation," *Appl. Opt.*, **14**, 3047–3052 (1975).

[27] D. E. Gray, ed., *American Institute of Physics Handbook*, 3rd Ed., McGraw-Hill, New York, 1972: (*a*) Section 6b, S. S. Ballard, J. S. Browder, and J. F. Ebersole, pp. 6-12–6-57. (*b*) Sect. 6c, S. S. Ballard, J. S. Browder, and J. F. Ebersole, pp. 6-58–6-94.

[28] D. T. Gillespie, A. L. Olsen, and L. W. Nichols, "Transmittance of optical materials at high temperatures in the 1 μ- to 12 μ-range," *Appl. Opt.*, **4**, 1488–1493 (1965).

[29] A. Smakula, "Synthetic crystals and polarizing materials," *Opt. Acta*, **9**, 205–222 (1962).

[30] D. E. McCarthy, "The reflection and transmission of infrared materials, . . . spectra from 2–50 microns." *Appl. Opt.*: I: **2**, 591–595; II: **2**, 596–603 (1963); III: **4**, 317–320; IV: **4**, 507–511 (1965); V: **7**, 1997–2000; VI: **7**, 2221–2225 (1968).

[31] A. Smakula, J. Kalnajs, and M. J. Redman, "Optical materials and their preparation," *Appl. Opt.*, **3**, 323–328 (1964).

[32] *Handbook of Plastic Optics*, U.S. Precision Lens, Cincinnati, OH, (1973).

[33] H. C. Raine, "Plastic glasses," in *Proc. London Conf. Opt. Instr. 1950*, Chapman & Hall, London, 1951, pp. 243–256.

[34] I. K. Pasco and J. H. Everest, "Plastic optics for opto-electronics," *Opt. Laser Technol.*, **10**, 71–76 (1978).

11

Spectral Filters

Spectral filters are devices used to modify the (spectral) transmittance of an optical system. They may be made of a homogeneous material with appropriate spectral absorption characteristics, or use some other optical phenomenon, such as multiple-beam interference or optical anisotropy, to effect the modification. The various types of filter are discussed in Sections 11.2–11.4. In Section 11.1, methods for classifying filtering characteristics are presented.

11.1 FILTERING CHARACTERISTICS

The optical properties of filters may be divided into integral and spectral characteristics. The spectral characteristics again fall into several subcategories:

1. Sharp-cutting (high-pass or low-pass).
2. Band-pass (wide or narrow).
3. Compensating.
4. Neutral density.

11.1.1 Integral Filter Characteristics

In integral terms the performance of a filter may be specified in terms of chromaticity and total transmittance. The chromaticity coordinates may be specified according to the C.I.E. tristimulus system, as described in Section 1.2.2.3, for any specific spectral distribution, $\Phi_\lambda(\lambda)$, of the incident flux, with the spectral transmittance, $\tau(\lambda)$, taking the place of R_λ in (1.22).

The total transmittance, τ_T, must be evaluated in terms of the flux spectral distribution, Φ_λ, in analogy with (18.5):

$$\tau_T = \frac{\displaystyle\int_0^\infty \tau(\lambda)\Phi_\lambda(\lambda)\,d\lambda}{\displaystyle\int_0^\infty \Phi_\lambda(\lambda)\,d\lambda}, \tag{11.1}$$

where obviously any variable directly proportional to Φ_λ may be substituted for it.

When the *effective* transmittance is desired, the spectral distribution of the incident flux must be weighted by that of the detector sensitivity $[S(\lambda)]$:

$$\tau_e = \frac{\displaystyle\int_0^\infty \tau(\lambda)\Phi_\lambda(\lambda)S(\lambda)\,d\lambda}{\displaystyle\int_0^\infty \Phi_\lambda(\lambda)S(\lambda)\,d\lambda}. \tag{11.2}$$

Occasionally, especially in photographic practice, this is called the *filter factor*.

For color photography practice, ANSI standards [2] specify that the filter factors of a filter or lens be presented as a series of numbers in the form:

$$100\tau_B - 100\tau_G - 100\tau_R. \tag{11.3}$$

Here τ_B, τ_G, and τ_R are, respectively, the filter factors evaluated relative to the blue-, green-, and red-sensitive layers of the color photographic emulsion. See Table 94 for a listing of these for a number of compensating filters.

In ophthalmic work, the *shade number* (N) is defined in terms of the luminous density (D) and the visual transmittance (τ_v) [3]:

$$N = 1 + (\tfrac{7}{3})D \qquad D = -\log_{10}\tau_v. \tag{11.4}$$

11.1.2 Sharp-Cutting Filter Characteristics

Sharp-cutting transmission filters pass flux, more or less completely, below a certain wavelength, and absorb (or reflect) the flux above this wavelength, or *vice versa*. Accordingly, such filters are classified as *short wavelength pass* or *long wavelength pass* (*short-pass* and *long-pass*), respectively. For filters used in reflection, this terminology may become confusing and should be made specific (e.g., "long-reflecting," etc.). Filters whose spectral characteristics change from transmitting to reflecting at a certain wavelength are called *dichroic*, that is, they are short-pass in transmittance and long-pass in reflection, or *vice versa* (see Section 11.3.3). In addition to the class of the sharp-cutting filter, the spectral position of the transition and the rate of transition are usually required. In the filter listing in Tables 90 and 91 [1a] the transition is taken at the wavelength where $\tau(\lambda) = 0.37$ and the rate of transition is specified as the spectral width, $\Delta\lambda$, of the transition from transparency to opacity, where transparency is assumed when $\tau > 0.7$ and opacity when $\tau < 0.05$. These values are quite arbitrarily chosen for convenience.

Since all long-pass filters absorb again in the far ir and all short-pass filters do so in the uv, the point where the transmittance drops again to 0.7 is listed as the "extent" of the filter in Tables 90 and 91, which cover, respectively, long- and short-pass filters of a number of manufacturers. Transmission spectra of many of these are given in Tables 96–98, which list these data for filters of Corning [4], [5] and Schott [6], respectively.

Of special interest are short-pass filters with the cutoff point just inside the ir. They are essentially clear in the visible but absorb heavily throughout the ir and serve as *heat absorbing filters.*

11.1.3 Band-Pass and Rejection Filters

A number of band-pass filters are listed in Table 92 [1a]. Their characteristics are often defined in terms of the half-peak transmittance points, that is, the wavelength values at which the transmittance has dropped to half its peak value. The midpoint between these defines the position of the passband and the spectral range between them, the *bandwidth.* To define the sharpness of the passband characteristic, the range between the 5% transmittance points, too, is sometimes given. In Table 92 it is given under *base bandwidth.* The *rejection range,* occasionally called the *free filter range,* is the spectral range, outside the *base bandwidth* range, over which the transmittance remains below a certain value, here taken as 1%. The filter efficiency may be defined in terms of the peak transmittance value.

Band-rejection filters are transparent over a wide spectral region and absorb radiation in a narrow range inside this region. They are defined in terms of the rejection band in a manner analogous to the definition of band-pass filters. Here the range of transparency replaces the rejection range in specifying their utility range. See Table 93 [1a] for a listing of band-rejection filters.

11.1.4 Compensating Filters

Filters having slowly changing transmittance spectra are generally classified as compensating filters. Their function is usually to modify somewhat the spectral luminance of a light source or the sensitivity spectrum of a detector.

When the purpose is to modify the color temperature of the luminous flux, the performance of the filter may be specified in terms of the change in color temperature, T'_c, measured in "mireds"[1]

$$T'_c = \frac{10^6}{T_c},\qquad(11.5)$$

where T_c is the color temperature measured in kelvin. The shift is then given by

$$\Delta T'_c = 10^6(T_1^{-1} - T_2^{-1}),\qquad(11.5a)$$

where T_1, and T_2 are the color temperatures of the flux, before and after the filtering, respectively. Accordingly, light blue filters, which raise the color temperature, exhibit a positive $\Delta T'_c$ and amber filters, a negative one. A number of filters, the shifts they produce, and their ANSI color-photographic density specifications (11.3) are presented in Table 94 [1a].

11.1.5 Neutral Density Filters

Neutral density filters may be specified in terms of their mean density, their range of neutrality, and the magnitude of the transmittance variation over this range; see Table 95 [1a] for listing of such filters. Note that, in addition to the usual dye filters (glass or gelatine), partially exposed and developed silver halide emulsions, evaporated metal films, crossed polaroid filters, and fine metal screens are occasionally convenient neutral-density filters.

11.2 ABSORPTION FILTERS [7], [8]

11.2.1 General Characteristics

Absorption filters are made by dissolving in glass dyes or metal ions that cause light to be absorbed over certain spectral regions. Dyes may also be dissolved in gelatine, which is subsequently formed into a thin sheet to act as a filter. When the absorption is uniform over the visible region of the spectrum, the filter appears neutral; if it varies over this region, it appears colored, in general.

---

[1] The term *mired* is a contraction of *mi*cro *re*ciprocal *d*egrees.

The absorption may be due to one of three causes:

1. For substances dissolved in the glass: absorption due to interatomic or molecular bonds.

2. For substances forming colloidal suspensions: scattering and subsequent absorption by the suspended particles.

3. For substances forming microscopic suspensions: absorption by the suspended crystals.

Heating the glass tends to strengthen the absorption bands and to shift them toward the longer wavelength region: the loosened interatomic bonds correspond to lower resonant frequencies and, therefore, longer wavelengths [7]. Subsequent quenching may freeze such a state into the glass. In addition, major color changes may be produced at elevated temperatures due to a shifting of the oxidation-reduction equilibria and changed electron mobility.

Irradiation can change the state of the color centers (sites of imperfections in the atomic network) in the glass and hence its absorption spectrum. When caused by sunlight, such changes are called solarization and typical processes responsible for them were discussed in Section 10.1.4.

The effect of *thickness* of the filter glass at any wavelength may be calculated from (10.15) and the effect of obliquity of the light passage from (10.16). Determination of the effect on the total transmittance and the chromaticity requires integration of these results over the active spectral region. Results for a number of filter glasses have been published [9]. These define the filter action in terms of the total luminous transmittance (Y) and the CIE chromaticity coordinates (x, y) of the transmitted flux—both for Standard Illuminant A. Such data are included in Ref. 10* for the Kodak Wratten filters.

11.2.2   Atomic and Molecular Absorption

Glasses may be colored by dissolving in them oxides of the transition elements or dyes. The oxides of the first group of transition elements (titanium, vanadium, chromium, manganese, iron, cobalt, nickel, and copper) yield slowly changing absorption spectra, while the dyes may yield very sharply defined ones. The higher transition element groups, primarily the rare earths, yield absorption spectra intermediate in sharpness.

Many elements when dissolved exhibit wide variations in their absorption spectra, depending on their state of oxidation and their bonding to the atoms in the glass matrix. Thus, not only the identity of the dissolved element, but also (1) its valence state and the atmosphere in which the glass is heated and (2) the constitution of the glass matrix and details of the annealing cycle can have strong effects on the resulting color.

Below we discuss briefly some of the more important colorants of this type.

Iron. The addition of ferric oxide (Fe_2O_3) to soda-lime-silicate glass tends to produce a green color due to the resulting ferric ions. See Curve (a) in Figure 11.1 [8] for the effect of 1.3% Fe_2O_3. Indeed, under the normal glass manufacturing conditions, it is difficult to produce glass entirely free from this ingredient, and it is responsible for the greenish appearance of most plate glass when viewed edgewise. The green tint is occasionally masked by the addition of Mn or Se, which produce the complementary color; this results in an even stronger absorption, but one that is visually neutral.

If the ions are reduced further (e.g., by adding carbon or by raising the melt to a higher temperature), the transmittance drops and the absorption curve shifts toward the blue. This also causes stronger absorption in the ir and Fe_2O_3 is, therefore, a popular constituent of green sunglasses.

The addition of ferric oxide to phosphate glasses results in only little absorption in the visible, while the absorption in the ir is even higher, yielding a "heat-absorbing" glass.

Manganese. The addition of Mn^{3+} ions to the glass gives rise to an absorption band in the 0.42–0.65-μm region, resulting in a purple appearance {see curve (a) Figure 11.2 [8]}. Under strong reducing conditions, an almost totally clear glass results [Curve (b) Figure 11.2].

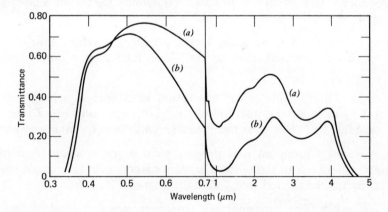

Figure 11.1 Effect of iron ions (1.3% Fe_2O_3) under (a) normal and (b) reducing conditions. Kreidl [8].

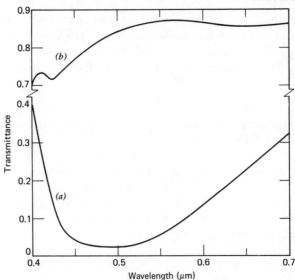

Figure 11.2 Effect of manganese ions on the transmittance spectrum of glass. (a) Mn^{3+} ions; (b) after reduction, yielding Mn^{2+} ions. Kreidl [8].

Cobalt. Cobalt, even in small amounts (ca. 0.01%) can produce a deep blue coloration, especially in potash-containing glasses {see Curve (a), Figure 11.3 [1b]}. Because of its strong coloring power, it is used in conjunction with copper to produce black glass. The blue color is due to an "unsaturated" state. In the "saturated" state, where each cobalt atom is associated with the maximum number of nearest neighbors, a red color is produced. It is obtained in glasses containing significant amounts of phosphoric or boric acid.

Nickel. In a soda-lime glass, nickel produces a yellow tint, characteristic of the saturated state, and in potash-lime glass, a purple coloration.

Copper. Cupric oxide causes absorption which increases progressively from 0.5 μm to beyond the end of the visible spectrum {see Curve (c), Figure 11.3 [1b]}. This gives the glass a greenish-blue (cyan) appearance.

Chromium. Chromium ions in glass yield a transmission band from approximately 0.5 to 0.65 μm. The transmission then rises again somewhat toward the far red {see Curve (b), Figure 11.3 [1b]}.

Neodymium. The transmittance spectrum of a typical didymium (neodymium-praseodymium) glass is shown in Figure 11.4 [4]. Its appearance is purple. When thin, the broad blue passband predominates; when

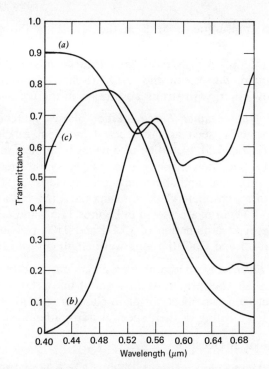

Figure 11.3 Transmittance spectra of passband filters. (*a*) Cobalt–blue; (*b*) chromium, copper, iron–green; (*c*) copper–cyan. Kreidl and Rood [1*b*].

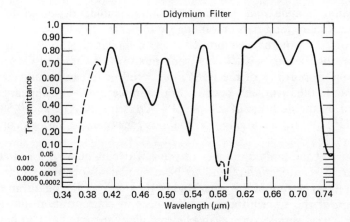

Figure 11.4 Transmittance spectrum of didymium glass (Corning No. 5120) [4].

thick the high transmittance peak in the red shifts the color in that direction.

Vanadium (V_2O_3), *Molybdenum, and Praseodymium*. All these tend to produce green glasses. In the V_2O_5 form, vanadium produces a brownish-yellow color, with strong absorption in the uv.

Synthesized Transmittance Spectra. Filter glasses matching specified transmittance spectra, such as the spectral luminous efficiency curves, $V(\lambda)$, $V'(\lambda)$ (see Fig. 1.5), would be of obvious usefulness in photometry and colorimetry. Such bell-shaped curves can be synthesized by combining additives with long- and short-cutting characteristics. The transmittance spectra of a phosphate glass matching the $V(\lambda)$ and a silicate glass matching the $V'(\lambda)$ curve are shown in Figures 11.5a and b, respectively. In both of these, a combination of CeO_2 and TiO_2 provided the short wavelength cut-off and CuO the long wavelength cut-off [10].

Dye-Based Filters. Dye-based filters are manufactured both in gelatine and glass matrices. The former, in the form of thin gelatine sheets, often sandwiched between two sheets of plate glass, are called *Wratten filters*. See Ref. 10* for a listing of their spectral transmittance values.

11.2.3 Colloidal Suspensions

Below we list some glass colorants that work as colloidal suspensions.

Gold. Gold was traditionally the means for obtaining a strong red tint in glass. When gold-containing glass is reheated to a high temperature, or "*struck*," gold crystals form and turn the glass ruby-red.

Cadmium Sulfide and Selenide. Cadmium sulfide dissolved in glass produces a sharp long-pass cutoff edge at about 0.525 μm, when "struck" and recooled. Cadmium selenide produces a similar edge, but in the ir, so that the glass becomes black. By using mixtures of CdS and CdSe, with increasing proportions of the latter, the cutoff edge can be moved from 0.525 μm to 0.65 μm, and beyond. For example, ruby-like appearance is obtained by the addition of 2% Se, 1% CdS, 1% As_2O_3, and 0.5% C.

See Table 99 for a summary of methods for obtaining various colors in glass.

Because of the limited number of ways of dyeing glass, colored glasses of different manufacturers tend to have similar absorption spectra. By plotting the spectrum of the logarithm of the density (the logarithm of the logarithm of the transmittance), differences in concentration and thickness appear as simple vertical displacements of the curves. The data for approximately 800 different glasses have been compared on this basis and summarized in a set of 44 graphs (each containing many curves) [11].

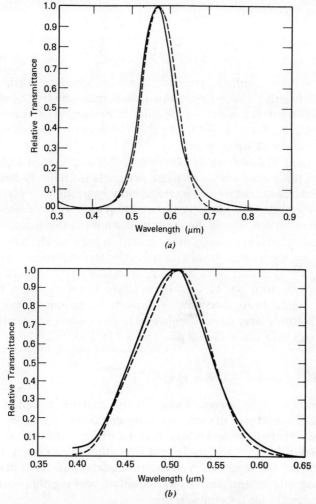

Figure 11.5 Comparison of transmittance spectra with spectral luminous efficiency curves. (a) Special phosphate glass transmittance (solid) and $V(\lambda)$ (broken); (b) special silicate glass transmittance (solid) and $V'(\lambda)$ (broken). Res *et al.* [10].

11.2.4 Absorption Filters for the Infrared

Semiconductor materials tend to have sharp-cutting long-pass transmittance characteristics; see Figure 10.13. These are due to the fact that they absorb very strongly all photons with energy in excess of the forbidden band gap energy (U_g). This energy controls the location, λ_a, of the absorption edge of the material. Analogously to (16.2), this wavelength is given by

$$\lambda_a = \frac{hc}{U_g}.$$ (11.6)

Beyond this wavelength the transmittance rises rapidly. At considerably longer wavelengths, it drops again due to absorption by free carriers. This form of transmittance spectrum permits using some semiconductors as filters in the ir. See Table 100 [12] for the cut-off wavelengths and refractive indices of some of these.

Due to their high refractive indices and the concomitant high surface reflectivity, the transmittance of these materials is relatively low even in the spectral region where their absorption is negligible; according to Fresnel's formulae (2.101–2.106) the reflectance losses from a plane parallel plate exceed 50% if the material has a refractive index of 3.36 or higher. Low-reflection coatings may be used to improve the transmittance of such materials; a single-layer film may raise the transmittance of Si and Ge plates ($n = 4$) to over 90% over half an octave of the ir spectrum [12].

Plastic sheets, too, may be used for ir filters. For instance, by removing hydrogen halide from successive groupings along the polymer chain, polyvinyl chloride may be made opaque in the visible, transmitting over only two bands in the ir (1–3.2 μm and 3.4–5.4 μm) [13].

11.3   THIN-FILM FILTERS [14]–[17]

Multiple-beam interference, especially as observed in a plane-parallel cavity (e.g., Fabry-Perot etalon), has been discussed in Section 2.4.4 and applications thereof in low- and high-reflection coatings in Sections 8.1.1.2 and 8.2.1.2, respectively. Such coatings are usually made by evaporation of the material in a vacum chamber. In all instances, the interference effects are wavelength sensitive, making thin-film devices potentially useful for spectral filtering.

By constructing multiple-layered films ("multilayers," in popular parlance) of dielectric materials, almost any desired spectral characteristic can be synthesized. Much work has gone into the development of methods for designing multilayer films with prescribed characteristics [18], but much more remains to be done. On the other hand, the far simpler analysis problem has been solved effectively and a number of methods have been developed [19], [20] for analyzing the multiple beam interference effects governing the spectral performance of multilayer films. One of these, based on matrix calculus, is described in Appendix 11.1. Here we discuss only certain special cases of particular interest.

The characteristics of a number of materials used in multilayer construction are presented in Table 101: dielectrics in Table 101*a* [16] and metallic films in Table 101*b* [21] (for the latter the transmittance values corresponding to a given reflectance value is given). Characteristics of the dielectric films are summarized in Figure 11.6 [22]. Filters for use in the uv and ir require special attention; see Refs. 23 and 22, respectively, for these.

11.3.1 Fabry-Perot Type Interference Filters

The popular simple interference filter is essentially a Fabry-Perot etalon, where the spacer is a thin layer of dielectric material deposited by evaporation between two layers of partially reflecting metallic coating. But, whereas in the Fabry-Perot interferometer the two surfaces must be extremely flat (deviations of $\lambda/40$ accumulate to half a wavelength after 10 reflections!), the requirements are far lower in the interference filter; here the evaporated layers follow the surface contour, so that the two surfaces are essentially parallel, tending to compensate mutually for any deviation from planeness.

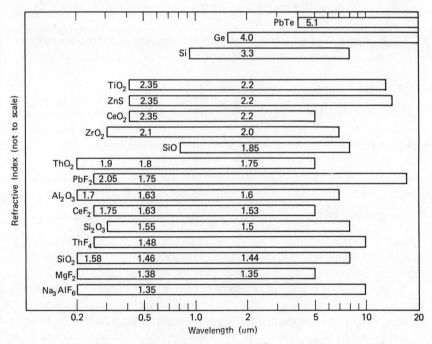

Figure 11.6 Dielectric thin-film materials: spectral ranges and refractive indices. Based on Hass and Ritter [22].

11.3.1.1   General Expression. The transmittance of such a system is given by (2.214) and (2.215) which here take the forms:

$$\boxed{\tau_T = \frac{\tau^2}{(1-\rho)^2 + 4\rho \sin^2 \delta}}$$ (11.7)

$$\delta = \phi + \frac{2\pi d \cos \theta}{\lambda} = \phi + \frac{2\pi d \sqrt{n^2 - \sin^2 \theta_0}}{\lambda_0},$$ (11.8)

$$= \phi + k_0 d \sqrt{n^2 - \sin^2 \theta_0},$$

where τ, ρ are transmittance and reflectance, respectively, of the two metallic layers, assumed identical,

   ϕ is the phase shift experienced by the wave on reflection at the two layers,

   d is the thickness of the dielectric layer,

   n is its refractive index,

   θ, θ_0 are the angles between the wavefront and the dielectric layer surface normals, inside and outside it, respectively, and

   $k_0 = 2\pi/\lambda_0$ is the magnitude of the wave vector outside the layer.

11.3.1.2   Peak Values. The transmittance is maximum and minimum when δ equals $m\pi$ or $(m+\frac{1}{2})\pi$, respectively, that is, when k has values k_p, k_t, such that

$$\phi + k_{p,t} d \sqrt{n^2 - \sin^2 \theta_0} = \begin{cases} m\pi \\ (m+\frac{1}{2})\pi \end{cases}, \quad m = 0, 1, 2 \ldots,$$
$$\text{(11.9a)}$$
$$\text{(11.9b)}$$

and hence, transmittance peaks occur for wavelengths

$$\lambda_p = \frac{2d \sqrt{n^2 - \sin^2 \theta_0}}{(m - \phi/\pi)}.$$ (11.10)

The maximum and minimum values of the transmittance are, then,

$$\tau_{mx} = \tau(k_p) = \frac{\tau^2}{(1-\rho)^2},$$ (11.11)

$$\tau_{mn} = \tau(k_t) = \frac{\tau^2}{(1+\rho)^2},$$ (11.12)

and the ratio between them, the filter *rejection ratio*, is

$$R_\tau = \frac{\tau_{mx}}{\tau_{mn}} = \frac{(1+\rho)^2}{(1-\rho)^2}.$$ (11.13)

The value of the peak transmittance is called the (optical) *efficiency of the filter*.

As stated in Section 11.1.3, the filter passband is defined by the *in vacuo* wavelengths (or wave vector magnitudes) λ_1, λ_2 (or k_1, k_2), where the transmittance has dropped to half its peak value, that is when (from 11.7)

$$\sin \delta = \pm \frac{1-\rho}{2\sqrt{\rho}}. \tag{11.14}$$

In conjunction with (11.8), this implies

$$\sin^{-1} \frac{1-\rho}{2\sqrt{\rho}} = \phi + k_1 d \sqrt{n^2 - \sin^2 \theta_0} \tag{11.15a}$$

$$\sin^{-1} \frac{1-\rho}{2\sqrt{\rho}} = -(\phi + k_2 d \sqrt{n^2 - \sin^2 \theta_0}). \tag{11.15b}$$

On adding these two equations, we find the passband width:

$$\delta k = k_1 - k_2 = \frac{2 \sin^{-1}[(1-\rho)/2\sqrt{\rho}]}{d\sqrt{n^2 - \sin^2 \theta_0}}. \tag{11.16}$$

From (11.9a) this may be written

$$\delta k = \frac{2 k_p \sin^{-1}[(1-\rho)/2\sqrt{\rho}]}{m\pi - \phi}. \tag{11.16a}$$

To find the equivalent expression in terms of the wavelength, we note that

$$\frac{dk}{d\lambda} = -\frac{2\pi}{\lambda^2}$$

and hence

$$\delta\lambda \approx \frac{\lambda_p^2}{2\pi} \delta k = \frac{2\lambda_p \sin^{-1}[(1-\rho)/2\sqrt{\rho}]}{m\pi - \phi}, \qquad \delta\lambda/\lambda_0 \ll 1. \tag{11.17}$$

Hence the resolving power (see 2.176) is

$$R = \frac{\lambda_p}{\delta\lambda} = \frac{m\pi - \phi}{2 \sin^{-1}[(1-\rho)/2\sqrt{\rho}]}. \tag{11.18}$$

11.3.1.3 Periodicity of Transmission Spectrum. From (11.7) and (11.8) it is apparent that the transmission spectrum is periodic in k_0. From (11.9) it is clear that the period equals the change, Δk, in k_0 which changes m by unity, that is,

$$\Delta k = \frac{\pi}{d\sqrt{n^2 - \sin^2 \theta_0}}$$

$$= \frac{k_p}{m - \phi/\pi}. \tag{11.19}$$

The corresponding wavelength change can be derived from this as

$$\Delta\lambda = \frac{\lambda_p}{m \mp 1 - \phi/\pi} \tag{11.20}$$

in the direction of increasing and decreasing wavelengths, respectively.

11.3.1.4 Effects of Angle of Incidence. Equation 11.10 also gives the dependence of the peak wavelength on the angle of incidence, θ_0, of the wavefront. The variation of λ_p with θ_0 effectively limits the usuable angular field of the interference filter. If we define the half-field as the angle, φ, at which k_p has shifted by $\frac{1}{2}\delta k$, we obtain from (11.9a), starting with normal incidence ($\theta_0 = 0$):

$$\tfrac{1}{2}\delta k = k_p(\varphi) - k_p(0) = \frac{(m\pi - \phi)}{d}\left[(n^2 - \sin^2 \varphi)^{-1/2} - \frac{1}{n}\right]. \tag{11.21}$$

Equating this to its value from (11.16) and multiplying through by nd, we obtain

$$\frac{\sin^{-1}(1-\rho)}{2\sqrt{\rho}} = (m\pi - \phi)\left[\left(\frac{1 - \sin^2 \varphi}{n^2}\right)^{-1/2} - 1\right]$$

and, hence,

$$\frac{\sin^2 \varphi}{n^2} = 1 - \frac{(m\pi - \phi)^2}{\{\sin^{-1}[(1-\rho)/2\sqrt{\rho}] + m\pi - \phi\}^2}, \tag{11.22}$$

as defining the usuable filed size.

With the value of the arcsine much less than $(m\pi - \phi)$, this approaches the value of $\delta\lambda/\lambda_p$ as given by (11.18). Thus, for a high resolving power interference filter, the maximum half-field angle, φ' (measured inside the dielectric), equals $1/\sqrt{R}$, approximately.

11.3.1.5 Practical Filter Parameters. For example, an interference filter whose reflecting layers each have a reflectance and transmittance of 90% and 6%, respectively, will yield a peak transmittance of 36% and a relative bandwidth of 1.7% when used in second order (3.4%, when used in the first order). Its usable field extends $\pm 7.5°$; when it is used near an image plane, the equivalent numerical aperture is 0.13. These values are quite representative of commercially available interference filters. According to (11.20), such a filter has additional passbands at twice the nominal wavelength and at $\frac{2}{3}$ of its wavelength. These may be eliminated by combining the filter with a second one, possibly an absorption filter, whose passband includes the primary passband, but not the secondary ones, of the original filter.

By increasing the order, at any given wavelength, the passband becomes narrower, but the undesirable adjacent passbands come closer.

11.3.1.6 All-Dielectric Filters. Higher efficiency, combined with higher rejection ratio, can be obtained by substituting multiple dielectric layers (see Section 8.2.1.2) for the metallic reflection layers [24]. The dielectric reflecting layers exhibit very low absorption and therefore facilitate higher reflectance values at a given transmittance level. A representative filter of this type has peak transmittance of 80% and a relative bandwidth of 1.25% in the first order [24].

11.3.2 Periodic Multilayer Interference Filters

11.3.2.1 General Considerations. The periodic multilayer film is relatively easy to analyze and yet permits considerable flexibility in synthesizing band-pass and sharp-cutting filters. This section is devoted to a discussion of its performance. The results given here are valid when the film is used at normal incidence; as discussed in Section 11.3.4.1, they may also be used at oblique incidence, provided appropriate modifications are made.

In general, a periodic multilayer film with a two-component period has a reflectance peak at the wavelength for which the total period is optically half a wavelength thick; the spectral width of the high-reflectance region increases with the ratio of the refractive indices of the two components. Thus, basically, such a multilayer film lends itself to blocking a spectral band in transmission. A typical application might be a safety filter for use with a laser operating at a fixed wavelength. Here we would like to have high transmission throughout the visible part of the spectrum, with only a narrow region, surrounding the reflection peak, blocked. This can be accomplished by a simple two-component periodic stack having a small refractive index difference between the two components [25]. Simultaneously, the same filter can be considered as a band-pass filter when it is used as a mirror in an optical system.

In spite of this basic restriction, such films can, in practice, be adapted to serve as short- and long-pass filters, as well. Cf. Section 11.3.3.2.

11.3.2.2 Quarter-Wave Stack. The simplest nontrivial periodic multilayer system to analyze is the one whose period consists of two layers of differing refractive indices (n_1, n_2), both having the same optical thickness

$$d_0 = d_j n_j, \qquad j = 1, 2, \tag{11.23}$$

where d_j is the geometrical thickness of the layer having refractive index n_j. Such a system has a reflectance maximum at that wavelength (say, λ_b) for which d_0 is one-quarter of λ_b, and is called a *simple quarter-wave stack*. The response of a number of typical quarter-wave stacks, consisting of lossless dielectric layers, is shown in Figures 11.7*a* and *b* [17] as a

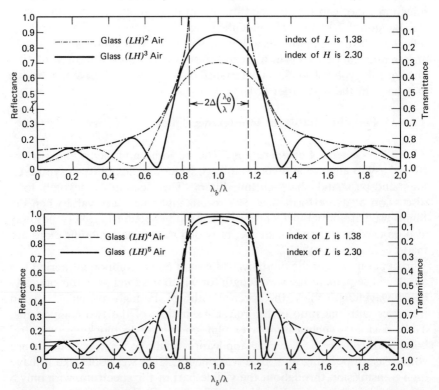

Figure 11.7 (*a, b*) Reflectance spectrum of some quarter-wave stacks with various numbers (*N*) of periods. Index ratio $n_1/n_2 = 0.6$. Dash-double-dot line is envelope. Baumeister [17].

function of the (normalized) reciprocal wavelength. This figure illustrates the fact that, as the number (N) of periods in the stack increases, the reflectance at λ_b approaches unity, the transition from reflection to transmission becomes sharper, and the number of oscillations in the region between 0 and 1 increases: indeed, the number of oscillations there equals N. As N approaches infinity ("infinite quarter-wave stack") the reflectance approaches unity over a band $\Delta(\lambda_b/\lambda)$ given by

$$\Delta\frac{\lambda_b}{\lambda} = \pi^{-1}\sin^{-1}\left|\frac{(1-n_1/n_2)}{(1+n_1/n_2)}\right|. \tag{11.24}$$

The reflectance at λ_b is given by

$$\rho_p = \frac{P+P^{-1}-2}{P+P^{-1}+2} \tag{11.25}$$

where

$$P = \left(\frac{n_2}{n_1}\right)^{2N}\frac{n_0}{n_s} \tag{11.26}$$

and $n_{s,0}$ are, respectively the refractive indices of the substrate and the medium above the stack (usually air). Clearly, the ratio n_2/n_1 controls both the width of the high-reflectance band and the rate at which the reflectance approaches unity.

Note that with the assumed losslessness, the transmittance of the stack is

$$\tau = 1-\rho. \tag{11.27}$$

Also note that the curves shown in Figures 11.7a and b are only one period of a periodic spectral response: additional peaks appear at λ as given by:

$$\frac{\lambda_b}{\lambda} = 1+2j, \qquad j = 1, 2, 3, \ldots. \tag{11.28}$$

11.3.2.3 The p:q Stack. If the period consists of two layers of unequal optical thickness

$$d_{o1} = pd_{oT}, \qquad d_{o2} = qd_{oT} = (1-p)d_{oT}, \tag{11.29}$$

the reflectance peak occurs at the wavelength, λ_b, at which the total optical thickness, d_{oT}, of the period equals half a wavelength:

$$d_{oT} = \frac{\lambda_b}{2}. \tag{11.30}$$

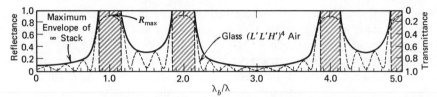

Figure 11.7   (c) Reflectance of some 2:1 quarter-wave stacks. Baumeister [17].

By way of illustration, the response for $p = \frac{1}{3}$ (the "2:1 stack") is shown in Figure 11.7c [17].

Reflectance maxima occur whenever

$$\frac{\lambda_b}{\lambda} = j, \qquad j = 1, 2, 3 \ldots, \tag{11.31}$$

except for those values of j where one of the layers has an optical thickness[2]

$$d_{o1}, d_{o2} = \frac{j}{2} \lambda. \tag{11.32}$$

As shown in Appendix 11.1, such a layer is effectively absent and is called an "absentee layer."

11.3.2.4   Modified Periodic Stacks.

If the individual layers deviate slightly from the $\lambda_b/4$ thickness, the peak reflectance is somewhat reduced and the transition from high to low reflectance becomes somewhat less sharp than the ideal curves shown in Figure 11.7. However, random deviations of up to 10% can be tolerated without serious effects.

It can be shown that for a nonperiodic quarter-wave stack, (11.25) still applies with the more general definition:

$$P = \prod_{j=0}^{J/2} \left(\frac{n_{2j+1}}{n_{2j}} \right)^2 \frac{n_o}{n_s}, \quad \text{when } J \text{ is even}$$

$$= \prod_{j=1}^{(J-1)/2} \left(\frac{n_{2j+1}}{n_{2j}} \right)^2 \frac{n_1}{n_o n_s}, \text{when } J \text{ is odd}, \tag{11.33}$$

where J is the total numbers of layers in the film. From (11.25) and

---

[2] In view of (11.29) and (11.30), the condition for absenteeism (11.32) can also be written

$$\frac{\lambda_b}{\lambda} = \frac{j}{p}, \qquad j = 1, 2, 3 \ldots,$$

accounting for the absence of all the even orders in the quarter-wave stack, as implied by (11.28).

(11.33) it is evident that P, and hence the peak reflectance, of a periodic quarter-wave stack can be significantly enhanced by adding a high-index layer to the top of the stack, so that the two extreme layers are of the high-index material.

The general possibilities of modifying the spectral characteristics of a multilayer stack are obviously unlimited. Below we list two illustrations:

1. Adding $\lambda_b/8$ thick layers at the extremes of a periodic quarter-wave stack has been used to reduce the secondary maxima at the short-wavelength side of the typical characteristic of Figure 11.7a and b.

2. The region of high reflectance may be widened by using two stacks in tandem. If these are constructed so that their regions of high reflectance are immediately adjacent to each other, the width of the high-reflectance band can be doubled.

11.3.3 Dichroic Beam Splitters: "Hot" and "Cold" Mirrors

It is frequently desirable to split the flux passing an optical system into long- and short-wavelength portions, without absorbing either portion. This can be accomplished by dielectric multilayer mirrors that operate as long- or short-wavelength pass filters, with both the transmitted and reflected portions of the flux continuing along their separate paths after reflection has taken place. Such mirrors are called dichroic.

11.3.3.1 Applications. Below we list some applications of dichroic mirrors.

1. In the projection of photographic slides and motion picture films, it is important to illuminate the transparency to be projected with the maximum luminous flux possible and to avoid illumination by flux outside the visible spectrum, primarily the ir flux, which only heats the transparency, possibly dangerously, without contributing to the visibility of the projected image. Often, in such systems, spherical (or ellipsoidal) mirrors are used to reflect an image of the source back onto itself and into the entrance pupil of the projection lens. It is then desirable to have the mirror reflect only the visible portion of the flux and to transmit the ir. This is preferable to having the mirror absorb the ir, being heated thereby, since the removal of heat from the immediate vicinity of the lamp is often a major problem in optical projection system design. Such a mirror is referred to as a "cold" mirror. For additional protection, an ir-reflecting, or "hot" mirror may be interposed between the light source and the transparency. Here, again, an ir-reflecting window is preferable to the more common ir-absorbing one for the reasons stated.

2. In colorimetry, it may be desired to split the flux into three spectral

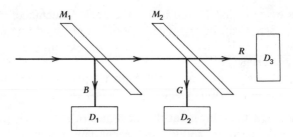

Figure 11.8 Schematic of dichroic mirror arrangement for splitting flux into trichromatic components.

bands to measure their relative flux content by means of three different detectors. This may be accomplished by means of two long- or short-pass filters, or a combination of these. A typical arrangement is shown in Figure 11.8. Here M_1 may be a mirror reflecting blue light and passing the flux at longer wavelengths, and M_2 may be a mirror transmitting red light and reflecting flux at shorter wavelengths. Then detectors D_1, D_2, and D_3 will receive flux corresponding to the blue, green, and red parts of the spectrum, respectively.

3. As an illustration of a more sophisticated task, we cite a radar display where it may be desired to view a phosphor display screen and, simultaneously, to photograph it. The phosphor should have a high radiance-time product in the part of the spectrum where the photographic film is sensitive (blue and uv) and should have a long persistence in the region of great visual efficacy (yellow-green). The flux from the phosphor screen may then be split by means of a dichroic mirror, which diverts most of the actinic flux to the photographic film without significantly reducing the luminance sensed by the observer.

11.3.3.2 Construction. The multilayer interference filters described in the preceding section lend themselves readily to the construction of dichroic mirrors. A broadband filter, possibly with its passband doubled by the technique described there, may be long- or short-pass over a considerable spectral range. The fact that it reverses its function and becomes transmitting again at very short (or long) wavelengths is irrelevant, if these wavelengths are outside the spectral range of the system as determined by the source and detector spectra and transmission spectra of the other components.

Since dichroic mirrors are often used at other than normal incidence (usually at 45° to the normal) they are frequently designed for that type of operation and this fact must be borne in mind when designing or specifying them. To apply the results of Section 11.3.2 to this situation, certain simple modifications must be introduced, as described presently.

11.3.4 Miscellaneous Characteristics

11.3.4.1 Angle of Incidence. When the wavefront traverses the dielectric multilayer obliquely, this affects both the pathlength difference introduced (11.8) and the reflectivity at the interfaces (2.101–2.105). The change in pathlength is readily accounted for by using an "equivalent" optical thickness

$$d_{oe} = d_j n_j \cos \theta_j \qquad (11.34)$$

instead of the actual optical thickness d_o as given by (11.23). The effect on the reflectivity, too, is often handled by introducing an "effective" refractive index. This index is the one that would result, at normal incidence, in the same reflectivity as the one actually observed at the given oblique incidence. The formula for the effective index is derived in Appendix 11.1 and is to be used in (11.24–11.26, 11.33); as shown in the appendix (11.76–11.77), it differs for the two directions of polarization, so that the filter response for these differs both in the width of the passband and in the values of the transmittance maxima and minima. When used with unpolarized light and in a system insensitive to the direction of polarization of the flux, the averages of the two transmittance and reflectance values can be used. Methods for minimizing these effects have been investigated [26].

When an interference filter is used at some distance from an image plane of a large-field object, or at some distance from the iris of an optical system operating at a large relative aperture, it receives rays over a large range of obliquity and, consequently, affects different parts of the ray bundle differently. Optimization of the filter design under such conditions has been analyzed for the task of detecting monochromatic radiation against a wide-spectrum background [27].

11.3.4.2 Aging. It has been found that the spectral characteristics of single and multilayer films tend to drift as the filter ages. Drifts of about 0.5% in the first year are typical. These drifts seem to be due to the slow crystallization of the amorphous material in the evaporated layers and is influenced primarily by conditions during manufacture, such as the temperature of the substrate, rate of evaporation, atmospheric pressure in the vaccum chamber, etc. [28].

11.3.4.3 Cascading. The transmittance of a series of absorption filters is well approximated by the product of the individual transmittance values; lumping reflection with the absorption losses does not introduce serious error there. In multilayer films, however, reflection is generally a major factor and, therefore, the particular sequence in which the component filters are arranged may influence the overall performance profoundly [29].

11.3.5 Special Thin-Film Systems

11.3.5.1 Induced Transmission Filters [22]. Over a certain thickness range, a thin layer of a highly conducting material will reflect a major part of the incident radiation, absorb a small part, and transmit the remainder. If such a layer is provided with a low reflection coating on both sides, the fraction of the reflected light can be drastically reduced—and the transmittance enhanced—over the spectral region in which the low-reflection coating is effective. If this region is narrow, the resulting device will act as a bandpass filter. Such devices have been made for use in the uv. One of these, having a bandwidth of 6.5 nm and peak transmittance of 55% at 0.265 μm, has been described [22].

11.3.5.2 Tunnel (FTIR) Layers [30]. In Section 2.2.2.2, it was shown that the total internal reflection may be frustrated by the presence, in the low-index region, of a higher index material in close proximity to the reflecting surface, and that the amount of transmitted light may be controlled by the spacing between the high-index media. Thus a very thin low-index layer sandwiched between higher index media can provide controlled reflectivity and replace the metal films, or dielectric thin-film stacks, in the Fabry-Perot-type interference filters of Section 11.3.1. Such layers are called *tunnel layers* in analogy to the "tunneling" of elementary particles through regions of high potential. See Figure 11.9 [30] for a diagram of a Fabry-Perot type interference filter using two tunnel layers (shown cross-hatched).

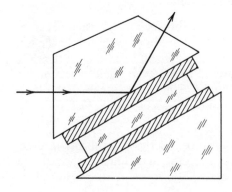

Figure 11.9 Tunnel layer (Fabry-Perot type) interference filter. Baumeister [30].

By itself, a tunnel layer can act as a long-pass filter, especially in the ir. Performance in this application can be enhanced considerably by cascading a series of tunnel layer.

Tunnel layers do not seem to be used widely [30]. This may be due to their high sensitivity to a change in the angle of incidence, which limits their useful angular field and/or the angular aperture with which they can be used. They are, in this respect, far more limited than the interference type multilayer films: differentiation of (11.34) shows that the rate of change of the effective optical thickness varies with the sine of the angle of incidence; this may be negligible at near-normal incidence, but is significant at the large angles at which the tunnel layers must be used. For the same reason, the undesirable polarization effects, discussed in the preceding section, are far more pronounced here.

11.4 MISCELLANEOUS FILTERS

11.4.1 Polarizing Filters

We have already noted (Section 8.3.5) that prisms of birefringent material may be used to produce polarized light. We have also noted (Section 8.1.1.1) that a stack of dielectric plates inclined to the wavefront at Brewster's angle can act as a polarizer.

More often, polarizer construction is based on another phenomenon. A series of long thin conductors, placed parallel to each other, acts as a polarizer, provided the length of the conductors is of the order of magnitude of the wavelength, at least, and their width (and spacing) is considerably less than that. When electromagnetic waves are incident on such a device, the component whose electric field oscillates parallel to the conductor length will set up currents in the conductor and be reflected or absorbed, whereas the transverse component will be affected only slightly.

Such a series of conductors, in the form of a wire grid, was used already by H. Hertz to determine the polarization of radio waves. Later such grids were used on microwaves, and even on radiation in the far ir. See Ref. 31 for a brief survey.

By using an ingenious technique (evaporating a thin metallic layer onto a blazed diffraction grating at grazing incidence), such conductor arrays have been obtained sufficiently fine to make them useful even in the near ir region. See Table 102 [31] for results obtained with aluminum and gold films evaporated onto a diffraction grating replica of 2160 lines/mm. Operation was effective from 0.7 μm to beyond 15 μm, but strong absorption bands, due to the substrate, blocked transmittance in the spectral bands 7.7–9.2 μm and 10–11 μm [31].

The thin, elongated regions of high conductivity need not be in the form of actual conductors. They may, instead, consist of strongly oriented molecules, whose associated electron clouds take the place of the conductors in absorbing radiation polarized parallel to their long axis. To be effective, the length of these "electron clouds" must be of the order of magnitude of the wavelength of the light and this is about a thousand times as large as the usual lattice spacing. Thus only needle-shaped crystals and long polymer chains have been considered for this purpose. (Such materials, whose absorption spectra vary with the direction of polarization of the traversing light, are called *dichroic*.

The popular Polaroid polarizing sheets are based on this principle. The so-called J-type sheet contains submicroscopic needles of herapathite (quinine sulfate periodide) crystals, about 1 μm thick. These are suspended in a matrix and coated onto a substrate by extrusion through a thin aperture, ensuring their orientation roughly parallel to each other.

The other types consist essentially of polyvinyl alcohol sheets with the alignment of the polymer chains accomplished by stretching. The most commonly used, the so-called H-type sheet, is obtained by causing iodine atoms to be absorbed in the stretched sheet. These atoms attach themselves at the appropriate sites along each polymer chain and thus form the elongated regions of electric conductivity required for the absorption of light [32]. By proper control of the iodine staining process, the polarizer can be made effective over a spectral region extending from 0.3 to 0.7 μm [33]. The K-type polarizer is made by dehydrating the stretched polyvinyl alcohol sheet; the resulting double-bonds, too, combine to act as elongated "conductors." These latter-type polarizers exhibit better stability at temperatures up to 90°C [32].

See Figure 11.10 [32] for the transmittance characteristic of typical Polaroid sheet polarizers. Figures 11.10a and b show the internal spectral transmittance for the parallel and transverse components, respectively. Overall transmittance values for commercial materials are shown in Table 103. Note that better extinction is invariably obtained at the cost of reduced peak transmittance.

Characteristics of polarizing sheets for use in the ir and uv are shown in Figures 11.11a and b, respectively [34]. Here and in Table 103, $\tau_{1,2}$ represent the transmittance values with the polarizers parallel and crossed, respectively. In general, with an angle θ between polarizing axes, the transmittance at a given wavelength, λ, is

cf. (2.90). $\tau(\lambda, \theta) = [\tau_1(\lambda) - \tau_2(\lambda)] \cos^2 \theta + \tau_2(\lambda),$ (11.35)

Polaroid polarizing sheets are also made sandwiched with a quarter-wave ($\lambda/4$) retardation sheet whose fast axis is oriented at 45° to the transmitting axis of the polarizer. The $\lambda/4$-film converts the output of the

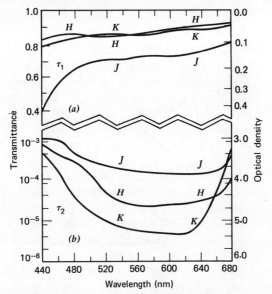

Figure 11.10 Transmittance spectra of Polaroid polarizing filters. (*a*) Parallel; (*b*) crossed. The right-hand ordinates give the equivalent density values. Land [32].

first layer into circularly polarized light. When such a combination is placed over a mirror, the linearly polarized light passes through the $\lambda/4$ sheet twice before returning to the polarizing sheet. Thus it arrives there polarized normal to the transmitting axis of the polarizer and is blocked (see Figure 11.12).

11.4.2 Birefringent Filters

In general, polarized light passing through birefringent material changes its state of polarization. The rate of change is proportional to the wave number of the light—inversely proportional to the wavelength—when dispersion effects are neglected. A number of filters based on this fact have been developed.

11.4.2.1 Lyot-Öhman (L—O) Filters. Basics [35–37]. In the L—O filter, a birefringent plate is placed between the two polarizers, whose polarization axes are parallel to each other and at 45° to the optic axis of the birefringent plate—all perpendicular to the normal to the plates (see Figure 11.13). If the birefringent material is a half-wave plate [i.e., $(n_o - n_e)d = \lambda/2$, see Section 2.2.3.1], or an odd multiple thereof, the linearly polarized light entering it will have its polarization direction turned through 90° and will be completely blocked by the second polarizer. This will happen for any wavelength (λ) for which the plate

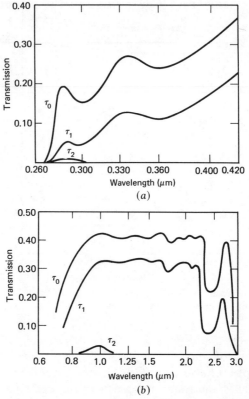

Figure 11.11 Polaroid polarizing sheets for the uv and ir: transmittance spectrum. (*a*) Type NHP'B (uv); (*b*) Type HR (ir). $\tau_{1,2}$ represent the transmittance spectra with the two polarizers parallel and crossed, respectively; τ_0 refers to a single filter. From Polaroid [34].

thickness (*d*) satisfies the condition

$$\lambda = \frac{2d\Delta n}{m}, \qquad m = 1, 3, 5, \ldots \tag{11.36}$$

where $\Delta n = |n_o - n_e|$ is the difference between the ordinary and extraordinary ray refractive indices. Indeed, it is readily shown that the transmittance of the sandwich will vary with the wavelength according to[3]

$$\tau = \tau_0 \cos^2 \frac{\pi d \Delta n}{\lambda}. \tag{11.37}$$

---

[3] On passing through successive layers of the birefringent plate, the linearly polarized light becomes elliptically polarized, with the eccentricity of the ellipse decreasing until, at a phase difference of $\pi/2$ (quarter-wave thickness) the light is circularly polarized. The relative length of the major axis of the ellipse is $|\cos{(\pi d\Delta n/\lambda)}|$ and this is the amplitude of the component in the direction of the original polarization. From this, (11.37) follows immediately.

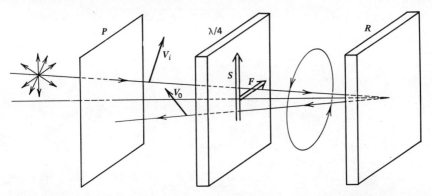

Figure 11.12 Circular polarizer blocking reflected light. A light ray, indicated by a solid arrowhead, passes linear polarizer (P), leaving it polarized as V_i. After passing through the quarter-wave retardation plate ($\lambda/4$), it is circularly polarized. It is then reflected from metallic reflector (R) and passes again through the quarter-wave retardation plate. It has then experienced a total retardation of π and is consequently again linearly polarized, but normal to the original direction, as shown by vector V_0. This is blocked by the polarizer, P. The combination ($P + \lambda/4$) is called circular polarizer. The fast and slow axes of the retardation plate are indicated by double arrows, marked F and S, respectively.

Neglecting dispersion effects,[4] the width, $\delta\lambda$, of the transmission band between half-intensity points can be found from the following considerations. Let m be the number of full-wave cycles represented by d at the peak wavelength λ_0:

$$\frac{d\Delta n}{\lambda_0} = m. \tag{11.38}$$

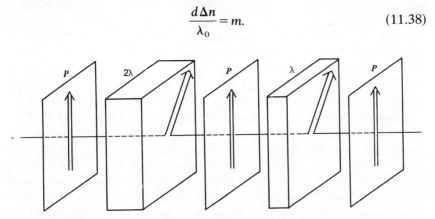

Figure 11.13 Orientation of axes in Lyot-Öhman filter. P are linear polarizers. The blocks are retardation plates, with retardation as indicated. The axes of all components are indicated by double arrows.

[4] In practice, the actual spectral spacings are about 10% less than this.

Then the increase in wavelength ($\delta\lambda/2$) required to reduce the transmittance by half is equivalent to a quarter cycle and given by

$$m - \frac{1}{4} = \frac{d \, \Delta n}{(\lambda_0 + \delta\lambda/2)}$$

$$\approx \frac{d \, \Delta n}{\lambda_0}\left(1 - \frac{\delta\lambda}{2\lambda_0}\right), \qquad \delta\lambda \ll 2\lambda_0. \tag{11.39}$$

In conjunction with (11.38), this shows that

$$\delta\lambda \approx \frac{\lambda_0^2}{2d \, \Delta n}. \tag{11.40}$$

Thus any desired passband width can be obtained by the appropriate choice of thickness

$$d = \frac{\lambda_0^2}{2 \, \delta\lambda \, \Delta n}. \tag{11.40a}$$

There remains, however, the problem of the additional transmittance peaks appearing, spaced 2 $\delta\lambda$, 4 $\delta\lambda$, etc. Since this spacing is only twice as large as the width of the passband, such a plate, in itself, would not usually be useful as a filter—the spectrum of the transmitted light would simply be crossed by a series of dark fringes (channel spectrum). The basic contribution of Lyot [35] was to eliminate all but the desired transmittance peak over a full octave. He noted that if the above filter is cascaded with another of half the thickness, the nearest transmittance peak will be twice as far away and a transmittance zero will appear at the locations of the originally neighboring peaks and every second such peak beyond these; the number of transmittance peaks has been halved and their spacing doubled, whereas the peak width has been very slightly reduced. If this is now followed by another such filter whose plate is half as thick as that of its predecessor, and the process is continued until a simple half-wave plate is reached, the first spurious transmittance peak will appear at $\lambda_0/2$. See Figure 11.14, where a shows the transmittance spectrum of a 16λ plate, b that of an 8λ plate, etc., with e showing that of a full-wave plate. [Note that the abscissae in Figure 11.14 are (λ_0/λ), the normalized wave number.] Figure 11.14f shows the transmittance spectrum of the combination: the product of the transmittance curves a–e, illustrating the peak at $\lambda = \lambda_0$ with a passband as given by (11.40) and the next transmittance peak at $\lambda = \frac{1}{2}\lambda_0$ ($\lambda_0/\lambda = 2$). In general, a filter consisting of N birefringent plates of thicknesses

$$d_1, 2d_1, 4d_1, \ldots 2^{N-1} d_1$$

will have a transmittance of the form

$$\tau_T(\lambda) = \tau_0 \prod_{k=1}^{N} \cos^2 2^k \phi = \frac{\tau_0}{4^N} \frac{\sin^2 2^N \phi}{\sin^2 \phi} \qquad (11.41)$$

where τ_0 is the transmittance of the filter when ϕ equals π, the second equality follows from a trigonometric identity,[5] and

$$\phi = \frac{\pi \lambda_0}{\lambda} = \frac{\pi d_1 \Delta n}{\lambda} \qquad (11.42)$$

is the retardation (phase difference between the two polarization components) introduced by the thinnest plate. The reader may recognize the last member of (11.41) as the formula for the diffraction pattern of a diffraction grating having 2^N infinitesimal slits; see (2.171) and Figure 2.18b.

For instance, a 10-stage filter of this type will be equivalent to a diffraction grating having 1024 slits and will have a passband of less than $10^{-3} \lambda_0$. The spectral resolution is thus limited only by the available thickness of the birefringent material and manufacturing accuracy. At present, commercial filters are made primarily of quartz ($\Delta n \approx 0.0093$) or calcite ($\Delta n \approx 0.17$) [37]. See Table 77 for the refractive indices of these and Table 74 for those of ADP, KDP, and MgF_2 crystals. See also Ref. 38 for extensive data on refractive indices of all of these.

---

[5] In the second member of (11.41), write the typical factor:

$$\cos 2^k \phi = \tfrac{1}{2}(e^{i 2^k \phi} + e^{-i 2^k \phi}).$$

Performing the indicated multiplication, we find

$$\sqrt{\frac{\tau_T}{\tau_0}} = \frac{1}{N'} \sum_{k=-N'/2}^{(1/2)N'-1} e^{i(2k+1)\phi},$$

where we have written

$$N' = 2^N.$$

This sum of a finite geometrial series may be evaluated in a closed form:

$$\sqrt{\frac{\tau_T}{\tau_0}} = \frac{e^{-i(N'-1)\phi} - e^{i(N'+1)\phi}}{N'(1 - e^{i2\phi})}$$

$$= \frac{(e^{iN'\phi} - e^{-iN'\phi})}{N'(e^{i\phi} - e^{-i\phi})}$$

$$= \frac{\sin N'\phi}{N' \sin \phi},$$

from which the last member of (11.41) follows immediately.

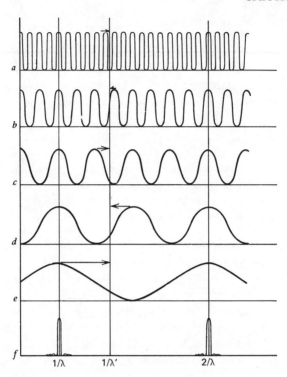

Figure 11.14 Transmittance spectra of successive plates in L-O filter (a–e) and its overall spectrum (f).

11.4.2.2 Effect of Angle of Incidence.

Clearly, the retardation varies with the angle of incidence of the wave. Specifically, the new retardation (ϕ) relative to that at normal incidence (ϕ_0) is [39]

$$\frac{\phi}{\phi_0} \approx 1 + c\theta^2(n_x \cos^2 \alpha - n_z \sin^2 \alpha), \qquad \theta \ll 1, \qquad (11.43)$$

where θ is the angle of incidence, with the surface normal assumed parallel to the y axis,

α is the azimuth angle of the plane of incidence relative to the x axis,

$n_{x,y,z}$ are the refractive indices along the axes of the index ellipsoid, in order of increasing magnitude, and

$$c = (n_x n_z - n_y^2)/2(n_z - n_x)n_x n_y^2 n_z. \qquad (11.44)$$

Taking, for instance, a uniaxial crystal with the optic axis coinciding with the x axis, ($n_y = n_z$) and in the plane of incidence ($\alpha = 0$), we have for the resultant retardation

$$\frac{\phi}{\phi_0} - 1 = \frac{\Delta\lambda}{\lambda_0} = \frac{\theta^2}{2n_z^2},\qquad(11.45)$$

where $\Delta\lambda$ here is the shift in the transmitted wavelength. Thus, to assure a shift of less than $10^{-4}\lambda_0$, the angle of incidence must be maintained below about one degree.

This rather severe restriction may be overcome by cutting each birefringent block into half, along a plane parallel to its surface, rotating the second half through 90°, and placing a half-wave retarding plate between them. This will cause the azimuth reference in the second half to be rotated through 90°, so that the change introduced by it will cancel that introduced by the first half—to the first-order approximation to which (11.43) is correct.

11.4.2.3 Temperature Effects. Temperature changes affect the Δn and the thickness of the birefringent plates. This results in a spectrum shift amounting to about -66 and -42 pm/°C for quartz and calcite, respectively [37]. Hence, for filter stability, the temperature must be controlled carefully. Alternatively, temperature control can be used for fine tuning of the filter over a very limited range.

11.4.2.4 Tunable Birefringent Filters [39]. In addition to the temperature control just mentioned, several other techniques are available for tuning an L—O filter. These permit tuning over very substantial spectral ranges.

If the retardation is changed by a fractional amount $\Delta\phi/\phi$, this will result in an equal relative spectral shift of the filter characteristic. For the thick component plates with large ϕ, this may call for a large $\Delta\phi$, which may not be practical. That requirement may, however, be eliminated by noting that the condition just mentioned is sufficient, but not necessary for producing the desired spectral shift. In the thickest plate, the shift $\Delta\phi$ corresponds to a shift of $2^N\Delta\phi_0$ rad or $2^{N-1}\Delta\phi_0/\pi$ cycles of the transmittance pattern (11.36). By shifting, instead, the nearest transmittance maximum to the desired new location, a considerably smaller shift, $\Delta\phi'$, suffices. This can be done in each plate so that nowhere will the required retardation shift be more than π [40]. See Figure 11.14, where $1/\lambda'$ illustrates the new position and the shifts are indicated by horizontal arrows originating at a peak.

Methods for actually obtaining the shift include the following:

1. The construction of the birefringent plates in the form of a pair of compensating wedges, with the effective plate thickness controlled by the amount of overlap (see Figure 11.15).

Figure 11.15 Wedges for variable effective thickness.

2. The addition, to each birefringent plate, of a supplementary plate, whose retardation can be controlled by mechanical stress (elastooptic effect) or voltage (electrooptic effect) [40].

3. The addition of a $\lambda/4$ retardation plate to each birefringent plate with the axis of the former at 45° to the latter [41].[6] By rotating the retardation plate-polarizer combination in its plane, the transmittance maximum can be shifted over a full cycle.

This system requires that the $\lambda/4$ plate be achromatic over the desired operating range. This can be accomplished by using a combination of quartz and magnesium fluoride, who differ significantly in the variation of the dispersion of Δn [38].

11.4.2.5 Laser-Tuning Applications. Birefringent filters have been used, in conjunction with a diffraction grating, for tuning a dye laser and spectral widths of about 1 pm have been obtained with electrooptic tuning [42]. When a birefringent filter is used inside a laser cavity, the resulting multiple passage of the wave effectively sharpens the transmittance spectrum. In such applications it may suffice to insert a birefringent plate at Brewster's angle, so that the crystal surfaces act as polarizers and the external polarizing plates may be dispensed with. Tuning may then be accomplished by simply rotating the plate in its plane [43].

11.4.2.6 Solč[7] Filter [44]. The Solč filter consists of a pile of N birefringent plates, each identical to the thinnest plate in the equivalent L—O filter, with the whole pile placed between two polarizers. This filter is roughly equivalent to a L—O filter whose thickest plate equals the pile thickness. Two versions of the Solč filter are used. In the *fan arrangement*, the two polarizers are positioned with their axes parallel and the bire-

---

[6] The effect of the $\lambda/4$ plate can be understood as follows. Linearly polarized light entering the birefringent plate will, in general, leave it elliptically polarized with the major (or minor) axis of the ellipse parallel to the original direction of polarization. This elliptically polarized light may be decomposed into two components parallel to the major and minor axes of the ellipse, respectively, and with amplitudes (a, b) equal to these and a $\pi/2$ phase difference between them. The addition of a $\lambda/4$ plate with its optic axis parallel to the major (or minor) ellipse axis, will cancel this phase shift—or bring it to π—and convert the light into linearly polarized light at an angle $(\tan^{-1} a/b)$ to the optic axis of the $\lambda/4$ plate. By aligning the analyzer following the $\lambda/4$ plate accordingly, it can be made to exhibit a transmittance maximum at any desired value of a/b.

[7] I. Solč (1953).

fringent plates are arranged so that the optic axis of the jth plate is oriented at an angle $(j-\frac{1}{2})\alpha$ relative to the polarizer axis, where α is a small angle, optimally equal to $\pi/2N$. In the *folded arrangement*, the polarizer axes are crossed and the plates are arranged with alternating optic axis orientations, so that the optic axis of the jth plate is oriented at $(-1)^j\alpha/2$ relative to the axis of the first polarizer.

The advantages of the Solč over the L—O filter are higher overall transmittance (due to the elimination of all, but two, of the polarizers) and economy in terms of material required. Its disadvantages are poor suppression of secondary maxima in the immediate neighborhood of the primary maximum and considerably greater difficulty in aligning the stack.

11.4.3 Residual Ray Filters

The phenomenon of *residual rays* (reststrahlen), too, can be used for filtering, especially in the ir. In crystals with almost ionic bonding, incident radiation tends to set up vibrations between pairs of neighboring positive-negative valence atoms. These vibrations exhibit resonance effects and at the resonant frequency a large portion of the incident wave field is absorbed in a thin surface layer and converted into mechanical vibrations. These vibrations, in turn, generate radiation (at the same frequency) which appears in the form of enhanced reflectivity. The heightened reflectivity does extend over a considerable wavelength band; but this band can be effectively narrowed by using multiple reflections. A device built on this principle has been described [45].

Table 104 [46] lists residual ray spectral peaks for a number of crystals. Reflectance curves for some of these have been published [47].

11.4.4 Christiansen Filters

If particles in a liquid suspension have a refractive index whose rate of change with wavelength differs from that of the liquid, and these refractive indices are equal at one particular wavelength, the suspension will transmit the light at that wavelength unscattered. At other wavelengths, the light will be scattered at the particle-liquid interfaces and the transmittance will be lowered. Such a suspension can therefore be used as a filter. It is named after its originator, Christiansen[8] and its performance has been analyzed on the basis of scattering theory [48]. Because of the high temperature sensitivity of the refractive index of the liquid, such a filter can be tuned by controlling its temperature.

As a representative example we describe here a filter made of optical

---

[8] C. Christiansen (1884).

crown glass particles suspended in ethyl silicate. As the temperature is raised from 22 to 57°C, the wavelength of peak transmittance drops from 0.573 to 0.433 μm. The passband is approximately Gaussian in shape and grows with the peak wavelength [48]. Detailed results for such a filter made of a 75-mm-thick cell, are listed in Table 105 [49].

Christiansen filters are also suitable for use in the ir. There, alkali halides and other materials have regions of anomalous dispersion, where their refractive index value passes through unity. At that point, the matrix liquid can be dispensed with, and the powder alone acts as a Christiansen filter. The locations of the transmission peaks of various materials used in this manner are listed in Table 104 [46].

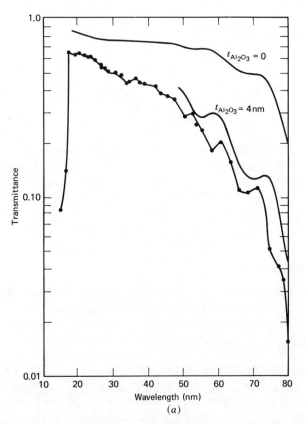

Figure 11.16 Transmittance spectra of thin metal films. (a) Calculated (8-nm-thick aluminum film, with and without Al_2O_3 protective coat) and observed (lowest curve). (b) Indium film, two thicknesses; dashed curves show calculated values. Hunter, Angel, and Tousey [50].

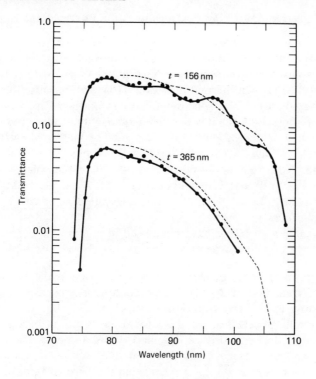

Figure 11.16 (*Continued*)

11.4.5   Thin Metal Films as Filters in the Ultraviolet

Thin metal films, opaque in the visible part of the spectrum, may have regions of high transparency in the far uv and may, therefore, be useful as filters there. Typical transmittance curves are shown in Figure 11.16*a* for an 80–nm thick aluminum film, with and without an aluminum oxide protective coat, and in Figure 11.16*b* for two different thicknesses of an indium film [50]. Note the typically rapid rise at the x-ray edge, followed by a slow drop, which becomes a sharp cutoff at the critical wavelength, corresponding to the electron eigen loss.

Since these filters are generally used below 0.1 μm, the substrate absorption becomes a major problem. Cellulose nitrate and related materials have been used as substrates, but with proper precautions [51] many metal films can be used totally unbacked.

11.4.6   Selective Absorbing Surfaces in Solar Energy Exploitation [52]

In designing the absorber for a solar energy collector, we are faced with seemingly contradictory requirements: the surface must be a good absorber to solar radiation and must have a low emissivity (ε) (i.e., poor absorptivity) to its own thermal radiation. However, because there is very little overlap between the spectra of the sun and a radiating body at a temperature below 400°C, it is possible to satisfy both these requirements simultaneously. We need only provide a surface that is essentially black ($\varepsilon \rightarrow 1$) below, say, 3 μm and essentially white ($\varepsilon \rightarrow 0$) above this wavelength. Note that only approximately 2% of the solar radiation is beyond 3 μm and that a blackbody at 400°K has only 0.2% of its radiation below 3 μm. Such surfaces are called selectively absorbing (*selective surfaces* for short).

A number of approaches have been used to prepare surfaces approaching the above emissivity spectrum.

1. Interference coatings may be designed to have the desired sharp reflectance transition at 2–4 μm. Such coatings are usually prepared by electroplating [53]. (The traditional evaporation techniques of Section 11.3 are totally impractical for the large areas required here.)

2. A high-reflectance surface (low absorption and emissivity) is coated with a layer that absorbs well over most of the solar spectrum, but becomes transparent in the infrared, so that the base surface, with its low emissivity, is active in that region. This approach may be combined with 1 above, to enhance its effectiveness [53]. Amorphous silicon on a highly reflective metal surface has been suggested for this purpose. Conversion efficiencies of close to 80% have been obtained with this [54].

3. The surface can be broken by small pits, which act as blackbodies in the spectral region in which they are large compared to the wavelength; in the long-wavelength region, however, they may be negligible compared to the wavelength, presenting effectively a "smooth" surface to the infrared radiation.

A deep saw-tooth structure will reflect a normally incident ray many times before reflecting it out again. This lowers the effective reflectance from the single surface reflectance ($\rho < 1$) to a value ρ^n, where n is the number of reflections. The absorbed fraction is then $(1 - \rho^n)$. At a shallow angle of incidence (or emittance), n will be small and therefore the effective reflectance relatively high. This corresponds to a major part of the thermally radiated energy [55].

A number of selective surfaces are listed in Table 106 [52]. Here α refers to the total absorbed fraction of solar radiation and ε to the total emissivity relative to the spectrum of a blackbody at about 400°K. In Table 106b [56] we present data on a number of absorber layers consisting of stacks.

It should be noted that similar considerations apply also to the surfaces of buildings, including the windows: for many purposes, conservation of energy is equivalent to its production.

The use of ir-reflecting mirrors, in conjunction with selectively absorbing surfaces, has been investigated [57]; but considerations of economics tend to make such arrangements impractical at this time.

APPENDIX 11.1 ANALYSIS OF MULTILAYER THIN FILMS [16, 17, 19]

1 Basic Formulation

Consider a plane wave, *in vacuo* wavelength λ_0, and of infinite extent, incident on a multilayer thin film. At the first surface, some of the energy will be reflected and some will be transmitted, as shown in Figure 11.17, where $\mathbf{k}$, $\mathbf{k}'$, and $\mathbf{k}''$ represent the vectors of the incident, refracted and reflected waves, respectively; the subscript indicates the surface number.

As shown in Figure 2.2, at any interface, the sum of the field components parallel to it (the *tangential* components) must be equal on the two sides of the interface. Consider now the axial point, A_j, of the jth interface and the electric field component $E_{\perp j}$ perpendicular to the plane of incidence. Since this component is tangential to the surface, the required equality yields, assuming no attenuation in the layer,

$$E_{\perp j} + E''_{\perp j} = E'_{\perp j} + E'''_{\perp j} = E_{\perp j+1}e^{-i\phi j} + E''_{\perp j+1}e^{i\phi j}. \qquad (11.46)$$

Here E'''_j is the electric field at A_j due to the wave reflected at the $(j+1)$th interface. Equation 11.46 is a slight generalization of (2.99) and the last equality follows from the identity of the wave pairs $(\mathbf{k}'_j, \mathbf{k}_{j+1})$ and $(\mathbf{k}'''_j, \mathbf{k}''_{j+1})$. $\mathbf{k}'''_j$ refers to the wave incident on surface j after reflection from surface

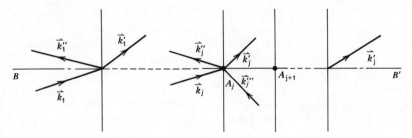

Figure 11.17 Ray paths in multilayer film.

$j+1$. The phase shift ϕ_j relates the wave at A_j to the corresponding one at A_{j+1} and is given by

$$\phi_j = k_j d_j \cos \theta_j = \frac{2\pi n_j d_j \cos \theta_j}{\lambda_0}, \qquad (11.47)$$

where $k_j = |\mathbf{k}_j|$

d_j is the thickness of layer following the jth surface,

n_j is its refractive index, and

θ_j is the angle between k_j' and the film normal, BB'.

The corresponding equations for the magnetic field vectors $(H_\parallel)$ are from (2.100)

$$n_{j-1} \cos \theta_{j-1}(E_{\perp j} - E_{\perp j}'') = n_j \cos \theta_j (E_{\perp j}' - E_{\perp j}''')$$
$$= n_j \cos \theta_j (E_{\perp j+1} e^{-i\phi j} - E_{\perp j+1} e^{i\phi j}), \quad (11.48)$$

appropriately generalized and in conjunction with the discussion following (2.106).

The reader will readily confirm that for the electric field components $(E_\parallel)$ parallel to the plane of incidence, the corresponding equations are

$$\cos \theta_{j-1} (E_{\parallel j} + E_{\parallel j}'') = \cos \theta_j (E_{\parallel j+1} e^{-i\phi j} + E_{\parallel j+1}'' e^{i\phi j}) \qquad (11.49)$$

and

$$n_{j-1}(E_{\parallel j} - E_{\parallel j}'') = n_j (E_{\parallel j+1} e^{-i\phi j} - E_{\parallel j+1}'' e^{i\phi j}). \qquad (11.50)$$

We can simplify the appearance of (11.47–11.50) somewhat by writing the tangential field component

$$E_t = E_\perp, \qquad E_\parallel \cos \theta_{j-1} \qquad (11.51)$$

for both the components perpendicular and parallel to the plane of incidence. In terms of E_t, then

$$E_{t_j} + E_{t_j}'' = E_{t_{j+1}} e^{-i\phi j} + E_{t_{j+1}}'' e^{i\phi j} \qquad (11.52)$$

and

$$E_{t_j} - E_{t_j}'' = c_j (E_{t_{j+1}} e^{-i\phi j} - E_{t_{j+1}} e^{i\phi j}), \qquad (11.53)$$

where

$$c_j = \frac{n_j \cos \theta_j}{n_{j-1} \cos \theta_{j-1}}, \qquad \frac{n_j \cos \theta_{j-1}}{n_{j-1} \cos \theta_j}, \qquad (11.54)$$

respectively for the perpendicular and parallel components.

Solving (11.52–11.53), we find

$$E_{t_j}, E_{t_j}'' = \tfrac{1}{2}[(1 \pm c_j)E_{t_{j+1}} e^{-i\phi j} + (1 \mp c_j)E_{t_{j+1}}'' e^{i\phi j}], \qquad (11.55)$$

where the upper and lower signs relate to the incident and the reflected waves, respectively.

To solve these equations for a given multilayer film consisting of a total of N layers (including substrate and surface medium), we compute the fields backwards, starting with

$$E_{tN} = A, \qquad E_N'' = 0, \tag{11.56}$$

since there is no reflection from the right-hand boundary of the last medium. On substituting these into (11.55), we obtain $E_{t_{N+1}}$, $E_{t_{N-1}}''$. From these, the preceding E_t's may be calculated, until E_{t_1}, E_{t_1}'' are found in terms of A. The film reflectance and transmittance, respectively, are then given by

$$\rho = \left| \frac{E_{t_1}''}{E_{t_1}} \right|^2 \tag{11.57}$$

and

$$\tau = \frac{\mathcal{S}'}{\mathcal{S}} = \left| \frac{A}{E_t} \right|^2 \frac{n'}{n}, \tag{11.58}$$

where n, n' are the refractive indices in the media preceding and following the multilayer film, respectively. Here we made use of (2.62) and (2.39), assuming μ/μ_0 equal to unity. Equation 11.58 is valid when the illumination is measured parallel to the wavefronts of the entering and exiting waves. When it is measured parallel to the film surfaces, its value is reduced by a factor $\cos \theta$, so that then

$$\tau_F = \left| \frac{A}{E_t} \right|^2 \frac{n' \cos \theta'}{n \cos \theta}, \tag{11.59}$$

where

$$\theta = \theta_1 \qquad \text{and} \qquad \theta' = \theta_N'. \tag{11.60}$$

Occasionally, especially when working with subgroups of multilayers, matrix notation provides a convenient shorthand formulation. In this notation (11.55) becomes

$$\left\| \begin{matrix} E_{t_j} \\ E_{t_j}'' \end{matrix} \right\| = M_j \left\| \begin{matrix} E_{t_{j+1}} \\ E_{t_{j+1}}'' \end{matrix} \right\| \tag{11.61}$$

where

$$M_j = \left\| \begin{matrix} (1+c_j)e^{-i\phi j} & (1-c_j)e^{i\phi j} \\ (1-c_j)e^{-i\phi j} & (1+c_j)e^{i\phi j} \end{matrix} \right\| \tag{11.62}$$

and

$$\left\|\begin{matrix} E_{t_1} \\ E_{t_1}'' \end{matrix}\right\| = \prod_1^{N-1} \mathbf{M}_j \left\|\begin{matrix} E_{t_N} \\ 0 \end{matrix}\right\|. \tag{11.63}$$

2 Total Tangential Field Formulation

Actual calculations may be simplified by rewriting (11.55) in terms of the total tangential fields at the interfaces:

$$E_{T_i} = E_{t_i} + E_{t_i}'' \tag{11.64}$$

$$H_{T_i} = H_{t_i} + H_{t_i}'' = c_i^*(E_{t_i} - E_{t_i}''), \tag{11.65}$$

where

$$c_j^* = \frac{n_{j-1}}{\cos \theta_{j-1}}, \quad \text{for the } E_\parallel \text{ component} \tag{11.66a}$$

and

$$= n_{j-1} \cos \theta_{j-1}, \quad \text{for the } E_\perp \text{ component.}[9] \tag{11.66b}$$

Hence, by dividing (11.65) by c_j^* and adding it to and subtracting it from (11.64), we find

$$E_{t_i}, E_{t_i}'' = \tfrac{1}{2}\left(E_{T_i} \pm \frac{H_{T_i}}{c_j^*}\right). \tag{11.68}$$

On substituting this into (11.55), we find

$$E_{T_i} = \frac{1}{2}\left[\left(E_{T_{i+1}} + \frac{H_{T_{i+1}}}{c_j^*}\right)e^{-i\phi j} + \left(E_{T_{i+1}} - \frac{H_{T_{i+1}}}{c_j^*}\right)e^{i\phi j}\right]$$

$$= E_{T_{i+1}} \cos \phi_j + i\frac{H_{T_{i+1}}}{c_j^*} \sin \phi_j \tag{11.69}$$

and

$$H_{T_i} = c_i^*(E_{t_i} - E_{t_i}'')$$

$$= \tfrac{1}{2}c_j^*[(1+c_j)E_{t_{i+1}}e^{-i\phi j} + (1-c_j)E_{t_{i+1}}''e^{i\phi j} - (1-c_j)E_{t_{i+1}}e^{-i\phi j}$$

$$\qquad\qquad\qquad\qquad\qquad\qquad - (1+c_j)E_{t_{i+1}}''e^{i\phi j}]$$

$$= \tfrac{1}{2}c_j^*c_j\left[\left(E_{T_{i+1}} + \frac{H_{T_{i+1}}}{c_{j+1}^*}e^{-i\phi j}\right) - \left(E_{T_{i+1}} - \frac{H_{T_{i+1}}}{c_{j+1}^*}e^{i\phi j}\right)\right]$$

$$= ic_{j+1}^*E_{T_{i+1}} \sin \phi_j + H_{T_{i+1}} \cos \phi_j, \tag{11.70}$$

[9] N.B. $c_j = c_{j+1}^*/c_j^*$.

$$\tag{11.67}$$

with the last step using the result (11.67).

The matrix equation corresponding to (11.69) and (11.70) is

$$\left\| \begin{matrix} E_{T_j} \\ H_{T_j} \end{matrix} \right\| = \left\| \begin{matrix} \cos \phi_j & (i/c_{j+1}^*) \sin \phi_j \\ ic_{j+1}^* \sin \phi_j & \cos \phi_j \end{matrix} \right\| \left\| \begin{matrix} E_{T_{j+1}} \\ H_{T_{j+1}} \end{matrix} \right\| \tag{11.71}$$

Here the 2×2 transformation matrix has a unity determinant and may be more convenient to work with than (11.62).

We now briefly consider two special cases.

3   Absentee Layers

When ϕ_j is an odd or even multiple of π, the 2×2 matrix of (11.71) takes the form

$$\left\| \begin{matrix} \pm 1 & 0 \\ 0 & \pm 1 \end{matrix} \right\|, \tag{11.72}$$

respectively. In the matrix product of the multilayer film, this has no effect, or only changes the sign of the existing wave. Hence, in terms of transmission and reflection, such a layer has no effect at all; it is called an *absentee layer*.

4   Absorbing Layers

So far we have discussed only nonabsorbing layers. In an absorbing medium, with absorption coefficient α and refractive index n_R, the field amplitude may be written

$$\frac{E(z)}{E(0)} = e^{-(1/2)\alpha z} d^{in_R k_0 z} = e^{i\hat{n} k_0 z} \tag{11.73}$$

where

$$\hat{n} = n_R + \frac{i\alpha}{2k_0} \tag{11.74}$$

is called the *complex refractive index*. Here the z axis is parallel to the (*in vacuo*) wave vector, $\mathbf{k}_0$, and the second member of (11.73) is based on (4.59). The use of $\hat{n}$ permits our results, (11.55) etc., to be applied to absorbing layers as well.

5   Oblique Incidence

The results given in Section 11.3.2 for periodic and quarterwave stacks are all based on normal incidence. However, by using an effective thickness and index, as explained in Section 11.3.4, they may be applied more generally. Here we derive the expression for the effective index.

The refractive index controls the reflectivity at the film interfaces according to Fresnel's equations, e.g., (2.101–2.105), the reflectivity being given by $|E''/E|^2$. At normal incidence, these equations take on the simple form

$$\frac{E''}{E} = \frac{n - n'}{n + n'}.$$

(11.75)

If we substitute into this, for n and n',

$$n_e = n \cos \theta, \qquad n_e' = n' \cos \theta'$$

(11.76)

or

$$n_e = \frac{n}{\cos \theta}, \qquad n_e' = \frac{n'}{\cos \theta'},$$

(11.77)

we obtain the general equations $(2.101a)$ and $(2.105a)$ for perpendicularly and parallel polarized light, respectively. Therefore, we may account for the effects of the obliquity by using the effective refractive indices (11.76–11.77).

REFERENCES

[1] R. Kingslake, ed., *Applied Optics and Optical Engineering*, Vol. 1, Academic Press, New York, 1965; (a) P. T. Scharf, "Filters," Chapter 3. (b) N. J. Kreidl and J. L. Rood, "Optical Materials," Chapter 5.

[2] ANSI Standard PH3.37, American National Standards Institute, Washington, D.C., 1969.

[3] For example I. M. Borish, *Clinical Refraction*, 2nd. Ed., Professional, Chicago, 1954, p. 461.

[4] "Corning color filter glasses," Corning Glass Works, Corning, New York (1970).

[5] R. G. Saxton, in *Handbook of Chemistry and Physics*, 58th Ed., Chemical Rubber Co., Cleveland (1977).

[6] "Schott color filter glass," Jenaer Glaswerk Schott & Gen., Mainz.

[7] W. A. Weyl, *Coloured Glasses*, Dawson's, London, 1959.

[8] N. J. Kreidl, "Optical properties;" Section 14 in *Handbook of Glass Manufacture*, Vol. 2, Ogden, New York, 1960, pp. 1–48.

[9] A. J. Werner, "Luminous transmittance, and chromaticity of colored filter glasses in CIE 1964 uniform color space," *Appl. Opt.* **7**, 849–855 (1968).

[10] M. A. Res, C. J. Kok, K. Kröger, F. Hengstberger, and M. Res, "Filter glasses for photometric application," *Appl. Opt.* **14**, 1017–1020 (1975).

[10*] "Kodak filters," Kodak Publ. No. B-3, Eastman Kodak Co., Rochester, 1976.

[11] J. A. Dobrowolski, G. E. Marsh, D. G. Charbonneau, J. Eng, and P. D. Josephy, "Colored filter glasses: an intercomparison of glasses made by different manufacturers," *Appl. Opt.* **16**, 1491–1512 (1977).

[12] J. T. Cox, G. Hass, and G. F. Jacobus, "Infrared filters of antireflected Si, Ge, InAs, and InSb," *J. Opt. Soc. Am.* **51**, 714–718 (1961).

[13] E. R. Blout, R. S. Corley, and P. L. Snow, "Infra-red transmitting filters, II. The region 1 to 6 μm." *J. Opt. Soc. Am.* **40**, 415–418 (1950).

[14] O. S. Heavens, *Optical Properties of Thin Solid Films*, Butterworth's, London, 1955; Dover, New York, 1965.

[15] A. Vasiček, *Optics of Thin Films*, North-Holland, Amsterdam, 1960.

[16] H. A. McLeod, *Thin-Film Optical Filters*, Hilger, London, 1969.

[17] P. Baumeister, "Applications of thin film coatings," MIL-HDBK-141, U.S. Dept. of Defense, 1962, Section 20.

[18] E. Delano and R. J. Pegis, "Methods of synthesis for dielectric multilayer films," in *Progress in Optics*, Vol. 7, E. Wolf, ed., North-Holland, Amsterdam, 1969, pp. 69–137.

[19] F. Abèles, "Optics of thin films," in *Advanced Optical Techniques*, A. C. S. Van Heel, ed., North-Holland, Amsterdam, 1967, pp. 143–188.

[20] K. Rabinovitch and A. Pagis, "Multilayer antireflection coatings: theoretical model and design parameters," *Appl. Opt.* **14**, 1326–1334 (1975).

[21] L. Holland, *Vacuum Deposition of Thin Films*, Chapman and Hall, London, 1970.

[22] E. Ritter, "Dielectric film materials for optical applications," in *Physics of Thin Films*, Vol. 8, G. Hass et al., eds., Academic, New York, 1975, pp. 1–49; also "Optical film materials and their applications," *Appl. Opt.* **15**, 2318–2327 (1976).

[23] P. W. Baumeister, V. R. Costich, and S. C. Pieper, "Bandpass filters for the ultraviolet," *Appl. Opt.* **4**, 911–914 (1965).

[24] H. D. Polster, "A symmetrical, all-dielectric interference filter," *J. Opt. Soc. Am.* **42**, 21–24 (1952).

[25] L. Young, "Multilayer interference filters with narrow stop bands," *Appl. Opt.* **6**, 297–315 (1967).

[26] K. Rabinovitch and A. Pagis, "Polarization effects in multilayer dielectric thin films," *Opt. Acta* **21**, 963–980 (1974).

[27] S. L. Linder, "Optimization of narrow optical spectral filters for nonparallel monochromatic radiation," *Appl. Opt.* **6**, 1201–1204 (1967).

[28] J. Meaburn, "The stability of interference filters," *Appl. Opt.* **5**, 1757–1759 (1966).

[29] P. Baumeister, R. Hahn, and D. Harrison, "The radiant transmittance of tandem arrays of filters," *Opt. Acta* **19**, 853–864 (1972).

[30] P. W. Baumeister, "Optical tunneling and its applications to optical filters," *Appl. Opt.* **6**, 897–905 (1967).

[31] G. R. Bird and M. Parrish, "The wire grid as a near-infrared polarizer," *J. Opt. Soc. Am.* **50**, 886–891 (1960).

[32] E. H. Land, "Some aspects of the development of sheet polarizers," *J. Opt. Soc. Am.* **41**, 957–963 (1951).

[33] A. S. Makas, "Film polarizer for visibile and ultraviolet radiation," *J. Opt. Soc. Am.* **52**, 43–44 (1962).

[34] "Polarized light," Polaroid Corp., Cambridge, Mass., 1970.

[35] B. Lyot, "Le filtre monochromatique polarisant et ses applications en physique solaire," *Ann. Astrophys.* **7**, 31–79 (1944).

[36] Y. Öhman, "A new monochromator," *Nature* **141**, 157–158 (1938).

[37] D. E. Gray, ed., *American Institute of Physics Handbook*, 3rd Ed., McGraw-Hill, New York, 1972, Section 6i, P. Baumeister and J. Evans, pp. 6–170 to 6–182.

[38] J. M. Beckers, "Achromatic linear retarders," *Appl. Opt.* **10**, 973–975 (1971); also **11**, 681–2 (1972).

[39] J. W. Evans, "The birefringent filter," *J. Opt. Soc. Am.* **39,** 229–242 (1949).

[40] B. H. Billings, "A tunable narrow-band optical filter," *J. Opt. Soc. Am.* **37,** 738–746 (1947).

[41] J. M. Beckers, L. Dickson, and R. S. Joyce, "Observing the sun with a fully tunable Lyot-Ohman filter," *Appl. Opt.* **14,** 2061–2066 (1975).

[42] H. Walther and J. L. Hall, "Tunable dye laser with narrow spectral output," *Appl. Phys. Lett.* **17,** 239–242 (1970).

[43] A. L. Bloom, "Modes of a laser resonator containing tilted birefringent plates," *J. Opt. Soc. Am.* **64,** 447–452 (1974).

[44] J. W. Evans, "Solč birefringent filter," *J. Opt. Soc. Am.* **48,** 142–145 (1958).

[45] J. Strong, *Procedures in Experimental Physics*, Prentice-Hall, New York, 1938, p. 383.

[46] R. B. Barnes and L. G. Bonner, "The Christiansen Filter Effect in the Infrared," *Phys. Rev.* **49,** 732–740 (1936).

[47] W. M. Sinton and W. C. Davis, "Far infrared reflectance of TlCl, TlBr, TlI, PbS, PbCl$_2$, ZnS, and CsBr," *J. Opt. Soc. Am.* **44,** 957–963 (1954).

[48] R. H. Clarke, "A Theory for the Christiansen filter," *Appl. Opt.* **7,** 861–868 (1968).

[49] H. S. Denmark and W. M. Cady, "Optimum grain size in the Christiansen filter," *J. Opt. Soc. Am.* **25,** 330–331 (1935).

[50] W. R. Hunter, D. W. Angel, and R. Tousey, "Thin films and their uses for the extreme ultraviolet," *Appl. Opt.* **4,** 891–898 (1965).

[51] W. R. Hunter, "The preparation and use of unbacked metal films as filters in the extreme ultraviolet," *Physics of Thin Films*, Vol. **7,** G. Hass, M. H. Francombe, and R. W. Hoffman, eds., Academic, New York, 1973, pp. 43–114.

[52] J. A. Duffie and W. A. Beckman, *Solar Energy Thermal Processes*, Wiley, New York, 1974, Section 5.6.

[53] H. Tabor, H. Weinberger, and J. Harris, "Surfaces of controlled spectral absorptance," Symposium on Thermal Radiation of Solids, NASA SP-55, 1965, pp. 525–530.

[54] D. E. Ackley and J. Tauc, "Silicon films as selective absorbers for solar energy conversion," *Appl. Opt.* **16,** 2806–2809 (1977).

[55] K. G. T. Hollands, "Directional selectivity, emittance and absorptance properties of vee corrugated specular surfaces," *Solar Energy* **7,** 108–116 (1963).

[56] D. E. Soule and D. W. Smith, "Infrared spectral emittance profiles of spectrally selective solar absorbing layers at elevated temperatures," *Appl. Opt.* **16,** 2818–2821 (1977).

[57] J. C. C. Fan and F. J. Bachner, "Transparent heat mirrors for solar energy applications," *Appl. Opt.* **15,** 1012–1017 (1976).

12

Optics of the Atmosphere

12.1 STRUCTURE OF THE ATMOSPHERE

Often optical observations are made through extended segments of the atmosphere; and its effects may then be significant. Before analyzing the effect of the atmosphere on optical signals, we must briefly review the structure of the atmosphere.

12.1.1 Air at Sea Level

12.1.1.1 General Description. The gaseous shell enveloping the earth, held to it by gravitational forces, is called the *atmosphere*. It is a mixture of many component gases, but nitrogen and oxygen together account for almost 99% of its mass. See Table 107 [1] for a listing of its permanent components, which are present in almost constant amounts, and of some variable components. In the absence of water vapor its molecular weight is

$$M_d = 28.97 \text{ g.}$$

12.1.1.2 Water Vapor. The major variable component of the atmosphere is water vapor. The atmospheric water content may be specified in terms of *mixing ratio*, which gives the ratio of water mass to the mass of the dry air in any given volume element of atmosphere. It may reach values up to 0.02 (20 g/kg) in the tropics. More often it is measured in terms of *absolute humidity*, the mass of water contained in a unit volume of air. This has an upper limit determined by the temperature; air containing the maximum amount of humidity is said to be *saturated*. The absolute humidity (H_s) of saturated air is tabulated in Table 108. The absolute humidity (H) of the air relative to that of air saturated with water vapor, is called *relative humidity*:

$$\mathcal{H} = \frac{H}{H_s}.$$

(12.1)

With decreasing temperature, H_s drops and hence, for a given body of air, $\mathcal{H}$ will rise, in general. At the temperature where the value of H_s has dropped to that of H, water will usually start to condense out of the air; this temperature is referred to as the *dew point*.

Relative humidity may also be specified in terms of the partial pressure (p_v) of the water vapor relative to the vapor pressure (p_w) of water at the ambient temperature:

$$\mathcal{H} = \frac{p_v}{p_w}.$$

(12.2)

Vapor pressure of water as a function of temperature is listed in Table 109 [3]. The absolute humidity may be calculated by the formula:

$$H = \frac{M_w}{V_0} \frac{p_v}{p_0} \frac{T_0}{T} = \frac{288.6 \, p_v^*}{T} \text{ g/m}^3$$

$$\boxed{= \frac{288.6 \, \mathcal{H} p_w^*}{T} \text{ g/m}^3}$$

(12.3)

where $M_w = 18$ is the molecular weight of water,

$V_0 = 0.0224136 \text{ m}^3$: the volume of a mole of gas at standard temperature and pressure,

p_v = the partial pressure of the water vapor in the atmosphere,

$p_0 = 1.01325 \times 10^5 \text{ N/m}^2 = 1.01325 \text{ bar} = 760 \text{ torr}$ (i.e., mm Hg) is the standard atmospheric pressure,

$T_0 = (0°C =) 273.16 \text{ K}$ is standard temperature,

T = the temperature of the atmosphere, and

p_j^* = the pressure, p_j, in mm Hg.

Under standard conditions and at 100% relative humidity

$$H = 4.839 \text{ g/m}^3.$$

(Note that $H \approx p_v^* \text{ g/m}^3$ at the usual ambient temperatures.)

For purposes of estimating the optical absorption effects of water vapor contained in the atmosphere, the humidity is often stated in *length of precipitable water*, which is the thickness of the layer of water that would result from the condensation of the atmospheric water vapor in a cylinder extending over the viewing range. If H^* is the absolute humidity in grams per cubic meter, a cylinder having a 1-cm² cross-sectional area and a

1-km length, will contain $0.1 H^*$ g of water. Assuming the water to condense to $1 \text{ cm}^3/\text{g}$, this implies a layer of thickness

$$H^* \text{ mm} = (\mathcal{H}H_s) \text{ mm}. \tag{12.4}$$

Thus the length of precipitable water in millimeters per kilometer equals the absolute humidity in (g/m^3).[1]

The *specific gravity* of humid air at atmospheric pressures, p^*, and absolute temperature, T, may be calculated from

$$
\begin{aligned}
\rho &= \frac{M_d(p - p_v) + M_w p_v}{V_0 p_0 T/T_0} \\
&= \frac{(T_0 M_d/V_0 p_0)[p - (1 - M_w/M_d)p_v]}{T} \\
&= \frac{0.46456 \times 10^{-3}(p^* - 0.3787 \, p_v^*)}{T}
\end{aligned}
\tag{12.5}
$$

where the volume V_0 is taken in cubic centimeters and the partial pressure, p_v^*, of the water vapor may be calculated from the relative humidity and p_w^* as given in Table 109:

$$p_v^* = \mathcal{H} p_w^*. \tag{12.2a}$$

For dry air at standard atmospheric conditions

$$\rho_0 = 1.2925 \times 10^{-3}$$

corresponding to

$$N = \frac{\rho_0 N_A}{M_d} = 2.688 \times 10^{25} \text{ molecules/m}^3 \tag{12.6}$$

where $N_A = 6.02217 \times 10^{23} \text{ mole}^{-1}$ is Avogadro's number. The molecular weight of humid air is then given by

$$M = 28.97\left(1 - \frac{0.3787 p_v}{p}\right) \text{ g}. \tag{12.7}$$

12.1.1.3 Distribution in Latitude.
The mean values of atmospheric pressure $(\bar{p})$, partial pressure of atmospheric water vapor $(\bar{p}_v)$, and the temperature $(\bar{T})$, and the temperature fluctuations (ΔT) vary with latitude. Table 111 [1] presents these values at sea level.

[1] Note that liquid water absorbs ir radiation far more strongly than does the same mass of water vapor. The absorption coefficient of liquid water is listed in Table 110, together with its refractive index and reflectance. For characteristics of seawater, see Ref. 42.

***12.1.1.4   Aerosols* [2b] [4a].** In addition to the molecular components of the atmosphere, there are various types of particles and droplets suspended in it. These *aerosols* absorb and scatter radiation. Typical aerosol particle counts are given in Table 112 for various locations and wind speeds.

The size distribution of aerosols over land and sea are presented in Figure 12.1 [4a], with the logarithmic derivative of the particle concentration

$$\frac{dN}{d(\log r)} = \frac{r\,dN}{dr} \tag{12.8}$$

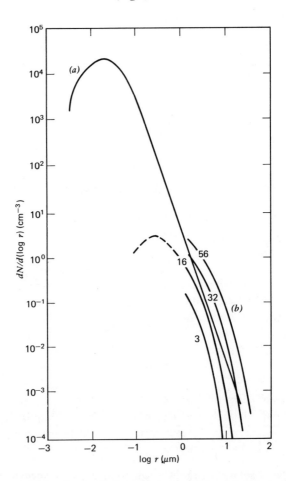

Figure 12.1   Aerosol size distribution in continental air (*a*) and maritime air (*b*). Parameter marked on (*b*) is wind speed in km/hr. After Manson [4a].

as ordinate. Here r is the particle radius, and the distribution over the oceans is given for various wind speeds (km/hr) as marked on the curves.

A major part of the atmospheric aerosols is in the form of water droplets, which constitute haze, fog, and clouds. These droplets form around small nuclei as they absorb water from the vapor content of the atmosphere. In haze, the droplets tend to be smaller than 1 μm in diameter, whereas in fog and clouds they tend to be larger than 3 μm in diameter. See Table 112 and the caption of Table 118 for representative droplet distributions.

12.1.2 Distribution in Altitude [1], [2a]

12.1.2.1 Temperature. In terms of the observed temperature gradient, the atmosphere is divided into four major shells. Up to an altitude of approximately 12 km, the temperature drops with increasing altitude, this lapse being approximately 6.5°C/km. This region is called the troposphere and its upper boundary, the *tropopause*. Its mean height, temperature, and pressure are listed in Table 111. In the *stratosphere*, extending from the tropopause to the *stratopause* at approximately 50 km, the temperature increases slowly. This is followed by the *mesophere*, the *mesopause*, and the *thermosphere*, in which the temperature gradient is negative, vanishes, and is positive, respectively. See Figure 12.2 [5a] for the course of atmospheric temperature.

12.1.2.2 Compositional Distribution. In the *homosphere*, extending to an altitude of about 100 km, the "permanent" constituents of the atmosphere are quite homogeneously distributed. In the *heterosphere*, above this altitude, the composition is governed by the differential diffusion rates. In the *exosphere*, above 1000 km, molecular collisions become negligible.

In the *ozonosphere*, coinciding roughly with the stratosphere, there is a relatively high concentration of ozone (O_3) (see Figure 12.3 [6]), which is significant in that it absorbs a major part of the uv radiation, protecting organisms near the earth's surface from the high levels of uv flux density found in the solar radiation above the troposphere.

The ozone results from the dissociation of molecular oxygen by the uv portion of the solar radiation: the subsequent combination of atomic and molecular oxygen forms O_3. Its concentration is high at an altitude of approximately 20 km and drops for both higher and lower altitudes: at higher altitudes because of the low concentration of oxygen there and at lower altitudes because of the low level of uv radiation, which penetrates the ozone layer but poorly. Its total atmosphere length is approximately

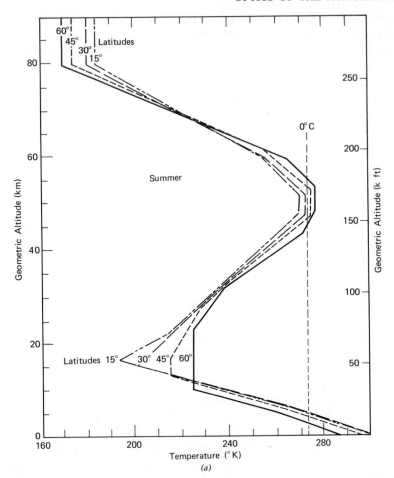

Figure 12.2 Temperature variation with altitude. (a) Summer; (b) winter; for tropics (15°), subtropics (30°), mid-latitudes (45°), subarctic (60°). Cole, Court, and Kantor [5a].

3 mm. (See Section 12.2.2.2 for the meaning of atmosphere length.) Constants for ozone absorption in the uv are tabulated in Tables 116 and 120. Ozone has another strong absorption band in the ir (9.5–10 μm).

The *ionosphere*, above 70 km, is important because of its high reflectivity to radio waves.

12.1.2.3 Density Distribution. At constant temperature and composition, the density of the atmosphere is proportional to the pressure and this, in turn, is given by the weight (w) per unit area of the air mass above

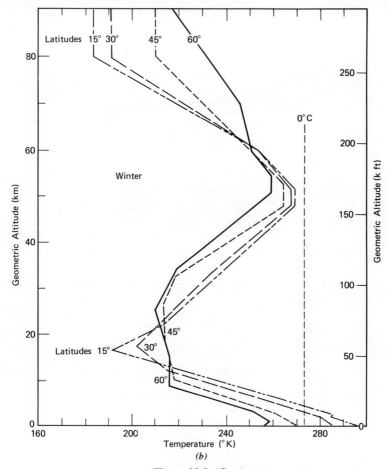

Figure 12.2 (Continued)

the point under consideration. The change in pressure (p) over a differential change in height (h) is given by

$$dp = dw = g \rho \, dh = k \, p \, dh \tag{12.9}$$

where

$$k = \frac{gM}{RT} \tag{12.10}$$

is a constant determined by the gravitational acceleration

$$g = 9.81 \text{ m/sec}^2,$$

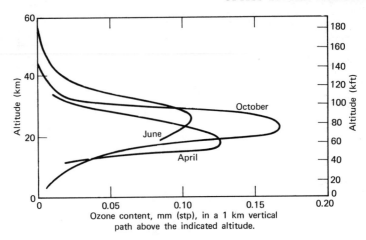

Figure 12.3 Vertical distribution of atmospheric ozone. Hudson [6].

the molecular weight (M) of the air, as given by (12.7), its temperature, T, and the gas constant

$$R = 8.3143 \ \text{J/K} \cdot \text{mole.}$$

Solving (12.9) we find the pressure at any height

$$p = p_0 e^{-kh} = p_0 e^{-h/h^*}, \tag{12.11}$$

where p_0 is the atmospheric pressure at sea level, and

$$h^* = \frac{RT}{gM}, \tag{12.12}$$

called the *scale height,* is the height difference over which the pressure drops by a factor of e. It also equals the *height of the (fictitious) homogeneous atmosphere,* that is, the height that the atmosphere would have if it were compressed to a state of constant pressure, p_0. Substituting their values for M, R, and g,, we find for the scale height

$$h^* = \frac{29.3 \ T}{1 - 0.378 p_v/p} \ \text{m} \tag{12.13}$$

$$\approx 8000 \ \text{m at } 0°C \text{ and low humidity.}$$

Note that the above derivation is based on the isothermal model, assuming a constant value of k and, hence, of T, an assumption which is, strictly speaking, certainly not justified. However, the results are sufficiently accurate to be useful for many purposes.

The actual atmosphere has variable scale heights, whose values are approximately 8.44 km and 10.4 km, respectively, for pressure and density [2a].

12.1.2.4 Humidity. The scale height for water vapor is considerably less than that for the permanent constituents of the atmosphere, approximately 2600 m. Therefore, the humidity in the atmosphere can usually be neglected above 10 km altitude.

12.1.2.5 Aerosols. Aerosol concentration drops roughly exponentially with altitude over the first 3–4 km:

$$N = N_0 e^{-\beta h} \qquad (12.14)$$

and then drops slowly (to approximately one half at 18 km) up to approximately 20 km. Above this altitude, the small-particle concentration again decreases exponentially, whereas the large-particle ($r >$ 0.1 μm) concentration increases somewhat.

Based on Ref. 7, representative values for the constants in (12.14) are as follows:

Visibility (km)	5	23
$N_0\,(m^{-3})$	13.78×10^9	2.828×10^9
$\beta\,(km^{-1})$	1	0.83

(The term "visibility" is defined in Section 12.4.3.)

12.1.2.6 Summary. See Tables 113 [1] and 114 [8a] for the altitude variations of pressure, temperature, densities, scale height, and mean free path. Table 114 is a standardized atmosphere adopted for use in the United States.

12.2 TRANSMISSION OF RADIATION THROUGH THE ATMOSPHERE

12.2.1 Atmospheric Refraction

12.2.1.1 Refractive Index of Air. The refractive index of dry air is approximately equal to 1.00028 in the visible spectrum. More precisely it is given by [9]

$$n = 1 + 10^{-6} c, \qquad (12.15)$$

where, at 15°C and 760 mm Hg pressure,

$$c_0 = 83.4213 + \frac{2.40603 \times 10^4}{130 - \lambda^{-2}} + \frac{159.97}{38.9 - \lambda^{-2}}$$

$$(= 277.82, \lambda = 0.55 \ \mu m) \tag{12.16}$$

and λ is the *in vacuo* wavelength in μm. At other temperatures and pressures,

$$c = \frac{1.388 \times 10^{-3} p^* c_0}{1 + 0.003671 T^*}$$

$$= \frac{1.388 \times 10^{-3} p^* c_0}{0.003671 T - 0.00277} \tag{12.17}$$

$$\approx \frac{0.38 p^* c_0}{T}$$

$$\left(\approx 105.6 \frac{p^*}{T}, \qquad \lambda = 0.55 \ \mu m \right) \tag{12.17a}$$

where T and T^* are, respectively, the air temperature values in K and °C, and p^* is the barometric pressure in mm Hg.

In the presence of water vapor at partial pressure, p_v^*, c is *reduced* by the amount

$$\left(0.05722 - \frac{0.000457}{\lambda^2} \right) p_v^*$$

$$(= 0.0557 p_v^*, \qquad \lambda = 0.55 \ \mu m). \tag{12.18}$$

For the highest accuracy in calculating the refractive index, fluctuations in the carbon dioxide content of the air must be allowed for (a mole of "standard air" contains 3×10^{-4} mole of CO_2). See Ref. 10 for details. The refractive index of liquid water is included in Table 110 [11].

It is interesting to note that the dispersion implicit in (12.16) implies a broadening of short light pulses transmitted through the atmosphere. For example, over a 9–km path, the limiting pulse widths are 1.6 psec and 0.44 psec at 0.84 μm and 10.6 μm, respectively, if a 30% broadening of the pulse is accepted [12].

12.2.1.2 Turbulence. Random local fluctuations in air temperature result in corresponding fluctuations of refractive index. On the assumption that these changes are fast, they may be treated as adiabatic. Then the connection between pressure and temperature is

$$p = p_1 T^{-\gamma/(1-\gamma)} \tag{12.19}$$

where

$$\gamma = c_p/c_v \approx 1.4 \qquad (12.20)$$

is the ratio of specific heats and p_1 is a constant.
   Hence

$$dp = \frac{-\gamma}{1-\gamma}\frac{p}{T}dT$$

and, based on this and the approximate form of (12.17a)

$$\frac{dc}{dT} = \frac{\partial c}{\partial T} + \frac{\partial c}{\partial p}\frac{dp}{dT}$$

$$= \frac{-0.379c_0}{T}\left(\frac{p^*}{T} - \frac{dp}{dT}\right)$$

$$= \frac{-0.379c_0 p^*}{(1-\gamma)T^2}. \qquad (12.21)$$

Hence, referring to (12.17),

$$\frac{\delta c}{c} = \frac{-1}{1-\gamma}\frac{\delta T}{T}, \qquad (12.22)$$

represents the index fluctuations due to temperature fluctuations in the turbulent atmosphere.

The effects of turbulence on imaging through the atmosphere are discussed in Section 12.4.7.

12.2.1.3   Atmospheric Refraction in Astronomic Observations. As a result of refraction by the atmosphere, the apparent elevation (ε') of an extraterrestrial object point is higher than its true elevation (ε). The difference (Δ) depends on the elevation itself, as well as on temperature, barometric pressure, and humidity. Atmospheric refraction angles (Δ_0) are tabulated in Table 115 for 10°C and 760 mm Hg pressure. For other conditions, the formula [1]

$$\Delta = \frac{\Delta_0 p^*}{760(0.962 + 0.0038T^*)} \qquad (12.23)$$

may be used.
   Again, a convenient rule of thumb:

$$\Delta \approx 0.28 \cot \varepsilon \text{ mrad}, \qquad \varepsilon > 5°. \qquad (12.23a)$$

12.2.1.4 *Refraction in Terrestrial Observations* [2c]. *Mirages and "Looming"* [13].

The horizon is located at the points where the lines of sight are tangent to the earth's surface. For a spherical terrestrial surface and from simple geometric considerations, its distance (r'_H) from the observer would be found by drawing, from the observer's eye, a line tangent to the earth's surface); see Figure 12.4:

$$r'_H = [(r_e + h)^2 - r_e^2]^{1/2} \approx \sqrt{2r_e h} = 3.57 \sqrt{h}\,\text{km} \qquad (12.24)$$

where $r_e = 6370$ km is the earth's radius

h = the observer's altitude above the earth's surface and the numerical result is valid for h in meters and $h \ll r_e$.

The horizon occurs then at an elevation angle

$$\varepsilon_H = -\tan^{-1}\frac{r'_H}{r_e} \approx -\frac{r'_H}{r_e}\,\text{rad}$$

$$\boxed{\approx -0.0321\sqrt{h}\,\text{degrees.}} \qquad (12.25)$$

Due to atmospheric refraction, r_H is somewhat larger, as illustrated by the dotted line in Figure 12.4. Under average atmospheric conditions, the observed horizon distance is

$$r_H \approx 3.84\sqrt{h}\,\text{km.} \qquad (12.26)$$

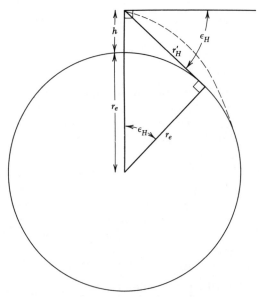

Figure 12.4 Deriving the location of the horizon. The line of sight with refraction is shown dotted.

The geometric visibility distance of a distant *elevated* object is simply the sum of its and the observer's horizon distances.

Under certain meteorologic conditions the variation of refractive index with altitude may exhibit abnormally large values in a given region. This will cause a pronounced bending of the rays passing through such a region, with rays bent toward the region of higher refractivity. For instance, on a sunny day a road surface may become very hot, especially if it has a dark color. It will then heat the layer of air next to it and the index gradient will be positive. This, in turn, will bend rays away from the surface of the earth and objects will appear located below ground level; see Figure 12.5a. The layer of air will act somewhat like a reflector, give the illusion of "water on the road," and cause the mirages and fata morgana reported by travelers especially in hot, desert regions. These, incidentally, can be observed in northern regions, also [14].

When the ground is appreciably colder than the atmosphere above it, the index gradient becomes negative, rays are bent toward the earth, and objects appear displaced upwards; see Figure 12.5b. This is called"looming" and, in connection with radio waves, "ducting."

In general, assuming only a horizontal stratification of the atmosphere,

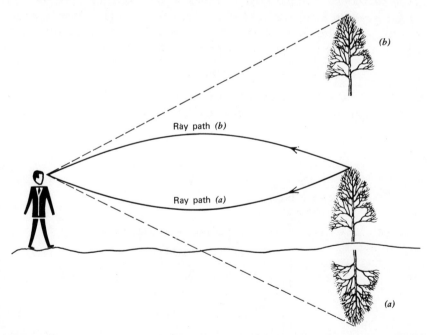

Figure 12.5 Mirage formation in atmosphere. (a) Positive index gradient; (b) Negative index gradient ("looming").

the curvature[2] produced by an index gradient, n', is given by

$$C = \frac{n'}{n} \cos \varepsilon, \qquad (12.30)$$

where ε is the angle the rays make with the horizontal. In air with near horizontal rays, this is approximately equal to n'. Alternatively, we note that, on the basis of Snell's law, the sine of the zenith angle at any level of refractive index, n, will be

$$\sin Z_1 = \frac{n_0}{n_1} \sin Z_0, \qquad (12.31)$$

where the ray originally had a zenith angle Z_0 in a layer of refractive index n_0. For mirages to occur, rays must be totally reflected in the atmosphere. This requires (12.31) to equal unity, that is,

$$\sin Z_0 = \frac{n_1}{n_0}. \qquad (12.32)$$

The maximum temperature differences actually observed over layers within a kilometer of the earth's surface, is approximately 30°C, or 10% of the absolute temperature [13]. This changes c in (12.15) by approximately 10% also and, hence, the refractive index by approximately 3×10^{-5}. Hence, the initial zenith angle of the ray must be no less than

$$Z \geqslant \sin^{-1} (1 - 3 \times 10^{-5}) \approx 89.5°, \qquad (12.33)$$

and the reflection will generally cause a deviation of 1°, at most.

Note, however, that on the basis of (12.31) and assumption (12.32), all

---

[2] Curvature is defined as

$$C = \frac{d\varepsilon}{ds} = \frac{1}{r_c}, \qquad (12.27)$$

where s = the ray segment length and
r_c = the radius of curvature.

Equation 12.27 is readily deduced from the general formula [15]:

$$n\mathbf{C} = \operatorname{grad} n - \mathbf{s}(\mathbf{s} \cdot \operatorname{grad} n), \qquad (12.28)$$

with

$$\operatorname{grad} n = n'\mathbf{y}^0, \qquad (12.29a)$$

$$\mathbf{s} = s \cos \varepsilon \mathbf{x}^0 + s \sin \varepsilon \mathbf{y}^0, \qquad (12.29b)$$

$$C = \sqrt{|\mathbf{C}|^2}, \qquad (12.29c)$$

and $\mathbf{x}^0$, $\mathbf{y}^0$ unit vectors in the x and y directions, respectively.

the rays would continue horizontally in an infinitesimally thin layer at $n = n_1$. This would create a severe nonuniformity of amplitude, which causes the geometrical approximation to break down, so that (12.30) must be used and total reflection occurs [16].

12.2.2 Extinction in General

12.2.2.1 Bouguer-Lambert and Beer Laws. As monochromatic radiation propagates through the atmosphere, its radiance is attenuated exponentially [Bouger-Lambert law (4.59)] due to absorption and scattering. The absorbed radiation flux is transformed into other forms of energy, usually into heat, whereas the scattered flux is redistributed angularly. Thus the apparent spectral radiance of an object drops with the distance, x, according to

$$L_\lambda(x) = L_\lambda(0) \exp\{-[\alpha_a(\lambda) + \alpha_s(\lambda)]x\}, \qquad (12.34)$$

where $\alpha_a(\lambda)$ and $\alpha_s(\lambda)$ are the absorption and scattering coefficients, respectively, at wavelength λ. These attenuation coefficients may be combined:

$$\alpha_T(\lambda) = \alpha_a(\lambda) + \alpha_s(\lambda); \qquad (12.35)$$

α_T is called the *extinction coefficient.*

Often the attenuation coefficients are proportional to the concentration, N, of the scattering and absorbing particles (Beer's law). Equation 12.30 then takes the form:

$$L_\lambda(x) = L_\lambda(0) \exp\{-N[\sigma_a(\lambda) + \sigma_s(\lambda)]x\}, \qquad (12.36)$$

where the σ's have the dimension of area and are referred to as absorbing and scattering cross sections, respectively. This terminology is based on the model according to which each particle acts as if it had cross-sectional area, σ, absorbing or scattering all flux striking it [see (4.61)].

A quantitative treatment of scattering in the atmosphere is presented in Section 12.2.3. The major absorption mechanisms in the atmosphere are molecular and act over bands, primarily in the ir region of the spectrum. These are discussed in Section 12.2.4.

12.2.2.2 Effective Pathlength. According to (12.36), the extinction exponent is proportional to the product of the concentration and the pathlength. This is conveniently expressed in terms of *atmosphere length*, which is analogous to the precipitable water thickness discussed in Section 12.1.1.2. The quantity (w) of a certain gas, in atmosphere length, is the thickness (l_e) of the gas layer resulting if all the gas in a cylinder extending over the pathlength, l were compressed to standard atmospheric pressure at $0°C$, while maintaining the cross-sectional area of the cylinder. It may be calculated from

$$l_e = l\left(\frac{T_0}{T}\right)\left(\frac{p_g}{p_0}\right) = 0.35942 p_g^* \frac{l}{T}, \tag{12.37}$$

where p_g^* is the partial pressure (in mm Hg) of the gas under consideration.

In the last column of Table 107, the reader will find the quantity of each of the permanent constituent gases in atmosphere-cm, when viewing through the total atmosphere. For horizontal viewing at sea level, the quantity of gas, per kilometer of viewing range, is obtained from this number by division by 8 km, the height of the homogeneous atmosphere.

For viewing through the whole atmosphere at an angle Z from the zenith, the atmosphere length must, to a first approximation, be divided by $\cos Z$.[3]

When treating radiation passing through the entire atmosphere, either from space to the earth's surface (as in astronomical observations) or from the earth's surface into space (as in satellite photography), it is convenient to refer to the number of "air masses" traversed by the radiation. Here one *air mass* is defined as the total atmosphere length of the air above the earth's surface at sea level. The number of air masses traversed by radiation from space reaching the earth's surface at various zenith angles is, again to a first approximation, proportional to sec Z. More accurate values are tabulated in Table 115. The extinction coefficient due to passage through the total atmosphere, as a function of wavelength, is tabulated in Table 116 [1].

12.2.3 Scattering [2d]

12.2.3.1 Basics of Atmospheric Scattering [17]. Scattering effects can be analyzed quite accurately on the basis of two formulations associated

---

[3] This approximation ignores the curvature of the earth and that of the rays in the atmosphere; the latter is due to refraction. At values of Z close to 90°, it breaks down, and more complex calculations are required. More accurately, the atmospheric pathlength is given by

$$l_A = \left(\frac{\Delta}{58.36''}\right) h^* \csc Z, \tag{12.38}$$

where Δ = the angle of atmospheric refraction and
$h^* \approx 8\ km$ is the scale height of the atmosphere.

Formula (12.38) is due to Laplace and is based on an isothermal atmosphere. More accurate values are included in Table 115.

When the optical path does not traverse the full atmosphere, the calculation of the atmosphere pathlength is facilitated by the use of Chapman's function (cf. Ref. 2c).

with the names of Rayleigh[4] and Mie,[5] respectively. Rayleigh scattering is applicable to particles small compared to the wavelength of the radiation (e.g., molecules); it varies according to

$$L(\theta) \sim 1 + \cos^2 \theta \qquad (12.39)$$

with the azimuth angle θ and inversely with the fourth power of the wavelength. This wavelength dependence accounts for the blue coloration of the sky (see Section 12.4.6). In the aerosol-free, dry atmosphere, the scattering coefficient due to Rayleigh scattering has the form

$$\alpha_s(\lambda) = 1.09 \times 10^{-3} \lambda^{-4.05} \text{ km}^{-1} \qquad (12.40)$$

with the term 0.05 in the exponent included to account for dispersion in the atmosphere; see (12.16).

When the dimensions of the scattering particles are comparable to the wavelength, the rigorous analysis of Mie scattering is required. This is mathematically complex and exhibits rapid fluctuations with wavelength. By way of illustration, the scattering efficiency of water droplets ($n = 1.33$) is shown in Figure 12.6 [17a] as a function of the Mie parameter

$$\mu = kr = \frac{C}{\lambda} \qquad (12.41)$$

where $k = 2\pi/\lambda$ is the magnitude of the wave propagation vector,
$\qquad r =$ the radius of the water droplet, and
$\qquad C = 2\pi r$ is its circumference.

The *extinction, absorption,* and *scattering efficiencies* (K_e, K_a, K_s) are defined, respectively, as the ratio of the extinction, absorption and

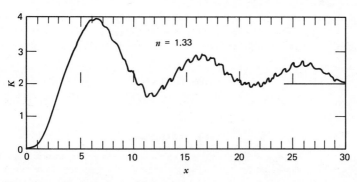

Figure 12.6 Mie scattering by nonabsorbing water droplets as function of the Mie parameter $x = kr$. Van de Hulst [17a].

[4] J. W. Strutt, Lord Rayleigh (1871).
[5] G. Mie (1908).

scattering cross sections to the geometric cross-sectional area. The rapid oscillations and slow convergence evident in Figure 12.6 illustrate the difficulty in making even approximate numerical calculations.

When the scattering particles are very much larger than the wavelength, the scattering cross section becomes independent of wavelength (clouds are white) and equal to twice the cross-sectional area of the particles. To explain this, we consider the action of a simple particle. It clearly reflects, refracts, and absorbs light with an effective area equal to its cross-sectional area. In addition to all this, it diffracts light. From the point of view of the diffraction process, the particle is an opaque disc in a transparent aperture. According to Babinet's[6] principle[7] this diffracts in a manner identical to a clear disc, of the same size and shape, inset in an opaque diaphragm, that is, according to (2.130). Thus an amount of flux equal to that reflected, refracted, and absorbed is also diffracted, making the effective extinction cross section equal to twice the geometrical cross-sectional area.

Plots of K_e and K_s as functions of water droplet radius, r, are shown in Figure 12.7 [18*] for three wavelengths.

Because the scattering efficiency, K, is almost constant for large Mie numbers, the scattering coefficient is then approximately proportional to the mean cross-sectional area $(\bar{A})$ and the number density (N):

$$\alpha = K \int A(r)N'(r)\, dr = \pi K \int r^2 N'(r)\, dr = \pi K \bar{r^2} N, \qquad (12.42)$$

where $N' = \partial N/\partial r$ is the number density per unit radius interval. However, when we analyze the scattering in terms of the liquid water concentration,

---

[6] A. Babinet (1837).

[7] Babinet's principle states that the diffraction effects due to an aperture transmittance distribution, $\tau(\mathbf{x})$, equal the diffraction effects of the complementary transmittance distribution:

$$\tau_c(\mathbf{x}) = 1 - \tau(\mathbf{x}).$$

This principle is readily demonstrated from the fact that the sum of the transmittance distributions,

$$\tau(\mathbf{x}) + \tau_c(\mathbf{x}) = 1$$

represents a clear aperture of infinite extent, whose diffraction effects vanish. Thus the sum of the diffraction amplitudes of $\tau(\mathbf{x})$ and $\tau_c(\mathbf{x})$ must vanish everywhere. For this to occur, their respective diffraction amplitudes must be equal and in phase opposition. Their intensities, proportional to the absolute-value-squared of their amplitudes, must therefore be identical.

Note that this equality does not include the undiffracted portion of the wave, which is a δ function for a finite opaque obstacle and zero for its complement.

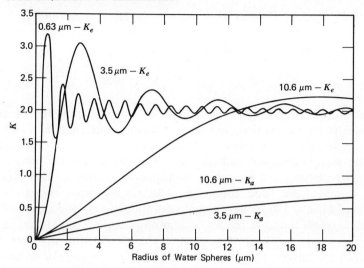

Figure 12.7 Mie extinction and absorption efficiencies of water droplets as a function of droplet radius, r, for the wavelengths indicated. At 0.63 μm the absorption efficiency is very small and has been omitted. Chu and Hogg [18*].

(w) (mass/unit volume), in the atmosphere, we find

$$w = \rho_w \int V(r)N'(r)\, dr = \tfrac{4}{3}\pi\rho_w \int r^3 N'(r)\, dr$$

$$= \frac{4\pi}{3}\,\rho_w \overline{r^3} N, \tag{12.43}$$

where $V = \tfrac{4}{3}\pi r^3$ is the volume of the droplet and $\rho_w = 10^6$ g/m^3 is the density of liquid water.

On substituting the value of N from (12.43) into (12.42),

$$\alpha = \frac{3Kw}{4\rho_w r^*}, \tag{12.44}$$

where $r^* = \overline{r^3}/\overline{r^2} \sim r$.

Thus for a given liquid water concentration, the scattering coefficient tends to vary *inversely* with the droplet radius. This explains why the visibility in rain is significantly higher than that in a fog with a much lower water concentration: in rain the drop radii are millimeters whereas in fogs they are micrometers in size.

See Table 117 [19] for representative absorption and reflective data on clouds.

Further details concerning Rayleigh and Mie scattering are presented in Appendix 12.1.

12.2.3.2 Typical Droplet Distributions in Haze, Fog, and Rain. The calculation of effective scattering cross section in the atmosphere requires, for each wavelength, the integration

$$\bar{\sigma}(\lambda) = \frac{1}{N} \int_0^\infty N'(r)\sigma(\lambda; r) \, dr \qquad (12.45)$$

of the number density and the scattering cross section (σ) of all the particles in the atmosphere. For various reasons, including the difficulty of treating the effects of multiple scattering and lack of information concerning the rapidly changing particle size distribution, such calculations appear to be impractical for actual atmospheric conditions. However, certain models have been found to yield results quite consistent with observed effects. Below we present such models for haze, fog, and rain.

The aerosols found in the fog-free atmosphere are called haze, even if the visibility is excellent. The particle size distribution in hazes differs from place to place—and from time to time in any place. For instance, the dense bluish haze often found in heavily industrial areas, is particularly rich in very small smoke particles. In general, however, the model distribution "C," as given in the caption of Table 118 [20], can be taken as representative of continental haze. The "M" distribution listed there is representative for haze in coastal regions.

The particle size distribution in fogs and clouds has been found to be representable as the product of an increasing power factor and a decreasing exponential of the radius,

$$N(r) = Ar^a \exp(-Br^b), \qquad (12.46)$$

where r is the radius of the droplet and A, B, a, and b, are constants.[8]

---

[8] In terms of N the total number of particles per unit volume, and r_p, the radius corresponding to the peak of the distribution, A and B are

$$A = \frac{a}{b} \frac{a+1}{b} \frac{b}{\Gamma\left(\dfrac{a+1}{b}\right) r_p^{a+1}} N \qquad (12.47)$$

$$B = \frac{a}{b r_p^b} \qquad (12.48)$$

and the total liquid water density in the atmosphere is:

$$w = \frac{4\pi}{3} \left(\frac{b}{a}\right)^{3/b} \frac{\Gamma\left(\dfrac{a+4}{b}\right)}{\Gamma\left(\dfrac{a+1}{b}\right)} r_p^3 N. \qquad (12.49)$$

Here Γ represents the gamma function.

Measurements indicate values of a between 1 and 6 and values of b between 0.5 and 1 [20].

The shape of the distribution curve of raindrop sizes tends to be determined primarily by the rate of rainfall. Representative distributions are shown in Figure 12.8 [21]. The radius of the median raindrop (in millimeters) is found to be given approximately by

$$\langle r \rangle = 0.62\, W^{0.182}, \qquad (12.50)$$

where W is the rate of rainfall in millimeters per hour [21].

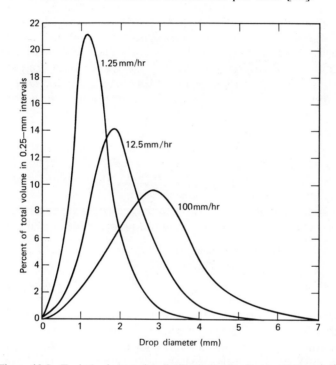

Figure 12.8 Typical raindrop size distribution. After Laws and Parsons [21].

12.2.3.3 Scattering in Haze, Fog, and Rain. The above distributions, together with a knowledge of the complex refractive index of water, permit the calculation of scattering and absorption in haze, fog, and rain. The total extinction coefficient (α_T) and the relative scattering coefficients (albedo),

$$\frac{\alpha_s}{\alpha_T} \qquad (12.51)$$

are tabulated in Table 118 for three hypothetical water aerosol distributions, including typical clouds and continental and coastal hazes. The assumed dependences of droplet concentration on droplet radius are given in the caption of that table.

The extinction and absorption coefficients for rain are shown in Figure 12.9 [19] as a function of the rate of rainfall. These are based on the above mentioned distribution models of Ref. 21.

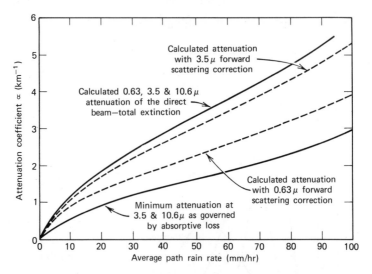

Figure 12.9 Extinction and absorption coefficients for rain. After Chu and Hogg [18*].

Incidentally, the scattering due to rain exhibits a surprising effect: the scattering is proportional to the areal drop number density, N_A (number of drops striking the ground per unit time per unit area), regardless of the size of these drops and the volume rate of rainfall. To explain this, we note that, for the large droplets usually constituting rain, (a) the scattering varies with the cross-sectional area of the droplet ($S = \pi r^2$) and according to Stokes' law, (b) the velocity of the falling droplets also varies with S. On the other hand, at a given volume rate of any flux, (c) the concentration varies inversely with the velocity [see (1.9)]. Hence the increased velocity of the larger drops cancels out exactly the effect of their larger size.

On substituting appropriate numerical values, this leads to an extinction coefficient [22a]

$$\alpha_T = 5.2 \times 10^{-8} \, N_A \, \text{m}^{-1} = 1.24 \times 10^{-8} \frac{z}{r^3} \, \text{m}^{-1}, \qquad (12.52)$$

where

$$N_A = \frac{z}{\frac{4}{3}\pi r^3} \tag{12.53}$$

is the number of drops striking the ground per square meter per second, when the rainfall is z m/sec. Transmittance values found under various rain conditions are tabulated in Table 119 [6].

Occasionally the aerosol content of the atmosphere is measured in terms of *turbidity*, which is here defined [23] as

$$B = \frac{\alpha_A}{\alpha_R}, \tag{12.54}$$

where $\alpha_{A,R}$ are the extinction coefficients due to aerosol and Rayleigh scattering, respectively.

12.2.3.4 Spectral Dependence of Scattering.

In the following, we give some generalizations concerning the spectra of atmospheric scattering coefficients, followed by certain illustrative examples that may be used as guidelines for estimating scattering effects.

For conditions of good visibility (meteorologic range[9] greater than 1 km, i.e., $\alpha_T < 3.9$ km^{-1}), radiation at shorter wavelengths tends to be scattered more, relative to that at longer wavelengths. For conditions of poor visibility $(\alpha_T > 3.9$ km$^{-1})$, the scattering is independent of wavelength. The wavelength independence of scattering at low visibility is due to the fact that this condition usually[10] is due to larger particles that scatter all wavelengths almost equally (see Figure 12.6).

The spectrum of the scattering coefficient, $\alpha_{sA}(\lambda)$ for aerosols at sea level on an "average clear day" is given in Table 120 [4b]. At altitude, h, other than zero, this coefficient must be multiplied by $[\exp-(h/1.2$ km$)]$ to account for the dilution of the aerosol [see (12.14)]. Also given there is the cross section (σ_R) for Rayleigh scattering; to obtain the corresponding contribution to α, this must be multiplied by the particle concentration, N, (see Table 113). For the total atmosphere effect, see Table 116.

Under usual haze conditions, with visibility range below 6 km, the following formula permits estimation of the scattering coefficient spectrum, $\alpha(\lambda)$[11]:

$$\alpha(\lambda) = \alpha_v \left(\frac{0.53}{\lambda}\right)^{\sqrt[3]{\alpha_v}}, \tag{12.55}$$

---

[9] "Meteorologic range" is a concept roughly equal to the visibility range. See Section 12.4.3 for its definition.

[10] The "industrial haze," mentioned above, is an exception; it scatters short wavelength flux more strongly.

[11] This is almost equivalent to (12.98). It is similar to the spectral dependence proposed in Ref. 24 for average haze conditions; there, however, the exponent is about 5% smaller.

where α_v is the "effective" scattering coefficient for the total visible radiation (in km^{-1}), which may be determined from the meteorologic range, r_M, by

$$\alpha_v = \frac{3.912}{r_M};$$ (12.56)

see Section 12.4.3. Clearly (12.55) implies that $\alpha(0.53\ \mu m)$ equals the overall value of α_v. On very clear days, scattering has been found to be of the form [22b]

$$\alpha_s(\lambda) = \frac{0.008}{\lambda^{2.09}}\ \text{km}^{-1}.$$ (12.57)

The wavelength values appearing in (12.55) and (12.57) are in micrometers.

12.2.3.5 Scattering in the Infrared. It is well known that ir radiation penetrates haze better than does visible light. Extrapolation of (12.55) and (12.56) would imply that this advantage increases with the visibility range. Indeed, under haze conditions, a meteorologic range of 0.85 km (optical density: 2/km), the visibility range was found to be 0.91 km at 0.55 μm and 20 km at 10 μm, an improvement by a factor of more 20. On the other hand, under fog conditions, a meteorologic range of 60 m (optical density: 29/km), the visibility range goes only from 68 to 125 m [25].

12.2.4 Absorption

Absorption of radiation in the atmosphere takes place at molecules having electric moments and at aerosols with complex (dissipative) refractive index. In general, the absorption per particle is increased by multiple scattering, which implies a multiplication of photon-particle collisions. Molecular absorption takes place primarily in bands in the ir due to molecules of water vapor, carbon dioxide, and ozone, and in the uv due to ozone.

The effects of absorption by aerosols were included in our discussion of scattering in the preceding section. The continuous absorption due to ozone can be calculated from the data given in Table 117 and in the last column of Table 120, which gives the absorption coefficient, A_0, per unit atmosphere length. When this is multiplied by the concentration of ozone in the path (in atmosphere cm/km), the ozone contribution to the absorption coefficient is obtained; see Figure 12.3 for ozone concentration data at various altitudes. For passage through the total atmosphere, the approximate atmosphere length of ozone is $3(\sec Z)$ mm, where Z is the angle the path makes with the vertical.

The total effect of molecular absorption for a horizontal path through the atmosphere is illustrated in Figure 12.10 [6], with the molecules responsible for the major absorption bands indicated below the graph. The regions of high transmittance, for example, 2–2.5, 3–4, 4.5–5, and 8–14 μm, are called atmospheric "windows."

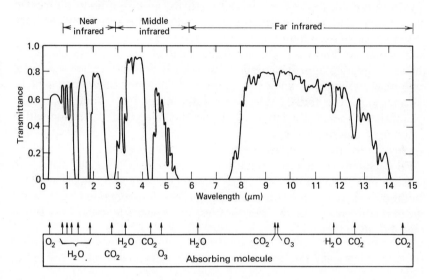

Figure 12.10 Representative atmospheric transmittance spectrum, horizontal path. Range: 1.83 km (6000 ft); humidity: 17 mm precip. water. After Gebbie et al., Proc. Roy. Soc. **A206**, 87 (1951). Molecules responsible for some of the absorption lines are indicated below the graph [6].

The calculation of absorption in the molecular absorption bands is very simple in principle. If the absorption coefficients, $\alpha(\lambda)$, of each constituent at all wavelengths are known, we simply substitute their values into an expression of the form (12.34) and integrate over all wavelengths. This method is not practical, however, because the absorption coefficients fluctuate very rapidly with wavelength. The usual absorption measurement, even if executed over a seemingly narrow spectral region, averages the transmittance measurement over some finite spectral band, which is large compared to the details of the absorption spectrum, so that the Bouguer–Lambert law does not apply (as mentioned in our discussion of the effect of thickness on the spectra of absorption filters in Section 11.2). If the strength and shape of each absorption line were known, its absorption exponent could be calculated for any atmosphere length, and the results summed. However, these lines are too numerous to make this "microscopic" approach practical at present.

Several approaches [26] have been developed to arrive at an approximation that is sufficiently accurate and yet avoids an unreasonable amount of computation. In the *line-by-line method*, the microscopic approach is used and computer time is saved by using sophisticated algorithms for minimizing the number of terms in the Legendre–Gauss quadrature used. In the band-model methods, groups of lines are treated as a whole and their mean effects calculated. Various forms of such groupings have been found to be appropriate for different spectral regions and a method known as the *aggregate method* has been developed to make reasonable approximations economically feasible. Other successful empirical methods are known as "Lowtran" and multiparameter analytical methods. The former relies on a large quantity of data empirically established, whereas the latter uses a complex analytical expression with a number of parameters whose values are chosen to make this expression applicable to the various situations and spectral regions.

To facilitate a coarse estimation of the transmission spectrum, we present here a far more compact, but far less accurate, "macroscopic" method [27]. It is based on empirical observations obtained over broad spectral absorption bands over various atmosphere lengths. It is found that the *absorption*

$$A = \int_{\bar{\nu}_2}^{\bar{\nu}_1} [(1 - e^{-\alpha(\bar{\nu})x})] \, d\bar{\nu} = 10^4 \int_{\lambda_1}^{\lambda_2} [1 - e^{-\alpha(\lambda)x}] \frac{d\lambda}{\lambda^2} \qquad (12.58)$$

can be approximated by one of two analytic functions of the atmospheric pressure (p) and atmosphere length (w), the choice of function depending on the value of A. Here $\bar{\nu}_{1,2}$ are the limits of the spectral range over which the band absorbs significantly. The spectrum is here represented in terms of the wave number

$$\bar{\nu}(\text{cm}^{-1}) = \frac{10^4}{\lambda(\mu\text{m})}. \qquad (12.59)$$

The empirical functions are

$$\begin{aligned} A &= c\sqrt{wp^q}, & A &< A_0 \\ &= C + D \log w + Q \log p, & A &> A_0. \end{aligned} \qquad (12.60)$$

Here A_0, c, q, C, D, and Q are constants differing for different absorbing substances and for the various bands of each absorbing substance. The values of these constants for the absorption bands of water vapor and carbon dioxide are given in Table 121.

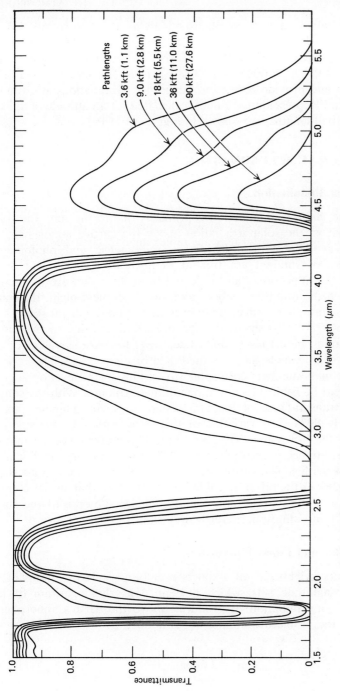

Figure 12.11 Representative atmospheric transmittance spectra, horizontal path at 4.6 km altitude; 2.5 mm precip. water per km. Hudson [6].

Extensive tables have been published giving the absorption due to water vapor and CO_2 for the spectral range $0.3(0.1)14 \, \mu m$ and for pathlengths ranging from 0.01 to 100 cm of precipitable water and pathlengths of 0.1 to 1000 km for CO_2 [6]. These are based on sea level observations and provide correction factors for observations at altitudes up to 30.5 km. Representative curves, calculated for an altitude of 4.6 km and various pathlengths, are shown in Figure 12.11 [6].

12.3 NATURAL ILLUMINATION

12.3.1 Solar Illumination

In daytime, natural illumination is primarily due to the sun. The solar constant, that is, the solar irradiation above the earth's atmosphere, is now taken at 1353 W/m^2 [28]. Other radiometric and photometric characteristics of sunlight are treated in Section 5.3. At night, light scattered in the atmosphere and reflected from the moon are the major contributors to natural illumination, and, on a moonless night, the stars, and various atmospheric effects may become the predominant sources of light. Figure 12.12 [29] shows the spectral irradiance due to the sun—above the atmosphere ($m = 0$) and at sea level for various values of air masses; this supplements the smoothed data of Figure 5.45. The various values of air mass numbers correspond to various solar elevation angles. They are listed in Table 115 in the penultimate column. With the sun at zenith, the transmittance of the atmosphere to solar luminous flux is approximately 80%. The variation of this value with solar elevation is tabulated in the last column of Table 115. The corresponding value of radiant transmittance is approximately 86%, for clean, dry air, but may drop to 60% with moist, dusty air. Solar illumination for various solar elevation angles (ε) is given in Table 126 [30]. Note that sunset corresponds to ($\varepsilon \approx -0.85°$) to account for the angular subtense of the sun's radius ($0.27°$) and atmospheric refraction ($0.58°$).

12.3.2 Stellar and Lunar Illumination

The apparent "brightness" of a star is usually measured in terms of its (apparent) magnitude (m), which is defined as the logarithm of the illumination (or irradiation) it produces above the earth's atmosphere. Specifically, the visible magnitude, m_v, is defined by

$$m_v = 2.5 \log_{10} \frac{E_{vo}}{E_{vs}}, \qquad (12.61)$$

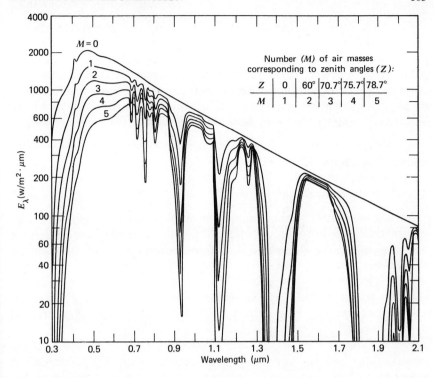

Figure 12.12 Solar spectral irradiance above the atmosphere and at sea level for various values of air masses (*M*). $p = 760$ mm Hg; 2 cm precip. water; 300 dust particles/cm³; 0.28 cm ozone. After Moon [29]. The data below 1.2 μm have been extrapolated into the region below 100 W/m² $\cdot$ μm by the present author.

where $E_{vo} = 2.54\ \mu$lux [1] and E_{vs} is the stellar illumination. The bolometric magnitude (m_b) is a measure of the total irradiation, E_{es}, appearing above the earth's atmosphere due to the star:

$$m_b = 2.5 \log_{10} \frac{E_{eo}}{E_{es}} \qquad (12.62)$$

where $E_{eo} = 24.8$ nW/m².

Other magnitudes have been defined to match specific detector spectral responses; see Refs. 1 and 31a, which list several such systems (U, G, R; IPg, IPv; UV, BG, R; Y, S, K; U, B, V; U_A, B_A, V_A).

The *color index* of a star is the difference between the magnitudes of the star as measured at two different wavelengths (or with two different detectors of different spectral response).

Some representative magnitude values are as follows:

Sun	$m_v = -26.74$
Moon (full)	$m_v = -12.73$
Venus (full)	$m_v = -4.22$
Sirius (brightest fixed star)	$m_v = -1.43$
Polaris	$m_v = +2.12$
Barely visible by unaided eye	$m_v = +6.$

Note also that a unity change in magnitude corresponds to an illumination change by a factor of

$$\frac{E_m}{E_{m+1}} = 10^{0.4} = 2.512. \tag{12.63}$$

There are two convenient rules of thumb:

1. The illumination due to a first magnitude star is

$$E_1 = \frac{2.54 \times 10^{-6}}{2.512} \approx 1 \ \mu\text{lux}$$

2. The ratio of the illumination value due to a first magnitude star to that of a sixth magnitude star is:

$$\frac{E_1}{E_6} = 100.$$

A more extensive listing of visual magnitude and color temperature of planets and stars is given in Table 122. Reference 5b contains a large compilation of data on stellar distribution and spectral characteristics. Spectral irradiation curves associated with the planets are shown in Figure 12.13a [32]. The total illumination from starlight is discussed in Section 12.3.3.

Illumination from moon light varies both with lunar elevation and phase. The intensity of the moon, as seen from the earth, varies with the phase (ϕ) according to the factor $k(\phi)$ as given in Table 123 [1]. The phase can be estimated from the formula:

$$\phi \approx \frac{180° t^{(d)}}{14.75}, \tag{12.64}$$

where $t^{(d)}$ is the time, in days, from the nearest full moon. Lunar illumination on a horizontal surface on earth is tabulated in Table 124.

It can also be estimated from the data of Figure 12.13b. This presents tabulated data of illumination due to the full moon at various elevation angles and also, in graph form, the attenuation factor due to the phase.

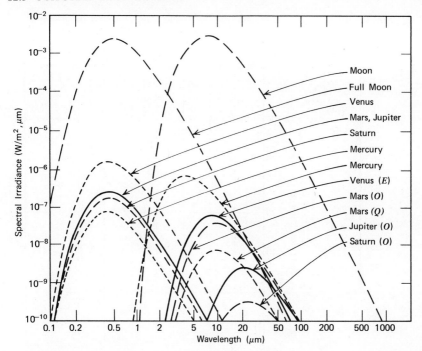

Figure 12.13a Calculated spectral irradiation above atmosphere due to moon and planets. The curves peaking near 0.5 μm are due to reflected solar radiation alone, when planet is at its brightest; the others are due to thermal emission. Parenthetical symbols: E-greatest elongation, O—opposition, Q—quadrature. Ramsey [32].

12.3.3 Sky Luminance

Sky luminance varies with solar elevation, with angular distance from the sun, and with meteorologic conditions. Representative values for the sky luminance near the horizon are listed in Table 125. The corresponding illumination values on the horizontal ground are also given there.[12] (Actual values may vary from those listed by as much as a factor of 10.) The spectral radiance of the daytime sky is shown in Figure 12.16a [33].

[12] Regarding the relationship between sky luminance (L_s) and the terrestrial illuminance (E) resulting from it, we note that with uniform sky luminance (e.g., under heavy overcast conditions) the illuminance on a horizontal surface is

$$E = \int_{\text{hemisph.}} L_s \cos Z \, d\Omega = \int_0^{\pi/2} L_s \cos Z (2\pi \sin Z) \, dZ = \pi L_s \qquad (12.65)$$

where Z is the zenith angle that is, the complement of the elevation.

Illumination
Due to Full Moon

ϵ	$\dfrac{L}{lx}$
0.1°	5.6×10^{-4}
10°	.0108
20°	.0426
30°	.100
40°	.162
50°	.227
60°	.286
70°	.328
80°	.355
90°	.372

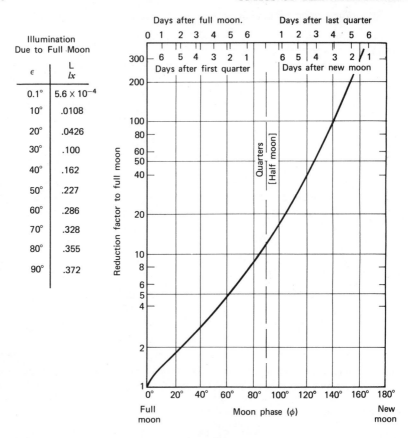

Figure 12.13b Table of illumination due to full moon on a surface normal to the incident flux and attenuation factor due to the moon's phase. The table is based on U.S. Govt. Publ. AD 402980 and the graph on R. O'B. Carpenter & R. M. Chapman (AD 417753) [37].

For the extinction spectrum of the atmosphere, see Table 116 [1].

Values of *twilight sky* luminance as a function of solar elevation angle are given in Table 127 [34] for various elevation and azimuth (relative to the suns meridian) angles. The last column of that table gives the illumination received by a horizontal surface at an altitude of 2.8 km. The course of sky luminance during twilight at sea level is plotted in Figure 12.14 [34] and zenith sky luminance, illumination on a horizontal surface, and its color temperature during twilight are listed in Table 128 [31b]. Figure 12.15 [35] shows the sky luminance at which stars of a given magnitude become visible under various conditions.

On a moonless *night*, the natural illumination consists primarily of the

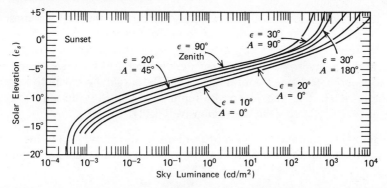

Figure 12.14 Sky luminance during twilight as observed at sea level. ε is elevation and A is azimuth angle relative to the sun's meridian. After Koomen *et al.* [34].

following:

1. Illumination due to visible stars is about 55 μlux on a horizontal surface.

2. Faint stars ($m_v > 6$) contribute approximately 70 μnit to the mean sky luminance, three times as much in the galactic plane (the "milky way") and less toward the galactic poles.

3. Zodiacal light is due to particles in the interplanetary space. It is strongest in the plane of the ecliptic, along the zodiac, and contributes about 122 μnit to the average sky luminance [36]—about half as much away from the zodiacal region [1].

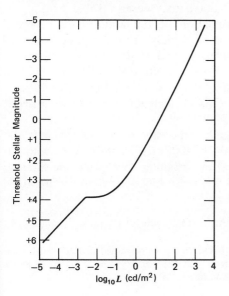

Figure 12.15 Threshold stellar magnitude as a function of sky luminance. Data for sea level; add 0.1 magnitude for each 300-m altitude. For viewing at zenith angle Z, the curve should be displaced upwards by $(0.127 \sec Z)$ magnitude units. The values are representative for visibility when the exact location of the star is known; if not, two units should be subtracted from the magnitude value. After Tousey and Koomen [35].

4. Airglow is due to the excitation of atoms in the upper atmosphere by ultraviolet solar radiation. It is concentrated primarily in a 2-km-thick layer approximately 95 km above the earth and, in the infrared, is due primarily to hydroxyl emission. In the visible part of the spectrum, there is a pronounced green continuum [37]. The airglow contributes approximately 40 μnit to the mean sky luminance [1], [36]. Its characteristics have recently been reviewed extensively [38].

Below we summarize the contributions of the above to the mean night sky luminance and the illumination on an unobstructed horizontal surface.

Factor	Luminance (μnit)	Illumination (μlux)	Fraction (%)
Visible stars	17.3	55	7
Faint stars	70	220	28
Zodiacal light	122	383	49
Airglow	40	125	16
Totals	~250	~785	100

Night sky luminance has been observed to go from 210 μnit at the zenith to about 280 μnit at a 15° elevation [1] and, on a clear moonless night the illumination on an unobstructed horizontal surface can be expected to be approximately 500–800 μlux. See also Table 128.

In the *infrared*, the thermal radiation of the atmosphere becomes important. Its spectral radiance rises rapidly from 0.02 at 3 μm to approximately 5 W/m^2 · μm at 6 μm. Beyond 15 μm it drops slowly. In the region between 7 and 14 μm, there is a dip (due to the atmospheric window there) which, at the zenith, amounts to about a factor of 10; it is about half as much at a 5° elevation and negligible toward the horizon; see Figure 12.16b [37].

When an object is viewed against the sky as a background, the "granularity" or "noise" of the sky radiance is an important factor in analyzing detection probability and in designing an appropriate filtering process. The Wiener spectrum of this noise has been found to approximate[13]

$$W(\nu) = a\nu^{-b} \text{(W/m}^2 \cdot \text{sr)}^2/\text{(cycle/rad)}^2, \tag{12.66}$$

[13] The daytime sky approximations are based on curves presented in Ref. 39 with the correction of a presumed typographical error in the slope value given for Figure 2a there. Night sky data from a private communication from R. C. Jones.

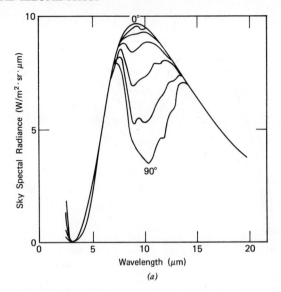

Figure 12.16 Sky spectral radiance for elevation angles: 0°, 1.8°, 3.6°, 7.2°, 14.5°, 30°, and 90°. (a) Summer, noon, sea level, light cirrus clouds, at 27°C; (b) Clear night, 3.35 km altitude, at 8°C. Bell *et al.* [33].

where the constants are, for various conditions, as follows

	daytime		nighttime	
clear	$a = 0.1$	$b = 3$	$a = 2 \times 10^{-6}$	$b = 3$
hazy	$a = 0.25$	$b = 2$	$a = 10^{-5}$	$b = 3$
cloudy	$a = 50$	$b = 2.7$	$a = 8 \times 10^{-3}$	$b = 3$

12.3.4 Coordinates of Sun and Stars [36]

Some technological applications, such as solar energy exploitation and guidance, require calculation of solar and stellar positions. We present here a brief and simplified introduction to spherical astronomy to enable the reader to do this to a good approximation.

12.3.4.1 The Celestial Sphere.
Figure 12.17a shows the celestial sphere with the earth at its center. The circles appearing as horizontal lines in the figure are the radial projections of the latitude circles on earth and mark the corresponding *declination* coordinate on the celestial sphere.[14] The

---

[14] More accurately, these circles mark off equal angles as subtended at the center of the celestial sphere, whereas the latitude circles are spaced at equal arc lengths as measured along the earth's surface, approximated by an oblate spheroid.

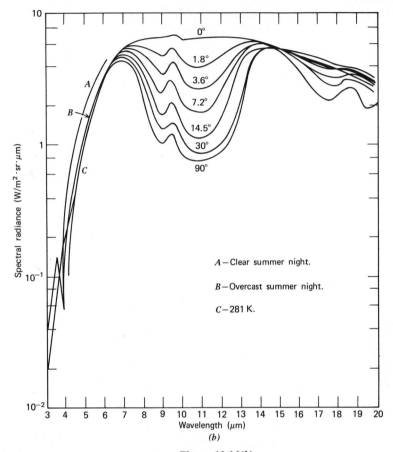

Figure 12.16(b)

central one, a great circle, is the *celestial equator* (Q) and the projections of the earth's poles are the *celestial poles* (P, P'). The lines crossing the declination circles and appearing elliptical in the figure, are analogous to the longtitude lines on earth, but differ from them in that they are fixed relative to the stars in the galaxy, rather than relative to the earth. They are called *hour circles* and indicate the right ascension, as discussed later. The local hour circle is called the *meridian*.

The straight line tangent to the earth, shown in the figure at 45° to the traces of the declination circles, represents the *horizon plane* of an observer at latitude 45°N indicated by a dot. Since the earth's size is infinitesimal relative to that of the celestial sphere, the horizon may be drawn through its center, as in the figure. To this observer his horizon plane appears horizontal and half the celestial sphere is redrawn in Figure

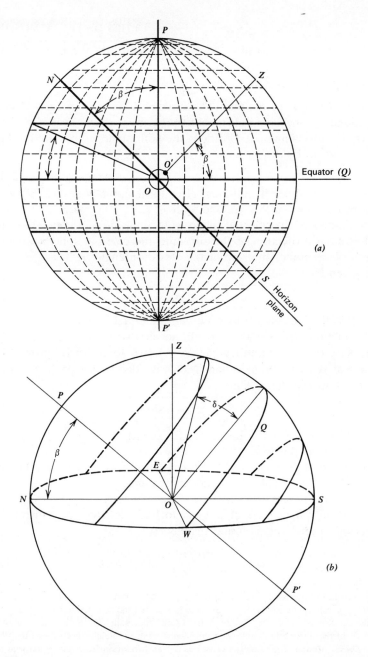

Figure 12.17 Celestial sphere. (*a*) Total view; (*b*, *c*) local view. *Q* and *P*, celestial equator and poles, respectively; *O*, observer; *C*, center of declination circle; *A*, sun; *Z*, zenith; *N*, *W*, *S*, *E*, the cardinal points; τ, the hour angle; β, latitude; ε, elevation, and ζ, azimuth of the sun.

111

12.17b as it appears to him. Also shown there are the cardinal points: north (N), west (W), south (S), and east (E), the celestial equator (Q), two declination circles at $\pm\delta$ and the celestial north pole (P).

12.3.4.2 *Position of the Sun.* The sun's path, on any one day, traverses approximately one complete declination circle, thus advancing $15°$ per hour or $1°$ every 4 min. Accordingly, the hour angle, τ, indicated in Figure 12.17c is given by

$$\tau = \frac{t}{4},$$

(12.67)

where t is the time, in minutes, after local noon. [See (12.72) for the relationship between this and clock time, t_c.]

If the radius of the celestial sphere is taken as unity, the height of Point A (the radial projection of the sun onto the celestial sphere) above the horizon plane equals the sine of the elevation angle, ε. This can be seen to be given by[15]

$$\boxed{\sin \varepsilon = \cos \tau \cos \beta \cos \delta + \sin \beta \sin \delta,}$$

(12.68)

where β is the latitude of the observer and equals the angle the planes of the declination circles make with the local vertical, see Figure 12.17b.

The sine of the azimuth angle, ζ, relative to north, is given by the distance of A from the meridian (NZS), divided by the projection of $\overline{OA}$ onto the horizon plane. Thus

$$\boxed{\sin \zeta = \frac{\cos \delta \sin \tau}{\cos \varepsilon},}$$

(12.69)

with west being positive.

The apparent declination (δ) of the sun associated with each day of the year is listed in Table 129 and shown graphically in Figure 12.18, together with the equation of time defined below.

Occasionally it may be of interest to find the time the sun reaches a certain elevation. This is found by solving (12.68):

$$\cos \tau = \frac{\sin \varepsilon}{\cos \beta \cos \delta} - \tan \beta \tan \delta.$$

(12.70)

[15] Referring to Figure 12.17c, this formula is readily derived on noting that the radius of the circle corresponding to declination δ is $\cos \delta$, and that the distance of that circle from the celestial equator is $\sin \delta$. The first term in (12.68) represents the height of A above the center, C, of the declination circle, and the second term the height of C above the horizon plane. These formulae are usually derived as a transformation from the equatorial to the horizon spherical coordinate system, using the spherical trigonometry formalism. For our purposes, however, the above suffices. Also note that the calculated value of ε must be corrected for atmospheric refraction (see Table 115) to make it agree with the observed value.

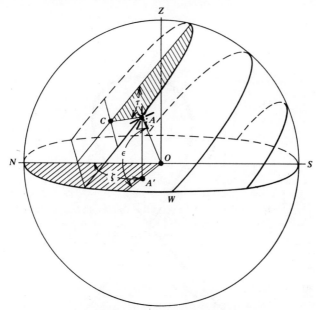

Figure 12.17 (c) (*Continued*)

12.3.4.3 Equation of Time and Standard Time. During the (tropical)[16]

year, the sun sets 365.2422 times, whereas the earth rotates about its axis 366.2422 times: one sunset per year is lost due to the earth's revolution around the sun. Starting with the sun at the meridian (noon): when the earth has completed one revolution, as observed by the return of a fixed star to the same meridian, it must turn another degree, approximately, for the sun to reach the meridian again. This is to compensate for the angle by which the earth has advanced in its orbit around the sun during the day. See Figure 12.19, where the earth is shown at A at noon for an observer at O. After completion of one rotation of $360°$, the earth is as shown at B, having advanced in its orbit by an angle, θ_1. A slight additional rotation, $\Delta\theta$, is missing from the new noon position. This additional rotation is given by

$$\Delta\theta = \theta_1 \sec \gamma \qquad (12.71)$$

where γ is the (mean) angle between the observer's velocity vector and the plane of the earth's orbit. Since γ varies with the season, so does $\Delta\theta$. θ_1 also varies seasonally, due to the variations in the earth's orbital velocity. Consequently the duration of an actual solar day generally differs from 24 hr; only the mean length of the solar day is 24 hr exactly.

---

[16] The tropical year is defined by the return of the sun to the same apparent declination.

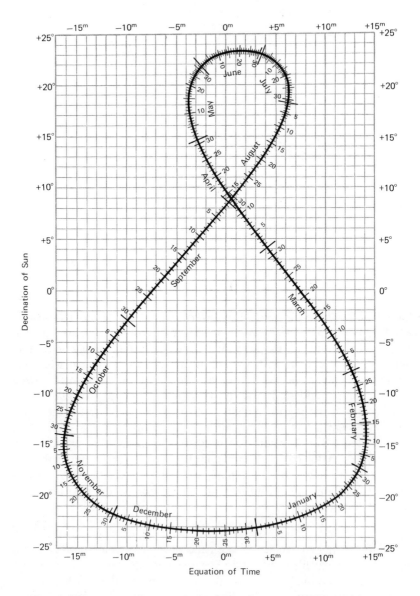

Figure 12.18 The equation of time (abscissa) and solar declination (ordinate).

114

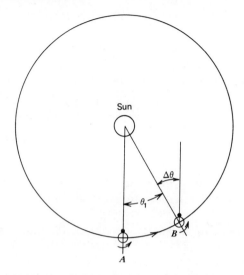

Figure 12.19 Illustrating the origin of the equation of time.

The slight differences in the length of a day accumulate periodically, so that the time (t_a) of the actual noon (sun at meridian) differs from the time (t_m) of local mean noon (12 o'clock on a correct time piece, set to minimize the difference between actual noon and 12 o'clock). The residual difference between actual and mean noon is called the *equation of time:*

$$\Delta t = t_a - t_m$$

and is tabulated in Table 130 and shown graphically in Figure 12.18, together with the daily values of the solar declination. The dates of the extrema and zeros of the equation of time are listed below (the value of the corresponding extremum in parentheses):

Extremum		Zero
11 Feb.	(-14.3^m)	16 April
14 May	$(+3.7^m)$	14 June
26 July	(-6.4^m)	2 Sept.
3 Nov.	$(+16.4^m)$	25 Dec.

If we would set our watches according to local mean time, we would have to reset them by 4 min every time we traveled 1° east or west, or about 3 sec every kilometer (at mid latitudes). This is obviously a rather inconvenient situation. Hence *standard time zones* have been instituted,

requiring us to reset our watches only after traveling approximately 15°
(ca. 1200 km) eastward or westward.

As a result of the two complications just discussed, the value of t_c from
our clocks must be corrected to obtain the value of t required for
calculating τ. Specifically, for t, t_c, and Δt in minutes:

$$\boxed{t = t_c + \Delta t + 4(L - L_0)} \qquad (12.72)$$

where L = the local longitude and
   L_0 = the longitude of the time zone, usually the nearest multiple of
   15°.

Here L and L_0 are in degrees and taken positive if they are east of
Greenwich and negative if they are west thereof.

12.3.4.4 Positions of Stars.
By definition, the sun always crosses the
meridian at noon. Because of the earth's orbit around the sun, the
"noon" of a star varies from day to day. Its occurrence can be calculated
from a knowledge of its right ascension (α) as follows.

The origin for the celestial system of hour circles (analogous to Greenwich in the terrestrial system of longitudes) is the *vernal equinox*, the
point where the sun crosses the celestial equator in Spring, on about 21
March. The *right ascension* of a star is the angular distance of its hour
circle, measured eastward,[17] from the vernal equinox or, equivalently, the
time (measured from noon) it crosses the meridian on the day of the
vernal equinox. See Table 122 for a listing of declination and right
ascension of the brightest stars.

To calculate the position of a star on any day, we may still use (12.68)
but replace t in (12.67) by

$$t_s = 60(\alpha_0 - \alpha_s) + t_c + 4(L - L_0) \qquad (12.73)$$

to obtain the time in minutes after actual local noon, for use in (12.68)
and (12.69). Here α_s is the right ascension of the star and α_0 is the right
ascension of the sun, both measured in hours; α_0 may be calculated
approximately from

$$\alpha_0 = (24/365.2422)(N - N_0) = 0.06571(N - 80), \qquad (12.74)$$

where N = the number of the day in the year [e.g., N (1 January) = 1,
   N(21 March) = 80] and
   $N_0 = N$ (vernal equinox) ≈ 80.[18]

---

[17] "Eastward" means the same sense as the earth's rotation and revolution that is,
counterclockwise, when viewed north-to-south, as shown, for example, in Figure 12.19.

[18] For dates 1 January to 21 March, use $N_0 = 80 - 365 = -285$.

Note that $(\alpha_0 - \alpha_s)$ represents the time (in hours after noon) when the star crosses the meridian.

12.3.5 Astronomical Data

In Table 131 we list some basic data of the solar system.

12.3.6 Reflectivity of Terrain and Water

Representative values of reflectivity of terrain features are listed in Table 132 [40]; reflectivity of water surfaces, as a function of the angle of incidence of the light, is listed in Table 133 [8]. See Table 110 for specular reflectivity of water and Table 134 [41] for its ir emissivity spectrum, normal to its surface.

Reflectance spectra of typical vegetation and soils are shown in Figure 12.20a and b [37]. The reflectance spectra of seawater are investigated in Ref. 42.

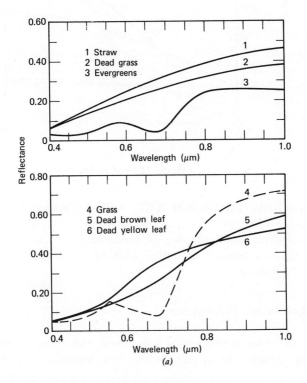

Figure 12.20 Typical reflectance spectra. (a) Vegetation; (b) Soil [37].

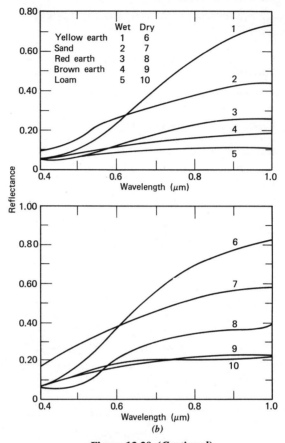

Figure 12.20 (*Continued*)

12.4 IMAGING THROUGH THE ATMOSPHERE[19]

The absorption and scattering of radiation, as it passes through the atmosphere, were treated in Section 12.2. Here we analyze their effects on image formation.

12.4.1 Luminance Propagation

As far as image formation is concerned, the effect of absorption is simply an attenuation of the image flux, which is exponential with

---

[19] Except for Section 12.4.7, the discussion here follows essentially that of Ref. 22.

distance at any one wavelength. The effects of scattering are, however, twofold:

1. It attenuates the direct (undeviated) flux, which alone contributes to the image formation.

2. It distributes part of the subtracted flux over the total image area, lowering the image modulation contrast.

The latter effect may be treated as a veiling process: the atmospheric layer redistributes part of the flux it receives from each object point into the directions corresponding to the total image area; thus it acts as if it had a luminance of its own, veiling the object luminance. Neglecting spectral variations of the absorption and scattering coefficients,[20] we have for the direct flux component an apparent luminance [see (12.30)]:

$$L_d(r) = L_0 e^{-(\alpha_a + \alpha_s)r} = L_0 e^{-\alpha_T r} \qquad (12.75)$$

where L_0 = the true object luminance,
$\quad r$ = the range, and
$\quad \alpha_{a,s,T}$ = are the absorption, scattering and extinction coefficients, respectively.

In the open atmosphere, the veiling effect is conveniently treated in terms of the *luminance density*, l_A, of the atmosphere, which is the luminance of a layer of atmosphere of unit thickness. This is determined by the illumination received by the atmosphere from all directions and by the *volume scattering function*, $\beta(\theta)$, which states what fraction of the incident flux is scattered through angle θ. To define $\beta(\theta)$, we note that for an illuminated volume element, δV, the effective intensity will be

$$\delta I(\theta) = E\beta(\theta)\,\delta V, \qquad (12.76)$$

where E = the illumination received (assumed incident from one direction only) and
$\quad \theta$ = the angle measured relative to the direction of the incident flux.

Expressing the volume element in terms of its cross-sectional area (δA), and its thickness (δd), we find

$$\beta(\theta) = \frac{\delta I(\theta)}{E\,\delta A\,\delta d} = \frac{\delta L_A(\theta)}{E\,\delta d}$$

$$= \frac{l_A(\theta)}{E}. \qquad (12.77)$$

[20] Ignoring the spectral dependence of the α's introduces inaccuracies. The difficulties caused by it, however, seem to be outweighed by the convenience afforded by the resulting simplification of the analysis implicit in (12.75) and subsequent formulae based on it [43]. The empirical equation (12.98) enables us to formulate a more accurate analysis.

Hence

$$l_A(\theta) = E\beta(\theta). \tag{12.78}$$

In the general situation treated here, the illumination is incident from all directions, requiring us to substitute the directional density of the illumination:

$$E_\Omega(\mathbf{u}) = \frac{\partial E}{\partial \Omega}\bigg|_{\mathbf{u}}$$

for E into (12.78) and to integrate over the sphere:

$$l_A(\mathbf{u}) = \int E_\Omega(u')\beta(\theta'')\,d\Omega \tag{12.79}$$

where θ'' is the angle between unit vectors $\mathbf{u}$ and $\mathbf{u}'$ corresponding, respectively, to the directions in which the luminance density and the illumination density are taken.

Now consider the passage of the flux through a differential lamina of the atmosphere. The resulting change in the luminance of the beam is

$$dL = [l_A - L(r)\alpha_T]\,dr, \tag{12.80}$$

where r is the range, the first term gives the flux added due to the veiling effect, and the second term the loss due to absorption and scattering.

To integrate this equation, we must establish the variation of α_T and l_A along the range. By assuming that α_T and l_A vary similarly, we obtain a rather simple result. This assumption is reasonable if the absorption effects are small compared to the scattering—a condition normally valid in the visible part of the spectrum. Describing this variation along the range by a dimensionless function $f(r)$, we have

$$\alpha_T(r) = \alpha f(r)$$
$$l_A(r) = l f(r) \tag{12.81}$$

and, on substituting this into (12.80) and integrating, we obtain

$$\int_{L_0}^{L_r} \frac{dL}{L(r)\alpha - l} = \int_0^r f(r')\,dr'$$

and, hence,

$$\log\frac{L_r\alpha - l}{L_0\alpha - l} = -\alpha\bar{r}$$

or

$$L_r\alpha - l = (L_0\alpha - l)e^{-\alpha\bar{r}}$$

and, finally,

$$L_r = \frac{l}{\alpha}(1 - e^{-\alpha \bar{r}}) + L_0 e^{-\alpha \bar{r}}, \tag{12.82}$$

where L_r is the apparent luminance as observed at range, r, and we have written $\bar{r}$ for the effective range given by the integral of $f(r)$. The first term of (12.82) is occasionally called the *path luminance*.

On letting r go to infinity in this equation, we note that the horizon luminance (L_h), corresponding to an infinitely distant object, is

$$L_h = \frac{l}{\alpha}. \tag{12.83}$$

(Note, however, that in general L_h is a function of the direction of viewing and may even be unobservable in that direction.) In terms of this luminance, we may now rewrite (12.82) as

$$L_r = (L_0 - L_h)e^{-\alpha \bar{r}} + L_h. \tag{12.84}$$

The apparent luminance increases and decreases, respectively, for $(L_0 < L_h)$ and $(L_0 > L_h)$; it always tends toward L_h, the so-called *equilibrium luminance*.

It has been shown that even for viewing through the total atmosphere, all the required α and l values can be obtained from relatively simple measurements made from the ground. The extinction coefficient is obtained from a measurement of the apparent luminance (or radiance) of the sun; when this is compared to its known value as observed above the atmosphere, the total specular (i.e., unscattered) transmittance of the atmosphere is obtained. The scattering coefficient—including its angular dependence—can be obtained from the luminance (or radiance) measurements near the horizon. This approach has been confirmed experimentally. [44].

12.4.2 Contrast Propagation

To find the effect of the atmosphere on the modulation contrast, we note that the luminance difference between an object and its background is given by

$$L_r - L_r' = (L_0 - L_0')e^{-\alpha \bar{r}}, \tag{12.85}$$

which we obtained by subtracting (12.84) for the background from the equivalent equation for the object. The prime denotes quantities relating to the background. Hence the expressions for the original and apparent contrasts are

$$C_0 = \frac{L_0 - L_0'}{L_0'},$$

$$C_r = \frac{L_r - L_r'}{L_r'}. \qquad (12.86)$$

On dividing C_r by C_0 from (12.86) and substituting for $(L_r - L_r')/(L_0 - L_0')$ from (12.85), we obtain a simple expression for the contrast reduction due to the atmosphere:

$$\frac{C_r}{C_0} = \frac{L_0'}{L_r'} e^{-\alpha \bar{r}}. \qquad (12.87)$$

Referring back to (12.75), we have

$$\frac{L_r'}{L_0'} = R(1 - e^{-\alpha \bar{r}}) + e^{-\alpha \bar{r}}$$

$$= R + (1 - R)e^{-\alpha \bar{r}}, \qquad (12.88)$$

where the *sky-background ratio* is

$$R = \frac{l}{\alpha L_0'} = L_h/L_0'. \qquad (12.89)$$

See Table 135 [22c] for some representative values of R. On substituting the reciprocal of (12.88) into (12.87) we find a more useful expression for the contrast reduction in the atmosphere:

$$\boxed{\frac{C_r}{C_0} = (1 - R + Re^{\alpha \bar{r}})^{-1}.} \qquad (12.90)$$

For viewing against the sky as a background ($R = 1$), this simplifies further to

$$\frac{C_r}{C_0} = e^{-\alpha \bar{r}}. \qquad (12.91)$$

12.4.3 Visual Range

The distance from which a large black object can be seen against the horizon is called the *visual range* (r_v) or *visibility* [not to be confused with the interference fringe visibility, defined in (2.187)]. Here "large" means sufficiently large so that only modulation contrast, and not the size, limits detection, that is, so that any increase in size would not enhance visual detectability.

Referring to (12.91), which applies here, and noting that, for a black object C_0 is unity, we find for the liminal contrast

$$C_L = e^{-\alpha r_v}. \tag{12.92}$$

We neglect here, and in the remainder of this section, the variation of $f(r)$ and hence the distinction between r_v and $\bar{r}_v$. The visual range based on C_L equal to 2% is called *meteorologic range*, r_M, and is, from (12.92):

$$r_M = -\frac{\log(0.02)}{\alpha} = \frac{3.912}{\alpha}, \tag{12.93}$$

or, since r_M is the observable,

$$\boxed{\alpha = \frac{3.912}{r_M}.} \tag{12.94}$$

This may be substituted into (12.84), (12.90), or (12.91) so that, for example, the contrast reduction is

$$C_r/C_0 = e^{-3.192r/r_M} = (0.02)^{r/r_M} \tag{12.95}$$

for viewing against the sky as a background.

The relationship of α to density ($\mathscr{D}$) per unit length is simply

$$\alpha = (\log 10)\mathscr{D} = 2.30\,\mathscr{D}' \tag{12.96}$$

and, hence, using (12.93)

$$r_M = \frac{1.702}{\mathscr{D}}. \tag{12.97}$$

See Table 136 [45] for an international scale of r_M.

The following formula has been suggested [40] for the evaluation of the spectral dependence of α:

$$\alpha(\lambda) = \frac{3.9}{r_M}\left(\frac{0.53}{\lambda}\right)^{0.62\sqrt[3]{r_M}} \text{km}^{-1} \tag{12.98}$$

where r_M is in kilometers and λ is in micrometers.

Visual threshold depends on object size and background luminance (see Section 15.3.3), so that we must consider these factors, in addition to object contrast and atmospheric conditions when estimating the distance from which a given object can be seen. Extensive nomograms are available to facilitate this estimation [22d].

On substituting (12.94) into (12.90), we obtain an expression for the atmospheric contrast attenuation against any background [22e]

$$\frac{C_r}{C_0} = (1 - R + 50^\rho R)^{-1} \qquad (12.99)$$

where $\rho = r/r_M$. This is illustrated in Figure 12.21 and tabulated in Table 137 as a function of ρ for various values of sky-background ratio, R.

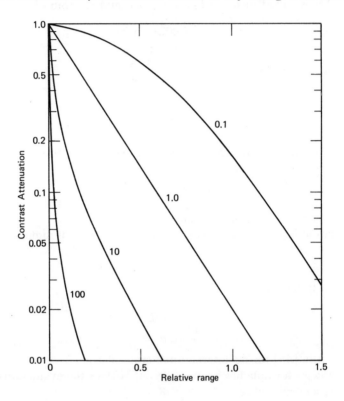

Figure 12.21 Contrast attenuation versus range for various values of sky-background luminance ratio. The range is relative to the meteorologic range r_M and the luminance ratio is indicated next to each curve. See Table 135 for typical values and Table 137 for more extensive data of contrast attenuation.

12.4.4 Small Light Sources

The illumination received through the atmosphere from a light source, intensity, I, at a distance, r is

$$E = \frac{Ie^{-\alpha_T r}}{r^2} = \frac{(0.02)^{r/r_M} I}{r^2}. \qquad (12.100)$$

The corresponding apparent luminance value is

$$L_r = (0.02)^{r/r_M} L_0.$$ (12.101)

In terms of the modulation contrast

$$C_0 = \frac{L_0}{L_b}, \qquad C_r = \frac{L_r}{L_b},$$

$$C_r = (0.02)^{r/r_M} C_0.$$ (12.102)

By substituting its threshold value, C_L, for C_r and solving for r, we can find the visibility range for any given light source. For the threshold contrast values as a function of ambient or background luminance, see Section 15.3.3.2.

12.4.5 Aided Vision

A telescope is capable of extending the range at which a small object can be detected; this is, essentially, its function. Indeed, if atmospheric effects—and the degradation due to the light scattering in the telescope— are negligible, the range is extended by a factor m^* exactly, where m^* is the angular magnification of the telescope.

However, when the objects are so large that their detectability is determined by the transmitted contrast, rather than by their size, a telescope can only *reduce* the range of detection by virtue of the contrast reduction introduced by it. In one survey of 18 different instruments, the observed contrast reduction factor ranged from 0.88 to 0.98 in daytime (0.75 to 0.96 at night); it was 0.96 and 0.92 for a standard 7×50 binocular for day and night use, respectively [46].

In summary, to predict the detection range for any object under given atmospheric conditions and with the use of a given instrument, the contrast rendition as calculated above must be reduced by the instrument contrast reduction factor and the object dimensions must be increased by m^*.

12.4.6 Color Effects in Atmospheric Imaging

We now consider the color changes that the atmosphere introduces into the appearance of objects. We treat successively a black, a grey (neutral), and a general object.

Black Object. For a black object $L_0 = 0$ and (12.84) takes the form

$$L_r = L_h (1 - e^{-\alpha r}).$$ (12.103)

We now recall that this equation is based on the assumption of an α

independent of wavelength and return to the monochromatic form of the equation in which alone it is valid. The spectral flux $(\Phi_{\lambda r})$ received at any given range (r) will be proportional to the apparent spectral luminance $(L_{\lambda r})$ there, so that we have

$$\Phi_{\lambda r} = kL_{\lambda h}[1 - e^{-\alpha(\lambda)r}], \tag{12.104}$$

where k is the constant of proportionality. On substituting this into (1.20), we find

$$X_r = k \int L_{\lambda h}[1 - e^{-\alpha(\lambda)r}]\bar{x}_\lambda \, d\lambda, \tag{12.105}$$

and so on, for X_r, Y_r, and Z_r. Equation 1.21 will then readily yield the chromaticity coordinates.

Neutral Object. To find a convenient expression for the apparent color attained by a neutral object, we express its luminance in terms of its contrast with the sky background (L_h):

$$C_0 = \frac{L_0 - L_h}{L_h}. \tag{12.106}$$

Substituting this into (12.84), we find

$$L_r = L_h(1 + C_0 e^{-\alpha r}), \tag{12.107}$$

and, following the argument applied in our derivation for a black object

$$X_r = k \int L_{\lambda h}[1 + C_0 e^{-\alpha(\lambda)r}]\bar{x}_\lambda \, d\lambda, \tag{12.108}$$

and so on.

General Object. Finally, we treat a general object having a spectral dependent reflectance, $\rho(\lambda)$. Hence its spectral luminance is

$$L_{0\lambda} = \frac{1}{\pi} E_\lambda \rho(\lambda), \tag{12.109}$$

where E_λ is the spectral illumination on the object. On substituting this into (12.84), we find for each spectral component

$$L_{\lambda r} = \left[\frac{1}{\pi} E_\lambda \rho(\lambda) - L_{\lambda h}\right] e^{-\alpha(\lambda)r} + L_{\lambda h} \tag{12.110}$$

and

$$X_r = k \int \left[\frac{1}{\pi} E_\lambda \rho(\lambda) - L_{\lambda h}\right][e^{-\alpha(\lambda)r} + L_{\lambda h}]\bar{x}_\lambda \, d\lambda, \tag{12.111}$$

and so on.

In Table 138 [22*f*] the reader will find tabulated values of the apparent luminance (relative to the horizon luminance), chromaticity coordinates, dominant wavelength, and excitation purity for neutral objects of various modulation contrast values and at various ranges, both for a theoretically aerosol-free atmosphere [Rayleigh scattering only see (12.40)] and for an actual very clear day, characterized by (12.57).

12.4.7 Atmospheric Turbulence

***12.4.7.1 Definitions* [47].** Changes of pressure and temperature in the atmosphere cause changes in its density and refractive index (see Section 12.2.1.2). These, in turn, if they are inhomogeneous, result in the distortion of the wavefront passing through the atmosphere from any object point. Such changes are always present in the natural atmosphere. They vary randomly and are called *turbulence*. In the visible part of the spectrum they are largely wavelength independent [48].

As a consequence of turbulence, an image formed through the atmosphere will, in general, exhibit scintillation, motion, and blurring effects. These are variously referred to as *shimmer, twinkling, atmospheric boil, dancing, wandering*, and *optical haze* [13], [22] and pose an ultimate limit on the image quality attainable through the atmosphere. In astronomical observation this effect is called *seeing* and represents the major limitation on resolution and geometrical accuracy. In optical communication over a light beam, it causes random displacements and phase distortion in the beam, so that communication over long distances requires a mechanical or optical conduit (see Chapter 13).

A thorough analysis of the problem through 1959 can be found in Ref. 49 and much of the work cited below is based on this. See Ref. 50 for more recent reviews.

Consider first imaging from an *incoherent source*, where the wave front is essentially unlimited in extent. Here the effects are conveniently divided into phase and amplitude disturbances.

The *phase disturbance*, in turn, may be divided into two components:

1. The rotation of the mean direction of the wavefront produces a shift in the image location.

2. The distortion of the wavefront around its mean produces blurring of the image.

The *amplitude disturbance* can be observed by viewing the illumination from a distant point source as a shadowgram of the intervening atmosphere. The illuminated region is generally crossed by dark bands and patches, indicating regions of destructive interference. These regions move across the illuminated region with a velocity that changes very slowly with time. This moving pattern is said to be "frozen in" the atmosphere, but changes as it moves along. While the aperture of the imaging system is covered by such a dark region, the image formed, too, will be dark; it lights up when a bright region moves across the aperture. The alternation between these two states, caused by the motion and shape variation of the interference pattern, is the source of the scintillation phenomenon.

The "frozen in" pattern is due to a tendency of the turbulence to break the wavefront into patches separated by large phase jumps. Under conditions of strong turbulence, these may be as small as 10 mm [47]. Various attempts have been made to overcome these phase shift problems, including dynamically shifting the local phase [51].

When a *coherent beam* of limited cross-sectional area traverses the atmosphere, two additional phenomena can be observed [52]:

1. *Beam steering.* The beam, as a whole, is displaced laterally. It may, consequently, move out of the receiving aperture, in part or totally, again causing scintillation in the received signal. This is caused by large-volume inhomogeneities.

2. *Beam spreading.* Small-volume inhomogeneities cause the spreading of the beam with a consequent lowering of the transmission efficiency. Note that all of the above variations are spatio-temporal in nature.

12.4.7.2 Blurring—Temporal Aspects.

12.4.7.2 Blurring—Temporal Aspects. In treating blurring due to turbulence, we should differentiate between short- and long-exposure observations. The blurring apparent in short-exposure observations includes only the wavefront distortion after subtraction of the mean; the rotation of the mean wavefront appears only as a shift in the image location. In long-exposure observations, on the other hand, the random rotation of the wavefront direction causes a "random walk" of the image point over the image plane and the blurring due to it must be included. In this case the turbulence effects may be treated with the theory of incoherent optical systems and an optical transfer function (otf) may be ascribed to the atmospheric layer traversed. In the short-exposure case, we can speak only of an otf of the atmosphere in combination with the lens.

The temporal effects of turbulence on imaging can be approximated by assuming the index perturbations to be "frozen" into the atmosphere, moving along with the speed of the wind.[21] If the exposure is so long that these perturbations move across the full aperture diameter during the

[21] To enable the reader to estimate wind speed values, we note that any speed less than 0.5 m/sec is considered "calm"; 2–3 m/sec, a slight breeze; 11–14 m/sec a strong breeze (just below gale force). The Beaufort scale of winds is given in Table 139 [53].

exposure, the result begins to approximate that of long-term exposure (see the results of Ref. 52).

Angular variations of approximately 10^{-4} rad may take place over an interval shorter than a millisecond [47].

12.4.7.3 Blurring—Spatial Aspects. Optical Transfer Function [47], [54].

The optical effects of turbulence are conveniently expressed in terms of a *coherence distance* [47], d_0. This represents the distance, measured across the aperture of the optical system, over which the wavefront from a point source remains essentially coherent after having passed through the atmosphere. Specifically, d_0 is defined by

$$d_0 = \frac{\sqrt{4.8\Gamma(1.2)}}{(A_1/2)^{3/5}} = \frac{3.18}{A_1^{0.6}}, \qquad (12.112)$$

where A_1 is defined by

$$A_1 = \text{var} (\phi) + \text{var} (\log \Delta u) \big|_{d=1}, \qquad (12.113)$$

the sum of the variances of ϕ and $\log \Delta u$ as measured over a unity displacement. Here ϕ is the phase shift and Δu is the electric field amplitude change relative to the mean value of the field—both random changes caused by the turbulence. On the basis of analytic work [55], it appears that the sum of the variances varies with the $\frac{5}{3}$ power[22] of the distance, d, over which it is measured:

$$A = \text{var} (\phi) + \text{var} (\Delta u) = A_1 \, d^{5/3}. \qquad (12.114)$$

It is called the *structure function* of the turbulent atmosphere.

The distance d_0 as defined in (12.112) is significant in a number of ways [57]:

1. Over a circle of diameter d_0, the wavefront distortion has an rms value almost exactly equal to one radian.

2. The maximum resolution[23] attainable with long exposures cannot exceed the resolution attainable in free space with diffraction-limited optics with an aperture diameter d_0. Hence the angular blur, with any aperture, is always approximately

$$\frac{\lambda}{d_0},$$

at least. With short exposures, the resolution is optimized by taking the aperture diameter equal to $3.87 \; d_0$.

[22] The validity of this form of dependence seems to be rather limited [50], [56].
[23] See following (12.121) for the quantitative definition of "resolution" here.

3. No matter how large its initial diameter, the mean relative energy density transmitted in a laser beam through the turbulent atmosphere is never more than that obtained in free space with a beam of diameter d_0.

4. The average antenna gain of a static optical heterodyne receiver can not exceed the free-space antenna gain of an antenna with diameter d_0.

For long exposures, the otf of the atmosphere is real and given by [47]

$$\boxed{T_l(\nu) = \exp\left(-3.44\,\nu_0^{5/3}\right)} \qquad (12.115)$$

where

$$\nu_0 = \nu\lambda b/d_0 = \nu_\theta\lambda/d_0 \qquad (12.116)$$

is the spatial frequency, normalized for an aperture diameter, d_0,

b = the image distance (for an infinitely distant object, b is the effective focal length of the optical system), and

ν_θ = the spatial frequency in angular (radian) measure.

Note that (12.115) is close to Gaussian in shape [56].

For short exposures the otf of the atmosphere-optical system combination is given by

$$T_c(\nu) = T_0(\nu)T_l(\nu)\exp\left(3.44\beta\nu_0^{5/3}\nu_r^{1/3}\right), \qquad \nu_r \leqslant 1, \qquad (12.117)$$

where T_0 = the otf of the optical system, aperture diameter, D,

$$\nu_r = \nu\lambda b/D \qquad (12.118)$$

is the normalized spatial frequency, and

$$\beta(\tfrac{1}{2} < \beta < 1) = \text{the fraction of } A$$

represented by var (ϕ). It depends on the length, r, of passage through the atmosphere and is unity for the *near-field* case:

$$r \ll \frac{d_e^2}{\lambda} \qquad (12.119a)$$

and equal to half for the *far-field* case:

$$r \gg \frac{d_e^2}{\lambda}. \qquad (12.119b)$$

Here d_e is the effective aperture diameter, d_0, or the actual diameter, D, whichever is smaller.

Referring back to (12.117) we note that by going to a short exposure, we can increase the otf by factors

$$\exp\left(3.44\nu_0^{5/3}\nu_r^{1/3}\right) \text{ and } \exp\left(1.72\nu_0^{5/3}\nu_r^{1/3}\right) \qquad (12.120)$$

for near- and far-field operation, respectively. These can be significant factors, especially as the spatial frequency approaches the cutoff frequency $(1/\lambda F)$ of the optical system. The improvement is, obviously, more significant in the near-field.

The resolution, W, as a function of optical system aperture diameter is shown in Figure 12.22 and Table 140 [47]. For normalization purposes, the resolution is taken relative to W_{mx} and the aperture diameter relative to d_0. Here

$$W_{mx} = \pi \left(\frac{d_0}{2\lambda b}\right)^2 \tag{12.121}$$

is the resolution obtainable with an infinitely large aperture, and "resolution" is defined as the integral of the otf over the total frequency range.

12.4.7.4 Determination of the Coherence Distance. The value of A_1, as used in the definition (12.112) of d_0, may be written [57]

$$A_1 = 2.91k^2 \int_0^r Q(x)C_n^2(x)\, dx \tag{12.122}$$

where $k = 2\pi/\lambda$,

$Q(x) =$ a factor depending on the shape of the wavefront:

$$Q(x) = \begin{cases} 1, \text{ for an infinite plane wave} \\ (x/r)^{5/3} \text{ for a spherical wave, radius, } r, \text{ and} \end{cases}$$

$C_n =$ the *index structure constant*, defined by

$$C_n^2 = \frac{\text{var}(n)}{d^{2/3}} = \text{constant}, \tag{12.123}$$

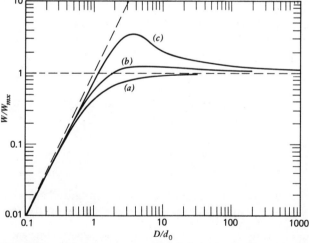

Figure 12.22 Resolution (W) versus system aperture (D). W_{mx} and d_0 are defined in the text. (a) Results for long exposure; (b) short exposure, far field; (c) short exposure, near field. Fried [47].

where var (n) signifies the mean squared value of the refractive index difference as determined over a distance d.

For a horizontal path, with C_n assumed constant [55],

$$A_1 = 2.91 k^2 r C_n^2 \qquad (12.124)$$

and hence, from (12.112)

$$d_0 = \frac{1.676}{(k^2 r C_n^2)^{0.6}}. \qquad (12.125)$$

The 6/5-power dependence on wavelength implied by (12.125) has been confirmed experimentally over the wavelength range 0.63–10.6 μm [58].

Representative values of C_n as a function of altitude are shown in Figure 12.23 [55]. Its variation during a typical clear[24] day near the equinox is shown in Figure 12.24 [59]. Also shown is the variance of the concomitant log-amplitude fluctuations. Note especially the typically low level of scintillations immediately after sunrise and sunset and the high level during mid-day.

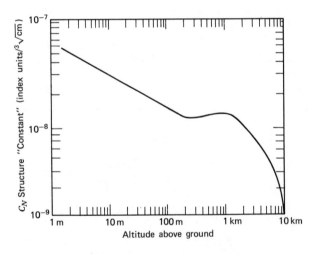

Figure 12.23 Typical values of refractive index structure constant, C_N, as a function of altitude. Hufnagel and Stanley [55].

For viewing extraterrestrial objects from an altitude of about[25] 2400 m above sea-level, an approximate median value of d_0 for night time

[24] There were some clouds after 16:30 causing alternating sunlight and shade.

[25] This result is based on extensive observations made by A. B. Meinel at the Kitt Peak National Observatory and by A. A. Hoag at the U. S. Naval Observatory at Flagstaff, Arizona, with the two sets of data in good agreement with each other.

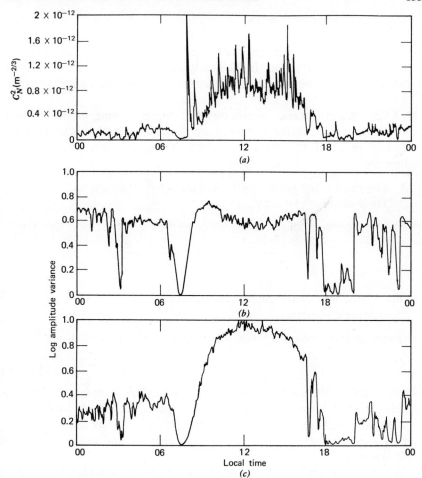

Figure 12.24 Refractive index structure constant at $d = 1$ cm (a); and log amplitude variance of laser beam scintillations over a 1000 m, (b); and a 490-m path (c). Lawrence, Ochs, and Clifford [59].

observation has been found to be given by

$$d_0 = 0.114 \left(\frac{\lambda}{0.55 \ \mu\mathrm{m}}\right)^{1.2} (\cos Z)^{0.6} \ \mathrm{m}, \qquad (12.126)$$

where Z is the zenith angle of the observation. The standard deviation from the value given by (12.126) is about 35% [57].

At an altitude of 4.2 km, the limit of good response was found to vary between 50 and 200 cycles/mrad [60] where "good response" is taken as the region where the otf is above 25%.

Modulation transfer function measurements made on an aircraft with a retro-reflector placed on the wing tip showed considerable deterioration in flight compared to measurements made on the ground. This seems to indicate that the turbulence immediately surrounding the aircraft may be a major factor in the atmosphere blurring of aerial photography [61].

12.4.7.5 *Scintillations.*

As discussed at the beginning of this section, scintillation of the image is caused by the changing illumination at the entrance pupil of the imaging system. These changes consist of the following:

1. The passage of interference patches across the aperture,
2. Their changing structure.

The time constant of the changes due to the former factor is given by

$$\tau \sim \frac{l}{v},\qquad(12.127)$$

where $l =$ the size of the interference patches and
$v =$ their velocity.

The size of these patches tends to be of the order of magnitude

$$l \sim \sqrt{\lambda r},\qquad(12.128)$$

where r is the distance to the turbulent region. This is equal to the radius of the central zone of a Fresnel zone plate (2.152) to which it is conceptually related. The velocity of the patch is due to the wind and is set by the wind velocity component transverse to the direction of light propagation.

When the aperture of the imaging system is considerably larger than the size of the interference patches (a few centimeters), the magnitude of the scintillations will be reduced; this effect is called *aperture smoothing.* Specifically, we would expect the standard deviation (σ_s) of the scintillation relative to the mean illumination ($\bar{E}$) to vary inversely with the square root of the mean number ($\bar{N}$) of patches in the entrance pupil, that is, the modulation index

$$\mathcal{M}^2 = \frac{\sigma_s^2}{\bar{E}^2} \sim \frac{1}{\bar{N}} \sim \frac{A_{\text{patch}}}{A_{\text{pupil}}} \sim \frac{1}{D^2},\qquad(12.129)$$

where $A_{\text{pupil,patch}}$ are the areas of the entrance pupil and the interference patches, respectively, D is the diameter of the entrance pupil, and we assumed A_{patch} to be constant.

Observations show, however, a dependence [62]

$$\mathcal{M}^2 \approx \frac{c}{D}.$$ (12.129a)

This may be due to the fact that the interference patches tend to take the form of long bands [144a], so that the number included in the aperture varies with its diameter rather than its area. The value of the constant, c, in (12.129a) has been found to be in the range 0.25–0.5 cm, depending on atmospheric conditions [62].

The temporal spectrum of scintillations is observed to consist of two components: one governing at low frequencies and dropping but slowly with frequency and another governing at higher frequencies and dropping more rapidly with frequency. An increase in aperture diameter shifts the crossover frequency (ν_{tc}) toward the lower frequencies and, for a diameter larger than 60 mm, the crossover phenomenon can no longer be observed. The spectrum can thus be approximated by a formula of the form [62]

$$S(\nu_t) = \frac{c'}{\nu_t^a + \beta \nu_t^b}$$ (12.130)

where c' = a constant varying with the aperture diameter and atmospheric conditions and

$a = 0.5$,
$b = 3 + D/10$,
$\beta = \nu_c^{a-b} = (105 - 9D)^{a-b}$, and
D = the aperture diameter in centimeters. (12.131)

These results were obtained by observations over a 300-m range parallel to the ground and may be contrasted with the spectra of stellar scintillations as shown in Figure 12.25a and b [63].

12.4.7.6 Laser Beam Propagation. For detailed discussions of turbulence effects on laser propagation and on communication via such a beam, the reader is referred to Refs. 62 and 64. Here not only the turbulence, but also the slowly changing atmospheric index gradients are important in grossly bending the laser beam (see Section 12.2.1).

See Ref. 65 for a calculation of the detection probability of an optical laser pulse transmitted through a turbulent atmosphere. It is interesting to note that scintillations actually improve the detection probability for pulses small compared to the detector noise level.

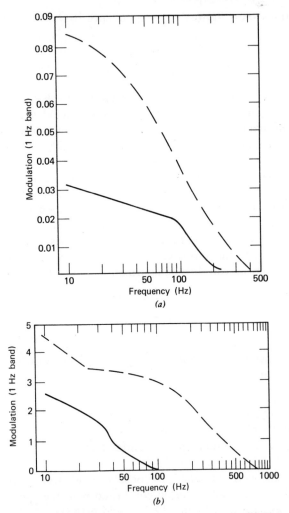

Figure 12.25   Stellar scintillation spectra. (a) Washington, D. C.: (b) Flagstaff, Ariz. Each on two days. After Mikesell *et al.* [63].

APPENDIX 12.1   RAYLEIGH AND MIE SCATTERING

1   Rayleigh Scattering. Refractive Index and Scattering Coefficient [66]

Here we establish the spectral and angular dependence of molecular scattering and its connection with the refractive index. Consider, first, scattering from a thin layer of a gas containing scattering particles small compared to the wavelength, λ, of a normally incident monochromatic, linearly polarized, plane wave. Let N be the concentration of scattering particles and δz the thickness of the layer; δz is assumed large compared to the particle size and small compared to the wavelength (see Figure 12.26a). Each particle in the layer will scatter a certain amount of power from the flux passing it. For this scattering action, we assume the following mechanism: the electromagnetic wave field sets up electric dipole oscillations in each particle, parallel to the direction of polarization of the wave. The resulting dipole oscillators reradiate the extracted energy in all directions with an amplitude proportional to the amplitude (a) of the incident wave and to $\sin \beta$, where β is the angle between the dipole axis and the direction of scattering. See Figure 12.26 where the y-axis is taken parallel to the direction of polarization and the z-axis parallel to the direction of wave propagation and normal to the scattering layer; Figure 12.26a shows a ray emanating from a point on the y-axis.

Consider now a point P at a distance r_0 from the layer and let O be the foot of the perpendicular from P to the layer. Any area element, dA, will then contribute to the amplitude at P an amount

$$d\hat{a}_s = caN \, \delta z \, dA e^{ikr} \frac{\sin \beta}{r}, \tag{12.132}$$

where $c =$ a constant of proportionality, the *scattering factor*,
$k = 2\pi/\lambda$ is the magnitude of the incident wave vector, and
$r =$ the distance from the area element to P.

This amplitude element will be inclined at an angle $(\tfrac{1}{2}\pi - \beta)$ to the y-axis and its component parallel to that axis will be

$$d\hat{a}_{sy} = caN \, \delta z \, dA e^{ikr} \frac{\sin^2 \beta}{r}. \tag{12.133}$$

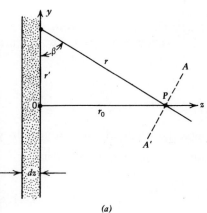

(a)

Figure 12.26 Illustrating molecular scattering in a differential layer.

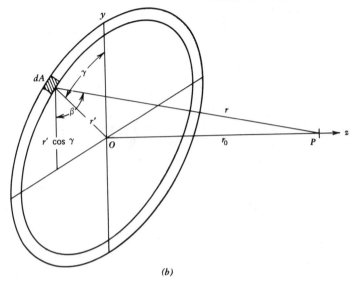

(b)

Figure 12.26 (Continued)

To obtain the total contribution of the scattering layer at P, we integrate $d\hat{a}_{sy}$ over the whole plane; we choose $d\hat{a}_{sy}$ instead of $d\hat{a}_{s}$, in view of the fact that the z components of the latter cancel out for symmetry reasons. We integrate first over an annular element, radius r, and then integrate the contributions of these elements over the plane. Since they form a right triangle, r, r_0, and r' are related according to

$$r^2 = r_0^2 + r'^2. \tag{12.134}$$

Here r' is the component of r in the $x-y$ plane. The area element in the annulus will be

$$dA = r'\, d\gamma\, dr', \tag{12.135}$$

where γ is the angle between the y axis and the vector from O to the area element. On differentiating (12.134) for constant r_0, it becomes apparent that

$$r\, dr = r'\, dr', \tag{12.136}$$

so that we may rewrite (12.135):

$$dA = r\, d\gamma\, dr. \tag{12.137}$$

For the annulus under consideration, the angle β will vary with γ. Specifically, it can be seen from Figure 12.26b and (12.134) that

$$\cos\beta = \frac{r'\cos\gamma}{r} - \frac{\sqrt{r^2 - r_0^2}}{r}\cos\gamma. \tag{12.138}$$

Hence

$$\sin^2 \beta = 1 - \cos^2 \beta$$

$$= \sin^2 \gamma + \left(\frac{r_0}{r}\right)^2 \cos^2 \gamma. \tag{12.139}$$

On substituting this and (12.137) into (12.133) and integrating first over the annulus and then over the plane, we find

$$\hat{a}_s = caN \, \delta z \int_{r_0}^{\infty} e^{ikr} \int_{-\pi}^{\pi} \left[\sin^2 \gamma + \left(\frac{r_0}{r}\right)^2 \cos^2 \gamma\right] d\gamma \, dr$$

$$= \pi caN \, \delta z \left[\int_{r_0}^{\infty} e^{ikr} \, dr + r_0^2 \int_{r_0}^{\infty} \frac{e^{ikr}}{r^2} \, dr\right]. \tag{12.140}$$

Consider the first integral. Formally it becomes:

$$\lim_{B \to \infty} \frac{e^{ikr}}{ik}\bigg|_{r_0}^{B} = \lim_{B \to \infty} \frac{e^{ikB}}{ik} - \frac{e^{ikr_0}}{ik} \sim \frac{ie^{ikr_0}}{k}, \tag{12.141}$$

where we have dropped the term due to the upper limit, since it oscillates about zero as B approaches infinity. The second integral is an exponential integral (E_2), which we bring into its standard form by substituting:

$$\frac{r}{r_0} = t.$$

Then

$$r_0^2 \int_{r_0}^{\infty} \frac{e^{ikr}}{r^2} \, dr = r_0 \int_{1}^{\infty} \frac{e^{ikr_0 t}}{t^2} \, dt = r_0 E_2(-ikr_0). \tag{12.142}$$

For very large kr_0, that is for r_0 very much larger than the wavelength, we can use the first term of the asymptotic expansion of $E_2(z)$:

$$E_2(z) \approx \frac{e^{-z}}{z}. \tag{12.143}$$

Hence, the second integral, too, equals

$$\frac{ie^{ikr_0}}{k}. \tag{12.144}$$

Consequently (12.140) becomes

$$\hat{a}_s = \frac{i2\pi caN \, \delta z}{k} e^{ikr_0}$$

$$= i\lambda caN \, \delta z e^{ikr_0}. \tag{12.145}$$

When we add to this the amplitude of the unscattered original radiation, we find for the total field at P

$$\hat{a}_p = \hat{a} + \hat{a}_s = \hat{a}(1 + i\lambda cN\,\delta z)e^{ikr_0}. \qquad (12.146)$$

The expression in parentheses may be viewed as the first two terms in the Taylor series expansion of an exponential and, if δz is sufficiently small, higher order terms may be neglected. We may then substitute the exponential for the sum in the parentheses, yielding

$$\hat{a}_p = a \exp i(kr_0 + \phi), \qquad (12.147)$$

the original wave with a phase shift

$$\phi = \lambda cN\,\delta z. \qquad (12.148)$$

If we view the assemblage of molecules in the layer as a medium with a gross refractive index, n, the phase shift introduced by the layer can be seen to be[26]

$$\phi = \frac{2\pi}{\lambda}(n-1)\,\delta z. \qquad (12.149)$$

On equating (12.148) and (12.149) we find, for the scattering factor

$$c = \frac{2\pi(n-1)}{\lambda^2 N}. \qquad (12.150)$$

We must now find the relationship between the factor c and the scattering coefficient α_s. Based on the definition of α_s (4.58), the total amount of flux scattered per unit area in the layer is

$$\delta E = E\alpha_s\,\delta z. \qquad (12.151)$$

This must also equal the total amount of flux scattered by all the particles included in this area element of the layer. This, in turn, is given by the integral:

$$\delta E = N\,\delta z \int_{\text{sphere}} (ca \sin \beta)^2 \, d\Omega, \qquad (12.152)$$

where $(N\,\delta z)$ is the total number of particles participating in the scattering process, $(ca \sin \beta)$ is the amplitude scattered into direction β, and we square the integrand because the flux is proportional to the square of the amplitude. Also

$$a^2 = E. \qquad (12.153)$$

[26] The axial shift introduced into the wavefront equals the transit time $[\delta z/(c/n)]$ multiplied by the reduction in phase velocity $(c - c/n)$. This shift multiplied by the magnitude of the wave propagation vector $(2\pi/\lambda)$ yields the phase shift (12.149).

Substituting this into (12.152) and taking the element

$$d\Omega = 2\pi \sin \beta \, d\beta, \tag{12.154}$$

we find

$$\delta E = 2\pi c^2 EN \, \delta z \int_0^\pi \sin^3 \beta \, d\beta$$

$$= \tfrac{8}{3}\pi c^2 EN \, \delta z. \tag{12.155}$$

On equating this to (12.151), we find

$$\alpha_s = \tfrac{8}{3}\pi c^2 N. \tag{12.156}$$

Substituting into this the value of c as given by (12.150):

$$\boxed{\alpha_s = \frac{32\pi^3}{3}\frac{(n-1)^2}{N\lambda^4}.} \tag{12.157}$$

This is Rayleigh's formula connecting the scattering coefficient, α_s, with the refractive index in a randomly structured medium, such as a gas. It also illustrates the inverse fourth-power dependence of the scattering on the wavelength. This, in turn, shows that the scattering in an aerosol-free atmosphere affects primarily flux at the shorter wavelengths, accounting for the blue color of the clear daylight sky and the red of the setting sun. Note that actual scattering in aerosol-free air is slightly greater than this, by a factor of approximately 1.06 due to the anisotropy of some of the component molecules.

The dependence of the scattered amplitude on the angle β (between the direction of polarization and the line of sight to the scatterer) shows that the intensities of the two polarization components will generally differ in any given direction. Indeed, the component polarized normal to the plane of the scattering will be scattered isotropically, whereas the component polarized parallel to that plane will exhibit an intensity varying according to

$$\sin^2 \beta = \sin^2 \theta,$$

where θ is the angle through which the wave is scattered. Hence, the light scattered at right angles to the direction of the incident light will be fully polarized—normal to the plane of scattering. For totally unpolarized incident light, the total intensity of Rayleigh scattering then varies with direction according to

$$L(\theta) = \tfrac{1}{2}L_0(1 + \cos^2 \theta), \tag{12.158}$$

where L_0 is the intensity of the forward (or backward) scattered light.

2 Mie Scattering [2f]

When the diameters of the scattering spheres are neither very large nor very small compared to the wavelength of the scattered radiation, Mie theory must be used to obtain an accurate prediction of the scattering. This theory represents an exact solution of Maxwell's equation for scattering spheres with complex refractive index $\hat{n}$. We present here only the final results and refer the reader to Ref. 17 for a complete derivation.

The scattered radiation is polarization dependent and we denote by $S_{1,2}(\theta)$ the normalized amplitudes of the flux scattered through angle θ; the subscripts 1 and 2 refer to flux polarized normal to the scattering plane, and parallel to it, respectively. The normalization is made relative to the amplitude incident on the sphere cross section.

According to Mie theory

$$S_1(\theta) = \sum_{m=1}^{\infty} \frac{2m+1}{m(m+1)} [a_m \pi_m(\cos \theta) + b_m \tau_m(\cos \theta)] \quad (12.159)$$

$$S_2(\theta) = \sum_{m=1}^{\infty} \frac{2m+1}{m(m+1)} [b_m \pi_m(\cos \theta) + a_m \tau_m(\cos \theta)], \quad (12.160)$$

where

$$\pi_m(\cos \theta) = \frac{1}{\sin \theta} P_m^1(\cos \theta) \quad (12.161)$$

$$\tau_m(\cos \theta) = \frac{d}{d\theta} P_m^1(\cos \theta), \quad (12.162)$$

P_m^1 is an associated Legendre polynomial, and the coefficients a_m and b_m are given by

$$a_m = \frac{\psi'_m(\hat{n}x)\psi_m(x) - \hat{n}\psi_m(\hat{n}x)\psi'_m(x)}{\psi'_m(\hat{n}x)\zeta_m(x) - \hat{n}\psi_m(\hat{n}x)\zeta'_m(x)} \quad (12.163)$$

$$b_m = \frac{\hat{n}\psi'_m(\hat{n}x)\psi_m(x) - \psi_m(\hat{n}x)\psi'_m(x)}{\hat{n}\psi'_m(\hat{n}x)\zeta_m(x) - \psi_m(\hat{n}x)\zeta'_m(x)}, \quad (12.164)$$

where

$$\psi_m(x) = \sqrt{\pi x/2} J_{m+(1/2)}(x) \quad (12.165)$$

$$\zeta_m(x) = \sqrt{\pi x/2} [J_{m+(1/2)}(x) + (-1)^m i J_{-m-(1/2)}(x)] \quad (12.166)$$

are the Riccati–Bessel functions and J_m are the Bessel functions of the first kind. For a sphere of radius, a, the Mie variable

$$x = ka = \frac{2\pi a}{\lambda} \quad (12.167)$$

is given by the ratio of the sphere circumference to the wavelength.

The extinction and scattering cross sections associated with each scattering sphere are given, respectively by

$$\sigma_e = \sigma_g K_e \qquad (12.168)$$

$$\sigma_s = \sigma_g K_s \qquad (12.169)$$

where σ_g is the geometric cross-sectional area and

$$K_e = \frac{2}{x^2} \sum_{m=1}^{\infty} (2m+1)\mathcal{R}[a_n + b_n] \qquad (12.170)$$

$$K_s = \frac{2}{x^2} \sum_{m=1}^{\infty} (2m+1)(a_m a_m^* + b_m b_m^*) \qquad (12.171)$$

are the extinction and scattering efficiencies, respectively.

Extensive evaluation of these can be found in Ref. 17 and for complex refractive indices in Ref. 67. For spheres whose refractive index is close to unity, and this includes water droplets, and whose Mie variable (x) exceeds unity, the following simplified formulae [17a] give reasonably accurate results:

$$K_e = 2 - 4y \exp(-\rho \tan \beta)$$
$$\times [\sin(\rho - \beta) - y \cos(\rho - 2\beta) + y \cos \beta], \qquad (12.172)$$

$$K_s = 1 + \frac{\exp(-4xn_I)}{2xn_I} + \frac{\exp(-4xn_I) - 1}{8x^2 n_I^2}, \qquad (12.173)$$

where $\rho = 2(n_R - 1)x$,
  $\tan \beta = n_I/(n_R - 1)$,
  $y = (\cos \beta)/\rho$, and
n_R, n_I = the real and imaginary parts of the refractive index of the scattering spheres [see (11.74)].

Evaluations of (12.172) and (12.173) for aerosols of constant radius water droplets are shown in Figure 12.7 [19].

REFERENCES

[1] S. W. Allen, *Astrophysical Quantities*, 3rd Ed., Athlone, London, 1973.

[2] E. J. McCartney, *Optics of the Atmosphere*, Wiley, New York, 1976: (a) Chapter 2; (b) Chapter 3; (c) Section 2.5.1; (d) Chapter 1; (e) Chapter 4; (f) Chapter 5.

[3] D. R. Stull, "Vapor pressure," in *American Institute of Physics Handbook*, 3rd Ed., D. E. Gray, ed., McGraw-Hill, New York, 1972, pp. 4–261 to 4–315.

[4] S. L. Valley, ed., *Handbook of Geophysics and Space Environment*, Air Force Cambridge Research Laboratory, Bedford, Mass., 1965: (a) J. E. Manson, "Aerosols," Section 5.5, (b) L. Elterman, "Atmospheric attenuation model," Section 7.1.

[5] W. L. Wolfe, ed., *Handbook of Military Infrared Technology*, Office of Naval Research, Washington, D. C. 1965: (a) G. N. Plass and H. Yates, "Atmospheric phenomena," Chapter 6, pp. 175–279; (b) R. Kanth, "Backgrounds," Chapter 5, pp. 95–173.

[6] R. D. Hudson, *Infrared System Engineering*, Wiley, New York, 1969, Chapter 4.

[7] R. A. McClatchey, R. W. Fenn, J. E. A. Selby, F. E. Volz, and J. S. Garing, "Optical properties of the atmosphere," 3rd Ed., Air Force Cambridge Research Laboratory, Bedford, Mass., 1972, Table 3.

[8] M. F. Harris, "Meteorological information," in *American Institute of Physics Handbook*, 3rd Ed., D. E. Gray, ed., McGraw-Hill, New York, 1972, pp. 2–133 to 2–147, (a) From Gvt. Prtg. Office, Cat. No. NAS 1.2:At6/962, Washington, D. C.

[9] B. Edlén, "The refractive index of air," *Metrologia* **2**, 71–80 (1966).

[10] J. C. Owens, "Optical refractive index of air: Dependence on pressure, temperature, and composition," *Appl. Opt.* **6**, 51–59 (1967).

[11] G. M. Hale and M. R. Querry, "Optical constants of water in the 200 nm to 200 μm wavelength region," *Appl. Opt.* **12**, 555–563 (1973).

[12] E. Brookner, "Atmosphere propagation and communication channel model for laser wavelengths," *IEEE Trans.* **COM-18**, 396–416 (1970).

[13] W. Viezee, "Optical mirage," in *Scientific Study of Unidentified Flying Objects*, E. V. Condon, ed., Vision, London, 1970

[14] M. Minnaert, *The Nature of Light and Color in the Open Air*, Dover, New York, 1954, Chapter 4.

[15] A. Sommerfeld, *Optics*, Academic, New York, 1954, Section 48A.

[16] C. V. Raman and S. Pancharatnam, "The optics of mirages," *Proc. Ind. Acad. Sci.*, **49**, 251–261 (1959).

[17] H. C. van de Hulst, *Light Scattering by Small Particles*, Wiley, New York, 1957, (a) Sections 11.22 and 11.23.

[18] Petr Chylek, "Asymptotic limits of Mie-scattering characteristics," *J. Opt. Soc. Am.* **65**, 1316–1318 (1975).

[18*] T. S. Chu and D. C. Hogg, "Effects of precipitation on propagation at 0.63, 3.5 and 10.6 microns," *Bell Syst. Tech. J.* **47**, 723–759 (1968).

[19] L. W. Carrier, G. A. Cato, and K. J. von Essen, "The backscattering and extinction of visible and infrared radiation by selected major cloud models," *Appl. Opt.* **6**, 1209–1216 (1967).

[20] D. Deirmendjian, "Scattering and Polarization Properties of Water, Clouds and Hazes in the Visible and Infrared," *Appl. Opt.* **3**, 187–196 (1964).

[21] J. O. Laws and D. A. Parsons, "The relation of raindrop-size to intensity," *Trans. Am. Geophys. Union* **24**, 452–460 (1943).

[22] W. E. K. Middleton, *Vision through the Atmosphere*, University of Toronto Press, Toronto, 1952: (a) Section 6.2; (b) Note p. 155, Table 8.3; (c) p. 73; (d) Figures 6.3–6.11; (e) After H. Siedentopf; (f) Table 8.3; (g) Table 5.2.

[23] L. Elterman, "Aerosol measurements since 1973 for normal and volcanic atmospheres," *Appl. Opt.* **15**, 1113–1114 (1976).

[24] F. Löhle, "Über die Lichtzerstreuung im Nebel," *Phys. Z.* **45**, 199–205 (1944).

[25] A. Arnulf, J. Bricard, E. Curé, and C. Véret, "Transmission of haze and fog in the spectral region 0.35 to 10 microns," *J. Opt. Soc. Am.* **47**, 491–498 (1957).

[26] A. J. La Rocca, "Methods of calculating atmospheric transmittance and radiance in the infrared," *Proc. IEEE* **63**, 75–94 (1975).

[27] J. N. Howard, D. E. Burch, and D. Williams, "Infrared transmission of synthetic atmospheres," *J. Opt. Soc. Am.* **46**, 186, 237, 242, 334, 452 (1956).

[28] M. P. Thekaekara, "Solar irradiance: Total and spectral and its possible variations," *Appl. Opt.* **15**, 915–920 (1976).

[29] P. Moon, "Proposed standard solar-radiation curves for engineering use," *J. Franklin Inst.* **230**, 583–617 (1940).

[30] L. A. Jones and H. R. Condit, "Sunlight and skylight as determinants of photographic exposure. I. Luminous density as determined by solar altitude and atmospheric conditions," *J. Opt. Soc. Am.* **38**, 123–178 (1948).

[31] Landolt-Börnstein, *Numerical Data and Functional Relationships in Science and Technology*, New Series, Vol. 6/1, H. H. Voigt, ed., Springer, Berlin, 1965: (*a*) E. Lamla, "Colors of the stars," Section 5.2.7; (*b*) H. Siedentopf and H. Scheffler, "Brightness of twilight and night sky," Section 1.5.3; (*c*) H. Siedentopf and H. Scheffler, "Astronomical refraction and extinction," Section 1.5.1; (*d*) F. Gondolatsch, "Mechanical data of planets and satellites," Section 4.2.1; (*e*) G. P. Kuiper, "Physics of planets and satellites," Section 4.2.2; (*f*) W. Gliese, "Stellar positions," Section 5.1.1.

[32] R. C. Ramsey, "Spectral irradiance from stars and planets, above the atmosphere, from 0.1 to 100 microns," *Appl. Opt.* **1**, 465–471 (1962).

[33] E. E. Bell, L. Eisner, J. Young, and R. Oetjen, "Spectral radiance of sky and terrain at wavelengths between 1 and 20 microns. II. Sky measurements," *J. Opt. Soc. Am.* **50**, 1313–1320 (1960).

[34] M. J. Koomen, C. Lock, D. M. Packer, R. Scolnik, R. Tousey, and E. O. Hulburt, "Measurement of brightness of the twilight sky," *J. Opt. Soc. Am.* **42**, 353–356 (1952).

[35] R. Tousey and M. J. Koomen, "The visibility of stars and planets during twilight," *J. Opt. Soc. Am.* **43**, 177–183 (1953).

[36] K. R. Lang, *Astrophysical Formulae*, Springer, Berlin, 1974, Chapter 5.

[37] H. V. Soule, *Electro-Optical Photography at Low Illumination Levels*, Wiley, New York, 1968.

[38] A. T. Vassy and E. Vassy, "La luminescence nocturne," in *Encyclopedia of Physics*, Vol. 49/5, S. Flügge, ed., (*Geophysics III/5*, K. Rawer, ed.) Springer, Berlin, 1976, pp. 2–116.

[39] G. F. Aroyan, "The technique of spatial filtering," *Proc. IRE* **47**, 1561–1568 (1959).

[40] H. S. Stewart and R. F. Hopfield, "Atmospheric effects," *Applied Optics and Optical Engineering*, Vol. 1, R. Kingslake, ed., Academic, New York, 1965, Chapter 4.

[41] M. A. Bramson, *Infrared Radiation*, Plenum, New York, 1968 [Original Russian Ed., 1964).

[42] L. W. Pinkley and D. Williams, "Optical properties of sea water in the infrared," *J. Opt. Soc. Am.* **66**, 554–558 (1976).

[43] A. Cohen, "Horizontal visibility and the measurement of atmospheric optical depth of lidar," *Appl. Opt.* **14**, 2878–2882 (1975).

[44] J. I. Gordon, J. L. Harris, and S. Q. Duntley, "Measuring space-to-earth contrast transmittance from ground stations," *Appl. Opt.* **12**, 1317–1324 (1973).

[45] E. O. Hulburt, "Optics of atmospheric haze," *J. Opt. Soc. Am.* **31**, 467–476 (1941).

[46] H. S. Coleman and W. S. Verplanck, "A comparison of computed and experimental detection ranges of objects viewed with telescopic systems from aboard ship," *J. Opt. Soc. Am.* **38**, 250–253 (1948).

[47] D. L. Fried, "Optical resolution through a randomly inhomogeneous medium for very long and very short exposures," *J. Opt. Soc. Am.* **56**, 1372–1379 (1966).

[48] G. K. Born, R. Bogenberger, K. D. Erben, F. Frank, F. Mohr, and G. Sepp, "Phase-front distortion of laser radiation in a turbulent atmosphere," *Appl. Opt.* **14**, 2857–2863 (1975).

[49] V. I. Tatarski, *Wave Propagation in a Turbulent Medium*, Dover, New York, 1967.

[50] E. Brookner, "Atmosphere propagation and communication channel model for laser wavelengths," *IEEE Trans.* **COM-18**, 396–416 (1970); also **COM-22**, 265–270 (1974).

[51] See The March 1977 issue of *J. Opt. Soc. Am.*, **67**, which, in its entirety, is devoted to "adaptive optics."

[52] J. J. Davis, "Consideration of atmospheric turbulence in laser system design," *Appl. Opt.* **5**, 139–147 (1966).

[53] S. Pettersson, *Introduction to Meteorology*, 2nd Ed., McGraw-Hill, New York, 1958, Table 2.

[54] C. Roddier and F. Roddier, "Influence of exposure time on spectral properties of turbulence-degraded astronomical images," *J. Opt. Soc. Am.* **65**, 664–667 (1975).

[55] R. E. Hufnagel and N. R. Stanley, "Modulation transfer function associated with image transmission through turbulent media," *J. Opt. Soc. Am.* **54**, 52–61 (1964).

[56] R. F. Lutomirski and H. T. Yura, "Wave structure function and mutual coherence function of an optical wave in a turbulent atmosphere," *J. Opt. Soc. Am.* **61**, 482–487 (1971).

[57] D. L. Fried and G. E. Mevers, "Evaluation of r_0 for propagation down through the atmosphere," *Appl. Opt.* **13**, 2620–2622 (1974); Corrections: **14**, 2567 (1975), **16**, 549 (1977).

[58] T. J. Gilmartin and J. Z. Holtz, "Focused beam and atmospheric coherence measurements at 10.6 μm and 0.63 μm," *Appl. Opt.* **13**, 1906–1912 (1974).

[59] R. S. Lawrence, G. R. Ochs, and S. F. Clifford, "Measurements of atmospheric turbulence relevant to optical propagation," *J. Opt. Soc. Am.* **60**, 826–830 (1970).

[60] J. C. Dainty and R. J. Scaddan, "Measurements of the atmospheric transfer function at Mauna Kea, Hawaii," *Mon. Not. R. Astr. Soc.* **170**, 519–532 (1975).

[61] D. Kelsall, "Optical 'seeing' through the atmosphere by an interferometric technique," *J. Opt. Soc. Am.* **63**, 1472–1484 (1973); "Interferometric evaluation of the imaging characteristics of laser beams propagated through the turbulent atmosphere," *Phot. Sci. Eng.* **21**, 123–129 (1977).

[62] B. N. Edwards and R. R. Steen, "Effects of atmospheric turbulence on the transmission of visible and near infrared radiation," *Appl. Opt.* **4**, 311–316 (1965).

[63] A. H. Mikesell, A. A. Hoag, and J. S. Hall, "The scintillation of starlight," *J. Opt. Soc. Am.* **41**, 689–695 (1951); (*a*) Figure 10.

[64] R. S. Lawrence and J. W. Strohbehn, "A survey of clear-air propagation effects relevant to optical communication," *Proc. IEEE* **58**, 1523–1545 (1970).

[65] R. W. McMillan and N. P. Barnes, "Detection of optical pulses: the effect of atmospheric scintillation," *Appl. Opt.* **15**, 2501–2503 (1976).

[66] J. W. Strutt, Lord Rayleigh, *Phil. Mag.* **47**, 375–384 (1899) [*Scientific Papers* **4**, 397–406, Dover, London 1964.]

[67] D. Deirmendjian, R. Clasen, and W. Viezee, "Mie scattering with complex index of refraction," *J. Opt. Soc. Am.* **51**, 620–633 (1961).

13

Integrated and Fiber Optics

13.1 INTEGRATED OPTICS [1]–[4]

13.1.1 General Description

In many applications light waves can replace electric waves ranging from radio to microwave frequencies. The potential advantage of this replacement lies in the great reduction in component dimensions that, in principle, is possible. Light waves are shorter than microwaves by a factor of 10^4, so that miniaturization by 10^8 in cross-sectional area—12 full decades in volume—is conceivably possible.

The advantages of miniaturization are manifold. We list some here.

1. Reduced materials costs (e.g., in monocrystalline materials).
2. Reduced manufacturing costs.
3. Reduced packaging costs. (In some cases, this may make feasible certain devices that otherwise would be totally impractical. See the illustration below.)
4. Reduced influence of the environment (such as vibration).
5. Speeding up of energy and information transmission due to the shorter distances. This may be important in its own right (as in computer applications) or in terms of the increased bandwidth made feasible thereby.
6. Reduced energy requirements.
7. Facilitating high-energy densities necessary in some applications

For these potential advantages to be realized, ways must be found to produce highly compact components:

1. Optical waveguides whose cross-sectional dimensions are indeed of the order of magnitude of an optical wavelength.

2. Light sources and detectors of similar dimensions.

3. Light modulation and steering devices of a size consistent with 1 and 2.

By way of illustration, a single, fully coherent laser beam could carry n telephone conversations:

$$n = \frac{\nu_t}{\Delta \nu_t} = 2 \times 10^{11}, \tag{13.1}$$

where ν_t the frequency of the laser radiation ($\nu_t = c/\lambda \approx 6 \times 10^{14}$ Hz)

$\Delta \nu_t$ the required bandwidth ($\Delta \nu_t \approx 3 \times 10^3$ Hz), and the numbers given correspond to 0.5 μm radiation and an assumed 3-kHz bandwidth requirement.[1]

This beam could be transmitted over a channel of approximately 0.5 μm diam; if it is 1 mm in diameter, 4 million times as many conversations could be transmitted simultaneously.

All this sounds very promising. But let us stop to think: how could all these conversations be fed into our thin waveguide? Allowing one tenth of a cubic centimeter for each feed-in connection, the resulting volume for the transmission unit would be 8×10^{10} cubic meters—80 times the size of a 1-km cube! On the other hand, with a cubic micrometer allotment to each junction, the volume is less than a 1-m cube.

Ideally, all the components of an integrated optics subsystem, light sources, light guides, modulators, couplers, and detectors, would be developed integrally on a single substrate, resulting in *monolithic integrated optics*. At present, gallium arsenide is the most promising "universal" substrate. On the other hand, it may be desirable to combine components of various substances in a *hybrid integrated optics* device. For example, light modulators would be made of a material chosen for its large electrooptic coefficient, and the guide, for its high transmittance. It may then be possible first to deposit the localized components on the substrate and then to form an overall guide layer over them. The light may be transferred from guide to component and back again by an appropriate choice of relative refractive index [5] or by the proper shaping of the microscopic interfaces [6].

Many of the techniques developed in the semiconductor electronics industry, such as photolithographic masking, crystal growing, ion implantation, etc., are suitable in the manufacturing of integrated optics components and systems, as well. These are the subject of the next section.

---

[1] It is also assumed that a carrier wave can be modulated with a signal whose bandwidth is half the carrier frequency and that two mutually normal polarization components can be used.

13.1.2 Manufacturing [7], [4]

The layers and channels of integrated optics may be produced either by the deposition of one material onto a substrate of some other material, or by modifying volume elements on the surface of the substrate—or inside it—by physical or chemical processes.

13.1.2.1 Sputtering [8]–[10] and Related Film Deposition Techniques.

Sputtering refers to the removal of material from a surface by means of ion bombardment. Frequently this term includes the deposition of the sputtered material onto another surface. Sputtering is a popular method for generating the thin films used in integrated optics components because the resulting films tend to exhibit high transparency.

Dielectrics, such as glass, as well as metals may be sputtered. The ions must strike the material with energy levels in excess of approximately 30 eV. They penetrate into the target and dislodge·atoms from the lattice. These atoms may then leave the target, if they are at the surface, or transfer their momentum to such atoms. The rate of sputtering is determined both by the characteristics (energy and rate) of the incident ions and by the ease with which atoms are removed from the target. This implies that the proportion of the components in a film sputtered from compound materials may differ from that of the target material. The composition of the resulting film is also affected by the structure of the substrate on which the film is formed: some atoms may be more readily accepted than others. For instance, it has been found that films sputtered from Corning 7059 Pyrex glass contain no boron oxide, although the original glass is 13% B_2O_3. Also, the atoms of the gas sustaining the discharge influence the sputtering rate: light atoms are less effective, but tend to carry a disproportionate share of the current, so that their presence should be minimized for fast sputtering.

Sputtering is a rather slow process. Films grow at a rate of approximately 2–3 nm/sec [10]; but their uniformity can readily be maintained to within 1% over a 20 cm area [8].

In plasma sputtering, the source of ions is a rf discharge in a vacuum chamber, with the pressure maintained at about 0.01 torr. Alternatively, the target may be maintained in a high vacuum and be bombarded with an ion beam generated by an ion gun. With this method, the ions may be made to strike the target obliquely, leading to improved sputtering efficiency. It is more convenient in other respects as well—especially when the substrate must be maintained in a vacuum during the sputtering—but film uniformity over large areas becomes more difficult to obtain [9].

Layers of metal oxides may be deposited by sputtering from metal targets if the atoms are permitted to react with oxygen atoms in the discharge. This technique is called *reactive sputtering*.

Complex polycrystalline layers can be deposited by *chemical vapor deposition*, by permitting two chemicals to react in the vapor phase, with the product forming a deposit on the substrate. For example, a layer of siliconoxynitride resulting from the reaction of NO and SiH_4 at 850°C has been used as a waveguide film [11]. Here the film composition, and hence the refractive index, can be controlled by controlling the proportion of the reacting gasses.

In vacuo evaporation, too, has been used in the construction of film waveguides [3a], but these techniques do not seem to have proven themselves.

In *plasma polymerization*, a discharge is passed through a gas containing an organic monomer, causing it to polymerize and settle on the electrodes. With proper precautions, this deposit may take the form of a clear film. Films with losses as low as 0.04 dB/cm at 0.633 μm have been obtained [12].

Films successfully used for this purpose include vinyl trimethyl silane (VTMS) and hexadimethyldisiloxane (HMDS). In a 200-W discharge, VTMS films grow at about 3–4 nm/sec. The characteristics of these films can be altered subsequent to their deposition by exposure to oxygen, nitrogen, chlorine, and nitric oxide. For example, heat treatment at 140°C in an oxygen atmosphere causes the refractive index of VTMS to drop from 1.53 to 1.48 in 1.5 hr. This effect offers a potential capability of "tuning" the index by controlled heat treatment [12].

13.1.2.2 Dipping and Spinning [8]. Varnishes and epoxy resins may be suitable media for waveguides in integrated optics. The substrates may be dipped into the material and withdrawn slowly. They may be withdrawn rapidly, but then the excess material must be permitted to run off. Spinning, too, may be used to spread the material more uniformly.

This technique is rapid and the film composition, including doping, may be readily controlled. However, it is difficult to obtain good uniformity. Polymethyl methacrylate and polystyrene films deposited from solution (in chloroform and toluene, respectively) exhibit birefringence due to stresses set up during the evaporation process. Photoresist, if properly processed [13], can also be used as an optical waveguide medium.

13.1.2.3 Epitaxial Crystal Films. Lightguiding films may also be grown epitaxially on compatible crystal substrates, with a somewhat lower refractive index. The required difference in refractive index may be due to one of the following:

1. A difference in composition and basic structure of the two crystals (heterostructure).

2. Free charge carriers, whose presence controls the refractive index (see Section 13.1.5.1 for a numerical illustration).

A typical example of heterostructure integrated optics is a gallium-aluminum arsenide layer grown onto a gallium arsenide substrate by liquid-phase epitaxy. In this technique, the substrate is passed over molten $Ga_{1-x}Al_xAs$. Slow cooling (0.1–0.5°C/min) permits the material to form a monocrystalline layer on the substrate.

The structure based on differences in free carrier concentration is illustrated by a layer of relatively pure, intrinsic gallium arsenide, grown onto a n-doped GaAs substrate from the vapor phase. The higher density of free charge carriers in the substrate lowers its refractive index, so that light in the intrinsic layer may be trapped and guided.

The junction between the two types of GaAs may also serve as a semiconductor junction laser in integrated optics.

13.1.2.4 Diffusion and Ion Implantation.

Turning now to methods for modifying existing layers in a substrate, we mention first the diffusion of donor-producing (n-type) atoms into a substrate doped with acceptor-producing (p-type) materials, or vice versa.

For instance, zinc may be diffused into n-type GaAs by placing the latter, together with Zn_3As_2 into an enclosure and heating it to approximately 750°C for several minutes. This produces a layer of p-type material at the surface, with a thin neutral layer below it followed by the n-type substrate, essentially unmodified. [7]

Alternatively, both semiconductor crystals and glasses can have their surface layers modified by bombarding them with protons or light ions. In crystals, such bombardment creates traps and hence reduces the density of free carriers; this increases resistivity and refractive index. The trap creation is thought to be due to displacement of lattice atoms disrupting the crystal structure [7]. By way of illustration, in gallium arsenide, protons penetrate approximately 1 μm for each 100 keV of energy.

Waveguides have also been produced by irradiating silica glass [14]. The resulting increase in refractive index may again be due to the creation of carrier traps or to the increased mass density. Conceivably, the chemical changes produced could also contribute to the change in refractive index. Results cited in Ref. 15 imply that protons penetrate silicate glass a distance:

$$d = (5W)^{1.75} \ \mu m, \tag{13.2}$$

their so-called range. Here W is the initial proton energy in MeV. As they penetrate into the material, the energy of the bombarding protons is absorbed in the form of ionization until the remaining energy approaches a critical value, at which the probability of atom-displacement-causing collisions becomes significant. This occurs approximately 0.1 μm from the end of the range, so that "monochromatic" (better: monoenergetic) protons produce a very thin layer of increased refractive index. Due to the range of energies represented by the proton beam, the region is considerably thicker than that and can be shaped as desired by scanning the proton beam through the appropriate energy range, at an appropriate rate.

Annealing the bombarded material at an elevated temperature tends to reduce drastically the effects of the bombardment. This is taken as evidence, that the major effect of the bombardment is the dislocation of lattice atoms, with annealing tending to eliminate such dislocations.

Both diffusion and ion bombardment produce refractive index changes by changing the free carrier concentration (see Section 13.1.5.1 for further discussion of this).

13.1.2.5   Use of Masks. Frequently it is desired to deposit a certain layer over a limited area or to remove material from a certain precisely defined region. This may be accomplished by masking, executed photographically or electrographically; this technique was developed for use in semiconductor electronics. The substrate is coated with a layer of photoresistive material. It is then exposed by an uv or optical image, or by an electron beam. This exposure changes the physical characteristics so that the subsequent processing removes only the exposed regions (positive) or the unexposed regions (negative material). After the deposition of the desired layer (or removal of the unwanted material), the remaining photoresist is removed (see Section 17.3.2).

In some processes, a metallic mask is required. In these, the substrate is first coated with a thin metallic film, for example, by evaporation in a vacuum, and the photoresist is deposited over the metal film. Subsequent exposure and processing leaves the metal film exposed, except in the regions to be covered by the final mask. Now the uncovered metal film can be removed, possibly by an acid treatment. The remaining photoresist is then removed, leaving the desired mask in the form of a thin metallic film.

When shaped individual channels are desired, the strip intended for the channel may be masked by a quartz fiber bent to the desired shape. The desired thickness of material is now removed by sputtering from the surface, remaining only in the region covered by the fiber, and the desired channel results [15].

13.1.3   Propagation in a Waveguide [3], [16]–[20]

13.1.3.1   Introduction. Guiding light within thin films and along small channels is a major function in integrated optics. These films and channels are then said to serve as optical waveguides, and we present here a brief survey of the basic theory underlying this process. Qualitatively, the phenomenon may be understood in terms of light rays that are totally reflected at the interface between the guide and the substrate or other matrix in which it is embedded. The condition for total reflection is simply that the angle of incidence of the ray on the interface equal, or exceed, the critical angle, θ_c, defined by

$$\boxed{\sin \theta_c = \frac{n'}{n}} \,, \qquad\qquad (13.3)$$

where n, n' are, respectively, the refractive indices of the media on the incidence side and outside of the interface. See Section 2.2.2.2.

When the thickness of the guiding layer far exceeds the wavelength of the light, and there is a substantial difference between the refractive indices, the performance can be analyzed accurately in terms of rays. When these conditions are not met, only certain rays (or field *modes*) are possible. These are found by means of Maxwell's equations with the appropriate boundary conditions. Most of the phenomena may, however, be explained in terms of rays, multiple beam interference (2.2.4)–(2.2.5), and Fresnel's equations, all of which already have been discussed. We chose this approach here because it more readily gives concrete insight into the processes.

13.1.3.2 Interference Effects

Origin of Modes. Consider a thin film of thickness, d, and of infinite extent laterally, in which a light ray propagates, making an angle

$$\alpha = \frac{\pi}{2} - \theta \qquad\qquad (13.4)$$

with the film surfaces. This is called a *slab waveguide*. If $\alpha \le (\pi/2) - \theta_c$, the ray will propagate zigzag fashion, as it bounces back and forth between the surfaces bounding the film. See Figure 13.1, which also shows the

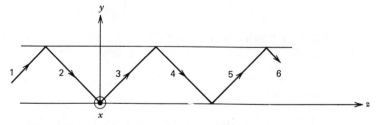

Figure 13.1 Ray path corresponding to a trapped mode in a waveguide.

coordinate system used in our analysis: the y axis is taken normal to the film surface, the z axis is parallel to the direction in which the light advances, and the x axis is normal to both. Note that the legs of the zigzag path have been numbered consecutively.

If we look upon the ray in the figure as representing an infinite wavefront normal to it, we must expect interference effects between the waves represented by the odd-numbered legs, since these propagate parallel to each other and the wavefronts have infinite extent. Similar effects apply, of course, to the even-numbered legs. At any given point inside the film, we have a superposition of waves of all the legs. At such a point, the path difference between two consecutive odd legs is

$$\Delta s = 2d \sin \alpha \qquad (13.5)$$

as implied by (2.209); note that our α is complementary there to θ. Considering the phaseshifts, ϕ_a, ϕ_b, introduced by reflection at the upper and lower surface, respectively, we find the total phase change introduced by a round trip (two legs of the zigzag path):

$$\Delta \phi = k_1 \Delta s - \phi_a - \phi_b = 2n_1 k_0 d \sin \alpha - \phi_a - \phi_b, \qquad (13.6)$$

where

$$k_1 = 2\pi/\lambda_1 = 2\pi n_1/\lambda_0 \qquad (13.7)$$

is the wave propagation constant in the film,

$n_1 =$ the refractive index there, and

$\lambda_1, \lambda_0 =$ the wavelengths in the film and *in vacuo*, respectively.

Note that ϕ_1, ϕ_2 represent phase retardation and therefore enter the phase change, $\Delta \phi$, negatively.

The condition on α representing a possible ray propagation angle is that the waves in every other leg interfere constructively, that is

$$\Delta \phi_m = 2\pi m, \qquad m = 0, 1, 2, 3 \ldots \qquad (13.8)$$

Substituting this into (13.6) we find the condition on α:

$$\boxed{\sin \alpha_m = \frac{2\pi m + \phi_a + \phi_b}{2n_1 k_0 d}}. \qquad (13.9)$$

A distinct mode is associated with each value of m. The wave associated with each such mode will have a relative phase factor

$$\exp\left[-ik_0 n_1(\pm y \sin \alpha_m + z \cos \alpha_m)\right], \qquad (13.10)$$

with the upper and lower signs associated with the upward and downward directed legs of the zig-zag, respectively. Thus the resulting wave will propagate in the z-direction with an effective propagation constant

$$k_m = k_0 n_1 \cos \alpha_m \equiv \beta. \tag{13.11}$$

In the literature, this propagation constant is often designated by the letter β, and the factor

$$N_m = n_1 \cos \alpha_m < n_1 \tag{13.12}$$

is called the *effective guide index.*

Note that from (13.3) and (13.4) the condition for total reflection throughout the path is

$$\cos \alpha_m \geqslant \frac{n_2}{n_1}, \tag{13.13}$$

where n_2 is the larger of n_a and n_b. On substituting this into (13.12), we find the lower limit on N as well and can summarize

$$n_2 < N_m < n_1. \tag{13.14}$$

At a surface where condition (13.13) is not fulfilled, part of the flux escapes upon each reflection at that surface and the corresponding modes are called "radiative." If the flux escapes only on the substrate side, but not into the superstrate (usually air), they are occasionally referred to as *substrate modes.*

Limits on Modes. Equation 13.14 implies an upper limit on the mode number, m. To find this, we solve (13.9) for m:

$$m = \frac{2n_1 k_0 d \sin \alpha_m - \phi_a - \phi_b}{2\pi} \tag{13.15}$$

$$= \frac{2n_1 k_0 d \sqrt{1 - \cos^2 \alpha_m} - \phi_a - \phi_b}{2\pi}. \tag{13.16}$$

Introducing the inequality (13.13), we find that, for trapped modes, the mode number can not exceed

$$m \leqslant \frac{k_0 d \sqrt{n_1^2 - n_2^2} - (\phi_a/2) - (\phi_b/2)}{\pi}$$

$$= \frac{V}{\pi} - \delta m. \tag{13.17}$$

Here we have introduced the very convenient quantity

$$\boxed{V = k_0 d \sqrt{n_1^2 - n_2^2} = 2\pi \frac{d}{\lambda} \sqrt{2\bar{n} \Delta n}}, \tag{13.18}$$

which is occasionally called *normalized film thickness*; it equals, approximately, π times the highest mode number the film can carry; also,

$$\delta m = \frac{\phi_a + \phi_b}{2\pi}, \tag{13.19}$$

the mode-number residue, which can not exceed unity: each ϕ vanishes at the critical angle and equals π at grazing incidence [see (13.35) and 13.36), below]. In (13.18), $\bar{n}$ is the mean of n_1 and n_2 and Δn is their difference.

Equation 13.17 implies that with

$$V < \pi(1 + \delta m) \tag{13.20}$$

no more than a single mode can be propagated and that for

$$V < \pi \, \delta m \tag{13.21}$$

no modes whatsoever can be carried. According to (13.18), the limiting value of V_1 as implied by (13.21) corresponds to a frequency:

$$\omega = k_0 c = \frac{V}{d\sqrt{n_1^2 - n_2^2}} = \frac{\pi \, \delta m}{d\sqrt{n_1^2 - n_2^2}}. \tag{13.22}$$

This is called the *cutoff frequency* of the guide.

In a symmetrical guide ($n_a = n_b = n_2$)

$$\phi_a = \phi_b = \phi, \tag{13.23}$$

say, and hence

$$\delta m = \frac{\phi}{\pi}. \tag{13.24}$$

For any given value of V, it is then possible to find an angle, α, such that

$$\phi < V,$$

so that condition (13.21) is not met and one mode is supported. In an asymmetrical guide, however, the index difference may be very small at, say, the lower interface and large at the upper. In that situation, ϕ_a may still be almost equal to π even when the critical angle has been reached ($\phi_b = 0$) at the lower surface. In that case,

$$\delta m = \tfrac{1}{2} - \varepsilon, \qquad 0 < \varepsilon \ll 1 \tag{13.25}$$

and there will be no mode propagated when

$$V < \frac{\pi}{2} - \varepsilon. \tag{13.26}$$

Similarly, condition (13.20) becomes

$$V < \pi, \qquad \frac{3\pi}{2} - \varepsilon, \qquad (13.27)$$

respectively, for symmetrical and asymmetrical arrangements.

Solving (13.17) and (13.18) for Δn, we find the condition for the existence of modes up to the mth:

$$\Delta n > \frac{(m + \delta m)^2 \lambda^2}{8 \bar{n} d^2}. \qquad (13.28)$$

In a symmetrical guide ($n_a = n_b$), the highest mode may be at the critical angle where δm vanishes. Then the condition simplifies to

$$\Delta n > \frac{(m\lambda/d)^2}{8\bar{n}}, \text{ (symmetrical).} \qquad (13.28a)$$

However, if the refractive index difference is small only at, say, the lower surface and is large at the upper surface, $\delta m = 0.5$, approximately and the condition becomes:

$$\Delta n > \frac{(m + \frac{1}{2})^2 \lambda^2}{8 \bar{n} d^2} \text{ (strongly asymmetrical).} \qquad (13.28b)$$

This inequality, with $m = 0$, gives us the condition for cut-off:

$$\Delta n < \frac{\lambda^2}{32 \bar{n} d^2}, \qquad (13.29)$$

or

$$\boxed{d < \frac{\lambda}{4\sqrt{2\bar{n}\Delta n}}}. \qquad (13.29a)$$

Transverse Distribution of Amplitude. In the y-direction, the zig-zagging wave represents a standing wave. From the y term in the exponent of (13.10), we can deduce the distribution of the wave across the film thickness. Note that the amplitude of the field does not change on reflection, so that the absolute phase at the upper surface must be

$$\pm m\pi + \frac{\phi_a}{2}; \qquad (13.30)$$

the absolute phase is similarly written for the lower surface, with the subscript b replacing a. Thus, taking the origin at the lower surface, we can write the y dependence of the field:

$$E = E_0 \exp\left(-ik_0 n_1 y \sin \alpha_m + \frac{i\phi_b}{2}\right), \qquad (13.31)$$

with the condition at the upper surface requiring

$$k_0 n_1 d \sin \alpha_m - \frac{\phi_b}{2} = m\pi + \frac{\phi_a}{2}. \tag{13.32}$$

On substituting $\sin \alpha_m$ from (13.9) into this, we obtain an identity, showing (13.30)–(13.32) to be consistent with (13.9). Equations 13.9 and 13.31 imply that the field amplitude varies sinusoidally across the film thickness, with m complete half-cycles included. Beyond the interface, the field amplitude drops exponentially, as shown in (2.110). In terms of the present parameters, the drop is according to

$$\exp\left(k_1 y \sqrt{\cos^2 \alpha - \frac{n_b^2}{n_1^2}}\right) = \exp\left(k_0 y \sqrt{n_1^2 \cos^2 \alpha - n_b^2}\right) \tag{13.33}$$

and

$$\exp-\left[k_1(y-d)\sqrt{\cos^2 \alpha - \frac{n_a^2}{n_1^2}}\right] \tag{13.34}$$

below the lower and above the upper surface, respectively. See Figure 13.2 for representative distributions corresponding to $m = 0$, 1, and 2.

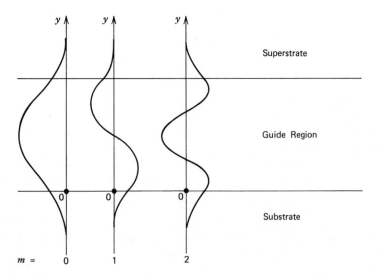

Figure 13.2 Transverse distribution of field amplitude in trapped modes.

13.1.3.3 *Phase Shifts.* TE *and* TM *Modes.* The phase shift introduced by the reflections at the interfaces can be deduced from Fresnel's equations (2.101a), (2.105a), after expressing θ' in terms of θ. This yields

$$\frac{E''_{\text{TE}}}{E_{\text{TE}}} = \frac{n_1 \cos\theta - n_2\sqrt{1 - (n_1/n_2)^2 \sin^2\theta}}{n_1 \cos\theta + n_2\sqrt{1 - (n_1/n_2)^2 \sin^2\theta}} \qquad (13.35)$$

and

$$\frac{E''_{\text{TM}}}{E_{\text{TM}}} = \frac{n_2 \cos\theta - n_1\sqrt{1 - (n_1/n_2)^2 \sin^2\theta}}{n_2 \cos\theta + n_1\sqrt{1 - (n_1/n_2)^2 \sin^2\theta}}. \qquad (13.36)$$

(In accordance with popular usage in the analysis of guided waves, we have substituted the subscripts TE and TM for $\perp$ and $\parallel$, respectively. TE and TM stand for "transverse electric" and "transverse magnetic," respectively.)

When $\sin\theta$ exceeds n_2/n_1, the second term in each factor becomes imaginary and the expressions may be written

$$\frac{E''_{\text{TE}}}{E_{\text{TE}}} = \frac{\cos\theta - i\sqrt{\sin^2\theta - (n_2/n_1)^2}}{\cos\theta + i\sqrt{\sin^2\theta - (n_2/n_1)^2}} \qquad (13.37)$$

and

$$\frac{E''_{\text{TM}}}{E_{\text{TM}}} = \frac{\cos\theta - i(n_1/n_2)^2\sqrt{\sin^2\theta - (n_2/n_1)^2}}{\cos\theta + i(n_1/n_2)^2\sqrt{\sin^2\theta - (n_1/n_2)^2}}. \qquad (13.38)$$

Note that in both of these the phase of the phasor associated with the numerator is the negative of that associated with the denominator and that the magnitudes of the two are equal. Hence the phasor associated with the fraction has unity magnitude and a phase equal to twice that of the phasor of the numerator alone, so that we can write immediately the tangent of half the phase shift introduced by the reflection: it is simply the ratio of the imaginary to the real part of the numerator, that is

$$\tan \tfrac{1}{2}\phi_{\text{TE}} = \frac{\sqrt{\sin^2\theta - (n_2/n_1)^2}}{\cos\theta}$$

$$= \frac{\sqrt{\cos^2\alpha - (n_2/n_1)^2}}{\sin\alpha} \qquad (13.39)$$

$$\tan \tfrac{1}{2}\phi_{\text{TM}} = \frac{\sqrt{\sin^2\theta - (n_2/n_1)^2}}{(n_2/n_1)^2 \cos\theta}$$

$$= \frac{\sqrt{\cos^2\alpha - (n_2/n_1)^2}}{(n_2/n_1)^2 \sin\alpha}, \qquad (13.40)$$

where we have made use of (13.4) and changed the sign so that the ϕ represent phase retardation.

Goos–Hänchen shift. So far we have considered plane wavefronts of infinite extent. It has been found that, when a narrow pencil of light is totally reflected at an interface between two dielectrics, it experiences a lateral displacement (Δz) parallel to the interface, in addition to the phase shift. This displacement is known as the Goos–Hänchen shift.[2] It can be interpreted in terms of a model in which the reflection does not take place at the interface itself, but rather at a virtual surface, parallel to the surface, and displaced a distance Δy into the lower-index region (see Figure 13.3). We present here a heuristic derivation, which, it is hoped, will provide insight into the origin of this phenomenon.

We simulate a limited bundle of rays by providing two infinite plane waves with a small angle between them, so that their propagation constants may be represented by

$$\beta + \tfrac{1}{2}\Delta\beta, \qquad \beta - \tfrac{1}{2}\Delta\beta,$$

respectively [see (13.11)]. This implies beats in the plane of the interface, with the spacing between peaks:

$$z_{j+i} - z_j = \frac{2\pi}{\Delta\beta}, \tag{13.41}$$

where z_j designates the location of the jth peak. [See the equation preceding (2.41) with $t = 0$ and $a = 1$ and a unity change in n; also note that $\Delta\beta$ here is analogous to $2\Delta k$ there.] More generally, a displacement Δz corresponds to a phase difference $\Delta\phi$:

$$\Delta z = \frac{\Delta\phi}{\Delta\beta}. \tag{13.42}$$

In this equation we may now interpret the phase shift $\Delta\phi$ as being due to

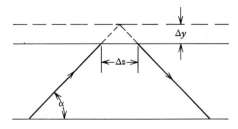

Figure 13.3 Illustrating the Goos–Hänchen shift, Δz.

[2] F. Goos and H. Lindberg–Hänchen (1943–1949).

the difference in ϕ_a of (13.39) and (13.40) due to the difference, $\Delta\beta$, in the propagation constants:

$$\Delta\phi = \phi_a(\beta + \tfrac{1}{2}\Delta\beta) - \phi_a(\beta - \tfrac{1}{2}\Delta\beta). \tag{13.43}$$

Δz is then the resulting shift in the location of the beat peak. In the limit, (13.42) becomes the expression for the Goos–Hänchen shift:

$$\Delta z = \frac{d\phi}{d\beta} \equiv \frac{\dfrac{d\phi}{d\alpha}}{\dfrac{d\beta}{d\alpha}} \tag{13.44}$$

$$\Delta z = \frac{2 \cot \alpha}{k_0 n_1 \sqrt{\cos^2 \alpha - (n_2/n_1)^2}}, \qquad \text{TE} \tag{13.44a}$$

$$= \frac{2 n_1 \cot \alpha}{k_0 n_2^2 \sqrt{\cos^2 \alpha - (n_2/n_1)^2}}, \qquad \text{TM}, \tag{13.44b}$$

where we differentiated (13.39) and (13.40) with respect to α, noting that, for the left side of the resulting equation

$$\frac{d}{d\alpha} \tan \tfrac{1}{2}\phi = \tfrac{1}{2}(1 + \tan^2 \tfrac{1}{2}\phi)\frac{d\phi}{d\alpha} = \frac{1 - (n_2/n_1)^2}{\sin^2 \alpha}\frac{d\phi}{d\alpha}, \tag{13.45}$$

and obtained from (13.11)

$$\frac{d\beta}{d\alpha} = -k_0 n_1 \sin \alpha. \tag{13.46}$$

From Figure 13.3 it is evident that the equivalent penetration depth Δy is given by

$$\Delta y = \tfrac{1}{2}\frac{\Delta z}{\cot \alpha}$$

$$= \frac{1}{k_0 \sqrt{n_1^2 \cos^2 \alpha - n_2^2}}, \qquad \text{TE} \tag{13.47a}$$

$$= \frac{n_1^2}{n_2^2 k_0 \sqrt{n_1^2 \cos^2 \alpha - n_2^2}}, \qquad \text{TM}. \tag{13.47b}$$

Comparison with (13.33) shows that the penetration depth for the TE mode is equal to the depth yielding attenuation by $1/e$. This shift is accompanied by a time delay [21]

$$\Delta t = n_1 \cos \alpha \frac{\Delta z}{c}. \tag{13.48}$$

It can be shown that the group velocity obtained from this ray picture is correct when the time delay, Δt, is included [21].

13.1.3.4 Channel Guides. The number of modes discussed in Section 13.1.3.2 referred to a single direction relative to the z axis. In an infinite slab, any other direction in the x–z plane (azimuth) is possible, so that there is an infinity of possible modes. When the waveguide is limited in two directions, it is called a channel waveguide. In such waveguides, modes are restricted in azimuth as well as in elevation. Considering a rectangular cross section, $d_x \times d_y$, the conditions on the modes are similar to (13.9), except that there are now two angles θ: θ_y and θ_x are the angle the wavevector makes with the y and x axes, respectively. Each of these, individually, must satisfy (13.9). The component in the direction of the guide axis is then, again, as in (13.11)

$$\beta = n_1 k_0 \cos \alpha,$$

with

$$\cos \alpha = \sqrt{1 - \cos^2 \theta_x - \cos^2 \theta_y}. \tag{13.49}$$

The permitted values of $\theta_{x,y}$ are then, analogously to (13.9),

$$\cos \theta_{x,y} = \frac{2\pi m + \phi_a + \phi_b}{2 n_1 k_0 d_{x,y}}, \tag{13.50}$$

with $\sin \theta_{x,y} > n_2/n_1$ for the ray to be trapped.

13.1.4 Coupling of Wave Energy [3], [4], [16], [22]

The transfer of light from one component to another is done via radiation coupling. This occurs when light from a radiation source is introduced into an optical waveguide and when it is transferred from the waveguide to the detector. It may also occur between light guides, transferring light from one guide to another. A number of mechanisms are available for accomplishing this.

13.1.4.1 End-On Coupling. The most obvious way of inserting light into a waveguide is by simple illumination of the guide edge, called "end fire" coupling. However, for this insertion to be efficient, the distribution of the illumination must match the amplitude-squared distribution in the guide, for example, as shown in Figure 13.2. Simultaneously, the illumination must be limited in angular distribution, again to match the modes to be excited in the guide. This limitation precludes substantial demagnification of any source not highly collimated. Consequently, efficient coupling is feasible only from lasers. Since the zero-order modes of slab and rectangular waveguides resemble the Gaussian distribution often obtained from laser sources, the coupling could, in principle, be made quite efficient. In practice, however, the method is very difficult, because the microscopic dimensions of the guides demand enormous precision on part of the associated optics.

Some interesting progress has been made in this area; microscopic lenses, both spherical and cylindrical, have been made of photoresist, directly on the optical fiber waveguides [23]. More recently, such lenses have been made by melting glass or epoxy onto a fiber ending that has been appropriately prepared [24]. Nevertheless, best results are obtained with the monolithic laser-guide arrangement described in Section 13.1.5.2 and the prism and related couplers described in the next section.

13.1.4.2 Tapered Film, Prism, and Grating Coupling. More efficient methods of coupling are based on techniques in which electromagnetic flux runs parallel to the guide and gradually leaks into it.

Tapered Film. The *tapered thin-film coupler* is illustrated in Figure 13.4, which shows a ray propagating along a slightly tapered film. If the refractive index, n_3, of the tapered film is greater than n_1, that of the receiving film below it, the ray will be totally reflected for large angles of incidence. As the ray propagates down the film toward the thin end, its angle of incidence on the film surfaces decreases by 2γ upon each second reflection. As soon as this angle falls below the critical angle of the interface ($\sin^{-1} n_1/n_3$), the flux starts leaking into the receiving film in significant quantities. When the tapered film has shrunk to the size (13.29a) below which it can no longer sustain any mode, all the flux will have been transferred to the receiving film, or reflected backward, toward the thick end of the tapered film. For small index differences ($n_3 - n_1$), this latter portion should be very small.

This technique can be used to feed laser light into a waveguide, as well as to transfer light from one guide to another in hybrid integrated optics

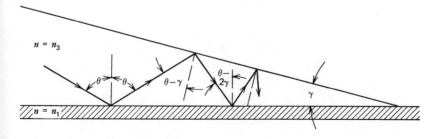

Figure 13.4 Tapered thin-film coupler.

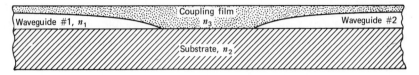

Figure 13.5 "Coupling film" in hybrid integrated optics system.

circuits; see Figure 13.5. Here the "coupling film" has an index, n_3, greater than that of the substrate. If, in addition, $n_3 > n_1$, the index of the first guide, the light will enter the "coupling film" immediately on striking it. Otherwise it will do so when the angle of incidence exceeds the critical angle. The light then passes along the "coupling film" until it enters the second waveguide [6].

Prism Coupler. The prism coupler is illustrated in Figure 13.6. It consists of a prism above, and in close proximity to, the waveguide layer, a thin film in our example. The light enters the prism and is partially reflected at the base, which runs parallel to the film surface, or almost so. Part of the incident flux, however, penetrates the airgap and enters the film to excite in it guide modes, which then propagate as described in Section 13.1.3. The excitation takes place over the extended region BB'. If the spatial frequency of the amplitude distribution along BB' matches that of the mode in the film, the mode will be reinforced as it advances from B to B', and substantial amounts of flux may be transferred in this manner. This frequency can be seen to be represented by the propagation constant

$$k_3 = k_0 n_3 \sin \theta_3, \qquad (13.51)$$

where n_3 is the refractive index of the prism material, and θ_3 is the angle of incidence on the prism-film interface. For phase matching this must

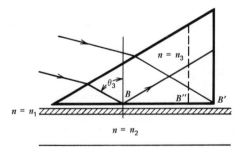

Figure 13.6 Prism coupler, coupling into a film.

equal one of the permissible mode propagation constants (13.11)

$$\beta_m = k_0 n_1 \sin \theta_1, \tag{13.52}$$

where $\theta_1 [= (\pi/2) - \alpha]$ is the angle the mode's wavevector makes with the film surface normal. Note that, in view of Snell's law, the matching condition [k_3 (13.51) equals β_m (13.52)] simply implies that the angle of refraction be that corresponding to the desired zig-zag path.

In the direction normal to the interface, the wave entering from the prism, will decay exponentially in the airgap and so will the wave trapped in the guiding film. The extent of overlap between these two field distributions in the airgap determines the coupling constant, K_{13}, which describes the rate at which flux is transferred across the gap. As the mode builds up in the film, some of its energy will be transferred back to the prism, so that the prism must be terminated at some point, B'', for optimum transfer to the film.

We can gain valuable insight on coupling optimization from analyzing the reverse process: coupling flux from the film into the prism, the process appropriate, for instance, at the detection stage. Here there is no transfer in the reverse direction to complicate the analysis (see Figure 13.7). Since the rate of amplitude transfer is proportional to the amplitude (A_1) in the film:

$$\frac{dA_1}{dz} = -K_{13}A_1, \tag{13.53}$$

the amplitude in the film will decay exponentially along z. (See the next section for a more thorough discussion of the coupling constant, K_{13}.) Here all the energy will, eventually, be transferred to the prism, provided it is made infinitely long. From the reversibility of optical rays, it follows that we could obtain complete transfer from the prism to the film simply by reversing the direction of all rays in Figure 13.7. This implies, of course, an exponentially increasing illumination onto the prism. If, instead, uniform illumination is used, it is found that the optimum transfer efficiency is approximately 81% and requires an illuminated region of length [16]

$$L = \frac{1.25}{K_{13}}, \tag{13.54}$$

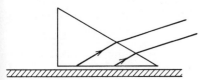

Figure 13.7 Prism coupler, coupling out of a film.

where the coupling constant

$$K_{13} = \frac{e^{-2pd}}{w_e} \sin \phi_{12} \sin \phi_{32} \tan \alpha, \qquad (13.55)$$

where $p = \sqrt{\beta^2 - k_0^2 n_2^2}$ (13.56)

d = the gap thickness

w_e = the effective film thickness including the Δy's of the Goos–Hänchen effect (13.47) and

ϕ_{12}, ϕ_{23} = the phase shifts due to the reflections as given by (13.39) and (13.40), with subscript 2 referring to the gap.

Instead of varying the illumination exponentially, the coupling constant may be varied by controlling the gap width, for example, by applying nonuniform pressure to the substrate carrying the film. The major remaining obstacle to the highest transfer efficiencies is the extremely sharp edge required of the prism at point B'' in Figure 13.6. There the radiation density is maximum, and any imperfection, such as chips or bevels, will cause significant radiation loss.

Grating Couplers. The grating coupler works on a principle similar to that of the prism. The light is incident at an angle such that the diffraction angle corresponds to that of one of the permitted zig-zag directions. Alternatively, the condition may be described in terms of the periodic field variations of the wavefront obliquely incident on the grating, as modified by the grating complex transmittance: for efficient flux transfer one of the harmonic components of this product must match the phase of a permitted mode of the film.

Grating couplers of this type can be constructed integrally with the film. An appropriate layer is deposited over the desired grating area. A mask is then formed over it by means of an evaporated metal film and photoresist, exposed by an interference pattern and processed. The resulting mask controls the subsequent sputtering by means of which parallel strips of the layer are deposited or removed, leaving the desired grating. Thick phase gratings may be made by the holographic techniques described in Sections 19.3.1.6 and 19.3.6.1. Such a grating, if it satisfies the Bragg condition (14.117), can be highly efficient, launching most of the flux into one mode. Unfortunately, it is there impossible to satisfy exactly the phase-matching and Bragg conditions simultaneously, so that a compromise is necessary [25].

13.1.4.3   Directional Coupling. The type of coupling described in our discussion of the prism coupler, occurs also whenever two optical waveguides are arranged so that their mode fields overlap. This is called *directional coupling.* For example, if two identical waveguide channels run parallel and in close proximity to each other, the wave propagating in one of them will slowly be transferred to the neighboring one. Upon completing the transfer to the second one, the wave will begin to be transferred back to the first: the energy will oscillate back and forth between the two guides. Specifically, the rate at which the amplitude (A_2) builds up in the second channel may be described by the differential equation:

$$\frac{dA_2}{dz} = -i\beta_c A_2(z) + K_{21} A_1(z) \tag{13.57a}$$

and the decay of the amplitude, A_1, in the first one by

$$\frac{dA_1}{dz} = -i\beta_c A_1(z) + K_{12} A_2(z), \tag{13.57b}$$

where

$$\beta_c = \beta - i\alpha_a/2 \tag{13.58}$$

is the complex mode propagation constant along z,

α_a = the absorption coefficient in the guides, and
K_{ij}, $i, j = 1, 2$ are the coupling constants.

The first term in each equation represents the loss of amplitude from the channel under consideration and the second term, the energy transferred to it from the opposite channel. The coupling constants here are imaginary and can be written

$$K_{12} = K_{21} = -iK, \tag{13.59}$$

where K is real.[3]

The solution of (13.57) can be seen to be:

$$A_1(z) = A \cos Kz \, e^{i\beta z} \tag{13.60a}$$

$$A_2(z) = A \sin Kz \, e^{i(\beta z - \pi/2)}. \tag{13.60b}$$

The flux in the two guides is then, respectively,

$$P_1 = A_1 A_1^* = A^2 \cos^2 Kz \, e^{-\alpha z} \tag{13.61a}$$

$$P_2 = A_2 A_2^* = A^2 \sin^2 Kz \, e^{-\alpha z}. \tag{13.61b}$$

According to these equations, if a wave is launched into one of the two channels, all the power will have been transferred to the other one after

---

[3] Note that the coefficient $(-i)$ signifies that the contribution to one guide, due to its neighbor, lags in phase $\pi/2$ radians behind the phase of the donor. Hence, any influence of the recipient, back to the donor, is in phase opposition and hence contributes only negatively. This explains why energy is transferred totally to the neighbor, before it starts building up again in the original "donor."

propagating over a distance

$$L = \frac{\pi}{2K}. \tag{13.62}$$

After twice that distance, it will have returned to the first channel, and a propagation distance $L/2$ is required for half the power to be transferred.

13.1.5 Components [3], [7]

Having surveyed the major processes utilized in integrated optical systems, we now turn to a discussion of the chief components used.

13.1.5.1 Waveguides

Substance. An optical waveguide consists of an extended region of relatively high refractive index and low attenuation surrounded by media of lower refractive index, so that electromagnetic waves can propagate in it with reasonably small absorption and radiation losses. As discussed in Section 13.1.2.3, the difference in refractive index may be obtained by making the guide of a chemical composition differing from that of its surroundings or by inducing a change in free carrier concentration. Specifically, the change in refractive index, due to a concentration of N free carriers per unit volume, is given by [7]

$$\Delta n = \frac{-Nq_e^2}{2\varepsilon_0 n\omega^2 m^*} = \frac{-Nq_e^2\lambda_0^2 \times 10^{-7}}{2\pi n m^*}, \tag{13.63}$$

where n = the refractive index of the intrinsic semiconductor,
$q_e (\approx 1.6 \times 10^{-19} C)$ is the free carrier charge,
$\varepsilon_0 = 10^7/4\pi c^2$ is the permittivity of free space, and
m^* = the effective free carrier mass.

For electrons (mass $\approx 9.1 \times 10^{-31}$ kg) the effective mass is about 6×10^{-32} kg.

Referring back to Section 13.1.3.2, we can determine the required increase (ΔN) in carrier concentration as determined by the desired maximum mode number, m. Thus, solving (13.63) for N and substituting for Δn from (11.28b):

$$\Delta N = \frac{10^7 \pi (m + \frac{1}{2})^2 m^*}{4q_e^2 d^2}. \tag{13.64}$$

By way of illustration, let us consider a waveguide constructed by growing an intrinsic epitaxial film of GaAs (free carrier concentration N_f) on a substrate of n type GaAs (free carrier concentration $N_s \gg N_f$). With a layer thickness of 1 μm, we find the required carrier concentration

$$N_s \approx N_s - N_f \approx 5 \times 10^{24} \text{ m}^{-3}$$

for a single-mode guide ($m = 0$). Returning to (13.63) and noting that the refractive index of GaAs is approximately 3.3, we find, for $\lambda_0 = 1 \mu$m, that the corresponding refractive index difference is approximately 0.01.

In epitaxially grown heterostructure waveguides, care must be taken to ensure that the lattice structure approximate closely that of the substrate. A popular form is a layer of gallium–aluminum arsenide ($Ga_{1-x}Al_xAs$) grown on gallium arsenide (GaAs). Such a layer has an index of refraction lower than that of the GaAs by an amount

$$\Delta n = 0.4x. \tag{13.65}$$

A waveguide requires a layer of higher index; this can be obtained by depositing another layer of GaAs over the layer of GaAlAs, with the latter acting as an insulation layer, isolating the waveguide from the substrate.

Form. Waveguides may be classified according to the shape of their cross section. The simplest such guides are in the form of films and are effectively limited only in one dimension. These are called *slab* or *planar guides*. When limited in width as well, they are called channel waveguides. Those without substrate are called fibers and are treated in Section 13.2.

As discussed in Section 13.1.2, channel waveguides may be deposited above a substrate; these are called *ridged* or *strip* guides. They may also be *embedded* in the substrate. A third alternative, the *strip-loaded planar guide*, is a hybrid of the ridged channel and the planar waveguide. If the thickness of the planar guide is below cutoff [see (13.21) and (13.22)], and the strip is sufficient to bring it above cutoff, by increasing n_2 and hence V [see (13.18)], waves can propagate in the planar guide only under the strip. Nevertheless, most of the energy may be propagating in the planar guide and not in the strip. This enables us to minimize drastically the scattering losses due to edge-roughness normally plaguing channel guides: the strip-loaded planar guide is an attempt to obtain the benefits of a channel guide without its disadvantages [26]. See Figure 13.8 for an illustration of the three types of channel guides and the associated energy distributions.

13.1.5.2 Light Sources. In principle, any light source can be used in integrated optics. In general, however, effective coupling is difficult (see Section 13.1.4). For efficient coupling, a small source area must be combined with directional concentration, and often monochromaticity,

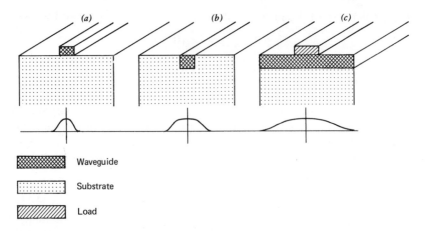

Waveguide

Substrate

Load

Figure 13.8   Channel guide configurations. (*a*) Ridged or strip; (*b*) embedded; (*c*) strip-loaded. The associated amplitude distribution is shown below each schematic.

too, is important. Lasers combine all these requirements and are, therefore, the most popular light sources used in integrated optics. Also, strong coherence is necessary if a single mode is to launched, and this, too, is relatively readily available from a laser. Light emitting diodes (LED) satisfy the small area and monochromaticity requirements and, therefore, are an occasional alternative to the laser.

The choice among the various types of laser available is dictated by the requirement of small size which is the *raison d'être* of integrated optics. This points to the solid-state and semiconductor lasers as prime candidates. For ease in pumping, the injection-pumped semiconductor laser appears to be the most important light source in integrated optics. Such a laser can be coupled to the waveguide by any of the techniques available for radiation coupling as described in Section 13.1.4.

If the light source is to be formed monolithically with the guide, GaAs appears to be the most promising material. For operation at room temperature, the double heterostructure laser, consisting of a layer of GaAs sandwiched between two layers of GaAlAs has been used. In practice, this sandwich itself is sandwiched between GaAs layers for better electric contacts, as required for the injection pumping. The layers of GaAlAs confine the radiation to the central layer, and the only remaining problem is the formation of the reflecting end faces required for the laser cavity. In lasers made separately, cleaving suffices to this end. However, for lasers formed monolithically on the GaAs substrate, other methods must be used. It has been possible to grow the required GaAs layers from the vapor phase in the form of mesa with faces sufficiently smooth and well oriented to form a laser cavity. Unfortunately, these lasers must be operated at liquid nitrogen temperatures; the heterostructure laser can not be grown in this manner so that the required cavity can not be obtained.

Fortunately, there exists an alternative to the reflecting surfaces: the so-called *distributed feedback*. With the proper choice of pitch, a periodic modulation of the waveguide surface will act as as a reflector. In principle, this can be understood as the action of a grating diffracting the wave impinging on it in its zig-zag progression along the guide. More quantitatively, it should be considered as a Bragg diffraction effect with normal incidence, as described by (14.117) with $\theta = \pi/2$. Hence the condition for reflection is that the pitch of the grating be an integral multiple of half-wavelengths. In a pumped laser cavity corrugated at that spatial frequency, a wave will build up, due to the population inversion, as it progresses along the cavity. It will then diminish, due to the scattering by the corrugations and as it diminishes, the wave progressing in the opposite direction builds up [27], and an effective cavity results.

13.1.5.3 Lenses and Prisms. Lenses and prisms in "two dimensional" versions can be incorporated into integrated optics systems. In one technique, the lens or prism is deposited on the substrate in the form of a thin cross-sectional slice of the prototype, generally via a mask (see Figure 13.9). A coupling film is then deposited over them. This permits coupling the wave into the prism or lens and from one into the other, via the coupling film. The edges may be either sharp or tapered. An isosceles right triangle, with the hypotenuse normal to the axis, has been suggested as a "two-dimensional corner-cube," possibly to serve as a reflector in an integrated optics laser cavity [5].

An electrooptically induced "prism" for use in integrated optics has been proposed. It is illustrated schematically in Figure 13.10. The N-shaped arrangement of metallic electrodes is deposited onto electrooptic

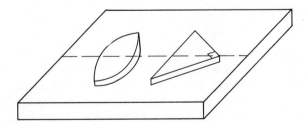

Figure 13.9 "Two-dimensional" lens and prism in an integrated optical system.

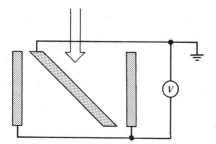

Figure 13.10 Electrooptic "prism" for integrated optics.

material (e.g., titanium-doped lithium niobate). When the diagonal electrode is grounded and a voltage is applied to the legs of the "N," fields of opposite polarity will appear above and below the diagonal. Furthermore, each one of these fields will run oblique to the direction of the light propagation (shown by a hollow arrow in the figure). This, in turn, will introduce a refractive index gradient into the electrooptic material, causing the deflection of the wavefront propagating through it. A reversal of the polarity, will reverse the direction of the deflection, so that this "prism" can be used as a double-throw switch [28].

13.1.5.4 Modulators [3], [7], [29]. The light passing through the integrated optical system may be modulated in accordance with an applied electrical signal by means of any of the basic techniques—electrooptic, magnetooptic, or acoustooptic—used for high-frequency modulation in the usual optical system.

Electrooptic Modulators. If an electric field is applied to an optical waveguide made of electrooptically susceptible material, birefringence is introduced, in general (see Section 14.3). Consider an electric field applied normal to the wavevector of the light. If the light is polarized in the direction of the applied field, or normal to it, the induced birefringence will appear simply as a change in refractive index. If the light is linearly polarized in any other direction, or elliptically polarized, its ellipticity will change as it passes through the guide. Either of these two effects may be used to produce amplitude modulation. Below we list some of the ways in which this can be done.

1. The increase in refractive index due to the electrooptic effect may suffice to convert a region of the crystal into a waveguide. Alternatively, a reduction in refractive index may bring an existing guiding region to cutoff, so that it will no longer guide the wave. In a related effect, the ability to confine certain modes may be impaired by the index reduction, so that some modes will no longer be guided, but radiated or absorbed in metallic substrates.

In this technique, the index change introduced by the electric field must be comparable in magnitude to that required for the guide to carry a single mode [see (13.29)]. For instance, in GaAs, the maximum index attainable (before electrical breakdown of the material occurs) is $\Delta n = 1.6 \times 10^{-3}$. This implies (see 13.29a) that the guide must have a thickness of at least

$$d \geqslant 2.5 \ \mu\text{m}, \quad \text{at} \quad \lambda = 1 \ \mu\text{m},$$

to sustain a single mode.

2. In a directional coupling arrangement, the index change may be used to create or destroy the coupling effect between two parallel guides and this effect, too, may be used for modulation.

3. Directional coupling may also be used to detect phase shifts by means of interference effects. The "balanced bridge modulator," is analogous to the Mach–Zehnder interferometer (Figure 2.32) and is illustrated in Figure 13.11. Here the regions marked "C" are coupling regions in which half the energy is transferred from the guide with the leading phase into its neighbor. At the points marked $\pm\phi$, phase shifts, as indicated, are introduced electrooptically. If now light flux Φ is introduced at Port 1, the flux received at Ports 3 and 4 will be, respectively,

$$\Phi = \Phi_1 \cos^2 \phi, \quad \Phi_1 \sin^2 \phi. \tag{13.66}$$

4. The induced birefringence may be used to convert the guide into a $\lambda/2$ retardation plate and thus to rotate the direction of polarization of linearly polarized light through 90°. In conjunction with a polarization analyzer, this can be converted into amplitude modulation. More generally, for linearly polarized light, a retardation ϕ will result in a transmittance factor

$$\tau = \cos^2\phi \quad \sin^2\phi \tag{13.67}$$

for parallel and crossed polarizer arrangements, respectively.

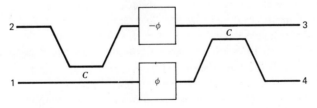

Figure 13.11 Balanced bridge modulator. After F. Zernike [29].

5. By evaporating a periodic metallic grating onto the electrooptic layer, and applying a voltage to it, a periodic index variation may be introduced into the layer. If the grating lines run parallel to the light propagation vector, or nearly so, the system will act like a phase diffraction grating — quite efficiently so, if the Bragg condition is satisfied.

The grating lines may also be used running perpendicular to the propagation vector. This may introduce, into the guided flux, new modes with propagation constants

$$k_m \pm \frac{2\pi}{\Lambda},$$ (13.68)

where Λ is the grating spacing.

Because of the enormously high levels of energy concentration readily obtainable in integrated optics systems, electrooptic light modulation can be made very *efficient*. This constitutes a major part of the promise held by integrated optics as such. Consider (14.194) according to which the power requirements for electrooptic light modulation are

$$\boxed{P = \frac{K\lambda^2 \phi_{mx}^2 \, \Delta\nu A_c}{L}},$$ (13.69)

where K is a constant including the optical characteristics of the material,
   ϕ_{mx} is the amplitude of the resulting phase modulation,
   L is the length of the passage through the material,
   A_c is the cross-sectional area of the beam, and
   $\Delta\nu$ is the modulation bandwidth.

Since the power requirements are proportional to the cross-sectional area of the region modulated by the electrooptic effect, we clearly reduce these requirements by a factor of one million when we go from millimeter to micrometer dimensions.

The time required for the light to pass through the modulated region forms a limit on the modulation *bandwidths* obtainable in these devices: if the modulating field has passed through π radians, i.e., has reversed its direction, before the modified lightwave has completed its passage through the modulated region, the field applied when the light reaches the end of the region will cancel out the effect of the field applied π radians earlier. *Traveling-wave* modulators, in which the modulating field moves along with the light, permit the extension of the modulating region without, thereby, limiting the attainable bandwidth.

In conclusion, we mention *electroabsorption*, which may be used to modulate the transmitted light flux directly by means of an electric field.

For light whose wavelength is near the bandgap of a semiconductor material, its transmittance can be modified substantially by an applied field, which shifts the bandgap. This phenomenon has been used to modulate light with power requirements of only 200 μW/MHz [30]. See Section 14.3.3 for additional discussion of this phenomenon.

Acoustooptic Modulators. The periodic index modulation described under 5. above, can also be obtained by means of acoustic waves. As surface waves, these modulate the thickness and as compressional waves the refractive index. Both of these form gratings coupling a frequency change into the lightwave propagating along the guide and, hence, may switch it into another mode. This new mode may be lost to the circuit, for example, by being unconfined, or it may be coupled to (or uncoupled from) a parallel guide. In planar slab guides, the grating may change the direction of propagation of the guided beam and hence switch it from one detector to another.

Magnetooptic Modulators. A magnetooptic effect, Faraday rotation (Section 14.4.2), too, has been used for modulating the flux in an integrated optics system. Specifically, a film of iron-garnet, grown epitaxially on a gallium-garnet substrate, permitted modulation at 80 MHz, at least. The modulation was effected by switching flux from the TE to the TM mode and *vice versa.*

13.1.5.5 Detectors [7]. The detector used in conjunction with an integrated-optics device may, of course, be external with the light coupled into it by any of the techniques described in Section 13.1.4. Ideally, however, the detector should be integral with the system as a whole. The coupling can then be end-on, preferably by having the detector butt the guide; or a coupling layer, with its tapered edge, may be used as described in Section 13.1.4.2 for use in hybrid integrated optics systems.

A number of detectors that have been built integrally in this manner are listed below.

1. Silicon p-i-n diodes have been diffused by the usual techniques onto a silicon substrate. The associated lightguide was a film of glass sputtered onto the substrate with a layer of SiO_2 (obtained by oxidizing the silicon) serving as an isolation layer. Coupling to the diode was via a tapered guide. Such a diode, with its intrinsic, light-sensitive region buried some micrometers below the surface, is ideal for use with embedded waveguides.

2. Schottky barrier diodes are more suitable for use with surface guides, in conjunction with a tapered coupling film.

3. Detectors may also be grown epitaxially into a hole etched into the waveguide region.

If the detector could be made of the same material as the guide, this would offer great advantages in manufacturing economy. The obvious difficulty here is that, as a waveguide, the material must be transparent and, as a detector, it must absorb light efficiently. Several techniques may be used to accomplish this.

1. If the material is used near the bandgap, an electric field may induce electroabsorption and this may create a photodiode in a heterostructure waveguide.

2. Ion implantation may change the guide material locally into a photodetector.

3. Diffused impurities may be used to convert a portion of the guide into a detector.

13.2 SINGLE FIBERS [31]–[34]

Fiber optics is closely related to integrated optics and may be considered the first of its branches to be developed. Since it is unrestrained by the need for a substrate, fiber technology is more flexible and has already found numerous important applications. The characteristics of single fibers are described in this section and those of fiber bundles in the next.

13.2.1 Manufacture and Materials

13.2.1.1 Cladding. Just as in integrated optics, the functioning of a fiber as a waveguide depends on total internal reflection. This implies, that the fiber must be surrounded by a medium of relatively lower refractive index. If the fiber is freely suspended in air, this condition is well satisfied. However, light will be lost at any point of contact with another material, especially if it is of equal or higher refractive index, and such contact is generally difficult to avoid. Contamination of the surface by dust, grease, etc. are likely to cause similar losses. Hence optical fibers are usually surrounded by some kind of coating of lower-index material. This may take the form of a well-defined sheath or cladding, resulting in a *stepped-index fiber*, or it may be accomplished by a gradual reduction of the index with distance from the fiber axis, resulting in a *graded-index fiber*. In a third form, the *single-material fiber*, the core is suspended by means of a thin wall-to-wall plate inside a tube. Some advantages in mode selection and transmittance seem to be attainable with *doubly-clad fibers*, consisting of a higher-index core surrounded by a lower index cladding, which, in turn, is surrounded by another cladding of intermediate index [35]. See Figures 13.12*a–d* for schematics of the various cross sections. Here the density of the stippling indicates the relative value of the refractive index.

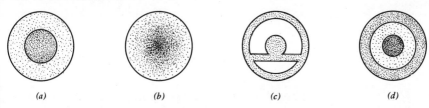

Figure 13.12 Types of fiber cladding. (a) Stepped-index; (b) graded-index; (c) single-material; (d) doubly-clad fiber.

Fibers for the transmission of illumination or images over short distances are generally made with large refractive index differences between core and cladding, and in such fibers the core is usually large, with a thin cladding sufficient to provide optical isolation. On the other hand, in fibers for long-distance, large-bandwidth optical communication, the number of modes must often be restricted. Such fibers therefore are made with very small refractive index differences (see Section 13.1.3.2); consequently the guided flux fields penetrate far into the cladding so that it must be made very thick, often many times as thick as the core.

13.2.1.2 Generalities and Stepped-Index Fibers [31a], [32a], [36]. A single fiber is readily drawn from a glass rod, when this is heated to an appropriate viscosity level. Clad fibers can similarly be drawn from a greatly scaled-up version. Thus, a stepped-index fiber can be drawn from a rod placed inside a tube and a graded-index fiber from a rod or tube prepared to have a similar gradient, as described below.

Drawing. For the drawing process, the initial piece (rod, rod-in-tube, or tube) is placed inside an oven and heated to the appropriate working temperature (see Figure 10.6). A fiber is then drawn from the tip of the piece, outside the oven, at a speed v_F, and wound onto a spool. Meanwhile the initial piece is slowly fed into the oven from the other side at speed v_I, where conservation of volume implies that

$$\frac{v_I}{v_F} = \frac{A_F}{A_I} = \frac{r_F^2}{r_I^2},$$ (13.70)

where A and r represent cross sectional area and radius, respectively, and the subscripts F and I refer to the fiber and initial piece, respectively. In a typical arrangement, a 30-mm-diam rod is fed into the oven at a speed of about 1 cm/sec and a 150 μm fiber is withdrawn at 400 m/sec, by means of a large diameter drum, rotating at a constant speed. Larger diameter fibers, which can not be bent around a drum, can be withdrawn by passing between pressure rollers. The fiber shape can be readily maintained to an accuracy of approximately 1% during the drawing.

In the rod-in-tube technique, special care must be taken to ensure that the outer surface of the rod and the inner surface of the tube be free of contamination, which might frustrate the total internal reflection in the final fiber.

Double Crucible Method. In an alternative approach, the glass is melted in two concentric crucibles (see Figure 13.13) and is withdrawn simultaneously from both. Here great care must be exercised to prevent contamination of the glass melt by the crucible material, which is generally platinum or a refractory ceramic. On the other hand, there is less difficulty in obtaining a clean interface, which is also crucial to low-loss light conduction.

Vapor Deposition Method [36], [37]. When the ultimate in glass purity is required, as in fibers for communication systems (see Section 10.1.6), the material may be formed directly from component reagents, which are mixed in a flame and deposited on a rotating glass mandrel in the form of a soot. This mandrel also reciprocates slowly until, after many passes, the desired thickness has been deposited. The deposited material is then removed in the form of a sleeve and consolidated, by sintering, into a fiber blank, which, in turn, can be drawn into a fiber of the desired diameter. Alternatively, the soot may be deposited *inside* a tube, which eventually becomes the cladding of the finished fiber.

Single-Material Fiber. The single-material fiber, illustrated in Figure 13.12c, is made from a thin rod fused onto a support plate, which is then inserted into a tube and fused in place. This assembly can now be drawn into a fiber.

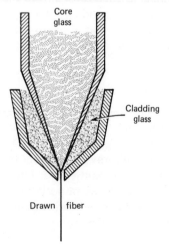

Core glass

Cladding glass

Drawn | fiber

Figure 13.13 Schematic of double-crucibel method for drawing clad fibers.

13.2.1.3 Graded-Index Fibers. Early graded-index fibers were made by a discontinuous ion-exchange technique: the glass rod, initially homogeneous, is immersed in a molten salt bath. The exchange of ions between it and the bath lowers the refractive index of the rod, primarily in its outer layers, less so, toward the center. This results in a varying refractive index, whose gradient may closely resemble the desired form.

More recently, such fibers are made in a continuous double-crucible technique, where the ion exchange between core and cladding takes place rapidly as the fiber cools [38]. For instance, Selfoc fibers are made of borosilicate glass, with the core doped with thallium and the cladding with sodium [38].

The vapor deposition techniques also lend themselves to making graded index fibers. The composition of the reagent mixture can be carefully controlled and is varied on successive passes to yield accurately the desired index gradient [37], [39].

13.2.1.4 Fibers with Plastics and with Liquid Cores. For many applications, silica fibers enclosed in low-index plastic sheaths are attractive. These are manufactured by extruding the plastic sheath simultaneously with the drawing of the fiber. Usually fluoroethylene-propylene copolymer (Teflon FEP) is used for the cladding. This material has the unusually low refractive index of 1.34. In such fibers, the jacket fits but loosely around the core, making contact at just one point on the cross section [40].

Fully plastic fibers, too, have been developed. They have the advantage of much greater flexibility, so that they can be used in significantly larger diameters. This lowers the cost of bundles of a given cross-sectional area and facilitates the insertion of the light flux. The efficient light coupling may more than compensate for their somewhat lower transmittance characteristics, especially in short lengths. While plastic fibers are more resistant to failure under flexing, they are far more sensitive to surface damage and high temperatures. They must generally be maintained below 80°C [41*a*], whereas glass fibers may be constructed for use at several hundred degrees celsius.

One popular type (Du Pont "Crofon") consists of a polymethyl methacrylate core with the cladding partly fluorinated, and has a nominal diameter of 0.25 mm [41*a*]. The extinction coefficient of these fibers is approximately $0.44 \, \text{m}^{-1}$ in the middle of the visible spectrum; at $0.9 \, \mu\text{m}$, however, they have a strong absorption line. See Table 141 [41*b*] for the coefficients at various wavelengths and Figure 13.14 [41*b*] for transmission spectra for various lengths. Such plastic fibers remain flexible to diameters of approximately 1.5 mm, whereas glass fibers must be limited to a tenth of this to retain their flexibility.

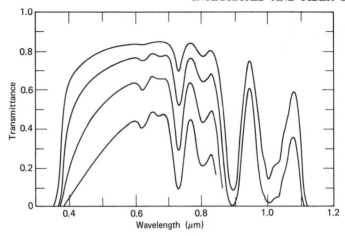

Figure 13.14 Transmittance spectrum of Crofon fiber bundles of lengths 30 cm, 60 cm, 120 cm, and 240 cm; calculated. Brown and Derick [41b].

Lately, plastic fibers have been developed with a graded index. These are especially interesting for imaging applications in medical examinations of internal organs. Resolving powers of over 5 mm^{-1} have already been obtained with fibers 3 mm in diameter and 150 mm long [42].

In an effort to develop low-loss fibers, tubes of silica glass have been drawn into fine capillaries and filled with higher-index liquids, such as tetrachloroethylene (C_2Cl_4), occasionally mixed with carbon tetrachloride (CCl_4). Such tubes have been tested in lengths up to a kilometer and showed promise at the time [43].

13.2.1.5 Materials Considerations

High Transmittance. As discussed in Section 10.1.2 and 10.1.5, optical glasses absorb relatively little in the visible part of the spectrum and near ir. Hence they are all suitable for optical fibers in lengths up to several meters. However, when lengths of hundreds of meters are required, as in optical communication systems, even their low absorption coefficients may be excessive and special precautions and techniques are required (see Section 10.1.6).

Uv Transmission. For uv transmission over shorter distances, silica glass fibers may be covered with a fluorocarbon resin of even lower refractive index such as the Teflon mentioned in the preceding section; these resins transmit well in the uv down to 0.2 μm. A 1-m length of such a fiber (70 μm core and 100 μm cladding diameter) transmitted almost 60% over the visible part of the spectrum and dropped to a quarter of that value at about 0.21 μm [44a].

Ir Transmitting. For use in the ir, up to approximately 8 μm, arsenic trisulfide glass can be used for the fibér. The cladding is then usually made with modified glass of the same basic type [32b].

Mechanical Strength. The mechanical strength of glass components is generally found to be considerably less than the theoretically predicted value (see Section 10.1.3.1). This is due to microscopic cracks and related defects of the surface, which tend to magnify locally any stress applied to the component: it is the surface that controls, to a great extent, the strength of the component as a whole. Consequently the tensile strength of glass rods and fibers (in terms of force per unit cross-sectional area) increases as the fiber diameter decreases, since the reduced surface area reduces the probability of fracture-inducing surface defects. The dependence of tensile strength on fiber diameter is illustrated in Figure 13.15 [32a]. The values shown there are mean values and actual strength values may deviate from these by an octave, up or down.

The tensile strength of optical glass fibers can be increased by prestressing them in compression. This is done by using for the cladding a material with a thermal coefficient of expansion that is slightly greater than that of the core material. On cooling this will produce the desired stress in the core.

Aging. Due to atmospheric effects, primarily chemical in nature, and also due to the mechanical damage likely to be induced whenever the fiber surface is touched, the strength of optical fibers tends to drop rapidly from its initial value. To minimize this effect, it seems to have become

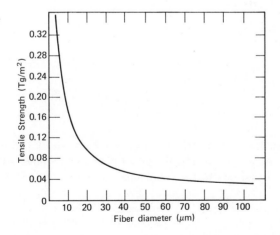

Figure 13.15.   Glass fibers: Variation of tensile strength, with diameter, Allan [32a].

common practice to provide a protective coating as soon as possible after drawing [40].

13.2.2 Geometrical Optics Analysis [31b], [32c]

We now proceed to the analysis of light propagation in an optical fiber. When the fiber diameter is large compared to the wavelength, diffraction effects may be neglected and the propagation may be treated in terms of ray optics. This is the subject of the present section. When the fiber diameter is comparable in size to the wavelength of the radiation it carries, the wave nature of light dominates and the analysis must be made in terms of physical optics. This is done in Section 13.2.3.

13.2.2.1 Trapped Rays. The effectiveness of the fiber as a waveguide is based on the phenomenon of total internal reflection (see Section 2.2.2.2). In terms of geometrical optics, this simply means that all the light inside a higher index (n_1) medium striking the interface with a lower index (n_2) medium, will be reflected without any losses, provided the sine of the angle of incidence (θ) exceeds the index ratio:

$$\sin \theta \geq \frac{n_2}{n_1}. \tag{13.71}$$

It should be noted that, in practice, the reflection will not be perfect due to contaminants, or nonuniformities, at the surface, destroying condition (13.71) Any ray violating (13.71) to a significant degree,[4] will be severely attenuated and may be presumed lost if this occurs more than once or twice. If (13.71) is satisfied, the ray may continue along the fiber, reflected back and forth from its surface, with only slight attenuation. Such a ray is *trapped*.

13.2.2.2 Propagation in a Uniform, Straight Cylindric Fiber. Let us now investigate the progress of a ray in a straight, circularly cylindric fiber, radius r_0. The fiber is assumed to have a uniform refractive index, n_1, to be surrounded by a medium of index, n_2, and to have its entrance face cut normal to the axis.

We first treat a meridional ray, that is, a ray passing through the fiber axis. Afterward we proceed to the general, skew, ray and to the effects of a surface cut obliquely to the fiber axis.

Meridional Ray. A meridional ray, entering the fiber at an angle α

---

[4] For example, for $n_1/n_2 = 1.5$, the critical angle for total reflection is 41.8°. At 40° incidence, each reflection will attenuate a ray to below 25% if it is unpolarized—to 10%, if it is polarized parallel to the plane of incidence.

with the axis, will strike the fiber envelope at an angle

$$\theta = \frac{\pi}{2} - \alpha', \tag{13.72}$$

where, according to Snell's law (2.94), the angle of refraction, α', is given by

$$\alpha' = \sin^{-1}\left(\frac{n_0}{n_1}\sin\alpha\right) \tag{13.73}$$

and n_0 is the refractive index of the medium from which the ray enters the fiber. See Figure 13.16, which shows a ray entering a fiber at A and striking its envelope at B. Clearly the condition for total reflection at B is, from (13.71)–(13.77),

$$\frac{n_2}{n_1} \leqslant \sin\theta = \cos\alpha' = \sqrt{1 - \sin^2\alpha'}$$

$$= \sqrt{1 - \left(\frac{n_0}{n_1}\right)^2 \sin^2\alpha}. \tag{13.74}$$

Solving for $\sin\alpha$, we find the condition

$$\sin\alpha \leqslant \boxed{\sqrt{n_1^2 - n_2^2}/n_0 \equiv \sin\alpha_{mx}} \ ; \tag{13.75}$$

for any value α less than α_{mx}, as defined in (13.75), the ray will be trapped. The corresponding angle inside the fiber is

$$\sin\alpha'_{mx} = n_0 \frac{\sin\alpha_{mx}}{n_1} = \sqrt{1 - \left(\frac{n_2}{n_1}\right)^2}. \tag{13.75a}$$

Accordingly, we may define the *numerical aperture* (na) of the fiber as

$$\boxed{A_0 = n_0 \sin\alpha_{mx} = \sqrt{n_1^2 - n_2^2}} \ . \tag{13.76}$$

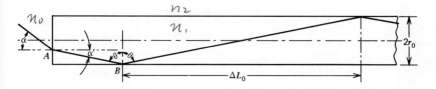

Figure 13.16. Meridional ray in a uniform cylindric fiber.

Note that a fiber may easily have a na in excess of unity. A fiber clad with a medium of index 1.41, must have a core index of 1.73 to have a unity na; an unclad fiber of index 1.414 has such a na.

When dealing with clad fibers whose indices n_1 and n_2 are very close to each other, it is convenient to introduce the mean value of the index

$$\bar{n} = \frac{n_1 + n_2}{2} \tag{13.77}$$

and the fractional index difference

$$\delta = \frac{n_1 - n_2}{\bar{n}}. \tag{13.78}$$

In terms of these, (13.76) becomes

$$\boxed{A_0 = \bar{n}\sqrt{2\delta}}. \tag{13.76a}$$

Returning to the ray in Figure 13.16, we note that it will be reflected at B, with the angle of reflection again equal to θ, so that it will again cross the axis at an angle α'. Hence its subsequent angle of incidence on the envelope will again be equal to θ, and this angle will be maintained throughout its travel.

The *total pathlength* of this ray, in a fiber of length L, will be

$$\dot{s} = L \sec \alpha' = \frac{n_1 L}{(n_1^2 - n_0^2 \sin^2 \alpha)^{1/2}} \tag{13.79}$$

and the interval between reflections, as measured along the fiber, will be

$$\Delta L_0 = 2r_0 \cot \alpha' = 2r_0 \frac{(n_1^2 - n_0^2 \sin^2 \alpha)^{1/2}}{n_0 \sin \alpha}. \tag{13.80}$$

Hence the number of reflections experienced by a ray traveling a distance L along the fiber will be

$$N_0 \approx \frac{L}{\Delta L_0} = L \frac{\tan \alpha'}{2r_0}$$

$$= \frac{L n_0 \sin \alpha}{2r_0(n_1^2 - n_0^2 \sin^2 \alpha)^{1/2}}, \tag{13.81}$$

with the exact value depending on the heights of entry and exit and differing from N_0 by less than unity.

Skew Rays. The analysis of skew rays is somewhat more complex and is presented in Appendix 13.1. The results, however, are quite simple. In the uniform, circularly cylindric fiber, a skew ray describes an angular

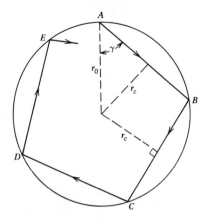

Figure 13.17 Path of a skew ray in a circular fiber: profile projection.

"helix" as it progresses along the fiber; see Figure 13.17 for a projection of its path onto a cross section of the fiber. Throughout, the ray segments remain at a constant distance, r_c, from the fiber axis; also the angle of incidence, θ, on the fiber surface remains constant:

$$\cos \theta = \sin \alpha' \cos \gamma = \frac{n_0}{n_1} \sin \alpha \cos \gamma , \qquad (13.82)$$

where

$$\sin \gamma = \frac{r_c}{r_0}. \qquad (13.83)$$

Hence the trapping condition on the angle of entry (α) becomes

$$n_0 \sin \alpha \leqslant A_0 \sec \gamma , \qquad (13.84)$$

where we solved (13.82) for $\sin \alpha$, substituted for θ from (13.71) and for $(n_1^2 - n_2^2)$ from (13.76). We may thus speak of a virtual numerical aperture of skew rays, which is

$$A = n_0 \sin \alpha_{mxs} = \sqrt{n_1^2 - n_2^2} \sec \gamma \geqslant A_0. \qquad (13.85)$$

This clearly implies that, for skew rays, the acceptance angle of the fiber increases with the ray's distance from the fiber. For two reasons this does not, however, yield a corresponding increase in real na.

1. For any such angle of entry ($\alpha > \alpha_{mx}$), the fiber will trap the ray only if it is incident on an area segment that is restricted both radially and in azimuth.

2. All of these rays are not fully trapped, although they may be

trapped, practically speaking. They are called tunneling rays and are discussed further in Section 13.2.3.2.

When we view the exit surface of a diffusely illuminated fiber head on, the surface will appear uniformly bright. If we now turn the axis until it makes an angle greater than α_{mx} with our line of sight, a black band will appear across the center of the exit surface. As we continue turning the axis away, the band will broaden until it covers the whole face when 90° is reached [45].

The total pathlength of a skew ray is also given by (13.79). The distance between reflections, as measured along the fiber, is given by

$$\Delta L = 2r_0 \cos \gamma \cot \alpha'$$
$$= 2 \frac{\sqrt{(n_1^2 - n_0^2 \sin^2 \alpha)(r_0^2 - r_c^2)}}{n_0 \sin \alpha}. \tag{13.86}$$

On comparing this with (13.80), we note that the number of reflections (N) is increased by a factor $\sec \gamma$:

$$N = N_0 \sec \gamma. \tag{13.87}$$

Neglecting losses, the totally reflected flux transmitted through such a fiber from a Lambertian source, luminance L, is

$$\Phi(\alpha_0) = \pi^2 r_0^2 L \sin^2 \alpha_0 \qquad \alpha_0 \leq \alpha_{mx}\left[= \sin^{-1} \frac{\sqrt{n_1^2 - n_2^2}}{n_0}\right]$$

$$\Phi_T = \pi^2 r_0^2 L \frac{(n_1^2 - n_2^2)}{n_0^2} + \Phi_s, \qquad \alpha_0 > \alpha_{mx}, \tag{13.88}$$

$$= \left(\frac{\pi r_0 A_0}{n_0}\right)^2 L + \Phi_s,$$

where α_0 is the half-angle subtended by the source (assumed circular) at the fiber entrance face,

Φ_s represents the flux in the tunneling rays, which enter at an angle larger than α_{mx} but are, due to their skewness, totally reflected, according to the ray picture.

For an infinite source

$$\Phi_s = \Phi(\alpha_{mx}), \tag{13.88a}$$

see Appendix 13.2.

Oblique Entrance Face. When the entrance face is cut oblique to the fiber axis, the cone of trapped rays, too, will be oblique to the fiber axis outside the fiber. There the axial ray will make an angle γ:

$$\sin \gamma = \frac{n_1}{n_0} \sin \beta \qquad (13.89)$$

with the axis. Here β is the angle between the fiber axis and the normal to the entrance face. For small angles β and α_{mx}, the trapped rays still form a near-circular cone, half-apex angle α_{mx}, around this ray.

The new direction for any specific ray is most conveniently found by means of the vector relationships (9.6 and (9.7).[5] The vector, $\mathbf{s}'$, representing the refracted ray is given by

$$\mathbf{s}' = \mathbf{s} + \{[n'^2 - n^2 + (\mathbf{n} \cdot \mathbf{s})^2]^{1/2} - (\mathbf{n} \cdot \mathbf{s})\}\mathbf{n}, \qquad (13.90)$$

where $\mathbf{s}$ is the vector representing the incident ray and vectors $\mathbf{s}, \mathbf{s}'$ have magnitudes equal to the refractive index of their respective media,

$\mathbf{n}$ is a unit vector normal to the surface, and

n, n' are the refractive indices of the two media.

13.2.2.3 Conical Fibers. When the fiber is conical, the angle of incidence of a trapped ray changes on successive reflections (see Figure 13.18). A ray of the type shown may eventually return to the entrance face. The fate of any meridional ray in such a fiber is conveniently

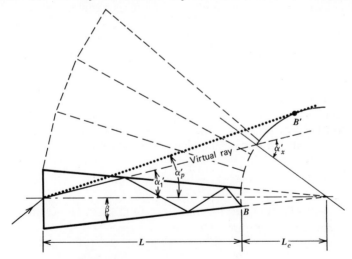

Figure 13.18. Ray path in a conical fiber.

[5] In Volume 1 (9.7) was printed in error. It should read:

$$K = \sqrt{n'^2 - n^2 + (\mathbf{n} \cdot \mathbf{s})^2} - \mathbf{n} \cdot \mathbf{s}.$$

analyzed by imaging the axial section of the fiber repeatedly in the tangential reflection plane. See the broken lines in Figure 13.18. The ray is then represented by a straight line as shown there and its angle of incidence at the jth reflection can be seen to be

$$\theta_j = \frac{\pi}{2} - \alpha_1' + (2j-1)\beta, \tag{13.91}$$

where α_1' is the angle the ray makes with the axis inside the fiber, before the first reflection, and β is the semi-apex-angle of the cone.

Let α_p' represent the angle subtended, at the point of entry, by the polygon BB', generated by the exit face in the imaging process just described; then, for a ray to reach the exit face, we have the condition

$$\alpha_1' \leqslant \alpha_p'. \tag{13.92}$$

If the ray does not meet this condition, it will eventually return to the entrance face.

In the situation usually of interest in fibers, β is very small so that the polygon may be approximated by a circle, and we use this approximation in the following. In the figure L represents the length of the cone segment included between the fiber end faces and L_c the length between the exit face and the cone apex. From similar triangles the ratio between the cross-sectional radii at exit and entrance faces is given by

$$\frac{r_2}{r_1} = \frac{L_c}{L + L_c} = \sin \alpha_p', \tag{13.93}$$

as evident from the figure, where the line making angle α_p' with the axis is now tangent to the circle of radius L_c.

A more general ray leaving the entrance face at angle α_1' and meeting condition (13.92) will strike the exit face at an angle α_x'. Inspection of Figure 13.18 shows that application of the sine-theorem yields:

$$\sin \alpha_x' = \frac{L + L_c}{L_c} \sin \alpha_0' = \frac{r_1}{r_2} \sin \alpha_0'. \tag{13.94}$$

For the ray to be trapped, this must satisfy (13.75). Hence the condition of the angle at the entrance face becomes

$$n_0 \sin \alpha_0 = n_1 \sin \alpha_0' = n_1 \frac{r_2}{r_1} \sin \alpha_x' \leqslant \sqrt{n_1^2 - n_2^2} \frac{r_2}{r_1}, \tag{13.95}$$

where we have applied consecutively: Snell's law, (13.94), and (13.75). Comparing this result with (13.76), we note that the na of the fiber has been reduced by the factor (r_2/r_1):

$$A_{oc} = \frac{A_0 r_2}{r_1}. \qquad (13.96)$$

We conclude that we may taper a fiber in order to obtain a greater spatial concentration of light, but this is accomplished at the cost of a greater angular divergence of the beam. Alternatively, we may increase the area illuminated by the beam, at the cost of decreasing the angle of the cone converging upon the illuminated region. This is another manifestation of the law which limits the product of area and solid angle in an optical system to its value at the entrance. [This product may be called "relative flux acceptance"; see note 9 in Appendix 13.1. Also cf. Lagrange's invariant (9.23) and (9.47)].

13.2.2.4 Bent Fibers. When a fiber is bent, certain rays previously trapped, may go into a radiating mode. However, for the large bending radii frequently encountered in practice, this loss may be negligible and the possibility of bending fibers without significant loss of radiation is a major advantage of fibers as light guides. Here we briefly analyze the restrictions on bending.

The bending of a fiber destroys the axial symmetry that we have assumed so far. The effect of bending is most pronounced for meridional rays in the plane of the curvature (viz. the plane osculatory to the curve), and we analyze here that particular ray to find the effects of bending. We may then conclude that the effects on other rays will be smaller.

Refer to Figure 13.19, which shows a fiber, cross-sectional radius, r_0,

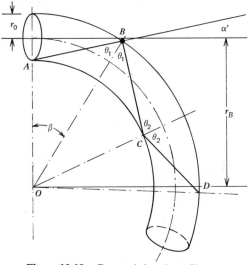

Figure 13.19. Ray path in a bent fiber.

bent into a circular arc, radius r_b, with the center of curvature at O. To simplify the derivation, we treat first a ray entering at a point on the envelope and generalize only later. A meridional ray enters at point A, making an angle α' with the axis. It strikes the envelope again at point B. Applying the law of sines to Triangle ABO, we find the angle of incidence at B:

$$\sin \theta_1 = \frac{r_b - r_0}{r_b + r_0} \sin BAO = \frac{r_b - r_0}{r_b + r_0} \cos \alpha' \qquad (13.97)$$

and the angle between cross-sectional planes through points of reflection:

$$\angle AOB \equiv \beta = \frac{\pi}{2} - \alpha' - \theta_1. \qquad (13.98)$$

The angle of incidence at the next reflection point, C, is clearly equal to that at A:

$$\theta_2 = \frac{\pi}{2} - \alpha'. \qquad (13.99)$$

The length of the passage in the fiber between two points of reflection is, again from the law of sines,

$$\Delta s = \overline{AB} = (r_b + r_0) \frac{\sin \beta}{\cos \alpha'}, \qquad (13.100)$$

corresponding to a fiber segment of length

$$\Delta L = \beta r_b. \qquad (13.101)$$

Thus the ratio of ray pathlength to fiber length is

$$\frac{s}{L} = \frac{r_b + r_0}{r_b} \frac{\sin \beta}{\beta \cos \alpha'}$$

$$= (1 + \rho) \frac{\sin \beta / \beta}{\cos \alpha'}, \qquad (13.102)$$

where we have abbreviated:

$$\rho = \frac{r_0}{r_b}. \qquad (13.103)$$

The ray will escape the fiber if

$$\frac{n_2}{n_1} < \sin \theta_1 = \cos \alpha' \frac{1 - \rho}{1 + \rho}, \qquad (13.104)$$

with the first step from (13.74) and the equality from (13.97). Hence, the na is reduced by the bending. The cone of rays that are captured by the fiber is now restricted to a semiapex angle, α_{mxb}, corresponding to a virtual numerical aperture which is, analogously to (13.76),

$$A_b = n_0 \sin \alpha_{mxb} = \left[n_1^2 - n_2^2 \frac{(1+\rho)^2}{(1-\rho)^2} \right]^{1/2}. \tag{13.105}$$

Regarding the meridional ray, therefore, the bending is equivalent to an increase of n_2 by a factor

$$\frac{(1+\rho)}{(1-\rho)} \approx 1 + 2\rho, \qquad \rho \ll 1. \tag{13.106}$$

When this quantity exceeds n_1/n_2, even a ray entering the bend parallel to the fiber axis will escape. Hence on equating (13.106) to this ratio, we obtain the critical bending ratio

$$\boxed{\rho_c = \frac{n_1 - n_2}{n_1 + n_2} = 2\delta} \tag{13.107}$$

for which even a ray entering parallel to the fiber axis will escape and only rays inclined to the axis in the direction of the bend are trapped. See (13.77) and (13.78) for the last step.

The reader will readily confirm the following. When a ray enters the bend at a distance, h, from the fiber axis $(-r_0 < h < r_0)$, the equation for θ_1 becomes

$$\sin \theta_1(h) = \frac{(r_b + h) \cos \alpha'}{r_b + r_0}, \tag{13.108}$$

with the remainder of the equations remaining unaffected (except for the obvious exceptions relating to the first lap of the ray path).

See Figure 13.20 for the dependence of the virtual na on the bending ratio for various values of index ratio. On each curve the index ratio is indicated. A more precise evaluation of losses due to bending requires a physical optics analysis [19a].

From (13.107) it is evident that for fibers with small index differences even slight bending can destroy the trapping effectiveness. Since the bend need not extend over a long segment of the fiber, even a very slight deviation from straightness, a "microbend," can impair trapping significantly and, if such microbends exist in larger numbers, they may dominate the attenuation in the fiber.

The following formula has been developed for estimating the microbend induced losses (in dB) in terms of practical fiber construction. It refers to a fiber encapsulated in a cable and distorted by the presence of

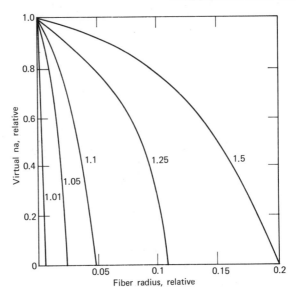

Figure 13.20. Reduction of na due to bending. Numbers on curves give the index ratio (core/cladding) of the fiber.

N bumps or particles, height h, with the encapsulation tending to restore the fiber beyond the disturbance. The attenuation is then [46]

$$a^* = \frac{0.9N\overline{h^2}r_0^4(kE_e/E_f)^{3/2}}{\delta^3 r_f^6},\qquad (13.109)$$

where N is the number of bend-causing disturbances,
    $r_{0,f}$ are the core and outside fiber radii, respectively,
    $E_{e,f}$ are the Young's moduli of the encapsulating and fiber materials, respectively,
    δ is the relative index difference of the fiber materials [see (13.78)] and

$$k = \frac{F(h)}{hE_e}\qquad (13.110)$$

is a dimensionless constant, of unity order of magnitude, relating the restoring force, F, to the displacement, h, causing it and to the modulus, E_e.

13.2.2.5 *Graded-Index Fibers.* When the refractive index drops away from the axis, rays passing along the fiber are turned toward the axis in such a way that, at any radial distance, r:

$$n(r) \sin \theta(r) = \text{constant}, \qquad (13.111a)$$

where $n(r)$ is the index at the distance r,

$\theta(r)$ is the angle that the ray at a point distance r from the axis, makes with the radius to that point.

If the ray is trapped, it will eventually return toward the axis.

Considering a meridional ray in a fiber whose refractive index drops monotonically away from the axis and measuring the angles, α', relative to the fiber axis, this becomes

$$n(r) \cos \alpha'(r) = n(0) \cos \alpha_0' = n(r_{mx}), \qquad (13.111)$$

where α_0' is the angle at which the ray crosses the fiber axis and

r_{mx} is the maximum radius reached by that ray.

The ray progresses as follows. It crosses the axis at angle α_0'; as it recedes from the axis, its angle of inclination to the axis drops, until it is parallel to the axis at a distance r_{mx} from it. As it proceeds, it starts approaching the axis, and eventually crosses it, to repeat the process on the other side of the fiber axis. All meridional rays, crossing the fiber axis at angle α_0', are confined to a cylindric region of radius r_{mx}, provided the fiber index drops to the value called for by (13.111). Equation 13.111 has a form similar to that of (13.74) and, hence, the effective na of the graded-index fiber is, analogous to (13.76),

$$A_0 = \sqrt{n^2(0) - n^2(r_0)}, \qquad (13.112)$$

provided the index decreases monotonically from the axis to radius r_0.

With proper design, such a fiber can act as a lens: it can image a luminance distribution at its entrance face into a similar distribution at its exit face. It can do this rigorously for meridional rays and approximately for all rays. The conditions for this capability are as follows:

1. The fiber length is such that each ray completes an integral number of cycles during its passage.

2. The optical pathlength per cycle is independent of the ray inclination α_0:

$$\hat{s} = \int_{\text{cycle}} n(r) \, ds = \text{constant for all rays.} \qquad (13.113)$$

It can be shown (see Appendix 13.3) that an index distribution

$$n(r) = n(0) \operatorname{sech} Kr = n(0)\left(1 - \tfrac{1}{2}K^2 r^2 + \frac{5}{24} K^4 r^4 - \cdots\right) \qquad (13.114)$$

leads to perfect focusing of meridional rays at a distance

$$\Delta L = \frac{\pi}{K} \tag{13.115}$$

or any integral multiple thereof. In other words, in a fiber with such an index distribution, any luminance distribution in one cross-sectional plane will give rise to a similar illumination distribution at any cross-sectional plane $m\Delta L$ away from it, where m is an integer, provided that only meridional rays are considered. Clearly, for $\Delta L \gg r$ $(Kr \ll 1)$, the above is quite accurately approximated by the parabolic distribution

$$\boxed{n(r) = (1 - \tfrac{1}{2}K^2 r^2)n(0)} \quad . \tag{13.116}$$

Although this perfect imaging does not include skew rays exactly [47], the aberrations are quite small, for small values of Kr. [48].

Actual imaging may be an interesting application for graded-index fibers, however the major impetus for their development came from the requirements of guided-wave communication systems (Section 13.3.3). There pulse spreading in transit is a major limitation on the pulse frequency and, hence, on the signal rates achievable. The two major factors in pulse spreading are dispersion (treated in the next section) and mode spreading (Section 13.2.3.3). In geometrical optical terms, this latter effect is evident in the dependence of the optical pathlength on the angle of incidence, α' [see (13.79)]. The resulting difference in transit time implies that an infinitesimal pulse entering the fiber will be spread out over a finite interval on arrival. The first components to arrive will be those launched parallel to the fiber axis and the last ones, those incident on the fiber surface at the critical angle. In general, the increase in the optical pathlength due to ray inclination at α, will be

$$\delta\hat{s} = n_1 L - \frac{n_1 L}{\sqrt{n_1^2 - n_0^2 \sin^2 \alpha}}$$

$$= n_1 L \left(1 - \frac{1}{\sqrt{n_1^2 - n_0^2 \sin^2 \alpha}}\right), \tag{13.117}$$

For the marginal rays, at the critical angle $(\alpha = \alpha_{mx})$:

$$\delta\hat{s} = \frac{n_1 L(n_2 - 1)}{n_2} \tag{13.118}$$

and the corresponding delay:

$$\delta t = \frac{\delta\hat{s}}{c} = \frac{n_1 L(n_2 - 1)}{cn_2}. \tag{13.119}$$

Clearly skew rays may be transmitted at larger angles, so that they may contribute even more to pulse spreading. All these differences are effectively eliminated in graded-index fibers satisfying (13.113).

13.2.2.6 Dispersive Pulse Spreading. In fibers used for communication at high information rates, very short pulses are transmitted over very long distances. Consequently, slight differences in the velocity of propagation may measurably affect the arrival time of a pulse. If the pulse itself is composed of various modes and wavelengths, the different components may propagate at different velocities, resulting in the spreading of the pulse. This, in turn, limits the bandwidth and information rate at which the system can operate. Because of the significant differences in the propagation velocities of the various modes, fibers for such communication systems are often constructed to operate in only a single mode.

To minimize the spread due to dispersion, monochromatic light may be used. However, even with monochromatic light, the wave packet making up the pulse necessarily covers a finite wavelength region. According to (3.49) this is approximately

$$\Delta \nu_t \approx \frac{1}{2\pi\tau}, \tag{13.120}$$

where τ is the pulse width. In terms of wavelength spread, we have (since $\nu_t = c/\lambda$; see 2.32):

$$\Delta\lambda = \frac{\Delta\nu_t}{d\nu/d\lambda} = -\frac{\lambda^2 \Delta\nu_t}{c} \approx -\frac{\lambda^2}{2\pi c\tau}. \tag{13.121}$$

The corresponding spread in transit time is

$$\Delta t = \Delta \frac{L}{v} = \Delta \frac{Ln}{c \cos \alpha'} = \frac{L\Delta n}{c \cos \alpha'}, \tag{13.122}$$

where L = the length of the fiber,
  $v = c/n \cos \alpha'$ is the velocity of propagation of the ray along the fiber axis,
  c = the *in vacuo* velocity of light propagation,
  α' = the angle the mode makes with the fiber axis, and
  $\Delta n = (dn/d\lambda) \Delta\lambda = (dn/d\lambda)\lambda^2/2\pi c\tau$ is the change in refractive index corresponding to the wavelength change $\Delta\lambda$.

For more accurate results, we note that the pulse propagates with the group velocity, v_g, given according to (2.44), by

$$v_g = v - \lambda \frac{dv}{d\lambda} = \frac{c(n+\lambda n_\lambda)}{n^2}, \tag{13.123}$$

where we substituted for the phase velocity

$$v = \frac{c}{n}$$

and denote

$$n_\lambda = \frac{dn}{d\lambda}.$$

We now evaluate the pulse spreading (Δt) in terms of the transit time

$$t = \frac{L}{v_g} \tag{13.124}$$

and the derivative of the group velocity

$$\frac{dv_g}{d\lambda} = \frac{c\lambda(nn_{\lambda\lambda} - 2n_\lambda^2)}{n^3}, \tag{13.125}$$

where

$$n_{\lambda\lambda} = \frac{d^2n}{d\lambda^2}.$$

Hence

$$\Delta t = \frac{dt}{d\lambda}\, \Delta\lambda = -\frac{L(dv_g/d\lambda)\,\Delta\lambda}{v_g^2}$$

$$= \frac{-t_0\lambda(nn_{\lambda\lambda} - 2n_\lambda^2)\,\Delta\lambda}{(n + \lambda n)^2}, \tag{13.126}$$

where

$$t_0 = \frac{nL}{c} \tag{13.127}$$

is the transit time of the axial mode.

By way of illustration, for silica glass the values of n_λ, $n_{\lambda\lambda}$ are $-12.6 \times 10^3 \, \text{m}^{-1}$ and $13 \times 10^9 \, \text{m}^{-2}$, respectively. Using (13.121) and (13.122), these yield for a 1-ps-pulse of pure 1 μm radiation, traveling down a 1 km long fiber:

$$\Delta t = 19 \, \text{ps}.$$

If, however, an aluminum–gallium arsenide LED, with a spectral width of $\Delta\lambda = 32 \times 10^{-9} \, \text{m}$ is used, we find

$$\Delta t = 1.2 \, \text{ns}. \cdot$$

These results illustrate both the inherent limitations and the additional importance of monochromaticity in the transmission of very short pulses.

In the usual optical glasses a rough estimate of the pulse spreading can be obtained from the following formulae. For a pulse of radiation covering a spectral range $\Delta\lambda$ [49]:

$$\Delta t \approx 0.17 \, \Delta\lambda \frac{L}{\lambda} \text{ ns.} \tag{13.128a}$$

For a monochromatic pulse, the pulse length, Δt, resulting in spreading, again equal to Δt, is given by:

$$\Delta t \approx 0.57\sqrt{L} \text{ ps.} \tag{13.128b}$$

Here L is the fiber length in meters [49].

For a discussion of the attenuation of the wave, see Section 13.2.3.4.

13.2.3 Physical Optics Considerations

Some important aspects of the guiding of light through fibers can not be explained by geometrical optics; especially when the fiber diameter is of the order of magnitude of a wavelength, mode structure becomes important and requires treatment in terms of physical optics. These matters are considered in the present section.

13.2.3.1 Mode Structure in Circular Fibers [50].
Just like slab and rectangular waveguides, fibers with a circular cross section, too, can support several modes. Qualitatively, the modes may be described in terms of radial variations of the field, with a maximum, or a minimum, at the axis and, possibly, additional maxima along the radius of the core. We denote by m the number of maxima along the radius. Bound modes are again characterized by a field that decays monotonically outside the core. Superimposed on this radial variation, there may appear an azimuthal variation: the field can vary cyclically around the circumference. The total circumference must correspond to an integral number (l) of cycles. If the light in the modes is linearly polarized (LP), the various modes are often designated by symbols of the form, LP_{lm}.

Maxwell's equations have been completely solved for waves propagating in a circular cylindric structure [18]–[20]. However, the results are quite cumbersome and we give here a greatly simplified solution, valid when the index difference between core and cladding is small. The error is of the order of magnitude $\Delta n/\bar{n}$.

Mode Fields. Just as in slab waveguides, here, too, the number of possible modes is determined by the value of V as given by (13.18)

$$V = k_0 r_0 \sqrt{n_1^2 - n_2^2} = 2\pi\left(\frac{r_0}{\lambda}\right)\bar{n}\sqrt{2\delta}, \tag{13.129}$$

where $k_0 = 2\pi/\lambda_0$ is the wave propagation constant *in vacuo*,
             r_0 = the radius of the core,
$n_{1,2}(n_1 > n_2)$ = the refractive indices of core and cladding, respectively, and
             $\bar{n}, \delta$ = their mean value and relative difference, respectively.

In the mode characterization presented here, V is decomposed into two orthogonal components, U and W, such that

$$V^2 = U^2 + W^2, \tag{13.130}$$

where U and W characterize the fields carried in the core and the cladding, respectively, and are defined by

$$U = r_0 k_0 \sqrt{n_1^2 - N_m^2} \tag{13.131}$$

$$W = r_0 k_0 \sqrt{N_m^2 - n_2^2}. \tag{13.132}$$

Here N_m is the effective guide index (not known initially). For trapped rays, this is, according to (13.14) limited: $n_2 < N_m < n_1$. Such rays therefore are characterized by real values of U and W.

The field amplitude due to any mode LP_{lm} is then described approximately by equations of the form

$$E_{x,y} = E_0 \cos(l\varphi) \frac{J_l(Ur/r_0)}{J_l(U)}, \qquad r < r_0 \tag{13.133a}$$

$$= E_0 \cos(l\varphi) \frac{K_l(Wr/r_0)}{K_l(W)}, \qquad r > r_0 \tag{13.133b}$$

where φ is the azimuth coordinate variable,
       E_0 is the amplitude of the field at the core surface, and
       J_l denotes the Bessel function and K_l the modified Hankel function.
For $l > 0$, additional modes are obtained when the cosines are replaced by sines.

U and W are determined by the boundary conditions at the core-cladding interface. These lead to the equation:

$$\frac{UJ_{l-1}(U)}{J_l(U)} = -\frac{WK_{l-1}(W)}{K_l(W)}. \tag{13.134}$$

This in conjunction with (13.130) determines the values of U and W. In fully trapped modes, U and W are real and $U \leqslant V$. When U attains its maximum value, V, W vanishes, as indicated by (13.130). Referring back to (13.134), we note that $W = 0$ corresponds to $J_{l-1}(U) = 0$, that is,

$$U = j_{l-1,m} = V, \tag{13.135}$$

the mth zero of the $(l-1)$th order Bessel function. See Table 142 [51] for the values of $j_{l,m}$. This value of V is referred to as the cutoff value for the mode LP_{lm}. For values of V below this, U becomes complex and the mode is no longer trapped.

The following approximations are useful for finding the value of U for trapped modes. For the lowest mode ($l = 0$, $m = 1$):

$$U = \frac{(1+\sqrt{2})V}{1+(4+V^4)^{1/4}}. \tag{13.136}$$

For higher modes, the following relationship holds approximately:

$$m = \frac{\sqrt{U^2 - l^2} - l\cos^{-1}(l/U)}{\pi}. \tag{13.137}$$

To facilitate the use of this equation to find U, we show in Figure 13.21 curves of l/V versus m/V for various values of U/V. These may be entered with any given set of l/V and m/V to find the corresponding U/V.

In addition to the two axial LP_{01} modes, a given fiber can carry a number of mode families as determined by the value of V. Specifically, the number of possible mode families is given by the number of Bessel function (J_l) zeros below V, considering all orders but excluding the zeros at the origin. Each of the mode families, with $l > 0$, consists of four modes, differing in the direction of polarization and/or the orientation of the azimuthal variation [see the remarks following (13.133)]. For large values of V, the total number (N) of modes is given approximately by

$$N \approx \frac{V^2}{2}. \tag{13.138}$$

Mode Propagation Parameters. Once the values of V, U, and W have been established, the *mode propagation constant*, k_{lm}, is readily evaluated from

$$k_{lm} = \frac{k\sqrt{U^2 n_2^2 + W^2 n_1^2}}{V} = \frac{(V/\delta - U^2)^{1/2}}{r_0} \tag{13.139}$$

and hence the phase velocity

$$v = \frac{ck}{k_{lm}} = \frac{cV}{\sqrt{U^2 n_2^2 + W^2 n_1^2}}. \tag{13.140}$$

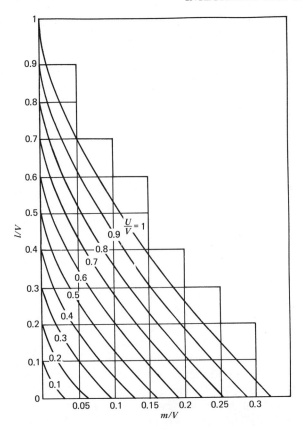

Figure 13.21   Plots of l/V versus m/V with U/V as parameter. According to (11.133).

The *attenuation* of any particular fully trapped mode is due to absorption and scattering in (a) the core, (b) the cladding, and (c) the interface. To estimate the expected value of overall attenuation, we must find the fraction (P_1/P) of the power (P) carried in the core, and the power density (p) at the interface. These are given by

$$\frac{P_1}{P} = \left(\frac{W}{V}\right)^2 (1 + \iota_l) \approx \frac{U^2}{V^2\sqrt{W^2 + l^2 + 1}},$$ (13.141)

where

$$\iota_l = \frac{J_l^2(U)}{J_{l+1}(U)J_{l-1}(U)}.$$ (13.142)

Also,

$$\frac{p}{P} = \frac{\iota_l W^2}{\pi r_0^2 V^2}. \tag{13.143}$$

Ray Analogs. It is instructive to consider the ray picture analogous to the field distributions just discussed. See Figure 13.22, which shows the wavevector associated with a certain plane wavefront. Inside the fiber this has the magnitude nk, where

$$n = n_1, n_2$$

in the core and the cladding, respectively. Working in the cylindric coordinate system (r, φ, z), the z component has the magnitude k_{ml} and that in the φ direction has the magnitude l/r [corresponding to $(2\pi/\Lambda)$ for the spatial period of length $(\Lambda = 2\pi r/l)$]. Since the square of the resultant (kn) equals the sum of the squares of the three orthogonal components, the radial component has the magnitude

$$q(r) = \left[k^2 n^2 - k_{ml}^2 - \left(\frac{l}{r}\right)^2 \right]^{1/2}. \tag{13.144}$$

On substituting $n = n_1$; from (13.139) for k_{lm}; from (13.130) for $(V^2 - W^2)$; and (13.129) for V, we find:

$$q(r) = \left(\frac{U^2}{r_0^2} - \frac{l^2}{r^2} \right)^{1/2}, \qquad r < r_0, \tag{13.145}$$

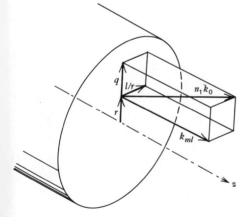

Figure 13.22 Wave vector of a trapped mode.

for the radial component in the core. On substituting $n = n_2$ and following a similar procedure, we find

$$q(r) = i \left(\frac{W^2}{r_0^2} + \frac{l^2}{r^2} \right)^{1/2}, \qquad r > r_0. \qquad (13.146)$$

for the radial component in the cladding.

Inspection of (13.145) shows that q is real in the region

$$\frac{lr_0}{U} < r < r_0 \qquad (13.147)$$

and imaginary outside this region. Since the wave propagation in the radial direction is represented by

$$Ae^{iqr},$$

it is evident that an imaginary value of q represents an exponential decay in the direction r, an evanescent wave [see (2.110)]. Since the wave is trapped, we expected this for $r > r_0$. The decay of the field for

$$r < r_c = \frac{lr_0}{U}, \qquad (13.148)$$

too, has a simple ray optic interpretation. Consider an end view of the fiber with our ray propagating as shown in Figure 13.23. The angle, γ, the ray projection makes with the fiber radius is the same as that which the transverse component of the wave vector makes with the radius. This is given by

$$\cos \gamma = q(r_0) \left[q^2(r_0) + \left(\frac{l}{r_0} \right)^2 \right]^{1/2}$$

$$= \sqrt{1 - \frac{l^2}{U^2}} \qquad (13.149)$$

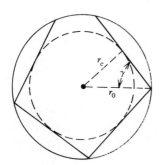

Figure 13.23 Skew ray propagating down a circular fiber; profile projection.

on substituting from (13.145). Hence

$$\sin \gamma = \frac{l}{U}.$$ (13.150)

In the ray picture, all rays with this value of γ are tangent to a circle with radius

$$\boxed{r_c = r_0 \sin \gamma = \frac{l r_0}{U}},$$ (13.151)

which is identical to the inner limit of the mode fields. No modes with these values of l and U and no rays with this value of γ exist inside the circle of radius r_c. This circle forms the envelope, or *caustic* of all rays corresponding to the mode family characterized by these values of l and U.

The angle, α', these wavevectors make with fiber axis is given by

$$\cos \alpha' = \frac{k_{lm}}{n_1 k_0} = \frac{(n_2^2 U^2 + n_1^2 W^2)^{1/2}}{n_1 V}.$$ (13.152)

On substituting this into the trigonometric identity

$$\sin^2 \alpha' = 1 - \cos^2 \alpha'$$

and eliminating W and V from the resulting expression by means of (13.130) and (13.129), respectively, we find

$$\sin \alpha' = \frac{U}{n_1 k_0 r_0}$$ (13.153)

and, by Snell's law, in the space outside the fiber ends:

$$\sin \alpha = \frac{U}{n_0 k_0 r_0}.$$ (13.154)

Since for trapped rays the maximum value of U is V, we find for the na of the fiber

$$A_0 = n_0 \sin \alpha_{mx} = \frac{V}{k_0 r_0} = \sqrt{n_1^2 - n_2^2},$$ (13.155)

in agreement with (13.76).

13.2.3.2 Tunneling Rays [52]. We have, so far, considered only real solutions of (13.130). In fact, complex values of U and W are possible, as well. These are found to correspond to rays whose angle α' with the fiber axis is greater than the complement to the critical angle

$$\alpha' > \alpha'_{mx}$$

as given by (13.75a). Such rays, if they are meridional, are radiative (i.e., not trapped); however, if they are skew rays, they may be trapped when considered in terms of geometrical optics. See (13.84) and (13.85) and the subsequent discussion. Such rays, with

$$\sqrt{n_1^2 - n_2^2} < n_0 \sin \alpha < \cos \gamma \sqrt{n_1^2 - n_2^2} \qquad (13.156)$$

are fully trapped only in the infinitesimal wavelength limit and do exhibit some residual radiation ("leakage"), leaving the fiber by a tunneling process.

In our mode-field terminology, such rays have a complex value of U and W. The field, as given by (13.133b) becomes oscillatory for large values of r, say,

$$r > r_L > r_0,$$

where r_L, the distance from the fiber axis at which the field becomes oscillatory, is given by

$$r_L = \frac{r_0(l-1)}{\sqrt{U^2 - V^2}}. \qquad (13.157)$$

The denominator here is iW, but we wrote it in the form of a radical to show that it is real: in these tunneling modes, the real part of U is greater than V. The cylindrical surface defined by r_L, forms a caustic of tunneling rays radiated from the fiber—tangent to the caustic.

The accurate calculation of the attenuation due to such tunneling is quite difficult. We present here a result that gives a reasonably close approximation. The relative power at a distance z from the entrance plane is given by

$$\frac{P(z)}{P(0)} = e^{-iza/r_0}, \qquad (13.158)$$

where the dimensionless attenuation constant

$$a \approx 2RB \tan \alpha' \exp \frac{-2VB^3}{3n_1(R^2 - 1)}, \qquad (13.159)$$

where

$$R = \alpha'/\alpha'_{mx} \approx \frac{\sin \alpha'}{\sin \alpha'_{mx}}, \qquad (13.160a)$$

and

$$B = (1 - R^2 \cos^2 \gamma)^{1/2}. \qquad (13.160b)$$

13.2.3.3 Multimode Pulse Spreading [34], [50]. The phase velocity of any mode LP_{lm}, in a fiber is given by

$$v_p = \frac{v_t}{k_{lm}} = \frac{c}{\lambda_0 k_{lm}}. \tag{13.161}$$

As evidenced by (13.129), any change in wavelength results in a change in V and this, in turn, in a change in k_{lm} and, hence, in the phase velocity. Clearly the presence of a multiplicity of wavelengths in a single-mode light pulse causes a spreading out of this pulse as it travels down the fiber. This is independent of any pulse spreading due to the dispersion in the material as discussed at length in Section 13.2.2.6. It is also usually negligible compared to the pulse spreading caused by the dispersion of the material (all this within a single-mode family LP_{lm}). When several mode families are represented in the pulse, the phase velocity differences between these will introduce further pulse spreading. It becomes apparent that these various causes of pulse spreading, with their interactions, become rather difficult to analyze. They may also become the limiting factor in long-distance, large-bandwidth communication by fiber.

For an upper-limit estimate of the pulse broadening introduced by mode differences, let us take the difference in transit time of two monochromatic waves traveling in the core and the cladding, respectively. This yields

$$\Delta t = \frac{L}{v_1} - \frac{L}{v_2} = \frac{L \, \Delta n}{c}. \tag{13.162}$$

For a 1-km fiber with $\Delta n = 0.03$, this predicts a transit time difference

$$\Delta t = 100 \text{ ns}.$$

Indeed, it can be shown [53] that the time delay experienced by any mode group, m, is given by

$$t \approx \frac{\left(n_1 + \Delta n \dfrac{m}{m_{mx}}\right)L}{c} \approx \left(n_1 + \pi \, \Delta n \frac{m}{V}\right) \tag{13.163}$$

if index dispersion is neglected. Let us compare this to a specific calculation. Again we consider a strictly monochromatic wave propagating in two different modes and calculate the difference in transit time of the two modes.

Consider a fiber with radius $50\lambda_0/\pi$ and indices 1.5 ± 0.015 in the core and the cladding, respectively. According to (13.129) this will have a value of $V = 30$. According to (13.130) and (13.136), this yields for the LP_{01} mode

$$U = 2.33633, \qquad W = 29.90889.$$

According to (13.140) the corresponding phase velocity, relative to that of light *in vacuo*, is

$$\frac{v_{10}}{c} = \frac{V}{\sqrt{n_2^2 U^2 + n_1^2 W^2}} = 0.660144.$$

For the LP_{63} mode, we find from (13.130) and (13.137)

$$U = 17.83, \qquad W = 24.12656,$$

and

$$\frac{v_{63}}{c} = 0.664685.$$

The difference in propagation velocity is, thus,

$$\frac{\Delta v}{c} = 0.00454.$$

Over a 1-km path, this results in a transit time difference

$$\Delta t = L \left(\frac{1}{v_{10}} - \frac{1}{v_{63}} \right) \approx \frac{L \, \Delta v}{v^2}$$

$$= 34 \text{ ns}.$$

Formula (13.163) yields for this case 20 ns.

In a *graded-index fiber* with perfect focusing, there would be no pulse spreading at all due to mode differences. It is found that by a proper choice of index distribution the pulse spreading can be reduced to [54], [34]

$$\Delta t = \frac{t_0 \delta^2}{8}, \tag{13.164}$$

where δ is the relative index difference (13.78) and
t_0 is the axial mode transit time (13.127).

The required index distribution corresponding to this value is approximately parabolic and is given at the end of Appendix 13.3. The spread (13.164) is approximately 0.25 ns/km for a 2% index difference. However, an index deviation of as little as 10^{-4} from the optimum profile, will broaden the pulse by double this amount. Measurements on actual fibers have shown rms pulse widths of 0.13 ns/km [55].

13.2.3.4 Beam Attenuation. The attenuation of a beam during its passage through a fiber is due primarily to one of the following:

1. Reflection at the entrance and exit faces (see Section 10.1.2).
2. Scattering and absorption in the material (see also Section 10.1.2).
3. Incomplete "total" internal reflection at the fiber surfaces.

If the extinction coefficient of the fiber material is α_T and the reflectance of its surface, under conditions of "total" internal reflection, is ρ_0, we have for the total attenuation inside the fiber

$$\frac{\Phi_{out}}{\Phi_{in}} = \rho_0^N e^{-\alpha_T s} = \exp\left(N \log \rho_0 - \alpha_T s\right).$$

On substitution from (13.79) and (13.81), this becomes

$$\frac{\Phi_{out}}{\Phi_{in}} = \exp\frac{(n_0 \log \rho_0 \sin \alpha/2r_0 - n_1\alpha_T)L}{\sqrt{n_1^2 - n_0^2 \sin^2 \alpha}},\tag{13.165}$$

$$= e^{-\hat{\alpha}L}$$

where we have combined the coefficient of L in the second member under the symbol $\hat{\alpha}$ in order to emphasize the dependence on L.

In general, the extinction coefficient will be different for core and cladding materials. The effective value is a combination of the two, weighted according to the fraction of the power carried by each; see (13.141).

Equation 13.165 does not include the losses due to reflections at the end faces of the fiber; these are given by (8.3) and (10.15). Also, it is exact only for meridional rays; for skew rays passing at a distance r_c from the axis, the total number of reflections increases according to

$$\sec \gamma = 1 \Big/ \sqrt{1 - \frac{r_c^2}{r_0^2}}\tag{13.166}$$

see (13.87).

The rays not satisfying (13.71) will be rapidly eliminated in the form of radiation losses. Waves in the tunneling modes of Section 13.2.3.2, too, will be attenuated more rapidly than the fully trapped waves. Consequently, the energy in the guided wave will tend to consist more and more exclusively of trapped modes, as the wave progresses over greater distances. Since the fraction of the power penetrating into the cladding is also less for the trapped modes, the significance of the extinction coefficient of the cladding material diminishes as well with increasing distance [56].

13.2.4 Applications of Single Fibers [32a]

At the moment, the most widespread potential use of optical fibers would seem to be in communication. There the fibers are combined into cables; this application is discussed in a special section (13.3.3). Here we treat other applications of single fibers. Applications of fiber bundles are treated in Section 13.3.4.

Single optical fibers are used as small apertures, especially in scanning applications. One end of a fiber may be made to scan an image, with the other end feeding into a photometric device. With useful fiber diameters down to below 10 μm, such a device may be similar in cost to a comparable pinhole and provide the additional convenience of a flexible connection to the output.

Vibratory scanning may be implemented, with the fiber itself providing the resonant mechanism. For example, a small bead of iron attached to a fairly thick fiber (0.25 mm diam) and placed between the poles of an electromagnet, provides a convenient scan when an alternating current is applied to the magnet at the mechanical resonance frequency of the fiber-bead system. Vibrational amplitudes of up to 1 cm have been obtained with such a system.

Single-point measurements in inaccessible areas, too, may be facilitated by a single fiber unit. Single fibers have also been used for mixing light, providing a uniform output from a nonuniform input illumination.

They have also been used for transporting radiative energy for heating purposes. Such a fiber, in conjunction with a 100-W tungsten lamp as input, has been used for soldering connections in delicate electronic systems.

When used to transmit laser radiation, for example, for retinal surgery, the extremely high-power levels may cause difficulty. Excessive heating may result if the medium absorbs a significant amount of energy, even though this may represent a very low percentage. In well-polished, high-transmittance glasses, however, the major problem seems to be solarization. For instance, at a power density of 15 kW/cm^2, an ordinary glass fiber, 1.5 m long, had its transmittance drop from 53 to 25% in 7 min. Glasses resistant to solarization have been developed exhibiting a drop of only 10% after 1 hr under the above conditions [44b].

In conclusion, we mention a proposed application of an entirely different nature: a very long fiber may be used as a spectroscope. If mode dispersion has been eliminated (e.g., in a graded index fiber), the material dispersion will spread a pulse temporally according to its spectral components. If a sufficiently fast detector is available, this permits the analysis of the pulse's spectral composition [57].

13.3 FIBER BUNDLES

A single fiber, if thin enough to be flexible, is severely limited in terms of the amount of energy and information it can transmit. Therefore fibers are often combined into cables or bundles. Such fiber bundles fall into two categories:

1. When the energy transported does not carry a signal, or if the signal in one fiber is unrelated to that in the other, the relative positioning of the various fibers in the bundle is unimportant and the far simpler *incoherent* bundles suffice. These are often referred to as *light guides.*
2. When pictorial information is to be transmitted, the relative positions of the fibers in the bundle must be controlled. Such bundles are then referred to as *coherent.*

Here we first discuss the incoherent bundles in Section 13.3.1 and then proceed to the coherent ones in the following section. Fiber cables for communication purposes are treated in Section 13.3.3 and applications in the last section.

13.3.1 Incoherent Bundles

13.3.1.1 Advantages of Fiber Bundles. The primary function of incoherent fiber bundles is the transmission of light from place to place. For this task, fibers have a number of advantages over other optical devices.

1. *Flexibility.* By making the component fibers of a bundle sufficiently thin, the bundle can be made flexible so that light can be guided around complicated paths without resorting to mirrors and prisms with the attendant cost and technical problems.
2. *Efficiency.* A fiber-optic transmission line may also have a higher overall transmittance than the equivalent mirror system. The fact that a fiber may have a na in excess of unity [see (13.76)] also may make it attractive means for bringing light to its target.
3. *Compactness.* The flexibility of fiber optic bundles may lead to a more compact optical system.
4. *Shaping a light beam.* The cross-sectional shape of the transmitted light beam may be changed by (a) tapering the component fibers or (b) rearranging the fibers in the bundle. Thus a circular beam may be transformed into an oblong one; a single beam may be split into a number of separate beams; various beams may be combined into one; or the cross-sectional area of the illuminated region may be changed.

13.3.1.2 Construction of Bundles. In the simplest method of manufacturing bundles, the fiber is drawn from a furnace onto a drum, as described in Section 13.2.1.2. After an appropriate number of drum revolutions has been completed, the fibers on the drum may be cut parallel to the drum axis, removed, and tied or fused together. This yields a bundle of a length roughly equal to the drum circumference.

The manufacture of bundles by a continuous process and without length limitations requires a multiplicity of openings through which the fibers are drawn individually, to be combined after they have hardened. The tying of the bundle as it is drawn together can, of course, be made part of the continuous process.

13.3.1.3   Optical Properties. For tightest packing, the fiber lattice takes on a hexagonal form as illustrated in Figure 13.24a. The reader will readily confirm that the circles in such an arrangement take up a fraction

$$f_h = \frac{\pi}{2\sqrt{3}} = 0.9069 \qquad (13.167)$$

of each lattice cell, and an equal fraction of the bundle cross section if edge effects are neglected. When cladding is allowed for, the fraction

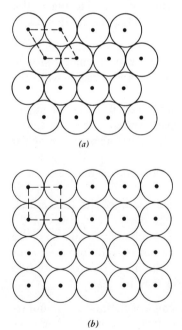

(a)

(b)

Figure 13.24   Packing lattices of fiber bundles. (a) Hexagonal; (b) Square. In each, a lattice cell is indicated by broken lines.

taken up by the cores is

$$f_{hc} = \frac{r_1^2}{r_2^2} f_h, \tag{13.168}$$

where r_1, r_2 are the outer radii of core and cladding, respectively. This is called the core packing fraction. If the reflection losses at the entrance and exit faces (10.15) and attenuation (13.165), too, are considered, we find the overall transmittance of a hexagonal bundle:

$$\tau = \frac{\pi}{2\sqrt{3}} \left[\frac{4n_1 r_1}{(n_1+1)^2 r_2} \right]^2 e^{-\hat{a}L} \tag{13.169}$$

$$\approx 0.677 \, e^{-\hat{a}L}, \qquad n_1 = 1.5, \qquad \frac{r_1}{r_2} = 0.9.$$

A typical value for the extinction coefficient of fibers in a bundle is $0.15 \, \mathrm{m}^{-1}$ in the visible portion of the spectrum. See Figure 10.3 and Table 80 for the extinction coefficient spectra of typical optical glasses and plastics.

Breakage of individual fibers occurs during manufacturing and tends to continue during usual handling. This reduces transmittance proportionally. The fraction of broken fibers in a new bundle should not exceed 1%.

The alternative to the hexagonal is the square lattice. Here the circles take up a fraction

$$f_s = \frac{\pi}{4} = 0.785. \tag{13.170}$$

The transmittance of such a bundle is therefore lower than that of a hexagonal bundle by a factor

$$\frac{\sqrt{3}}{2} = 0.866.$$

See Figure 13.24b.

13.3.1.4 Mechanical Properties. Glass fibers are quite flexible up to diameters of approximately 0.15 mm, plastic fibers up to 1.5 mm diam. When a fiber bundle is bent, the fibers located on the outside of the bend are stressed in tension and those on the inside, in compression. Such stresses tend to break those fibers that already suffer from internal stresses and, as a result, reduce the transmittance of the bundle somewhat. However, in glass fibers, the transmittance tends to reach a stable value only 1 or 2% below the original value after approximately 100 such bendings. In plastic fibers, small cracks ("crazing") tend to develop and to lower the transmittance value further.

The mechanical strength of a straight fiber bundle equals the combined strengths of the individual fibers that have been discussed in Section 13.2.1.5. However, when a bent fiber bundle is loaded, the loading differs for different parts of the bundle, the tensile load being greater for fibers on the outside of the bend. Differences in loading are transmitted by the friction between fibers in the bundle. This friction itself, in conjunction with the relative motion induced by the bending, also tends to introduce mechanical damage, weakening the fibers. Figure 13.25 [32d] shows the effect of lubrication on the strength of a bundle bent at a right angle around a 5-mm-diameter mandrel. It appears that proper lubrication can increase the strength of the bundle by a factor of more than four under these conditions.

The operating temperature range of a glass fiber bundle is usually set by the characteristics of the sheathing material and cement used to bind it. Bundles have been made for operation up to 400°C. Plastic fiber bundles are limited by the temperature characteristics of the plastic material as discussed earlier (Section 13.2.1.4).

13.3.2 Coherent Bundles

Since each fiber in a bundle can be made to carry a certain signal flux, unaffected by the flux in the neighboring fibers, it can serve to carry information from a given object element to an image plane. A bundle of such fibers can be used to transfer an image from one location to another.

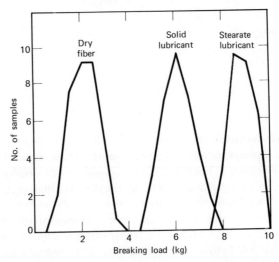

Figure 13.25 Strength of bent fiber bundles: effect of lubrication. Bundles bent through 90° around a 5-mm-diam mandrel. Allan [32d].

In this application, however, the location of each fiber relative to the others in the bundle must be carefully controlled. As noted earlier, bundles with such controlled location are called coherent.

13.3.2.1 Construction of Coherent Fiber Bundles [32e]. Before proceeding to the discussion of their optical characteristics in the next section, we describe here some of the manufacturing techniques used in constructing coherent bundles.

Monofilament Fiber Bundles. A coherent fiber bundle may be made by winding a single fiber onto a drum, essentially as described in Section 13.3.1.2, taking care, however, to position successive turns of the helix in close proximity and with no overlap. When the desired width has been reached, the next layer may be wound onto the first, upon simply reversing the sense of the helix. Unfortunately, this simple technique leads to a number of difficulties and the so-called *hoop technique* may yield better results. Here the single layer is consolidated over a narrow region along its length, for example, by cementing, and the layer is then removed from the drum in the form of a hoop. When the required number of single-layer hoops have been prepared, these are combined into a two-dimensional array by cementing them together at their consolidated section. This may then be cut through the combined "consolidated" region, resulting in a coherent fiber bundle, which is completed by grinding and polishing the two end faces.

Since fibers thinner than approximately 20 μm are very difficult to handle individually, multiple fibers are used when thinner fibers are required, for example, because of resolution requirements.

Multiple Fibers. A *multiple fiber* is a coherent fiber bundle that was fused and again drawn, so that it takes on the outward form of a single fiber, but is, in fact, a multiplicity of even thinner fibers. Such multiple fibers may then be stacked again into larger fiber bundles. By this technique, bundles with cross sections of several centimeters may be constructed, consisting of fibers with diameters as small as 5 μm.

Note that in fused fiber bundles, such as these, the effective cladding thickness is doubled, since the claddings of two neighboring fibers are always fused into one continuum.

On occasion, it may be convenient to fuse and draw bundles of multiple fibers, subsequently to combine them into a *two-stage multiple fiber*.

In general, multiple fibers are more economical when large diameter bundles are needed. Their main disadvantage seems to be that any fiber breakage affects a large area element of the image field, which may be more objectionable than many scattered area elements constituting the same total area. Due to the required fusing, multiple fibers are also less flexible than monofilament fiber bundles.

Optical Isolation. Whenever the cladding has a finite thickness, a small amount of the "trapped" flux leaves its fiber. Scattering in the fibers and at their surfaces, too, transfers some of the trapped-mode flux into radiative modes. Both of these phenomena contribute to the passage of light from one fiber to another. In incoherent bundles, this leads, at worst, to a light loss, which is generally negligible. But in coherent bundles, it results in a lowering of the contrast in the final image, possibly a far more serious defect. To reduce this passage of light from fiber to fiber, the component fibers may be coated with a metallic layer. This, however, precludes the drawing of multiple fibers.

In a more flexible approach, each fiber is coated with a layer of opaque glass. This, in turn, has the disadvantage of increasing significantly the overall fiber diameter for a given core diameter, thereby reducing the core packing fraction.[6] Since the opaque layer generally used need be no more than a micrometer thick, this is not a serious problem for the thicker fibers; but in thin fibers, the loss may be significant.

In fibers with a diameter of less than 20 μm, it may be better to place small absorbing fibers into the interstices between the circular guiding fibers. In hexagonal and square lattices, such interstitial fibers must have a radius of less than

$$\left(\frac{2}{\sqrt{3}} - 1\right)r_0 = 0.1547r_0 \quad \text{and} \quad (\sqrt{2} - 1)r_0 = 0.4142r_0,$$

respectively. In bundles including a large number of fibers, any radiated ray is very likely to strike one of the absorbing fibers before reaching the image.

For ease in manufacturing, the absorbing fibers are made of a glass similar to that constituting the guiding fibers. It is rendered opaque by the addition of metal oxide ions (see Section 11.2.2).

Rigid Bundles. In applications where flexibility in operation is not required, the bundle itself may be fused into a single rigid unit with the desired path shape. Such bundles are called *image conduits.* Often the bundle length is short compared to its cross-sectional width; it is then

---

[6] The increase in fiber diameter is due (a) to the thickness of the absorbing material required to ensure a sufficiently low transmittance and (b) to the fact that the cladding thickness must also be increased if absorption of the guided flux is to be avoided: the two neighboring layers of cladding are no longer in contact, so that each one of them individually must have the required thickness.

called a *fused fiber plate*. Such plates are used primarily as field flatteners and for transferring images from a phosphor screen in a vacuum tube to an accessible surface.

Splicing. Splicing large bundles of fibers is generally not considered a serious problem, even with coherent bundles. The end faces are simply polished and then butted. As long as the core packing density is reasonably large, there will be some losses at the junction, but these will not be much larger than implied by the packing fraction.

In fibers for communication, where the core may be very much smaller than the overall fiber dimension and, in addition, each fiber may have to be matched to its mate, the problem takes on a completely different dimension. This is discussed below (Section 13.3.3.1).

Arrays Formed in Plastic Sheets. Single-layer plastic fiber bundles may be made by casting or pressing. In one system, a plate with a series of saw-tooth ridges (100/mm) was cast of a material with relatively high refractive index (polycarbonate, $n = 1.59$). Subsequently, a lower-index monomer (methyl acrylate, $n = 1.4$) was permitted to diffuse into it. Finally the diffused monomer was polymerized by γ radiation [58].

In another system, ridges are embossed onto a plastic sheet and filled with a liquid that is then solidified [59]. Light guiding arrays have also been formed photographically in a polymer film [13].

Woven Fiber Bundles. Recently weaving has been developed as a method for bundling fibers [60]. The fibers are fed through a jacquard loom. Plastic fibers are used and, to minimize fiber bending, the shed (consisting of alternately raised and lowered warp threads) is formed only of binder threads interposed between each pair of fibers. Multiple-ply ribbons can be woven and subsequently stacked to form the final bundle. The technique is especially useful when a complex, but repetitive, fiber distribution pattern is required. For instance, a three-color crt display, operating from three separate monochrome displays, was implemented by this technique [60].

13.3.2.2 Optical Properties: Spread Functions

General. Optical properties of coherent optical fiber bundles are similar to those of the incoherent bundles, except that the inclusion of isolation means may reduce, somewhat, the effective numerical aperture by causing increased attenuation of the more oblique rays.

In addition, the effective spread function of the bundle is here of interest, and we devote the present section to an analysis of this. A brief discussion of light leakage and imaging defects in fiber bundles is presented in the next section.

Analyses of the image transmission characteristics have been published [61], [62]. Here we present a simple step-by-step analysis that will, hopefully, give both insight into the limitations and sufficient information for most system design needs.

Static Fiber Imaging. Because of the lack of isoplanaticism, the Fourier analytic approach, including the concept of otf, does not apply here. To illustrate this, let us consider the image modulation resulting from placing a tightly packed bundle of square fibers over an object whose luminance varies sinusoidally in one dimension. We place the bundle so that a side of each square is parallel to the object fringes. The object may then be represented by

$$L(x) = \bar{L}(1 + M_0 \cos 2\pi\nu x), \tag{13.171}$$

where $\bar{L}$ = the mean luminance of the object,
  M_0 = its modulation,
  $\nu = 1/\Lambda$ is its spatial frequency, and
  Λ = the pitch of the object fringes.

This implies taking the x axis normal to the object fringes, with its origin at a luminance maximum. If we place over this a square aperture (here a fiber cross section), width a, with its center at x_j, the integrated luminance at the input will be

$$I_j = \int_{x_j - a/2}^{x_j + a/2} L(x)\, dx. \tag{13.172}$$

On substituting from (13.171) and integrating, we find the mean luminance over the aperture, relative to the mean luminance over the object

$$\hat{L}_j = \frac{I_j}{a^2 \bar{L}} = 1 + M_0 \cos 2\pi\nu x_j \, \mathrm{sinc}\, \frac{a}{\Lambda}, \tag{13.173}$$

where $\mathrm{sinc}\, u = \sin(\pi u)/\pi u$, represents the so-called sinc function. If we assume perfect mixing of flux in the fiber, we may take this also as the normalized luminance of the fiber exit face. Clearly, the output luminance of the fiber at the maximum ($|x_j| < a/2$) varies not only with relative pitch of the bundle (a/Λ), but also with the position of the maximum-luminance fiber (x_j) relative to the luminance maximum. This illustrates the lack of isoplanaticism referred to above.

To illustrate its effect on modulation let us proceed to the bundle and find the modulation transfer:

$$T = \frac{\text{image modulation}}{\text{object modulation}}$$

$$= \frac{M}{M_0} = \frac{\hat{L}_{mx} - \hat{L}_{mn}}{M_0(\hat{L}_{mx} + \hat{L}_{mn})}$$

$$= \frac{(\cos 2\pi \nu x' + \cos 2\pi \nu x'') \operatorname{sinc} a/\Lambda}{2 + M_0 (\cos 2\pi \nu x' - \cos 2\pi \nu x'') \operatorname{sinc} a/\Lambda}, \tag{13.174}$$

where x' and x'' are the distances of the nearest fiber center from the luminance maximum and minimum, respectively. This has a maximum value when both x' and x'' vanish. Then

$$(M/M_0)_{mx} = \operatorname{sinc} a\nu = \operatorname{sinc} a/\Lambda. \tag{13.175}$$

It has its minimum value when

$$x' = x'' = \frac{a}{2}. \tag{13.176}$$

Then

$$\left(\frac{M}{M_0}\right)_{mn} = \cos \pi a\nu \operatorname{sinc} a\nu$$

$$= \cos \pi \frac{a}{\Lambda} \operatorname{sinc} \frac{a}{\Lambda}. \tag{13.177}$$

Note that for $a/\Lambda \geq \frac{1}{2}$, the modulation in the image may vanish entirely.

In an actual fiber bundle, the observed modulation will vary between the values (13.175) and (13.177), depending on the fiber position relative to the luminance cycle. Clearly, if this is random, the effect constitutes a multiplicative noise with a variance:

$$\overline{N^2} = \overline{\cos^2 \pi \nu x'} - \overline{\cos \pi \nu x'}^2 = \frac{\Lambda}{\pi a} \int_0^{\pi a/\Lambda} \cos^2 u \, du - \left(\frac{\Lambda}{\pi a} \int_0^{\pi a/\Lambda} \cos u \, du\right)^2$$

$$= \frac{1}{2} \left(1 + \operatorname{sinc} \frac{2a}{\Lambda} - 2 \operatorname{sinc}^2 \frac{a}{\Lambda}\right). \tag{13.178}$$

Dynamic Fiber Imaging. The noise just described may be effectively eliminated by moving the fiber bundle[7] relative to the object, resulting in *dynamic fiber imaging.* If the detector response is sufficiently slow, or the motion is sufficiently fast, so that the output is integrated over the time

---

[7] Note that both the input and the output faces must be moved in unison.

interval required for the bundle to move the distance $a/2$, the detected, or observed, maximum will be given by (13.173), averaged over $0 < x_j < a/2$:

$$\bar{L}_{mx} = 1 + M_0 \operatorname{sinc}^2 a\nu. \tag{13.179}$$

The same expression, with a minus sign replacing the plus sign, gives the resulting luminance observed at the minimum. Hence, in analogy to (13.174) the modulation transfer function becomes:

$$\mathsf{T}(\nu) = \frac{M}{M_0} = \operatorname{sinc}^2 a\nu. \tag{13.180}$$

This is now independent of fiber location and hence (a) the noise has been eliminated and (b) isoplanaticism has been established, justifying us to identify (13.180) with the mtf.

It can be shown that if the fiber bundle lattice is rotated through an angle β, this expression becomes [62]

$$\mathsf{T}_\beta(\nu) = \operatorname{sinc}^2(a\nu \cos \beta) \operatorname{sinc}^2(a\nu \sin \beta). \tag{13.181}$$

At $\beta = 0$, $\pi/2$, this becomes identical to (13.180). It attains its minimum value

$$\mathsf{T}_{\pi/4}(\nu) = \operatorname{sinc}^4 \frac{a\nu}{\sqrt{2}}$$

at $\beta = \pi/4$.

Circular Fibers. In the more important case of circular fibers, the derivation of the imaging performance follows essentially the same steps, but is mathematically more difficult. It is presented in Appendix 13.4 and yields the result:

$$\boxed{\mathsf{T}_c(\nu) = \left[\frac{J_1(2\pi\nu r_0)}{\pi\nu r_0} \right]^2} . \tag{13.182}$$

13.3.2.3. Optical Properties: Stray Light and Defects

Effect of Stray Light. In general, the isolation between fibers is not perfect as assumed in the preceding discussion. A fraction, b, of the light passes from fiber to fiber and is eventually distributed, more or less evenly, over the image, resulting in a background luminance $b\bar{L}$. Consequently, from (13.179)

$$\bar{L}_{mx,mn} = 1 \pm \left(\frac{1-b}{1+b} \right) M_0 \sin a\nu \tag{13.183}$$

and hence the dynamic mtf becomes

$$T(\nu) = \left(\frac{1-b}{1+b}\right) \text{sinc}^2 a\nu. \tag{13.184}$$

When the illumination at the input to the fiber bundle is held to within a cone of semiapex angle,

$$\alpha \leq \alpha_{mx}$$

(see 13.75), stray light is due only to one of the following:

1. Light entering the cladding and interstices directly.
2. Frustration of total internal reflection.
3. Scattering in the fiber and at its surfaces,
4. Bending effects, when applicable.

The fraction of light entering the interstices is given by

$$b_i = 1 - f_{hc}, \tag{13.185}$$

where f_{hc} is the core packing fraction (13.168). Except for this, all the above should be negligible in a properly constructed and used fiber bundle.

However, when the incident light is not limited to the acceptance cone of the fibers, there may be considerably more unguided light. For instance, with a Lambertian source at the entrance to the fiber, the density of the flux trapped in the fiber is

$$E = \pi L \sin^2 \alpha_{mx} \tag{13.186}$$

out of the total emittance (πL) [see (1.12) with the integration carried only to α_{mx} and where we have divided both sides by the cross-sectional area of the fiber]. Thus the fraction of unguided light, from this source alone, is

$$b_\alpha = 1 - \sin^2 \alpha_{mx} = \cos^2 \alpha_{mx}. \tag{13.187}$$

Crosstalk Power. The amount of crosstalk between fibers can be calculated more accurately from the field distributions in the cladding [18a], [19b], [20a]. For fibers carrying a large number of modes, the crosstalk power (P_c) relative to the entering power (P_0) is given by [63]

$$\frac{P_c}{P_0} \approx \frac{1 - \sin L^*/L^*}{2}, \tag{13.188}$$

where the normalized length

$$L^* = \frac{4z\alpha_i^2 e^{-V(d/r_0 - 2)}}{\alpha'_{mx}\sqrt{2\pi V r_0 d}}, \tag{13.189}$$

and α'_{mx} ($\ll 1$) and $V(\gg 1)$ are given by (13.75) and (13.129), respectively,

α_i is the half-apex angle of incident illumination cone, assumed uniform,

d is the center-to-center spacing between the fibers, and it is assumed that the illumination is concentrated on the fiber axis, so that only meridional modes are excited.

Imaging Defects. Invariably, manufacturing deficiencies also cause defects in the image. These include dark spots, due to broken or cracked fibers, and image distortions due to fiber misalignment. If the fibers in a fused plate are not accurately normal to the fiber axis, the image will appear displaced laterally.

13.3.3 Fiber Cables for Communication Systems [64]

Communication systems have grown at an enormous rate over the past decades. By way of illustration: the carrier network in the Bell System had only 50 million circuit kilometers in 1954; 20 years later it had a billion, with about one third of it consisting of electrical wires and cables, and the remainder of radio channels [65]. Such a volume warrants a careful examination of material costs and information capacity of the channels; here it appears that the potential of communication by light waves carries considerable promise. This has given rise to a tremendous effort to develop integrated optics and fiber waveguides in that direction.

Most of the relevant component considerations have already been covered in the earlier parts of this chapter. There remain the special construction problems of fiber cables for communication and system considerations. These are the subjects of the present section.

13.3.3.1 Construction of Fiber Cables [64a]

Protective Measures. The approach to the cabling of fibers for communication systems is governed by the severe enviromental conditions to which these cables may be exposed, both in handling and in location. This effect is acerbated by the inherent fragility of glass fibers. Consequently, the major portion of the cable volume may be taken up by components designed to protect the fibers. Such components must be provided to protect the fibers against (a) abrasion and contamination, (b) tensile stresses, and (c) excessive bending and gross impact stresses.

To protect their surfaces from abrasion, contamination, and chemical attack, the fibers are usually coated with some plastic material. This may be extruded simultaneously with the drawing of the fiber or the fiber may be coated subsequently, depending on the workability and solubility of the coating material. Among the materials used are tetrafluoroethylene

(Teflon), perfluoronated ethylene-propylene (FEP), perfluoroalkoxy resin (PFA), and polyurethane.

Protection against tensile stresses may be provided by including fibers of materials that have a high Young's modulus, such as steel. With proper design, these can be made to absorb almost all of the tensile stress applied to the cable.

In one system, the guide fibers are wrapped helically around a central strengthening member. This introduces into the guide fibers a variety of stresses and care must be taken to minimize these, for example, by rotating the fibers appropriately during the wrapping process, so as to minimize the torsional stress. Alternatively, the strengthening members may be placed around the guide fibers, so that they contribute to the protection against crushing. To protect the cable against gross impact stresses and crushing, the cable is usually covered with a strong protective sheath.

Since the refractive index differences usually used there are small, communication systems fibers are quite sensitive to bending. Even very small bends and slight wrinkles, so-called *microbends*, are a major factor determining the guidance losses in such fibers (see Section 13.2.2.4). This calls for special precautions in manufacturing. The protective sheath can also be designed to prevent bending beyond the permissible curvature.

Extensive test results obtained on glass fibers under rough service conditions have been published [66]. Extensive data on the mechanical, chemical, and thermal strength of plastic fiber cables, too, have been compiled [41a].

Coupling. The coupling requirements of fiber cables fall into two major classes: connect-disconnect junctions and permanent splices. The former can use optical components to provide efficient coupling between fiber ends (see Section 13.1.4.1). The splices, however, must be made by connecting individual fiber ends. Here both the finishing of the broken fiber ends and their accurate alignment pose difficult technological problems, especially if they are to be executed in the field by maintenance personnel.

An ingenious technique of combining bending with tension has been shown to lead to a clean break [67].

Fiber alignment can be facilitated by providing forms into which the fiber ends may be fitted, so that the end faces of matching fiber pairs match. Some cementing or fusion process may then be applied. If necessary, the form may, of course, remain as a permanent part of the junction. Successful splicing of one-dimensional fiber arrays, with losses of only 0.1 dB, have been demonstrated with multimode fibers (120 μm outside and 80 μm core diameters) [68].

13.3.3.2 System Considerations [34]. Since the function of a communication system is the transmission of information, such systems must be evaluated and compared in terms of channel capacity at a given error rate. At present, the major limitations on the achievable channel capacity is pulse spreading (due to material and modal dispersions) and the attainable power levels, with the attendant signal-to-noise levels.

Limits on Power Levels. Due to the small cross-sectional areas used in optical fibers, even moderate amounts of power correspond to extremely high values of energy density, so that nonlinear effects in the materials must be considered. At high energy density, stimulated Raman and Brillouin scattering will attenuate the signal. Assuming that we wish to limit the stimulated scattering everywhere to below the received signal level, it has been shown [69] that the critical power level is set primarily by Brillouin scattering and is given by

$$P \approx 20 \frac{\alpha_T A}{\gamma}, \qquad (13.190)$$

where α_T is the extinction coefficient
A is the effective fiber cross-sectional area, and
γ is the gain coefficient of the backward Brillouin process.

Assuming values of $0.005/m$ and 3×10^{-11} m/W for α_T and γ, respectively, we find the limit on the permissible power to be approximately 0.75 W for a $10 \, \mu$m-diam fiber. Raman and Brillouin scattering are discussed in Sections 4.3.3 and 14.2.3.5, respectively.

Choice of Modulation. Since the error rate is determined by the signal-to-noise ratio and this is, ultimately, limited by quantum noise, it is advantageous to have the instantaneous signal level as high as possible. [Note that the instantaneous signal-to-quantum-noise ratio is proportional to the square root of the signal (3.98).] Therefore some form of pulse modulation is often preferred over simple analog amplitude modulation, even though bandwidth is thereby sacrificed.

Choice of Fiber Type. With a laser source, single mode fibers, or multimode graded index fibers, give the highest information rate, since they minimize pulse spreading. With LED sources, the fiber must be multimode to permit coupling useful amounts of flux into it. A good rule of thumb is that the pulse spacing should be at least $4\Delta t_{rms}$; this limits the modulation loss to about 1 dB. Table 143 [34] lists the information rate, in M bits · km/sec for various fiber profiles, as limited by this condition.

Choice of Wavelength and Detector. The ultimate lower limit on the attenuation is set by Rayleigh scattering, which varies inversely with λ^4.

This points to the choice of longer wavelength in long-distance communication. On the other hand, glasses have absorption bands at 2.7, 0.95, and 0.72 μm due to OH^- ions, so that the neighborhoods of these bands should be avoided.

In the wavelength range generally used ($\lambda < 1.2\,\mu$m), p-i-n diodes (Section 16.3.2.2) offer the simplest solution. With these, dark current noise tends to be the limiting factor (at low signal levels). This limitation can be overcome by introducing gain into the detection process, as provided by avalanche diodes (Section 16.3.2.3), for instance. Here the gain process itself introduces excess noise, so that there is an optimum gain for each signal level.

Figure 13.26 [34] shows the number of signal photoelectrons required at the detector of an optical fiber communication system for an error rate of 10^{-9}. This is shown as a function of the information rate. The stippled bands show the operating regions with and without detector gain. See the figure caption for a description of additional data presented there.

13.3.4 Other Applications of Fiber Bundles [32f]

We survey here some applications of fiber bundles. This may provide the reader with ideas on how fiber bundles could solve some of the problems confronting him.

13.3.4.1 Illumination. Optical fibers have several advantages as adjuncts in illumination.

1. They permit separating the light source from the area to be illuminated. This is particularly important in medical scopes, which are inserted into various body openings for visual inspection of internal organs. In such applications, illumination must be provided and the need for inserting a lamp complicates further an already very difficult task, carries risks in terms of electrical connections, and, because of the concomitant heating, limits the amount of illumination feasible. All of these difficulties and limitations can be avoided by introducing the illumination via a thin fiber bundle.

2. Fiber bundles permit miniaturization, which may be crucial in applications using many light sources. Even though very small incandescent filaments are feasible, and commercially available (see Section 5.1.2.3), they invariably come in much larger envelopes that preclude tight packing.

3. When used for instrument illumination, fiber optics are particularly useful in conjunction with panels including many instruments. In such applications, the multiplicity of lamps increases the frequency of lamp

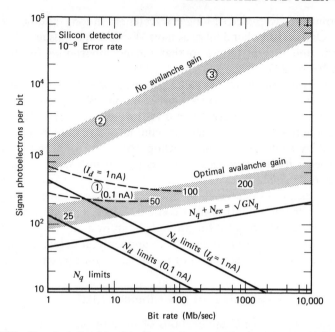

Figure 13.26 Required signal level in fiber optic communication systems. Average primary signal photoelectrons per bit interval required to achieve 10^{-9} error rate versus bit rate for high-impedance integrating front-end optical receivers employing direct detection. Silicon photodetectors and FET first-stage amplifiers are assumed. The ordinate is also proportional to the required minimum received average power divided by the bit rate, the proportionality constant being $(\eta/h\nu)$. The dotted bands indicate the expected performance based on current device parameters. The numbers (25, 50, 100, 200) shown in the middle of the lower band are the calculated values of optimal avalanche gain for the respective bit rates. The dotted lines show degradation of performance due to primary dark current (I_d) in the presence of optimal avalanche gain. The solid horizontal line gives the quantum noise (N_q) limit without gain and represents the ultimate in receiver performance. The two solid lines with negative slope set the limit of performance for no gain if dark current noise (N_d) were the only noise source. The solid line with positive slope represents the limit due to quantum and excess noise (N_q+N_{ex}) with optimal gain but negligible dark current. The circles represent experimentally achieved values by Goell: ① corresponds to -62 dBm, ② corresponds to -52 dBm, and ③ corresponds to -29 dBm. The curves are based on calculations of Personick. From Miller, Li, and Marcatili [34].

failure. This, in conjunction with the difficulty of lamp replacement, makes the traditional method of illumination undesirable. The use of a fiber bundle, illuminated by a single lamp and split into the required number of branches, can eliminate both these difficulties. It also facilitates the provision of a stand-by light source in devices where lamp failure carries serious risks.

4. In optoelectronic devices, electrically controlled light sources are

used, in conjunction with photosensitive components, to make and break electrical connections. This method offers major advantages in facilitating electrical isolation from high voltages and in providing immunity from radio interference. Full exploitation of these potential advantages often requires the transportation of the light over some distances, conveniently accomplished by a fiber optics bundle.

5. The reliable monitoring of light sources located in inaccessible places, too, may be facilitated by fiber optics bundles. In one application, a diagram of the automobile's outline was displayed on its dashboard, with fiber bundles carrying light from the various lamps to their analogous position on the diagram. This enabled the driver to check, at a glance, the condition of all his exterior lamps.

13.3.4.2   Redistribution of Light. The fact that the shape of the fiber bundle cross section can easily be changed, opens the way to many applications. We include in this category the option of splitting the original fiber into a number of branches (already mentioned under 3 in the preceding section) or the combining of a number of branches into a single trunk (see 5 in the preceding section). The only basic restriction to be observed is that the overall illuminated area remain constant, if cylindrical fibers are used.

1. Whenever the shape of the area element to be illuminated differs greatly from the shape of the available light source and has an aspect ratio much larger than unity, it becomes difficult to avoid major inefficiencies in the utilization of the available light. The illumination of the entrance slit of a spectroscope or a single line on a document, are examples of this type of problem. In such cases, a fiber bundle going from a circular to an oblong outline offers a convenient solution.

2. The same approach can solve the inverse problem posed by semiconductor lasers, which constitute oblong light sources with very large aspect ratios, making it very difficult to collimate their light. By using a fiber bundle going from a linear to a circular cross section, the largest dimension of the source can be reduced by, say, a factor of 10, and, consequently, the volume of the collimating system by a factor of 1000!

3. When the light from a scanning slit is to be brought to a detector, which generally has an aspect ratio close to unity, the inefficiencies may be even more serious than in the analogous illumination problems. Here the bundle of 2 may be used to boost the utilization of the available light flux.

4. In punched card and tape readers, the light from a single source is efficiently brought to a large number of small area elements. This is much

more efficient than the equivalent illumination of a much larger area of inconvenient dimensions from the same source, without the use of fibers. It is also significantly superior to the numerous small lamps required by the alternative approach of providing an individual source for illuminating each row.

5. High-luminance display of characters, or a limited selection of shapes, too, can be executed efficiently with incoherent fiber bundles. Here the use of fibers may greatly reduce the number of lamps required and, consequently, maintenance costs. Colored filters are, of course, readily incorporated into the input section of the optical system.

In traffic control signs, which must often operate at very high levels of ambient illumination, the appearance of the undisplayed alternative signal ("phantom") may pose a serious difficulty. This, too, has been eliminated by means of display systems using incoherent fiber bundles.

In one system for implementing road signs [44c], hundreds of sub-bundles are combined into a single bundle for joined illumination. The subbundles may then be plugged into a peg board for displaying the desired information.

6. Fiber optics have even been used in multichannel correlators, where subbundles from various locations are combined for signal summing. [44d].

13.3.4.3 Scanning Applications.
Branched fiber bundles (y guides) may be used in scanning applications, where the branch carrying light to the object to be scanned is joined to another branch carrying the reflected light from the object to the detector. Since bundles with submillimeter cross sections are easily implemented and can be operated close to the object, imaging optics may be eliminated altogether. The amount of light spreading from the illuminating bundle to the object, can be controlled by limiting the na (13.76) of the fibers. The fibers that pick up the reflected light are conveniently placed in a circular annulus around the aferrent fiber bundle.

The fraction of emitted light from a plane circular Lambertian radiator received by a parallel circular disc concentric with it at a distance, d, is given by [70]

$$f(r, r', d) = \frac{r^2 + r'^2 + d^2 - \sqrt{(r^2 + r'^2 + d^2)^2 - 4r^2 r'^2}}{2r^2}, \quad (13.191)$$

where r, r' are the radii of the radiating and receiving disks, respectively. We may apply this formula to the system just described, with r now the radius of the illuminating fiber bundle and r' that of the scanned object element. If we now add r'' as the outer radius of the annular bundle of pickup fibers, we obtain for the efficiency of the system:

$$\eta = \rho f(r, r', d)[f(r', r'', d) - f(r', r, d)], \qquad (13.192)$$

where ρ is the reflectivity of the object element scanned. This formula neglects the possibility of confining the radiated energy to a cone of semiapex angle α; when this is considered, the actual efficiency may be considerably higher.

In addition to converting object luminance (or reflectance) distributions into temporal signal distribution, such fibers can be used as sensing elements in monitoring the lateral position of a moving strip. Because of the strong dependence of η on d, it can be used to monitor axial displacement of an object, as well.

See Section 13.3.4.5.(3) for a scanning application using a coherent fiber bundle.

13.3.4.4 Coherent Bundles—Simple Imaging. The simplest function of coherent bundles, is the transfer of images from one location to another. There are a number of applications of this function.

Fiberscopes. It is generally impossible to place the fiber bundle in contact with the object. Optical means, such as an objective lens, must then be used to image the object on the entrance face of the bundle. Furthermore, it is often desirable to enlarge the image received at the exit face of the bundle; this can be done by means of a magnifier, or eyepiece, lens. The resulting combination (objective—coherent fiber bundle— eyepiece) is essentially a telescope, or microscope, with the intermediate image plane extended axially by means of a fiber bundle. Such a device is called a *fiberscope.*

Fiberscopes have important applications in medicine and in industrial inspection tasks, where access to internal surfaces may be severely restricted. The need for accurate focusing of the objective can be minimized by using a short-focal-length lens. However, resolution requirements limit this and, with short object distances, some focusing may still be required. This may be accomplished by mounting the fiber bundle end so that it can be moved (by means of a control cable) in guide rails, relative to the objective lens. Steering capability may be provided for by mounting the bundle in a flexible wire helix, which tends to maintain it straight but allows for bending by means of tension on a guide wire running along the helix.

Image Conduits. In some applications the desired path of the bundle is constant, so that it may be made rigid. Such an image conduit provides greater stability and, frequently, greater convenience in handling.

Fiber Plates. In some applications, the required bundle may be very short, with its cross-sectional diameter large compared to its length. Such fiber plates are often used as faceplates in crt's. They may provide increased contrast by eliminating internal reflections and halos (see Section 6.3.1.3). In addition, they bring the image to an accessible surface so that contact photography of the display becomes feasible. Thirdly, the fiber plate can be used to eliminate the curvature often needed in the crt screen to maintain focusing throughout the scan.

Fiber plates have also found important application in multistage image intensifier and converter tubes. There the image must be transferred from the phosphor screen of one stage to the photocathode of the next. Since these are usually strongly convex toward each other (see Section 16.4.2.2), not only is it impossible to place them in contact with each other, even optical imaging becomes very difficult. The use of fiber optic faceplates solves this problem, while eliminating the need for imaging by lenses.

13.3.4.5 Coherent Bundles—Complex Imaging. In coherent fiber bundles, the arrangement of the fibers at the output face must bear a closely controlled relationship to that at the input face, but it need not be identical to it. This permits the implementation of a number of useful techniques.

Large-Scale Displays. In Section 13.2.2.3 we have already described conical fibers that, when used in coherent bundles, will magnify or demagnify images. In practice, however, the magnification factor achieved is limited. But image magnification can be achieved with simple cylindrical fibers also, if the fiber spacing (pitch) at the output is larger than that at the input. In such displays, designed for large viewing distances, the lack of contiguity in the displayed image may not be objectionable. An ingenious method for obtaining such increased spacing has been described [30g].

Image Reversion. In some applications, such as simple telescopes and proximity-focused image intensifier tubes using microchannel plates (see Section 16.4.2.3), a reversed image is obtained, requiring either a compound prism system, or an additional imaging stage for its reversion. A fiber optic plate may accomplish this more simply [31c, 44e]. It is only necessary to rotate one end of the bundle through 180° relative to the other, before fusing the bundle.

Raster Converters. A one-dimensional fiber bundle, where fibers are arranged along a straight line at one end and in a closed circle at the other, can be used to convert a continuous circular scan into a constant velocity linear scan with negligible flyback time. A constant speed circular scan is readily obtained on a crt by applying equal and purely sinusoidal signals to the two sets of deflection plates, 90° out of phase with each other. Alternatively, a rotating fiber can be used to provide the circular scan at the input.

Similarly, a two-dimensional array can be changed into a single line, as useful in streak camera recording for high-speed photography [31d].

APPENDIX 13.1 SKEW RAYS IN FIBERS

To analyze the progress of a general skew ray through a cylindrical fiber, consider Figure 13.27, which depicts a segment of such a ray between two reflections, at points A and B. Point C is the foot of the perpendicular from B to the cross-sectional plane, P, passing through A. Note that $\overline{BC}$ is parallel to the axis $(\overline{OO'})$. Now consider triangle ABC. Evidently the angle at B equals α', the angle the ray makes with the axis. Hence the angle at A is

$$\varepsilon = \frac{\pi}{2} - \alpha'. \tag{13.193}$$

Let O be the intercept of the axis with plane P. Then radius, $\overline{OA} = r_0$, is the normal to the cylinder envelope at A. The angle

$$\sphericalangle OAC = \gamma = \sin^{-1} \frac{r_c}{r_0}, \tag{13.194}$$

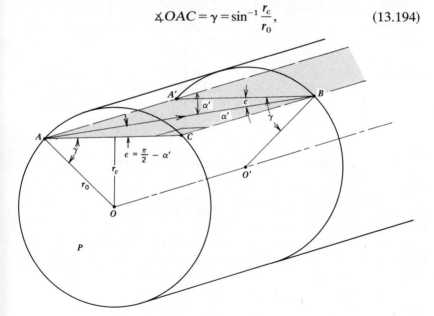

Figure 13.27 Skew ray in circular fiber.

where r_c is the distance of the ray from the axis. We can now write an expression for the angle of reflection $\sphericalangle OAB = \theta$:

$$\cos \theta = \cos \varepsilon \cos \gamma = \sin \alpha' \cos \gamma. \qquad (13.195)$$

This follows from the fact that θ is compounded of ε and γ, which lie in mutually perpendicular planes.[8] Note that both ε and γ, and hence θ, at B are equal to the analogous angles at A. Clearly the angle θ is invariant throughout the passage of the ray, regardless of the number of reflections.

For the ray to be trapped, θ must satisfy the inequality (13.74). This implies

$$\cos \theta \leq \sqrt{1 - \left(\frac{n_2}{n_1}\right)^2} \qquad (13.196)$$

and hence, from (13.195),

$$\sin \alpha' = \frac{\cos \theta}{\cos \gamma} \leq \frac{\sqrt{1 - (n_2/n_1)^2}}{\cos \gamma}. \qquad (13.197)$$

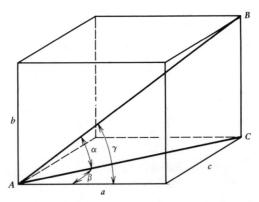

Figure 13.28 Compound angle in parallelepiped.

[8] Consider a parallelepiped with unity diagonal, $\overline{AB}$, and sides a, b, and c. Let $\overline{AC} = \sqrt{a^2 + c^2}$ be the diagonal of the base, as shown in Figure 13.28. Let α be the angle between it and $\overline{AB}$, β the angle between it and side a, and γ the angle between $\overline{AB}$ and side a. It then follows from inspection of the figure that

$$\cos \alpha = \overline{AC} = \sqrt{a^2 + c^2}$$

and

$$\cos \beta = \frac{a}{\sqrt{a^2 + c^2}}.$$

Hence

$$\cos \gamma = a = \cos \alpha \cos \beta.$$

QED

In analogy to (13.75), for any given value of γ, a ray will be trapped only if its α satisfies the condition

$$\sin \alpha \leqslant \sin \alpha_{mxs} \equiv \frac{\sqrt{n_1^2 - n_2^2}}{n_0 \cos \gamma} = \frac{\sin \alpha_{mx}}{\cos \gamma}. \tag{13.198}$$

Also, in analogy to (13.76), the na is

$$A_{skew} = \sqrt{n_1^2 - n_2^2} \sec \gamma, \tag{13.199}$$

which is equivalent to (13.85)

APPENDIX 13.2   RELATIVE FLUX-ACCEPTANCE[9] OF AN OPTICAL FIBER

We now seek an expression for the flux transmitted by a fiber illuminated by an extended Lambertian source of uniform luminance L, adjacent to the fiber entrance face. We consider only the portion of the flux totally reflected in the fiber ($\theta \geqslant \theta_c$) and neglect reflection and absorption losses. The value of the flux is obtained by integrating the luminance over the area, S, of the entrance face and over the angular region, Ω_R, of total internal reflection:

$$\Phi = L \int_S \int_{\Omega_R} \cos \alpha \, d\Omega \, dS. \tag{13.200}$$

We divide the trapped flux into two parts: the part (Φ_0), which satisfies (13.75), and the part (Φ_s), which does not and therefore consists exclusively of skew rays.

The integral covering Φ_0 is readily evaluated, since every area element, dS, is equivalent. We therefore simply substitute the total area (πr_0^2) for the area integral. For the integral over the solid angle, we use annular area elements of the form ($2\pi \sin \alpha \, d\alpha$), with α going from zero to α_{mx}. Hence, here (13.200) takes the form

$$\Phi_0 = L(\pi r_0^2) 2\pi \int_0^{\alpha_{mx}} \sin \alpha \cos \alpha \, d\alpha$$

$$= \pi^2 r_0^2 L \sin^2 \alpha_{mx} = \pi^2 r_0^2 L \frac{n_1^2 - n_2^2}{n_0^2}, \tag{13.201}$$

where we have made use of (13.75) in the last step.

[9] The terms "throughput," "étendue," and others have been proposed for this quantity [71].

To evaluate Φ_s, we divide the area into concentric annular elements

$$dS = 2\pi r\,dr,$$

integrating between the limits 0 and r_0. For any point, B, on such an annulus, we choose, out of its radiation cone (see Figure 13.29) a radial sector at azimuth angle β with width $d\beta$; these sectors go from $\beta = -\pi/2$ to $\beta = \pi/2$ and are taken twice, since they extend to both sides of the radiation cone axis. All the rays radiated into such a sector have a common value of γ, as shown in the figure. This sector, in turn, is broken up into elements of length $d\alpha$ (and width $\sin\alpha\,d\beta$). The integration over the sector is performed over the limits α_{mx} and α_{mxs}, as given by (13.75) and (13.85), respectively. The value of γ associated with any value β can be found from the projection $(\overline{BD})$ of the element $(B'D')$ at β onto the cross section plane. From triangle OCD, we find

$$\sin\gamma = \frac{r\sin\beta}{r_0}. \tag{13.202}$$

Here C is the point where the projection of element β intersects the circumference to the fiber cross section.

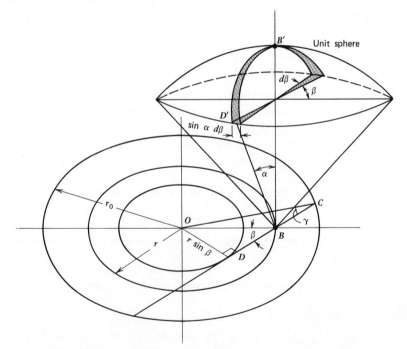

Figure 13.29   Deriving the relative flux-acceptance of a circular fiber.

Hence we rewrite (13.200), for Φ_s:

$$\Phi_s = 2\pi L \int_0^{r_0} r \int_{\Omega_s} \cos \alpha \, d\Omega \, dr, \qquad (13.203)$$

where the integral over Ω_s is

$$\int_{\Omega_s} \cos \alpha \, d\Omega = 2 \int_{-\pi/2}^{\pi/2} \int_{\alpha_{mx}}^{\alpha_{mxs}} \sin \alpha \cos \alpha \, d\alpha \, d\beta$$

$$= \int_{-\pi/2}^{\pi/2} (\sin^2 \alpha_{mxs} - \sin^2 \alpha_{mx}) \, d\beta.$$

Substituting from (13.198) and factoring out $\sin^2 \alpha_{mx}$, and then substituting from (13.202), we find for the Ω_s integral

$$\sin^2 \alpha_{mx} \int_{-\pi/2}^{\pi/2} \left(\frac{1}{\cos^2 \gamma} - 1 \right) d\beta$$

$$= \sin^2 \alpha_{mx} \int_{-\pi/2}^{\pi/2} \left[\left(1 - \frac{r}{r_0} \sin \beta \right)^{-1} - 1 \right] d\beta$$

$$= \pi \sin^2 \alpha_{mx} \left[\frac{1}{\sqrt{1 - \left(\frac{r}{r_0} \right)^2}} - 1 \right]. \qquad (13.204)$$

On substituting this into (13.203), we find:

$$\Phi_s = 2\pi^2 \sin^2 \alpha_{mx} L \left[\int_0^{r_0} \frac{r \, dr}{\sqrt{1 - (r/r_0)^2}} - \int_0^{r_0} r \, dr \right]$$

$$= 2\pi^2 r_0^2 L \sin^2 \alpha_{mx} \left(\int_0^1 \frac{u \, du}{\sqrt{1 - u^2}} - \int_0^1 u \, du \right)$$

$$= \pi^2 r_0^2 L \sin^2 \alpha_{mx}, \qquad (13.205)$$

where we have substituted the variable of integration

$$u = \frac{r}{r_0}.$$

Equation 13.205 confirms (13.88a).

APPENDIX 13.3 GRADED-INDEX FIBERS: FOCUSING CONDITION

We derive here the index distribution in a fiber that will cause perfect imaging for any axial point of the fiber, that is, satisfies (13.113) for meridional rays. We use cylindrical coordinates (r, φ, z) coaxial with the fiber.

Let us first evaluate the derivative

$$\left(\frac{dr}{dz}\right)^2 = \tan^2 \alpha' = \sec^2 \alpha' - 1 = \frac{n^2(r)}{n^2(0)\cos^2 \alpha_0'} - 1, \qquad (13.206)$$

with α' the angle the path makes with the z-axis, and the last step following from (13.111).

From this we can now derive an expression for the length of a ray element:

$$ds^2 = \left[1 + \left(\frac{dz}{dr}\right)^2\right] dr^2$$

$$= \frac{n^2(r)\, dr^2}{n^2(r) - n^2(0)\cos^2 \alpha_0'}. \qquad (13.207)$$

On substituting the root of this into (13.113), we find:

$$\hat{s} = 4 \int_0^{r_{mx}} \frac{n^2(r)\, dr}{[n^2(r) - n^2(0)\cos^2 \alpha_0']^{1/2}} = \text{constant}. \qquad (13.208)$$

Following Ref. 47 we now show that

$$n(r) = n(0)\, \text{sech}\, Kr \qquad (13.209)$$

satisfies (13.208). Here

$$K = \frac{2\pi}{\Lambda} \qquad (13.210)$$

and Λ is the length of one cycle of the ray path. On substituting (13.209) into (13.208) and then substituting

$$u = \tanh Kr$$

$$du = K\, \text{sech}^2 \, Kr\, dr \qquad (13.211)$$

and noting that

$$\text{sech}^2 \, Kr = 1 - \tanh^2 Kr = 1 - u^2, \qquad (13.212)$$

we find, after applying a simple trigonometric identity,

$$\hat{s} = \frac{4n(0)}{K} \int_0^{\tanh Kr_{mx}} \frac{du}{[\sin^2 \alpha_0' - u^2]^{1/2}}$$

$$= \frac{4n(0)}{K} \sin^{-1} \frac{\tanh Kr_{mx}}{\sin \alpha_0'}. \qquad (13.213)$$

But, from (13.111) and (13.209),

$$\sin^2 \alpha_0' = 1 - \cos^2 \alpha_0' = 1 - \frac{n^2(r_{mx})}{n^2(0)}$$

$$= 1 - \text{sech}^2 Kr_{mx} = \tanh^2 Kr_{mx}. \qquad (13.214)$$

On substituting this into (13.213), we find that the arc sine equals $\pi/2$ and, in conjunction with (13.210), the integral

$$\int_{\text{cycle}} n(r)\, ds = n(0)\Lambda, \qquad (13.215)$$

which is independent of α_0'. This justifies the choice of (13.209), which is identical with (13.114).

The effectiveness of graded-index profiles has been analyzed in terms of the general form [54]

$$n(r) = n(0) \left[1 - 2\delta \left(\frac{r}{r_0} \right)^b \right]^{1/2}, \qquad r < r_0,$$

$$= n(0)\sqrt{1 - 2\delta}, \qquad r \geq r_0. \qquad (13.216)$$

where

$$\delta = 2\,\frac{n(0) - n(r_0)}{n(0) + n(r_0)}$$

is the relative index difference, assumed much less than unity, and b is an exponent that may be chosen arbitrarily. It is found that there is a pronounced optimum for

$$b = 2(1 - \delta) \approx 2. \qquad (13.217)$$

Note that the first two terms of the series expansion of (13.216) with (13.217) are

$$n(r) = n_0 \left[1 - \delta \left(\frac{r}{r_0} \right)^2 + \ldots \right]. \qquad (13.218)$$

Comparing this to the expansion of (13.209)

$$n(r) = n(0) \left[1 - \frac{K^2 r^2}{2} + \ldots \right], \qquad (13.219)$$

we find that the two distributions are equal to a second-order approximation, provided

$$K = \frac{\sqrt{2\delta}}{r_0}, \qquad (13.220)$$

which is implied, to the same approximation, by (13.214) in conjunction with (13.76a).

APPENDIX 13.4. TRANSFER FUNCTIONS OF CIRCULAR FIBER BUNDLES

To derive the mtf for a bundle of circular fibers, we follow the general procedure outlined in Section 13.3.2.2. We again refer to a sinusoidal object (13.165)

$$L(x) = \bar{L}(1 + M_0 \cos 2\pi\nu x).$$ (13.221)

Over this object we place a circular fiber, centered at x_j (see Figure 13.30). Now the luminance integrated over the aperture will be

$$I_j = \int_{A_F} L(x)\, dA,$$ (13.222)

where $L(x)$ is given by (13.221) and the area element, as indicated in the figure is

$$dA = 2[r_0^2 - (x - x_j)^2]^{1/2}\, dx,$$ (13.223)

and the limits of integration are $x_j \pm r_0$. On substituting these and (13.221) and (13.223) into (13.222) and introducing the dimensions variable

$$u = \frac{x - x_j}{r_0},$$ (13.224)

we obtain the relative image luminance

$$
\begin{aligned}
\hat{L}(x_j) &= \frac{I_j}{\pi r_0^2 \bar{L}} = \frac{2}{\pi} \int_{-1}^{1} \left(1 + M_0 \cos 2\pi\nu r_0 \frac{u + x_j}{r_0}\right) \sqrt{1 - u^2}\, du \\
&= \frac{2}{\pi} \left[\int_{-1}^{1} \sqrt{1 - u^2}\, du + M_0 \cos 2\pi\nu x_j \right. \\
&\quad \times \int_{-1}^{1} \cos(2\pi\nu r_0 u)\sqrt{1 - u^2}\, du \\
&\quad \left. - M_0 \sin 2\pi\nu x_j \int_{-1}^{1} \sin(2\pi\nu r_0 u)\sqrt{1 - u^2}\, du \right].
\end{aligned}
$$

The first integral in the brackets equals $\pi/2$; the second equals

$$\frac{\pi J_1(2\pi\nu r_0)}{2\pi\nu r_0};$$

the third one vanishes due to its antisymmetric integrand. Thus,

$$\hat{L}(x_j) = 1 - 2M_0 \cos 2\pi\nu x_j \frac{J_1(2\pi\nu r_0)}{2\pi\nu r_0}.$$ (13.225)

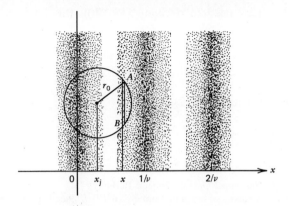

Figure 13.30   Deriving the transfer function of a bundle of circular fibers.

This is the relative luminance of the fiber cross section. When we average this over an image line $(x = x_j)$, we must weight $\hat{L}$ by the fraction of the line covered by the fiber cross section. Referring to Figure 13.30, we readily see that the normalized weighting is given by

$$\left(\frac{4}{\pi}\right)\left(\frac{\overline{AB}}{2r_0}\right) = \frac{4[1-(x-x_j)^2/r_0^2]^{1/2}}{\pi}, \tag{13.226}$$

with the factor $4/\pi$ required to make the mean weight equal to unity. Consequently, we have for the line spread function:

$$L(x) = 1 + \left(\frac{8}{\pi}\right)M_0\left[\frac{1-(x-x_j)^2}{r_0^2}\right]^{1/2}\cos 2\pi\nu x_j\,\frac{J_1(2\pi\nu r_0)}{2\pi\nu r_0}. \tag{13.227}$$

This, again, is not only a function of x, but also depends on x_j, the position of the fiber relative to the object maximum; it is nonisoplanatic.

Again introducing motion-averaging, we obtain for the luminance values corresponding to a maximum and a minimum, respectively,

$$\frac{1}{2r_0}\int_{-r_0}^{r_0}\left\{1\pm 8M_0\left[1-\frac{(x-x_j)^2}{r_0^2}\right]^{1/2}\cos 2\pi\nu x_j\,\frac{J_1(2\pi\nu r_0)}{2\pi^2\nu r_0}\right\}dx_j$$

$$= 1 \pm 4\,\frac{M_0}{r_0}\cdot\frac{J_1(2\pi\nu r_0)}{2\pi^2\nu r_0}\cdot r_0\int_{-1}^{1}\sqrt{1-u^2}\cos 2\pi\nu r_0 u\,du$$

$$= 1 \pm \frac{4M_0 J_1^2(2\pi\nu r_0)}{(2\pi\nu r_0)^2}. \tag{13.228}$$

Following the steps used in deriving (13.180), this leads to a relative modulation

$$\frac{M}{M_0} = 4\left[\frac{J_1(2\pi\nu r_0)}{(2\pi\nu r_0)}\right]^2, \tag{13.229}$$

which is identical with (13.182).

REFERENCES

[1] T. Tamir, Ed., *Integrated Optics*, Springer, Berlin, 1975.

[2] S. E. Miller, "A survey of integrated optics," *IEEE J. Quant. Electr.* **QE-8,** 199–205 (1972).

[3] W. S. C. Chang, M. W. Muller, and F. J. Rosenbaum, "Integrated optics," in *Laser Applications*, Vol. 2, M. Ross, ed., Academic, New York, 1974, pp. 227–334; (a) Section III B2

[4] H. F. Taylor and A. Yariv, "Guided wave optics," *Proc. IEEE* **62,** 1044–1060 (1974).

[5] P. K. Tien, S. Riva-Sanseverino, R. J. Martin, and G. Smolinsky, "Two-layered construction of integrated optical circuits and formation of thin-film prisms, lenses, and reflectors," *Appl. Phys. Lett.* **24,** 547–549 (1974).

[6] J. A. McMurray and C. R. Stanley, "Taper-coupling between 7059-glass and CdS films and phase modulation in the composite waveguide structure," *Appl. Phys. Lett.* **28,** 126–128 (1976).

[7] E. Garmire, "Semiconductor components for monolithic applications," Ref. 1, pp. 243–304.

[8] F. Zernike, "Fabrication and measurement of passive components," in Ref. 1, pp. 202–241.

[9] H. L. Garvin, E. Garmire, S. Somekh, H. Stoll, and A. Yariv, "Ion beam micromachining of integrated optics components," *Appl. Opt.* **12,** 455–459 (1973).

[10] W. J. Coleman, "Evolution of optical thin films by sputtering," *Appl. Opt.* **13,** 946–951 (1974).

[11] M. J. Rand and R. D. Standley, "Silicon oxynitride films on fused silica for optical waveguides," *Appl. Opt.* **11,** 2482–2488 (1972).

[12] P. K. Tien, G. Smolinsky, and R. J. Martin, "Thin organosilicon films for integrated optics," *Appl. Opt.* **11,** 637–642 (1972).

[13] E. A. Chandross, C. A. Pryde, W. J. Tomlinson, and H. P. Weber, "Photolocking—A new technique for fabricating optical waveguide circuits," *Appl. Phys. Lett.* **24,** 72–74 (1974); H. P. Weber, R. Ulrich, E. A. Chandross, and W. J. Tomlinson, "Light-guiding structures of photoresist films," in *Digest of Technical Papers* Opt. Soc. Am., Topical Meeting, Integr. Opt. etc., Feb. 1972. Paper TuA8.

[14] E. R. Schineller, R. P. Flam, and D. W. Wilmot, "Optical waveguides formed by proton irradiation of fused silica," *J. Opt. Soc. Am.* **58,** 1171–1176 (1968).

[15] J. E. Goell and R. D. Standley, "Sputtered glass waveguide for integrated optical circuits," *Bell Syst. Tech. J.* **48,** 3445–3448 (1969).

[16] P. K. Tien, "Light waves in thin films and integrated optics," *Appl. Opt.* **10,** 2395–2413 (1971).

[17] H. Kogelnik, "Theory of dielectric waveguides," in Ref. 1, pp. 15–81.

[18] N. S. Kapany and J. J. Burke, *Optical Waveguides*, Academic, New York, 1972.

[19] D. Marcuse, *Light Transmission Optics*, Van Nostrand Reinhold, New York, 1972; (a) Chapter 9; (b) Chapter 10.

[20] J. A. Arnaud, *Beam and Fiber Optics*, Academic, New York, 1976; (a) Section 5.13.

[21] H. Kogelnik and H. P. Weber, "Rays, stored energy, and power flow in dielectric waveguides," *J. Opt. Soc. Am.* **64**, 174–185 (1974).

[22] T. Tamir, "Beam and waveguide couplers," in Ref. 1, pp. 84–137.

[23] L. G. Cohen and M. V. Schneider, "Microlenses for coupling junction lasers to optical fibers," *Appl. Opt.* **13**, 89–94 (1974).

[24] C. C. Timmermann, "Highly-efficient light coupling from GaAlAs lasers into optical fibers," *Appl. Opt.* **15**, 2432–2433 (1976).

[25] H. Kogelnik and T. P. Sosnowski, "Holographic thin film couplers," *Bell Syst. Tech. J.* **49**, 1602–1608 (1970).

[26] H. Furuta, H. Noda, and A. Ihaya, "Novel optical waveguide for integrated optics," *Appl. Opt.* **13**, 322–326 (1974).

[27] H. Kogelnik and C. V. Shank, "Coupled-wave theory of distributed feedback lasers," *J. Appl. Phys.* **43**, 2327–2335 (1972).

[28] I. P. Kaminow and L. W. Stulz, "A planar electrooptic-prism switch," *IEEE J. Quant. Electr.* **QE-11**, 633–635 (1975).

[29] I. P. Kaminow, "Optical waveguide modulators," *IEEE Trans.* **MTT-23**, 57–70 (1975).

[30] F. K. Reinhart, "Electroabsorption in Al_yGa_{1-y}—Al_xGa_{1-x} As Double Heterostructures," *Appl. Phys. Lett.* **22**, 372–374 (1973).

[31] N. S. Kapany, *Fiber Optics*, Academic, New York, 1967; (a) Chapter 5; (b) Chapter 2; (c) Section 13.4; (d) Chapter 10.

[32] W. B. Allan, *Fiber Optics*, Plenum, New York, 1973; (a) Chapter 3; (b) Section 10.1; (c) Chapter 2; (d) p. 70; (e) Chapter 7; (f) Chapters 5 and 8; (g) Section 8VC.

[33] R. D. Maurer, "Glass fibers for optical communications," *Proc. IEEE* **61**, 452–462 (1973).

[34] S. E. Miller, E. A. J. Marcatili, and T. Li, "Research toward optical-fiber transmission systems," *Proc. IEEE* **61**, 1703–1751 (1973).

[35] S. Kawakami and S. Nishida, "Characteristics of a doubly clad optical fiber with a low-index inner cladding," *IEEE J. Quant. Electr.* **QE-10**, 879–887 (1974); also "W-type optical fiber: Relation between refractive index difference and transmission bandwidth," *Appl. Opt.* **15**, 1121–1122 (1976).

[36] Optical Society of America, *Optical Fiber Transmission*, New York, 1975; (a) M. D. Rigterink, "Materials, systems and fiber fabrication processes," TuA1; (b) W. G. French and G. W. Tasker, "Fabrication of graded index and single mode fibers with silica cores," TuA2.

[37] J. B. MacChesney, P. B. O'Connor, and H. M. Presby, "A new technique for the preparation of low-loss and graded-index fibers," *Proc. IEEE* **62**, 1280–1281 (1974).

[38] K. Koizumi, Y. Ikeda, I. Kitano, M. Furukawa, and T. Sumimoto, "New light-focusing fibers made by a continuous process," *Appl. Opt.* **13**, 255–260 (1974).

[39] W. G. French, G. W. Tasker, and J. R. Simpson, "Graded index fiber waveguides with borosilicate composition: fabrication techniques," *Appl. Opt.* **15**, 1803–1807 (1976).

[40] P. Kaiser, A. C. Hart, and L. L. Blyler, "Low-loss FEP-clad silica fibers," *Appl. Opt.* **14**, 156–162 (1975).

[41] (a) T. C. Hager, R. G. Brown, and B. N. Derick, "Plastics fiber optics," *Soc. Plast. Eng. J.* **23**, 36–42 (1967). (b) R. G. Brown and B. N. Derick, "Plastic fiber optics. II: Loss measurements and loss mechanisms," *Appl. Opt.* **7**, 1565–1569 (1968).

[42] K. Iga and N. Yamamoto, "Plastic focusing fiber for imaging applications," *Appl. Opt.* **16**, 1305–1310 (1977).

[43] J. Stone, "Optical transmission in liquid-core quartz fibers," *Appl. Phys. Lett.* **20**, 239–240 (1972).

[44] "Fiber Optics Comes of Age," *Proc. SPIE* **31** (1972); (a) H. Buyken and W. Kriege, "Properties of uv-fibers," pp. 37–42; (b) F. C. Unterleitner and H. L. Sowers, "High power transmission through fiber optics," pp. 45–47; (c) M. X. FitzPatrick, "Fiber optic road signs," pp. 75–76; (d) N. A. M. Mackay and M. G. Dohler, "A fiber-optics matrix for a multi-channel correlator," pp. 51–53; (e) W. P. Siegmund and H. B. Cole, "Fiber optic image inverters," pp. 99–104.

[45] R. J. Potter, E. Donath, and R. Tynan, "Light collecting properties of a perfect circular optical fiber," *J. Opt. Soc. Am.* **53**, 256–260 (1963).

[46] R. Olshansky, "Distortion losses in cabled optical fibers," *Appl. Opt.* **14**, 20–21 (1975).

[47] S. Kawakami and J. Nishizawa, "An optical waveguide with the optimum distribution of the refractive index with reference to waveform distortion," *IEEE Trans.* **MTT-16**, 814–818 (1968).

[48] E. G. Rawson, D. R. Herriott, and J. McKenna, "Analysis of refractive index distributions in cylindrical graded-index glass rods (GRIN rods) used as image relays," *Appl. Opt.* **9**, 753–759 (1970).

[49] M. Di Domenico, "Material dispersion in optical fiber waveguides," *Appl. Opt.* **11**, 652–654 (1972).

[50] D. Gloge, "Propagation effects in optical fibers," *IEEE Trans.* **MTT-23**, 106–120 (1975).

[51] M. Abramowitz and J. A. Stegun, *Handbook of Mathematical Functions*, U. S. National Bureau of Standards, Washington, D.C., 1964, Table 9.5.

[52] A. W. Snyder and D. J. Mitchell, "Leaky rays on circular optical fibers," *J. Opt. Soc. Am.* **64**, 599–607 (1974).

[53] D. Gloge, "Dispersion in weakly guiding fibers," *Appl. Opt.* **10**, 2442–2445 (1971).

[54] D. Gloge and E. A. J. Marcatili, "Multimode theory of graded-core fibers," *Bell Syst. Tech. J.* **52**, 1563–1578 (1973).

[55] L. G. Cohen, "Pulse transmission measurements for determining near optimal profile gradings in multimode borosilicate optical fibers," *Appl. Opt.* **15**, 1808–1814 (1976).

[56] C. Pask and A. W. Snyder, "Multimode optical fibers: Interplay of absorption and radiation losses," *Appl. Opt.* **15**, 1295–1298 (1976).

[57] T. A. Orofino and F. C. Unterleitner, "Optical fibers for dispersion in the time domain," *Appl. Opt.* **15**, 1907–1909 (1976).

[58] H. Nomura, T. Okada, S. Oikawa, and J. Shimada, "Fiber optic sheets of ridged polymer films," *Appl. Opt.* **14**, 586–588 (1975).

[59] R. Ulrich, H. P. Weber, E. A. Chandross, W. J. Tomlinson, and E. A. Franke, "Embossed optical waveguides," *Appl. Phys. Lett.* **20**, 213–215 (1972).

[60] A. C. Schmidt, J. S. Courtney-Pratt, and E. A. Ross, "Woven fiber optics," *Appl. Opt.* **14**, 280–287 (1975).

[61] R. Drougard, "Optical transfer properties of fiber bundles," *J. Opt. Soc. Am.* **54**, 907–914 (1964).

[62] H. Ohzu, "Image transmission characteristics of fiber bundles," in Ref. 31, pp. 357–371; M. Vanwormhoudt and W. De Kinder, "Space variant imagery in fiber optics," *ibid*, pp. 397–409.

[63] A. W. Snyder and P. McIntyre, "Crosstalk between light pipes," *J. Opt. Soc. Am.* **66**, 877–882 (1976).

[64] M. K. Barnoski, ed., *Fundamentals of Optical Fiber Communications*, Academic, New York, 1976; (a) J. E. Goell, "Optical fiber cable," pp. 59–81.

[65] S. J. Buchsbaum, "Lightwave communications—An overview," *Phys. Today* **29**[5], 23–25 (May 1976).

[66] R. L. Lebduska, "Fiber optic cable test evaluation," *Opt. Eng.* **13,** 49–55 (1974).

[67] D. Gloge, P. W. Smith, D. L. Bisbee, and E. I. Chinnock, "Optical fiber end preparation for low-loss splices," *Bell Syst. Tech. J.* **52,** 1579–1588 (1973).

[68] P. W. Smith, D. L. Bisbee, D. Gloge, and E. L. Chinnock, "A molded-plastic technique for connecting and splicing optical-fiber tapes and cables," *Bell Syst. Tech. J.* **54,** 971–984 (1975).

[69] R. G. Smith, "Optical power handling capacity of low loss optical fibers as determined by stimulated Raman and Brillouin scattering," *Appl. Opt.* **11,** 2489–2494 (1972).

[70] J. W. T. Walsh, "Radiation from a perfectly diffusing circular disc," *Proc. Phys. Soc.* **32,** 59–71 (1920).

[71] See W. H. Steel, "Luminosity, throughput or étendue? Further comments," *Appl. Opt.* **14,** 252 (1975); J. J. Horan, "The $A\Omega$ Question," *ibid,* p. 2033 (1975).

14

Stress Optics:
Elasto-, Electro-, and Magnetooptic Effects

The refractive index and other optical characteristics of a medium may be changed by the application of a force field, which may be mechanical, electrical, or magnetic. This change may then be used to modify the light flux passing through it. The present chapter treats these techniques and some of their applications.

14.1 STATIC ELASTOOPTIC EFFECT—PHOTOELASTICITY

The application of mechanical stress to any material changes its refractive index. When the stress is anisotropic, it will cause birefringence, in general. These effects have application both in static arrangements, where the phenomenon is referred to as photoelasticity, and at high frequencies, where the effects of stress waves traveling through the medium are called acoustooptic. The present section is devoted to the static effects and Section 14.2 to the dynamic effects.

Before proceeding to a discussion of the elastooptic effects, we present a brief introduction to the analysis of stress and strain in solids and fluids.

14.1.1 Stresses in the General Solid [1]

14.1.1.1 Stress. Stresses are defined as forces per unit area acting on a surface. ("Surface" is here used in its mathematical sense and may be located inside a homogeneous body.) The stress on a surface may be represented by a vector with components T_x, T_y, T_z. The stress components perpendicular to the surface are called *normal* (tensile or compressive) and those parallel to it are called *shear*.

The stress at any point is a function of the surface orientation. In general, it may be represented in terms of components of the form

$$T_{ij}, \qquad i,j = x, y, z$$

representing the j component of a stress acting on a plane normal to the i direction. Here

$$T_{xx}, \qquad T_{yy}, \qquad T_{zz}$$

represent normal stresses and the other components, shear stresses. Under static conditions, equilibrium implies that

$$T_{ij} = T_{ji}. \tag{14.1}$$

Hence there are, in general, only six stress components to be specified. These are conventionally written

$$T_{xx} = T_1, \qquad T_{yy} = T_2, \qquad T_{zz} = T_3,$$
$$T_{yz} = T_{zy} = T_4, \qquad T_{xz} = T_{zx} = T_5, \qquad T_{xy} = T_{yx} = T_6. \tag{14.2}$$

The stress in any general direction is given as the sum of the components (in that direction) of the component stresses.

As explained further in Appendix 14.1.1, there exists a principal coordinate system in which the components with mixed subscripts vanish: the stresses on the three principal surfaces are all normal stresses.

14.1.1.2 Strain. Any real material suffers a distortion, or *strain*, as a result of the applied stress. We consider, first, a two-dimensional case, with two orthogonal line segments, $\overline{OA}$ and $\overline{OB}$, of infinitesimal lengths, X and Y, respectively, embedded in the solid (see Figure 14.1). As a result of the stress, they will be, in general, displaced, changed in length, and rotated, so that they will appear as shown by the broken lines, $\overline{OA'}$, $\overline{OB'}$ in the figure. (The figure does not show the displacement of the origin because we consider it fixed in the volume element depicted.) The points A and B have experienced displacements $\mathbf{U}$ and $\mathbf{V}$, respectively. The components of the displacement, $\mathbf{u}$, of a general point, at x of the segment $\overline{OA}$, can be written from similar triangles:

$$u_x = \frac{U_x}{X}x = S_{xx}x, \qquad u_y = \frac{U_y}{X}x = S'_{xy}x \tag{14.3}$$

and, similarly, for a point at y of segment $\overline{OB}$, the components of the displacement, $\mathbf{v}$, are

$$v_x = \frac{V_x}{Y}y = S'_{yx}y, \qquad v_y = \frac{V_y}{Y}y = S_{yy}y, \tag{14.4}$$

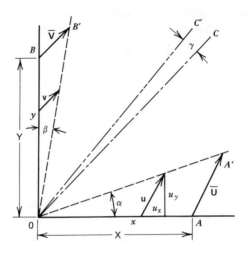

Figure 14.1 Illustrating strain.

where the strain coefficients, S_{ij}, are defined as follows:

$$S_{xx} = \frac{U_x}{X} = \frac{\partial u_x}{\partial x} \qquad (14.5)$$

is (the x component of) the relative elongation of an element originally parallel to the x axis,

$$S_{yy} = \frac{V_y}{Y} = \frac{\partial v_y}{\partial y} \qquad (14.6)$$

is the analogous elongation for an element parallel to the y direction,

$$S'_{xy} = \frac{U_y}{X} = \frac{\partial u_y}{\partial x} \approx \alpha, \qquad (14.7)$$

the angle through which the segment $\overline{OA}$ has been rotated and

$$S'_{yx} = \frac{V_x}{Y} = \frac{\partial v_x}{\partial y} \approx \beta, \qquad (14.8)$$

the corresponding angle for segment $\overline{OB}$.

We are justified in writing the ratios as derivatives because we chose the original elements to be of infinitesimal length. The approximations in (14.7) and (14.8) are based on the assumption that all the partial derivatives are much smaller than unity.

Since we are here concerned only with the *relative* displacements within the volume element, we must subtract the overall rotation experienced by the volume element. This is given by γ, the angle between the angle bisectors $\overline{OC}$, $\overline{OC'}$, before and after the straining, respectively. By inspection we see that required rotation is

$$\gamma = \tfrac{1}{2}(\alpha - \beta) \approx \frac{1}{2}\left(\frac{\partial u_y}{\partial x} - \frac{\partial v_x}{\partial y}\right). \tag{14.9}$$

When we apply this to the definition of our strain coefficients, we obtain

$$S_{xy} = S'_{xy} - \gamma = \frac{\partial u_y/\partial x + \partial v_x/\partial y}{2} \tag{14.10a}$$

$$S_{yx} = S'_{yx} + \gamma = \frac{\partial u_y/\partial x + \partial v_x/\partial y}{2} = S_{xy}. \tag{14.10b}$$

We readily generalize these results to three dimensions and write for the six strain components

$$S_1 \equiv S_{xx} = \frac{\partial u_x}{\partial x} \qquad (x \text{ elongation}) \tag{14.11}$$

$$S_2 \equiv S_{yy} = \frac{\partial v_y}{\partial y} \qquad (y \text{ elongation})$$

$$S_3 \equiv S_{zz} = \frac{\partial w_z}{\partial z} \qquad (z \text{ elongation})$$

$$S_4 \equiv S_{yz} = \frac{\partial v_z/\partial y + \partial w_y/\partial z}{2} \qquad (\text{rotation about } x \text{ axis})$$

$$S_5 \equiv S_{xz} = \frac{\partial u_z/\partial x + \partial w_x/\partial z}{2} \qquad (\text{rotation about } y \text{ axis})$$

$$S_6 \equiv S_{xy} = \frac{\partial u_y/\partial x + \partial v_x/\partial y}{2} \qquad (\text{rotation about } z \text{ axis}).$$

14.1.1.3. *Elasticity.* In many materials, the strains are proportional to the stresses over a limited range. These are then said to obey Hooke's law and to be elastic over that range. The strain components may then be written in terms of the stress components:

$$S_1 = s_{11}T_1 + s_{12}T_2 + s_{13}T_3 + s_{14}T_4 + s_{15}T_5 + s_{16}T_6$$

$$S_2 = s_{21}T_1 + s_{22}T_2 + s_{23}T_3 + \ldots \ldots \ldots$$

$$\cdot \; \cdot \; \cdot \; \cdot \; \cdot \; \cdot \; \cdot \; \cdot \; \cdot \; \cdot \; \cdot \; \cdot \; \cdot \; \cdot \; \cdot \; \cdot$$

$$\cdot \; \cdot \; \cdot \; \cdot \; \cdot \; \cdot \; \cdot \; \cdot \; \cdot \; \cdot \; \cdot \; \cdot \; \cdot \; \cdot \; \cdot$$

$$S_6 = s_{61}T_1 + s_{62}T_2 + s_{63}T_3 + \ldots \ldots \ldots \tag{14.12}$$

The coefficients s_{ij} relating strain to stress are called *elastic compliances* (or elastic constants).

If (14.12) are solved for the stresses, we obtain a similar set of equations with a set of coefficients c_{ij}, called *elastic stiffness coefficients* (or moduli of elasticity):

$$T_1 = c_{11}S_1 + c_{12}S_2 + c_{13}S_3 + c_{14}S_4 + c_{15}S_5 + c_{16}S_6. \tag{14.13}$$

$$\cdot \quad \cdot \quad \cdot \quad \cdot \quad \cdot \quad \cdot \quad \cdot \quad \cdot \quad \cdot \quad \cdot \quad \cdot \quad \cdot \quad \cdot \quad \cdot \quad \cdot \quad \cdot$$

$$\cdot \quad \cdot \quad \cdot \quad \cdot \quad \cdot \quad \cdot \quad \cdot \quad \cdot \quad \cdot \quad \cdot \quad \cdot \quad \cdot \quad \cdot \quad \cdot \quad \cdot \quad \cdot$$

See Appendix 14.1.2 for the relationship between the s_{ij} and c_{ij}. We discuss there also the application of this general form to crystals of different symmetries. Here we restrict ourselves to isotropic materials.

14.1.2   Elasticity in Isotropic Solids and Fluids

In isotropic solids, such as glasses, most of the elastic constants vanish and the remaining constants are related by equalities, so that there are only two independent constants:

$$s_{11} = s_{22} = s_{33} \tag{14.14a}$$

$$s_{12} = s_{13} = s_{21} = s_{23} = s_{31} = s_{32} \tag{14.14b}$$

$$s_{44} = s_{55} = s_{66} = 2(s_{11} - s_{12}). \tag{14.14c}$$

The stiffness coefficients are similarly related, except that

$$c_{44} = c_{55} \doteq c_{66} = \tfrac{1}{2}(c_{11} - c_{12}). \tag{14.15}$$

We now determine the value of these elastic coefficients in terms of the more common engineering parameters. In a rod-shaped isotropic material, an axial stress, say T_{xx}, produces a corresponding axial strain

$$S_{xx} = \frac{T_{xx}}{E}, \tag{14.16}$$

and, in addition, a smaller transverse strain

$$S_{yy} = S_{zz} = -\sigma S_{xx}. \tag{14.17}$$

E is known as *Young's modulus* and

$$\sigma \leqslant 0.5$$

as *Poisson's ratio*. More generally, the strains in such a material can be summarized by the following two equations:

$$S_{ii} = \frac{T_{ii} - \sigma(T_{jj} + T_{kk})}{E} \qquad (14.18)$$

$$S_{ij} = \frac{T_{ij}}{G}; \qquad i, j, k = x, y, z; \qquad i \neq j \neq k. \qquad (14.19)$$

G is called modulus of rigidity or shear modulus.
On solving for the stresses, these become:

$$T_{ii} = (\lambda + 2\mu)S_{ii} + \lambda(S_{jj} + S_{kk}) \qquad (14.20)$$

and

$$T_{ij} = \mu S_{ij}, \qquad (14.21)$$

where

$$\lambda = \frac{\sigma E}{(1+\sigma)(1-2\sigma)} \qquad (14.22)$$

and

$$\mu = G = \frac{E}{2(1+\sigma)} \qquad (14.23)$$

are known as Lamé's constants.
The reader will readily confirm that

$$E = (3\lambda + 2\mu)(1 - 2\sigma). \qquad (14.24)$$

On comparing (14.20) with the appropriate equation derived from (14.13),

$$T_{ii} = c_{11}S_{ii} + c_{12}(S_{jj} + S_{kk}), \qquad (14.25)$$

we find that

$$c_{11} = \lambda + 2\mu = \frac{E(1-\sigma)}{(1+\sigma)(1-2\sigma)} \qquad (14.26)$$

and

$$c_{12} = \lambda = \frac{\sigma E}{(1+\sigma)(1-2\sigma)}. \qquad (14.27)$$

Similarly from (14.21) we find

$$c_{44} = \mu. \qquad (14.28)$$

For future reference, we define two more moduli:

1. The volume strain is readily determined from (14.18):

$$S_V = S_{xx} + S_{yy} + S_{zz}$$
$$= \frac{(T_{xx} + T_{yy} + T_{zz})(1 - 2\sigma)}{E}. \tag{14.29}$$

If the pressure is hydrostatically applied,

$$T_V = T_{xx} = T_{yy} = T_{zz}, \tag{14.30}$$

and the volume strain becomes

$$S_V = \frac{3T_V(1 - 2\sigma)}{E}. \tag{14.31}$$

This permits us to define the *bulk modulus*

$$K = \frac{T_V}{S_V} = \frac{E}{3(1 - 2\sigma)} = \lambda + \frac{2\mu}{3}. \tag{14.32}$$

2. In an infinite plate, in compression or tension, T_{ii}, in thickness, the lateral strains, S_{jj}, S_{kk}, vanish and (14.18) becomes

$$T_{ii} = (\lambda + 2\mu)S_{ii}. \tag{14.33}$$

Hence the *plate modulus* is given by

$$K' = \frac{T_{ii}}{S_{ii}} = \lambda + 2\mu = \frac{(1 - \sigma)E}{(1 + \sigma)(1 - 2\sigma)}. \tag{14.34}$$

14.1.3   Elastooptic Effect

14.1.3.1   Isotropic Solids. The stresses just discussed produce changes in the refractive index of the material; in general, these will be anisotropic—the material becomes (temporarily) birefringent. Let us first consider isotropic materials. In these the stress produces an index ellipsoid with the index change

$$n_i - n_0 = aT_i + b(T_j + T_k), \qquad i, j, k = \text{I, II, III} \tag{14.35}$$

where $T_{\text{I,II,III}}$ are the principal stresses, that is, the stresses in the direction of the principal axes, x', y', z', respectively,

n_0 is the refractive index of the unstressed material, and

n_i is the index experienced by a light wave polarized in the direction of principal axis i.

By taking the differences of Equations 14.35 in pairs, we find

$$n_i - n_j = (a - b)(T_i - T_j) = B(T_i - T_j), \qquad (14.36)$$

where

$$B = a - b$$

is called the *stress-optic coefficient*. It is usually measured in brewsters (Br):

$$1 \, \text{Br} = (10^{12} \, \text{N/m}^2)^{-1}.$$

Values for a number of glasses, plastics, and other materials are listed in Table 144.

If a single tensile or compressive stress is applied, the resulting index ellipsoid is one of revolution and the birefringence is similar to that in an uniaxial crystal. If two or more nonparallel stresses are applied, the birefringence may become like that of a biaxial crystal.

By way of illustration, consider a plate, thickness d, placed parallel to the x–y plane and stressed parallel to its plane by a tensile stress with components T_x and T_y ($T_I = T_x$, $T_{II} = T_y$, $T_{III} = 0$). The resulting index ellipsoid is illustrated in Figure 14.2. From (14.36) we obtain the index differences

$$n_x - n_y = B(T_x - T_y) \qquad (14.37a)$$

$$n_x - n_z = BT_x \qquad (14.37b)$$

$$n_y - n_z = BT_y. \qquad (14.37c)$$

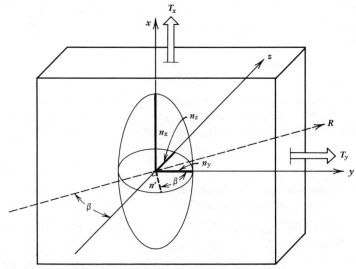

Figure 14.2 Index ellipsoid in strained plate. The three ellipsoid axes are indicated by heavy lines and the strain forces by hollow arrows.

A general polarized electromagnetic wave incident along the z axis can be broken up into components whose electric vectors vibrate parallel to the x and y directions, respectively. These will experience a relative retardation (in length)

$$D_{xy} = (n_x - n_y) \, d = B(T_x - T_y) \, d. \tag{14.38}$$

Ray R, in the y–z plane, making an angle β with the z axis, can be similarly split. The component vibrating parallel to the x axis will still experience the index

$$n_x = n_z + BT_y.$$

But the component whose electric field vibrates parallel to the y–z plane will experience an index n', as indicated in the figure. This can be computed from the equation of the ellipse cut from the index ellipsoid by the y–z plane. This equation is

$$\frac{y^2}{n_y^2} + \frac{z^2}{n_z^2} = \left(\frac{n' \cos \beta}{n_y}\right)^2 + \left(\frac{n' \sin \beta}{n_z}\right)^2 = 1. \tag{14.39}$$

Hence

$$n' = \left(\frac{\cos^2 \beta}{n_y^2} + \frac{\sin^2 \beta}{n_z^2}\right)^{-1/2}$$

$$= \left[\frac{\cos^2 \beta}{(BT_y + n_z)^2} + \frac{\sin^2 \beta}{n_z^2}\right]^{-1/2}. \tag{14.40}$$

14.1.3.2   Crystalline Materials. In crystalline materials, the index ellipsoid is given by

$$a_1 x^2 + a_2 y^2 + a_3 z^2 + 2a_4 yz + 2a_5 xz + 2a_6 xy = 1. \tag{14.41}$$

The distance from the origin to any point on this ellipsoid represents the refractive index experienced by light polarized in that direction (see Section 2.2.3.1). Note that a_1, a_2, and a_3 represent, respectively, n_x^{-2}, n_y^{-2}, and n_z^{-2}. The change in the values of the a's is given by

$$\Delta(a_p) = \sum_q p_{pq} T_q, \tag{14.42}$$

where T_q are the applied stresses and
   p_{pq} are the stress-optic coefficients.

See Table 145 for representative data and Appendix 14.1.4 for the usual matrix formulation.
   Noting that

$$\Delta\frac{1}{n^2} \approx -\frac{2\Delta n}{n^3}, \tag{14.43}$$

we find for the first three terms of (14.41)

$$\Delta n_j = n_j^3 \sum_q p_{jq} \frac{T_q}{2}, \qquad j = 1, 2, 3.$$

14.1.3.3   Liquids.

In liquids, the Lorentz-Lorenz equation (10.3) implies[1] that the change in refractive index is related to the strain (S) according to

$$\Delta n = \frac{S_V(n^2 - 1)(n^2 + 2)}{6n}$$

$$= \frac{T_V(n^2 - 1)(n^2 + 2)}{6nK}, \tag{14.47}$$

where T_V is the hydrostatic pressure (stress) and K is the bulk modulus.

For instance, for water ($n = 1.333$, $K = 2.3 \times 10^9 \, \text{N/m}^2$):

$$\Delta n = 0.367 \, S = 0.16 \times 10^{-9} \, (\text{N/m}^2)^{-1} \, T.$$

---

[1] To derive this result, we note that the relative change in density equals the volume strain:

$$\frac{\Delta \rho}{\rho} = S_V. \tag{14.44}$$

From (10.3) we readily confirm that

$$(n^2 - 1)(n^2 + 2) = \frac{9C}{(1 - C)^2} \tag{14.45}$$

and from (10.4) that, since C is proportional to ρ,

$$\Delta C = \frac{C\Delta\rho}{\rho} = S_V C, \tag{14.46}$$

with the last step from (14.44).

On differentiating (10.3), we find

$$2n\Delta n = \Delta C \frac{2(1 - C) + (1 + 2C)}{(1 - C)^2}$$

or

$$\Delta n = \frac{3\Delta C}{2n(1 - C)^2}.$$

On substituting into this successively from (14.45) and (14.46), this becomes

$$\Delta n = \frac{3S_V C}{2n(1 - C)^2}$$

$$= \frac{S_V(n^2 - 1)(n^2 + 2)}{6n}. \tag{14.47a}$$

This is identical to (14.47) in the text.

14.1.4   Polariscopy [2] and Photoelastic Modulators

14.1.4.1   General Description of the Polariscope.
The major application of the static elastooptic effect is in stress analysis. Here a transparent model is prepared and put under stress. The resulting birefringence pattern is viewed between crossed polarizers. A device for doing this is called polariscope.

Let us consider what happens if the plate of Figure 14.2 is illuminated along the z axis by light linearly polarized with the direction of polarization making an angle θ with the x axis. The phase retardation of the y component relative to the x component will be

$$\phi_{xy} = \frac{2\pi}{\lambda} D_{xy}, \tag{14.48}$$

where D_{xy} is given by (14.38). Hence the components of the electric field vector can be represented by the phasors

$$\hat{E}_x = E_0 \cos \theta$$
$$\hat{E}_y = E_0 \sin \theta e^{-i\phi_{xy}}.$$

If the analyzer makes an angle θ' with the x axis, the field passing it will be the sum of the phasor components parallel to the analyzer axis:

$$\hat{E}_r = E_0(\cos \theta \cos \theta' + \sin \theta \sin \theta' e^{-i\phi_{xy}}). \tag{14.49}$$

This corresponds to a flux density

$$M = |\hat{E}_r|^2 = E_0^2[(\cos \theta \cos \theta' + \sin \theta \sin \theta' \cos \phi_{xy})^2$$
$$+ (\sin \theta \sin \theta' \sin \phi_{xy})^2]$$
$$= M_0 \frac{\cos 2\theta \cos 2\theta' + \sin 2\theta \sin 2\theta' \cos \phi_{xy}}{2}, \tag{14.50}$$

where we have written

$$M_0 = E_0^2$$

for the incident flux density.

In the special case of crossed polarizers $[\theta' = (\pi/2) + \theta]$, the transmitted flux density is

$$M_\perp = \tfrac{1}{2} M_0 (1 - \cos^2 2\theta - \sin^2 2\theta \cos \phi_{xy})$$
$$= M_0 \sin^2 2\theta \sin^2 \left(\frac{\phi_{xy}}{2}\right). \tag{14.51}$$

14.1.4.2 Isoclinic and Isochromatic Fringes. The flux density (14.51) can be seen to vanish in two distinct cases:

$$\textbf{1} \quad \theta = 0, \pi/2 \qquad\qquad (14.52a)$$

$$\textbf{2} \quad \phi_{xy} = 2\pi m, \qquad m = 0, 1, 2 \cdots . \qquad (14.52b)$$

The former corresponds to the elements for whom the principal stresses run parallel or normal to the polarizer axis. Since these generally form a continuum in a stressed body, they give rise to dark fringes called *isoclinic*. The second case corresponds to locations yielding a retardation of an integral number of wavelengths, regardless of its orientation. If monochromatic light is used, they, too, will form a set of dark fringes. If white light is used, however, they appear as fringes where a certain spectral region is attenuated from the transmitted light. Therefore, they appear as fringes of a given color and are accordingly called *isochromatic* fringes.

Even in monochromatic light it is possible to distinguish between these two sets of fringes: When the two polarizers are rotated in unison, the isoclinic fringes move, while the isochromatic fringes remain unaffected. The converse is true, when the magnitude of the loading is changed while maintaining the load distribution similar.

The spectrum of the light transmitted at any given retardation, D, is complicated by the fact that for large values of D, several wavelength bands may be missing, corresponding to various orders [various values of m in (14.52b)]. In Table 146 [2] we list a description of the colors obtained with various values of D, when the illuminating flux is white.

When the polarizers are oriented parallel, the transmitted flux is the complement of that given in (14.51):

$$M_{\parallel} = M_0[1 - \sin^2(2\theta) \sin^2(\tfrac{1}{2}\phi_{xy})]. \qquad (14.53)$$

14.1.4.3 Polariscopy in Circularly Polarized Light. Occasionally two quarter-wave plates are inserted between the polarizers and the stressed object, with their axes rotated 45° from those of the polarizers. This causes the object illumination to be circularly polarized, so that the transmitted flux becomes independent of the polarizer orientation. It thus takes the form

$$M = \tfrac{1}{2}M_0 \sin^2 \phi_{xy} \qquad (14.54)$$

and the isoclinic fringes are eliminated.

14.1.4.4 Phase Modulators. Some methods used in studying the optical properties of materials require phase modulation of light. This may be obtained by means of elastooptic light modulators, consisting of a piezoelectric transducer in conjunction with an acoustic medium [3]. These are similar to the acoustooptic modulators of the next section, but operate at much lower frequencies and, therefore, with much longer acoustic wavelengths.

14.1.4.5 Photoplasticity. In addition to elasticity, solids also exhibit plasticity, a nonlinear and irreversible process far more complex than elasticity. It, too, is accompanied, in general, by optical birefringence. The observation of this birefringence is an important tool in studying plasticity both in man-made polymers, the modern "plastics," and in structural materials, such as metals and polycrystalline aggregates, near their elastic limit [4].

14.2 DYNAMIC ELASTOOPTIC EFFECT—ACOUSTOOPTICS

At frequencies in the audio range, that is, below, say, 15 kHz, electromechanical devices, such as galvanometer mirrors or loud speaker motors, can be used to modulate light. But at frequencies significantly above this, only elasto-, electro-, and magnetooptic techniques are applicable. Here we discuss the first of these.

Acoustic waves in a transparent medium introduce periodic refractive index changes and these turn the medium effectively into a diffraction grating. This grating is amenable to modification by the simple expedient of changing the waves fed into the medium. This provides convenient and powerful means for modulating and deflecting light.

14.2.1 Generation of Acoustic Waves [5]

The only effects presently used in the generation of pressure waves at high frequencies are the magnetostrictive and inverse piezoelectric (electrostrictive) effects and, at frequencies above a megahertz, only the latter is useful. Since in most optical applications we are concerned with higher frequencies, we discuss here only the piezoelectric transducers and refer the reader to Ref. 5a for details of magnetostrictive acoustic generators.

14.2.1.1 Formulation of the Piezoelectric Effect [5b, c]. In the (inverse) piezoelectric effect, a mechanical distortion is produced in a crystal by the application of an electric field. Any crystal having a polar axis[2] is piezoelectric, and the distortion can take the form of expansion, contraction, shear, or torsion. In its elementary aspects, the effect can be understood with the aid of the diagrams shown in Figure 14.3, which

---

[2] A polar axis is an axis such that no plane normal to it is a plane of symmetry.

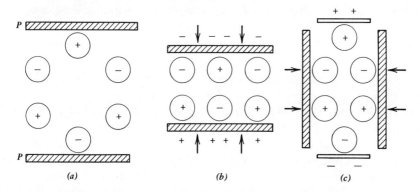

Figure 14.3 Illustrating the origin of the piezoelectric effect.

shows a typical neutral lattice cell consisting of three positive and three negative ions—(a) in its unstrained state and (b) after a stress is applied. In (a) no free charges will appear on the plates, P; in (b) free charges will appear due to the redistribution of the ions. Conversely, the application of an electric field will distort the cell as shown in (b). Figure 14.3c illustrates the effect of a transverse compression giving rise to free charges on the surfaces parallel to the direction of the stress.

By convention, a set of rectangular coordinates, x, y, z, are assigned to each crystal. See Ref. 6 for details. For instance, in a quartz crystal, the axis analogous to the vertical direction in Figure 14.3 is the x axis, and the axis normal to the plane of the figure, the z axis. In piezoelectric ceramics [poled polycrystalline materials (see Section 14.3.1.3)], the poling axis, which is an axis of infinite symmetry, is taken as the z axis [7].

14.2.1.2 Static Versus Dynamic Effects. For a quantitative description of the piezoelectric effect in static terms, see Appendix 14.1.3. In acoustooptic applications, the dynamic characteristics are dominant. These are controlled by the mechanical resonance characteristics of the piezoelectric transducer, which is usually in the shape of a plate, whose deformation in thickness or shear is controlled by the electric potential applied across its faces. If the transducer is excited by an alternating voltage near its mechanical resonance frequency, relatively large vibrational amplitudes can be excited. The resonance frequence, ν_{tR}, is given by the reciprocal of the transit time of an acoustic wave across a transducer and back:

$$\nu_{tR} = \frac{v_a}{2d},$$ (14.55)

where d is the thickness of the transducer and
    v_a is the acoustic velocity (14.62).

The width, $\Delta \nu_t$, of the resonance curve at half-amplitude can be calculated from the Q factor [see (7.36)], which represents the ratio of the energy stored in the transducer to the energy dissipated per cycle:

$$\Delta \nu_t = \frac{\nu_{tR}}{Q} \qquad (14.56)$$

[see (7.36) and (7.37)]. The energy dissipated per cycle includes frictional losses in the transducer and loading, that is, energy radiated, both into the acoustic medium and into the general environment. The amount of loading is controlled by the matching of the transducer impedance to that of the acoustic medium (see the next section) and by the coupling factor, which represents the ratio of radiated-to-stored energy under optimum matching conditions.

In Table 147 [8] we list a number of piezoelectric transducers of various cuts and modes, including their density, coupling factors, dielectric constants, frequency constants ($v_a/2$) and mechanical impedance.

14.2.2   Propagation of Acoustic Waves

14.2.2.1   The Wave Equations. To develop the equations governing acoustic waves, we consider a volume element, length Δx (see Figure 14.4), one edge of which has been displaced by an amount, D. The displacement of the other end will be

$$D + \frac{\partial D}{\partial x} \Delta x.$$

The corresponding change in element length is the difference between these:

$$\Delta D = \frac{\partial D}{\partial x} \Delta x,$$

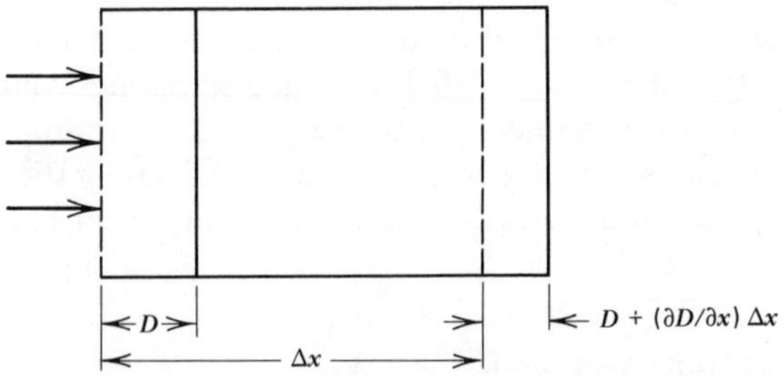

Figure 14.4   Illustrating acoustic wave propagation.

and the strain, that is, the relative elongation, is

$$S = \frac{\Delta D}{\Delta x} = \frac{\partial D}{\partial x}. \tag{14.57}$$

On the other hand, the net lateral force on the element is

$$F = A\left[\left(T + \frac{\partial T}{\partial x}\Delta x\right) - T\right] = A\frac{\partial T}{\partial x}\Delta x = V\frac{\partial T}{\partial x}, \tag{14.58}$$

where T is the stress, that is, the force per unit area,
A is the cross-sectional area of the element, and
V is its volume.

According to Newton's second law, the volume element will be accelerated by an amount

$$\frac{\partial^2 D}{\partial t^2} = \frac{F}{m} = \frac{V(\partial T/\partial x)}{V\rho}$$

$$= \rho^{-1}\frac{\partial T}{\partial x}, \tag{14.59}$$

where m is the mass of the volume element and
ρ is its mass density.

We now recall the relationship between stress and strain and thus find from (14.13) and (14.57)

$$T = cS = c\frac{\partial D}{\partial x}. \tag{14.60}$$

On substituting this into (14.59), we obtain the differential equation for elastic waves

$$\frac{c}{\rho}\frac{\partial^2 D}{\partial x^2} = \frac{\partial^2 D}{\partial t^2}. \tag{14.61}$$

Comparing this with the analogous equation for electromagnetic waves (2.18), and noting (2.21), we find that the phase velocity (v_a) is given by the square root of the coefficient on the left:

$$v_a = \sqrt{\frac{c}{\rho}}. \tag{14.62}$$

Here c is the appropriate stiffness coefficient. For a compressional wave in a thin rod, it is Young's modulus (14.16); in an extended medium, the plate modulus (14.34); for shear waves, the rigidity modulus, G; in a liquid, the bulk modulus (14.32), or the reciprocal of the adiabatic compressibility; and, finally, in gases, it is effectively (γp), the product of the ambient pressure, p, and the specific heat ratio, γ [see (12.20)].

A plane, simple harmonic acoustic wave, radian frequency Ω, may therefore be written in terms of the maximum particle displacement, D_{mx}, as follows. The particle displacement is

$$D = D_{mx} \cos \Omega \left(\frac{x}{v_a} - t\right). \tag{14.63}$$

The particle velocity is derived therefrom by differentiation:

$$v_p = D = \Omega D_{mx} \sin \Omega \left(\frac{x}{v_a} - t\right) \tag{14.64}$$

and the particle acceleration

$$a_p = \dot{v}_p = -\Omega^2 D_{mx} \cos \Omega \left(\frac{x}{v_a} - t\right). \tag{14.65}$$

The corresponding wavelength is

$$\Lambda = \frac{2\pi v_a}{\Omega}. \tag{14.66}$$

From (14.57) the strain is

$$S = \frac{\partial D}{\partial x} = -\frac{\Omega D_{mx}}{v_a} \sin \Omega \left(\frac{x}{v_a} - t\right), \tag{14.67}$$

and the maximum displacement in terms of the maximum value, S_{mx}, of the strain:

$$D_{mx} = \frac{v_a S_{mx}}{\Omega}. \tag{14.68}$$

From (14.59), the stress gradient equals the density times the particle acceleration:

$$\frac{\partial T}{\partial x} = \rho a_p = -\rho \Omega^2 D_{mx} \cos \Omega \left(\frac{x}{v_a} - t\right) \tag{14.69}$$

and hence the stress itself is:

$$T = -v_a \rho \Omega D_{mx} \sin \Omega \left(\frac{x}{v_a} - t\right) = \rho v_a^2 S. \tag{14.70}$$

The ratio of stress to particle velocity,

$$Z = \left|\frac{T}{v_p}\right| = \rho v_a = \sqrt{c\rho} \tag{14.71}$$

is known as the *acoustic impedance* and determines the transmission of acoustic energy across a boundary separating different media.

14.2.2.2 Acoustic Energy. The potential energy density in an acoustic wave is given by half the product of displacement and stress gradient

$$u_p = \tfrac{1}{2}D\frac{\partial T}{\partial x}$$

$$= \tfrac{1}{2}\rho\Omega^2 D_{mx}^2 \cos^2 \Omega \left(\frac{x}{v_a} - t\right). \tag{14.72}$$

The kinetic energy density,

$$u_k = \tfrac{1}{2}\rho v_p^2$$

$$= \tfrac{1}{2}\rho\Omega^2 D_{mx}^2 \sin^2 \Omega \left(\frac{x}{v_a} - t\right). \tag{14.73}$$

Clearly, the energy density oscillates back and forth between these two states and its total, the sum of these two, is constant:

$$u = \tfrac{1}{2}\rho\Omega^2 D_{mx}^2 = \tfrac{1}{2}\rho v_a^2 S_{mx}^2, \tag{14.74}$$

with the last step following from (12.68).

In a plane wave, energy crosses a unit area, normal to its propagation vector, at the rate

$$E = u v_a = \tfrac{1}{2}\rho v_a^3 S_{mx}^2 = \frac{T_{mx}^2}{2Z}. \tag{14.75}$$

in analogy with (1.8). Here E is known as the acoustic *flux density* (sometimes *intensity*), measured in W/m^2. When the acoustic intensity is measured in decibels (dB), this is often taken relative to the reference intensity[3]

$$E_0 = 1 \text{ pW/m}^2, \tag{14.76}$$

---

[3] Sometimes sound *pressure* level is also expressed in decibels. It is then given by

$$p^* = 20 \log_{10}\frac{p}{p_0} \text{ dB},$$

where p is the pressure (in Pa) and

$$p_0 = 20 \text{ Pa};$$

occasionally

$$p_0 = 0.1 \text{ Pa}.$$

For the above definition to agree with (14.76) in air, p_0 should equal 29 Pa.

NOTE: The pascal (Pa) is the SI unit for pressure: $1 \text{ Pa} \equiv 1 \text{ N/m}^2$.

approximately the threshold of audibility. Thus the acoustic intensity in decibels is

$$J = 10 \log_{10} \frac{E}{E_0}. \tag{14.76a}$$

From (14.75), we can readily calculate the total acoustic flux in a plane wave of cross-sectional area, A:

$$W = AE = \tfrac{1}{2}\rho v_a^3 S_{mx}^2 A, \tag{14.77}$$

and, hence, the maximum strain resulting from a given amount of acoustic power

$$S_{mx} = \left(\frac{2W}{\rho v_a^3 A} \right)^{1/2}. \tag{14.78}$$

14.2.2.3   Reflection and Refraction [9]. An acoustic wave incident on an interface between two different media is, in general, partly reflected and partly refracted. The refraction is governed by the acoustic analog of Snell's law,

$$\frac{1}{v_a} \sin \theta = \frac{1}{v_a'} \sin \theta', \tag{14.79}$$

where the unprimed and primed variables refer to the incidence and refraction medium, respectively and θ is the angle the wavevector makes with the normal to the interface. In analogy with Fresnel's equations, the reflected and refracted amplitudes, respectively, are given by

$$E_a'' = \frac{Z' \cos \theta - Z \cos \theta'}{Z' \cos \theta + Z \cos \theta'} E_a \tag{14.80}$$

$$E_a' = \frac{2\rho v_a' \cos \theta}{Z' \cos \theta + Z \cos \theta'} E_a, \tag{14.81}$$

where Z is the acoustic impedance.

The reflectance and transmittance can be obtained from $(E_a''/E_a)^2$ and $(E_a'/E_a)^2$, respectively. At normal incidence, these are

$$\rho_0 = \frac{(Z' - Z)^2}{(Z' + Z)^2}, \tag{14.82}$$

$$\tau_0 = \frac{4\rho^2 v_a'^2}{(Z' + Z)^2}. \tag{14.83}$$

Conditions for the vanishing of reflection are clearly, in general,

$$\frac{Z}{\cos \theta} = \frac{Z'}{\cos \theta'},$$ (14.84)

and at normal incidence

$$Z = Z'.$$ (14.84a)

Analogously to low reflection coatings in optics, matching layers can be used between transducer and medium, or between one medium and the next, to maximize the transfer of acoustic flux.

14.2.2.4   Absorption. As the plane acoustic wave is propagated down the medium, its amplitude diminishes due to energy losses. In a liquid, these losses are primarily due to friction (proportional to the viscosity of the medium) and also due to heat conduction from the higher temperature regions (in the compressional portion of the wave) to the lower temperature regions (in the dilated portion). For a given wave amplitude, both of these losses are proportional to the square of the frequency, so that the attenuation coefficient, too, is proportional to the square of the frequency. The ratio of amplitudes, at two points, spaced by x, along the wave propagation, is

$$\frac{A(x_1 + x)}{A(x_1)} = e^{-\alpha^* \Omega^2 x},$$ (14.85)

where α^* is a constant for a given medium.

In solids there are a number of additional contributions to the attenuation, and the frequency dependence is more complicated.

14.2.3   Optical Characteristics of Acoustic Waves [10]

Just like static stresses, periodic stress waves are accompanied by refractive index changes. Table 148 [11] lists the elastooptic constants of isotropic and crystalline materials used in acoustooptics. The scattering of light by acoustic waves was first predicted by Brillouin[4] and is named after him, especially when the acoustic frequencies are very high.

14.2.3.1   Qualitative Considerations. Consider now the simple case of a uniform plane wave of monochromatic light (wavelength, λ) incident upon an acoustic wave field consisting of sinusoidal index variations (wavelength, Λ) traveling at right angles to the light wave. See Figure 14.5, where the alternating solid and broken straight lines indicate

---

[4] L. Brillouin (1922).

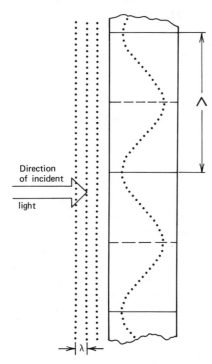

Figure 14.5 Illustrating acoustooptic interaction.

acoustic wave peaks and troughs, respectively, and the progressing optical wave front is indicated by dotted lines.

As the light enters the acoustic field, the portions of the wavefront entering the region near a pressure peak will experience a higher refractive index and will therefore advance with a lower phase velocity. On the other hand, those portions entering regions of dilation will exhibit a higher phase velocity, and the wavefront will soon take on a corrugated appearance as shown by the wavy line. Since the acoustic velocity is much less than that of the light wave, we ignore it for the moment and consider the acoustic wave as if "frozen" in the medium.

Since the wave elements advance essentially normal to the local wavefront, this corrugation implies changes of direction for almost all wave elements. This, in turn, leads to a redistribution of the light flux, which tends to concentrate near regions of compression. The superposition of this amplitude redistribution must be considered when considering deep acoustic fields and complicates the analysis of the optical effects in these. Significant simplifications, however, are possible by using approximations valid in various regions of operation.

1. When the acoustic wavelength is very much larger than the optical wavelength, diffraction effects can be ignored and a geometrical optical analysis is valid.

When the acoustic wavelength becomes comparable in size to the optical wavelength, a physical optics analysis must be used. There are, then, two distinct cases, which may be classified as shallow or deep acoustic field.

2. In the shallow field, the amplitude redistribution may be ignored.

3. In the deep acoustic field, it may not. However, here the situation rapidly approaches one where the phenomenon of Bragg diffraction sets in, and here, again, the analysis becomes more tractable.

Below we discuss each of these and conclude the section with a brief discussion of the quantum theoretical aspects, which become especially significant at high power densities.

14.2.3.2 Geometric Optic Analysis. When a ray travels through a medium of nonuniform refractive index, it is bent, unless it travels parallel to the index gradient [see (12.30) and Appendix 13.3].

Consider a series of rays, parallel to the z axis, as shown in Figure 14.6

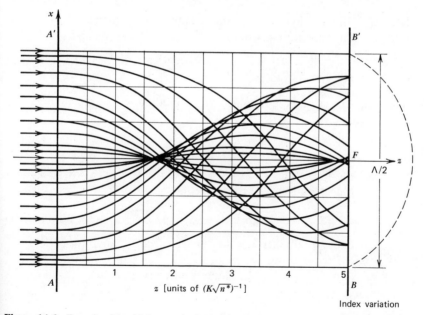

Figure 14.6 Rays in sinusoidal acoustic field. Dotted line represents the index variation. After Lucas and Biquard [12].

[12], incident on an acoustic field with sinusoidal index variations:

$$n = n_0 \left(1 + n^* \cos \frac{2\pi x}{\Lambda} \right),$$

$$= n_0(1 + n^* \cos Kx), \tag{14.86}$$

where n^* is the amplitude of the index variation, relative to the mean value, n_0, of the index, and

$$K = \frac{2\pi}{\Lambda}$$

is the propagation constant of the acoustic wave. Here the field is confined to the space between the planes AA', BB', and the index variations are indicated graphically along BB' as an axis. There are no variations assumed normal to the plane of the figure. In the plane of a pressure maximum, shown as the z axis, the index is constant at

$$n(0) = n_0(1 + n^*), \tag{14.87}$$

the index gradient vanishes, and the rays remain straight.

From (13.111) we have for the angle, α_0, at which the ray crosses the axis:

$$\cos \alpha_0 = \frac{n(x_0)}{n(0)}, \tag{14.88}$$

where x_0 is the x value at the point of entry, where the ray is parallel to the z axis. On substituting from (14.86) and (14.87) we find

$$\cos \alpha_0 = \frac{1 + n^* \cos Kx_0}{1 + n^*}. \tag{14.89}$$

This cosine has a minimum value, corresponding to the maximum value $(\alpha_{0\,mx})$ of α_0:

$$\cos \alpha_{0\,mx} = \frac{1 - n^*}{1 + n^*}. \tag{14.90}$$

On expanding both sides of (14.89) and (14.90) in a Taylor series and comparing the second-order terms, we find

$$\alpha_0 \approx [2(1 - \cos Kx_0)n^*]^{1/2} \le \alpha_{0\,mx} = 2\sqrt{n^*}, \qquad n^* \ll 1, \tag{14.91}$$

as the maximum angle of refraction experienced by any given ray.

To obtain a more complete picture of the ray paths, we refer back to (13.206). According to this, the slope of any ray at a general distance, x, from the axis is given by

$$\frac{dx}{dz} = \left[\frac{n^2(x)}{n^2(0)} \cos^2 \alpha_0 - 1 \right]^{1/2}. \tag{14.92}$$

Again from (13.111), we find

$$n(0) \cos \alpha_0 = n(x_0) = n_0(1 + n^* \cos Kx_0). \tag{14.93}$$

On substituting this and (14.86) into (14.92), and, expanding, we find

$$\frac{dx}{dz} = \frac{[2n^*(\cos Kx - \cos Kx_0) + n^{*2}(\cos^2 Kx - \cos^2 Kx_0)]^{1/2}}{(1 + n^* \cos Kx_0)}$$

$$\approx [2n^*(\cos Kx - \cos Kx_0)]^{1/2}, \tag{14.94}$$

where we have retained terms in n^* only to the first order. To obtain the ray path, we must integrate this:

$$z = \int_{x_0}^{x} \frac{dx}{[2n^*(\cos Kx - \cos Kx_0)]^{1/2}}. \tag{14.95}$$

For paraxial rays, where we may substitute

$$\cos Kx = 1 - \tfrac{1}{2}K^2x^2,$$

this takes the form

$$z_p = \frac{1}{\sqrt{n^*}K} \int_{x_0}^{x_p} \frac{dx}{\sqrt{x_0^2 - x^2}} = \frac{\sin^{-1}(x_p/x_0) - \pi/2}{\sqrt{n^*}K} \tag{14.96}$$

or

$$x_p = x_0 \cos(\sqrt{n^*}Kz). \tag{14.97}$$

This has zeros, that is, nodes or foci, at

$$z_0 = \frac{(2m-1)\pi}{2K\sqrt{n^*}} = \frac{(2m-1)\Lambda}{4\sqrt{n^*}}. \tag{14.98}$$

To obtain the nonparaxial solution, we put (14.95) into the form of an elliptic integral [13a]

$$z = (\sqrt{n^*}K)^{-1} \int_{Kx_0}^{Kx} \frac{du}{\sqrt{\cos u - b}},$$

$$= (\sqrt{n^*}K)^{-1} \left[F(\varphi, a) - F\left(\frac{\pi}{2}, a\right) \right], \tag{14.99}$$

where

$$F(\varphi, a) = \int_0^{\varphi} \frac{du}{\sqrt{1 - a^2 \sin^2 u}} \tag{14.100a}$$

is the elliptical integral of the first kind, and

$$\varphi = \sin^{-1}\left[\frac{(1-\cos u)}{(1-b)}\right]^{1/2}$$ (14.100b)

$$a = \sqrt{\frac{(1-b)}{2}}$$ (14.100c)

and

$$b = \cos Kx_0.$$ (14.100d)

Plots of representative ray paths are shown in Figure 14.6, with foci at $\frac{1}{2}\pi K\sqrt{n^*}$ and $\frac{3}{2}\pi K\sqrt{n^*}$.

If we fix our attention on a narrow bundle represented by a single ray in the figure, we find that, as the acoustic wave travels along the x axis, the ray oscillates angularly in the x–y plane with a maximum excursion which will not exceed $\alpha_{0\,mx}$ as given by (14.91). Alternatively, observation at a point fixed in space, say in plane BB', will reveal intensity fluctuations as the "focal points," such as F, pass it.

14.2.3.3 Physical Optics Analysis: Shallow Field (Raman–Nath [91] approximation).
To cover acoustic fields whose wavelength is comparable to the optical wavelength, we now turn to a physical optics analysis. Here the complex amplitude transmittance of the acoustic field may be represented as a phasor:

$$\hat{\tau} = \sqrt{\tau}e^{i\phi(x)},$$ (14.101)

where τ is the flux transmittance and
$\phi(x)$ is the phase shift given by

$$\phi(x) = \phi_0 \cos\frac{2\pi x}{\Lambda},$$ (14.102a)

$$\phi_0 = kd\,\Delta n$$ (14.102b)

$$k = \frac{2\pi n_0}{\lambda_0}$$ (14.102c)

is the optical wave propagation constant in the medium, and Δn is the amplitude of the index fluctuation.

The Fraunhofer diffraction pattern corresponding to (14.101) is obtained as the Fourier transform of the amplitude transmittance, with the variable

$$x^* = \frac{x}{\lambda}$$ (14.103)

replacing x. Since $\hat{\tau}$ is periodic in x^*, we obtain a Fourier series

$$A(\theta) = \sqrt{\tau} \sum_{n=-\infty}^{\infty} c_n e^{i 2\pi n x^* \lambda / \Lambda}, \tag{14.104}$$

where[5]

$$c_n = \int_{-\Lambda/2\lambda}^{\Lambda/2\lambda} \cos \frac{2\pi n x^* \lambda}{\Lambda} \exp\left(i\phi_0 \cos \frac{2\pi x^* \lambda}{\Lambda}\right) dx^*$$

$$= i^n \frac{\Lambda}{\lambda} J_n(\phi_0), \qquad n = \cdots -1, 0, 1, 2 \cdots \tag{14.107}$$

and the exponential in (14.104) represents a wave propagating in the direction

$$\theta_n = \frac{n\lambda}{\Lambda}. \tag{14.108}$$

We conclude that the Fraunhofer diffraction pattern consists of sharp lines in the directions θ_n. Since the flux density is proportional to the square of the amplitude, the intensities of these diffraction lines are

$$\boxed{\eta_n = J_n^2(\phi_0)} \tag{14.109}$$

relative to the incident intensity. This implies that the zeroth order vanishes when

$$\phi_0 = j_{01} = 2.4048 \tag{14.110}$$

and that the first orders attain their maximum value

$$J_{1mx}^2(\phi_0) = (0.58187)^2 = 0.33857, \tag{14.111}$$

when

$$\phi_0 = 1.84118. \tag{14.111a}$$

These results were based on the Fourier series approximation, which is

[5] The integral may be evaluated on substituting the variable of integration

$$u = \frac{2\pi x^* \lambda}{\Lambda}. \tag{14.105}$$

This yields [13b]

$$c_n = \frac{\Lambda}{2\pi\lambda} \int_{-\pi}^{\pi} \cos nu \, e^{i\phi_0 \cos u} \, du \tag{14.106}$$

$$= i^n \frac{\Lambda}{\lambda} J_n(\phi_0). \tag{14.107}$$

strictly valid only when the illuminated acoustic field is infinitely long. With limited apertures, the lines in the diffraction pattern spread out into bands (compare Figure 2.18a and b), whose width is controlled by the effective[6] length, L_e, of the acoustic field, just as the length of a diffraction grating controls the width of each order. Specifically, according to (2.175), the half-width of each diffraction band will be

$$\delta\theta = \frac{\lambda}{L_e}.$$ (14.112)

When the shallow acoustic field makes an angle α with the optical wave vector, each ray passes through a range $(\Delta\phi)$ of phases of the acoustic wave. Consequently, the maximum phase shift experienced by the optical wave is reduced. But even with the optical wave incident normally on the acoustic wave, the effects of refraction and diffraction will soon bend it, so that it passes through regions of various acoustic phases. This effect limits the range of validity of the shallow field analysis.

This range of validity can be estimated from the condition that a beam limited to a positive half-cycle of the acoustic field will not be spread substantially to the neighboring negative half-cycle, either by refraction or diffraction. These two requirements lead to two conditions that must be satisfied for the shallow-field approximation to be valid. The condition based on refraction is

$$\boxed{Q_1 = \sqrt{\frac{\Delta n}{n_0}}\frac{d}{\Lambda} < \frac{1}{2}}$$ (14.113)

and that obtained from the limitations on diffraction effect is

$$\boxed{Q_2 = \frac{2\pi\lambda}{\Lambda^2}\frac{d}{} \ll 1}.$$ (14.114)

See Appendix 14.2 for the derivation of these conditions and for other requirements.

14.2.3.4 Physical Optics Analysis: Deep Field. When conditions (14.113) and (14.114) are not met, the analysis becomes very complex [14]. But when $Q_2 \geqslant 1$, it can again be simplified considerably: under this condition, the acoustic field acts very much like a thick diffraction grating, that is, a grating made up of planes rather than lines. In such a grating,

---

[6] We use here the qualifier "effective" to call attention to the fact, that only the regions where the acoustic and electromagnetic waves overlap are to be considered. Indeed, any nonuniformity, such as attenuation in the acoustic beam, affect the effective length.

the basic equation governing the diffraction angle is still valid. However, a significant amount of diffraction takes place only around certain angles of incidence and results in just one diffraction order; for other angles of incidence, there is effectively no diffraction. This effect is called Bragg[7] diffraction and we analyze it briefly. Consider Figure 14.7a. This shows a few layers of a three-dimensional diffraction grating oriented normal to the plane of the diagram. A plane wavefront is incident from the upper left and is scattered on the planes (traces shown as broken lines). Significant amounts of light will emerge only in directions in which the interference is constructive. For this to occur in an infinite grating, two conditions must be fulfilled simultaneously:

1. Energy scattered throughout a given grating plane must arrive at the new wavefront in phase.

2. Energy scattered at successive grating planes must arrive at the new wavefront with a path difference which equals an integral number of wavelengths.

The first of these conditions is identical to that for reflection from a smooth surface and yields

$$\theta_d = \theta_i. \tag{14.115}$$

The second condition is the same as that of a plane diffraction grating[8]

$$\sin \theta_i + \sin \theta_d = \frac{m\lambda}{D}, \qquad m = 0, 1, 2, 3 \cdots, \tag{14.116}$$

where λ is the wavelength of the light and
    D is the spacing between layers.

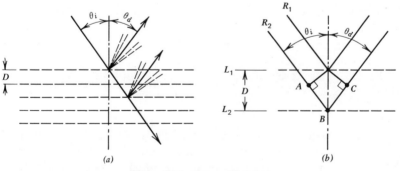

(a)   (b)

Figure 14.7 Bragg diffraction.

[7] W. L. Bragg (1912).

[8] The reader can readily confirm this equation by reference to Figure 14.7b, which shows two rays (R_1, R_2) scattered at two successive layers (L_1, L_2), respectively. The path difference $(\overline{AB} + \overline{BC})$ is clearly given by: $D(\sin \theta_i + \sin \theta_d)$. For fully constructive interference, this must equal an integral number of wavelengths $(m\lambda)$.

The simultaneous fulfillment of both conditions (14.115) (14.116) implies

$$\sin \theta_i = \sin \theta_d = \frac{m\lambda}{2D};$$ (14.117)

the diffraction will be similar to that obtained with a plane diffraction grating, but only for special angles of incidence: the angle of incidence must equal the angle of diffraction.

In acoustooptic applications, only the first order is generally used, and D equals the acoustic wavelength, Λ. Hence the equation for the so-called Bragg angle, θ_B, becomes here

$$\sin \theta_B = \frac{\lambda}{2\Lambda}.$$ (14.117a)

The diffraction efficiency of a thick phase grating, like the one under consideration here, and its dependence on the exact angle of incidence, can be calculated in a number of ways [15]. It is found [10a] that the diffraction efficiency at the Bragg angle is

$$\boxed{\eta = \sin^2 \frac{\phi}{2}},$$ (14.118)

where

$$\phi = 2\pi \, \Delta n \, \frac{d}{\lambda \cos \theta_i}$$ (14.119)

and

$$\sin \theta_i = \frac{\lambda}{2D}.$$ (14.120)

Note that here the efficiency can theoretically reach 100%, namely when $\phi = \pi$, $3\pi \cdots$. This, because the flux is here diffracted into but one order, in contrast to the shallow field case where we saw the maximum efficiency in a single order to be less than 34%.

The efficiency varies also with the deviation ($\Delta\theta$) of the angle of incidence from the Bragg angle (θ_B) as follows:

$$\eta = \frac{\sin^2 \sqrt{\phi^2/4 + \delta^2}}{1 + 4 \, \delta^2/\phi^2},$$ (14.121)

where

$$\delta = \Delta\theta K d \, \frac{\cos \theta_B}{2 \cos \theta_d}$$

$$\approx \frac{K d \, \Delta\theta}{2}, \qquad \Delta\theta \ll \theta_B.$$ (14.122)

We do not cover here the difficult intermediate case of

$$0.1 \lesssim Q_2 \lesssim 1. \tag{14.123}$$

The reader can find extensive numerical results covering that range in Ref. 16.

14.2.3.5 Quantum Optical Effects.

From the quantum theoretical point of view, the diffraction of light at an acoustical wave may be considered as a transfer of momentum. A quantum of acoustic wave energy, a phonon, has a momentum given by

$$p_{\text{phon}} = \frac{h}{\Lambda}, \tag{14.124}$$

just as a photon has a momentum

$$p_{\text{phot}} = \frac{h}{\lambda}, \tag{13.125}$$

[see (4.73)].[9] Scattering takes place when a photon absorbs a phonon, acquiring its momentum. Figure 14.8 shows the vector relationship

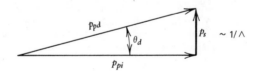

Figure 14.8 Scattering of photon by phonon: vector diagram. p are momentum vectors; subscripts p refer to photon, s to phonon, i to incident, and d to diffracted light.

involved when a horizontally incident photon absorbs a phonon traveling vertically upward. In addition to increasing its momentum, the photon also increased its energy by that of the phonon. This increase in energy corresponds to an increase in frequency [see (4.2)], so that the scattered light will have a new frequency, ω', according to

$$\hbar\omega' = \hbar\omega + \hbar\Omega$$

[9] In the first printing, the h in the last member of this equation was erroneously printed with a bar.

or

$$\omega' = \omega + \Omega, \tag{14.126}$$

where Ω is the radian frequency of the acoustic wave.[10]

This frequency shift may be observed by combining the diffracted with the undiffracted wave and observing an optical modulation frequency, Ω, corresponding to the beats (see Figure 2.1). It has also been observed by using the resonance absorption of a gas as an optical filter (for light generated by a discharge in the same gas species). Such a "filter" may be detuned by means of the Zeeman effect (Section 14.4.1). The detuning required to absorb the diffracted wave is found to equal Ω [17].

Analogously, the wave diffracted in the direction opposing the acoustic wave propagation, corresponds to the creation of phonons, with the momentum subtracted from that of the photon. The energy of the photon, too, is reduced by $\hbar\Omega$ and its frequency by Ω.

---

[10] This change in frequency can be explained also in terms of physical optics. As explained at length in Section 19.3.5.4, a displacement, δx, of a grating aperture, introduces a phase change

$$\delta\phi = \frac{2\pi\,\delta x}{\Lambda}. \tag{14.127}$$

For a unity phase change, the displacement is $(\Lambda/2\pi)$. This happens Ω times every second in our acoustic grating, so that Ω is added to the optical frequency.

The observed frequency shift has also been explained as a Doppler shift. Light from a real source and reflected at a surface is equivalent to light originating at a source located at the image of the actual source. When the reflecting surface moves normal to its plane, the image will have a velocity twice that of the surface. Light reflected from such a surface will therefore exhibit the Doppler shift observed from a moving source (2.51):

$$\omega_r = \left(\frac{1+v/c}{1-v/c}\right)^{1/2}\omega,$$

$$\approx \left(1+\frac{v}{c}\right)\omega, \qquad v \ll c. \tag{14.128}$$

Since the effective source velocity is here twice the acoustic velocity, v_a, its component in the direction of the optical wave propagation is

$$v_e = 2v_a \sin\theta_B = v_a \frac{\lambda}{\Lambda}. \tag{14.129}$$

Hence the shift in frequency is from (14.128):

$$\Delta\omega = v_e \frac{\omega}{c} = v_a \frac{\lambda\omega}{c\Lambda} = \Omega, \tag{14.130}$$

with the last step following from (2.32) applied to both the electromagnetic and the acoustic waves.

The probability of photon-phonon interaction is proportional to the product of their respective densities. Hence it should be possible to transduce significant amounts of acoustic power into optical power, by letting very intense acoustic fields interact with weaker optical waves. Conversely, it should be possible to transmute optical into acoustic power by using very intense light beams. Such intense beams will generate phonons in the manner just discussed. This phenomenon is called stimulated Brillouin scattering and has been observed in the laboratory [18], [10b].

14.2.3.6   Anisotropic Media [19]. Returning to the momentum diagram of Figure 14.8, we note that the phonon vector must be normal to the bisector of $\mathbf{k}_i$ and $\mathbf{k}_d$, if the equality

$$|\mathbf{k}_i| = |\mathbf{k}_d| \tag{14.131}$$

is to be maintained. (The frequency shift is generally negligible relative to the momentum change.) This is another explanation of the Bragg diffraction phenomenon in acoustooptics.

This result, however, is valid only when the refractive index is identical for the two waves. In acoustic waves in solid materials, especially when shear waves are involved, the diffracted wave may be polarized normal to the polarization of the incident wave. In birefringent media, the rotation of the plane of polarization may result in a significant change in refractive index and, hence, in wavelength and momentum. Consequently, the associated momentum vector diagram is no longer symmetric and may appear as shown in Figure 14.9a, with θ_d significantly different from θ_i. Indeed $\mathbf{K}$ may form an obtuse angle with $\mathbf{k}_i$, as shown in Figure 14.9b. As we decrease K, while maintaining a constant difference between k_i and k_d, the direction of $\mathbf{k}_d$ approaches that of $\mathbf{k}_i$, until $\mathbf{k}_i, \mathbf{k}_d$, and $\mathbf{K}$ become collinear when

$$K = \pm(k_d - k_i),$$

as illustrated in Figure 14.9c. In view of the relationships:

$$k = n\frac{\omega}{c}, \quad K = \frac{\Omega}{v_a}, \tag{14.132}$$

this corresponds to an acoustic frequency:

$$\Omega_{mn} = \frac{v_a\omega}{c}(n_d - n_i) = v_a k_0(n_d - n_i). \tag{14.133}$$

This is the minimum frequency at which efficient diffraction will occur in

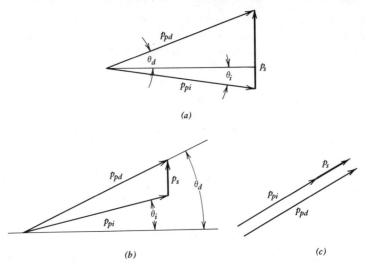

Figure 14.9 Acoustooptic scattering in birefringent medium: vector diagram. Symbols defined in Fig. 14.8.

the deep field situation. Similarly, an upper limit is reached when

$$K = k_d + k_i,$$ (14.134)

that is, when

$$\Omega_{mx} = v_a k_0 (n_d + n_i).$$ (14.135)

Returning to Figure 14.9a, we see, more generally, that the conditions for efficient diffraction—the "generalized Bragg conditions"—are

$$k_i \sin \theta_i + k_d \sin \theta_d = K$$ (14.136)

and

$$k_i \cos \theta_i = k_d \cos \theta_d.$$ (14.137)

On solving these equations for $\sin \theta_i$ and $\sin \theta_d$, and substituting

$$k_i = \frac{2\pi n_i}{\lambda}, \qquad k_d = \frac{2\pi n_d}{\lambda},$$ (14.138)

we obtain, after some routine algebraic and trigonometric manipulation[11]

$$\sin \theta_i = \frac{\lambda_0}{\Lambda} \frac{1 + (n_i^2 - n_d^2)\Lambda^2/\lambda_0^2}{2n_i}$$ (14.139)

$$\sin \theta_d = \frac{\lambda_0}{\Lambda} \frac{1 - (n_i^2 - n_d^2)\Lambda^2/\lambda_0^2}{2n_d}.$$ (14.140)

[11] Square (14.137), express the cosines in terms of sines and solve for $\sin^2 \theta_d$. Now solve (14.136) for $\sin \theta_d$ and square. Equate the two results and solve for $\sin \theta_i$.

On substituting

$$\frac{\Lambda^2}{\lambda_0^2} = \frac{k_0^2 v^2}{\Omega^2},$$ (14.141)

the right-hand members take the form

$$\frac{1}{2}\frac{\lambda}{\Lambda}\left[1 \pm \frac{k_0^2 v^2(n_i^2 - n_d^2)}{\Omega^2}\right].$$ (14.142)

The first terms in (14.139) and (14.140) represent the ordinary Bragg condition and the second terms the deviation therefrom due to the birefringence. From (14.142), the two contributions are seen to be equal for a frequency

$$\Omega_1 = k_0 v \sqrt{|n_i^2 - n_d^2|}.$$ (14.143)

For acoustic frequencies considerably higher than this, the deviation from the usual Bragg condition becomes negligible. But for frequencies lower than Ω_1, the effects due to birefringence dominate. See Table 149 [19] for some representative values of $\nu_{mn,1,mx}(=\Omega_{mn,1,mx}/2\pi)$.

14.2.3.7 Surface Waves. In conclusion, we mention but briefly acoustic surface waves. These, too, can be used in light diffraction, usually working in reflection. Here they follow the shallow-field equation with

$$\phi_0 = 2ka \cos \theta_i,$$ (14.144)

where a is the wave amplitude. See Ref. 20 for reviews of work in the acoustooptics of surface waves.

14.2.4 Applications of Acoustooptics [11]

In addition to being an important tool in the investigation of mechanical characteristics of materials, acoustooptics have found a number of applications in technology. We review some of these here.

14.2.4.1 Acoustooptic Light Modulation. Passage through an acoustic wave field in itself, does not affect the total flux. However, the accompanying refractive effects can readily be utilized to modulate the light intensity. A basic arrangement for illustrating the "diffraction grating" effect, is shown in Figure 14.10. Ignoring, for the present, the broken lines, we see a point source of light, S, whose flux is collimated by lens, L_1, at whose focal point S is located. This lens is followed by a cell, C, carrying acoustic waves and having a piezoelectric transducer at one end and an absorber at the other. It is followed by lens, L_2, which focuses the

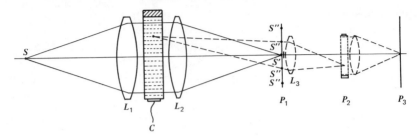

Figure 14.10   Acoustooptic light diffraction. S, slit light source; L_1, collimating lens; C, ultrasonic cell; L_2, focusing lens; P_1, plane of Fraunhofer diffraction pattern; P_2, plane of cell image; P_3 conjugate to P_1.

collimated light in plane P_1, imaging S at S'. Due to the acoustic waves, diffraction images of S may appear at either side of S'. These are indicated on the figure by S'' and are spaced approximately $(f_2\lambda/\Lambda)$ from S' and from each other, where f_2 is the focal length of L_2. The illumination at S'' is controlled by the amplitude of the waves in the cell, according to (14.109): the illumination at S'' is modulated by means of modulating these waves. The modulation bandwidth achievable with this technique is limited by the time interval

$$\Delta t = \frac{L}{v_a} \qquad (14.145)$$

required to clear the cell, length L, of the acoustic wave.[12] However, we

---

[12] It is instructive to note the consistency of this result with the requirements of Bragg diffraction [21]. The required angle of incidence there is

$$\theta_B = \frac{\lambda}{2\Lambda} = \frac{\lambda v_t}{2v_a},$$

where v_t is the temporal frequency of the acoustic wave. The change in angle required for a frequency change Δv_t is, then

$$\Delta\theta_B = \frac{\lambda\,\Delta v_t}{2v_a}.$$

For effective modulation, light diffracted by the central frequency and the changed frequencies, $v_t \pm \Delta v_t$, must overlap. Hence the optical and acoustic waves must both have an angular spread of $2\Delta\theta_B$, at least. According to the laws of Fraunhofer diffraction (2.157), this spread is given by

$$\Delta\theta = \frac{\lambda}{L}.$$

This, in conjunction with the preceding equation, yields a bandwidth

$$\Delta v_t = \frac{v_a}{L},$$

equivalent to (14.145).

can achieve much greater modulation bandwidths by the more sophisticated arrangements, which will be described shortly.

We may now block the light of S', as shown by the short bar at P_1 in
the figure, and place a lens, L_3, immediately thereafter, to image the cell
in plane P_2. This image will be formed only of light diffracted out of S' by
the acoustic wave. Consequently, only those portions of the cell, C, which
diffract light to points S'', will appear illuminated in the image at P_2. Also
only a fraction of the flux illuminating any point in the cell will reach the
corresponding point in the image. This fraction will be controlled by the
acoustic amplitude at that point. In other words, the cell image at P_2 will
be a rendition of the acoustic wave field in the cell. For instance, if the
voltage applied to the transducer consists of a carrier wave, frequency Ω,
with its amplitude modulated by the signal corresponding to one line of a
television display, this line will be displayed at P_2—at least as much of it
as the cell can hold, with the cell capacity given by (14.145).

We seem to have here a method for displaying video signals. There is,
however, one major difficulty: the image at P_2 does not stand still; it
travels (downward, in the figure) with a speed

$$v' = mv_a, \qquad (14.146)$$

where m is the magnification of the cell image at P_2, relative to the
original cell.

This motion would obviously blur the display. To eliminate it, a
rotating mirror may be placed after P_1. If the rotational speed of the
mirror can compensate exactly for the motion of the wave images, these
images will remain stationary (see Figure 14.11). From the geometry, it is
evident that the required mirror angular velocity, ω_m, is given by

$$2\omega_m D = mv_a, \qquad (14.147)$$

where D is the distance of the rotating mirror, M, from plane P_2.

With this arrangement, the image of any wave packet will remain

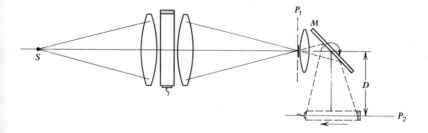

Figure 14.11   Stabilized acoustooptic display. M, rotating mirror.

stationary in P_2 and the cell image will appear as a window scanning across the wave packet image. This technique clearly also makes the length of the displayed wave train entirely independent of the cell length. Such devices have been used for television display [22], radar mapping [23], and could be used for video recording, as well [24].

The same device can produce a full-color display in an amazingly simple manner [25]. If the source, at S, is polychromatic and sufficiently small, the images at S'' will represent the spectrum of S. Note the wavelength dependence of the angle of diffraction (14.108). If now we place a narrow slit at P_1 of Figure 14.11 at the position of, say, the first-order images S'', only a narrow spectral region will pass on to form the cell image. The particular spectral region selected can be controlled by means of the carrier frequency of the acoustic field. In brief, we have here a display device in which the display luminance is controlled by amplitude modulating and its color, by frequency modulating the electrical carrier wave.

14.2.4.2 Beam Deflectors [11], [22].

As indicated by (14.108), the angle of diffraction experienced by the diffracted light varies inversely with the wavelength, and hence directly with the frequency, of the acoustic wave. Hence the position of S'' can be scanned over an appreciable distance by changing the frequency applied to the transducer. The position of the spot is given by

$$x = f_2 \frac{\lambda}{\Lambda} = \frac{2\pi f_2 \lambda \Omega}{v_a} \tag{14.148}$$

and consequently the length of the scan:

$$\Delta x = \frac{2\pi f_2 \lambda \, \Delta\Omega}{v_a}. \tag{14.149}$$

The number of resolvable spots is one of the fundamental parameters of any deflector. This is here given by

$$N = \frac{\Delta x}{\delta x} = \frac{2\pi \, \Delta\Omega L}{v_a}$$

$$= \Delta\nu_t \, \Delta t, \tag{14.150}$$

where $\delta(x) = f_2 \, \delta\theta$,

$\delta\theta$ = the angular half-width of the spot as given by (14.112), and $\Delta\nu_t = \Delta\Omega/2\pi$ is the width of the frequency sweep.

We now consider a few limitations and problems and indicate briefly how they can be overcome. Large values of N depend on large frequency scans, and these, in turn, imply short acoustic wavelengths and therefore operation in the Bragg diffraction domain. As we have seen, this operation requires a carefully controlled angle of incidence, (14.117a), which varies with the acoustic wavelength:

$$\Delta\theta_B \approx \frac{d\theta}{d\Lambda}\Delta\Lambda = \frac{\lambda\,\Delta\Lambda}{2\Lambda^2}. \tag{14.151}$$

Hence, to implement a large frequency shift, the angle of incidence must be changed along with the frequency. The required angular change can be approximated by replacing the transducer by a series of individual transducers arrayed along the z axis. These are mounted in a staggered, staircase-like arrangement, each successive transducer advanced by $\Lambda_0/2$ relative to its predecessor [22]. If now the signals applied to these are also successively delayed by half a cycle, a plane acoustic wavefront, parallel to the z axis, will be generated at the frequency corresponding to Λ_0. When the wavelength is increased by $\Delta\Lambda$, the wave due to one transducer, relative to its predecessor, will be advanced by a distance

$$\delta\Lambda = \tfrac{1}{2}\Lambda_0 - \tfrac{1}{2}(\Lambda_0 + \Delta\Lambda) = -\frac{\Delta\Lambda}{2}. \tag{14.152}$$

This effectively tilts the wavefront to a slope

$$\frac{\Delta\Lambda}{2d_1},$$

where d_1 is the spacing between transducers. This will be equal to the required tilt (14.151) provided

$$d_1 = \frac{\Lambda^2}{\lambda}. \tag{14.153}$$

Alternatively, the transducers may be driven with relative phase shift introduced electrically so as to tilt the resulting wave front as required by the Bragg condition [21]. A method using a single transducer with shaped and appropriately interconnected electrodes has also been described [26].

The time taken to complete the transfer of a spot from one position to another equals the time Δt [see (14.145)] required to fill the cell. According to (14.150), the number of resolvable spots is proportional to Δt. Hence, with a given frequency sweep, we can increase N only at the expense of a lengthened access time.

When a continuous sweep is required, the acoustic frequency may be changed continuously. The simultaneous presence, in the cell, of a range of wavelengths, implies that the diffraction angle changes continuously along the cell. This is equivalent to the action of a cylindric lens and can be canceled by the insertion of the inverse cylindric lens. The required focal length can be seen to be

$$f_c = \frac{\delta x}{\delta \theta},$$ (14.154)

where $\delta \theta$ is the change in diffraction angle due to the frequency change observed for a displacement:
δx along the cell.

Here the factors can be evaluated from

$$\delta x = v_a \, \delta t$$ (14.155)

and

$$\delta \theta = \frac{\delta t \Theta}{t_s},$$ (14.156)

where Θ is the total angular sweep and
t_s is the sweep time.

On substituting these into (12.154), we find the required focal length:

$$f_c = \frac{v_a t_s}{\Theta}.$$ (14.157)

Since this is constant throughout the scan, it can be compensated for by a single lens with the inverse focal length—negative if the scan is in the same direction as the wave propagation and *vice versa*.

Acoustic scanners can be used, for instance, in display devices, facsimile recorders, flying spot scanners, computer memories, optical printers, and for read-out in a holographic information storage system.

14.2.4.3 *Acoustooptic Filters.* At a given angle of incidence, efficient diffraction takes place over a narrow spectral region only, in deep acoustic fields. This fact can be used to construct tunable spectral filters. A possible arrangement is shown in Figure 14.12, which illustrates a *collinear* filter arrangement in which the incident and diffracted optic waves travel collinearly. As shown in Figure 14.9c this is possible in birefringent media in which the magnitude of the wave vector of the ordinary wave differs from that of the extraordinary one. If the acoustic wave vector is equal to that difference, and the diffraction switches the optical wave from ordinary to extraordinary (or *vice versa*), the "generalized Bragg condition" will be fulfilled for the collinear case. (See the discussion of birefringent media in Section 14.2.3.6.)

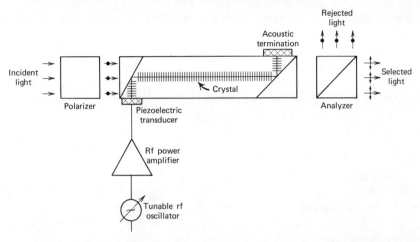

Figure 14.12   Tunable acoustooptic filter. After S. E. Harris, from Chang [11].

It has been shown that such a filter using a 5-cm-long $LiNbO_3$ crystal could yield filter tuning from 0.4 to 0.7 μm, with a relative bandwidth (i.e., reciprocal of resolving power) of $1.3 \times 10^{-4} \lambda^*$ and requiring about 1.4 W of acoustic power [27]. Here λ^* is the wavelength in micrometers.

Collinearity is not essential to an acoustooptic filter: several successful noncollinear filters have been reported [28], [29].

14.2.4.4 Optoelectronic and Signal Processing Applications.

The frequency shift introduced by the diffraction process may be used to produce the desired modulation by a *superheterodyne* process [30]. When the diffracted beam is added to the directly transmitted beam, the mixture will produce beats at the acoustic frequency. Denote the amplitudes of the diffracted and transmitted beams by a_d and a_t, respectively, taken relative to the incident amplitude, so that

$$a_d^2 + a_t^2 = 1.$$

Then the modulation of the mixed beam will be

$$M_m = \frac{(a_t + a_d)^2 - (a_t - a_d)^2}{(a_t + a_d)^2 + (a_t - a_d)^2} = 2a_t a_d. \tag{14.158}$$

This should be compared with the modulation of the transmitted beam by itself:

$$M_t = \frac{1 - a_t^2}{1 + a_t^2} = \frac{a_d^2}{2 - a_d^2}$$

$$\approx \frac{a_d^2}{2}, \qquad a_d \ll 1. \tag{14.159}$$

We see that, for small values of a_d, the heterodyning process yields a gain of $4a_t/a_d$ in modulation—a gain that may be substantial.

An acoustooptical arrangement may also be used as a *variable delay line:* the electrical signal is converted into an acoustic wave traveling down the medium. It may be subsequently recovered at any point by the diffraction of an incident optical beam, introducing a time delay equal to

$$t_d = \frac{x}{v_a}, \tag{14.160}$$

where x is the distance traveled by the wave from the transducer to the pick-up point [31].

An arrangement similar to that shown in Figure 14.10 can be used to perform correlation or convolution integrals at megahertz frequencies [31], [32]. Assume that we place into plane P_2 a transparency whose transmittance is given by

$$\tau(x) = F_1(x).$$

If we now modulate the carrier wave entering C by another signal $F_2(v't)$, it will appear at P_2 as a similar illumination distribution. Here we have written the time function as a function of $v't$, simply by multiplying by the acoustic velocity v' as it appears in P_2. If all the flux transmitted by the transparency is collected and detected, it will yield a signal

$$C(t) = \int_0^L F_1(x)F_2(v't-x)\, dx, \tag{14.161}$$

which is the convolution of F_1 and F_2. By reversing the orientation of the transparency, this can be changed into a correlation.

This process can be made dynamic, that is, the reference function F_1 can be controlled independently by placing another cell, similar to C, but oppositely oriented, in plane P_2. If the function $F_1(v't)$ is fed into it as a time function, its effective transmittance, too, can be made to follow F_1 and the total flux appearing at the aperture in plane P_3 (Figure 14.10) will be given by

$$C(t) = \int_0^L F_1(x-v_a t)F_2(v't-x)\, dx. \tag{14.162}$$

Other useful signal processing operations, too, can be executed by acoustooptics [31].

14.3   ELECTROOPTIC EFFECTS

When an electric field is applied across an optical medium, the distribution of electrons within it is distorted, changing the polarizability (α) of the material and hence its refractive index [see (10.3)] nonisotropically: even if the material was originally isotropic, it becomes birefringent due to the applied field. This is the electrooptic effect.

We distinguish between the linear and quadratic electrooptic effects. The linear effect is treated in Section 14.3.1. It occurs only in crystals lacking a center of symmetry—the same crystals that exhibit the piezoelectric effect. The quadratic effect occurs in these, as well as in isotropic substances, and even in liquids; this is discussed in Section 14.3.2.

The electrooptic effect is, in part, due to the elastooptic (or piezooptic) effect, whenever this occurs; the applied field causes a strain in the medium and this strain, in turn, causes a change in the refractive index. This is referred to as the secondary electrooptic effect. The effect observed when the crystal is clamped, or at frequencies that prevent any substantial strain from occurring, is called primary.

In addition to the electrooptic effect proper, as just described, there are other electrooptic effects, such as electrooptic absorption, scattering, and spectral effects. These, too, are discussed here briefly, in Sections 14.3.3 to 14.3.5, respectively. Electrooptic effects in liquid crystals are treated in Sections 14.5.1.4 and 14.5.2.2. The electrooptic effects have found many technological applications. These are reviewed in Sections 14.3.6–14.3.8.

14.3.1 First-Order Electrooptics—Pockels Effect

14.3.1.1 Basic Formulation **[33a], [34].** In its most general form, the electrooptic effect is conveniently expressed as a change in the coefficients of the index ellipsoid (14.41). Denoting the principal axes of this ellipsoid by (x, y, z), we can write the field-free ellipsoid

$$\frac{x^2}{n_x^2} + \frac{y^2}{n_y^2} + \frac{z^2}{n_z^2} = 1.$$ (14.163)

When an electric field $\mathbf{E}$ is applied, this changes to

$$x^2\left(n_x^{-2} + \sum r_{1j}E_j\right) + y^2\left(n_y^{-2} + \sum r_{2j}E_j\right) + z^2\left(n_z^{-2} + \sum r_{3j}E_j\right)$$

$$+ 2yz \sum r_{4j}E_j + 2zx \sum r_{5j}E_j + 2xy \sum r_{6j}E_j = 1, \qquad j = x, y, z, \quad (14.164)$$

where $E_{x,y,z}$ are the components of $\mathbf{E}$,

$\quad r_{ij}$ are the first-order electrooptic coefficients, and higher order effects are neglected.

The first-order electrooptic birefringence is called Pockels[13] effect.

---

[13] F. Pockels (1906).

Except in triclinic crystals (which do not seem to be very useful in electrooptics), symmetry conditions demand that many of the electrooptic coefficients vanish and that others be mutually dependent. See Table 150b for a listing of the interrelations for all crystal symmetry classes.

The coefficients r_{1j}, r_{2j}, r_{3j} do not affect the orientation of the ellipsoid axes, they only cause a change in the respective index by an amount Δn_i, such that

$$(n_i + \Delta n_i)^{-2} = n_i^{-2} - 2n_i^{-3} \Delta n_i + \cdots = n_i^{-2} + \sum_j r_{ij} E_j. \qquad (14.165)$$

Thus, for small index changes, Δn_i can be approximated by

$$\Delta n_i = -\tfrac{1}{2} n_i^3 \sum r_{ij} E_j. \qquad (14.166)$$

The coefficients r_{4j}, r_{5j}, r_{6j}, on the other hand cause a rotation of the ellipsoid axes, in addition to affecting their lengths. These changes can be determined by transforming the ellipsoid to a new coordinate system, whose axes coincide with the principal axes of the new ellipsoid. A method for doing this is given in Appendix 14.1.6. The coefficients of the ellipsoids in the rotated coordinate systems, and the angles of rotation, for all symmetry classes, are listed in Table 151 [35].

14.3.1.2 Materials [36], [37]. Any transparent crystal lacking an inversion center of symmetry, exhibits a first-order electrooptic effect. However, to be useful, such crystals must not only have sizeable electrooptic coefficients, but must also be available in substantial sizes, with good optical quality and at a reasonable cost. Many of the technologically useful materials are listed in Table 152 together with the point group symmetry,[14] dielectric constants, refractive indices, electrooptic coefficients, and their Curie temperature, which is the upper limit for the given structure.

Some brief comments on a few of the more important listed materials follow. KDP and ADP are available in large sizes, up to 50 mm, at relatively low cost; but they are water soluble and fragile. If deuterium (D) is substituted for the hydrogen in their manufacture, their electrooptic properties are greatly enhanced. See KD_2PO_4 in the table. $BaTiO_3$,

---

[14] The point group refers to the symmetry of each crystal lattice point. It is described by listing (a) axes of symmetry (in terms of the multiplicity of the symmetry about the axis) and (b) mirror symmetry planes (denoted by the letter m). A bar over a number, say N, indicates that the symmetry involves an inversion, that is, that the original constellation is repeated upon rotation by $(2\pi/N)$ radians *plus* changing all displacement vectors $\mathbf{r}$ into $-\mathbf{r}$.

The reader is referred to any modern text covering crystallography for a more detailed explanation of the nomenclature [33b], [38].

$LiNbO_3$, and $LiTaO_3$ are readily available in sizes up to 1 cm. They are ferroelectric and require special discussion; see the next section.

14.3.1.3   Linear Effects in Ferroelectrics

Ferroelectric Crystals. Ferroelectricity refers to the spontaneous electrical polarization observed in some crystals below a certain temperature, the Curie temperature, at which the crystal structure changes and the ferroelectricity vanishes. This polarization should result in the appearance of an electric potential across the crystal, but is not normally observable macroscopically because the crystal tends to break up into polarization domains. The polarization direction is constant within each domain, but varies from domain to domain, so that the various contributions to the overall polarization tend to cancel out. In general, this situation can be changed by the application of a strong electric field. In such a field, the domains polarized in the direction of the field tend to grow at the expense of the other domains, so that for large fields practically the whole crystal is polarized in one direction. This process is called *poling* and is especially easy near the Curie temperature.[15] For use in electrooptic devices, ferroelectric materials should be poled.

In addition to preparing the crystal for electrooptic modulation, poling induces static electrooptic birefringence in the crystal. This is a steady state and opens the way to electrooptic memories, which require no external field for their maintenance. A small external field, superimposed on the polarization, will add vectorially and modify the birefringence. A large field may reverse the poling.

Polycrystalline Ferroelectrics. Because of technical difficulties in growing them, crystals often become very expensive in large sizes. The possibility of using polycrystalline materials therefore is interesting. Among such materials developed, lanthanum-doped lead zirconate-tantalate ($Pb_{1-\varepsilon}La_\varepsilon Zr_{1-x}Ti_xO_3$), popularly known as PLZT, has aroused special interest. This material is hot-pressed into polycrystalline disks whose surfaces accept an optical quality polish. When an electric field is applied, the material becomes birefringent, with the index change proportional to the square of the applied voltage. This effect is due to poling. At fields beyond E_s, where the polarization saturates, the response becomes linear.

---

[15] Some crystals are highly resistant to poling. In some the domain structure remains unaffected until the field is strong enough to break down the crystal. In such crystals any net polarization, introduced perhaps at the Curie temperature, will soon be neutralized by free charges accumulated from the environment and can be observed only when the temperature is changed. This changes the amount of polarization and a net electric potential will appear across the crystal. Such crystals are called *pyroelectric* and may serve as radiation detectors. See Section 16.5.1.3.

The coercive field (required to destroy the polarization) and the remanent polarization (observed in the absence of an electric field) are shown in Figures 14.13*a* and *b*, respectively, as a function of temperature, for a number of compositions of PLZT [39].

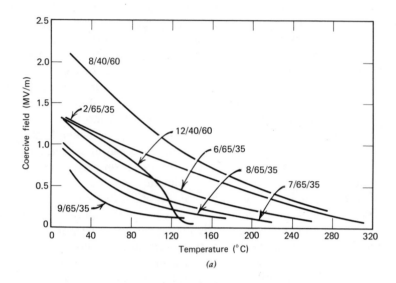

(a)

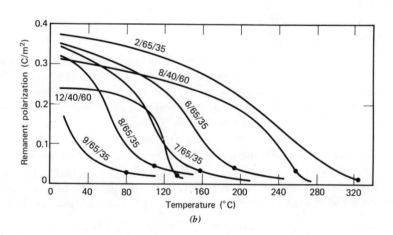

(b)

Figure 14.13 Ferroelectric properties of PLZT polycrystalline plates, as a function of temperature. (*a*) Coercive field; (*b*) Remanent polarization (dots indicate the temperature for the maxima of the dielectric constants). The composition is indicated next to each curve, in the form: lanthanum/zirconium/titanium percent. Haertling [39].

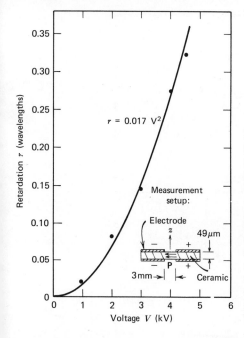

Figure 14.14   Electrooptic birefringence in PLZT (2/65/35). Increase in retardation with applied voltage for a poled 2/65/35 ceramic plate operated between the zero-field saturation remanence state and the field-on saturation remanence states of the ferroelectric hysteresis loop. The actual retardation at zero-field saturation remanence $(V = 0)$ is 0.66 wavelengths. Voltage was applied as shown in the inset in order to produce the ceramic poling direction **P**, and light of wavelength 0.656 μm was propagated in the z direction perpendicular to **P**. Thacher and Land [40].

This material (with $\varepsilon = 0.02$, $x = 0.35$) has been investigated extensively [40]. When fully poled, it exhibits an index difference

$$\Delta n = n_0 - n_e = 0.0087,$$

and with $\varepsilon = 0.06$, values of Δn to 0.015 have been observed [41]. Figure 14.14 [40] shows the response of a plate of such material with dimensions and electrode placement as shown in the inset. The transmittance spectra for plates of various thicknesses are shown in Figure 14.15a [40]. Some of the transmitted radiation is scattered and hence the transmittance depends on the acceptance angle of the detector; the curves shown are for a 4°-detector aperture and for a plate poled normal to the surfaces. Figure 14.15b shows the transmittance spectrum in the visible and near-ir in greater detail.

The major potential applications of this material seem to lie in information storage, memory or display. These are discussed further in Section 14.3.7.2.

14.3.1.4   Illustrations of Pockels Effect

KDP. By way of illustrating the Pockels effect, let us consider a KDP crystal, point group $\overline{4}2m$. Since the crystal is uniaxial, its index ellipsoid

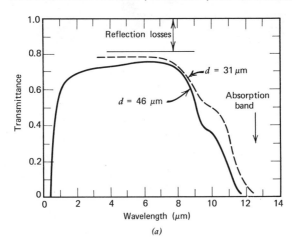

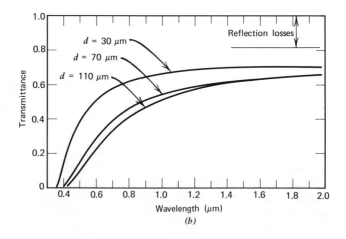

Figure 14.15 Transmittance spectra of PLZT plates of various thicknesses (d). Plates 2/65/35 and poled normal to plate surface; light measured over a 4° aperture. Thacher and Land [40].

has the form

$$\frac{x^2}{n_0^2}+\frac{y^2}{n_0^2}+\frac{z^2}{n_e^2}=1. \tag{14.167}$$

Its elevation and profile projections are shown by the solid lines in Figure 14.16a and b, respectively. When a general electric field is applied, the

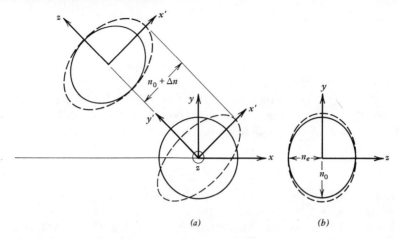

(a)   (b)

Figure 14.16 Index ellipsoids of KDP, schematic. Solid lines: field-free; broken lines: on application of an electric field in the z-direction.

ellipsoid becomes

$$\frac{x^2}{n_0^2}+\frac{y^2}{n_0^2}+\frac{z^2}{n_e^2}+2r_{41}(E_x yz+E_y xz)+2r_{63}E_z xy = 1, \qquad (14.168)$$

since for this symmetry class $r_{52}=r_{41}$ and all other coefficients, except r_{63}, vanish (see Table 150b). To simplify the illustration, let us assume that the field is applied in the z direction ($E_x=E_y=0$). The new ellipsoid may then be written:

$$\frac{x^2+y^2}{n_0^2}+\frac{z^2}{n_e^2}+2r_{63}xy = 1. \qquad (14.169)$$

This represents an ellipsoid whose axes are rotated about the z axis and from symmetry considerations the rotation is clearly through an angle of 45°. In terms of the rotated coordinate system,

$$x = x' \cos 45° - y' \sin 45°,$$
$$y = x' \sin 45° + y' \cos 45°, \qquad (14.170)$$

and the ellipsoid takes the form

$$(n_0^{-2}+r_{63}E_z)x'^2+(n_0^{-2}-r_{63}E_z)y'^2+\frac{z^2}{n_e^2} = 1.$$

This can be written in the form:

$$\frac{x'^2}{(n_0+\Delta n)^2}+\frac{y'^2}{(n_0-\Delta n)^2}+\frac{z^2}{n_e^2} = 1, \qquad (14.171)$$

where

$$\Delta n \approx \tfrac{1}{2}n_0^3 r_{63} E_z \qquad (14.172)$$

[see (14.166)]. This new ellipsoid, with its principal axes, is shown in Figure 14.16 by dotted lines.

In practice the effect may be observed as follows. In the absence of an electric field, polarized light entering parallel to the z axis will experience no retardation. When the field E_z is applied, however, it will experience a retardation

$$\phi = 2\Delta n d k, \qquad (14.173)$$

where d is the length of the light path in the stressed crystal and k is the wave propagation constant.

$LiTaO_3$. As a second illustration, we take $LiTaO_3$, which is also uniaxial and of symmetry $3m$. With the field again applied along the z axis, the only electrooptic coefficients involved are

$$r_{23} = r_{13} \quad \text{and} \quad r_{33}.$$

Its field-free index ellipsoid is of the form (14.167). When the field is applied in the z direction, it takes the form

$$\frac{(x^2+y^2)}{(n_0+\Delta n_0)^2} + \frac{z^2}{(n_e+\Delta n_e)^2}$$
$$= (n_0^{-2}+r_{13}E_z)(x^2+y^2)+(n_e^{-2}+r_{33}E_z)z^2 = 1. \qquad (14.174)$$

This ellipsoid is shown in Figure 14.17 before and after application of the electric field.

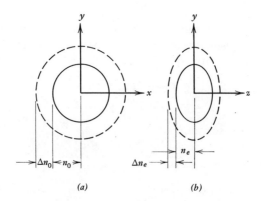

(a)   *(b)*

Figure 14.17   Index ellipsoids of $LiTaO_3$, schematic. Solid lines: field-free; broken lines: on application of an electric field in the z-direction.

Here, light entering along the z axis will experience no retardation, independent of the applied field. However, light incident along the x or the y axis will experience a retardation

$$\phi_0 = (n_e - n_0)kd, \qquad (14.175)$$

$$\phi = (n_e + \Delta n_e - n_0 - \Delta n_0)kd, \qquad (14.176)$$

respectively, before and after the field is applied. Thus the retardation is changed by

$$\Delta\phi = (\Delta n_e - \Delta n_0)kd$$
$$\approx \tfrac{1}{2}(n_e^3 r_{33} - n_0^3 r_{13})E_z kd. \qquad (14.177)$$

14.3.2 Second Order Electrooptics—Kerr Effect

14.3.2.1 Kerr Effect in Isotropic Materials [42]. The lowest order electrooptic effect in isotropic media is quadratic and is called electric Kerr[16] effect. This is evident from symmetry considerations, which prevent the sense of the applied field from affecting the result, so that a field **E** must have the same effect as a field $-$**E**. Since any formula, such as (14.164), in which E appears raised to an odd power implies a change with sense, such powers are impossible here. There are, however, a number of substances, including some liquids, that exhibit a substantial Kerr effect.

In isotropic substances, the Kerr effect can be viewed as introducing birefringence: when an electric field is applied to an isotropic liquid, its index ellipsoid changes from spherical to spheroidal, similar to that of a uniaxial crystal, with the axis of symmetry parallel to the applied field. Polarized light incident normal to this axis will experience a retardation given by

$$\phi = \Delta ndk, \qquad (14.178)$$

where

$$\Delta n = n_\| - n_\perp$$

and $n_\|$, $n_\perp$ are, respectively, the refractive indices for oscillations parallel and perpendicular to the applied field (analogous to the extraordinary and ordinary refractive indices, respectively, of uniaxial crystal),

d is the pathlength in the stressed medium, and

k is the *in vacuo* wave propagation constant.

---

[16] J. Kerr (1875). In addition to the second-order electrooptic effect, there is also a magnetooptic effect named after Kerr. Here we shall be referring to the former, except when the attribute "magnetic" is used in conjunction with the name.

We now define the Kerr constant:

$$K = \Delta n \frac{k}{E^2}, \tag{14.179}$$

where E is the magnitude of the applied field. In terms of this constant, the retardation is given by

$$\phi = KE^2 d \text{ rad.} \tag{14.180}$$

The values of K for a number of liquids are listed below.

	Kerr Constant [pm/V^2]	
Substance	Ref. 43a	Ref. 44a
Water	0.33	0.28
Nitrobenzene	15.4	22.8
Chloroform	−2.51	—

These values were obtained near 20°C and at wavelength 0.589 μm.

In fluids, two factors may contribute to the Kerr effect: the polarization of molecules and the alignment of molecules having a dipole moment. In fluids whose molecules do not have a dipole moment, only the first contribution exists. In these substances, the modulating field and the field of the electromagnetic radiation cooperate and hence the Kerr effect is always positive. In dipolar fluids, the Kerr effect is also positive, provided the dipole axis is parallel to the axis of maximum polarizability [α of (10.4)]. However, if the dipole axis is essentially normal to the axis of maximum polarizability, the molecular-alignment contribution will subtract from the polarization contribution and, if it outweights the latter, the Kerr effect becomes negative. See chloroform in the table shown above.

To understand the temperature dependence of the Kerr effect, we note that the polarization and alignment of the molecules represent a transfer of energy from the applied electric field to the molecules. This energy acts to overcome their thermally induced randomness, and its effectiveness in this must be evaluated relative to the thermal energy.

Detailed analysis shows that the polarizability and alignment contributions vary inversely with the absolute temperature and its square, respectively.

14.3.2.2 Second Order Effects in Crystals

General Form. To include second-order effects, we must rewrite (14.164) as follows:

$$\sum_{i=1}^{3} \sum_{j=1}^{3} \sum_{k=1}^{3} \sum_{l=1}^{3} (n_{ij}^{-2} + r_{ijk}E_k + R_{ijkl}E_kE_l)x_ix_j = 1, \qquad (14.181)$$

where R_{ijkl} are the second-order electrooptic coefficients and x_1, x_2, x_3 are x, y, z, respectively.

Occasionally the second order effect is expressed in terms of the polarization components (P_i) instead of the field components, where the polarization vector

$$\mathbf{P} = \mathbf{D} - \varepsilon_0\mathbf{E} = \varepsilon_0(\epsilon - 1)\mathbf{E}, \qquad (14.182)$$

with components

$$P_i = \varepsilon_0 \sum_j (\varepsilon_{ij} - 1)E_j \qquad (14.183)$$

and

$$\mathbf{D} = \varepsilon_0\epsilon\mathbf{E} \qquad (14.184)$$

is the electric displacement (flux) vector, ε_0 is the permittivity of free space (see footnote 2, Chapter 2) and ϵ is the dielectric constant tensor with elements ε_{ij}.

Terms of the form ($g_{pq}P_kP_l$) are then used in place of ($R_{ijkl}E_kE_l$), where p and q go from 1 to 6 and represent a pair of subscripts each, as shown in the following tabulation.

p (or q)	1	2	3	4	5	6
ij (or kl)	11	22	33	23	13	12

See the subscripts of the a's in (14.41).

Second-Order Effect in Ferroelectrics [45], [37]. Certain ferroelectrics, having a cubic structure above the Curie temperature, have an inversion center, and, therefore, do not show a first-order effect. They do, however, exhibit a second-order effect and, by means of a bias polarization, this may effectively be converted into a strong first order effect. We may describe this in terms of the bias polarization, P_b, and an applied signal polarization

$$P_s \ll P_b.$$

The change in refractive index may then be written

$$\Delta(n_{ij}^{-2}) = \sum_k \sum_l g_{ijkl}P_kP_l, \qquad i, j, k, l = 1, 2, 3$$

$$\approx \sum_k \sum_l g_{ijkl}P_{bk}P_{bl}\left(1 + \frac{P_{sk}}{P_{bk}} + \frac{P_{sl}}{P_{bl}}\right), \qquad (14.185)$$

with the P_b constant. This is linear in the signal polarization, P_s. Alternatively, we may say that the bias polarization destroys the symmetry, so that the crystal develops a first-order effect.

The process just described is particularly important near the Curie point (T_c), where the dielectric constant, and hence, the polarization induced by a given field, varies with the temperature, T, as $(T - T_c)^{-1}$. Since the index change varies with the polarization [see (10.3)], large changes can be induced here by reasonably small fields.

But operation close to the Curie point carries with it difficulties of its own. Due to the steep temperature dependence of the modulation, the demands on temperature stability of, and uniformity in, the crystal are very high. By using a self-compensatory crystal arrangement, it is possible to relax the stability requirement, but the uniformity requirement remains a serious handicap. Also, as the Curie point is approached, a greater portion of the applied high-frequency signal power tends to go into heating the crystal. This not only makes the operation less efficient, but also aggravates the difficulty in maintaining the required temperature stability. In practice, the designer must choose a temperature that balances the conflicting requirements of large response and limited temperature sensitivity.

14.3.3   Electrooptic Absorption—Franz-Keldysh Effect

The absorption edges of optical media (see Section 10.1.2.3) are sensitive to applied electric fields. Specifically, the application of an electric field shifts the absorption edge toward the longer wavelengths. If the absorption edge is sharp, the field may shift the edge across the wavelength of the light passing through the crystal, making it effectively opaque. This is called the Franz-Keldysh effect [46], [47]. It has been observed in CdS crystals [48] and a similar effect in pn-junction diodes [49]. The transmittance spectra of a CdS crystal, with and without applied voltage, are shown in Figure 14.18 [48].

14.3.4   Electrooptic Scattering [50]

In addition to the birefringence induced in poled ferroelectric polycrystalline materials by an electric field (Section 14.3.1.3), an electrooptic

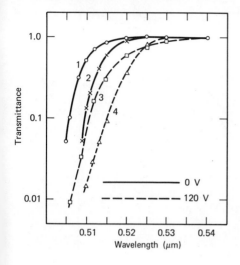

Figure 14.18 Transmittance spectra of Cds crystal, with and without applied electric field. 1 and 2: zero field; 3 and 4: 120 V applied; 1 and 3 extraordinary ray; 2 and 4: ordinary ray. Williams [48].

scattering effect is also observed there. The degree of scattering and depolarization of the transmitted light is a function of the poling direction: scattering is for more pronounced when the material is poled transverse to the light propagation. Figure 14.19 [50] shows the transmittance values observed with various relative apertures, in unpoled material and materials poled parallel and normal to the surface of the plate. Note that for a numerical aperture of 0.1, indicated by the broken line in the figure, the transmittance ratio is 10:1. This scattering effect becomes dominant and swamps the birefringence in coarse-grained materials whose crystals are larger than approximately 20 μm wide.

14.3.5 Stark Effect [44b]

When an electrical discharge is subjected to a strong electric field (ca. 10^7 V/m) and its spectrum is observed transverse to the field, the lines can be observed to be split into a symmetrical line pattern, with some of the lines polarized parallel and some normal to the field. When viewed parallel to the field, only the latter components are observed, and these appear unpolarized.

This effect is due to the modification of the energy levels in the atoms making the radiating transitions: depending on the orientation of the atom relative to the field, the transition may be more or less energetic than in the field-free discharge, resulting, respectively, in a decrease or increase in the radiated wavelength. To a first approximation, the change

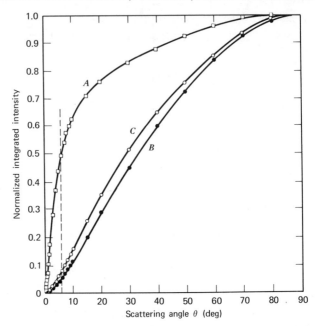

Figure 14.19 Relative transmittance of coarse grained lead zirconate-tantalate (PZT) plate as a function of the angle (θ) subtended by the optical aperture. Plate poled normal (Curve A) and parallel (Curve B) to the surface. Curve C is for a randomly polarized plate. PZT is 65/35 with 2% bismuth, 80 μm thick and nominal grain size of 5 μm. The abscissa corresponding $na = 0.1$ is indicated by a broken line. Land and Thacher [50].

in energy is directly proportional to the applied field; higher order terms, however, are also present. If we write the energy change in terms of the frequency shift, we have

$$\Delta U = h \, \Delta \nu_t = h(a_1 E + a_2 E^2 + a_3 E^3 + \cdots), \qquad (14.186)$$

where, for hydrogen,

$$a_1 = 1.92 \times 10^4 \, \text{Hz} \cdot \text{m/V}, \qquad a_2 = 1.56 \times 10^{-9} \, \text{Hz} \cdot \text{m}^2/\text{V}^2$$

and

$$a_3 = 4.56 \times 10^{-21} \, \text{Hz} \cdot \text{m}^3/\text{V}^3.$$

The corresponding change in wavelength is approximately

$$\Delta \lambda = -\frac{\lambda^2 \, \Delta \nu_t}{c};$$

see (4.37).

Thus an electric field may be used to "tune" a spectral line and, in conjunction with a very narrow filter, to modulate the transmitted optical flux.

14.3.6 Applications: Light Modulation [10c], [37], [45]

In electrooptic birefringence, the application of an electric field modulates the refractive index, and this, in turn, modulates the phase of the transmitted light—an effect that is not directly observable. Clearly, interference effects may be used to convert the phase modulation into an amplitude modulation. Alternatively, since the effects depend on the direction of polarization, this fact can be used to convert the modulation into one of amplitude.

If the transit time, t_t, of the light through the cell is short compared to the period of the modulation frequency, ω_m

$$t_t = \frac{d}{c} \ll \frac{1}{\omega_m}, \qquad (14.187)$$

the modulator action can be analyzed in terms of lumped constants. When the modulation frequency is higher than that, the analysis must be made in terms of waves traveling down the cell; see Section 14.3.6.5.

14.3.6.1 Lumped Modulators. All the electrooptic effects can be used for light modulation. By way of illustration, consider the liquid Kerr cell. The arrangement is shown in Figure 14.20. Light from a source, S, passes through a polarizer, P_1, that polarizes the light at 45° to the horizontal, as indicated by the hollow arrow. It then passes through a trough, the Kerr cell, K, which has optical windows, W, for this purpose. The analyzer, P_2, with its polarization direction normal to that of P_1, follows the cell, so

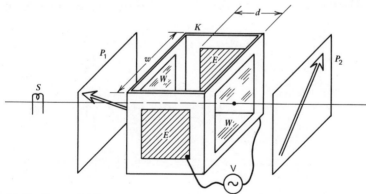

Figure 14.20 Kerr cell (K) used as a light modulator. P_1 and P_2, crossed polarizers; E, electrodes; W, windows.

that normally no light passes it. Now a transverse electric field is applied to K, by means of two electrodes, E, connected to a voltage source. Due to the applied voltage, the horizontal polarization component of the incident wave, the component parallel to the electric field, will experience an index different from that experienced by the vertical, transverse, component. This introduces, into the light, a retardation, which is given by (14.180):

$$\phi = KE^2 d = \frac{KV^2 d}{w^2},$$
(14.180a)

where V is the applied voltage and
      w is the spacing between the electrodes,

so that the light emerges elliptically polarized, in general. With $\phi = \pi$, however, it emerges linearly polarized, with the direction of the electric field vector turned through 90°; under this condition the emerging direction is parallel to the transmission axis of the analyzer, P_2, and the light passes it fully. More generally, the transmittance of the system is given by

$$\tau = \tau_0 \sin^2 (\phi/2),$$
(14.188)

where τ_0 is the system transmittance observed when P_1 and P_2 are aligned with their axes parallel and we have neglected scattering of light and any residual transmittance with the polarizers crossed. Kerr shutters have been built with time constants of only a few picoseconds [50*].

As a second illustration, consider a Pockels cell light modulator using a KDP crystal subjected to an electric field applied along the optic axis (z). As discussed in detail in Section 14.3.1.4, this results in birefringence experienced by light passing parallel to this axis. An arrangement utilizing this effect is illustrated in Figure 14.21. Here the modulating field is

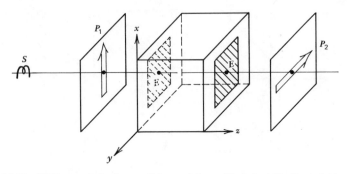

Figure 14.21 KDP crystal used as a light modulator (Pockels cell). P_1 and P_2, crossed polarizers; E, electrodes.

applied parallel to the wave propagation vector of the light to be modulated. This can be done by means of transparent electrodes, as indicated at E in the figure. The change in the index is given by (14.172) and the phase change by (14.173):

$$\phi = n_0^3 r_{63} E_z \, dk = n_0^3 r_{63} V k. \tag{14.173a}$$

Here a change in voltage

$$V_\pi = \frac{\pi}{n_0^3 r_{63} k} = \frac{\lambda}{2 n_0^3 r_{63}} \tag{14.189}$$

will yield a phase change of π and hence will bring the transmittance from near zero to maximum. When the spacing (w) between the electrodes differs from the light path length (d) through the stressed crystal, we must multiply (14.189) by (w/d). Equation 14.189 is also readily generalized to any desired phase shift, ϕ, by multiplication by (ϕ/π). Thus we may write

$$V(\phi) = \frac{w\lambda\phi}{2\pi n_0^3 r_{63} \, d}. \tag{14.189a}$$

14.3.6.2 Energy Considerations. The high field levels required for modulators of the type just described, imply that a high energy density must be attained in the electrooptic crystal. This density is given by (2.55) and may be written:

$$u = \tfrac{1}{2}\varepsilon_0 \varepsilon_k E^2 \cos \beta = \varepsilon_0 \varepsilon_k V^2 \frac{\cos \beta}{2w^2}, \tag{14.190}$$

where ε_k is the dielectric constant of the material and
β is the angle between the electric field ($\mathbf{E}$) and flux ($\mathbf{D}$) vectors.

For an electrode area, A, the energy stored in the electrooptic material will be

$$W = uwA. \tag{14.191}$$

When an alternating voltage is applied at radian frequency ω, the power dissipated in the medium will be [see (7.31)]

$$P = \frac{uwA\omega}{Q}, \tag{14.192}$$

where the Q factor is indicated by the desired bandwidth, $\Delta\omega$, according to (7.35):

$$Q = \frac{2\omega}{\Delta\omega}. \tag{14.193}$$

On substituting (14.189a), (14.190) and (14.193) into (14.192), we find the power dissipation to be

$$P = \varepsilon_0 \varepsilon_k V^2 \cos \beta \frac{A \, \Delta\omega}{4w} = \frac{\varepsilon_0}{16\pi^2} \cdot \lambda_0^2 \cdot \frac{\varepsilon_k \cos \beta}{n^6 r^2} \cdot \frac{wA}{d^2} \cdot \phi^2 \, \Delta\omega, \quad (14.194)$$

where the first factor is a constant, the second is set by the optical wavelength, the third by the material, the fourth by the geometry of the system, and the last by the modulation and bandwidth desired.

14.3.6.3 *Geometrical Considerations.* We see from (14.194) that, in order to reduce the power dissipation, we must minimize the cross-sectional area (A) and maximize the optical pathlength (d).

Assuming that we wish to modulate a laser beam with a Gaussian cross section, we have from (19.41) that the maximum attainable pathlength is

$$d_{mx} = \frac{\pi W^2}{4\lambda} = \frac{\pi n W^2}{4\lambda_0}, \quad (14.195)$$

if the beam is to be confined to a width, W. Here λ_0 is the *in vacuo* wavelength of the light to be confined.

Since the width W is defined at the point where the beam has dropped to $1/e$ of its peak luminance, considerable flux falls beyond this point. For this reason, and to provide room for mechanical tolerances, it may be desirable to provide an electrooptic crystal of a diameter somewhat larger than this, say by a factor C. This leads to an electrode area

$$A = C^2 W^2 = \frac{4dC^2\lambda_0}{\pi n}, \quad (14.196)$$

for a longitudinally applied field. Assuming $w = d$, the minimum value for the geometrical factor in (14.194) is then

$$\left(\frac{wA}{d^2} \right)_{mn} = \frac{4C^2\lambda_0}{\pi n}. \quad (14.197)$$

If the field is applied transversely,

$$A = CWd \quad (14.198)$$

and the electrode spacing is

$$w = CW. \quad (14.199)$$

Hence the geometrical factor is again as given by (14.197). We may therefore write as a general expression for the minimum power dissipation

$$P_{mn} = \frac{\varepsilon_0}{2\pi^2} \lambda^3 \left(\frac{\varepsilon_k \cos \beta}{n^7 r^2} \right) C^2 \phi^2 \Delta \nu_t, \qquad (14.200)$$

where

$$\Delta \nu_t = \frac{\Delta \omega}{2\pi}$$

is the required bandwidth in hertz.

To overcome the limitation implied by (14.200), we must find an alternative to beam confinement by focusing. Confinement within a light guiding structure, as discussed in Section 13.1.5.4, is such an alternative. For a guide with dimensions $w \times W$, the electrode area will be as in (14.198) and the geometrical factor becomes

$$\frac{wA}{d^2} = \frac{LCwW}{d} = \frac{C^2 A_g}{d}, \qquad (14.201)$$

which can be made arbitrarily small by increasing d. Here A_g is the cross-sectional area of the beam in the guide.

14.3.6.4 Fabry–Perot Modulator. Another method for reducing the energy requirements is based on the Fabry–Perot interferometer. Here the waves bounce back and forth as they gradually leak out of the cavity. The pathlength of the flux in such a cavity, measured to the point where it has decayed to $1/e$ of its initial value, is approximately

$$d_{eff} = \frac{-d}{\log \rho} \approx \frac{d}{1-\rho}, \qquad 1-\rho \ll 1, \qquad (14.202)$$

where ρ is the reflectivity of the mirrors making up the cavity. With this arrangement, however, care must be taken that both carrier and sidebands of the modulated wave fall into the Fabry–Perot interferometer passband as given by (2.220) and (2.221).

14.3.6.5 Traveling Wave Modulator [37]. To overcome the restriction of (14.187), the modulating signal can be applied transversely in the form of a wave traveling along the electrodes with a velocity equal to the group velocity of the optical signal propagating through the modulator. The optical wave will then experience a constant refractive index as it passes through the modulator and much higher modulating frequencies are possible.

In materials whose polarization is primarily electronic,[17] $n \approx \sqrt{\varepsilon}$, and therefore the synchronization of the two waves can readily be approximated by constructing the waveguide around the electrooptic material. In most electrooptic materials, however, the lattice contribution is significant, so that at the signal frequencies $n < \sqrt{\varepsilon}$. The desired synchronization may, however, still be achieved by including a substantial area filled with air in the electronic waveguide cross section.

Instead of speeding up the electrical modulating wave, the modulated optical wave may be slowed down by letting it propagate obliquely to the waveguide axis, so that it progresses through the material along a zig-zag path.

14.3.7 Imaging Devices and Spatial Light Modulators

By applying the electric field locally, the effective transmittance of an electrooptic plate (sandwiched between polarizers) can be modulated spatially, so that such a plate can be used to present an optical image. Such devices are occasionally called *spatial light modulators.* In ferroelectric materials there is also an obvious storage possibility; but even in other electrooptic plates it may be possible to store, on the plate surfaces, electric charges inducing an electric field and, hence, birefringence. Both mono- and polycrystalline materials have been used in such imaging devices.

If the electrooptic crystal is photoconductive, the charge distribution may be obtained by applying a voltage to electrodes sandwiching the crystal and separated from it by thin insulating layers. The irradiation induces free carriers that will, partially neutralize the applied field.

Alternatively, the charge distribution may be generated by covering the crystal with a layer of photoconductive material whose surface is initially covered with a uniform charge. Exposure of the image causes a selective leakage of the charge through the photoconductor and onto the crystal surface, generating the required field.

Potentially even more interesting is the phenomenon called "optical damage" or "photorefraction." Here the electrons excited in the illuminated region diffuse toward the unilluminated regions, there to be trapped. These trapped electrons create an electric field across the illuminated region and, via the electrooptic effect, a change in the refractive index.

14.3.7.1 *Single-Crystal Devices.* Devices based on the inherent photoconductivity have been made of single crystals of ZnS, ZnSe, and

---

[17] Note that n is the square root of the dielectric constant [see (2.39)] *at the optical frequency.* When the polarization is electronic, it remains constant up to very high frequencies.

$Bi_{12}SiO_{20}$ in sizes up to 25 mm diameter, and have been dubbed PROM (*Pockels readout optical modulator*). The resolution is typically 100 mm^{-1}, and frame rates in the kilohertz range are possible [51], [51*].

The "Titus" (*tube image à transparence variable spatio temporelle*) is an image-forming cathode ray tube, producing a transparent image from an electric video signal input. "Phototitus" is a similar device, in which an optical image input is converted into the transparency, with a photoconductive layer substituting for the scanning electron beam. In both, the heart of the device is a deuterated KDP crystal, cut normal to its z axis and operated near its Curie temperature, at approximately $-53°C$. Operation at this point, takes advantage of the very high electrooptic sensitivity there, so that a voltage of only 150 V is required for full modulation, when a reflecting electrode is used.

Near the Curie temperature, the normal dielectric constant is in excess of 600, whereas the transverse one is only one-tenth of this value. This permits a resolution well in excess of the plate thickness. The crystal is maintained at this low temperature by means of a Peltier effect cooler [52].

Experimental models of this type of device have been built with a resolution of 3000 elements across the image and with luminance ratios as high as 1000:1 [52*].

Clearly, a photoemissive detector can be used in place of the photoconductor. If the photoemissive current is amplified by means of a microchannel plate (cf. Section 16.4.2.3), a highly sensitive spatial light modulator may be obtained [52**].

14.3.7.2 Polycrystalline Devices [41].

Polycrystalline materials have been used as image-forming devices in a manner similar to "phototitus" just described, but at room temperature.

A representative device consists of a plate of PLZT ferroelectric material coated on both sides with a layer of polyvinyl carbazole (PVK), each of these layers, in turn, being topped with a transparent electrode. The panel is initially poled parallel to its surface and the transparent electrodes are charged by means of a voltage applied across them. As in "phototitus," the voltage appears primarily across the photoconductive layers, except in areas exposed to light, where it appears across the ferroelectric and induces poling normal to the plate. In fine-grained materials, this causes the plate's birefringence to disappear, at least partly, in the illuminated region. With the plate placed between polarizers, this can cause the illuminated area to turn light or dark, depending on the relative orientation of the polarizer axes. Such a device has been called "ferpic" [41]. In a coarse-grained plate (called "cerampic" [53]) the normal poling reduces the scattering in the illuminated region, causing it to appear clear when viewed through an imaging system with a limited aperture.

To erase the image and prepare the plate for reuse, a field may be applied parallel to the panel surface by means of closely spaced narrow electrode strips, interdigitally placed on the PLZT plate as shown in Figure 14.22. Alternatively, the desired poling can be induced by tensile or compressive stress, applied, for instance, by bending the plate slightly. The resulting strain causes poling in the direction of the applied stress.

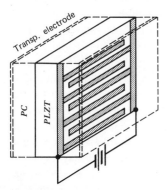

Figure 14.22   Ferroelectric image panel.

14.3.7.3   Photorefraction [53*]. Ferroelectric materials, which are both photoconductive and electrooptically sensitive, may exhibit *photorefraction*. This is due to the redistribution of electrons photoconductively excited and subsequently trapped. The resulting internal electric fields locally produce birefringence. Such materials can be used as three-dimensional, reversible optical storage media, potentially capable of very high resolution and reasonably high sensitivity. The major materials used presently are lithium and strontium-barium niobates ($LiNbO_3$ and "SBN"), barium titanate ($BaTiO_3$), and the polycrystalline PLZT discussed in Section 14.3.1.3.

Erasure is effected by illuminating the sample uniformly. To avoid erasure by the read-out light, the latter must be of a wavelength to which the photoconductor is insensitive. Alternatively, the charge pattern can be fixed in the crystal by thermally activated ion conductivity, transforming the electronic charge distribution into a more stable ion charge pattern. In some materials, field-assisted reversal of ferroelectric domains can be used to accomplish fixing.

These materials have been used widely as holographic recording media.

14.3.8 Other Applications

Here we describe very briefly additional applications. Space restrictions force us to limit the discussion and the interested reader is referred to the cited literature for further details.

Scanning and Switching. In Section 14.3.6.1 we described optical systems in which the application of a voltage V_π will turn the direction of polarization of the transmitted optical flux through 90°. In conjunction with polarizing prisms of the type discussed in Section 8.3.5, such a switching arrangement can be used to turn a beam through 90°—or any other desired angle. Digital light beam scanners can be constructed on this principle [54].

Deflection of light beams by means of diffraction gratings, electrooptically induced in crystals, were discussed in Section 13.1.5.4(5).

Frequency Shifting. The Pockels effect has also been used to shift the frequency of monochromatic flux. In one system [55] a rotating electric field is applied to a crystal having a threefold z axis, with the axis of rotation of the electric field parallel to that crystal axis. The field distorts the initially circular cross section of the index ellipsoid, so that light traversing the crystal parallel to the z axis experiences a rotating ellipsoid. If this light is circularly polarized, it will experience a shift of frequency equal to the rotational frequency of the field; the shift will be up or down, depending on the relative sense of the rotations of the modulating field and the electromagnetic field vector.

In an alternative technique, the system parameters are arranged so that the light pulses traversing the crystal experience a constantly rising or falling refractive index. This is equivalent to an optical pathlength changing in time and, hence, introduces a Doppler shift, down or up, respectively, in frequency [56].

Both these modulation techniques permit the frequency shift to be varied continuously and to obtain theoretically close to full conversion!

Pulse Compression. In conclusion, we note an ingenious laser pulse compression technique. The pulse is electrooptically modulated with a linearly varying frequency ("chirp") and then passed through an appropriately dispersive element, which delays the leading edge by an amount approximately equal to the pulse width [57].

14.4 MAGNETOOPTIC EFFECTS

The presence of a magnetic field may also affect the optical processes in a medium. In this section we discuss individually the major magnetooptic effects, pointing out parallels to the analogous electrooptic effects, as we go along.

14.4.1   Zeeman Effect [44b], [58]

When a magnetic field is applied to an excited gas, the spectral lines are split up in a characteristic manner. This is called the *Zeeman*[18] *effect.* Its origin can be explained as follows.

Any elliptical orbital motion of an electron can be resolved into three orthogonal components, each of which is a simple harmonic. We take the z component parallel to the applied magnetic field, and resolve the resultant of the other linear components into *two* circular components, noting that a linear simple harmonic motion can be considered the resultant of two oppositely directed circular motions (see Figure 2.9). The application of the magnetic field raises the frequency of one of these circular components (the one obeying the right-hand rule of motor action) and slows down the other component. Consequently, when the electrical discharge is viewed parallel to the applied field, its spectrum will be observed to consist of two lines, symmetrically displaced, on the two sides of the original line. These spectral lines will be observed to consist of circularly polarized light, the sense of the polarization differing in the two lines. The observed change in optical frequency can be shown to be[19]

$$\Delta\omega = \frac{q_e B}{2m_e} = 87.94 \times 10^9 \ B \ \text{rad/sec} \qquad (14.207)$$

$$\Delta\nu_t = 13.996 \times 10^9 \ B \ \text{Hz},$$

[18] P. Zeeman (1896).

[19] For a particle moving in a circular orbit, radius, r, with angular frequency ω_0, the acceleration is $(\omega_0^2 r)$ so that the required centripetal force is

$$F_0 = m\omega_0^2 r. \qquad (14.203)$$

When a magnetic field of flux density B is now applied normal to the plane of the orbit, the particle will experience a total force:

$$F = F_0 \pm qBv = m\omega_0^2 r \pm qB\omega_0 r, \qquad (14.204)$$

if it carries a charge q. According to (14.203), this corresponds to a new angular frequency

$$\omega^2 = \frac{F}{mr} = \omega_0^2 \pm \frac{qB\omega_0}{m}. \qquad (14.205)$$

Hence

$$\omega^2 - \omega_0^2 = (\omega + \omega_0)(\omega - \omega_0) = \pm \frac{qB\omega_0}{m}$$

and the corresponding frequency change

$$\Delta\omega = \omega - \omega_0 = \frac{qB}{m(\omega/\omega_0 + 1)} \approx \frac{qB}{2m}. \qquad (14.206)$$

This is called the Larmor precession frequency (J. Larmor, 1897).

where $q_e/m_e = 1.7588 \times 10^{11}$ C/kg is the ratio of the electronic charge to its mass and

B = the flux density of the magnetic field (in tesla, i.e. Wb/m^2).

The z component is unaffected by the magnetic field. It will not radiate in the z direction and will, therefore, not be visible when viewed in that direction.

When the discharge is viewed normal to the direction of the field, the two shifted lines will again appear, but will now be linearly polarized normal to the applied field since the rotating vectors are viewed parallel to their plane of rotation. The unaffected z component will appear at the original location of the line and will be polarized parallel to the direction of the field.

In practice, the observed line splitting is often more complicated due to quadratic and quantum effects.

The *inverse Zeeman effect* refers to the resonance absorption lines exhibited by nonexcited gases which are also shifted according to (14.207) when a magnetic field is applied.

The Zeeman effect is the magnetic analog to the electrooptic Stark effect.

14.4.2 Faraday and Magnetooptic Kerr Effects

14.4.2.1 Faraday Effect [42b]. When plane polarized light passes through a substance subjected to a magnetic field, its plane of polarization is observed to be rotated by an amount proportional to the magnetic field component parallel to the direction of the light propagation. This is known as the Faraday[20] effect and is closely related to the axially observed Zeeman effect. The electromagnetic wave passing through the medium can again be decomposed into two circularly polarized components, of opposite sense. Each of these will set up corresponding oscillations in the electron clouds of the medium. However, as discussed in the preceding section, these oscillations will be oppositely affected by the presence of the magnetic field. Hence the two components will experience different refractive indices—n_r and n_l for the right and left circularly polarized components, respectively. As explained in our discussion of optical activity (Section 2.2.3.2), this results in a rotation of the polarization vector as the wave propagates through the medium.

There is, however, an essential difference between the Faraday effect and optical activity. In the former, the rotation is independent of the sense of the light passage, so that, if the light is reflected at the end of its passage and passes through the medium once more, it will be rotated

[20] M. Faraday (1845).

through twice the single-passage rotation. In optical activity, on the other hand, the sense of the rotation is fixed relative to the direction of light propagation, so that if it is reflected back through the medium, no net rotation will remain, much as the rotation of a screw as it passes through a nut will be canceled out by the rotation required to extract it from that nut.

This asymmetry makes it possible to build a truly one-way optical channel. If light passes successively through an optically active plate and a Faraday cell, each with a $\pi/4$ rotation, the two rotations will cancel in one direction and add to a 90° rotation for light passing in the opposite sense. If this combination is placed between parallel or crossed polarizers, it will act as a one-way channel [59].

14.4.2.2 Faraday Effect in Nonmagnetic Materials. In para- and diamagnetic materials, the permeability differs but little from that of free space (μ_0) so that the product of μ_0 with the magnetic intensity (H) may be substituted for the magnetic flux density. The rotation of the plane of polarization is then given by

$$\theta = VBd \approx \mu_0 VHd \qquad (14.208)$$

where B is the magnetic flux component parallel to the light propagation,
  d is the length of the light passage through the stressed medium, and
  V is Verdet's constant, specific to the material.

See Table 153 [44a], [60] for a listing of Verdet's constants and their variation with wavelength for number of liquids and solids.
  In terms of the refractive indices n_r and n_l.

$$\theta = \frac{k_0[n_r^{(R)} - n_l^{(R)}]d}{2}, \qquad (14.209)$$

where k_0 is the *in vacuo* propagation constant of the light and the superscripts indicate that only the real component of the index is to be taken.

14.4.2.3 Faraday Effect in Ferromagnetic Materials [61]. In ferromagnetic materials the relative permeability,

$$\mu^* = \frac{\mu}{\mu_0}, \qquad (14.210)$$

is much larger than unity and must be considered. In such materials

$$B = \mu H = \mu_0(H + M) \approx \mu_0 M, \qquad (14.211)$$

where M is the magnetization (magnetic dipole moment per unit volume, analogous to the electric polarization, P, in a material subjected to an electric field). The permeability is here constant for $M \leq M_s$, the saturation magnetization. In a saturated material, we may approximate

$$B \approx \mu_0 M_s \tag{14.212}$$

and (14.208) takes the form

$$\theta = \mu_0 V M d$$
$$= K'Md \leq K'M_s d, \tag{14.213}$$

where K' is Kundt's constant.

Values of the saturation magnetization (M_s) and the specific saturation rotation,

$$F_s = K'M_s, \tag{14.214}$$

for some ferromagnetic materials are given in Table 154 [61].

As implied by (14.209), the specific rotation is given by

$$F = \frac{k_0[n_r^{(R)} - n_l^{(R)}]}{2}. \tag{14.215}$$

14.4.2.4 Kerr Rotation. When plane polarized light is incident on the surface of a material subjected to a magnetic field, the reflected beam is found to have its plane of polarization rotated, provided that the wave propagation vector has a component parallel to the field. This is known as *Kerr rotation* or as the magnetooptic Kerr effect (to distinguish it from the electrooptic Kerr effect discussed in Section 14.3.2). It is readily understood as the reflective equivalent of the Faraday effect. The phenomenon of reflection can be explained as the result of electron currents set up in the material by the incident electromagnetic wave (see Section 2.2.1). Under the influence of the magnetic field, these currents experience Larmor precession (14.207) and this, in turn, causes the observed rotation of the plane of polarization.

For normally incident light, the Kerr rotation is found to be given by

$$\theta_K = -\mathcal{I}\left[\frac{n_r - n_l}{n_r n_l - 1}\right], \tag{14.216}$$

where $\mathcal{I}$ indicates that the imaginary part is to be taken.

In practice, the largest values of Kerr rotation are obtained from ferromagnetic materials at saturation magnetization. With such materials the rotation rarely exceeds 1° in uncoated materials. Note, however, that coating can increase θ_K significantly by reducing the amount of unrotated light reflected from the surface [61].

14.4.3   Cotton-Mouton and Voigt Effects [41c]

When a fluid is subjected to a magnetic field, it is found to become birefringent with the optic axis parallel to the magnetic field. This phenomenon is called Cotton-Mouton[21] effect and is related to the transverse Zeeman effect, just as the Faraday effect is related to the axial Zeeman effect. It is of second order only, compared to the Faraday effect, generally much smaller and is proportional to the square of the magnetic flux.

The phase difference introduced by the passage through the stressed medium is given by

$$\phi = (n_e - n_0)kd = CH^2d, \qquad (14.217)$$

where C is the Cotton-Mouton constant. Representative values of this are given below [44a]:

	Temperature, C	
Liquid	°C	$(T^2m)^{-1}$
Acetone	20.2	0.0376
Benzene	26.5	0.0075
Chloroform	17.2	−0.0658
Carbon disulfide	28.0	−0.004

These were measured at 0.58 μm.

In a gas near an absorption line, a first-order magnetooptic birefringence can be observed. This is called the *Voigt effect*.

14.4.4   Summary of Electro- and Magnetooptic Effects

For convenient reference the electro- and magnetooptic effects are summarized in Table 155. The table is arranged to emphasize the analogies between them. An extensive bibliography on the magnetooptic effects has been published [62].

[21] A. Cotton and H. Mouton (1907).

14.4.5 Computer Memories [61] and Other Applications of Magnetooptics

Because electric fields are generally easier to generate than magnetic fields, electrooptic are popularly preferred over magnetooptic devices, except in the investigations of the structure of materials and in applications requiring spatial modulation, where the domain structure of ferromagnetic materials offers a pronounced advantage. Such applications are the subject of this section. A magnetooptic modulator for use in integrated optics was described in Section 13.1.5.4.

14.4.5.1 Advantages. At this time, the magnetooptics application receiving most attention appears to be in large capacity computer memories. As computer memories expand, the search for more compact and cheaper memories becomes more urgent. Such memories should not only be capable of storing very large amounts of information in a relatively small area, but should permit very rapid read-out of the information, as well—serially or, preferably, with random access. The usual magnetic memories are limited in the minimum element size attainable primarily because of the mechanical clearance required between the recording head and the magnetic material to be switched. The high speed at which the magnetic material must be passed under the pickup head puts a lower limit on this clearance and, hence, also on the element size. In addition, the required mechanical motion limits the speed at which a large memory can be read out.

In both these areas, optical techniques can overcome the limitation, and hence the research into magnetooptic memories. In these the element size could be of the order of magnitude of the wavelength of the light used and, because beam deflection is essentially inertialess, access speeds, in principle, could approach the speed of light.

14.4.5.2 Reading. The magnetooptic memories developed so far seem to be based on the Faraday or Kerr effects, as observed in a magnetized ferro- or ferrimagnetic material. To read out the information, linearly polarized light is passed through (or reflected from) the magnetized film and then split into two and passed through a pair of analyzers. These are followed by detectors; comparison of their outputs permits the detection of any rotation of the polarization direction caused by the magnetized film.

14.4.5.3 Writing. In films initially magnetized in one direction, writing consists of reversing the direction of polarization. This may be accomplished by raising the temperature of the element above the Curie point, usually by means of illumination with a pulse of laser radiation. The local rise in temperature will destroy the magnetization at the illuminated area element and, on cooling, it will acquire a reversed magnetization under the influence of the closure flux exerted by the neighboring elements. The write beam must be significantly more intense than the read beam, since the latter must not raise the element to the Curie temperature.

Note that writing need not necessarily be done at the Curie temperature. The magnetic coercivity of some materials varies strongly with temperature, even below the Curie point, so that writing can be done in conjunction with an auxiliary magnetic field at lower temperatures.

Experimentally determined irradiation levels required for reliable writing are shown in Figure 14.23 [63] for a number of materials. All materials at present considered for magnetooptic memories require more than 10^{-10} J/bit writing energy [61].

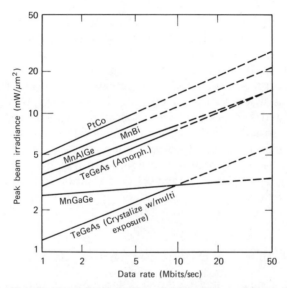

Figure 14.23   Measured beam irradiation for reliable writing as a function of the data rate. The dotted extensions of the lines are extrapolations of doubtful validity. Brown [63].

14.4.5.4   Erasure.
Erasure consists of restoring to the element the original direction of magnetization. This can be accomplished by heating it above the Curie temperature and simultaneously applying an external magnetic field returning the element to the original polarity.

14.4.5.5   Error Rate.
If we assume the noise to have a Gaussian distribution, a sufficiently low error rate ε, requires a signal-to-noise ratio,

R, such that

$$\tfrac{1}{2}\text{erfc}\,\frac{R}{\sqrt{2}}, \qquad \text{erfc}\,\frac{R}{\sqrt{2}} \leqslant \varepsilon \tag{14.218}$$

for binary and multilevel signals, respectively. If we take[22] $\varepsilon = 10^{-14}$, we find the required signal-to-noise ratios to be 7.65 and 7.74, respectively.

The signal-to-noise ratio is primarily controlled by the quantum noise that varies with the square root of the signal level. This implies a lower limit to the acceptable signal energy. At the high signal rates considered here, high flux levels are required and the maximization of the signal level becomes one of the major design parameters, which dictates, among others, the film thickness of the magnetic material. For high light transmittance, the film must be thin; on the other hand, the amount of Faraday rotation is directly proportional to the film thickness, and the signal actually detected varies with this rotation. Hence

$$M_F = \frac{F}{\alpha} \tag{14.219}$$

becomes a useful merit function for the magnetic material used in transmission. Here F is the Faraday rotation (at saturation) per unit thickness and α is the extinction coefficient. M_F may be interpreted as the rotation obtainable from a film with a thickness such that it attenuates the transmitted flux by a factor of $1/e$. Values of M_F are included in Table 154.

In addition to the signal quantum noise just discussed, there are other important noise sources that must be considered [64].

1. The optical flux reaching the detectors is only partly signal flux. Background flux is due to the spreading of the read-out spot beyond the memory element, depolarization of the light due to scattering effects, and stray light. All of these contribute to the quantum noise.

2. Nonuniformities in the film surface and film granularity cause noise effects, too.

3. The noise contributed by the detectors and the associated circuitry must be considered, as well.

14.4.5.6 System Considerations. Magnetooptic memories can be constructed either for direct reading and writing or for holographic access. In direct access memories a reciprocal relationship exists between the data

---

[22] This figure is based on one error per year, assuming three readings each microsecond, noting that there are 3.1536×10^7 sec in an ordinary calendar year.

The noise amplitude is here taken as its standard deviation and the signal level as half the interval between successive reference levels.

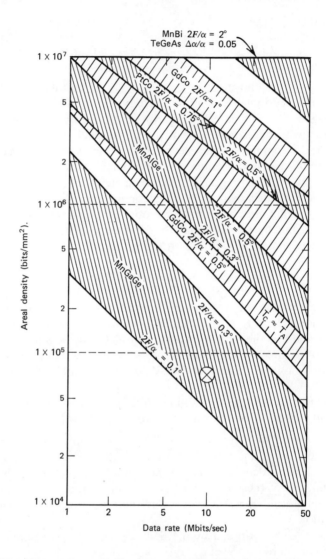

Figure 14.24 Bit density versus data rate for various magnetooptic films, based on a 17-dB read-out signal-to-noise ratio. To cover various material compositions, the bands are shown broad; note especially GdCo, whose band straddles those of PtCo and MnAlGe, because of the wide range of read-out beam irradiation accepted by various compositions of this material. On the basis of experimental evidence, it is assumed that the read-out irradiation levels are one-third those of the writing levels shown in the preceding figure. X indicates an experimentally verified detection-limited point. Brown [63].

314

rate and the storage density. For a given signal-to-noise ratio, as dictated by the required limit on the error rate, a given amount of energy must reach the detector. The higher the data rate, therefore, the higher must be the power level of the read-out beam. On the other hand, the irradiation of the memory element must be maintained well below the writing and erasure thresholds. Irradiation, i.e., power per unit area, being thus limited, the element area must be increased with the required power level. The increased element area obviously implies a reduced element density. See the shaded bands in Figure 14.24 [63] for the dependence of storage density on data rates for six magnetooptic materials. The irradiation level permissible for read-out is a major factor determining this relationship and this, in turn, is strongly influenced by both detailed composition (hence the broad bands in the figure) and by the thermal characteristics of the substrate. All the materials represented in Figure 14.24 are assumed to be deposited on a silica glass substrate, and MnBi and GdCo to be overcoated with SiO to prevent oxidation. In interpreting the data of this figure, note that areal densities above 10^6 mm^{-2} may not be practical for optical reasons. Thus, for a number of materials, performance is limited by purely optical and electronic considerations, rather than material and thermal characteristics. Despite their potential significance, noise sources, other than signal quantum noise, have been neglected in calculating the data for Figure 14.24. In MnGaGe and MnAlGe, for instance, the granularity limits the storage density to approximately 10^5 mm^{-2}. Lack of domain stability similarly limits the storage density in GdCo [63]. Granularity data are included in Table 154.

To overcome some of the granularity limitations, data can be stored holographically, as described briefly in Section 19.3.6.2, on the magnetooptic films. Here a large block of data is read out at one time and therefore the noise problems are reduced. However, the laser power levels required (ca. 10^3 W) in conjunction with the pulse generation requirements (10 nsec pulses at megahertz frequencies) pose serious technical problems.

Pertinent data of magnetooptic materials used in data recording are listed in Table 154.

14.5 LIQUID CRYSTALS IN OPTICS

14.5.1 General Description of Liquid Crystals [65], [66]

14.5.1.1 Phases

Classifications. The liquid crystal state is a mesophase, a state intermediate between crystal and liquid. Many organic substances pass through the liquid crystal phase in making the transition from crystal to liquid. While in this phase, often over a considerable temperature range, they are liquid in the usual sense, but exhibit extensive ordering of their molecules, giving them optical properties usually found only in crystals.

Liquid crystals formed when a crystal is melted, are called *thermotropic*; those formed as a crystal is dissolved, are called *lyotropic*. Primarily the former are used in optical applications, and only they are discussed here.

All liquid crystals consist of elongated molecules that tend to line up with their long axes parallel. When this is the only regularity evident, the phase is described as *nematic* (threadlike). See Figure 14.25a. In these crystals, the crystallinity expresses itself only in terms of the orientation of the molecules, and their centers of gravity do not conform to any lattice.

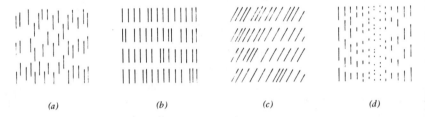

(a)   (b)   (c)   (d)

Figure 14.25   Molecular alignment patterns in liquid crystals. (a) Nematic; (b) smectic A: (c) smectic C; (d) cholesteric.

When the molecules are not only parallel to each other, but are also arranged in layers, they are called *smectic* (greaselike); here all the molecule ends form a set of parallel, equispaced surfaces, often plane. Thus, in addition to exhibiting orientational ordering, such crystals form a regular lattice in one dimension. See Figures 14.25b and c for two illustrations.

Smectic crystals are further classified, according to the ordering within the layers, as Types A–H. In Type A the molecules are arranged with their long axes normal to the layer surface and in Type C, inclined to this surface, but still parallel to each other. In neither of these is there any ordering within each layer. In Type B, domains form within each layer, giving the crystal the appearance of a mosaic. Other types, denoted D to H have been identified, but this division does not seem to be fully established.

Nematic crystals consisting of optically active substances, that is those whose molecules do not exhibit a mirror symmetry plane, exhibit a further peculiarity. In these, the molecules are still lined up with their axes parallel to a fixed plane and, within any one layer parallel to that plane, they also have a common direction in azimuth; but this common direction varies continuously as we progress from layer to layer: a unit vector parallel to the direction of the alignment describes a helix as it is displaced normal to the plane of the layer. See Figures 14.25*d* and 14.26. This phase is called *cholesteric*, because it is most often observed in cholesterol derivatives. It is sometimes considered to be a third form, distinct from the nematic [65*a*].

Figure 14.26   Illustrating molecular alignment in a cholesteric crystal.

A given substance may pass successively through both smectic and nematic or smectic and cholesteric phases as it is heated from the crystalline to the isotropic liquid state.

Detailed Structure. The pictures presented in Figure 14.25 represent great oversimplifications. In any real liquid crystal (a) the orientation of individual molecules deviates slightly from the local mean value and (b) over substantial distances, the local mean can vary appreciably. According to the "swarm" hypothesis, nematic crystals consist of domains (swarms) of approximately 10^5 molecules each, in which the local mean is constant; it varies only from swarm to swarm. These swarms were thought to be embedded in a less constrained liquid-crystal matrix in which the molecule orientation varied slowly to make the transition from the orientation of one swarm to that of its neighbor.

Due to their highly asymmetrical structure, the electrical polarizability of liquid crystals is strongly anisotropic. This gives rise to birefringence and to refractive index fluctuations accompanying the orientational variations. These, in turn, scatter incident radiation. The swarm hypothesis was formulated, in part, to account for this scattering; but continuous slow variations, too, can account for the observed scattering, provided significant changes in orientation take place over distances comparable to the wavelength [66*a*].

Viscosity. Nematic crystals are not very viscous (typically $0.01 \, N \cdot s/m^2$) although significantly more so than water (ca. $0.001 \, N \cdot s/m^2$). Smectic crystals, on the other hand, tend to be highly viscous, due to their higher degree of ordering.

14.5.1.2   Textures [65b], [66b]. Because liquid crystals consist of strongly elongated molecules and are, consequently, birefringent, their structure can effectively be observed between polarizers. When such observations are made under a microscope, various patterns, or textures, are observed.

When large solid crystals of smectic or nematic materials are formed between parallel microscope slides and then are slowly heated, the melt appears constituted of homogeneous regions corresponding to the areas of the original crystals. Within any such region, the birefringence is constant and these represent domains within which the molecules are aligned parallel to each other, much as illustrated in Figure 14.25. Also, the orientation of the long axis must be parallel to the layer, or at least partly so, to account for the observed birefringence. This texture is called *homogeneous* (see Figure 14.27*a*).

When a slight relative displacement is imparted to the slides, the liquid crystal may undergo a change such that birefringence is no longer observed and the crystal appears iso- or homeotropic. This indicates that the long axes of the molecules are now normal to the glass plates, resulting in the so-called *homeotropic* texture (see Figure 14.27*b*).

When no special precautions are taken and the crystals are formed either by heating an aggregate of crystals or by cooling the isotropic liquid, the *focal-conic* texture is usually obtained. This consists of microscopic domains. The molecule orientation is ordered within each of these, but varies randomly from domain to domain. The name of this texture is derived from the fact that fine black lines are observed in it, corresponding to the elliptical sections of cones and the hyperbolic loci of the apices of the cones compatible with the elliptical sections. (The locus of the apices passes through a focus of the ellipse.) These conic sections are due to the formation, within the crystal, of domains in which molecules follow conical arrangements.[23]

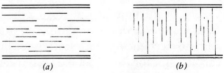

(a)   (b)

Figure 14.27   (*a*) Homogeneous and (*b*) homeotropic molecular alignment patterns in liquid crystal cells.

[23] Space limitations permit no more than this rather incomplete description. Alternatively the arrangement of the molecules could have been described as doughnutlike, with the doughnut generally both elliptical and of uneven thickness around its circumference. See Ref. 65*c* and especially Ref. 66*c* for a picturesque illustration.

The cholesteric crystals, too, tend to form a focal-conic texture. But, on being disturbed, they tend to go over into the *Grandjean*[24] planar texture, in which the axis of the helix is normal to the glass plates, so that the region of parallel molecules extends throughout the layer, instead of being broken up into domains, as in the focal-conic texture, where the helix axes are parallel to the layer and, hence, may vary in azimuthal orientation.

14.5.1.3 Surface Effects [67]. So far, the absolute direction of the molecule axes has been left undetermined; only the relative orientation is fixed by intermolecular forces. The surfaces of the container may be used to fix this absolute direction, especially in liquid crystal cells, thin layers of liquid crystals sandwiched between flat plates. The surfaces of these plates tend to favor one alignment direction over all others, depending on their physical and molecular structure. In general terms, a solid surface that is readily wetted by the liquid crystal (critical surface tension of the solid greater than the surface tension of the liquid) will tend to produce an alignment of molecules axes parallel to the surface, that is, a homogeneous alignment (see Figure 14.27a). A surface with poor wetting characteristics will tend to induce a molecule alignment perpendicular to the surface, that is, homeotropic (see Figure 14.27b). The surface characteristics may be changed drastically by the adsorption of surfactants (active surfacing substances) that interact with the liquid crystal molecules to produce one or the other of the above arrangements. Lecithin seems to be a popular coating material for inducing homeotropic alignment. Such materials may also be supplied as additives to the liquid crystal, from which they are then selectively adsorbed at the surface. Some liquid crystals, such as MBBA (see Section 14.5.1.5) may themselves form a monomolecular adsorption layer on the solid, so that homeotropic alignment results even on surfaces that have a high critical surface tension. Note that the ambient relative humidity may have a substantial influence on the critical surface tension. For instance, the critical surface tension of soda-lime glass drops from about 0.075 N/m to 0.038 N/m as the relative humidity rises from 0.001 to 10%.

Surface irregularities may also influence the molecule alignment. Rubbing the contact surface in one direction, is found to induce homogeneous alignment in that direction. This appears to be due to microscopic grooves produced in the surface. If the two plates of the nematic liquid crystal cell are grooved in directions perpendicular to each other, this will induce, across the cell, a quarter turn twist in the molecule axis alignment, so that the material appears similar to a cholesteric crystal, but with accurately controlled pitch.

[24] F. Grandjean (1921).

14.5.1.4   Optical Characteristics

Transparency. In thin layers (tens of micrometers thick) liquid crystals are transparent when undisturbed, because of the relatively large optically homogeneous regions. When the structure is disturbed, however, as by an electric current passing through the crystal, the dependence of the refractive index on molecule orientation induces optical inhomogeneity; the liquid scatters light and becomes effectively opaque.

Birefringence. Liquid crystals act as uniaxial crystals with the optic axis parallel to the long axis of the molecule. They are strongly birefringent to light passing them in a direction normal to the long axis.

Optical activity. Cholesteric crystals are optically highly active and may rotate the plane of polarization of incident light through thousands of radians per millimeter thickness. The same effect may be obtained in nematic crystals. We noted in the preceding section that a twist may be introduced into a layer of ordinary nematic crystals by using "rubbed" surface plates, rotated relative to each other, to bound the layer. Such a *twisted nematic crystal* will exhibit optical activity similar to that of a cholesteric crystal.

Circular dichroicism. Over certain narrow spectral regions, cholesteric crystals are circularly dichroic, that is, they are transparent to light circularly polarized in one sense and scatter light circularly polarized in the opposite sense.

Diffraction. The periodicity of its helical structure gives cholesteric crystals some of the characteristics of a three-dimensional diffraction grating, obeying the Bragg diffraction rule approximately [see (14.117)]. Thus a layer of cholesteric crystal in the Grandjean texture may appear in a saturated color when viewed in white light. The pitch of the helical structure, here constituting the grating constant, is highly sensitive to external factors such as temperature and mechanical stress: even a small change in these can produce a large relative change in pitch and hence in the color observed.

Thin layers of such crystals, on a black backing, have been used for sensitive temperature measurements, especially when the temperature is to be mapped over an extended area.

14.5.1.5   Materials.
Many materials exhibit a liquid crystal phase; but they do not, necessarily, have the characteristics desirable in terms of applications. For instance, a device that is to be used at ambient temperature conditions without supplementary heating or cooling requires a material remaining in the desired phase over the anticipated temperature

range. In other words, the solid-liquid transition temperature must be at the lower end of the anticipated range (or below it) and the nematic-isotropic transition temperature at the upper end (or above). This requirement may pose some difficulty. However, by mixing two or more liquid crystal species, the solid-liquid transition may be lowered considerably below that of either one in the pure state; the nematic-isotropic transition temperature of the mixture tends to be intermediate to those of the components.

Table 156 [68], [69] lists some of the more popular materials including, where applicable, their usual abbreviated names. Some commercially available liquid crystals are also listed. Some cholesteric crystals, and the colors attainable from them, are listed in Table 157 [70]. Cholestryl p-nitrobenzoate is of special interest; it exhibits brilliant colors ranging from violet to red as the temperature falls. It remains in the liquid crystal phase between 186° and 258°C.

Certain electrooptic applications of liquid crystals require a given level of conductivity—very low, if currents are to be minimized, higher, if currents are desired. The conductivity of liquid crystal materials is determined, to a great extent, by the contaminants or dopants present, so that extremely high purity may be required for some applications.

14.5.2 Stress-Optical Effects in Liquid Crystals

14.5.2.1 Elastooptic Effects **[71].** Although liquid crystals do not exhibit a shear elastooptic effect, optical effects can be induced by bending, in some instances. Consider molecules that have the form of an elongated wedge and are confined to a thin film. When this film is bent, the layer nearest the center of curvature will be compressed and the layer on the opposite surface will be expanded. Consequently, there will be a tendency for the molecules to line up with their thin ends pointing toward the center of curvature. If these molecules carry an electric dipole moment parallel to their long axis, this alignment will produce an electric polarization. This effect has been called "piezoelectric" or "flexoelectric" [66d]. A similar effect can be imagined for arc-shaped (banana-shaped) molecules, which will tend to line up with the arc formed by the bent film. If they carry a transverse electric dipole moment, here too the bending will induce polarization. Because of the birefringent character of the molecules, this polarization implies certain optical effects.

14.5.2.2 Electrooptic Effects **[68], [72].** Liquid crystals exhibit many electrooptical effects, and we discuss below briefly those that appear most important in terms of technological application.

Homogeneous Birefringence. Because of the dielectric anisotropy inherent in the molecules, the orientation, and hence the birefringence, of a liquid crystal may be controlled by an externally applied electric field. Specifically, let us denote the difference in dielectric constants by

$$\Delta\varepsilon = \varepsilon_{\parallel} - \varepsilon_{\perp}, \tag{14.220}$$

where $\varepsilon_{\parallel}$, $\varepsilon_{\perp}$ are the dielectric constants parallel and perpendicular, respectively, to the long axis of the molecule. A molecule carrying an electric dipole moment parallel to its axis will exhibit a positive value of $\Delta\varepsilon$ and, in an applied electric field, will tend to line up with its axis parallel to the field. Conversely, a molecule carrying a dipole moment transverse to its long axis, will have a negative $\Delta\varepsilon$ and align itself transversely to an applied electric field. Liquid crystals are classified as positively or negatively anisotropic according to the sign of $\Delta\varepsilon$ as defined in (14.220).

Now consider a liquid crystal cell provided with transparent electrodes and placed between parallel polarizers. See Figure 14.28. If the crystal is

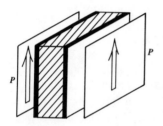

Figure 14.28 Liquid crystal cell for modulating light by tunable birefringence. The electrodes are indicated by heavy lines. The liquid crystal is shown shaded. The polarizers are indicated by P and their axes by hollow arrows.

homogeneously ordered, it will be birefringent to light passing the cell normally. If, furthermore, the cell thickness (d) is chosen to provide a half-wave retardation

$$k\,\Delta nd = \pi \tag{14.221}$$

and is oriented at 45° to the polarizer axes, essentially no light will pass. If the material has a positive $\Delta\varepsilon$ and a strong electric field is applied across the electrodes, the molecular arrangement will approach the homeotropic (except for thin layers adjacent to the containing plates) and the birefringence will no longer be observed for the normally incident light: the light will pass the sandwich. The extent of the molecular rotation will depend on the field strength applied, so that the transmittance of the sandwich can be controlled by the applied voltage.

In a material with a negative $\Delta\varepsilon$, we may start with a homeotropic ordering changed to a homogeneous one[25] when the field is applied. Such an arrangement will perform similarly to the former, if the polarizers are crossed [73]. These phenomena are referred to as *tunable birefringence* or *Freedericksz*[26] *transitions*. The last one is occasionally called "deformation of alignment phase" (DAP).

Color Modulation. To obtain maximum transmittance in the arrangement just described, the film thickness (d) must be such as to cause a retardation, which is an odd multiple of $\pi/2$, and this value $\pm\pi/4$ for half-intensity:

$$\frac{d\,\Delta n}{\lambda}\begin{cases}=m+\frac{1}{2}\\ =m+\frac{1}{4},\\ =m+\frac{3}{4}\end{cases} \quad m=0,1,2,3\ldots, \tag{14.222}$$

for the three points, where Δn is the difference in the refractive indices associated with the two directions and is controlled by the applied field. Solving (14.222) for λ, we find the wavelength for the peak transmittance to be

$$\lambda_p = \frac{2d\,\Delta n}{2m+1} \tag{14.223a}$$

and that half this intensity will be observed at

$$\lambda_1 = \frac{4d\,\Delta n}{4m+3} \tag{14.223b}$$

and

$$\lambda_2 = \frac{4d\,\Delta n}{4m+1}. \tag{14.223c}$$

If we define the passband by these half-intensity points, it will extend up and down from the peak wavelength by

$$\delta\lambda_1 = \frac{\lambda_p - \lambda_1}{\lambda_p} = \frac{1}{4m+3} \tag{14.224a}$$

$$\delta\lambda_2 = \frac{\lambda_2 - \lambda_p}{\lambda_p} = \frac{1}{(4m+1)}. \tag{14.224b}$$

[25] In order to fix the azimuth of the homogeneous texture induced, we may start with a modified homeotropic orientation: instead of being aligned exactly normal to the film surfaces, a small tilt may be induced in the molecular orientation, e.g., by surface treatment (cf. Section 14.5.1.3). The azimuth of this tilt will then fix that of the induced homogeneous texture.

[26] V. Freedericksz (1927).

For $m = 0$, this means that the passband will extend 100% above and 33% below the peak wavelength. Specifically, it covers the whole visible spectrum (say from 0.4 to 0.7 μm) for λ_p ranging from 0.35 to 0.6 μm. In the zero order, therefore, significant intensity modulation can be obtained with practically no color effects. By operating at higher voltages, $m = 2$, for example, the passband will be far narrower (-9% and $+11\%$) and the color of the transmitted light can be controlled by means of the applied field.

A display device combining simultaneous color and black and white capability has been developed based on this principle [74].

Optical Activity. Optical activity in liquid crystal devices, especially that in twisted nematics as described in Section 14.5.1.4, may be used for light modulation. As discussed there, a quarter-turn twist may be induced into the ordering of a nematic crystal by orientating the faceplates of the retaining cell with their grooves running perpendicular to each other. Now consider such a cell consisting of positively anisotropic material and placed between two parallel polarizers. When plane polarized light passes through it, the plane of polarization, too, is turned through 90° and no light will pass. If a strong electric field is now applied across the cell, the molecules will take on an essentially homeotropic ordering, the optical activity will vanish and the arrangement becomes transparent [75].

Scattering Effects. When a steady electric field is applied across a negatively anisotropic crystal, small domains form in it, somewhat analogously to the domains in ferroelectric materials. These are called *Williams domains* after their discoverer [76]. They are microscopic, elongated, and appear separated by domain walls appearing as dark lines. The pattern is somewhat reminiscent of fingerprint patterns.

At higher voltages (ca. 0.5 V/μm), the induced currents produce turbulence and the crystal becomes highly scattering. This effect is called *dynamic scattering* [77] and can be used to switch a liquid crystal cell from a clear to an opaque state. The same effect is observed under ac excitation, provided the frequency is sufficiently low (presumably the applied voltage must have a period longer than the transit time of an ion). Upon removal of the field, the crystal reverts to the ordered state spontaneously in approximately 0.1 sec. The required voltage is of the order of 3 V/μm and rises with increasing frequency. At a certain threshold frequency, the turbulence takes on a new character and decays rapidly, in a matter of milliseconds [78]. This threshold depends on the resistivity of the liquid crystal and on the cell thickness. For instance, in a 12.5-μm-thick cell, it dropped from 500 to 50 Hz as the resistivity rose from 2 to 20 GΩ-cm.

As the frequency rises further, a second threshold is reached. Above this, the applied field no longer induces turbulence. On the contrary, it tends to restore order, and even to prevent the formation of Williams domains [79].

Storage Effects [80]. Liquid crystal cells in which self-sustained opacity may be induced and removed at will are made by using a mixture of nematic and cholesteric crystals. The applied dc or low-frequency field emulsifies the mixture and the disordered crystals scatter light; they turn milky. This milky state decays very slowly and in a matter of hours the crystal clears. It may, however, be cleared rapidly, in less than a second, by the application of a field at a frequency above the second threshold.

Field-Induced Phase Transitions. As mentioned in Section 14.5.1.4, cholesteric crystals tend to be highly opaque unless the Grandjean texture is induced. Such crystals may, however, be converted into a clear nematic phase by a strong electric field. This effect offers an additional electrooptical modulation technique [81].

Dichroic Switching. Liquid crystals have also been used as orientors for dichroic molecules that could not be oriented individually by an external field. By mixing such dichroic molecules (e.g., methyl red) with a nematic crystal (e.g., *p-n*-butoxy benzoic acid), it was found that the orientation of the dichroic molecules becomes controllable by the external field, so that at 4 V/μm the mixture in a 12-μm-thick cell was switched from a reddish orange to yellow in 5 ms; 100 ms were required for the mixture to return to its original color [82].

14.5.2.3 Magnetooptic Effects [66e]. The molecules of a number of the important liquid crystals contain aromatic rings. A magnetic field normal to the plane of such a ring tends to set up a current in the ring, generating a counter magnetic field. As a result, there is a tendency for such rings to align themselves parallel to an applied magnetic field. This effect is so weak that it is generally not detectable in isotropic liquids. However, in liquid crystals its force is multiplied due to the cooperation between individual molecules that extend over large numbers, of the order of magnitude of 10^5. In such crystals, therefore, a magnetic field can be used to align the molecules with their axes parallel to the field. A 100-μm-thick crystal requires a field of approximately 1000 Oe to align it. Comparison to alignment by means of an electric field shows that 1 G is roughly equivalent to 1 V/cm [66f].

14.5.3   Liquid Crystal Optical Devices

In addition to the thermometric application mentioned in Section 14.5.1.4, the main consumer applications for liquid crystals seem to be in the area of displays. The passive liquid crystal display is superior to self-luminous displays in applications of high ambient illumination. The conversion of an opaque original to a transparency is often desirable for efficient projection or coherent optical processing. This, too, is an important potential application of liquid crystals.

The liquid crystal cell can be used in transmission or can be backed with a high-reflectance surface for use in reflection. In conjunction with a scattering mode, the backing may be an absorbing black layer upon which the scattering liquid will appear white.

For use in image-forming devices, liquid crystal films can be made to modulate the system transmittance locally using any of the electrooptic effects discussed in the preceding section. The image can be formed by means of a matrix consisting of discrete elements or it can be formed in a continuous manner. Below we discuss techniques used in the two approaches. In all cases, we deal with a liquid crystal film sandwiched between electrodes, at least one of which is transparent.

***14.5.3.1   Matrix Operated Displays* [83].** If one of the two electrodes is broken up into elements, each provided with its own electrical connection, the individual elements can be activated, and the desired display generated, at will, simply by energizing the appropriate elements.

This scheme is suitable for a numeral display, consisting of seven bars arranged in the form of a figure 8 and, perhaps, even for a 7×5 matrix by means of which alphanumeric characters may be displayed. But with all its conceptual simplicity, this scheme has serious limitations: it is obviously not suitable for displays with large information content. A display of 1000×1000 elements would require a million individually controlled elements.

The coordinate-addressed matrix arrangement, illustrated in Figure 14.29 is a useful alternative. Here, all the lower electrodes in each row are connected to one lead, and similarly, all the upper electrodes in each column. If, now, a voltage $+V/2$ is applied to the ath row and $-V/2$ to the pth column, only the element (a, p) will experience the voltage V, sufficient to activate it. Thus the number of required leads is reduced from $(m \times n)$ to $(m + n)$, where m is the number of rows and n the number of columns. There are, however, some difficulties with this scheme also. If all the unactivated leads are at the intermediate voltage—ground potential in our example—half the operating voltage will appear

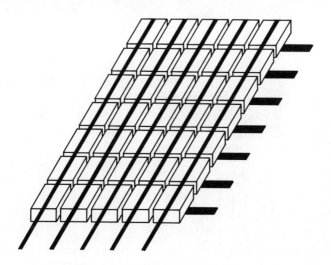

Figure 14.29 Schematic of coordinate-addressed matrix display.

across the undesired elements in row a and in column p. This scheme is called *half-selection*. Here threshold activating voltage (V_t) must satisfy

$$\frac{V}{2} < V_t \leqslant V \tag{14.225}$$

to permit proper operation. In *third-selection*, a voltage $-V/6$ would be applied to all unactivated rows and $+V/6$ to all such columns and $+V/2$ and $-V/2$ to the activated row and column, respectively. Consequently, a potential difference of only $V/3$ appears across all unactivated elements. Clearly, these methods give good results only with mechanisms having a more or less well-defined threshold and are quite unsuitable to linear effects. The frequency dependence of the electrooptic effects in liquid crystals has been used to enhance the speed and contrasts obtainable in matrix displays [79], [84], [85].

When elements in different rows and columns are to be activated, this can generally not be done simultaneously: the activation of elements (a, p) and (b, q) would, simultaneously, activate elements (a, q) and (b, p), also. Here elements must be activated sequentially, possibly a column at a time.

14.5.3.2 Continuous Imaging Devices [86], [87]. By applying the operating voltage to the liquid crystal film via a photoconductor, a light-activated continuous display can be obtained. The cell arrangement is illustrated in Figure 14.30 [74]. The cell consists of a photoconductor

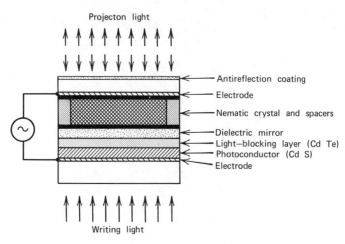

Projecton light

Figure 14.30   Light-activated liquid crystal cell for converting opaque to transparent copy, schematic. Solid black indicates insulating layers and clear white, glass plates. After Grinberg et al. [74].

layer, which together with the liquid crystal film, is sandwiched between two transparent electrodes. The liquid crystal film is separated from the photoconductor by a specular reflecting film and an optical isolator. In the dark, the voltage applied across the sandwich appears primarily across the photoconductor. When the image to be converted is projected onto the surface of the photoconductor, the induced conduction transfers the surface charges partly to the interface between the photoconductor and the liquid crystal, so that the voltage across the latter rises, activating it. The liquid crystal film is viewed in reflected light and the optical isolator serves to prevent any of the viewing light from activating the photoconductor. Such a device—combining color and black and white modulation—has already been referred to above [74].

14.5.3.3   Thermally Activated Display Devices. Smectic crystals have been used in optical information display [88]. CBOA crystals melt into a nematic phase at 67°C and become isotropic at 72°C. When they are cooled slowly, they take on the smectic A phase at 61°C and become solid at 35°C. On·the other hand, when they are cooled rapidly, they freeze into a highly scattering polycrystalline state, unless an alternating electric field of appropriate amplitude is applied during cooling.

In the device, a panel of such a crystal is heated locally by means of a scanned laser beam, goes into the isotropic liquid phase and becomes opaque on cooling. To erase the spot, it is again melted and an alternating electric field is applied during the cooling process. Such a device has been demonstrated resolving 3500 elements on a side, on a 35×35 mm useful area [89].

APPENDIX 14.1   MATRIX TREATMENT OF ELASTIC AND FIELD EFFECTS IN SOLIDS

1   Stress

The state of stress of an element in a solid can be expressed in terms of the stresses on three mutually orthogonal surfaces, normal, say, to the x, y, and z axes of a Cartesian coordinate system. As stated in Section 14.1.1.1, the components of these stresses are written in the form

$$T_{ij}, \qquad i, j = x, y, z \qquad (14.226)$$

representing the j component of the stress on the surface normal to the i axis.

The stress vector acting on an arbitrary surface, designated by its unit normal vector, n, can then be written as the matrix product[27]:

$$\begin{Vmatrix} T_x \\ T_y \\ T_z \end{Vmatrix} = \begin{Vmatrix} T_{xx} & T_{yx} & T_{zx} \\ T_{xy} & T_{yy} & T_{zy} \\ T_{xz} & T_{yz} & T_{zz} \end{Vmatrix} \begin{Vmatrix} n_x \\ n_y \\ n_z \end{Vmatrix}. \qquad (14.227)$$

Note that

$$n_x = \cos \alpha, \qquad n_y = \cos \beta, \qquad n_z = \cos \gamma, \qquad (14.228)$$

where α, β, and γ are the angles the normal makes with the x, y, and z axes, respectively.

The matrix $\|T_{ij}\|$ represents a two-dimensional tensor, It may be diagonalized, that is, there is a Cartesian coordinate system (x', y', z'), rotated with respect to the original coordinate system, in which all but the diagonal elements vanish. In this coordinate system, this matrix takes the form:

$$\begin{Vmatrix} T_I & 0 & 0 \\ 0 & T_{II} & 0 \\ 0 & 0 & T_{II} \end{Vmatrix} \qquad (14.229)$$

and the stress on the general surface has components:

$$T_{x'} = T_I \cos \alpha', \qquad T_{y'} = T_{II} \cos \beta', \qquad T_{z'} = T_{III} \cos \gamma', \qquad (14.230)$$

---

[27] To avoid confusion, we denote matrices here by means of double vertical lines, rather than by parentheses as was done in the first volume.

respectively. A method for finding the new coordinate system is given below in Section 6 of this appendix.

2 Elasticity

In matrix formulation, the relationship between stress and strain takes the form

$$\|S_p\| = \|s_{pq}\| \, \|T_q\|$$
$$\|T_p\| = \|c_{pq}\| \, \|S_q\|, \qquad p, q = 1, 2, \ldots, 6, \qquad (14.231)$$

where $\|S_p\|$ and $\|T_p\|$ are two-dimensional tensors, representable by six-element column matrices and $\|s_{pq}\|$ and $\|c_{pq}\|$ are four-dimensional tensors, representable by six-by-six matrices. $\|c_{pq}\|$ is the matrix inverse to $\|s_{pq}\|$. Its elements may be calculated from s_{pq} according to:

$$c_{rs} = \frac{|s_{pq}|^{(rs)}}{|s_{pq}|}, \qquad (14.232)$$

where the denominator is the determinant of the matrix $\|s_{pq}\|$ and the numerator the minor of element (rs), that is, $(-1)^{r+s}$ times the determinant obtained from the matrix $\|s_{pq}\|$ after deleting the rth row and the sth column.

In a cubic crystal, these matrices take the form

$$
\begin{Vmatrix}
s_{11} & s_{12} & s_{12} & 0 & 0 & 0 \\
s_{12} & s_{11} & s_{12} & 0 & 0 & 0 \\
s_{12} & s_{12} & s_{11} & 0 & 0 & 0 \\
0 & 0 & 0 & s_{44} & 0 & 0 \\
0 & 0 & 0 & 0 & s_{44} & 0 \\
0 & 0 & 0 & 0 & 0 & s_{44}
\end{Vmatrix}. \qquad (14.233)
$$

In an isotropic solid we have, in addition,

$$s_{44} = 2(s_{11} - s_{12}). \qquad (14.234)$$

The c matrix has here the same form, except that in an isotropic solid

$$c_{44} = \frac{c_{11} - c_{12}}{2}. \qquad (14.235)$$

In other crystal lattice symmetries, other elements may vanish and other equalities may appear. Details of their forms are listed in Table 150a.

3   Piezoelectricity

The piezoelectric effect is usually described by the piezoelectric strain (d_{ip}) or stress (e_{ip}) constants:

$$d_{ip} = \frac{\partial S_p}{\partial E_i}\bigg|_T = \frac{\partial D_i}{\partial T_p}\bigg|_E \qquad (14.236)$$

$$e_{ip} = \frac{\partial T_p}{\partial E_i}\bigg|_S = \frac{D_i}{S_p}\bigg|_E . \qquad (14.237)$$

Here S_p and T_p are the strain and stress components

E_i and D_i are the electric field and displacement (flux density) components, and, throughout this appendix,

S, T, D, E as subscripts or superscripts indicate that the corresponding quantity is held constant during the differentiation, and,

$$i, j = 1, 2, 3; \qquad p, q, r = 1, 2, \ldots, 6.$$

The constants d_{ip}, e_{ip} form 6×3 matrices, whose forms in terms of zeros and internal equalities, are governed by crystal symmetries (see Table 150b for details). These matrices connect vectors ($\mathbf{E}, \mathbf{D}$, i.e., one-dimensional tensors) with two-dimensional tensors ($\mathbf{T}$ and $\mathbf{S}$); they represent three-dimensional tensors. In terms of these constants, the piezoelectric strain (in an unstressed crystal) is given by

$$\|S_p\| = \|d_{ip}\| \, \|E_i\| \qquad (14.238)$$

and the short-circuited, induced flux density

$$\|D_i\| = \|d_{ip}\| \, \|T_p\| . \qquad (14.239)$$

Since the reader may come across some additional piezoelectric constants in his perusal of the extensive literature on the subject, we present here definitions of other constants in use:

$$h_{ip} = -\frac{\partial T_p}{\partial D_i}\bigg|_S = -\frac{\partial E_i}{\partial S_p}\bigg|_D \qquad (14.240)$$

$$g_{ip} = \frac{\partial S_p}{\partial D_i}\bigg|_T = -\frac{\partial E_i}{\partial T_p}\bigg|_D . \qquad (14.241)$$

We can now write the general equations for stress, strain, electrical displacement and field, combining elastic and piezoelectric effects:

$$T_p = c_{pq}^E S_q - e_{ip}E_i = c_{pq}^D S_q - h_{ip}D_i \qquad (14.242)$$

$$S_p = s_{pq}^E T_q + d_{ip}E_i = s_{pq}^D T_q + g_{ip}D_i \qquad (14.243)$$

$$D_i = e_{ip}S_p + \varepsilon_{ij}^S E_j = d_{ip}T_p + \varepsilon_{ij}^T D_j \qquad (14.244)$$

$$E_i = -g_{ip}T_p + \beta_{ij}^T D_j = -h_{ip}S_p + \beta_{ij}^S D_j \qquad (14.245)$$

and the relationships between these various constants can be seen to be

$$d_{ip} = \varepsilon_{ji}^T g_{ip} = e_{iq}S_{qp}^E \qquad (14.246)$$

$$g_{ip} = \beta_{ji}^T d_{jp} = h_{iq}S_{qp}^D \qquad (14.247)$$

$$e_{ip} = \varepsilon_{ji}^S h_{jp} = d_{iq}c_{qp}^E \qquad (14.248)$$

$$h_{ip} = \beta_{ji}^S e_{jp} = g_{iq}c_{qp}^D. \qquad (14.249)$$

Here it is understood that each product is summed over the repeated index.

ε_{ij} are the dielectric constants relating D to E:

$$\varepsilon_{ij} = \frac{\partial D_i}{\partial E_j} \qquad (14.250)$$

and $\|\beta_{ij}\|$ is the inverse of $\|\varepsilon_{ij}\|$; see (14.232) for the formula relating one to the other. Clearly, it suffices to know one (a) elasticity matrix (e.g., $\|c_{pq}\|$), (b) dielectric matrix (e.g., $\|\varepsilon_{ij}\|$), and (c) piezoelectric matrix (e.g., $\|d_{ip}\|$) to permit the calculation of all other constants.

4   Elastooptics

The elastooptic constants most commonly used are

$$p_{pq} = \frac{\partial \Delta_p}{\partial S_q}, \qquad p, q = 1, 2, \ldots, 6, \qquad (14.251)$$

where Δ_p is the change in the pth index ellipsoid coefficient, the six coefficients being, respectively,

$$\frac{1}{n_x^2}, \quad \frac{1}{n_y^2}, \quad \frac{1}{n_z^2}, \quad 2a_4, \quad 2a_5, \quad 2a_6.$$

In terms of the p_{pq}, the change in the index ellipsoid coefficients is given by

$$\Delta_p = \sum_q p_{pq}S_q. \qquad (14.252)$$

Since both Δ and S are two-dimensional tensors, the p tensor is four-dimensional: the form of the p-matrix is similar to that of the elasticity matrix, except that diagonal symmetry is no longer necessary (see Table 150c).

5 Electrooptics

The most commonly used electrooptic constants are

$$r_{pi} = \frac{\partial \Delta_p}{\partial E_i}. \tag{14.253}$$

In terms of these, the changes in the index ellipsoid coefficients are

$$\Delta_p = \sum_i r_{pi} E_i. \tag{14.254}$$

Here the Δ_p are components of a two-dimensional tensor and $\mathbf{E}$ is a vector. Hence r is a three-dimensional tensor: the form of the r matrix is similar to that of the piezoelectric constants (see Table 150b).

6 Principal Axes Transformation and Mohr's Circle [33c]

Many quantities in physics are representable by three-dimensional ellipsoids. Among these are stress, strain, and refractive index. When this is done, it is often of interest to find the principal Cartesian coordinate system, that is, the coordinate system in which the axes of the system coincide with the principal axes of the ellipsoid. In such a system, the cross terms vanish and the ellipsoid takes the simple form:

$$\frac{x^2}{A_1^2} + \frac{y^2}{A_2^2} + \frac{z^2}{A_3^2} = 1 \tag{14.255}$$

where A_1, A_2, and A_3 are the principal semiaxes.

The principal axes can be found from the expression of the general ellipsoid, say

$$a_1 x^2 + a_2 y^2 + a_3 z^2 + a_4 yz + a_5 xz + a_6 xy = 1 \tag{14.256}$$

by writing the coefficients in the form of a 3×3 matrix,

$$\begin{Vmatrix} a_1 & a_6 & a_5 \\ a_6 & a_2 & a_4 \\ a_5 & a_4 & a_3 \end{Vmatrix}$$

and finding its eigenvalues (λ_i) and normalized eigenvectors $\mathbf{V}_i$. The lengths of the three principal semiaxes are then simply given by

$$A_i = \frac{1}{\sqrt{\lambda_i}},$$ (14.257)

and their directions are those of the corresponding eigenvectors V_i.

The eigenvalues can be found by equating to zero the following determinant:

$$\begin{vmatrix} a_1 - \lambda & a_6 & a_5 \\ a_6 & a_2 - \lambda & a_4 \\ a_5 & a_4 & a_3 - \lambda \end{vmatrix}$$ (14.258)

and solving for λ. Since this determinant yields an expression which is cubic in λ, the equation has three roots, which may be arbitrarily denoted by λ_1, λ_2, and λ_3, respectively.

The eigenvectors are then found from the matrix equation

$$\begin{Vmatrix} a_1 - \lambda_i & a_6 & a_5 \\ a_6 & a_2 - \lambda_i & a_4 \\ a_5 & a_4 & a_3 - \lambda_i \end{Vmatrix} \begin{Vmatrix} V_{ix} \\ V_{iy} \\ V_{iz} \end{Vmatrix} = 0.$$ (14.259)

This yields three simultaneous equations for V_{ix}, V_{iy}, and V_{iz}, the components of V_i. From these the ratios

$$V_{ix} : V_{iy} : V_{iz}$$

can be found. From the normalization condition

$$V_{ix}^2 + V_{iy}^2 + V_{iz}^2 = 1,$$ (14.260)

the absolute values are obtained, and these represent the direction cosines of the ith principal axis.

In the two-dimensional case, the rotation of the coordinate system can be analyzed relatively simply. The matrices of the rotation through θ, and its inverse, are

$$\begin{Vmatrix} \cos\theta & -\sin\theta \\ \sin\theta & \cos\theta \end{Vmatrix} \quad \begin{Vmatrix} \cos\theta & \sin\theta \\ -\sin\theta & \cos\theta \end{Vmatrix},$$ (14.261)

whereas the stress tensor in the principal coordinate system is

$$\begin{Vmatrix} T_{\mathrm{I}} & 0 \\ 0 & T_{\mathrm{II}} \end{Vmatrix}.$$ (14.262)

After pre- and postmultiplying this with the above two matrices, respectively, we obtain the stress matrix in the rotated coordinate system:

$$\left\| \begin{matrix} T_I \cos^2 \theta + T_{II} \sin^2 \theta & (T_I - T_{II}) \sin \theta \cos \theta \\ (T_I - T_{II}) \sin \theta \cos \theta & T_I \sin^2 \theta + T_{II} \cos^2 \theta \end{matrix} \right\|. \tag{14.263}$$

This transformation can be represented by a simple geometrical construction, called Mohr's circle,[28] illustrated in Figure 14.31. T_I and T_{II} are layed off on the abscissae axis. With their midpoint, C, as a center, draw a circle though these. Each point on this circle represents the stresses on a surface through the point considered. Specifically, to find the stresses on a plane whose normal is inclined at θ to the principal coordinate system, draw a radius through C at 2θ to the abscissae axis. This intersects the circle at A. The coordinates of A relative to the origin, O, represent the normal (T_n) and shear (T_s) stresses relative to the plane. The reader can readily confirm that these correspond, respectively, to the terms in the top row of (14.263).

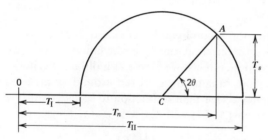

Figure 14.31 Mohr circle for two-dimensional stress analysis. See text for explanation.

With T_n, T_s given for two planes oblique to each other, two points, such as A and B can be found and Mohr's circle constructed. This gives immediately the principal stresses (as the intercepts on the abscissae axis) and permits readily the determination of the stresses on any other plane.

By using a two-step process, this method can be used to solve the three-dimensional case as well (see Ref. 2).

APPENDIX 14.2 MEASURES OF ACOUSTIC FIELD DEPTH

In Section 14.2.3.3, we presented a very simple and convenient approximation, which permits the diffraction efficiency to be calculated as the square of a Bessel function. But as shown there, this is valid only for

---

[28] O. Mohr (1835–1918).

"shallow" acoustic fields. We must now define a measure for this shallowness.

Since the Bessel function approximation is based on the assumption that the optical wave amplitude remains uniform across the wavefront, we may define a depth, d_0, which represents the depth of the field in which a ray is refracted through half a cycle of the acoustic wave, that is, a depth that permits a normally incident ray to traverse completely one compression half-cycle. From Figure 14.6 we see that this is twice the distance to the first zero, which, for the paraxial mode is given by z_0. Hence

$$d_0 \approx 2z_0 = \frac{\Lambda}{2\sqrt{n^*}}.$$ (14.264)

Assuming that the shallow field approximation is valid for any depth $d < d_0$, we can write its range of validity

$$Q_1 = \frac{d}{d_0} = 2\sqrt{n^*}\frac{d}{\Lambda} < 1.$$ (14.265)

This is essentially the second condition stated in Ref. 90.

If we base our restriction on light deviation caused by diffraction, rather than refraction, effects, we arrive at a different criterion. Assuming an effective aperture ranging over ± 1 rad around an acoustic wave peak, we find the distance d'_0 for light diffraction to spread over one cycle of the acoustic wave [from (2.157) with $2x = \Lambda$, $n = 1$, $f = d'_0$, $a = \Lambda/2\pi$]:

$$d'_0 = \frac{(1/2\pi)\Lambda^2}{\lambda}$$ (14.266)

and, hence, a second criterion

$$Q_2 = \frac{d}{d'_0} = \frac{2\pi\lambda_0 d}{n_0\Lambda^2} \ll 1.$$ (14.267)

This is the first condition stated by Ref. 90.

It has been found [10] that the Bragg approximation is quite accurate for

$$Q_2 \geqslant 7.$$

Under these conditions, 95% of the radiation is diffracted into the Bragg order when Condition 14.120 is fulfilled.

For higher diffraction orders (m), another condition has been stated [14a], [91]:

$$Q_3 = \frac{m\lambda}{\sqrt{n_0 \Delta n}} < 1.$$ (14.268)

REFERENCES

[1] I. S. Sokolnikoff, *Mathematical Theory of Elasticity*, 2nd ed., McGraw-Hill, New York, 1956.

[2] A. Kuske and G. Robertson, *Photoelastic Stress Analysis*, Wiley, New York, 1974.

[3] J. C. Cheng, L. A. Nafie, S. D. Allen, and A. I. Braunstein, "Photoelastic Modulator for the 0.55–13-μm Range," *Appl. Opt.* **15**, 1960–1965 (1976).

[4] J. Javornicky, *Photoplasticity*, Elsevier, Amsterdam, 1974.

[5] O. E. Mattiat, ed., *Ultrasonic Transducer Materials*, Plenum, New York, 1971; (a) Y. Kikuchi, "Magnetostrictive metals and piezomagnetic ceramics as transducer materials," pp. 1–61; (b) D. Berlincourt, "Piezoelectric crystals and ceramics," pp. 63–124; (c) A. H. Meitzler, "Piezoelectric transducer materials and techniques for ultrasonic devices operating above 100 MHz," pp. 125–182.

[6] "IEEE Standard of Piezoelectricity," IEEE Std. 176–1978, IEEE, New York, (1978).

[7] "IRE standards on piezoelectric crystals: Measurement of piezoelectric ceramics, 1961," *Proc. I.R.E.* **49**, 1161–1169 (1961).

[8] N. Uchida and N. Niizeki, "Acoustooptic deflection materials and techniques," *Proc. IEEE* **61**, 1073–1092 (1973).

[9] J. W. Strutt, Lord Rayleigh. *The Theory of Sound*, 2nd Ed., Dover, New York, 1945, Section 270.

[10] A. Yariv, *Quantum Electronics*, 2nd Ed., Wiley, New York, 1975; (a) Sections 14.8–14.11; (b) Section 18.6; (c) Sections 14.3–14.6.

[11] I. C. Chang, "Acoustooptic devices and applications," *IEEE Trans.* **SU-23**, 2–22 (1976).

[12] R. Lucas and P. Biquard, "Propriétés optiques des milieux solides et liquides soumis aux vibrations élastiques ultrasonores," *J. Phys. Rad.* (Series 7) **3**, 464–477 (1932).

[13] I. S. Gradshteyn and I. M. Ryzhik, *Tables of Integrals, Series, and Products*, Academic, New York, 1965; (a) 2.571(4); (b) 3.915(2).

[14] M. V. Berry, *The Diffraction of Light by Ultrasound*, Academic, New York, 1966; (a) Chapter 2.

[15] H. Kogelnik, "Coupled wave theory for thick hologram gratings," *Bell Syst. Tech. J.* **48**, 2909–2947 (1969).

[16] W. R. Klein and B. D. Cook, "Unified approach to ultrasonic light diffraction," *IEEE Trans.* **SU-14**, 123–134 (1967).

[17] L. Ali, "Über den Nachweis der Frequenzänderung des Lichtes durch Doppler-Effekt bei der Lichtbeugung an Ultraschallwellen," *Helv. Phys. Acta* **8**, 502–505 (1935).

[18] R. Y. Chiao, C. H. Townes, and B. P. Stoicheff, "Stimulated Brillouin scattering and coherent generation of intense hypersound waves," *Phys. Rev. Lett.* **12**, 592–595 (1964).

[19] R. W. Dixon, "Acoustic diffraction of light in anisotropic media," *IEEE J. Quant. Electr.* **QE-3**, 85–93 (1967).

[20] R. M. O'Connell and P. H. Carr, "New materials for surface acoustic wave (SAW) devices," *Opt. Eng.* **16**, 440–445 (1977).

[21] E. I. Gordon, "A review of acoustooptical deflection and modulation devices," *Proc. IEEE* **54**, 1391–1401 (1966).

[22] A. Korpel, R. Adler, P. Desmares, and W. Watson, "A television display using acoustic deflection and modulation of coherent light," *Appl. Opt.* **5**, 1667–1675 (1966); also *Proc. IEEE* **54**, 1429–1437 (1966).

[23] L. Levi, "Light modulator records airborne radar display," *Electronics* **31**, 80–83 (1 Aug. 1958).

[24] L. Levi, "High fidelity video recording using ultrasonic light modulation," *J. Soc. Mot. Pict. TV Eng.* **67**, 657–661 (1958).

[25] A. H. Rosenthal, "Color control by ultrasonic wave gratings," *J. Opt. Soc. Am.* **45**, 751–756 (1955).

[26] G. H. Alphonse, "Broadband acoustooptic deflectors: New results," *Appl. Opt.* **14**, 201–207 (1975).

[27] I. C. Chang, "Tunable acousto-optic filters: An overview," *Opt. Eng.* **16**, 455–460 (1977).

[28] S. E. Harris and R. W. Wallace, "Acousto-optic tunable filter," *J. Opt. Soc. Am.* **59**, 744–747 (1969).

[29] N. Uchida and S. Saito, "Acoustooptic tunable filter using TeO_2," *Proc. IEEE* **62**, 1279–1280 (1974).

[30] R. W. Dixon and E. I. Gordon, "Acoustic light modulators using optical heterodyne mixing," *Bell Syst. Tech. J.* **46**, 367–389 (1967).

[31] R. W. Damon, W. T. Maloney, and D. H. McMahon, "Interaction of light with ultrasound: Phenomena and applications," in *Physical Acoustics*, Vol. 7, W. P. Mason and R. N. Thurston, eds., Academic, New York, 1970, pp. 273–366.

[32] R. A. Sprague, "A review of acousto-optic signal correlators," *Opt. Eng.* **16**, 467–474 (1977).

[33] J. F. Nye, *Physical Properties of Crystals*, Clarendon, Oxford, 1957; (a) Chapter 13; (b) Appendix B; (c) Chapter 2.

[34] I. P. Kaminow, *An Introduction to Electrooptic Devices*, Academic, New York, 1974.

[35] O. G. Vlokh and I. S. Zheludev, "Changes in the optical parameters of crystals caused by electric fields (the linear electrooptical effect)," *Kristallografia* **5**, 390–402 (1960), *Soviet Physics. Crystallography* **5**, 368–380 (1960).

[36] R. Bechmann, R. F. S. Hearmon, and S. K. Kurtz in Landolt–Börnstein, *Numerical Data and Functional Relationships In Science and Technology*, New Series Vol. III/1, K. H. Hellwege, ed., Springer, Berlin, 1966; Vol. III/2, 1969.

[37] I. P. Kaminow and E. H. Turner, "Electrooptic light modulators," *Appl. Opt.* **5**, 1612–1628 (1966); also *Proc. IEEE* **54**, 1374–1390 (1966).

[38] C. Kittel, *Introduction to Solid State Physics*, 4th Ed., Wiley, New York, 1971, Chapter 1.

[39] G. H. Haertling, "Improved hot-pressed electrooptic ceramics in the (Pb, La) (Zr, Ti)O_3 system," *J. Am. Ceram. Soc.* **54**, 303–309 (1971).

[40] P. D. Thacher and C. E. Land, "Ferroelectric electrooptic ceramics with reduced scattering," *IEEE Trans.* **ED-16**, 515–521 (1969).

[41] J. R. Maldonado, D. B. Fraser, and A. H. Meitzler, "Display applications of PLZT ceramics," in *Adv. Image Pickup Display*, Vol. 2, B. Kazan, ed., Academic, New York, 1975, pp. 65–168; also "Image storage and display devices using fine-grain, ferroelectric ceramics," *Bell Syst. Tech. J.* **49**, 953–967 (1970).

[42] M. Born, *Optik*, Springer, Berlin, 1933; (a) Par. 80; (b) Par. 78; (c) Par. 79.

[43] W. R. Cook and H. Jaffe, "Magneto-, electro-, and elasto-optic constants," Section 6m in *American Institute of Physics Handbook*, D. E. Gray, ed., 3rd Ed., McGraw-Hill, New York, 1972; (a) From International Critical Tables; (b) D. W. Saunders, J. F. Rudd, and R. D. Andrews; (c) R. W. Dixon, N. F. Borelli, R. A. Miller, and R. S. Krishnan.

[44] E. V. Condon and H. Odishaw, eds., *Handbook of Physics*, 2nd Ed., McGraw-Hill, New York, 1967; (a) E. V. Condon, "Molecular optics," Part 6; (b) E. V. Condon, "Atomic structure," Part 7.

[45] F. S. Chen, "Modulators for optical communications," *Proc. IEEE* **58**, 1440–1457 (1970).

[46] W. Franz, "Einfluss eines elektrischen Feldes auf eine optische Absorptionskante," *Z. Naturforsch.* **13A**, 484–489 (1958).

[47] L. V. Keldysh, "The effect of a strong electric field on the optical properties of insulating crystals," *J. Exp. Theoret. Phys.* **34**, 1138–1141 (1958); *Sov. Phys. JETP* **7**, 788–790 (1958).

[48] R. Williams, "Electric field induced light absorption in CdS," *Phys. Rev.* **117**, 1487–1490 (1960).

[49] G. Racette, "Absorption edge modulator utilizing a $p-n$ junction," *Proc. IEEE* **52**, 716 (1964).

[50] C. E. Land and P. D. Thacher, "Ferroelectric ceramic electrooptic materials and devices," *Proc. IEEE* **57**, 751–768 (1969).

[50*] M. A. Duguay, "The ultrafast optical Kerr shutter," in *Progress in Optics*, Vol. 14, E. Wolf, ed., North-Holland, Amsterdam, 1976, pp. 161–193.

[51] J. Feinleib and D. S. Oliver, "Reusable optical image storage and processing device," *Appl. Opt.* **11**, 2752–2759 (1972).

[51*] B. A. Horwitz and F. J. Corbett, "The PROM—theory and applications for the Pockels readout optical modulator," *Opt. Eng.* **17**, 353–364 (1978).

[52] G. Marie, J. Donjon, and J. P. Hazan, "Pockels effect imaging devices and their applications," in *Adv.Image Pickup Display*, Vol. 1, B. Kazan, ed., Academic, New York, 1974, pp. 225–302.

[52*] D. Casasent, "E-beam DKDP light valves," *Opt. Eng.* **17**, 344–352 (1978); and "Photo DKDP light valve: a review," *ibid.*, 365–370.

[52**] C. Warde, A. D. Fisher, D. M. Cocco, and M. Y. Burmawi, "Microchannel spatial light modulator," *Opt. Let.* **3**, 196–198 (1978).

[53] C. E. Land, "Optical information storage and spatial light modulation in PLZT ceramics," *Opt. Eng.* **17**, 317–326 (1978).

[53*] D. L. Staebler, "Ferroelectric crystals," in *Holographic Recording Materials*, H. M. Smith, ed., Springer, Berlin, 1977, pp. 101–132.

[54] H. Meyer, D. Riekmann, K. P. Schmidt, U. S. Schmidt, M. Rahlff, E. Schröder, and W. Thust, "Design and performance of a 20-stage digital light beam deflector," *Appl. Opt.* **11**, 1752–1756 (1972).

[55] J. P. Campbell and W. H. Steier, "Rotating-waveplate optical-frequency shifting in lithium niobate," *IEEE J. Quant. Electr.* **QE-7**, 450–457 (1971).

[56] M. A. Duguay and J. W. Hansen, "Optical frequency shifting of a mode-locked laser beam," *IEEE J. Quant. Electr.* **QE-4**, 477–481 (1968).

[57] M. A. Duguay and J. W. Hansen, "Compression of pulses from a mode-locked He–Ne laser," *Appl. Phys. Lett.* **14**, 14–16 (1969).

[58] F. A. Jenkins and H. E. White, *Fundamentals of Optics*, 4th Ed., McGraw-Hill, New York, 1976, Chapter 32.

[59] J. W. Strutt, Lord Rayleigh, "On the constant of magnetic rotation of light in bisulfide of carbon," *Philo. Trans.* **176**, 343–366 (1885).

[60] U. Cappeller, "Faraday-Effekt von Atomen, Ionen und Molekeln," Section 1328 in Landolt-Börnstein. (see Ref. 36) 6th Ed., Vol. I/1; G. Joos, A. Eucken and K. H. Hellwege, eds., Springer, Berlin, 1950, pp. 405–432.

[61] D. Chen, "Magnetic materials for optical recording," *Appl. Opt.* **13**, 767–778 (1974).

[62] E. D. Palik and B. W. Henvis, "A bibliography of magnetooptics of solids," *Appl. Opt.* **6**, 603–630 (1967).

[63] B. R. Brown, "Optical data storage potential of six materials," *Appl. Opt.* **13**, 761–766 (1974).

[64] B. R. Brown, "Readout performance and analysis of a cryogenic magnetooptical data storage system," *IBM J. R&D* **16**, 19–26 (1972).

[65] G. W. Gray, *Molecular Structure and the Properties of Liquid Crystals*, Academic, New York, 1962; (*a*) p. 40; (*b*) Chapter 2; (*c*) pp. 23–31.

[66] P. G. de Gennes, *The Physics of Liquid Crystals*, Clarendon, Oxford, 1974; (*a*) Section 3.4.3; (*b*) Chapters 4, 6, and 7; (*c*) P. 274; (*d*) Pp. 97–101; (*e*) Section 3.2; (*f*) Section 3.3.

[67] F. J. Kahn, G. N. Taylor, and H. Schonhorn, "Surface-produced alignment of liquid crystals," *Proc. IEEE* **61**, 823–828 (1973).

[68] L. T. Creagh, "Nematic liquid crystal materials for displays," *Proc. IEEE* **61**, 814–822 (1973).

[69] G. H. Brown and J. W. Doane, "Liquid crystals and some of their applications," *Appl. Phys.* **4**, 1–15 (1974).

[70] J. L. Fergason, N. N. Goldberg, and R. J. Nadalin, "Cholesteric structure. II. Chemical significance," *Mol. Cryst.* **1**, 309–323 (1966).

[71] R. B. Meyer, "Piezoelectric effects in liquid crystals," *Phys. Rev. Lett.* **22**, 918–921 (1969).

[72] W. Helfrich, "Electric alignment of liquid crystals," *Mol. Cryst. Liq. Cryst.* **21**, 187–209 (1973).

[73] M. F. Schiekel and K. Fahrenschon, "Deformation of nematic liquid crystals with vertical orientation in electrical fields," *Appl. Phys. Lett.* **19**, 391–393 (1971).

[74] J. Grinberg, W. P. Bleha, A. D. Jacobson, A. M. Lackner, G. D. Myer, L. J. Miller, J. D. Margrum, L. M. Fraas, and D. D. Boswell, "Photoactivated birefringent liquid-crystal light valve for color symbology display," *IEEE Trans.* **ED-22**, 775–783 (1975).

[75] M. Schadt and W. Helfrich, "Voltage-dependent optical activity of a twisted nematic liquid crystal," *Appl. Phys. Lett.* **18**, 127–128 (1971).

[76] R. Williams, "Domains in liquid crystals," *J. Chem. Phys.* **39**, 384–388 (1963).

[77] G. H. Heilmeier, L. A. Zanoni, and L. A. Barton, "Dynamic scattering: A new electrooptic effect in certain classes of nematic liquid crystals," *Proc. IEEE* **56**, 1162–1171 (1968).

[78] G. H. Heilmeier and W. Helfrich, "Orientational oscillations in nematic liquid crystals," *Appl. Phys. Lett.* **16**, 155–157 (1970).

[79] P. J. Wild and J. Nehring, "Turn-on time reduction and contrast enhancement in matrix-addressed liquid-crystal light valves," *Appl. Phys. Lett.* **19**, 335–336 (1971).

[80] G. H. Heilmeier and J. E. Goldmacher, "A new electric field controlled reflective optical storage effect in mixed liquid crystal systems," *Proc. IEEE* **57**, 34–38 (1969).

[81] T. Ohtsuka and M. Tsukamoto, "Ac electric-field-induced cholesteric-nematic phase transition in mixed liquid crystal films," *Jap. J. Appl. Phys.* **12**, 22–29 (1973).

[82] G. H. Heilmeier and L. A. Zanoni, "Guest-host interactions in nematic liquid crystals. A new electrooptic effect," *Appl. Phys. Lett.* **13**, 91–92 (1968).

[83] C. H. Gooch, "Liquid crystal matrix displays," in *Progress in Electro-Optics*, E. Camatini, ed., Plenum, New York, 1975, pp. 87–107.

[84] C. R. Stein and R. A. Kashnow, "A two-frequency coincidence addressing scheme for nematic-liquid-crystal displays," *Appl. Phys. Lett.* **19**, 343–345 (1971).

[85] A. Alimonda and V. Meyer, "A method for reducing the decay time of a liquid crystal," *IEEE Trans.* **ED-20**, 332 (1973).

[86] G. Labrunie, J. Robert, and J. Borel, "Nematic liquid crystal 1024 bits page composer," *Appl. Opt.* **13**, 1355–1358 (1974).

[87] W. P. Bleha, L. T. Lipton, E. Wiener-Avnear, J. Grinberg, P. G. Reif, D. Casasent, H. B. Brown, and B. V. Markevitch, "Application of the liquid crystal light valve to real-time optical data processing," *Opt. Eng.* **17**, 371–384 (1978).

[88] F. J. Kahn, "Ir-laser-addressed thermo-optic smectic liquid-crystal storage displays," *Appl. Phys. Lett.* **22**, 111–113 (1973).

[89] F. J. Kahn, D. Maydan, and G. N. Taylor,'"Performance and characteristics of smectic liquid crystal storage displays," in *Liquid Crystal Devices*, T. Kallard, ed., Optosonic, New York, 1973.

[90] R. Extermann and G. Wannier, "Théorie de la diffraction de la lumière pour les ultrasons," *Helv. Phys. Acta* **9**, 520–532 (1936).

[91] C. V. Raman and N. S. N. Nath, "The diffraction of light by high frequency sound waves. Part IV. Generalized theory," *Proc. Ind. Acad. Sci.* **3**, 119–125 (1936).

15

Vision

15.1 ANATOMY AND PHYSIOLOGY OF THE VISUAL SYSTEM [1a], [2a,b], [3a], [4a,b]

15.1.1 Anatomic Factors in Vision

The visual process may be divided into four functions:

1. Formation of the image on the retina.
2. Conversion of the optical image into nerve pulses.
3. Interaction between these pulses and their conduction to the brain.
4. Interpretation of these pulses in the brain.

The optical media of the eye perform the first function. The retina performs the second function and also contains some of the interconnections between nerve channels. The optic nerve conducts the nerve pulses to the lateral geniculi in the brain. From these, the information is transferred to the visual centers in the cortex of the brain for interpretation.

The eye, which performs all but the last function, at least in part, is shown in Figure 15.1 in schematic form. It is almost spherical in shape, with a radius of curvature of approximately 12 mm, and enclosed in a tough sheath, the *sclera*, the white of the eye. In the front, a segment of the sclera is replaced by a bulge of clear tissue, the *cornea*. The front surface of the cornea has a radius of approximately 8 mm, and it is there that most of the refraction in image formation takes place. A short distance behind the cornea, we find a variable diaphragm, the *iris*, the colored part of the eye. On its outer rim, this is attached to a ring-shaped, *ciliary*, muscle. The near-circular opening in the center of the iris is the *pupil*. Immediately behind the iris is a flexible body, which, because of its shape and function, is called the *lens*. This consists of layers of tissue

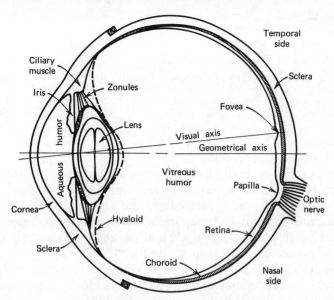

Figure 15.1   Eye cross section, schematic.

enclosed in a tough capsule. It is suspended from the ciliary muscle by means of the *zonule fibers*.

The space between the cornea and the rear of the lens (anterior chamber) is filled with a liquid, the *aqueous humor*, whereas the rest of the globe (posterior chamber) is filled with the *vitreous humor*, which has a jellylike consistency and is divided from the aqueous humor by the *hyaloid diaphragm*.

At the rear of the eye, the inside of the sclera is lined by the *retina* on which the image is formed. The retina is backed by the *choroid layer*, which absorbs unused radiation.

The retina contains the photosensitive elements that convert the incident illumination into nerve impulses. It is penetrated by the optic nerve at the *papilla*, corresponding to the blind spot of vision. It also exhibits a small depression, approximately 1.5 mm diam, the *fovea*. Only here is vision of fine detail possible. The line joining the center of the fovea to the center of the pupil is called the visual axis.[1] This does not coincide with the optical axis, the axis of (approximate) symmetry of the eye. The angle (α) between these two axes is approximately 5° in the horizontal plane.

---

[1] Strictly, the visual axis is the line joining the center of the fovea to the rear nodal point. The nodal points are defined by a ray leaving the eye parallel to the direction of its entry. Its intersections with the optical axis are the nodal points.

15.1.2 Optical System

The optical media of the eye form a reversed[2] image of the external object on the retina. The size of the image (h) is proportional to the angle (φ) subtended by the object at the pupil. Specifically, it is given approximately by

$$h = 16.7\varphi \text{ mm.} \tag{15.1}$$

Although most (42 diopter) of the eye's power is concentrated at the cornea, the lens, too, contributes power. When the eye is in the relaxed state, the lens power is about 22 diopters and, in the *emmetropic* (normal) eye, the images of distant objects are then formed on the retina.

In this relaxed state, the zonules are tensed and the lens is relatively flat. Contraction of the circular muscles inside the ciliary body relaxes the zonules and permits the lens to thicken at the center. This increases its power by anywhere up to 10 diopters (in the 20 year old). By means of this *accommodation* process, the emmetropic eye can cause the images of closer objects, too, to be formed on the retina. The closest object position that may still be accommodated to yield focusing on the retina is called the *near point*. In the normal adult eye, this varies from about 8 cm in the 15 year old to 2 m in the 60 year old. Beyond the age of 70 years, the lens has usually hardened so that no more accommodation is possible. This loss of accommodation power, caused by aging, is called *presbyopia*. (As a rule of thumb, accommodation drops at the rate of 0.25 diopter/year from 14 diopter at the age of 8 years to 6 diopter at 40 years, then at 0.5 diopter/year to 1 diopter at the age of 50 years.)

Other forms of *ametropia* (ocular focusing abnormalities) include hypermetropia and myopia. In *hypermetropia* (far-sightedness) the image of a distant object is formed behind the retina, when the eye is relaxed and accommodation is required to focus such an image, leaving insufficient accommodation for near objects. This condition may be compensated for by a positive (spectacle) lens. In *myopia* (near-sightedness) the image of a distant object is formed in front of the retina, and only a closer object position will yield a distinct retinal image. The object position

---

[2] When first thinking of the reversion of the retinal image, we might wonder how we see upright images. In fact, of course, there is no direct connection between the retinal image and the "image" in our consciousness. By correlating the perceived image with the information of other senses, we slowly associate the corresponding regions of the visual field with "up," "down," etc. In an interesting experiment, G. M. Stratton wore, during his waking hours, eye glasses which reversed the perceived image. After about a week, he found that he saw upright, despite the glasses. But after removing them, he saw everything reversed and it required almost as long until he saw upright again. [*Psych. Rev.* **4**, 341–360, 463–481 (1897).]

yielding a focused image when the lens is relaxed is called the *far point.* We may compensate for myopia by preposing a lens of appropriate negative power.

From the above description it might appear that the normal eye is focused for a distant object, whenever accommodation is not activated: the neutral point of accommodation is at infinity. This appears to be widely accepted as correct. There is, however, strong evidence that the neutral point is at a distance of approximately 50 cm (the precise value varies between observers and ranges from 20 cm to 2 m) [5]. The evidence comes from the phenomena called: empty field, dark, night, twilight, and instrument myopia. In all of these, the focusing cues are either absent or reduced and, simultaneously, accommodation tends toward the above neutral point [6]. Extensive measurements of the neutral point in total darkness have shown a mean distance of 36 cm—2.76 diopter, with a standard deviation that varied greatly from observer to observer, but averaged 0.36 diopter [6*]. For distant objects the effect is similar to myopia; it simulates hypermetropia for near objects.

The actual eye may be approximated by an optical system with interfaces of specified radii and media of specified refractive indices and thickness. Such "schematic eyes" have been proposed. The parameters of two versions (due to Le Grand, 1946) are listed in Table 158 [3a]. The second of these is a simplified version of the first in that the cornea is taken uniform with the region of the aqueous and vitreous humors and the lens is taken to be thin. These approximations may suffice for many applications.

15.1.3 Retina [4a,b], [7a]

The retina is the screen on which the ocular image is formed. It also contains the photosensitive elements that transduce the illumination into nerve impulses. In addition to this, a considerable amount of "image processing" takes place in the retina; that is to say, the nerve pulses traveling from the eye to the brain along the optic nerve are not simply a replica of the illumination distribution on the retina. Instead, this distribution is operated upon within the retina. Operations may include the formation of weighted sums, differencing of adjacent illumination values, and so on. This explains the rather complicated structure of this wonderful organ.

15.1.3.1 *Retinal Structure.* Let us start with a description of the retinal area. We have already mentioned the foveal depression where alone detailed vision is possible. It corresponds to a field of approximately 5° diameter in the object space. The central region of the fovea, the *foveola,*

has a diameter of approximately 1.4° and, at its center, the *central island*, has a diameter of approximately 0.2°. This is the region of highest visual acuity. The fovea is covered by a layer of yellow pigment, the *macula*.

The fovea is surrounded by zones called *parafovea* (8.6°), *perifovea* (19°), and the *peripheral* regions. These zones exhibit progressively lower acuity. Approximately 5 mm to the nasal side of the central island we find the papilla, the region where the optic nerve exits from the retina. There are no receptors here, and it therefore gives rise to the *"blind spot"*—a gap in the visual field—displaced horizontally, toward the temple, about 16° from the point of fixation. It is oval in shape, extending approximately 5.5° horizontally and 7.5° vertically.

In depth, the retinal structure is dictated by its functions. Essentially, the retina contains the receptors, which transduce the illumination into nerve impulses, and three cascaded layers of nerves[3]:

1. The nerves associated with the individual receptors,
2. The ganglion cells, which are extensions of the optic nerve fibers,
3. Between them, the bipolar nerves mediate between the receptor nerves and the optic nerve.

The thickness of the retina varies from 130 μm at the center of the fovea to 370 μm in the parafovea. This structure is shown in more detail in Figure 15.2 [2*b*]. We may distinguish between the following layers

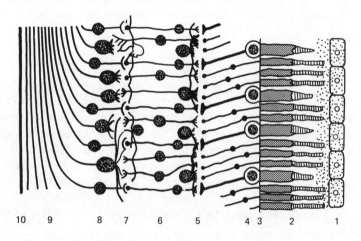

10   9   8   7   6   5   4   3   2   1

Figure 15.2   Retinal layers, schematic.

[3] For convenience we use here the term nerve in place of the technically more correct *neuron*. The neuron is a single nerve cell, whereas the term *nerve* is usually used for a bundle of neurons, held together in a common sheath.

(note that the light enters from the left, while our discussion starts from the right):

1. The *pigment epithelium*, the rearmost layer, consists of heavily pigmented cells that absorb unused radiation.

2. The *receptor-cell layer* contains the rods and cones in which the actual transducing of illumination to nerve impulses takes place.

3. This is followed by the *external limiting membrane*, which separates the receptor-cell layer from the following layers,

4. The *outer nuclear layer*, contains the nerves associated with each of the receptors.

5 & 7. The *outer* and *inner plexiform* (or synaptic) layers containing the connections (synapses) between the various cell types.

6. The *inner nuclear layer* contains the bipolar cells, mediating between receptor and optic nerves.

8 & 9. The layers of the ganglion cells (8) and (9) the nerve fibers,

10. The *internal limiting membrane* adjoining the hyaloid membrane which encloses the vitreous humor.

15.1.3.2 Retinal Receptors. There are in the retina two types of receptors, named, because of their shapes, rods and cones; see Figure 15.3 for

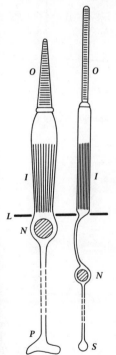

Figure 15.3 Cone and rod. *O*, outer segment; *I*, inner segment; *N*, nucleus; *L*, limiting membrane; *P*, pedicle of cone; *S*, spherule of rod.

representative forms. The cones are concentrated in the foveal and parafoveal region, whereas the rods are absent in the foveola and become more dense in the peripheral region; see Figures 15.4 and 15.5 [8], [9]. In the foveola the cones are very densely packed and may be as small as 1 μm in diameter; as the periphery is approached, the cone diameter increases to approximately 3 μm. The diameter of the rods, too, ranges from approximately 1–2.5 μm.

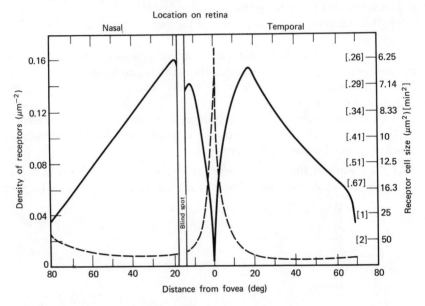

Figure 15.4 Density of rods and cones in retinal regions. After Østerberg [9].

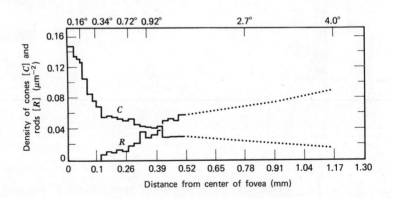

Figure 15.5 Density of rods and cones in the foveal region. Pirenne [8] after Østerberg [9].

The human retina has approximately 120 million rods and 6.8 million cones (110,000 of these in the fovea and 25,000 in the foveola). On the other hand, there are less than a million nerve fibers in the optic nerve, implying a great deal of data reduction in the retina. As a matter of fact, some highly sophisticated signal processing takes place in the visual system (see Section 15.4.4), much of it in the retina itself. This facilitates image sharpening, contrast enhancement, brightness constancy largely independent of ambient illumination, edge detection, etc.

This signal processing relies, in part, on interconnections between the various nerves in the retina. These are illustrated in Figure 15.6 [10] representing an enlarged section of Figure 15.2. (The numbers along the lower edge of the figure permit the reader to match the various regions with the corresponding ones in Fig. 15.2.)

This figure shows, in layer 6,

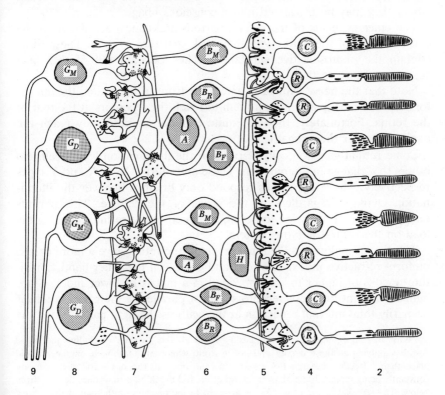

Figure 15.6 Detail of retinal layers, showing cones (C); rods (R); bipolars: midget (B_M), rod (B_R), and flat (B_F); ganglion cells: diffuse (G_D) and midget (G_M); as well as horizontal (H) and amacrine (A) cells. After Dowling and Boycott [10].

1. Rod bipolars (B_R), each forming synapses with several rod detectors (R),

2. Flat bipolars (B_F) each synapsing with several cone detectors (C).

In layer 8 it shows the diffuse ganglion cells (G_D) synapsing with several bipolars. The midget bipolars (B_M) and midget ganglion cells (G_M) synapse with but a single cone nerve and bipolar, respectively.

In addition we see in the outer plexiform layer, 5, horizontal cells (H), which interconnect cone detectors over long distances. The amacrine cells (A) in the inner plexiform layer, 7, seem to perform a similar function for the bipolar cells, connecting these with each other and with ganglion cells.

Each ganglion cell exits in a single fiber, its *axon*. All the axons run to the papilla, where they are joined into the optic nerve, leaving the retina and extending to the lateral geniculate body in the thalamus portion of the brain. From there the signals are transferred to the brain cortex, where they may be translated into a conscious image.

The *outer segments* of the receptors (rods and cones) are shown at the extreme right of the figure with their disk structure evident. These disks contain the photosensitive substances. They develop out of the *inner segments*, shown to their left.

Note that the nerve axons run across the front of the retina; the image-forming flux must pass through them to reach the receptors at the rear of the retina. Fortunately, they are quite transparent until they reach the papilla to enter their sheath.[4]

Some retinal ganglion cells respond to the illumination level itself and others to a change in illumination. These are referred to as X and Y types respectively. The Y-type cells respond only briefly whenever the illumination changes, regardless of the sense of the change. This permits temporal image processing to be implemented in the retina, reducing the load on the central nervous system.

The tissues making up the inner layers of the retina have a more or less uniform refractive index. The rods and cones, on the other hand, constitute regions of higher index embedded in a matrix of lower index. Since light is incident on these receptors almost parallel to their axes, there is primarily total internal reflection at the walls as the light passes along the

---

[4] Why are the detector elements placed behind all the "wiring," including the blood vessels supplying all these nerves, and not in front, where any reasonable engineer would place them? Recent research has shown that the disks in the outer rod segments are constantly being replaced: the disks at the tip of the rod break off as new ones develop from below. The removal of the dead disks is assigned to the pigment epithelium at the rear of the retina. This is why the detectors must be located there. It also explains the shape of the cones. The first disks produced in the fetus are always smaller than the later ones. In the cones these remain; in the rods they are soon replaced [11].

receptor. Thus these receptors act as optic fibers, funneling almost all the incident flux along for absorption in any photosensitive materials within their body. As the diameter of these "fibers" is comparable to the wavelength of light involved, the action is similar to that in a waveguide [12]. Flux entering through the outer zones of the pupil deviates significantly from this parallelism and is therefore less completely funneled. This accounts, at least in part, for the decline in effectiveness of light entering through the outer regions of the pupil, the Stiles-Crawford effect discussed in Section 15.1.4.4.

15.1.3.3 Duality of the Retina—Photopic and Scotopic Vision. For many purposes the eye may be considered to have two interleaved retinas, the photopic and scotopic, each having its own spectral response curve. (See Figure 1.5 and Tables 159 [3b] and 160 for the spectral response of the normal observer.[5]) This difference accounts for the *Purkinje*[6] *effect* (or *shift*), the relative luminances of a blue and red object change when the level of illumination changes from daylight to twilight. Thus, a red patch, which appears brighter in daylight may appear darker than an adjoining blue patch when viewed at low illumination levels.

The photopic retina is concentrated in the foveal region and has a sensitivity threshold of approximately 10^{-3} cd/m^2 (30 mTd), discriminates between colors and is capable of high resolution. The scotopic retina is practically absent in the fovea, has a threshold of approximately 10^{-6} cd/m^2 (35 μTd), cannot discriminate between colors or resolve fine detail. Each of these retinas seem to have its distinct photochemical detection process accounting for the differing adaptation rates, as discussed in Section 15.5.1.4.

Even though the scotopic retina is more sensitive, the differential sensivity of the photopic retina is greater at higher luminance levels, so that its response governs there. Above approximately 3 m cd/m^2 (0.1 Td), the cones are more sensitive than the rods. Indeed, severe saturation of the rod response starts at approximately 10 cd/m^2 (100 Td) and is almost total at 300 cd/m^2 (200 Td) [15].

The discontinuities apparent in the curves of threshold contrast (Figure 15.18), dark adaptation thresholds (Figure 15.35), and stereoscopic resolution threshold (Figure 15.54) can all be accounted for by the duality of

---

[5] Table 159 is an updated version of Table 2 in Volume 1. Table 160 was computed from the logarithmic values adopted by the C.I.E. in 1951 [13]. Even though these tables cut off at 0.83 and 0.79 μm, respectively, the sensitivities extend well into the ir. Beyond 0.65 μm, the foveal sensitivity drops by a factor of 10 for each wavelength increase of approximately 36.5 nm, up to 0.95 μm, and somewhat more slowly thereafter. Measurements have been reported to 1.064 μm [14] see also Note 26.

[6] J. E. Purkinje (1825).

the retina, with the rods and cones responsible for scotopic and photopic vision, respectively.

The transition from scotopic to photopic response takes place at intermediate luminances, from 1 m cd/m^2 to 1 cd/m^2, approximately. Vision at these levels is called *mesopic*. As expected, the mesopic spectral sensitivity curve is intermediate between the scotopic and photopic. The following empirical formula has been found to represent the luminance (L_o) observed at mesopic levels, in terms of the luminances (L_s and L_p) calculated on the basis of the scotopic and photopic spectral luminous efficiency curves, respectively:

$$L_o = \frac{L^* L_s + L_p^2}{L^* + L_p},$$

(15.2)

with $L^* = 0.0143$, 0.0628, 0.123 cd/m^2 for field sizes: 5°, 15°, and 45°, respectively [16].

Note that the peripheral retina has a spectral response similar to the scotopic, even at photopic levels [17].

15.1.3.4 Physical Processes in Visual Light Detection. The actual processes responsible for converting the absorbed light flux into the electrical signal of the nerve impulse are photochemical. In the rods, a photosensitive substance, *rhodopsin* or *visual purple* is bleached by the absorbed light. Its spectral absorption curve is close to the scotopic luminous efficiency curve and presumably, it provides at least a first step in the light sensation transduction process. See also Note 25 in this chapter.

When the illumination is removed, the retina regenerates the rhodopsin, requiring about an hour for substantially total regeneration after strong bleaching.[7] This, too, is in good agreement with the rate at which the sensitivity is restored (see Section 15.5.1.4). However, the bleaching, as such, cannot account for the decrease in sensitivity: it becomes optically measurable only at approximately 10^4 cd/m^2, far above the point where rod response saturates.

Both subjective threshold measurements on humans and electroretinography on animals indicate that the *logarithm* of the retinal sensitivity varies directly with the bleaching fraction of the rhodopsin, with the underlying process as yet unknown [4c].

Cones, too, contain photosensitive molecules somewhat similar to rhodopsin, but with different response spectra and considerably faster regeneration rates.[7] Three different sensitivity spectra have been identified, responsible for the red, green, and blue sensations (see Section 15.7.2 for further details).

---

[7] In the dark, rhodopsin regeneration follows an exponential course, with a time constant of about 7.5 min (shorter, if the bleaching was due to a short flash) [18]. The regeneration of cone pigment is similar, but with a time constant of 2 min [19].

Mechanical pressure on the eye, or its electrical stimulation, can excite the retinal nerves and cause visual sensations. These are called *phosphenes* [2c], [20a], [21]. Strong pressure may stop the blood circulation in the retina, causing temporary blindness [20b].

15.1.4 Retinal Illumination

15.1.4.1 Geometrical Factors. According to (9.124), the illumination in an image is given by

$$E = L_o \tau \frac{S_A}{b^2} \left(\frac{n'}{n}\right)^2$$

$$= 0.0025 L_o S_A, \text{ for the retina } (S_A \text{ in mm}^2), \qquad (15.3)$$

where L_o is the object luminance,
  τ is the over-all transmittance of the media,
  S_A is the area of the exit pupil,
  b is the image distance from the exit pupil, and
  n, n' are the refractive indices of the object and image media, respectively, and we have substituted the approximation

$$\Omega'_p = \frac{S_A}{b^2} = \pi \sin^2 \alpha' \qquad (15.3a)$$

for the solid angle subtended by the exit pupil at the image. The third member of (15.3) results from substituting appropriate values ($n = 1$, $n' = 1.336$, $\tau = 0.75$, and $b = 22.29$) for the ocular constants.

The amount of light reaching the retina from a small source may be increased significantly by imaging the source on the pupil with a high-aperture lens. This arrangement is called Maxwellian view.[8]

15.1.4.2 Transmission Factors. The transmittance of the eye seems to be controlled primarily by that of the lens and by reflection at the cornea, absorption in the other media being relatively negligible. Representative curves of the transmission spectrum of the eye at various ages are shown in Figure 15.7 [2a]. The density spectrum, as obtained by combining results from various techniques, is shown in Figure 15.8 [22]. In applying such data, the reader should keep in mind the large variations between individuals, even in the same age group and variations over the pupil diameter, which are due to the variation in lens thickness.

The cutoff of visibility at the ultraviolet seems to be due to absorption in the lens, rather than insensitivity of the retinal receptors. Thus

---

[8]J. C. Maxwell (1860).

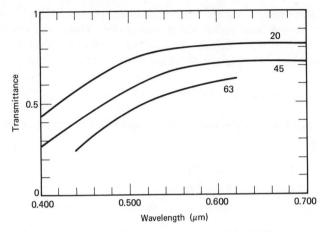

Figure 15.7 Eye transmittance spectra for various ages. The age, in years, is marked next to the respective curve. After Said and Weale [2a].

aphakics, i.e., persons from whose eyes the lenses have been removed, can see ultraviolet down to approximately 0.31 μm. The fact that this appears to them as blue rather than violet may be explained by the fact that the retina fluoresces with a greenish color and sensation of this, when mixed with the basic violet sensation, appears blue. Incidentally, the lens, too, fluoresces when illuminated by ultraviolet light.

At the infrared end, transmission is substantial up to about 1.2 μm and at 1 μm almost half the flux is transmitted. The cutoff here is due to the insensitivity of the retinal detectors.[5]

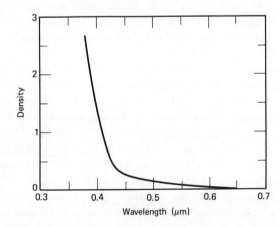

Figure 15.8 Density spectrum of eye. After van Norren and Vos [22].

15.1.4.3   Scattering. In addition to being absorbed, light is also scattered in the ocular media, especially in retinal layers before reaching the receptors. The amount of light scattering may be measured with a small bright source being imaged on the retina. With such an image, the retinal illumination due to scattered light is found to drop with distance from the image. The illumination outside the image proper tends to veil any other images formed there. The ratio of equivalent veiling luminance (L_v) to the illumination (E) at the pupil can be estimated by the formula:

$$\frac{L_v}{E} = \frac{10}{\theta^2}, \tag{15.4}$$

where θ is the angular distance (in degrees) from the image. This is discussed further in conjunction with the eye's spread function in Section 15.4.1.

15.1.4.4   Directional   Sensitivity—Stiles-Crawford   Effect. The magnitude of the internal visual stimulus is determined primarily by the amount of flux reaching the retina and its spectral composition. In addition to this, the angle of incidence, too, is significant. The drop in effectiveness of illumination with increasing angle of incidence on the retina is known as the Stiles-Crawford effect [23]. Note that light entering through the center of the pupil is incident almost normally on the retina and that increasing the distance of the point of entry from this center implies an increasing angle of incidence: in cone vision, flux entering at 3 mm from the pupil center is only one third as effective as the flux entering at the center. The following formula approximates the relative effectiveness (η) of the flux entering at a distance (r) from the pupil center:

$$\eta = 10^{-pr^2}, \tag{15.5}$$

where, for cone vision, $p = 0.05$ mm^{-2}. In rod vision, the effect is less pronounced. It reduces response significantly only for entry more than 3 mm from the center of the pupil, and then only by approximately 35%—compared to an 80% drop in cone response [24].

15.1.4.5   Unit of Retinal Stimulation—the Troland. As indicated by (15.3), the retinal stimulation is proportional to both field luminance (L) and pupil area (S). Indeed, the product

$$I = LS \tag{15.6}$$

is taken as the magnitude of the stimulus, the unit being the *troland* (Td), when L is in cd/m^2 and S in mm^2. I may be interpreted as the *intensity* of the radiation leaving the pupil, the troland being equivalent to a microcandela.

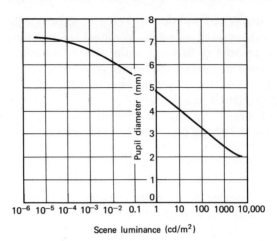

Scene luminance (cd/m²)

Figure 15.9 Pupil diameter versus field luminance, average. After de Groot and Gebhard [30].

To convert object luminance into retinal illumination under free viewing conditions, the pupil size must be known. Assuming this to be given by the curve of Figure 15.9, the relationship between Td and cd/m² is given in Figure 15.10, which may be used for conversion from one to the other.

Occasionally, efforts have been made to include the effect of the Stiles-Crawford effect in computing the effective retinal stimulation, but

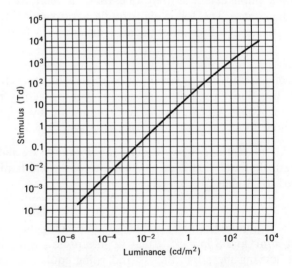

Luminance (cd/m²)

Figure 15.10   Relating troland to cd/m².

this does not seem to have found wide acceptance. Also, absorption and scattering losses in the ocular media are generally ignored when computing the troland value, which is therefore referred to as the *conventional* visual stimulus.

15.1.5 Interocular Summation

We now briefly consider the interaction between the two eyes. The image information from each eye is passed along its optic nerve to the optic *chiasma* located behind the midpoint between the two eyes. There some of the fibers from the two eyes cross over, so that all the fibers pertaining to the right-hand side of the retinas of both eyes come to the right-hand side of the brain and the others to the left-hand side. (Due to the optical reversion of the retinal image, this means that the left-hand side of the field comes to the right-hand side of the brain.)

Normally, information from the two eyes is combined in the brain in such a way, that a single image only is sensed. This is called *fusion* and is possible only if corresponding points in the images in the two eyes fall at approximately equal displacements from the center of the fovea. For an infinitely distant object, this requires the two optical axes to be parallel. For an object at a finite distance, it requires the eyes to converge on the object. This *convergence* on the viewed object is generally an automatic and unconscious process.

In Section 15.8.1 we shall see how slight relative displacements of corresponding image points can serve as important cues in depth perception.

Regarding adaption, the two eyes seem to act almost independently. On the other hand, the irises in the two eyes respond in unison, and the illumination values received by the two eyes seem to carry equal weight in this respect. See Figure 15.11 [25], which shows pupil diameter as a function of luminance with various ratios of right eye to left eye luminance.

15.1.6 Pupil Reflex

In the healthy eye, the diameter of the pupil is variable and influenced by many factors. Among the conditions causing contraction are accommodation and convergence (required for near objects), and eye irritation. Strong emotions and fatigue tend to dilate the pupil. Drugs, too, may be used to cause pupillary contraction or dilation. But the most important factor in pupil control is the illumination striking the retina. Here the pupil reflex obviously tends to stabilize the retinal illumination; but when we note that a change of luminance by a factor of 10^6 is required to

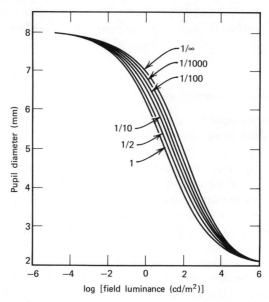

Figure 15.11 Pupil diameter vs. (right eye) field luminance, for various ratios of right eye/left eye luminance. Bartleson [25].

produce a change of 15 in aperture area, it becomes just as obvious that we must look a little deeper to appreciate the significance of this wonderful device.

In terms of acuity performance, we find that the pupil reflex tends to optimize performance [26], [27]. At low light levels, where quantum noise and absolute signal level are the primary limits on perception (see Section 15.6.2) the large pupil optimizes performance. At higher light levels, where the imaging performance of the eye is the controlling factor, the pupil contracts to the size that is optimum from the point of view of image formation: at approximately 2 mm diameter, diffraction effects and aberrations are balanced (see Figure 15.21).

A further important function of the pupil reflex is its indirect contribution to dark adaptation. The contraction of the pupil in high ambient luminance reduces the retinal illumination and, consequently, speeds up dark adaptation dramatically (see Figure 15.35). A 10-fold improvement in effective sensitivity has been observed [28].

Both rods and cones are effective in controlling pupil size, each at its level of effectiveness. Below 1 Td the rods are the major factor and above 100 Td, the cones primarily are in control. It has been shown that below 1 Td, cone stimulation may cause the *expansion* of the pupil, somewhat contracted due to rod stimulation applied simultaneously [29].

Pupil diameter at a given luminance can vary widely—by more than half an octave—between individuals [30]. Average values based on a large number of observers and obtained with a uniformly luminous field are shown in Figure 15.9 [30].

When retinal illumination is nonuniform [2d], visual and pupil-lomotoric effectiveness are quite similar in terms of threshold, including variations with stimulus size and duration and adaption level [31].

When a light-adapted observer enters a dark environment, the pupil expansion takes place in three stages. First there is a rapid dilation, which is completed in less than 10 sec; this is followed by an equally rapid partial contraction, which, in turn, is followed by a slow dilation, requiring approximately $\frac{1}{2}$ hr for its completion. This last stage parallels closely the rate of rhodopsin regeneration [32]. The pupillary contraction on returning to a well-lit environment is essentially complete after approximately 10 sec; see Figure 15.12, which represents pupil reaction following application and removal of a 320-cd/m² field [3a].

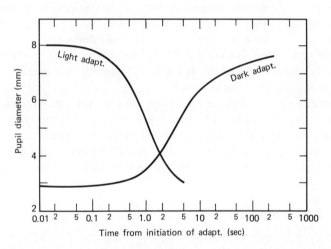

Figure 15.12 Pupil response following application and removal of a 320 cd/m² field. After Reeves [3a].

15.2 PSYCHOPHYSICAL MEASUREMENT: SOME CONCEPTS [33]

When the visual system is considered as a component in a communication system, transfer, spatial, and temporal characteristics are the fundamental parameters of interest. The "noise" of the system is next in importance; for, in conjunction with the other parameters, it sets the limits on system performance. In the following we devote a section to each.

Because the study of these parameters is based on psychophysical concepts and methods, we preface our discussion with a brief presentation of some of these.

In physical systems, the transfer characteristic can be determined by measuring the output corresponding to any input. When human sensation is involved, direct measurement of the output is not feasible and some indirect method must be employed.

15.2.1 Absolute Methods and Interocular Matching

The most obvious method of "measuring" sensation is to ask the subject to assign a numerical value to the sensation level. This is called magnitude estimation and is generally guided by providing a reference stimulus that is arbitrarily taken as, say, 100. Alternatively, two stimuli, straddling the test stimulus, may be provided, and the subject asked to adjust the test stimulus until it is, say, at the midpoint between the reference stimuli. This is called *ratio scaling*.

On the basis of such absolute methods, a brightness scale, with the *bril* as a unit, has been established [33*]. *Brightness* is the psychological correlate of luminance and the bril is the brightness of a microlambert as sensed under standard conditions.[9]

Interocular matching is an important tool in studying retinal effects. Here the sensation obtained by one eye is compared to that obtained by the other as the two views are exposed alternatingly. The effectiveness of the method is based on the observation that processes taking place in the retina of one eye do not affect the functioning of the retina of the other eye.

15.2.2 Threshold Techniques

An alternative approach uses observed thresholds to establish a scale. On the assumption that the sensed magnitude of a barely detectable change in signal is constant, independent of the signal level, a curve relating sensation to stimulus magnitude can be developed. For instance, we may increase luminance from zero until it is just detected. This threshold luminance corresponds, then, to a unit brightness. The luminance is now increased until the change is barely detectable. The required luminance change then again corresponds to a unit change in brightness, and so forth.

---

[9] The standard conditions are: the observer is fully dark adapted, and the object is a 5°-diam disk and is presented for 5 sec.

As means for establishing a scale, this method is rather unreliable, because the underlying assumption (constant sensation difference at threshold) is highly questionable. Nevertheless, more generally threshold measurements are very convenient and provide a powerful tool in the investigation of psychophysical processes.

The reader should not be misled by the term "threshold" as used in this context. In its usual sense, the term implies that an input signal will fail to produce any output until it reaches a certain level, that is, the threshold level. This is not the sense in which it is used here. The detection process in human senses seems to be similar to a physical detection process limited by noise. As the stimulus (input signal) approaches threshold, it is more and more likely to be detected or, in other words, it is detected more frequently. Consequently, we must define the threshold as the stimulus value that is detected with a given probability, say 50%. This point is discussed further in Section 15.6.1.

15.2.3 The "Laws" of Weber and Fechner

In some human senses, such as sight, hearing, and touch, it is found that the threshold (ΔL) varies directly with the background level (L):

$$\frac{\Delta L}{L} = \text{constant} \tag{15.7}$$

over large ranges of background level. This phenomenon is known as Weber's[10] law and the ratio (15.7) as Weber's fraction. If the sensation (x) is assumed to be a logarithmic function of the signal

$$x(L) = a + b \log L, \tag{15.8}$$

then the sensation change (Δx) due to small change (ΔL) in the stimulus will be given by

$$\Delta x \approx \Delta L \frac{dx}{dL} = b \frac{\Delta L}{L}, \tag{15.9}$$

which is observed to be constant. Thus the assumption of a constant Δx at threshold implies a logarithmic response. This is called Fechner's[11] law, but its validity has been challenged repeatedly [34].

15.2.4 Signal Detection Theory [35], [36]

The decision of an observer whether, or not, a signal was present, is a complex process. Signal detection theory is an effort to analyze it in terms

---

[10] E. H. Weber (1834).
[11] G. T. Fechner (1860).

of signal theory by means of an assumed underlying model. In this theory, the sensation obtained in the absence and presence of the signal are both presumed to be disturbed by "noise" (see Figure 15.13). Here the dashed curve represents the probability density of sensation received from the background, that is, in the absence of signal. This curve is centered on x_b. The solid curve represents the density in the presence of signal which, on the average, results in a sensation level x_s.

The observer now observes values x and must decide whether they are due to background or signal. In the case depicted in the figure, large values of $x(x > x_2)$ are clearly due to a signal; the background alone is very unlikely to yield such a value. Similarly small values of $x(x < x_1)$ are clearly due to background alone. However, any value of x between x_1 and x_2 could be due to either curve and, consequently, leaves the observer uncertain concerning the signal. To cover all cases, the observer may choose a decision level, x_d, and opt for "signal" whenever $x > x_d$ and "background" otherwise.

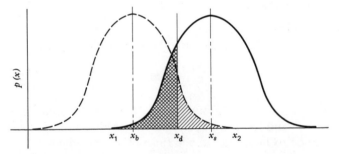

Figure 15.13   The probability density $p(x)$, of a sensation level x. Solid curve: in the presence of signal. Dashed curve: background only. x_b and x_s are the mean sensation values in the absence and presence, respectively, of the signal. x_d is the presumed decision level.

There is a certain error probability inherent in this procedure, but it is the best he can do under the circumstances. The error probabilities are indicated in the figure. The shaded area to the right of x_d represents the probability of deciding "signal" when, in fact, the background was responsible $[P(S' | B)]$. This type of error is called a *false alarm*. The cross-hatched area to the left of x_d, represents the probability of a "background" decision when, in fact, a signal was present $[P(B' | S)]$. This error is called a *miss*. The area under the solid curve and to the right of $x_d [P(S' | S)]$ represents the *hit* probability. Clearly, the hit and error rates depend on the location of x_d, and the observer will change this location depending on the probability of signal and background and on the relative cost of false alarms and misses.

The plot of hit rate versus false alarm rate, as the decision level is varied, is called the receiver operating characteristic (ROC). On this curve, purely random behavior appears as a 45° line through the origin. The area between this line and the ROC curve is a measure of the discrimination between the two alternatives under study. When noise is Gaussian, with standard deviation σ, the value

$$d' = \frac{x_s - x_b}{\sigma} \tag{15.10}$$

is called the *index of sensitivity* and is a measure of this discrimination.

15.2.5 Specific Experimental Techniques [1b]

In conclusion, we describe briefly a number of experimental techniques used to determine stimulus thresholds or matches.

Method of Adjustment. The observer may be asked to adjust a stimulus until he just perceives it or until it matches perfectly a comparison stimulus. This is the method of adjustment and is often the simplest and fastest to use. It is, however, highly susceptible to individual differences and suffers from poor repeatability [37].

Forced Choice. In the forced choice method, the observer is asked to provide some information concerning the signal and, if he does not have this information, to make his best guess. He may, for instance, be asked which of two signals was brighter, in which quadrant of his field did he observe a flash, or whether the lines in a barely resolved grating were horizontal or vertical. In this method, pure guessing is part of the response, and this must be allowed for in the analysis of the data. On the other hand, even subliminal (i.e., below threshold) perception can influence the response to some extent and, consequently, far higher sensitivity values are obtained [37]. This method is, however, much more time consuming than is the method of adjustment.

Method of Limits. The experimenter may raise the signal level in a series of predetermined steps, starting from below threshold, until the observer responds positively. The threshold is then presumed to fall between this level and the preceding one. The procedure is then repeated, starting from above threshold and proceeding until the observer fails to detect the signal. The threshold is estimated as the average obtained from a number of such cycles. This is the method of limits.

Method of Constant Stimuli. In the method of constant stimuli, a number of fixed levels straddling the threshold are presented to the observer many times and in random order. He is asked each time whether he detected it (or whether it differed from a comparison signal). Statistics are compiled on the detection probability as a function of the signal level. A plot of the results permits the determination of a threshold—at 50% detection, for instance. If the probability is normally distributed and the results are plotted on "probability graph" paper, the points should fall on a straight line. This fact permits a more accurate estimate, obtained by optimizing the fit of a straight line drawn through the points.

15.3 TRANSFER CHARACTERISTIC

15.3.1 Luminance Range of Human Vision

Human vision ranges from an absolute threshold of approximately $1 \mu cd/m^2$ $(40 \mu Td)$ to approximately $30 mcd/m^2$ $(10^8 Td)$ where the rhodopsin becomes just about fully bleached after 1 sec. At approximately 30 times this level, coagulation of the receptor substance commences. See Figure 15.14 [38] where threshold and saturation points are indicated. This implies that human vision can accommodate a luminance range of 14 decades, probably unmatched in our technology. At the same time, human brightness sensation ranges over approximately three decades only. Matching to this an input range a hundred billion times as large would seem to pose a major engineering problem.

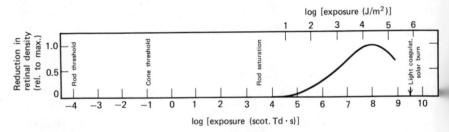

Figure 15.14 Normalized retinal density change as a function of log exposure. Threshold and saturation levels, too, are indicated. After Weale [38].

The solution to this problem is essentially twofold.

1. The instantaneous operating range is limited to about four decades, with the operating point moving along the input scale.

2. Within the instantaneous operating range, the response is nonlinear, with the sensed brightness rising relatively slowly with the stimulus luminance.

There are essentially four processes involved in setting the operating point.

1. *The duality of the retina* (see Section 15.1.3.3). At levels below 3 mcd/m^2 (0.1 Td), the rod output controls and the cones are effectively inoperative (scotopic vision). The rods thus cover a range of only approximately 3.5 decades.

Above this level the cone output tends to control, leaving a total of over 10 decades to these.

2. *Pupil reflex* (see Section 15.1.6). Over a decade is added to the visual range by the pupil reflex, which can change the ratio of retinal illumination to object luminance by a factor of approximately 15.

3. At very high levels, the bleaching of the visual pigments is bound to reduce the sensitivity of the detector units in the retina. This mechanism is significant only in cone vision.

4. The remaining shifting of the operating point, the lion's share, is due to various adaptation processes, as discussed below in Sections 15.3.2 and further in Section 15.5.1.4.

15.3.2 Brightness Function

The transfer characteristic of the visual system, that is, the dependence of the brightness on the stimulus luminance, is called the *brightness function*. It is strongly influenced by the adaption level, by the size of the object field, and by its duration. It is little influenced by the wavelength and even the location on the retina [39]. In this section we consider only the steady-state effects; the effects of signal duration are treated in Section 15.5.

15.3.2.1 Large Fields. At the lowest luminance levels, to about $2 \text{ } \mu\text{cd/m}^2$, the response is close to linear:

$$B = 5000 \, L \text{ bril}, \tag{15.11}$$

where L is in cd/m^2. At higher levels, the response varies roughly with the $\frac{1}{3}$ power. For the dark-adapted eye, it can be approximated by:

$$B = 0.68 \sqrt[3]{L} \tag{15.12}$$

and more generally by [33*]

$$B = k(L - L_0)^\beta, \qquad 10^{-4} < L < 10^4, \tag{15.13}$$

where k, L_0, and β are constants determined by the adaptation luminance. The form of this function is quite universal, applying to very small fields [39] as well as to a *ganzfeld* [40], where the total field of view is uniform. Only the values of the "constants" vary from case to case [41]. For intermediate object size (5°) and steady exposure, the values of the constants are as given in Figure 15.15 [34]. The compression of the response $(\beta < 1)$ is due to processes at both plexiform layers [39]. At reasonably low illumination levels, the bleaching of the photopigment can be taken as directly proportional to the retinal exposure. In conjunction with the inhibitory responses generated and mediated by the horizontal cells [42], [43], this produces at the bipolar cells a response varying with the 0.7 power [39]. At the ganglion cells, the response is found to be proportional to the square root of this [44], i.e. directly proportional to the sensed brightness.

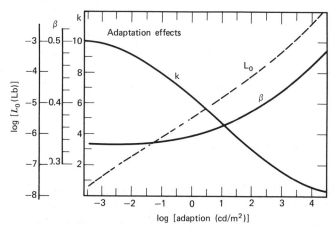

Figure 15.15 Constants for use in (15.13). From Stevens and Stevens [33]. Note: The values of L_0 are in lamberts and the values of k too, are appropriate for L-values in lamberts.

15.3.2.2 Effect of Field Size. When the luminous object subtends a small field, close to the resolution limit of the eye, the brightness function is found to have a square root, rather than a cube root form [39]. This change in response is presumably due to the absence of inhibitory influence from the surrounding detectors.

The power-law form of the brightness function has also been confirmed by means of measurement of the pupil reflex threshold. For the dark adapted eye, the exponent was found to go from 0.28 to 0.65 as the object diameter dropped from 10° to 0.6°. For a 5°-diam object, it rose from 0.35 to 0.6 as the luminance rose from 0 to 35 kTd. [41]

15.3.2.3 Fully Adapted State. The brightness sensed when the eye is fully adapted to a uniform field of luminance L appears to follow the form [45]:

$$B = \frac{B_\infty I^\beta}{I^\beta + K} \tag{15.14}$$

with an exponent $\beta = 0.32$ and

$$B_\infty = 37.5 \text{ bril}, \qquad K = 19. \qquad (15.15)$$

This implies a maximum sensed brightness of 37.5 bril in the adapted state and accounts partly for the observed phenomenon of brightness constancy: at photopic levels the sensed brightness changes little when the ambient illumination changes greatly, as discussed in Section 15.3.4. Neglecting the exponent, (15.14) corresponds both to a simple automatic gain control and to elementary photochemical models [46].

15.3.2.4 Dazzle. When the adaption level is much lower than the object luminance, a type of visual discomfort called *dazzle* may be experienced. This occurs for instance when viewing a very bright small object against a dark background. (To avoid dazzle, projection slides should consist of dark lines on a bright background rather than vice versa.)

15.3.3 Luminance Thresholds

The detection of a luminance difference is one of the basic functions of vision. The luminance that can just be distinguished from total darkness is called the absolute luminance threshold. Both absolute and difference thresholds depend, among other factors, on object size and, to the extent that this is true, they can be treated as acuity thresholds (see Section 15.4.3). We treat them here, however, since the emphasis is primarily on the luminance level and the object size is viewed as a secondary parameter.

15.3.3.1 Absolute Threshold [2e]. The absolute threshold is influenced by many factors, including the part of the retina on which the image appears; the dimensions of the luminous patch; the time interval over which it is viewed; the spectral composition of the light; and the history of the retina over the preceding hours.

Receptor Type. At low luminance levels detection can be obtained only by means of the rods and therefore extrafoveal vision must be employed. The switchover is often unconscious and appears on threshold curves as a discontinuity of the slope at approximately 2.5 mcd/m^2 (see Figures 15.18 and 15.35). Under optimum conditions the absolute thresholds for cones and rods are approximately 1 mcd/m^2 and 1 μcd/m^2, respectively, with values as low as 0.3 μcd/m^2 reported occasionally.

Object size. The dependence of absolute threshold on size is shown in Figure 15.16 [2e]. Note that below 40 min the slope of the curve is almost

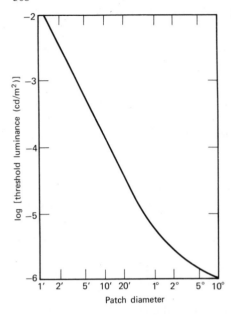

Figure 15.16 Absolute threshold as a function of angular diameter. After Le Grand [2e].

exactly -2, that is, the threshold luminance varies inversely with the solid angle. This is an expression of Ricco's[12] law: at threshold

$$L\Omega = \text{constant},$$

$$\approx 10^{-9}\ \text{nit} \cdot \text{sr, according to Figure 15.16.} \quad (15.16)$$

For large object size it approaches unity, implying that

$$L\sqrt{\Omega} = \text{constant, at threshold.} \quad (15.17)$$

This is called Piper's[13] formula. (Piéron's[14] formula states that $L\sqrt[3]{\Omega}$ is constant.) See Appendix 15.1.2 for the theoretical bases of (15.16) and (15.17).

Viewing Time. Since the signal energy depends on the product of flux (Φ) and time (t), these show a reciprocal relationship

$$\Phi t = \text{constant, at threshold,} \quad (15.18)$$

provided that t does not exceed the integration time of vision. This is known as Bloch's[15] law. The integration time is approximately 0.1 sec at low levels and drops at higher luminance levels (see Section 15.5.1.2).

[12] A. Ricco (1877).
[13] H. Piper (1903).
[14] H. Piéron (1920).
[15] A. M. Bloch (1885).

Spectral Dependence. The spectral dependence of threshold, for both foveal and extrafoveal vision, is presented in Figure 15.17 [47]. As expected, these follow the respective luminous efficiency curves, with a significant deviation occurring only in the photopic curve below 0.47 μm.

Adaptation. Effects of earlier exposures and subsequent dark adaptation are summarized in Table 161 (see Section 15.5.1.4).

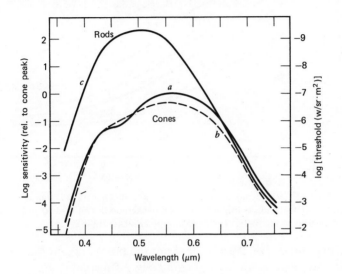

Figure 15.17 Dependence of absolute threshold on wavelength. (*a*) Foveal cones; (*b*) Cones, 8° above fovea; (*c*) Rods, 8° above fovea. After Wald [47].

15.3.3.2 Modulation Contrast Threshold.

When the object is seen against a luminous background, the threshold must be expected to rise. According to Weber's law, the rise should be directly proportional to the background adaptation level, so that the modulation contrast[16] threshold should remain constant. In fact, the increase is less than that, at least at the lower luminance levels. There, the increase in luminance raises the signal-to-quantum-noise ratio [see (3.98)] and therefore lowers the contrast threshold. This is discussed further in Section 15.6.2.

Results from an extensive set of tests are shown in Figure 15.18 [48]. This shows clearly the drop of contrast threshold with increasing background adaptation luminance. It also shows how the threshold drops with

[16] Modulation contrast is defined as the ratio of the luminance increase to the background luminance. For brevity, we simply write "contrast" in the remainder of this chapter. In the present context, this should not lead to any ambiguity.

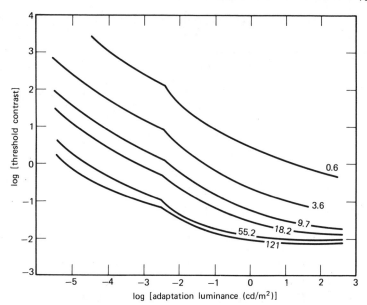

Figure 15.18 Threshold modulation contrast as a function of adaptation (background) luminance for circular disks brighter than the background, with angular disk diameter (in minutes) marked at the end of each curve. After Blackwell [48].

increasing object size. Also evident are the breaks due to the changeover from rod to cone vision at approximately 2.5 mcd/m².

15.3.4 Brightness Constancy and Simultaneous Contrast

15.3.4.1 Brightness Constancy [7c]. When we look at a piece of white paper in a relatively dimly lit room, it appears just as white to us as it appears when viewed in bright sunlight. This is true even though its physical luminance may be lower by a factor of 1000 and a sheet of paper having the same luminance, when viewed in sunlight, would appear black. This is an illustration of the psychophysical phenomenon called *brightness constancy*: two objects with equal reflectivity will be judged equally bright even though they are viewed under widely differing illumination conditions, so that they produce widely different retinal illumination values. It is as if the viewer could sense the reflectivity directly when, in fact, all he can sense reasonably directly is the luminance. In general, brightness constancy must be based on an evaluation of the illumination level; each area element is sensed relative to some average of the luminance of its environment. Adaptation is a major factor contributing to brightness constancy. Here we discuss some additional contributors.

15.3.4.2 Simultaneous Contrast. The brightness sensed at any point is strongly influenced by the luminance of its immediate environment. This effect is called *simultaneous contrast* or *spatial induction*. When negative it is occasionally called *masking*. If two equally luminous patches A and B are viewed in a uniform background field, they will, of course, appear equally bright. If now a relatively bright object is introduced near A, A will appear less bright than B. If a dark patch is introduced, instead, A's brightness will be enhanced. The effectiveness of this induction appears to follow the form [49]

$$B_t = a\frac{L_t^n - kL_i^n}{1 - k^2},$$ (15.19)

where B_t is the observed brightness of the test patch,

L_t, L_i are the luminance values of the test patch and the inducing field, respectively,

$n(\approx \frac{1}{3})$ is the exponent of the brightness function, and

k is the induction coefficient, which has been found to be 0.1 and 0.6, respectively, for adjacent and surrounding inducing fields.

A test patch will appear darker when it is surrounded by a narrow, more luminous region. As the width of this region increases, the darkening effect, too, increases up to a width of approximately 3 min. Any further increase in its width reduces the effect; the additional area has an inhibitory effect on the induction. This effect is related to the spatial summation of retinal receptor outputs and has been studied extensively. Results for various retinal locations are shown in Figure 15.19 [50]. A small test area of fixed luminance was flashed at the center of a circular inducing field. As the inducing field area was increased, its luminance had to be reduced to prevent it from masking the test flash. However, beyond a certain point, the inducing field luminance could be raised. The curves of Figure 15.19 show the luminance of the inducing field as a function of its area for various eccentricities.

Induction means, in effect, that the brightness of each area element is sensed relative to the average of its immediate environment; this is obviously an important factor in brightness constancy. In this context it is occasionally called *simultaneous contrast*.

This phenomenon can also be illustrated by the well-known optical illusion where a grey patch in a black surround will appear much brighter than an identical patch surrounded by a white region.

A related illusion [7c] is based on the fact that the visual apparatus fails to detect a gradual change in luminance, provided the gradient is sufficiently low. Consider two identical rectangular patches adjacent to each other as shown in Figure 15.20a. A density trace across this pair is shown

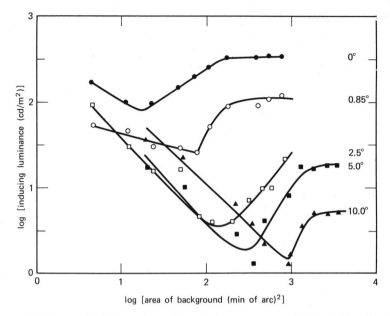

log [area of background (min of arc)²]

Figure 15.19 Inducing field luminance just sufficient to mask a small flash (2' diam, 0.015 sec duration) of fixed luminance, as a function of its solid angle, with retinal eccentricity as a parameter. Westheimer [50].

in Figure 15.20b. Because of the small slope of the density pattern, each of the patches appears uniform. At the boundary, however, there is a sharp step, readily detected by the observer. This step is automatically interpreted as showing that the left-hand patch is more dense [51]. The illusion can be destroyed by eliminating the sharp transition, for example, by placing a pencil over the boundary [52]. Contrasts as high as 50%

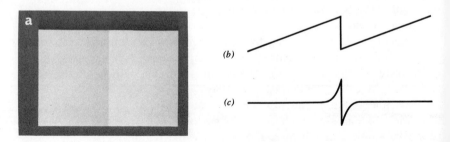

Figure 15.20 (a) Two identical patches exhibiting an apparent brightness difference. Note that any small break, such as a pencil placed across the boundary, destroys the illusion. Land & McCann [52]. (b) A density trace across the pattern of (a). (c) Density trace for Cornsweet illusion pattern.

are obtained with this phenomenon, and, for contrasts up to 20%, the perceived contrast equals that implied by the step height [52].

Closely related to the above is the *Cornsweet illusion* [7a]. Here the two patches are both identical and uniform but are separated by a "spur" (see Figure 15.20c for the corresponding density trace). Here, again, the steep transition is perceived as such and the slow transition is ignored, so that, again, the left-hand patch appears darker.

15.3.4.3 Retinex Theory [52]. These phenomena are fundamental to the retinex theory, according to which the relative brightness of two separate patches is estimated by summing all the boundary steps between them. This permits sensible brightness constancy in a scene, even if substantial nonuniformities of illumination are introduced, provided these nonuniformities are gradual.

On the assumption that there is a triad of independent receptor arrays responsible for color vision, retinex theory applied to each of these individually can account for color constancy and induction effects [53].

15.4 SPATIAL CHARACTERISTICS

The visual system is conveniently divided into four stages, each having its own impact on the system imaging performance.

1. *Optical Stage.* This refers to the imaging performed by the eye itself, including the focusing action of cornea and lens and the scattering of light in retina and lens.

2. *Retinal Receptor Stage.* This includes the transduction of the electromagnetic energy into a neural signal.

3. *Retinal Neural Stage.* This includes the signal processing taking place in the retina, which is a major determinant of the transfer characteristic. In view of the extensive transverse interconnections among retinal receptors and nerves, we expect this stage to have a significant impact on the spread function as well.

4. *Central Neurological Stage.* Some image processing appears to take place in the brain also. This can generally be identified by the fact that the effects of stimulation on one eye will affect the other eye only if the process is central.

Each of these stages influences the spatial performance of vision. The effects of the first stage can be treated in terms of geometrical and physical optics and that of the second stage in terms of sampling theory (Section 3.3). The latter two stages are less accessible to quantitative analysis and must, at present, be investigated empirically.

15.4.1 The Optical Stage

Just as in any optical system, three primary factors contribute to the spreading of the light in the retinal image of a point source: diffraction, aberrations, and scattering. As noted above (Section 15.1.3.1) the receptor density permits high acuity only in the fovea, especially in its central region over a diameter of a fraction of a degree. Only in this region is the optical performance of the eye a limiting factor. With such a small field, the only primary aberrations that need be considered are spherical and chromatic aberrations.

Spherical aberration is found to be considerably lower than expected on the basis of the spherical surfaces of the Le Grand model [54] due to the flattening of the cornea away from the apex and the lowering of the refractive index of the lens toward its edge. For the unaccommodated eye the spherical aberration can be approximated by

$$\Delta v = 0.16 r^2 \text{ diopter}, \tag{15.20}$$

where r is the radial distance of the pupil zone in millimeters [54]. During accommodation, the spherical aberration tends to vanish and even to change sign [55].

Chromatic aberration has been found to amount to a difference in power of about 2.5 diopter between 0.4 and 0.7 μm. This corresponds to a difference in focal length of approximately 0.7 mm [56].

The spread function of the ocular optics has been measured by observing, through the pupil, the retinal image of a point source. Assuming the reflection at the retina to be essentially diffuse, the observed distribution will be the result of passing the ocular optics twice and, therefore, should represent the autoconvolution of the actual spread function. In terms of the derived transfer function, the single-passage modulation transfer function (mtf) should be the square root of the mtf derived from the observation. Considerable data have been accumulated in this manner [57], [58]. Representative mtf curves, calculated from such data, are shown in Figure 15.21 [58]. For a 2-mm pupil diameter, the total diameter of the point spread function for white light is found to be approximately 1 min, about equal to that of the Airy disc.

For the smallest pupil diameters, the mtf is set primarily by diffraction effects, so that performance could be improved by increasing the pupil diameter. At higher spatial frequencies this can indeed be observed: beyond 30 cycles/deg, pupil diameters of 3 and 3.8 mm permit improved performance. But over most of the spatial frequency spectrum, the 2-mm pupil diameter permits better performance because the reduced aberration effects more than compensate for the increased diffraction. Mtf

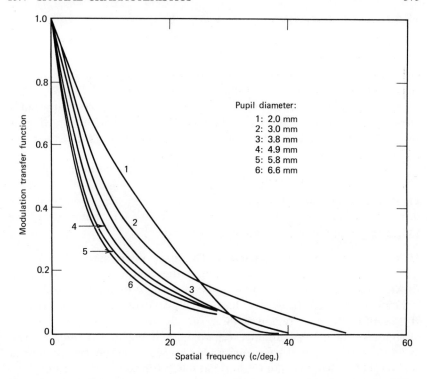

Figure 15.21 Mtf of ocular optics for various pupil diameters. Campbell and Gubish [58].

curves have also been calculated on the basis of measured aberrations [54].

With eccentricity, the spread function deteriorates only slightly up to 20° and quite rapidly beyond. At 30° it is more than double its width in the foveal region. [58*]

Light scattering in the eye becomes significant when a very bright spot is located in the field of view. Here scattered light may mask image details, even if the scattering is relatively small. Such masking, including both physical and psychological factors, is called *glare*. Its effect has been found to be representable by an equivalent veiling luminance (L_v) given approximately by the following formulae [59]:

$$L_v \approx \frac{29E}{(\theta + 0.13)^{2.8}} \text{ cd/m}^2, \qquad 0.15° < \theta < 4° \qquad (15.21a)$$

$$\approx \frac{10E}{\theta^2} \text{ cd/m}^2, \qquad 4° < \theta < 100°. \qquad (15.21b)$$

Here E is the illumination at the pupil (in lux), and
   θ is the angular distance from the point image and the factor 10
   tends to increase to approximately 25 with age.

Data covering the total spread function for foveal imaging, at various
pupil diameters, are given in Table 162 [59].

Approximately half the image flux falls within a radius of 1.2 min of a
point source image when the pupil diameter is 2 mm. This radius is
doubled for a 6.6-mm pupil diameter. A representative good retinal line
spread function (at 80 cd/m²) can be taken as [60]

$$L(x) = 0.29e^{-3.3x^2} + 0.33e^{-0.93|x|}, \qquad (15.22)$$

where x is in minutes of arc.

15.4.2 System Spread Function

15.4.2.1 Shape of Spread Function. The total system spread function is
far more complex because of the nonlinearities and receptor interactions;
also its determination is far more difficult because of the psychophysical
techniques required. In general terms, the point spread function consists
of a central circular region with a sharply peaked positive response
surrounded by a wide annular region with a negative response. The
response at high spatial frequencies is governed primarily by the central
peaked region, whereas the response at low frequencies is determined, to
a great extent, by the characteristics of the negative annulus.

Determinations in the foveal and parafoveal regions have shown that
the line spread function there can be approximated by a difference
between two Gaussian distributions: a larger positive one, with a small
standard deviation, combined with a smaller negative one, with a large
standard deviation. Specifically, the line spread function can be written

$$L(x) = A\frac{e^{-x^2/2a^2} - ce^{-x^2/2b^2}}{\sqrt{2\pi}\,a}, \qquad (15.23)$$

where a, b are the standard deviations of the two Gaussians,
   A, B are their areas, and
   c is short for $c = aB/bA$.

For normalization $A - B = 1$. Extensive measurements on three observers
have shown that

$$\frac{b}{a} = 1.5$$

throughout the region checked (to 2.5° from the fovea). At a luminance of 30 cd/m^2, the ratio A/B ranges from approximately 1.2 at the center of the fovea to approximately 1.35 at a point 2.5° away from it. The values of a are approximately 0.038° at the center and 0.044° at 2.5° [61]. Thus

$$L(x) = 1.11e^{-0.096x^2} - 0.61e^{-0.041x^2} \qquad (15.24)$$

is a representative foveal system spread function. Here x is again in minutes of arc.

Another investigator [62] found the following to be a good approximation to the visual point spread function at 1300 Td:

$$P(r) = (6 - 9r^2)(r^2 + 1)^{-7/2} \qquad (15.25)$$

with r again in minutes of arc.

Clearly, none of these approximations should be relied upon too heavily.

Another limitation is illustrated by the fact that the shape of the spread function changes with the intensity. Because of the nonlinearity of the retinal response as discussed in Section 5.2.3, Fourier transform is not applicable. Furthermore, the retina is far from homogeneous (see Figures 15.4 and 15.5), so that the lack of isoplanaticity, too, makes the application of Fourier theory questionable. Nevertheless, for limited areas on the retina and small values of modulation, the response approximates linearity [63] with an effective gain of γ, the fractional differential gain at the operating point (see Section 18.4.1.1). In view of the power-law response of the visual system, γ is constant. Experimental evidence indicates that at spatial frequencies below 3 cycles/deg, even high modulation values may be treated as linear if a logarithmic response is allowed for [63].

15.4.2.2 Mtf Measurements. Below we summarize the results of spatial frequency response measurements. In this discussion, we use the term "mtf" in quotes indicating that the data are valid only for sinusoidal signals.

Extensive measurements of the spatial frequency threshold spectrum have been made; these are discussed in Section 15.4.3.3. On the very questionable assumption of constant signal at threshold, the mtf can be taken as the reciprocal of these.

Spatial frequency response measurements have also been made on the retina-brain section alone. By coherently illuminating two narrow slits in front of the pupil with monochromatic light, Young's interference fringes are generated on the retina [64]. These are but little affected by aberrations and focusing errors in the optical system. A representative threshold curve obtained with this technique is shown in Figure 15.22 [65]. This and other techniques for establishing the spatial frequency response of the visual system have been reviewed in Refs. 66 and 67.

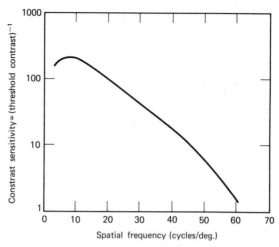

Figure 15.22   Spatial frequency response of retina-brain section of human visual system. Campbell [65].

Suprathreshold mesurements of the visual "mtf" have been made by matching the apparent luminances of a sinusoidal luminance pattern (at peak and trough) with a probe luminance adjacent to the pattern. Results of such measurements are shown in Figure 15.23 [68]. It should be noted, however, that the effects of induction, which are essential in forming the "mtf," will also affect the brightness of the probing luminance; therefore the results should be taken as no more than qualitative. Suprathreshold measurements obtained with the absolute method, too, have been published [69]. Others [70] have found the mtf to drop off as $e^{-\nu}$ at high

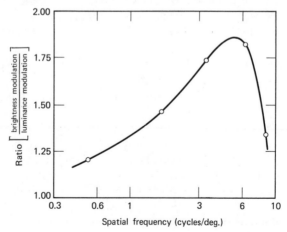

Figure 15.23   Suprathreshold spatial frequency response. Bryngdahl [68].

frequencies and the frequency of peak response to drop from 5 cycles deg (at $17 \, cd/m^2$) to 1.5 cycles/deg (at $17 \, mcd/m^2$); it is below 0.5 cycles/deg at $1.7 \, mcd/m^2$.

15.4.2.3 Effects of Inhibition. Mach Bands.

A common feature of all these mtf curves is a dropping off of the response toward very low spatial frequencies, generating a response peak in the region between 1 and 6 cycles/deg. This is a consequence of the negative annulus surrounding the positive disk of the point spread function (see end of Appendix 18.1) and can be ascribed to the lateral inhibition in the retina, to insensitivity to low luminance gradients, and to adaptation (all interrelated effects). These become less pronounced at lower luminance levels and, hence, the peak of the response spectrum shifts to lower frequencies as the luminance drops [67a].

A second consequence of the negative annulus is an increase in the brightness of small luminous objects especially if they are smaller than the positive central disk of the spread function. Since it is the spatial analog of the temporal Broca-Sulzer effect (Section 15.5.1.2) that name has, on occasion, been applied to this phenomenon [71].

At sharp transitions between two luminance levels, the negative annulus should lead to an overshoot phenomenon, that is, a bright band should appear in the higher luminance region, adjacent to the transition, and a dark band similarly in the lower luminance region. Such bands, called after Mach,[17] are indeed observed. They become more pronounced up to a point as the transition becomes steeper: for transitions taking place within a resolution element, they almost disappear. This disappearance points to nonlinear processes in the visual system [72] such as truncation or "clipping," following the spatial interactions. These processes do not seem to have been identified as yet [72].

A practical consequence of the Mach phenomenon is a significant enhancement of luminance transitions in the field of view. Such transitions may lead to brightness steps, and brightness slopes, considerably greater than the underlying luminance step and slope, making the transition more readily detectable. This form of response emphasizes contours. These are, perhaps, the major perceptual characteristics of the viewed scene, so that the emphasis may be a major advantage.

In addition, this characteristic compensates partly for the blurring introduced by the diffraction and aberrations introduced by the optical stage. These cause a descending mtf, which is partly compensated for by the "mtf" of the retina-brain section, which rises at low spatial frequencies.

[17] E. Mach (1865).

The Mach phenomenon extends over approximately 0.1° and varies with retinal illumination and locality [73]. Another phenomenon, the brightness slope appearing across the width of a density step in a density staircase, extends over much larger angles (approximately 3°) and its origin has been termed Hering-type[18] inhibition [74]; see also Section 15.3.4.2.

15.4.2.4 Variation with Luminance and Viewing Time. From the point of view of optimizing performance, it would be desirable to have the system spread function vary with luminance: the spread function should be as narrow as possible at high luminance levels and broaden at lower levels to permit greater integration of the limited number of photons received from each element. The Designer of man does not disappoint us on this score, either. There are strong indications that the region of integration drops with the square root of the luminance, just as called for by these considerations.

The determining quantity here is the total number of quanta received from the element. This is set by both spatial and temporal integration intervals and the temporal integration interval, too, drops roughly with the square root of the luminance (L). Thus the total integration volume [75]

$$\int_x \int_y \int_t \frac{P}{P(0)}\, dx\, dy\, dt \sim L^{-1/4}. \qquad (15.26)$$

15.4.2.5 Variation with Wavelength and Retinal Location. The effect of wavelength on imaging performance can be accounted for almost completely in terms of diffraction effects and chromatic aberration [66], [76].

The variation of the system spread function with retinal location has been studied in terms of acuity and is treated in greater detail in the next section. Qualitatively, it is found that in the central region, up to a radius of approximately 2°, the visual system performance approaches that implied by the density of receptors and that optical effects are significant up to a radius of approximately 4.5°. Beyond that, the continued drop in acuity seems to be due only to the neural interconnections in the retina, with very many receptors pooling their output into a single optic nerve fiber.

15.4.3 Resolution and Acutance [1c]

15.4.3.1 Definitions. Spatial threshold performance of an optical system is usually stated in terms of *resolution*: the ability to separate, or

[18] E. Hering (1920).

resolve, two objects, that is, to recognize them as two. In vision, the term *acuity* is often used instead. Strictly, this term refers to sharpness, that is, to the sharpness of the brightness slope; but here it is used as a measure of the ability to recognize detail. Specifically, acuity is defined as the reciprocal of the angle subtended, at the eye, by the resolved object detail. When the acuity angle is measured in minutes of arc, the reciprocal is called *decimal acuity* (*A**) [66]. *Visual efficiency* has been defined as [66]

$$\eta_v = 0.836^{(1/A^*)-1} \qquad (15.27)$$

as a measure of visual capacity in terms of general functioning.

Below we list the types of object frequently used for visual acuity tests, with the defining detail given in brackets. They are illustrated in Figure 15.24:

1. Single line [line width].
2. Vernier arrangement [misalignment].
3. Bar pattern [space between bars].
4. Landolt ring [gap width].
5. Snellen letters [thickness of stroke].
6. Sinusoidal pattern [half-period].

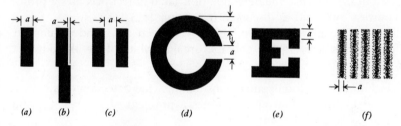

(a)   (b)   (c)   (d)   (e)   (f)

Figure 15.24   Acuity test objects (*a*) Single bar; (*b*) vernier; (*c*) bar pair; (*d*) Landolt "C"; (*e*) Snellen "E"; (*f*) sinusoidal bar pattern. In each the defining detail is shown as *a*.

The relationships between the various measures of acuity and spatial frequency are given below for an object testing for a decimal acuity *x*:

Acuity: $x[\text{min}^{-1}] = 60x\,[\text{deg}^{-1}] = \dfrac{1080}{\pi}\,x\,[\text{rad}^{-1}] = \dfrac{1.08}{\pi}\,x\,[\text{mrad}^{-1}]$

Spatial frequency: $0.5x\,[\text{cycles/min}] = 30x\,[\text{cycles/deg}]$

$$= \dfrac{540}{\pi}\,x\,[\text{cycles/rad}] = 90x\,[\text{cycles/mm}]; \qquad (15.28)$$

the last value refers to the spatial frequency on the retina assuming for the eye an effective focal length of 17 mm.

Threshold performance with patterns 1–5 is discussed in the next section and that with the sinusoidal bar pattern in Section 15.4.3.3.

15.4.3.2 Acuity Limits for Various Objects

Bright Line. The acuity of a bright line on a dark background is limited only by its luminance: an infinitely luminous line will yield an infinite acuity. This becomes obvious when we think in terms of the retinal image formed. Due to diffraction and aberration effects, an infinitesimal point will be imaged as a disk with a luminance distribution similar to the point spread function of the eye and independent of the actual shape of the infinitesimal object. The image of any real object will be almost identical to this, provided its angular subtense is very small compared to the angular subtense of the central maximum of the point spread function, approximately 1 min. Thus the visual system can not possibly differentiate between a small circular disk of area A and luminance L, and another of area $A/2$ and luminance $2L$. The sensed parameter is the intensity:

$$I = \int_{area} L \, dA = LA, \qquad (15.29)$$

where the last member applies when the luminance is uniform. When viewed in these terms, the detection of a small object becomes a problem in contrast threshold as discussed in Section 15.3.3.2.

The reciprocal relationship (15.29) between area and luminance is called Ricco's law (15.16) and the area at which it breaks down, the *critical area.*

Dark Line. In a totally analogous manner, the detection of a dark line on a bright background also becomes a problem of contrast threshold. Here Ricco's law takes the form:

$$wC = \text{constant, at threshold.} \qquad (15.30)$$

Here w is the line width and C is the modulation contrast. Now, however, there is a definite limit to the observed acuity: the contrast, being negative, can never exceed unity and hence there is a lower limit to the detectable width. Under favorable circumstances, dark lines can be detected down to a width of $0.5''$, a feat that is truly amazing in view of the fact that the eye's spread function is a hundred times as wide.

The fact that the line detection task is essentially one of contrast threshold implies that acuity will rise with luminance. This dependence is shown in Figure 15.25 [77]. The total number of photons absorbed per element can also be increased by increasing the viewing time up to the limit imposed by the integration time (see Section 15.5.1.2). As in Bloch's

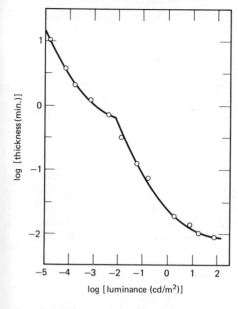

Figure 15.25 Dark-line acuity dependence on luminance. Hecht and Mintz [77].

law (15.18), viewing time and threshold line width are reciprocally related up to approximately 0.1 sec.

Localization. The Vernier arrangement (Figure 15.24b) is an example of localization acuity. In such tasks, visual acuity is again amazingly good. Displacements of only 2″ can occasionally be detected. Observers asked to align two abutting bars accomplished this with a standard deviation of 5–6″ [78]. When monochromatic light is used, the optimum spectral region varies from observer to observer [78]. Image motion may leave acuity performance unaffected, even if the image traverses tens of cones during the integration period of 0.1 sec [79]. In another localization task, where a thin line is to bisect a small square, the acuity angle was found to be 4″ [60].

Acuity performance on single lines and localization are evidently better, by an order of magnitude, than implied by the size of the receptor element. Such performance has been termed *hyperacuity*. In connection with line detection, we have explained it quite simply as a contrast discrimination task. In a localization task, we note analogously, that the localization of the maximum of a smooth curve, too, is a contrast discrimination task.

Resolution. Resolution targets of the form of Figure 15.24c typically yield an optimum acuity angle of 20″ [60] and patterns with many bars, one of 25″. Both the number of lines in the pattern and their length influence acuity. As expected from photon noise considerations, acuity at a given contrast rises with background luminance. This is illustrated in Figure 15.26 [80]. The dependence on viewing time is here similar to that described in the preceding paragraph.

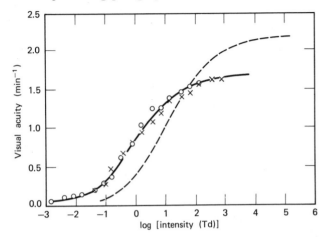

Figure 15.26 Dependence of acuity on luminance for grating (solid line) and Landolt C (broken line). Shlaer [80].

When the pattern of Figure 15.24c is formed of two bright bars on a dark background, the acuity is only approximately 1 min. Here the "dip" is identical to that obtained with the dark bars, but it appears on a bright background and is therefore lower in terms of modulation contrast.

The Landolt C yields results similar to those of the double bar pattern. The rise of acuity with luminance for this pattern is included in Figure 15.26.

Effect of Retinal Location. Published results for the variation of acuity with eccentricity on the retina vary widely [81]. A representative result gives a threshold angle of 1.5′ at the center of the fovea, increasing by 1.5′ for every degree of eccentricity, up to 20° [81].

15.4.3.3 Sinusoidal Gratings. The sinusoidal grating has been studied extensively [67]. Threshold contrast as a function of spatial frequency at various intensities is shown in Figure 15.27 [82]. The decrease in threshold with increasing luminance is clearly evident, as is its rise at lower spatial frequencies. Note that this rise vanishes at 3 Td and becomes progressively more pronounced at higher intensities.

Of special interest are the curves showing threshold modulation as a function of intensity at a given spatial frequency. Such curves obtained with monochromatic light are shown in Figure 15.28 [76]. These clearly

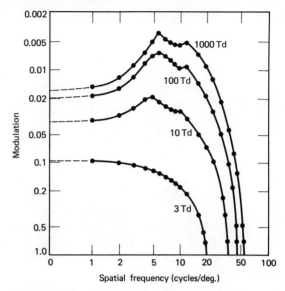

Figure 15.27 Threshold modulation versus spatial frequency for sinusoidal gratings at various intensities. Patel [82].

demonstrate the quantum-noise-limited regions, where the required modulation varies inversely with the square root of the flux density (slope $-\frac{1}{2}$ on the log-log plot) and the higher level noise-limited region, where the threshold modulation becomes independent of intensity (horizontal lines). Also evident is the upward shift with spatial frequency of the transition intensity. It has already been mentioned (Section 15.1.6) that the pupil reflex tends to optimize visual performance at any luminance level [26], [27].

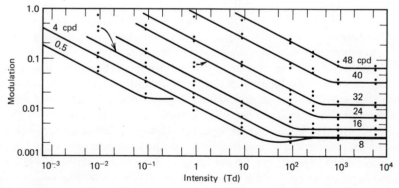

Figure 15.28 Threshold modulation versus intensity for sinusoidal gratings at various spatial frequencies, with 0.525 μm light. Pupil diameter 2 mm. Van Nes and Bouman [76].

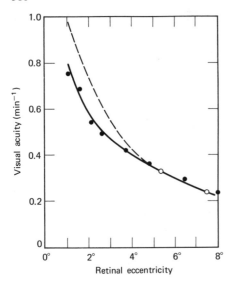

Figure 15.29 Visual acuity dependence on retinal position for sinusoidal fringes. For total visual system (solid line) and retina-brain section (broken line). Green [83].

Variation of sine-wave acuity with eccentricity (distance from foveola) is shown in Figure 15.29 [83], both for a high-contrast target (solid lines) and for interference fringes formed on the retina (broken lines).

Orientation, too, is an important determinant of threshold. Threshold has been found to rise fourfold at 45° to the vertical compared to horizontally or vertically orientated gratings [4d], [84], [85].

15.4.4 Spatial Coding in the Visual System [4d]

In Section 15.4.2.3 we noted that the lateral summation structure of the retina, including both positive and negative elements, leads to a significant brightening of small point or line sources; we may say that the system is "wired" to detect such sources. There is strong evidence, both psychophysical and purely physiological, that there are additional and more complex interconnections geared to the detection of edges and even of sinusoids of specific frequencies [86]. Some parts of the underlying processes seem to be located in the retina and other parts more centrally [7d]. These processes also give rise to various optical illusions and also to the fact that the visual acuity is orientation dependent.

Especially the indication of specific channels sensitive to individual spatial frequencies has given rise to much experimental work. Such channels would have obvious implications in size estimation [87]. They seem to have a bandwidth of approximately 1 cycle/deg and to act independently of each other [88]. They may be located in the retina alone [62].

15.5 TEMPORAL CHARACTERISTICS

A number of temporal effects are involved in vision. Here we treat the response of the light detection process proper in the first section, spatiotemporal interactions in the second and involuntary eye motion in the last section. Temporal aspects of the pupil reflex have already been discussed (Section 15.1.6).

15.5.1 Time Response of the Visual Process [2f], [4e]

15.5.1.1 General Description. We can gain good insight into the temporal characteristics of the visual process by studying its step-function response, that is, its response to a constant stimulus as observed from the time of its onset. A representative response of this type is shown schematically in Figure 15.30, which shows the response (solid line) to a constant stimulus (broken line) initiated at *O*.

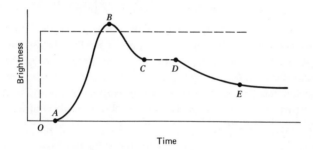

Figure 15.30 Step function response of visual system (solid line) to a constant stimulus (broken line) initiated at time *O*.

There is a delay in the response (*OA*) called the *latency period*, followed by a rapid rise (*AB*) and a similarly rapid, but relatively minor, drop (*BC*) to an almost steady plateau. In fact, there is a continuous slow decline, visible (with the contracted time scale) over the interval *DE*.

Let us now briefly consider the processes underlying the response features.

1. *Latency* (*OA*). The exact time interval between onset of luminance and the sensing of the brightness is very difficult to establish by psychophysical techniques. Electroretinograms show that an electrical signal (the early receptor potential) is generated a bare 25 μsec after incidence of the stimulus, whereas full response may require 100 msec to peak. Other factors in latency are the speed with which the signals propagate along the optic nerve. These range from 3 to 20 m/sec, rising with the nerve fiber diameter.

Reaction time, the interval between a visible signal and a manual response, may be shorter than 200 msec. It increases with decreasing luminance (L) as $(1/\sqrt{L})$ [89]. In one test it doubled to 400 msec when the luminance was lowered to 2 cd/m^2 [90]. Latency varies over the retina inversely with receptor density [91].

2. *Onset and Partial Decline of Response* (ABC). Once the response is initiated, it rises rapidly to a peak in approximately 10 msec. This portion of the response curve is hardly affected by the level of adaptation [92]. The period of descent is estimated to be 30 and 100 msec in the light and dark adapted eyes, respectively [92], [93].

3. *Light Adaptation* (DE). The plateau at D is followed by a slow decline toward the final brightness corresponding to an eye adapted to the new luminance level. The duration of this process depends on the luminance level and requires 2–3 min for attainment of maximum sensitivity.

15.5.1.2   Impulse Response of the Visual System

General Form. If the visual process were linear, the impulse response could be derived from the curve of Figure 15.30 by differentiation and, although this condition is not met, the general shape of the impulse response can be deduced. It consists of a rapid rise in brightness, followed by a rapid decline that carries it to below zero (inhibition); it then slowly rises to zero, possibly with some small oscillations about the axis.

Pulse Techniques. The impulse response curve can be studied by observing the interactions of two equal light flashes presented in close succession. The variation of this interaction with their spacing reveals the impulse response. If their spacing is very close (short relative to AB), we obtain simple summation (Bloch's law); when measured in terms of threshold, two closely spaced brief flashes are as detectable as a single flash of twice the intensity.

As the spacing between the pulses increases, the effectiveness of the second pulse drops until, at very large spacing, there is no interaction and the contribution of the second pulse is only due to the increase in detection probability of two pulses over one, a lowering of the threshold by approximately 25%. At some spacing less than this, the contribution of the second pulse vanishes almost completely, indicating that it fell into the region of inhibition generated by the first pulse [92]. The data given under 2 in the preceding section were obtained in this manner.

Pulse resolution is another problem. It has been found that the mere task of differentiating between two very closely spaced pulses and a more widely spaced pair, the difference in spacing had to be at least 35 ms and the longer spacing had to be at least 45 msec absolute [93].

Meta- and Paracontrast. The masking of the second flash by the first may occur even if they are not at the same retinal location exactly. This is sometimes called *paracontrast.* In a parallel, although paradoxical, phenomenon, called *retroactive* or *metacontrast,* the larger, later flash masks an earlier pulse in its vicinity [94]. Of course, the second pulse only appears to be acting backward in time. The explanation is simply that the signal due to the stronger flash is perceived more quickly, than is the weaker pulse; this is due to differences in the reaction time, as discussed in the preceding section: the stronger pulse thus "overtakes" the latter and blocks its perception.

Square Pulse Response. The shape of the impulse response can also be studied by observing the brightness of a square pulse as a function of its length. It is found that the observed brightness first increases with pulse length paralleling the brightness function; this is again simply a manifestation of Bloch's law. It then levels off and drops to a steady level, so that over some range a shorter pulse is brighter than a longer one of the same intensity. This is the Broca-Sulzer[19] effect. At high luminance levels, the perceived brightness is greatest for 50 ms pulses; at 0.1 cd/m² the pulse must be half a second long to be its brightest [39]. The maximum brightness grows with the square root of the pulse energy, whereas the long-pulse brightness grows only with its cube root. This corresponds to the vanishing of the Broca-Sulzer effect at low luminance levels.

Critical Time. Neglecting the overshoot, we can use a logarithmic plot of threshold energy as a function of pulse duration to establish a *critical time* (t_c), separating the Bloch-law region (characterized by constant energy[20]) from the steady-state region (characterized by constant power, i.e., linearly rising energy): a straight line is fitted to each of the two regions and the abscissa of their intersection is taken as t_c. This time is found to shrink with increase in both area and luminance of the object. The luminance dependence can be approximated by [95]:

$$t_c = 0.065 \, L^{-0.15}, \tag{15.31}$$

where L is the background luminance in cd/m². See also Refs. 39 and 96.

The transition between the two regions is described by the Blondel–Rey[21] formula:

$$L = L_\infty \frac{t + t_c}{t}, \tag{15.32}$$

[19] A. Broca and D. Sulzer (1902).
[20] The energy is here usefully measured in lumen-seconds, often called *talbot.* Another unit, *lumerg,* equals 10^{-7} talbot.
[21] A. Blondel and J. Rey (1911).

where L is the luminance required for the pulse to match a steady-state luminance, L_∞.

15.5.1.3   Periodic Signals

Flicker Threshold. The response of the eye can also be studied by means of periodic signals. The frequency at which the flicker ceases to be perceptible is called the critical flicker frequency (CFF). Over a wide luminance range CFF is directly proportional to the logarithm of the luminance (Ferry–Porter[22] law) and to the logarithm of the stimulus area (Granit–Harper[23] relation). See Figure 15.31 [2f]. The dependence of CFF on wavelength, stimulus size, and retinal location is also indicated there. At frequencies above the critical, the brightness equals that of a steady signal with a luminance equal to the mean luminance of the flickering signal (Talbot–Plateau[24] law). This is closely related to Bloch's law (15.18).

The flicker phenomenon is often described by means of curves showing the threshold modulation as a function of CFF. Such curves are called de-Lange [97] curves when plotted log–log with increasing modulation

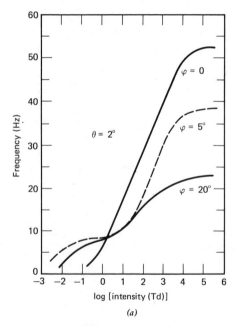

(a)

Figure 15.31   Critical flicker frequency as function of luminance for various stimulus diameters (θ), eccentricities (φ) and wavelengths (λ) [in nm]. Hecht [2f].

[22] E. S. Ferry (1892), T. C. Porter (1898).
[23] R. Granit and P. Harper (1930).
[24] F. Talbot and J. Plateau (1834).

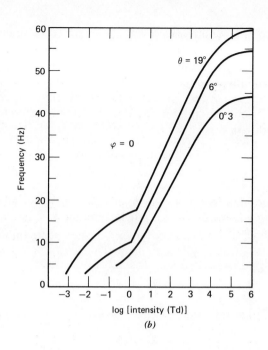

(b)

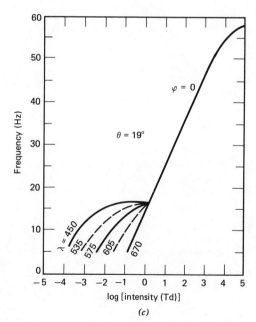

(c)

391

plotted downward. At intensity levels above 0.5 Td these curves start at a threshold modulation of 0.6–0.7 at low frequencies, rise to a peak somewhere between 5 and 20 Hz and then drop rapidly. At lower luminances, the low-frequency threshold is higher and the sensitivity rise with frequency is far less pronounced. See Figure 15.32 [98]. When, instead, the absolute amplitude (ΔL) of the luminance oscillations, rather than the modulation:

$$M = \frac{\Delta L}{L}$$

is plotted, the high-frequency portions of the curves of all the luminance values tend to coincide into a single asymptote. This is illustrated in Figure 15.33 [99].

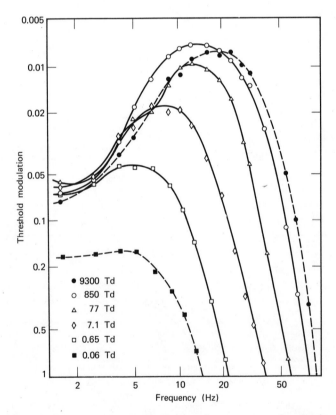

Figure 15.32 Flicker modulation threshold versus frequency at various luminance levels. Kelly [98].

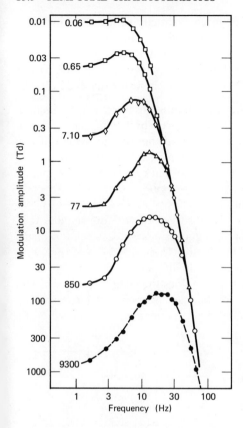

Figure 15.33 Threshold flicker amplitude versus frequency at various luminance levels. Kelly [99]. The intensity (in Td) is marked next to each curve.

Underlying Mechanism. The flicker sensitivity characteristics can be understood in terms of two frequency response mechanisms. One mechanism, perhaps a diffusion process located in the receptor units, controls the high frequency response and is independent of the luminance level. It varies as $e^{-\sqrt{\omega/2}}$, where ω is the frequency. Another mechanism, involving lateral summation and inhibition effects, is presumably located in the plexiform layers, is slower and controls the low-frequency response; it is strongly influenced by the luminance level [99].

Note that the sensation of flicker should not be taken as equivalent to temporal resolution. At the higher frequencies, the observer does not sense the individual pulses and has no quantitative awareness of their frequency. He simply senses the stimulus as flickering.

Spectral Effects. Although the spectral distribution does have some effect on the critical flicker frequency, the three color mechanisms all do respond to flicker in a manner similar to the white light response [100].

When the alternating fields differ in chromaticity rather than in luminance, the CFF is considerably lower. This is the basis for flicker photometry (see Section 1.4.1 and Ref. 101). It has also been found that a phase shift may appear between the two fields, presumably due to the difference in critical times of the two color mechanisms involved [102].

Suprathreshold Performance. The modulation observed for luminance oscillations above threshold have been measured by having the observer match a comparison field, first to the peak and then to the trough of the oscillations [103]. In another experiment the modulation of test signals at various frequencies was compared directly to that of a reference signal at a fixed frequency [104]. The results are in poor agreement with each other and with the threshold results.

The Broca–Sulzer effect can also be observed in periodic signals. It is then called simply (although rather ambiguously) *brightness enhancement.* Representative data are shown in Figure 15.34 [105]. At very low frequencies, the pulses are seen individually and according to their luminance (1.0). As the frequency rises, so does the brightness, due to the Broca–Sulzer effect. At yet higher frequencies, averaging begins to set in and the brightness drops toward the mean luminance (0.5) according to the Talbot–Plateau law.

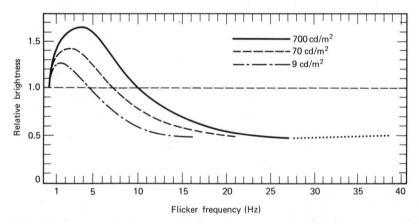

Figure 15.34 Relative brightness (in terms of a matching steady luminance) of an oscillating luminance as a function of frequency at three luminance levels. Field size 3.8°. After Rabelo and Grüsser [105].

15.5.1.4 Adaptation [4c]

General Description. At the beginning of Section 5.3 we pointed out that the enormous dynamic range of the visual system is accommodated, in part, by means of a shifting of the operating range, called adaptation. This can be seen as setting the system gain.

The adaptation process consists of two components. One of these is controlled directly by the neural response to the stimulus and acts rapidly. The other is mediated by the state of bleaching of the photosensitive substances and acts with the same time constant as these do: 7.5 and 2 min for rods and cones, respectively. The state of bleaching acts as an equivalent background in lowering the perceived brightness and in raising the threshold [18]. (See Section 15.6.1 for an explanation of the mechanism in terms of light quanta absorbed.) Specifically, it has been found that the threshold intensity change is given by [18]

$$\Delta I = I_0 e^{ab}, \tag{15.33}$$

where I_0 is the absolute threshold intensity,
  b is the bleached fraction, and
  a is a constant. $a = 45$ and 7 for rods and cones, respectively, for large objects.

The summation of the signals from different receptors, as mediated by the horizontal and amacrine cells (Figure 15.6), also plays a role in adaptation: at low luminances, detector output is *effectively* summed over a wider retinal area. This does not imply that the signal paths change with luminance; it is more likely that the nature of the signal received from the more distant receptor changes from inhibition to summation at low light levels.

Light Adaptation. When an observer enters a well-lit environment after an extended stay in the dark, he is instantaneously blinded: the gain of his visual system is set too high so that it saturates. Minimum sensitivity is attained after approximately 2–3 min. After this interval, the sensitivity rises slightly and reaches its steady state after approximately 10 min [106].

Dark Adaptation. Dark adaptation is a far slower process and often requires an hour for its completion. Two distinct phases can be identified (see Figure 15.35 [107], Curve 1). This shows log threshold as a function of dark adaptation time, following preadaptation to 4×10^5 Td. The threshold drops rather quickly during the first 4 min, to reach a plateau just below 1 Td. Only after 12 min in the dark, is the process resumed with a rather rapid drop for 10 min, followed by a much slower rate lasting for about an hour. The two phases can be identified with dark adaptation of cones and rods, respectively, the former being far more rapid than the latter. The first plateau indicates completion of cone adaptation, and there is no further increase in sensitivity until the rod sensitivity attains

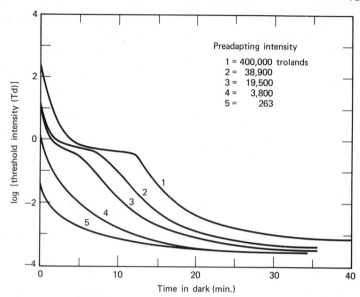

Figure 15.35 Dark adaptation curves: intensity threshold versus time in darkness, following preadaptation to various levels. Hecht, Haig, and Chase [107].

the absolute sensitivity of the cones. At that point, rod sensitivity begins to control and the threshold continues dropping, following the rod adaptation curve. The same effects are noticeable in Curves 2 and 3, corresponding to preadaptation levels of 4×10^4 and 2×10^4 Td, respectively. Curves 4 and 5 are fully controlled by rod adaptation.

Although even low ambient luminance levels lower rod sensitivity significantly, the thorough destruction of a state of dark adaptation requires prolonged exposures at high luminance levels. Dark adaptation curves for 5-sec exposures at various luminance levels are shown in Figure 15.36 [108] and those following various exposures to 1060 cd/m^2 in Figure 15.37 [109]. Typical results are listed in Table 161 [2e].

Due to the difference in dark adaptation rates for cones and rods, the size and location of the test field influence the measured adaptation field. Also because of this duality, care must be taken in evaluating the state of adaptation and its destruction. This must be done in terms of the receptor class involved. For instance, if we are concerned with scotopic dark adaptation, we must evaluate the preadapting light also in scotopic units, even though the luminance may be at a high level. Similarly, when it is necessary to work at mesopic luminance levels while protecting the state of dark adaptation, it may be advantageous to use red illumination, which is relatively more effective for cones, leaving the rods relatively well dark adapted [110].

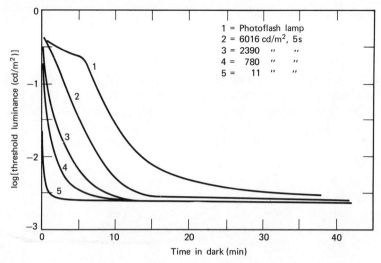

Figure 15.36 Dark adaptation curves: luminance threshold versus time in dark, following exposure to a photoflash and various 5-sec exposures. Wald and Clark [108].

Stabilized Images [7b]. In normal vision, images move across the retina sporadically due to ocular motion (see Section 15.5.3). This motion can, however, be eliminated by applying to the eye a contact lens holding a mirror. The image to be viewed is projected onto a screen in front of the eye and the contact lens mirror is incorporated into the optical system

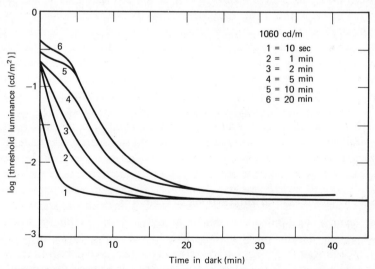

Figure 15.37. Dark adaptation curves: luminance threshold versus time in dark after exposure to 1060 cd/m² for various times. Baker [109].

of the projection arrangement in such a way that, when the eye rotates, the mirror motion compensates for the eye motion, so that the image remains stabilized on the retina [111]. It is found that for exposures up to 0.2 sec stabilization does not affect acuity [112].

With longer periods of stabilization, the perceived image slowly fades and finally disappears. It may reappear thereafter, only to fade again [111]. This disappearance seems to be an effect of adaptation. The effect increases with spatial frequency [113]. Relative motion between the image and the retina, or any substantial change in luminance, can cause the image to reappear.

After-Images [4f]. When an observer enters a dark environment, or closes his eyes immediately after fixating on a bright object, he will continue seeing it for some time. This sensation is called *after-image* and it can be understood in terms of retinal areas that are light adapted to various degrees. A high correlation has been established between the brightness of the after-image and the degree of retinal bleaching [114]. The periodic disappearance and reappearance of the after-image seems to be related to the similar behavior of real images in stabilized vision. The temporal course of after-images can be described by an exponentially decaying brightness with a damped sinusoidal variation superimposed. The period of the sinusoid is about 4 sec [115].

15.5.1.5 Temporal Effects on Acuity. Just like sensitivity, acuity increases with viewing time. In general, the required time increases with the difficulty of the task. For sinusoidal gratings at intermediate spatial frequencies, the threshold contrast drops linearly with increasing viewing time up to 50 msec and more slowly thereafter, up to 1 sec, whereupon it remains constant [116].

On this basis it is possible to establish a "critical time" of acuity from a plot of threshold contrast versus viewing time. Straight lines are fitted to the short- and long-viewing-time portions of the curve; their intersection determines the "critical time." This has been found to fall into the range 150–200 msec at 3 cycles/deg and to rise for both lower and higher spatial frequencies [117].

15.5.2 Spatiotemporal Interactions in Vision

The general visual signal is four dimensional and, if we limit ourselves to monocular vision, three dimensional: the stimulus varies in the temporal and in two spatial dimensions. For convenience, we often treat the spatial and temporal variations individually; but there are important interactions so that we can not always maintain the isolation.

Brightness Function. For instance, the brightness function varies linearly with the product of time and area when both area and time are small and with its cube root when both are large. When the area is large and the viewing time short, or *vice versa*, the variations follow the square root of the luminance [39].

Frequency Response. Signals varying sinusoidally both in time and space have been studied at threshold and above it. Suprathreshold experiments showed that at low frequencies (2 Hz) the perceived modulation of the sinusoidal luminance pattern was more than doubled. At approximately 5 Hz it appeared equal to that of the steady pattern, and at 10–20 Hz it had about half this value [103].

The thresholds for combined spatial and temporal oscillations, at 50 Td, are shown by the solid curve in Figure 15.38 [118]. The intercept on the temporal frequency axis is the CFF and that on the spatial frequency axis is the acuity limit. The area under this curve is divided roughly into three regions. At high spatial frequencies, the thresholds vary with the square root of the luminance; they are controlled by the quantum noise and obey the Rose regimen (see Section 15.6). At high temporal frequencies, the threshold values are quite independent of

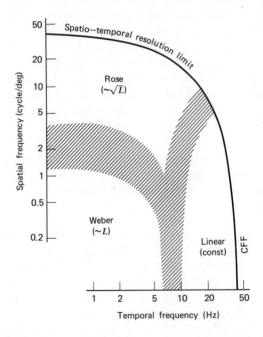

Figure 15.38 Spatiotemporal threshold map for the 50-Td level. Kelly [118].

luminance (see Figure 15.33). When both spatial and temporal frequencies are low, the contrast threshold is independent of luminance: vision here obeys Weber's law [118], at least at high luminance values.

Ocular Motion. When we view a bar pattern, any motion of the eye resulting in image motion across the retina will translate the spatial pattern into a temporal one, from the point of view of the individual receptor unit. This point is the subject of the next section.

15.5.3 Ocular Motion and Its Effects on Acuity

Image motion does not seem to have a significant effect on acuity, even for substantial velocities. Even in Vernier acuity tests, where acuities of approximately 5″ are attained, it is found that velocities up to 4 deg/sec do not affect performance. An exposure time of 0.2 sec was employed there, and hence any target point traveled across 90 receptors during the exposure while a misalignment corresponding to 0.2 receptor diameters was detected [79].

This insensitivity to image motion is rather fortunate in view of the fact that the living human eye is stationary for only very brief periods. Even during intense fixation, continuous scanning takes place. This scanning occurs in various types of motion ranging from rapid *saccades* covering a fraction of a minute of arc, to long-term drift, which may range up to half a degree [119].

The effect of motion of the image across the retina has been evaluated by the obvious method of oscillating the image. Frequencies below 10 Hz with amplitudes above 10′ have been found to be advantageous [120]. More sophisticated studies have been made by attaching a reflector to the eye (usually to a contact lens placed over the cornea), permitting optical stabilization of the retinal image, independent of ocular motion. With a slight change, this arrangement can be used to enhance the movement of the image on the retina [111].

These spontaneous ocular movements may be serving three important functions:

1. The eye's motion prevents the fading of the image, which quickly disappears when it is stabilized on the retina (see Section 15.5.1.4).

2. More complex, and potentially also very important, is the coupling between spatial and temporal effects introduced by this motion. For instance, noting that visual response to luminance fluctuations is optimum at some frequency (depending on mean luminance, see Figure 15.32), a lateral oscillation of the image at ν_t will stimulate a retinal receptor unit at its optimum frequency, if the receptor is located at the boundary

between regions of differing luminance; it will leave unaffected the response of receptors within a region of uniform luminance. Thus, ocular scanning may contribute to "edge enhancement," which is one of the more important spatial visual effects.

3. Ocular scanning has also been invoked in efforts to account for visual acuity which is, as already noted, far in excess of the receptor element size. It is difficult to explain such performance when a relatively coarse-grained receptor pattern is assumed superimposed on an image already blurred by diffraction effects, so that it is only marginally resolvable. Rapid, small-amplitude image motion could provide a smoothing effect, similar to the effect of motion when using a ground glass viewing screen (Section 19.1.3) or a fiber bundle (Section 13.3.2.2). These explanations are, however, far from proven.

In conclusion we describe in more detail the involuntary motion during normal visual fixation. There are essentially four types of such motion [119]:

1. Small, rapid (saccadic) movements with an amplitude of approximately 18″ and at a frequency of 30–70 Hz.

2. Slow, irregular movements with amplitudes of 1–5′ and at a frequency of 2–5 Hz.

3. Slow drift of less than 5′.

4. Rapid jerks with amplitudes ranging from 2 to 25′, although they rarely exceed 10′. These jerks are spaced 0.2–4 sec, and occasionally tend to compensate for slow drift.

The total drift rarely exceeds 10′ but does occasionally reach half a degree.

Extensive statistics of eye movements have been compiled [121]. These can be summarized in terms of the median value of the eye deflection. This value grows linearly with observation time at a rate of approximately 3.6′/sec, at least for fixation periods no longer than 1 sec.

15.6 NOISE EFFECTS IN VISION

15.6.1 The Nature of Visual Threshold

Thresholds have played a major role in our discussion of vision, brightness perception, and spatial and temporal resolution. It therefore behooves us to take a close look at the nature of these thresholds.

There are basically two types of threshold. In one type, a subthreshold input will have no effect at all; it will not combine with another similar input to produce an output. This is how we described, for instance, the threshold photon energy in our discussion of the photoelectric effect (Section 4.2). In linear systems, the nature of the threshold is altogether different. Here it is determined by the random fluctuations (called "noise") exhibited by the output in the absence of an input. A given small input signal may well produce a correspondingly small output; but if this signal output is indistinguishable from the random output obtained in the absence of a signal, it will not be detectable as a signal. To be detectable as a signal, it must rise to a certain level, roughly equal to the noise level. The input level required to produce such an output is the threshold signal. In such a situation, the threshold can not be deterministically defined. As the input signal increases, the *probability* of detecting it increases. The probability level accepted as detection must be defined here.

What is the nature of the visual thresholds? As a step toward answering this question, let us consider in some detail an early experiment [122]. It was found that at absolute threshold $2-6 \times 10^{-17}$ J must be delivered to the pupil of the eye for 60% detection probability. This referred to a circular object, $10'$ in diameter exposed for 1 msec with 0.51μm light. Considering a 50% transmittance for the ocular media and also that the retina absorbs at most 20% of the incident flux, we find that each flash correspond to $2-6 \times 10^{-18}$ J delivered to the retina. At $\lambda = 0.51 \mu$m, the energy of each quantum is [see (4.2)]:

$$E = \frac{hc}{\lambda} \approx 0.4 \times 10^{-18} \text{ J},$$

so that for each threshold flash the retina absorbed only 5–15 quanta. Now, a $10'$-diam image implies that these quanta were distributed over a retinal area of approximately 2×10^{-3} mm^2. But, at the point of the retina used ($20°$ temporal eccentricity) the receptor density is approximately 150×10^3 mm^{-2} (Figure 15.8). Hence our quanta were distributed over 300 receptors: only one out of each 30 received just one quantum. If the spontaneous firing rate of receptors in the dark is close to this, this would imply that vision threshold is limited strictly by the "dark current"!

Clearly one quantum suffices to yield a detectable output from a receptor.[25] It seems, however, that at least five such receptors must be activated and their outputs pooled, before the subject has a sensation of light. The first type of threshold has not yet been precluded by this experiment. More recent experiments, however, using signal detection theory, have shown that with some observers, a single quantum absorbed increases their "seen" response probability [123].

---

[25] This does not imply that a single photon can directly initiate neural activity. There is apparently a gain mechanism in the receptor whereby the absorption of a single photon permits the flow of a large number of charges to take place. The exact mechanism has apparently not yet been identified.

We conclude that visual thresholds are strictly statistical in nature, except, of course, for the absolute threshold of one photon implicit in the quantum nature of light.

How is this performance affected by adaptation? It is found that the quantum efficiency of the receptors is maintained at its very high level. Only the subsequent decision level is raised, perhaps a thousandfold [18].

15.6.2 Noise Sources in the Visual System [45]

On the basis of physical theory, we must ascribe to the visual system at least three noise sources.

1. *Detector noise.* By whatever mechanism the eye converts radiation into a neural impulse, we must expect this mechanism to operate occasionally even in the absence of incident radiation, and, the more sensitive the detector, the more likely it is to be triggered spontaneously. This prediction has been confirmed experimentally and the effect has been called *"dark light"* and *eigengrau* [18].[26]

2. *Sensation and Neural Noise.* At the other end of the neural pathway, we must expect spontaneous stimulation of sensation, even in the absence of neural activity in the optic nerve; the process that translates neural impulses into sensation must be expected to take place occasionally even in the absence of neural impulses, resulting in sensation noise.

In addition to this "sensation noise" spurious pulses must be expected to occur at all neural terminals, such as in the plexiform layers in the retina and in the lateral geniculate body. We shall not discuss these further, however, because at the present state of knowledge they may be treated together with either of the preceding two noise sources.

3. *Radiation Noise.* In addition to the noise sources internal to the visual system, there is noise superimposed on the entering radiation at the time it enters the eye. This noise is due to the quantum nature of light: the fact that light arrives in the form of individual quanta. This noise has been analyzed most extensively and is often referred to as "quantum noise" [125]–[127].

---

[26] It is often pointed out that the sensitivity spectrum of the eye matches the solar radiation energy spectrum amazingly well. It seems that the visual system was well designed. The question has been raised, however, that, since the eye is a quantum detector, its sensitivity should be compared with the photon-spectrum of the sun, and this yields a very poor match; the sun's spectrum is rich in ir photons to which the eye is barely sensitive. But further analysis shows that sensitivity to ir photons requires a lowering of the work function of the photosensitive material and this implies an increase in "dark light"—the net result would be a reduction in absolute sensitivity [124].

These noise effects determine the threshold contrast that the visual system can detect. Specifically, we assume that, to be detected, the signal must be some multiple, say b, of the noise. If we now take the luminance difference (ΔL) between object and background to be the signal, we find for the threshold contrast:

$$C_0 = \frac{\Delta L}{L_b} = \frac{bL_n}{L_b}, \qquad (15.34)$$

where L_b is the background luminance and
$\quad L_n$ is the rms value of the luminance noise.

Turning to the various noise sources in the visual system, we note that they differ in their dependence on the background luminance.

The *dark noise* should remain constant, so that the threshold contrast observed due to it should vary inversely with the background luminance

$$C_d = \frac{k_d}{L_b}, \qquad (15.35)$$

where k_d is b times the dark noise.

The *quantum noise*, on the other hand, varies with the square root of the number of quanta detected, that is, according to $\sqrt{L_b}$, assuming a Poisson distribution and a low contrast. Hence the contrast due to it will be

$$C_q \sim \frac{\sqrt{L_b}}{L_b},$$

$$= \frac{k_q}{\sqrt{L_b}}, \qquad (15.36)$$

where k_q is a constant determined by the size of the quanta. Following Ref. 126, we show in Figure 15.39 the threshold contrast as a function of spatial frequency at various illumination levels, as determined by quantum noise alone.

The *neural noise* can be expected to be independent of the luminance. However, to refer it to the input stage, it must be divided by the gain introduced by the visual system, and this varies greatly with the adaptation level, L_b. Consequently, the threshold contrast due to it takes the form:

$$C_\psi \approx k_\psi (1 + KL^{-\beta}). \qquad (15.37)$$

See Appendix 15.1.4 for a derivation of this expression.

To obtain the total threshold contrast, as given by (15.34), we must

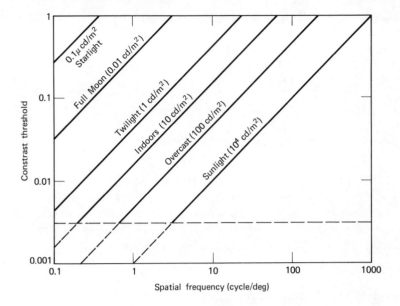

Spatial frequency (cycle/deg)

Figure 15.39 Threshold contrast versus spatial frequency for the indicated values of luminance, as implied by quantum noise. For contrast thresholds below ca. 0.003 (broken line), other noise factors limit detection.

obtain the overall noise level by summing the squares of the individual noise contributions. In terms of the individual threshold contrasts, we obtain the expression:

$$C_0^2 = C_d^2 + C_q^2 + C_\psi^2$$

$$\boxed{= \frac{k_d^2}{L_b^2} + \frac{k_q^2}{L_b} + k_\psi^2(1 + KL_b^{-\beta})^2,} \quad \beta < 1 \qquad (15.38)$$

(see Appendix 15.1.1).

It is immediately evident that at the lowest levels of luminance, the first term dominates, so that there the threshold contrast varies inversely with the luminance. Equation 15.38 also implies that at very high levels of luminance the last term dominates and the threshold luminance approaches a constant value. In this region vision would obey Weber's law. It is shown in Appendix 15.1.5 that there is an intermediate region in which the middle term dominates and the threshold contrast varies inversely with $\sqrt{L_b}$.

Specifically, when (15.38) is written in terms of the troland value, I_b, instead of L_b, the values of the coefficients are roughly:

$$k_d = \frac{4 \times 10^{-8}}{\Omega}, \qquad \alpha < 0.6° \quad [\text{see}(15.16)]$$

$$= \frac{4.5 \times 10^{-6}}{\sqrt{\Omega}}, \qquad \alpha > 0.6° \text{ (see Figure 15.16)}$$

$$k_q = \frac{6 \times 10^{-4}}{\sqrt{\Omega}}, \qquad [\text{see } (15.70)]$$

$$k_\psi = \frac{0.014}{\sqrt{\Omega}}, \qquad \alpha < 0.6° \text{ [see Ref. 45]}$$

$$= 0.0012, \qquad \alpha > 0.6°, \tag{15.39}$$

and $K = 19$, $\beta = 0.32$. Here Ω is the object size in steradian and α is the object diameter.

15.6.3   Noise Amplitude Distributions

The noise amplitude distribution may be studied from the statistics of the error rates as the signal amplitude is varied as implied by Figure 15.13. In principle, such investigations could yield details of the combined noise distributions. In practice, a very large number of decisions is required for accurate results. This makes the measurement very difficult[128].

15.7   COLOR VISION

The major characteristics of color perception have already been surveyed in Section 1.2 We restrict ourselves here to (a) a discussion of visual performance in color discrimination, (b) a brief survey of theories of color vision, and (c) of color vision defects.

15.7.1   Color Discrimination and Chromaticness Stability [3d]

We wish to know by how much a color must be changed, so that the change becomes noticeable to a normal observer. Due to three-dimensionality of color sensation, such changes may be in any of the three parameters: hue, saturation, and luminance, or any combination of these. We present in this section information concerning visual sensitivity to changes in hue and saturation separately. Sensitivity to luminance changes is discussed in Section 15.3 in conjunction with the transfer characteristic of vision. Regarding multidimensional color changes, we restrict ourselves to general comments: the study of this field is complex and in a state of flux.

15.7.1.1 Hue Difference. In a pure spectral color, the just detectable wavelength change varies considerably along the spectrum. Even a casual view of the spectrum as generated, for instance, by a prism shows regions where the color changes rapidly (e.g., in the yellow region) and other regions where the color seems sensibly constant (e.g., in the red end of the spectrum). Measurements of the threshold of wavelength differences are usually made with a split field of view, with the two parts of the field illuminated with fluxes differing slightly in wavelength. The change in wavelength required to make the difference just noticeable may be plotted as a function of wavelength. A representative curve of such measurements is shown in Figure 15.40 [129]. It exhibits pronounced minima (maxima of sensitivity) at 0.49 μm (blue-green) and at 0.59 μm (orange-yellow); a secondary minimum appears at 0.44 μm. At the two ends of the spectrum, the curve rises sharply, indicating a rapid decline of sensitivity to wavelength change.

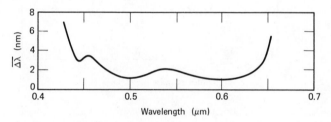

Figure 15.40 Wavelength discrimination spectrum: wavelength difference threshold as a function of wavelength. 2° field, 70 Td. Wright and Pitt, *Proc. Phys. Soc.* (London) **46**, 459 (1934).

Similar measurements can be made using patterns consisting of bars illuminated with flux of one wavelength, alternating with bars illuminated with flux of another. The required wavelength difference varies then not only with the spectral region but also with the spatial frequency. A set of representative curves is shown in Figure 15.41 [130]. The threshold value of wavelength difference is shown as a function of the center wavelength for various spatial frequencies (in cycles per degree) of the bar grating. The minima at 0.49 μm and 0.59 μm are discernible here, too, at the lower spatial frequencies.

The sensitivities of the eye to luminance and wavelength differences, respectively, are compared in Figure 15.42 [130]. One may note there that the rise in threshold at low spatial frequencies is pronounced in the luminance difference but absent in the wavelength difference curves.

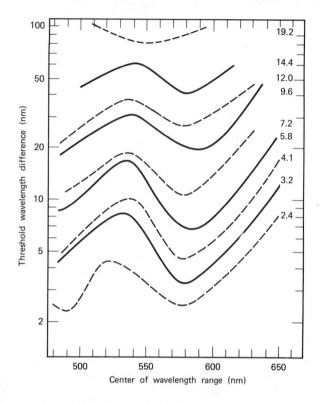

Figure 15.41 Wavelength discrimination spectra at various spatial frequencies. The spatial frequency, in cycles/deg is marked next to each curve. Hilz and Cavonius [130].

It should be interesting to determine the highest spatial frequencies detectable with equiluminance gratings that vary only in hue. Presumably this maximum is attained with a pair of complementary colors. For instance, a yellow-blue pair of complementaries occurs at 0.475 μm and 0.575 μm, respectively and are consequently spaced by 100 nm. Entering Figure 15.41 with their center wavelength (525 nm) we find that the frequency of 19.2 cycles/deg corresponds almost to this wavelength difference. We conclude, that such a grating can be resolved to approximately 20 cycles/deg. A red-green grating should be resolvable at a slightly higher spatial difference (by extrapolation estimated at 23 cycles/deg). This should be compared to a threshold frequency of approximately 60 cycles/deg for fully modulated monochromatic [26] and achromatic gratings [82].

Extensive experiments have also been made with sinusoidally varying, equiluminous patterns, using pairs of complementary colors and varying

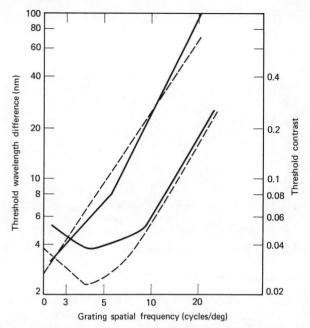

Figure 15.42 Comparison of wavelength discrimination (upper curves) with luminance discrimination at 0.5 μm (solid curves) and 0.6 μm (broken curves). Hilz and Cavonius [130].

their purity to determine the threshold [131], [132]. Figures 15.43 and 15.44 [131] show the threshold purity as a function of spatial frequency for various luminance levels, the two figures referring to red-green and to yellow-blue modulation, respectively. Moving patterns have been used to check the temporal effects [131].

15.7.1.2 Saturation Difference. When the two fields are identical in hue and luminance and differ only in saturation (i.e., chroma or purity), one may measure, for instance, the least colormetric-purity step (Δp_c) away from white, which is still detectable. Representative results of such measurements are shown in Figure 15.45 [129], where the negative logarithm of Δp_c is plotted as a function of wavelength. A striking feature is the pronounced minimum in the yellow (Δp_c large), which implies that even spectrally pure yellow is sensed as relatively unsaturated.

The size of the minimum discernible step varies also with the purity itself and peaks at intermediate purities. See Figure 15.46 for the threshold values Δp_c as a function of purity for red light (0.65 μm). From curves such as this, the number of discernible saturation steps at any wavelength can be estimated. The results of such an estimation are given

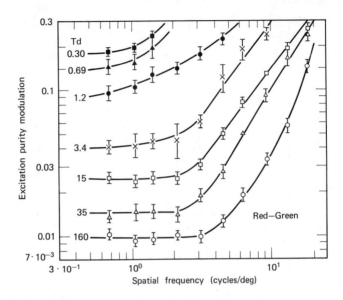

Figure 15.43 Threshold purity modulation for complementary colors as a function of spatial frequency at various values of intensity. Red (610 nm); green (492 nm). Excitation purity is expressed with respect to dominant wavelength 492 nm. Horst and Bouman [131].

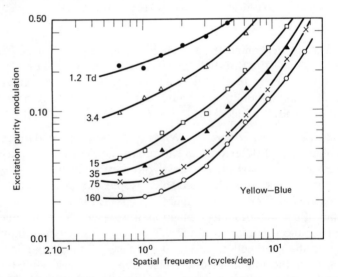

Figure 15.44 Threshold purity modulation for complementary colors as a function of spatial frequency at various values of intensity. Yellow (573 nm), blue (463 nm). Excitation purity is expressed with respect to dominant wavelength of 573 nm. Horst and Bouman [131].

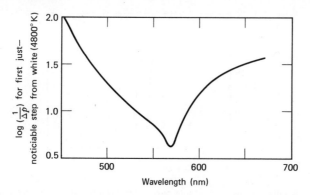

Figure 15.45 Purity discrimination spectrum near the neutral point. Wright [129].

in Figure 15.47. By this criterion, again, pure spectral yellow is the least saturated color.

For a review of the various threshold saturation differences, see Ref. 133 which covers especially the thresholds for almost pure spectral lights.

15.7.1.3 General Color Differences and Change in Color Temperature. In general, when two colors are compared, hue, purity, and luminance all differ. Much work has been done in an effort to determine how changes in these various dimensions combine. Some of it is reviewed in Section 1.2.3.3. It is aimed at distorting the CIE chromaticity diagram in such a way that equal displacements in it correspond to equally noticeable color changes. Results of such efforts are called *uniform color scales.* Some of these, too, are reviewed in that section. It should be noted, however, that the nonisotropy of visual sensitivity is such that a truly uniform color scale could be represented only on a warped surface even when we restrict our efforts to the two-dimensional chromaticity (see Section 6.7. of Ref. 3 for an extensive review of such efforts).

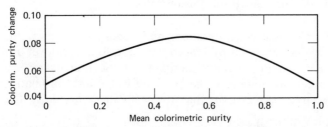

Figure 15.46 Purity discrimination as a function of purity. 2° field, 30 Td, "white" is 4800°K, dominant wavelength is 0.65 μm. After Martin, Warburton, and Morgan [3d].

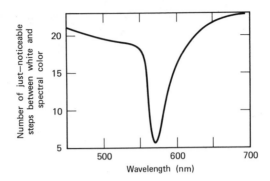

Figure 15.47 Number of just-perceptible purity steps as a function of wavelength. 2° field, "white" is 4800°K. After Martin, Warburton, and Morgan [3*d*].

The minimum detectable change in color temperature is about 5.5 mired, independent of the color temperature itself. [See (11.5) for the definition of the mired unit.]

15.7.1.4   Chromaticness Stability [3e]. Already in Section 1.2.3.2 we pointed out that the hue of a color changes with luminance even though the wavelength is maintained constant. This phenomenon is called the Bezold–Brücke[27] effect and is illustrated in two different forms in Figures 1.16 and 15.48 [3*e*]. From these figures it appears that there are three wavelengths, 0.47 μm, 0.51 μm, and 0.575 μm that exhibit hues that are stable with changing luminance. These correspond to "psychologically simple" blue, green, and yellow, respectively. In fact, there is a fourth stable hue in the nonspectral region (reddish purple). The analogous change of hue with changing saturation is called *Abney effect*[28] and is illustrated in the curvature of the constant-hue lines in Figures 1.14, and 1.15.

It is also found that the sensed saturation of a color changes with changing luminance, although its chromaticity coordinates remain fixed. This is sometimes called the Purdy[29] effect.

15.7.1.5   Chromatic Adaptation and Color Constancy. In Section 15.5.1.4 we noted that a prolonged exposure to a high level of illumination lowers, temporarily, the sensitivity of the retinal area exposed, a process called adaptation. If the exposure is due to a chromatic stimulus, it is reasonable to assume that this will affect primarily the sensors

[27] W. v. Bezold (1873) and E. T. v. Brücke (1878).
[28] W. de W. Abney (1913).
[29] D. McL. Purdy (1931).

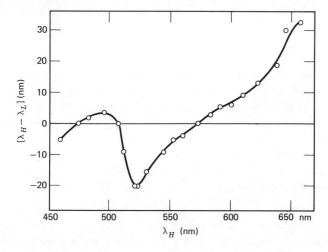

Figure 15.48 Bezold–Brücke effect: apparent change of dominant wavelength when the intensity is raised from 100 to 1000 Td. λ_L and λ_H are, respectively, the dominant wavelengths at these levels. After Purdy [3d].

sensitive to the spectral region of the stimulus. (It has been found that each type of receptor saturates independently and controls its own summation pool [18].) On subsequent exposure to other chromatic stimuli, the response of the adapted sensors will be reduced relative to that of the others. As a result, the color sensed will be different from that sensed by an unadapted eye. Some of the results can be approximated by shifting the chromaticity point away from the chromaticity of the adapting flux, if this leads to a reduced saturation. This is not necessarily true, when this shift would lead to an increased saturation of the complementary color.

Tests with adaptation fields of color temperatures from 2000 to 6500 K have indicated very little dependence of color discrimination on adaptation field color. The thresholds are only slightly higher at low color temperatures [134].

Color constancy is, in a sense, the psychological counterpart of adaptation. It describes the observation that objects do not change their sensed (reflectance or transmittance) colors with changing illumination chromaticity, as would be required by simple psychophysical theories. Rather, the visual system enables us to cancel out the effects of the chromaticity of the illuminating flux, distorting the chromaticity diagram by shifting the neutral point toward the chromaticity of the overall illumination: the chromaticity of an object known to reflect neutrally is classified as neutral and the sensation of other colors is judged relative to it [see also the retinex theory (Section 15.3.4.3)].

15.7.2 Theories of Color Vision [1d, 2g]

To measure color precisely, as described in Section 1.2, is one matter; to understand how the human eye does it, is another. We noted in Chapter 1 that, although it is mediated by spectra that are essentially infinity dimensional,[30] color sensation is only three-dimensional. Hence the color sensing mechanism in the retina need not contain a large number of resonant circuits (as does the ear): a set of three receptor types suffices to account for the observed color sensation. Here we survey the major theories of color vision.

***15.7.2.1 Tristimulus Theory* [135].** According to the Young (-Helmholtz) theory,[31] the retina contains three types of receptors, each with its own sensitivity spectrum, analogous to the tristimulus value curves of the CIE system (Figure 1.13 and Table 7). Indeed, three types of cone have been identified in the retina. These differ in the photosensitive pigment they contain. Their absorption spectra are shown in Figure 15.49 [135]. We may assume that the sensitivity spectra are similar.[32] Light at any given wavelength may excite all three cone types, each according to its sensitivity. For instance, as indicated on the figure, light at wavelength λ excites the red (R), green (G), and blue (B) sensitive cones in the ratio:

$$r(\lambda):g(\lambda):b(\lambda).$$

This ratio determines the sensed chromaticity, analogous to (1.20) with r, g, b) taking the places of $(\bar{x}_\lambda, \bar{y}_\lambda, \bar{z}_\lambda)$.

The photosensitive pigments carried by the red, green, and blue sensitive cone types are similar to the rhodopsin of the rods and are referred to as *erythrolab*, *chlorolab*, and *cyanolab*, respectively.

To determine the spatial characteristics of the three cone systems, their threshold modulation has been measured as a function of spatial frequency. The red and green systems have been found to be quite similar. The threshold for the blue system is considerably higher—3 times at 0.25 cycles/deg and 70 times at 10 cycles/deg [136].

---

[30] In principle, the spectral intensity at each wavelength is independent of that at any other wavelength. In fact, however, line broadening (see Section 4.3.2) reduces this dimensionality.

[31] T. Young (1802) and H. von Helmholtz (1852).

[32] Each cone type responds in a given manner to each absorbed photon, regardless of its wavelength; only the fraction of photons absorbed varies with wavelength (principle of univariance).

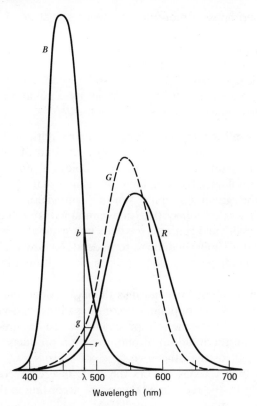

Figure 15.49 Sensitivity spectra of red, green, and blue (*R, G, B*) sensitive cones. Normalized for equal areas. Rushton [135].

15.7.2.2 Dominator–Modulator Theory. The activity of the sensors in the retina may be measured by means of microelectrodes. Such measurements indicate that the retina is composed of two major types of sensor units:

1. Approximately 95% of the units have a wide spectral response. These units are called *dominators*.
2. There are other neural units having very narrow responses. These are called *modulators*. The spectral responses are scattered, but there are three major types.

The Granit theory,[33] based on these findings, postulates that the dominators are responsible for the sensation of brightness, while the

[33] R. Granit (1930).

modulators superimpose a more or less strong sensation of hue over this brightness sensation.

Difficulties with this theory arise from the narrow absorption spectrum of the modulators and from their excessive variety. It is therefore thought that the electrical measurements of neural response measure, not the primary effects, but the result of interaction, by summation or inhibition, of various primary receptor types.

15.7.2.3 Opponent Color and Zone Theories. Although the Young-Helmholtz theory accounts well for the physical end of the color vision problem, it is less satisfactory at the psychological end. For instance, while it is psychologically acceptable to claim that the simultaneous excitation of the green and blue sensors will yield the sensation of blue-green, it is less so to say that the simultaneous excitation of green and red sensors will yield yellow. One does not, generally, sense yellow as a "red-green"! It is similarly difficult to understand how the simultaneous excitation by complementary colors yields the achromatic sensation of white.

Because of these objections Hering's theory[34] gained much support. It postulates three pairs of opponent color sensations: red-green, yellow-blue, white-black. The same sensor produces red or green sensation, depending only on the build-up of photoreaction products within it. The same is postulated for the other two types of sensors.

Color vision theory today, favors the *zone theory*. According to this, detection of light by the receptors takes place according to the trichromatic hypothesis of Young and Helmholtz. At a higher level, or "zone," in the nervous system, these sensations are converted into the three pairs of opponent colors and, therefore sensed in that form at the conscious level [2g].

15.7.3 Color Vision Defects [4g]

Significant defects in color vision occur in about 8% of males and 0.4% of females. It may manifest itself in a deficiency, or absence, of one or more of the color mechanisms.

Observers with all three color mechanisms are called *trichromats*, those with one mechanism missing, *dichromats* and those with but one color mechanism, *monochromats*. The *anomalous trichromat* possesses all three mechanisms, but one or more of these is defective.

Dichromatism. Dichromats with the red, green, or blue mechanism missing are called *protanopes, deuteranopes,* and *tritanopes,* respectively.[35]

---

[34] E. Hering (1872).

[35] Protos = first; deuteros = secondary; tritos = third. An = negative; opsis = vision.

Protanopia and deuteranopia are the most common color vision defects. People suffering from these, confuse red and green, apparently seeing both as yellow. The protanope has a strongly reduced sensitivity in the red; but the deuteranope has a spectral luminance response close to normal. His deficiency in the green is not so significant in this respect because the red mechanism is quite sensitive there, too. Tritanopia is very rare and leads to a confusion of green with blue and pink with orange.

Dichromats can be diagnosed by means of their *neutral locus*, that is the locus (on the CIE chromaticity diagram) of colors that to them appear neutral or white. The intersection of this with the locus of the spectral colors is called the *neutral point*. It is at 0.4955 μm and 0.5 μm for protanopes and deuteranopes, respectively. For the tritanope, it is in the yellow. Other colors that appear, to the dichromat, identical in hue fall along *confusion loci*. These are shown in Figures 15.50 and 15.51 [137] for prota- and deuteranopes, respectively.

Prota- and deuteranopia are due to a lack of erythrolab and chlorolab,

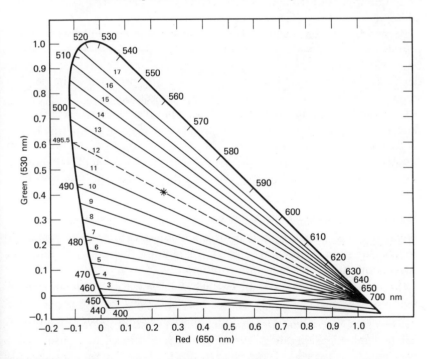

Figure 15.50 Confusion loci for protanopes. The broken line represents the neutral locus and the asterisk that of Standard Illuminant C. The loci are spaced to correspond to threshold chromaticity changes. The chromaticity diagram is based on 0.65, 0.53, and 0.46 μm as primaries. Pitt [137].

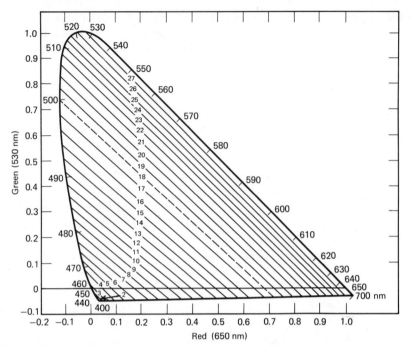

Figure 15.51 Confusion loci for deuteranopes. The broken line represents the neutral locus. The loci are spaced to correspond to threshold chromaticity changes. The chromaticity diagram is based on 0.65, 0.53, and 0.46 μm as primaries. Pitt [137].

respectively and there is strong evidence that tritanopia is due to a lack of cyanolab.

Note that the *normal* observer is effectively tritanopic in the central island of the fovea. This appears to be due to both the macula and a deficiency of cyanolab [137*].

Chromatic Anomaly. Anomalous trichromats are called, respectively, *protanomalous, deuteranomalous,* or *tritanomalous* (protans, deutans, and tritans, for short) if their red, green, or blue mechanisms are defective. Anomalous trichromacy varies continuously from normal vision to the respective dichromacy. The physiological basis for these defects is still unclear.

Tests for Color Vision Defects. Detailed analysis of color vision defects can be made by asking the observer to match metameric hues displayed in two halves of his field. The device to do this is called *anomaloscope*. In another test, the observer is asked to sort color chips according to hue and saturation. For rapid screening, test charts have been developed. These display colored discs which appear of different hues to the normal observer, but will be confused by the common types of dichromats.

15.8 STEREOSCOPY AND SPACE PERCEPTION [1e]

One may well ask how we are able, by means of the two-dimensional retinal image, to get an amazingly complete picture of the three-dimensional space around us. There are, in fact, many cues that we utilize in space perception, but only two of these, stereoscopy and monocular parallax, are sufficiently quantitative for use in instrumentation. It is on these, therefore, that our discussion here is concentrated.

15.8.1 Stereoscopy

We noted earlier that fusion of the images from the two eyes is a prerequisite for good binocular vision and that this requires that a given object point be imaged at approximately equal displacements from the centers of the foveas in the two eyes. Small differences in these displacements do not interfere with fusion, but rather may provide important cues for object distances. The process of providing object distance cues in this form is called *stereoscopy*.

15.8.1.1 Geometrical Relationships. As we just noted, the stereoscopic effect is based on the difference in retinal image point displacements. To calculate this, let us refer to Figure 15.52. Here L and R denote the centers of rotation of the left and right eye, respectively, and we assume that these are sufficiently close to the respective nodal points to permit us to take them as coincident. $\overline{LR}$ is the interocular axis. It subtends an angle θ at the object point Q. If the eyes do not converge (optical axes parallel), the image point displacements on the two retinas will differ by θ. If, however, the eyes converge by an angle θ_0, the difference in displacement will be

$$\Delta\theta = \theta - \theta_0. \tag{15.40}$$

In the figure, the two optical axes are represented by $\overline{P_0P_0'}$ and $\overline{P_0P_0''}$, respectively, and the displacement difference of the image points is the sum of the arcs

$$\Delta\theta = P_0'Q' + Q''P_0''. \tag{15.41}$$

Note that an object point at the point of convergence, P, will exhibit no relative displacement and that this will be true for any point P on the circle through L, R, and P_0. (According to Euclidean geometry, an arc segment, α, of a circle subtends and angle $\frac{1}{2}\alpha$ at every point on the circle.) This circle is called *horopter*. We may approximate

$$\Delta\theta = \theta - \theta_0 \approx \frac{R_0 - R}{RR_0} B_0, \tag{15.42}$$

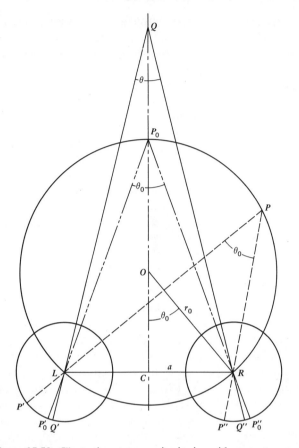

Figure 15.52   Illustrating stereoscopic viewing with convergence.

where RR_0 are the distances to the two object points and
  B_0 is the interocular distance.

A more exact relationship is derived in Appendix 15.2, but the above approximation is valid for conditions generally met in instrumentation.

15.8.1.2   Observer Performance. For the small values of $\Delta\theta$ met with in stereoscopy, the brain fuses the two image points and interprets their displacement difference as due to a difference in object distance. Under good conditions of observation, the normal observer is sensitive to $\Delta\theta$ as small as a few seconds of arc: most observers have a stereoscopic threshold below 10 sec. Taking the interocular distance as 65 mm and zero convergence, we find that for an object at $R = 1$ km

$$\Delta\theta = 65 \times 10^{-6} = 13.4''$$

such an object will normally still provide a noticeable stereoscopic effect as contrasted with a much more distant background.

We list here a number of factors influencing stereoscopic performance. As expected, angular separation between the two object points raises the stereoscopic threshold. Experimental results are shown in Figure 15.53. Object size, too, is found to influence performance, and an object diameter of 2.4′ seems to permit optimum discrimination.

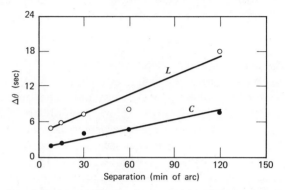

Figure 15.53 Stereoscopic acuity ($\Delta\theta$) as a function of angular separation of compared object points. After Matsubayashi [1d].

Performance is better for more distant objects. High object luminance, too, improves stereoscopic performance. Representative data are shown in Figure 15.54 [138]. On the other hand, differences in visual acuity between the two eyes does not seem to have much effect on stereoscopic discrimination up to a reduction to one fourth of the resolving power of

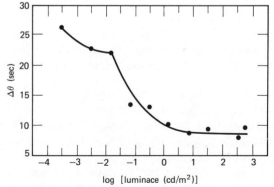

Figure 15.54 Stereoscopic acuity ($\Delta\theta$) as a function of object luminance. Mueller and Lloyd [138].

one eye relative to that in the other. See Ref. 139 for a dramatic illustration.

15.8.1.3 Stereoscopic Instruments.

The stereoscopic effect may be enhanced by magnification, m^*, of the angular subtense of the object, as in microscopy and telescopy. It may also be enhanced by separating the two points of view and bringing the images to their respective eyes by means of mirrors or prisms; see Figure 15.55. The resulting displacement difference is then still given by (15.42) with B_0 replaced by the effective baseline

$$B_e = m^*B, \tag{15.43}$$

where B is the actual baseline of the instrument.

The stereoscopic effect may be simulated by presenting to the two eyes different images corresponding to views of the same object from two positions differing in azimuth angle. (Here azimuth angle is an angle measured in a plane containing the interocular axis.) In aerial photography, for instance, the terrain may be photographed from two widely separated aircraft (i.e., camera) positions. When the two developed photographs are presented to the two eyes, respectively (with the direction corresponding to the camera displacement parallel to the interocular axis) pronounced stereoscopic effects may be observed in the terrain region common to the two photographs.

Several techniques can be used for presenting stereoscopic image pairs to an observer. In the simplest arrangement, the two images are spaced by the interocular distance and the observer views them without convergence. This may require some practice since the eyes tend to converge to a degree appropriate to the object distance as signaled by the accommodation of the lenses of the eyes. (Here a myopic observer may have an advantage: he simply views the pair without his eyeglasses at a distance corresponding to his far point.) Again, a separate viewing lens may be provided for each of the two images, with the images placed near the focal points of these lenses. If the image spacing differs significantly from the interocular distance, this may be compensated for by making the lenses prismatic and thus deviating the line of sight [see (8.37)]. If the required prism angle would be too large, an arrangement similar to that of Figure 15.55 may be used.

Occasionally, especially when free viewing from many positions is needed, it is desirable to superimpose the two images on the same overall area. The images to be presented to the two eyes may then be separated by forming the images, respectively, by light (a) in different wavelength regions or (b) linearly polarized in different directions. In both techniques, the viewer uses a pair of optical filters in front of his eyes to

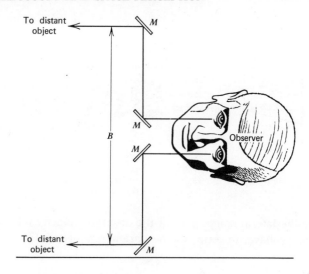

Figure 15.55 Enlargement of stereoscopic base.

eliminate from each eye the unwanted image. In (a), these are spectral filters, opaque to the wavelength range of the unwanted image, and in (b) they are linear polarizers crossed with the direction of polarization of the light in the unwanted image.

In another technique, the image for each eye is formed through a uniform bar grating that blocks alternating bands. The two images are then printed interleafed. Bars, placed a small distance in front of the mixed print, block from each eye's view the unwanted portion of the print (see Figure 15.56).

For a review of instruments used in stereoscopy, see Ref. 140.

15.8.2 Monocular Parallax

In stereoscopic viewing we found that generally the relative position of two images, as seen by the two eyes, differed, with the difference depending on the distance to the object (more correctly, with the radius of the corresponding horopter.) This effect may be used to estimate relative object distances even with a single eye: the observer moves his eye laterally and notes the changing relative image positions—the parallax.

This technique may be used to locate image positions in space by checking the parallax motion relative to a pin superimposed on the image in space. When there is no motion of the pin relative to the image, the pin is in the image plane.

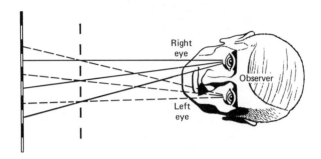

Figure 15.56 Interleaved stereoscopic image system.

Denoting the lateral component of the observer velocity relative to the object by v, we have, in terms of the parameters of (15.42):

$$v = \frac{dB_0}{dt} \qquad (15.44)$$

and for the angular velocity of the relative image positions

$$\delta\omega = \frac{d\Delta\theta}{dt} = \frac{d\Delta\theta}{dB} \cdot \frac{dB}{dt} \approx \frac{R_0 - R}{RR_0} v = \frac{\Delta R}{R} \omega, \qquad (15.45)$$

where $\Delta R = R - R_0$, and
  $\omega = v/R_0$ is the angular velocity of the observer.

The sensitivity is higher for smaller velocities and representative data are shown in Figure 15.57 [141]. The dependence of the threshold on luminance is shown in Figure 15.58 [141].

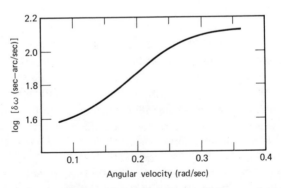

Figure 15.57 Threshold of differential angular velocity ($\delta\omega$), as a function of angular velocity. Graham, Baker, Hecht, and Lloyd [141].

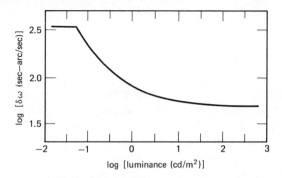

Figure 15.58 Threshold of differential angular velocity ($\delta\omega$) as a function of luminance. Graham, Baker, Hecht, and Lloyd [141].

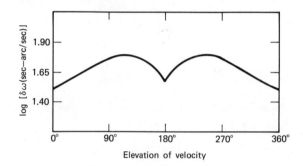

Figure 15.59 Threshold of differential angular velocity ($\delta\omega$) as a function of movement direction. Direction measured from horizontal. Note that the threshold is minimum for horizontal and maximum for vertical movement. Graham, Baker, Hecht, and Lloyd [141].

Sensitivity is also found to be greatest for horizontal movement and least for vertical movement. Threshold angular velocity as a function of direction of velocity is shown in Figure 15.59 [141].

15.8.3 Other Cues in Space Perception

Here we list only briefly other cues in space perception. Some of these depend on physiological factors, such as convergence and accommodation. The observer senses, usually on the subconscious level, the amount of convergence and accommodation required to see the object clearly and this provides him with information concerning its distance from him.

In other instances, the information is conveyed based on his presumed knowledge of the actual dimensions and apparent size of the object: we can estimate quite well the distance to another person based on our knowledge of his actual size. Similarly, we can estimate distance along a road based on perspective, the convergence of its edges in the distance. Here our estimate is based on the assumption of uniform road width. The appearance of texture of known fineness, too, can give us a clue concerning object distances [142].

At larger distances, the effect of atmospheric haze is usually such as to lower object contrast and to change the hue toward the blue (see Section 12.4.6). Such changes, too, can then serve as cues to object distance. The interested reader can get details and relevant references to these in Ref. 1*e*.

APPENDIX 15.1 THRESHOLD CONTRAST AND NOISE IN THE VISUAL SYSTEM

1 Threshold Contrast as a Function of Noise

At the detection threshold, the signal-to-noise ratio equals some minimum value, say b:

$$\frac{Q_s}{Q_n} = b. \tag{15.46}$$

The threshold contrast, on the other hand, is defined as the luminance difference divided by the background luminance:

$$C = \frac{\Delta L}{L_b} = \frac{Q_s}{Q_b} = \frac{bQ_n}{Q_b}, \tag{15.47}$$

where we have taken the signal to be given by the luminance difference and assumed the detected energy to bear a fixed ratio to the luminance. The last step of (15.47) follows from (15.46).

When several noise sources contribute to the system noise, their total mean-square value equals the sum of their mean-square values. Referring all of them to the input stage, we must divide each by the signal gain (G_j) up to that stage:

$$Q_n^2 = \sum_j \frac{Q_{nj}^2}{G_j^2}. \tag{15.48}$$

Hence, from (15.47):

$$C_\theta^2 = \left(\frac{b}{Q_b}\right)^2 \sum \left(\frac{Q_{nj}}{G_j}\right)^2$$

$$= \sum \left(\frac{C_j}{G_j}\right)^2, \tag{15.49}$$

where we have written the noise energy in terms of the partial contrasts:

$$C_j = b\frac{Q_{nj}}{Q_b}. \tag{15.50}$$

In other words, we obtain the threshold contrast squared as the sum of the partial contrasts squared, referred to the input.

2 Threshold Due to Dark Noise

The magnitude of the dark noise can be estimated from the absolute threshold measurements. At absolute threshold, the luminance is presumably b times the dark noise. Hence k_d in (15.35) is identical to the absolute threshold. From the data of Figure 15.16 we find this to be (in cd/m²)

$$L_0 \approx \frac{10^{-9}}{\Omega} \text{ cd/m}^2, \tag{15.51}$$

for objects well below 1° in diameter. Here we have taken the object size in steradians:

$$\Omega = \pi\alpha^2 = \left(\frac{\pi}{180\times 60}\right)^2 \pi\alpha^{*2}, \tag{15.52}$$

where α, α^* are the angular diameter of the object in radians and minutes of arc, respectively. To obtain the absolute threshold in trolands, we must multiply L_0 by the pupil area in mm², that is, by approximately 40. Thus we find

$$k_d = I_0 = \frac{4\times 10^{-8}}{\Omega} \text{ Td},$$

for small objects, as given by (15.39), and the corresponding threshold contrast

$$C_d = \frac{k_d}{I_b}. \tag{15.53}$$

This point of view also carries implications concerning the variation of dark light with object area. For small objects, the dark light will be summed over a fixed area, independent of the smaller object area. This implies a constant noise level, independent of the object size. Since the signal does vary directly with the object area, the absolute threshold varies inversely with object area. This accounts for Ricco's law (15.16). For larger objects, the signal is summed over the object area, so that its fluctuations vary with the square root of that area. Since the signal varies directly with the area, the threshold intensity varies inversely with the square root of the area. This accounts for Piper's law (15.17); see, also (3.102).

3   Quantum Limited Thresholds

To determine thresholds set by quantum effects, let us first calculate the noise due to a (quasi-)monochromatically illuminated object and background, with radiance values L_{eo} and L_{eb}, respectively. The signal energy is, then:

$$Q_s = \eta(L_{eo}A_0 - L_{eb}A_b)t\frac{A_p}{d^2}, \qquad (15.54)$$

where η is the (quantum) efficiency with which flux is detected, relative to the flux incident on the pupil,

A_0, A_b, A_p are, respectively, the areas of the object, the background segment over which the eye integrates, and the pupil,

t is the integration time of the visual process, and

d is the distance of the eye from the object.

We take the variance of the signal energy as the noise squared (Q_n^2). This equals the sum of the variances of the object and the background. This, in turn, equals the sum of the quanta detected from both, multiplied by the energy-per-quantum-squared. Hence

$$Q_n = \sqrt{n}\, Q_q = \sqrt{Q^+ Q_q} \qquad (15.55)$$

where $n = Q^+/Q_q$ (15.56)

is the number of photons detected,

$Q_q = hc/\lambda$ (15.57)

is the energy per photon, and

Q^+ = the total received energy involved in the detection process. It equals (15.54) with the minus sign replaced by a plus sign.

Hence the signal-to-noise ratio is

$$\frac{Q_s}{Q_n} = \frac{Q_s}{\sqrt{Q^+ Q_q}}. \qquad (15.58)$$

At the detection threshold, this must equal b. To obtain an expression in terms of contrast:

$$C = \frac{L_{e0} - L_{eb}}{L_{eb}}, \tag{15.59a}$$

we substitute:

$$L_{e0} - L_{eb} = CL_{eb} \tag{15.59b}$$

and

$$L_{e0} + L_{eb} = (2 + C)L_{eb} \tag{15.59c}$$

into (15.58).

Since we do not know the value of A_b, we assume it equal to A_0 and in order to obtain a simple expression we limit ourselves to low contrast values ($C \ll 2$). Then, on substituting (15.54) and (15.59) into (15.58), we find:

$$\left(\frac{Q_s}{Q_n}\right)^2 = \frac{\eta A_p A_0 t}{d^2} \frac{L_{eb} C^2}{2Q_q}$$

$$= \frac{\eta I_{eb}}{2Q_q} \Omega t C^2, \tag{15.60}$$

where we have written the solid angle subtended by the object:

$$\Omega = \frac{A_0}{d^2}, \tag{15.61}$$

and, in anticipation of our conversion to trolands:

$$I_e = L_e A_p. \tag{15.62}$$

On equating this to b^2 and solving for C^2, we find:

$$C^2 = \frac{2Q_q b^2}{\eta t I_{eb} \Omega}. \tag{15.63}$$

We must now find the corresponding expression when polychromatic light, with an intensity distribution characterized by $I_{e\lambda}$, is used. For such radiation, the coefficient of (15.60) must be integrated over the spectrum. It can be shown [45a] that the quantum efficiency is, in terms of the standard visibility factor, V:

$$\eta(\lambda) = (\lambda_0/\lambda) V \eta_0, \tag{15.64}$$

where we have written η_0 for $\eta(\lambda_0)$. When we substitute this into the coefficient of (15.60) and integrate, we obtain

$$\frac{1}{2} \int \left(\frac{\lambda_0}{\lambda} V \eta_0 \frac{I_{eb}}{Q_q} \right) d\lambda = \frac{\eta_0}{2Q_{q0}} \int VI_{eb}\, d\lambda, \tag{15.65}$$

where

$$Q_{q0} = hc/\lambda_0 \tag{15.66a}$$

is the photon energy at λ_0 and we noted that

$$Q_q = hc/\lambda. \tag{15.66b}$$

Now note that the integral of the right-hand member of (15.65), when multiplied by $10^6 K_0$ equals the intensity in trolands [see (1.7) and the subsequent discussion as well as Section 15.1.4.5]. On substituting this (I_b) into (15.65) and the result into (15.60), we find the signal-to-noise ratio squared:

$$\left(\frac{Q_s}{Q_n} \right)^2 = \left(\frac{\eta_0}{2Q_{q0}} \right) \left(\frac{I_b}{10^6 K_0} \right) tC^2 \tag{15.67}$$

and the corresponding expression for the contrast:

$$C^2 = \frac{2 \times 10^6 Q_{q0} b^2}{\eta_0 t I_b \Omega}. \tag{15.68}$$

Based on this, Rose wrote his fundamental relationship:

$$C^2 \Omega I_b = \text{constant.} \tag{15.69}$$

Comparing (15.68) to (15.35), we see that

$$k_q = \sqrt{2} \times 10^3 b \sqrt{\frac{K_0 Q_{q0}}{\eta_0 t \Omega}}$$

$$\approx \frac{0.56 \times 10^{-3}}{\sqrt{\Omega}}, \tag{15.70}$$

on substituting the following estimates:

$$b = 3, \qquad K_0 = 680\,\text{lm/W}, \qquad Q_{q0} = 0.36 \times 10^{-18}\,\text{W},$$

$$\eta_0 = 0.07,^{36} \qquad \text{and} \qquad t = 0.2\,\text{sec.}$$

4 Neural Threshold

Even assuming the neural noise to be constant, its effect on the threshold contrast will vary with the adaptation luminance, because this

---

[36] This value is based on the assumption of 50% transmission through the ocular media at λ_0 [1f], 20% of the energy incident on the retina absorbed in the rhodopsin [1f] and a 70% quantum efficiency within the rhodopsin [143].

controls the gain in the visual system. Since the visual transfer characteristic is nonlinear and we have limited our analysis to low contrasts, we may use the differential gain, and the neural noise must be divided by this to yield the noise-equivalent trolands.

Corresponding to (15.14), the brightness function is

$$B = kI^\beta \qquad (15.71)$$

with

$$k = \frac{B_\infty}{I_a^\beta + K}, \qquad (15.72)$$

where I_a is the adaptation intensity. From this we readily find the differential gain:

$$G' = \frac{dB}{dI} = \frac{\beta k I^\beta}{I} = \frac{\beta B}{I}. \qquad (15.73)$$

Hence the neural threshold contrast is

$$C_\psi = \frac{b B_n / G'}{I_b} = \frac{b B_n I}{\beta B I_b}. \qquad (15.74)$$

From (15.71) and (15.72) the value of B is

$$B = \frac{B_\infty}{(1 - K I_b^{-\beta})}, \qquad (15.75)$$

at the adaptation intensity

$$I = I_a = I_b.$$

Accordingly, the expression for the threshold contrast becomes:

$$C_\psi = \frac{b B_n}{\beta B_\infty} (1 + K I_b^{-\beta})$$
$$= k_\psi (1 + K I_b^{-\beta}). \qquad (15.76)$$

The value of k_ψ can be determined from threshold contrast data at high intensities, where the other noise components are negligible. The values of k_ψ given at the end of Section 15.6.2 are based on Blackwell's data; see Figure 15.18 [45], [48].

5 Summary

On substituting (15.53), (15.68), and (15.76) into (15.49) we obtain, analogous to (15.38):

$$C_\theta^2 = C_d^2 + c_q^2 + C_\psi^2$$
$$= \frac{k_d^2}{I^2} + \frac{k_q^2}{I} + k_\psi^2 (1 + K I^{-\beta})^2, \qquad (15.77)$$

where I is the background intensity. In Figure 15.60 we see plots of the component contrasts and the resultant overall contrast, corresponding to a 2° diameter circular disk object [45].

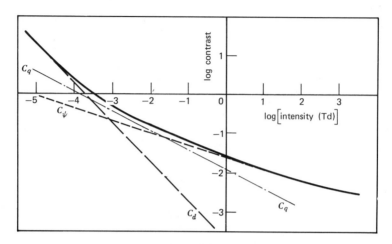

Figure 15.60   Component contrast thresholds (C_d, C_q, and C_ψ) and the resultant contrast threshold (solid curve), as a function of background intensity. Note that C_d is a straight line with negative unity slope; C_q is a straight line of slope -0.5, and C_ψ has a slope of -0.32 at low intensities.

APPENDIX 15.2   STEREOSCOPIC ANGULAR DIFFERENCES

Here we derive the displacement differences between the two retinal images due to object distance. Refer to Figure 15.52.

The semiinterocular axis,

$$\overline{CR} = a,$$

subtends an angle θ_0 at the center of the horopter circle, where θ_0 is the angle the total interocular axis subtends at its circumference. Thus the angle (θ_0) subtended by the interocular axis at object point P_0 is given by

$$\sin \theta_0 = \frac{a}{r_0} \qquad (15.78)$$

and for any other point by

$$\sin \theta = \frac{a}{r}, \qquad (15.78a)$$

where r, r_0 are the radii of the corresponding horopter circles. The difference in relative displacement is

$$\Delta\theta = \theta - \theta_0 \approx \sin\theta - \sin\theta_0 = \frac{a(r_0 - r)}{rr_0}. \tag{15.79}$$

The radii of the horopter circles may be calculated by reference to Figure 15.61. Applying the cosine theorem to Triangle COP we find that

$$r^2 = \overline{OC}^2 + R^2 - 2\overline{OC}\,R\cos\varphi, \tag{15.80}$$

where R is the distance to the object point and
  φ is its azimuthal position.

From the right triangle COR we have

$$\overline{OC}^2 = r^2 - a^2, \tag{15.81}$$

so that

$$r^2 = r^2 - a^2 + R^2 - 2R\sqrt{r^2 - a^2}\cos\varphi \tag{15.82}$$

and hence

$$r^2 = a^2 + \left(\frac{R^2 - a^2}{2R\cos\varphi}\right)^2. \tag{15.83}$$

With $a \ll R$ and $\varphi \ll 1$, we have, approximately,

$$r = \tfrac{1}{2}R. \tag{15.84}$$

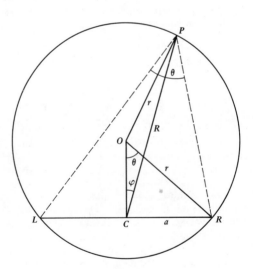

Figure 15.61   Finding the horopter radius.

On substituting this into (15.79) we find

$$\Delta\theta = \frac{R_0 - R}{RR_0} \, (2a).$$

This is the approximation cited in (15.42).

REFERENCES

[1] C. H. Graham, ed., *Vision and Visual Perception*, Wiley, New York, 1965; (a) J. L. Brown, "The structure of the visual system," pp. 39–59; (b) C. H. Graham, "Some basic terms and methods," pp. 60–67; (c) L. A. Riggs, "Visual acuity," pp. 321–349; (d) C. H. Graham, "Color: Data and theories," pp. 414–451; (e) C. H. Graham, "Visual space perception," pp. 504–507; (f) N. R. Bartlett, "Threshold as dependent on some energy relations and characteristics of the subject," pp. 154–184.

[2] Y. Le Grand, *Light, Color and Vision*, 2nd ed., Chapman & Hall, London, 1968; (a) Chapters 3 and 5; (b) Chapter 16; (c) Chapter 18; (d) p. 101; (e) Chapter 10; (f) Chapter 13; (g) Chapter 19.

[3] G. Wyszecki and W. S. Stiles, *Color Science*, Wiley, New York, 1967; (a) Section 2; (b) Table 3.3; (c) Table 2.7; (d) Section 6.6; (e) Section 7.4.

[4] H. Davson, ed., *The Eye, Vol. 2A: Visual Function in Man*, 2nd Ed. Academic, New York, 1976; (a) H. Ripps and R. A. Weale, "*The visual photoreceptors*", Chapter 1; (b) G. B. Arden and A. L. Holden, "The retina", Part II; (c) H. Ripps and R. A. Weale, "Visual adaptation," Sec. IV A, Chapter 3; (d) H. Ripps and R. A. Weale, "Contrast and border phenomena: Visual resolution," Chapter 4; (e) H. Ripps and R. A. Weale, "Temporal analysis and resolution," Chapter 5; (f) H. Ripps and R. A. Weale, "After-images," Chapter 6, (g) F. H. C. Marriott, "Abnormal color vision," Chapter 12.

[5] R. T. Hennessy, "Instrument myopia," *J. Opt. Soc. Am.* **65**, 1114–1120 (1975).

[6] C. A. Johnson, "Effects of luminance and stimulus distance on accommodation and visual resolution," *J. Opt. Soc. Am.* **66**, 138–142 (1976).

[6*] R. J. Miller, "Temporal stability of the dark focus of accommodation," *Am. J. Optom. Physiol. Opt.* **55**, 447–450 (1978).

[7] T. N. Cornsweet, *Visual Perception*, Academic, New York, 1970; (a) Chapters 5–7; (b) pp. 399–410; (c) Chapters 11 and 13; (d) Chapters 15 and 16.

[8] M. H. Pirenne, *Vision and the Eye*, 2nd Ed., Chapman & Hall, London, 1967.

[9] G. A. Østerberg, "Topography of the Layer of Rods and Cones in the Human Retina," *Acta Ophthalmol.* Suppl. 6 (1935).

[10] J. E. Dowling and B. B. Boycott, "Organization of the primate retina: Electronmicroscopy," *Proc. Roy. Soc. (Lond.)* **B166**, 80–111 (1966).

[11] R. W. Young, "Visual Cells," *Sci. Amer.* **223**, 80–91 (Oct. 1970).

[12] E. Snitzer and H. Osterberg, "Observed Dielectric Waveguide Modes in the Visible Spectrum," *J. Opt. Soc. Am.* **51**, 499–505 (1961).

[13] International Commission on Illumination, *J. Opt. Soc. Am.* **41**, 734–738 (1951).

[14] D. H. Sliney, R. T. Wangemann, J. K. Franks, and M. L. Wolbarsht, "Visual sensitivity of the eye to infrared laser radiation," *J. Opt. Soc. Am.* **66**, 339–341 (1976).

[15] A. Aguilar and W. S. Stiles, "Saturation of the rod mechanism of the retina at high levels of stimulation," *Opt. Acta* **1**, 59–65 (1954).

[16] D. A. Palmer, "Standard observer for large-field photometry at any level," *J. Opt. Soc. Am.* **58,** 1296–1299 (1968).

[17] I. Abramov and J. Gordon, "Color vision in the peripheral retina. I. Spectral sensitivity," *J. Opt. Soc. Am.* **67,** 195–202 (1977).

[18] W. A. H. Rushton, "Visual adaptation," *Proc. Roy. Soc. (Lond.)* **B162,** 20–46 (1965).

[19] W. A. H. Rushton and G. H. Henry, "Bleaching and regeneration of cone pigments in man," *Vision Res.* **8,** 617–631 (1968).

[20] S. Duke-Elder ed., *System of Ophthalmology,* Vol. IV, Kimpton, London, 1968, (a) "Non-photic stimuli," pp. 465–468; (b) "The nutrition of the retina," pp. 393–395.

[21] G. Oster, "Optical art," *Appl. Opt.* **4,** 1359–1369 (1965), Sec. V.

[22] D. van Norren and J. J. Vos, "Spectral transmission of the human ocular media," *Vision Res.* **14,** 1237–1243 (1974).

[23] W. S. Stiles and B. H. Crawford, "The luminous efficiency of rays entering the eye pupil at different points," *Proc. Roy. Soc. (Lond.)* **B112,** 428–450 (1933).

[24] J. A. van Loo and J. M. Enoch, "The scotopic Stiles-Crawford effect," *Vision Res.* **15,** 1005–1009 (1975).

[25] C. J. Bartleson, "Pupil diameters and retinal illuminances in interocular matching," *J. Opt. Soc. Am.* **58,** 853–855 (1968).

[26] F. W. Campbell and A. H. Gregory, "Effect of size of pupil on visual acuity," *Nature* **187,** 1121–1123 (1960).

[27] J. M. Woodhouse, "The effect of pupil size on grating detection at various contrast levels," *Vision Res.* **15,** 645–648 (1975).

[28] J. M. Woodhouse and F. W. Campbell, "The role of the pupil light reflex in aiding adaptation to the dark," *Vision Res.* **15,** 649–653 (1975).

[29] J. ten Doesschate and M. Alpern, "The effect of photoexcitation of the two retinas on pupil size," *J. Neurophysiol.* **30,** 562–576 (1967).

[30] S. G. de Groot and J. W. Gebhard, "Pupil size as determined by adapting luminance," *J. Opt. Soc. Am.* **42,** 492–495 (1952).

[31] E. Alexandridis, "Räumliche und zeitliche Summation pupillomotorisch wirksamer Lichtreize beim Menschen," *Albrecht v. Graefes Arch. klin. exp. Ophthal.* **180,** 12–19 (1970).

[32] M. Alpern and F. W. Campbell, "The behavior of the pupil during dark adaptation," *J. Physiol. (Lond.)* **165,** 5P–7P (1963).

[33] G. A. Gescheider, *Psychophysics, Method and Theory,* Erlbaum, Hillsdale, N. J., 1976.

[33*] J. C. Stevens and S. S. Stevens, "Brightness Function: Effects of Adaptation," *J. Opt. Soc. Am.* **53,** 375–385 (1963).

[34] S. S. Stevens, "To honor Fechner and repeal his law," *Science* **133,** 80–86 (1961).

[35] W. W. Petersen, T. G. Birdsall, and W. C. Fox, "The theory of signal detectability," *Trans. IRE* **PGIT-4,** 171–212 (1954).

[36] D. M. Green and J. A. Swets, *Signal Detection Theory and Psychophysics,* Wiley, New York, 1966.

[37] D. H. Kelly and R. E. Savoie, "A study of sine-wave contrast sensitivity by two psychophysical methods," *Percept. Psychophys.* **14,** 313–318 (1973).

[38] R. A. Weale, "Limits of human vision," *Nature* **191,** 471–473 (1961).

[39] R. J. W. Mansfield, "Brightness function effect of area and duration," *J. Opt. Soc. Am.* **63,** 913–920 (1973).

[40] R. B. Barlow and R. T. Verrillo, "Brightness sensation in a ganzfeld," *Vision Res.* **16,** 1291–1297 (1976).

[41] F. Thoss and S. Bougrina, "The influence of adaptation and field area on the exponents of Stevens' power functions at the light reaction of the human pupil," *Vision Res.* **16**, 317–320 (1976).

[42] R. A. Normann and F. S. Werblin, "Control of retinal sensitivity. I. Light and dark adaptation of vertebrate rods and cones," *J. Gen. Physiol.* **63**, 37–61 (1974).

[43] F. S. Werblin, "Control of retinal sensitivity. II. Lateral interaction at the outer plexiform layer," *J. Gen. Physiol.* **63**, 62–87 (1974).

[44] R. J. W. Mansfield, "Visual adaptation: Retinal transduction, brightness, and sensitivity," *Vision Res.* **16**, 679–690 (1976).

[45] L. Levi, "Types of noise in the visual system," to be published (*a*) Appendix.

[46] L. Levi, "Automatic gain control model for vision," *Nature* **233**, 396–397 (1969).

[47] G. Wald, "Human vision and the spectrum," *Science* **101**, 653–658 (1945).

[48] H. R. Blackwell, "Contrast thresholds of the human eye," *J. Opt. Soc. Am.* **36**, 624–643 (1946).

[49] A. M. Marsden, "An elemental theory of induction," *Vision Res.* **9**, 653–663 (1969).

[50] G. Westheimer, "Spatial interaction in human cone vision," *J. Physiol. (Lond.)* **190**, 139–154 (1967).

[51] G. van den Brink and C. J. Kreemink, "Luminance gradients and edge effects," *Vision Res.* **16**, 155–159 (1976).

[52] E. H. Land and J. J. McCann, "Lightness and retinex theory," *J. Opt. Soc. Am.* **61**, 1–11 (1971).

[52*] R. P. Dooley and M. I. Greenfield, "Measurements of edge-induced visual contrast and a spatial-frequency interaction of the Cornsweet illusion," *J. Opt. Soc. Am.* **67**, 761–765 (1977).

[53] E. H. Land, "The retinex theory of color vision," *Proc. Roy. Inst. Gr. Brit.* **47**, 23–58 (1974).

[54] A. van Meeteren, "Calculations on the optical modulation transfer function of the human eye for white light," *Opt. Acta* **21**, 395–412 (1974).

[55] A. Ivanoff, "About the spherical aberration of the eye," *J. Opt. Soc. Am.* **46**, 901–903 (1956).

[56] W. N. Charman and J. A. M. Jennings, "Objective measurements of the longitudinal chromatic aberration of the human eye," *Vision Res.* **16**, 999–1005 (1976).

[57] G. Westheimer, "Optical and motor factors in the formation of the retinal image," *J. Opt. Soc. Am.* **53**, 86–93 (1963).

[58] F. W. Campbell and R. W. Gubish, "Optical quality of the human eye," *J. Physiol. (Lond.)* **186**, 558–578 (1966).

[58*] J. A. M. Jennings and W. N. Charman, "Optical image quality in the peripheral retina," *Am. J. Optom. Physiol. Opt.* **55**, 582–590 (1978).

[59] J. J. Vos, J. Walráven, and A. van Meeteren, "Light profiles of the foveal image of a point source," *Vision Res.* **16**, 215–219 (1976).

[60] G. Westheimer, "Spatial frequency and light-spread descriptions of visual acuity and hyperacuity," *J. Opt. Soc. Am.* **67**, 207–212 (1977).

[61] M. Hines, "Line spread function variation near the fovea," *Vision Res.* **16**, 567–572 (1976).

[62] D. H. Kelly, "Spatial frequency selectivity in the retina," *Vision Res.* **15**, 665–672 (1975).

[63] M. Davidson, "Perturbation approach to spatial brightness interaction in human vision," *J. Opt. Soc. Am.* **58**, 1300–1308 (1968).

[64] Y. Le Grand, "Sur la mesure de l'acuité visuelle au moyen de franges d'interférence," *C. R. Acad. Sci.* **200**, 490–491 (1935).

[65] F. W. Campbell, "The human eye as an optical filter," *Proc. IEEE* **56**, 1009–1014 (1968).

[66] G. A. Fry, "The optical performance of the human eye," in *Progress in Optics*, Vol. 8, E. Wolf, ed., North-Holland, Amsterdam, 1970, pp. 53–131.

[67] L. Levi, "Vision in communication," in *Progress in Optics*, Vol. 8, E. Wolf, ed., North-Holland, Amsterdam, 1970, pp. 345–372, (a) Figure 5.

[68] O. Bryngdahl, "Characteristics of the visual system: psychophysical measurements of the response to spatial sine-wave stimuli in the photopic region," *J. Opt. Soc. Am.* **56**, 811–821 (1966).

[69] O. Franzen and M. Berkley, "Apparent contrast as a function of modulation depth and spatial frequency," *Vision Res.* **15**, 655–660 (1975).

[70] R. L. de Valois and H. Morgan, "Psychophysical studies of monkey vision. III. Spatial luminance contrast sensitivity test of macaque and human observers," *Vision Res.* **14**, 75–81 (1974).

[71] K. E. Higgins and E. J. Renalducci, "Suprathreshold intensity-area relationships: A spatial Broca-Sulzer effect," *Vision Res.* **15**, 129–143 (1975).

[72] M. Davidson and J. A. Whiteside, "Human brightness perception near sharp contours," *J. Opt. Soc. Am.* **61**, 530–536 (1971).

[73] A. Remole, "Effect of retinal illuminance and eccentricity on border enhancement extent," *Vision Res.* **16**, 1323–1327 (1976).

[74] G. von Békésy, "Mach- and Hering-type lateral inhibition in vision," *Vision Res.* **8**, 1483–1499 (1968).

[75] H. B. Barlow, "Temporal and spatial summation in human vision at different background intensities," *J. Physiol. (Lond.)* **141**, 337–350 (1958).

[76] F. L. Van Nes and M. A. Bouman, "Spatial modulation transfer in the human eye," *J. Opt. Soc. Am.* **57**, 401–406 (1967).

[77] S. Hecht and E. V. Mintz, "The visibility of single lines at various illuminations and the retinal basis of visual resolution," *J. Gen. Physiol.* **22**, 593–612 (1939).

[78] J. A. Foley-Fisher, "Measurements of Vernier acuity in white and coloured light," *Vision Res.* **8**, 1055–1065 (1968).

[79] G. Westheimer and S. P. McKee, "Integration regions for visual hyperacuity," *Vision Res.* **17**, 89–93 (1977).

[80] S. Shlaer, "The relation between visual acuity and illumination," *J. Gen. Physiol.* **21**, 165–188 (1937).

[81] F. W. Weymouth, "Visual sensory units and the minimum angle of resolution," *Am. J. Ophthalmol.* **46**, 102–113 (1958).

[82] A. S. Patel, "Spatial resolution by the human visual system. The effect of mean retinal illuminance," *J. Opt. Soc. Am.* **56**, 689–694 (1966).

[83] D. G. Green, "Regional variations in the visual acuity for interference fringes on the retina," *J. Physiol. (Lond.)* **207**, 351–356 (1970).

[84] A. Watanabe, T. Mori, S. Nayata, and K. Hiwatashi, "Spatial sine-wave responses of the human visual system," *Vision Res.* **8**, 1245–1263 (1968).

[85] C. W. Taylor and D. E. Mitchell, "Orientation differences for perception of sinusoidal line stimuli," *Vision Res.* **17**, 83–88 (1977).

[86] C. Blakemore and F. W. Campbell, "On the existence of neurones in the human visual system selectively sensitive to orientation and size of retinal images," *J. Physiol. (Lond.)* **203**, 237–260 (1969).

[87] A. Pantle and R. Sekuler, "Size-detecting mechanisms in human vision," *Science* **162**, 1146–1148 (1968).

[88] M. B. Sachs, J. Nachmias, and J. G. Robson, "Spatial-frequency channels in human vision," *J. Opt. Soc. Am.* **61**, 1176–1186 (1971).

[89] T. Veno, "Reaction time as a measure of temporal summation at suprathreshold levels," *Vision Res.* **17,** 227–232 (1977).

[90] L. E. Hufford, "Reaction time and the retina area-stimulus intensity relationship," *J. Opt. Soc. Am.* **54,** 1368–1373 (1964).

[91] M. Lichtenstein and C. T. White, "Relative visual latency as a function of retinal locus," *J. Opt. Soc. Am.* **51,** 1033–1034 (1961).

[92] T. Vetsuki and M. Ikeda, "Study of temporal visual response by the summation index," *J. Opt. Soc. Am.* **60,** 377–381 (1970).

[93] T. H. Nilsson, "Two-pulse-interval vision thresholds," *J. Opt. Soc. Am.* **59,** 753–756 (1969).

[94] L. A. Lefton, "Metacontrast: A review," *Percept. Psychophys.* **13,** (Suppl. 1B), 161–171 (1973).

[95] R. M. Herrick, "Foveal luminance discrimination as a function of the duration of the decrement or increment of luminance," *J. Comp. Physiol. Psychol.* **49,** 437–443 (1956).

[96] J. M. Anglin and R. J. W. Mansfield, "On the brightness of short and long flashes," *Percept. Psychophys.* **4,** 161–162 (1968).

[97] H. de Lange, "Research into the dynamic nature of the human fovea-cortex system with intermittent and modulated light. I. Attenuation characteristics with white and colored light," *J. Opt. Soc. Am.* **48,** 774–784 (1958).

[98] D. H. Kelly, "Visual response to time-dependent stimuli," *J. Opt. Soc. Am.* **51,** 422–429 (1961).

[99] D. H. Kelly, "Theory of flicker and transient response. I. Uniform fields," *J. Opt. Soc. Am.* **61,** 537–546 (1971).

[100] C. R. Cavonius and O. Estévez, "Sensitivity of human color mechanism to gratings and flicker," *J. Opt. Soc. Am.* **65,** 966–968 (1975).

[101] C. V. Truss, "Chromatic flicker fusion frequency as a function of chromatic difference," *J. Opt. Soc. Am.* **47,** 1130–1134 (1957).

[102] P. L. Walraven and H. J. Leebeek, "Phase shift of sinusoidally alternating colored stimuli," *J. Opt. Soc. Am.* **54,** 78–82 (1964).

[103] O. Bryngdahl, "Effects of spatiotemporal sinusoidally varying stimuli on brightness perception," *J. Opt. Soc. Am.* **56,** 706–707 (1966).

[104] F. Veringa, "On some properties of nonthreshold flicker," *J. Opt. Soc. Am.* **48,** 500–502 (1958).

[105] C. Rabelo and O. J. Grüsser, "Die Abhängigkeit der subjektiven Helligkeit intermittierender Lichtreize von der Flimmerfrequenz (Brücke-Effeckt, "brightness enhancement"): Untersuchungen bei verschiedener Leuchtdichte und Feldgrösse," *Psychol. Forsch.* **26,** 299–312 (1961).

[106] H. D. Baker, "The course of foveal light adaptation measured by the threshold intensity increment," *J. Opt. Soc. Am.* **39,** 172–179 (1949).

[107] S. Hecht, C. Haig, and A. M. Chase, "The influence of light adaptation on subsequent dark adaptation of the eye," *J. Gen. Physiol.* **20,** 831–850 (1937).

[108] G. Wald and A. B. Clark, "Visual adaptation and the chemistry of the rods," *J. Gen. Physiol.* **21,** 93–105 (1937).

[109] H. D. Baker, "The instantaneous threshold and early dark adaptation," *J. Opt. Soc. Am.* **43,** 798–803 (1953).

[110] S. Hecht and Y. Hsia, "Dark adaptation following light adaptation to red and white lights," *J. Opt. Soc. Am.* **35,** 261–267 (1945).

[111] L. A. Riggs, F. Ratliff, J. C. Cornsweet, and T. N. Cornsweet, "The disappearance of steadily fixated visual test objects," *J. Opt. Soc. Am.* **43,** 495–501 (1953).

[112] V. T. Keesey, "Effects of involuntary eye movements on visual acuity", *J. Opt. Soc. Am.* **50,** 769–774 (1960).

[113] L. A. Arend, "Temporal determinants of the form of the spatial contrast threshold MTF," *Vision Res.* **16,** 1035–1042 (1976).

[114] H. B. Barlow and J. M. B. Sparrock, "The role of after images in dark adaptation," *Science* **144,** 1309–1314 (1964).

[115] G. Ekman and R. Lindman, "Note on measurement of visual after image intensity as a function of time," *Vision Res.* **4,** 579–584 (1964).

[116] V. Tulunay-Keesey and R. M. Jones, "The effect of micromovements of the eye and exposure duration on contrast sensitivity," *Vision Res.* **16,** 481–488 (1976).

[117] J. L. Brown and J. E. Black, "Critical duration for resolution of acuity targets," *Vision Res.* **16,** 309–315 (1976).

[118] D. H. Kelly, "Adaptation effects on spatio-temporal sine-wave thresholds," *Vision Res.* **12,** 89–101 (1972).

[119] F. Ratliff and L. A. Riggs, "Involuntary motion of the eye during monocular fixation," *J. Exp. Psychol.* **40,** 687–701 (1950).

[120] J. Krauskopf, "Effect of retinal image motion on contrast thresholds for maintained vision," *J. Opt. Soc. Am.* **47,** 740–744 (1957).

[121] L. A. Riggs, J. C. Armington, and F. Ratliff, "Motion of the retinal image during fixation," *J. Opt. Soc. Am.* **44,** 315–321 (1954).

[122] S. Hecht, S. Shlaer, and M. H. Pirenne, "Energy, quanta, and vision," *J. Gen. Physiol.* **25,** 819–840 (1942).

[123] B. Sakitt, "Counting every quantum," *J. Physiol. (Lond.)* **223,** 131–150 (1972).

[124] N. Ben Yosef and A. Rose, "Spectral response of the human eye," *J. Opt. Soc. Am.* **68,** 935–936 (1978).

[125] A. Rose, "The sensitivity performance of the human eye on an absolute scale," *J. Opt. Soc. Am.* **38,** 196–208 (1948).

[126] A. Rose, *Vision: Human and Electronic*, Plenum, New York, 1973, Chapters 1 and 2.

[127] O. H. Schade, "Optical and photoelectric analog of the eye," *J. Opt. Soc. Am.* **46,** 721–739 (1956).

[128] H. R. Blackwell, "Studies on the form of visual threshold data," *J. Opt. Soc. Am.* **43,** 456–463 (1953).

[129] W. D. Wright, *Researches on Normal and Defective Colour Vision*, Kimpton, London, 1946.

[130] R. Hilz and C. R. Cavonius, "Wavelength discrimination measured with square-wave gratings," *J. Opt. Soc. Am.* **60,** 273–277 (1970).

[131] G. J. C. van der Horst and M. A. Bouman, "Spatiotemporal chromaticity discrimination," *J. Opt. Soc. Am.* **59,** 1482–1488 (1969).

[132] E. M. Granger and J. C. Heurtly, "Visual chromaticity-modulation transfer function," *J. Opt. Soc. Am.* **63,** 1173–1174 (1973).

[133] P. K. Kaiser, J. P. Comerford, and D. M. Bodinger, "Saturation of spectral lights," *J. Opt. Soc. Am.* **66,** 818–826 (1976).

[134] M. R. Pointer, "Color discrimination as a function of observer adaptation," *J. Opt. Soc. Am.* **64,** 750–759 (1974).

[135] W. A. H. Rushton, "Pigments and signals in colour vision," *J. Physiol. (Lond.)* **220,** 1–31P (1972).

[136] C. R. Cavonius and O. Estévez, "Contrast sensitivity of individual color mechanisms of human vision," *J. Physiol. (Lond.)* **248,** 649–662 (1975).

[137] F. H. G. Pitt, "The nature of normal trichromatic and dichromatic vision," *Proc. Roy. Soc. (Lond.)* **B132,** 101–117 (1944).

[137*] B. R. Wooten, K. Fuld, and L. Spillman, "Photopic spectral sensitivity of the peripheral retina," *J. Opt. Soc. Am.* **65,** 334–342 (1975).

[138] C. G. Mueller and V. V. Lloyd, "Stereoscopic acuity for various levels of illumination," *Proc. Natl. Acad. Sci.* **34,** 223–227 (1948).

[139] B. Julesz, "Texture and visual perception," *Sci. Amer.* **218,** 38–48 (Feb. 1965).

[140] L. P. Dudley, "Stereoscopy," in *Applied Optics and Optical Engineering*, R. Kingslake, ed., Vol. 2, Academic, New York, 1965.

[141] C. H. Graham, K. E. Baker, M. Hecht, and V. V. Lloyd, "Factors influencing thresholds for monocular movement parallax," *J. Exp. Psychol.* **38,** 205–223 (1948).

[142] U. Neisser, "The processes of vision," *Sci. Amer.* **221,** 204–214 (Sept. 1968).

[143] A. Knowles and H. J. A. Dartnall, *The Photobiology of Vision*, Vol. 2B of Ref 4, 1977.

16

Photoelectric and Thermal Detectors

16.1 PHENOMENA EMPLOYED IN PHOTOELECTRIC DETECTION

16.1.1 Fundamentals and Classification [1a], [2a]

To detect light we may, in principle, use any action it exerts on a substance. In this chapter we are primarily concerned with phenomena based on the excitation of electrons in solids due to absorbed radiation. Only the last section is devoted to thermal effects of radiation as used in detection.

It has been mentioned earlier (Section 4.8.2) that in a solid the electrons are distributed in energy bands separated by forbidden gaps. The phenomena of interest to us occur between the valence band where electrons are almost immobile, the conduction band where they are quite mobile, and the semi-infinite "band," which we may call "freedom," where electrons are no longer bound to the original solid. This band exists only beyond the surface and overlaps the conduction band, at least in part. (See Fig. 16.1a).

If the electron receives an amount of energy sufficient to raise it to the *vacuum level* (the bottom of the "freedom" band), it may escape the solid. Once freed, the electron may be transported to another structure for detection. The freeing of electrons from a solid by incident radiant flux is called the *photoemissive effect.*[1]

When the absorbed radiation transfers the electron only into the conduction band, this is called the *photoconductive effect.* The direct use of

---

[1] It is often called simply the photoelectric effect. However, here we shall use the more specific term "photoemissive" and apply the more general "photoelectric" to all the optical electron excitation phenomena.

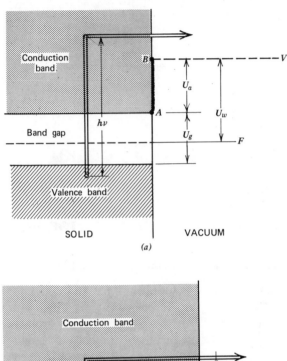

(a)

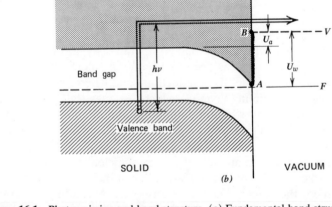

(b)

Figure 16.1 Photoemission and band structure. (a) Fundamental band structure. (b) Band bending. F and V are the Fermi and vacuum levels, respectively. $\overline{AB}$ is the surface barrier. The band gap energy (U_g), electron affinity (U_a), and work function (U_w) are indicated. The hollow arrows illustrate a photoemission process due to a photon of energy $h\nu$. Note that with band bending the required photon energy is lowered.

442

this effect is important, but frequently it is used in conjunction with semiconductor junctions that permit the manipulation of the motion of the freed carriers.

The excitation of an electron is a rapid process requiring only about 10^{-12} sec [3]. In photoemission, this excitation, together with the subsequent rapid electron transport, essentially completes the detection process. Consequently, the response times of photoemissive detectors are generally very short (of the order of a nanosecond) making them suitable for high-frequency operation and the observation of short-duration events. In photoconductive mechanisms, the operation may depend critically on carrier life times, which are quite long, often measured in milliseconds.

Not all incident photons participate in the above processes. Some are reflected or transmitted, others are absorbed in phonon processes [see Section 16.1.6.2 (b)] and still others have the electrons, freed by them, absorbed. The fraction of photons that participate in any detection process is called the quantum efficiency of the process.

16.1.2 Photoemissive Effect

In photoemission an electron must absorb an amount of energy at least equal to the difference between its energy and the vacuum level. Since, at the usual temperatures, the unexcited electron can not rise far above the Fermi level [cf. (4.83)], this amount of energy equals the difference between the Fermi and vacuum levels. It is called *work function. U_w.* If the photons constituting the incident flux do not have this energy, they will not free electrons, no matter how intense the incident flux.[2] Planck's relationship (4.2) relates the wavelength (λ) of a photon to its energy:

$$U_p = h\nu_t = \frac{hc}{\lambda}. \tag{16.1}$$

Here $h = 6.6262 \times 10^{-34}$ J · sec is Planck's quantum of action, ν_t is the frequency of the photon, and $c = 2.997925 \times 10^8$ m/sec is the *in vacuo* velocity of light.

Thus, if the incident photon is to have an energy of at least U_w, its wavelength may not exceed the threshold value

$$\lambda_0 = \frac{hc}{U_w}, \tag{16.2}$$

obtained by setting $U_p = U_w$ in (16.1).

---

[2] We exclude here multiple-photon (nonlinear optic) effects, which become significant only at extremely high flux densities.

If the incident photons have an energy in excess of that amount, they will impart this, too, to the electron, which may then leave the solid with some net excess energy $(U_p - U_w)$, unless it loses some of this energy on the way to the surface. However, one does not rely on this energy for operation in practical detection devices. In these, the released electrons are accelerated by an external electric field, with the total electric current indicating the value of the incident radiant flux.

The work function consists of two parts.

1. Part of the band gap energy, U_g.
2. The *electron affinity*, U_a, representing the difference between the bottom of the conduction band and the vacuum level.

These are indicated in Figure 16.1a. In the first step of photoemission, the electron is transported into the conduction band. It must then diffuse to the surface of the solid. If it reaches this surface with kinetic energy at least equal to the electron affinity, it may escape. Obviously, the ideal photoemitter will have a low electron affinity. At first sight this is limited by the surface barrier, $\overline{AB}$. However, it has been found that the application of certain coatings, principally Cs and Cs_2O, will bend the bands downward at the surface, lowering the electron affinity; see Figure 16.1b. With such *band bending*, even negative values of electron affinity can be obtained. Here, any conduction band electron reaching the surface will have sufficient energy to escape [4].

These considerations control the basic spectral response of photoemissive materials. Consider a decrease in the wavelength of incident flux. When this wavelength reaches the value λ_0, photoemission begins to occur. As the wavelength is reduced further, the photoemissive efficiency increases rapidly. However, as the discrepancy between the incident photon energy and the work function grows, the efficiency, with which the incident photons are absorbed, drops. In addition, because of the excess energy, such electrons are more likely to be excited deep inside the semiconductor and to lose their energy on the way to the surface. Both these effects tend to lower the quantum efficiency with the reduction of wavelength.

The angular distribution of the released electrons usually follows a cosine (Lambert) law distribution [5a].

The existence of an electric field at the emitting surface modifies the shape of the barrier and may enhance the emission of electrons significantly (Schottky effect[3]).

[3] W. Schottky (1938).

16.1.3 Photoconduction [1a]

When the energy of the incident flux suffices to carry electrons across the bandgap into the conduction band, photoconductive detection techniques become possible. In the simple photoconductive effect, the increased conductivity of the material is utilized to measure the amount of incident flux. Since the conductivity of the solid increases directly with the number of electrons in the conduction band and the number of "holes" in the valence band (the "*carriers*"), a rather simple relationship is implied. This is complicated somewhat by the fact that the freed electrons have a tendency to drop back (or to be trapped at the donor sites) and, further, by the existence of some carriers even in the absence of irradiation. Nevertheless, the change in conductivity (γ) is simply [1a]

$$\frac{\Delta\gamma}{\gamma} = k\,\Delta n, \tag{16.3}$$

where the increase in carrier concentration, at equilibrium, is given by

$$\Delta n = g\tau \tag{16.4}$$

and the proportionality factor by

$$k = \frac{b+1}{bn+p}. \tag{16.5}$$

Here g is the carrier generation rate per unit volume; τ is the effective carrier lifetime; n, p are, respectively, the concentration of negative and positive carriers (conduction band electrons and valence band holes), and b is the ratio of electron to hole mobility. See Section 16.1.6 for additional details.

These expressions are based on an assumed state of equilibrium. Because of the relativity long carrier lifetimes in semiconductors, the attainment of equilibrium may be a slow process and photoconductors generally have a relatively poor response to high-frequency flux changes.

With doping, that is, the introduction of impurity atoms of a different valence (see Section 4.8.3), it is possible to reduce the amount of energy required to generate a carrier. Semiconductors whose carriers originate in donor (or acceptor) sites, are called *extrinsic*, in contrast to *intrinsic* semiconductors where such contributions are, at most, of secondary importance.

The basic spectral response considerations here are similar to those of the photoemissive effect [1a].

16.1.4 Junction Effects

The joining of two extrinsic semiconductors of opposite polarity provides powerful means of current control and radiation detection. In the negatively doped (n-type) semiconductor, free electrons predominate; these are electrically balanced by positively charged donor sites. Conversely, in the positively doped (p-type) semiconductor, free "holes" predominate, electrically balanced by negative acceptor sites. On being joined, therefore, the two sections are electrically neutral. As carriers diffuse back and forth across the junction, however, an electrical polarization is quickly established since initially there are more electrons to diffuse from the n-type into the p-type, than *vice versa*, and conversely for the holes. The introduction of excess electrons into the p-type section creates a negatively charged region there; similarly, the diffusion of holes into the n-type material creates a positive region.

This arrangement is called a p-n junction and is illustrated schematically in Figure 16.2a. The steady-state charge distribution is shown in b. Since the divergence of the electric field is proportional to the charge

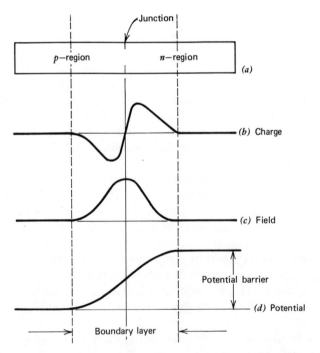

Figure 16.2 Schematic of a p-n junction and the associated charge, field, and potential distributions.

density, the electric field will be as shown in Figure 16.2c. Again, the electric field is the negative of the gradient of the electric potential, which is shown in Figure 16.2d. Thus a potential step, as shown there, develops in the neighborhood of the junction. Clearly, as this barrier develops, it will slow down the drift of electrons into the p-region and that of holes into the n-region.

In addition, even initially, there will be some negative carriers in the p-type material generated by thermal effects, which suffice to carry some electrons into the conduction band. For obvious reasons, these are called *minority carriers*, in contrast to the more numerous holes that constitute the *majority carriers*. The converse situation applied in n-type materials.

The electric field distribution illustrated in Figure 16.2d will tend to accelerate the diffusion of minority carriers across the gap. The slowing down of majority carrier diffusion and the acceleration of minority carrier diffusion continues until the two diffusion rates become equal and equilibrium is reached.

Next we must consider the behavior of the junction under the influence of an externally applied field. To understand this, we note an essential difference between the diffusion of majority and minority carriers: The minority carriers *drop* across the junction, and even those with almost no initial energy can make the crossing. The majority carriers, on the other hand, must have an initial energy equal to the height of the barrier if they are to cross it. Thus all minority carriers are available for diffusion across the junction, while the fraction of available majority carriers depends strongly on the height of the barrier. Specifically, the availability of electrons in the n-doped section varies with the potential, U/q_e, of the conduction band in the p-region according to the Fermi distribution (4.83). For reasonably high barriers ($U \gg kT$), this implies, effectively, an exponential distribution. Here q_e is our symbol for the charge of the electron.

This difference between minority and majority carrier behavior has important consequences for the behavior of the junction under the influence of an externally applied electric field. Consider, for instance, what happens when the potential of the (low-potential) p-region is raised relative to that of the n-region. This will clearly lower the height of the barrier and may increase significantly the number of majority carriers available to cross the junction. Indeed, this number, and therefore the resulting current, will grow exponentially with the applied potential. On the other hand, if we reverse the applied potential, no majority carriers will be drawn across the junction, but all the available minority carriers will cross the junction. Increasing the reversed potential will then have little effect on the current. This is the rectifying effect of the p-n junction

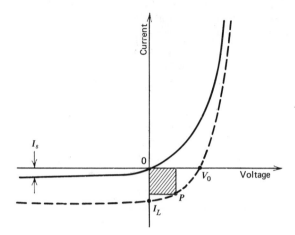

Figure 16.3 Current-voltage characteristic in a p-n junction in the dark (solid line) and under illumination (broken line). I_s is the reverse dark current; V_0 and I_L are the open circuit voltage and short circuit current, respectively.

and is illustrated by the representative current-voltage characteristic shown in curve A of Figure 16.3, exhibiting an exponential rise in current at a positive applied potential and an almost constant small current at negative applied potentials.

The condition of reverse bias is of special interest in connection with radiation detection. In this condition, the current is a direct measure of the concentration of minority carriers. In the preceding section we saw that radiation creates carrier pairs, including minority carriers, so that the number of exciting photons absorbed can be measured by means of a reverse-biased p-n diode. (The newly created majority carriers cannot flow through the circuit across the junction in opposition to the externally applied potential.) The quantum efficiency of the p-n junction may be quite high, well over 50%. Once this value is known, the diode may be used as an absolute "quantum counting" device.

The electron-hole pairs created by the incident light will be separated by the voltage gradient at the junction, with the electrons flowing into the n-region and the holes into the p-region. Thus a current will flow as long as radiation is incident, even with no external voltage applied. This displaces the current-voltage characteristic of the diode downward, to the form shown in Figure 16.3, Curve B. The current just mentioned is known as the short-circuit current, I_L. Alternatively, at open circuit, this photocurrent will cause voltage to build up to a value V_0, at which any further current is prevented from flowing. The creation of voltage V_0 is known as the *photovoltaic effect*. I_L and V_0 are shown in Figure 16.3.

The underlying physical phenomenon, the generation of conduction-band carriers, is the same as in photoconduction, so that the same spectral considerations apply.

The carrier lifetimes in junctions are much shorter than they are in photoconductors and the former therefore lend themselves to work at much higher frequencies. However, the two regions of charge distribution create a capacitive effect that limits the frequency to below about 100 kHz. This limitation may be overcome by interposing an intrinsic (undoped) layer between the p- and n-regions. The resulting *pin* (p-i-n) diodes have cut-off frequencies as high as 10^9 Hz and above [6].

The detection sensitivity of semiconductor junction detectors may be further enhanced by using a double junction arrangement: p-n-p or n-p-n, with the two outer sections far more strongly doped than the center section. Such an arrangement is called a transistor; the former version, which we discuss here as an illustration, is shown in Figure 16.4. According to the usual nomenclature, the sections are labeled emitter, base, and collector, respectively. The applied bias will constitute a forward bias for the first (emitter-base) junction, so that there will initially be a strong hole current into the base. Because the base section is very thin, this current will continue on into the p-type collector section without heavy losses; the energy gained by passing from emitter to base suffices to carry the holes across the barrier from base to collector. At the same time, there will be a strong electron current from the base into the emitter, but this will not be balanced by a matching flow from the collector, which has very few carrier electrons. The base will thus quickly become positively charged, canceling, across the emitter-base junction, the effect of the external bias, so that most of the current ceases.

If exciting radiation is now permitted to fall onto the base, the resulting holes will be drawn directly into the collector, much as in an n-p diode.

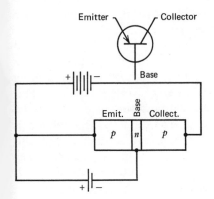

Figure 16.4 Schematic of a p-n-p transistor.

In addition to this, however, the base potential is lowered as a result of the hole exodus, thus reactivating the emitter-base junction. Because of the relatively much stronger doping of the emitter, many more holes will flow into the base and thence into the collector than the photoelectrons that flow from the base back into the emitter and photoholes flowing from the base into the collector, which constitute the current in a simple photodiode. Thus the photocurrent in our phototransistor will be much stronger than that in a simple diode.

Here again, the spectral response is governed by the same considerations as in photoconduction.

16.1.5 Photodiffusion and Photon Drag Effects [1a]

Consider a homogeneous slab of photoconductor material, as shown in Figure 16.5, with radiation (wavelength $\lambda < \lambda_0$) incident on one side. This radiation will create electron-hole pairs, which will diffuse away from the surface. This diffusion current has been used in two ways to measure the irradiation.

1. By immersing the photoconductor in a magnetic field, an electric potential is generated across it, as illustrated in Figure 16.5. As the electron-hole pairs created at the upper surface diffuse downward, the magnetic field will tend to deflect them at right angles to both the field and their direction of propagation. Since the electrons and holes are deflected in opposite directions, this results in a potential difference across the slab. This potential difference can be taken as a measure of the incident flux. This is called the *photoelectromagnetic (PEM) effect.*

2. Without the magnetic field, the above arrangement will generate a potential difference in the direction of the diffusion. This is due to the difference in mobility of electrons and holes. The more mobile electrons will diffuse further away from the irradiated surface, than will the holes; the

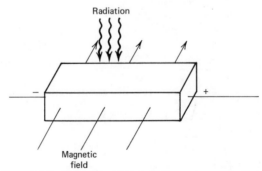

Figure 16.5 Illustrating the photoelectromagnetic effect.

resulting charge gradient results in a potential difference. This is known as the *Dember effect.*[4]

Photons incident on a semiconductor transfer some of their momentum to the free charge carriers in the material. This *photon drag effect* establishes an electromotive force, and hence a potential difference, in the direction of the momentum of the incident photons.

Elongated germanium slabs have been used as radiation detectors in this manner. Such detectors may retain a linear response to irradiation values as high as $400 \, \text{GW/m}^2$ [7] and are therefore especially useful in measuring laser radiation. They may have a linear dynamic range of 10^{10}. Response times may be as short as tens of nanoseconds [8].

16.1.6 Charge Carriers in Semiconductors

In Section 4.8 we gave an introduction to electons in solids, sufficient for most topics covered so far. In connection with our treatment of semiconductor detectors, however, we must delve somewhat deeper into the life story of carriers, electrons, and holes, in a semiconductor material.

We must distinguish between two types of semiconductors. The *intrinsic* semiconductors are essentially similar to insulators, except that the gap separating the conduction from the valence band may be crossed thermally at usual temperatures. The *extrinsic* semiconductors have either chemical impurities or other crystal irregularities imposed on them, so that they exhibit a number of permitted energy levels inside the band gap as described in Section 4.8.3.

In this section we discuss the relationship of conductivity to carrier concentration and the factors affecting, and affected by, the lifetime of carriers.

16.1.6.1 Conductivity and Carrier Mobility [1b]. In the absence of a net electric field, carriers move through the lattice until they collide with a lattice imperfection or a vibrating lattice atom. They then continue in a new direction. These collisions clearly control the average velocity of the carriers and are called *scattering*. With this type of motion, a uniform distribution of electrons will tend to remain uniform.

In addition to this Brownian-type motion, there are two others.

1. If there is a sudden increase in carrier concentration in one region, for example, due to current injection or the photogeneration of hole–electron pairs, the random motion will tend toward a uniform distribution

---

[4] H. Dember (1931).

(i.e. the carriers will have a net velocity away from the region of concentration). This type of motion is called *diffusion*.

2. If there is an overall electric field applied to the material, the carriers will experience a certain acceleration between collisions. This acceleration, in turn, will produce a net velocity in the direction of the field. The resulting motion is called *drift*. At low field strengths, the drift velocity is proportional to the applied field and Ohm's law is obeyed. As the field is increased, a certain limiting terminal velocity is reached so that the current no longer increases proportionally to the applied voltage (velocity saturation). At still higher fields, the carriers accumulate, between collisions, enough energy to ionize the atoms with which they collide, increasing the current by carrier multiplication rather than velocity rise (avalanche breakdown). See Sections 16.2.2.1 and 16.3.2.3.

In an isotropic crystal, Ohm's law (2.11) may be written in the one-dimensional form

$$J_x = \gamma \mathscr{E}_x, \tag{16.6}$$

where J_x and $\mathscr{E}_x$ are current density and field, respectively, in the x direction, and γ is the conductivity. In terms of the electron density, n, and the average velocity v_x, this may be written

$$J_x = nqv_x = \gamma \mathscr{E}_x, \tag{16.6a}$$

where q is the charge on the electron.

Ohm's law thus implies that the velocity (v) of an electron in a conductor is proportional to the applied field. The constant of proportionality, μ, is called *mobility*:

$$v = \mu \mathscr{E}. \tag{16.7}$$

Thus

$$\gamma = nq\mu. \tag{16.8}$$

Since holes, too, contribute to the current, and do so with a mobility different from that of electrons, we distinguish between electron (μ_e) and hole mobility (μ_h). If both electrons and holes participate in the conduction, with concentrations n and p, respectively, the conductivity is

$$\gamma = (n\mu_e + p\mu_h)q = q\mu_h(nb + p), \tag{16.9}$$

where

$$b = \frac{\mu_e}{\mu_h} \tag{16.10}$$

is the *mobility ratio*.

We can thus rewrite (16.6) in vector form

$$\mathbf{J} = q\mu_h(nb + p)\vec{\mathscr{E}}.$$ (16.11)

It can be shown that the mobility (μ_I) as controlled by scattering at ionized impurities increases approximately with the $\frac{3}{2}$ power of the temperature and varies inversely with the density of ionized impurities. Scattering due to lattice vibrations (μ_L), on the other hand, varies inversely with the $\frac{3}{2}$ power of temperature. The total mobility may be calculated from these according to

$$\frac{1}{\mu} = \frac{1}{\mu_I} + \frac{1}{\mu_L}.$$ (16.12)

Thus, at low temperatures, the mobility is controlled by μ_I and at high temperatures by μ_L, and the total mobility passes through a maximum at some intermediate temperature.

16.1.6.2 Carrier Generation, Recombination, and Lifetime [9a]. Carrier pairs are generated by thermal and photoeffects and are annihilated by recombination. There are a number of different recombination processes.

1. The simplest process involves the chance meeting of an electron with a positively ionized atom (a hole); the electron "drops into the hole," emitting a photon; the carrier pair has vanished. This is called *direct* radiative recombination and, generally, is a rare occurrence.
2. If the electron drops into the hole in a number of smaller steps, it releases its energy as quantized lattice vibration packets, called *phonons*.
3. The energy released when an electron combines with a hole may also be absorbed by a third excited carrier, lifting it to a higher energy level. This is called *Auger recombination.*

In practice, most recombinations take place at *recombination centers*, crystal imperfections tending to bind minority carriers, making them easy targets for majority carriers in search of a partner. For a recombination center to be effective, it must lie near the center of the forbidden band in order to be easily accessible to both types of carrier. If it strongly favors one type of carrier, it becomes a *trap*, delaying recombination rather than facilitating it.

The carrier concentration is determined primarily by three processes: (a) Thermal generation at a constant rate A; (b) photogeneration at a rate proportional to the incident illumination, E; and (c) recombination at a rate proportional to the concentration of carriers, n.[5] [In practice, both negative (electrons) and positive (holes) carriers will contribute. However, we derive here only the simpler relationship, valid when the contribution of one of these is negligible.]

The general expression for the rate of change of carrier concentration is, therefore

$$\dot{n} = A + \alpha E - \frac{n}{\tau}, \qquad (16.13)$$

where the constant τ is called the carrier lifetime. At equilibrium, $\dot{n}$ vanishes, so that we have for the carrier concentration with illumination

$$n_i = (A + \alpha E)\tau = n_0 + \Delta n, \qquad (16.14)$$

where $n_0 = A\tau$ is the equilibrium concentration in the absence of light and

$$\Delta n = \alpha E \tau \qquad (16.15)$$

is the increase in equilibrium concentration under illumination. More generally we have, on integration of (16.13),

$$n = n_2(1 - e^{-t/\tau}) + n_1 e^{-t/\tau} = n_2 - \Delta n e^{-t/\tau} = n_1 + \Delta n(1 - e^{-t/\tau}), \quad (16.16)$$

where n_1 is the concentration existing when the present state of illumination was initiated at time $t = 0$, n_2 is the equilibrium concentration for the present state of illumination, and $\Delta n = n_2 - n_1$. [The reader may readily confirm this solution by noting that (16.13) may be written

$$\dot{n} = \frac{n_i - n}{\tau}$$

and substituting from (16.16) for n.]

Thus, when illumination is initiated, the carrier density varies as

$$n = n_i - \alpha E \tau e^{-t/\tau} \qquad (16.17)$$

and, when it is turned off, as

$$n = n_0 + \alpha E \tau e^{-t/\tau}. \qquad (16.18)$$

Note that the lifetime, τ, is also a time constant of the detector, the time required until the process comes within $1/e$ of its final carrier concentration. In photoconductors, the time constant varies from less than 10^{-8} sec in zinc-doped germanium to 0.05 sec in cadmium sulfide.

---

[5] This is an assumption of questionable validity. In reality, the recombination rate is also influenced by the availability of the carrier partner and this, in turn, varies with n also. However, when it suffices to explain the observed phenomena approximately, these subtleties are often sacrificed on the altar of convenience. We, too, follow this custom in the service of pedagogy.

In terms of conductivity change, the increase due to Δn carriers is [from (16.9) and (16.15) and assuming that $\Delta n = \Delta p$]

$$\Delta\gamma = q\mu_h\,\Delta n(b+1) = q\mu_h\alpha E\tau(b+1), \qquad (16.19)$$

and the fractional increase

$$\frac{\Delta\gamma}{\gamma} = \frac{b+1}{nb+p}\,\alpha E\tau$$

$$= \rho q\mu_h\eta(b+1)\frac{E\tau}{h\nu d}, \qquad (16.20)$$

where we have substituted its value $(\eta/h\nu d)$ for α and the resistivity $[\rho = 1/\gamma]$ from (16.9). Here η is the quantum efficiency and d is the thickness of the layer. (N.B. this form is valid only for thin layers, where η varies directly with d.) Equation 16.20 is identical with (16.3) with the generation rate αE replacing g. Clearly, the increase in conductivity is directly proportional to the lifetime, τ.

As mentioned in the previous section, the presence of traps tends to prevent recombination, leaving the counterpart carrier to contribute to conductivity. This lengthens the effective lifetime and thus contributes to increasing the sensitivity of the photoconductor.

We note, for future reference, that there is a mean distance over which a carrier will diffuse before recombining. This distance is called *diffusion length*.

16.1.6.3 Types of Contact, Gain, and Bandwidth [10a]

Ohmic and Blocking Contacts. The photoconductor may function as a photon detector delivering a carrier for each photon absorbed or as a transducer of illumination to electrical conductivity. The form of operation depends on the nature of the contact between the electrodes and the photoconductor.

If there is essentially no potential barrier at the electrode contacts (i.e., the barrier is lower than the thermal energy kT) any carrier drawn from the photoconductor at one electrode, will be immediately replaced at the other electrode in order to maintain overall charge neutrality in the photoconductor. Such contacts are called *ohmic*, because they permit the detector to obey Ohm's law: the current is proportional to the applied voltage.

When such contacts are provided, the concentration of free carriers is maintained as given by (16.14) and is best treated in terms of change in conductivity, as given by (16.18) and (16.19). In that event, many

electrons may be drawn from the photoconductor for each photon absorbed, the actual number depending on the applied voltage.

When there is a high potential barrier at the electrode contacts, a
photocarrier drawn from the semiconductor cannot be replaced. Thus the
maintenance of charge neutrality, and steady currents, is possible only
when the carrier is generated at the contact or, for volume detectors,
when carrier pairs are generated. In either event, here the number of
carriers, or carrier pairs, collected cannot exceed the number of photons
absorbed (except if secondary ionization occurs). See Section 16.3.2.3.

Gain. The gain of a detector may be defined as the ratio

$$G = \frac{N_c}{N_p},$$ (16.21)

where N_c is the number of photocurrent carriers collected at the electrodes and N_p is the number of photon-generated carriers.

If the lifetime of carriers (τ) in the photoconductor is shorter than the
transit time (τ_t), some of the photocarriers will be lost (by recombination)
before leaving the photoconductor, We may then write for the gain

$$G = \frac{\tau}{\tau_t},$$ (16.22)

where the transit time is given in terms of the conduction path length (l)

$$\tau_t = \frac{l}{v} = \frac{l}{\mu(V/l)} = \frac{l^2}{\mu V}.$$ (16.23)

Compare (16.7) with V/l replacing the field strength $\mathscr{E}$.

This expression for gain is valid also when the transit time is less than τ,
assuming ohmic contact. The gain may then be substantially greater than
one.

The current density is given by the carrier charge concentration (nq)
multiplied by the carrier velocity (see the discussion preceding Figure
1.3). The velocity, in turn, is given by the product of electric field strength
(V/l) and mobility (μ). Thus the current for an area S is

$$I = \frac{nqS\mu V}{l}$$ (16.24)

and the resistance

$$R = \frac{V}{I} = \frac{l}{\mu nqS}.$$ (16.25)

Space-Charge Limited Current. Since the transit time varies inversely
with the applied voltage, the gain may be increased by increasing this

voltage. The amount of voltage that may usefully be applied to the photoconductor is limited, however, by the condition where the charge applied across the detector (viewed as a capacitor) equals the photogenerated charge within the photoconductor volume:

$$Q = VC = nq(lS) \qquad (16.26)$$

or

$$V = \frac{nqlS}{C}. \qquad (16.27)$$

Here C is the capacitance. On substituting this into (16.23) we find the corresponding transit time to be

$$\tau_t = \frac{lC}{\mu nqS}. \qquad (16.28)$$

We note that this is identical to the RC-constant (τ_{RC}) of the detector, obtained by multiplying (16.25) by C:

$$\tau_t = \tau_{RC}. \qquad (16.29)$$

Thus, from (16.22) and (16.29) we may write the maximum useful gain

$$G_{mx} = \frac{\tau}{\tau_{RC}}. \qquad (16.30)$$

We now recall that the detector bandwidth (i.e., the rate of change of signal to which it can respond) is limited by the carrier lifetime

$$\Delta \nu = \frac{1}{2\tau}. \qquad (16.31)$$

(See sampling theorem, Section 3.3.) Hence we may write for the gain-bandwidth product

$$G\Delta\nu = \frac{1}{2\tau_{RC}}. \qquad (16.32)$$

When the voltage exceeds the limit set by (16.27), the illumination-independent (dark) current dominates. This current is limited by the fields generated by the space charge distributed throughout the conductor. It is given by

$$I_S = \frac{Q}{\tau_t} = \frac{VC}{l^2/\mu V} = \frac{\mu V^2 C}{l^2}, \qquad (16.33)$$

where we have used (16.23) and (16.26). In contrast with ohmic current, it varies with the square of the applied voltage.

Sensitizing [11*a*]. To increase the gain, the permissible voltage limit must be raised. This can be done by providing both trapping *and* recombination centers. The introduction of impurities to yield such traps is called *sensitizing*, since it raises the gain capability of the detector. Alternatively sensitizing can be understood in terms of the lengthening of the lifetime of one of the carrier species [10*b*].

16.1.6.4 Spectral Response.

The response of a photoconductive detector is generally limited to a relatively narrow spectral region. A long-wavelength limit is obviously given by the cutoff wavelength, λ_0 [see (16.2)]. At short wavelengths, the incident flux is absorbed, and the carriers are generated near the detector surface, where recombination is very likely. Consequently, carriers resulting from such short-wavelength flux tend to have short lifetimes resulting in a low responsivity [see (16.20)].

16.1.6.5 Frequency Response.

Reference to (16.16) shows that the photoconductor response decays exponentially in time. This corresponds to a frequency response of the form[6]

$$S(\nu) \sim [1 + (2\pi\nu\tau)^2]^{-1}. \tag{16.34}$$

This should be the frequency response of a homogeneous photoconductor, where the carrier lifetime limits the gain. In photodiodes, where the photocarriers are swept out quickly, that is, in an interval short compared to the carrier lifetime, the frequency response will be considerably higher; many photodiodes respond to microwave frequencies—well into the GHz range [12].

16.1.7 Photoionization [13]

In concluding this section, we briefly mention that photoionization (effectively photoconductivity) in gases, too, can be used for light detection. By applying a bias voltage near the breakdown value for the gas, a highly effective detector may be obtained, especially for use in the uv. Indeed, D^* values above $10^{11} \, (\text{cm} \cdot \text{Hz}^{1/2}/\text{W})$ have been obtained at 300 nm (see also Ref. 14).

---

[6] The autocorrelation function of an exponentially decaying signal $(e^{-t/\tau})$ has the form $\tau e^{-|t|/\tau}$. the power spectrum of such a signal is the Fourier transform of this and has the form (16.34).

16.2 PHOTOTUBES

16.2.1 Photocathodes and Vacuum Photodiodes

16.2.1.1 Cathode Materials **[5b], [15a].** To obtain useful photoemitting cathodes, a relatively low work function must, of course, be provided. Clean metal surfaces are relatively poor photoemitters. On the other hand, when monomolecular layers of alkali metal ions are absorbed on such bases as silver or tungsten, a far better response is obtained. This is, presumably, due to the formation of an electrical double layer on the metal surface. The electrical field associated with this double layer facilitates the escape of photoelectrons. Further improvements may be obtained by partial oxidation of the alkali layer.

However, to obtain high efficiencies, a low work function does not suffice; the cathode must be such that useful photoexitation can take place also within the material, up to the depth of several tens of nanometers. Once they have been excited into the conduction band, electrons must be able to reach the surface with a major portion of their energy intact. This is much more likely in a dielectric or a semiconductor than in a metal. On the other hand, there must be enough electrical conductivity to permit the neutralization of the charge accumulated due to the photoemission; otherwise, the positive charge on the cathode will inhibit the escape of photoelectrons, resulting in fatigue effects (Section 16.2.1.7).

Efforts to satisfy all these requirements have resulted in a number of highly efficient photocathodes with various efficiencies and spectral responses. The spectral responses of most commercially available photocathodes have assigned to them an *S number* by which they can be specified. Here, too, we present data on photocathodes according to these S numbers. See Tables 163 [16] and 164 [17] and Figure 16.6 for extensive data. (S number curves have been standardized for many photoconductive materials, as well; these are shown in Table 170.)

Originally, the two most notable photocathodes were the silver–oxygen–cesium (Ag–O–Cs) and the antimony–cesium (Sb–Cs) cathodes. Comparing these, we note that the Ag–O–Cs [S1] cathode has the lowest work function of any commercially available cathode and can, consequently, be used well into the infrared, to beyond 1.1 μm. (Specially prepared cathodes of this type have been used up to 1.7 μm.) Its peak sensitivity, on the other hand, is only one-tenth of that attainable in the more popular Sb–Cs cathodes. Due to its low work function, its dark emission is several thousand times as high as that of the Sb–Cs cathode.

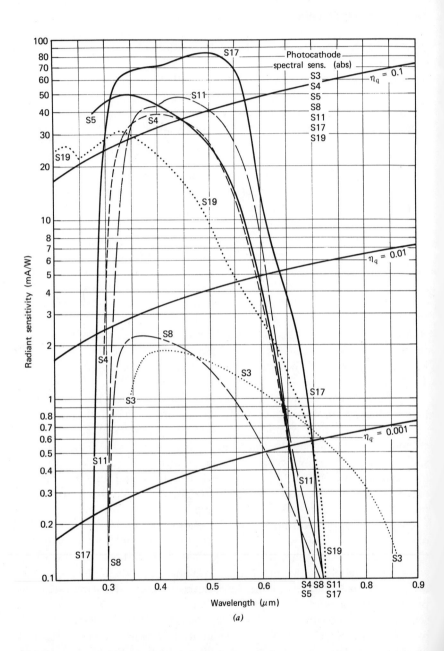

Figure 16.6 Photocathode sensitivity spectra. (a), (b) S curves. (c) RCA curves, from *Photomultiplier Tubes*, RCA Electronic Components, Harrison, N. J. 1971, with permission.

460

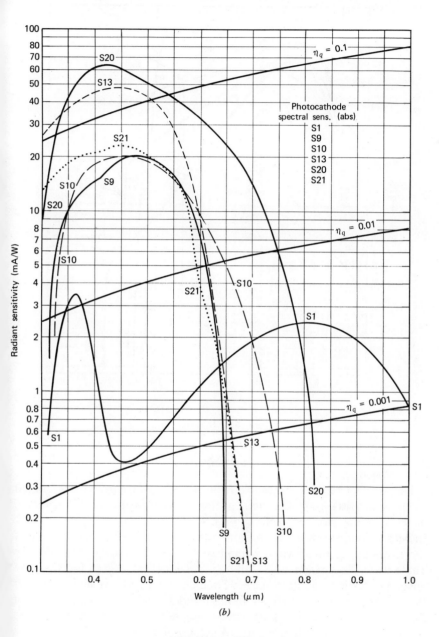

Figure 16.6 (*Continued*).

461

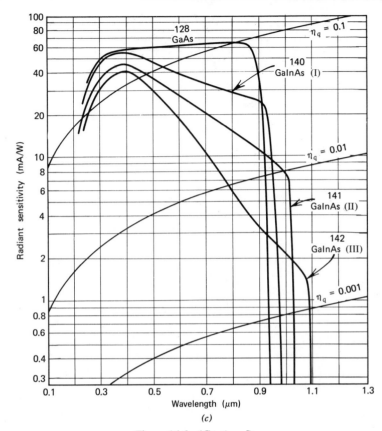

Figure 16.6   (Continued).

These two factors limit its usefulness at very low illumination levels (see Section 16.2.4.6).

The Sb–Cs cathodes are usable from the ultraviolet (usually as limited by the transmittance of the tube envelope) to approximately 0.7 μm and, in the visible region, have the highest sensitivity commercially available. In detail, their response depends on thickness and on the transmittance of the envelope, so that several S numbers have been assigned to them (see Table 163). For example, S4 and S5 are identical except for the tube envelope; S9 is a semitransparent Sb–Cs cathode; S11 is similar, except that it consists of a thinner antimony layer. It, therefore, absorbs less of the red radiation, while short-wavelength radiation is favored, because those photoelectrons that are freed escape more readily. (This explains the shift of the S11 peak toward the blue.) For work in the visible part of the spectrum, S4 and S11 are probably the most popular.

More recently, multialkali cathodes with very desirable characteristics became commercially available (e.g., S20). These extend the high sensitivity into the near infrared, without significant increase in dark current.

Lately, further significant improvements were obtained with materials compounded of elements from Columns III and V of the periodic table. Chief among these is GaAs and tertiary compounds such as InGaAs. With a monocrystalline structure and provided with a layer of Cs_2O, these combine high sensitivity in the visible with good response in the ir. GaAs has a quantum efficiency in excess of 0.1 up to 0.85 μm and some InGaAs cathodes a quantum efficiency above 10^{-3} to almost 1.1 μm. With such cathodes, luminous sensitivities well in excess of a mA/lm have been obtained—more than ten times the sensitivity of the typical multialkali cathode [4].

16.2.1.2 Cathode Construction. The manufacturing processes of these cathodes have been described in the literature (see, e.g., Refs. 4 and 5b). They are, however, so sensitive to many variables that reproducibility is often poor and selection from within production batches must often be employed. For example, the presence of oxygen at pressures as low as 5×10^{-10} torr will reduce a multialkali (S20) cathode's sensitivity by 10% in 90 sec [18].

Photocathodes may be made either opaque or partially transparent. The latter form is usually used with light incident on one side of the cathode and electrons emitted on the other. This permits a tube construction with better optical coupling characteristics. On the other hand, the cathode thickness becomes critical—if the cathode is too thin, it will not absorb enough photons; if too thick, the photoelectrons will not be able to escape for collection at the anode. Also, the required electrical conduction is difficult to obtain with the very thin transparent cathodes, so that conducting tin-oxide layers are occasionally provided under the cathode layer. In absolute terms, the opaque cathodes are superior [19a], but with careful design the transparent cathodes may come close to them [20]. Methods have been described for increasing absorption in thin photocathodes by using total internal reflection (see Section 16.2.3.9).

The cathode spectral response is somewhat dependent on the angle of incidence. Especially in the thinner cathodes, such as S11, increased obliquity increases the absorption of red radiation resulting in a slight shift of the response toward the long wavelengths.

16.2.1.3 Photodiode Construction. The simplest device employing photoemission is the vacuum photodiode. This consists of a photoemissive cathode and an electron-collecting anode, enclosed in an evacuated

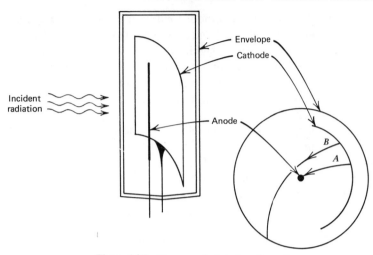

Figure 16.7 Vacuum photoiode, schematic.

envelope that permits the entry of the radiation to be measured. Some representative diode types are listed in Table 165.

Generally, the photocathode is made large to facilitate the collection of a maximum of light flux. In photodiodes having opaque cathodes, the emission of electrons is toward the incident light. The collecting anode must then be on the same side. It is, therefore, generally made in the form of a relatively thin wire to minimize the blockage of the incident light. See Figure 16.7, which illustrates schematically a typical photodiode. On the other hand, if the anode is made too thin, the tangential component of the original momentum of the electrons will cause these to miss the anode and continue on to return to the cathode or to strike the envelope (Path B in Figure 16.7). Not only are such electrons lost to the output current, but they may cause even more serious difficulty through the consequent charge accumulation on the envelope. [This accumulation may be positive or negative, depending on the secondary emission ratio (see Section 16.2.3.1)].

16.2.1.4 Responsivity and Transfer Characteristics. Cathode sensitivity in commercial tubes range from about 0.002 A/W (for S1 and S3) to 0.085 A/W (for S17 and some multialkali cathodes) at the peak of the response curve. See Table 164. [These values should be compared to the limit:

$$\frac{q_e \lambda}{hc} \approx 0.8\lambda \, (\lambda \text{ in } \mu m)$$

corresponding to unity quantum efficiency.]

The linearity of photocathode response, output current versus input flux, is excellent up to approximately 50 μA/cm² [based on Ref. 21a]. In practice, manufacturers recommend a limit of approximately 4 μA/cm². This corresponds to approximately 0.2 and 0.08 lm/cm² in S1 and S4 photocathodes, respectively, when a 2870 K light source is used. Beyond this point, fatigue effects and thermal decomposition may begin to set in.

16.2.1.5 *Current-Voltage Characteristics.* In photodiodes, the current usually rises rapidly with voltage up to approximately 20 V. As the voltage is increased further, the current continues to rise, but far more slowly (see Figure 16.8 [5c]). This response can be explained as follows:

The photoelectrons, upon emission, will have some slight momentum,

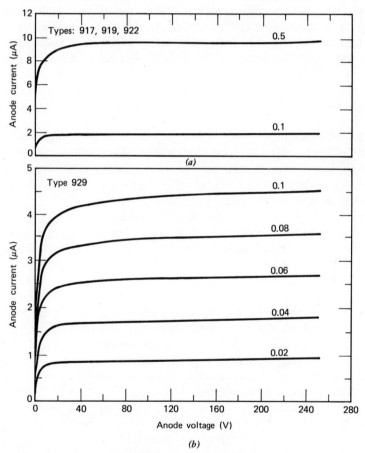

(a)

(b)

Figure 16.8 Current-voltage characteristics of vacuum photodiodes. The incident flux (in lumen) is indicated above the corresponding curve. (*a*) Ag—O—Cs; (*b*) Sb—Cs cathode. Tube type numbers shown on the figure. Zworykin and Ramberg [5c].

which may suffice to carry some of them to the anode, even without an externally applied field. This momentum is the sum of their initial momentum and the momentum supplied to them by the absorbed photon (modified by possible collisions on the way to the surface). The shorter the wavelength of the photon, the greater its energy and the greater the momentum of the emitted electron. However, in the absence of an external field, the exit momentum must be directed precisely toward the anode if the electron is to appear in the external circuit. Since the exit momentum distribution follows a cosine-law and the anode is generally quite small, the observed current will be very small. As the anode potential is raised relative to the cathode, an increasing number of photoelectrons will be drawn to it and the current rises rapidly until a voltage of approximately 20 V is applied, at which point most of them will be collected. Beyond this point, the further increase in collection efficiency will cause the current to continue to rise, but generally at a much lower rate. Also, the large electrostatic fields that may appear at the complex cathode surface may facilitate electron emission (Schottky effect) and thus contribute to the continued rise in current.

16.2.1.6 Temporal Response. The transit time of photoelectrons from cathode to anode varies directly with their spacing and inversely with the square root of the accelerating voltage. In the usual configurations, it is a few nanoseconds. The delay time of the photoemissive effect is considerably less than that—of the order of a picosecond.

Usually, however, it is not the absolute value of the delay that is of interest, but rather variations in this delay. These variations limit the frequency with which a light source may be modulated if this modulation is to be detected—the interval between two light pulses, if these pulses are to be resolved temporally. At high voltages, where the initial electron momentum is relatively unimportant, it is primarily the variation of distance to the anode, for various points on the cathode, that limits the frequency response. But appreciable differences in initial momentum, as caused for instance by the presence of short wavelengths, can also affect this response. If high-frequency response is crucial, it may be advisable to narrow the spectrum of the incident light as much as possible [3].

It should be noted, however, that in a vacuum diode it will usually be the circuit components that will limit the high-frequency response. The interelectrode capacitance, which is likely to amount to a number of picofarads, coupled with the stray capacitance of the associated circuitry, would require an extremely low load resistance, if the high-frequency potential of the diode is to be realized. This will generally be impractical except for very intense and short-duration pulses, where large current values become feasible.

16.2.1.7 Fatigue and Life Characteristics. When the photocathode is operated at relatively high current levels, temporary *fatigue effects* must be expected due to the positive charge accumulated on the cathode by the electron depletion. This effect can be minimized by improving the conductivity of the photocathode as mentioned in Section 16.2.1.2.

Sensitivity also tends to drop slowly with use. Representative life curves for an S1 and an S4 cathode, respectively, are shown in Figure 16.9 [21b]. Both were operated at an illumination that yielded approximately 3 μA/cm^2 initially.

One Sb–Cs cathode, operated at 10 μA/cm^2 (at 300 V) exhibited a 70% drop in sensitivity after 2 hr of operation. After 20 days of recovery time, sensitivity had returned to only 50%. On the other hand, after 5 hr of operation at 1 μA/cm^2 this type of tube (6094) suffered only a 10% drop in sensitivity [22].

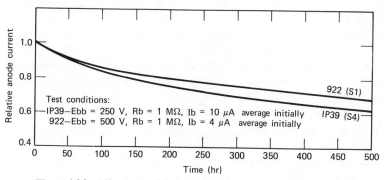

Figure 16.9   Life curves of S-1 and S-4 photocathode. Engstrom [21].

16.2.2   Gaseous Amplification and Gas Photodiodes [5d]

To increase the current output of a photodiode without adding a significant amount of noise, internal current amplification must be employed. Two methods are used to accomplish this. The use of gaseous current amplification is the subject of the present section. The other approach, employing secondary electron emission, is discussed in Section 16.2.3.

16.2.2.1 Gaseous Current Amplification. Consider an electron, accelerated by an electrostatic field, traveling through a gas. There is a certain probability that is will collide with a gas molecule. If, at the time of the collision, the electron has enough energy to ionize the molecule, this collision may produce a second free electron (plus a positive ion). We now have two electrons being accelerated by the electrostatic field; each of these may participate in another ionizing collision to double, again, the number of free electrons in the current. This phenomenon is known as *gaseous current amplification* and is used to increase the output of photodiodes.

To a first approximation, the field ($\mathscr{E}$) and gas density should be so related that the electron gains an amount of energy equal to the ionizing energy in the distance it travels, on the average, between collisions; this is the *mean free path* (l_0); in other words, the product $l_0\mathscr{E}$ should equal the ionizing energy in electron volts. The mean free path is simply related to the density (n) of molecules in the gas. At the temperatures here considered, the motion of the molecules will be slow compared to that of the electrons, so that we may consider them stationary. Assume, first, that the molecules are circular with radius r. As the electron (negligible radius) travels a distance l through the gas, it will collide with any molecule whose center is within the distance r of the electron's path. The electron thus sweeps out a cylinder, with cross-sectional radius r and length l, colliding with any molecule whose center is within this cylinder. It will thus collide with a total of $\pi r^2 l n = S l n$ molecules, where S is the cross-sectional area of the molecule. The mean distance between collisions will be the value of l for which there will be, on the average, one collision:

$$l_0 = \frac{1}{Sn}. \tag{16.35}$$

A little thought will convince the reader that this relationship will hold, independently of the assumed circularity of molecular cross section.

The electrostatic field is given by

$$\mathscr{E} = \frac{V}{d}, \tag{16.36}$$

where V is the potential applied across the diode and d is the separation of the electrodes. Thus the condition for an accumulation of sufficient energy (U_i) for ionizing a molecule between collisions is

$$U_i = l_0\mathscr{E} = \frac{V}{dnS} \text{ eV}. \tag{16.37}$$

This may be used to estimate the required n (and hence pressure) to yield operation at a suitable V.

To obtain the current amplification, we must use, instead of l_0, the distance l_i, the mean distance between ionizing collisions. More usually its reciprocal, α, is used where α is the mean number of ionizing collisions per unit distance traveled; it is known as *Townsend's first coefficient* [23]

We have then for the change in current (i) during an infinitesimal travel distance (dx)

$$di = i\alpha\, dx, \qquad (16.38)$$

Hence, the anode current will be

$$i = i_o e^{\alpha d'}, \qquad (16.39)$$

where i_0 is the initial cathode current and
  d' is the interelectrode distance minus the distance required for the electron to require energy U_i:

$$d' = d - \frac{U_i d}{V}. \qquad (16.40)$$

Here α is proportional to n (and pressure) as long as condition (16.37) is satisfied.

16.2.2.2 Gain Limitations. Since (16.39) implies that the amplification factor $G(=i/i_0)$ can be raised arbitrarily simply by an increase in pressure and a compensating increase in V, it must be mentioned that the amount of amplification that can practically be obtained from a gas diode is limited by other considerations. The ions that are produced by the electrons will drift to the cathode and may there produce secondary electrons by impact. (These ions need not be positively charged; it has been found that excited argon atoms in a metastable state may be responsible for a major portion of the secondary electrons [24].) The secondary electrons thus released are added to the cathode current and amplified as they travel from cathode to anode. Assuming that each electron has a probability, γ, of causing (via an ion) the release of a secondary electron, we have for the anode current, including the secondary electrons, as amplified,

$$i = i_0 e^{\alpha d'}(1 + \gamma e^{\alpha d'}).$$

Since these secondary electrons also produce ions, and therefore another generation of secondary electrons, and so on, we have

$$i = i_0 e^{\alpha d'}[1 + \gamma e^{\alpha d'} + (\gamma e^{\alpha d'})^2 + \cdots]$$

$$= \frac{i_0 e^{\alpha d'}}{1 - \gamma e^{\alpha d'}}. \qquad (16.41)$$

It is evident from (16.41) that, as α approaches $(-1/d')\log\gamma$, the current tends to become very large and self-sustaining, independent of the initial cathode current. This condition is called glow discharge and must be avoided for stable operation. In addition, large cathode currents tend to shorten tube life. As a result, amplification factors much in excess of 10 are rarely recommended.

16.2.2.3 Gas Photodiode Characteristics. A typical transfer characteristic of a gas phototube is shown in Figure 16.10 [5d]. The slight nonlinearity evident is due to the initially low field at the cylindrical cathode. At substantial current levels the field there is increased due to the slow-moving positive ions, and, as a result, the field is more uniform and gas amplification more efficient [21c].

A typical series of current-voltage characteristics is shown in Figure 16.11 [5d].

Since gas ions contribute significantly to the total current, both as positive ions and via secondary electron generation, and since these ions move very slowly compared to electron velocities, the response of gas diodes is considerably slower than that of vacuum diodes. This, in turn, results in a more limited frequency response as illustrated in Figure 16.12 [5d], where curves A and B refer to S1 and S4 cathodes, respectively. Some representative gas diodes are included in Table 165.

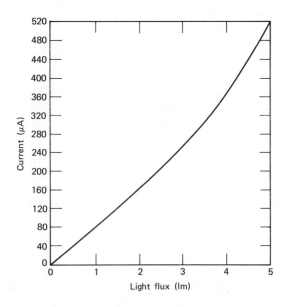

Figure 16.10 Gas photodiode transfer characteristic. Zworykin and Ramberg [5d].

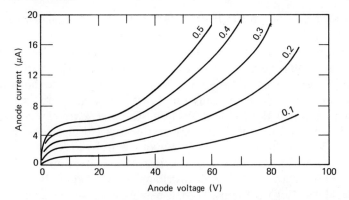

Figure 16.11 Gas photodiode (Type 868), current-voltage characteristic for various flux values (in lumen). Zworykin and Ramberg [5d].

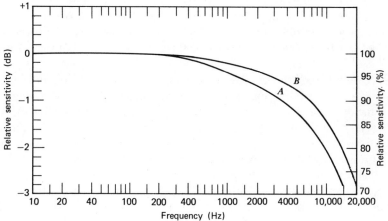

Figure 16.12 Gas photodiode, frequency response. A : Ag—O—Cs; B : Sb—Cs cathode. Zworykin and Ramberg [5d].

16.2.3 Electron Multiplication and Multiplier Phototubes (MPTs)

Often the output of a phototube must be amplified for measurement and detection. Invariably, such amplification adds noise to the signal. The amount of noise added, however, depends on the method of amplification (see Section 16.2.4). A relatively low-noise amplification process is that of electron-multiplication, and therefore this is frequently used in phototubes. Tubes employing such electron multipliers are called *multiplier phototubes*; this section is devoted to a description of such tubes. Electron multipliers are also used in certain image tubes (see Section 16.4.2.3).

***16.2.3.1 Electron Multiplication* [15b].** When an energetic electron strikes a solid, it excites electrons internal to the solid and some of these excited electrons may escape. The number of electrons thus freed depends on the energy of the incident electron and on the material bombarded; the energy of the true secondary electrons is approximately 2 eV and does not change significantly either with the bombarded material or with the energy of the incident electron. The *secondary emission ratio* is defined as the ratio of the total number of emitted electrons (including scattered incident ones) to the number of electrons incident.

The energy distribution of the electrons leaving the material is illustrated in Figure 16.13 [25]. This distribution is conveniently divided into three regions. Region I consists of a narrow pulse at the energy of the incident electrons. It represents those of the incident electrons that are reflected elastically. Region II is a low probability density region, dropping gradually from the foot of Pulse I toward the origin. It represents the remainder of the incident electrons, which are inelastically scattered, giving up some of their energy to the material bombarded. The true secondary electrons constitute Region III, which has the form of a somewhat broader pulse, usually peaking in the region between 1.4 and 2.2 eV.

As mentioned above, the secondary emission ratio depends on the energy of the incident electrons and on the material. The dependence on the incident electron energy is illustrated in Figure 16.14. At low electron energies, the ratio drops from unity, as more and more of the incident electrons are absorbed. As the bombarding electrons become more energetic, they begin to excite electrons in the material and the secondary emission ratio rises, reaching unity at the "first cross-over energy level," U_{cr1}. It continues to rise until, at U_{peak}, the bombarding electrons penetrate so deeply into the cathode that increasing numbers of the excited

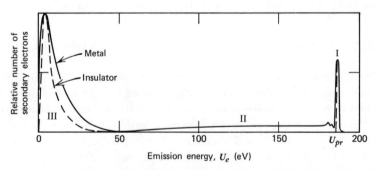

Figure 16.13 Secondary electrons, typical energy distribution. Hachenberg and Brauer [25].

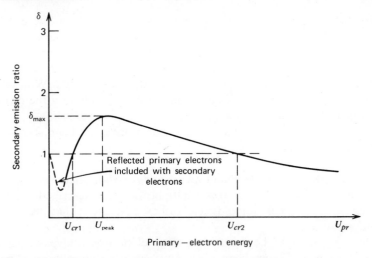

Figure 16.14 Secondary emission ratio, versus energy of primary electrons.

electrons lose too much energy in collisions on the way to the surface and fail to escape. Further increase in the energy of the bombarding electrons reduces the secondary emission ratio until, at the "second cross-over energy level," U_{cr2}, it again drops below unity. (The two energy levels, $U_{cr1,2}$, contain between themselves the energy region in which the bombarded material will assume a positive potential as a result of secondary electron emission in excess of unity.) Note that point U_{cr1} is unstable.

Turning to the other factor, the electrode material, we find that the best electron multiplier materials are intermetallic compounds and insulators [25] presumably because of the ease with which excited electrons drift to the surface in these. Indeed, good photoelectric emitters are, as a rule, also good electron multipliers and Sb–Cs is one of the most efficient electron multiplier materials, with yields up to 10 times. Although used fairly often, this material has certain shortcomings in that it is sensitive to many factors, such as exposure to air, to heat in excess of 75°C, and to high current densities (in excess of $0.1 \, \text{mA/cm}^2$). For these reasons, magnesium oxide, in a silver matrix, is a popular secondary emission material. It is capable of secondary emission ratios in the neighborhood of nine. With careful deposition and high bombarding voltages, a ratio of 12 can be obtained [26]. The conducting silver matrix is essential for avoiding the buildup of charges and the accompanying disturbing field effects.

Even though the emitted secondary electrons, too, follow a cosine-law distribution, the angle of incidence of the bombarding electrons may have a significant effect on the secondary emission ratio. When the energy of these exceeds U_{peak}, the ratio rises with increasingly oblique incidence. This is readily understood if we recall that, at these energy levels, a significant portion of the excited electrons fail to reach the surface because of the distance below it at which they are excited. With oblique incidence, this distance is reduced and a greater fraction of the electrons can reach the surface.

The time delay in a secondary emission process has been shown to be less than 30 ps [27].

16.2.3.2 Multiplier Electrodes and Channel Multipliers.

Popular electrode arrangements for electron multiplication are described in the following section. These employ electrodes that are either flat or only slightly curved.

One exception is the *continuous channel electron multiplier*, commercially known as Channeltron [28]. This consists, typically, of a narrow glass tube, approximately 1 mm in diameter and 50–100 diameters long, with an electron-multiplying coating applied to the inside wall. A large voltage (typically approximately 2 kV) is applied across the length of the tube so that electrons are accelerated along it. Such electrons will generally have a transverse velocity component and thus strike the coated wall along the way. When this happens, secondary electrons are emitted and accelerated along the tube. These, too, will have a transverse velocity component and may, in turn, strike the wall to release additional electrons. Gains of approximately 10^6 are thus attained. At a gain of approximately 10^7, the process begins to saturate and output pulses all tend to have equal heights. This saturation effect may be useful in certain applications [29] and has been found to be due, probably, to space charge effects [30]. A slight curving of the channel tube has been found to reduce ion feedback significantly [29].

The use of smaller channels, tens of microns in diameter, may give excellent frequency response to 5.6 GHz [31]. Such channels may also be assembled into microchannel plates used in image intensification (see Section 16.4.2.3).

16.2.3.3 Multiplier Phototube Construction [5e], [21d].

Multiplier phototubes consist of an evacuated envelope containing an electron-emitting cathode, a number of electron-multiplying electrodes called *dynodes* (or a channel electron multiplier) and an electron collecting anode. The construction must allow the light to strike the photocathode and permit the collection of most of the photoelectrons onto the first dynode at an energy appropriate for the desired multiplication ratio.

Successive dynodes and the anode must then be placed and electrified such that they similarly collect the electrons emitted by the preceding dynode.

In the simplest arrangements, the electrons are made to strike dynodes as they are accelerated from the cathode toward the anode. Perhaps the most popular of these types is the "venetian-blind" dynode system, illustrated in Figure 16.15a. Electrons strike one of a series of obliquely

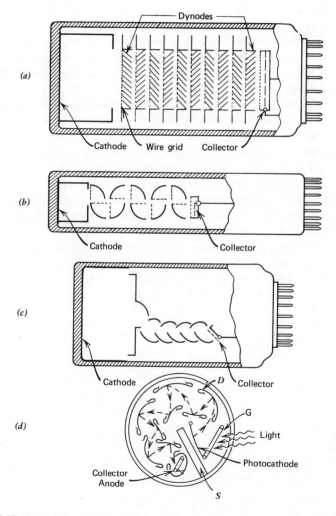

Figure 16.15 Dynode arrangements: (a) venetian blind; (b) box type; (c) linear focused; (d) circular cage focused.

placed dynode slats at each dynode stage, and the resulting secondary electrons are drawn on to the next stage by a thin wire grid placed over the succeeding dynode. In another dynode arrangement, the box-type illustrated in Figure 16.15*b*, the electrons follow a castellate path. In both these types, little electron-focusing is used, and a large dynode structure is needed to obtain an acceptable collection efficiency. This difficulty is overcome in the Rajchman linear dynode structure, illustrated in Figure 16.15*c*, where the dynode surfaces are shaped to produce the desired electrostatic field configuration. Because of its good electron-focusing characteristics, such an arrangement is suitable for use with a large number of dynode stages. A more compact form of this, the circular-cage arrangement, is shown in Figure 16.15*d*. Here the incident light passes through a thin wire grill (G) to strike the cathode. The ejected electrons are accelerated toward the first dynode (D), and thence to successive dynodes in turn. A shield (S) ensures that no electrons pass directly from cathode to anode, bypassing the multiplier arrangement, or from anode to cathode, increasing the dark current.

In the unfocused forms (*a*) and (*b*), the electrostatic field at the dynode surfaces is rather low, and, consequently, the transit time varies significantly among electrons. Therefore one of the focused types should be used if a high-frequency response is required. In addition, the electron collection efficiency tends to be somewhat lower in the venetian-blind arrangement.

Data on some representative MPTs are included in Table 165. For operation at microwave frequencies ($>10^{10}$ Hz), special constructions have been devised. See Ref. 12 for a review of these.

16.2.3.4 Gain Characteristics. The main superiority of the MPT over the photodiode lies in its overall multiplication factor or gain (G). This gain can be very large. For instance, if 10 dynode stages ($N = 10$) are used and each exhibits an average gain-per-stage (g) of 4, the overall gain will be

$$G = g^N \approx 10^6. \tag{16.42}$$

Since the gain per stage depends strongly on the interstage voltage, the overall gain varies even more strongly with the voltage. Gains run roughly from 2.5 to 4/stage for voltages of 60–120 V/stage. Representative overall gains versus voltage characteristics are shown in Figure 16.16 [32].

An additional consideration becomes evident when we differentiate (16.42) to find the relative rate of change of G with V:

$$\frac{\Delta G}{\Delta V} \bigg/ \frac{G}{V} \approx \left(\frac{V}{G}\right) g' \bigg/ \frac{dG}{dg} = \frac{VNg'}{g}, \tag{16.43}$$

where $g' = dg/dV$ is the rate of change of the gain per stage with voltage.

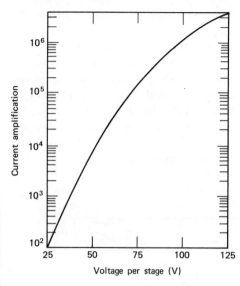

Figure 16.16 MPT gain versus voltage characteristic (Tube Type 931A). Engstrom [32].

Since the gain versus voltage curve is almost linear ($Vg'/g \approx 1$) up to voltages of 100 V/stage, the fractional change of gain with voltage approximates N in that range. This, in turn, would imply that the voltage control and stability must be N times as fine as the desired output stability (e.g., if we desire a 1% output stability, we must provide a 0.1% voltage stability in a 10-stage tube). In practice, the situation is only slightly better than that.

Even though multiplication factors of 10/stage are possible at 400–500 V/stage, at continuous operation such voltages are impractical because of regenerative ionization effects; that is, the MPT begins to act like a gas tube because of the residual gas that inevitably remains. However, when pulsed operation is acceptable, such voltages may be used advantageously. In addition to the increased gain, they result in reduced pulse spreading (dark current pulses in such operation have been measured at 0.5 ps between half-amplitude points) [33]. If pulse duration is limited to about 2.5 μs, it becomes possible to operate at the peak of the multiplication characteristic, so that voltage control is no longer critical [33].

16.2.3.5 Responsivity and Transfer Characteristics. The responsivity of a MPT equals the product of the cathode responsivity and the overall gain, which includes both electron collection efficiency and electron multiplication factors. Both cathode responsivity (Section 16.2.1.4) and gain (in the preceding section) have already been discussed.

The excellent linearity of the photocathode, already mentioned in connection with the photodiode, becomes even more significant in the MPT. In the diode the output current is limited by the permissible cathode current, which is usually limited to a few $\mu A/cm^2$. In the MPT, on the other hand, the upper limit is set by the anode current, which may be two decades higher than this, usually several tenths of milliamperes. Thus the range of linear operation is extended upward. With photon-counting techniques (see Section 16.2.4.7) the linear operating range may be extended downward, so that it extends over almost 10 decades (see Figure 16.17) [32].

It must be noted that the circuits associated with the MPT also may affect linearity, as discussed in Section 16.2.3.9.

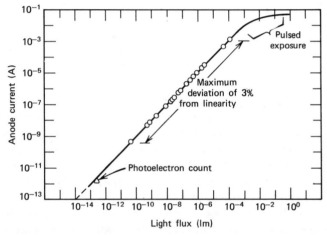

Figure 16.17 Typical MPT transfer characteristic. 100 V/stage saturation at 45 mA. Data at high current levels were taken with pulsed light, 1% duty cycle. Engstrom [32].

16.2.3.6   Temporal Response.

Because of delays in the emission of electrons and their transit, there is a delay from the time of light incidence until the current appears at the anode. This delay, however, is not usually significant. What is more significant, is the spread in arrival time between secondary electrons generated by the same photoelectron. As a result of this spread, for instance, two closely spaced light pulses might fuse in the output.

Because photoemission and secondary electron emission are extremely fast (measured in picoseconds), the frequency response of MPTs is limited primarily by spread in the electron transit times. Because pathlengths and initial velocities differ for different electrons, the current pulse produced by a very short light pulse may, in practice, be much wider than the latter.

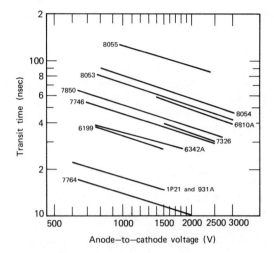

Figure 16.18 MPT transit times versus voltage for tube types indicated. Engstrom [21].

With signal frequencies below 100 MHz, these effects are usually negligible; but above this point, special efforts must be made if the optimum response is desired. We have already noted that the focused MPTs have a better uniformity of transit times. But, beyond this, some tubes have been designed especially to optimize frequency response [34].

As expected, transit velocities vary with $\sqrt{V}$, the square root of the applied voltage, since the accelerating force is directly proportional to this voltage. It is interesting to note that the spread in arrival time of electrons, too is proportional to $1/\sqrt{V}$. This would indicate that differences in pathlength are the principal factor responsible for the spread. For a typical 10-stage venetian-blind tube (8053) at 100 V/stage, the transit time is 70 ns and the time spread (between half-amplitude points) 30 ns. For a typical circular-cage MPT (931A), the transit time is approximately 17 ns and the pulse rise time (10%–90% full amplitude) is less than 2 ns. [21e]. Transit and rise time data for a number of MPTs are shown in Figures 16.18 and 16.19, respectively [21e].

16.2.3.7 Fatigue and Life Characteristics. Fatigue and life characteristics of photocathodes are discussed in Section 16.2.1.7. When a MPT is operated at anode currents approaching 100 μA/cm^2, the responsivity tends to drop progressively. When the current is removed, the sensitivity tends to return but slowly. This effect is more pronounced in Sb–Cs dynodes and may be due to the sublimation of cesium. Silver–magnesium dynodes tend to increase in sensitivity by approximately 45% (for the first 200 hr at 2 mA) and then to decrease slowly [21f].

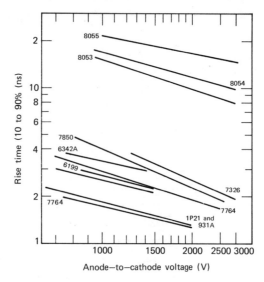

Figure 16.19 MPT rise times versus voltage for tube types indicated. Engstrom [21].

One 10-stage MPT (6094) with Sb–Cs venetian-blind dynodes and started at 2 mA anode current exhibited a declining output that stabilized at 1.3 mA after 10 min [22]. One author recommends running MPTs below 10^{-7} A for constant and linear operation in astronomical applications [19a].

16.2.3.8 Miscellaneous Effects. *Temperature* affects primarily the dark current level in phototubes; this is discussed in Section 16.2.4.3. However, a slight increase in sensitivity, too, is often observed with increasing temperature. In one Sb–Cs tube (1P21), this was measured at 0.3%/°C [32]. This increase occurs primarily at the longer wavelengths, whereas at shorter wavelengths there is a slight drop in sensitivity. For Sb–Cs cathodes (S4 and S11) the crossover occurs at approximately 0.59 μm [35]. One practical result of this is that changes in temperature cannot be compensated for unless the spectral composition of the radiation is known [35]. Therefore, when MPTs are cooled for low light level use, they should be used only at shorter wavelengths: below 0.5 μm for Sb–Cs, 0.65 μm for trialkali; S1 can be used at least up to 1 μm [36].

The presence of *magnetic fields* may interfere with the electrostatic focusing in the MPT and may, therefore, lower its sensitivity. Representative values of magnetic flux parallel to the tube axis causing a 50% drop in sensitivity are [21g]

931A	(circular-cage dynode arrangement, axis parallel to field	20 gauss
8053	(venetian-blind dynodes)	15 gauss
7029	(circular-cage, axis transverse to field)	4 gauss
6342A	(electrostatically focused, longitudinal)	0.7 gauss

16.2.3.9 Installation Considerations. A number of factors should be considered when installing a MPT. In view of the effects described in the preceding section, magnetic shielding may be required for maintaining full tube sensitivity. When very low light levels are to be detected, cooling may be used to lower the dark current, and the noise associated with it, considerably.

The coupling of the light source to the phototube, too, should be carefully considered. If the light source image moves, but there is a stationary aperture through which all the light passes, it may be advantageous to image this aperture on the photocathode. The image should not be too small, lest the cathode be overloaded locally; but on the other hand, it should be confined to the active part of the cathode.

Ocassionally, the fraction of the light absorbed in a semitransparent cathode may be increased significantly by exploiting total internal reflection in the face plate. To this end the light may be introduced into the face plate at an angle exceeding the critical angle, by means of a prism in optical contact with the face plate [37].

It is common practice in MPT circuit design to use a very well regulated and filtered dc power supply, with the positive terminal grounded and the negative terminal connected directly to the cathode. A series of $N+1$ resistors, where N is the number of dynodes (approximately 20 kΩ each), are connected in series across the power supply, and the dynodes are connected to the successive junctions between resistors. The anode is connected to ground via the load or current-measuring device (see Figure 16.20).

It should be noted that when the cathode is illuminated and the phototube conducts, two potential nonlinearities are introduced [38], [39].

1. Current flows through the load resistance to the anode, lowering its potential relative to the last dynode. This tends to lower the tube gain with increasing illumination.

2. (Positive) current flows out of the dynodes into the resistors, as shown by the arrows, when the dynodes supply electrons to the tube current. This additional current increases the bias of the dynodes with

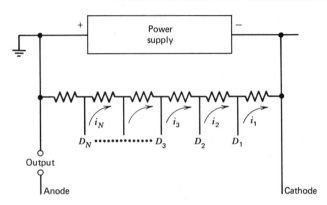

Figure 16.20 MPT circuit, schematic. D_i is the ith dynode.

respect to each other and, therefore, tends to raise the tube gain with increasing illumination.

These effects should be considered when designing a MPT installation in which linearity is crucial.

16.2.4 Noise and Dark Current in Phototubes

Since the ultimate limitation of phototube performance is set by its noise characteristics, we must now consider these in detail. A number of more or less independent sources of noise must be considered when dealing with phototube circuits. These are here first treated individually; their interactions are then considered in Section 16.2.4.6.

16.2.4.1 Johnson Noise [9b]. The photocurrent is usually measured or detected by passing it through a resistance, actual or virtual. Such a resistance contains charged particles, which, as a result of their temperature, exhibit random motion. Since they are charged, these random motions represent random currents, and these are called *Johnson noise*. The Johnson noise is "white"; that is, its power is uniformly distributed over the frequency spectrum up to very high frequencies (probably up to 10^{13} Hz).

It can be shown that the energy generated in any conductor as a result of Johnson noise has a mean value of

$$\bar{W} = 4kT\,\Delta\nu_t$$
$$\approx (1.6 \times 10^{-20}\,\text{J}) \times \Delta\nu_t, \quad \text{at room temperature.} \quad (16.44)$$

Here $k = 1.38054 \times 10^{-23}$ J/K is Boltzmann's constant,
  T = the absolute temperature, and
  $\Delta\nu_t$ = the frequency band width over which the noise is measured.

In a resistor of resistance r, this corresponds to a noise current

$$I_{NJ}^2 = 4kT \frac{\Delta\nu_t}{r}. \qquad (16.44a)$$

This must be compared to the anode signal current, i_{sa}. Equation 16.44 is known as Nyquist's[7] formula.

16.2.4.2 Shot Noise. A second source of noise is due to the quantum nature of the charge generated at the cathode. This charge corresponds to the number, n, of electrons. If n is assumed to be Poisson distributed, the variance of n will be $\bar{n}$, the mean number of electrons (see Section 3.4.4). If the charge is measured by accumulating the cathode current, i_c, over a period of time τ, the mean value of the charge will be $i_c\tau$. The fluctuation in the number of electrons will have a standard deviation (or rms value) of

$$\text{sd}\,(n) = \sqrt{n} = \sqrt{\frac{i_c\tau}{q_e}}.$$

This corresponds to a current fluctuation

$$\sqrt{\overline{i_{NC}^2}} = \frac{\sqrt{\bar{n}}q_e}{\tau} = \sqrt{\frac{i_c q_e}{\tau}}, \qquad (16.45)$$

where $q_e = 1.6021 \times 10^{-19}$ C is the charge of the electron. We obtain the corresponding expression in terms of bandwidth, by replacing τ^{-1} by $2\Delta\nu_t$[8]:

$$\boxed{\bar{i}_{NC}^2 = 2q_e i_c \Delta\nu_t.} \qquad (16.46)$$

This is the usual expression for shot noise. In gas or multiplier phototubes, this current is multiplied by the overall gain, G, so that the output noise power due to cathode shot noise is

$$I_{NC}^2 = 2q_e i_c \Delta\nu_t G^2. \qquad (16.47)$$

16.2.4.3 Dark Current and Its Noise. Even when no illumination enters the phototube, some current flows to the anode. This is called *dark current*. It may be generated by a number of processes.

[7] H. Nyquist (1928).

[8] This interval thus corresponds to a half-cycle of the passband. See Section 18.2.2.3 for a quantitative derivation.

1. *Thermal Emission of Electrons from the Cathode and Dynodes.* Due to the low work function of the photocathode, some electrons acquire enough energy thermally, to escape. This effect can be greatly reduced by cooling. Thus the thermal dark current in a MPT has been observed to drop from 10^{-8} A at 28°C to 10^{-11} A at −20°C [19a]. This current seems to be significant only above −20°C [40].

2. *Field Effects.* When sharp discontinuities in conducting surfaces are present, very high electrostatic fields may appear locally at the surface of the conductor. These fields may facilitate the emission of electrons via the Schottky effect.

3. *Ion Feedback.* There is always some residual gas remaining inside the envelope (if only evaporated electrode coating materials, such as cesium), and positive ions do occur. At high voltages these can cause secondary electron emission contributing to the dark current. Also metastably excited atoms drifting to the cathode can produce emission of secondary electrons.

4. *Ohmic Leakage.* Even when the tube construction itself provides for extremely high electrical resistance between electrodes, contamination of the tube envelope surfaces, inside or out, may cause leakage current. Especially on the outside, the mere presence of humidity will increase the flow of leakage current. Therefore, if the ultimate in dark current reduction is desired, the tube must be mounted in a dessicated housing [19a]. Occasionally, a moisture-resistant coating on the base may be helpful.

5. *Fluorescence.* The relatively high currents appearing at the MPT anode due to the electron multiplication, result in some fluorescent emission and some of this light may strike the sensitive cathode to increase the dark current. In addition, high-energy particles striking the envelope also cause fluorescence and thus contribute to the dark current. The radioactive K^{40} usually present in the glass will often contribute to this [19a].

The dark current, to the extent that it is due to thermionic emission, can be minimized by cooling. Dark current sources 2 and 3 are reduced at lower interelectrode voltages beyond the obvious reduction due to the lower gain. Thus the noise current in one MPT (1P21) was found to increase from 2 to 80 nA as the voltage was increased from 80 to 110 V/stage, but only to 18 nA when the envelope was maintained at cathode potential [41]. Hence dark current may be significantly reduced by coating the envelope and maintaining it at cathode potential. In general, the minimum acceptable voltage should be used at low signal levels.

The roles of various sources of dark current have been studied extensively on the basis of the output pulse height distribution [42]. The dark current itself is, however, not usually a limiting factor. As long as it remains constant it can readily be subtracted from the output, leaving only the signal current. The trouble caused by the dark current is due to its fluctuations, which are primarily due to shot noise. Following (16.46), the shot noise of the cathode dark current (i_d) is

$$\overline{i_{Nd}^2} = 2q_e i_d \Delta \nu_t. \tag{16.48}$$

The dark current noise is usually given in terms of the noise equivalent input (NEI) (see Section 3.4.5). This is the amount of input signal producing an increase in output equal to the (noise) output in the absence of signal. This is a useful way of specifying noise, because it is relatively independent of gain setting and intertube variations [43]. Obviously, it is useful only when the signal-to-noise ratio is dark current limited, that is, at very low signal levels.

In the absence of signal, dark current and Johnson noise (Section 16.2.4.1) will normally be the only significant noise sources, except that, in MPTs, the dark current noise is increased by noise contributions from the dynodes as follows.

16.2.4.4 Dynode Shot Noise. In a MPT the shot noise generated at the cathode is, of course, amplified along with the signal current. But the dynodes contribute shot noise even beyond this, as may be seen by considering a noninteger mean gain factor per stage, say $g = 3.5$. With such a gain factor, some electrons will give rise to three and others to four secondary electrons. But even when g is an integer—say 3—the actual gain will occasionally be only 2, balanced by an occasional gain of 4, etc. These random fluctuations will contribute to the noise.

The calculation of the output pulse height distribution resulting from a single electron is quite complicated, even when a Poisson distribution is assumed at each dynode. Computer-generated results are shown in Figure 16.21 [44].

When it is important to know this distribution, an approximation may be of interest. When the gain per stage (g) is high enough to make negligible the probability (P_0) of zero output for r electrons at the cathode, a relatively simple approximation is, indeed, possible. According to this [45] the probability of n electrons at the anode for r electrons at the cathode is

$$P(n) = \frac{\beta^\alpha n^{\alpha-1} e^{-\beta n}}{(\alpha-1)!}, \tag{16.49}$$

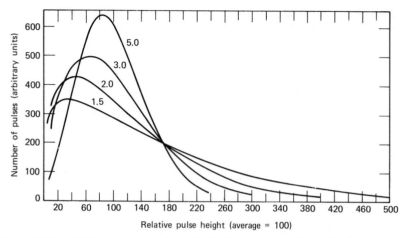

Figure 16.21 MPT output pulse height distribution, assuming Poisson-distributed electron multiplication. The mean value of gain per dynode is indicated above the corresponding curve. Lombard and Martin [44].

where[9] $\alpha = r(1 - g^{-1})$

$\beta = (1 - g^{-1})/G.$

The value of P_0 may be calculated from [45]

$$P_0 = [p_N(0)]^r \qquad (16.50a)$$

by means of the recurrence relationship

$$P_K(0) = \exp\{g[P_{K-1}(0) - 1]\}, \qquad (16.50b)$$

where

$$P_1(0) = e^{-g}. \qquad (16.50c)$$

In many applications, however, the rms value of the shot noise suffices. This may be calculated quite readily. We simply assume that each stage contributes an amount of shot noise as given by (16.46) (with i_c replaced by the mean value of the dynode current) and that each such contribution is amplified by the gain factors of all the subsequent stages. Assuming a gain g at every stage, we note that the mean current at stage j is $i_c g^j$. We can then write for the amplified contribution of that stage:

$$I_{Nj}^2 = (2q_e i_c g^j \Delta \nu_t)(g^{N-j})^2 = [2q_e i_c \Delta \nu_t G^2] g^{-j},$$

where $N =$ the number of stages and

$G = g^N$ is the overall gain.

[9] The expressions for α and β in [45] are both larger than ours by a factor of g^2, presumably due to a typographical error.

Comparison with (16.47) shows that the bracketed factor is simply the cathode-generated shot noise, I_{NC}^2. The sum (I_{NG}^2) of the contributions of all the stages is the sum of I_{Nj}^2 over all j; this involves only the sum of a geometric progression and can therefore be written

$$I_{NS}^2 = I_{NC}^2 + I_{NG}^2 = \sum_{j=0}^{N} I_{Nj}^2 = \frac{1 - g^{-(N+1)}}{1 - g^{-1}} I_{NC}^2 = \frac{g - G^{-1}}{g - 1} I_{NC}^2. \quad (16.51)$$

It follows that the dynode generated noise alone is

$$I_{NG}^2 = \left(\frac{g - G^{-1}}{g - 1} - 1 \right) I_{NC}^2 = \frac{1 - G^{-1}}{g - 1} I_{NC}^2$$

$$\approx \frac{1}{g - 1} I_{NC}^2, \quad (16.52)$$

when the overall gain is high.

Some investigators [46] have confirmed the correctness of the factor $(g - 1)^{-1}$, while others have found that it should be [47]

$$\frac{B}{g - 1},$$

where $B = 1.54$ for a 10-stage, end-on S11 tube (5819). It has been suggested that discrepancies may be due to nonuniformities in the gain factor g over the dynode surface [48].

Based on (16.51), the signal-to-noise power ratio is reduced by a factor

$$F^* = \frac{R_{in}^2}{R_{out}^2} = \frac{I_{NS}^2}{I_{NC}^2} = \frac{g - G^{-1}}{g - 1} \approx \frac{g}{g - 1}. \quad (16.53)$$

This equals the reciprocal of the detective quantum efficiency (see Section 3.4.4) and is closely related to the *noise factor* used in the theory of electronics systems. In a number of tubes it has been measured to range between 4 and 1.7, its value according to (16.53) at $g = 2.5$ [49] (see Table 165).

16.2.4.5 Background Noise. Often the radiation received at the cathode consists of both signal and nonsignal (or background) radiation. For example, in astronomical radiometry, the background radiation may be due to starlight scattered by the atmosphere. Although much of this may be excluded by means of a field stop, such exclusion is limited by "seeing" the random motion of the star image due to atmospheric turbulence: if the field aperture is too small, it will exclude the signal (star) radiation part of the time.

Since the shot noise is generated by the total cathode current, whether it originates in the signal or elsewhere, such nonsignal current contributes to the shot noise and therefore lowers the detection probability or the accuracy to which the signal may be measured. This condition may be considered as an increase in noise, or we may consider both signal and background as a more complex "signal" and demand a larger signal-to-noise ratio for detection. Here we choose the second approach because it simplifies the following discussion considerably.

16.2.4.6   Signal-to-Noise Ratio.

Having diagnosed the various noise sources, we are now in a position to estimate the key parameter in detection—the signal-to-noise power ratio

$$R^2 = \left(\frac{I_S}{I_N}\right)^2 = \frac{G^2 i_S^2}{(I_{NC}^2 + I_{NG}^2 + I_{NJ}^2)}, \tag{16.54}$$

where i_S and I_S are the signal current at the cathode and anode, respectively. The cathode current shot noise, I_{NC}, includes the contribution of the cathode-generated dark current, if i_c in (16.47) is taken as the sum of the signal and this dark current.

This expression can generally be simplified since some of the noise terms are usually negligible. To this end we first calculate the magnitude of the shot noise relative to the Johnson noise. Reference to (16.44a) and (16.47) shows immediately that this ratio is

$$\rho = \frac{I_{NC}^2}{I_{NJ}^2} = \frac{q_e i_c G^2 r}{2kT} = \frac{q_e V_0 G}{2kT} \tag{16.55}$$

$$\approx 20 V_0 G, \qquad \text{at room temperature,}$$

where $V_0 = i_c r G$ is the voltage (in volts) across the output resistor.

Thus for output voltages not much less than 1 V, Johnson noise is negligible, even in diodes. In MPTs, it is necessary only to set the gain well in excess of

$$G \gg \frac{0.05}{V_0} \qquad \text{or} \qquad G \gg (20 i_c r)^{-1/2} \tag{16.56}$$

to insure that the Johnson noise be negligible.

Consider (16.55) further as it applies to a diode. Here it may be written

$$\rho = \frac{q_e i_c r}{2kT}. \tag{16.57}$$

We may now consider r to be limited by the desired bandwidth ($\Delta \nu_t$) and the stray capacitance (C) of the diode, where the relationship may be approximated by

$$\Delta \nu_t = (2\pi r C)^{-1}. \tag{16.58}$$

On substituting this for r into (16.57), we obtain

$$\rho = \frac{q_e i_c}{4\pi kTC \, \Delta\nu_t},$$

$$\approx \frac{i_c^*}{6\Delta\nu_t}, \tag{16.59}$$

at room temperature and $C = 20 \, pfd$; here i_c^* is in pico-amperes and $\Delta\nu_t$ in hertz.

Generally, then, when working with diodes, we must consider shot noise and Johnson noise, unless $\rho \gg 1$. In MPTs we need consider only shot noise (including total cathode current and dynode effects). Thus we may write (16.54) for diodes:

$$R^2 = \frac{i_S^2}{2[q_e(i_S + i_d) + 2kT/r]\Delta\nu_t} \tag{16.60}$$

and for MPTs:

$$R^2 = \frac{i_S^2}{[2q_e(i_S + i_d) \, \Delta\nu_t g/(g-1)]}$$

$$\approx i_S \frac{1 - g^{-1}}{2q_e \, \Delta\nu_t}$$

$$\approx \frac{i_S}{2q_e \, \Delta\nu_t} \approx 3 \times 10^{18} \frac{i_S}{\Delta\nu_t}, \tag{16.61}$$

where the first approximation is valid for a signal current well in excess of the dark current. If the gain per stage may be considered to be large, we obtain the very simple expression resulting from the second approximation.

Note that it is not the high gain of the MPT, but rather its cathode sensitivity (determining i_S*) which controls the signal-to-noise ratio attainable*; the gain must only suffice to make Johnson noise negligible. It has been found experimentally also, that the signal-to-noise ratio is improved only slightly if at all, by the applied voltage, at least beyond 100 V/stage [50].

16.2.4.7 Photon Counting [51], [52]. Instead of measuring the total current output from the MPT, we may "count photons" by exploiting the very large gains of which the MPT is capable. These enable us to detect the current pulse initiated by a single electron leaving the cathode. We may, therefore, count these pulses instead of measuring the anode current. Such counting essentially eliminates the noise contributed by the dynode system, which is manifested by variations in pulse heights. It also makes the signal-to-noise ratio independent of bandwidth [52].

In addition to this, photon counting at low light levels enables us to discriminate significantly against those portions of the dark current that do not originate at the cathode. This ability is based on the fact that noncathode dark current appears in the form of many pulses of very small height. By setting the detection threshold of the counter above the height of these pulses, we eliminate a major portion of the dark current noise, while losing only very few of the signal pulses of interest to us [32], [19b], [53]. By this approach illumination levels as low as 2×10^{-16} lumens were detected over 30 years ago [32]. This detection required liquid air cooling ($-190°C$) and counting approximately one photon/sec over a 5 min period! Dark current pulse rates of some tubes have been measured and are listed in Table 166 [40], [43]. Assuming that these counts are Poisson distributed, the standard deviation from this count will be $\sqrt{mt}$, where m is the number of pulses per unit time and t is the counting time. For a reasonable detection probability, the number of signal pulses to be counted during this interval must exceed this number by some factor, say 2, and must be at least unity. It should be noted that erratic bursts of dark current are occasionally reported [43], so that the assumption of a Poisson distribution is questionable.

16.3 PHOTOCONDUCTIVE DETECTORS

Photoconductive detectors fall into two major categories.

1. Homogeneous detectors, where a carrier, freed anywhere inside the material is drawn out by an externally applied field.

2. Junction detectors, where the carriers are generated in the neighborhood of a p-n junction to be separated by an internally generated field.

In the junction detectors, with the exception of the avalanche diodes, there is no such gain. On the other hand, very short time constants are attainable (see Table 167 [54a]).

We present in Section 16.3.1 and 16.3.2 brief descriptions of the more important photoconductive detectors, homogeneous and junction, respectively. These are followed, in Sections 16.3.3 and 16.3.4, by a discussion of noise and operating characteristics. The characteristics of these detectors are summarized in Tables 168 [55a], [1c] and 169 [55a]; those for use in the ir, also in Figure 16.22 [55a]. A number of homogeneous semiconductors have been assigned S curves and these are presented in Table 170, where the tolerance range, too, is indicated [16b]. For construction techniques, refer to Ref. 11b.

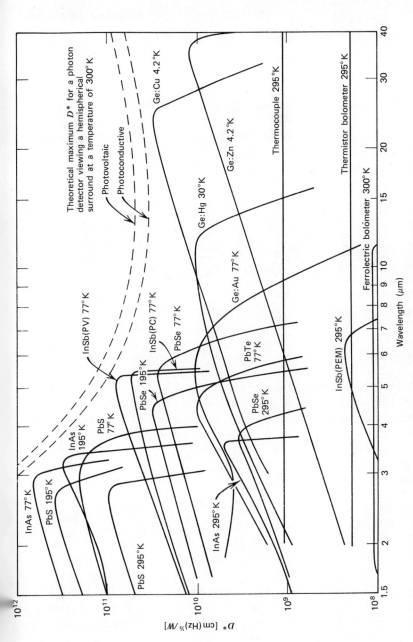

Figure 16.22 D^* of various infrared detectors when operated at the indicated temperature. Chopping frequency is 1800 Hz for all detectors except InSb(PEM) 1000 Hz; ferroelectric bolometer 100 Hz; thermocouple 10 Hz. Each detector is assumed to view a hemispherical surround at a temperature of 300 K. Hudson [55a].

16.3.1 Homogeneous Photoconductive Detectors [1c]

Before discussing specific photoconductors, we note here that the efficacy of a homogeneous photoconductor may be specified in terms of (1) responsivity, (2) detectivity, (3) gain, or (4) specific sensitivity.

1. *Responsivity* in its usual sense (see Section 3.1.3) depends on both load and applied voltage. To obtain a general quantity, it is here usually defined as change (Δg) in conductance per unit incident power (P):

$$k_r = \frac{\Delta g}{P}. \tag{16.62}$$

2. *Detectivity* is usually given as D^* (see Sections 3.4.5 and 16.3.3.2).
3. *Gain* is here essentially quantum gain (see Section 16.1.6.3).
4. *Specific sensitivity* or *photoresponse* [11c] is defined as the responsivity multiplied by the square of the electrode spacing (l)

$$S = \frac{\Delta g l^2}{P} \, \mathrm{m^2/\Omega W}, \tag{16.63}$$

where the irradiation is assumed uniform. Whereas responsivity is a characteristic of a specific detector unit, specific sensitivity is a characteristic of a photoconductive layer, independent of size and electrode configuration. (The spacing is squared to compensate for the increase in incident power when the electrode spacing is increased. The increase in Δg when the electrode is increased, is compensated for by the corresponding increase in power.)

In defining the specific sensitivity, the thickness of the photoconductive layer is taken as an intrinsic factor, whereas the length and width (electrode configuration) is not. This is done because the layer thickness affects the detector response in a very basic way: the absorption characteristics, and even the spectral characteristics, are strongly dependent on the layer thickness. Electrode configuration, on the other hand, affects only the electrical characteristics, and these in a strictly proportional manner. This configuration is therefore generally used to match the detector to the available auxiliary circuits, such as power supply and amplifier. Because of the great flexibility in this regard, detector characteristics are often given in a form making them independent of the electrode configuration. The resistance may then be given in "*ohms per square*," that is, ohms for a square detector area. For a fixed thickness (t), the resistance is then independent of the size of the square.[10]

For the present purpose it is convenient to divide the homogeneous detectors into *intrinsic* and *extrinsic* types. The extrinsic detectors are more difficult to make, because of the critical nature of the doping levels. Also, their responsivity is lower because of their lower absorption. On the other hand, they provide far greater flexibility.

16.3.1.1 Intrinsic Detectors [9c]. The most commonly used intrinsic photoconductors are cadmium sulfide (CdS) in the visible region and lead salts and indium antimonide (InSb) in the near infrared (ir). Both CdS and the lead salts are used as polycrystalline films, either sintered (CdS), evaporated in a vacuum (PbS), or chemically deposited (other lead salts) on a suitable substrate. Electrodes may be in the form of evaporated gold or painted on graphite. Sensitization may be applied by heating the film in an oxygen atmosphere. The finished detector is protected by means of a cover of suitable material such as glass or fused quartz. The InSb detectors are generally made from single crystals, sliced, lapped, and etched to the desired thickness.

In certain applications, photoconductors with very high resistivity are required as, for instance, when a photoconductively generated charge image is to be stored on the surface of the photoconductor. For such applications selenium and antimony trisulfide (Sb_2S_3) are useful. In these applications, conduction takes place primarily through the layer rather than along it, and the useful layer thickness is dictated by considerations of the mean free path of the carriers and the depth of penetration of the incident radiation.

One such material, zinc oxide, is used primarily in a resinous matrix for electrophotography. Its characteristics are, therefore, discussed in that connection in Section 17.4.1.1.

Let us now consider briefly each of the commonly used photoconductors.

Cadmium Sulfide and Selenide [56]–[58]. Cadmium sulfide is the most sensitive photoconductive material. Its sensitivity may extend throughout

---

[10] This is readily shown on writing the resistance in terms of resistivity (ρ), length (l), width (w):

$$R = \frac{l\rho}{wt}. \tag{16.64}$$

For a square configuration ($l = w$), we have then

$$R_\square = \frac{\rho}{t}, \tag{16.65}$$

independent of the size of the square.

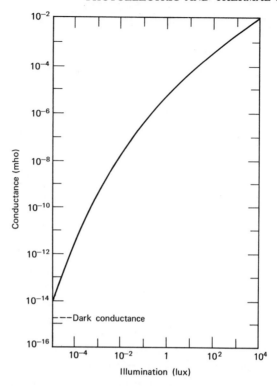

Figure 16.23 Transfer characteristic of CdS detector [58].

the visible spectrum and beyond. The light-to-dark conductance ratio has been given as high as 10^{11} [58] (at 10 lux illumination), although 10^6 is more usually quoted [57]. This material is normally doped with copper and chlorine and has both intrinsic and extrinsic response characteristics. Photocells may be prepared by evaporation, sintering or applying the CdS powder in a binding matrix, such as polystyrene [56], [57]. Transfer characteristic and spectral response for a representative type are shown in Figures 16.23 and 16.24 [58].

The time constant of such cells depends heavily on the illumination level. For instance, the above cell has a time constant of 0.07 sec at 100 lux Also, there are long-term effects, presumably due to impurity trapping. Due to these effects, the cell will require about 10 hr in darkness before it reaches its normal dark conductance. The response to illumination is considerably faster; rise-time data for a representative CdS cell are shown in Figure 16.25a and decay characteristics in Figure 16.25b [58].

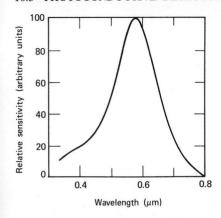

Figure 16.24 Spectral response of CdS [58].

Pure CdS has been found to have a specific sensitivity of 10^{-5} cm^2/$\Omega \cdot$ W and an electron lifetime of a microsecond. The corresponding values for sensitized CdS are 10^{-1} cm^2/$\Omega \cdot$ W and 10 msec, respectively [11d].

CdSe is also used for photodetection. Although its responsivity is lower, its sensitivity extends somewhat further into the ir [1c].

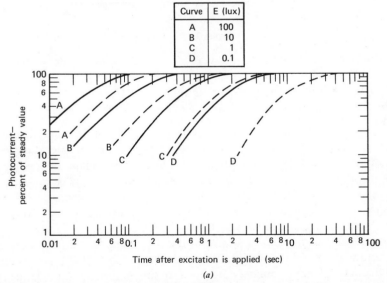

Curve	E (lux)
A	100
B	10
C	1
D	0.1

(a)

Figure 16.25 Temporal response of CdS cell. Ambient temperature: 25°C; color temperature: 2870 K. (a) Rise times. Solid and broken lines for 5 sec and 5 min dark storage, respectively; illumination values listed. (b) Decay times. 22.5 V applied; illumination (lux) shown next to each curve [58].

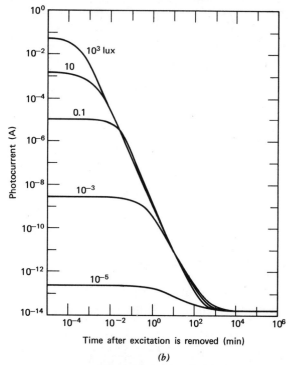

Time after excitation is removed (min)

(b)

Figure 16.25 (Continued).

Thallous Sulfide [59]. In the very near ir, thallous sulfide is the most sensitive detector material, having a peak D^* of 2×10^{12} cm · $Hz^{1/2}/W$ at $0.9 \mu m$. Its usefulness, however, is limited to below $1.3 \mu m$ and it is, therefore, not very popular. The time constant is 0.53 msec.

Lead Sulfide [60]. In the near infrared ($1.2–4.0 \mu m$) lead sulfide detectors are the most sensitive photoconductors at room temperature. The resistance of the films varies between 10^4 and $10^8 \Omega/$square, depending on doping and heat treatment. The time constant may be anywhere between 20 and $500 \mu sec$ and tends to vary inversely with the resistance. The responsivity is maximum at about $5 \times 10^6 \Omega/$square and is approximately $300 \, Vcm^2/W$ at $2.5 \mu m$ wavelength.

Although this detector is usable at room temperature, its detectivity is optimum near 195 K (solid CO_2). Cooling shifts the peak from a wavelength of approximately $2 \mu m$ at room temperature to $2.5 \mu m$ at 195 K and below. It is usable from 0.6 to $3.0 \mu m$.

Lead Selenide [61] *and Telluride* [62]. Thin film lead selenide detectors are similar to lead sulfide. The sensitizing heat treatment may be used to optimize operation at various temperatures (dry ice or liquid nitrogen[11]). It is most useful in the 4–6 μm region, although it has a strong competitor in InSb there. Lead telluride is also used in that region but has a considerably lower detectivity.

Indium Antimonide [63] *and Arsenide* [64]. In contrast to the poly-crystalline lead salt photoconductors, indium antimonide detectors are usually made of single crystals. Detector elements range from square 0.5 to 8.0 mm on the side. The larger detectors can, however, not be made so thin and therefore exhibit a lower detectivity. The major manufacturing problem seems to be that of purity. Arrays of crystals can be used. The energy band gap is 0.18 eV and the material is an excellent photoconductive detector in the wavelength range between 3.5 and 7.0 μm.

The peak of the spectral response curve moves from 7 to 5.5 μm as the detector is cooled from room temperature to 77 K. In the same interval, the detectivity $D^*(\lambda, 800, 1)$ moves from 2.5×10^8 to 5×10^{10} cm/Hz$^{1/2} \cdot$ W and the time constant from 5×10^{-8} to 5×10^{-6} sec. The responsivity at the optimum bias is about 0.3 V/W at room temperature and 500 V/W and 2×10^4 V/W at 190 and 77 K, respectively, for a square crystal 1 mm on the side and a 500 K blackbody source.

Selenium [65]. Selenium was the first photoconductive material discovered. Originally the gray, hexagonal form, which is semiconducting, was used. Later it was found that the reddish, amorphous form is both highly resistive and a sensitive photoconductor. It is usually prepared by evaporative deposition. Because of its high dark-resistivity (in excess of 10^{13} Ω/cm) it is widely used when storage times of several seconds are required, as, for instance, in electrography. Its sensitivity extends from the far uv through the green and peaks at approximately 0.4–0.45 μm. The range of the carriers (holes) is approximately 10 μm (at 5 V/μm) and layers up to this thickness are useful. (Electrons have a range of only $\frac{1}{10}$ of this.) Quantum efficiencies close to unity can be obtained.

Alloying with a few percent of tellurium extends the sensitivity well into the near ir [66]. Evaporation at elevated temperatures causes the development of small crystals inside the amorphous matrix of the film and this, too, results in enhanced sensitivity in the near ir [66].

The response to illumination is almost linear ($\gamma = 0.9$)[12] at low illumination levels and high applied electric fields. If the field is not maintained sufficiently high, commensurate with the illumination, saturation due to space charge limiting is observed.

---

[11] See Table 172 for temperatures corresponding to various coolants.
[12] See Section 18.3.1.1 for the definition of gamma (γ).

At low voltages, the dark current obeys Ohm's law, but when the field reaches approximately 4 V/μm, a rapid rise of dark current is observed.

The time constant of selenium is less than 50 μsec, but its measurement is complicated by severe fatigue effects, which become pronounced within a fraction of a second.

Lowering the temperature greatly reduces the conductivity; going from 20 to 10°C may reduce it by a factor of five. Heating, on the other hand, initiates recrystallization to the metallic form with much lower resistivity. At 80°C deterioration is marked already after 15 min. At 30–35°C this occurs after several hundreds of hours.

Antimony Trisulfide [67]. In Sb_2S_3 the range of the carriers is of the order of only a micrometer so that in the usual screens photoconduction requires penetration by the illumination. Since Sb_2S_3 absorbs shorter wavelength (blue) more readily than longer wavelength (red) radiation, the sensitivity peaks in the red for thicker layers. As the layer is made thinner, the sensitivity peak moves toward the blue end of the spectrum. In Figure 16.26 [67], the spectral sensitivity of a typical layer is compared

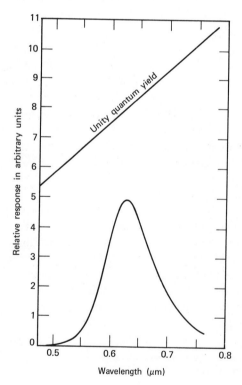

Figure 16.26 Spectral response of an Sb_2S_3 target, Forgue *et al.* [67].

to that corresponding to unity quantum efficiency. The responsivity is occasionally as high as 1 mA/lm; it drops with increasing illumination, and the transfer characteristic exhibits a gamma of approximately 0.5. The time constant, too, is a function of the illumination level. At 100 lx it is of the order of milliseconds, and at 10 lx it is approximately 50 msec. Faint residual photoconductivity may persist for seconds after removal of the illumination.

When an Sb_2S_3 screen is heated, both sensitivity and dark current rise rapidly—decades between 0 and 200°C. However, within this range changes seem to be reversible. Baking a layer of this material in air lowers the dark current and increases the responsivity; but on operating it in a vacuum, it tends to revert to its initial characteristics.

Mercury–Cadmium Telluride (CMT, MCT). For use in the ir, intrinsic detectors are superior to extrinsic ones in that they require less cooling; they can operate at approximately $1000 \text{ K} \cdot \mu\text{m}/\lambda_c$, where extrinsic detectors must be cooled to approximately $500 \text{ K} \cdot \mu\text{m}/\lambda_c$, where λ_c is the cutoff frequency of the detector [68]. Especially for work further out in the ir, it would be highly desirable to have an intrinsic detector available. Unfortunately, among the simple compounds none is known intrinsically suitable for the important 8–14 μm atmospheric window region. There are, however, some that can be alloyed together to yield the desired bandgap. Best known among these are the alloys of HgTe and CdTe. These compounds have band gaps of -0.3 and 1.6 eV, respectively.[13] When they are mixed, the resulting alloy exhibits a bandgap anywhere between these limits, depending almost linearly on the respective proportions of the two components.

The variation of the bandgap with temperature is positive for HgTe and negative for CdTe (both approximately 5×10^{-4} eV/K), so that alloying reduces the temperature dependence.

It has been found that the bandgap of $Hg_{1-x}Cd_xTe$ is very closely ($x = \pm 0.005$) approximated by the following expression:

$$V_g = [1.59x - 0.25 + 5.233 \times 10^{-4} \, T(1 - 2.08x) + 0.0327x^3] \, \text{eV},$$

$$(16.66)$$

$$0.17 < x < 0.6$$

$$20 < T < 300 \text{ K}$$

where x is the mole fraction of CdTe and
T is the temperature in K [69].

---

[13] The negative bandgap implies that the valence band extends beyond the bottom of the conduction band. These bands do not overlap in momentum space, however, so that the material is still a semiconductor (or semimetal).

The corresponding cutoff wavelength:

$$\lambda_c = 1.24 \ \mu\text{m/eV} \cdot V_g$$

is about 10% larger than the wavelength at peak response and the response there is down to one half its peak value.

Detectivities $[D^*(10.6 \ \mu\text{m}, \ 1800 \ \text{Hz}, \ 1 \ \text{Hz})]$ can be in excess of $10^{10} \ \text{cm}\sqrt{\text{Hz}}/\text{W}$ [70]. D^* rises from this value to 10^{12}, as the detector is cooled from 170 to 77 K [71].

Pb/SnTe is another promising alloy, having some advantages over Hg/CdTe [68]. For Hg/CdTe the response times are approximately 1 nsec and 1 μsec for the photovoltaic and the most sensitive photoconductive versions, respectively. The most sensitive Pb/SnTe (LTT) detectors have response times below 50 nsec (see Ref. 70).

All the above can be operated at liquid nitrogen temperatures (77 K) as contrasted with 30 K for the equivalent mercury-doped germanium detector.

16.3.1.2 Extrinsic Detectors [72].

To obtain sensitivity extending further into the ir, carriers with lower excitation energies are required. These may be provided by means of "doping"—the substitution of atoms of different valence for some of the matrix atoms as discussed in Section 4.8.3. In principle, both silicon and germanium are suitable, and both positive and negative doping may be used. In practice, however, positively doped germanium is used almost exclusively.

When materials with very low excitation energy are used, there is a strong tendency for thermal excitation to prevail and to mask the photo-excitation of interest. This can be prevented only by cooling; the cooling requirements will be the stricter the further into the ir the sensitivity extends.

The lower spectral limit of such detectors is set by the spectral transmissivity characteristic of the matrix material and is approximately 1.8 μm for germanium. Further limitations are inherent in the necessarily low concentration of the active atoms, which limits the efficiency with which photons are absorbed. Also, as a consequence of the severely limited dopant concentration, the effects of accidental and unavoidable impurities become significant. These are especially harmful when they produce traps so shallow as to be thermally excited. In that event, they provide recombination centers, which may severely limit the lifetime, and hence the detector sensitivity. This problem is usually treated by providing shallow donor centers, which free electrons to fill, and inactivate, the shallow acceptor levels. This *compensation* lengthens the lifetime and, hence, increases the sensitivity of the detector.

The following are representative doping materials.

1. *Gold* doping with concentrations as high as $10^{-13}/mm^3$ and excitation energy of 0.16 eV (corresponding to 7.75 μm) has a response peak at 5.4 μm and is usable to about 9 μm. This detector should be operated at 60 K; operation at the temperature of liquid nitrogen (77 K) lowers the detectivity by a factor of three.

2. *Mercury* doping produces a detector with a response peak at 11 μm and usability up to 14 μm. It must be operated below 30 K.

3. *Copper* doping produces acceptor sites with an excitation energy of 0.04 eV. It yields detectors with a peak at 24 μm and a cutoff at 30 μm. These must be operated below 15 K so that liquid helium cooling is used.

4. *Zinc* doping of germanium produces "*zip*" (zinc impurity photoconductor) detectors exhibiting an acceptor level at 0.033 eV. These peak at 36 μm and remain usable even beyond 40 μm. They must be operated at liquid helium temperatures (4 K).

5. *Tin* doping permits operation to 110 μm, but operating temperatures must be maintained below 4 K.

16.3.2   Photodiodes and Triodes

Photodiodes may be used to overcome the limitations of photoconductors as inherent in their long time constants. In addition to this advantage, photodiodes provide the option of operating in a photovoltaic mode: they may be used as active power sources, where the power generated is proportional to the incident radiation. When used in this configuration, they may be used to convert electromagnetic flux, such as sunlight, directly into electrical power. Diodes constructed for this purpose are called *solar cells* (Section 16.3.2.1). The highest detectivities are attained with reverse bias, and diodes in detection applications are generally operated with such a bias (Section 16.3.2.2).

To overcome the lack of internal gain in the ordinary photodiode, it may be operated at a voltage sufficiently high to provide avalanche breakdown (Section 16.3.2.3) or a dual junction (transistor) form may be used (Section 16.3.2.4).

Diode junctions may be of the *p-n* type as described in Section 16.1.4. Alternatively, they may consist of a metal-semiconductor interface called the Schottky barrier or the classical point-contact junction [54b]. All of these may exhibit both photovoltaic and rectifying effects.

16.3.2.1   Photovoltaic Cells [54c].   We refer here as "photovoltaic" to those diodes that are operated in an active mode, i.e., those operated in

the fourth quadrant of the current-voltage curve; see Figure 16.3. The shaded area there shows the power output corresponding to operation at Point P.

The positive current in an illuminated diode rises exponentially with the external voltage (V) and drops (or rises negatively) with increasing illumination. It can be written in the form

$$I = I_s(1 - e^{\beta V}) + I_L, \tag{16.67}$$

where I_s is the *saturation current* observed in the dark ($L_L = 0$) with
$\qquad V = -\infty$,

$\qquad I_L$ is the *short circuit current* observed due to the illumination, and
$\qquad \beta = q_e/kT$ [see following equations (16.44) and (16.45) for definition
$\qquad$ of symbols]; Note that I_s and I_L are both negative.

For operation in the visible spectral region, photovoltaic cells are generally made of silicon and, in the near ir, indium arsenide is often used. For use as a *solar cell*, the silicon junction has a large area and is usually backed with a solid conducting layer, whereas the front surface, receiving the radiation, is provided with thin "fingers" of conducting coating, all connected to a strip of conducting coating running along the edge of the detector [73] (see Figure 16.27).

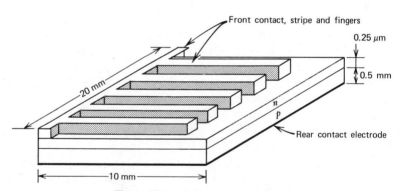

Figure 16.27 Solar cell, schematic.

16.3.2.2 Depletion Layer Photodiodes [54d].
Photovoltaic operation relies on the internally generated electric field; to the extent that this is limited, the carrier drift velocities and, therefore, the response speed, too, are limited. Due to the lengthened transit time, recombination may reduce the number of free carriers reaching the external circuit. Thus, both high-frequency response and responsivity suffer from the electric field limitations. To overcome these limitations, we may apply an external

field to supplement the internal one. This field must be applied in the "reverse" direction, that is, in the direction that permits only minority carriers to flow. The fact that carriers generated in the neighborhood of the junctions are quickly swept out of it creates a *depletion layer*, which becomes the active region of the photodetector. Photovoltaic cells designed for operation in this manner are called *depletion layer photodiodes*.

p-i-n Diodes. Photodiode junctions exhibit a capacitive effect. To reduce this capacitance a very lightly doped (intrinsic) region may be grown between the positively and negatively doped regions [6]. The resulting diode is referred to as *p-i-n* diode and should be usable to 10^{10} Hz. Actual tests have shown no drop in response at 2×10^9 Hz in a Ge diode. This consisted of a Ga-doped *p*-base covered by a very lightly positively doped (π) layer 2.5 μm thick and topped by a Sb-doped *n* layer. The *n* layer is provided with a number of thin gold electrode-strips (see Figure 16.28). This method of construction has the additional advantage of permitting freer control of the depletion layer geometry.

By way of illustration, silicon *p-i-n* diodes have a high quantum efficiency (approximately 0.65) in the spectral region from 0.5–0.8 μm and

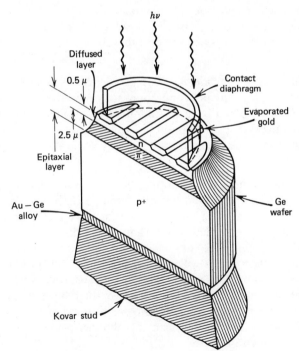

Figure 16.28 *p-i-n* diode section, schematic. Sze [54].

are still about 20% efficient at 0.4 and 1.0 μm. They operate linearly over 10 decades of flux magnitude and, even at room temperature, they have

$$D^*(0.8\ \mu\text{m}, 100, 6) = 4 \times 10^{12}\ \text{cm}\sqrt{\text{Hz}}/\text{W}$$

These data are for typical diodes with an active region having a diameter of approximately 0.5 mm [74].

Heterojunction Diodes. When good response at high frequencies is required, the designer faces two conflicting requirements [12]. On the one hand, absorption in the active (depletion) region must be good, and this calls for a thick layer. On the other hand, thick layers lengthen response time and, therefore, lower response at high frequencies. This dilemma may be resolved by the use of a *heterojunction*, that is, a junction in which the *p* and *n* layers consist of different materials. The top layer consists of a material transparent to the radiation of interest, whereas a material highly opaque to it is used for the lower level. Consequently, the radiation is absorbed almost completely in the immediate neighborhood of the junction.

Such a diode, in which the *n* doped top layer consisted of GaAs and the *p* doped lower layer of Ge, has been described. Here the 10-μm-thick *n* layer absorbed less than 0.1% of 0.845-μm radiation, whereas 90% of the radiation was absorbed in the first micrometer of the *p* layer [75].

Edge-Illuminated Photodiodes. The problems posed by the thin layers required for high-frequency operation may be overcome by another approach. Instead of receiving the illumination through the top layer, the *p-n* junction may be sliced and illuminated from the side [76]. This leads to long, narrow detector regions, but in many applications this may not be objectionable.

Point-contact Photodiodes. Another technique for obtaining good high-frequency operation is by the use of a point contact junction. Here the junction is between a *p*- (or *n*-) type material and a metal electrode in the form of a thin wire provided with a sharp point through which contact is made. Such junctions have long been used for rectification, but can also be used in photodetection, where they are capable of operation at very high frequencies—perhaps to 50 GHz [77]. They have a serious disadvantage in that the incident flux, to be effective, must be concentrated into a very narrow region in the immediate neighborhood of the junction. This difficulty is aggravated by the need for oblique incidence to avoid light blockage by the wire electrode.

This latter difficulty may be avoided by extreme thinning of the semiconductor layer in the neighborhood of the contact and illuminating from the other side [78] (see Figure 16.29). This device, when made of silicon, has a quantum efficiency close to unity, and well in excess of unity (2–3) when made of germanium. It is interesting to note that the manner of forming the contact has a significant influence on the current-voltage characteristic of the diode. The usual current pulse-formed contacts result in diodes exhibiting a photovoltaic effect similar to that of p-n diodes. When formed by illumination with an intense laser beam, the dynamic range of the photodiode is increased, but the photovoltaic effect is lost [78].

Schottky Barrier Photodiode. In addition to the depletion layer and point-contact junctions already mentioned, metal-semiconductor junctions known as Schottky barrier junctions, also exhibit rectifying and photovoltaic effects. Thus these, too, can be used in photodiodes.

A very thin (approximately 0.01 μm) metal layer is evaporated onto the semiconductor. The metal layer serves as one electrode; the other is provided by a metallic backing on the semiconductor.

The major shortcoming in this construction lies in the high reflectance of the metallic coating. To enable a major portion of the incident flux to reach the semiconductor region to generate carriers, the metallic film has a low reflection coating evaporated onto it. If the detector is designed for monochromatic radiation at a fixed wavelength, these low-reflection coatings can be made very efficient. Thus an n-type diode with an evaporated gold film and a ZnS low-reflection coating has exhibited a quantum efficiency of 70% with good response to pulses of 0.5 ns when operated at a 50 V reverse bias [79]. In addition to silicon, ZnS and GaAs

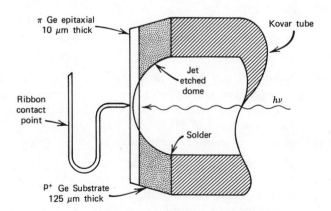

Figure 16.29 Point contact diode with thinning. Di Domenico *et al.* [78].

have been used for metal-semiconductor diodes. (See Figure 16.31 for the quantum efficiency spectra of various high-speed photodiodes, many of them of the Schottky barrier type [80]).

16.3.2.3 Avalanche Photodiodes [54d], [80], [81]. As noted earlier, photodiodes are superior to photoconductors in terms of speed. On the other hand, they are incapable, as a rule, of gains in excess of unity, which are readily obtained from the photoconductors. There is, however, one very successful technique that permits large gains to be obtained from fast photodiodes. This technique is analogous to electron multiplication in gases (Section 16.2.2): as it is accelerated by the electric field, the electron excites additional electrons on colliding with them and these secondary electrons join the current.

Just as in gas diodes, operation is limited by breakdown. Special care must be taken to ensure that this occurs uniformly over the whole active area. Therefore the active area is usually held quite small (approximately 150 μm in diam) and special attention is given to the uniformity of the crystal. Since breakdown tends to occur first at the edges, *guard rings*, enclosing the active region, are grown into the diode. These are regions with a slow, uniform rate of change of doping that prevent the buildup of very large voltage gradients so that breakdown will not occur there.

The avalanche multiplication process clearly must increase the noise level in the signal. Careful control of this process, by appropriate shaping of the doping profile, may be used to reduce the noise generation. The response time, too, may be reduced by such shaping [82]. Silicon avalanche diodes based on this principle have exhibited a gain-bandwidth product of 150 GHz and an excess noise factor of 5; this at a quantum efficiency of 0.66 at 0.83 μm [82]. Signal and noise power dependence on the multiplication factor are illustrated in Figure 16.30 [82].

Avalanche diodes made of silicon are useful up to 1.1 μm and those of germanium up to 1.55 μm. With cooling to 77 K, InSb diodes are good to 5.5 μm.

16.3.2.4 Phototransistors [10c], [83]. Another method for increasing the gain in junction photoconductors is by use of the amplication effects of transistors as described in Section 16.1.4. By this method gains of 1000 may be obtained, but gains of 50 are more usual.

The gain of a p-n-p phototransistor may be shown to be

$$G = \frac{\gamma_p L_{np}}{\gamma_n w}, \tag{16.68}$$

where $\gamma_{p,n}$ are the conductivities in the p and n regions, respectively,
L_{np} is the electron diffusion length in the p region and
w is the width of the n region.

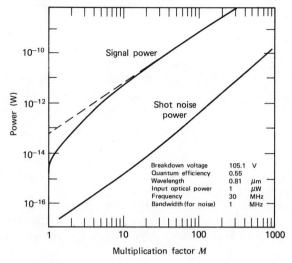

Figure 16.30 Signal and shot-noise power versus multiplication factor for avalanche diode. A GaAlAs LED, radiating at 0.81 μm was used as a signal. The broken line corresponds to $P \sim M^2$. Kanbe *et al.* [82].

The gain-bandwidth product may be written, analogously to (16.25)

$$G \cdot \Delta \nu = \frac{1}{2\pi RC},$$ (16.69)

where the resistance R and the capacitance C refer to the emitter-base space.

16.3.2.5 Conclusions. A large variety of junction photodetectors have been described. The choice from among them will be governed by such factors as frequency response, impedance requirements, and, above all, by the wavelength range to be detected. Since the opacity, that is the rate of absorption of radiant flux, tends to vary inversely with wavelength, short wavelength flux tends to be absorbed close to the surface of the detector, so that the Schottky barrier detectors, with their very thin metallic layer, are ideal. For radiation in the visible range of the spectrum, diffused p-n and p-i-n junctions are indicated. Infrared radiation tends to penetrate quite deeply into the semiconductor so that speed of response must generally be sacrificed unless the edge-illumination junction [76] is used.

The spectral quantum efficiencies of various diodes are compared in Figure 16.31. Also shown in this figure are lines of constant responsivity in mA/W [80]. See Table 171 for a listing of the characteristics of many

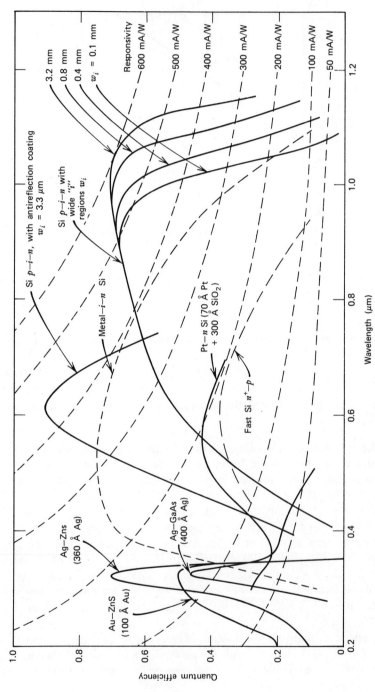

Figure 16.31 Spectral response of some high-speed photodiodes. Melchior *et al.* [80].

high-speed diodes. This is taken from Ref. 80, where sources are given for each entry.

The choice of homogeneous photoconductor will be dictated primarily by considerations of spectral range and detectivity. Figure 16.22 (and Table 168) may conveniently be used to guide this choice.

16.3.3 Noise in Photoconductors

The noise characteristics of photoconductors are quite similar to those of multiplier phototubes. The differences that do exist are discussed in the first subsection. The usual approach to noise analysis in photoconductors is, however, different. Here the accepted approach seems to be to consider background radiation—and the photon noise associated with it—as a noise source. This background noise is often considered to be beyond the control of the designer so that there exists a design target called background limited infrared photodetection (BLIP) operation [84]. The second subsection is devoted to a discussion of this.

16.3.3.1 Noise Characteristics. The noise characteristics of homogeneous photoconductors differ from those of the multiplier phototube primarily because of the substantial lifetime (τ) of the carriers in the photoconductor. We have already noted (16.34) that this limits the frequency response of the detector according to $[1+(2\pi\nu\tau)^2]^{-1}$; it has a similar effect on those noise components which are inherent in the photoconductor.

Johnson Noise. Johnson noise, often called *thermal noise* in this context, has the same form here as in multiplier phototubes (16.44). It can be controlled only by means of cooling and such cooling is frequently utilized to lower the thermal noise level well below the background noise, making it negligible.

Shot Noise. In photoconductors, shot noise is called *generation-recombination* (gr) *noise* because it is caused by the statistical fluctuations in both the generation and recombination rates which initiate and terminate, respectively, the carrier existence. Thus there are two random occurrences here, in contrast to phototube operation where there is no random process analogous to recombination and therefore only one random variable. This doubles the expression for gr noise power. In addition, the substantial carrier lifetime applies a frequency-dependent factor to the gr noise spectrum, as mentioned at the beginning of the section. Thus, instead of (16.46), we arrive at an expression of the form

$$I_{gr}^2 = \frac{4q_e I_0 \,\Delta\nu G^2}{1+(2\pi\nu\tau)^2}, \qquad (16.70)$$

where G is the gain as defined in (16.21).

I_0 is the total time rate at which carrier charge is generated, by the incident radiation:

$$I_0 = I_p + I_T = \eta_p q_e \Phi_p + I_T, \qquad (16.71)$$

Here I_p, I_T are the rates at which charge is generated by the incident photons and thermal fluctuations, respectively,

η_p is the conversion efficiency, that is the number of carriers generated by each incident photon, and

Φ_p is the incident photon flux.

On substituting (16.71) for I_0 into (16.70), we obtain

$$I_{gr}^2 = \frac{4q_e(q_e \eta_p \Phi_p + I_T)\,\Delta\nu G^2}{1 + (2\pi\nu\tau)^2}. \qquad (16.72)$$

The frequency spectrum given by (16.72) is almost flat for low frequencies ($\nu \ll 1/2\pi\tau$). At $\nu = \nu_1 \equiv 1/2\pi\tau$, it has dropped to half its original value and at frequencies yet higher, it drops rapidly.

In photodiodes the photo-carriers are immediately swept apart due to the high field at the junction. Thus the gain may be close to unity and the effective lifetime is short (usually $\tau \ll 1/\nu$). Hence (16.46) may be applied to photodiodes.

Current Noise [9d]. In metal conductors the frequency-independent Johnson noise is constant regardless of signal current. In all other conductors, however, the presence of current generates a noise having a spectrum which varies inversely with the frequency. This is called *current* or *flicker noise* (occasionally "1/f noise"). It is found to vary approximately with the square of the current. It may thus be written as

$$I_C^2 = A I_p^2\,\Delta\nu/\nu, \qquad (16.73)$$

where A is a constant, usually found to be in the neighborhood of 10^{-11}. This formula has been found to be valid even at frequencies as low as 10^{-3} Hz and for thin films. Apparently the cause of this noise has not yet been definitely established but certain carrier trapping effects have been suggested as a possible source [85].

Detailed analysis shows that the noise power varies with the square of current density. Therefore it may be reduced by using large-diameter conductors.

The effects of current noise can be minimized by operating at high frequencies. When the quantities to be measured are constant or vary but slowly, it is often found advisable to modulate—or "chop"—the incident flux by a rotating sectored wheel, vibrating tuning fork or other light-modulating device, to bring operation into a frequency range where the current noise is negligible.

Conclusion. On reviewing the major sources of noise and their variation with frequency, we find a general spectrum shape as shown in Figure 16.32.

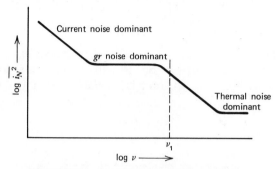

Figure 16.32 Semiconductor noise spectrum.

16.3.3.2 Background-Limited (BLIP) Operation [84].

As indicated in the preceding discussion, all noise factors, except gr noise, can be effectively eliminated by cooling the detector and "chopping" the incident flux. Such cooling can effectively eliminate the thermal term in the expression for gr noise (16.72), leaving only the contribution of the incident flux consisting of signal and background photons. Operation at these conditions, combined with unity conversion efficiency ($\eta_p = 1$) is referred to as BLIP operation. It is often instructive to consider this, since it represents an upper limit to the performance and also since it can serve as a design goal.

The noise under BLIP conditions is given by (16.72) with

$$\eta_p = 1 \quad \text{and} \quad I_T = 0. \tag{16.74}$$

Under these conditions the signal current is

$$I_S = \frac{q\Phi_p GM}{\sqrt{1+(2\pi\nu\tau)^2}}, \tag{16.75}$$

where M is the signal modulation, that is the ratio of signal to total flux—a measure of signal-to-background flux ratio. From (16.75) and

(16.72) (as modified by 16.74), we can readily calculate the BLIP signal-to-noise ratio:

$$R_{\mathrm{BLIP}}^2 = \frac{I_s^2}{I_N^2} = \frac{1}{4}\left(\frac{\Phi_p}{\Delta \nu}\right).\tag{16.76}$$

Note that this *signal-to-noise ratio is independent of frequency and time constant.* Thus in principle we can, under these conditions, ignore the influence of the time constant on the operation bandwidth. On the other hand, responsivity will drop at higher frequencies when τ is increased. Although this can generally be compensated for by amplification, this amplification may add noise.

As discussed in Section 3.4.5, a useful detectivity quality factor is given by D^* [cf. (3.102)] which may be defined as the signal-to-noise ratio referred to unit flux received, unit bandwidth, and unit detector area. Specifically

$$D^* = \frac{R\sqrt{\Delta \omega}}{E\sqrt{S}},\tag{16.77}$$

where R is the signal-to-noise ratio when the detector receives homogeneous irradiation E, and
  S is the detector area.
  Note that

$$E = \frac{\Phi_p Q_p}{S},\tag{16.78}$$

where Φ_p is the number of photons incident per unit time,

$$Q_p(\lambda) = \frac{hc}{\lambda}\tag{16.79}$$

  is the energy of a photon at wavelength λ,
  S is the detector area
  h is Planck's constant, and
  c is the velocity of light [cf. the values following (16.1)].

On substituting into (16.77) the value of R from (16.76) and that of E from (16.78), we find the D^* for BLIP operation:

$$D_{\mathrm{BLIP}}^*(\lambda) = \frac{1}{2}\sqrt{\frac{\Phi_p}{\Delta \nu} \cdot \frac{\Delta \nu}{S} \cdot \frac{S}{\Phi_p Q_p}} = \frac{1}{2Q_p(\lambda)\sqrt{E_p}},\tag{16.80}$$

where we have introduced the photon irradiation

$$E_p = \frac{\Phi_p}{S}.\tag{16.81}$$

Curves of $D^*_{\mathrm{BLIP}}(\lambda)$ are included in Table 169.
The *noise factor* of a detector is then defined as

$$F = \left(\frac{R_{\mathrm{BLIP}}}{R}\right)^2 \geqslant 1. \qquad (16.82)$$

It approaches unity as BLIP operation is approached.

For a hemispherically radiating blackbody field at temperature T and with unity conversion efficiency at all wavelengths[14] (e.g. thermal detectors)

$$D^*_{\mathrm{BLIP}} = (16c_3 kT^5)^{-1/2} = \frac{2.8258 \times 10^{14}}{T^{5/2}}$$

$$= 1.81 \times 10^8 \qquad \text{at room temperature (300 K).} \qquad (16.83)$$

Here c_3 and k are the Stefan–Boltzmann and Boltzmann constants, respectively. See (4.16) and (4.41) for the values and interrelationship of these constants.

Photovoltaic detectors respond to the number of absorbed photons ($\eta\lambda/hc$), up to $\lambda = \lambda_0$, and not at all beyond. Under these conditions,

$$\boxed{D^*_{\mathrm{BLIP}} \approx \left(\frac{\lambda\lambda_c}{2c}\right)\sqrt{\pi h k T}\, e^{hc/2\lambda_0 kT}.} \qquad (16.83a)$$

(For the photoconductive mode, this must be divided by $\sqrt{2}$.)

The ultimate limitation in detection capabilities of BLIP operation detector systems, and its interrelationship with cost, are analyzed in Ref. 88.

[14] An approximation to this may be readily derived by writing the expression for $(D^*_{\mathrm{BLIP}})^{-2}$ from (16.80), substituting $M_{e\lambda} = E_p Q_p$ and, from (16.79), for the remaining Q_p. The expression for $M_{e\lambda}$ is taken from (4.47). The resulting expression may then be integrated over all λ. To this end substitute the new variable of integration

$$u = \frac{c_2}{\lambda T}, \qquad d\lambda = -\frac{c_2\, du}{u^2 T}$$

for λ, and note the definite integral [86a], [86*a]

$$\int_0^\infty \frac{u^n\, du}{e^u - 1} = n!\, \zeta(n+1),$$

where $\zeta(5) = 1.03693$. If the expressions for c_1, c_2, and c_3 are substituted [see (4.39) and (4.41)], we obtain (16.83) with 15.33 in place of 16.

The discrepancy is due to the fact that we have assumed that photons are distributed randomly whereas they obey Bose–Einstein statistics. This fact adds a factor of $[1 - e^{-u}]^{-1}$ to the integrand. When this is considered, (16.83) is obtained [87].

16.3.4 Operating Conditions

16.3.4.1 Electrical Considerations

Homogeneous Photoconductor. For most purposes the homogeneous photoconductor with ohmic contacts may be represented by a resistor whose conductance varies with illumination—often almost linearly. More generally, it may be useful to approximate the conductance of a photoconductor in a given region of illumination values (around, say, E_0) by the formula

$$G = G_0 \left(\frac{E}{E_0}\right)^\gamma, \tag{16.84}$$

where G_0 is the conductance at E_0,

       E is the illumination on the active detector area, and

       γ is measure of linearity often called "gamma." (See footnote 12.)

An equivalent circuit of the photoconductor is shown in Figure 16.33. The stray capacitance (C_s) is shown in broken lines and is not included in the following analysis because it is significant only at high frequencies where homogeneous photoconductors are generally not used.

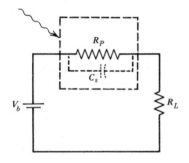

Figure 16.33 Photoconductor detector circuit. The photoconductor equivalent circuit is enclosed in broken lines.

It is readily seen that the voltages across the photoconductor and the load are, respectively,

$$V_p = \frac{V_b R_p}{R_p + R_L} \tag{16.85a}$$

$$V_L = \frac{V_b R_L}{R_p + R_L} \tag{16.85b}$$

and that the current is

$$I = \frac{V_b}{R_p + R_L}, \tag{16.85c}$$

where R_p and R_L are the resistance values of the detector and the load, respectively. Thus the power to the load is

$$P_L = V_L I = \frac{V_b^2 R_L}{(R_p + R_L)^2} \, . \tag{16.86}$$

For a given value of V_b, the load power can readily be shown to be maximized when R_L equals R_p. This is the optimum condition when Johnson noise is the limiting factor.

The power dissipation in the photoconductor is subject to some upper limit, say P_0. For any given value of R_L, this implies an upper limit, V_{max}, on the voltage that may be applied. This limit can be obtained from $(P_p = V_p I)$ and is found to be

$$V_{max} = (R_p + R_l) \sqrt{\frac{P_0}{R_p}}. \tag{16.87}$$

If, for any given value of V_b, the load resistance is to be chosen so as to maximize the minimum load power at any illumination level, it must be set equal to the maximum value consistent with the condition $R_L = R_p$, that is it must be set equal to the dark resistance, R_0, of the detector. Under these conditions,

$$V_{max} = 2\sqrt{P_0 R_0} \tag{16.88}$$

and

$$P_L = \frac{4P_0}{(a+1)^2}, \tag{16.89}$$

where

$$a = \frac{R_p}{R_0} \le 1 \tag{16.90}$$

is the detector resistance relative to its dark resistance.

Frequently we are not interested in maximizing the load power but rather in detecting a small change in the received flux, that is, we wish to detect a small change in detector resistance. In order to minimize the effect of noise in the output circuit of the detector, we must maximize the change in load power due to a small change in detector resistance. This change may be approximated by

$$\Delta P_L = A \, \Delta R_p, \tag{16.91}$$

where the coefficient

$$A = -\frac{dP_L}{dR_p} = \frac{2V_b^2 R_L}{(R_p + R_L)^3} = \frac{2P_L}{(R_p + R_L)} \qquad (16.92)$$

is the "differential gain" of the detector circuit.

To maximize ΔP_L, we must maximize A. To find the load resistance which will maximize A, for any given V_b, we equate to zero the derivative of A with respect to R_L:

$$\frac{dA}{dR_L} = 2V_b^2 \frac{(R_p + R_L)^3 - 3R_L(R_p + R_L)^2}{(R_p + R_L)^6} = 0. \qquad (16.93)$$

Thus

$$R_L = \tfrac{1}{2}R_p. \qquad (16.94)$$

On substituting this into (16.92) we find for the maximum value of A

$$A_{max} = \frac{8}{27} \cdot \frac{V_b^2}{R_p^2}. \qquad (16.95)$$

If we include the photoconductor responsivity

$$r_p = -\frac{dR_p}{d\Phi}, \qquad (16.96)$$

we may write for the maximum value of the detector device responsivity

$$r_{d\,max} = \frac{dP_L}{d\Phi} = \left(\frac{dP_L}{dR_p}\right)_{max} \cdot \frac{dR_p}{d\Phi} = \frac{8}{27} \cdot \frac{V_b^2}{R_p^2} r_p. \qquad (16.97)$$

(Note that in ir work the responsivity is occasionally defined as $dV_L/d\Phi$.)

Photovoltaic Mode. The power drawn from a semiconductor junction operating in a passive, photovoltaic mode is [see (16.67)]

$$P = -IV = -V[I_s(1 - e^{\beta V}) + I_L]. \qquad (16.98)$$

This is represented by the shaded area in Figure 16.3.

The power output may be maximized by differentiating (16.98) with respect to V and equating the result to zero. This yields

$$I_L = [(1 + V_p\beta)e^{\beta V_p} - 1]I_s. \qquad (16.99)$$

This may be solved for the voltage (V_p) providing peak power. We may also substitute the value of I_L from (16.99) into (16.67) to obtain the current for peak power and the peak power itself:

$$I_p = \beta I_s V_p e^{\beta V_p} \qquad (16.100)$$

and

$$P'_p = -I_p V_p \equiv \beta I_s V_p^2 e^{\beta V_p}. \tag{16.101}$$

The corresponding load resistor is

$$R_L = \frac{V^2}{P_p} = \frac{-e^{-\beta V_p}}{\beta I_s}. \tag{16.102}$$

The short-circuit responsivity of a photovoltaic detector is, from (16.67),

$$r_s = \frac{-I(V=0)}{\Phi} = \frac{-I_L}{\Phi}. \tag{16.103}$$

If we substitute for these their values in terms of photon flux, Φ_p,

$$-I_L = \eta_q q_e \Phi_p \tag{16.104}$$

$$\Phi = h\nu \Phi_p, \tag{16.105}$$

we find

$$r_s = \frac{\eta_q q_e}{h\nu}. \tag{16.106}$$

Biased Photodiode. When operated as a photodiode, the junction device may be represented by two parallel current sources, i_d and i_p, representing dark and photocurrent, respectively. These are shunted by an internal resistance (R_i) and capacitance (C_i), as shown in Figure 16.34). The manufacturer usually provides data concerning i_d and the open circuit time constant

$$\tau_0 = R_i C_i. \tag{16.107}$$

Information concerning i_p is usually given in terms of ampere/lumen, ampere/nit, or the radiant analogs of these. Maximum signal power out is

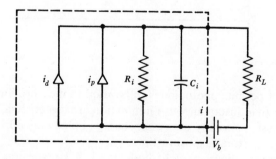

Figure 16.34 Biased photodiode circuit. The diode equivalent circuit is enclosed in broken lines.

again obtained with a matched load. Another important consideration in chosing R_L is, however, often the bandwidth requirement. The time constant of the detector circuit is

$$\tau = R_{iL}C_i,\qquad(16.108)$$

where the combined resistance

$$R_{iL} = \frac{R_iR_L}{R_i + R_L}$$

$$\approx R_L,\qquad R_L \ll R_i.\qquad(16.109)$$

Thus, operation at frequency ν_t, limits τ to $1/2\pi\nu_t$ and R_L to

$$R_L < \frac{1}{2\pi\nu_tC_i}.\qquad(16.110)$$

Light Integrating Mode. Since the dark current, i_d, can be made extremely small and the internal resistance R_i is very high, any voltage stored across the capacitor C_i will decay but slowly in the absence of i_p. See Figure 16.35 for a representative decay characteristic [89]. On the other hand, when the diode is illuminated, the stored charge will decay at the rate of (i_p/C) volt/sec.

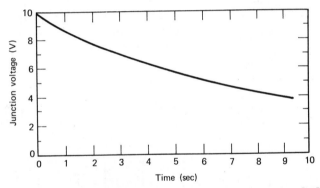

Figure 16.35 Dark decay of open-circuit photodiode. Weckler [89].

This fact permits photodiodes to be used to measure integrated exposure rather than instantaneous illumination. This is especially significant in scanning systems, where each element is sampled only during a very small fraction of the total exposure time.

A schematic diagram of a diode used in storage mode is shown in Figure 16.36. The diode, represented by its equivalent circuit enclosed in broken lines, is put in series with a voltage supply V_b, a load resistor R

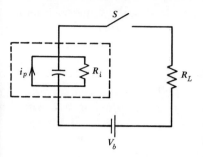

Figure 16.36 Photodiode in integrating mode, equivalent circuit.

($R_L \ll R_i$), and a switch S. When the switch is closed, the capacitor C_i quickly charges up to V_b. The switch is then opened for a fixed interval, Δt, during which the capacitor discharges by an amount essentially equal to $i_p \Delta t$ (assuming $i_p \gg i_d$, V_b/R_i). When the switch is closed again, there will be a current surge through R_L encompassing $i_p \Delta t$ coulombs—equal to the integrated charge, photo-generated during this interval. Representative discharge curves are shown in Figure 16.37 [89].

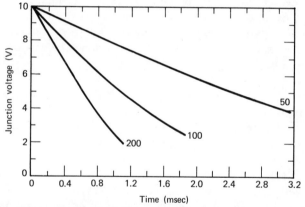

Figure 16.37 Decay of open-circuit photodiode, illuminated. The illumination values, in lux, are indicated next to the corresponding curves. Weckler [89].

16.3.4.2 Optical Considerations.

The detected signal-to-noise power ratio varies with the signal amplitude when shot noise limited and with the square of the signal amplitude when thermally, dark, or background limited. The efficiency of light collection is therefore a crucial design factor in any system involving detection near the limit of acceptable signal-to-noise ratio. The optical system should then be designed to maximize the image flux (divided by the square root of the detector area). When the detector area is limited, this implies that the image illumination be maximized and this, in turn, implies maximizing the numerical aperture of the imaging system in the image domain [see (9.134)].

In addition to using a large relative aperture, the numerical aperture can be increased by providing the image region with a medium having a high refractive index. This fact stimulated the use of "immersion optics" designs [90]. In these, the detector is mounted in optical contact with a flux-transmitting medium of high refractive index, n. (When ir radiation is to be detected, germanium with $n = 4$ may be used [1c].) If the front surface of this medium is made hemispherical, there results an increase in detectivity (D^*) by a factor of n. Instead of being placed at the center of the sphere, the detector may be placed at the third aplanatic point [cf. (9.52)] at a distance r/n beyond the center, where r is the radius of the hemisphere. This permits the use of an immersion medium of hyperhemispherical form and results in an increase of the effective area of the detector and an increase in D^* by a factor of n^2 [90]. These increases are valid only for detection limited by noise other than background; the limitations resulting from background radiation are not alleviated by this technique.

Extrinsic detectors frequently suffer from a low detectivity resulting from a low flux absorption coefficient. The detectivity of such detectors may be significantly increased by passing the flux to be detected through the photoconductive medium several times. Hence, such detectors are often placed inside integrating spheres (see Section 1.4.1) with high reflectance walls.

When the absorption is poor due to high surface reflectance, low reflection coatings may be provided [79] as discussed in connection with metal-semiconductor diodes in Section 16.3.2.2. Absorption may, in such instances, be increased also by the use of multiple total internal reflection in a manner similar to that discussed in connection with multiplier phototubes (Section 16.2.3.8).

Alternatively, the surface structure geometry may be modified, for example, by etching, to induce multiple reflections and therefore a higher absorption [91].

16.3.4.3 Mechanical Considerations. In the simplest arrangement, the photoconductive film is placed on a substrate between two evaporated (or painted-on) conducting electrodes (Figure 16.38a). To reduce the resistance of the photoconductor, it may be desirable to provide a very short path between two long electrodes. To accomplish this compactly, the electrodes may be formed as two sets of interleaved fingers (Figure 16.38b).

In photodiodes the direction of current flow is usually perpendicular to

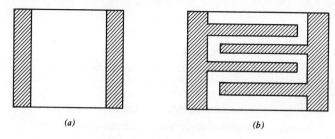

(a)   (b)

Figure 16.38 Electrode arrangements: (a) simple; (b) interleaved.

the plane of the detector. Here the substrate will be a conductor serving as one electrode, whereas the other electrode must be either transparent or cover only a small fraction of the photosensitive surface. In large-area diodes, the upper electrode may be of the form shown in Figure 16.27 to reduce the mean required current pathlength.

16.3.4.4 Thermal Considerations [55b]. We have seen in the preceding sections that many photoconductive detectors, especially those intended for the infrared region of the spectrum, must be operated at very low temperatures if the desired detectivities are to be obtained. Systems designed to provide the required cooling are called *cryogenic*.

To provide such cooling efficiently, the coolant is placed inside a Dewar[15] flask shielding it from heating by the environment. The detector is then placed in thermal contact with the coolant. A Dewar of this type usually provides a heat load of 0.1 W. For operation at liquid helium temperatures, more complex systems, such as a dual Dewar must be used. Here an outer Dewar maintains the outer wall of the inner Dewar at some intermediate temperature (see Figure 16.39b).

For operation in the far infrared, where the background radiation generated by the mount itself is likely to be significant, all components within the acceptance field of the detector must be cooled also. This implies that the detector is mounted inside a cooled enclosure that has an aperture just sufficient to accommodate the desired field of view of the detector.

The cooling may be provided in an open cycle where the coolant is discarded or in a closed cycle system where the coolant is recycled after

---

[15] Named after its originator, J. Dewar (1842–1923).

A Dewar flask, "Dewar" for short, is a double-walled jar with the region between the two walls evacuated to avoid thermal conduction and convection across the wall. The surfaces facing the vacuum are often metallized for high reflectivity, to minimize heating by radiation (see Figure 16.39a). The reader may recognize his thermos bottle as a Dewar.

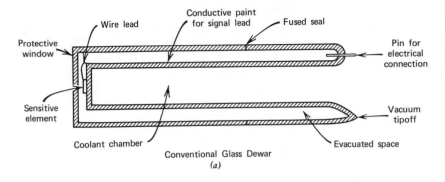

Conventional Glass Dewar
(a)

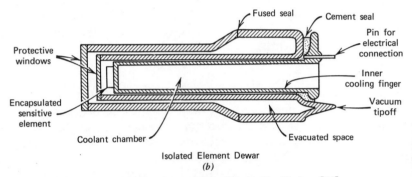

Isolated Element Dewar
(b)

Figure 16.39   Dewar flasks: (a) simple; (b) double. Hudson [55].

use. Characteristics of the more popular coolants are listed in Table 172 [55b].

In open cycle systems, as the coolant is heated and evaporated, the Dewar may simply be refilled as needed from a ready supply of coolant. In the more sophisticated *Joule–Thompson refrigerator* (or *cryostat*), the coolant gas is introduced at very high pressure, to be "throttled," that is, permitted to expand freely in the vicinity of the detector. This throttling process cools the gas to the desired temperature. On its way out, the throttled gas, still very cold, cools the tube carrying the high-pressure entering gas.

When a device must operate unattended for extended periods of time, a *solid refrigerant* may be used. A metal rod, acting as a thermal conductor, carries the heat from the detector, the "heat load," to the coolant, which is placed inside a well-insulated container. The temperature is controlled by controlling the pressure inside the container. It has been estimated that a 12-kg package can maintain a 0.1 W heat load at 88 K for a full year [92].

In the closed-cycle refrigeration system, the spent coolant is recompressed for reuse, as in conventional home refrigerators. The Joule–Thompson system lends itself to such use. More efficiently, the *Claude refrigerator*, expands the compressed gas in a working mode. In the *Stirling refrigerator*, two piston-cylinder sets are used, alternating as compressors and expanders. Closed-cycle refrigerators are usually rated in terms of the *coefficient of performance* (COP) which is defined as the ratio of the refrigeration power to the total required input power.

Detectors used in space vehicles, may be cooled by means of *radiative-transfer coolers*. These consist of a good thermal conductor ending in an efficient radiator, usually hemispherical in shape and blackened for high emissivity. When the radiator is pointed at empty space and shielded from radiation sources (including the vehicle on which it is mounted) it can provide effective cooling. Such coolers have been used to maintain detectors at 195 K, although considerably lower temperatures should be obtainable.

Cooling to 195 K by means of the *Peltier effect*, too, has been used. The Peltier effect is the inverse of the Seebeck effect on which the thermocouple is based. In the Peltier effect, a current is passed through a junction, usually a bismuth-telluride p-n junction, cooling or heating it, depending on the direction of current. When used for cooling, a maximum temperature differential of

$$\Delta T_{\max} = 0.5 z T_j^2 \tag{16.111}$$

can be obtained. Here T_j is the temperature of the cooled junction and z is the *thermoelectric figure of merit*

$$z = \frac{\alpha^2}{\rho k}, \tag{16.112}$$

where α is the thermoelectric coefficient (in V/K).
  ρ is the resistivity (in Ω/cm), and
  k is the thermal conductivity (in W/K $\cdot$ cm).

For bismuth-telluride, z equals 3×10^{-3}/K. For cooling to low temperatures, several Peltier junctions must be cascaded and the high electric power input requirements soon become prohibitive, especially in terms of removing the heat resulting from this input.

16.4 IMAGE DETECTORS [93], [94]

Up to this point we have discussed photoelectric detectors that measure total flux and do not, of themselves, respond usefully to changes in flux distribution. In conjunction with scanning devices, mechanical or optical, such detectors are capable of investigating the distribution of flux in an image. There are, however, other photoelectric detection devices that are capable of transforming an input irradiation distribution into a new one or into a sequence of electrical signals. The former type of device is usually called an "image tube" and the latter an "image pick-up tube", (television) camera tube, or "signal generating tube." (The latter nomenclature implies that only electrical signals are signals—a rather narrow definition which the reader, hopefully, will at this point no longer endorse.) A more appropriate nomenclature might be optical-optical and optical-electrical image transducers [15].

In many of these devices, the optical image is first converted into an electron image, that is, an electron cloud whose density distribution follows that of the optical irradiation distribution. The first subsection here is therefore devoted to a brief introduction to electron optics. The second is devoted to image tubes (optical-optical transducers); these have already been introduced in conjunction with luminescence (Section 6.32), but this chapter offers a welcome opportunity to report on some more recent trends. Cathode ray image tubes (optical-electrical) based on photoemissive and photoconductive principles are discussed in Sections 16.4.3 and 16.4.4, respectively, and solid-state image transducers in Section 16.4.5. Noise and resolution characteristics are treated in the final section, 16.4.6.

16.4.1   Electron Optics

Many of the image-forming detectors employ electron optical lenses and deflecting fields and therefore we review here briefly the elements of this discipline, especially as it applies to image tubes. To avoid unnecessary complications we neglect relativity effects throughout.

16.4.1.1   *Motion of Charged Particles in Electric and Magnetic Fields* [95]

Uniform Electrostatic Field. A charged particle in an electric field experiences a force given by [see (2.2)]

$$\mathbf{F} = q\mathbf{E}, \tag{16.113}$$

where $\mathbf{E}$ is the electric field strength and q is the charge on the particle. The resulting acceleration is given by

$$\dot{\mathbf{v}} = \frac{\mathbf{F}}{m} = \frac{q\mathbf{E}}{m}, \tag{16.114}$$

where $\mathbf{v}$ is the velocity and m the mass of the particle. A particle falling from rest through a potential difference, V, acquires an amount of kinetic energy given by

$$\tfrac{1}{2}mv^2 = qV. \tag{16.115}$$

Thus it acquires a velocity

$$v = \sqrt{\frac{2qV}{m}}. \tag{16.116}$$

To find the trajectory of the particle, we resolve its velocity into two components, parallel and perpendicular to the field, respectively. The former changes at a rate given by (16.114) and the latter remains constant. Thus, if we choose Cartesian coordinates, with the field in the z direction, the velocity in the x–z plane, and the origin at the electron position at time $t = 0$, we arrive at the following set of differential equations:

$$\ddot{x} = 0 \tag{16.117}$$

$$\ddot{z} = \frac{qE}{m}, \tag{16.118}$$

with boundary conditions

$$x(0) = z(0) = 0 \tag{16.119a,b}$$

$$\dot{x}(0) = v_{xo}, \qquad \dot{z}(0) = v_{zo}, \tag{16.119c,d}$$

where v_{xo} and v_{zo} are the original x and z components of the velocity. Solution of these equations yields

$$x = v_{xo}t, \tag{16.120}$$

$$v_z = v_{zo} + \frac{q}{m}Et, \tag{16.121}$$

$$z = v_{zo}t + \frac{1}{2}\frac{q}{m}Et^2. \tag{16.122}$$

We obtain the form of the trajectory, by eliminating t from (16.120)–(16.122):

$$z = \frac{v_{zo}}{v_{xo}}x + \frac{q}{m}\frac{E}{v_{xo}^2}x^2, \tag{16.123}$$

a parabola.

Uniform Magnetic Field. In a magnetic field, the force is normal to both the field and the velocity and can be written as

$$\mathbf{F} = q\mathbf{v} \times \mathbf{B}, \tag{16.124}$$

where $\mathbf{B}$ is the magnetic flux density in units of tesla (T), in the MKSA system, which we have adopted here.

Consider first velocities perpendicular to the magnetic field. The magnitudes of the force and acceleration are then given by

$$F = qvB, \tag{16.124a}$$

$$\dot{v} = \frac{q}{m} vB. \tag{16.125}$$

They will be directed normal to the velocity and hence will not affect the kinetic energy of the particle. Force will be constant and centripetal, causing the particle to move in a circular path with constant speed. Such a particle experiences an acceleration

$$\dot{v} = \frac{v^2}{r} \tag{16.126}$$

normal to both velocity and field vectors. On equating (16.125) and (16.126) we find the radius of the trajectory

$$r = \frac{v}{B} \frac{m}{q} \tag{16.127}$$

and the time for a single revolution

$$t_r = \frac{2\pi r}{v} = \frac{2\pi m}{qB}. \tag{16.128}$$

For an electron ($q/m = 1.758796 \times 10^{11}$ C/kg), this leads to

$$t_r = 3.57243 \times 10^{-11} B^{-1} \text{ sec}, \tag{16.128a}$$

where B is in tesla.

Note that the period is independent of particle velocity.

In the more general case, where particle velocity is not perpendicular to the magnetic field, it may be resolved into two components: v_z parallel to the field and v_s perpendicular to it. Then (16.128) still apply, if we substitute v_s for v. Since the velocity component v_z is unaffected by the field, it remains constant. Thus in a uniform magnetic field, in the absence of an electric field, the general particle follows a path shaped like a circular helix with constant pitch

$$Z = v_z t_r \tag{16.129}$$

(see Figure 16.40a).

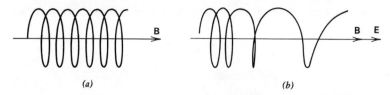

Figure 16.40 Electron path in magnetic field: (a) without electric field; (b) with axial electric field.

Parallel Electric and Magnetic Fields. When magnetic and electrostatic fields operate simultaneously, the particle motion can be analyzed by resolving the velocity and the electric field each into two components, parallel, and transverse to the magnetic field, respectively. The arrangement of special interest to us here is that of the parallel electrostatic and magnetic fields. Here, the component transverse to the fields remains unaffected by the electric field, while the parallel component, v_z, unaffected by the magnetic field, experiences a constant acceleration,

$$\dot{v}_z = \frac{F}{m} = \frac{qE}{m},$$ (16.130)

due to the electric field.

Thus the particle path describes a circular helix with a constantly changing pitch, given by [see (16.121)].

$$Z = v_z(t)t_r = \left(v_{zo} + \frac{q}{m} Et\right)t_r,$$ (16.131)

where t is the time elapsed since the particle had z-velocity, v_{zo} (see Figure 16.40b).

16.4.1.2 The Electron Optics Analogy [96a]. According to the Maupertuis principle,[16] any particle under the influence of a force field follows the path of least action, in going from point P_1 to point P_2:

$$\delta\left[\int_{P_1}^{P_2} p \, ds\right] = 0,$$ (16.132)

where p is the momentum,
  s is the displacement, and
  δ denotes the variation due to an infinitesmal change in the particle path.

[16] P. de Maupertuis (1747).

Fermat[17] had previously formulated a similar principle concerning a ray of light, whose path is such as to make the transit time an extremum:

$$\delta t = \delta \left[\int_{P_1}^{P_2} \frac{n}{c} \, ds \right] = 0, \tag{16.133}$$

where n is the refractive index and
c is the speed of light *in vacuo*.

If we now consider the path of an electron in an electric field and define as zero the potential at which the electron velocity vanishes, we find the velocity as given by (16.116) and the electron momentum

$$p = mv = \sqrt{2qmV}, \tag{16.134}$$

equal to a constant times $\sqrt{V}$.

Without affecting the validity of (16.132) and (16.133), we may drop the constant factors in their integrands and write

$$\delta \left[\int_{P_1}^{P_2} \sqrt{V} \, ds \right] = 0, \tag{16.132a}$$

$$\delta \left[\int_{P_1}^{P_2} n \, ds \right] = 0. \tag{16.133a}$$

Comparison of these equations shows that the path of a charged particle in an electric field is identical with that of a ray of light in a variable medium, with the refractive index of the medium proportional to the square root of the potential of the electric field.

Thus an electrostatic field with axial symmetry is to an electron, what a lens is to a ray of light. The former is therefore often called an "electron lens." Usually the potential changes continuously so that the electron lens is analogous to an optical lens with continuously changing refractive index.

In Appendix 16.1 we show that paraxial imaging requires that at every transverse plane each paraxial ray be deviated through an angle proportional to the angle it makes with the axis and that this does, indeed, occur in an axially symmetric electrostatic field. Thus, any such field images a paraxial electron source point into an image point that may be real (all electrons cross the axis at one point) or imaginary (the electron paths converge into a single point when projected backward).

If the potential at the object differs from that at the image, this corresponds to a lens with object and image in spaces with differing refractive indices; such a lens is referred to as an *immersion lens*.

[17] P. de Fermat (1657).

The imaging in the paraxial region being Gaussian, every electron lens can be characterized by three pairs of planes. The planes where collimated beams of electrons entering the lens are focused are called *focal planes*; there are two of these corresponding to the two directions in which the beam may enter. There exists another pair of planes that are imaged on each other at unity linear magnification; these are called *principal planes*. The two conjugate planes where the angular magnification is unity, are called *nodal planes* (see footnote 1, Chapter 15). The distance from a principal plane to the corresponding focal plane is the *focal length* of the lens. When the spacing between the principal planes is negligible compared to the focal length, the electron lens may be treated as a thin lens.

The axially symmetric fields discussed so far are useful in imaging. For deflection purposes, transverse fields are employed. A potential gradient that is essentially transverse to the electron beam corresponds to a refractive index gradient transverse to the beam, so that such a field acts much like an optical prism deflecting the beam.

16.4.1.3 Electron Optical Aberration. Just as in light optics, electron rays passing an electron lens at a finite distance from the axis generally experience a focal length different from that of the paraxial rays. This is called spherical aberration and may be described by

$$y = a_3 \alpha^3 + a_5 \alpha^5 + \ldots, \qquad (16.135)$$

where y is the blur radius, that is, the displacement from the axis of the ray intercept in the paraxial image plane, for a ray from an axial object point,

α is the angular aperture (for a collimated beam entering, the aperture radius), and

a_i is the ith order aberration coefficient.

Often only the first term need be considered. The other Seidel aberrations: coma, field curvature, astigmatism, and distortion, all have their obvious analogs in electron optics also, when off-axis object points are involved.

To find the analog to chromatic aberrations, we note that the "refractive index" of the electron lens was proportional to $\sqrt{V - V_0}$, where V_0 is the potential at which the electron velocity would vanish. When the electron beam contains electrons of various initial velocities, this implies different values of V_0 and, therefore, of "refractive index." This is fully analogous to the causes of chromatic aberrations in light optics. Hence degradations resulting from such velocity differences are called chromatic aberrations.

As shown in Section 16.4.1.6, these effects play an important role in magnetic imaging, but occasionally they must be considered also in electrostatic lenses. For instance, when the electron passes through a transverse electrostatic field, it accumulates a certain transverse momentum component. The magnitude of this component is directly proportional to the time the particle spends in the field and this, in turn, is inversely proportional to its axial velocity. Thus electrons with different velocities will experience different amounts of "refraction."

16.4.1.4 Electron Optics in Image Tubes [97]. In image tubes including, among others, cathode ray display tubes, image intensifier tubes, and television "camera" tubes, electron optics are used for three essentially different functions.

1. *Imaging.* An electron density distribution in one plane is reproduced in another plane. Here the exact distribution of electrons is significant.

2. *Beam Concentration.* The electrons in a beam are concentrated into a small cross section, with no regard to the exact distribution within this cross section.

3. *Deflection.* The electron beam is deflected and guided to various desired locations.

Imaging is the essential function in image intensifier tubes and in the "write" stage of television camera tubes. Beam concentration, in conjunction with deflection, is used in cathode ray tubes and in the "read" stage of television camera tubes.

All of these functions can be performed either electrostatically or magnetically and in the following sections we briefly survey electron-optical techniques used in such image tubes.

16.4.1.5 Electrostatic Lenses

Imaging Lenses. Since the velocity of electrons is relatively low at the time of their emission, photocathodes acting as sources of electrons are generally at a relatively low potential, in view of our convention referring all potentials to the zero-velocity potential as zero. This corresponds to an object in a medium of near zero refractive index and poses special problems.

One form of such an electron-optical system, consisting of three concentric spherical surfaces containing, respectively, cathode, anode, and image surface has been analyzed exhaustively [98]. However, efforts to approximate in practice the required transparent anode have met with serious difficulties so that another form, as shown in Figure 16.41 [99] is usually used. This figure shows the photocathode at zero potential and the

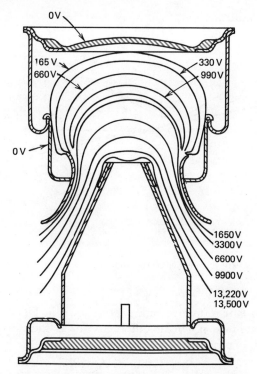

Figure 16.41 Type 6914 ir image converter tube, potential distribution. Wreathall [99].

equipotential surfaces at progressively higher potentials approximating
spherical surfaces. The resulting electron ray paths and field curvatures,
both tangential and sagittal, are illustrated in Figure 16.42 [99].

We gain the ability to focus without affecting the potential of cathode
and anode, by insulating the middle section of the electron-optical
structure as shown in Figure 16.43 [100], and controlling its potential
independently. Also shown there are the image surfaces, again both
tangential and sagittal, obtained with various values of potential on the
focusing structure. The ability to control image magnification in this
manner is also evident in this figure; this only requires relocating the
screen for each change in focusing potential. Instead, we may control the
screen potential independently and achieve true varifocal ("zoom") action
[100].

Symmetrical Two-Tubes Lens. A number of electrostatic lenses may be
used in the "read" section of image tubes. One of these consists of two
identical coaxial tubes separated by but a small gap and maintained at
two different potentials V_1 and V_2. The equipotential lines, indicating

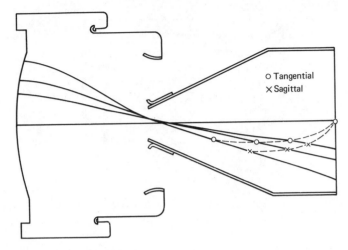

Figure 16.42 Type 6914 ir image converter tube, chief rays and field surfaces. Wreathall [99].

surfaces of constant refractive index, are shown in Figure 16.44 [96b], where the percentage potential difference (d) is indicated. The actual potential at the line marked "d%" is

$$V(d) = \frac{(100-d)V_1 + dV_2}{100}.$$ (16.136)

We may consider this analogous to an onion-like optical lens formed in layers of differing refractive index. The lens is convex and has a relative refractive index larger than unity to the left of the mid-plane and a relative refractive index less than unity to the right of it, if V_2 is larger than V_1. (The index is taken relative to the space in which the lens is

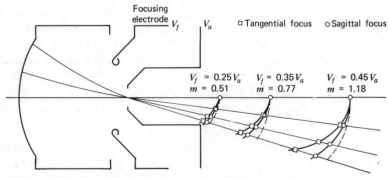

Figure 16.43 Triode image tube. Vine [100].

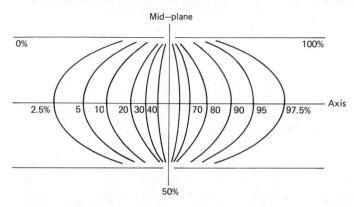

Figure 16.44 Symmetrical two-tube lens, potential distribution. Klemperer [96].

immersed, that is, the potential of the enclosing tube.) Thus the left half has positive and the right half has negative focusing power. The net effect of the lens, however, is a positive focusing action. The locations of the focal (F_1, F_2) and principal (P_1, P_2) planes are indicated in Figure 16.45 [96c] for the case of $V_1 = 1000$ V and $V_2 = 5000$ V. The distances of these planes from the midplane and focal lengths (f_1, f_2) are tabulated in Table 173 [96d] for various voltage ratios V_2/V_1. Note that there the values subscripted "1" relate to electrons being decelerated, that is, passing from right to left in Figure 16.44. Those subscripted "2" are for electrons passing in the direction of acceleration, from left to right in that figure. For the determination of magnification, the positions of the nodal planes (N_1, N_2) are significant. They can be found from the data of Table 173 [96d] by means of the simple relations [96e].

$$N_1 F_1 = P_2 F_2, \qquad N_2 F_2 = P_1 F_1. \qquad (16.137)$$

Three-Tube Lenses. The popular three-tube lens, too, is often used in crt's and in the read section of camera tubes. It is illustrated in Figure 16.46 [97]. Two equal coaxial tubes, diameter D_1, are separated by

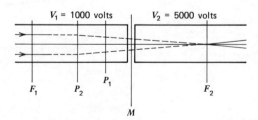

Figure 16.45 Symmetrical two-tube lens, Gaussian planes. Klemperer [96].

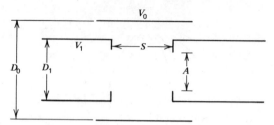

Figure 16.46   Three-tube lens, schematic.

distance S. This gap is enclosed by the outer tube, diameter D_o, coaxial with the inner tubes. On the inner tubes, the ends facing each other, are partly closed off, leaving a circular aperture of diameter A. Both inner tubes have the potential V_1 applied to them, while the potential V_o is applied to the outer tube. Both focal length, f (relative to D_o) and transverse spherical aberration, Δy, (relative to f) are given in Figure 16.47a and b [97] as functions of A (relative to D_o). All these data are for the configuration:

$$D_1 = S = 0.6D_o.$$

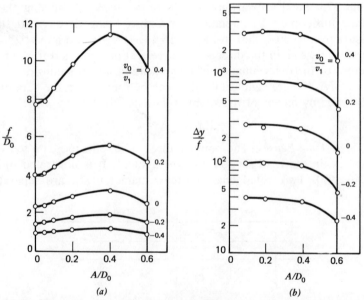

Figure 16.47   Three-tube lens: (a) focal length (relative to outer tube diameter); (b) spherical aberration (relative to focal length), both as functions of relative aperture. Vine [97].

Note that the abscissa $(A/D_o = 0.6)$ corresponds to completely open inner tubes.

To understand the action of this lens, we return to our analogy of $n \sim \sqrt{V}$. In the gap between the inner tubes, the influence of V_o, primarily, is felt, whereas inside the inner tubes, their potential, V_1, controls. Thus, as long as $V_o < V_1$, we generate, in the gap region, a lens having a refractive index less than that of the surrounding medium. The equipotential surfaces bounding the gap bulge into it, so that the surfaces, V, in the gap $(V_o < V < V_1)$ form a biconcave lens having a relative refractive index below unity. Such a lens exhibits a positive focusing action.

This lens permits focusing control, by means of V_o, such focusing leaves unaffected the electrostatic field outside the lens.

16.4.1.6 Magnetic Focusing. Focusing by a magnetic field is based on the result (16.128) that an electron traveling in a uniform magnetic field returns to the original flux line after a fixed period, t_r, regardless of its original momentum. (The original flux line is the line, parallel to the magnetic field, passing through the initial electron position.) Note that this is true even for an electron accelerated in a uniform electrostatic field parallel to the magnetic field.

This implies that the electrons emitted from a given point will all return to a second point, provided they are placed in a uniform magnetic field with no transverse field components and provided that the component (v_z) of their velocity parallel to the field is identical for all the electrons. Indeed, this focusing repeats itself periodically, as long as the field conditions apply. The image points are displaced from the original object point by a distance [see (14.129)]

$$z_N = v_z t_r N, \qquad N = 1, 2, 3, \ldots, \tag{16.138}$$

where N is the number of loops completed by the electron before reaching the image point.

If an electrostatic field accelerates the electron along the direction of the magnetic field, this becomes, using (16.122) and (16.128) with $t = Nt_r$:

$$z_N = Nt_r \left(v_{zo} + \frac{qENt_r}{2m} \right)$$

$$= \frac{3.57 \times 10^{-11} N}{B} \left(v_{zo} + \frac{\pi EN}{B} \right). \tag{16.139}$$

Chromatic Aberration. Inspection of (16.139) shows that, with constant electric and magnetic fields, only variations in v_{zo} can cause incomplete focusing, that is variations in z_N. Such variations do, as a rule, exist and the resulting defocusing is called chromatic aberration, as mentioned in Section 16.4.1.3. Such differences exist for two reasons:

1. Differences in the initial velocity at which the particles are emitted from the cathode,

2. Differences in the direction at which the electrons enter the field. Neglecting differences of Type (1), all electrons have the same velocity (v) on arriving at the field boundary, since all started with the same initial velocity and all have passed through the same potential difference. The component parallel to the field, however, is given by

$$v_{zo} = v_o \cos \theta,$$

where θ is the angle the electron velocity makes with the fields. Thus v_{zo} depends on θ and this, in turn, gives rise to chromatic aberration of amount

$$\delta z_N = N t_r \, \delta v_{zo} = 2\pi \frac{m}{q} \frac{N}{B} (\cos \theta_1 - \cos \theta_2) v_o$$

$$\approx 1.786 \times 10^{-11} \frac{N}{B} v_o (\theta_2^2 - \theta_1^2), \qquad \text{for} \qquad \theta_1, \theta_2 \ll 1. \qquad (16.140)$$

Taking the parallel electron as reference and assuming small angles θ, this yields

$$\delta z_N \approx 1.786 \times 10^{-11} \frac{N v_o \theta^2}{B}. \qquad (16.140a)$$

16.4.1.7 Electron Prisms. When an electric or a magnetic field is transverse to the electron beam, it will bend it, much as a prism bends a light beam. Such fields are therefore used to deflect electron beams.

Electrostatic Deflection. Assume an electron that has dropped from rest through a potential V and consequently acquired a velocity

$$v_o = \sqrt{2 \frac{q}{m} V}. \qquad (16.116)$$

If this electron now enters a uniform electric field, E, transverse to the velocity, its trajectory will be given by

$$\dot{v}_x = \frac{qE}{m}, \qquad \dot{v}_z = 0,$$

where we have placed the x axis parallel to the field and the z axis parallel to the initial velocity. This is identical to the situation we treated in Section 16.4.1.1, starting with (16.117), except that we interchange x and z and that v_{z_o} there vanishes. Thus we find [see (16.123)] that

$$x = \frac{1}{2} \frac{q}{m} \frac{E}{v_o^2} z^2 = \frac{Ez^2}{4V}, \qquad (16.141)$$

where we have used (16.116) to eliminate v_o.

If we now denote by Z the extent of the deflecting field in the z direction, we find the exit time and velocity

$$t_e = \frac{Z}{v_o}, \qquad (16.142)$$

$$v_{xe} = \frac{q}{m} \frac{EZ}{v_o}, \qquad v_{ze} = v_o. \qquad (16.143)$$

Thus the particle velocity is deflected through an angle θ given by

$$\tan \theta = \frac{v_{xe}}{v_{ze}} = \frac{qEZ}{mv_0^2} = \frac{EZ}{2V} = \frac{\Delta x}{\frac{1}{2}Z}, \qquad (16.144)$$

where we have again used (16.116) to eliminate v_o and (16.141) to eliminate E and V (see Figure 16.48a). This implies that the electron will leave the deflecting field as if coming from a point on the extension of the entrance path, at the midpoint of the field region.

Magnetic Deflection. If a uniform magnetic field is transverse to it, the electron beam will follow, as long as it is within the field, a circular path with a radius given by (16.127) (see Figure 16.48b). It is immediately

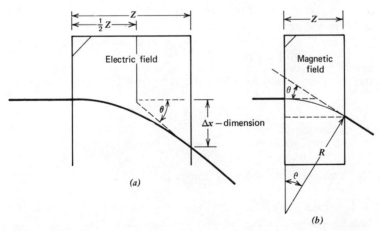

Figure 16.48 Electron trajectories: (a) in electric field; (b) in magnetic field.

evident from this figure that the deflection angle θ is given by

$$\sin \theta = Z/R = \frac{q}{m}\frac{B}{v}Z, \qquad (16.145)$$

where Z is the z component of the electron path inside the magnetic field.

It should be noted that actual fields are never uniform and that, if edge effects of the fields are considered, the path will be considerably more complicated, both in electric and magnetic deflection.

16.4.1.8 Proximity "Focusing." When the source of the electrons is close to the receiving screen and the electrostatic field there is very high, it may be possible to dispense with focusing altogether. Paradoxically this procedure is called *proximity focusing*.

For high acuity, large fields must be applied. This increases the sensitivity of the photocathode but, simultaneously, increases the dark current, both due to the Schottky effect (Section 16.1.2). In practice, a suitable compromise must be found [101].

When a Maxwell distribution is assumed for the electron energy and the cathode radiates in a Lambertian manner, the mtf resulting from proximity focusing can be found in a closed form (see Ref. 103):

$$\mathsf{T}(\nu) = \exp - \left(\frac{\nu}{\nu_0}\right)^2, \qquad (16.146)$$

where

$$\nu_0 = \frac{\sqrt{q_e V/kT}}{2\pi d}, \qquad (16.147)$$

   q_e is the charge of the electron,
   k is Boltzmann's constant,
   V, d are the cathode-screen voltage and spacing, respectively, and
   T is the cathode temperature.

Actual mtf's are found to follow the form [103]

$$\mathsf{T}(\nu) = \exp - \left(\frac{\nu}{\nu_0}\right)^{1.8}; \qquad (16.148)$$

see (16.153).

16.4.2 Optical–Optical Image Tubes [93c], [102]

As described in Section 6.3.2, (optical–optical) image tubes consist of a photoemissive cathode, an electron optical system, and a phosphor screen; they are called *image intensifier* tubes if they serve to increase the image luminance and *image converter* tubes if they serve primarily to change the wavelength of the image flux. The basic structure can be seen from Figure 16.42.

16.4.2.1 Single-Stage Image Tubes

Proximity Focused Tubes. The simplest image tubes are constructed with proximity focusing (see Section 16.4.1.8). The phosphor screen is placed close to the photocathode and a very high field is applied between them. As given there, the mtf due to such imaging is almost Gaussian, and a representative mtf curve for such a tube is shown in Figure 16.49 [103], [104].

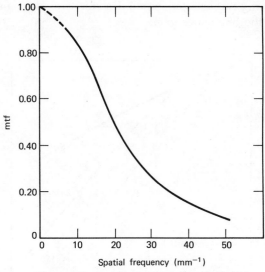

Spatial frequency (mm⁻¹)

Figure 16.49 Mtf of proximity-focused image intensifier. Light: white; cathode: K_2CsSb; field: 10 kV over 1.25 mm gap. Goodson *et al.*, Garfield *et al.* [103], [104].

Production of proximity focused tubes is beset by many difficulties, such as activating the cathode while it is in close proximity to the phosphor screen and preventing sputtering of the aluminum backing of the phosphor screen under the high electrostatic field used (ca. 10 kV/mm). However, these seem to have been overcome to a great extent. Useful diameters up to 40 mm and, perhaps, 75 mm are possible. A representative tube, operated at 10 kV/mm has a luminance gain of 50, resolves 50 mm⁻¹, and has a background luminance corresponding to 0.2 μlux. It has an S25 photocathode and a P20 phosphor screen [104].

Single-Stage Electrostatic Tubes. The mtf of a representative electro-statically focused tube, together with those of the screen and electron optics individually, is shown in Figure 16.50 [105]. These curves were obtained with red light (0.645 μm) and at a point at one third of full field angle. The strong image field curvature is not matched by that of the screen. Thus on-axis the performance is limited by the screen, but off-axis by the electron optics [105].

With such tubes luminance gains of up to 200 times are feasible with simple one-to-one imaging. By producing on the phosphor an electron image smaller than the original one at the cathode, the electron current density, and hence the screen brightness, may be increased. This requires, however, either an increase in the active cathode diameter or special viewing optics at the output if it is to yield significant advantages.

In the following two subsections we describe two additional techniques for increasing the luminance gain.

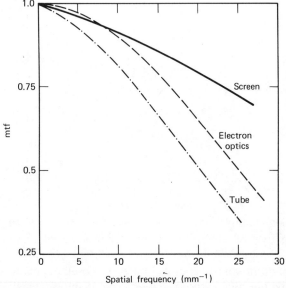

Figure 16.50 Mtf curves of screen, electron optics, and complete tube for electrostatically focused image tube, 4 mm from center of screen. Stark *et al.* [105].

16.4.2.2 Cascaded Tubes. Additional luminance gain can be obtained by cascading a number of image tubes, the output of one serving as the input to the next. Rather than actually arranging tubes in such a manner, it may be advantageous to build a single tube containing a number of intermediate stages. These, then, contain intensifier screens, or dynodes, which are a sandwich of a phosphor and a photocathode. The gain of such a dynode may be 150 (see Figure 16.51 [106]) and several of these may be cascaded. Considerations of gain and resolution conflict in defining the optimum phosphor thickness. With two such dynodes, an overall gain of close to 10^6 may be obtained, so that individual photoelectrons produce visible flashes [106]. Although, with this method, there is some loss in resolution and detective quantum efficiency, the increased responsivity may make it worthwhile.

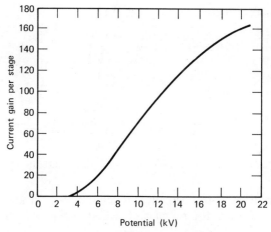

Figure 16.51 Dynode current gain as function of incident electron energy. McGee, *et al.* [106].

The mtf of such a tube (and that of some of its components) is shown in Figure 16.52 [107]. This tube has a 40 mm clear diameter and uses an S11 photocathode and a P11 phosphor. Its temporal frequency response drops from 0.7 to 0.1 as the frequency rises from 430 to 8300 Hz [107].

This tube is magnetically focused; with the more convenient electrostatic focusing, difficulties arise from the fact that there the object and image fields are inherently concave toward each other. This prevents a simple matching of the phosphor screen of one stage to the photocathode of the next. Fiber optics are used successfully to overcome this difficulty. Airtight fiber face plates, concave on the inside and flat on the outside, are used for a modular construction in which the user can assemble a number of stages according to his needs. See Figure 16.53 [108] for a two-stage tube of this type. The luminance gain and transfer function of this device are shown in Figures 16.54 and 16.55 [108], respectively, and resolving power, as a function of illumination, for three values of object modulation, in Figure 16.56 [108]. In this unit, successive images have diameters

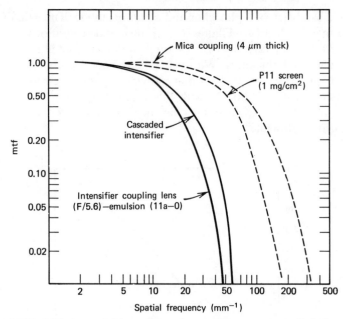

Figure 16.52 Mtf of cascaded intensifier and some of its components. Delori *et al.* [107].

of 40, 25, and 16 mm, respectively, yielding an overall magnification of 0.4. Cathode materials for initial and dynode stages are S25 and S20, respectively.

In another cascading technique, electron multiplication screens are used as dynodes at intermediate image planes. Each electron incident on one side of the very thin screen gives rise to several secondary electrons

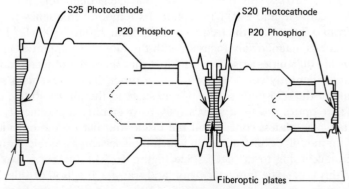

Figure 16.53 Fiber-coupled, two-stage intensifier with demagnification, Collings *et al.* [108].

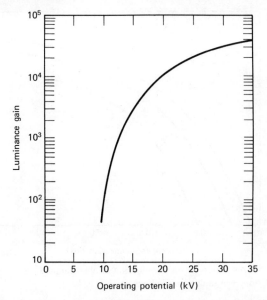

Figure 16.54 Luminance gain vs. potential of fiber-coupled, two-stage intensifier. Collings *et al.* [108].

emitted from the other side. Such screens, too, cannot be concave toward both sides simultaneously, precluding electrostatic focusing. But even with magnetic focusing, the primary electrons traversing the screen have velocities much higher than those of the secondary electrons and are therefore not focused satisfactorily ("chromatic aberration"). This method does not seem to be practical.

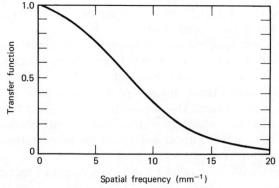

Figure 16.55 Transfer function (square wave) of fiber-coupled, two-stage intensifier. Collings *et al.* [108].

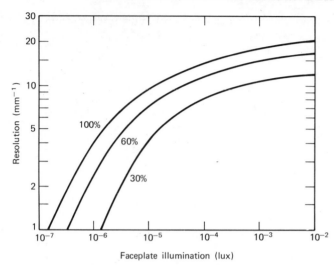

Faceplate illumination (lux)

Figure 16.56 Input resolution versus illumination of fiber-coupled, two-stage intensifier for various values of object modulation. Collings *et al.* [108].

16.4.2.3 The Microchannel Plate in Image Tubes. The channel electron multiplier (Section 16.2.3.2) can be used effectively in image forming tubes to amplify the current to each image element. A large number of such channels, each drawn to a very small diameter (approximately 50 μm) are assembled into a microchannel plate (mcp). When a high voltage is applied across such a plate, a single input electron may yield thousands of output electrons.

Manufacture. The mcp's are generally made of a silica glass, made conducting by the addition of lead and bismuth [109]. The dependence of the secondary electron yield on the angle of incidence and the glass composition, has been studied [110].

The channels are usually drawn in two stages: first individual tubes are drawn to a diameter of approximately 1 mm. These are then assembled and fused into groups of several hundred and further drawn to the final diameter (20–50 μm). These groups are then assembled to the desired final plate area and fused. This assembly is then sliced to the desired thickness, usually approximately 50 times the channel diameter. Soluble glass cores may be used to avoid collapse of the tubes in the outer zone; this is etched out after completion of the plate. The channels are connected electrically in parallel by the evaporation of metal electrodes on the two faces of the plate [111], [112], [112*].

In an alternative method, channels are etched into very thin glass platelets (e.g., microscope cover slips) and these are then stacked, fused, and sliced to the desired plate thickness [113].

Transfer Characteristic. When a high voltage is applied to a mcp, entering electrons produce secondary electrons on striking the wall. The field within the channels is maintained uniform by the leakage current flowing along the channel walls. For signal current values small compared to the leakage current, the gain is quite constant and a function of the applied voltage; see Figure 16.57 [111]. As the output current rises, the electron current flowing along the wall is reduced by the heavy secondary emission near the output end of the plate. Thus the field along the channels becomes nonuniform and the gain drops. This phenomenon becomes significant when the channel current attains approximately 10% of the leakage current.

As in all high-voltage vacuum devices, here, too, positive ions are produced due to the collisions of electrons with residual gas atoms. These ions are drawn to the input end of the plate and may (a) produce additional secondary electrons, increasing the gain (as in gas diodes, see Section 16.2.2) or (b) reach the photocathode and cause serious damage

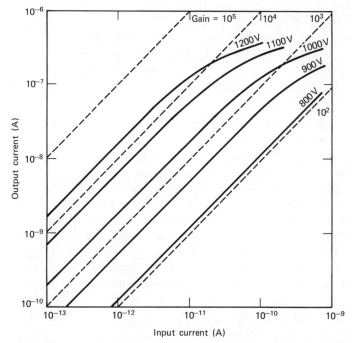

Figure 16.57 Transfer characteristics of a channel plate. Manley *et al.* [111].

there [114]. When tube geometry permits it, these ions may be drawn off before reaching the cathode. Other techniques for minimizing ion damage involve curving the channels or applying the field obliquely.

Noise Level. Noise genesis in a channel multiplier is similar to that in a multiplier phototube, except that the variable number of multiplication stages increases the noise. Experimentally, the output pulses for single input electrons have been found to be distributed according to a negative exponential [111]. This yields a signal variance of $\bar{G}^2$ for each input electron, where $\bar{G}$ is the mean value of the gain (see Section 18.2.2.2). That is, the specific noise[18] of the mcp is

$$R_p^{-2} = \frac{\eta n_i \bar{G}^2}{(n_i \eta \bar{G})^2} = \frac{1}{\eta n_i}, \tag{16.149}$$

where n_i is the number of electrons incident on an image element and
   η is the efficiency with which these enter the mcp, as limited by the effective channel area (electrons striking the wall cross section portion of the plate, do not contribute to the output).

Since the specific noise at the input (assuming a Poisson distribution) is also

$$R_i^{-2} = \frac{1}{\eta n_i}, \tag{16.150}$$

the resulting signal-to-noise ratio is [see (18.24)]

$$R_0^2 \doteq [R_i^{-2} + R_p^{-2}]^{-1} = \tfrac{1}{2}\eta n_i \tag{16.151}$$

—only one half of that at the input. The input signal-to-noise ratio already had been reduced by η, which is also of the order of magnitude of 0.5. In practice, the signal-to-noise power ratio at the output is only one quarter of that at the input [111], [115].

By using ratios length/diameter larger than 50, saturation effects become evident earlier and the noise factor may be reduced [115].

The noise factor may also be reduced by coating the channels with a material having a high secondary emission ratio. It is necessary only to coat a short section at the input end, where the first multiplications take place [116].

The noise factor $(R_0/R_i)^2$ is shown in Figures 16.58 and 16.59 [111] as a function of plate voltage and input electron energy, respectively.

---

[18] By "specific noise" we refer to the ratio of noise power to signal power—$1/R^2$, where R is the signal-to-noise ratio. See Section 18.2.2.1. Note also that the noise power of n carriers is n times the variance of the gain.

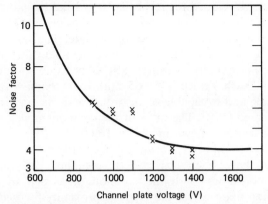

Figure 16.58 Noise factor versus channel plate voltage, calculated. Experimental points are indicated by x's. Manley *et al.* [111].

Mcp Image Tubes. In contrast to the earlier *first-generation* image tubes, tubes using mcp's are called *second-generation.*

In practical second-generation tubes, the entrance plane of the mcp may be covered with a thin aluminum film; this serves a number of functions.

1. It scatters the entering electrons, effectively eliminating the loss in gain due to traveling parallel to the channel axis.

2. It protects the cathode from positive ion bombardment.

3. Electrons back-scattered from the interchannel glass are prevented from reentering the mcp [117].

A variety of image tubes incorporating mcp's have been constructed.

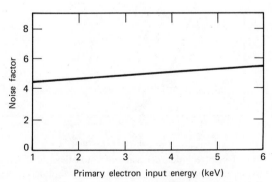

Figure 16.59 Noise factor versus input electron energy, with 1 kV on channel (measured). Manley *et al.* [111].

One device was developed for recording very brief luminous events such as fast crt traces. This consists simply of a sandwich of a photocathode, a mcp, a phosphor screen, and a fiber optic plate. The recording material (photographic film) is placed in contact with the fiber optic plate and only very small gaps (proximity focusing) exist at the input and output ends of the mcp. One such device (45×65 mm) uses channel spaces 50 μm center-to-center and is capable of recording a single trace at 5 mm/nsec from a P11 screen [118]. The mtf's of the component stages and that of the overall system are shown in Figure 16.60 [118]. Such tubes may be superior to electron-optically focused tubes [119].

Other tubes have been constructed essentially like first-generation image intensifier tubes, with the addition of a mcp in front of the phosphor screen upon which the mcp is "proximity-focused." The cross section of such a tube is shown in Figure 16.61 [120] and mtf's of several in Figure 16.62 [120]. In these tubes, electrostatic focusing, rather than proximity focusing, is used at the input to the mcp, despite the resulting increase in size and distortion, (a) because the image inversion obtained in first-generation tubes is required in some applications and (b) to avoid the blurring due to the proximity-focusing at the input. (Because field

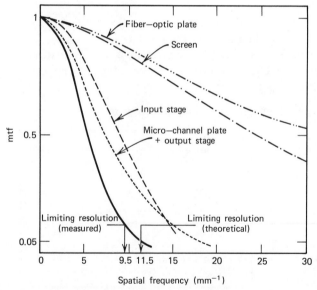

Figure 16.60 Mtf's of mcp recorder, calculated. Photocathode and screen: S20 and P11, respectively. Fiber optics fiber diameter: 6 μm. Potential differences: Photocathode—channel plate, 250 V. Across channel plate, 800 V. Channel plate–screen: 7 kV. Graf et al. [118].

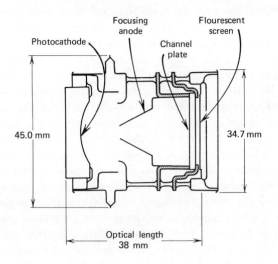

Figure 16.61 Mcp image intensifier tube, schematic. Emberson and Holmshaw [120].

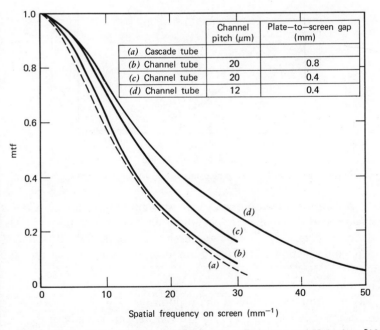

Figure 16.62 Mtf's of various intensifier tubes. Emberson and Holmshaw [120].

emission effects at the photocathode limit the permissible field there, the blurring due to proximity-focusing at the input to the mcp is far more serious than that at the output.) Such tubes, with 1 kV across a 2-mm-thick mcp have a current gain of several thousand and an overall luminance gain of almost a million [121].

The saturation effects referred to above, and evident in Figure 16.57, may be useful to provide some limiting of the luminance in the bright regions of the image, without reducing the gain in the darker regions where it may be more important. In this respect mcp tubes are superior to cascaded image intensifier tubes. Other advantages are compactness and ease of gain control independent of other imaging parameters [120].

Mcp tubes have also been used for intensification in roentgenographic recording. Here the image itself is so unsharp that the additional blurring due to the proximity-focusing at the input may not be objectionable. This permits the construction of large compact screens. The performance of units up to 124 mm in diameter has been reported. The mtf of such a unit is shown in Figure 16.63 [112].

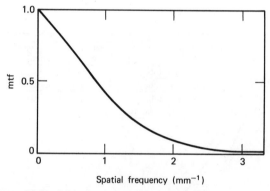

Figure 16.63   Mtf of x-ray mcp intensifier. Millar *et al.* [112].

16.4.2.4   Background Luminance. A major limitation of image tubes is due to background luminance, the analog of dark current in photodiodes. This includes all components of screen luminance not directly attributable to incident illumination. The more important among these are the following:

1. Thermal dark current.
2. Field emission, especially at sharp points.
3. Secondary electron emission due to ion impact on cathode.
4. Scintillations and secondary electrons from envelope.
5. Scattered output light reaching the photocathode.

Careful tube design and construction may reduce all of these significantly.

16.4.3 Optical–Electrical Image Transducers: Photoemissive [15c]

The basic operation of photoemissive image transducers is illustrated by the schematic shown in Figure 16.64. The optical image is formed on a photoemissive screen K. In the *write section*, the photoelectrons liberated thereby are focused onto the leading surface of the target plate or storage screen T. Through some process, this causes a change in the electrical potential on the other side of that plate. This is then scanned by an electron beam from the down-light portion of the tube, which is called *read section* and is similar to the usual crt (see Section 6.3.1). This causes the change in potential to be sensed in some manner. The choice of process by which the deposited charge controls the potential on the other side of the plate and the manner in which the change of potential is sensed gives rise to a number of different tube types, the most important of which are discussed in the following subsections.

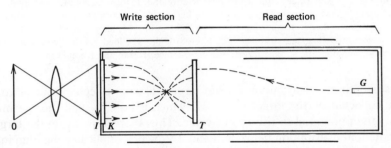

Figure 16.64 Photoemissive image tube (optical-electrical), schematic. O, object; I, optical image; K and T, photoemissive and target screens, respectively; G, electron gun; the broken lines represent electron paths.

16.4.3.1 *Image*[19] *Orthicon* [122] *and Isocon* [123]. The image orthicon is illustrated in Figure 16.65 [15c]. We discuss the writing and reading phases separately.

Writing. The previous scanning of the target plate by the electron beam has brought the rear surface of the target plate (T) to the potential of the reading gun, usually ground potential. Because of leakage current through the target plate, the front surface, too, is at that potential.

The optical image formed on the photocathode (PE) causes the emission of photoelectrons. These are accelerated toward the target due to the potential of the cathode, which is low (-500 V) relative to the target. They are focused there by the focusing coil F. Thus electrons strike the target in a pattern whose density duplicates the illumination at the photocathode, causing the emission of secondary electrons and hence a

---

[19] The prefix "image" signifies an electron-optical image intensifier section.

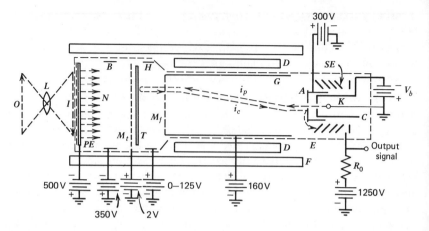

Figure 16.65 Image orthicon. A, gun anode and first dynode; B, accelerator anode; C, control grid of gun; D, deflection coils; E, tube envelope; F, focusing coil; G, wall anode; H, decelerator anode; I, optical image; i_c, return-beam collected current; i_p, primary beam; K, electron-gun cathode; L, imaging lens; M_F, field mesh; M_t, target mesh; N, pattern of photoelectrons; O, external object; PE, photocathode; R_o, load resistor; SE, secondary-electron multiplier; T, storage target; V_b, bias voltage. Kazan and Knoll [15c].

positive charge distribution on the target. The slow secondary electrons are collected by the target mesh, M_t, which is at $+2$ V and thus permits the target potential to rise to this value. Thus there appears on the target a potential distribution between 0 and 2 V, which duplicates the illumination in the image at the photocathode. By capacitive coupling, this potential distribution is transferred to the read side of the target.

Reading. During reading the target is scanned by an electron beam, i_p, that charges the surface to the electron gun (ground) potential. The remaining electrons constitute the collected current, i_c, and reflect the variations in the target potential distribution. This current returns along a path close to the primary electron beam and strikes the electron gun anode (A). This anode acts simultaneously as the first dynode of an electron multiplier section (SE) arranged around the electron gun. After passing all the dynodes, the multiplied electron beam is collected and produces a voltage signal across the load resistor R_0.

The charge deposited on the rear surface of the target brings the front surface, too, to ground potential by means of a small leakage current.

Note that the read beam must be normally incident to bring the target to the electron gun potential. This follows from elementary mechanics considerations: To land on a target at gun potential, the electron must have used up all its kinetic energy in overcoming the potential barrier; if it has any transverse velocity component, it will reach zero axial velocity before having used up all its kinetic energy. To ensure the required normal incidence, a field mesh (M_f), at anode potential, is often provided to produce a strong, uniform field near the target.

We now proceed to some of the important design characteristics.

Target Mesh. The transfer characteristic (output current versus illumination) is quite linear in the image orthicon, up to the illumination corresponding to the target mesh potential. At this point the transfer characteristic exhibits a *knee*. When the illumination exceeds this value, the excess secondary electrons are repelled by the target mesh and redistribute themselves over the target. This causes a dark border to appear around bright areas in the final displayed image. The resulting "sharpening" of the image [analogous to the inhibition effects in vision (Section 15.4.2.3) and the adjacency effects in photography (Section 17.2.2.2)] may be desirable, so that image orthicons are often operated above the knee. This effect also tends to speed up the return of the target to ground potential, giving the tube a better temporal frequency response, an important consideration when it is used with changing or moving objects.

The spacing between mesh and target controls the capacitance and, hence, the amount of charge corresponding to the knee. By increasing this spacing, the range of the tube is lowered, but its sensitivity is increased. Wide spacing (ca. 4 mm) is therefore used in low light level applications, whereas for the usual applications the spacing is about 0.05 mm.

Noise Characteristics and Target Frequency Response [124]. The multiplier section typically provides a gain of approximately 500, so that the amplifier noise is negligible [see (16.56) with $V > 10^{-3}$ V]. On the other hand, the shot noise in the read beam must be added, and it is therefore important to keep this to a minimum (see Section 16.4.6.2 for further detail).

A typical general purpose tube (No. 5820) has a resolution of 600–700 lines/frame when operated at the knee. Larger tubes have a resolution of 800 lines/frame. At low light levels the resolution may be considerably lower; see (16.172) and Figure 16.66 [123].

Lag Effect. When considering the frequency response of an image orthicon, we must distinguish between the bandwidth of the read-out circuit and the limitations imposed by the time constant of the target. Thus it may be possible that a single frame be read out in a millisecond at a bandwidth of, say, 150 MHz, but that the target will not be ready for a new frame until a full second has elapsed. This latter effect is called *lag*

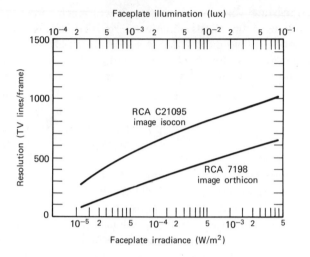

Figure 16.66 Resolution versus irradiance for image orthicon and isocon. Bandwidth; 12 MHz; modulation: 100%; irradiance spectrum: 2850 K; aspect ratio: 3×4. Musselman [123].

and depends on the level of illumination. This dependence can be explained as follows. The electrons in the read beam have various small energies at the time they are emitted from the cathode. Those having axial velocity components at the time of emission will also have such components at the time of landing on the target (assuming electrostatic operation). At low illumination levels, where the target remains essentially at electron gun potential, they will be the only ones able to land and will cause a negative potential at the target, which then becomes the reference potential causing a slowing of the discharge rate [125]. Lag effects cause the blurring of moving images, so that the resolving power of the device becomes dependent on image velocity.

Size. The typical image orthicon has an aperture 75 mm in diameter and is 375 mm long, but commercially available apertures range 35–115 mm. The electron optics are typically purely magnetic, although electrostatic deflection has been found useful too [126].

Image Isocon. A major limitation of the image orthicon is due to the shot noise introduced by the beam current collected after reflection from the target. The lower the illumination, the higher will be the collected current and, therefore, the noise. This severely limits operation at low light levels. The isocon read system was developed to overcome that limitation. In this system only the electrons scattered at the target are collected, so that the collected current varies directly with the illumination and the shot noise is least at low illumination values.

The selection of the scattered electrons is accomplished at the anode of the return beam, which consists of a plate surrounding the electron gun. Due to the magnetic focusing field, the specularly reflected electrons in the return beam describe helices with relatively little spread due to differences in radial velocity. The scattered electrons, on the other hand, are spread over a wide region. At the separator plate, the specularly reflected electrons are blocked and only the scattered ones are passed on to the first dynode of the multiplier section.

The number of scattered electrons is proportional to the number striking the target and this, in turn, is proportional to the illumination.

Image Intensifier Orthicon. In an alternative method for increasing the sensitivity of the image orthicon, an image intensifier tube is preposed. This may consist of a front section, built integrally with the image orthicon, to act as an image intensifier and to provide a much higher level of illumination at the orthicon cathode.

16.4.3.2 Conductivity Modulated Tubes [15d].

The sensitivity of the image orthicon may also be increased by providing for electron multiplication in the target plate. Three approaches to this have been developed. In two of these, the target plate consists of a dielectric mounted on a thin metallic backing facing the photocathode and maintained at some positive voltage (e.g., 20 V). See Figure 16.67 [15d] for an illustration of such a type. The read beam, similarly to that in the image orthicon, returns the other side of the target plate to ground potential, transforming the target effectively into a charged condenser. Photoelectrons generated by the image formed on the cathode are accelerated by a high potential difference (e.g., 20 kV) toward the target plate into which they penetrate, causing an increase in conduction. The local change in conduction permits a corresponding discharging of the condenser so that a charge-image of the optical image appears on the far side of the target plate. During the read portion of the cycle, this side of the target plate is scanned by an electron beam that returns it to ground potential by depositing an amount of charge proportional to the voltage rise generated by the previous conduction through the target. This current is capacitatively conducted to the metallic backing and thence flows out through the load resistor R_0 connected to this backing, providing a signal that essentially follows the distribution of photoelectrons and, hence, illumination.

The increase in electron conduction referred to may be electron bombardment conductivity (EBC, Ebicon), secondary electron conductivity (SEC), or conduction through a reverse-biased diode as in the silicon diode array target tube (variously called EBS, IDAC, SEM, SiEBIR, SIT).

Electron Bombardment Conductivity (EBC) [15e]. In EBC, the penetrating electrons create electron-hole pairs much as do the photons in photoconductors. But, whereas photons give rise to no more than one carrier pair (they have less than 10 eV energy for wavelengths above 0.124 μm), the bombarding electrons, which may have energies of thousands of electron volts, may each generate hundreds of carrier pairs.

EBC targets are frequently made of As_2S_3, approximately 1 μm thick. With such targets, gains of 500 at 20 kV are typical. They are supported by an aluminum oxide film (on the write beam side) followed by a thin film of aluminum (0.2 μm thick) with the insulating layer facing the read beam side.

Secondary Electron Conductivity (SEC) [127], [128]. In SEC tubes, as shown in Figure 16.67, the screen consists of porous material and the charge carriers are secondary electrons freed by the bombarding electrons into the vacuum in the interstices. There they are drawn along by the applied field generating "tertiary" electrons on the way. This latter effect makes the target gain dependent on the bias voltage across the target. The electrons passing through the target plate and drawn to the field mesh, M_f, also contribute to the charging of the target.

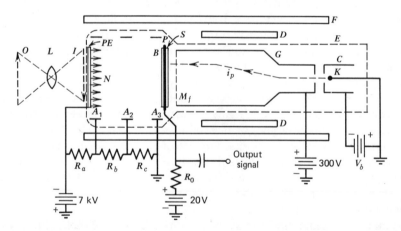

Figure 16.67 Secondary-emission-conductivity (SEC) camera tube. $A_1–A_3$, accelerating anodes; B, target-support layer; C, control grid of electron gun; D, deflection coils; E, tube envelope; F, focusing coil; G, electron-gun anode; I, input optical image; i_p, reading beam; K, electron-gun cathode; L, lens; M_f, field mesh; N, pattern of photoelectrons; O, external object; P, metallic target blackplate; PE, photocathode; $R_a–R_c$, voltage divider; R_o, load resistor; S, porous storage layer; V_b, bias voltage. Kazan and Knoll [15d].

When compared with the EBC, it is found that the greater thickness of the SEC screen makes for a lower capacitance and the greater electron mobility, too, for a more rapid response; both these factors give the SEC screen a reduced lag in reading. Another advantage of the SEC screen is its higher penetrability, which permits the use of considerably lower voltages in the write section, 7 kV being a representative value. The considerably higher resistivity makes for storage capabilities extending over many hours. This makes the SEC tube useful both for integrating at low light levels and for storing images of transient phenomena.

The target generally consists of a layer of porous KCl (ca. 20 μm thick density ca. 0.02). The resistivity of such a layer is approximately $10^{17} \, \Omega \cdot$ cm. It has a gain of approximately 200 with 30–40 V across it. (When operating above this voltage, conduction electrons become significant and operation approaches EBC characteristics.)

These tubes have a sublinear response due to the drop in bias voltage with increasing signal. At high levels (10^{-1} lux) the resolution is approximately 800 lines/frame; at low levels (10^{-4} lux), it drops to 100 lines. The noise level is close to the input shot noise for exposures around 1000 photoelectrons/element with picture element size up to $10^4 \, \mu$m^2 [129].

As noted earlier, the lag effect in SEC tubes is relatively low. Nevertheless, at very low illumination levels it may become significant. Normally only 5% of the signal remains after 50 msec; however, at 4×10^{-3} lux, 15% may remain. Thus a tube that resolves 550 lines/frame on a stationary image, drops to 460 lines when this crosses the field in 10 sec [127].

SEC targets have been made with diameters up to 30 mm. The (square wave) transfer function of such a tube is shown in Figure 16.68 [130]. The linearity of response can be judged from the transfer characteristic shown in Figure 16.69 [130].

SEC tubes, too, have been used with a preposed image intensifier section, yielding useful performance levels (200 lines/frame for full contrast, stationary image) at 10^{-5} lux.

Tubes with SEC targets have also been designed with proximity focusing in the read section. This permits a more compact tube and also allows for flat face plates. One such tube having a 25 mm diameter target has been made for use in the ultraviolet [131].

Silicon Diode Array Target Tubes (SiEBIR etc.) [132]. The silicon diode array target tube is similar, in operation, to the electron bombardment conductivity tubes described earlier in this section. Here, however, an array of silicon diodes replaces the target plate of that tube.

The target plate consists of a wafer of n-type silicon, one surface of which has been oxidized. The layer of SiO_2 contains an array of windows

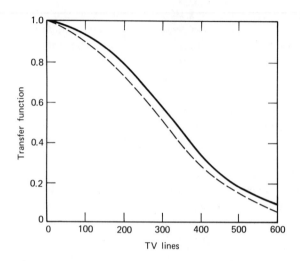

Figure 16.68 Transfer function (square wave) of SEC tube, at the center (solid) and near corners (broken). "Corner" measurements on diagonal, at 80% points photocathode voltage: $-8\,\text{kV}$; signal plate voltage: 11 V; suppr. screen voltage: 30 V. Sato and Takahashi [130].

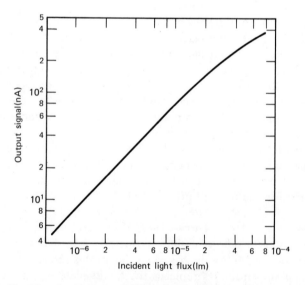

Figure 16.69 Transfer characteristic of SEC tube. Sensitivity: $166\mu\text{A/lm}$; photocathode voltage: $-8\,\text{kV}$; signal plate voltage: 11 V; suppr. screen voltage: 30 V. Sato and Takahaski [130].

through which boron has been diffused to form p-type islands. This wafer, with the oxidized surface facing the read gun, functions as the target plate. See Figure 16.70 [133].

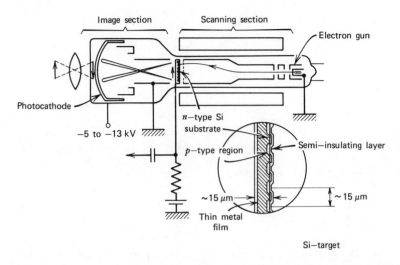

Figure 16.70 Diode array target tube, schematic with target detail. Miyashiro and Shirouzo [133].

In operation, the photo-electrons are greatly accelerated and focused on the target plate. These electrons penetrate into the silicon layer, each of them generating thousands of carriers on their way. These carriers drift to the p-type island which has previously been charged to the read gun potential, several volts negative relative to the write surface. There they accumulate during the read scan period until they are neutralized by the read gun. The amount of charge accumulated is detected during this neutralization.

Tubes built on this principle have exhibited excellent linearity over a range of 1000:1, starting at illumination levels below 0.5 $\mu W/m^2$. Typical electron gain is 2000 at 10 kV between photocathode and target. These tubes are also relatively rugged mechanically and optically.

Si-diode targets have been made with diameters up to 25 mm and the transfer function of a tube with that diameter is compared with that of a 16-mm-diam target tube in Figure 16.71 [134]. As in other tubes of this type, the target is not fully neutralized during the read cycle, so that some after-image (lag) remains for a number of frames, lowering resolution of moving targets. Limiting resolution for Si-diode array target tubes is

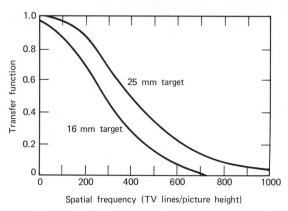

Spatial frequency (TV lines/picture height)

Figure 16.71 Transfer function (square wave) of EBS-operated Si diode array target tubes at 400 nA signal current. Upper and lower curves are for 25-mm and 16-mm diagonal targets, resp. Santilli and Conger [134].

compared with that of SEC tubes for both stationary and moving targets in Figure 16.72 [134].

16.4.4 Optical–Electrical Image Transducers: Photoconductive [15f], [93d]

A simpler and more rugged image tube results when the photoconductive effect is used for primary detection. Both homogeneous (vidicon) and

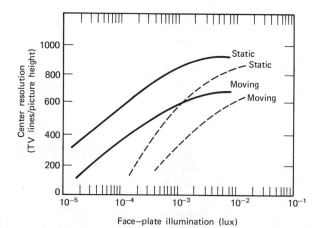

Face–plate illumination (lux)

Figure 16.72 Resolution versus illumination of Si diode array target tubes, for stationary and moving targets. Solid lines: 25-mm EBS target (27 mA/lx sensitivity). Broken lines: 32-mm SEC target (1.2 mA/lx sensitivity). Santilli and Conger [134].

junction (plumbicon, silicon diode array) screens are used, and we discuss these now.

16.4.4.1 *Vidicon*. The vidicon, due to its simple, economical, and compact construction, is today the most popular image pickup tube. This is especially true since, at normal daytime light levels, its performance does not lag far behind that of the other, more expensive tubes.

A schematic diagram of a typical vidicon is shown in Figure 16.73 [15*f*]. The inner surface of the faceplate is first coated with a transparent conducting film (*B*), which acts as the backplate of the target. This is covered with a photoconductive layer (*PC*). As in the EBC, SEC, and IDAC tubes, the backplate is maintained at a positive voltage, say 20 V, while the other side of the photoconductive layer is brought to ground potential by the read beam. This layer thus acts as a charged capacitor. When an optical image is formed on this layer, the induced photoconductivity permits the condenser to start discharging locally at a rate proportional to the level of illumination. The vacuum side of the photoconductor then assumes a positive voltage distribution, also proportional to the illumination.

During reading, the inner photoconductor surface is scanned by an electron beam depositing charge sufficient to return this surface to ground

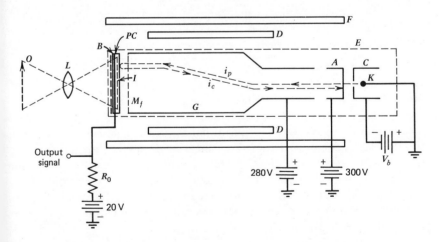

Figure 16.73 Camera tube with "homogeneous" photoconductive target (vidicon). *A*, accelerating anode; *B*, transparent backplate electrode; *C*, control grid; *D*, deflection coils; *E*, tube envelope; *F*, focusing coil; *G*, wall anode; *I*, optical image; i_c, return beam; i_p, primary beam; *K*, electron gun cathode; *L*, lens; M_f, field grid; *O*, external object; *PC*, photoconductive layer; R_0, load resistor; V_b, bias voltage. Kazan and Knoll [15*f*].

potential. This change of potential is capacitatively transferred to the backplate, producing a signal current, which flows out through the load resistor R_0, across which the signal appears.

Target Characteristics. Target thickness and resistivity must be such that charge can be maintained over a frame period (τ). In practice a resistivity of approximately $10^{12} \, \Omega \cdot cm$ and a photoconductor response time no longer than τ is required. The typical vidicon screen consists of a layer of Sb_2S_3 several microns thick. This material has peak response in the extreme red portion of the spectrum, but by evaporating in a (low-pressure) gaseous atmosphere, a porous layer is obtained, peaking near the human visual response peak. Its response is somewhat slow, so that a slight blurring of moving targets must be accepted, especially at lower illumination levels. For uv and X-ray images, screens of amorphous selenium and lead oxide are used. See the discussion of plumbicons for screens usable in the ir.

Both upper and lower limits on target capacitance must be observed. The amount of charge, and hence the output signal level, varies directly with the capacitance. Thus, to obtain an acceptable signal-to-noise ratio, the screen capacity may not be far below a nanofarad. On the other hand, the higher the capacitance, the more difficult it becomes to neutralize the target during reading with the limited beam current available. This is true especially at low light levels where lag effects (see Section 16.4.3.1) become pronounced. Target capacitance is, therefore, usually maintained in the range between 1 and 4 nF. On the other hand, for storage applications, layers with higher capacitance (0.5 μm thick) are used.

Performance Parameters. Vidicon performance may be controlled via two readily adjustable parameters: backplate voltage and read beam current (ca. 1 μA). Increasing the backplate voltage increases the responsivity; but, above a certain level, the conductivity rises rapidly (space charge limited operation) and performance drops. If the read current is too low, lag becomes noticeable and smearing of moving images results[20]; if too high, resolution is lost. Also, care must be taken that no point reaches a voltage above the first secondary-emission crossover point (E_{cr1}, Figure 16.14), lest the read beam increase its potential instead of neutralizing it. When this does happen, special procedures are required to return the screen to operating condition.

Response Time. The vidicon is usually used with a frame scan time of approximately 30 msec. With an Sb_2S_3 screen and at high illumination

---

[20] Scanning rapidly during the flyback period with a defocused "read" beam has been investigated as a method for overcoming this difficulty [135].

levels, this time suffices to ensure that the photoconductivity decay to below 10% during a frame period. However at intermediate light levels, yielding $4 \, \text{mA/m}^2$ screen current (corresponding to an illumination of approximately 20 lux), the photoconductor response is so slow that the signal decays to only 50% during a frame period and five cycles are required for decay to 10%. At even lower levels, such as $0.25 \, \text{mA/m}^2$, three frame cycles are required for a decay to even just 50% [67]. This is a serious limitation of the vidicon at lower light levels and with rapidly moving images.

Operating Parameters. Typical vidicon operating parameters are as follows. The transfer characteristic is sublinear ($\gamma \approx 0.7$). Signal-to-noise ratios of 100 are obtainable at sufficient illumination, which usually falls in the range of 1–10 lux. The detective quantum efficiency is thus 10^{-3}, that is, one-tenth of that of the image orthicon. Although commercially available tube diameters range from 12 to 115 mm, 25 mm is the most commonly used, with a tube length of approximately 150 mm. With such a tube, resolution of 1000 lines/frame is attainable; under usual operating conditions, however, 600 lines are the limit.

16.4.4.2 Tubes with Junction Targets. We have noted that vidicon sensitivity is limited by the permissible voltage across the target and that, at lower light levels, its usefulness is impaired by its long photoconductor response time. Both these limitations are overcome by using a diode structure for the photoconductor layer. This permits the application of relatively high (reverse) voltage across the photoconductor, ensuring that almost all carriers generated by the illumination be swept out of the junction to be read by the read beam (quantum efficiency close to unity). It also results in a much faster response, so that operation at lower light levels becomes possible. Indeed, it has been shown [136] that in many applications, the performance of such tubes, despite their relative simplicity, is comparable to that of the image orthicon. This form of operation also results in an almost linear transfer characteristic.

An additional advantage of such tubes is their suitability to ir images. Ir sensitivity demands a gap below 1.5 eV.[21] On the other hand, a small band gap is inconsistent with high resistivity at room temperature and the above-mentioned required resistivity of $10^{12} \, \Omega \cdot \text{cm}$ cannot be attained with band gaps below 1.7 eV. With p-n junction screens, however, band gaps of 1 eV may still have sufficiently high resistance.

Plumbicon [136]. The plumbicon was the first tube of this type. Its

---

[21] Recall that the cutoff wavelength is given by $(1.24 \, \text{eV}/\Delta V) \, \mu\text{m}$, where ΔV is the width of the band gap in eV (see the last line of Table 19).

screen consists primarily of a film of PbO, deposited on a transparent conducting plate. The layer in contact with this plate is n-type and the layer facing the read beam is p-type. Sensitivities range from 300 to 400 μA/lm. The transfer characteristic of a typical plumbicon (P) is compared to those of a number of vidicons (V_I, V_{II}) and image orthicons (O_I, O_{II}, O_{III}) in Figure 16.74 [136]. The Xs mark the optimum operating points.

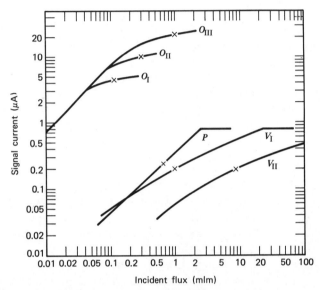

Figure 16.74 Transfer characteristics of various camera tubes. The signal current I_s is plotted as a function of the luminous flux on the photosensitive layer, the illumination being uniform. P applies to a "Plumbicon" of average sensitivity (350 μA lm), O_I to a very sensitive 3 in. image orthicon, O_{II} to a 3 in. image orthicon with the knee at a higher illumination, and O_{III} to one of the types of 4.5 in. image orthicon often used whenever the highest picture quality is required.

V_I applies to an Sb_2S_3 vidicon at a sensitive setting (signal plate voltage V_s = about 40 V, dark current $I_d = 0.02$ μA), and V_{II} to a similar vidicon with a lower signal plate voltage (V_s = about 15 V, $I_d = 0.005$ μA) and increased speed of response.

The operating point for optimum setting is given on each curve by an x. The illumination required for this setting can be read on the abscissa. Van Doorn [136].

The resolution of the plumbicon is limited by the transverse electrical conductivity of the target layer. By making this layer very thin (10–20 μm) and not doping it too heavily, the typical tubes have an mtf that drops from 0.45 to 0.2 as the spatial frequency goes from 400 to 600 lines/frame.

For operation at very low light levels, an image intensifier may be placed before the input to the plumbicon [137]. This permits operation to well below a microlux. The resolution values attained with plumbicons with a preposed n-stage ($n = 0, 1, 2, 3$) image intensifier are shown in Figure 16.75 [137].

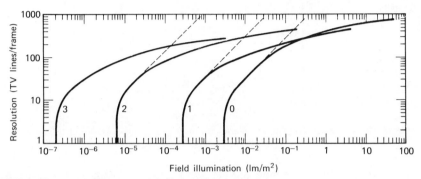

Figure 16.75 Resolution of plumbicon, with and without image intensifier preposed. The number of preposed intensifier stages is indicated next to each curve. Taylor *et al.* [137].

Silicon Diode Array Camera Tube (SIDAC) [138]. The silicon diode array plate described as part of a photoemissively operated tube in Section 16.4.3.2 can also be used as the photodetector in a vidicon-type arrangement. When used in this manner, it has the advantages of the broad spectral response of silicon and of ruggedness. Also the short carrier lifetimes reduce the undesirable lag effects just as in the plumbicon. Spatial resolution, here, is determined by the array structure. Representative mtf curves for such a camera tube are shown in Figure 16.76 [138].

16.4.5 Solid-State Arrays

The functions of optical–optical and optical–electrical image transduction can also be accomplished by using photoconductive detectors in conjunction with phosphors or electric switching circuitry, respectively. Position-sensitive detectors, too, can be used.

16.4.5.1 Solid-State Image Intensifiers and Converters (Optical–Optical) [15g].
Solid-state image intensifiers and converters generally employ combinations of photoconductive (PC) and electroluminescent (EL) materials. The basic principle is illustrated in Figure 16.77. An

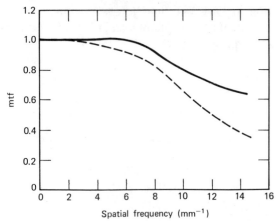

Figure 16.76 Mtf of diode array tube for magnetic (solid curve) and electrostatic (broken curve) focus. 0.55 μm flux. Crowell and Lebuda [138].

alternating voltage is connected across PC and EL layers in series. In the dark, the voltage drop is primarily across the PC layer, so that there is no significant EL excitation. When the PC is illuminated, the resistance drops and significant voltage appears across the EL screen, which consequently luminesces. Early versions of such devices have already been described (Section 6.3.3); we mention here only some more recent work along this line.

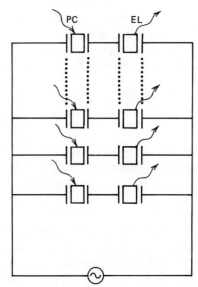

Figure 16.77 Solid-state image intensifier, general schematic.

A major problem is the attainment of sufficiently high resistance across the PC layer, which must be thin enough to be penetrated by the input image radiation. Various approaches to overcome this difficulty have been developed. Compare the grooved and pedestaled layers shown in Figures 6.19 and 6.20 and the embedded electrodes of Ref. 139.

A more recent technique [140] employs nonmatching electrodes evaporated on opposite sides of the photoconductive layer: the circular electrode on the far surface is concentric with a larger window in the near surface electrode. This increases the resistance between electrodes and reduces the capacitance that would otherwise tend to short circuit the voltage applied across it. As a result, one may use layers that are quite thin (50 μm) and readily penetrated by the light. Figures 16.78 and 16.79 [140] show spectral response and transfer characteristic of such a panel, using CdSe for the PC. The corresponding operating efficiency was about 10 l/W.

With X-ray images, the thickness of the PC layer is not so limited because of the greater penetrating power of the X-rays. The transfer characteristic of one storage panel for use with X-rays is shown in Figure 16.80 [141] and that of another—with a preposed SEC tube—in Figure 16.81 [142].

Solid-state image converters sensitive to ir would seem to require photoconductors sensitive to the longer wavelengths. Instead of this, quenching of photoconductivity by ir may be used. For example, in CdS the photoconductivity is destroyed by ir illumination. Image converter

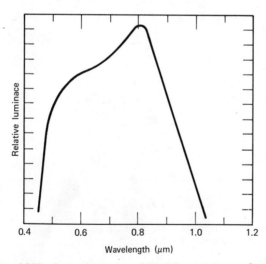

Figure 16.78 Spectral response of EL–PC panel. Stewart [140].

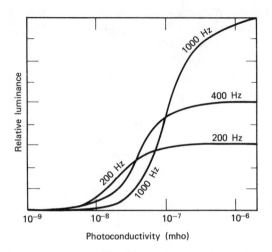

Figure 16.79 Transfer characteristic of El–PC panel, frequency dependence. 100 V applied. Stewart [140].

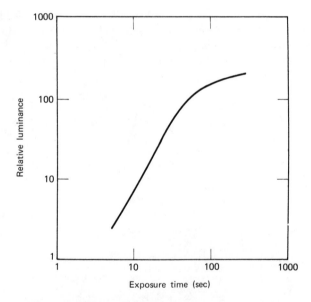

Figure 16.80 Transfer characteristic of X-ray image storage panel, 100-kV, 5-mA X-rays, with 12.5-mm aluminum filter. Ranby and Ellerbeck [141].

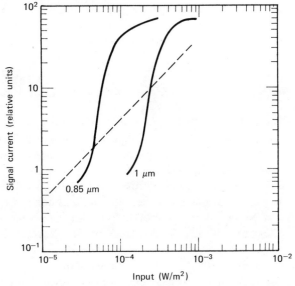

Input (W/m²)

Figure 16.81 Transfer characteristic of PC–EL converter with SEC-tube, at two wavelengths, compared to SEC-tube alone, at 0.85 μm (broken line). Szepesi and Novice [142].

panels based on this principle have been developed in two versions. In the simpler version, the arrangement is as shown in Figure 16.82 [143]. The photoconductivity is provided by "bias" illumination in the spectral region of the CdS photoconductive sensitivity and quenched by the ir flux incident from the object. In contrast to the usual PC–EL panels, the PC here is not shielded from the EL flux; instead, this flux serves as additional bias light so that the bright spots become brighter and a bi-stable, two-tone image results [143].

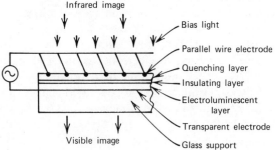

Figure 16.82 Image converter using ir quenching of photoconductivity, schematic. Kohashi *et al.* [143].

In a second version, the photoconductive (CdS) layer contains also phosphor material (ZnS), which is ir quenchable. The quenching of this phosphor then controls the photoconductivity and hence the voltage applied across the EL layer.

16.4.5.2   Solid-State Optical–Electrical Image Transducers [144]

Switched Devices. The function of a television camera tube, too, can be accomplished with solid-state techniques. Such image panels generally consist of an array of diodes whose elements are read sequentially by switching. For instance, the collector layer may be diffused in rows and base and emitter regions may be diffused discretely into the, say, silicon substrate. Copper strips may then be applied to connect the emitters in columns. By switching to the desired column and applying the read-out signal to the desired row, any desired element may be read [145]. In such devices it is often desirable that the incident flux be integrated over a long time. This is possible with the above arrangement.

In another arrangement, the scanning pulse, which controls read-out, travels automatically along the row at the scanning speed, so that only the transverse scan requires switching [146].

Excellent storage and efficient read-out have been made possible by the use of the metal-oxide-silicon (MOS) capacitor, underlying also many of the charge transfer devices discussed in the next subsection. This device consists of a doped semiconductor substrate covered by an insulating oxide layer, which, in turn, is topped by a metal electrode. Consider such a device formed on p-type silicon. When a positive potential is applied to the electrode, the region below it is depleted of free majority carriers and any minority carriers generated there will accumulate on the underside of the oxide layer, which acts as a capacitor. The depleted region is limited laterally by forming a heavily doped (p^+) conducting layer below the oxide layer, except under the electrode.

Such capacitors, connected in pairs and arranged in a two-dimensional array make promising solid-state image detectors, when operated as charge-injection devices (CID) [147]. Here the two electrodes of each capacitor pair are connected to electrodes running along each row and column, respectively. The application of a potential to, say, the row-connected capacitors permits the photogenerated charge to accumulate in it. Any particular capacitor in the row may be read out by transferring the charge to its partner (by energizing the corresponding column electrode) and then injecting it into the substrate, from which it may be read out. Alternatively, rows may be read out sequentially, yielding the equivalent of a line scan.

Charge Transfer Devices (*CTD*) [148]. In a more recently developed method, the array is read out by transferring the charge, accumulated at any one element, from element to element along the row—to be detected when it reaches the last element in the row. The charge is held in a potential well until its transfer to the neighboring element is initiated by a pulse from a "clock."

Two types of such devices have been developed. In the *bucket-brigade device* (BBD), the elements can be represented as capacitors alternating with transistor gates. The capacitor is charged by photoconductive currents generated by the incident flux in the adjacent material. Now consider a sequence of such units in which charge has accumulated in the odd-numbered capacitors. The even-numbered capacitors are maintained at reference potential by means of a line (*B*) connected to them (see Figure 16.83*a* [146]). When it is desired to read out the accumulated

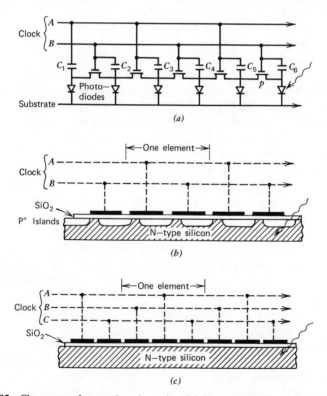

Figure 16.83 Charge transfer panels, schematics: (*a*) "bucket brigade" sensor equivalent circuit; (*b*) "bucket brigade" sensor cross section schematic; (*c*) charge-coupled sensor cross section schematic. Weimer *et al.* [146].

charge, line A is activated by a clock pulse, permitting the transistor following the charged (odd-numbered) condensers to transfer the accumulated charge to the following (even-numbered) condenser. This continues until the original condensers have reached reference potential, at which time all the charge has been transferred to the even-numbered condensers. This process is repeated, with clock pulses alternatingly activating even- and odd-numbered transistors, until the charge accumulated on the first capacitor (C_1) has reached the end of the line (following C_6 in Figure 16.83).

Physically, the device consists of p^+ islands diffused into an n-type substrate, covered by an insulating layer of SiO_2, on which conducting patches have been evaporated (see Figure 16.83b).

In the alternative, *charge-coupled device* (CCD) illustrated in Figure 16.83c, the semiconductor layer is uniform and the clock pulses cause the formation of a depletion layer under the appropriate conducting patches. This depletion layer serves as a potential well to hold the charge, while the neighboring patches on both sides are neutral. When a neighboring patch is brought to an even lower potential, the charge is transferred into the depletion layer under it. The original well is then eliminated and the second well brought to the intermediate depth, to prepare for transfer to the next patch. Thus, at any instant a triplet of patches is involved in the handling of one batch and three transmission lines of clock pulses, appropriately staggered, are used to ensure the transfer of charge in the forward direction only. This requirement can be avoided by the introduction of an asymmetry into the element structure.

In the devices just described, the charges are transferred through channels running along the underside of the oxide layer. In these *surface charge-coupled devices* (SCCD), a considerable amount of noise is introduced by interface effects. This can be eliminated by using a channel running some distance below the interface. Such a *buried channel* can be produced by converting the upper layer of the p-type substrate into an n-type layer, causing the junction to take over the function of the channel. These *bulk*—or, buried—*channel charge-coupled devices* (BCCD) exhibit considerably lower noise. Also, the increased fringing of the applied fields aids in charge transfer; in such devices, inefficiencies can be held to below 10^{-4} up to 25 MHz.

In both of the above techniques, illumination detected during the read-out period causes cross talk or "smearing," that is, the appearance of photogenerated charge belonging to one element on the readings obtained from subsequent elements. This effect may be negligible if the duration of the read-out period is very short compared to the total exposure time. Otherwise, exposure must be interrupted during read-out.

Alternatively, two parallel lines of elements may be used. Exposure and the accumulation of photogenerated charge occurs in the write line. When read-out is desired, the charge accumulated in the elements of the write line is transferred to the elements in the read line—by charge-transfer—each element in the write line transferring its charge to its partner in the read line, which is shielded from incident flux. Read-out then takes place there, unaffected by illumination, while the photosensitive elements in the write line can continue accumulating photogenerated charge.

CCD's can have element spacings as small as 30 μm. Linear arrays have been produced with 1728 sensor elements and having inefficiencies less than 5×10^{-5} at 2.5 MHz.

In two-dimensional arrangements of charge-coupled devices, a whole two-dimensional array of accumulated photogenerated charges can be transferred at once from the write to a storage array. To initiate read-out, the first row in the storage array is shifted into the read array consisting of a single row of charge-coupled elements and all other rows are shifted down to the elements of the neighboring row in the storage array. Upon completion of read-out of the first row, all rows are again shifted by one and the next row enters the read array. This process is continued until all lines have been read [149].

Area arrays have been produced with 320×512 elements, to match U.S. TV broadcast standards. In the BCCD version, they can have noise equivalent signal values of less than 50 electrons, when cooled to $-20°C$ to reduce the dark current, which tends to be relatively large in these; useful images have been obtained with CCDs at irradiation levels as low as starlight (10^{-5} W/m^2, with 2854°K radiation).

Ir Devices [148]. In conclusion we note briefly that CCDs have been adapted for work in the ir, as well, by using doped (In, Ga) silicon or Schottky-barrier [150] diodes. These are usable in the 3.5-μm window and are capable of 1°C resolution [150].

16.4.5.3 *Position-Sensitive Detectors.*

When the image consists of only a point, or a line, whose location is to be determined, a single position-sensitive detector may be employed. Such a detector may consist of a reverse-biased large area rectifying junction, where one of the layers has a relatively large conductivity. The incident flux generates carrier pairs that drift to opposite junction layers. The carriers in the conducting layer will diffuse rapidly and distribute themselves almost uniformly over the junction area. Those in the other layer, however, will remain in the neighborhood of the illuminated region (see Figure 16.84). Thus a potential difference will arise between the region of the illuminated spot

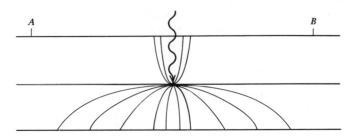

Figure 16.84 Illustrating a position-sensitive detector.

and regions remote from it—a potential difference that grows with the distance. Hence, by measuring the potential difference between two points, such as A and B, it is possible to determine the location of the illuminated spot, I. Specifically, a null for this difference indicates that the illuminated spot is at the midpoint between A and B. If, now, an external bias field is applied across the device, parallel to the junction, the null will correspond to a different location of the illuminated region. By applying a saw-tooth voltage, the null-yielding location can be scanned across the diode and the location of the illuminated spot determined by the timing of the null [151]. Such a device is illustrated by a Schottky-barrier diode, using a 10-mm spacing of lateral electrodes with a pulse width below 0.1 μsec [152].

The *scanistor* [153] is another area device capable of locating an illuminated spot and of measuring illumination distribution. It consists of pairs of bucking junction diodes arranged in a row. For such a pair to conduct, one of the diodes must be forward-biased and the other one illuminated. A bleeder resistor, running from voltage V_B to ground is connected to one side of the diode-pairs and a constant voltage bus (at V_0) to the other (see Figure 16.85). The voltage across the diode pairs therefore varies according to

$$V(x) = \frac{x}{X} V_B - V_0, \qquad (16.152)$$

where V_B and V_0 are the voltages applied to the bleeder resistor and the bus, respectively,

x is the distance along the bleeder resistor measured from the grounded terminal, and

X is the total length of the bleeder resistor.

The value, x_0, of x for which V vanishes is called the null point. All diodes at $x > x_0$ will not conduct, whereas those with $x < x_0$ will conduct,

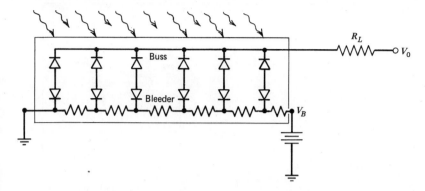

Figure 16.85 Illustrating the scanistor principle.

to an extent proportional to the illumination received by them. By varying V_0 (or V_B), the null point can be scanned along the row: the resulting current flow will be a measure of the total flux in the region $(0 < x < x_0)$. Thus the illumination distribution can be found. For example, the illumination along the row may be read in the form of a linear scan by applying a saw-tooth voltage at V_0 and differentiating the resulting voltage appearing across the load resistor R_L.

16.4.6 Noise and Bandwidth Considerations

We have previously met with both temporal and spatial (granularity) noise individually. In image tubes these appear together. The usual spatial Fourier formalism may be used to analyze these with the simple expedient of adding another dimensions to allow for temporal variation [154]. Since the detailed formulation of the noise characteristics varies according to the type of image tube, we discuss them individually.

We mention briefly an empirical relationship that has been found to describe accurately the mtf of a large number of imaging devices. It is of the form [155]

$$T(\nu) = \exp - \left(\frac{\nu}{\nu_0}\right)^a, \tag{16.153}$$

where ν_0 and a are constants typical for the device. The values of these constants for a number of devices are listed in Figure 16.86 and Table 174 [155]. In this discussion we utilize a number of results of system analysis theory; the interested reader is referred to Chapter 18 for their derivations.

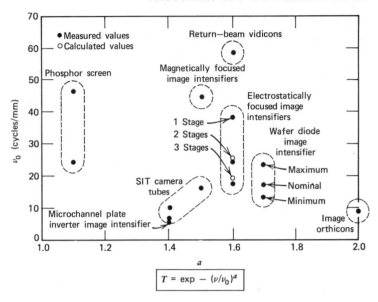

Figure 16.86 Ranges of constants in empirical mtf formula (as given below the figure) for electron optical devices. From Johnson [155]. See that paper for detailed references.

16.4.6.1 Optical–Optical Image Tubes

Temporal Noise. In considering the temporal noise in image tubes, our primary concern is again shot noise, and the relevant formulae developed for phototubes are valid, with the following differences:

1. The effect of the luminescence on the noise must be determined.
2. The signal is in the form of current density, rather than current.
3. The effective bandwidth is limited by the temporal response of the phosphor and the observer.

Let us discuss these individually.

1. The action of the phosphor may be described as producing an average number, g_p, of photons for each incident electron. If we assume the process to be Poisson-distributed, this will also be the variance of the number of photons resulting from a single electron. The specific noise, R_2^{-2}, of the luminescence process is obtained by multiplying this by the number of electrons (n_e) incident on a phosphor area element (see Note 17):

$$R_2^{-2} = n_e g_p. \tag{16.154}$$

The specific noise, R_1^{-2}, of the incident electrons, also assumed Poisson distributed, is n_e. Noting that the specific noise at the output equals the sum of the specific noise values of the constituent processes, we have for this

$$R_{out}^{-2} = R_1^{-2} + R_2^{-2} \qquad (16.155)$$

and, for the noise factor of the phosphor screen:

$$F = \frac{R_{out}^{-2}}{R_1^{-2}} = \frac{R_1^{-2} + R_2^{-2}}{R_1^{-2}} = 1 + \frac{R_2^{-2}}{R_1^{-2}} = 1 + \frac{1}{g_p}. \qquad (16.156)$$

2. We multiply the current density by the effective element area, $\bar{A}$, in order to convert to current.

3. The effective time constant, τ, can be found from the equivalent bandwidth.

Referring to (16.61) (with $\tau = \frac{1}{2}\Delta\nu$) and to (16.156), we can now summarize our results to obtain for the squared temporal signal-to-noise ratio:

$$R_t^2 = \frac{j\bar{A}\tau}{q_e F} = \frac{(E\eta_c)\bar{A}\tau g_p}{q_e(1 + g_p)}, \qquad (16.157)$$

where the electron current density (j) is given as the product of the incident irradiation (E) and the cathode sensitivity (η_c).

Introducing the value of q_e ($1.602 \times 10^{-19}\,C$) and assuming that the phosphor gain g_p is much larger than unity and that the eye response ($\tau \approx 0.1$ sec) controls, we can simplify (16.157) to

$$R_t^2 = 0.624 \times 10^{18}\, E\eta_c\bar{A}. \qquad (16.158)$$

To treat the system at levels of E so low that the dark current noise can not be neglected, we write signal s and noise N separately. We then find

$$R_t^2 = \frac{s^2}{N^2} = \frac{(j_s\bar{A}\tau)^2 g_p}{(j_s + j_d)\bar{A}\tau q_e(1 + g_p)} = \frac{j_s\bar{A}\tau g_p}{(1 + \rho)q_e(1 + g_p)}$$

$$= \frac{E_s\eta_c\bar{A}\tau g_p}{(1 + \rho)q_e(1 + g_p)} \approx 0.624 \times 10^{18}\, \frac{E_s\eta_c\bar{A}}{1 + \rho}, \qquad (16.159)$$

where E_s is the signal irradiation and

$$\rho = \frac{j_d}{j_s} \qquad (16.160)$$

is the ratio of dark to signal current density, and the last approximation is based on the assumptions of (16.158).

When the current density, j_b, generated by the background irradiation

(E_b) is to be treated as a source of noise, the same result is obtained, except that we must then define

$$\rho = \frac{j_d + j_b}{j_s} \tag{16.161}$$

$$\approx \frac{j_b}{j_s} = \frac{E_b}{E_s} \tag{16.162}$$

when dark current is negligible.

On substituting this latter value into (16.159) we find

$$R_t^2 = \frac{E_s^2 \eta_c \bar{A} \tau g_p}{(E_s + E_b) q_e (1 + g_p)}. \tag{16.163}$$

Since the noise consists of pulses of the form $l(t)$, its Wiener spectrum has the form $|\mathscr{F}[l(t)]|^2$.

Spatial Noise. Spatial noise may be treated in terms of two components. When dealing with long observation times, the effects treated in the preceding discussion become negligible and only the variations due to phosphor nonuniformities remain significant. These may be referred to as *screen granularity.* When the observation time element is short, the spatial noise introduced by the temporal fluctuations, too, must be considered. Its value is again given by (16.157)–(16.159), and its Wiener spectrum by $|\mathscr{F}[P(x, y)]|^2$.

Representative results of measurements permitting the estimation of screen granularity are shown in Figure 16.87 [156]. These represent

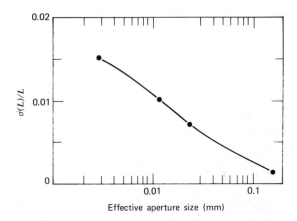

Figure 16.87 Phosphor screen granularity versus scanning aperture diameter. The ordinates are the standard deviation of luminance relative to the luminance. Haseyawa [156].

the specific standard deviation of intensity as a function of scanning aperture diameter.

Total Noise. The combined specific variance is obtained as the sum of the specific variance values of the components and the resultant signal-to-noise ratio is

$$R_T^2 = \frac{R_t^2 R_s^2}{R_t^2 + R_s^2}.$$ (16.164)

Temporal noise equals spatial noise when

$$\frac{E\eta_c \bar{A}\tau g_p}{(1+g_p)q_e} = R_s^2,$$ (16.165)

that is, when the irradiation equals

$$E = \frac{q_e(1+g_p)R_s^2}{\eta_c \bar{A}\tau g_p} \approx \frac{1.6 \times 10^{18}}{\eta_c \bar{A}\tau} R_s^2.$$ (16.166)

If the irradiation is substantially less than this value, the time-dependent noise (16.157)–(16.159) controls; if the irradiation is much higher than this, only screen granularity need be considered.

Blurring due to target motion results in a signal attenuation that increases with increasing spatial frequency. Specifically, it can be shown that the effect of uniform motion can be represented as an mtf of the form [157]

$$T_m(\nu) = \frac{1}{\sqrt{1 - \left(\dfrac{2\pi\nu\upsilon}{\alpha}\right)^2}},$$ (16.167)

where υ is the image velocity on the phosphor screen and the result is valid for a phosphor decaying according to

$$L(t) = L(0)e^{-\alpha t};$$ (16.168)

see Section 18.4.1.5.

Liminal Resolving Power [102]. Occasionally it is convenient to rate image tubes in terms of the size of the resolution element. Such an element size may be computed on the basis of an assumed threshold signal-to-noise ratio, required for element detection. Since the display size is often readily variable, the resolution is often rated as the number of elements (N_h) in the image height. Denoting the ratio of image width (w) to height (h) by

$$\alpha = \frac{w}{h},$$ (16.169)

we find the total number of image elements

$$N_T = \alpha N_h^2 \qquad (16.170)$$

and the element area, A_E, in terms of the total display area, A_T, as

$$A_E = \frac{A_T}{N_T} = \frac{A_T}{\alpha N_h^2}, \qquad (16.171)$$

where we assume the resolution element dimensions to be equal in the h and w directions.

We substitute this for $\bar{A}$ into (16.163) and that, in turn, into (16.164); we may, then, equate (16.164) to the threshold S/N ratio and solve for N_h. The result is relatively simple if we limit our consideration to low irradiation levels. We then simply substitute (16.171) into (16.163) and solve for N_h to find

$$N_h = \sqrt{\frac{\eta_c A_T \tau g_p}{\alpha (E_b + E_s) q_e (1 + g_p)} \frac{E_s}{R_t}}, \qquad (16.172)$$

where we now interpret R_t as the assumed threshold signal-to-noise ratio.

16.4.6.2   Optical–Electrical Image Tubes.
As described in Sections 16.4.3 and 16.4.4, in many optical–electrical image tubes the electron image is accumulated on a target plate and then read by means of a scanning beam. This method increases the signal-to-noise ratio due to the accumulation of the charge over the complete frame period. On the other hand, additional shot noise is added by the read beam.

Noise in Image Orthicons and Isocons. Noise in the image orthicon write section may be calculated analogously to that in a vacuum photo-diode (see Section 16.2.4). Assuming that the photoelectric emission approximates Poisson statistics, we have for the signal-to-shot-noise power ratio

$$R_w^2 = n_E = \frac{q_E}{q_e} = \frac{i_E \tau_f}{q_e} = \frac{\eta_c E A_E \tau_f}{q_e}, \qquad (16.173)$$

where n_E is the number of electrons accumulated in one signal element, q_E and i_E are the charge and current, respectively, corresponding to n_E,
  τ_f is the frame period, the time over which charge is accumulated before reading,
  q_e is the charge of an electron,
  η_c is the cathode sensitivity (in A/W),
  E is the mean image irradiation (in W/m^2), and
  A_E is the effective element area.

In the read beam, the shot-noise-current-squared will be [see (16.45)]

$$i_N^2 = \frac{i_c q_e}{\tau_E} = \frac{(i_r - q_E/\tau_E)q_e}{\tau_E} = \frac{q_e}{\tau_E}\left(i_r - \frac{i_E \tau_f}{\tau_E}\right), \qquad (16.174)$$

where i_r is the current in the incident read beam,
$\quad\;\; i_c$ is the current in the returning read beam, and
$\quad\;\; \tau_E$ is the time for reading out one element:

$$\tau_E = \frac{\tau_f}{N_T}, \qquad A_E = \frac{A_T}{N_T}, \qquad (16.175)$$

where $\;\;N_T$ is the total number of resolution elements in the raster and
$\quad\;\; A_T$ is the total area of the raster.

On the other hand, the signal current is

$$i_s = \frac{q_E}{\tau_E} = \frac{i_E \tau_f}{\tau_E} = i_E N_T. \qquad (16.176)$$

Thus the signal-to-noise power ratio for the read section is

$$R_r^2 = \frac{i_s^2}{i_N^2} = \frac{i_E^2 N_T^2 q_e}{\eta_c E A_E \tau_f} = \frac{i_E^2 N_T^3 q_e}{\eta_c E A_T \tau_f}. \qquad (16.177)$$

The combined signal-to-noise ratio may be calculated according to

$$R_T^2 = \frac{R_w^2 R_r^2}{R_w^2 + R_r^2}. \qquad (16.178)$$

In the ideal *isocon* scan, the return beam current is identical with the signal current, which, in turn, is a fraction (say k) of the average current ($N_T i_E$) supplied to the element being read. Thus

$$i_s = i_c = k N_T i_E \qquad (16.179)$$

and

$$i_N^2 = \frac{k N_T i_E q_e}{\tau_E}. \qquad (16.180)$$

Thus the signal-to-noise power ratio in the read section is here

$$R_T^2 = \frac{k N_T i_E \tau_E}{q_e}. \qquad (16.181)$$

By way of illustration, a general purpose image orthicon tube (No. 5820), operated at 4.5-MHz bandwidth, has a signal-to-noise ratio that rises from 10 to 45 as the illumination rises from 10^{-2} lux to the knee. This corresponds to a detective quantum efficiency of 0.025.

Noise in Vidicons. Noise in the vidicon is essentially similar to that in the analogous photoconductor (see Section 16.3.3) with the effective bandwidth, $\Delta\nu$, replaced by $1/2\tau_f$.

Because of the relatively long frame time usually employed, the time constants of the detector are likely to have a negligible effect on the bandwidth of the image tube. This will rather be determined by the spatial frequency transfer function—with the spatial frequency multiplied by the scanning velocity to yield the corresponding temporal frequency.

16.5 THERMAL DETECTORS

The detectors discussed up to this point are activated in a quantum mode; each absorbed quantum, on the average, contributes a certain amount to the signal output. The thermal detectors, to be discussed in this section, respond due to a change in temperature which is a function of the absorbed energy and independent of its wavelength; their spectral response is determined solely by their spectral absorptivity and is therefore generally much more uniform than that obtained with a simple photon detector.

Thermal detectors are mostly elemental, that is, they measure the total radiation incident on their sensitive region, regardless of its distribution (Section 16.5.1). Thus scanning must be used to observe an irradiation distribution. However, image-forming thermal detectors, too, have been developed. Three of these are described in Section 16.5.2. Noise characteristics are discussed in Section 16.5.3.

16.5.1 Elemental Thermal Detectors

In principle, any substance having a measurable physical characteristic that changes in a predictable manner with temperature can be used as a thermal radiation detector. In practice, there are, however, primarily only four types of such detectors in use: (a) the bolometer, (b) the thermocouple, (c) the pyroelectric, and (d) the pneumatic detector. Here a subsection is devoted to each of these.

The temperature change experienced by such a detector element due to energy incident at the rate Φ may be found by equating the rate of temperature change (multiplied by the heat capacity $\mathscr{C}$) to the net rate at which power is supplied to the detector element[22]:

$$\mathscr{C}\frac{d\,\Delta T}{dt} = \varepsilon\Phi - \mathscr{G}\,\Delta T, \tag{16.182}$$

---

[22] Here we neglect temperature-dependent power generation internal to the detector and energy losses by radiation. Convection losses are included in $\mathscr{G}$.

where ΔT is the difference between the detector and ambient temperature,

    ε is the detector element emissivity, elsewhere designated by e and

    $\mathcal{G}$ is the thermal conductivity from the element to its environment.

Under equilibrium conditions, the members of this equation vanish, and

$$\Delta T = \frac{\varepsilon \Phi}{\mathcal{G}}. \tag{16.183}$$

With a sinusoidally varying Φ (using complex notation, see Chapter 2, Note 4):

$$\Phi = \Phi_0 e^{i\omega t}, \tag{16.184}$$

the temperature, too, will vary sinusoidally with, say, amplitude ΔT_0. If changes in the values of the coefficients can be neglected,

$$\Delta T_0 (\mathcal{G} + i\omega \mathcal{C}) = \varepsilon \Phi_0. \tag{16.185}$$

Thus

$$|\Delta T_0| = \frac{\varepsilon \Phi_0}{\sqrt{\mathcal{G}^2 + \omega^2 \mathcal{C}^2}}. \tag{16.186}$$

16.5.1.1 Bolometers [1d], [9e].

General Description. The operation of the bolometer is based on the change of resistance exhibited by a substance when its temperature is changed. When a current is passed through this substance, this change in resistance can be detected as a change in voltage. Circuit arrangements optimizing the detector performance have been investigated [158]. Frequently a circuit as shown in Figure 16.88 [1d] is used. The flux incident

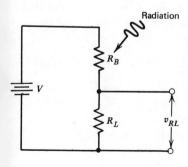

Figure 16.88 Typical bolometer circuit.

on the bolometer element, R_B, changes its temperature and, consequently, its resistance by an amount ΔR_B. As a consequence, the current, and hence the voltage (V_L) across the load resistor, R_L, changes.

Specifically, this voltage is

$$V_L = \frac{VR_L}{R_L + R_B} = \frac{V}{1 + R_B/R_L},\qquad (16.187)$$

where V is the applied voltage. The signal is the negative change, $-\Delta V_L$, in this voltage:

$$-\Delta V_L \approx \frac{-dV_L}{dR_B}\Delta R_B = \frac{VR_L\,\Delta R_B}{(R_L + R_B)^2}$$

$$= \frac{V_L\,\Delta R_B}{R_L + R_B}.\qquad (16.188)$$

In terms of fractional changes

$$\delta V_L = \frac{\Delta V_L}{V_L}\quad\text{and}\quad \delta R_B = \frac{\Delta R_B}{R_B} \approx \frac{dR_B}{dT}\frac{\Delta T}{R_B} = \alpha\Delta T,\quad (16.189)$$

this becomes

$$-\delta V_L = \left(1 + \frac{R_L}{R_B}\right)^{-1}\delta R_B = \frac{\alpha\,\Delta T}{1 + R_L/R_B},\qquad (16.190)$$

where we have denoted the specific temperature dependence of the resistance by

$$\alpha = \frac{1}{R_B}\frac{dR_B}{dT}.\qquad (16.191)$$

We must now consider the change in temperature and resistance resulting from a given radiant flux, Φ, absorbed by the bolometer element.

Consider first the temperature attained by R_B under steady-state conditions with no flux incident. Then the heat power (W) generated in the bolometer element by the current must equal the flow to the environment

$$W = I^2 R_B(T_i) = \mathscr{G}_{10}(T_1 - T_0) \approx \mathscr{G}(T_1 - T_0),\qquad (16.192)$$

where $\mathscr{G}_{10}$ is the effective thermal conductance from the element (at temperature T_1) to the environment (at temperature T_0), including both radiative and conductive terms. If the temperature difference is small, $\mathscr{G}_{10}$ may be approximated by the conductance $\mathscr{G}$ at T_1, as indicated in the last member of (16.192).

If the element is now irradiated, the temperature of the element will change by an amount ΔT, which may be evaluated by equating the rate of change of internal energy ($\mathscr{C}\,\Delta T$) to the sum of (a) the absorbed flux ($\varepsilon\Phi$), (b) the change in ohmic heating (ΔW), and (c) the change in negative energy losses to the environment:

$$\frac{\mathscr{C}\,d\,\Delta T}{dt} = \varepsilon\Phi + \Delta W - \mathscr{G}\,\Delta T. \tag{16.193}$$

Here ε is the emissivity which is identical to the fractional absorption. The change ΔW, in turn, is found by approximating

$$\Delta W = \frac{dW}{dT}\Delta T = \frac{d}{dT}(I^2 R_B)\,\Delta T = \frac{d}{dT}\left[\frac{V^2 R_B}{(R_L + R_B)^2}\right]\Delta T$$

$$= \frac{V^2 R_B}{(R_L + R_B)^2}\frac{R_L - R_B}{R_L + R_B}\alpha\,\Delta T = \frac{R_L - R_B}{R_L + R_B}W\alpha\,\Delta T. \tag{16.194}$$

Substituting for W from (16.192) into this equation and the result for ΔW into (16.193), we obtain

$$\mathscr{C}\frac{d\,\Delta T}{dt} + \mathscr{G}_e\,\Delta T = \varepsilon\Phi, \tag{16.195}$$

where we have written for short

$$\mathscr{G}_e = \mathscr{G} - \frac{\Delta W}{\Delta T} = \mathscr{G}\left[1 - (T_1 - T_0)\alpha\frac{R_L - R_B}{R_L + R_B}\right]. \tag{16.196}$$

The solution of this is

$$\Delta T = \frac{\varepsilon\Phi}{\mathscr{G}_e}(1 - e^{-\mathscr{G}_e t/\mathscr{C}}) \tag{16.197}$$

for a constant Φ. With the incident flux sinusoidal, say

$$\Phi = \Phi_0 e^{i\omega t}, \tag{16.184}$$

the solution is

$$\Delta T = \frac{(1 - e^{-\mathscr{G}_e t/\mathscr{C}})\varepsilon\Phi_0 e^{i\omega t}}{\mathscr{G}_e + i\omega\mathscr{C}}. \tag{16.198}$$

This may be substituted into (16.190) to obtain the instantaneous voltage change observed. After steady-state conditions are attained, the real exponential will vanish, provided $\mathscr{G}_e$ is positive. We have then the voltage change for constant signal and the amplitude of the sinusoidal voltage fluctuations, respectively,

$$\delta V_{L_0} = \frac{\alpha \, \Delta T}{1 + R_L/R_B} = \frac{\alpha \varepsilon \Phi}{\mathscr{G}_e(1 + R_L/R_B)} \tag{16.199}$$

and

$$|\delta V_{L_s}| = \frac{\alpha \, |\Delta T|}{(1 + R_L/R_B)} = \frac{\alpha \varepsilon \Phi_0}{(\mathscr{G}_e^2 + \omega^2 \mathscr{C}^2)^{1/2}(1 + R_L/R_B)}. \tag{16.200}$$

These results are particularly useful when

$$\Delta T \ll T_1 - T_0.$$

Inspection of (16.196)–(16.198) indicates that the bolometer element will heat up without limit, and hence burn itself out, unless $\mathscr{G}_e$ is positive. The condition for this may be written from (16.196) as

$$(T_1 - T_0)\alpha \frac{R_L - R_B}{R_L + R_B} < 1. \tag{16.201}$$

In the usual arrangement, $R_L \gg R_B$ to maximize ΔV_L. Then (16.201) simplifies to

$$\alpha(T_1 - T_0) < 1. \tag{16.202}$$

Occasionally, the responsivity, k (in volts/incident watt) is of interest. We readily find, for the circuit of Figure 16.88,

$$k_0 = \frac{\Delta V_L}{\Phi} = \frac{V_L \delta V_{L_0}}{\Phi} = \frac{\alpha \varepsilon V}{\mathscr{G}_e(R_L + R_B)^2} \tag{16.203}$$

and

$$k_s = \frac{\Delta V_L}{\Phi_0} = \frac{\alpha \varepsilon V}{(\mathscr{G}_e^2 + \omega^2 \mathscr{C}^2)^{1/2}(R_L + R_B)^2}, \tag{16.204}$$

for steady and sinusoidal signals, respectively, where we have substituted from (16.187) and (16.199).

Metal Bolometers. Metal bolometers are made in the form of wires or films. The resistance of a metal conductor varies with temperature according to

$$R = R_0[1 + \gamma(T - T_0)], \tag{16.205}$$

where R_0 is the resistance at temperature T_0. The constant, γ, generally has a value in the neighborhood of $0.005°C^{-1}$. Accordingly, on substituting (16.205) into (16.191), we have

$$\alpha = \left(\frac{1}{\gamma} + T - T_0\right)^{-1} \tag{16.206}$$

and the condition (16.202) is always satisfied for positive γ, as long as the assumptions, such as ($\mathscr{G}_{10} = \mathscr{G}$), are satisfied.

Thermistor Bolometers. Semiconductors exhibit a larger temperature dependence of resistivity. Some materials, consisting of sintered oxides of manganese, cobalt, and nickel, were especially designed for use as thermally sensitive resistors, *thermistors*, for short. The thermistor is generally made in the form of a flake approximately 10 μm thick and mounted on a heat absorbing substrate, the thermal sink. Frequency response depends on the thermal conductance between the flake and this sink. The flake may also be gas-backed, yielding higher sensitivity at the cost of slower response.

The temperature dependence of resistance of semiconductor materials has the form

$$R = R_0 e^{\beta/T}, \tag{16.207}$$

so that its thermal coefficient of resistivity is, from (16.191),

$$\alpha = \frac{-\beta}{T^2}. \tag{16.208}$$

Since this is negative, (16.202) is always satisfied. However, if $R_B > R_L$ inspection of (16.201) shows that spontaneous burnout becomes a possibility and should be considered in setting the operating range.

Superconducting Bolometer. The superconducting bolometer is based on the rapid transition from normal conductivity to superconductivity exhibited by certain materials when cooled to near absolute zero. By means of placement in a Dewar flask (see Section 16.3.4.4) the resistor element is maintained near the bottom of the transition range; the incident radiation will then cause a small rise in temperature, which will result in a large rise in resistance.

Operation at low temperatures also reduces the thermal noise [see Section 16.5.3 (16.236)] and the heat capacity of the detector element. However, the required cooling to the temperature of liquid helium and maintenance of the temperature to an accuracy of approximately 10^{-4} K pose major technical problems.

16.5.1.2 Thermocouples and Thermopiles [1e], [9f].

The thermocouple is another thermal detector. It consists, essentially, of two wires of dissimilar materials joined to each other at both ends. If the two junctions are maintained at different temperatures, a thermoelectric voltage is generated in the circuit and, as a result, a current will flow in it. If we maintain one junction at the ambient temperature, T_0, and expose the other to incident radiation, the resulting temperature rise will cause a thermoelectric current that may be used to measure the temperature rise and, hence, the incident power.

The generated voltage is found to be given by

$$V = P_{AB} \Delta T, \qquad (16.209)$$

where ΔT is the temperature difference between the two junctions, and

$$P_{AB} = P_A - P_B$$

is the difference between the *thermoelectric power* of the two materials A and B from which the thermocouple is constructed. Representative values of P are given in Table 175 [9f]. From there it can be seen, for example, that P_{AB} for a bismuth-antimony thermocouple is 100 $\mu V/°C$.

To evaluate the temperature rise resulting from a given amount of incident radiation, we must consider not only the thermal conductance, $\mathcal{G}$, but also the *Peltier effect*, which contributes to the cooling of the "hot" junction, by removing energy at the rate of

$$\Delta W = P_{AB} i T, \qquad (16.210)$$

where i is the current flowing through the junction, and
T is the absolute temperature of the junction.

Hence, at equilibrium, the absorbed fraction, ε, of the incident energy flux, Φ, must equal the sum of the losses due to heat conduction and Peltier effect:

$$\varepsilon \Phi = \mathcal{G} \Delta T + P_{AB} i T, \qquad (16.211)$$

where i, in turn, is given by the generated voltage, V, divided by the total circuit resistance, R. Hence

$$\varepsilon \Phi = \mathcal{G} \Delta T + \frac{P_{AB} T V}{R} = \Delta T (\mathcal{G} + T P_{AB}^2 / R), \qquad (16.212)$$

where we have made use of (16.209).

The corresponding responsivity is now readily found as

$$k_0 = \frac{V}{\Phi} = \frac{\varepsilon P_{AB}}{\mathcal{G} + P_{AB}^2 T / R}. \qquad (16.213)$$

As the circuit resistance, R, approaches infinity, the Peltier cooling becomes negligible, and we find

$$k_0' = \frac{\varepsilon P_{AB}}{\mathcal{G}}. \qquad (16.214)$$

This result shows that both a large value of P_{AB} and a small value of thermal conductance are required to ensure good responsivity. Therefore thermocouples are often placed in evacuated enclosures to reduce the cooling by conduction and convection through the air.

Thermocouples may be cascaded; this arrangement is called a *thermopile*. Clearly, alternating junctions in the resulting chain must be held at the reference temperature, while the others are exposed to the incident flux. The resulting voltage is then the sum of the voltages generated by the component thermocouples. However, since the incident flux is divided among these junctions, the responsivity is identical with that of a single thermocouple.

16.5.1.3 Pyroelectric Detectors [159]-[165].

Below their Curie temperature, ferroelectric materials are constituted of microscopic domains with oppositely oriented polarization. If we denote by f the fraction of domains oriented in one direction, we see that a spontaneous bound charge density

$$q' = [f - (1-f)]P_s = (2f-1)P_s \qquad (16.215)$$

will appear on the surface of the material. This is eventually neutralized by free charges, which flow through the material at a rate determined by its resistivity. The constant, P_s, is called spontaneous polarization and *varies with temperature*. This phenomenon is called *pyroelectricity* and may be used to detect changes in temperature.

Because of the spontaneous charge neutralization, pyroelectric detectors are used with alternating incident flux, that is, the flux is modulated by means of a chopper.

Since the signal generation here takes place across a capacitor, rather than a resistor, useful high-frequency response values are possible. Pyroelectric detectors can be operated in the megahertz range and, with heterodyne techniques, noise equivalent input power (NEP) of 2×10^{-15} W/Hz$^{1/2}$ has been obtained [161b].

The temperature dependence is especially pronounced as we approach the temperature at which ferroelectricity disappears, the *Curie point*. Figure 16.89 [159] shows the temperature dependence of P_s for triglycine sulfate (TGS); note especially the large slope as the Curie temperature is approached at 48°C.

The change in polarization (ΔP), or charge density, due to a temperature change ΔT, is

$$\Delta P = K_p \, \Delta T, \qquad (16.216)$$

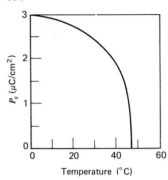

Figure 16.89 Spontaneous polarization (P_s) of TGS versus temperature in the neighborhood of the Curie point. Hadni [159].

where

$$K_p = \frac{\partial P_s}{\partial T} \qquad (16.217)$$

is the thermal dependence of polarization. Hence the charge, q, on the detector is

$$q = C\,\Delta V = A\,\Delta P = AK_p\,\Delta T, \qquad (16.218)$$

where C is the capacitance,
  ΔV is the change in voltage, and
  A is the surface area of the detector plate.

Electrically, the action of the detector may be described by a current source, I, in parallel with a detector capacitance, C, an internal leakage resistance, R_i, and a load resistance, R_L (see Figure 16.90). Assuming a sinusoidally varying current

$$I = I_0 e^{i\omega t},$$

the output voltage, too, will be sinusoidal with amplitude v_0 given by

$$v_0 = \frac{I_0}{\sum_j Z_j^{-1}} = \frac{I_0}{i\omega C + 1/R_i + 1/R_L}$$

$$= \frac{I_0}{i\omega C + G}. \qquad (16.219)$$

Here Z_j is the impedance of the jth component.
Hence

$$|v_0| = I_0(G^2 + \omega^2 C^2)^{-1/2}, \qquad (16.219a)$$

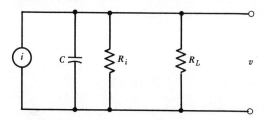

Figure 16.90 Pyroelectric detector, equivalent circuit.

where we have set the equivalent conductance

$$G = \frac{1}{R_i} + \frac{1}{R_L}. \tag{16.220}$$

The current can be derived from (16.218) as

$$i = \frac{dq}{dt} = AK_p \frac{d\,\Delta T}{dt}. \tag{16.221}$$

For sinusoidal variations of incident flux, we have [see (16.186)]

$$|\Delta T| = \varepsilon \Phi_0 e^{i\omega t} [\mathscr{G}^2 + \omega^2 \mathscr{C}^2]^{-1/2} \tag{16.222}$$

and hence

$$\frac{d\,|\Delta T|}{dt} = \varepsilon \Phi_0 \omega [\mathscr{G}^2 + \omega^2 \mathscr{C}^2]^{-1/2} e^{i\omega t}. \tag{16.223}$$

On substituting this into (16.221) we find for I_0,

$$I_0 = AK_p \varepsilon \Phi_0 \omega [\mathscr{G}^2 + \omega^2 \mathscr{C}^2]^{-1/2}. \tag{16.224}$$

Hence, using (16.219a), we find for the responsivity

$$k_s = \frac{|v_0|}{\Phi_0} = AK_p \varepsilon (\mathscr{G}^2 + \omega^2 \mathscr{C}^2)^{-1/2} (G^2 + \omega^2 C^2)^{-1/2}. \tag{16.225}$$

Since $\mathscr{G}/\mathscr{C}$ is of the order of magnitude of 1 Hz, usually $\omega \gg \mathscr{G}/\mathscr{C}$; we may then write

$$k_s = \frac{AK_p \varepsilon}{\omega \mathscr{C} G} [1 + (\omega \tau_e)^2]^{-1/2}, \tag{16.226}$$

where $\tau_e = C/G$ is the electrical time constant of the circuit.

Note that the heat capacity is given by

$$\mathscr{C} = c'Ad, \tag{16.227}$$

where c' is the specific heat capacity and
  d is the thickness of the detector element.

The electrical conductance is given by

$$G = \varepsilon_0 \varepsilon'' \omega, \tag{16.228}$$

where ε_0 is the permittivity of space and
  ε'' is the imaginary component of the dielectric constant.

A number of materials useful as pyroelectric detectors have been investigated intensively. We list their relevant characteristics in Table 176. Because triglycine sulfate (TGS) and triglycine selenate (TGSe) have high pyroelectric coefficients (K_p) near room temperature, they are the most popular pyroelectric materials. The quality factor, Q, listed in Table 176, is defined in (16.240). More recently, it has been found that doping TGS with alanine improves its stability considerably and its detectivity somewhat [161b].

16.5.1.4 Pneumatic Detectors [1f], [9g]. Photoacoustic Effect. In the pneumatic detector, the absorbed flux is used to heat the gas in an enclosure. The resulting increase in gas pressure causes the deformation of a flexible mirror that is part of the enclosure surface. This mirror is illuminated via a grid of alternating clear and opaque strips. After reflection from the mirror, when quiescent, these are imaged upon themselves in such a way that the opaque strips are imaged on the clear spaces and *vice versa*, so that the passage of light is blocked. When the mirror bulges due to the absorbed flux, some light passes and is detected. This detector is known as the *Golay cell*.[23]

This detector potentially approaches an ideal detector in performance. It deviates from this due to reflections from the window, incomplete absorption of the incident flux, and thermal losses to the housing. Its response time is approximately 20 msec. Its fragility is a limiting factor in some applications.

In commercially available cells, the pneumatic connections to the atmosphere external to the gas cell are constructed to minimize the effects of fluctuations in ambient temperature, on the one hand, and, on the other hand, to respond to reasonably rapid fluctuations in the incident flux. See Figure 16.91. The cell is usually used in conjunction with a chopper at about 10 Hz. Normal detectivity is about 10^{10} W^{-1}.

Only in passing we note that the absorption of pulsed optical flux generates pressure waves in solids, liquids, and gases. These waves can

[23] M. J. E. Golay (1947).

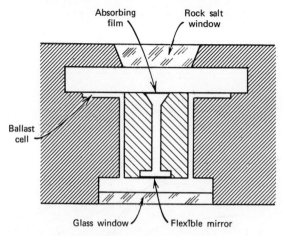

Figure 16.91 Golay cell, section.

then be detected by a "microphone" such as piezoelectric crystal. This effect is called *photoacoustic* and has been used to measure absorption spectra [165*] and in microscopy [165**].

See Figure 16.92 [161*b*] for a comparison of the detectivities of various uncooled thermal detectors.

16.5.2 Thermal Image Detectors [165]

We now describe briefly two systems of thermal detection capable of displaying an image directly and one thermal optical-electrical image transducer.

16.5.2.1 Evaporograph [1g], [166]. Already in 1840, Sir John Herschel used differential evaporation of moisture to make an infrared image visible. More recently, this principle was developed into a much more sophisticated and sensitive device, the *evaporograph* [166], [167].

This device is illustrated in Figure 16.93 [166]. It is based on the differential evaporation of a layer of oil from a reflecting surface, with the rate of evaporation, and therefore the layer thickness, controlled by the flux density of the image formed on the surface. Due to interference effects, the thickness of the layer of liquid at any point, in turn, controls the color appearing there.

The evaporograph cell (Figure 16.94 [166]) consists of two chambers, both connected to a vacuum pump, but separated by a thin plastic membrane. The chamber facing the viewer is provided with a heater and is filled with oil vapor, which tends to condense on the membrane. The

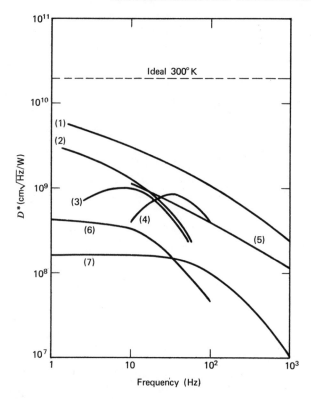

Figure 16.92   Performance of some uncooled thermal detectors.
(1) TGS: alanine (1.5 mm × 1.5 mm, 10 μm thick);
(2) spectroscopic thermopile (0.4 mm²);
(3) golay cell; (4) TRIAS cell;
(5) TGS, ruggedized (0.5 mm × 0.5 mm);
(6) evaporated film thermopile (0.12 mm × 0.12 mm);
(7) immersed thermistor (0.1 mm × 0.1 mm). Putley [161b]. For specific references, see that article.

membrane is provided with a black layer that absorbs the flux incident on it from the object side. The resulting change in temperature of the membrane controls the rate of oil condensation and, therefore, the thickness of the layer.

In the system illustrated in Figure 16.93, the image is formed by a germanium lens through an infrared filter and the cell input window. The interference image is made visible by illuminating the oil film from the viewing side via a beam splitter.

An instrument built on this principle has attained a "sensitivity" of 1°C

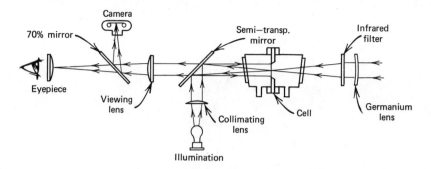

Figure 16.93 Evaporograph, schematic. McDaniel and Robinson [166].

under favorable conditions. With a raster varying between 20°C and 30°C, it resolved 10 lines/mm on the membrane. Response time varies between 15 sec (for the image of a man) and a fraction of a second for an object at 300°C.

16.5.2.2 Absorption Edge Image Converter [1h], [168]. The spectral location of the absorption edge of semiconductors varies with temperature. In conjunction with the rapid change of transmittance with wavelength in the absorption edge region, this implies a rapid change of transmittance with temperature for flux whose wavelength falls into this region. Figure 16.95 illustrates how a small shift, $\Delta\lambda$, in the absorption edge results in a large shift, from a to b, in density.

A practical unit built on this principle is illustrated in Figure 16.96 [168]. The detector is a 1-μm-thick film of amorphous selenium (with a

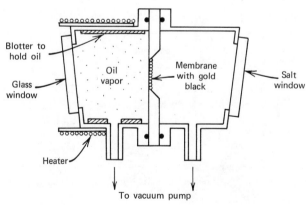

Figure 16.94 Evaporograph cell, cross-sectional schematic. McDaniel and Robinson [166].

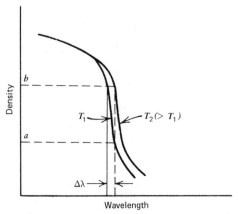

Figure 16.95   Absorption edge shift. Kruse *et al.* [1*h*].

film of chromium to enhance absorption). The detector is illuminated with light from a sodium arc lamp, whose flux wavelength lies in the center of the absorption edge of the selenium. The concave mirror forms an infrared image of the object on the detector, which is viewed through an aperture in that mirror. The time constant of this device is of the order of magnitude of a second, and objects 10°C above ambient can certainly be observed

16.5.2.3   Pyroelectric Vidicons [169], [170]. With a target plate of pyroelectric material, a vidicon can be made to operate in the 8–14 μm band. Presently such devices can detect temperature differences as small as 0.2°C and 1°C at 0 and 3 mm^{-1} spatial frequencies, respectively.

Since the pyroelectric detector senses only the change in temperature, objects at a constant temperature produce no signal unless some artificial

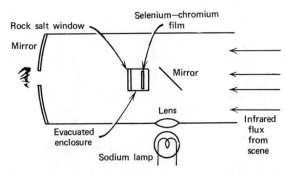

Figure 16.96   Absorption edge image converter. Kruse *et al.* [1*h*].

change is introduced. The required changes can be produced by chopping, as in the elemental detector, or by scanning ("panning") the image across the pyroelectric plate. The latter appears to yield better results [170]. Since this alternating component of the output signal is bipolar, depending on the sense of the temperature change, a pedestal signal must be provided to permit reading out by the electron beam, which can accommodate only positive signals.

The resolution of these devices is generally limited to spatial frequencies below 3–4 mm^{-1}. The mtf is anisotropic because of the scanning raster employed and becomes even more so, when the panning mode is used to introduce the required temperature fluctuations [171].

Like the elemental pyroelectric detectors, the targets in these tubes are usually made of TGS. But important advantages (e.g., operability in a vacuum) are gained by using lower resistivity materials, such as lead germanate [172]. Pyroelectric vidicons hold promise in medical, industrial, and surveillance applications [173].

16.5.3 Noise in Thermal Detectors and Their Comparison

Noise. In addition to Johnson noise, temperature noise (i.e., random fluctuations in the detector temperature) plays an important role in thermal detectors. These fluctuations are caused by the random absorption and emission by the detector of energy in the form of photons. The resulting fluctuations are analogous to shot noise, but differ in that energy carried by photons, rather than the number of photons, is the significant fluctuating quantity.

By analogy with (16.46), the fluctuations in the monochromatic radiant flux $\Phi(\lambda)$ emitted from a radiator are

$$\Phi_n^2 \approx 2 \frac{hc}{\lambda} \Phi \, \Delta \nu_t, \qquad (16.229)$$

where hc/λ is the energy per photon at the wavelength, λ, of the photon and

$\Delta \nu_t$ is the signal band width over which the detector follows temperature fluctuations.

This expression is based on an assumed Poisson distribution. Because the photons obey the statistics of indistinguishable particles (Bose–Einstein statistics), an additional factor of

$$\frac{e^{c_2/\lambda T}}{e^{c_2/\lambda T} - 1} \qquad (16.230)$$

must be supplied [174].

If we consider a grey body radiator (defined in Section 4.4.1), emissivity ε, and quasimonochromatic radiation of spectral width $\delta\lambda$, we have [see (4.47)]

$$\Phi = \frac{\varepsilon c_1 A \,\delta\lambda}{\lambda^5 (e^{c_2/\lambda T} - 1)}, \tag{16.231}$$

where A is the radiating surface area. Then, substituting this into (16.229) with factor (16.230) applied, we obtain for the fluctuation summed over the total spectrum

$$\Phi_n^2 = 2\varepsilon hcc_1 A \,\Delta\nu_t \int_0^\infty \frac{e^{c_2/\lambda T} \,d\lambda}{\lambda^6 (e^{c_2/\lambda T} - 1)^2}. \tag{16.232}$$

This integral may be evaluated on substituting the new variable of integration

$$u = \frac{c_2}{\lambda T}, \qquad d\lambda = -c_2 \frac{du}{u^2 T} \tag{16.233}$$

and noting the definite integral [86b], [86$^*b$]

$$\int_0^\infty \frac{u^4 e^u \,du}{(e^u - 1)^2} = 4! \,\zeta(4), \tag{16.234}$$

where the Riemann zeta function

$$\zeta(4) \equiv \sum_{n=1}^\infty n^{-4} = \frac{\pi^4}{90}. \tag{16.235}$$

This leads to the result

$$\Phi_n^2 = \tfrac{8}{15}\pi^4 ecc_1 c_2^{-5} hAT^5 \,\Delta\nu_t.$$

On substituting the values of c_2 [from (4.39b)] and c_1 in terms of the Stefan–Boltzmann constant c_3 [from (4.41)], this becomes

$$\Phi_n^2 = 8c_3 k\varepsilon AT^5 \,\Delta\nu_t. \tag{16.236}$$

The detectivity, D, is defined as the reciprocal of the noise-equivalent input power (NEP) and [see (3.102)]

$$D^{*2} = D^2 A \,\Delta\nu_t = (8c_3 k\varepsilon T^5)^{-1}. \tag{16.237}$$

When Johnson noise is the controlling factor, the noise voltage is [see (16.44)]

$$V_{nJ}^2 = \frac{W_{nJ}}{G} = \frac{4kT \,\Delta\nu_t}{G}, \tag{16.238}$$

where G is the conductance of the detector.

Considering, for instance, a pyroelectric detector in which the Johnson noise controls, we have

$$D^* = k_s \frac{\sqrt{A \, \Delta \nu_t}}{V_{nJ}}, \qquad (16.239)$$

where k_s is the responsivity in V/W. We substitute into this the values of the various factors from (16.226), (16.228), and (16.238); we then find that the parameters peculiar to the material constitute a "quality" factor

$$Q = \frac{K_p}{c' \sqrt{\varepsilon''}}. \qquad (16.240)$$

This, too, is listed in Table 176.

When the photon noise of the background radiation is considered, it is found that the detectivity takes the form [175]

$$D^* = 4 \times 10^{16} \left(\frac{\varepsilon}{T^5 + T_b^5} \right)^{1/2} \text{cm} \sqrt{\text{Hz}}/\text{W}, \qquad (16.241)$$

where ε is the detector emissivity and
T_b is the background temperature.

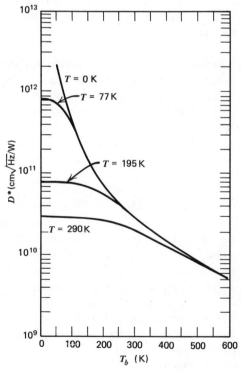

Figure 16.97   Thermal detector detectivity, theoretical limits. Limperis [175].

See Figure 16.97 [175] for the dependence of D^* on T_b for various values of detector temperature.

The values of D^* attainable with thermal detectors (excepting super-conducting bolometers) generally do not exceed $10^9 \, \text{cm}\sqrt{\text{Hz}}/\text{W}$ for a 500 K blackbody (see Table 168).

Comparison of Thermal Detectors. The choice of the thermal detector to be used will be dictated primarily by considerations of required sensitivity and operating frequency. The highest sensitivities are attainable only with bolometers cooled to approximately 2 K. This makes operation inconvenient and expensive. For operation at room temperature and at very low frequencies (<10 Hz), the thermocouple and pneumatic cell are optimum; the bolometer may be used at higher frequencies, whereas above 1 kHz the pyroelectric detectors must be used.

16.6 COMPARISON OF DETECTORS

We have already discussed considerations for choosing radiation detectors within specific groups (Section 16.3.2.5 for photoconductors and Section 16.5.3 for thermal detectors). Here we briefly compare detectors from different categories.

Thermal versus Photon Detectors. When they cover the required spectral range, photon detectors will be found, as a rule, to be more sensitive than thermal detectors and will therefore be preferred, except when they must be rejected for specific reasons. Such reasons may be the price, the cooling, which is required for all the photon detectors in the far infrared or, possibly, a limitation in the dynamic range. When a very broad, or uniform, spectral response is required, the thermal detectors may be superior.

Photoemissive versus Photoconductive. Photoemissive (vacuum tube) detectors are limited almost exclusively to the near ir range of the spectrum and below. Because of the feasibility of electron multiplication, their gain and overall detectivity is there generally superior to those obtainable with photoconductive detectors. However, when operation is required only at higher levels of illumination, where these advantages are not effective, the more convenient photoconductive detectors may be preferable even in this spectral range. Especially in applications where it is desirable to operate without an external power supply, the photovoltaic or photoelectromagnetic modes of operation of the photoconductor will make it the preferred device.

Photodiodes versus Homogeneous Photoconductors [83], [176]. It is found that the photodiode is superior to the homogeneous photoconductor at low light levels (milliwatts). On the other hand, because of the high intrinsic capacitance of the diode, its output circuit must be matched to the operating frequency and, with broad-band operation, the efficiency will be lower in diodes than in the homogeneous photoconductors.

In terms of responsivity, homogeneous photoconductors are superior because of the high gains of which they are capable, whereas the highest gain-bandwidth products are obtained in avalanche diodes.

APPENDIX 16.1 CONDITIONS FOR PERFECT IMAGING IN ELECTRON OPTICS

To derive the condition for perfect paraxial imaging in a radially symmetric optical system, refer to Figure 16.98 where O and I denote an object point and the corresponding image point, respectively, and P, P' the principal planes. Perfect imaging refers to the condition where all rays emanating from O converge into I. This, in turn, implies that the total angular refraction[24]

$$\Delta\alpha = \alpha_i' - \alpha_i = R_i\left(\frac{1}{l'} - \frac{1}{l}\right) = \frac{R_i}{f}$$ (16.242)

is proportional to the ray height R_i on the principal planes, where the constant of proportionality

$$\frac{1}{l'} - \frac{1}{l} = \frac{1}{f}$$

is the system power, which is defined as the reciprocal of the focal length. (The symbols used here are defined in Section 9.2.1.1.)

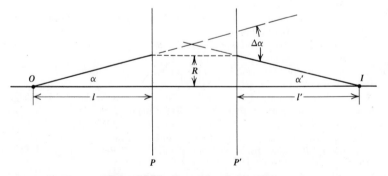

Figure 16.98 Paraxial optical imaging.

[24] Note that in Figure 16.98 both α and l are negative according to the convention of Section 9.2.

It can readily be shown that a radially symmetric electrostatic field satisfies requirement (16.242) paraxially. Consider a thin short cylindrical volume element coaxial with the field. The sum of the flux entering it must vanish (flux leaving it is considered negative). Thus we find that the flux entering through the walls of the cylinder must equal the excess of the flux leaving through the far end over that entering at the near end:

$$2\pi RZE_r = \pi R^2 \frac{dE_z}{dz} Z$$

or

$$E_r = \tfrac{1}{2} R \frac{dE_z}{dz}, \tag{16.243}$$

where R and Z are radius and length of the cylindric element, and εE is the electric flux density [see (2.3)], with subscripts r and z indicating the radial and axial components, respectively. Since the amount of deflection ($\Delta\alpha$) is proportional to E_r [see (16.144)], and this, in turn, is here proportional to R as evidenced by the last equation, we conclude that a radially symmetric field does satisfy requirement (16.242) for paraxial imaging.

Note also that according to (16.243) a voltage gradient in the direction of the axis implies a radial field component.

REFERENCES

[1] P. W. Kruse, L. D. McGlauchline, and R. B. McQuistan, *Elements of Infrared Technology*, Wiley, New York (1962); (a) Section 8.3; (b) Section 6.7; (c) Chapter 10; (d) Sections 8.4.1, 8.4.2, and 9.3.4; (e) Section 8.4.3; (f) Section 8.4.4; (g) Section 8.4.5; (h) Section 8.4.6.

[2] T. S. Moss, G. J. Burrell, and B. Ellis, *Semiconductor Opto-Electronics*, Butterworths, London, 1973, Chapter 5.

[3] W. E. Spicer and F. Wooten, "Photoemission and photomultipliers," *IEEE Proc.* **51,** 1119–1126 (1963).

[4] H. R. Zwicker, "Photoemissive detectors," in *Optical and Infrared Detectors*, R. J. Keyes, ed., Springer, Berlin, New York, 1977, Chapter 5.

[5] V. K. Zworykin and E. G. Ramberg, *Photoelectricity*, Wiley, New York, 1949; (a) Chapter 6; (b) Chapter 3; (c) Figure 6.6b; (d) Chapter 7; (e) Chapter 8.

[6] R. P. Riesz, "High speed semiconductor photodiodes," *Rev. Sci. Instr.* **33,** 994–998 (1962).

[7] P. J. Bishop, A. F. Gibson, and M. F. Kimmitt, "The performance of photon-drag detectors at high laser intensities," *IEEE J. Quant. Electr.* **QE–9,** 1007–1011 (1973).

[8] B. S. Patel, "Photon-drag Ge detectors for CO_2 lasers," *Proc. IEEE* **61,** 795–796 (1973).

[9] R. A. Smith, F. E. Jones, and R. P. Chasmar, *The Detection and Measurement of Infra-Red Radiation*, 2nd Ed., Clarendon, Oxford, 1968; (a) Sections 4.7, 4.8, and 3.2–3.6; (b) Section 5.3; (c) Sections 4.9–4.12; (d) Section 5.5; (c) Section 7.2; (f) p. 156; (g) p. 161.

[10] A. Rose, *Concepts in Photoconductivity and Allied Problems*, Interscience, New York, 1963; (a) Chapters 2, 4, and 8; (b) Section 3.11; (c) Sections 5.4 and 5.5.

[11] R. H. Bube, *Photoconductivity of Solids*, Wiley, New York, 1960; (a) Section 6.5; (b) Chapter 4; (c) Section 3.3; (d) "Photoconductors," in *Photoelectronic Materials and Devices*, S. Larach, ed., Van Nostrand Reinhold, New York, 1965, Chapter 2.

[12] L. K. Anderson and B. J. McMurtry, "High-speed Photodetectors," *Appl. Opt.* **5**, 1573–1587 (1966).

[13] N. S. Kopeika, R. Gellman, and A. P. Kushelevsky, "Improved detection of ultraviolet radiation with gas-filled phototubes through photoionization of excited atoms," *Appl. Opt.* **16**, 2470–2476 (1977).

[14] R. B. Green, R. A. Keller, G. G. Luther, P. K. Schenck, and J. C. Travis, "Galvanic detection of optical absorption in a gas discharge," *Appl. Phys. Lett.* **29**, 727–729 (1976); also "Use of an opto-galvanic effect to frequency-lock a carrier wave dye laser," *IEEE J. Quant. Electr.* **QE–13**, 63–64 (1977).

[15] B. Kazan and M. Knoll, *Electronic Image Storage*, Academic, New York, 1968; (a) Section I C; (b) Section I A; (c) Section V B; (d) Section V C; (e) Section I B; (f) Section V D; (g) Section VI B.

[16] Electronic Industries Association, Washington, D.C., (a) "Relative spectral response data for photosensitive devices," JEDEC Publ. No. 50, 1964; (b) "Relative spectral response curves for semiconductor infrared detectors," JEDEC Publ. No. 78, (1969).

[17] R. W. Engstrom, "Absolute spectral response characteristics of photosensitive devices," *RCA Rev.* **21**, 184–190 (1960).

[18] D. McMullan and J. R. Powell, "Residual gases and the stability of photocathodes," in Ref. 94d, pp. 427–439.

[19] W. A. Hiltner, ed., *Astronomical Techniques*, University of Chicago Press, Chicago, 1962; (a) A. Lallemand, "Photomultipliers," pp. 126–156; (b) W. A. Baum, "The detection and measurements of faint astronomical sources," pp. 1–33.

[20] J. R. Sizelove and J. A. Love III, "Analysis of translucent and opaque photocathodes," *Appl. Opt.* **5**, 1419–1422 (1966); also *ibid.* **6**, 356–357 (1967).

[21] R. W. Engstrom, in *RCA Phototubes and Photocells*, RCA, Lancaster, PA, 1963; (a) Figure 22; (b) Figure 24; (c) p. 35; (d) pp. 38–44; (e) pp. 62–63; (f) pp. 67–69; (g) p. 66; (h) p. 51.

[22] J. P. Keene, "Fatigue and saturation in photomultipliers," *Rev. Sci. Instr.* **34**, 1220–1222 (1963).

[23] L. B. Loeb, *Basic Processes of Gaseous Electronics*, University of California Press, Berkely, 1955.

[24] R. W. Engstrom and W. S. Huxford, "Time-lag analysis of the Townsend discharge in argon with activated caesium electrodes," *Phys. Rev.* **58**, 67–77 (1940).

[25] O. Hachenberg and W. Brauer, "Secondary electron emission from solids," in *Advances in Electronics and Electronic Physics*, Vol. 11, L. Marton, ed., Academic, New York, 1959, pp. 413–499.

[26] P. Wargo, B. V. Haxby, and W. G. Shepard, "Preparation and properties of thin film MgO secondary emitters," *J. Appl. Phys.* **22**, 1311–1316 (1956).

[27] G. Diemer and J. L. H. Jonker, "On the time delay of secondary emission," *Phil. Res. Repts.* **5**, 161–172 (1950).

[28] G. W. Goodrich and W. C. Wiley, "Continuous channel electron multiplier," *Rev. Sci. Instr.* **33**, 761–762 (1962).

[29] D. S. Evans, "Low energy charged-particle detection using the continuous-channel electron multiplier," *Rev. Sci. Instr.* **36,** 375–382 (1965), see also Ref. 109.

[30] K. C. Schmidt and C. F. Handee, "Continuous channel electron multiplier operated in pulse saturated mode," *IEEE Trans.* **NS–13,** 100–111 (1966).

[31] K. Oba and H. Maeda, "Impulse and frequency response of channel electron multipliers," in Ref. 94d, pp. 123–139.

[32] R. W. Engstrom, "Multiplier photo-tube characteristics: Application to low light levels." *J. Opt. Soc. Am.* **37,** 420–431 (1947).

[33] R. F. Post, "Performance of pulsed photomultipliers," *Nucleonics* **10,** 46–50 (1952).

[34] R. W. Engstrom and R. M. Matheson, "Multiplier phototube development at RCA Lancaster," *IRE Trans.* **NS–7**², 52–57 (1960).

[35] A. R. Boileau and F. D. Miller, "Changes in spectral sensitivity of multiplier phototubes resulting from changes in temperature," *Appl. Opt.* **6,** 1179–1182 (1967).

[36] A. T. Young, "Temperature effects in photomultipliers and astronomical photometry," *Appl. Opt.* **2,** 51–60 (1963).

[37] T. Hirschfeld, "Improvements in photomultipliers with total internal reflection sensitivity enhancement," *Appl. Opt.* **7,** 443–449 (1968).

[38] D. J. Baker and C. L. Wyatt, "Irradiance linearity corrections for multiplier phototubes," *Appl. Opt.* **3,** 89–91 (1964).

[39] P. L. Land, "A discussion of the region of linear operation of photomultipliers," *Rev. Sci. Instr.* **42,** 420–425 (1971).

[40] J. P. Rodman and H. J. Smith, "Test of photomultipliers for astronomical pulse-counting applications," *Appl. Opt.* **2,** 181–186 (1963).

[41] H. J. Marrinan, "The use of photomultipliers in Raman spectroscopy," *J. Opt. Soc. Am.* **43,** 1211–1215 (1953).

[42] J. A. Baicker, "Dark current in photomultipliers," *IEEE Trans.* **NS–7,** 74–80 (1960).

[43] E. H. Eberhardt, "Threshold sensitivity and noise ratings of multiplier phototubes," *Appl. Opt.* **6,** 251–255 (1967).

[44] F. J. Lombard and F. Martin, "Statistics of electron multiplication," *Rev. Sci. Instr.* **32,** 200–201 (1961).

[45] H. J. Gale and J. A. B. Gibson, "Methods of calculating the pulse height distribution at the output of a scintillation counter," *J. Sci. Instr.* **43,** 224–228 (1966).

[46] R. F. Tusting, Q. A. Kerns, and H. K. Knudsen, "Photomultiplier single-electron statistics," *IEEE Trans.* **NS–9**³, 118–123 (1962).

[47] R. W. Engstrom, R. G. Stroudenheimer, and A. M. Glover, "Production testing of multiplier phototubes," *Nucleonics* **10,** 58–62 (1952).

[48] G. C. Baldwin and S. I. Friedman, "Statistics of single-electron multiplication," *Rev. Sci. Instr.* **36,** 16–18 (1965).

[49] E. H. Eberhardt, "Noise factor measurements in multiplier phototubes," *Appl. Opt.* **6,** 359–360 (1967).

[50] M. Jonas and Y. Alon, "Dependence of signal-to-noise ratio on operating voltage in photomultipliers," *Appl. Opt.* **10,** 2436–2438 (1971).

[51] R. Foord, R. Jones, C. J. Oliver, and E. R. Pike, "The use of photomultiplier tubes for photon counting," *Appl. Opt.* **8,** 1975–1989 (1969).

[52] E. H. Eberhardt, "Multiplier phototubes in quantum counters," *Appl. Opt.* **6,** 161–162 (1967).

[53] E. H. Eberhardt, "Multiplier phototubes for single electron counting," *IEEE Trans.* **NS–11**³, 48–55 (1964).

[54] S. M. Sze, *Physics of Semiconductor Devices,* Wiley, New York, 1969; (a) Table 12.1; (b) Chapter 8; (c) Section 12.3; (d) Section 12.4.

[55] R. D. Hudson, *Infrared System Engineering,* Wiley, New York, 1969; (a) Sections 7.5 and 10.4; (b) Chapter 11.

[56] S. Rothschild, "Large-screen, high-current photoconductive cells using zinc sulfide cadmium sulfide, or cadmium sulfide, or cadmium selenide phosphors," *J. Opt. Soc. Am.* **46**, 662–663 (1956).

[57] F. H. Nicoll and B. Kazan, "Large area high-current photoconductive cells using cadmium sulfide powder," *J. Opt. Soc. Am.* **45**, 647–650 (1955).

[58] Anonymous, in Ref. 21, pp. 71–76.

[59] R. J. Caskman, "Film-type infrared photoconductors," *IRE Proc.* **47**, 1471–1475 (1959).

[60] J. N. Humphrey, "Optimum utilization of lead sulfide infrared detectors under diverse operating conditions," *Appl. Opt.* **4**, 665–675 (1965).

[61] D. E. Bode, T. H. Johnson, and B. N. McLean, "Lead selenide detectors for intermediate temperature operation," *Appl. Opt.* **4**, 327–331 (1965).

[62] W. Beyen, P. Bratt, H. Davis, L. Johnson, H. Levinstein, and A. MacRae, "Cooled photoconductive infrared detectors," *J. Opt. Soc. Am.* **49**, 686–692 (1959).

[63] F. D. Morten and R. E. J. King, "Photoconductive indium antimonide detectors," *Appl. Opt.* **4**, 659–663 (1965).

[64] C. Hilsum, "Photoelectric effects in InAs at room temperature," *Phys. Soc. Proc.* **70B**, 1011–1012 (1957).

[65] P. K. Weimer and A. D. Cope, "Photoconductivity in amorphous selenium," *RCA Rev.* **12**, 314–334 (1951).

[66] P. H. Keck, "Photoconductivity in vacuum coated selenium films," *J. Opt. Soc. Am.* **42**, 221–225 (1952).

[67] S. V. Forgue, R. R. Goodrich, and A. D. Cope, "Properties of some photoconductors, principally antimony trisulfide," *RCA Rev.* **12**, 335–349 (1951).

[68] T. S. Moss, "Infrared detectors," *IR Phys.* **16**, 29–36 (1976).

[69] J. L. Schmit and E. L. Stelzer, "Temperature and compositional dependences of the energy gap of $Hg_{1-x}Cd_xTe$," *J. Appl. Phys.* **40**, 4865–4869 (1969).

[70] C. Vérié and M. Sirieix, "Gigahertz cutoff frequency capabilities of CdHgTe photo-voltaic detectors at 10.6 μ," *IEEE J. Quant. Electr* **QE-8**, 180–184 (1972).

[71] M. A. Kinch, S. R. Borrello, and A. Simmons, "0.1 eV HgCdTe photoconductive detector performance," *IR Phys.* **17**, 127–135 (1977).

[72] H. Levinstein, "Extrinsic detectors," *Appl. Opt.* **4**, 639–647 (1965).

[73] The April 1977 issue of *IEEE Trans.* **ED-24** is, almost in its entirety, devoted to silicon solar cells.

[74] G. J. Deboo and C. N. Burrous, *Integrated Circuits and Semiconductor Devices*, McGraw-Hill, New York, 1971, Section 6.5.

[75] R. H. Rediker, T. M. Quist, and B. Lax, "High speed heterojunction photodiodes and beam-of-light transistors," *IRE Proc.* **51**, 218–219 (1963).

[76] M. I. Grace and D. E. Sawyer, "UHF photoparametric amplifier," *IEEE Trans.* **ED-13**, 901–903 (1966).

[77] L. U. Kibler, "A high-speed point contact photodiode," *IRE Proc.* **50**, 1834–1835 (1962).

[78] M. DiDomenico, W. M. Sharpless, and J. J. McNicol, "High speed photodetection in germanium and silicon cartridge-type point-contact photodiodes," *Appl. Opt.* **4**, 677–682 (1965).

[79] M. V. Schneider, "Schottky barrier photodiodes with antireflection coating," *Bell Syst. Tech. J.* **45**, 1611–1638 (1966).

[80] H. Melchior, M. B. Fisher, and F. R. Arams, "Photodetectors for optical communication systems," *IEEE Proc.* **58**, 1466–1486 (1970).

[81] G. E. Stillman and C. M. Wolfe, "Avalanche photodiodes," in *Semiconductors and Semimetals*, Vol. 12, R. K. Willardson and A. C. Beer, eds., Academic, New York, 1977, Chapter 5, pp. 291–393.

[82] H. Kanbe, T. Kimura, Y. Mizushima, and K. Kajiyama, "Silicon avalanche photodiodes with low multiplication noise and high-speed response," *IEEE Trans.* **ED-23**, 1337–1343 (1976).

[83] R. H. Bube, "Comparison of solid-state photoelectronic radiation detectors," *Metal. Soc. AIME Trans.* **239**, 291–300 (1967).

[84] R. L. Petritz, "Fundamentals of infrared detectors," *IRE Proc.* **47**, 1458–1467 (1959).

[85] A. Van der Ziel, "Flicker noise in semiconductors: not a true bulk effect," *Appl. Phys. Lett.* **33**, 883–884 (1978).

[86] I. S. Gradshteyn and I. M. Ryzhik, *Tables of Integrals, Series, and Products*, Academic, New York, 1965; (a) 3.411 (1); (b) 3.423 (2).

[86*] H. B. Dwight, *Tables of Integrals and Other Mathematical Data*, 4th Ed., Macmillan, New York, 1961; (a) 860.39; (b) 860.518.

[87] R. C. Jones, "Performance of Detectors for Visible and Infrared Radiation," in *Advances in Electronics*, Vol. 5, L. Marton, ed., Academic, New York, 1953, Section II5.

[88] J. A. Jamieson, "Passive infrared sensors: Limitations on performance," *Appl. Opt.* **15**, 891–909 (1976).

[89] G. P. Weckler, "Operation of *p-n* junction photodetectors in a photon flux integrating mode," *IEEE Trans.* **SC-2**, 65–73 (1967).

[90] R. C. Jones, "Immersed radiation detectors," *Appl. Opt.* **1**, 607–613 (1962).

[91] F. Restrepo and C. E. Backus, "On black solar cells or the tetrahedral texturing of a silicon surface," *IEEE Trans.* **ED-23**, 1195–1197 (1976).

[92] U. E. Gross and A. I. Weinstein, "A cryogenic-solid cooling system," *IR Phys.* **4**, 161–169 (1964).

[93] L. M. Biberman and S. Nudelman, eds., *Photoelectronic Imaging Devices*, Plenum, New York, 1971; (a) Vol. 1; (b) Vol. 2; (c) Vol. 2, Chapters 5–8; (d) Vol. 2, Chapters 13–18.

[94] J. D. McGee, D. McMullen, E. Kahan, B. L. Morgan, and R. W. Airey, eds., *Photo-Electronic Image Devices, Adv. Electronics Electron Devices*, Academic, New York; (aa) Vol. 12, 1960; (a) Vol. 22, 1966; (b) Vol. 28, 1969; (c) Vol. 33, 1972; (d) Vol. 40, 1976.

[95] K. R. Spangenberg, *Fundamentals of Electron Devices*, McGraw-Hill, New York, 1957, Chapter 4.

[96] O. Klemperer, *Electron Optics*, 3rd ed., Cambridge University Press, Cambridge, 1971; (a) Section 1.2; (b) Figure 3.5; (c) Figure 2.1; (d) Table 4.1; (e) p. 21.

[97] J. Vine, "Electron optics," Ch. 10 in Ref. 93a.

[98] P. Schagen, H. Bruining, and J. C. Francken, "A simple electrostatic electron optical system with only one voltage," *Philips Res. Rept.* **7**, 119–130 (1952).

[99] W. M. Wreathall, "Aberrations of diode image tubes," Ref. 94a, pp. 583–590.

[100] J. Vine, "The design of electrostatic zoom image intensifiers," Ref. 94b, pp. 537–543.

[101] J. A. Cochrane and R. F. Thumwood, "The effects of high electric fields on photocathodes," Ref. 94d, pp. 444–448.

[102] P. Schagen, "Electronic aids to night vision," *Roy. Soc. Phil. Trans.* **269**, 233–263 (1971).

[103] J. Goodson, A. J. Woolgar, J. Higgins, and R. F. Thumwood, "The proximity focused image intensifier," Ref. 94c, pp. 83–92.

[104] B. R. C. Garfield, R. J. F. Wilson, J. G. Goodson, and D. J. Butler, "Developments in proximity focused diode image intensifiers," Ref. 94d, pp. 11–20.

[105] A. M. Stark, D. L. Lamport, and A. W. Woodhead, "Calculation of the modulation transfer function of an image tube," Ref. 94b, pp. 567–575.

[106] J. D. McGee, R. W. Airey, M. Aslam, J. R. Powell, and C. E. Catchpole, "A cascaded image intensifier," Ref. 94a, pp. 89–104.

[107] F. C. Delori, R. W. Airey, J. D. McGee, "Further research on the Imperial College cascade image intensifier," Ref. 94c, pp. 99–116.

[108] P. R. Collings, R. R. Beyer, J. S. Kalafut, and G. W. Goetze, "A family of multi-stage direct-view image intensifiers with fiber-optic coupling," Ref. 94b, pp. 105–118.

[109] V. P. Boutot, G. Eschard, R. Polaert, and V. Duchenois, "A microchannel plate with curved channels: An improvement in gain, relative variance and ion noise for channel plate tubes," Ref. 94d, pp. 103–111.

[110] G. E. Hill, "Secondary electron emission and compositional studies on channel plate glass surfaces," Ref. 94d, pp. 153–165.

[111] B. W. Manley, A. Guest, and R. T. Holmshaw, "Channel multiplier plates for imaging applications," Ref. 94b, pp. 471–486.

[112] I. C. P. Millar, D. Washington, and D. L. Lawport, "Channel electron multiplier plates in x-ray image intensification," Ref. 94c, pp. 153–165.

[112*] A. R. Asam, "Advances in microchannel plate technology and applications," Opt. Eng. 17, 640–644 (1978).

[113] W. Baumgartner and V. Zimmermann, "A high gain channel electron multiplier (CEM) array and some of its operational characteristics," Ref. 94c, pp. 125–131.

[114] W. M. Sackinger and G. A. Gislason, "Ion feedback noise in channel multipliers," Ref. 94c, pp. 175–182.

[115] V. Chalmeton and G. Eschard, "Reduction of the relative variance of the single-electron response at the output of a microchannel plate," Ref. 94c, pp. 167–174.

[116] H. Pollehn, J. Bratton, and R. Feingold, "Low noise proximity focused image intensifiers," Ref. 94d, pp. 21–23.

[117] R. Ward, "Noise measurements on image intensifiers," Ref. 94d, pp. 553–564.

[118] J. Graf, M. Fouassier, R. Polaert, and G. Savin, "Characteristics and performance of a microchannel image intensifier designed for recording fast luminous events," Ref. 94c, pp. 145–152.

[119] G. Eschard, J. Graf, and R. Polaert, "Signal to noise and collection efficiency measurements in microchannel wafer image intensifiers," Ref. 94d, pp. 141–152.

[120] D. L. Emberson and R. T. Holmshaw, "Some aspects of the design and performance of a small high-contrast channel image intensifier," Ref. 94c, pp. 133–144.

[121] C. B. Johnson, C. E. Catchpole, and C. C. Mathe, "Microchannel plate inverter image intensifiers," IEEE Trans. ED–18, 1113–1116 (1971).

[122] R. W. Redington, "The image orthicon," Ref. 93b, Chapter 9, pp. 93–202.

[123] E. M. Musselman, "The new image isocon—Its performance compared to the image orthicon," Ref. 93b, Chapter 10, pp. 203–215.

[124] A. D. Cope and H. Borkan, "Isocon scan—a low-noise, wide-dynamic-range camera tube technique," Appl. Opt. 2, 253–261 (1963).

[125] P. D. Nelson, "The development of image isocons for low light level applications," Ref. 94b, pp. 209–227.

[126] S. Miyashiro and S. Shirouzu, "Electrostatically scanned image orthicon," Ref. 94b, pp. 191–207.

[127] G. W. Goetze and A. H. Boerio, "SEC camera-tube performance characteristics and applications," Ref. 94b, pp. 159–171.

[128] A. Choudry, "Characteristics of an optically scanned SEC device," Ref. 94d, pp, 253–262.

[129] P. Zucchino, "Photometric statistical performance of the SEC target," Ref. 94d, pp. 239–251.

[130] K. Sato and M. Takahashi, "A magnetically focused SEC camera tube," Ref. 94c, pp. 241–251.

[131] P. R. Collings, L. G. Healy, A. B. Laponsky, and R. A. Shaffer, "A proximity focused ultraviolet-sensitive SEC camera tube," Ref. 94c, pp. 253–261.

[132] G. W. Goetze and A. B. Laponsky, "Early stages in the development of camera tubes employing the silicon-diode array as an electron-imaging charge-storage target," Ref. 93b, pp. 253–262.

[133] S. Miyashiro and S. Shirouzo, "A supersensitive camera tube incorporating a silicon electron-multiplication target," Ref. 94c, pp. 207–218.

[134] V. J. Santilli and G. B. Conger III, "TV camera tubes with large silicon diode array targets operating in the electron bombardment mode," Ref. 94c, pp. 219–228.

[135] J. H. T. van Roosmalen, "Adjustable saturation in a pick-up tube with linear light transfer characteristic," Ref. 94b, pp. 281–288.

[136] A. G. van Doorn, "The 'plumbicon' compared with other television camera tubes," Philips Tech. Rev. 27, 1–14 (1966).

[137] D. G. Taylor, C. H. Petley, and K. G. Freeman, "Television at low light levels by coupling an image intensifier to a plumbicon," Ref. 94b, pp. 837–849.

[138] M. H. Crowell and E. F. Labuda, "The silicon diode array camera tube," Bell Syst. Tech. J. 48, 1481–1528 (1969), Ref. 93b, Chapter 15.

[139] T. Sasaki, T. Nakamura, and S. Goto, "Experiments on a wire-electrode type image intensifier using electroluminescence," Adv. Electr. Electron Phys. 16, 621–631 (1962).

[140] R. D. Stewart, "A solid state image converter," IEEE Trans. ED–15, 220–225 (1968).

[141] P. W. Ranby and R. P. Ellerbeck, "A new type of thorn image storage panel," J. Phot. Sci. 19, 77–82 (1971).

[142] Z. Szepesi and M. Novice, "Solid-state radiographic amplifiers and infra-red converters," Ref. 94b, pp. 1087–1098.

[143] T. Kohashi, T. Nakamura, S. Nakamura, and K. Miyaji, "Recent developments in solid-state infra-red image converters," Ref. 94b, pp. 1073–1086.

[144] C. H. Séquin and M. F. Tompsett, Charge Transfer Devices, Advances in Electronics and Electron Physics, Suppl. 8, Academic, New York, 1975.

[145] P. K. Weimer, G. Sadasiv, J. E. Meyer, L. Meray-Horvath, and W. S. Pike, "A self-scanned solid-state image sensor," IEEE Proc. 55, 1591–1602 (1967).

[146] P. K. Weimer, M. G. Kovac, F. V. Shallcross, and W. S. Pike, "Self-scanned image sensors based on charge transfer by the bucket-brigade method," IEEE Trans. ED–18, 996–1010 (1971).

[147] H. K. Burke and G. J. Michon, "Charge-injection imaging: Operating techniques and performance characteristics," IEEE Trans. ED–23, 189–195 (1976).

[148] The February 1976 issue of the above Transactions is, in its entirety, devoted to CTDs. Note here especially: D. F. Barbe, "Charge-coupled device and charge-injection device imaging," pp. 177–182.

[149] C. H. Séquin, D. A. Sealer, W. J. Bertram, M. F. Tompsett, R. R. Buckley, T. A. Shankoff, and W. J. McNamara, "A charge-coupled area image sensor and frame store," IEEE Trans. ED–20, 244–252 (1973).

[150] E. S. Kohn, "A charge-coupled infrared imaging array with Schottky-barrier detectors," IEEE Trans. ED–23, 207–214 (1976).

[151] J. W. Wallmark, "A new semiconductor photocell using lateral photoeffect," IRE Proc. 45, 474–483 (1957).

[152] T. C. Carr, J. C. Richmond, and R. G. Wagner, "Position-sensitive Schottky barrier photodiodes: Time-dependent signals and background saturation effects," IEEE Trans. ED–17, 507–513 (1970).

[153] J. W. Horton, R. V. Mazza, and H. Dym. "The Scanistor—A solid-state image scanner," IEEE Proc. 52, 1513–1528 (1964).

[154] L. Levi, "On combined spatial and temporal characteristics of optical systems," *Opt. Acta* **17,** 869–872 (1970).

[155] C. B. Johnson, "A method for characterizing electro-optical device modulation transfer functions," *Photogr. Sci. Eng.* **14,** 413–415 (1970); also "Classification of electron-optical device MTF's," Ref. 94c, pp. 579–584.

[156] S. Hasegawa, "Resolving power of image tubes," Ref. 94b, pp. 553–565.

[157] L. Levi, "Motion blurring with decaying detector response," *Appl. Opt.* **10,** 38–41 (1971).

[158] M. Smith, "Optimization of bias in thermistor radiation detectors," *Appl. Opt.* **8,** 1027–1033, 1213–1216 (1969).

[159] A. Hadni, *Essentials of Modern Physics Applied to the Study of the Infrared,* Pergamon, Oxford, 1967; Section. II5.

[160] J. Cooper, "Minimum detectable power of a pyroelectric thermal receiver," *Rev. Sci. Instr.* **33,** 92–95 (1962).

[161] E. H. Putley, "The pyroelectric detector," in *Semiconductors and Semimetals,* R. K. Willardson and A. C. Beer, Eds., Academic, New York; (a) Vol. 5, 1970, Chapter 5, pp. 259–285; (b) Vol. 12, 1977, Chapter 7, pp. 441–449.

[162] A. M. Glass, "Ferroelectric $Sr_{1-x}Ba_xNb_2O_6$ as a Fast and Sensitive Detector of Infrared Radiation," *Appl. Phys. Lett.* **13,** 147–149 (1968).

[163] A. Hadni, "Improvements in the detectivity of infrared pyroelectric detectors," *Opt. Commun.* **1,** 251–253 (1969).

[164] P. A. Jansson, "Hot-pressed TGS for pyroelectric detector applications," *Appl. Opt.* **13,** 1293–1294 (1974).

[165] J. M. Lloyd, *Thermal Imaging Systems,* Plenum, New York, 1975.

[165*] M. M. Farrow, R. K. Burnham, M. Anzanneau, S. L. Olsen, N. Purdie, and E. M. Eyring, "Piezoelectric detection of photoacoustic signals," *Appl. Opt.* **17,** 1093–1098 (1978).

[165**] H. K. Wickramasinghe, R. C. Bray, V. Jipson, C. F. Quate, and J. R. Salcedo, "Photoacoustics on a microscopic scale," *Appl. Phys. Lett.* **33,** 923–925 (1978).

[166] G. W. McDaniel and D. Z. Robinson, "Thermal imaging by means of the evaporograph," *Appl. Opt.* **1,** 311–324 (1962).

[167] M. Czerny, "Über Photographie im Ultraroten," *Z. Phys.* **53,** 1–12 (1924).

[168] W. R. Harding, C. Hilsum, and C. D. Northrup, "A new thermal image converter," *Nature* **181,** 691–692 (1958).

[169] M. F. Tompsett, "A pyroelectric thermal imaging camera tube," *IEEE Trans.* **ED-18,** 1070–1074 (1972).

[170] R. Watton, C. Smith, B. Harper, and W. M. Wreathall, "Performance of the pyroelectric vidicon for thermal imaging in the 8–14 micron band," *IEEE Trans.* **ED-21,** 462–469 (1974).

[171] S. E. Stokowski, "Thermal modulation transfer function analysis of pyroelectric device characteristics," *Appl. Opt.* **15,** 1767–1774 (1976).

[172] R. Watton, G. R. Jones, and C. Smith, "Pyroelectric materials for operation in a hard vacuum pyroelectric vidicon," Ref. 94d, pp. 301–312.

[173] T. Conklin and E. H. Strupp, "Applications of the pyroelectric vidicon," *Opt. Eng.* **15,** 510–515 (1976).

[174] R. C. Tolman, *The Principles of Statistical Mechanics,* Oxford University Press, London, 1938; p. 512.

[175] T. Limperis, "Detectors," in *Handbook of Military Infrared Technology,* W. L. Wolfe, ed., Office of Naval Research, Washington, D.C., 1965.

[176] M. DiDomenico and O. Svelto, "Solid-state photodetection: A comparison between photodiodes and photoconductors," *IEEE Proc.* **52,** 136–144 (1964).

17

Photography

Techniques of storing exposure information for extended periods are called *photography*. Except for electrophotography, almost all photographic techniques rely on photochemical changes in layers containing photosensitive materials. The most popular of these techniques is silver halide photography and the first part of this chapter is devoted to it, with later sections treating other photochemical techniques (Section 17.3) and electrophotography (Section 17.4); the final part of the chapter is devoted to color photography (Section 17.5).

17.1 SILVER HALIDE PHOTOGRAPHY: MATERIALS AND PROCESSES

The most important photographic system today is the one using a gelatin-silver halide emulsion on a substrate. Its popularity is due to its potentially very high responsive quantum efficiency, as described in Section 17.1.3.1. (Its detective quantum efficiency is not very high; see Section 17.2.4.3.) In this section we treat its structure and operation. The optical characteristics of the photographic record are described in the next section.

17.1.1 Structure of the Photographic Materials

17.1.1.1 Photographic Emulsions [1a], [2a], [3a]. The silver halide emulsion consists of a suspension of silver chloride or bromide in gelatin. The silver halide is the photosensitive component and the gelatin serves primarily as the matrix. In addition, the gelatin absorbs the released halides and may also contribute to the formation of electron traps at the crystal surfaces.

Of the two halides used, bromide is the more sensitive and its sensitivity is usually enhanced further by the incorporation of some iodide.

Such photographic emulsions are produced by mixing, in the gelatin, solutions of an alkali halide and silver nitrate with a resulting precipitation of submicroscopic silver halide crystals. This is followed by a slow physical *ripening* process during which the larger crystals grow at the expense of the smaller ones, the transfer taking place by convection of a solvent or actual coagulation of crystals. One function of the gelatin is the inhibition of this ripening so that the gelatin concentration determines the ultimate size and uniformity of the ripened crystals or *grains*. In practical emulsions these vary from spherical particles with a diameter of approximately 0.05 μm (Lippmann emulsions) to 3-μm particles in the fast emulsions. The larger particles usually take the shape of triangular platelets. The statistics of crystal sizes vary for different emulsions. Representative values of mean projected area and number of particles per unit volume are shown in Table 177. Physical ripening is followed by washing, which removes the excess halide, and by *chemical ripening*, which permits sensitizing substances to react with the crystals to increase their photosensitivity.

Because silver halide materials are sensitive only in the ultraviolet and blue end of the spectrum, spectral sensitizers are usually added to the emulsion to extend its sensitivity throughout the visible spectrum. These are generally dyes that are adsorbed at the crystal surfaces. They selectively absorb longer wavelength radiation and transfer the absorbed energy to the crystal, boosting its sensitivity to such radiation.

***17.1.1.2 Substrates and Mechanical Stability* [2b], [3b], [4a].** The emulsion is then coated onto a substrate—a plastic film or glass for transparency, or a paper for reflection photography. The main function of the substrate, or base, is to impart dimensional stability. In addition, it must transmit or reflect efficiently the radiation passing through the emulsion.

Substrate dimensions are affected by temperature, humidity, plastic flow, chemical changes, and stresses introduced during processing. Representative dimensional changes for popular substrates are given in Table 178.

Dimensional changes due to variations in temperature and humidity are generally reversible. The others, due to the release of mechanical stresses and evaporation of plasticizers and residual solvents, are irreversible. Ordinarily cellulose triacetate is used as the film base. When dimensional stability is very important, polyethylene terephthalate films are recommended.

Because of the considerable systematic dimensional changes exhibited by film bases, the random changes, too, are appreciable. Published data indicate fractional standard deviations of 17 and 4×10^{-5} for Kodak Plus-X aerial films on cellulose acetate butyrate and polyethylene terephthalate (Estar) bases, respectively. These changes were obtained over 15-mm distances on a large film and are superimposed on any linear dimensional changes [5].

The paper substrates require special preparation for desirable mechanical characteristics (strength maintained during wetting and aging) and optical ones (reflectivity high, uniform, and diffuse). The desirable optical properties are usually obtained by coating the paper with a layer of *baryta*, barium sulfate in a gelatin matrix, before depositing the photographic emulsion. The baryta may contain some yellow-absorbing or blue-fluorescing dyes to equalize the reflection spectrum, normally somewhat attenuated in the blue by the paper and the baryta. The size of the crystals in the baryta layer determines the glossiness of the resulting photographic paper. Any desired texture, too, may be embossed onto this layer.

In addition to the dimensional variations introduced by the substrate, the *gelatin effects* may contribute significantly to image distortion. To appreciate their importance, it should be noted that the gelatin is strongly hygroscopic, swelling to five to eight times its dry thickness during development. Also, the removal of unexposed crystals during fixing decreases the emulsion thickness by approximately 20% in the unexposed areas, whereas exposed and developed grains show a slight increase in volume. In addition, the developer, together with the reaction products, have a tanning effect on the neighboring emulsion so that the gelatin there absorbs less water. These phenomena, together with slight nonuniformities in the gelatin, clearly call for slow, even processing if distortions are to be minimized (see Ref. 1b).

When due precautions are taken, the overall distortion of isolated star images on photographic plates can be maintained below 2 μm except for points within approximately 15 mm from the edge of the plate [1b], [5]. Near this edge, distortions may be considerable (up to 50 μm at 1 mm from the edge). The residual random distortion that remains after an appropriate quadratic radial distortion is allowed for, is less than a micron. In addition to the random distortions, certain predictable distortions are often present. It has been found that, due to gelatin effects, exposed image points have a tendency to approach each other during development by 1–2 μm. This *Ross effect*[1] is quite constant for spacings up to approximately 100 μm and clearly tends to cancel the Kostinsky effect (Section 17.2.2.3). It has also been found that the tanning of the

[1] F. E. Ross (1920).

gelatin in exposed regions has a pronounced shrinking effect on such images when surrounded by unexposed emulsions.

17.1.1.3 Halation Control. When the emulsion is coated on a transparent base, an antihalation device too is generally provided. *Halation* is due to the reflection of light from the rear surface of the base, after having passed through the emulsion and the base. This causes exposure of emulsion areas outside the image area. Light scattered at the critical angle and beyond can be especially harmful since it is totally reflected at the outside surface of the substrate. Around any image point, this light forms an annulus, or halo, whose inside radius is

$$r_H = 2d \tan \theta_c, \qquad (17.1)$$

where the critical angle

$$\theta_c = \csc^{-1} n \qquad (17.2)$$

is set by the refractive index (n) of the base material (see Figure 17.1).

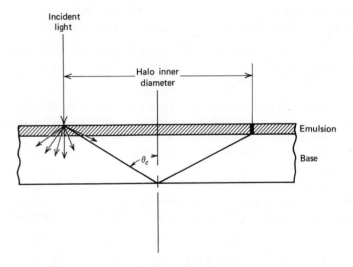

Figure 17.1 Illustrating the halation ring inner diameter.

To reduce it, the back of the base is usually coated with an opaque dye of low reflectivity, which absorbs most of the radiation transmitted by the emulsion. This coating must be removed during processing. Alternatively, the base may be impregnated with an absorbing dye through which the halation light must pass twice before causing the objectionable exposure.

Since this dye remains in the base after processing, its density may not be very high if the required read out illumination levels are to be maintained at reasonably low values. The higher this density, the more stringent, too, are the requirements on the dye uniformity.

17.1.1.4 Dimensional Standards. Dimensions of photographic films and plates have been standardized [6a]. In Table 179 we summarize the basic parameters of the most popular films.[2]

In motion picture recording and display as well as in miniature cameras and when accurate film control is important, sprockets are often used for film transport. Films to be used with these are provided with sprocket holes appropriately shaped [6a], [7].

17.1.2 Latent Image Formation [1c], [2c], [3c]

The usual photographic process in silver halide emulsions takes place in two basic steps: the formation of the latent image and its subsequent development. The *latent image* consists of changes in the crystal structure at the atomic level. These changes are not directly observable but influence subsequent chemical behavior, permitting the conversion of the submicroscopic latent image into the gross photographic one.

Gurney-Mott Mechanism. The following basic process, known as the Gurney-Mott[3] mechanism, is now generally accepted as underlying the formation of the latent image.

1. The absorbed photon raises an electron into the conduction band.
2. This electron is trapped at the site of a crystal defect, here called the *sensitivity speck*, which consequently becomes negatively charged.
3. This site attracts, and binds, a mobile interstitial silver ion, converting it into atomic silver and hence neutralizing it. This prepares the site for the trapping of an additional ion.
4. Steps 1–3 may then be repeated again and again, causing the build-up of a small speck of metallic silver at the sensitivity speck.

This silver speck is the unit of the latent image that can act as a catalyst permitting the subsequent developing process to convert the whole crystal into metallic silver. It is called a *development center.*

The build-up of metallic silver (Step 4) imparts stability to the silver speck. The first silver ion trapped is not very tightly bound to the sensitivity site, and it is likely to escape unless another is attracted within

---

[2] The *Journal of the Society of Motion Picture and Television Engineers* publishes new standards as they are issued.

[3] R. W. Gurney and N. F. Mott (1938).

a short time. Apparently two atoms suffice for stability and as few as three may be enough to make the crystal developable. (This stability is only relative and some fading of the latent image occurs with aging, especially in finer grain emulsions; humidity and oxygen in the atmosphere speed this process [8].)

Print-Out Silver. Extensive continuation of silver build-up can yield silver specks visible under the microscope. These are called print-out silver.

Solarization. The halogens released by the accumulation of silver atoms are generally absorbed by the gelatin. At extremely high exposures, the resulting density may be less than that obtained at lower exposures. This effect is called *solarization* and is due to the overproduction of halogens, which can no longer be absorbed by the gelatin. These may attack the previously formed silver specks, reducing the developability of some of the crystals.

Any complete analysis of the latent image formation must include consideration of the holes generated by the photoexcitation of the electrons; but their role appears to be secondary [9].

17.1.3 Processing the Photographic Record [1d], [2d], [3d]

17.1.3.1 General Considerations

Gain. The function of the development process is to convert into metallic silver all those crystals that have acquired a development center due to exposure. This represents an enormous gain, since the original 20 odd photons responsible for the formation of the development center eventually cause the conversion of, say, 10^9 silver ions into metallic silver. This accounts for the high sensitivity of these emulsions. It is also evident that the gain is directly proportional to the crystal size. Hence, for a given crystal type, higher gain can be obtained only by accepting coarser grains.

Chemical and Physical Development. Basically the development takes place by means of a reagent that converts silver halide into metallic silver, with the silver speck acting as a catalyst. This process is called *direct* or *chemical development.* Simultaneously with it, silver ions in the developer solution may be deposited on the developing silver specks. This process is called *physical development.*

Fog. During the manufacture of the emulsion, some metallic silver specks are likely to occur in such magnitude that they make their grains developable without any exposure. Also, the developer occasionally may reduce an unexposed silver halide grain. These and similar phenomena producing metallic silver from unexposed grains are termed chemical *fog.*

PHOTOGRAPHY

17.1.3.2 Classical Processing. In the usual processing procedure, the emulsion is treated successively in four baths: developer, stop bath, fixer, wash.

Developer. The actual development occurs at the first stage. It takes place at the silver–silver halide interface, which is initially very small, so that the development proceeds very slowly at first, accelerating as the silver specks grow in size. In addition to the actual development reagent, the developer solution contains a preservative compound that binds by-products that would otherwise cause rapid oxidation of the developer, and antifogging agents (generally soluble bromides) to minimize the undesirable development of unexposed crystals. In addition, a buffer is required to maintain the alkalinity required for the development reaction.

The length of time that the development is permitted to proceed strongly influences the nature of the resulting photographic image. Fog level, gamma, and speed, all tend to increase with development time (the latter terms are explained below). Development time must therefore be carefully controlled. Several minutes are usually required.

Developing solutions containing no significant amounts of solvents for the silver halide can act only on the grain surface so that the latent image speck must be located there to be effective. Such *surface development* differs in the effects obtained from *volume development*, which occurs in the presence of solvents and causes the development of crystals with internal development centers, as well.

The products of development cause three-dimensional cross linking in the gelatin. This, in turn appears as hardening or *tanning*—the gelatin becomes stronger mechanically, its melting point is raised, and it becomes more compact. The process is occasionally encouraged by the addition of special tanning agents, such as formaldehyde, since the resulting strength makes the emulsion less vulnerable to mechanical damage.

Stop Bath. After completion of the development process, the emulsion is transferred to a weak acid, usually dilute acetic acid, to neutralize the alkalinity and stop the development process. This is the stop bath.

Fixing and Wash Baths. For a stable record, the remaining silver halide crystals must be removed from the emulsion. This is generally done in two stages. First the crystals are converted into soluble salts and dissolved in the fixing bath. Sodium thiosulfate (or hyposulfate — "hypo", for short) is the usual agent used. The remaining soluble crystals and fixing solution are then removed in a wash bath.

If only a short-term record is required, these stages can be bypassed by means of a process called stabilization, whereby the silver halide crystals are converted into compounds insensitive to light and left in place.

17.1.3.3   Monobaths and Activators [1e], [2d]. When simplicity of processing is more important than dynamic range and long-term stability, the fixing agent may be incorporated into the developing solution. When such a "monobath" solution is used, the dissolution of the unexposed silver halide crystals starts simultaneously with the development process. Unfortunately, exposed crystals are dissolved almost as readily as unexposed ones. Nevertheless dense images are obtained by a process of physical development in which silver ions from the solution build up on the silver specks originating in the exposure, so that satisfactory results can often be obtained.

In an alternative technique, the development agent is incorporated into the emulsion itself in a dry, inactive form. To initiate the development, an *activator liquid*, usually an alkaline solution, is applied. This process is generally combined with the use of a stabilizer to avoid the need for the lengthy fixation process.

17.1.3.4   Rapid Processing. The development time, normally 5 to 10 min, can be greatly reduced by processing at high temperature. This tends to have adverse effects on fog level and granularity, but may still be desirable in certain applications. A process requiring only one third of a second from exposure to projection has been described [10], as have been other processes lasting about 30 sec, but providing better results [11]. In these techniques the method of applying the solution [12] and preparing the emulsion need special attention.

17.1.3.5   Processing for Phase Modulation. In our description of development, we noted that tanning tends to be an automatic by-product of this process. By taking special steps (e.g., eliminating sulfites) it can be encouraged and lead to pronounced surface deformations following the exposure distribution.

The tanning increases the dry thickness of the emulsion and decreases its thickness while wet. The change in thickness varies almost linearly with density. The transfer function of this process starts near zero at low spatial frequencies, peaks at a frequency given by

$$\nu_p \approx \frac{1}{6d},\tag{17.3}$$

where d is the unexposed emulsion thickness, and then drops off with the optical transfer function of the emulsion [13].

Now the silver image can be bleached, leaving a pure phase-modulated image, consisting of both thickness and index modulation. Some of the relevant processes are discussed in conjunction with holography (Section 19.3.4.2).

17.1.3.6 Silver Transfer and One-Step Photography. The silver transfer process is a technique providing a positive copy from a photographic negative in a material that is not photosensitive. This is accomplished by developing the primary negative photographic image, dissolving the undeveloped silver halide crystals in the photosensitive sheet, and permitting the freed silver ions to diffuse into the matrix coated onto the receiving sheet. This matrix contains nuclei on which silver ions accumulate, forming metallic silver crystals by physical development, principally in the areas next to the unexposed emulsion areas. This results in a positive image. Here the processing solution must provide rapid development followed by the dissolution of the undeveloped silver halide. This process is called *diffusion transfer* [2d]. It is used in one-step photography in which the exposed photograph is processed automatically and quickly [3e].

17.1.4 Nonoptical Photography [1f]

Photographic emulsions are also used for recording nonoptical radiation, such as fast electrons, X- and γ-rays. The emulsion response to such radiation differs from that governing optical exposure.

17.1.4.1 Exposure by Electrons

Effects of Electrons. As energetic electrons pass through an emulsion, they lose energy both in the grains and in the gelatin. The loss in the grains is partly used to excite electrons into the conduction band and such electrons may make the grain developable, just as the photoelectrons did during optical exposure. A major difference lies in the fact that an optical photon excites at most one such electron in one grain, whereas the energetic electron may excite hundreds of electrons in tens of grains: the responsive quantum efficiency of the emulsion to such exposure may be much larger than unity.

The penetration range (r) of electrons with energy values (W) between 5 and 100 keV follows approximately a power law [1g]

$$r = \left(\frac{W}{9\,\text{keV}}\right)^{1.77} \mu\text{m}. \qquad (17.4)$$

Since, as a rule, one exposing particle suffices to make the grain developable, there is here no failure of reciprocity.

The distribution of exposure due to a single energetic electron is governed by the rate at which it generates conduction electrons. This is low at high energy levels and becomes high as the electron approaches the end of its range, where it generates a roughly spherical region of exposure (see Figure 6.6, which depicts a similar situation in cathodoluminescence).

The resulting density grows monotonically with electron energy until the resulting range exceeds the emulsion thickness. At that point there occurs a sudden drop in the resulting density. Simple geometric considerations show that the effective thickness of the emulsion varies directly with the secant of the angle of incidence.

Electronography. Electronography is a photographic technique developed for recording images at very low light levels especially in astronomy [14]. The equipment is essentially an image intensifier tube (Section 16.4.2) with photographic film replacing the phosphor. A sensitive photocathode generates an electron image. After substantial acceleration, this is reimaged on the emulsion to produce a photographic image. The preferred materials are nuclear track emulsions, which are fine grained and also have a high density of silver halide.

17.1.4.2   Exposure by X- and γ-Rays

Characteristic Curves and Spectral Sensitivity. The passage of highly energetic electromagnetic radiation, such as X- and γ-rays, causes electron excitation by the photoelectric effect and Compton scattering and, at very high energies, by the generation of electron-positron pairs.

Because, typically, a fixed number of grains is made developable by each photon, the number of exposed grains, and hence the density, grows linearly with the exposure [see Nutting's formula [17.61]] up to densities of one or two. With X-rays of low and medium energy, each photon will generally make no more than a single grain developable. The energy cutoff depends on the emulsion type and generally occurs between 0.3 and 4.5 nm (0.3–4 keV).

At higher energy levels the mean number (η_g) of grains rendered developable rises with the photon energy (W) according to

$$\eta_g \approx 0.08 \ W, \tag{17.5}$$

where W is in keV (based on data of Ref. 15).

The linear density (D) versus exposure (H) curve implies a characteristic curve of slope[4]

$$\gamma^* = \frac{dD}{d \log H} = (\log 10) \ H \frac{d(cH)}{dH} = (\log 10) \ cH = (\log 10) \ D, \tag{17.6}$$

where we have substituted [on the basis of Nutting's formula (17.61)] $(cH = D)$.

---

[4] See Section 17.2.1.2 for the definition of the characteristic curve and Note 24 for the notation γ^*.

Since the small signal gain of the film equals γ^*, it is advisable to work at high values of D. Here, too, reciprocity is maintained as a rule.

In terms of roentgens[5] required to achieve a given density, the spectral sensitivity curves peak at 30–40 keV; they have dropped approximately 15 dB at 200 keV and remain quite constant beyond that.

Because of the greater penetration of the more energetic photons, the proportion of internal latent image specks is greater in X- and γ-ray than in optical exposure. Therefore the contrast[6] obtained with such recordings can be increased by using an internal developer.

Radiography. The best known, and still the most important, application of X-ray photography is in radiography—the photographic recording of X-ray shadows, primarily in medicine, but also in nondestructive testing. In this application, the performance limitations may be quite different from those in optical photography. Here, for instance, the granularity may be controlled by the photon noise rather than the characteristics of the material. Especially in medical applications, where the amount of radiation must be strictly limited, the number of X-ray quanta per image element generally determines the granularity. With limited exposure, this number can be increased only by enhancing the absorption efficiency. This is usually accomplished by placing roentgenoluminescent panels, usually on a cardboard substrate, in contact with the photographic film.

The resolution may be limited by the size of the X-ray source rather than by the turbidity of the photographic material. When luminescent panels are used, the spreading of the light between them and the emulsion may further degrade resolution (see Section 6.3.4 for further details).

Dosimetry [16]. Personnel working in the proximity of radiation hazards must be protected from overexposure. Despite reasonable precautions, leaks and other inadvertent exposures must still be reckoned with. To keep track of such events, these people often carry pieces of photographic film that may be processed (and replaced) periodically. If various types of radiation are involved, such radiation dosimetry is complicated by the fact that emulsion sensitivity may vary from type to type, in a manner different from the biological sensitivity [16a].

---

[5] The roentgen was defined as the amount of radiation generating one esu of charge in one cm^3 of dry air at standard conditions (0°C, 760 mm Hg). This is equivalent to 2.58×10^{-4} C/kg.

[6] In photography, the term contrast is conventionally used for the slope of the characteristic curve, the quantity symbolized by γ. This is defined in Section 17.2.1.

17.2 OPTICAL CHARACTERISTICS OF PHOTOGRAPHIC EMULSIONS

In this section we discuss the optical characteristics of photographic materials. This includes transfer characteristics, spatial response (spread function), and noise characteristics (granularity). We shall refer specifically to silver halide materials, but much of the discussion will be applicable to other materials as well. Their specific characteristics are treated in their respective sections.

17.2.1 Transfer Characteristics

17.2.1.1 Photographic Density **[1b], [3f], [17].** From the systems point of view, the output of a photographic record should be measured in terms of transmittance, τ, (or reflectance), since this will determine the signal power at the output. However, traditionally, the output is measured in terms of its logarithm, the density

$$D = -\log_{10} \tau, \qquad (17.7)$$

which corresponds more closely to the physiologically detected quantity.

Both transmittance, and the resulting density need careful definition. In general terms transmittance is defined as the ratio of the transmitted to the incident flux. But this ratio depends on the method of measurement. If we illuminate the film by means of a collimated beam, we obtain one reading; if we illuminate it diffusely, we will increase the transmittance observed in any one direction, but decrease the total transmittance. The dependence of observed transmittance on the angle over which the transmitted light is collected is illustrated in Figure 17.2a [4b] [18] for collimated light incident normally. Figure 17.2b [4b] illustrates the equivalent dependence when the illumination is totally diffuse.

When the transmittance is determined with collimated light incident and only the unscattered light is measured, the resulting transmittance is referred to as *specular*. It may be measured, for instance, by imaging a point source on a pinhole and taking the ratio of the fluxes passing this pinhole, with and without the film interposed.

When a collimated incident beam is used, but all the transmitted flux is measured, we obtain the *diffuse transmittance*. To measure the *doubly-diffuse transmittance*, the emulsion is diffusely illuminated and all the transmitted flux is measured.

Except at low densities, the ratio of specular to diffuse density is quite

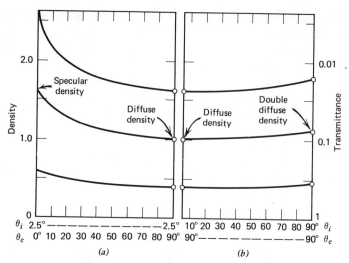

Figure 17.2 Density of high-speed negative film. (a) As a function of the collection angle (θ_c) with collimated light incident. (b) As a function of the illumination angle (θ_i) with perfectly diffuse collection. $\theta_{c,i}$ are half the apex angles of the cones concerned.

constant for any given material. It is called the Callier[7] Q-factor

$$Q = \frac{D(\text{specular})}{D(\text{diffuse})} \tag{17.8}$$

and can be used as a measure of granularity (see Section 17.2.3.2).

Other factors affecting the measured density are the spectral composition of the radiation (absorption and scattering characteristics of the emulsion generally vary with the wavelength) and the type of diffuser used: interreflections between the emulsion surface and the diffuser can distort the density measurement significantly. When the beam traverses the emulsion obliquely, the measured density increases directly with the internal pathlength:

$$D_\theta = D_0 \sec \theta$$

where θ is the angle at which the beam traverses the emulsion [18].

17.2.1.2   The Characteristic Curve. The input to a photographic emulsion is usually measured in terms of *exposure*, which is the time integral of illumination. In the visible region the units are usually in lux-seconds. In general, radiant units (J/m²) are more useful.

The response of the photographic emulsion to exposure and subsequent

[7] A. Callier (1909).

processing is measured in terms of *opacity*, the ratio of incident to transmitted light, or, more frequently, by *density*, the common logarithm of the opacity. This response is usually presented in the form of the *characteristic curve*, which gives density as a function of log (exposure). It is frequently referred to as the H-D curve (after Hurter and Driffield[8]).[9]

The characteristic curve may conveniently be divided into three regions. At low exposures it starts with a horizontal section followed by a region of increasing slope, the *toe* region. This is followed by a *linear* section with positive slope, which then begins to level off into the *shoulder* region. The constant slope of the linear section is referred to as the *"gamma"* (γ). In this section the opacity (O) may thus be represented in terms of the exposure (H) by

$$O = \left(\frac{H}{H_i}\right)^{\gamma}. \tag{17.10}$$

Clearly, gamma is a measure of the linearity of the emulsion response to exposure, unity gamma indicating that opacity is proportional to exposure. The constant H_i, which represents the ($D = 0$) intercept of the linear section (K in Figure 17.3), is called emulsion *inertia* and the value of D for zero exposure is called *fog*.

In silver halide emulsions, the gamma and, to a much lesser extent, the fog increase as development progresses, whereas the inertia tends to remain constant. Development time thus offers a practical method for controlling gamma over a considerable range. A typical characteristic curve is shown in Figure 17.3.

Since the output of a photographic system is usually in terms of transmitted (or reflected) light, the relationship between exposure and *transmittance*, the reciprocal of opacity, is of primary concern. In negative processes, a linear relationship between exposure and transmittance is most readily obtained by cascading two photographic processes, such that the product of their gammas equals unity. Note that in pictorial photography the final gamma is usually made approximately 1.6 to compensate psychologically for deficiencies in the process. Alternatively, one of the

---

[8] F. Hurter and V. C. Driffield (1890).

[9] Another parameter that has been found useful is the increase of opacity above fog level:

$$\Delta O = \frac{1}{\tau} - \frac{1}{\tau_0} = \frac{\tau_0/\tau - 1}{\tau_0}, \tag{17.9}$$

where τ_0 is the transmittance of the unexposed emulsion. In contrast to the usual H-D curve a log–log plot of ΔO versus exposure often approximates a straight line, down to very low density values [19]. [The value discussed in Ref. 19 is identical to the numerator of (17.9), that is, it differs from (17.9) by a constant factor only.]

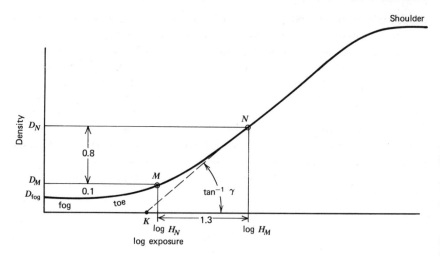

Figure 17.3 Representative characteristic $(H \& D)$ curve. Points M and N defined. $K = \log H_i$.

reversal processes may be used. Some of these are briefly described in Section 17.2.1.7. They are occasionally referred to as *autopositive*.

We now survey briefly how emulsion parameters influence the shape of the characteristic curve. The maximum density, and the gamma, which is proportional to it, are roughly proportional to the total cross-sectional area of the silver grains per unit emulsion area, with the thickness of the grains irrelevant. Consequently, a given amount of silver in an emulsion will produce a greater maximum density when it consists of smaller grains and when the emulsion is thin, so that there is less grain overlap.

A simple analysis based on a Poisson distribution shows that very large gammas would result if all grains were of equal sensitivity, that is, if they would require the same number of quanta of radiant energy to make them developable [see Appendix 17.2.2)]. Thus, in addition to the maximum density attainable, it is the uniformity of sensitivity of the grains that is responsible for a high gamma [20].

An effect analogous to grain size dispersion is observed when microscopic illumination variations disturb the exposure; these, too, lower the γ [21].

17.2.1.3 Sensitivity and "Speed." If the various luminance values of the object are to be distinguishable in the photograph, the corresponding exposure values must fall into the range between toe and shoulder of the characteristic curve and the exposure must be timed accordingly. In many applications it is not practical to determine this range. A single average luminance measurement may then be made and matched to the sensitivity of the emulsion, with allowance for a reasonable spread of luminance values. Qualitatively, this sensitivity is represented by the position of the characteristic curve along the abscissa. Quantitatively there are primarily two ways used to specify the sensitivity. Both are referred to as the *speed* of the emulsion and are based on the abscissa, $\log H_M$, of a point, M, on the characteristic curve (see Figure 17.3). H_M is defined as the exposure required to raise the emulsion 0.1 density units above the fog level.[10] The speeds are then given by

$$S_{ASA} = 0.8/H_M \qquad \text{[ANSI, ASA] } [6b], \qquad (17.11)$$

$$S_{DIN} = -10 \log_{10} H_M \qquad \text{[DIN (No. 4512)]}, \qquad (17.11a)$$

where H_M is expressed in lux-seconds. The formulae connecting the speed values are, clearly:

$$S_{DIN} = 10 \, (\log_{10} S_{ASA} - \log_{10} 0.8) \approx 10 \log_{10} S_{ASA} + 1 \qquad (17.12)$$

$$S_{ASA} \approx 10^{(S_{DIN}-1)/10}. \qquad (17.12a)$$

Photographic speed values are traditionally based on luminous units.

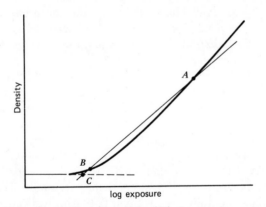

Figure 17.4   Illustrating contrast index (CI). See text for explanation.

[10] The exposure is to be made with a standard source and the development is such than an exposure ($H_N = 20H_M$) raises the density to 0.9 units above fog level [6b] (see Figure 17.3).

Occasionally a number called *contrast index* is substituted for the gamma. It represents the slope of the characteristic curve averaged over a segment starting at a point roughly equivalent to M in Figure 17.4 and ending at a point 2 log-units away from it [22]. Specifically, the contrast index (CI) is based on a line intersecting the characteristic curve at points A and B and the extended fog level line at C such that $\overline{AB} = 2$ and $\overline{BC} = 0.2$ logarithmic units (see Figure 17.4). The slope of this line is the contrast index.

As the use of artificial nonthermal light sources increases, these values become less relevant. Sensitivity figures based on the energy spectrum of specific sources have been proposed [23], [24]. The theory underlying this approach is treated in Section 18.1.4.1.

17.2.1.4   Spectral Sensitivity. Since the speed is defined in luminous units and the sensitivity spectrum differs from that of the normal observer, the effective speed depends on the spectral composition of the exposing radiation. To evaluate the effective speed for any particular exposure spectrum, the sensitivity spectrum of the photographic emulsion must be known (see Section 18.1.4).

Spectral sensitivity curves are usually given in terms of the reciprocal of the exposure (in J/m^2) required at any given wavelength to produce unit density or some other convenient density level. Representative curves are given in Appendix 17.1.

Undyed emulsions are generally uv and *blue-sensitive*, meaning that their sensitivity curves drop sharply in the green region of the spectrum and that they exhibit extremely low sensitivities at the longer wavelengths. When dyes are added to sensitize the emulsion throughout the visible spectrum, the emulsion is referred to as *panchromatic*. (Intermediate sensitization produces an *orthochromatic* emulsion.) Special ir emulsions are also available with significant sensitivity extending to approximately 1.24 μm. [25a]. Sensitivity at even longer wavelengths, where black body radiation at room temperature has significant components, makes storage very difficult.

In the far uv, that is, at wavelengths less than 0.25 μm, where the gelatin absorbs significantly, extremely thin emulsions must be used. These are named after Schumann and consist of uncovered silver halide crystals glued to the base by gelatin. They are useful down to 0.0075 μm. Alternatively, ordinary photographic emulsions may be coated with fluorescent materials [25a].

The efficiency with which radiation is absorbed depends largely on the transmittance spectrum of the silver halide crystals. These have uv absorption bands extending to 0.42 and 0.49 μm, respectively, for silver chloride and bromide (see Figure 10.13). Silver bromide with 3 mole% iodide, has a band extending to approximately 0.52 μm [1i]. These absorption bands control, on the one hand, the sensitivity spectrum of the unsensitized crystal and, on the other hand, the depth at which the radiation is absorbed. The less penetrating, shorter wavelength radiation creates essentially only surface development centers, whereas the longer wavelength radiation can cause such centers also internal to the crystals. Also, with short-wavelength radiation the crystal's cross-sectional *area* determines its effectiveness, whereas the crystal *volume* is the determinant at longer wavelengths. Consequently, the emulsion characteristic varies with the spectrum of the exposing radiation. For instance, the gamma usually increases with wavelength [1h]. It is thus possible that a red pattern on a blue background appear either dark on light or light on dark, depending on the exposure level.

Also the maximum density attained before solarization sets in depends on the spectral composition of the exposing radiation.

17.2.1.5   Reciprocity Law [1j], [2e], [3c]

Definition. According to the Bunsen-Roscoe principle,[11] the amount of photochemical action is directly proportional to the energy absorbed, independent of the time interval over which it is applied. For a given photographic density, the areal energy density, or exposure

$$H = Et = \text{constant.} \tag{17.13}$$

This implies a reciprocal relationship between irradiation (E) and exposure time (t) and is often called the *reciprocity law*.

Reciprocity Law Failure [26]. This "law" applies quite well at intermediate exposure times, but both very short and very long exposure times are found to require more energy for the same effect. This is referred to as reciprocity law failure and can be readily explained in terms of the Gurney–Mott theory. At very long exposure times (low irradiance) the captured silver ion may escape before it is joined by a comrade to stabilize it. This clearly lowers the quantum efficiency. At very short exposure times, on the other hand, we are faced with a very high irradiance and the second photoelectron is likely to arrive at the sensitivity spot before the first one succeeds in capturing a silver ion. It will consequently be repelled, to be trapped elsewhere or to return to the ground state. This, too, clearly lowers the efficiency with which the irradiation builds up latent image specks. At even shorter exposure times the generation of photoelectrons takes place effectively instantaneously so that, in this range, a change of exposure time has no effect: the reciprocity law again applies. The same is true for exposure by ionizing radiation (such as X- and γ-rays) where each photon may excite many electrons, but does so essentially instantaneously.

By way of illustration, when Eastman Kodak 649-F plates are exposed with ruby laser radiation ($\lambda = 0.6943\ \mu$m), the exposures required to attain a density of 2.0 are 10, 23, and 71 J/m^2 for exposure times of 60 s, 250 μs, and 15 ns, respectively [27].

---

[11] R. W. Bunsen and H. E. Roscoe (1862).

Figure 17.5 [26] shows some typical curves of reciprocity law failure. Conventionally, these are shown as a log–log plot of the required exposure as a function of the irradiance. In such a plot loci of constant exposure times appear as straight lines at unity slope. (From the point of view of revealing the underlying physical process, it would seem preferable to plot the exposure as a function of the exposure time.)

Because reciprocity law failure at short exposure times is controlled primarily by the mobility of the silver ions, it increases as the temperature is lowered. At liquid nitrogen temperature, the ion mobility is negligible and capture takes place only upon warming for subsequent processing. Hence (a) the sensitivity is much lower and (b) the reciprocity law holds. On the other hand, spectral composition of the light has no effect on reciprocity: when required exposure is plotted as a function of exposure time, the curves corresponding to various wavelengths are identical in shape and are merely displaced vertically, corresponding to the spectral sensitivity of the emulsion.

Intermittency Effects. When the exposure is intermittent, the reciprocity law failure depends on the frequency of the intermittence. At low

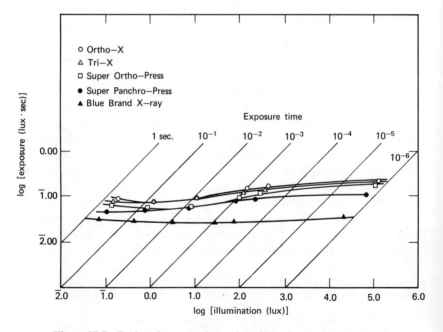

Figure 17.5   Reciprocity curves for various films. Castle and Webb [26].

frequencies, the effect of the exposure corresponds to the actual irradiance—at high frequencies to the mean radiance. This leads to the following results.

1. At intermediate irradiance levels, where the reciprocity law holds, the resultant exposure is independent of frequency,

2. At very high levels of irradiance, the efficiency improves with frequency,

3. At very low levels, the efficiency decreases with increasing frequency.

If each crystal receives, on the average, only a single photon per cycle, the additional interruption due to the modulation of the radiation is obviously not significant and the process proceeds as if there were a continuous exposure at the mean irradiance value. The frequency at which this occurs is called the critical frequency and is given by

$$\nu_c = E_p A_c, \tag{17.14}$$

where E_p is the irradiance in photons per unit area per second and
A_c is the effective absorption cross section of the average crystal in the emulsion.

This frequency divides, roughly, between the regions governed by the actual and the mean irradiance.

17.2.1.6 Hypersensitization and Latensification [1c,j], [3c]. When an exposure at moderate intensity is followed by a low-intensity exposure, the resulting density is generally higher than the density obtained when the same exposures are applied in reverse order. This phenomenon is occasionally called the *sequence effect*. It is readily explained on the basis of the Gurney–Mott hypothesis according to which reciprocity law failure at low intensities is associated with the first stages of latent image speck formation. After these stages have been passed, for instance, by means of the moderate-intensity exposure, reciprocity law failure no longer takes place. Thus low-intensity failure may be overcome by means of preexposure. This technique is called *hypersensitization*.

Similarly, since high-intensity failure is due, at least in part, to inefficiency in the growth of stable subimage specks to a developable size, it may be alleviated by a postexposure. This effectively intensifies the latent image and is often called *latensification*, for short. Latensification can also be accomplished by chemical means, for example, by exposing the latent image to mercury vapor [1d].

17.2.1.7 Miscellaneous Multiple Exposure Effects [1k], [2e], [3c]. A brief high-intensity local exposure may desensitize the areas of the emulsion thus exposed so that, on a subsequent normal exposure, these areas exhibit less density than the remainder. This is known as the *Clayden effect*[12] (or "black lightning") and can be accounted for by assuming that the too-rapid formation of free electrons causes the formation of internal latent image specks, which, during the subsequent exposure, compete with the surface specks. Since, with the usual developers, internal latent image specks are less accessible, the resulting image will exhibit reduced density. Mechanical pressure or slight bending of the film may have a similar effect [2*e*].

The *Villard effect*[13] following an X-ray exposure is similar, although there is some evidence that there the released halogens, too, play a role [2*e*].

A blue-sensitive emulsion is exposed to blue light and then an area on it is exposed to red light (to which it is not primarily sensitive). On development it is found that the area subsequently exposed to red light exhibits less density than the rest of the emulsion. This phenomenon, known as the *Herschel effect*,[14] can be accounted for by the freeing of electrons from the metallic silver specks by the longer wavelength radiation. This permits the silver ions to diffuse. In weak latent images, this may destroy developability. On the other hand, strong original exposures may eliminate the Herschel effect.

A displacement, rather than the destruction, of the latent image specks seems to be involved; but, to the extent that the new specks are internal, they are less developable. This can be shown by destroying the external latent image specks by treating the emulsion with chromic acid, in which case the red exposure increases the density although the emulsion is not primarily sensitive to it. This is referred to as the *Debot effect*.[15]

The fact that treatment with chromic acid desensitizes an exposed emulsion more than an unexposed one is referred to as the *Albert effect*[16] and can be used to obtain *reversal* (a positive rather than a negative image); it is merely necessary to treat the emulsion with the acid and then give it an overall exposure before developing.

The *Sabattier effect*[17] refers to the fact that exposure in conjunction with development seems to desensitize the area thus treated. It is accounted for only in part by the screening effect of the metallic silver and adjacency effects (Section 17.2.2.3). The reversal obtained when both

[12] A. W. Clayden (1899).
[13] P. Villard (1899).
[14] J. F. W. Herschel (1840).
[15] R. Debot (1941).
[16] E. Albert (1899).
[17] M. Sabattier (1860).

development and acid treatment are applied between the exposures is referred to as the classical reversal process.

The above effects are tabulated in Table 180 for convenient reference.

17.2.2 Spatial Characteristics [1b], [2f], [3g], [28a]

From consideration of effects of exposure level and time, we proceed to consider the spatial response of photographic emulsions.

17.2.2.1 Spread and Transfer Functions.

Light incident at one point of an emulsion is scattered into neighboring areas—primarily due to *turbidity*, that is, refractive nonuniformities within the emulsion. This, in conjunction with other mechanisms such as finite grain size and halation (reflection from the rear surface of the base, see Section 17.1.1.4), causes blackening over a larger emulsion area for such a point exposure. Similarly, an exposure by a long line of light causes blackening over a relatively broad band. The extents of these blackenings are the true point and line spread function, respectively, of the emulsion.

Virtual Exposure. Due to the nonlinearity of the characteristic curve, the course of this blackening changes with the exposure of the point or line and the background and also with development technique. This prevents the application of Fourier techniques and leads to obvious analytic difficulties when more complex exposure patterns are considered. To overcome these nonlinearities, the spread functions are generally expressed in terms of exposure distribution rather than density. This is done by using the characteristic curve of the emulsion to translate the measured density into the corresponding exposure value, which may properly be called *virtual exposure*. This procedure leads to characteristics, which are close to linear when development is carefully controlled to eliminate adjacency effects (Section 17.2.2.2) although significant deviations are present [29]. Hence the practical point and line-spread functions are defined as the virtual exposure resulting from an aerial point or line image formed on the emulsion.

Transfer Functions. The Fourier transform of the line spread function is the optical transfer function (otf). Since the spread functions are symmetrical, a cosine transform may be used and, since it is real, the otf is identical to the mtf and the relationships between these and the spread functions are as derived in Appendix 18.1.

When multiplied by the aerial image modulation, this mtf yields the virtual exposure modulation. The nonlinear response of the emulsion, as given by the characteristic curve of (17.10) must be used to determine the actual transmittance modulation. For low modulation values, however, the response can be approximated as linear with a gain factor equal to the gamma (γ) of the emulsion, as shown in Section 18.1.4.1. Hence the modulation (M_p) of the photographic image is

$$M_p \approx \gamma M_a, \tag{17.15}$$

where M_a is the modulation of the virtual exposure.

Mtf curves for a number of photographic emulsions are included in the data of Appendix 17.1. The theoretically predicted Gaussian spread is not found; but an empirical formula of the form [30].

$$T(\nu) = g \exp - \left(\frac{\nu}{\nu_g}\right)^n, \qquad \nu > 5 \text{ mm}^{-1}, \tag{17.16}$$

has been found to fit a large number of emulsions quite accurately (± 0.05). Here g, ν_g, and n are constants; their values for Panatomic-X film are listed together with the mtf curve in Appendix 17.1.

For some purposes, the exponential mtf [(17.16) with $n = 1$] is sufficiently accurate. It then takes the form:

$$T(\nu) = e^{-b\nu}, \tag{17.17}$$

where $b = 4.45$, 4.45, 4.2, and 2.6 μm, respectively, for Royal-, Tri-, Plus-, and Panatomic-X films [31c]. Data for Agfa emulsions (IFF, IF, ISS, IU) can be found in Ref. 32.

For high-resolution plates, the mtf has been found to be almost constant up to frequencies of 1500 mm^{-1}, at least. Representative values for a number of these are listed in Table 181 [33].

Phase Modulation. Density variations in the emulsion are accompanied by changes in the emulsion thickness, because of the chemical changes in the crystals and the tanning effects in the gelatin (see Section 17.1.3.2). In addition, these changes modify the refractive index of the emulsion, so that the density change in the emulsion is accompanied by a phase change. This makes the (amplitude) spread function complex; this is especially significant in coherent optical systems, such as spatial filtering and holography. In such systems, the effects of the induced thickness variations may be eliminated by immersing the film into a cell filled with an index-matching liquid and provided with optical grade windows: a "liquid gate"; see Section 19.3.3.5.

17.2.2.2. Adjacency Effects [1b], [2f], [34]. The exposure at one point of an emulsion may influence the image at neighboring points due to effects on processing. These *adjacency effects* are primarily due to the diffusion of by-products of development, spent developing solution, etc., across the emulsion area. (The gelatin effect also belongs into this general category of processing effects but has been treated already in Section 17.1.1.3.)

The diffusion process responsible for the adjacency effects takes place laterally through the emulsion and, with insufficient agitation, also along the emulsion surface. The diffusion taking place through the emulsion itself cannot readily be eliminated and may account in part for the deviation of the spread functions from the expected Gaussian form. On the other hand, the diffusion outside the emulsion, which accounts for the grosser effects, may be eliminated by constant brushing of the surface during development.

When this diffusion is not eliminated, it may result in various distortions of the photographic image. Thus, at a sharp transition from a dense to a light region, fresh developer diffuses from the light into the dense region causing overdevelopment there near the boundary. Similarly, exhausted developer and halide ions diffusing from the dense into the light region retard development there. As a result, there may appear a band of increased density just inside the dense region and a band of decreased density just inside the region of low density, giving the density trace across such a transition the appearance of "overshoot." These phenomena are referred to as the *border* and *fringe effects*, respectively. When the image consists of a small area of highly exposed emulsion on a background of low exposure, similar reasoning accounts for the density obtained, which may exceed greatly the density expected on the basis of the characteristic curve. This is known as the *Eberhard effect*.[18] The apparent increase in the width of the low density gap between two neighboring areas of high density is known as the *Kostinsky effect*.[19] The effect of the density difference between image and background, *for a given image size*, is called *background effect*. Some of these effects are illustrated in Figure 17.6, which shows schematically density traces across a bar exposure for various widths of bar. With the widest bar (1), fringe and border effects are visible at both edges and a plateau is evident in the middle. As the bar is narrowed, the border effects overlap and the image center becomes progressively more dense (2–4) (Eberhard effect). At (4), the bar is comparable in width to the line spread function of the material and any further reduction in bar width results in a lowering of density (5).

All these effects correspond qualitatively to a point spread function having a strong positive peak, surrounded by a shallow negative annulus. This has been called the *chemical spread function* [29] and is similar to the negative part of the spread function found in vision (see Section 15.4.2). It implies also an mtf that rises away from the origin, and such an

[18] G. Eberhard (1912).
[19] S. Kostinsky (1906).

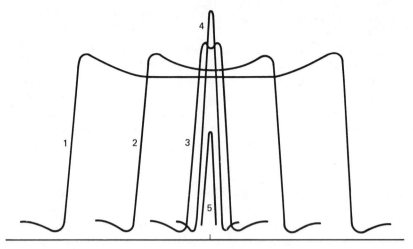

Figure 17.6 Illustrating adjacency effects: schematics of density traces across a series of progressively narrower exposure bands.

increase is, indeed, observed: mtf's tend to peak in the region between 10 and 25 cycles/mm, especially when processed with good agitation. The dependence of diffusion on the developer composition is dramatically illustrated in terms of the resulting transfer function in Figure 17.7 [34]. Note the 22% rise of the transfer function at $15 \, \text{mm}^{-1}$ with one developer, compared to a negligible rise with the other. Also note the excess of about 30 percentage points of the former over a large part of the spatial frequency range. This illustrates the image enhancement obtainable by an appropriate choice of processing. Alternatively, the adjacency effects, if known, may be removed in image processing [35].

With insufficient agitation, and especially with the photographic material stationary, gravity causes the released bromides—or fresh developer—to diffuse primarily in one direction, resulting (a) in light *bromide streaks* below dense areas and (b) in dark *developer streaks* below light areas. If the material is horizontal during processing, the analogous effects may cause a mottling, especially evident in uniformly exposed areas.

Under certain circumstances, development products at one point in an emulsion stimulate development at a neighboring point. This is called *infectious development* or *fogging*, depending on whether, or not, the affected point was a latent image point.

17.2.2.3 Effects of Radiation Spectrum and Other Factors. The nature of turbidity indicates that it should be sensitive to the spectral composition of the exposing radiation. Thus we expect shorter wavelength

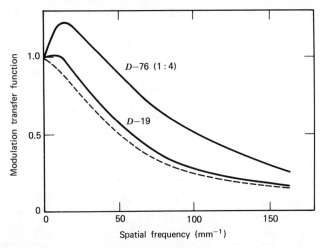

Figure 17.7   Mtf curves of Kodak Panatomic-X film, illustrating the effect of the developer and the concommitant adjacency effects. The type of developer used is marked next to each curve. The broken curve is obtained on subtracting adjacency effects. Barrows and Wolfe [34].

radiation to be both scattered and absorbed more than the longer wavelength radiation, two effects that tend to cancel. Some data along this line have been published showing that over most of the range the transfer function due to the longer wavelengths is higher, whereas for very high spatial frequencies there seems to be a tendency to reverse this condition [32]. Other factors influencing the mtf are aging, which tends to lower the transfer function, and surface development, which tends to raise it [32].

17.2.3   Granularity [1b], [2f], [3g], [28b]

We now turn to the spatial equivalent to noise in a photographic image. When a uniformly exposed photographic emulsion is processed, it is found to have a uniform density, macroscopically speaking. On closer examination, however, it is found that density fluctuates considerably from one area element to the other. An objective measure of these fluctuations (statistical in nature) is referred to as *granularity* as opposed to *graininess*, which measures the subjective sensation of nonuniformity.

17.2.3.1   Measures of Granularity

Selwyn's Granularity Constant. When the uniformly exposed and processed photographic emulsion is scanned with an aperture large enough to cover many grains, it is found that the observed density fluctuations follow a Gaussian distribution approximately, so that their statistics can be described by a single constant. We may write for the probability density:

$$p(\Delta D_0) = \sqrt{\frac{A_s}{\pi G^2}} \exp \frac{-A_s \, \Delta D_0^2}{G^2} \, , \qquad (17.18)$$

where A_s is the area of the scanning aperture,

G is Selwyn's[20] granularity constant, and

$\Delta D_0 = D_0 - \bar{D}$ is the deviation of the observed density, D_0 from its mean value, $\bar{D}$.

The dependence on the scanning aperture area is due to the fact that the observed density is averaged over a number of grains that varies directly with the area. See Appendix 17.2.3 for a derivation of this equation. It has been found that the constant G is quite independent of the area, for aperture diameters between 4 and 50 μm, and that it tends to vary with the mean density according to a power law

$$G(\bar{D}) \approx G(1)\bar{D}^{0.23}. \qquad (17.19)$$

Granularity increases also with development time [36]. See Figure 17.8

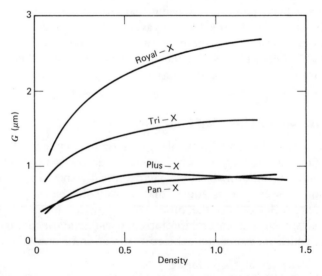

Figure 17.8 (Selwyn) Granularity versus density for four Kodak films. Based on data by Jones [31a].

[20] E. W. H. Selwyn (1935).

[31a] for values of G and their dependence on the density for a number of American emulsions, and Ref. 36 for data on some British films (Ilford Pan F, FP4, HP4). A nomenclature scale, as used by one manufacturer, is given in Table 182 [25b].

The standard deviation of the density fluctuation, in terms of G, can be written by inspection from (17.18)

$$\sigma_{D_0} = \frac{G}{\sqrt{2A_s}}. \tag{17.20}$$

Correlogram. As shown in Appendix 17.2.3, (17.18) is based on a simple model and is valid for apertures much larger than the intergrain spacing. When this model is not applicable, and when the apertures are small, spatial aspects of granularity become more complex. A more general description is obtained from the density autocorrelation function[21]

$$C_D(x, y) = \lim_{X, Y \to \infty} \frac{1}{4XY} \int_{-Y}^{Y} \int_{-X}^{X} \Delta D(x', y') \, \Delta D(x'-x, y'-y) \, dx \, dy$$

$$= \Delta D \odot \Delta D. \tag{17.21}$$

When normalized at the origin, it has been called *correlogram*:

$$C^*(x, y) = \frac{C(x, y)}{C(0, 0)}. \tag{17.22}$$

Wiener Spectrum. By the Wiener-Khintchine Theorem (3.44) the Fourier transform of the autocorrelation (17.21) yields the Wiener spectrum of the granularity

$$W_D(\nu_x, \nu_y) = \lim_{X, Y \to \infty} \left| \frac{1}{4XY} \int_{-Y}^{Y} \int_{-X}^{X} \Delta D(x, y) e^{i2\pi(\nu_x x + \nu_y y)} \, dx \, dy \right|^2$$

$$= \int\!\!\int_{-\infty}^{\infty} C_D(x, y) \cos 2\pi(\nu_x x + \nu_y y) \, dx \, dy. \tag{17.23}$$

Note that, from (17.21), at the origin the autocorrelation function equals the mean squared value of the density fluctuations:

$$C_D(0, 0) = \overline{\Delta D^2(x, y)} = \sigma_D^2. \tag{17.24}$$

---

[21] Here, and in the remainder of this section, we assume ergodic statistics so that we can substitute spatial averages for ensemble averages.

From (17.23) we note that

$$W_D(0,0) = \int\int C_D(x,y)\,dx\,dy: \qquad (17.25)$$

the Wiener spectrum at the origin equals the volume under the autocorrelation function. Similarly, from the inverse Fourier transform of the Wiener spectrum, which is equal to the autocorrelation function, we find

$$C_D(0,0) = \int\int W_D(\nu_x, \nu_y)\,d\nu_x\,d\nu_y = \sigma_D^2. \qquad (17.26)$$

Since the grains in a uniformly exposed silver-halide emulsion are quite unrelated to each other, we expect the autocorrelation to have significant values only for

$$x, y < d_g,$$

where d_g is the diameter of the largest grains, or, more generally, the extent of the influence of a grain on its neighborhood. Referring back to (17.23), we note that for

$$\nu_x, \nu_y \ll \frac{1}{d_g},$$

the Wiener spectrum of a silver-halide emulsion will be quite constant (see Figure 17.9a [37]). This is not true for other types of emulsions; see Figure 17.9b [38]. Note the rapid drop in the spectrum of X-ray film, whose granularity is due to quantum effects in the exposure rather than

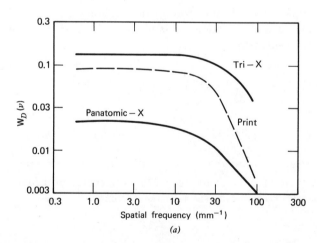

Figure 17.9a Wiener spectra of a fast (Tri-X) and a slow (Panatomic-X) emulsion. The broken curve shows the reduction in granularity obtained when the fast emulsion is printed onto the slow (fine grained) one. Higgins [29].

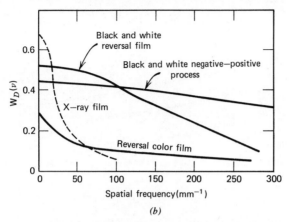

Figure 17.9b Comparison of Wiener spectra of various film types. After Klein and Langner [38].

the film grains. Additional spectra for high-resolution materials are shown in Figure 17.10 [39]. (These are actually measurements of light scattering corresponding to unit two-dimensional bandwidth and taken relative to incident flux, as explained in the next section.)

For apertures much larger than d_g, the constant Wiener spectrum implies a simple relationship between its value and the Selwyn granularity constant:

$$\boxed{W(0, 0) = \tfrac{1}{2}G^2.}$$
(17.27)

This is derived in Appendix 17.2.4.

Fluctuations in Transmittance. The transmittance change ($\Delta\tau$) corresponding to a density change (ΔD) is readily found from the definition of density:

$$\Delta D = D - \bar{D} = \log_{10}(\bar{\tau} + \Delta\tau) - \log_{10}(\bar{\tau}) = \log_{10}\left(1 + \frac{\Delta\tau}{\bar{\tau}}\right)$$

$$\approx \frac{\Delta\tau}{(\log 10)\,\bar{\tau}}, \qquad \Delta\tau \ll \bar{\tau},$$
(17.28)

with the last member following from the series expansion of the logarithm.

For the corresponding transmittance modulation, we have, then

$$M_\tau = \frac{\Delta\tau}{\bar{\tau}} = (\log 10)\,\Delta D.$$
(17.29)

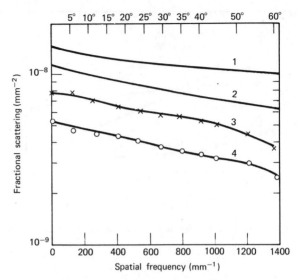

Figure 17.10 Light scattering spectra of some high resolution emulsions. (1) and (2): Kodak 649 F. (3) and (4): Agfa–Gaevert Scienta 10E70. Transmittance values are: (1): 0.5; (2): 0.36 and 0.78; (3): 0.11; (4): 0.07. After Vilkomerson [39].

Specifically, the modulation corresponding to the standard deviation of the density fluctuations is, from (17.20):

$$M_\tau(\sigma_D) = (\log 10)\,\sigma_D = \frac{(\log 10)G}{\sqrt{2A_s}} = \frac{1.628G}{\sqrt{A_s}}. \tag{17.30}$$

The transmittance Wiener spectrum is related to the density Wiener spectrum as

$$\frac{W_\tau}{W_D} = \left(\frac{\Delta\tau}{\Delta D}\right)^2 = [(\log 10)\bar{\tau}]^2 \tag{17.31}$$

and

$$W_\tau = \tfrac{1}{2}[(\log 10)\bar{\tau}G]^2 \tag{17.32}$$

when (17.27) applies.

17.2.3.2 Measurement of Granularity. The granularity may be measured by scanning the emulsion with a small aperture to determine the standard deviation (σ_D) of the density and hence G via (17.20) and $W(0, 0)$ via (17.27). One manufacturer has standardized this procedure with an aperture having a 48 μm diam, giving the numerical value of the rms granularity as $1000\,\sigma_D$.[22] The range of values obtained are included in Table 182.

Alternatively, the light scattering characteristics of the emulsion may be used to determine its granularity. In an appropriate optical arrangement, we may consider the grainy emulsion as an aperture and the light scattered by it, as its Fraunhofer diffraction pattern. For instance, with the arrangement of Figure 19.10, and the emulsion at P_0, the Fraunhofer diffraction pattern would appear at P_F. Equation 2.139 implies that this illumination pattern follows the square of its Fourier transform:

$$|\mathscr{F}[\tau]|^2 \sim \mathbf{W} \qquad (17.33)$$

and hence is similar to the transmittance Wiener spectrum.

The Callier coefficient (17.8), measuring, in effect, the fraction of light scattered by the emulsion, was the earliest quantitative measure of granularity. The above result (17.33), taken in conjunction with (17.26), shows why this was indeed appropriate.

17.2.3.3 Graininess. Although graininess is the psychological correlate of granularity, it is influenced by other factors, as well [40]. Especially magnification is a major determinant. To obtain a physical quantity correlating more closely with graininess, the *syzygetic* density difference has been defined as the mean value of the difference in density as sensed by two adjacent retinal receptors. Clearly, the sensed graininess depends also on the visual acuity and hence on the luminance and mean density.

17.2.4 Performance Criteria

17.2.4.1 Resolving Power. The classical quality factor for optical systems is resolving power. Basically, resolving power refers to the reciprocal of the separation of two object elements that are barely "resolved," that is, recognized as two. It is not surprising to find that the value obtained for resolving power depends heavily on the type of test pattern used as well as on the modulation [45]. Although many patterns have been used for resolving power determination, the periodic bar (or columnar) pattern, consisting of a series of bars and gaps of equal width, seems to be by far the most popular. This popularity may be only partly due to the fact that this pattern—consisting of three bars—appears to give values of resolving power higher than those yielded by any other of the conventional test patterns for high-contrast targets [41]. Certain methods for testing photographic resolving power have been standardized [6c].

[22] The procedure is further standardized by developing the material to unity density and measuring with an optical system measuring diffuse density as specified in ANSI Std. PH 2.19 [25b].

In the following we present an analysis of resolving power when a sinusoidal test pattern is used.

Visual Resolution. When resolving power is determined visually, the viewing magnification of the photographic image is of critical importance. To a first approximation, the retinal cone elements as projected onto the image may be considered as the scanning elements. At low magnification, the cone diameter may not be negligible compared to the period of the sinusoidal variations and the modulation will consequently be reduced significantly. On the other hand, with excessive magnification, the resulting small area elements will cause the granularity effects to be very large.

We obtain an estimate of the resolving power as the spatial frequency at which the standard deviation of the fluctuations equals the image modulation amplitude. This estimate may be optimistic because a unity signal-to-noise ratio may not suffice for detection; it may be pessimistic because visual integration effects over several retinal receptors are neglected.

The image modulation is approximately

$$M_i = \gamma M_0 T_i T_p ,$$ (17.34)

where M_0 is the original object modulation,

T_i, T_p are the mtf's of the imaging system and the photographic material, respectively, and we have made use of (17.15).

According to (17.30), the granularity modulation is

$$M_\tau = \frac{(\log 10)G}{\sqrt{2A_s}} ,$$ (17.35)

where A_s is here the emulsion area corresponding to the receptive field of one retinal receptor.

Assuming that optimum magnification is used and that this is such that the receptor diameter corresponds to one half cycle:

$$A_s = \frac{\pi}{4}\left(\frac{1}{2\nu}\right)^2 ,$$ (17.36)

where ν is the spatial frequency on the emulsion. On substituting this into (17.35) and equating the latter to (17.34), we find the following expression governing the threshold frequency (see Ref. 42):

$$\boxed{\gamma M_0 T_i(\nu)T_p(\nu) = \frac{4(\log 10)G\nu}{\sqrt{2\pi}} = 3.67\,G\nu = 0.0037\,G^*\nu^* ,}$$ (17.37)

where G^* is the Selwyn granularity constant in density micrometers and ν^* is the threshold spatial frequency in mm^{-1}.

The corresponding optimum magnification is

$$m_0 = \frac{d_r'}{1/2\nu} = \frac{2\nu d_r b_0}{f_e} = 0.04\nu^*, \tag{17.38}$$

where d_r is the effective cross-sectional diameter of a foveal cone (assumed $1.3\,\mu m$),

d_r' is its projection at the standard viewing distance, b_0, $b_0 = 250\,mm$, and

$f_e = 17\,mm$ is the effective focal length of the eye.

Scanner Resolution. When the sinusoidal test pattern is to be detected photometrically, it is obvious that the scanning slit (parallel to the test pattern lines) should be as long as possible to reduce granularity effects. Its width, however, must be maintained small in order to reduce the modulation degradation. On the other hand, too narrow a slit will result in pronounced granularity effects. To optimize resolution, the ratio of slit modulation transfer coefficient to granularity fluctuations must be maximized. This occurs when this width is $1.16556\ldots$[23] radians of the image cycle. When the slit width is thus optimized, the resolving power, ν, is found to be given by the equation [43]:

$$\nu = c \left[\frac{\gamma M_a}{G \ \mathrm{erf}^{-1}(1-2P)} \right]^2 b, \tag{17.39}$$

where $c = 87.0\ldots$,

M_a is the modulation of the aerial image,

P is the allowed probability of the density at a trough exceeding the adjacent peak density measurement, and

b is the slit length.

Once a confidence level has been chosen, the equation simplifies to

$$\nu = c' \left(\frac{\gamma M_a}{G} \right)^2 b, \tag{17.40}$$

where $c' = 106, 32, 18$ for

$P = 0.1, 0.01, 0.001$, respectively.

Note that γ, M_a, and G all are normally of unity order of magnitude, with G in micrometers.

[23] This is the real root of the equation: $2y \cos y - \sin y = 0$.

17.2.4.2   Information   Capacity   [2g],   [28c],   [44]. Another important
performance criterion for photographic materials is their spatial information capacity: how much information can be stored on them per unit area.
In a linear system under the usual conditions and normal signal and noise
distributions, the information capacity per unit area is given by (3.90),
which here takes the form:

$$H = \int\int\limits_{-\infty}^{\infty} \log_2 \left(1 + \frac{T^2 W_s}{W_G}\right) dv_x \, dv_y, \qquad (17.41)$$

where W_s, W_G are the signal and granularity Wiener spectra, respectively.

In the case of photographic emulsions, this calculation is complicated
by the nonlinear character of the system response and especially by the
limitations on signal amplitudes imposed by the limited density range of
the photographic emulsion ($2\Delta D = D_{mx} - D_{mn}$). This limitation makes the
assumption of a normal signal distribution unrealistic. An additional
complication arises from the fact that the noise level (W_G) varies with the
density level in photographic emulsions. An approximate solution has
been worked out assuming an exponential spread function and a constant
W_G [45], [46]. Based on this approximation, the estimated values of H
for various emulsions are given in Table 183a.

When the problem is viewed in terms of independent information cells
reliably recordable, there are two distinct cases, depending on the ratio of
granularity to spread function diameter. When this ratio is large, granularity sets a lower limit on the element size and this occurs with binary
recording. When the ratio is small, the spread function sets the lower limit
to the cell size. In both cases the information capacity is given by (3.8).
Results for a number of Kodak films are shown in Table 183b [2g].

17.2.4.3   Detectivity   [2h],   [31],   [28d]

Fundamentals. Up to this point we considered only the spatial performance of the medium, without regard to its sensitivity. We now consider
its ability to detect small amounts of flux. As shown previously (Sections
3.4.5 and 16.2.4), this is limited by the "dark" noise and the efficiency
with which the quanta activate the detector. In photographic emulsions,
this is limited not only by the efficiency with which photons are absorbed,
but also by the existence of a threshold—more than one quantum must be
absorbed to make the grain developable. As noted earlier (Section
17.2.1.6), this can be overcome, in part, by preexposure.

The efficiency with which photons are usefully absorbed is limited not
only by the characteristics of the grains themselves, but also by their

density and their binary nature: once a grain has become developable it is no longer sensitive and any additional photon absorbed by it is wasted. This causes the quantum efficiency to decline with preexposure. There should, therefore, be an intermediate value of preexposure, which optimizes the detectivity.

Let us now determine the emulsion *noise equivalent energy* (NEE), that is, the amount of energy yielding a density increase equal to the density fluctuations of the granularity [31], conventionally represented by its standard deviation:

$$\Delta D = \sigma_D. \tag{17.42}$$

Large Area Elements [31a,b] Let us first consider area elements large compared to the point spread function. The noise equivalent energy is then given by

$$Q_N = \Delta D \frac{dQ}{dD} = \sigma_D A_e \frac{dH}{dD}, \tag{17.43}$$

where we wrote the energy (Q) in terms of the exposure (H):

$$Q = HA_e \tag{17.44}$$

and A_e is the element area. The derivative (dH/dD) is closely related to the gamma, γ^{*24}:

$$\gamma^* = \frac{dD}{d (\log_{10} H)} = (\log 10) \, H \frac{dD}{dH}. \tag{17.45}$$

This may be substituted into (17.43). For σ_D we substitute from (17.20). Thus we find:

$$Q_N = \frac{(\log 10) \, H \sqrt{A_e} G}{\sqrt{2} \gamma^*}. \tag{17.46}$$

The energy detectivity is the reciprocal of this:

$$D_e = \frac{\sqrt{2/A_e} \gamma^*}{(\log 10) \, HG} \tag{17.47}$$

and the specific energy detectivity [see (3.102)]:

$$D_e^* = \sqrt{A_e} D_e = \frac{\sqrt{2}}{\log 10} \frac{\gamma^*}{HG} = \frac{0.6142 \gamma^*}{HG}. \tag{17.48}$$

---

[24] We use the asterisk to distinguish the local value of the slope from its value, γ, at the linear portion of the characteristic curve.

The signal-to-noise ratio resulting from an exposure of Q watts is:

$$R = \frac{Q}{Q_N} = (HA_e)D_e = \frac{\sqrt{2A_e}\gamma^*}{(\log 10)\, G}. \tag{17.49}$$

The detective quantum efficiency (DQE) is given by the ratio of signal-to-noise-squared at the output to that at the input (see Section 3.4.4):

$$\kappa_d = \frac{R^2}{R_{in}^2} = \frac{R^2}{n_q}, \tag{17.50}$$

where we have assumed the input signal-to-noise ratio to be governed by Poisson statistics, so that it equals the square root of the number of incident quanta, n_q. Taking R from (17.49) and writing the number of quanta:

$$n_q = \frac{HA_e}{Q_0}, \tag{17.51}$$

we find the detective quantum efficiency

$$\kappa_d = \frac{2\gamma^{*2}Q_0}{(\log 10)^2\, HG^2}, \tag{17.52}$$

where Q_0 is the average energy per quantum. As a rule κ_d values are lower than the ideal by a factor of approximately 100, due primarily to the recombination of electrons and holes. This is required to be a likely event, in order to permit a reasonable shelf life. Other factors responsible for the low values of κ_d are the incomplete absorption of light, the dispersion of grain sizes and locations, etc. [47].

In Figures 17.11 and 17.12 [31] we present, respectively, the energy detectivity (D_e), and the detective quantum efficiency (κ_d) as a function of exposure for four Eastman Kodak films. Detectivity data for some British films (Ilford Pan F, FP4 and HP4) have been published in Ref. 36.

Other related parameters, such as the number of noise equivalent quanta (NEQ), are discussed in Appendix 17.3.

Application of Theory. What is the practical significance of all of these functions when we wish to detect weak or low-modulation signals?

1. If we wish to detect a weak signal of limited duration (or, equivalently, when the permitted exposure time is limited), it is the energy detectivity (17.47) that must be maximized. To optimize detection, we must choose the emulsion having the highest value of D_e and provide the optimum bias exposure, that is, bring the emulsion to the optimum operating point, by pre- or postexposing it. (When there is some background, this may serve to contribute to the required bias exposure.)

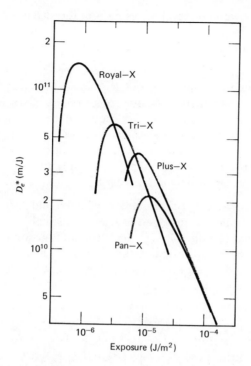

Figure 17.11 Specific energy detectivity of some Kodak films, as a function of exposure. Exposure wavelength: 0.43 μm. After Jones [31b].

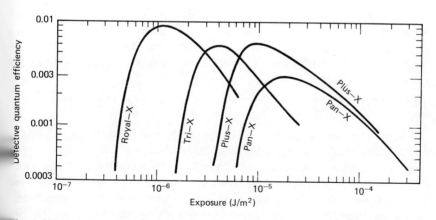

Figure 17.12 Detective quantum efficiency of some Kodak films, as a function of exposure. Jones [31a].

647

2. When there are no time limitations, but the signal is of low contrast, we must maximize the signal-to-noise ratio of the recording process, and time the exposure accordingly.

a. Assuming that we are limited to a single exposure, the emulsion should be chosen for its high value of R_{mx} and exposed until it attains this value, approximately at the inflection point of the characteristic curve.

b. If several exposures may be used and the results combined to reduce the effects of noise, the signal-to-noise ratio may be improved indefinitely by using more and more independent exposures. If, however, the total exposure time, t_T, is limited, we can optimize its utilization by timing each exposure to maximize κ_d [48].

Small Area Element [31c]. Equation 17.47 would seem to imply that the detectivity could be increased indefinitely, simply by decreasing A_e. Unfortunately A_e is limited by the point spread function of the emulsion. Specifically, if the aerial image to be detected is small compared to the point spread function, we must substitute its equivalent area, A_0, for A_e. The equivalent area is

$$A_0 = \frac{2\pi \left[\int_0^\infty r P(r)\, dr \right]^2}{\int_0^\infty r P^2(r)\, dr}. \tag{17.53}$$

Equation 17.47 then takes the form:

$$D_{e0} = \frac{D_e^*}{\sqrt{A_0}} = \frac{(0.6142/\sqrt{A_0})\gamma^*}{HG}, \tag{17.54}$$

where the coefficient in the parentheses is independent of the exposure. A_0 has been evaluated as 500, 500, 490, and 156 μm^2 for Eastman Kodak Royal-X, Tri-X, Plus-X, and Pan-X films, respectively [31c].

17.3 OTHER PHOTOSENSITIVE MATERIALS [3h], [49]

Many other photosensitive materials can be used instead of the silver halides. These generally lack the high gain obtainable with the latter, but they may have compensating advantages, such as economy or convenience, especially in reprography (photocopying) where high sensitivity is often not essential. In this section we describe the more popular of these materials. Summaries of papers presented at conferences on "unconventional" photographic systems have been published periodically [50], as well as a complete monograph [50*] and a bibliography of reviews of such systems covering the period 1960–1969 [51].

17.3.1 Diazotype and Other Dye-Forming Processes [49a], [52]

17.3.1.1 General Characteristics. Many aromatic diazo compounds[25] are broken down by the action of light. In the presence of water molecules, this photolysis changes the diazo into a phenol or similar molecule, with the release of nitrogen. The sensitivity spectra of these materials are concentrated primarily in the uv; a representative spectrum is shown in Figure 17.13a [52a]. Certain diazo compounds have sensitivities extending further into the visible part of the spectrum, as shown by Figure 17.13b; these are commerically referred to as "superfast" and can be exposed effectively by tungsten and fluorescent lamps, which are too weak in the uv for use with the unsensitized diazo materials.

Photography using diazo materials is inherently capable of resolving power higher than that of silver halide materials. However, the diazo materials generally exhibit no quantum gain and, therefore, have a much lower sensitivity.

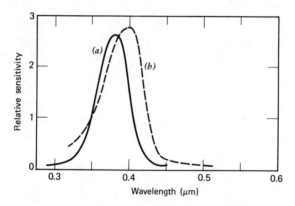

Figure 17.13 Sensitivity spectra of typical diazotype papers. (*a*) Regular; (*b*) "superfast." Based on Tyrrell [52a].

17.3.1.2 Positive Processes. The photosensitive diazo materials also readily combine with certain couplers to produce azo dyes. If now a layer containing such diazo compounds is partially exposed to light and then brought into contact with an appropriate coupler, the layer will remain

---

[25] In these a group containing *two nitrogen* atoms is substituted for one of the hydrogen atoms attached to the benzene ring. Hence the name di-azo—*azote* is French for nitrogen.

clear in the exposed areas and become dyed in the unexposed areas. This effect forms the basis of diazo photography. Its sensitometry has been analyzed theoretically and compared with experimental results [53].

The required coupler may be applied to the material after exposure in a moist ("semi-wet") process, or it may be incorporated into the original photosensitive layer, yielding a dry process. Here an acid environment is established to prevent coupling before the image exposure. The acidity may be neutralized, and coupling initiated, by subjecting the film or paper to an alkali, such as ammonia vapors; this corresponds to development. Alternatively, the alkalinity may be produced by heating.

17.3.1.3 Negative Processes. In the negative processes, the dyes may be formed by providing a coupler that reacts with the photolytic products. In some systems, the phenolic photolysis product itself acts as a dye-forming coupler. It is present only in exposed areas and only there will dyes be formed.

Other diazo compounds (diazo-sulfonates) will not couple to the dye-forming component until they are activated by light, decomposing them into diazonium and sulfite ions. These too form negative images.

In all of these processes, fixing is essential, that is, the remaining diazo material must be removed after developing.

17.3.1.4 Vesicular Process. Dye formation is not essential to diazo photography. In the vesicular process, the diazo material is embedded in a thermoplastic matrix. When exposed and then heated, the nitrogen gas released by the photolysis forms microscopic bubbles or vesicles. These scatter the read-out light incident on them. The simplicity of processing, which requires only moderate heating (approximately 80°C minimum, 115°C optimum), combined with a high modulation transfer function makes these materials useful despite their relatively low sensitivity. They are commercially available under the name *Kalvar.*

The vesicles have diameters ranging from 0.25 to 2 μm. One commercially available material has an mtf of 0.5 at approximately 100 mm^{-1} and 0.2 at 380 mm^{-1} [54]. These materials have a sensitivity spectrum peaking at 0.385 μm and, at that wavelength, require about 2 MJ/m^2 for maximum density. Since the vesicles occupy a volume much larger than the molecules responsible for them, we may here speak of a quantum gain.

Fixing requires the exposure of all the remaining diazo molecules after the vesicular image has been developed. Heating the matrix permits the resulting nitrogen gas to escape; this heating must be held below the temperature at which vesicles form.

The fact that these materials rely on scattering, rather than absorption, for their response, makes their transmittance highly dependent on read-out conditions. For instance, the same material that exhibits a density range of 0.9 when viewed with diffuse light will have a density range well over three when illuminated and viewed with f/16 lenses (see Figure 17.14 [55]). The corresponding gammas are approximately 0.5 and 3. This illustrates, incidentally, the extent to which the contrast of these materials can be controlled by the read-out technique. More flexibility is inherent in the fact that exposure increases density when the image is read out by transmitted light and decreases density when the reflected light is used for read-out. In the latter system, a low reflectance background must, of course, be provided. Kalvar material is available on a black paper base for autopositive recording.

Reversal exposures, too, are possible. Here the nitrogen released in the original exposure is permitted to diffuse. This is followed by a total exposure and heating to vesiculation temperature.

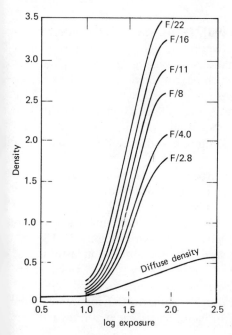

Figure 17.14 Characteristic curves of vesicular material (Kalvar 7BTC), illustrating their dependence on the relative aperture of the viewing system. The F/number is noted next to each curve. The exposure units are arbitrary. Nieset [55].

17.3.1.5 Other Dye-Forming Processes.
Dyes, besides diazo compounds, may be used in photographic systems. Those that are negative working form dyes on exposure to light by breaking up into molecular fragments that either are themselves colored or react with other

molecules to form colored products. A number of these have been reviewed in Ref. 49*b*.

When the dye is destroyed by the radiation, the process is positive working. Such materials have been reviewed in Ref. 49*c*.

17.3.2 Photopolymerization and Related Techniques

In certain substances, the action of light changes the gross physical characteristics, making the material harder, less soluble, etc. Such materials may be used simply to form visible images; but they are especially attractive for certain manufacturing processes in which the mechanical changes are used directly, such as in the manufacture of printing plates. In the form of photoresists, these materials may provide a protective layer over a substrate to control its processing, such as etching. Thus, they are a powerful tool in a number of industrial processes such as micro-machining the production of printed and integrated circuits, integrated optics, and related technologies.

17.3.2.1 Dichromated Colloids **[49d], [52b], [56a]** *and Other Photo-resists* **[52c], [56b], [57].** Upon activation by light, dichromates tend to induce cross-linking in colloidal matrices such as gelatin, fish glue, and shellac and in synthetic materials such as polyvinyl alcohol. This cross-linking reduces the solubility of these materials, so that a photographic system may consist of a film of dichromated colloid which is developed by soaking or rinsing in water or some other solvent. When clear, the resulting record will be a phase photograph and if colored, it will be a (negative) absorption photograph.

At elevated temperatures, these processes have a tendency to take place in the dark as well. Maintaining the substance below $-7°C$ is found to inhibit this "dark reaction" effectively.

The most commonly used dichromates are the ammonium and potassium forms. A typical sensitivity spectrum is shown in Figure 17.15 [58], illustrating that the sensitivity is governed by the absorption spectrum of the dichromate, as predicted by the Grotthus-Draper law,[26] which states that only absorbed radiation can be responsible for chemical action. In general, the quantum efficiency is roughly equal to 0.3 for wavelengths shorter than 0.41 μm and drops with increasing wavelength beyond that point.

Special techniques have been developed to maximize the index change resulting from a given exposure. Useful effects have been obtained with this method with exposures as low as 10 J/m^2 [58].

---

[26] T. C. J. D. Grotthus (1818) and J. W. Draper (1841).

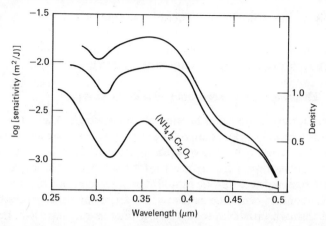

Figure 17.15 Sensitivity spectra of two commercial photoresists, compared to the absorption spectrum of ammonium dichromate (lowest curve and right-hand ordinates.) The sensitivity is here the reciprocal of the exposure (J/m^2) required to produce unity density. O'Brien [56].

Because of its high resolution capability, dichromated gelatin is used in holography (see Section 19.3.4.2).

Polyvinyl cinnamate forms cross linkages under exposure to uv radiation and is a popular substance for photoresists [59]. The characteristic curves of a number of commercial photoresists have been measured. Their exposure requirements are listed in Table 184 [59*].

Insolubility is the most popular change exploited in photopolymer photography, but others are used as well. See Table 185 [60] for a listing of the physical states used and their applications.

***17.3.2.2 Photopolymerization* [49e], [52c].** Photopolymerization is a process closely related to the above. Here the photosensitive material consists of a monomer which, upon exposure to the appropriate radiation, will polymerize. This process is initiated when a relatively small monomer molecule absorbs energy enabling it to form a link to is neighbor. Once this link is formed, the absorbed energy is available for forming an additional link and this process may progress, forming very long chains constituting giant molecules, called polymers. As a result of polymerization, the substance becomes harder and less soluble and this fact may be used to form an optical image. The chain reaction process implies a large quantum gain.

In principle, the polymerization reaction can be initiated directly by the absorption of radiation, but the process is made much more efficient by including a substance, called *initiator*, which absorbs the energy more efficiently and transfers it to the monomer. The spectral sensitivity of the material is then determined by that of the initiator.

Recent results with electronically exposed polymethyl methacrylate (PMMA) resists indicate a resolution of 2×10^4 lines/mm, at least [60*].

17.3.3 Thermographic Recording Processes [49f]

In thermography, the local heating of a film produces a change in color by inducing a chemical change (e.g., by physical contact induced by melting); by fusing a bright, opaque film covering a dark substrate; by selectively evaporating a solvent; or by some other processes. This is often used in *reflex exposure* applications, where the original copy is placed in contact with the sensitive material and the sandwich is irradiated through either layer. The dark regions of the original absorb a greater portion of the radiation, experience a greater termperature rise, and may, therefore, activate the thermographic material.—Note that a mirror image is obtained if the image-forming layer is placed in face-to-face contact with the original. In such cases, a transfer process may follow the thermographic process.

Thermal printing uses contact with a matrix whose elements may be selectively heated [61]; but this is outside our present scope.

17.3.4 Photochromism [49g], [52d], [56c] and the Photorefractive Effect [56d]

Many substances are *phototropic*, that is, they change their molecular structure reversibly under the influence of radiation in such a way that their absorption spectra are modified. When a visible color change is produced, this phototropism is referred to as *photochromic* and can be used for temporary photographic recording. (Often the changes can be made permanent, but their unique reversibility is generally the main reason for using photochromics.)

In inorganic materials, the color change may be due to the formation of F-centers (see Section 6.4) or the filling of electron traps at impurity sites [62]. In organic materials, the molecular change may be a switch between *cis*- and *trans*-forms or between an open or a closed ring [49g].

Reversion may be induced by heating, by irradiation with a wavelength beyond the actinic spectrum, or may take place spontaneously by phosphorescence.

Calcium fluoride doped with certain rare earths (La, Ce) and strontium or calcium titanate doped with transition metals (Ni and/or Mo), seem to be the most promising for information storage applications [57c].

Glasses have been made phototropic by dissolving in them cerium or europium salts or silver halide crystals. Such glasses have been used, for instance, in sunglasses that automatically adapt their transmittance to the ambient illumination level and in protective goggles to guard the wearer against retinal damage from sudden intense light flashes in his field of view [63]. Many other potential uses, including optical computer memories, have been suggested and investigated [64].

One representative photochromic glass consists of silver halide crystals approximately 10 nm wide, spaced about 60 nm on the average [64]. Representative absorption spectra, before and after darkening, are shown in Figure 17.16 and darkening and bleaching curves in Figure 17.17 [65]. Thermal fading curves are shown in Figure 17.18 for two representative materials [65].

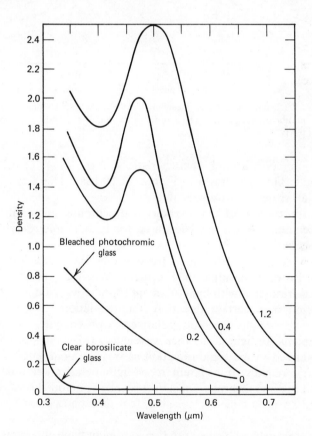

Figure 17.16 Density spectra of silver halide photochromic glass (Corning Type 5, 8 mm thick). The darkening, in terms of density increase relative to the unexposed glass, is indicated next to each curve. Megla [65].

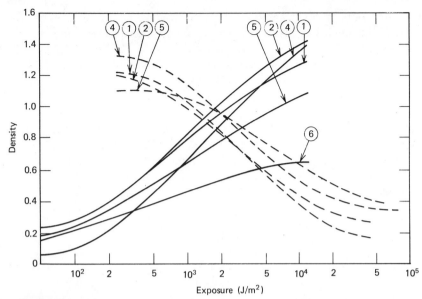

Exposure (J/m^2)

Figure 17.17 Darkening (solid) and bleaching (broken) curves for some representative silver halide photochromic glasses. The Corning-type number is indicated next to each curve. Megla [65].

Absorption spectra and characteristic curves for a representative photochromic film are shown in Figure 17.19 [66]. The active material there was alkyl mercury dithizonate.

As an illustration of an organic photochromic material we cite salicylideneaniline, which has been used for holographic recording [67]. See also Section 19.3.4.2.

M-centers, too, have been used for erasable photographic recording. These consist of two adjacent F-centers. When a material carrying M-centers is irradiated with light of an appropriate wavelength (M_F-band) and polarized in a certain direction, the oscillations of the electrons trapped at the M-site become polarized. Consequently the material becomes opaque to light at another wavelength (M-band)—provided this light is polarized in the same direction as was the polarizing (M_F) light. The material remains transparent to M-light polarized normal to that direction: it became dichroic. This effect is called photodichroism [67*]. It is potentially well suitable for erasable, real-time photographic recording [68].

Color centers excited by electron bombardment have also been used for information storage [68*]; but these are outside the scope of the present discussion. See however, Section 6.4.

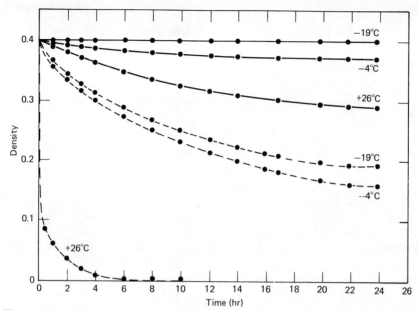

Figure 17.18 Fading curves of two silver halide photochromic glasses at various temperatures, as indicated next to each. The solid curves refer to slow-fading glass, Type 2; the broken curves to fast-fading glass, Type 6. Megla [65].

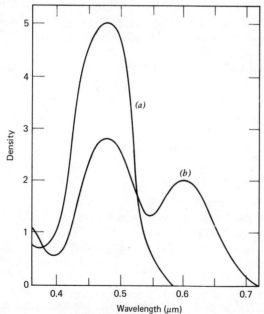

Figure 17.19 Density spectra of photochromic film. (*a*) Unexposed; (*b*) exposed. American Cyanamid Type 63–071 (discontinued). Film thickness: 12.7 μm. Baldwin [66].

657

The photorefractive effect in ferroelectric materials has been used extensively in hologram recording. See Section 14.3.7.3 for a brief description.

17.4 ELECTROPHOTOGRAPHY [3i], [69]

Electrophotography includes all optical image recording techniques involving externally applied electrical processes. The most widespread of these is *xerography*[27] in which a photoconductive layer is used to convert an illumination pattern (the optical image) into a similar electrical charge distribution, which, in turn, is made visible by one of many "development" techniques. Other electrophotographic techniques are based on persistent internal polarization (PIP), persistent photoconductivity, and a number of processes in which development occurs simultaneously with the exposure. Here we present the most important xerographic techniques and survey briefly some of the other electrophotographic processes that are not, at the moment, used commercially.

17.4.1 Xerography

A wide variety of xerographic techniques is possible, differing in the manner in which the latent image is formed and in the development technique used. For instance, the static charge could be formed by a photoemissive process. However, in the practical systems a photoconductive layer is used and this is the technique discussed here. Specifically, the electrostatic image is formed by charging the surface of the photoconductor uniformly, for example, by passing it under a corona discharge, and then forming an optical image on this surface. The response of the photoconductor will permit the charge to leak off at the exposed points, leaving the required electrostatic image.

17.4.1.1 Toner Xerography

Basic Systems. In toner-developed xerography the electrostatic image is developed by permitting colored particles (the *toner*) to be deposited at the charged regions. This is probably the most popular form of electrophotography and is used in two versions.

1. The photoconductive layer is coated on a reusable plate from which the toner is transferred to the paper, film or other expendable base.
2. The expendable base itself carries the photoconductor.

[27] Xeros is Greek for "dry."

Examples of these are found in the Xerox and Electrofax processes, respectively.

The Xerox process uses a selenium-coated plate as the electrostatic image carrier and a resinous developing toner. The completed powder image is electrostatically transferred to the paper and fused there to form a permanent image.[28] Electrofax uses zinc oxide coated onto the paper directly. Here the powder image need not be transferred.

Spectra. Sensitivity spectra of representative xerographic materials are shown in Figure 17.20 [70]. The apparent quantum efficiency spectrum of selenium, and its field dependence, are shown in Figure 17.21 [3i].

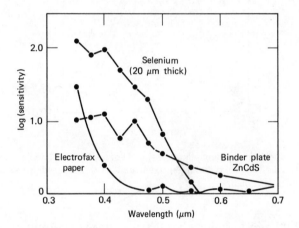

Figure 17.20 Sensitivity spectra of representative xerographic materials. The sensitivity is here the reciprocal of the exposure (J/m^2) required to discharge the plate surface potential by half. Chapman and Stryker [70].

It should be noted that the spectral response of the Electrofax paper is naturally limited almost exclusively to the ultraviolet, but can be extended by adding dyes to the photoconducting layer or by using a colored zinc oxide. An example of this is included in Figure 17.20.

Toner Application. The "developing" powder is usually charged by using the phenomenon of triboelectricity (transfer of electric charge by friction between dissimilar materials). It may be applied as an aerosol, a liquid suspension, or it may be brushed onto the electrostatic image

---

[28] Instead of the powder image, the original charge itself may be transferred. This is called transfer of *e*lectro-*s*tatic *i*mages (tesi).

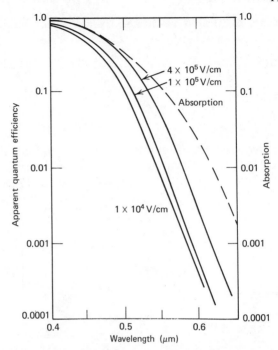

Figure 17.21 Apparent quantum efficiency spectra of a selenium plate for various values of applied field (solid, left-hand ordinate), compared to the absorption spectrum (broken, right-hand ordinate). Lehmbeck [3i].

directly. This latter technique has the advantage of permitting a threshold for the electrostatic field to be set. Due to the above-mentioned triboelectric effect, the brush material takes on a charge with a polarity opposite to that of the developing powder. Thus the field at the plate surface must exceed the field between the brush and the powder before development takes place. In another version, granules carrying the toner are permitted to cascade across the electrostatic charge image. This is referred to as *cascade development*.

The commercial versions of the Electrofax-type process appear to be using toners primarily in liquid suspensions. Here the liquid passes between the paper carrying the charge image and a *developing electrode* plate. This plate is maintained at a potential so as to cause the desired rate of deposition of toner onto the paper. After drying, the fine toner particles adhere to the coated paper surface, requiring no further fixing. Aerosol type development requires 1–10 sec, whereas electrophoretic deposition can yield development times as short as 0.1 sec.

Large-Area Effects and Characteristic Curve. Since the electrostatic field, rather than the charge, at the plate surface controls development, there is a tendency to develop only narrow areas and edges of extended areas where the changing charge density induces high fields. This tendency can be overcome by using a half-tone screen to break up the large areas in the image. Alternatively, image fidelity can be maintained by placing a conducting *"developing" electrode* above the surface of the photoconductor, leaving a small gap. If this plate is maintained at the same potential as the back of the xerographic plate (usually ground potential) the field will be close to normal to the plate and proportional to the residual charge. Since exposure removes charge, "development" tends to vary inversely with exposure, yielding an "autopositive"-type image. A response curve for a selenium plate is shown in Figure 17.22 [71].

The effect on the mtf, of the spacing between a developing electrode and the charged surface is illustrated in the amplitude response curves of Figure 17.23 [72]. These do not include the effects of the mtf of the charge image itself. For the latter see Ref. 72*.

Gamma Control. The above discussion assumed that the developer particles are charged with polarity opposite to the charge on the photoconductor surface. If they are charged with the same polarity and the

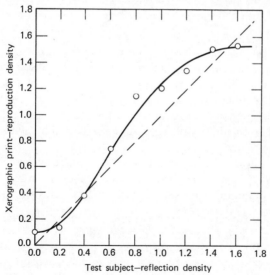

Figure 17.22 Characteristic curve of xerographic plate. Dessauer, Mott, and Bogdanoff [71].

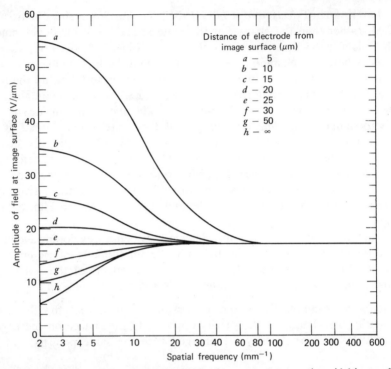

Figure 17.23 Field variation amplitude resulting from exposure to a sinusoidal image, for various spacings of the development electrode; calculated. Schaffert [72].

developing electrode also has the same polarity but a potential higher than the highest residual potential remaining on the photoconductor, the developer will be attracted to the exposed portions of the xerographic plate, forming a negative image (or a positive if a negative was used for the exposure). The contrast (gamma) of the system can be controlled by means of either the initial charging potential or that of the developing electrode.

Because latent image growth is not constant and, specifically slows down at high exposure levels, the total exposure must be carefully controlled to maintain contrast, at least for low contrast objects.

Sensitivity. Sensitivity data for a number of photoconductive materials as used in xerography are listed in Table 186.

The dark resistance of the photoconductor decreases with increasing applied electric field. This leads to an upper limit, when the charge leakage exactly cancels the rate at which the charge can be supplied. This limit is called the *acceptance potential.*

Density Range. Noise. The density range attainable with xerography is limited at the lower end by the reflectivity of the paper base, and has an upper limit set by the reflectivity of the toner. This range rarely exceeds one density unit.

Contrast is further reduced by the occasional unavoidable adhesion of toner particles in fully exposed regions. When the particle are microscopic, this basically lowers the background lightness; if they are visible, they lead to mottling of the background [73]. See [73*] for noise factors.

17.4.1.2 Deformational Xerography

Thermoplastic Xerography [56e] [74] [76]. Another active field in electrophotography is thermoplastic xerography (photothermoplastic recording). Here the latent charge image is developed as a deformation of the thermoplastic film and made visible by an appropriate optical system.

The basic structure generally has the form shown in Figure 17.24. Before the exposure, a charge is applied to the upper surface, causing a mirror charge to appear at the electrode. Exposure causes the conductivity of the photoconductor to rise and the charges to flow from the electrode to the plastic-photoconductor interface: the field across the plastic remains unchanged, but the potential at the upper surface drops. On recharging the upper surface to restore the original potential, the charge, and hence the potential, is increased in the exposed areas. (This procedure, incidentally, makes the recording insensitive to further exposure.) On heating the plastic to its softening point, the field-induced forces compress the plastic layer in the exposed regions, deforming it in accordance with the exposure distribution. The deformations can be made visible in a Schlieren-type optical system (similar to dark-field microscopy). When the deformations are large, the resulting scattering, by itself, may suffice to make the deformation visible in an ordinary projection or viewing system. The heating can be accomplished by a stream of hot air, by irradiation from a lamp, or by sending a current pulse through the electrode.

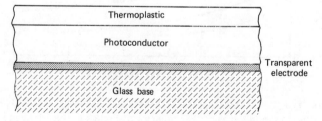

Figure 17.24 Typical thermoplastic xerographic material; schematic.

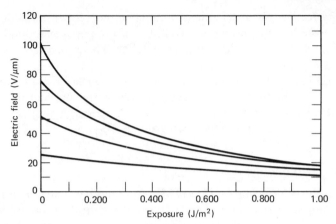

Figure 17.25   Field-exposure characteristics of PVK dyed with brilliant green, for various initial field values. Layer 4 μm thick. Bergen [74].

A material commonly used for the photoconductive layer is polyvinyl carbazole (PVK), which may be dyed to obtain the desired sensitivity spectrum. The field-exposure characteristic of a representative layer is shown in Figure 17.25 [74] and dark decay in Figure 17.26 [74].

As in silver halide photography, here, too, the sensitivity may be increased by supplementary overall exposures.

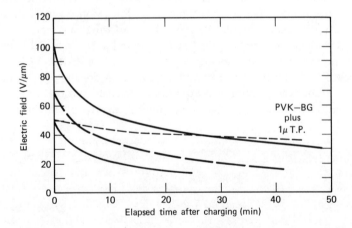

Figure 17.26   Dark-decay characteristics of PVK dyed with brilliant green, for various initial field values. Layer 4 μm thick. The broken curve refers to layer covered with 1 μm layer of thermoplastic. Bergen [74].

Frost versus Screened Recording [75]. The basic process, as described above would produce surface gradients, and the associated refraction, only at points of illumination gradient. Uniformly exposed areas would appear dark. In fact, on heating, such areas tend to form a wrinkled surface structure. This is called *frost*; it constitutes noise and, simultaneously, serves as a signal carrier: the noise amplitude is controlled by the signal, much as in a granular silver halide photograph. A representative spatial Wiener spectrum of frost is shown in Figure 17.27 [75]. Note the peak near 190 mm^{-1}. This represents a "natural" frequency at which the layer is most readily deformed; it varies inversely with the layer thickness [see (17.3)]. This frequency is unaffected by the exposure.

In an alternative approach, *screened recording*, the image is modulated by a regular grating. This modulation then serves as a carrier and largely eliminates the frost effect, especially if its frequency is close to the natural frequency of the layer. This method results in a considerably higher resolution and signal-to-noise ratio. Spatial frequencies as high as 4000 mm^{-1} have been reported [76].

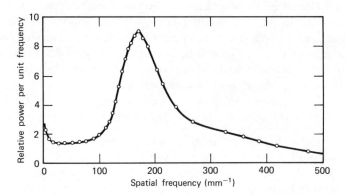

Figure 17.27 Wiener spectrum of frost deformation. Urbach [75].

Other Deformational Techniques. The heating required in thermoplastic xerography can be avoided by substituting an elastomer for the thermoplastic. No heating is then required, but the forces must be maintained so that, in practice, charges must be applied continuously. Such devices have been named ruticons and classified according to the method of supplying charges. In the α-ruticon, the charges are supplied by a glow discharge maintained above the elastomer layer; in the β-ruticon by means of a conductive liquid, such as mercury; and in the γ-ruticon by means of a metallic film coated onto the upper surface of the elastomer [77]. The latter appears to be the most promising form [78].

Other deformable media used to display electrostatic images are oil [79] and cholesteric liquid crystals [80].

17.4.2 Other Forms of Electrophotography

17.4.2.1 Persistent Internal Polarization [69b]. When an electric field is applied across a photoconductive plate while it is being exposed to an optical image, the plate becomes electrically polarized, with the polarization persisting for substantial periods of time after the field and image are removed. This phenomenon is referred to as *persistent internal polarization* and the plate as a *photoelectret.* This effect is due to the trapping of electrons and holes in the photoconductor, and the persistence depends on the depth of the available traps. The persistence can be extended by short-circuiting the two electrodes, eliminating the external electric field. The image can be developed by removing one of the electrodes and using one of the usual xerographic development techniques.

17.4.2.2 Persistent Conductivity [69c]. Often the increase in conductivity observed in photoconductors persists for a considerable time even in the dark. This effect is due to carriers in shallow traps in the bandgap. Their release can be accelerated by heating, ir irradiation, or application of a strong electric field. This persistent conductivity can be used to generate a charge image, as in xerography. It can also be used directly in an electrolytic arrangement to yield a metallic deposit proportional to the exposure or in an electrophoretic system to capture charged toner particles.

17.4.2.3 Other Forms. In addition to the techniques already described, numerous other electrophotographic systems have been suggested; some of them have been confirmed experimentally. Some use a suspension of photoconductive particles that become charged during exposure and migrate through a softened medium during development [81]. Others use conductivity changes due to heating [82]. The interested reader is referred to Ref. 69d for a survey of these.

17.4.3 Nonoptical Electrography

17.4.3.1 Electronic Electrography. The charge image that forms the basis of the xerographic process can, of course, be produced directly without using a photoelectric effect. For instance, it can be formed on a suitable dielectric base by means of an electron beam or by means of a voltage-modulated stylus passing in close proximity [83]. The development can then be by toner or the deformation of a liquid (Eidophor) [84] or a heated thermoplastic film [85] (e.g., the lumatron tube [86]).

A number of electrographic printing techniques, too, have been developed. See Ref. 69e for a review.

17.4.3.2 Electroradiography [69f]. In conclusion, we note that the xerographic plates are sensitive to X-rays as well. Indeed, their edge-enhancing characteristics may be of special interest in such applications. This technique is called *xeroradiography* and has been used in both medical applications and nondestructive testing.

17.5 COLOR PHOTOGRAPHY

17.5.1 Color Photography as a System Component

We have so far treated black-and-white photographic emulsions as two-dimensional communication channels, carrying one-dimensional signals. If we now wish to consider color films, we find ourselves confronted with a three-dimensional signal; ideally, the three color coordinates can vary independently at each point on the emulsion.

The concept of a multidimensional signal does not pose a major complication. The color photograph is equivalent to three independent, but simultaneous and coordinated, one-dimensional channels.

Color materials were developed primarily to record a visual stimulus and for esthetic purposes. They are less important as information channels. In that application, increased processing difficulties and the unavoidable residue of cross talk between these channels make color emulsions less attractive than the equivalent black-and-white emulsions. But in many cases their advantages may well outweigh these objections. When storage volume considerations or registry of simultaneous area elements are critical, color emulsions may be the answer. It is for this reason that a brief section on these is included here.

Color processes can be divided into three groups: subtractive, additive, and "spectral." [29] Of these, only the two former methods are of practical significance today and are treated below.

---

[29] One of the "spectral" techniques is based on dispersion; the image is broken up, by means of a line grating, into narrow strips with relatively large gaps between them. A dispersing device (glass prism or diffraction grating) is placed near the imaging lens aperture, so that each element along each line is broken up into a spectrum transverse to the line. This spectrum is recorded on auto-positive black-and-white film. When the film is processed, replaced in its "taking" position and illuminated with white light, a color image will appear when viewed through the line grating. The other technique, due to G. Lippmann (1891), is based on interference, a reflecting surface being provided at the back of the very fine-grained emulsion, so that standing electromagnetic waves are set up in the emulsion. At the nodes no blackening will occur. Light reflected from the processed emulsion will exhibit constructive interference at the wavelength of the exposure.

17.5.2   Subtractive Color Films [1k], [2i], [3j]

Most color photography today depends on a subtractive system. The emulsion consists of three layers, each sensitive to a different region of the spectrum. In processing, each of these is dyed to absorb (subtract) the corresponding spectral region of the incident light. The extent of this absorption at any emulsion element will depend on the amount of exposure received by the element.

Basic Structure. The following seems to be the most widely used emulsion arrangement. The uppermost layer is blue sensitive only. The middle emulsion layer is sensitive in the blue-to-green region, while the last layer is sensitive in the blue and red regions, with negligible sensitivity in the green. A thin (yellow) layer to absorb the residual blue light transmitted by the uppermost layer is interposed between it and the middle layer, so that it becomes effectively green sensitive only. Simultaneously the last layer becomes effectively red sensitive only.

Each of these emulsion layers is provided with a *coupler*, which combines with the oxidation product of the developer to produce, respectively, yellow (minus blue), magenta (minus green), and cyan (minus red) dyes. Each of these absorb, then, from the transmitted read-out light the spectral region which exposed it (see Figure 17.28).

Positive versus Negative Emulsions. When a negative image is desired the coupler is applied immediately upon development of the photographic image. In the reversal processes, on the other hand, the photographic image is first developed without application of the couplers to yield a negative black-and-white image. The emulsion is then given a *"reversal exposure"* exposing all of the emulsion uniformly. Subsequent development will affect primarily the originally unexposed portions of the emulsion. During this development the color developer and couplers are incorporated in the processing solution. Consequently the dyes form primarily in the originally unexposed areas and a positive color image results.

Processing. In some emulsions the coupler is added during processing (e.g., Kodachrome), whereas in others (e.g., Ektachrome) it is included in the appropriate emulsion layer encapsulated in globules to prevent diffusion.

During fixing, the silver grains and the above-mentioned yellow layer are removed, so that only the dye images remain. This accounts for the considerably reduced granularity in subtractive color films.

Cross Talk. Ideally (from the recording point of view) the spectral regions of sensitivity in the three layers should have no overlap and the

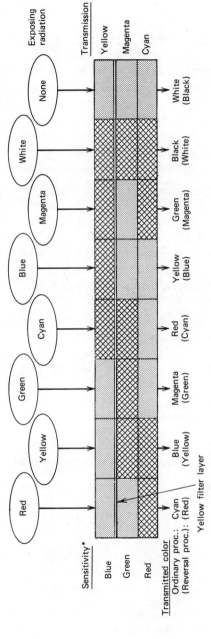

Figure 17.28 Schematic of subtractive color film. Each layer is sensitive to the color marked at the left and takes on, after processing, the color marked on the right. Cross hatching represents the areas exposed due to light of various colors, as indicated. With negative processing, these will be the dense, colored areas. The colors shown immediately below each patch refer to the light transmitted after such processing.

When the emulsion is reversal (positively) processed, the stippled areas will be the colored ones and the transmitted light will have the color shown in parentheses.

669

same should be true of the regions of absorption. But real emulsions must be expected to deviate from this ideal and therefore some cross talk between the three channels is to be expected both in recording and in read-out.

The effects of the cross talk due to the dyes can be substantially reduced by making the "raw" magenta forming coupler slightly yellow, that is, blue absorbing. Its blue-absorbing ability is then left unchanged during coupling, that is, independent of exposure. Similarly, the cyan forming coupler is made slightly red, to leave its blue and green absorption independent of exposure. The undesired blue absorption of the magenta layer and green and blue absorption of the cyan layer can then be compensated for during read-out. This method is used in negative color materials and accounts for their orange overcast appearance.

Characteristic Curves. The sensitivity and density spectra of representative color films are shown in Appendix 17.1. The sensitivity curves represent the reciprocal of the exposure value (in J/m^2) required—for each layer—to yield an overall unity neutral density. The density spectra represent the density at each wavelength, to yield unity neutral density. The characteristic curves are shown for the individual layers. If the gammas differ in the different layers, the resulting image hue will change with the exposure.

ANSI speed [6d]. The ANSI speed of color films is based on a mean value of a log-exposure value of the three colors:

$$S_A = \frac{1}{V},$$ (17.55)

where

$$V = \frac{\log_{10} H_B + \log_{10} H_G + \log_{10} H_R}{3}$$ (17.56)

and $H_{B,G,R}$ are the reference exposure values for the blue-, green-, and red-light densities of the film, when exposed to white light.

The reference exposure is defined as the exposure that raises the film a certain amount, ΔD, above the fog level, when all density measurements are made with spectral light of the corresponding color.[30]

[30] The value of ΔD is

$$\Delta D = \frac{D_N - D_M}{5},$$ (17.57)

where $H_{M,N}$ are defined as in Figure 17.4 and $D_{M,N}$ are the corresponding density values, when all measurements are made in green light (0.5461-μm line of the mercury arc). Incidentally, the blue and red density measurements referred to in the text, are made at the 0.4358-μm mercury line and the 0.6438-μm cadium line, respectively [6e].

One-Step Color Photography [3e]. In one-step color photography, a negative and a positive develop successively in a single multi-layer sandwich using a dye-transfer process. The lower section is similar to the subtractive color films described above, except that each sensitive layer has an inactivated layer of dye-developer associated with it. In each of these, the dye absorption spectrum is similar to the absorption spectrum of the associated silver halide layer, as in Figure 17.28.

After exposure, an activator brings the dye-developer materials into the associated exposed layers, where it initiates development, forms a negative image, and is absorbed to an extent proportional to the exposure. The remaining quantity of the dye-developer is complementary to the exposure level and diffuses into the upper layer, to form a positive image.

Meanwhile, an opaque white layer forms between the negative and positive images, covering the negative from view and, simultaneously, protecting it from additional exposure during the processing interval.

17.5.3   Additive Color Films [3k]

Additive color films differ fundamentally from their subtractive counterparts in that the photosensitivity and color generation functions may be separated. Such films may consist of strictly black-and-white emulsions, with the concomitant ease and flexibility in processing. On the other hand, they do not offer any saving in storage volume. From the information point of view, they are thus of interest only when the registry of simultaneous area elements is highly critical.

Perhaps the most elementary three-color photography system is the Dufay process. Here a *reseau* of blue, green, and red area elements is printed onto the base, below the regular black-and-white emulsion. The emulsion is exposed through this reseau and developed with reversal. The result is viewed by transmitted light. Because of the reversal processing, the photographic emulsion acts as a gate in front of each filter element: light will pass the filter element only to the extent that its passband spectrum was present in the exposing light.

In most additive systems, recording and display are accomplished by means of a special optical system. A representative form of this type of color film is illustrated in Figure 17.29. The film consists of an ordinary black-and-white emulsion coated on a lenticulated base, whose other side is filled with spherical bosses or "lenticules" as shown. The curvature of these lenticules is such as to put their focal points approximately at the emulsion. During recording, each lenticule will record one image element in the form of an image of the exit pupil of the objective with a density

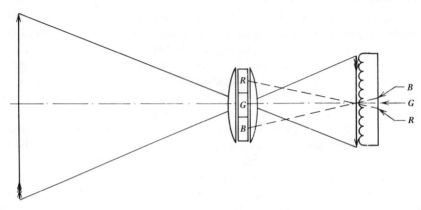

Figure 17.29　Illustrating lenticulated color film.

corresponding to the mean brightness of the corresponding area element of the object. We now place at the exit pupil three filters, transmitting only the red, green, and blue regions of the spectrum, respectively, as shown. The lenticules will then record separately the light transmitted by the three regions of the exit pupil; that is, the red, green, and blue components of the object are recorded separately. If the emulsion is now developed with a reversal process and projected by means of an objective with a similar filter arrangement, a more or less true color rendition of the object will result. Evidently the lenticules need not be spherical; cylindric ridges, parallel to the filter boundaries, suffice.

Incidentally, such lenticulated films need not be used in conjunction with color filters. Consider, for instance, a system where the recording can be done with high-resolution optics, while read-out must use a larger scanning spot. Here the three channels can be recorded directly at their appropriate locations, conceivably from the back of the film, without passing through the lenticules. During read-out, three photoelectric detectors can be placed at the locations corresponding to the three filters in Figure 17.28. Each of these will then read the data corresponding to one channel.

The limitations of this system are evident. Resolution is strictly limited by the width of the lenticules; in a system involving scanning, these can be made to correspond to the scan lines. Cross talk between channels is given by the line spread function of the emulsion as compared to the (channel) element size, which is one third the width of a lenticule. Such films used to be marketed with the following characteristics [87]: relative system aperture (also lenticule aperture), F/2.3; lenticule spatial frequency 25/mm; channel separation about one density unit.

In an alternative system, three linear grids are superimposed on the film during recording, modulating, respectively, the exposure in the red, green, and blue regions of the spectrum. These grids are oriented at 120° from each other (see Figure 17.30a). If the exposed and processed recording is placed into a spatial filtering arrangement (P_0, Figure 19.10) and is illuminated by collimated white light, the modulated recording will form, in the diffraction transform plane (P_F of Figure 19.10), diffraction images falling onto the circumference of a circle; see Figure 17.30b. Each of these images will be formed by density variations caused by object light of the corresponding color and each will receive white light proportional to the exposure at that part of the spectrum. Filters with the appropriate transmittance spectra are placed at the location of these diffraction patterns, as indicated in Figure 17.30b and the undiffracted light is blocked. At the image plane, a positive color reproduction of the exposing image will then be formed [88].

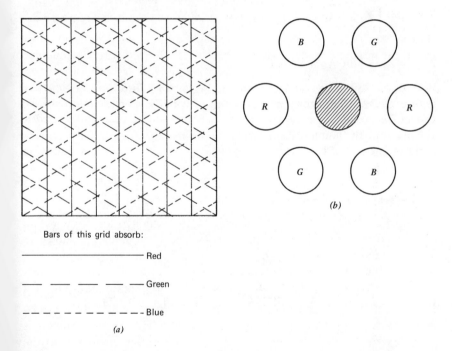

Bars of this grid absorb:

——————————————— Red

—— —— —— —— —— Green

— — — — — — — — — — Blue

(a)

Figure 17.30 Illustrating the three-grid additive color system. (a) Grid pattern used during recording. (b) Filter pattern used during reconstruction.

APPENDIX 17.1 PERFORMANCE DATA OF SOME PHOTOGRAPHIC FILMS

Some performance data on silver halide photographic films, both monochrome and color, are presented here to the extent that the manufacturers made them available for publication.

1 Panatomic-X Film

Figure 17.31 shows data on Eastman Kodak Panatomic-X film:

1. Characteristic curves, with development time as a parameter.
2. Sensitivity spectrum (in the visible) for two density levels,
3. Mtf curve. [Constants for use in (17.16) are: $g = 1.2$, $\nu_g = 81$, $n = 1.3$.]

The reciprocity law is obeyed between 1 and 100 msec. At 10 μsec and 1 sec exposures, the exposure should be doubled. At 10 sec and 100 sec, it should be taken as four and eight times, respectively, the value indicated by the exposure meter. For granularity data, see Figure 17.8.

2 Agfapan and Agfaortho Films

Data on Agfa-Gevaert Agfapan and Agfaortho films are presented in Figure 17.32. Characteristic curves (a), sensitivity spectra (relative) (b), and mtf curves (c) are shown.

Note that the numbers appearing in the names of these films correspond to their ANSI speeds.

3 Ektachrome and Gevacolor

Data for Eastman Kodak Ektachrome 64 daylight color positive film are shown in Figure 17.33. Characteristic and mtf curves are shown in parts (a) and (b), respectively, and the sensitivity and density spectra of each of the three dye layers in (c) and (d), respectively. The rms granularity value is 12; cf. Note 22. The analogous data for Agfa-Gevaert Color negative film (Type 680) are shown in Figure 17.34a–d.

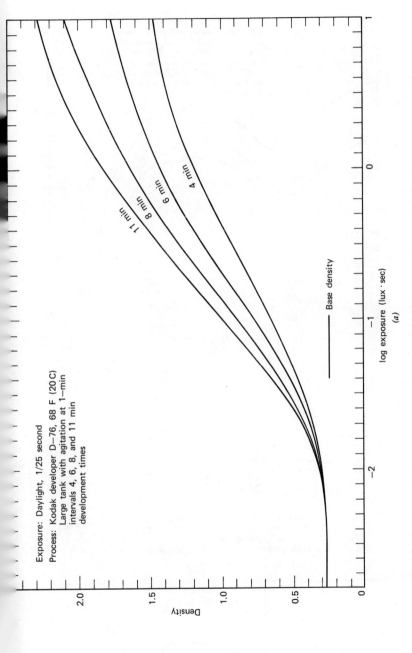

Figure 17.31 Kodak Panatomic-X film. (a) Characteristic curves, development time indicated next to each curve; (b) Sensitivity spectra, reference density indicated next to each curve; (c) mtf. Courtesy and copyright of the manufacturer. Note Notice on graph 31b; this applies to all data of this figure and Figures 17.33.

675

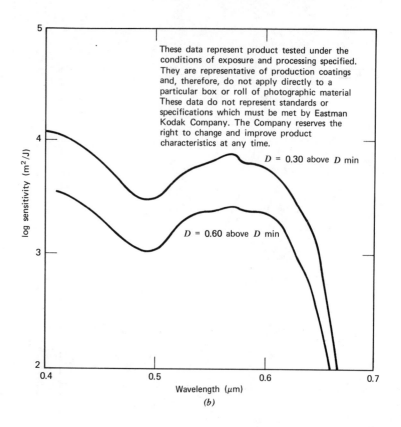

These data represent product tested under the conditions of exposure and processing specified. They are representative of production coatings and, therefore, do not apply directly to a particular box or roll of photographic material These data do not represent standards or specifications which must be met by Eastman Kodak Company. The Company reserves the right to change and improve product characteristics at any time.

D = 0.30 above D min

D = 0.60 above D min

(b)

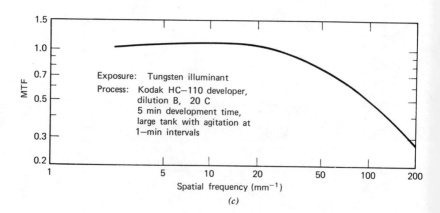

Exposure: Tungsten illuminant
Process: Kodak HC–110 developer,
         dilution B,   20 C
         5 min development time,
         large tank with agitation at
         1–min intervals

(c)

Figure 17.31 (*Continued*).

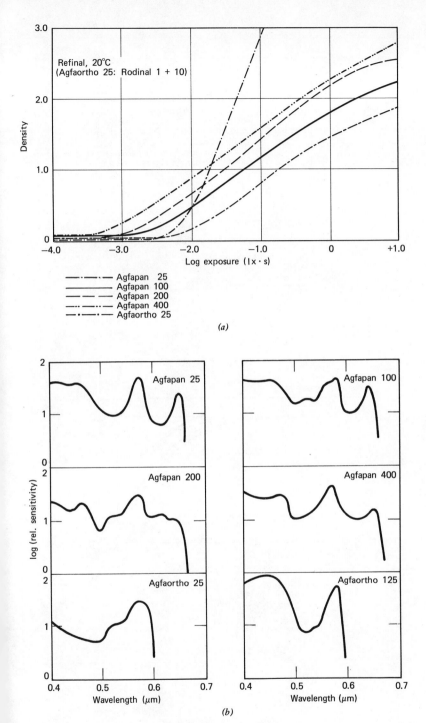

Figure 17.32 Agfapan and Agfaortho films. (*a*) Characteristic curves; (*b*) sensitivity spectra; (*c*) mtf. Courtesy and copyright of the manufacturer.

677

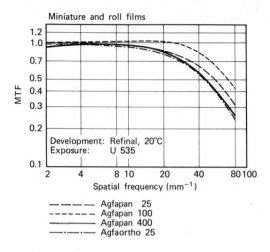

Miniature and roll films

Development: Refinal, 20°C
Exposure: U 535

Spatial frequency (mm⁻¹)

————— Agfapan 25
------- Agfapan 100
————— Agfapan 400
—·—·— Agfaortho 25

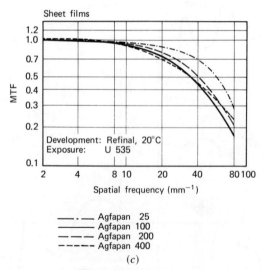

Sheet films

Development: Refinal, 20°C
Exposure: U 535

Spatial frequency (mm⁻¹)

—·— Agfapan 25
——— Agfapan 100
——— Agfapan 200
------- Agfapan 400

(c)

Figure 17.32 (Continued).

678

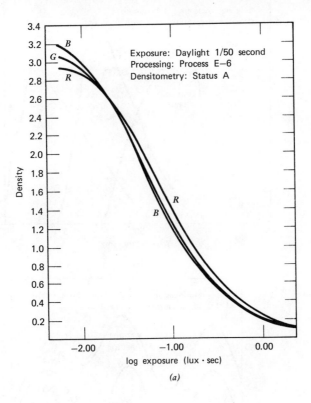

(a)

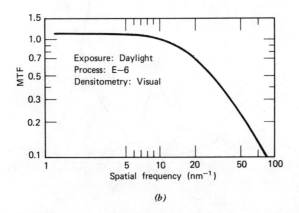

(b)

Figure 17.33 Kodak Ektachrome 64 color positive daylight film. (*a*) Characteristic curves of the three dye layers; (*b*) mtf, total for daylight exposure and photopic densitometry; (*c*) sensitivity spectra; and (*d*) density spectra, with dyes normalized for unity neutral density. Courtesy and copyright of the manufacturer.

679

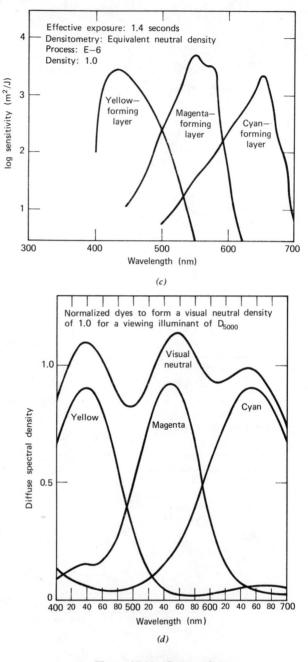

Effective exposure: 1.4 seconds
Densitometry: Equivalent neutral density
Process: E—6
Density: 1.0

Yellow—
forming
layer

Magenta—
forming
layer

Cyan—
forming
layer

log sensitivity (m²/J)

Wavelength (nm)

(c)

Normalized dyes to form a visual neutral density
of 1.0 for a viewing illuminant of D_{5000}

Visual
neutral

Yellow

Magenta

Cyan

Diffuse spectral density

Wavelength (nm)

(d)

Figure 17.33 (*Continued*).

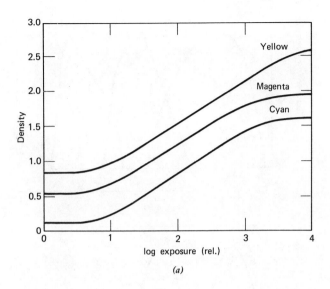

(a)

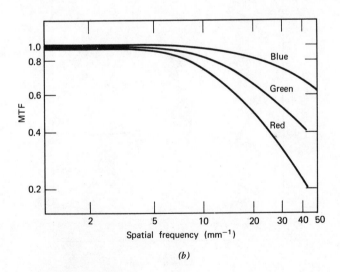

(b)

Figure 17.34 Agfa–Gevaert Gevacolor, Type 680, color negative film. (*a*) Characteristic curves; (*b*) mtf; (*c*) sensitivity spectra; and (*d*) density spectra—all for each of the three dye layers. Courtesy and copyright of the manufacturer.

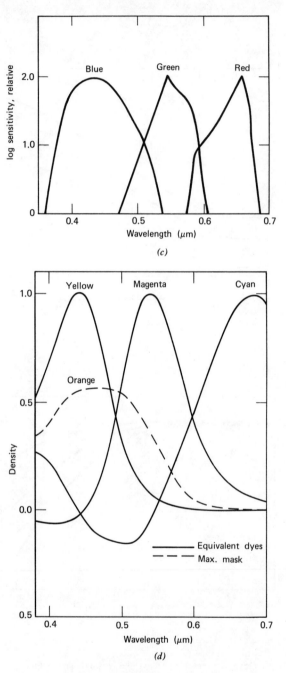

Figure 17.34 (Continued).

APPENDIX 17.2 SIMPLE MODEL FOR OPTICAL CHARACTERISTICS OF SILVER HALIDE EMULSIONS [28]

Here we derive on the basis of a simple model the optical characteristics of silver halide emulsions. Although the results are not accurate quantitatively (due to the oversimplified model), they are highly instructive in terms of the mechanisms responsible for the optical characteristics.

1 Transmittance in Terms of Grain Density

A thin photographic emulsion has a gross transmittance:

$$\tau_1 = 1 - \frac{A_1}{A_T},\qquad (17.58)$$

where A_1 is the sum of the cross-sectional areas of all the grains in the
emulsion and
A_T is the total emulsion area.

This, on the assumption that the grains are totally opaque, the matrix is totally transparent, and there is no overlap of grains.

When the emulsion is thicker, and overlapping of the grains may occur, the resulting transmittance can be estimated by considering the transmittance of each thin layer of the emulsion individually and multiplying these together. Thus

$$\tau = \prod_{j=1}^{J} \left(1 - \frac{A_j}{A_T}\right).\qquad (17.59)$$

Here J is the assumed number of thin layers; its value should not affect the final results.

By taking logarithms of each side, we convert the product into a sum:

$$-D = \log_{10}\tau = \sum \log_{10}\left(1 - \frac{A_j}{A_T}\right).\qquad (17.60)$$

Since A_j/A_T are very small, we can approximate the logarithm by the first term of its series expansion:

$$(\log 10)\, D = \frac{\sum A_j}{A_T} = \frac{A_G}{A_T} = N\bar{A}_g,\qquad (17.61)$$

where A_G is the total grain cross-sectional area;

$$A_G = N\bar{A}_g\qquad (17.62)$$

$\bar{A}_g$ is the mean value of the individual grain cross-sectional area, and N is the number of grains per unit emulsion area. Equation 17.61 is Nutting's formula.[31]

2 Density versus Exposure

We now derive the dependence of density on exposure, assuming that a grain will become developable if it receives at least n_0 photons, but not otherwise. We consider a thin emulsion, having all grains of the same size, cross-sectional area A_g, N_0 grains per unit area, receiving an effective exposure[32] of H_q quanta (photons) per unit area. We estimate the fraction of grains receiving at least n_0 photons as the probability of this happening. On the basis of Poisson statistics (3.94) this will be

$$\frac{N}{N_0} = \sum_{n=n_0}^{\infty} \frac{\bar{n}^n e^{-\bar{n}}}{n!} = 1 - \sum_{n=0}^{n_0-1} \frac{\bar{n}^n e^{-\bar{n}}}{n!}, \qquad (17.63)$$

where

$$\bar{n} = H_q A_g \qquad (17.64)$$

is the mean number of photons absorbed by each grain. By Nutting's formula, the resulting density will be:

$$D = A_g N_0 \frac{1 - \sum_{n=0}^{n_0-1} \bar{n}^n e^{-\bar{n}}/n!}{\log 10}. \qquad (17.65)$$

The corresponding slope of the characteristic curve will be

$$\gamma^* = \frac{dD}{d \log H_q} = H_q \frac{dD}{dH_q} = \bar{n} \frac{dD}{d\bar{n}}, \qquad (17.66)$$

where we have substituted for H_q from (17.64). On substituting into this the value of D from (17.65), we find

$$\gamma^* = A_g N_0 \bar{n} \left[1 + \sum_{n=1}^{n_0-1} \frac{(\bar{n}-n)\bar{n}^{n-1}}{n!} \right] e^{-\bar{n}}$$

$$= A_g N_0 \frac{\bar{n}^{n_0}}{(n_0-1)! \log 10}. \qquad (17.67)$$

---

[31] P. G. Nutting (1913).

[32] By *effective exposure* we mean $H_q = \int H_\lambda \alpha(\lambda) \, d\lambda$, where H_λ is the spectral exposure and $\alpha(\lambda)$ is the efficiency with which photons of wavelength λ are absorbed.

For instance, assuming $n_0 = 2$, we find

$$D_2 = 0.43 A_g N_0 (1 - e^{-\bar{n}} - \bar{n} e^{-\bar{n}})$$
$$= 0.43 A_g [1 - (1 + A_g H_q) e^{-A_g H_q}] \qquad (17.68)$$

and

$$\gamma_2^* = 0.43 A_G (A_g H_q)^2 e^{-A_g H_q}. \qquad (17.69)$$

These formulae lead to abnormally high values of γ^* at high exposure levels. When they are extended to cover thick emulsions and nonuniform grain sizes, lower values of γ^* are obtained [20]. See also Ref. 89.

3 Granularity

Let us now estimate the observed granularity. Here we measure the density over an aperture area A_a. The number of grains contained in this element will have a mean value

$$\bar{N}^* = \bar{N} A_a. \qquad (17.70)$$

Assuming this to be Poisson-distributed, its variance, too, will be equal to this value. The corresponding standard deviation of grain density is

$$\sigma(N) = \frac{\sqrt{\bar{N} A_a}}{A_a} = \sqrt{\frac{\bar{N}}{A_a}}. \qquad (17.71)$$

Hence the standard deviation of the density will be by Nutting's formula:

$$\sigma(D) = A_g \frac{\sqrt{N/A_a}}{\log 10}. \qquad (17.72)$$

Substituting into this the value of N from (17.61), we find

$$N = \frac{(\log 10) D}{A_g}$$

and

$$\sigma^2(D) = \frac{D A_g}{A_a (\log 10)}. \qquad (17.73)$$

This demonstrates the inverse dependence of the density variance on the aperture area. On substituting this value of $\sigma(D)$ into (17.20) we find for Selwyn's granularity constant:

$$G = \sqrt{\frac{2 A_g D}{\log 10}} = 0.932 \sqrt{A_g} \sqrt{D}. \qquad (17.74)$$

When allowance is made for variation in grain size, $\overline{A_g^2}/\overline{A}_g$ must be substituted into this for A_g. This yields Siedentopf's formula[33]:

$$G = 0.932 \sqrt{\frac{\overline{A_g^2}}{\overline{A}_g}} \, D^{1/2}. \tag{17.75}$$

Experimentally, G has been found to vary with the 0.23-power of D. This discrepancy is due to the change of the coefficient with density: the mean grain size decreases with increasing density because, at low exposures, predominantly the larger grains become developable. The smaller grains absorb the required number of quanta only as higher densities are attained [90].

4 Wiener Spectrum and the Granularity Constant

Here we demonstrate the approximate relationship (17.27) between Selwyn's granularity constant (G) and the Wiener spectrum that is constant at low spatial frequencies. Although we make the derivation in terms of density fluctuations, we see from (17.31) that these are directly proportional to transmittance fluctuations, with a proportionality factor $(\bar{\tau} \log 10)$, which is here a constant.

We derive the relationship by assuming a measuring aperture area, A_a, imposed on the emulsion. From isotropy considerations we see that the results are independent of the particular shape of the aperture; for convenience we take a square aperture $b \times b$. When this is imposed on the emulsion, the measured density deviation from the mean will be

$$\Delta D_0(x, y) = \int\limits_{-b/2}^{b/2}\!\!\int \Delta D(x' - x, y' - y) \, dx' \, dy' = \Delta D \odot S_b, \tag{17.76}$$

where the symbol S_b denotes the aperture function

$$S_b(x, y) = 1, \qquad -\frac{b}{2} < x, y < \frac{b}{2}$$

$$= 0, \qquad \text{elsewhere.} \tag{17.77}$$

The observed autocorrelation is, then,

$$C_0 = \overline{\Delta D_0 \odot \Delta D_0} = (\Delta D \odot S_b) \odot (\Delta D \odot S_b)$$

$$= \overline{C \odot S_b \odot S_b}, \tag{17.78}$$

where C is the true autocorrelation of the density fluctuations as given by (17.21).

[33] H. Siedentopf (1937).

The observed Wiener spectrum can be found as the Fourier transform of (17.78):

$$W_0 = \mathcal{F}[C]\{\mathcal{F}[S_b]\}^2$$

$$= W(\nu_x, \nu_y)\left(\frac{\sin \pi b\nu_x}{\pi b\nu_x} \cdot \frac{\sin \pi b\nu_y}{\pi b\nu_y}\right)^2. \tag{17.79}$$

The observed autocorrelation at the origin is

$$C_0(0, 0) = \left[\iint \Delta D \, dx \, dy\right]^2 = \sigma_{D_0}^2, \tag{17.80}$$

the variance of the observed density. From (17.20), this equals

$$C_0(0, 0) = \sigma_{D_0}^2 = \frac{G^2}{2A_a} = \frac{G^2}{2b^2}. \tag{17.81}$$

According to (17.26), it is also equal to the integral of W_0:

$$C_0(0, 0) = \int\limits_{-\infty}^{\infty}\!\!\int W(\nu_x, \omega_y)\left(\frac{\sin \pi b\nu_x}{\pi b\nu_x} \cdot \frac{\sin \pi b\nu_y}{\pi b\nu_y}\right)^2 d\nu_x \, d\nu_y. \tag{17.82}$$

For silver halide emulsions, the Wiener spectrum is quite constant at low frequencies (see Figure 17.8), so that for sufficiently large values of b, the sinc-functions in (17.79) will make the observed Wiener spectrum negligible except in the region of its constancy. We may therefore substitute $W(0, 0)$ for $W(\nu_x, \nu_y)$ and remove it from under the integral. In conjunction with (17.81), this yields

$$C_0(0, 0) = W(0, 0) \int_{-\infty}^{\infty}\left(\frac{\sin \pi b\nu_x}{\pi b\nu_x}\right)^2 d\nu_x \cdot \int_{-\infty}^{\infty}\left(\frac{\sin \pi b\nu_y}{\pi b\nu_y}\right)^2 d\nu_y$$

$$= \frac{W(0, 0)}{b^2} = \frac{G^2}{2b^2}. \tag{17.83}$$

Hence

$$W(0, 0) = \tfrac{1}{2}G^2,$$

confirming (17.27).

APPENDIX 17.3   ADDITIONAL TOPICS IN DETECTIVITY [28]

1   Noise Equivalent Quanta

On the basis of Poisson statistics alone, the effective number (n_e) of quanta detected equals the square of the signal-to-noise ratio at the output. Hence from (17.49):

$$n_e = R^2 = \frac{2\gamma^{*^2}A_e}{(\log 10)^2 G^2} = \frac{0.3772 A_e \gamma^{*^2}}{G^2}. \qquad (17.84)$$

Alternatively, we may define this number in terms of the detective quantum efficiency, κ_d, and n_q, the number of incident quanta:

$$n_e = n_q \kappa_d. \qquad (17.85)$$

[On substituting into this from (17.51) and (17.52), it can be seen to be identical with (17.84).]

This number is referred to as the *noise equivalent quanta* (NEQ). This should be interpreted as the number of input quanta, which would cause the observed noise on their own, assuming perfect detection. It should not be confused with the term "noise equivalent input" (Q_N/Q_0) which refers to an input whose resulting output equals the noise level observed at the output.

When taken per unit area, the NEQ is:

$$N_e = \frac{n_e}{A_e} = \frac{0.377 \gamma^{*^2}}{G^2}. \qquad (17.86)$$

2 Quantum Detectivity

Quantum detectivity (D_q) refers to the energy detectivity expressed in number of quanta:

$$D_q = Q_0 D_e, \qquad (17.87)$$

where Q_0 is the energy of the average quantum. It has also been defined as

$$D_q = \frac{R}{n_q} = \frac{\sqrt{n_e}}{n_q} = \frac{\sqrt{\kappa_d}}{n_q}, \qquad (17.88)$$

where we have used successively (17.84) and (17.85).

We obtain a third form by writing R (in the above definition) in terms of the signal and noise quanta, n_s, n_n, respectively, at the output. Then

$$D_q = \frac{n_s/n_n}{n_q} = \frac{n_s/n_q}{n_n} = \frac{\kappa_r}{n_n}, \qquad (17.89)$$

where κ_r is the responsive quantum efficiency and the last member is clearly the number of quanta that, at the input, would give rise to the observed number of output noise quanta.

On substituting (17.88) into (17.50) and then (17.49) and (17.87) into this, we find also

$$\kappa_d = RD_q = QQ_0D_e^2. \tag{17.90}$$

3 Noise Equivalent Contrast

The signal-to-noise ratio has been termed the *contrast detectivity* [31b]. Its reciprocal is the *noise equivalent contrast* (NEC).

REFERENCES

[1] *The Theory of the Photographic Process*, 3rd Ed., C. E. K. Mees and T. H. James, eds. Macmillan, New York, 1966; (a) C. R. Berry and R. P. Loveland, Chapter 2, and J. Pouradier, Section 3III; (b) F. H. Perrin, Chapter 23; (c) J. F. Hamilton and F. Urbach, Chapter 5; (d) W. E. Lee, R. B. Pontius, and T. H. James, Chapters 13–16; (e) G. T. Eaton, Chapter 18; (f) J. F. Hamilton and G. M. Corney, Chapter 10; (g) J. Spence, Chapter 9, data of Bizzetti and Della Corte; (h) J. L. Tupper, Chapter 19; (i) F. Moser and R. S. van Heyningen, Section 1 IV; (j) J. F. Hamilton, P. J. Hillson, and E. A. Sutherns, Chapter 7; (k) A. Weissberger, Chapter 17.

[2] P. Kowaliski, *Applied Photographic Theory*, Wiley, New York, 1972; (a) Section 5.2; (b) Section 5.1; (c) Sections 6.1 and 6.2; (d) Chapter 7; (e) Section 6.3; (f) Chapter 2; (g) Section 4.1; (h) Section 4.2; (i) Chapter 3.

[3] J. M. Sturge, ed., *Neblette's Handbook of Photography and Reprography*, 7th Ed., Van Nostrand Reinhold, New York, 1977; (a) F. W. H. Mueller, Chapter 2; (b) P. Z. Adelstein, G. G. Gray, and J. M. Burnham, Chapter 6; (c) W. West, Chapter 3; (d) R. W. Heun, Chapter 5; (e) E. H. Land, H. G. Rogers, and V. K. Walworth, Chapter 12; (f) H. Todd and R. Zakia, Chapter 8; (g) M. Abouelata, Chapter 9; (h) R. D. Murray, Chapter 15; (i) D. R. Lehmbeck, Chapter 13; (j) H. J. Bello, Chapter 14; (k) C. B. Neblette, Chapter 1.

[4] W. Thomas, ed., *SPSE Handbook of Photographic Science and Engineering*, Wiley, New York, 1973; (a) Section 8; (b) Figure 15.5.

[5] P. Z. Adelstein and D. A. Leister, "Non-uniform dimensional changes in topographic aerial films," *Photogram. Eng.* **29**, 149–161 (1963).

[6] American National Standards Institute, New York; (a) Standards No. PH 1 and 22; (b) PH 2.5 and 2.29; (c) PH 2.33; (d) PH 2.27; (e) PH 2.1.

[7] A. J. Miller and A. C. Robertson, "Motion picture film—its size and dimensional characteristics," *J. Soc. Mot. Pict. TV Eng.* **74**, 3–11 (1965).

[8] R. H. Armistead and F. B. Galimba, "Latent-image fading of three commercially available fine grained emulsions," *Photogr. Sci. Eng.* **17**, 42–46 (1973).

[9] J. P. Galvin, "Properties of holes in silver halides: A review," *Photogr. Sci. Eng.* **16**, 69–78 (1972).

[10] C. M. Tuttle, F. M. Brown, and N. R. Tuttle, "Equipment and methods for high speed photographic processing," *Photogr. Eng.* **3**, 65–77 (1952).

[11] R. P. Mason, "An advanced technique for short delay processing of photographic emulsions," *Photogr. Sci. Eng.* **5**, 79–86 (1961).

[12] S. L. Hersh and F. Smith, "Rapid processing: Present state of the art," *Photogr. Sci. Eng.* **5**, 48–54 (1961).

[13] H. M. Smith, "Photographic relief images," *J. Opt. Soc. Am.* **58**, 533–539 (1968).

[14] C. I. Coleman and S. P. Worswick, "Electronographic photometry," *Sci. Progr.* (Oxford) **63**, 265–292 (1976); see also *Adv. Electr. Electr.-Phys.* **40B**, Morgan, Airy, and Mullan eds., Academic, New York (1976), Section 2, pp. 613–692, which is devoted to electronographic devices.

[15] D. Bromley and R. H. Herz, "Quantum efficiency in photographic x-ray exposures," *Proc. Phys. Soc.* (London) **63B**, 90–106 (1950).

[16] K. Becker, *Photographic Film Dosimetry*, Focal, London, 1966, (*a*) pp. 142–144.

[17] C. S. McCamy, "New approaches in densitometry," *Photogr. Sci. Eng.* **21**, 103–108 (1977).

[18] K. S. Weaver, "Measurement of photographic transmission density," *J. Opt. Soc. Am.* **40**, 524–536 (1950).

[19] M. Margoshes, "Remarks on linearization of characteristic curves in photographic photometry," *Appl. Opt.* **8**, 818 (1969); also G. de Vaucouleurs, "Comments on several published and unpublished reactions to a paper on photographic characteristic curves," *Appl. Opt.* **8**, 818–819 (1969).

[20] E. F. Haugh, "Theory for optical density and developed silver as functions of exposure," *Photogr. Sci. Eng.* **6**, 370–375 (1962).

[21] C. T. Chang and J. L. Bjorkstam, "Effect of non-uniform irradiance, and irradiance fluctuations, upon the response of photographic film," *J. Opt. Soc. Am.* **65**, 1495–1501 (1975).

[22] C. J. Niederpruem, C. N. Nelson, and J. A. C. Yule, "Contrast index," *Photogr. Sci. Eng.* **10**, 35–41 (1966).

[23] L. Levi, "Detector response and perfect-lens-MTF in polychromatic light," *Appl. Opt.* **7**, 607–616 (1969).

[24] J. H. Altman, F. Gram, and C. N. Nelson, "Photographic speeds based on radiometric units," *Photogr. Sci. Eng.* **17**, 513–517 (1973).

[25] Kodak Publications, Eastman Kodak Co., Rochester, New York, 1965; (*a*) "Kodak plates and films for scientific photography," Publ. P-315; (*b*) "Understanding graininess and granularity," Publ. F-20; (*c*) "Sensitometry and image structure for Kodak color films," Publ. E-78.

[26] J. Castle and J. H. Webb, "Results of very short-duration exposures for several fast photographic emulsions," *Photogr. Eng.* **4**, 51–59 (1953).

[27] M. Hercher and B. Ruff, "High-intensity reciprocity failure in Kodak 649-F plates at 6943 Å," *J. Opt. Soc. Am.* **57**, 103–105 (1967).

[28] J. C. Dainty and R. Shaw, *Image Science*, Academic, New York, 1974; (*a*) Chapter 7; (*b*) Chapter 8; (*c*) Chapter 10; (*d*) Chapters 4 and 5.

[29] G. C. Higgins, "Methods for analyzing the photographic system, including the effects of nonlinearity and spatial frequency response," *Photogr. Sci. Eng.* **15**, 106–118 (1971).

[30] C. B. Johnson, "MTF parameters for all photographic films listed in Kodak pamphlet P-49," *Appl. Opt.* **15**, 1130–1131 (1976).

[31] R. C. Jones, (*a*) "On the quantum efficiency of photographic negatives," *Photogr. Sci. Eng.* **2**, 57–65 (1958); (*b*) "On the minimum energy detectable by photographic materials. Part II, Results for four current Kodak films," 191–197; (*c*) "On the minimum energy detectable by photographic materials. Part III. Energy incident on a microscopic area of the film," 198–204.

[32] L. O. Hendeberg, "Contrast transfer function of the light diffusion in photographic emulsions," *Ark. Fys.* **16**, 417–456 (1960).

[33] S. Johansson and K. Biedermann, "Multiple-sine-slit . . . 2: MTF data and other recording parameters of high resolution emulsions for holography," *Appl. Opt.* **13**, 2288–2291 (1974).

[34] R. S. Barrows and R. N. Wolfe, "A review of adjacency effects in silver photographic images," *Photogr. Sci. Eng.* **15,** 472–479 (1971).

[35] M. B. Silevitch, R. A. Gonsalves, and D. C. Ehn, "Prediction and removal of adjacency effects from photographic images," *Photogr. Sci. Eng.* **21,** 7–13 (1977).

[36] R. Shaw and A. Shipman, "Practical factors influencing the signal-to-noise ratio of photographic images," *J. Photogr. Sci.* **17,** 205–210 (1969).

[37] E. C. Doerner, "Wiener-spectrum analysis of photographic granularity," *J. Opt. Soc. Am.* **52,** 669–672 (1962).

[38] E. Klein and G. Langner, "Relations between granularity, graininess, and the Wiener-spectrum of the density deviations," *J. Photogr. Sci.* **11,** 177–185 (1963).

[39] D. H. R. Vilkomerson, "Measurement of the noise spectral power density of photosensitive materials at high spatial frequencies," *Appl. Opt.* **9,** 2080–2087 (1970).

[40] T. S. Huang and O. R. Mitchell, "Subjective effects of pictorial noise," *Photogr. Sci. Eng.* **21,** 129–136 (1977).

[41] F. H. Perrin and J. H. Altman, "Studies in the resolving power of photographic emulsions. VI. The effect of the type of pattern and the luminance ratio in the test object," *J. Opt. Soc. Am.* **43,** 780–790 (1953).

[42] E. W. H. Selwyn, "Visual and photographic resolving power II." *Photogr. J.* **88B,** 46–57 (1948).

[43] L. Levi, "Photometric resolving power," *Photogr. Sci. Eng.* **7,** 26–28 (1963).

[44] A. E. Saunders, "On the application of information theory to photography," *J. Photogr. Sci.* **21,** 257–262 (1973).

[45] R. C. Jones, "Information capacity of photographic films," *J. Opt. Soc. Am.* **51,** 1159–1171 (1961).

[46] C. N. Nelson, "Photographic system as a communication channel," *Appl. Opt.* **11,** 87–92 (1972).

[47] G. R. Bird, R. C. Jones, and A. E. Ames, "The efficiency of radiation detection by photographic films: State-of-the-art and methods of improvement," *Appl. Opt.* **8,** 2389–2405 (1969).

[48] R. Shaw, "The photographic process as a photon counting device," *J. Photogr. Sci.* **20,** 174–181 (1972).

[49] J. Kosar, *Light-Sensitive Systems: Chemistry and Application of Nonsilver Halide Photographic Processes,* Wiley, New York, 1965; (a) Chapter 6; (b) Sections 8.1 and 8.2; (c) Section 8.4; (d) Chapter 2; (e) Chapter 5; (f) Chapter 9; (g) Section 8.3.

[50] Society of Photographic Scientists and Engineers, Symposia on Unconventional Photographic Systems 1964, 1967, 1971, 1975; Seminar on "Novel Imaging Systems," 1969.

[50*] E. Brinckman, G. Delzenne, A. Poot, and J. Willems, *Unconventional Imaging Processes,* Focal, London, 1978.

[51] P. P. Hanson, "Unconventional photographic systems—A bibliography of reviews," *Photogr. Sci. Eng.* **14,** 438–442 (1970).

[52] A. Tyrrell, *Basics of Reprography,* Focal, London, 1972; (a) Chapter 10; (b) Chapter 8; (c) Chapter 11; (d) Chapter 12.

[53] C. E. Herrick, "The sensitometry of the positive diazotype process," *J. Opt. Soc. Am.* **42,** 904–910 (1952).

[54] M. E. Rabedeau, "The microimage characteristics of a Kalvar film," *Photogr. Sci. Eng.* **9,** 58–62 (1965).

[55] R. T. Nieset, "The basis of the Kalvar system of photography," *J. Photogr. Sci.* **10,** 188–195 (1962).

[56] *Holographic Recording Materials,* H. M. Smith, ed., Springer, Berlin (1977); (a) Chapter 3, D. Meyerhofer, "Dichromated gelatin." Pp. 75–99; (b) Chapter 7, R. A. Bartolini, "Photoresists." Pp. 209–227; (c) Chapter 5, R. C. Duncan, "Inorganic

photochromic materials." Pp. 133–160; (*d*) Chapter 4, D. L. Staebler, "Ferroelectric crystals." Pp. 101–132; (*e*) Chapter 6, J. C. Urbach, "Thermoplastic hologram recording." Pp. 161–207; (*f*) Chapter 8, J. Bardogna and S. A. Keneman, "Other materials and devices." Pp. 229–244 (p. 237).

[57] See *IEEE Trans.* **ED-22** No. 7 (July 1977) which is devoted to lithographic techniques. Note especially the group of 4 papers by F. H. Dill *et al.*, pp. 440–464.

[58] B. O'Brien, "Spectral sensitivity of photosensitive acid resists," *J. Opt. Soc. Am.* **42**, 101–105 (1952).

[58*] R. K. Curran and T. A. Shankoff, "The mechanism of hologram formation in dichromated gelatin," *Appl. Opt.* **9**, 1651–1657 (1970).

[59] L. M. Minsk, J. G. Smith, W. P. van Deusen, and J. F. Wright, "Photosensitive polymers. I. Cinnamate esters of poly (vinyl alcohol) and cellulose," *J. Appl. Polymer Sci.* **2**, 302–307 (1959).

[59*] M. R. Goldrick and L. R. Plauger, "Evaluation of exposure parameters for photoresist materials," *Photogr. Sci. Eng.* **17**, 386–389 (1973).

[60] P. Walker, "Aplications of photopolymers—A preface," *Proc. SPSE Seminar on Applications of Photopolymers*, Soc. Photogr. Sci. Eng. (1970).

[60*] A. N. Broers, J. M. E. Harper, and W. W. Molzen, "250—Å linewidths with PMMA electron resists," *Appl. Phys. Let.* **33**, 392–394 (1978).

[61] A. Olivei, "The optical properties of thermal prints," *Optik* **40**, 469–479, 497–517 (1974).

[62] Z. J. Kiss, "Photochromics," *Phys. Today* **23**, 42–49 (1970).

[63] W. R. Dawson and M. Windsor, "An eye protective panel for flash-blindness protection, using triplet state photochromism," *Appl. Opt.* **8**, 1045–1050 (1969).

[64] G. K. Megla, "Exploitation of photochromic glass," *Opt. Laser Tech.* **6**, 61–68 (1974).

[65] G. K. Megla, "Optical properties and applications of photochromic glass," *Appl. Opt.* **5**, 945–960 (1966).

[66] G. D. Baldwin, "Behavior of photochromic film under high power laser excitation," *Appl. Opt.* **8**, 1439–1446 (1969).

[67] D. S. Lo, "Photochromic salicylideneaniline as storage medium for interferometry and high speed recording," *Appl. Opt.* **13**, 861–865 (1974).

[67*] I. Schneider, M. Marrone, and M. N. Kabler, "Dichroic absorption of *M* centers as a basis for optical information storage," *Appl. Opt.* **9**, 1163–6 (1970).

[68] F. Caimi, "The photodichroic alkali-halides as optical processing elements," *Opt. Eng.* **17**, 327–333 (1978).

[68*] M. R. Tubbs and G. E. Scrivener, "Color center materials for holography, image recording, and data storage," *J. Phot. Sci.* **22**, 8–16 (1974).

[69] R. M. Schaffert, *Electrophotography*, Focal, London, Wiley, New York, 2nd Ed., 1975; (*a*) Chapter 2; (*b*) Chapter 3; (*c*) Chapter 4; (*d*) Chapter 5; (*e*) Chapter 9; (*f*) Chapter 8.

[70] D. W. Chapman and F. J. Stryker, "A binder-type plate for charge-transfer electrophotography," *Photogr. Sci. Eng.* **11**, 22–29 (1967).

[71] J. H. Dessauer, G. R. Mott, and H. Bogdanoff, "Xerography today," *Photogr. Eng.* **6**, 250–269 (1955).

[72] R. M. Schaffert, "The nature and behavior of electrostatic images," *Photogr. Sci. Eng.* **6**, 197–215 (1962).

[72*] J. C. Witte and J. F. Szezepanik, "Application of transfer function methods to the characterization of xerographic development," *J. Appl. Phot. Eng.* **4**, 52–56 (1978).

[73] G. T. Bauer, "Noise on nonimage areas in electrophotography," *Appl. Opt.* **13**, 1053–1059 (1974).

[73*] R. N. Goren and J. F. Szezepanik, "Image noise of magnetic brush xerographic development," *Photo. Sci. Eng.* **22,** 235–239 (1978).

[74] R. F. Bergen, "Characterization of a xerographic thermoplastic holographic recording material," *Photogr. Sci. Eng.* **17,** 473–479 (1973).

[75] J. C. Urbach, "The role of screening in thermoplastic xerography," *Photogr. Sci. Eng.* **10,** 287–297 (1966).

[76] W. S. Colburn and B. V. Chang, "Photoconductor-thermoplastic image transducer," *Opt. Eng.* **17,** 334–343 (1978).

[77] N. K. Sheridon, "The ruticon family of erasable image recording devices," *IEEE Trans.* **ED–19,** 1003–1010 (1972).

[78] D. Kermish, "Image formation mechanism in the γ-ruticon," *Appl. Opt.* **15,** 1775–1786 (1976).

[79] R. B. Gethman, "Light triggered oil waves," *Proc. IRE* **50,** 2381–2382 (1962).

[80] W. Haas, J. Adams, and J. Wysocki, "Imagewise deformation and color change of liquid crystals in electric fields," *Appl. Opt., Suppl. 3, Electrophotography,* pp. 196–198 (1969).

[81] W. L. Goffe, "Photographic migration imaging—A new concept in photography," *Photogr. Sci. Eng.* **15,** 304–308 (1971).

[82] V. Dresner and R. B. Comizzoli, "A thermo-electrophotographic device," *Photogr. Sci. Eng.* **16,** 43–46 (1972).

[83] R. A. Fotland and E. B. Noffsinger, "Contrography, a new electronic imaging technology," *Photogr. Sci. Eng.* **15,** 431–436 (1971).

[84] C. H. Evans, "Television optics," in *Applied Optics and Optical Engineering,* R. Kingslake, ed., Academic, New York, 1965 Vol. 2, pp. 305–306.

[85] W. E. Glenn, "Thermoplastic recording: A progress report," *J. Soc. Mot. Pict. TV Eng.* **74,** 663–665 (1965).

[86] R. J. Doyle and W. E. Glenn, "Lumatron: A high-resolution storage and projection display device," *IEEE Trans.* **ED–18,** 739–747 (1971).

[87] W. R. J. Brown, C. S. Combs, and R. B. Smith, "Densitometry of an embossed kinescope recording film," *J. Soc. Mot. Pict. TV Eng.* **65,** 648–651 (1956).

[88] P. F. Mueller, "Color image retrieval from monochrome transparencies," *Appl. Opt.* **8,** 2051–2057 (1969).

[89] M. Kawasaki, S. Fujiwara, and H. Hada, "Analysis of characteristic curves for monodisperse mono-grain layer emulsions and the minimum size of latent image specks," *Phot. Sci. Eng.* **22,** 290–295 (1978).

[90] E. F. Haugh, "A structural theory for the Selwyn granularity coefficient," *J. Photogr. Sci.* **11,** 65–68 (1963).

18

Imaging Systems: Analysis and Evaluation

18.1 CONCEPTS IN IMAGING SYSTEMS ANALYSIS

In this chapter we develop methods for analyzing optical and electro-optical imaging systems. Such systems vary much and an effort is made to present methods in a form sufficiently general for broad applicability.

To permit a reasonably self-contained presentation, it was necessary to include here a number of concepts that have been presented already in Chapter 3. Such duplication is, however, kept to a minimum.

18.1.1 Definition of Basic Concepts

By "imaging system" we refer to a system that maps an input signal (called "object") into an output signal (called "image") in such a way that each included[1] object coordinate (space and time) point has an image coordinate point associated with it. It is assumed that the object signal value at a given point influences primarily the image signal value at the corresponding point in image space. The signal *value* is specified in terms of suitable variables, usually energy or power density and the form in which the energy appears.

The primary characteristics of the imaging process are four: magnification, transfer characteristic, spread function, and noise. Essentially the imaging process involves changing scales: the change in scale of the

---

[1] Not every object point need be mapped. For instance, in a scanning raster, which translates spatial coordinates into temporal coordinates, a one-dimensional selection of points from the two-dimensional object space is mapped one-to-one into a one-dimensional image space, as the electrical signal.

independent variable is called *magnification* and that of the dependent variable, *transfer characteristic*. The image obtained when only these changes are applied is called the "ideal image" or the "(scaled) object." Actual images generally differ from the ideal ones primarily due to two factors. First, the signal from one object point contributes to the signal at image points corresponding to other object points. The measure of this contribution is called *spread function*. In addition, the system always suffers from random fluctuations superimposed on the signal; these are called *noise*. In principle, these fluctuations can occur in any of the preceding parameters; in practice, however, we are usually concerned only with those of the transfer characteristic and other random fluctuations superimposed on the signal value.

Viewing it as a communication channel, an ideal imaging system has an information capacity that is infinite in two ways: an infinite number of independent signal elements can be crowded into any finite image space and each signal element may have any of an infinite number of different values. The spread function limits the number of independent elements that can be crowded into a finite space (see sampling theorem, Section 3.3.1) and noise limits the number of levels of value that can be distinguished reliably.

18.1.2   Discussion of Basic Concepts

18.1.2.1   The Optics-Electronics Analogy. The basic concepts involved here have been developed in connection with electronic communication systems. When we apply them to electro-optical systems, however, several differences appear, which we list here at the outset.

1. *Magnification.* In an electronic system, the time scales in the input and output are generally identical. In optical systems, however, scale changes are usual in the spatial coordinates.

2. *Dimensionality.* Whereas the electronic system has normally only one independent variable, time, the optical imaging system usually has at least two spatial coordinates and, when temporal changes must be included, even three independent variables [1].

Similarly, in electronic systems the signal value is generally one-dimensional (voltage, charge, resistance, etc.), whereas in optical systems, the signal is often in the form of polychromatic light with different wavelength bands carrying independent information (color) and with the transfer and spread functions varying with wavelength.

3. *Realizability of Spread Function.* A spread function (or impulse response) extending into the region of negative time would imply an effect preceding its cause and is, therefore, impossible. Hence such

functions can not occur in electronic systems, having time as the independent variable. Such functions, with spatial coordinates replacing time, are quite common, however, in optical systems.

We now discuss the relevant parameters individually.

18.1.2.2 Magnification. We defined magnification as the independent-variable scale factor of the imaging process, that is, the ratio of the size of an infinitesimal image element to the corresponding object element. When object and image space dimensions are identical, the magnification is dimensionless. But when they differ, the magnification will have corresponding dimensions. For instance, if the object space dimension is time and the image space dimension is length, the magnification has the dimension of velocity and generally will correspond to the scanning velocity employed in translating the temporal into the spatial variations, as in a crt display.

When the magnification varies over the object space, this is called (geometrical) distortion. In a one-dimensional system, it is a scalar function of the independent variable. In a two-dimensional imaging system it is, in general, like mechanical strain, a tensor transforming one vector into another. Many optical systems, however, are radially symmetric; in these the magnification can be represented by a scalar as a function of the distance from the center of symmetry.

A system with constant magnification over the whole object space is distortionless.

18.1.2.3 Transfer Characteristic. The transfer characteristic describes the relationship between image and object values and may be represented by means of a graph or an equation. The ratio of image to object signal values at any point is called the *system gain*. When object and image energy are of the same nature (say, light with similar spectrum) or both are one dimensional (e.g., voltage, optical density, monochrome luminance), this *magnitude transfer characteristic* suffices to specify the transfer characteristic as a whole. However, when only one of the signals is in the form of polychromatic light, or both are, but differ in spectral composition, the definition of transfer characteristic becomes more difficult. It may then be convenient to divide this definition into two parts: (a) the magnitude transfer characteristic and (b) a method to account for the spectral nature of the values involved.

We consider here four forms of spectral transfer:

1. Both object and image are in the form of electromagnetic radiation and, at every wavelength, the magnitude of the image signal bears a fixed ratio to that of the object signal, with the ratio a function of wavelength

only. The spectral relationship between image and object may then be specified in terms of this ratio—the *spectral transmittance* characteristic. In that event the magnitude transfer characteristic is usually specified for unity spectral transmittance.

2. The object value is in the form of light and the "image" value is in the form of a one-dimensional variable (photographic density; electric charge, as in the photoelectric effect; etc.). Here the *spectral sensitivity* curve is used to describe the effectiveness of each wavelength region in producing the image value. This may be taken relative to the effectiveness at some reference wavelength and the magnitude transfer function is then evaluated at that reference wavelength, usually coinciding with the peak of the spectral sensitivity curve.

3. When the object signal value is one-dimensional (current, monochromatic radiation, photographic density, etc.) and the image is in the form of polychromatic radiation, as in cathodoluminescent phosphors, the spectral emittance curve can be used to describe the transfer characteristic. Here usually the total energy transfer characteristic is used to specify the magnitude transfer.

4. Components translating the spectral nature of the signal value, as in photo- and roentgenoluminescence, may be treated as two cascaded stages of type (2) and (3) respectively, with a virtual one-dimensional signal of arbitrary magnitude intermediate to input and output.

When the system gain is constant, the transfer characteristic is a straight line and the system is called *linear* (*in gain*).

18.1.2.4 Spread Function. The spread function $S(x)$ is the impulse response in an electronic system, the line and point spread function in one- and two-dimensional optical systems, respectively, and the three-dimensional "point-impulse response" [1] in a time-dependent optical system such as an image intensifier or a cathode ray tube.

If the system is linear (in gain), the signal value at any image point may be obtained as the sum of the values contributed by all the object points at that image point and the spread function is independent of the signal value. The image signal can then be obtained as the convolution of the (scaled) object with the spread function; see Section 3.2.1. As discussed there, the need for convolution can be avoided by working with the Fourier transforms of object, image, and spread functions; these are called object and image spectra and *transfer function*, respectively: the image spectrum is obtained as the product of object spectrum and transfer function.

In general, we define the spread function to be normalized such that the integral over the function equals unity.

The transfer function $T(v)$ is normalized to unity at a reference frequency that is usually taken at zero. It may be viewed as a gain factor, specific to a given spatial frequency and relating to the modulation as the signal, rather than to the absolute signal level. It may therefore be called modulation gain:

$$M_{out} = TM_{in}. \tag{18.1}$$

In contrast to the absolute gain, it is relative to the gain at the reference frequency.

The arguments x and v are vectors having dimensionality equal to that of the image space: two for an ordinary two-dimensional image; three, if variations in time, too, must be considered.

Note that the above manner of normalizing the spread function differs from that used in Section 3.2; it permits writing the relationships between spread and transfer functions in a simpler form. In Appendix 18.1 we summarize these relationships and in Table 187 we tabulate a number of useful sets of such functions.

In general, the spread function varies over the object space. Any region over which it is constant is called *isoplanatic*.

18.1.2.5 Noise. Noise, being essentially a random phenomenon, can be defined only in statistical terms. Often it is the squared value of the noise fluctuation $(n - \bar{n})^2$ that is significant [see (3.85)]. Note that the mean value of this is the variance of the noise and we often use this as a measure of noise magnitude.

Note also the fundamental limitation inherent in the random nature of the noise; whereas any deviation from the ideal image can, *in principle*, be restored,[2] if the system characteristics are known, the effect of noise can not be fully removed. Noise is therefore the essential cause of information loss in any system; the spread function can only amplify the information destroying power of the noise.

18.1.3 Illustrations

To clarify the meaning of the concepts defined above and to demonstrate their scope, we now present three illustrations.

1. *Photography by Camera.* Magnification is the ratio of the size of any object in the final print to that of the original object, the magnification in the usual sense. The magnitude gain is the ratio of print reflectance to object luminance. The spectral factor is the object spectral luminance correlated with the spectral transmittance of the lens system, including

---

[2] For evidence that this is true even at points where the transfer function vanishes, see Section 18.5.4.

filters, if any, and with the spectral sensitivity of the film. (The exact manner of combining these is discussed in Section 18.1.4.)

The transfer function is that of the lens, multiplied with those of the film and the printing process. Noise will be due primarily to the granularity of the film and paper, but in principle also to atmospheric turbulence and photon fluctuations.

2. *Television Camera Tube.* The object is the image formed on the photocathode and the image is the signal current. Magnification is the reciprocal of the scan velocity and magnitude gain is the cathode sensitivity at the peak of its response curve multiplied by the volt/ampere responsivity of the read and write portions of the tube. The spectral factor is given by the spectral sensitivity of the photocathode.

The spread function is governed differently in different coordinates (see Section 18.4.3.). In the line scan direction, for instance, it is given by

$$S(t) = J(vt) \circledast S_e(t), \tag{18.2}$$

where $J(x)$ is the current density function across the beam diameter, $S_e(t)$ is the impulse response of the associated electronic circuitry, and v is the read beam scan velocity.

Noise will include, among others, write current and read beam shot noise, Johnson noise in the output resistor, and cathode nonuniformities.

3. *Visual System.* In analyzing the performance of the visual system, the same concepts may be used, although they are here psycho-physical in nature, so that their accurate measurement is difficult.

Here the image is the mental image perceived by the viewer, and the magnification is its size relative to the object size. This is influenced both by geometrical and psychological factors (cf. "size constancy"). The gain is the ratio of the perceived brightness relative to the object luminance. The noise of the system includes the photon noise, the fluctuations in "dark light" and the random pulses generated in the neurological pathways and the brain.

18.1.4   System Characteristics in Polychromatic Light

If the imaging system discriminates between light values at various wavelengths, each of these constitutes another dimension of the dependent variable. To avoid the unpleasantness of infinity-dimensional gains, transfer functions, etc., we may use a much simpler device: the *spectral matching factor* (χ) [2] (sometimes called *spectral system efficacy* [3]) and the polychromatic transfer function.

Specifically, we wish to treat an optical system containing poly-chromatic optically active components. Such a component may change a polychromatic light signal into an electrical signal or *vice versa*, or it may change light of one spectral composition into light, generally of a different spectral composition, where energy at one wavelength at the input influences the energy at another wavelength in the output.

If our system contains several such components, it is convenient to divide it into subsystems, each commencing with an energy source and ending with a detector and to determine the spectral matching factor and polychromatic transfer function of each. In this scheme, a component changing light from one spectral composition to another may be treated as a detector followed by a light source.

18.1.4.1   The Spectral Matching Factor

Definition. Consider a subsystem with a light source spectral composi-tion $e(\lambda)$, a gain g_s, a spectral transmittance $\tau(\lambda)$, and a detector spectral sensitivity

$$k(\lambda) = k_0 \kappa(\lambda), \qquad (18.3)$$

where $k_0 = k(\lambda_0)$ is the absolute sensitivity at the reference wavelength λ_0, and

$\kappa(\lambda)$ is the relative sensitivity at any wavelength λ.

We can then write the ratio of image signal s_i to object signal s_0

$$\frac{s_i}{s_0} = g_s \frac{\displaystyle\int_0^\infty e(\lambda)\tau(\lambda)k(\lambda)\, d\lambda}{\displaystyle\int_0^\infty e(\lambda)\, d\lambda}$$

$$= g_s k_0 \chi, \qquad (18.4)$$

where the spectral matching factor is defined as

$$\boxed{\chi = \frac{\displaystyle\int_0^\infty e(\lambda)\tau(\lambda)\kappa(\lambda)\, d\lambda}{\displaystyle\int_0^\infty e(\lambda)\, d\lambda}} \qquad (18.5)$$

and we have used (18.3).

Thus we can calculate the output signal as if the total input signal were concentrated at wavelength λ_C, provided we multiply our result by the spectral matching factor:

$$\boxed{s_i = g_s k_0 \chi s_0.}$$ (18.6)

Object Signal in Luminous Terms. If the $\kappa(\lambda)$ in (18.5) is the spectral luminous efficiency of the standard observer (relative luminosity, standard visibility factor), the resulting spectral matching factor, χ_v, may be used to convert signal from luminous to radiant units. Specifically,

$$s_e = \frac{s_v}{\chi_v K_{mx}},$$ (18.7)

where s_e, s_v are signal values given in radiant (W) and luminous (lm) units, respectively, and

K_{mx} is the peak luminous efficacy (taken as 680 lm/W for photopic vision and 1746 "scotopic lumen"/W for scotopic vision).

Thus, if the object signal (s_{ov}) is given in luminous units, we find for the image signal, on substituting (18.7) into (18.6),

$$s_i = g_s k_0 \frac{\chi s_{ov}}{\chi_v K_{mx}}.$$ (18.8)

Spectral Matching Factor Values. Spectral matching factors must be calculated individually for each source-transmittance-detector combination—usually by means of numerical integration techniques. A number of such factors are given in Table 188. Matching factors for vision and the standardized S-curve [4a] detectors are given in Table 188a for blackbody radiators—for standardized (P-curve [4b]) phosphor screens in Table 188b [5]. The blackbody factors are calculated for both spectrally uniform transmittance [$\tau(\lambda) = 1$] and for cutoff below $\lambda = 0.35 \ \mu$m [$\tau(\lambda) = 0, \lambda < 0.35 \ \mu$m; $\tau(\lambda) = 1, \lambda > 0.35 \ \mu$m]. The latter are, however, given only when they differ from the former by at least 1% at 2854 K and are marked "filtered." Some additional factors are listed in Table 188c.

Spectral Matching Factor in Nonlinear Systems [3]. The technique just described is not immediately applicable in nonlinear systems because the transmittance, gain, or sensitivity depends on the signal level. However, by means of a simple modification, the spectral matching factor approach may be used in nonlinear systems also. We discuss here photographic emulsions as an example of a nonlinear detector to describe the technique. The reader should then be able to apply this approach to other nonlinear systems as well.

Equation 18.8 is not directly applicable to a system with a photographic detector, because the reference sensitivity, k_0, depends on the signal level. We therefore choose a reference signal level, say density level D_0, at which the sensitivity is known. This modified reference sensitivity, say $\hat{k}_0(D_0)$, has here the dimension of reciprocal exposure, for example, $(\text{J/m}^2)^{-1}$ or $(\text{lux} \cdot \text{sec})^{-1}$; it represents the reciprocal of the exposure required, at the reference wavelength, to bring the emulsion to the reference density level, D_0.

We calculate the exposure coefficient, h, that is, the multiple of the reference exposure constituting the effective exposure, H_e, applied:

$$h = H_e \hat{k}_0. \tag{18.9}$$

We then enter the characteristic curve of the detector, applying factor h to the exposure at the reference density to determine the actual density obtained. The exposure H_e, in turn, is calculated as before. Here it takes the form

$$H_e = \chi g_s L t, \tag{18.10}$$

where the spectral matching factor χ is calculated according to (18.5), L is the object radiance, and t is the exposure time. Referring to (18.9) we thus have

$$\boxed{h = \chi g_s L t \hat{k}_0 .} \tag{18.11}$$

This appears identical to (18.6) for the image signal. It differs only in the dimensionality of $\hat{k}$ (object signal dimension)$^{-1}$ versus that of k (image/object signal dimension).

Table 188c includes data for three types of film spectral response. When the modified reference sensitivity $\hat{k}_0$ is not known, it may be estimated from the ANSI (ASA) speed value (k_x) by the following relationship

$$\hat{k}_0(1) = \frac{12.5 k_x}{\chi_{ce}}$$
$$= 53 k_x, 18.4 k_x, 24.3 k_x \tag{18.12}$$

respectively, for blue-sensitive, orthochromatic, and panchromatic emulsions. Here χ_{ce} is the matching factor of the emulsion with the ANSI standard source. See Appendix 18.2 for the derivation of this formula.

18.1.4.2 Polychromatic Transfer Function. In polychromatic light, the system will form independent images at the component wavelengths. If the transfer function varies from wavelength to wavelength, an overall resultant transfer function (T_p) can be found by forming a weighted average of all the transfer functions at every spatial frequency ν:

$$T_p(\nu) = \frac{\int_0^\infty e(\lambda)\tau(\lambda)T(\nu, \lambda)\kappa(\lambda)\,d\lambda}{\int_0^\infty e(\lambda)\tau(\lambda)\kappa(\lambda)\,d\lambda}. \tag{18.13}$$

Generally, $T(\nu, \lambda)$ must be calculated or measured individually for each combination of ν and λ. One interesting case that may be stated in a more analytic form is that of the aberrationless lens with circular aperture. Its otf is real and of the form (see Tables 66–67):

$$T_0(\nu) = F(y) = \frac{2}{\pi}[\cos^{-1} y - y\sqrt{1-y^2}], \qquad y = \lambda F_e \nu = \frac{\lambda_o \nu}{2A}, \tag{18.14}$$

where F_e is the effective F/number, the ratio of the
   Gaussian image distance to the exit pupil diameter,
λ_0 is the wavelength in vacuum, and
A is the numerical aperture of the lens.

We define the specific spatial frequency

$$\nu_s = \frac{\nu n}{2A} \approx F_e \nu, \tag{18.15}$$

where n is the refractive index in the space under consideration. In this case, (18.13) becomes

$$T_{op}(\nu_s) = \frac{\int_0^\infty e(\lambda)\tau(\lambda)F(\lambda\nu_s)\kappa(\lambda)\,d\lambda}{\int_0^\infty e(\lambda)\tau(\lambda)\kappa(\lambda)\,d\lambda}, \tag{18.16}$$

where F is given by (18.14). This perfect-lens-otf in polychromatic light has been evaluated for a number of representative light sources and detectors with

$$\tau = 0, \qquad \lambda < 0.35 \ \mu\text{m}$$
$$\tau = 1, \qquad \lambda > 0.35 \ \mu\text{m}.$$

The results are presented in Table 189 [3].

18.2 LINEAR SYSTEMS ANALYSIS

For several reasons, linear processes are usually much easier to analyze than nonlinear ones. First, even if the functions are complicated, they can often be broken up into simpler components which may then be treated individually, and the final results added. Thus linearity permits the use of Fourier transform techniques. It is also of fundamental importance in statistical analysis where it facilitates averaging. In addition, linearity permits several operations readily to be combined into a single resultant operation as will be illustrated in the remainder of this section.

This section is therefore devoted to linear systems. Specifically, we answer the question: given an imaging system composed of a number of components, each of which may be treated as linear, (among other things this implies incoherent imaging) and assuming that we know magnification, gain, transfer function, spectral, and noise characteristics of each, how may we determine these characteristics for the system as a whole?

18.2.1 The Deterministic Characteristics

18.2.1.1 Magnification. The system magnification (M) is found by multiplying together the magnification values (m_i) of all the n components:

$$M = \prod_{i=1}^{n} m_i. \tag{18.17}$$

18.2.1.2 Gain. The system gain, G, too, is obtained as the product of all the component gain values, g_i, multiplied by the product of all spectral matching factors, χ_m, which were defined in Section 18.2.1:

$$G = \prod \chi_m \prod_{i=1}^{n} g_i. \tag{18.18}$$

18.2.1.3 Transfer Functions. The system transfer function, too, is obtainable as the product of the component transfer functions, except that here the arguments (frequencies) of all the transfer functions must be mapped into a common reference region by means of the intermediate magnification. The reference region will usually be either the object or the image region. In the following, we assume that the multiplication is to be made in the object space, where the scale is presumably prescribed and not subject to arbitrary changes by the designer. The system transfer function is then given by

$$T_s(\boldsymbol{v}) = \prod_{i=1}^{n} T_i\left(\frac{\boldsymbol{v}}{M_i}\right), \tag{18.19}$$

where the *cumulative magnification*, M_i, is defined as

$$M_i = \prod_{j=1}^{i} m_j \qquad (18.20)$$

and ν is the frequency in the object space.

18.2.2 System Noise

18.2.2.1 General Noise Distribution [6]

System Noise Power. The noise contribution of the various components may be combined by performing the convolution of the probability densities of the various noise terms. This yields the probability density of the system noise.

The mean noise power of the system is merely the sum of the mean noise power of all the contributing noise sources after mapping into the reference region. If the frequency dependence of the noise must be considered, the Wiener spectra of the noise terms must be summed, after mapping into the reference region. This time the mapping involves both the magnification and the gain of the intervening components. Hence the noise Wiener spectrum has the following form in the object space

$$\mathbf{W}_{n_s}(\nu) = \sum_{i=0}^{n} \frac{\mathbf{W}_{n_i}(\nu/M_i)}{G_i^2(\nu)}, \qquad (18.21)$$

where $\mathbf{W}_{n_i}(\nu)$ is the Wiener spectrum of the noise contributed by the ith component,[3] and the definition of the *cumulative gain*, G_i, is similar to that of the cumulative magnification, M_i, except that here the transfer function, too, must be considered:

$$G_i(\nu) = [\Pi^* \chi_m k_{0_m}] \prod_{j=1}^{i} g_j T_j\left(\frac{\nu}{M_j}\right); \qquad (18.22)$$

M_0 and $G_0(\nu)$ are taken as unity. The asterisk on the Π indicates that only factors pertaining to detectors preceding component $(i+1)$ are included.

When the spectrum is uniform, the variance (σ^2) may be substituted for the Wiener spectrum; both represent the mean value of the squared

---

[3] $\mathbf{W}_{n_i}$ is measured at the output of the ith stage, i.e., if we denote by $\mathbf{W}'_{n_i}, \mathbf{W}''_{n_i}$ the noise appearing at the input and output, respectively, of the ith stage, then

$$\mathbf{W}_{n_i} = \mathbf{W}''_{n_i} - g_i^2 \mathbf{W}'_{n_i}.$$

fluctuation. Equation 18.21 then becomes

$$\sigma_s^2 = \sum_{i=0}^{n} \frac{\sigma_i^2}{G_i^2} \ . \tag{18.23}$$

Specific Noise. Often the signal-to-noise power ratio (R^2) is more convenient to work with. We call its reciprocal *specific noise*—the ratio of mean noise power to mean signal power. The specific noise of the system is

$$R^{-2}(\nu) = \frac{W_{n_s}(\nu)}{S^2(\nu)} = S^{-2}(\nu) \sum_{i=0}^{n} \frac{W_{n_i}(\nu/M_i)}{G_i^2(\nu)}$$

$$\boxed{= \sum_{i=0}^{n} R_i^{-2}\left(\frac{\nu}{M_i}\right),} \tag{18.24}$$

where R_i^{-2} is the specific noise originated at stage i,
    $S(\nu)$ is the object spectrum.

We write S^2 instead of $|S|^2$, and we have made use of the fact that the signal energy at stage i is given by $S^2 G_i^2(\nu)$. In other words, *the specific noise of the system equals the sum of the specific noise contributions of all the components.* Neither signal, nor noise, nor gain need be known individually; the signal-to-noise ratios alone suffice for determining the system signal-to-noise ratio. This is particularly significant for components whose absolute noise level is difficult to determine, but whose signal-to-noise ratio may readily be estimated from threshold measurements, such as our most important optical detector—the human visual system.

The specific noise approach to analyzing system noise enables us to compare readily the contributions of the various noise sources to the system noise and, hence, may guide us in improving the system. Once the specific noise of a certain component has been made small compared to the others, not much is gained by reducing it further, until other noise sources have been treated. The use of image intensifiers for viewing low-luminance targets, such as fluoroscopic screens and starlight scenes, is a classical example of this situation. Here direct-view detectivity is limited by the noise in the visual system. With image intensification, the visual specific noise is reduced and quantum noise becomes the limiting factor. At this point no significant benefit accrues from intensifying the image further. Indeed, the visual detectivity may drop at higher luminance levels due to the reduced spatial integration domain in the retina at these levels; see Section 15.4.2.4.

By way of illustration, let us use (18.24) to calculate the specific shot noise of a multiplier phototube. There we have, from (3.98):

$$R_i^{-2} = \bar{n}_i^{-1} = (g^i \bar{n})^{-1},$$ (18.25)

where $\bar{n}_{0,i}$ are, respectively, the average number of electrons leaving the cathode and the ith dynode. When we substitute this into (18.24) and sum the resulting geometric series, we find:

$$R^{-2} = \sum_{i=0}^{n} \frac{1}{\bar{n}_0 g^i} = \frac{(1 - 1/g^{n+1})}{\bar{n}_0(1 - 1/g)}$$

$$= \frac{(g - 1/G)}{\bar{n}_0(g - 1)}.$$ (18.26)

The detective quantum efficiency is then

$$n_{qd} = \frac{R_0^{-2}}{R^{-2}} = \frac{(g - 1/G)}{(g - 1)},$$ (18.27)

in agreement with (16.53).

Noise Factor [7a]. The above formulation is convenient for analytic work. In experimental work, it is generally impossible to separate, at the output of a component, the portion of the noise due to the input, from that contributed by the component itself. For such work, and with specific reference to Johnson noise, the noise factor, F, has been defined as the ratio of the signal-to-noise power ratio at the input to that at the output:

$$F = \left(\frac{R_0}{R}\right)^2 = \frac{R^{-2}}{R_0^{-2}} = \frac{R_0^{-2} + R_a^{-2}}{R_0^{-2}},$$ (18.28)

so that

$$R_a^{-2} = (F - 1)R_0^{-2},$$ (18.29)

where $R_{a,0}^{-2}$ are the specific noise values of amplifier and source, respectively. For a cascaded amplifier system, we then obtain a formula similar to our (18.24). This is known as Friis' formula.[4]

18.2.2.2 Representative Noise Sources

Binomial and Poisson Distributions [6]. Since quantum noise sources play a major role in electro-optical systems, we discuss these here in some more detail to illustrate the above result (18.24). The statistics of many such noise sources may be represented by the binomial or by the Poisson distribution. The binomial distribution applies when a quantum is either

[4]H. T. Friis (1944).

simply transmitted or lost. The Poisson distribution seems to apply when the quantum is amplified by an integral factor, the gain, which varies around a fixed mean value, g.

In this context, the binomial process may be viewed as having a gain that varies randomly, having the value unity with a probability g and the value zero with a probability $(1-g)$. The mean value of the gain, too, is then g and the value of the variance $g(1-g)$. Note that in a Poisson process the variance equals the mean (3.95). Hence the variance of the gain, that is, the variance of the number of output quanta for a single input quantum, at stage i with gain g_i, is

$$\sigma_{1i}^2 = g_i, \qquad g_i(1-g_i) \tag{18.30}$$

for a Poisson and a binomial process, respectively.

To obtain the variance (σ_i^2) with a number n_{i-1} of identical input quanta and the corresponding

$$n_i = g_i n_{i-1} \tag{18.31}$$

output quanta at stage i, we use the fact that, for independent random variables, the variance of the sum equals the sum of the variances [8]. Since the variances are equal for the n_{i-1} incident quanta, the variance of the output number equals n_{i-1} times the variance for a single input quantum. Thus

$$R_i^{-2} = \frac{\sigma_i^2}{n_i^2} = \frac{n_{i-1}\sigma_{1i}^2}{n_i^2} = \frac{\sigma_{1i}^2}{g_i n_i}$$

$$= \frac{\sigma_{1i}^2}{g_i n_0 G_i}, \tag{18.32}$$

where we have made use of (18.31), n_0, n_m are the number of quanta at the system input, and output, respectively, and G_i is the gain through stage i [see (18.22)]. Combining (18.30) and (18.32), we find

$$R_i^{-2} = \frac{1}{G_i n_0} = n_i^{-1}, \tag{18.33}$$

$$R_i^{-2} = \frac{1-g_i}{G_i n_0} = \frac{1-g_i}{n_i} \tag{18.34}$$

for Poisson and binomial processes, respectively.

Specifically, if there is a series of m consecutive components that may be treated as following binomial statistics, their combined specific noise may be represented in terms of their total gain $(G = G_m)$, as

$$R_T^{-2} = \frac{1}{n_0} \sum_{i=1}^{m} \frac{1-g_i}{G_i} \equiv \frac{G}{n_0 G} \sum_{i=1}^{m} \frac{1-g_i}{G_i} = \frac{G}{n_m} \sum_{i=1}^{m} \left(\frac{1}{G_i} - \frac{1}{G_{i-1}} \right), \tag{18.35}$$

where $n_m = Gn_0$ is the number of quanta at the system output.

Now we observe that in the sum of the last member of (18.35), each term $(1/G_i)$ is canceled by the following $(1/G_{i-1})$-term, except for the first $(1/G_{i-1})$ and the last $(1/G_i)$. These are, respectively,

$$\frac{1}{G_0} = 1, \qquad \frac{1}{G_m} = \frac{1}{G}. \qquad (18.36)$$

Thus (18.35) becomes

$$R_T^{-2} \text{ (binom)} = \frac{1-G}{n_m} : \qquad (18.37)$$

the series of components may be treated as a single component with a gain equal to the product of the component gains.

If, in addition, the series is headed by a component obeying Poisson statistics, this eliminates the G term in the numerator, so that simply

$$R_T^{-2} = n_m^{-1}. \qquad (18.38)$$

Obviously, this approximation is valid also whenever $G \ll 1$ and when the initial component is binomially distributed but has $g \ll 1$. This fact broadens the applicability of (18.38). To illustrate, light from a distant object entering an objective lens (binomial, $g \ll 1$) which images it at the input surface of a fiber bundle (binomial), conducting the light to a photocathode (binomial) constitute such a series, and their total noise is given by (18.38). This accounts for the fact that the photon noise does not seem to appear in the total output noise of the phototube, a fact that seems to have caused some puzzlement [9a].

If the above series is followed by a phosphor with quantum gain g_p, the temporal specific noise for the total image intensifier system will be simply

$$R^{-2} = \frac{g_p + 1}{n_m}, \qquad (18.39)$$

where n_m now is the number of photons emitted by a single image element during the integration period of the detector.

Exponential Distribution. Some devices exhibit noise which appears as a gain fluctuation having a negative exponential probability distribution (see mcp's, Section 16.4.2.3):

$$p(g) = \alpha e^{-\alpha g}. \qquad (18.40)$$

For this distribution the mean value of the gain is

$$\bar{g} = \int_0^\infty g p(g) \, dg = \alpha \int_0^\infty g e^{-\alpha g} \, dg = \frac{1}{\alpha}. \qquad (18.41)$$

The variance of the gain is

$$\text{var}(g) = \overline{(\bar{g} - g)^2} = \overline{\bar{g}^2} - \bar{g}^2 = \int_0^\infty g^2 p(g)\, dg - \bar{g}^2$$

$$= \alpha \int_0^\infty g^2 e^{-\alpha g}\, dg - \frac{1}{\alpha^2} = \frac{1}{\alpha^2} = \bar{g}^2. \qquad (18.42)$$

18.2.2.3 Noise Equivalent Bandwidth and Interval. Often the noise spectrum may be considered white over the region passed by the system, while the signal is concentrated in a relatively narrow spectral region where the system transfer function may be considered constant. For purposes of noise calculations, it may then be convenient to represent the system frequency response in terms of an *equivalent passband*, which is a square transfer function, unity over a band, $\Delta\nu$, and vanishing outside it. This band is centered on the center of the signal spectrum. If $\Delta\nu$ is chosen to yield the same signal-to-noise ratio as the actual system, it is called the *(noise) equivalent bandwidth.*(We have already referred to this name in Section 9.4.2.)

To determine the equivalent bandwidth, assume the input signal power (P_i) to be concentrated at frequency ν_s. Let the system transfer function (not necessarily normalized) be represented by $G(\nu)$, and the noise power spectrum by $W_n(\nu)$, where

$$G(\nu) \approx 0, \qquad \nu > \nu_0$$

and

$$W_n(\nu) \approx W_{n0} = \text{constant}, \qquad \nu < \nu_0. \qquad (18.43)$$

Then the amount of signal and noise power passed by the system will be, respectively

$$P_s = P_i G^2(\nu_s) \qquad (18.44)$$

$$P_n = \int_{-\infty}^\infty W_n(\nu) G^2(\nu)\, d\nu = W_{n0} \int_{-\infty}^\infty G^2(\nu)\, d\nu. \qquad (18.45)$$

Here and in the remainder of this section we write G^2 for $|G|^2$.

The resulting specific noise is

$$R^{-2} = \frac{P_n}{P_s} = W_{n0} \frac{\displaystyle\int_{-\infty}^\infty G^2(\nu)\, d\nu}{P_i G^2(\nu_s)}. \qquad (18.46)$$

With a system having the equivalent passband, this value would have been

$$R^{-2} = \frac{2W_{n0}\Delta\nu}{P_i}; \qquad (18.47)$$

the factor two is included to account for the fact that the noise spectrum extends over both sides of the origin so that the equivalent passband passes two bands, each of width $\Delta\nu$, symmetrically placed about the origin. On solving (18.46) and (18.47) for $\Delta\nu$, we find

$$\Delta\nu = \frac{1}{2} \frac{\displaystyle\int_{-\infty}^{\infty} G^2(\nu)\, d\nu}{G^2(\nu_s)}. \tag{18.48}$$

Let us now consider an important special case: shot noise sumperimposed on a low frequency signal. Consider a uniform object. Due to quantum effects, it will appear as composed of randomly spaced small dots. Each of these may be represented by $a_j\, \delta(\mathbf{x}-\mathbf{x}_j)$, where the a_j are random variables with mean value $\bar{a}$, and the x_j are uniformly distributed over the domain with mean density n, so that the actual number of dots in any area element obey Poisson statistics. The object is imaged through a system having an (unnormalized) spread function $g(\mathbf{x})$. This will result in an image that is similar to the object, except that each small dot is replaced by a wider one of form $a_j g(\mathbf{x}-\mathbf{x}_j) = a_j g_j$, for short. If the system is an electronic circuit, $g(t)$ will be the impulse response. If the object is a two-dimensional optical one, $g(x, y)$ may be the point spread function of the system or the transmittance distribution in a scanning aperture over which all the signal values are instantaneously integrated. In either case, the spread function may be replaced by an equivalent "square" *interval* or window, that is, a spread function that is constant over an interval $\Delta\mathbf{x}$ and vanishes outside it, which is chosen to yield the same specific noise as the actual spread function.

To determine $\Delta\mathbf{x}$, we first calculate the specific noise obtained in the above system. The signal density will be the mean value at any point since the object was uniform ("dc"). Hence the signal power density will be:

$$P_s = \bar{s}^2 = \left[\sum a_j \int g(\mathbf{x})\, d\mathbf{x} \right]^2$$

$$= n^2 \bar{a}^2 \left[\int g(\mathbf{x})\, d\mathbf{x} \right]^2. \tag{18.49–54}$$

The noise, i.e. the signal variance, is

$$P_n = n(s^2 - \bar{s}^2) = n\bar{a}^2 \int g^2(\mathbf{x})\, d\mathbf{x} - P_s. \tag{18.55}$$

Hence the specific noise is

$$R^{-2} = \frac{P_n}{P_i} = \frac{\alpha \int_{-\infty}^{\infty} g^2(\mathbf{x}) \, d\mathbf{x}}{n\left[\int_{-\infty}^{\infty} g(\mathbf{x}) \, d\mathbf{x}\right]^2} - 1, \qquad (18.56)$$

where $\alpha[= \overline{a^2}/\bar{a}^2]$ is the specific variance of the amplitude. On the other hand, with the equivalent aperture $\Delta\mathbf{x}$, we have a mean number $n\,\Delta\mathbf{x}$ of Poisson-distributed quanta constituting the signal, so that the specific noise is [from (18.33)]

$$R^{-2} = \frac{1}{n\,\Delta\mathbf{x}}. \qquad (18.57)$$

The equivalent interval is now found by solving (18.55) and (18.57) for $\Delta\mathbf{x}$

$$\Delta\mathbf{x} = \frac{\left[\int_{-\infty}^{\infty} g(\mathbf{x}) \, d\mathbf{x}\right]^2}{\alpha \int_{-\infty}^{\infty} g^2(\mathbf{x}) \, d\mathbf{x}}. \qquad (18.58)$$

This represents a (noise) equivalent scanning aperture or spread function width. When all the a_i are equal, $\alpha = 1$ and may be dropped from (18.58).

Note that here the signal frequency is essentially zero, so that the noise equivalent bandwidth is, according to (18.48)

$$\Delta\boldsymbol{\nu} = \frac{1}{2} \frac{\int_{-\infty}^{\infty} G^2(\boldsymbol{\nu}) \, d\boldsymbol{\nu}}{G^2(0)},$$

$$= \frac{1}{2} \frac{\int_{-\infty}^{\infty} g^2(\mathbf{x}) \, d\mathbf{x}}{\left[\int_{-\infty}^{\infty} g(\mathbf{x}) \, d\mathbf{x}\right]^2}, \qquad (18.59)$$

where the change in the numerator is based on Parseval's theorem and that in the denominator on the fact that it represents Fourier transform of the spread function at the origin. Multiplying together (18.58) and (18.59), we find

$$\Delta\mathbf{x} \cdot \Delta\boldsymbol{\nu} = \frac{1}{2\alpha}. \qquad (18.60)$$

18.2.3 Summary

We have attempted to provide very general rules for analyzing linear systems and have presented a number of important simplifications. We summarize the main rules here

System magnification:

$$M = \prod_{i=1}^{n} m_i \qquad (18.17)$$

System gain:

$$G = \prod_{j} \chi_j \prod_{i=1}^{n} g_i \qquad (18.18)$$

System transfer function:

$$\mathsf{T}(\boldsymbol{\nu}) = \prod_{i=1}^{n} \mathsf{T}_i\left(\frac{\boldsymbol{\nu}}{M_i}\right) \qquad (18.19)$$

System specific noise:

$$R^{-2}(\boldsymbol{\nu}) = \sum_{i=0}^{n} R_i^{-2}\left(\frac{\boldsymbol{\nu}}{M_i}\right), \qquad (18.24)$$

where

$$\chi_j = \frac{\displaystyle\int_0^{\infty} e(\lambda)\tau(\lambda)\kappa(\lambda)\,d\lambda}{\displaystyle\int_0^{\infty} e(\lambda)\,d\lambda} \qquad (18.5)$$

for the jth source-detector combination,

$$M_i = \prod_{j=1}^{i} m_j, \qquad (18.20)$$

and

$$R_i^{-2}(\boldsymbol{\nu}) = \frac{\mathsf{W}_{n_i}(\boldsymbol{\nu})}{\mathsf{W}_{s_i}(\boldsymbol{\nu})}. \qquad (18.61)$$

To clarify further the principles developed in this section, an illustrative example of a system analysis is worked out in Appendix 3. The system parameters are somewhat idealized there to permit concentrating on the essential concepts. For an illustration of a more detailed system analysis, the reader is referred to Ref. 9, treating a television camera system.

Detailed analyses for spectral matching have been published, for example, for infrared radiometry [10] and for the photography of crt displays [11].

18.3 NONLINEAR AND NONISOPLANATIC SYSTEMS

As mentioned earlier, nonlinear systems are, as a rule, much more difficult to treat than linear ones. Straightforward methods may, however, be used when the spread function has negligible width (zero-spread process). In this section we discuss primarily such processes and only in Section 18.3.4. are finite spread nonlinear systems briefly treated.

Zero spread nonlinearities can be treated analytically or graphically in the spatial domain by techniques we describe in this section. Some efforts have been made to treat them in the frequency domain, also, by means of "describing function"; but these seem to be of rather limited usefulness [12], [13], [7b].

18.3.1 Concepts in Nonlinear Systems Analysis

18.3.1.1 Differential Gain. We define the gain of a system as the ratio of the output to the input signal

$$g = \frac{s'}{s}. \tag{18.62}$$

This is a useful concept in linear systems where the gain is constant. In a nonlinear system, the gain varies with the signal and the *differential gain* is important. This is defined as the ratio of the output change to the input change, as the latter approaches zero:

$$g^* = \lim_{\Delta s \to 0} \frac{\Delta s'}{\Delta s} = \frac{ds'}{ds}. \tag{18.63}$$

The *relative differential gain, often called "gamma"* (γ), is a dimensionless quantity defined as the differential gain evaluated relative to the gain proper:

$$\gamma = \frac{g^*}{g} = \frac{s}{s'} \frac{ds'}{ds} \approx \frac{\Delta s'/s'}{\Delta s/s}. \tag{18.64}$$

Alternatively, γ may be defined as a logarithmic derivative of the transfer characteristic, that is, the slope of the transfer characteristic when plotted on a log–log scale:

$$\gamma = \frac{d(\log s')}{d(\log s)} = \frac{ds'/s'}{ds/s} \approx \frac{\Delta s'/s'}{\Delta s/s}. \tag{18.65}$$

Thus, to find the change in output signal due to a small change in input signal, we multiply the latter by the differential gain. If we deal with fractional changes, we do the same, using the relative differential gain. For small values of modulation, the output modulation is obtained from the input modulation by multiplication with the differential gain:

$$M' = \gamma M. \tag{18.66}$$

To show this, consider a signal consisting of small variations, $\pm \Delta s$, superimposed on a mean level s_0. Its modulation is, then

$$M = \frac{s_{mx} - s_{mn}}{s_{mx} + s_{mn}} = \frac{(s_0 + \Delta s) - (s_0 - \Delta s)}{(s_0 + \Delta s) + (s_0 - \Delta s)} = \frac{\Delta s}{s_0}. \tag{18.67}$$

Similarly, the modulation of the output signal is

$$M' = \frac{\Delta s'}{s_0'} \tag{18.68}$$

and their ratio:

$$\frac{M'}{M} = \frac{\Delta s'/s_0'}{\Delta s/s_0} = \gamma. \tag{18.69}$$

In systems, where the effects of finite spread must be taken into account—by using the otf—the signal is conveniently expressed in relative, or modulation, terms. The above result then permits the application of linear analysis methods even to nonlinear components: the gamma, just like the transfer function, becomes simply another modulation gain factor.

18.3.1.2 Dynamic Range. The dynamic range, too, is an important concept applicable primarily to nonlinear systems, including systems linear over a limited range. Since all real systems are limited in the range of output signals they can provide effectively, it is a concept of great generality.

Lower limits are occasionally set by the presence of a *toe* in the transfer characteristic: below a certain value of input signal, the differential gain is too small to be useful and the system responds effectively only above a certain threshold input signal. If there is no toe, noise will set a lower limit to the useful input signal.

Upper limits are usually set by a shoulder in the transfer characteristic: the differential gain is very low above a certain value of input signal. Occasionally components may be damaged by input signals above a certain value.

The ratio of useful upper to lower limit of input signal is called the *dynamic range* of the system.

18.3.2 Some Nonlinear Transfer Characteristics

We now discuss a number of nonlinear transfer characteristics that are of special importance. They are illustrated in Figure 18.1 and Table 190.

18.3.2.1 Power-Law Response. A power-law response has the form:

$$s' = as^b. \tag{18.70}$$

Its relative differential gain is, then,

$$\gamma = \frac{s}{s'}\frac{ds'}{ds} = b, \qquad \text{a constant.}$$

Note that the "gamma" of a photographic emulsion may be defined as the slope of the curve of log (output) versus log (input), if we consider opacity as the output variable; see Section 17.2.1.2. For a photographic emulsion with a power-law response, we have, then [see (18.65)]:

$$\text{"gamma"} = \gamma. \tag{18.71}$$

The "gamma" of photographic theory is thus identical with the relative differential gain and constant gamma corresponds to a power-law response with exponent γ. However, the term "gamma" is used even in nonphotographic systems in this sense.

Unity gamma corresponds to linear response, so that the gamma value may serve as an indicator of linearity. Occasionally, responses with γ smaller or larger than unity have been called *sublinear* and *superlinear*, respectively. See Figure 18.1e for illustrations of these.

18.3.2.2 Logarithmic Response. A logarithmic response has the general form:

$$s' = a \log s + c. \tag{18.72}$$

Its differential gain is

$$g^* = \frac{ds'}{ds} = \frac{a}{s}. \tag{18.73}$$

Thus

$$\Delta s' \approx a\frac{\Delta s}{s}, \tag{18.74}$$

that is, a given fractional change in input signal will yield a constant absolute change in output signal, regardless of the signal level. For example, a 1% change in input will always yield a unity change in output, if $a = 100$.

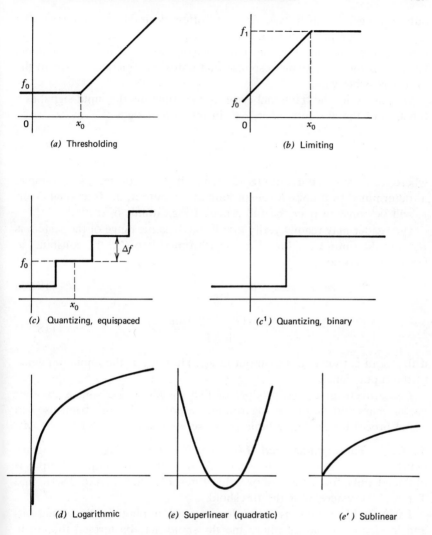

Figure 18.1. Representative nonlinear processes.

For instance, the observation of a constant luminance contrast $\Delta s/s$ at visual threshold (Weber's law) led to the assumption of a logarithmic sensory response (Fechner's law); this is based on the tacit assumption that the detectable output change is constant over the range of input signal levels (see Section 15.2.3).

In system design, the logarithmic response is useful when a large dynamic range signal is to be accommodated by a system with a small output range. In such systems, it is often desirable to minimize the fractional error in estimating the input signal, rather than the absolute error. If the noise level at the output is constant—independent of signal level—it is the logarithmic response that limits this fractional error to the lowest possible value.

We may write the fractional error, ε, in estimating the input signal as a function of the absolute error, $\Delta s'$, in the output signal:

$$\varepsilon = \frac{\Delta s}{s} \approx \frac{1}{s} \Delta s' \frac{ds}{ds'} = \frac{\Delta s'}{a}, \tag{18.75}$$

where we have made use of (18.73). Thus, if $\Delta s'$ is constant, for example, as determined by a noise level constant at the output, the fractional error, ε, will be constant over the full range of logarithmic operation.

The reader may readily verify that if the dynamic range of the system is R_D and the output is allowed to range from s'_{mn} to s'_{mx}, the constants in (18.72) should be

$$a = \frac{s'_{mx} - s'_{mn}}{\log R_D} \tag{18.76}$$

$$c = \frac{s'_{mn} - (s'_{mx} - s'_{mn}) \log s_{mn}}{\log R_D}, \tag{18.77}$$

if the input is to match the output range. Here s_{mn} is the minimum value of the input signal.

The conversion of the product of two signals into a sum is another useful characteristic of the logarithmic response. This has been used in signal processing. A logarithmic response is illustrated in Figure 18.1d.

18.3.2.3 Thresholding and Limiting. In *thresholding*, input signals below a certain level (the threshold) do not affect the output. Output is obtained only for signals above this threshold. See Figure 18.1a and Table 190a, where x_0 is the threshold.

This type of response is found, for instance, in photographic emulsions and in electronic diodes where the threshold may be termed the cut-in signal value, which is only very slightly above zero (see Figure 16.3).

This type of response may be used to eliminate low-level noise, if its level is below that of the significant signal. For example, graininess in the dark areas of a photographic transparency may be eliminated by printing this transparency such that the exposure due to the dark regions falls below the toe region of the characteristic curve of the copy material. Also, when multiplier phototubes are used in the photon-counting mode, thresholding may be used to prevent low-level noise pulses from affecting the count (see Section 16.2.4.7).

In *limiting*, all signals above a certain level (the limit) appear in the output as if they were input signals at that level. Saturation effects are a typical example of limiting. See Figure 18.1*b* and Table 190*b*.

Limiting may be used to eliminate undesired small signal components superimposed on a large signal, assuming that we are not interested in signal variations above the limit. It may also be used to prevent damage to subsequent components that might be harmed by an excessive output signal.

Thus limiting may be used to eliminate graininess in the light region of a photographic transparency by copying it so that the exposure of these regions is above the shoulder of the characteristic curve. Also, note that the target mesh, used in image orthicons to protect the target screen, does this by limiting the signal voltage, attainable by the latter.

18.3.2.4 Signal Quantization. Quantization converts a continuous input signal into a discontinuous output signal: all input signals falling into a finite neighborhood are not differentiated and are given a certain fixed value in the output; those in the next neighborhood, another output value, and so on, for the whole input range. See Figure 18.1*c* and Table 190, Item *c*.

Two-level quantization, is called binary. Binary quantization divides all input signals into two classes ("on"–"off," "black"–"white," etc.), depending on whether the input signal is below or above a certain value, called the *switching level.*

Quantization may be used (a) to prepare continuous signals for input to a digital computer. It may also be used (b) to reduce the amount of information that must be transmitted over an information channel. Binary quantization may be used (c) to eliminate noise from a signal that is essentially binary in nature.

Illustrations for these applications are as follows:

1. The spatial frequency spectrum of a photograph is to be found by means of a digital computer. This picture may be fed into the computer after quantization. (Note that this requires spatial quantization as well as signal level quantization.)

2. In a "video-phone" a recognizable television picture is to be transmitted over a telephone line that has a relatively small information rate capacity. Rather than blurring the picture excessively by limiting the spatial frequency, it may be advisable to limit the number of luminance levels.

3. In the photon-counting technique (Section 16.2.4.7) the output pulses are to be counted by an electronic "integrator." To eliminate spurious effects due to varying pulse heights, binary quantization of pulse

heights is indicated. (In addition, all pulse widths must be brought to the same value.)

18.3.3 Graphical Techniques in Nonlinear Systems Analysis

When the transfer characteristic falls into none of the above categories and cannot be conveniently approximated by an analytic expression, graphical methods may be useful. These are extensively used, for instance, in photography. A graph representing the transfer characteristic is entered, on the axis of the abscissa, with a given input value, and the corresponding ordinate yields the associated output. When the system consists of a number of nonlinear components in cascade, the graphical technique can be extended to include them sequentially. This procedure is illustrated in Figure 18.2 which represents the *tone-reproduction diagram* [14a] of a photographic system including camera (Quadrant 1), film (Quadrant 2), and paper (Quadrant 3). The overall system transfer characteristic is then obtained, in Quadrant 4, as the locus of the intersections of

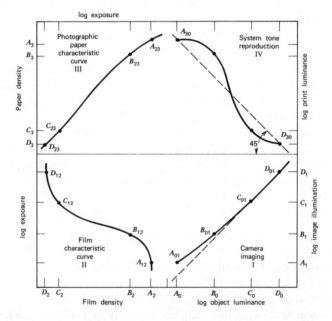

Figure 18.2. Tone reproduction diagram for camera system. Curves I, II, and III are known. Starting with given points A_0, B_0, C_0, and D_0, we find by means of Curve I the points $A_{01}, B_{01} \cdots$ and, thence, points $A_1, B_1 \cdots$. These, in turn, lead us to points $A_{12}, B_{12} \cdots$ on Curve II and, thence, to points $A_2, B_2 \cdots$. This process is repeated in Quadrant 3, yielding points $A_3, B_3 \cdots$. These, in conjunction with the original $A_0, B_0 \cdots$, form the system tone reproduction curve in Quadrant 4, in the upper right hand portion of the figure.

the original object luminance values, as abscissae with the corresponding reflectance values in the final print as ordinates.

Refer to Figure 18.2. Starting, on the horizontal axis, with points A_0, B_0, C_0, D_0 corresponding to four object luminance values we obtain, via Graph I, the corresponding image illumination values $(A_1, \ldots, D_1)$ on the vertical axis. The characteristic curve, rotated through $90°$, is placed into Quadrant 2, its vertical position there being determined by the exposure time. From it, resulting film density values $(A_2, \ldots, D_2)$ are obtained. The characteristic curve of the paper is then placed in the third quadrant, its horizontal position determined by the exposure time and illumination used in the printing process. From this curve the resulting density values $(A_3, \ldots, D_3)$ can be obtained. These, when modified by the viewing illumination, yield the resulting luminance values $(A'_3, \ldots, D'_3)$. The intersections between horizontal lines through these, with vertical lines through points $(A_0, \ldots, D_0)$ yield four points on the system transfer function, Curve IV.

18.3.4 Finite-Spread Nonlinearities

Finite-spread nonlinearities are generally very difficult to analyze. But suggestions have been made to treat them in stages, each of which is either linear, although finite-spread, or nonlinear, but zero-spread [15].

For instance, useful results have been obtained in analyzing the performance of the visual system by first transforming the signal nonlinearly, for example, logarithmically, and then operating on it linearly [16]. Similarly, the procedure for analyzing the performance of photographic emulsions by means of an "equivalent exposure" (Section 17.2.2.1) illustrates this procedure.

One important nonlinear degradation occurs when the noise is multiplicative, rather than additive. This may happen, for instance, when a scene is illuminated nonuniformly, introducing low-frequency noise into the object. The noise spectrum may then be far different from the signal spectrum, and yet can not be separated because of the nonlinearity of the process. It has been shown [17] that in this situation logarithmic amplification, converting multiplication into addition, makes linear filtering techniques applicable, so that the noise may be removed.

18.3.5 Nonisoplanatic Systems

When the spread function of an imaging system varies over the field, the system is said to be nonisoplanatic.[5] This condition is inconvenient

---

[5] This condition is occasionally called linear shift variant (LSV) in contrast to the linear shift invariant (LSI) isoplanatic case.

because, again, Fourier techniques are not directly applicable. Fortunately, this difficulty can often be circumvented by dividing the field into patches, within each of which the spread function can be treated as constant.

When this is not feasible, for example, if more precise results are required, it may be possible to distort the image by a zero spread process in such a manner that the spread function becomes uniform over the image. Linear techniques may then be applied, followed by a reversal of the (zero-spread) distortion. The treatment of one class of such inhomogeneously blurred imaging processes—"comalike" imaging—has been treated in detail [18]. Here the spread function grows proportionally to the distance from the center and the restoration is done in polar coordinates: Fourier transform is applied to the angle-coordinate and Mellin transform to the radial coordinate.

The inhomogeneous blurring caused by nonuniform relative motion between the image and the image recording medium, too, has been analyzed [19].

Discrete imaging, as occurs in fiber optic imaging and scanning, is also highly nonisoplanatic. This is treated in the next section.

18.4 MOTION EFFECTS; DISCRETE IMAGING SYSTEMS

18.4.1 Motion Effects [20], [21]

When the image is stored on a recording medium and moves relative to this medium during the exposure, blurring results. In a linear system, this blurring may be represented by a spread function and, hence, if it is uniform over the image,[6] by a transfer function. Thus the motion itself can be treated as a pseudocomponent with its own characteristics.

Consider an image, with coordinate system ξ fixed in it, moving over a recording medium with coordinate system x, such that the origin of the ξ system is at $x'(t)$ at any time t. Then

$$\xi = x - x'. \tag{18.78}$$

Let us assume that the relative flux distribution in the image is constant throughout the exposure, but that the total image flux, $\Phi(t)$, varies with

---

[6] If the motion is nonuniform over the image, for example, rotatory, this results in a spread function that varies over the image. This situation has been analyzed thoroughly [22], but is not discussed here. In the preceding section, the reader will find indications as to how such blurring may be corrected after the fact.

time. We may then normalize the total flux and write for the instantaneous image "intensitity":

$$A(t) = \frac{\Phi(t)}{\int_{-\infty}^{\infty} \Phi(t)\, dt}, \tag{18.79}$$

so that

$$\int_{-\infty}^{\infty} A(t)\, dt = 1. \tag{18.80}$$

The function $A(t)$ now also represents the normalized temporal variations of the illumination at any image point. We can then write the recorded image signal, $s'(\mathbf{x})$, in terms of the instantaneous image $s(\boldsymbol{\xi})$:

$$s'(\mathbf{x}) = \int_{-\infty}^{\infty} s(\boldsymbol{\xi}) A(t)\, dt = \int_{-\infty}^{\infty} s[\mathbf{x} - \mathbf{x}'(t)] A(t)\, dt. \tag{18.81}$$

The equivalent transfer function can be derived from this by means of the Fourier transform. Thus the spectrum of the recorded image is

$$S'(\boldsymbol{\nu}) = \int s'(\mathbf{x}) e^{i2\pi \boldsymbol{\nu} \cdot \mathbf{x}}\, d\mathbf{x}$$

$$= S(\boldsymbol{\nu}) \int_{-\infty}^{\infty} e^{i2\pi \boldsymbol{\nu} \cdot \mathbf{x}(t)} A(t)\, dt. \tag{18.82}$$

Clearly

$$\mathsf{T}(\boldsymbol{\nu}) = \int_{-\infty}^{\infty} e^{i2\pi \boldsymbol{\nu} \cdot \mathbf{x}(t)} A(t)\, dt \tag{18.83}$$

is the transfer function that transforms the input signal spectrum into that of the recorded image.

We can also derive the spread function of the motion directly from (18.81) by substituting a point-object:

$$s(\mathbf{x}) = \delta[\mathbf{x} - \mathbf{x}'(t)] \tag{18.84}$$

for the object. This yields

$$\mathsf{P}(\mathbf{x}) = \int_{-\infty}^{\infty} \delta[\mathbf{x} - \mathbf{x}'(t)] A(t)\, dt$$

$$= \int_{-\infty}^{\infty} \frac{\delta(t - t_x) A(t)}{v(t)}\, dt$$

$$= \frac{A[t(\mathbf{x})]}{v[t(\mathbf{x})]}, \tag{18.85}$$

where we have made use of the identity[7]

$$\delta[\mathbf{x} - \mathbf{x}'(t)] = \frac{\delta(t - t_x)}{v}, \tag{18.86}$$

where $t_x = t(x)$ is the time at which $\mathbf{x}' = \mathbf{x}$ and

$$v = \left|\frac{d\mathbf{x}}{dt}\right| = \sqrt{\dot{x}^2 + \dot{y}^2} \tag{18.87}$$

is the absolute value of the relative velocity. We now apply this result to some examples.

18.4.1.1 Linear Motion, Constant Flux. Let us first consider systems where the motion is in a straight line in the x direction:

$$\mathbf{x}'(t) = x'(t) = \int_0^t \dot{x}' \, dt \tag{18.88}$$

and

$$A(t) = \frac{1}{T}, \qquad |t| < \tfrac{1}{2}T$$
$$= 0 \qquad |t| \geqslant \tfrac{1}{2}T.$$

The transfer function may then be written

$$\mathsf{T}(\nu) = \int_{-(1/2)T}^{(1/2)T} \exp[i2\pi\nu x'(t)] \, dt. \tag{18.89}$$

For instance, if the velocity is constant,

$$x'(t) = vt \tag{18.90}$$

and

$$\mathsf{T}(\nu) = \frac{1}{T} \int_{-(1/2)T}^{(1/2)T} e^{i2\pi\nu vt} \, dt = \frac{\sin \pi\nu X}{\pi\nu X}, \tag{18.91}$$

where we have written

$$X = vT. \tag{18.92}$$

---

[7] This identity may be derived from the identity

$$\int_{-\infty}^{\infty} F(x) \, \delta(x) \, dx = \int_{-\infty}^{\infty} F[x(t)] \, \delta(t - t_0) \, dt = F(0),$$

where t_0 is the value of t where $x = 0$. This indicates that $\delta(x) \, dx = (-) \, \delta(t - t_0) \, dt$. In the event that t increases as x decreases, the path of the second integral is the reverse of that of the first, so that a minus sign must be included. Hence we include the absolute value of dx/dt in (18.86).

If the velocity is sinusoidal, with total excursion X,

$$x'(t) = \frac{1}{2} X \frac{\sin 2\pi t}{T}$$

and, if the exposure is limited to an integral number of half-cycles,

$$
\begin{aligned}
\mathsf{T}(\nu) &= \frac{1}{T} \int_{-(1/2)T}^{(1/2)T} \exp\left[i\pi\nu X \sin\left(2\pi t/T\right)\right] dt \\
&= \frac{1}{2\pi} \int_{-\pi}^{\pi} e^{i\pi\nu X \sin\theta}\, d\theta \\
&= J_0(\pi\nu X).
\end{aligned}
\tag{18.93}
$$

[For evaluation of this integral, see the remark preceding (18.164).]

18.4.1.2 Circular Motion, Constant Flux. If the motion is circular, with uniform speed and radius r_0, the point spread function has radial symmetry and is given by

$$\mathsf{P}(r) = \frac{\delta(r - r_0)}{\pi r_0}. \tag{18.94}$$

Hence, from (18.164),

$$\mathsf{T}(\nu) = 2\pi \int_0^\infty r\mathsf{P}(r) J_0(2\pi r\nu)\, dr = J_0(2\pi\nu r_0). \tag{18.95}$$

18.4.1.3 Random Motion. If the image experiences random displacements, as occurs, for instance, in imaging through a turbulent atmosphere, our deterministic approach can be used only if the variations are stationary and the exposure is sufficiently long so that the system can be approximated as having passed through all possible states (that is, close enough to them) with a frequency proportional to the probability of these states. If we make these assumptions, we can write the point spread function

$$\mathsf{P}(\mathbf{x}) \sim \iint \frac{p(\mathbf{x})}{v} p(v \mid \mathbf{x}) A p(A \mid \mathbf{x})\, dv\, dA, \tag{18.96}$$

where $p(\mathbf{x})$, $p(v)$, and $p(A)$ are the probability densities of $\mathbf{x}$, v, and A, respectively.

In the special case where the $\mathbf{x}$ are distributed in a Gaussian manner and the velocity and flux level are independent of $\mathbf{x}$ $[p(v \mid \mathbf{x}) = p(v), p(A \mid \mathbf{x}) = p(A)]$, we obtain the Gaussian point spread function

$$\mathsf{P}(r) = (2\pi\sigma^2)^{-1} e^{-r^2/2\sigma^2} \tag{18.97}$$

and

$$\mathsf{T}(\nu) = e^{-2(\pi\nu\sigma)^2}. \tag{18.98}$$

See Table 187.

18.4.1.4   *Linear Motion, Sinusoidal Light Modulation.* If the illumination varies sinusoidally:

$$A(t) = k[1 + M \cos 2\pi v_t t]$$

and the motion is constant

$$x' = vt,$$

(18.83) becomes

$$T(v) = \int_{-\infty}^{\infty} e^{-i2\pi v v t}(1 + M \cos 2\pi v_t t)\, dt$$

$$= \delta(v) + \frac{M}{2}\left[\delta\left(v - \frac{v_t}{v}\right) + \delta\left(v + \frac{v_t}{v}\right)\right]; \qquad (18.99)$$

the system transmits only the mean level and the spatial frequency components

$$v = \frac{v_t}{v}. \qquad (18.100)$$

Indeed, this arrangement has been used to determine the spectral content of a spread function—the mtf [23].

18.4.1.5   *Decaying Detector Response and CRT Scan Spot.* If the detector memory is imperfect, that is, the weight of an exposure contribution drops according to some function, $g(t)$, the equation for the output signal, becomes

$$s'(\mathbf{x}, t) = \int_{-\infty}^{t} [\mathbf{x} - \mathbf{x}'(t')]A(t')g(t - t')\, dt'. \qquad (18.101)$$

Again, Fourier transform with respect to $\mathbf{x}$ shows that the corresponding transfer function is

$$T(\mathbf{v}, t) = \frac{\displaystyle\int_{-\infty}^{t} e^{i2\pi \mathbf{v} \cdot \mathbf{x}'(t')} A(t')g(t - t')\, dt'}{\displaystyle\int_{-\infty}^{t} A(t')g(t - t')\, dt'}, \qquad (18.102)$$

where we have introduced the denominator for normalization at the $\mathbf{v}$ origin.

For the special case of uniform illumination over a period T and an exponential decay rate

$$g(t) = e^{-t/\tau}, \qquad t \geqslant 0,$$

this becomes

$$T(\nu) = \frac{\int_{-(1/2)T}^{(1/2)T} e^{i2\pi\boldsymbol{\nu}\cdot\mathbf{x}'(t)} e^{t/\tau}\, dt}{2\tau \sinh{(T/2\tau)}}. \tag{18.103}$$

With linear motion, speed v, this leads to a mtf of the form

$$|T(\nu)|^2 = \frac{1 + (\sin \pi\nu v T/\sinh T/2\tau)^2}{1 + (2\pi\nu v\tau)^2}. \tag{18.104}$$

Sinusoidal motion leads to an expression of the form of an infinite sum of Bessel functions [21].

Decaying "detector" response with linear motion occurs also in crt scanning, where the instantaneous luminance distribution is composed of the sum of the remnants of luminance remaining from earlier spot positions. This, therefore, corresponds to (18.102) with constant A and the upper limit of the integral vanishing. This leads to an otf of the form:

$$T(\nu) = [1 + i2\pi\nu v\tau]^{-1} \tag{18.105}$$

with the mtf

$$|T(\nu)| = \frac{1}{\sqrt{1 + (2\pi\nu v\tau)^2}}. \tag{18.106}$$

This otf must be applied in addition to the otf corresponding to the scanning spot function, when analyzing the imaging in a scanning system.

18.4.2 Discrete Imaging

In some optical systems, the object is sampled at only discrete points and the image is reconstructed from these samples. This situation occurs, for instance, in halftone printing with mesh screens (Benday screens), in imaging via fiber optics and microchannel plates, and in imaging by means of a raster scan. Such systems may still be linear, but are highly nonisoplanatic. However, when the sampling is close enough, they may be treated as isoplanatic systems with a superimposed "noise." The noise is the result of the lattice structure, which introduces a false signal. Since the value of the false signal depends on the chance positioning of the object relative to the sampling grid, it may properly be called noise.

18.4.2.1 Discrete Functions and their Fourier Transforms. We define the periodically discrete function, f_d, associated with the continuous function, f, as

$$f_d(x) = f(x) \sum_{j=-\infty}^{\infty} \delta(j\Delta x - x). \tag{18.107}$$

(In the following we omit the limits of summation, it being understood that all summations are to be from minus to plus infinity.) Its Fourier transform is readily found to be

$$\mathscr{F}[f_d] = F_D(\nu) = \int_{-\infty}^{\infty} f(x) \sum \delta(j\Delta x - x) e^{i2\pi\nu x} \, dx$$

$$= \sum f(j\Delta x) e^{i2\pi\nu j\Delta x}$$

$$= \sum F\left(\nu + \frac{j}{\Delta x}\right). \tag{18.108}$$

To prove the last step, note that $\sum F(\nu + j/\Delta x)$ is periodic with period $N = 1/\Delta x$. Hence it may be represented by a Fourier series

$$\sum_k F\left(\nu + \frac{k}{\Delta x}\right) = \sum_i c_j e^{i2\pi j\Delta x\nu}, \tag{18.109}$$

where

$$c_j = \int F(\nu) e^{-2\pi j\Delta x\nu} \, d\nu = f(j\,\Delta x). \tag{18.110}$$

On substituting this into (18.109) and changing the dummy index from k to j, we obtain the last equation of (18.108).

We have taken the summation from negative to positive infinity, although actual objects will be limited in extent. This is no limitation, however, since we may define our object function over the whole plane, simply assigning it zero value outside the range of the actual object.

18.4.2.2 Analysis of Discrete Imaging Systems

Derivation of the Basic Result. Consider a one-dimensional object, $s(x)$, sampled at intervals Δx by means of an aperture having a transmittance (or illumination) distribution $g_0(x)$; then the signal value entering the system from a typical aperture location, $j\,\Delta x$, will be

$$s'(j\,\Delta x) = \int s(x) g_0(x - j\,\Delta x) \, dx. \tag{18.111}$$

See Figure 18.3, where the transmittance at point $P(x)$ is given by $g_0(x - j\,\Delta x)$, when the origin of the aperture coordinate system is located at $j\,\Delta x$. We now assume that the signal value s' is redistributed over the image plane, around point $j\,\Delta x$, according to some spread function $g_1(x)$. Then the total signal, $s''(x'')$ at any point x'' will be equal to the sum of the contributions made by each of the sampling apertures:

$$s''(x'') = \sum s'(j\,\Delta x) g_1(x'' - j\,\Delta x)$$

$$= \sum g_1(x'' - j\,\Delta x) \int s(x) g_0(x - j\,\Delta x) \, dx. \tag{18.112}$$

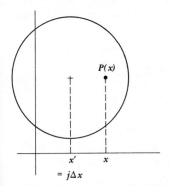

Figure 18.3. Discrete imaging—input.

To obtain the image signal spectrum, we calculate the Fourier transform of this and find

$$S''(\nu) = \int_{-\infty}^{\infty} e^{i2\pi\nu x''} \sum g_1(x'' - j\,\Delta x) \int s(x)g_0(x - j\,\Delta x)\, dx\, dx''.$$

$$(18.113)$$

Changing to a new variable of integration

$$x'' = u + j\,\Delta x, \qquad du = dx'',$$

we find

$$S''(\nu) = G_1(\nu) \sum e^{i2\pi\nu j\,\Delta x} \left[\int s(x)g_0(x - j\,\Delta x)\, dx \right]. \qquad (18.114)$$

Here, as usual, the capital letter denotes the Fourier transform of the function denoted by the corresponding lower case letter.

If we denote the cross correlation of s and g_0 by f, we find the bracketed factor to be $f(j\,\Delta x)$ and hence the sum as in (18.108). Hence

$$S''(\nu) = G_1(\nu) \sum F\!\left(\nu + \frac{j}{\Delta x}\right), \qquad (18.115)$$

where

$$F = \mathscr{F}[s \odot g_0] = SG_0^* \qquad (18.116)$$

is the Fourier transform of the cross correlation of s and g_0. Substituting this into (18.115) and separating out the $(j = 0)$ term, we find

$$\boxed{S''(\nu) = G_1(\nu)G_0^*(\nu)S(\nu) + G_1(\nu) \sum^{0} G_0^*\!\left(\nu + \frac{j}{\Delta x}\right) S\!\left(\nu + \frac{j}{\Delta x}\right),}$$

$$(18.117)$$

where the superscript zero indicates that the $j = 0$ term is excluded from the summation.

We may, therefore, represent the system as having a transfer function

$$T(\nu) = \frac{G_1(\nu)G_0^*(\nu)}{G_1(0)G_0(0)} = \frac{G_1(\nu)G_0^*(\nu)}{\int g_1(x)\, dx \int g_0(x)\, dx} \qquad (18.118)$$

and as adding noise of the form given by the sum in (18.117).

In Appendix 18.4 it is shown that this noise has a vanishing ensemble average, if the object position is equally likely at any lattice phase.

Discussion of Results. Inspection of the major result (18.117) shows that there will be no noise within the region of the object spectrum, if this is limited to

$$\nu < \frac{1}{2\,\Delta x}. \qquad (18.119)$$

The same is true, if the input transfer function satisfies this condition. Compare the sampling theorem (Section 3.3.1). When (18.119) is satisfied, the sums in (18.115) and (18.117) will have only one nonvanishing term at any spatial frequency ν. Indeed (18.115) shows that the image spectrum is a periodic version of the fundamental image spectrum

$$G_1 G_0^* S,$$

repeated at intervals $1/\Delta x$ and attenuated by a factor

$$\frac{G_1(\nu - j/\Delta x)}{G_i(\nu)}.$$

When condition (18.119) is satisfied, these repeated spectra do not overlap (see Figure 18.4). Such overlapping, when it does occur, is called *aliasing*.

The one-dimensional result here derived enables us to analyze two-dimensional systems also. If the two-dimensional sampling lattice is periodic and infinite, there will be an equivalent Δx corresponding to any direction so that the effective otf (18.118) can be derived for it. On the other hand, the spectrum of any two-dimensional object consists of a continuum of spectral components of all frequencies and directions, so that knowledge of all the corresponding otf's suffices to predict the resulting image.

The effect of a sampling aperture, integrating the signal over a finite area, is analyzed in Ref. 23*.

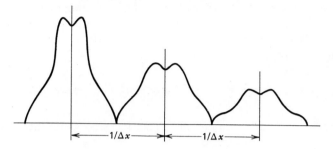

Figure 18.4. Discrete imaging spectrum.

18.4.3 Scanning Systems

The linear raster scan, continuous in one direction and discrete transverse to this, is a frequent case of discrete imaging, and we discuss it here in some detail. It was originally treated over 45 years ago [24]; optimization conditions, too, have been determined for it [25].

18.4.3.1 General Expression. We choose the object plane coordinate system so that the x axis coincides with the scan direction and assume a scanning aperture with transmittance (or illumination) distribution, $g_0(x, y)$, traversing the object with a speed v in raster lines spaced at Δy. The instantaneous signal will then be

$$s'(x; j) = \iint s(x', y') g_0(x' - x, y' - j \, \Delta y) \, dx' \, dy', \qquad (18.120)$$

where x is the position of the scanning aperture origin, and
$\quad j$ is the raster line number counted from the one passing through the origin of the object coordinate system.

To obtain this signal, $s'_t(t)$, as a function of time, we substitute into (18.20)

$$x = \frac{t^*}{v} \quad \text{and} \quad j = \left[\frac{t}{\Delta t}\right], \qquad (18.121a,b)$$

where Δt is the scan line period,

$$t^* = t - \Delta t \left[\frac{t}{\Delta t}\right] \qquad (18.121c)$$

and the square brackets indicate that only the integer portion of the fraction is to be taken. If the signal now passes through a subsystem with impulse response $g_t(t)$, the resulting signal will be given by

$$s''(t) = s'_t \circledast g_t. \qquad (18.122)$$

This signal now controls the illumination in the imaging plane, where the image is reconstructed with a synchronized raster scan, velocity v', spacing $\Delta y'$. The accumulated exposure at any point (x'', y'') in the image plane will then be

$$s'''(x'', y'') = \sum_j \int_{j\,\Delta t}^{(j+1)\Delta t} s''(t) g_1(x'' - t^* v', y'' - j\,\Delta y')\,dt, \quad (18.123)$$

where $j\,\Delta t, (j+1)\,\Delta t$ represent the times the scan starts and ends, respectively,

$t^* v' = (t - j\,\Delta t)v'$ represents the x-coordinate of the position of the scanning spot origin at time t, and

$g_1(x, y)$ represents the spread function of the image scan spot.

When the object scanning spot is formed on a medium with finite memory, such as a phosphor screen, g_0 includes the spreading of this beam due this memory; see (18.105) and (18.106).

18.4.3.2   Transfer Function with No Overlap. We may now evaluate the corresponding transfer functions in the x and y directions individually. We do this on the assumption of no overlap of raster lines. This enables us to dispense with the summation over j and permits us to work with the origin-raster line as representative.

We first assume an object varying in the x-direction only:

$$s(x, y) = s_x(x)$$

and introduce the line spread functions

$$g_{ix}(x) = \int g_i(x, y)\,dy, \quad i = 0, 1. \quad (18.124)$$

To avoid the inconvenience of changing to time and back we treat the case where the imaging raster is identical to the object scanning raster. (When these differ by a scale factor, this is readily applied to the result.) We also use the spatial equivalent of the impulse response

$$g_{tx}(x) = g_t\!\left(\frac{x}{v}\right). \quad (18.125)$$

We may then write for the one-dimensional image

$$s'''(x'') = \int s'''(x'', y'')\,dy'' = s_x \odot g_{0x} \circledast g_{tx} \circledast g_{1x}. \quad (18.126)$$

This corresponds to an otf of the simple form

$$\mathbf{T}_x = G_{0x}^* G_{tx} G_{1x}. \quad (18.127)$$

To determine the otf in the y direction, we define the analogous line spread functions

$$g_y(y) = \int g(x, y)\, dx \qquad (18.128)$$

and then find, referring to (18.117),

$$T_y = G_{0y}^* G_{1y} \qquad (18.129)$$

with an added noise term as in (18.117).

18.5 SYSTEM EVALUATION AND OPTIMIZATION [26]

System evaluation is an important phase of system analysis. We treat this in the present section. In this context we view the imaging system as an information channel and evaluate its performance in those terms. We concentrate on linear techniques and tacitly assume that these are applicable.

18.5.1 System Quality Criteria

Let us first list quality criteria that have been proposed.

1. *Information Capacity.* The expression for information capacity has a very simple form when the noise and signal power are both normally distributed and stationary over a sufficiently large area (see Section 3.4.3.1). These conditions are frequently satisfied and the information capacity may then be written

$$H = 2A_i \int_0^\infty \log\left(1 + \frac{W_0}{W_n}\right) d\mathbf{v}, \qquad (18.130)$$

where A_i is the image area, and
W_0, W_n are the Wiener (power) spectra of the object and the noise, respectively.

The information capacity has been used as a quality criterion [27], especially for detectors [28].[8]

---

[8] Many of the expressions developed in this section require the manipulation of mtf's, such as their integration or multiplication. These procedures may be inconvenient with the usual mtf's in their empirically obtained form. A rather general analytic approximation of the form

$$T(\nu) = e^{-(\nu/\nu_1)^a}$$

or more generally,

$$T(\nu) = T_1 e^{-(\nu/\nu_1)^a} + T_2 e^{-(\nu/\nu_2)^b}, \qquad T_1 + T_2 = 1$$

has been found to fit a large number of cases. Here T_i, ν_i, a, b are constants. Their values for a number of components have been published and their use may facilitate the analysis of system performance. See Refs. 16.155 and 17.30 for illustrative values.

2. *Linfoot's Figures of Merit.* Linfoot proposed three quality criteria for imaging systems [29], [30]. These have already been described in Section 9.4.2, and we briefly restate them here:

a. *Correlation quality* measures correlation between image and object:

$$Q_1 = \frac{\int \overline{s_0 s_i}\, dx}{\int \overline{s_0^2}\, dx} = \frac{\int W_0 T\, d\nu}{\int W_0\, d\nu}. \tag{18.131}$$

The subscript i refers to the image parameter.

b. *Relative structural content* measures the image power relative to the object power:

$$Q_2 = \frac{\int \overline{s_i^2}\, dx}{\int \overline{s_0^2}\, dx} = \frac{\int W_0 T^2\, d\nu}{\int W_0\, d\nu}. \tag{18.132}$$

c. *Fidelity* measures the closeness of correspondence of image with object:

$$Q_3 = 1 - \frac{\int \overline{(s_i - s_0)^2}\, dx}{\int \overline{s_0^2}\, dx} = 1 - \frac{\int W_0(1-T)^2\, d\nu}{\int W_0\, d\nu}. \tag{18.133}$$

Note that the correlation quality is the mean between the other two qualities:

$$Q_1 = \tfrac{1}{2}(Q_2 + Q_3). \tag{18.134}$$

In equations (18.131)–(18.133) we have made use of Parseval's theorem which implies that the integral of a function squared equals the integral of its Fourier transform squared. See (3.118).

3. *Signal-to-Noise Ratio.* The signal to noise ratio

$$R = \frac{\int \overline{s_i^2}\, dx}{\int \overline{n_i^2}\, dx} = \frac{\int W_0 T^2\, d\nu}{\int W_n\, d\nu}. \tag{18.135}$$

too, has been used as a quality indicator of an image [31]–[33]. Here n_i is the noise signal as measured at the image.

4. *Mean Squared Deviation.* Another criterion is the square of the deviation of the output, signal plus noise, from the ideal output signal, integrated over the total image:

$$E = \int \overline{[s_0 - (s_i + n_i)]^2} \, d\mathbf{x} = \int \overline{[S_0 - (S_0 T + N_i)]^2} \, d\mathbf{v}. \qquad (18.136)$$

Assuming the noise to be independent of the signal, this becomes:

$$E = \int [W_0(1 - T)^2 + W_n] \, d\mathbf{v}. \qquad (18.137)$$

This quantity, too, is very useful in image evaluation [34], [35].

5. *Transcorrelation.* The correlation between the ideal and the actual image, too, has been used as a quality criterion. Specifically, the *transcorrelation* has been defined [36][9]

$$C = \frac{\dfrac{1}{A} \int s_i' s_0 \, d\mathbf{x} - \dfrac{1}{A^2} \int s_i' \, d\mathbf{x} \int s_0 \, d\mathbf{x}}{\left\{ \left[\dfrac{1}{A} \int s_0^2 \, d\mathbf{x} - \left(\dfrac{1}{A} \int s_0 \, d\mathbf{x} \right)^2 \right] \left[\dfrac{1}{A} \int s_i'^2 \, d\mathbf{x} - \left(\dfrac{1}{A} \int s_i \, d\mathbf{x} \right)^2 \right] \right\}^{1/2}}, \qquad (18.138)$$

where s_i' is primed to indicate that it includes the noise in the image $s_i' = s_i + n_i$, and all the integrals are averaged.

On writing the signal as the sum of its mean (μ) and the deviation (δ):

$$s_0 = \mu_0 + \delta_0, \qquad s_i' = \mu_i + \delta_i, \qquad (18.139a)$$

with

$$\int \delta_0 \, d\mathbf{x} = \int \delta_i \, d\mathbf{x} = 0, \qquad (18.139b)$$

we readily find that equation (18.138) can be written:

$$C = \frac{\displaystyle\int \delta_i \delta_0 \, d\mathbf{x}}{\sqrt{\displaystyle\int \delta_i^2 \, d\mathbf{x} \int \delta_0^2 \, d\mathbf{x}}}. \qquad (18.140)$$

From Parseval's theorem this can be written in terms of the spectra:

$$C = \frac{\displaystyle\int S_0 S_i'^* \, d\mathbf{v}}{\sqrt{\displaystyle\int |S_0|^2 \, d\mathbf{v} \int |S_i'|^2 \, d\mathbf{v}}}. \qquad (18.141)$$

---

[9] Because we do not consider temporal variations here, I have dropped the time dependence from the original expression.

Noting that $S_i' = TS_0 + N_i$, and again assuming the noise independent of the signal, this yields:

$$C = \frac{\displaystyle\int W_0 T \, d\boldsymbol{v}}{\displaystyle\sqrt{\int W_0 \, d\boldsymbol{v} \int (W_0 T^2 + W_n) \, d\boldsymbol{v}}}. \tag{18.142}$$

18.5.2 Evaluation of Criteria

Having presented the "candidates," let us now evaluate them in conjunction with their task, communication. The general communication process has as its purpose the transmission of information. It can be broken up into three fundamental elements: (a) generation and coding; (b) transmission; and (c) reception, including decoding of the received signal. Once information has been generated, it can be lost, but not increased, by the other components of the system. (The alternative would imply that the later components generate information concerning the source, which contradicts their definition.) Thus, the best we can hope to do in phase (b) is to maintain the information constant or, if physical restrictions prevent this, to maximize the transmitted information. At the receiving end, too, information may be lost; but there we have two distinct tasks, detection and decoding, which we must separate conceptually even though they may actually be combined in a single physical process. Part of the confusion in connection with image enhancement seems to stem from this combination of tasks: the detector must maximize *information* detected—the decoder must maximize *fidelity*. To evaluate systems in respect to these tasks we must first have quantitative definitions of them. The definition of quantity of information given in Section 3.1.2 seems to be satisfactory, having passed all the relevant tests. (We must, of course, be careful to distinguish between significance and quantity of information; the scribblings of a madman may carry as much "information" as an encyclopedia.)

An appropriate criterion for fidelity is less obvious. The decoding process depends on the range of input signal possibilities. On the one hand, if only a small number of binary decisions are necessary, as in character recognition, high fidelity may be possible simply on the basis of these decisions. In that case, a weighted area integral of the total output will provide a number on which a decision concerning the input signal can be based, a different weighting function being required for each choice. (The resolution of spectral lines or double stars falls into this category also.) Here the output may still be optical but fidelity may no longer be defined in the general image sense—only the parameter of interest must enter the judgment of fidelity.

However, when the possible input signals are largely unrestricted, or continuously distributed, an attempt must be made to approximate them in the decoding, at least within some practical transformation. In such situations, the least-squared deviation is a good candidate for a measure of fidelity. It yields the best estimate of the signal in the following sense: assuming that nothing is known about signal and noise, except their mean Wiener spectra, and the fact that they have a Gaussian distribution, then the output filtered for least-squared deviation represents the signal more likely responsible for the received output than any other signal. (The *a posteriori* input signal probability density is a maximum at the signal represented by the filtered output.) This criterion has been criticized for giving equal weight to one large deviation and to many small ones, but this criticism does not seem to be obviously weighty.

Thus we may conclude that in transmission and in pure detection Criterion 1 is appropriate and in decoding of continuous-range signals Criterion 4 is appropriate. To understand the significance of the other criteria, we note that at the threshold of detectability (small signal-to-noise ratio) information capacity, as given by (18.78), approaches the signal-to-noise ratio, so that, in practice, the latter, Criterion 3, may be used as a criterion of detectability.[10]

When the system noise spectrum is uniform or totally unknown, Linfoot's "relative structural content" (Criterion 2b) becomes a measure of signal-to-noise ratio. When noise is negligible, his "fidelity" (Criterion 2c) becomes equivalent to the mean squared deviation (E) criterion. More specifically, his "fidelity" is

$$Q_3 = 1 - \varepsilon, \tag{18.143}$$

where ε is the integrated squared deviation relative to the total input signal "power."

The specific significance of transcorrelation is less obvious. It does become unity in an ideal system and it is clearly a measure of how closely the system approaches the ideal. But it is not clear which aspect of the system is optimized when we maximize transcorrelation rather than, say, minimize squared deviation. Note, however, that Linfoot's "correlation quality" (Criterion 2a), which is the noise-free equivalent of transcorrelation, is simply the arithmetic mean between "fidelity" and "relative structural content"; this indicates for transcorrelation a role intermediate to optimizing information content and fidelity. It might, therefore, be appropriate for systems that are to be used in both types of application [26].

---

[10] It should be noted that this optimization is valid only if the noise is strictly additive. If it varies with the signal value, other criteria must be applied [37].

18.5.3 Image Optimization Theory

Once criteria for evaluating a given image are available, optimization is the next step.

18.5.3.1 Optimization of Magnification and Gain. The component magnification, gain, and spectral characteristics may be used to optimize system performance. In other words, these component characteristics may be used to match the object characteristics to those of subsequent components, such as the detector, so that a maximum of information can be transmitted.

Magnification may be used to match transfer functions. The magnification must be such that the transfer function of a subsequent component does not attenuate the signal unduly. With the usual transfer function, which descends monotonically, this implies a sufficiently high magnification at an early stage. Compare (18.19), where the choice of M_i can be seen to control the effective point on the transfer function and, therefore, the effect of T_i on the system imaging process as described by (18.22).

Because of the importance of matching the spectral characteristics of the detector to those of the object, we must consider not only the sensitivity of the detector but the product of this with the spectral matching factor, in choosing a detector.

Signal-to-noise considerations dictate the choice of gain factors. For instance, by choosing g_i sufficiently high, the contributions of all subsequent noise sources can be made negligible. [Note the effect of G_i on the contribution of σ_i^2 in (18.23).] On the other hand, once the contributions of later stages are negligible, nothing is gained by increasing the gain g_i, as discussed in conjunction with specific noise in Section 18.2.2.1.

18.5.3.2 Spatial Frequency Filtering. The reader may note that the otf (T) entered into each of the image quality criteria listed in Section 18.5.1. It therefore seems reasonable to use modifications of the otf as a method of optimizing the system. Such modification is possible by the choice of the system components according to their otf. But, assuming that these have been chosen in an optimum manner, there is still the option of inserting an additional component whose specific function is the modification of the system otf. Such a component is called a (spatial) frequency filter (see Section 19.2).

Linear filtering has been treated thoroughly in the theory of electrical signals and analogous, two-dimensional techniques have been discussed extensively with many of the results from electric signal theory directly applicable [38].

Thus, just as in the theory of electrical signals, the filter maximizing the signal-to-noise ratio at an image point $\mathbf{x}_1$, is simply the matched filter [39], [40], [7c]

$$T_M(\mathbf{v}) \sim S_0^*(\mathbf{v})\, e^{-i2\pi\mathbf{v}\cdot\mathbf{x}_1}, \qquad (18.144)$$

provided the noise spectrum is white. If not, the right-hand member of (18.144) must be divided by W_n to yield the optimum filter. Note that this filter maximizes the signal-to-noise ratio at only one point, making it useful for the detection of a signal at a known location.

The filter minimizing the square deviation is the Wiener[11] filter [7c], [34], [35]

$$T_W(\mathbf{v}) = \frac{W_0(\mathbf{v})\, T^*(\mathbf{v})}{W_0\,|\,T(\mathbf{v})|^2 + W_n(\mathbf{v})} = \frac{W_i}{T(W_i + W_n)}. \qquad (18.145)$$

Derivations of these filter functions are given in Appendix 18.5.

In the absence of noise ($W_n = 0$), the Wiener filter (18.145) becomes

$$T_W = T^{-1} \qquad (18.146)$$

the reciprocal of the initial system otf. This simplified version has served as the basis for a number of image restoration schemes [41]–[43].

The filter (18.145) can be shown to be optimum even when the filter is not required to be linear [44]. However, when the original system transfer characteristic is not linear, the squared deviation of the final, distorted, image should be minimized, and this leads to other optimization requirements, which may be rather complex [16], [17].

18.5.4 Analytic Continuation of Signal Spectrum [45]

Since the spatial frequency response (otf) of any real optical system vanishes beyond a certain cutoff frequency (v_c), one might suspect that information in the portion of the spectrum beyond v_c is irretrievably lost. However, this is not necessarily true. Generally, the object size is limited, that is, the object vanishes beyond a certain limit, $|x| < \tfrac{1}{2}X$, say. This implies that its spectrum can be viewed as a band-limited signal. (Note that the spectrum of the spectrum is the original signal in reverse.) Such a signal is described by an analytic function. Like every such function, exact knowledge of the function values in an arbitrarily small neighborhood defines the function everywhere [46].

We may, for instance, represent the signal by a sum of functions (such as the prolate spheroidal wave functions), which are orthogonal in the object region. The coefficients of these orthogonal functions can be computed on the basis of the known signal values ($v < v_c$); the resulting

---

[11] N. Wiener (1949).

sum then defines the signal spectrum at all frequencies, including $v > v_c$ [47]–[49]. By means of such analytic continuation, or extrapolation, the complete signal can be recovered, provided there is no noise. The presence of even a small amount of noise seriously limits the capability of this method [50] and the signal-to-noise ratio must be at least 1000 before useful results can be obtained [51], [52].

In another method, one approaches the unknown ideal image function by successive approximation, using the known nonnegativity of the object function. This, too, seems to have met with some success [53].

APPENDIX 18.1 RELATIONSHIPS BETWEEN LINE AND POINT SPREAD FUNCTIONS AND THE OTF

We summarize here first the relationships between the line spread function, L, and the otf,[12] T. L is assumed normalized so that

$$\int_{-\infty}^{\infty} L(x)\, dx = 1. \tag{18.147}$$

In view of this, we can rewrite (3.31) and its inverse:

$$T(v) = \int_{-\infty}^{\infty} L(x) e^{i2\pi vx}\, dx \tag{18.148}$$

$$L(x) = \int_{-\infty}^{\infty} T(v) e^{-i2\pi vx}\, dv. \tag{18.149}$$

From (18.147) and (18.148):

$$T(0) = 1. \tag{18.150}$$

Since $L(x)$ is real,

$$T(-v) = T^*(v). \tag{18.151}$$

If $L(x)$ is symmetrical,

$$T(v) = \int_{-\infty}^{\infty} L(x) \cos 2\pi vx\, dx \tag{18.152}$$

and hence

$$T(-v) = T(v) \tag{18.153}$$

and

$$L(x) = \int_{-\infty}^{\infty} T(v) \cos 2\pi vx\, dx = 2 \int_{0}^{\infty} T(v) \cos 2\pi vx\, dv. \tag{18.154}$$

---

[12] In Section 3.2 we used the subscript c to distinguish the optical (complex) from the modulation transfer function. Here, however, we are concerned only with the otf, so that the subscript may safely be omitted.

The point spread function $P(x, y)$ may also be normalized:

$$\iint P(x, y) \, dx \, dy = 1. \tag{18.155}$$

Then the line spread function

$$L(x) = \int_{-\infty}^{\infty} P(x, y) \, dy. \tag{18.156}$$

When the point spread function is radially symmetric, as in photographic emulsions and on the axis of an axially symmetric optical system, it can be written as a function of just one variable

$$r = \sqrt{x^2 + y^2} \tag{18.157}$$

and we find [see (3.18) and (3.19)]

$$L(x) = \int_{x}^{\infty} P(r)(r^2 - x^2)^{-1/2} r \, dr \tag{18.158}$$

and

$$P(r) = -\frac{1}{\pi} \frac{d}{dr} \int_{r}^{\infty} \frac{r}{x} L(x)(x^2 - r^2)^{-1/2} \, dx. \tag{18.159}$$

To find the relationship between the symmetrical point spread function and the otf, let us first evaluate the two-dimensional Fourier transform, $\mathscr{F}^{(2)}(\nu_x, \nu_y)$, of a symmetrical function $f(r)$, $r = \sqrt{x^2 + y^2}$. From basic symmetry considerations, we note that this Fourier transform, too, will be symmetrical, a function of

$$\nu = \sqrt{\nu_x^2 + \nu_y^2} \, . \tag{18.160}$$

We may then write, using vector notation $\mathbf{x}$ for (x, y) and $\boldsymbol{\nu}$ for (ν_x, ν_y)

$$F(\nu) = \iint_{-\infty}^{\infty} f(r) e^{i 2\pi \boldsymbol{\nu} \cdot \mathbf{x}} \, d\mathbf{x}. \tag{18.161}$$

Going to polar coordinates (r, θ) for $\mathbf{x}$ and (ν, ϕ) for $\boldsymbol{\nu}$, we note that the differential area element

$$d\mathbf{x} = r \, d\theta \, dr \tag{18.162a}$$

and the scalar product

$$\boldsymbol{\nu} \cdot \mathbf{x} = \nu r \cos(\theta - \phi) \tag{18.162b}$$

so that the transform (18.161) may be written

$$F(\nu) = \int_0^\infty \int_{-\pi}^\pi f(r) e^{i2\pi r\nu \cos(\theta-\phi)} \, r \, d\theta \, dr$$

$$= \int_0^\infty rf(r) \left[\int_{-\pi+\phi}^{\pi+\phi} e^{i2\pi r\nu \cos\theta'} \, d\theta' \right] dr. \qquad (18.163)$$

Now, the integral of $e^{ia\cos\theta}$ over a complete cycle of θ equals $2\pi J_0(a)$ [54]. Hence

$$F(\nu) = 2\pi \int_0^\infty rf(r) J_0(2\pi r\nu) \, dr. \qquad (18.164)$$

Since $J_0(a)$ is symmetrical, we conclude that the inverse is identical in form:

$$f(r) = 2\pi \int_0^\infty \nu F(\nu) J_0(2\pi r\nu) \, d\nu. \qquad (18.165)$$

This transform, with the zero-order Bessel function as a kernel, is known as the Hankel (sometimes Fourier-Bessel) transform [55], [56].

Applying this result to the symmetrical point spread function, we obtain

$$\mathsf{T}(\nu) = 2\pi \int_0^\infty r\mathsf{P}(r) J_0(2\pi r\nu) \, dr \qquad (18.166a)$$

and

$$\mathsf{P}(r) = 2\pi \int_0^\infty \nu\mathsf{T}(\nu) J_0(2\pi r\nu) \, d\nu. \qquad (18.166b)$$

Equations 18.148, 18.149, 18.158, 18.159, and 18.166 permit us to pass from any one of the subject functions to any other, at least for the symmetrical P. The reader will find a listing of many Hankel transform pairs in Ref. 55a and a listing of some transfer and spread function sets in Table 187.

In conclusion we note that a line spread function that is everywhere positive implies that nowhere will the otf exceed unity (its value at the origin). Since the otf is the Fourier transform of the line spread function (L), an otf exceeding its value at the origin implies:

$$\left| \int \mathsf{L}(x) \exp(i2\pi\nu x) \, dx \right| > \left| \int \mathsf{L}(x) \, dx \right|. \qquad (18.167a)$$

The triangle inequality, however, shows that, with an L that is everywhere real and positive, this is impossible:

$$\left| \int L(x) \exp (i2\pi \nu x) \, dx \right| \leqslant \int |L(x) \exp (i2\pi \nu x)| \, dx$$

$$= \int L(x) \, | \exp (i2\pi \nu x)| \, dx$$

$$= \int L(x) \, dx. \qquad (18.167b)$$

We conclude that we must introduce negative regions into a line spread function, if we are to obtain an otf rising away from the origin. We have, indeed, met with such cases in vision (Section 15.4.2.3) and in photography (Section 17.2.2.2).

APPENDIX 18.2   REFERENCE SENSITIVITY OF PHOTOGRAPHIC EMULSIONS [3]

The determination of photographic exposure with radiation whose spectrum differs from that of the standard sources, is an important practical question [57]. Eq. 18.11 gives a simple answer, using the matching factors as given, for instance, in Tables 188. The film sensitivity needed there is given in (18.12), derived below, as a function of the ASA speed.

The arithmetic speed of a photographic emulsion is defined by [58a]

$$k_x = \frac{0.8}{H_m}, \qquad (18.168)$$

where H_m is the luminous exposure (in luxes) required to produce a density of 0.1 units above the fog level, the slope of the characteristic curve being such that $10^{1.3} H_m (\approx 20 H_m)$ is required to produce a density of 0.9 above fog level. Assuming that the fog level is approximately 0.1 density units, we find that the exposure (H_s) of $20 H_m$ brings the film to unity density:

$$H_s(1) = 20 H_m = \frac{16}{k_x}.$$

Denoting the flux at the reference wavelength and that of the standard sensitometric light by subscripts 0 and s respectively, and radiant exposure by an asterisk, we have:

$$\frac{1}{k_0} = H_0^*(1) = \chi_{se} H_s^*(1) = \frac{\chi_{se} H_s(1)}{K_0 \chi_{sv}}$$

$$= \frac{16 \chi_{se}}{K_0 \chi_{sv} k_x}, \qquad (18.169)$$

where χ_{se}, χ_{sv} are the matching factors for the standard light with the emulsion and visual response, respectively, and $K_0 = 680\,\text{lm/W}$ is the peak photopic efficacy. For daylight ratings, the appropriate standard sensitometric light is that defined in Ref. 58b. From Table 188c, the corresponding matching factors are

$$\chi_{sv} = 0.37$$
$$\chi_{se} = 0.297, 0.854, 0.648,$$

respectively, for blue-sensitive, orthochromatic, and panchromatic emulsions. Hence

$$\hat{k}_0 = \frac{K_0 \chi_{sv}}{16\chi_{se}} k_x = 15.7 k_x / \chi_{se}$$
$$= 53 k_x,\ 18.4\,k_x,\ 24.3\,k_x, \tag{18.170}$$

respectively, for these emulsions.
These are the values for use in (18.11).

APPENDIX 18.3   ANALYSIS OF SAMPLE OPTICAL SYSTEM

1   System Description

The system to be analyzed is shown in Figure 18.5. It consists of a dc amplifier whose output controls the electron beam current in a crt. The crt is scanned with a rectangular raster and is photographed. The following are details of the components:

1. Amplifier has gain of 10 and its frequency response is controlled by its output resistor $(R = 500\,\Omega)$ and capacitance $(C = 20\,\text{pFd})$. It has a transfer function of the form

$$\mathsf{T}_A(\nu_t) = [1 + (2\pi\nu_t RC)^2]^{-1/2}. \tag{18.171}$$

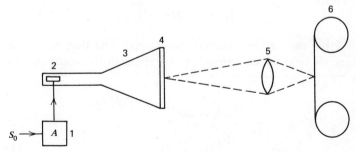

Figure 18.5. Sample compound optical system.

2. Electron-gun grid has gain of $5 \mu A/V$.

3. Crt has accelerating voltage of 5 kV and a P16 phosphor screen. Its efficiency may be assumed to be as given in Table 53. Its mtf is Gaussian with a standard deviation of 7.07 cycles/mm.

4. Raster parameters are 80 mm sweep length, 40 μsec sweep time, and 60 mm raster height. The raster may be assumed to be dense without overlap.

5. Lens has efl of 100 mm and a clear aperture of 20 mm diameter. It images the sweep length on the film at a 16 mm length, without aberrations. It has 10 uncoated air-glass surfaces.

6. Film is EK Tri-X film.

7. Input signal consists of a 200 mV sinusoidal voltage, modulated onto a 400 mV bias. The exposure is such as to yield a unity density on the film for the bias input alone.

2 Task

(a) Find the system gain (absolute and modulation), magnification, and transfer function. Evaluate the latter for a 10 MHz signal. (b) Find the required exposure time. (c) Find the modulation in the final image. (d) Find contributions to the noise in the final aerial image due to (1) Johnson noise at the amplifier, (2) shot noise in the electron beam, (3) photon noise in the aerial image. (e) Find specific noise and signal-to-noise ratio in the final transmittance image, including effects of granularity. (f) How would you effect a significant improvement in this signal-to-noise ratio, with minimum change in system parameters?

3 Solution—Preliminaries

We divide the system into a series of components as listed in the second column of Figure 18.6. The required parameters, not given explicitly, are evaluated below and then entered into the figure. Results for a 10-MHz signal are given in the last two columns and, below, in braces.

1. Amplifier [from (18.171)]:

$$T_1 = \frac{1}{\sqrt{1+[2\pi\nu_0(500)(20\times10^{-12})]^2}}$$
$$= [1+4\pi^2 10^{-16}\nu_0^2]^{-1/2}$$
$$\{ = [1+0.04\pi^2]^{-1/2} = 0.8467\}.$$

2. Electron gun. The gain of the electron gun $(5\times10^{-6} A/V)$ is given. Its transfer function may be taken as unity over the spectral range of interest.

Stage	Magnification		Gain		Mtf	Bandwidth	s'_{dc}	10 MHz	
								ν'	Mtf
	m	M	$g\{\chi\}$	G	T	$\Delta\nu$(MHz)	Gs_0	ν_0/M	T(ν')
0 Object	1	1	1	1		∞	0.4 V	10 MHz	
1 Amplifier	1	1	10 V/V	10	$(1+4\pi^2 10^{-16}\nu_0^2)^{-1/2}$	25	4 V	10 MHz	0.8467
2 Electron gun	1	1	5×10^{-6} A/V	50×10^{-6} A/V	1	∞	20 μA	10 MHz	1
3 Electron beam	1	1	5×10^3 W/A	0.25 W/V	1	∞	0.1 W	10 MHz	1
4 Crt raster	2×10^6 mm/sec	2×10^6 mm/sec	$3.23 \dfrac{\text{W/m}^2\cdot\text{sr}}{\text{W}}$	$0.8075 \left[\dfrac{\text{W/m}^2\cdot\text{sr}}{\text{V}}\right]$	$e^{-2.5\times10^{-15}\nu_0^2}$	12.53	0.323 W/m²·sr	$5\dfrac{\text{cyc}}{\text{mm}}$	0.7788
5 Lens	0.2	4×10^5 mm/sec	$0.0131 \dfrac{\text{W/m}^2}{\text{W/m}^2\cdot\text{sr}}$	$0.01057 \left[\dfrac{\text{W/m}^2}{\text{V}}\right]$	Eq. 18.16 Table 189	47.8	4.228 mW/m²	$25\dfrac{\text{cyc}}{\text{mm}}$	0.9254
6 Film exposure	1	4×10^5 mm/sec	59×10^{-3} J/W $\{0.969\}$	$6\times10^{-4} \left[\dfrac{\text{J/m}^2}{\text{V}}\right]$	$e^{-0.7\times10^{-7}\nu_0}$	7.143	0.155 mJ/m²	$25\dfrac{\text{cyc}}{\text{mm}}$	0.4966
7 Film processing	1	4×10^5 mm/sec			(0.65)		$\tau=1.0$		(0.65)
8 System						5.975			0.3×0.65 ≈0.020

Figure 18.6. Sample system parameters.

3. Electron beam. The gain here is represented by the accelerating voltage $(5 \times 10^3 \text{ V})$ which, in effect, converts amperes into watts. Here, too, the transfer function may be taken as unity over the range of interest.

4. Crt raster

$$m = v_{\text{scan}} = \frac{l_{\text{scan}}}{t_{\text{scan}}} = \frac{80}{40 \times 10^{-6}} = 2 \times 10^6 \text{ mm/sec}$$

$g' = 0.0487$ W/W (from Table 53).

Note: In gain dimensions, the numerator represents the dimensions of the output signal and the denominator that of the input. Clearly, the output of one stage is identical with the input of the next stage. Hence, the input to the lens stage has the dimensions: $\text{W/m}^2 \cdot \text{sr}$. This must be the numerator of the crt raster stage gain. To obtain this from the above g' we must divide by π [assuming a Lambertian screen; see (1.12)] and by the raster area: $A_r = 0.06 \times 0.08 = 48 \times 10^{-4} \text{ m}^2$. Thus

$$g = \frac{g'}{\pi A_r}$$

$$= \frac{0.0487}{\pi 48 \times 10^{-4}} = 3.23 \frac{\text{W/m}^2 \cdot \text{sr}}{\text{W}}$$

$$T(\nu_0) = e^{-\nu_4^2/2\sigma^2} = e^{-\nu_0^2/2M_4^2\sigma^2} = e^{-2.5 \times 10^{-15}\nu_0^2} \qquad (18.172)$$

$\{T(10^7) = e^{-0.25} = 0.7788\}$.

5. Lens:

$$m = \frac{l_{\text{image}}}{l_{\text{scan}}} = \frac{16}{80} = 0.2$$

$$F_e = \frac{f}{D}(1+m) = \frac{100}{20}(1+0.2) = 6. \qquad (18.173)$$

Assuming a 5% reflectance (ρ) at each air-glass surface, lens transmittance:

$$\tau = (1-\rho)^n = 0.95^{10} = 0.6. \qquad (18.174)$$

From (9.134)

$$g = \frac{E}{L} = \frac{\pi\tau}{4F_e^2} = \frac{\pi(0.6)}{4 \times 6^2} = 0.01309. \qquad (18.175)$$

To calculate the mtf, we first find the spatial frequency, ν_5, and the corresponding specific frequency, ν_s:

$$\nu_5 = \frac{\nu_0}{M} = \frac{10^7}{4 \times 10^5} = 25.$$

Hence

$$\nu_s = \nu_5 F_e \{ = 25 \times 6 = 150 \}.$$

From Table 188c: $\{T(150) = 0.9254\}$. [Alternatively, approximating by means of peak wavelength $(0.38 \, \mu\text{m})$ of P16 phosphor, we find the reduced spatial frequency

$$\nu_r = \lambda \nu_s = 0.38 \times 10^{-3} \times 150 = 0.057.$$

Then, from Table 53: $T(0.057) = 0.9275$.]

6. Film. (a) Gain. EK Tri-X film has an ANSI(ASA) speed of 400. We estimate the required exposure from (18.12) with $k_x = 400$ [see (18.9), with $g_3 = h = 1$]:

$$Lt = [\chi \hat{k}_0(1)]^{-1} = [22 k_x \chi_s]^{-1} = [22 \times 400 \times 0.454]^{-1} = 0.25 \text{ mJ/m}^2,$$

where the matching factor is obtained from Table 188c. On the other hand, the illumination at the film plane equals the mean input signal level $(0.4 \, \text{V})$ times the product of the gains of all the preceding stages $(0.01057 \, \text{W/m}^2 \cdot \text{V})$, i.e., $E = 4.228 \times 10^{-3} \, \text{W/m}^2$. Hence the required exposure time is

$$g_6 = H/E = \frac{0.25 \times 10^{-3}}{4.228 \times 10^{-3}} = 0.059 \text{ sec}.$$

The system gain, G_6, up to this stage is obtained by multiplying together:

$$G_6 = G_5 g_6 \chi_6 = 0.01057 \times 0.059 \times 0.969 = 0.6 \times 10^{-3}.$$

(b) MTF. From Section 17.2.2.1 we see that the mtf may be approximated by an exponential, $\exp(-2\pi b \nu_6)$, with $b = 4.45 \times 10^{-3}$ mm. Thus

$$T(\nu_6) = \exp \frac{-2\pi b \nu_0}{M_6} = e^{-7 \times 10^{-8} \nu_0} \qquad (18.176)$$

$$\{ = e^{-0.7} = 0.4966 \}.$$

The granularity figures we take from Figure 17.8:

$$G = 1.6 \, \mu\text{m}.$$

7. Film processing. Converting the virtual exposure to transmittance is

a nonlinear process. In view of the low signal modulation, however, we may approximate the *differential gain* by the gamma of the film [see (18.63)]. Referring to Figure 17.4 defining the ANSI(ASA) film speed, and assuming that the film will in fact be processed according to this, we estimate the gamma

$$\gamma = \frac{0.8}{1.3} = 0.615, \qquad \gamma \approx 0.65.$$

Since this acts on the modulation only, we treat it as an mtf constant at that value.

4 Answers to Questions

We can now answer the original questions.

(*a*) The system magnification (4×10^5 mm/sec) and gain (0.6 mJ/m$^2 \cdot$ V) can be read from the last entries of the M and G columns, respectively, where the gain refers to the linear portion (stages 1–6) of the system only.

The system transfer function is obtained as the product of all the entries in column "Mtf." In view of entry (5) this can not be expressed in analytic form. It can be approximated in such form, however, if we substitute from (18.14) for $T_5(\nu_0)$, with

$$y = \lambda_p F_e \nu_5 = \frac{\lambda_p F_e \nu_0}{M_5} = 0.38 \times 10^{-3} \times 6 \times 0.25 \times 10^{-5} \nu_0$$

$$= 0.57 \times 10^{-8} \nu_0.$$

We then obtain

$$T_T(\nu_0) = (1 + 4\pi 10^{-16} \nu_0^2)^{-1/2} e^{-2.5 \times 10^{-15} \nu_0^2} F(0.57 \times 10^{-8} \nu_0)(0.65),$$

$$(18.177)$$

where $F(y)$ is given by (18.14).

At 10^7 Hz it is given by the product of all entries in the last column (except that of Line 7, representing the nonlinearity):

$$T(10^7 \text{ Hz}) = 0.303 \approx 0.3.$$

(*b*) The required exposure time is the gain $g_6 = 59$ msec.

(*c*) The modulation in the final image is given as the product of the object modulation (0.5), the system mtf (0.3), and the film gamma (0.65). It is 0.0975.

(*d*) Noise power density contributions in the final image may be calculated from the specific noise at every stage multiplied by the signal power at the final image, or from the noise directly, multiplied by the intervening gain. The noise in a resolution element is then obtained by multiplying this with the area (A_e) of the resolution element

$$s_{ni}'^2 = \left(\frac{Gs_{ni}}{G_i}\right)^2 A_e.$$

To this end, we must first determine the noise equivalent bandwidth. This is a complicated procedure. To obtain an estimate, we take the component bandwidths and combine them according to their reciprocal squares. This is correct for Gaussian passbands, but, for lack of a practical alternative, we apply it to all the passbands:

$$\Delta \nu^{-2} \approx \sum \Delta \nu_i^{-2}. \tag{18.178}$$

The bandwidths may be obtained by means of the results in Table 187:

$$\Delta \nu_1 = \frac{1}{4} \times 10^8 = 25 \times 10^6, \qquad \Delta \nu_2 = \Delta \nu_3 = \infty,$$

$$\Delta \nu_4 = \sqrt{\frac{\pi}{8 \times 2.5 \times 10^{-15}}} = 12.53 \times 10^6$$

$$\Delta \nu_5 \approx 0.2724 \nu_c = \frac{0.2724 M_5}{\lambda F_e} = \frac{0.2724 \times 4 \times 10^5}{0.38 \times 10^{-3} \times 6} = 47.79 \times 10^6,$$

$$\Delta \nu_6 = \frac{1}{2 \times 0.7 \times 10^{-7}} = 7.143 \times 10^6$$

$$\Delta \nu_T^{-2} = 2.8 \times 10^{-14}, \qquad \Delta \nu_T = 5.975 \times 10^6.$$

We now estimate the area of the resolution element. From (18.61) with $\alpha = 1$,

$$A_e = \Delta x \, \Delta y = \frac{1}{2 \, \Delta \nu_T} \frac{1}{2 \, \Delta \nu_6} M^2 = \frac{400^2}{(11.95 \times 10^6)(14.29 \times 10^6)}$$

$$= 0.9369 \times 10^{-9}. \tag{18.179}$$

(d1) Johnson noise voltage [see (16.44)]

$$s_{nJ} = V_J = \sqrt{(4kT) \Delta \nu_T r} = (1.6 \times 10^{-20} \times 5.975 \times 10^6 \times 500)^{1/2}$$

$$= 6.914 \times 10^{-6} \text{ V}. \tag{18.180}$$

Hence,

$$s_{nJ}' = \frac{A_e G V_J}{G_1} = \frac{0.9369 \times 10^{-9} \times 0.6 \times 10^{-3} \times 6.914 \times 10^{-6}}{10}$$

$$= 3.89 \times 10^{-19} \text{ J}. \tag{18.181}$$

(d2) Shot noise current in electron beam ($i_B = 20\ \mu$A) [see (16.46)]

$$s_{ns} = i_s = \sqrt{2q_e i_B\ \Delta\nu} = (2 \times 1.602 \times 10^{-19} \times 20 \times 10^{-6} \times 5.975 \times 10^6)^{1/2}$$
$$= 6.188 \times 10^{-9}\ \text{A}.$$

Hence,

$$s'_{ns} = \frac{A_e G i_s}{G_2} = \frac{0.9369 \times 10^{-9} \times 0.378 \times 10^{-3} \times 6.188 \times 10^{-9}}{50 \times 10^{-6}}$$
$$= 6.957 \times 10^{-17}\ \text{J}. \tag{18.182}$$

(d3) Photon noise, H_p in the aerial image, where there are

$$n_p = \frac{HA_e}{hc/\lambda}\ \text{photons/resolution element:}$$

$$H_p = \frac{\sqrt{n_p}\,hc}{\lambda} = \sqrt{\frac{HA_e hc}{\lambda}}, \tag{18.183}$$

where H is the exposure, $H = s_5 = 0.25 \times 10^{-3}\ \text{J/m}^2$, and
hc/λ is the photon energy [see (4.2) with $\omega = 2\pi\nu_t = 2\pi c/\lambda$]
$h = 6.626 \times 10^{-34}\ \text{J sec},\ c = 3 \times 10^8\ \text{m/sec},\ \lambda = 0.38 \times 10^{-6}\ \text{m}.$

Hence $hc/\lambda = 0.5231 \times 10^{-18}\ \text{J}$, and

$$s'_{np} = H_p = (0.25 \times 10^{-3} \times 0.9369 \times 10^{-9} \times 0.5231 \times 10^{-18})^{1/2}$$
$$= 0.35 \times 10^{-15}\ \text{J}.$$

The three noise signals combine as the square root of the sum of their squares, so that at the image plane

$$s'_{nt} = \sqrt{s'^2_{nJ} + s'^2_{ns} + s'^2_{np}} = 0.357 \times 10^{-15}\ \text{J}.$$

The signal there is

$$s' = s_6 A_e = 0.25 \times 10^{-3} \times 0.9369 \times 10^{-9} = 0.234 \times 10^{-12}\ \text{J}.$$

Hence the specific noise of the linear portion of the system is

$$R_L^{-2} = \left(\frac{0.357 \times 10^{-15}}{0.234 \times 10^{-12}}\right)^2 = 2.328 \times 10^{-6}.$$

(e) The specific noise of the system is found as the sum of the specific noise values of the components

$$R_J^{-2} = \left(\frac{V_J}{V_1}\right)^2 = \left(\frac{6.914 \times 10^{-6}}{4}\right)^2 = 2.988 \times 10^{-12}$$

(18.184)

$$R_s^{-2} = \frac{1}{n_2} = \left[\frac{i_B \, \Delta t}{q_e}\right]^{-1} = \frac{2q_e \, \Delta \nu}{i_B}$$

$$= \frac{2 \times 1.602 \times 10^{-19} \times 5.975 \times 10^6}{20 \times 10^{-6}} = 9.572 \times 10^{-8}$$

(18.185)

$$R_p^{-2} = \frac{1}{n_6} = \left[\frac{HA_e}{hc/\lambda}\right]^{-1} = \frac{0.5231 \times 10^{-18}}{0.25 \times 10^{-3} \times 0.9369 \times 10^{-9}}$$

$$= 2.333 \times 10^{-6}.$$

(18.186)

Accordingly, the specific noise of the linear portion of the system equals the sum of these

$$R_L^{-2} = 2.429 \times 10^{-6}, \qquad R_L = 642,$$

in agreement with the result obtained under (d).

From analogy with (18.62), the relative transmittance noise contributed by the linear portion of the system is R_L^{-1} multiplied by gamma:

$$\left(\frac{\Delta \tau}{\bar{\tau}}\right)_L = \frac{\gamma}{R_L} = \frac{0.65}{655} = 0.992 \times 10^{-3}.$$

(18.187)

On the other hand, the fractional fluctuation in transmittance due to granularity is according to (17.30)

$$\left(\frac{\Delta \tau}{\bar{\tau}}\right)_g = \frac{1.628G}{\sqrt{A_e \times 10^{12}}} = \frac{1.628 \times 1.6}{\sqrt{936.9}} = 0.0851,$$

with the factor 10^{12} required to convert the area to $(\mu m)^2$, to be consistent with the usual dimensions of G. This is considerably larger than the contribution of the linear portion of the system, which is therefore negligible. Thus the overall system specific noise is

$$R_s^{-2} = \left(\frac{\Delta \tau}{\bar{\tau}}\right)_g^2 = 7.242 \times 10^{-3} = 0.0851^2.$$

Comparing the transmittance noise modulation (0.0851) with the final signal modulation (0.0975) as found under (c), we find a signal-to-noise ratio

$$R(10 \text{ MHz}) = \frac{0.0975}{0.0851} = 1.15$$

at the signal frequency. Such a low value indicates marginal detectability.

(f) If Panatomix-X film is used instead of the Tri-X film, the value of b in the exponent of the mtf-approximation is 2.6 μm and hence

$$T_6'(10 \text{ MHz}) = e^{-2\pi b\nu_0/M} = 0.6647.$$

This leads to an overall system mtf (including the effect of the gamma):

$$T_6'(10 \text{ MH}_z) = 0.264,$$

and, in view of the 50% input signal modulation, to an output signal modulation of 0.132.

Simultaneously, the Selwyn granularity constant is now $G = 0.85$ and the granularity modulation:

$$\left(\frac{\Delta\tau}{\bar{\tau}}\right)_g = \frac{1.628 \times 0.85}{\sqrt{936.9}} = 0.0452.$$

The signal-to-noise ratio at 10 MHz is now

$$R = \frac{0.132}{0.04} = 2.92.$$

If, in addition, the processing time is increased until the gamma becomes unity, we can obtain a signal-to-noise ratio of 4.5 and this should yield a readily visible image.

APPENDIX 18.4 "NOISE" IN DISCRETE IMAGING SYSTEMS—MEAN VALUE

In our discussion of the discrete imaging system, we chose our origin to coincide with one of the scan lines. When this is not done, the components of the noise term in (18.17) will each be multiplied by a complex factor of unity magnitude. This will not affect the noise power. However, when the relationship between the lattice and the object is changed during the exposure, averaging will take place and, if the change occurs uniformly over a full cycle of the lattice, the noise term will vanish (offering an attractive method for eliminating noise).

To prove this, we generalize slightly our definition of a discrete function:

$$f_d(x) = f(x) \sum \delta(x_0 + j \Delta x - x), \tag{18.188}$$

where x_0 is the displacement of the raster-zero from the object origin and we will eventually average over $0 < x_0 < \Delta x$. From a derivation analogous to that of (18.108), we find that the Fourier transform of f_d is then

$$F_D = \sum_j f(x_0 + j \Delta x) e^{i 2\pi (x_0 + j \Delta x)}$$

$$= \sum_j e^{i 2\pi j x_0 / \Delta x} F\left(\nu + \frac{j}{\Delta x}\right). \tag{18.189}$$

If we again consider $f(x)$ to be the cross correlation of the object signal with the object aperture transmittance, we see that (18.117) now takes the form

$$S''(\nu) = G_1(\nu) G_0^*(\nu) S(\nu) + G_1(\nu) \sum^0 e^{i 2\pi j x_0 / \Delta x} G_0^*\left(\nu + \frac{j}{\Delta x}\right) S\left(\nu + \frac{j}{\Delta x}\right). \tag{18.190}$$

On averaging this over $0 < x_0 < \Delta x$, we note that only the "noise" terms are affected and that there the coefficient of each component vanishes on averaging:

$$\bar{c} = \int_0^{\Delta x} e^{i 2\pi j x_0 / \Delta x} \, dx_0 = \frac{e^{i 2\pi j} - 1}{i 2\pi j / \Delta x} = 0. \tag{18.191}$$

APPENDIX 18.5   THE MATCHED AND LEAST-SQUARED DEVIATION FILTERS

1   The Matched Filter

Denote the irradiance in the scaled object and in the final image by $s_0(x)$, $s_f(x)$ and their amplitude spectra by $S_0(\nu)$, $S_f(\nu)$, respectively, the noise irradiance and its amplitude spectrum by $n(x)$ and $N(\nu)$, the mean Wiener spectra corresponding to s_0 and n by $W_0(\nu)$ and W_n, and the filter function by $T(\nu) = T_F(\nu) \exp[it(\nu)]$, where T_F and t are both real. Note that

$$W_0 = \overline{S_0^2}, \qquad W_n = \overline{N^2}. \tag{18.192}$$

We now seek the filter function maximizing the signal-to-noise ratio at point x_1. To do this we write the signal level as the Fourier transform of the signal spectrum and, assuming the noise to be stationary, the noise-level-squared proportional to the integrated noise spectrum:

$$R = \frac{s_f(x_1)^2}{n(x_1)^2} = \frac{\left| \int \int S_0 T \exp(i 2\pi \nu \cdot x_1) \, d\nu \right|^2}{k \int W_n T_F^2 \, d\nu}. \tag{18.193}$$

We do this for the case where $W_n(v) = W_n$ is constant over the range of T_F. Then the filter function maximizing R will be identical with the one maximizing

$$\frac{\left| \int S_0 T \exp(i2\pi v \cdot x_1) \, dv \right|^2}{\left[\int W_0 \, dv \int T_F^2 \, dv \right]}. \tag{18.194}$$

Now, combining the triangle and Schwarz inequalities:

$$\left| \int f(x)g(x) \, dx \right|^2 \leq \left[\int |f(x)g(x)| \, dx \right]^2 \leq \int |f(x)|^2 \, dx \int |g(x)|^2 \, dx, \tag{18.195}$$

with the left-hand member attaining its maximum value, namely equality with the right-hand member, when $f(x) = cg^*(x)$. Applying this result to (18.194) we find that R is maximum when

$$T = S_0^* \exp(-i2\pi v \cdot x_1) \tag{18.196}$$

and that this is the filter function maximizing the signal to noise ratio at x_1.

2   The Least Squared Deviation Filter

The filter minimizing the squared deviation can be derived readily even for a general noise spectrum. Note that

$$S_f = (S_0 + N)T. \tag{18.197}$$

Therefore the integrated mean squared deviation may be written:

$$E = \int \overline{(s_0 - s_f)^2} \, dx = \int \overline{(S_0 - S_f)^2} \, dv$$
$$= \int \overline{|S_0(1 - T) - NT|^2} \, dv. \tag{18.198}$$

If signal and noise are totally uncorrelated:

$$E = \int (\overline{|S_0|^2} \, |1 - T|^2 + \overline{|N|^2} \, |T|^2) \, dv$$
$$= \int [W_0(1 + T_F^2 - 2T_F \cos t) + W_n T_F^2] \, dv.$$

For minimum E, $\cos t$ must equal unity and

$$E = \int [(W_0 + W_n)T_F^2 - 2W_0 T_F + W_0] \, dv.$$

Completing the square of the terms involving T_F:

$$E = \int \left[\left(\sqrt{W_0 + W_n}\, T_F - \frac{W_0}{\sqrt{W_0 + W_n}} \right)^2 + \left(W_0 - \frac{W_0^2}{W_0 + W_n} \right) \right] d\nu$$

$$= \int \left[\left(\sqrt{W_0 + W_n}\, T_F - \frac{W_0}{\sqrt{W_0 + W_n}} \right)^2 + \frac{W_0 W_n}{W_0 + W_n} \right] d\nu. \qquad (18.199)$$

The term in parentheses, the only part of the integrand depending on T, cannot be negative, therefore E will be a minimum when that term vanishes, yielding for the desired filter function:

$$T_F = \frac{W_0}{W_0 + W_n}. \qquad (18.200)$$

When this condition is met, the mean squared deviation will have its minimum value as given by the last term of the integral of equation (18.199):

$$E_{min} = \int \frac{W_0 W_n}{W_0 + W_n}\, d\nu. \qquad (18.201)$$

In this discussion, W_0 was the object to the Wiener filter; but, simultaneously, it was the image of the original system with otf $T_i(\nu)$. Hence, in terms of the original object Wiener spectrum, W_i,

$$W_0 = W_i\, |T_i|^2. \qquad (18.202)$$

Furthermore, the filter function, T_F, compensates optimally only for the noise addition, as given in (18.197). To compensate for the degradation due to T_i, we must divide this restored image further, by T_i. Hence the corrected Wiener filter function is, with substitution of (18.202) into (18.200)

$$T_{F'} = \frac{T_F}{T_i} = \frac{1}{T_i} \frac{W\,|T_i|^2}{W\,|T_i|^2 + W_n} = \frac{W\,T_i^*}{W\,|T_i|^2 + W_n}, \qquad (18.203)$$

which is identical with (18.145).

REFERENCES

[1] L. Levi, "On combined spatial and temporal characteristics of optical systems," *Opt. Acta* **17**, 869–872 (1970).
[2] E. H. Eberhardt, "Source-detector spectral matching factors," *Appl. Opt.* **7**, 2037–2047 (1968).
[3] L. Levi, "Detector response and perfect-lens-MTF in polychromatic light," *Appl. Opt.* **7**, 607–616 (1969).

[4] Electronic Industries Association, Washington, D. C., (a) JEDEC Publ. No. 50, "Relative spectral response data for photosensitive devices," 1964; (b) JEDEC Publ. No. 16, "Optical characteristics of cathode ray tube screens," 1960.

[5] A. Blonder, Eng. Assoc. Thesis, Jerusalem College of Technology, Jerusalem, 1974.

[6] L. Levi, "On noise analysis of electro-optical systems," *Opt. Commun.* **9**, 325–326 (1973).

[7] W. B. Davenport and W. L. Root, *An Introduction to the Theory of Random Signals and Noise*, McGraw-Hill, New York, 1958; (a) Section 10.2; (b) Chapters 12 and 13; (c) Chapter 11.

[8] H. Cramér, *Mathematical Methods of Statistics*, Princeton University Press, Princeton, 1946, Equation 15.6.2.

[9] O. H. Schade, Sr., "The resolving power functions and quantum processes of television cameras," *RCA Rev.* **28**, 460–535 (1967); (a) Note, p. 460.

[10] M. Menat, "Applied filter radiometry," *Infrared Phys.* **11**, 133–146 (1971).

[11] L. Beiser, "A unified approach to photographic recording from the cathode-ray tube," *Photogr. Sci. Eng.* **7**, 196–204 (1963).

[12] E. Inglestam, "Attempts to treat non-linear imaging devices," *Jap. J. Appl. Phys.* **4**, Suppl. 1, 15–22 (1965).

[13] D. Graham and D. McRuer, *Analysis of Nonlinear Control Systems*, Wiley, New York, 1961.

[14] C. E. K. Mees and T. H. James, *The Theory of the Photographic Process*, 3rd Ed., Macmillan, New York, 1966; (a) e.g., Figures 22.10, 22.11, 22.16–22.18;

[15] R. Roehler, "Möglichkeiten zur Beschreibung der Abbildungseigenschaften einiger nichtlinearer optischer Systeme," *Optik* **22**, 174–187 (1965).

[16] T. G. Stockham, "Image processing in the context of a visual model," *Proc. IEEE* **60**, 828–842 (1972).

[17] A. V. Oppenheim, R. W. Schafer, and T. G. Stockham, "Nonlinear filtering of multiplied and convolved signals," *Proc. IEEE* **56**, 1264–1291 (1968).

[18] G. M. Robbins and T. S. Huang, "Inverse filtering for linear shift-variant imaging systems," *Proc. IEEE* **60**, 862–872 (1972).

[19] A. A. Sawchuk, "Space-variant image motion degradation and restoration," *Proc. IEEE* **60**, 854–861 (1972).

[20] A. Lohmann, "Aktive Kontrastübertragungstheorie," *Opt. Acta* **6**, 319–338 (1959).

[21] L. Levi, "Motion blurring with decaying detector response," *Appl. Opt.* **10**, 38–41 (1971).

[22] A. A. Sawchuk, "Space-variant system analysis of image motion," *J. Opt. Soc. Am.* **63**, 1052–1063 (1973).

[23] E. Inglestam, E. Djurle, and B. Sjögren, "Contrast-transmission functions determined experimentally for asymmetrical images and for the combination of lens and photographic emulsion," *J. Opt. Soc. Am.* **46**, 707–714 (1956).

[23*] W. Schneider and W. Fink, "Integral sampling in optics," *Opt. Acta* **23**, 1011–1028 (1976).

[24] P. Mertz and F. Gray, "A theory of scanning and its relation to the characteristics of the transmitted signal in telephotography and television," *Bell Syst. Tech. J.* **13**, 464–515 (1934).

[25] L. G. Callahan and W. M. Brown, "One- and two-dimensional processing in line scanning systems," *Appl. Opt.* **2**, 401–407 (1963).

[26] L. Levi, "On image evaluation and enhancement," *Opt. Acta* **17**, 59–76 (1970).

[27] F. B. Fellgett and E. H. Linfoot, "On the assessment of optical images," *Phil. Trans. Roy. Soc. (Lond.)* **A247**, 369–407 (1955).

[28] R. C. Jones "Information capacity of radiation detectors, II," *J. Opt. Soc. Am.* **52,** 1193–1200 (1962).

[29] E. H. Linfoot, "Transmission factors in optical design," *J. Opt. Soc. Am.* **46,** 740–752 (1956).

[30] E. H. Linfoot, "Quality evaluations of optical systems," *Opt. Acta* **5,** 1–14 (1958).

[31] J. L. Harris, "Resolving power and decision theory," *J. Opt. Soc. Am.* **54,** 606–611 (1964).

[32] P. G. Roetling, E. A. Trabka, and R. E. Kinzly, "Theoretical prediction of image quality," *J. Opt. Soc. Am.* **58,** 342–346 (1968).

[33] R. E. Kinzly, M. J. Mazurowski, and T. M. Holladay, "Image evaluation and its application to lunar orbiter," *Appl. Opt.* **7,** 1577–1586 (1968).

[34] D. Slepian, "Linear least-squares filtering of distorted images," *J. Opt. Soc. Am.* **57,** 918–922 (1967).

[35] C. W. Helstrom, "Linear restoration of incoherently radiating objects," *J. Opt. Soc. Am.* **62,** 416–423 (1972).

[36] R. S. Macmillan and G. O. Young, "Optimization of information processing optical systems by transcorrelation-function optimization," *J. Opt. Soc. Am.* **58,** 346–356 (1968).

[37] J. C. Dainty, "Detection of images immersed in speckle noise," *Opt. Acta* **18,** 327–339 (1971).

[38] A. van der Lugt, "A review of optical-data processing techniques," *Opt. Acta* **15,** 1–33 (1968).

[39] G. L. Turin, "An introduction to matched filters," *IRE Trans.* **IT–6,** 311–329 (1960).

[40] A. Kozma and D. L. Kelly, "Spatial filtering for detection of signals submerged in noise," *Appl. Opt.* **4,** 387–392 (1965).

[41] J. L. Harris, "Image evaluation and restoration," *J. Opt. Soc. Am.* **56,** 569–574 (1966).

[42] B. L. McGlamery, "Restoration of turbulence-degraded images," *J. Opt. Soc. Am.* **57,** 293–297 (1967).

[43] G. W. Stroke, "Image deblurring and aperture synthesis using à posteriori processing by Fourier-transform holography," *Opt. Acta* **16,** 401–422 (1969).

[44] B. R. Frieden, "Optimum nonlinear filtering of noisy images," *J. Opt. Soc. Am.* **58,** 1272–1275 (1968).

[45] T. S. Huang, W. F. Schreiber, and O. J. Tretiak, "Image processing," *Proc. IEEE* **59,** 1586–1609 (1971), Section IIC.

[46] A. Wolter, "On basic analogies and principal differences between optical and electronic information," in *Progress in Optics,* Vol. 1, E. Wolf, ed., North-Holland, Amsterdam, 1961, Sections 4.5 and 4.6, pp. 155–210.

[47] D. Slepian and H. O. Pollak, "Prolate spheroidal wave functions, Fourier analysis, and uncertainty, I," *Bell Syst. Tech. J.* **40,** 43–63 (1961).

[48] C. W. Barnes, "Object restoration in a diffraction-limited imaging system," *J. Opt. Soc. Am.* **56,** 575–578 (1966).

[49] B. R. Frieden, "Evaluation, design and extrapolation methods for optical signals, based on use of the prolate functions," in *Progress in Optics,* Vol. 9, E. Wolf, ed., North-Holland, Amsterdam, 1971, pp. 311–407.

[50] C. K. Rushforth and R. W. Harris, "Restoration, resolution and noise," *J. Opt. Soc. Am.* **58,** 539–545 (1968).

[51] J. L. Harris, "Diffraction and resolving power," *J. Opt. Soc. Am.* **54,** 931–936 (1964).

[52] J. L. Harris, "Information extraction from diffraction limited imaging," *Pattern Recognition* **2,** 69–77 (1970).

[53] Y. Biraud, "A new approach for increasing the resolving power by data processing," *Astron. Astrophys.* **1,** 124–127 (1969).

[54] H. B. Dwight, *Tables of Integrals and Other Mathematical Data*, 4th Ed., Macmillan, New York, 1961; Equation 866.03

[55] A. Papoulis, *Systems and Transforms with Applications in Optics*, McGraw-Hill, New York, 1968; Chapter 5; (*a*) pp. 144–145.

[56] J. W. Goodman, *Introduction to Fourier Optics*, McGraw-Hill, New York, 1968; pp. 11–13.

[57] J. H. Altman, F. Gram, and C. N. Nelson, "Photographic speeds based on radiometric units," *Photogr. Sci. Eng.* **17,** 513–517 (1973).

[58] Standards of the American National Standards Institute, New York; (*a*) ANSI PH 2.5–1972, "Method for determining speed of photographic negative materials (monochrome, continuous-tone)"; (*b*) USA Std. PH 2.29–1967, "Simulated daylight source for photographic sensitometry."

19

Coherent Optical Systems

19.1 IMAGING IN FULLY AND PARTIALLY COHERENT SYSTEMS

19.1.1 Limits of the Incoherent and Fully Coherent Approximations

When the object radiates incoherently, the optical imaging system is generally linear in intensities (except when the power levels of nonlinear optics are attained); when it radiates in a fully coherent manner, it may be treated as additive in complex amplitude (although here some unexpected phenomena may upset the apple cart, see Sections 19.1.3 and 19.1.4). Problems arise when we try to analyze a system that is partially coherent. In such systems, simple Fourier techniques are not applicable, so that convolution integrals must be evaluated in each case individually. In the next section we present the general formulation for imaging in partially coherent illumination and some basic illustrations.

To decide whether coherent, incoherent, or partially coherent analysis is in place, we must compare the width of the region of coherence [e.g., (2.195)] with the width of the spread function of the imaging system [e.g., (2.166)]. If the region of coherence is sufficiently wide so that coherence is essentially unity throughout the spread function, the system may be considered coherent. In the reverse situation, where the spread function varies but slightly over the (small) region of coherence, the system may be considered incoherent. If the two regions are of comparable size, the system must be treated as partially coherent when examining its performance at the limit of resolution.

19.1.2 Imaging in Partially Coherent Light [1]

General Expression. Quantitatively, the irradiation in the image of a partially coherent imaging system can be shown to have the form

$$E(\mathbf{x}) = K \int\!\!\int \gamma(\mathbf{x}_1' - \mathbf{x}_2')[u(\mathbf{x}_1')\mathbf{p}(\mathbf{x} - \mathbf{x}_1')][u^*(\mathbf{x}_2')\mathbf{p}^*(\mathbf{x} - \mathbf{x}_2')]\, d\mathbf{x}_1'\, d\mathbf{x}_2'. \quad (19.1)$$

Here $\mathbf{p}(\mathbf{x})$ is the "amplitude point spread function," that is,

$$|\mathbf{p}|^2 = \mathbf{P}, \quad (19.2)$$

γ is the complex degree of coherence [see (2.193)], where

$$\gamma_{12} = \gamma(\mathbf{x}_1' - \mathbf{x}_2') \quad (2.193a)$$

assuming spatially stationary coherence, and
$u(\mathbf{x})$ is the complex amplitude of the field at point $\mathbf{x}$, that is

$$|u|^2 = \mathbf{L}, \quad (19.3)$$

the radiance.

This integral may be understood as follows. The integrals over $u(\mathbf{x}_1')\mathbf{p}(\mathbf{x} - \mathbf{x}_1')$ and $u(\mathbf{x}_2')\mathbf{p}(\mathbf{x} - \mathbf{x}_2')$ sum the contributions of all points $\mathbf{x}_1'$, and $\mathbf{x}_2'$ to the amplitude at image point $\mathbf{x}$. If these contributions had a fixed phase relationship, all these contributions would contribute their full weight: the amplitude would be the sum of the contributions (with due regard to phase) and the power density would be obtained by squaring this sum, yielding all the cross-product terms with equal weight. In fact, however, the phase relationship is random, diminishing the weight of the cross-product terms by the factor $\gamma(\mathbf{x}_1' - \mathbf{x}_2')$, which is a measure of the randomness of the phase difference.

When γ is very narrow compared to $\mathbf{p}$, it may be replaced by a δ function and the equation reduced to

$$E(\mathbf{x}) = K \int |u(\mathbf{x}')|^2\, |\mathbf{p}(\mathbf{x} - \mathbf{x}')|^2\, d\mathbf{x}'$$

$$= K \int L(\mathbf{x}')\mathbf{P}(\mathbf{x} - \mathbf{x}')\, d\mathbf{x}', \quad (19.4)$$

the convolution of the object luminance distribution with the (energy) point spread function, equivalent to (3.20), which is the expression appropriate for incoherent imaging.

On the other hand, when $\mathbf{p}$ is very narrow compared to γ, the product $\mathbf{p}(\mathbf{x} - \mathbf{x}_1')\mathbf{p}^*(\mathbf{x} - \mathbf{x}_2')$ vanishes, except when $\mathbf{x}_1' - \mathbf{x}_2'$ is very small, that is, when $\gamma(\mathbf{x}_1' - \mathbf{x}_2')$ is close to unity. In that event, we may drop γ from the integrand of (19.1) to obtain two separable identical integrals and

$$E(\mathbf{x}) = K \left| \int u(\mathbf{x}')p(\mathbf{x} - \mathbf{x}') \, d\mathbf{x}' \right|^2, \qquad (19.5)$$

which is the form appropriate for coherent imaging.

Note that (19.1) contains convolutions so that it becomes tempting to apply Fourier transform. It can be shown that this leads to the image spectrum

$$S(\boldsymbol{\nu}) = \int_{-\infty}^{\infty} U(\boldsymbol{\nu}')U^*(\boldsymbol{\nu} - \boldsymbol{\nu}')$$

$$\times \left[\int \Gamma(\boldsymbol{\nu} - \boldsymbol{\nu}'' - \boldsymbol{\nu}')t(\boldsymbol{\nu} - \boldsymbol{\nu}'')t^*(\boldsymbol{\nu}'') \, d\boldsymbol{\nu}'' \right] d\boldsymbol{\nu}', \quad (19.6)$$

where U is the complex object amplitude spectrum, the Fourier transform of u,

Γ is the Fourier transform of γ, and

t is the "amplitude" transfer function of the imaging system, the Fourier transform of p.

The bracketed expression is called the "transmission cross coefficient." It is analogous to the otf and is occasionally called the "generalized transfer function," although the analogy is hardly close enough to warrant this nomenclature.

To develop a feeling for the significance of (19.1), let us consider some typical examples.

Sinusoidal Objects. When an incoherent sinusoidal object is imaged through a linear optical system, the image, too, will be sinusoidal (see Section 3.2.2). However, if the image is partially or totally coherent, harmonics will appear in the image. If one neglects these, one may speak of an "equivalent mtf," which is defined as the ratio of the image modulation (at the fundamental frequency) to the object modulation. The equivalent mtf is a function of the ratio of the coherence region width to the point spread function width. Figure 19.1 [2] shows results for an object illuminated by a slit-shaped incoherent source (width A) and imaged by an aberrationless system with a slit-shaped aperture (width B), for various values of the ratio $\varepsilon = \alpha/\beta$. Here

$$\alpha = \frac{A}{\lambda a}, \qquad \beta = \frac{B}{\lambda b}, \qquad \varepsilon = \frac{bA}{aB}, \qquad (19.7)$$

where a is the distance from the source to the object and

b is that from the lens to the image.

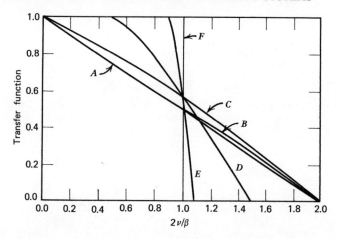

Figure 19.1 Equivalent mtf for slit aperture for various degrees of coherence. The curves correspond to the following values of ε. (A) $\varepsilon \gg 1$ (totally incoherent illumination; (B) $\varepsilon = 2$; (C) $\varepsilon = 1$; (D) $\varepsilon = 0.5$; (E) $\varepsilon = 0.1$; (F) $\varepsilon \ll 1$ (totally coherent illumination). Swing and Clay [2].

Note that ε goes from zero to infinity as the illumination goes from coherent to incoherent. The equivalent mtf is also affected by the value of the object modulation, but to a much lesser extent.

Knife-Edge Image. When an incoherently illuminated sharp transition from dark to light is imaged by an aberrationless optical system, the transition will appear somewhat smoothed in the image. When imaged under coherent illumination, the transition will be sharper, but spurious dark and light fringes will appear in the neighborhood of the transition; like the analogous phenomenon in acoustical systems, this is called "ringing." Results both experimental (coherent and incoherent) and theoretical (for various degrees of coherence) are shown in Figures 19.2 [6] and 19.3 [3], respectively. The latter refer to a circular source with various values of the incoherence parameter ε; see (19.7).

Two-Point Object. When the object in (19.1) consists of two δ functions located at b and $-b$, respectively, and the point spread function is that of an aberrationless ("diffraction-limited") imaging system with circular aperture, radius a, the resulting image can be shown to have the form

$$E(x) = 4a^2 L_0 \{ V^2[K(x-b)] + V^2[K(x+b)]$$
$$+ 2\gamma_{12} V[K(x-b)] V[K(x+b)] \}, \quad (19.8)$$

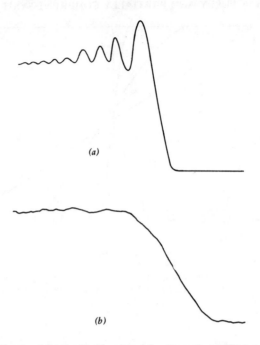

Figure 19.2 Diffraction images of knife-edge: (*a*) coherently and (*b*) incoherently imaged. Considine [6].

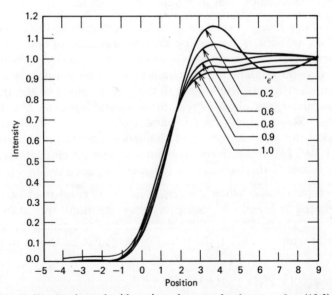

Figure 19.3 Knife-edge imaged with various degrees of coherence. See (19.3) for the definition of the incoherence parameter, ε. (N.B. incoherence increases with ε). Kinzley [3].

where we have used the abbreviated notations [see (2.165)]

$$V(y) = \frac{2J_1(y)}{y},$$

and $K = 2\pi a/\lambda f$. Plots of this equation for values of γ_{12} between zero and unity are shown in Figure 19.4a–c [4], where the curve with the highest center value corresponds to unity γ_{12} and the one with the deepest dip to zero γ_{12}. The three figures correspond to various point separations: $\delta = 3.2$, 4, 4.8, respectively, where

$$\delta = \frac{4\pi ab}{\lambda f} \tag{19.9}$$

is a dimensionless measure of the separation, $2b$.

Here the most striking phenomenon is the loss of resolution with increasing coherence. Whereas points spaced $\delta = 3.2$ are resolvable when incoherently illuminated, they must be separated by $\delta = 4.8$ if they are to be resolved under coherent illumination. We also note that increasing coherence shifts the point images toward each other (assuming that the point image appears to be located at the illumination peak). This is especially noticeable in Figure 19.4c.

19.1.3 Speckle Effect [5]

The Phenomenon. When a diffusing surface is illuminated by light that is coherent, at least in part, this gives rise to the *speckle effect*: when viewed from a distance, the surface luminance appears granular, even if the surface is uniformly illuminated, and, when the scattered light falls onto a distant surface, the illumination there, too, is granular in appearance. This "speckling" is due to interference effects and may very seriously degrade imaging with highly coherent light; the random luminance fluctuations superimposed on the image attain high modulation values.

To eliminate this effect, the coherence of the light must be destroyed. This cannot be done simply by introducing a scattering medium, such as a ground glass; the nonuniformities would have to vary the pathlength by an amount equal to the coherence length of the light. However, there are a number of techniques available to accomplish this [5a]. For instance the scattering medium may be moved so that the integration of the speckle pattern over the time constant of the detector (in visual observation—the integration time of the visual system) will reduce the modulation of the granularity to the desired level. Rotating ground glass screens and dilute aqueous solutions of milk have been used for this purpose [6]. In the

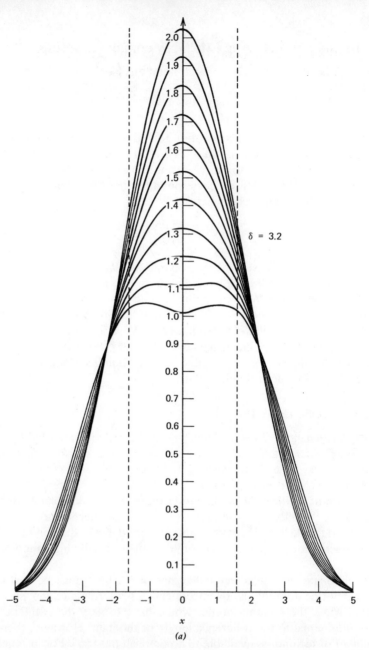

$\delta = 3.2$

x

(a)

Figure 19.4 Image of two point sources with various degrees of coherence. δ is a measure of the separation between points; see (19.9) for its definition. Curves are for various values of γ (0.0 to 1.0 in steps of 0.1). The abscissae (x) are measured in units of $\lambda F_e/2\pi$, where λ is the average wavelength and F_e is the effective F/number. The three figures are for the image point separation values, δ, indicated. δ is measured in the same units as x. The image point locations are indicated by dotted lines. Grimes & Thompson [4].

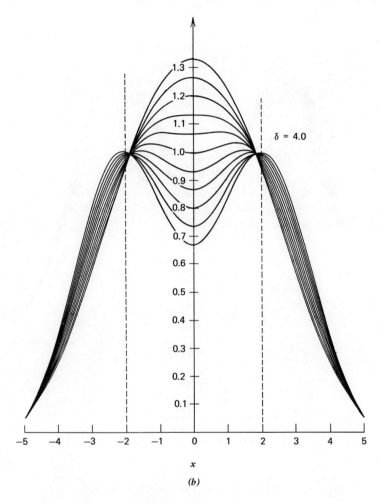

δ = 4.0

x

(b)

Figure 19.4 (*Continued*)

latter, the Brownian motion of the scattering particles sufficed to destroy the coherence of the scattered light. The moving ground glass screen can be made far more effective by combining it with a stationary one [7].

Statistics [5b,c], [8]. The detailed distribution of flux in coherent light scattered from a randomly scattering diffusing screen is clearly unpredictable. However, its second-order statistics may be calculated quite accurately.

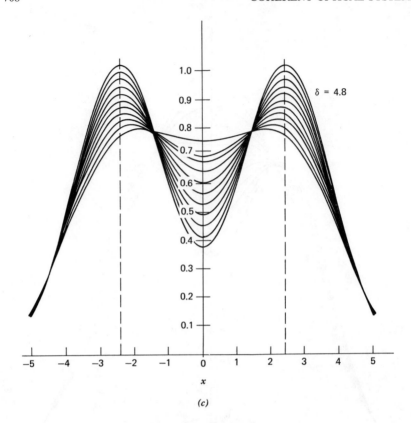

(c)

Figure 19.4 (Continued)

Let us first determine the illumination in the far field. Let

$$L(x, y) = k\rho(x, y)E_0(x, y) = |U(x, y)|^2 \qquad (19.10)$$

be the luminance at Point (x, y) on the diffusing screen.

Here $E_0(x, y)$ is the incident illumination,

  $\rho(x, y)$ is the reflectance,

    k is a directionality factor, including the diffusivity of the screen, assumed constant over the solid angle considered ($k = 1/\pi$ for a Lambertian reflector), and

  $U(x, y)$ is the complex amplitude of the field radiated into this solid angle.

We now divide the diffusing screen into very small equal area elements ($\Delta x \, \Delta y$) by means of a rectangular grid. The grid elements must be large enough so that the mean phase of the electromagnetic field reflected from one element is independent of that of radiation reflected from a neighboring element. We then account for the roughness of the screen by assuming that the typical element, i, introduces a random phase shift, φ_i, where $(\varphi_i, \mathrm{mod}\, 2\pi)$ is uniformly distributed over $0 < \varphi_i < 2\pi$.

We now determine the field at a point (X, Y) in the plane, P, at a distance z from the screen (see Figure 19.5). We limit ourselves to small angles so that Huygens' principle may be applied without a directionality factor. We then obtain the complex field at point (X, Y) by summing the contributions from all the screen elements:

$$V(X, Y) = e^{ikr_0} \sum_{i=1}^{n} \frac{U_i}{r_i} e^{i\varphi_i} e^{ikd_i} \sqrt{\Delta x\, \Delta y}, \qquad (19.11)$$

where $U_i = U(x_i, y_i)$,

(x_i, y_i) are the coordinates of the center of the ith element in the plane of the diffusing screen,

r_i is the distance from (x_i, y_i) to (X, Y)

r_0 is the distance from the origin in the diffusing screen plane to point (X, Y) and

$d_i = r_i - r_0$ is the pathlength difference for point (x_i, y_i).

To account for the square root sign over the element area, note that it is $U^2 \sim L$, which is proportional to $\Delta x\, \Delta y$.

Since we limit ourselves to distances

$$x_i, y_i, X, Y \ll z,$$

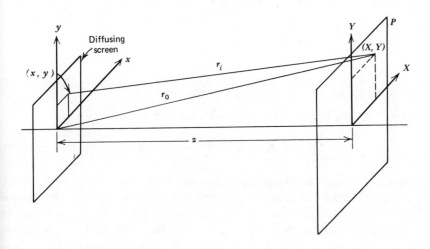

Figure 19.5 Geometry of speckle genesis.

we may approximate d_i as follows

$$d_i = r_i - r_0 = \sqrt{(x_i - X)^2 + (y_i - Y)^2 + z^2} - \sqrt{X^2 + Y^2 + z^2}$$

$$= z\left[\sqrt{1 - 2\frac{x_i X + y_i Y}{z^2} + \frac{x_i^2 + y_i^2}{z^2} + \frac{X^2 + Y^2}{z^2}} - \sqrt{1 + \frac{X^2 + Y^2}{z^2}} \right]$$

$$\approx -\frac{x_i X + y_i Y}{z} + \frac{x_i^2 + y_i^2}{2z}. \tag{19.12}$$

On substituting this into (19.11), lumping the last term in d_i with the random phase φ_i, we obtain

$$V(X, Y) = e^{ikr_0} \sum_i \frac{U_i}{r_i} e^{i\varphi_i'} e^{-ik(x_i X + y_i Y)/z} \sqrt{\Delta x\, \Delta y}, \tag{19.13}$$

where we have written

$$\varphi_i' = \varphi_i + \frac{k(x_i^2 + y_i^2)}{2z}. \tag{19.14}$$

Note that $\exp i\varphi_i'$ is uniformly distributed around the unity circle, just as is $\exp i\varphi_i$.

To obtain the illumination at (X, Y) we write

$$E(X, Y) = V(X, Y)V^*(X, Y) = \left[\sum_i \frac{U_i}{r_i} e^{i\varphi_i'} e^{-ik(x_i X + y_i Y)/z} \right]$$

$$\left[\sum_i \frac{U_i^*}{r_i} e^{-i\varphi_i'} e^{ik(x_i X + y_i Y)/z} \right] \Delta x\, \Delta y$$

$$= \sum_i \sum_j \frac{U_i U_j^*}{r_i r_j} e^{i(\varphi_i' - \varphi_j')} \exp\{ik[(x_j - x_i)X + (y_j - y_i)Y]/z\} \Delta x\, \Delta y. \tag{19.15}$$

Note that the portion of the double sum, for which $i = j$, yields

$$\bar{E}(X, Y) = \sum_i \frac{|U_i|^2}{r_i^2} \Delta x\, \Delta y. \tag{19.16}$$

This is precisely the illumination that would have been obtained from the same screen and illumination had the flux been incoherent. The portion including $i \neq j$ represents the fluctuations, $\tilde{E}$, in $E(X, Y)$, and may be analyzed as follows. We simplify the presentation by treating a uniformly illuminated screen ($U_i = U_j = U$) and neglect the variations in $r_i (r_i = r_j = r)$, which are relatively small since we are limiting our analysis to small angles. If we now combine terms in pairs—(i, j) with (j, i)—we obtain a sum of $n(n-1)/2$ terms, each of the form

$$2\frac{|U|^2}{r^2}\cos\theta_i.$$

In this sum θ_i is a random variable uniformly distributed between 0 and 2π. We may therefore write the sum

$$\tilde{E} = 2\left(\frac{|U|}{r}\right)^2 \sum_1^{2\,n(n-1)/2} \cos\theta_i \Delta x\,\Delta y. \tag{19.17}$$

Its mean value vanishes with $\overline{\cos\theta_i}$. Its variance, however, is [see (19.292)],

$$\overline{\tilde{E}^2} = \left(\frac{|U|}{r}\right)^4 (n^2 - n)\Delta x\,\Delta y. \tag{19.18}$$

On the other hand, under the same assumptions concerning U_i and r_i,

$$\bar{E}^2 = n^2\left(\frac{|U|}{r}\right)^4 \Delta x\,\Delta y \approx \tilde{E}^2, \qquad n \gg 1. \tag{19.19}$$

Thus the specific variance of the speckling is close to unity, when the light is fully coherent.

To obtain the second-order statistics of the spatial fluctuations, we find the autocorrelation of $E(X, Y)$:

$$C(\xi, \eta) = \iint E(X, Y)E^*(X-\xi, Y-\eta)\,dX\,dY$$

$$= \iint \left[\sum_i \sum_j \frac{U_i U_j^*}{r_i r_j} e^{i(\varphi_i' - \varphi_i')} \exp - ik\frac{X(x_i - x_j) + Y(y_i - y_j)}{z}\right]$$

$$\times \left[\sum_k \sum_l \frac{U_k^* U_l}{r_k' r_l'} r^{-i(\varphi_k' - \varphi_l')} \exp ik\frac{(X-\xi)(x_k - x_l) + (Y-\eta)(y_k - y_l)}{z}\right]$$

$$\times \Delta x^2\,\Delta y^2\,dX\,dY$$

$$= \iint \sum_i \sum_j \frac{|U_i|^2\,|U_j|^2}{r_i r_i' r_j r_j'} \exp\left[ik\frac{\xi(x_i - x_j) + \eta(y_i - y_j)}{z}\right]\Delta x^2\,\Delta y^2\,dX\,dY$$

$$+ \iint \sum_i \sum_j \sum_k \sum_l \cdots dX\,dY, \tag{19.20}$$

where we have collected all the quadratic terms in the first integral and denoted by prime marks distances measured to $(X-\xi, Y-\eta)$.

Note that the sum under the second double integral consists of $n^2(n^2-1)/2$ terms, each of magnitude of about $(|U|/r)^4$ and multiplied by a factor $\cos\theta_i$, with θ_i again randomly distributed between 0 and 2π. This sum, therefore, has a vanishing mean value. Also note that the first integrand is independent of the variables of integration, so that it may be removed from under the integral signs. Thus we may write for the mean value of the autocorrelation

$$\bar{C}(\xi, \eta) = (\Delta x\,\Delta y)^2 A \sum_i \sum_j \frac{|U_i|^2\,|U_j|^2}{r_i r_i' r_j r_j'} \exp ik\frac{\xi(x_i - x_j) + \eta(y_i - y_j)}{z}, \quad (19.21)$$

where we have written

$$A = \iint dX\,dY$$

for the area of the observation plane under consideration. The terms in (19.21) may be factored into two sums, one of which is the complex conjugate of the other. We may thus rewrite it in terms of a single sum:

$$C(\xi, \eta) = (\Delta x\,\Delta y)^2 A \left|\sum_i \frac{|U_i|^2}{r_i r_i'} e^{ik(\xi x_i + \eta y_i)/z}\right|^2$$

$$\approx \left(\frac{A}{z^2}\right) \left|\iint U^2(x, y)e^{ik(\xi x + \eta y)/z}\,dx\,dy\right|^2$$

$$= \frac{A}{z^2} \left|\iint L(x, y)e^{ik(\xi x + \eta y)/z}\,dx\,dy\right|^2, \quad (19.22)$$

where we have approximated the $r_i \approx z$ and the sum by an integral over the luminous region of the screen.

Note that the autocorrelation of the speckle pattern equals, within a constant factor, the absolute-value-squared of the Fourier transform of the screen luminance, with the transform variable (ν) replaced by the spatial coordinates (ξ, η) according to

$$\nu_x = \frac{\xi}{\lambda z}, \qquad \nu_y = \frac{\eta}{\lambda z}.$$

This gives a measure of the size of the speckle patches, which are then of the order of magnitude of $(z\lambda/a)$, where a is the width of the luminous region.

Alternatively we conclude that the spatial Wiener spectrum of the speckle pattern equals the autocorrelation of the diffusing screen luminance. We arrive at this conclusion on the basis of the Wiener-Khintchine theorem (Appendix 3.3) and the theorem according to which the Fourier transform turns the product of two functions into the correlation of their Fourier transforms [see (9.92) which follows a mathematically identical procedure].

When the diffusing screen is viewed, or otherwise imaged, at a distance z, its luminance also appears speckled. This can be accounted for by the variation in interference effects within light reflected from individual area patches. These patches cover regions whose light is summed coherently by the imaging system; that is, they correspond to the amplitude spread function of the imaging system. Thus the result (19.22) is valid if we multiply the luminance function in the integrand by $P(x, y)$, the point spread function of the eye or other imaging system.

Conclusions. From (19.19) it is evident that speckling may introduce a very serious disturbance into the imaging system. The speckle effect essentially superimposes a noise on the signal—with a signal-to-noise ratio of unity! That the signal is discernible in the noise, nevertheless, is primarily due to the fact that the noise level varies along with the signal level. On the other hand, speckling can be exploited for metrological and related purposes [5d–f].

19.1.4 Other Noiselike Effects and Overall Performance

When the object in a coherent imaging system is clear or specular, other disturbances become noticeable. For instance, any small nonuniformity, such as a speck of dust, somewhere in the illuminated region, may cause a conspicuous diffraction pattern. It generates a spherical wavefront that will interfere with the single wavefront of the illumination. (In an incoherently illuminated system, each point on the source is responsible for a separate wavefront, causing its own diffraction pattern; the multitude of different diffraction patterns, when superimposed, almost completely cancel out any net effect.) Also, if the illumination is spectrally quite pure (high temporal coherence) any transparent plate or film will tend to set up interference fringes that may, again, be very pronounced.

If we compare the performance of coherently illuminated systems with that of incoherently illuminated ones in terms of overall performance, we notice several disadvantages in the coherent system.

1. The cutoff spatial frequency is lower and resolving power is poorer in the coherent system.

2. The coherent system introduces spurious edge effects (ringing), which may be more objectionable than the mere blurring caused by incoherent systems.

3. The coherent system is far more prone to spurious "noiselike"

disturbances, which are analogous to "ghosts" and "haze" in incoherent systems, but are here far more pronounced and often more finely structured. An especially serious form of this appears in the speckling effect, which plagues any system having a diffuse component in the beam before the final image.

The advantages of the coherent optical systems are in that they permit far more freedom in processing the image, such as filtering, correlation procedures and holography. Highly coherent light sources (lasers) are also important because of the extremely high intensities attainable at relatively low power (see Table 54 and Section 7.4).

In imaging they may provide some advantage in the reduced blurring at an edge and the larger transfer function values in the lower half of the spatial frequency spectrum.

Hence, in imaging systems, the user will generally attempt to have his illumination as incoherent as possible, except if he requires the advantages of the coherent systems just listed. Here some caution is necessary as the following illustrations indicate.

The reader may recall that two types of illumination are used in microscopy: (a) Köhler illumination, in which the light source is imaged in the entrance pupil of the objective lens, and (b) "critical" illumination, in which it is imaged on the object. If the source is incoherent, we might be tempted to assume that the latter technique yields truly incoherent illumination. This is, however, not true. Since the wave from each source point is distributed over an object region of the size of the condenser spread function, the source image is partially coherent. Indeed, it can be shown that the coherence values obtained in the two types of illumination are identical [9a].

In photometric instrumentation dealing with high-resolution optics (e.g., microdensitometers), the effects of phase differences can distort measurements. These effects can be minimized by illuminating no more than the object field to be measured and using maximum numerical aperture [10].

19.1.5 Gaussian Cross-Sectional Beams

19.1.5.1 Description of the Gaussian Beam **[11].** In Section 2.1.2.1 we saw that a plane wave of infinite extent propagates parallel to itself and does not change its shape as it propagates. Real waves are, however, not infinite, and the limitation on their extent causes diffraction and, hence, changes in their form as they advance. Often, especially in confocal resonant cavities as used in lasers, the amplitude (U) drops off in a

Gaussian manner away from the axis of symmetry:

$$U = U_0 e^{-(r/w)^2}, \tag{19.23}$$

where r is the distance from the axis of symmetry and w, called the spot radius, is the distance from the axis at which the amplitude has dropped to $1/e$ from its peak value.

When a plane wave of this cross section propagates in a homogeneous and isotropic medium, it acquires curvature (radius, R) and expands (w increases). Its shape is given by

$$U(z, r) = U_0(w_0/w) \exp - \left[i(kz + \phi) + r^2 \left(\frac{1}{w^2} + \frac{ik}{2R} \right) \right], \tag{19.24}$$

where z is the coordinate normal to the wave front at its center, measured from the point where the wave front is plane. See Figure 19.6. The z axis coincides with the axis of symmetry,

w_0 is the spot radius at $z = 0$,

ϕ is a phase shift,

$k = 2\pi/\lambda$, is the magnitude of the wave vector (2.32).

ϕ, w, and R are given by

$$\tan \phi = \frac{z}{c} \tag{19.25}$$

$$w^2 = w_0^2 \left[1 + \left(\frac{z}{c} \right)^2 \right] \tag{19.26}$$

$$R = z \left[1 + \left(\frac{c}{z} \right)^2 \right], \tag{19.27}$$

where

$$c = \frac{\pi w_0^2}{\lambda}.$$

R is positive when the z coordinate, z_0, of the center of curvature is less than z. On solving (19.26)–(19.27) for w_0 and z, we find

$$w_0^2 = \frac{(wR\lambda)^2}{\pi^2 w^4 + R^2 \lambda^2} \tag{19.28}$$

and

$$z = \frac{\pi^2 w^4 R}{\pi^2 w^4 + R^2 \lambda^2}. \tag{19.29}$$

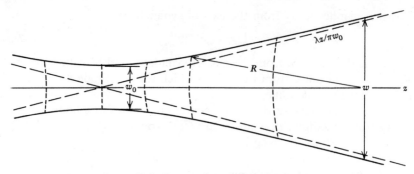

Figure 19.6 Propagation of Gaussian beam.

Equation 19.24 can be derived from Maxwell's equations [12], the scalar wave equation [11], or from Huygens' principle [13]. It shows that for such a wave, the wavefront is always concave toward the origin and the spot radius increases monotonically as the wave propagates away from the origin. The wave envelope exhibits a neck at the origin and the wave front is plane there. The envelope is a hyperbola with its focus at $r = \sqrt{w_0^2 + c^2}$. At large distances from the origin, the spot radius asymptotically approaches the value

$$w \simeq \frac{\lambda z}{\pi w_0},\tag{19.30}$$

which represents the Fraunhofer diffraction pattern of a Gaussian transmittance aperture, analogous to (9.109) (see Figure 19.6).

19.1.5.2 Effect of Lenses [11], [12]. To determine the effect of a lens—or other optical component—on the wave, we introduce the *complex radius, q*, where

$$\frac{1}{q} = \frac{1}{R} - \frac{i2}{kw^2}.\tag{19.31}$$

In terms of this complex radius (19.24) becomes

$$U(z, r) \sim \left(\frac{w_0}{w}\right) \exp\left[-ik\left(z + \frac{r^2}{2q}\right)\right].\tag{19.24a}$$

By substituting (19.26) and (19.27) into (19.31), the reader will readily confirm that

$$q(z) = q(0) + z = \tfrac{1}{2}ikw_0^2 + z = ic + z,\tag{19.32}$$

that is, the real part of q equals z.

If a Gaussian beam passes through a lens of focal length, f, and coaxial with it, the resulting transformation of the beam may be treated exactly as that experienced by a spherical wave emanating from an axial object point a distance, say a, in front of the lens. Its radius, b, on leaving the lens is given by (9.28a). Here we must merely replace a and b by q and q', respectively. Thus

$$\frac{1}{q'} = \frac{1}{q} - \frac{1}{f} = p_r + ip_i, \tag{19.33}$$

where we have written $p_{r,i}$ for the real and imaginary parts of $(1/q')$. Comparison with (19.31) shows

$$R' = \frac{1}{p_r} \tag{19.34}$$

and

$$w'^2 = \frac{2}{kp_i} \tag{19.35}$$

will be curvature and spot radii for the transmitted wave. By means of (19.28) and (19.29), we can find from these w_0, the spot radius at the new neck and its relative location $(-z')$ relative to the lens.

By way of illustration, if a positive thin lens of focal length f is placed at the neck of a Gaussian beam, the radius of curvature and spot radius on exiting the lens will be

$$R' = -f \tag{19.36}$$

$$w' = w_0. \tag{19.37}$$

The neck will be found at a distance

$$z' = \frac{f}{1 + (f\lambda/\pi w_0^2)^2} \tag{19.38}$$

and will have a spot radius

$$w_0' = \frac{w_0 \lambda f}{\sqrt{\pi^2 w_0^4 + \lambda^2 f^2}}. \tag{19.39}$$

Other schemes for analyzing effects of optical components on a Gaussian beam have been developed [11], [14].

19.1.5.3 Beam Confinement.

Occasionally we wish to confine a Gaussian beam to a certain spot radius, r_{mx}, over long distances. This can be accomplished by confining fibers (see Chapter 11) or by a "distributed lens," that is, a medium whose refractive index drops away from the axis in an axially symmetrical manner. (In terms of geometrical optics, all rays propagating away from the axis will be bent inward toward the axis.)

A third approach uses periodically spaced lenses to reconcentrate the diverging Gaussian beam whenever it has expanded to $w = r_{mx}$. Equations 19.38 and 19.39 enable us to calculate the required focal length and spacing, D (see Figure 19.7). Noting that for maximum spacing the spot radius at each lens must equal r_{mx} and that, due to the symmetry of the envelope (19.26), the neck must be at the midpoint between the lenses, we conclude that

$$D = -2z'_{mx},$$

with z'_{mx} obtained from maximizing z' as given by (19.38), with f as the independent variable. This readily leads to

$$f = \frac{\pi w_0^2}{\lambda} \tag{19.40}$$

and also,

$$D = -2z' = \frac{\pi w_0^2}{\lambda}. \tag{19.41}$$

Since in our system the lens is not placed at the neck, but at a distance z' from it, the power of the lens must be double the amount given by (19.40) and, hence,

$$f = \frac{\pi w_0^2}{2\lambda}. \tag{19.42}$$

As a numerical example assume that we wish to confine the beam of a helium-neon laser to a spot radius of 1 mm. To this end we must space the lenses no more than 5 m apart. On the other hand, if a 10 cm spot radius suffices, the spacing requirement becomes 50 km.

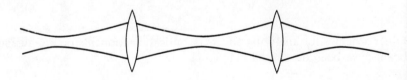

Figure 19.7 Confinement of Gaussian beam by lenses.

19.2 COHERENT SPATIAL FILTERING [15]–[18] AND IMAGE PROCESSING

Many operations performed on optical images are most simply expressed in terms of their spatial frequency spectrum. We have already seen (Sections 3.2, 9.3, 9.4, 18.1–18.2) that the modification of the spatial frequency spectrum is a major factor in the imaging process; hence, compensating for such modifications, too, may best be viewed in these terms. Even the improvement of signal-to-noise ratio can often be treated most conveniently in the spatial frequency domain.

Filtering of the temporal frequency spectrum is quite widespread in electronic systems, and it would be desirable to have a similar process available for optical images and their spatial spectra. Such a technique is indeed available and is the subject of the present section.

19.2.1 The Fourier Transform Plane

Coherent spatial filtering is based on the fact than an aberrationless lens can be used to obtain the Fourier transform of a transparent object. This is closely related to the fact that the amplitude distribution in the Fraunhofer diffraction pattern is similar to the Fourier transform of the complex transmittance of that aperture; see Section 2.3.1.2. Here we derive the conditions when a lens is used.

Let us first express the action of an aberrationless lens in wave-optical terms: The lens converts an incident plane wavefront into a spherical one, with radius f, where f is the effective focal length of the lens. Referring to Figure 19.8, we can see that this is equivalent to the introduction of an optical pathlength difference.

$$d = \sqrt{r^2 + f^2} - f = f\left[\frac{1}{2}\frac{r^2}{f^2} - \frac{1}{8}\frac{r^4}{f^4} + \cdots\right]. \qquad (19.43)$$

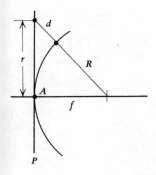

Figure 19.8 Illustrating action of perfect lens on plane wavefront. See text.

This, in turn, corresponds to a phase shift

$$\Delta\phi = -kd = -\frac{1}{2}kf\left[\frac{r^2}{f^2}-\frac{r^4}{4f^4}+\cdots\right]\approx\frac{-kr^2}{2f}, \qquad r\ll\sqrt[4]{\lambda f^3}. \quad (19.44)$$

Here r is the distance of the wavefront point from the axis,
  λ is the wavelength of the radiation involved, and
  $k = 2\pi/\lambda$ is the wavevector magnitude.

Thus the presence of an aberrationless lens, focal length, f, modifies the field distribution in its plane by a factor

$$e^{-ikr^2/2f}. \qquad (19.45)$$

In the remainder of this section we demonstrate that such a lens is capable of forming the Fourier transform of a wave amplitude distribution, and determine the conditions under which it does this.

Consider an object wavefront distribution, $\mathbf{u}(x, y)$ in the x, y plane, an aberrationless thin[1] lens in plane (X, Y) at a distance a from the former, and a "diffraction plane," x', y' at a distance b beyond the lens whose focal length is f. See Figure 19.9, where we have denoted by s the distance from any point, $P(x, y)$, in the object plane to any point (X, Y) in the lens plane, and similarly, by s' the analogous distance from point (X, Y) to any point, $P'(x', y')$ in the diffraction plane.

By applying Huygens' principle to the wavefront distribution $\mathbf{u}(x, y)$, we obtain the amplitude distribution at the plane of the lens:

$$\mathbf{v}(X, Y) = C\int\!\!\!\int_{-\infty}^{\infty}\mathbf{u}(x, y)e^{iks}\,dx\,dy$$

$$\approx Ce^{ika}\int\!\!\!\int\mathbf{u}(x, y)e^{i(k/2a)[(x-X)^2+(y-Y)^2]}\,dx\,dy. \quad (19.46)$$

Here we have used the approximation:

$$s = [a^2+(x-X)^2+(y-Y)^2]^{1/2}$$

$$= a+\frac{1}{2a}[(x-X)^2+(y-Y)^2]-\frac{1}{8a^3}[(x-X)^2+(y-Y)^2]^2+\cdots$$

$$\approx a+\frac{1}{2a}[(x-X)^2+(y-Y)^2], \qquad [(x-X)^2+(y-Y)^2]\ll\sqrt{\lambda a^3}.$$

$$(19.47)$$

[1] We have specified the lens as thin simply as a matter of convenience. The effect is identical if the lens is thick or compound, except that we must then refer to the location of its principal planes, wherever we now refer to its own location.

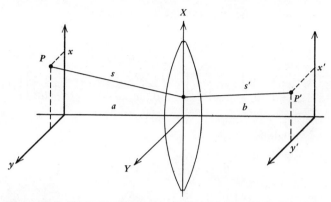

Figure 19.9 Illustrating illumination in Fourier transform plane.

The lens now introduces a phase shift as given by (19.45) so that the wavefront leaving it has the form:

$$\mathbf{v}'(X, Y) = Ce^{ika}e^{-ik(X^2+Y^2)/2f} \int\limits_{-\infty}^{\infty}\!\!\int \mathbf{u}(x, y)e^{ik[(x-X)^2+(y-Y)^2]/2a} \, dx \, dy,$$

$$X^2 + Y^2 < R^2,$$

$$= 0, \qquad X^2 + Y^2 > R^2, \tag{19.48}$$

where R is the radius of the lens aperture. In the following, we assume that the lens aperture is made sufficiently large so that the wavefront energy $|\mathbf{v}(X, Y)|^2$ is negligible for $(X^2 + Y^2 > R^2)$. This permits us to drop the restrictions on the first expression (19.48) without introducing a significant error.

Applying Huygens' principle to $\mathbf{v}'$, we find then the amplitude distribution $\mathbf{u}$ in the diffraction plane.

$$\mathbf{u}(x', y') = \int\limits_{-\infty}^{\infty}\!\!\int \mathbf{v}'(X, Y)e^{iks'} \, dX \, dY,$$

where, in analogy with (19.47),

$$s' = b + \frac{(X-x')^2+(Y-y')^2}{2b} \tag{19.47a}$$

and hence,

$$\mathbf{u}(x', y') = Ce^{ika} \int\limits_{-\infty}^{\infty}\!\!\int e^{-ik(X^2+Y^2)/2f}\left\{\int\limits_{-\infty}^{\infty}\!\!\int \mathbf{u}(x, y)e^{ik[(x-X)^2+(y-Y)^2]/2a} \, dx \, dy\right\}$$

$$e^{ikb}e^{ik[(X-x')^2+(Y-y')^2]/2b} \, dX \, dY,$$

$$[(X-x')^2+(Y-y')^2] \ll \sqrt{\lambda b^3}. \tag{19.49}$$

We first evaluate the integrals over X and Y:

$$I = \int \exp\left\{i\tfrac{1}{2}k\left[X^2\left(\frac{1}{a}+\frac{1}{b}-\frac{1}{f}\right)-2X\left(\frac{x}{a}+\frac{x'}{b}\right)\right]\right\} dX$$

$$\cdot \int \exp\left\{i\tfrac{1}{2}k\left[Y^2\left(\frac{1}{a}+\frac{1}{b}-\frac{1}{f}\right)-2Y\left(\frac{y}{a}+\frac{y'}{b}\right)\right]\right\} dY. \quad (19.50)$$

Note that we may complete the squares in the exponents; and substitute the new variables:

$$p = i\sqrt{\frac{ikc}{2}}\left[X-\left(\frac{x}{a}+\frac{x'}{b}\right)\Big/c\right] \qquad (19.51)$$

$$q = i\sqrt{\frac{ikc}{2}}\left[Y-\left(\frac{y}{a}+\frac{y'}{b}\right)\Big/c\right], \qquad (19.52)$$

where we have written for brevity:

$$c = \frac{1}{a}+\frac{1}{b}-\frac{1}{f}. \qquad (19.53)$$

On making this substitution, we obtain

$$I = \exp\left\{-i\left(\frac{k}{2c}\right)\left[\left(\frac{x}{a}+\frac{x'}{b}\right)^2+\left(\frac{y}{a}+\frac{y'}{b}\right)^2\right]\right\}$$

$$\cdot \frac{\int_{-\infty}^{\infty}e^{-p^2}\,dp}{i\sqrt{ikc/2}}\frac{\int_{-\infty}^{\infty}e^{-q^2}\,dq}{i\sqrt{ikc/2}}. \quad (19.50a)$$

Noting that the infinite Gaussian integral equals $\sqrt{\pi}$, we may rewrite $(19.50a)$:

$$I = \left(\frac{i2\pi}{kc}\right)\exp\left\{\left(\frac{-ik}{2c}\right)\left[\left(\frac{x}{a}+\frac{x'}{b}\right)^2+\left(\frac{y}{a}+\frac{y'}{b}\right)^2\right]\right\}$$

$$= \left(\frac{i2\pi}{kc}\right)\exp\left\{\frac{-ik}{2c}\left[\frac{(x^2+y^2)}{a^2}+\frac{2(xx'+yy')}{ab}+\frac{(x'^2+y'^2)}{b^2}\right]\right\}. \quad (19.50b)$$

We now substitute this into the integral (19.49) and find after removing from the integral all factors containing only x' and y':

$$\mathbf{u}(x',y') = \left(\frac{i2\pi C}{kc}\right)e^{ik/2c}e^{ik(a+b)}\exp\left[\left(\frac{ik}{2b}\right)(x'^2+y'^2)\left(1-\frac{1}{cb}\right)\right]$$

$$\cdot \iint \mathbf{u}(x,y)\exp\left[\left(\frac{ik}{2a}\right)(x^2+y^2)\left(1-\frac{1}{ca}\right)\right]$$

$$\exp\left[\left(\frac{ik}{abc}\right)(xx'+yy')\right]dx\,dy. \quad (19.54)$$

The integral remaining in (19.54) is a Fourier integral, provided the exponent involving $(x^2 + y^2)$ vanishes. The condition for this is that

$$\frac{1}{a} = c \equiv \frac{1}{a} + \frac{1}{b} - \frac{1}{f},$$

that is,

$$b = f. \tag{19.55}$$

The integral then takes the form

$$J = \iint \mathbf{u}(x, y) \exp \frac{ik(xx' + yy')}{f} \, dx \, dy, \tag{19.56}$$

which is a Fourier transform with the frequencies

$$\nu_x = \frac{x'}{\lambda f}, \qquad \nu_y = \frac{y'}{\lambda f}. \tag{19.57a}$$

In other words, the illumination amplitude in the plane (x', y') maps the spectrum of the amplitude in the plane (x, y), with frequency (ν_x, ν_y) plotted at point (x', y') where

$$x' = \lambda f \nu_x \qquad y' = \lambda f \nu_y. \tag{19.57b}$$

Alternatively, we may view x', y' as the Fourier transform variables, which permits us to view any spatial frequency (ν'_x, ν'_y) in the diffraction plane as generated by, and proportional to, a certain point (x, y) in the object plane, where

$$x = \lambda f \nu'_x \qquad y = \lambda f \nu'_y. \tag{19.58}$$

Referring back to (19.54) we note that the distribution in the diffraction plane contains, in addition to the Fourier transform (19.56), also a coefficient in the form of an imaginary exponential varying according to the distance $\sqrt{x'^2 + y'^2}$ of the point (x', y') from the axis. To eliminate this dependence, we must satisfy the further condition [see (19.55)]

$$\frac{1}{b} = c = \frac{1}{a} \frac{1}{f} \tag{19.59}$$

or

$$a = f. \tag{19.59a}$$

When this, too, is satisfied, (19.54) takes the form of a Fourier transform with a constant coefficient. We may summarize these conclusions as follows: *With the arrangement of Figure 19.9, we obtain at a distance f after the lens an amplitude distribution which is similar to the Fourier transform of the amplitude distribution in the object plane, except for a phase factor which varies with the square of the distance from the axis. This phase factor may be eliminated by placing the object at the front focal plane of the lens.* Equations 19.57 relate the position (x', y') to the spatial frequency (ν_x, ν_y) being mapped there.

The (x', y') plane is called the *Fourier transform plane* with respect to the object plane (x, y).

If the object is not at the front focal plane of the lens $(a \neq f)$, there remains a factor:

$$\exp\left[\left(\frac{ik}{2b}\right)(x'^2 + y'^2)\left(1 - \frac{1}{cb}\right)\right] = \exp\frac{ikr'^2(f-a)}{2f^2}, \qquad (19.60)$$

where we have made use of (17.55) and set

$$x'^2 + y'^2 = r'^2. \qquad (19.61)$$

On comparing (19.60) with (19.45) we see that the effect of displacing the object is equivalent to placing in the Fourier transform plane a thin lens with focal length

$$f_0 = \frac{-f^2}{f-a}. \qquad (19.62)$$

The effect of the displacement can thus be compensated for by a lens of focal length $(-f_0)$ placed in the Fourier transform plane. The reader will readily confirm that this is precisely the lens required to place at infinity the image of the object as viewed from the Fourier transform plane; that is, this lens simulates the condition of the object located at the front focal plane.

19.2.2 Principles of Spatial Filtering

The illumination at the Fourier transform plane may obviously be used to investigate the spatial frequency spectrum of the object wavefront: According to (19.57) every coordinate point (x', y') in this plane, corresponds to a spatial frequency (ν_x, ν_y) in the original object; the amplitude and phase of the wave there is a measure of the spectrum value and phase for the spatial frequency (ν_x, ν_y).

However, the importance of the Fourier transform plane goes far beyond this. It permits "filtering" images and correlating them with other images for the purpose of detecting similarities. To illustrate this, we cascade two transform stages of the type shown in Figure 19.9. Such a cascaded arrangement is shown in Figure 19.10, where the lens placed in

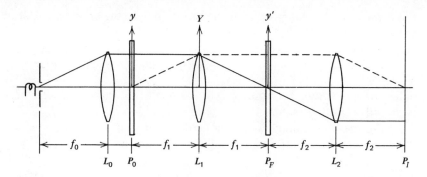

Figure 19.10 Schematic of spatial filtering arrangement.

the Fourier-transform plane images the first lens on the last and is required to avoid vignetting (see Section 9.5.1.3). Since the second stage is similar to the first, we obtain in the second Fourier-transform plane the Fourier transform of the Fourier transform of the object—i.e., simply an inverted reproduction of the object wavefront. Each position (x'', y'') in this last plane corresponds to a spatial frequency (ν'_x, ν'_y) in the first Fourier transform plane, where by analogy with (19.57b) and using (19.58)

$$x'' = \lambda f_2 \nu'_x = \left(\frac{f_2}{f_1}\right)x; \qquad y'' = \left(\frac{f_2}{f_1}\right)y. \tag{19.63}$$

Note that this is exactly what we would have expected on the basis of geometrical optical principles: the region between the two lenses is collimated with respect to the object points and the image will be an inverted rendition of the object, magnified by (f_2/f_1). Similarly, any spatial frequency in this image plane will be given by

$$\nu''_x = \frac{x'}{\lambda f_2} = \frac{f_1}{f_2}\,\nu_x, \qquad \nu''_y = \frac{f_1}{f_2}\,\nu_y. \tag{19.64}$$

We may now attenuate any frequency component of the object spectrum by placing in the Fourier-transform plane, P_F, an attenuator, at the location corresponding to that frequency. A device implementing such attenuations is called a *spatial filter*. For instance, we may eliminate all spatial frequencies whose value exceeds, say, ν_1, by placing in plane P_2 a circular aperture that is totally clear up to radius

$$r = \lambda f \nu_1 \tag{19.65}$$

and totally opaque beyond this radius. Alternatively, we may eliminate the mean, or "dc" component of the image, by placing an opaque dot at the axial point of P_F. If we wish to eliminate any other spatial frequency, ν_{x1}, say, we place a narrow strip at a point

$$x = \pm\lambda f\nu_{x1}.$$

(This may be used, for instance, to eliminate the raster pattern of a recorded television picture.)

Again, the object luminance may be differentiated by the introduction of the appropriate filter. Note that [19]

$$\mathcal{F}\left[\frac{df}{dx}\right] \sim \nu\mathcal{F}[f(x)]. \tag{19.66}$$

Hence, if we introduce at P_1 an absorber whose "transmittance" varies proportionally to x',

$$\tilde{\tau}(x') = ax', \tag{19.67}$$

we obtain at plane P_2 an image whose illumination is proportional to $\partial f/\partial x$, the x component of the gradient in the object plane. The transmittance referred to here is amplitude transmittance and its magnitude is equal to the square root of the ordinary (flux) transmittance [see (2.62)]:

$$\tau = \tilde{\tau}^2. \tag{19.68}$$

For future reference, we list here the relationships between optical density (D), absorption coefficient (α), and the flux transmittance (τ) and the amplitude transmittance $(\tilde{\tau})$. Let d be the pathlength of the flux in the absorbing medium considered. Then, from (19.68) and (4.59),

$$\tilde{\tau} = \sqrt{\tau} = e^{-(1/2)\alpha d}. \tag{19.69}$$

From (17.7)

$$D = \log_{10}\tau = (\log_{10}e)\log_e\tau = (\log_{10}e)\alpha d = -2\log_{10}\tilde{\tau}. \tag{19.70}$$

The filter functions derived in Appendix 18.5 as optimizing the image, too, may be implemented by this approach. Here, however, we meet with a practical difficulty. The filter functions discussed here so far were real. When the desired filter function is complex, its implementation requires the introduction of a phase shift at the Fourier-transform plane, and that may not be very easy. Indeed, the negative values of $\tilde{\tau}$ required in the negative half of the Fourier-transform plane according to (19.67) also implies the requirement of a phase shift of π there. Proposed techniques for such phase shifts include evaporated layers of dielectrics [20], controlled chemical action on a uniform clear dielectric film, controlled photopolymerization or deformation of a thermoplastic layer (Sections 17.3.2

and 17.4.1.2), ultrasonic wave fields (Section 14.2.4.4), polarization effects in special photographic materials [21] and an appropriately exposed, developed and, then, bleached silver halide emulsion [22]. Except for the latter, none of these techniques seem to have practical significance.

One relatively simple approach, the offset filter due to vander Lugt [23] is based on the same approach as the offset technique, which has made holography into a highly useful technique (Section 19.3.3.1). It is discussed in Section 19.2.4.

19.2.3 Image Processing by Spatial Filtering

An alternative approach is to view the spatial filter action as a multiplication of the Fourier transform of the object function with the filter function, where the filter function itself may be considered as the Fourier transform of some virtual "amplitude spread function." (The corresponding flux spread function is the absolute-value-squared of the "amplitude spread function.") According to the convolution theorem of Fourier theory [19], we obtain, then, in the image plane, P_I, the convolution of the object (amplitude) function with this "amplitude spread function." When the object is a point and its function a δ function, we obtain in the image plane this same "spread function" with its location determined by that of the δ function. (If the object consists of several δ functions, the image will consist of an equal number of copies of the "spread function.") Thus, in principle, it should be possible to generate any image by placing at the Fourier-transform plane a filter whose Fourier transform is the desired image.

Let us check the effect of a lateral displacement of the object. Using a basic result of Fourier theory (3.119), we note that a displacement of $\Delta x'$ will result in the Fourier transform in a factor $e^{i2\pi\nu_x''\Delta x'}$, that is, the resulting field distribution in the image plane will experience only a phase shift, an effect not detectable in direct viewing or recording. Thus such image formation will be insensitive to lateral object displacement.

More generally, when the coherent object is complex, we obtain in the image plane, P_I, a correlation or convolution of the object and the image underlying the filter. This effect can be used to detect the presence of such an image within an extensive object, and also to show its location there. It may be used in pattern recognition to pick out pattern details of interest, as in aerial photography, or element displacements in successive recordings [15]. Its application in character recognition is described in Section 19.3.6.3.

Since convolution effectively compresses an extended structured pattern into a narrow signal pulse, such correlation can also be used for pulse compression in coded-pulse radar systems [24].

As indicated in the preceding section, filtering can be used to restitute and, more generally, enhance images [24]. Considerable work along this line has been done also on the improvement of electron microscope images [25].

The use of spatial filtering in image processing is not limited to Fourier transform and convolution. Many transformations are possible and even nonlinear and nonisoplanatic processing can be implemented [26].

19.2.4 Offset Spatial Filtering [23]

Consider a filter with a transmittance varying sinusoidally at frequency ν_0:

$$\tilde{\tau} = \tau_0 (1 + M \cos 2\pi\nu_0 x'), \qquad (19.71)$$

placed in the Fourier transform plane. Its Fourier transform is the sum of three δ functions placed at

$$x'' = \pm\lambda f_2 \nu_0$$

and on the axis, respectively. In the image plane, P_I, we obtain then the object function convoluted with these. In other words, we obtain three versions of the object function; one on the axis and the others displaced by $\pm\lambda f_2\nu_0$—the sinusoid in the Fourier-transform plane has caused the object function to be offset. This sinusoid may also be multiplied (modulated) by any other real filter function to cause the offset version of the image to be filtered accordingly. The sinusoid (19.71) may thus be viewed as spatial "carrier" for the filter function, in complete analogy to the temporal carriers used in electronic communication systems.

Now consider the significance of displacing the sinusoid (19.71) in the x' direction through a distance Δx. As noted at the end of the preceding section, this will only cause the Fourier transform to be multiplied by a factor $\exp[i2\pi\nu_0 \Delta x]$. The image will still be at the same location; the filter displacement has only resulted in a phase shift of

$$\Delta\phi = 2\pi\nu_0 \Delta x \qquad (19.72)$$

in the flux arriving at the image plane. The displacement is thus equivalent to a phase shift introduced in the filter plane. Of course, we need not displace the whole carrier; we may displace any small segment of it in the neighborhood of, say, x'. This will then be equivalent to introducing, at x' on the filter, a phase shift, $\Delta\phi$.

The introduction of the carrier enables us to implement complex filter functions:

$$\hat{f}(x', y') = f(x', y')e^{i\phi(x',y')}, \qquad (19.73)$$

with purely real transmittance values. We write the carrier (19.71) in the form

$$\tilde{\tau}(x, y) = \tilde{\tau}_0\{1 + M(x', y') \cos 2\pi\nu_0[x' + \Delta x(x', y')]\} \qquad (19.74)$$

and, at every point (x', y'), we set the modulation

$$M(x', y') = \frac{cf(x', y')}{f_{mx}} \qquad (19.75)$$

and the shift term

$$\Delta x(x', y') = \frac{\phi(x', y')}{2\pi\nu_0}. \qquad (19.76)$$

Here c is a real constant

$$0 < c \leqslant 1 \qquad (19.77)$$

and f_{mx} is the maximum value of $f(x', y')$.

Since the image will appear around the points:

$$x'' = 0, \qquad \pm\lambda f_2\nu_0,$$

the carrier frequency, ν_0, must be chosen to be at least equal to twice the spatial frequency bandwidth of the object, if overlapping of the images is to be avoided. Methods for constructing such filters, using computer-controlled plotters, have been developed. Refer to Section 19.3.5.4 for an example of such a method.

If the Fourier transform of the desired filter function, $f(x', y')$, is available in the form of a transparency, or other object, it may be used to construct an offset filter photographically. In this technique, light from an off-axis point source, coherent with the object illumination, is permitted to interfere with light scattered from the object. The resulting interference pattern when recorded on the photographic material, serves as the filter. This is essentially a hologram of the object; see Section 19.3.1.1 for a mathematical treatment of the process involved.

Practical techniques have been described for producing offset spatial filters that perform such operations as differentiation [27], Hilbert transform [28], and image restoration [29]. An extensive bibliography on spatial filtering was published in 1972 [30].

19.2.5   Nonlinear Filtering

The half-tone screens of the graphic arts industry can be used to modulate the object transparency periodically. This results in a multiplicity of the object spectra in the Fourier-transform plane. The nonlinear growth of the halftone dots with object luminance—together with the nonlinear dependence of the diffracted flux on dot size—can be used to produce nonlinear coherent processing of the image [30*].

19.2.6 Other Coherent Image Processing Techniques

Coherent optical techniques can also be used to compare two objects and to synthesize images from two or more component objects. One basic approach is based on the same idea as the offset spatial filter: a diffraction grating introduces an angular displacement in the imaging of the radiation source. Here a diffraction grating is placed in the Fourier-transform plane and the two component objects, represented by complex transmittance $\hat{f}_1$ and $\hat{f}_2$, are placed in positions corresponding to the positive and negative first diffraction orders. The image corresponding to the negative first diffraction order of the first object and that of the positive first order of the second object will be superimposed at the location corresponding to zero order of the basic system. This results in the complex sum $\hat{f}_1 + \hat{f}_2$ [31].

If a phase shift of π radians is introduced into the path of one of the component objects, object subtraction $(\hat{f}_1 - \hat{f}_2)$ is obtained [31]. This has obvious application in detecting differences in two objects.

The basic grating used in the Fourier-transform plane may itself be a spatial filter generating a third image, $\hat{f}_3$. If this is combined with the above, a convolution

$$(\hat{f}_1 \pm \hat{f}_2) * \hat{f}_3$$

is obtained [32].

Another coherent optical method has been used to compare objects, such as fingerprints and to test them for identity [33]. It has the advantage of working directly in the Fourier-transform plane. With the need for an inverse Fourier transform eliminated, many of the difficulties of coherent optical systems are avoided. The technique may be understood as a Young's interference experiment (Section 2.3.3.1) with the two objects serving as the two slits required there; only if they are almost identical will interference fringes be obtained.

19.3 HOLOGRAPHY [34]

Holography is the process of recording the wavefront emanating from an irradiated object and then reconstructing it. The reconstruction of the wavefront should simulate the wavefield as it would be were the object in place, and in optical holography it should give rise to the same visual impression as the original wavefront. In terms of the spatial filtering theory of the preceding section, the hologram is a spatial filter whose spread function approximates the desired image, the hologram construction generally[2] being made by means of interference effects using the radiation emanating from the object whose image is to be formed.

Holography was invented by D. Gabor [35] in 1948 and developed from an obscure laboratory curiosity into a highly useful technique due to the development of the laser—and the invention of offset holography by Leith and Upatnieks [36] in 1962.

A major difficulty in recording the wavefront is posed by the fact that photographic techniques record only the total irradiation; due to averaging over time intervals long compared to the period of the wave (ca. 2×10^{-15} sec) all information concerning phase relationships is lost. In the usual holography, the wave scattered by the object is made to interfere with another wave coherent with it, so that the degree of constructiveness is a measure of the phase. The interference thus converts phase into irradiance which, in turn, may be recorded photographically. The photographic interference record is called *hologram*. When it is appropriately irradiated, it tends to give rise to a wavefield similar to that which emanated from the original object. We demonstrate this fact in the following section.

19.3.1 Basic Concepts in Holography

19.3.1.1 A Plane Hologram—An Illustration. To illustrate the basic concepts involved in holography, we now present a detailed quantitative analysis of a holographic process. To permit concentrating the discussion on essentials we (a) use the scalar wave equation, (b) assume a plane reference wave, (c) choose a photographic "thin" hologram (i.e., a photographic film or plate whose emulsion thickness is small compared to the size of the structural details recorded across the emulsion surface), and (d) assume an ideal photographic process.[3]

In Figure 19.11, the x–y plane (P) represents the plane of the holographic record. The plane wavefront, F, is incident on it at an angle θ_1. It

[2] Synthetic holograms (Section 19.3.5.4) are an exception.

[3] (a) Polarization effects are treated in Section 19.3.1.7, (b) spherical reference waves in Section 19.3.2, (c) "thick" holograms in Section 19.3.1.6, and (d) the effects of nonlinearity in Section 19.3.3.2.

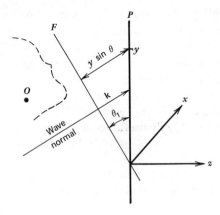

Figure 19.11 Illustrating the basic holographic recording process.

may be represented by[4]

$$\hat{A}_1(x, y, z) = A_1 \exp i\mathbf{k} \cdot \mathbf{s}, \tag{19.78}$$

where the caret denotes a phasor
  $A_1 = |\hat{A}_1|$ is the amplitude of the wave
  $\mathbf{k}$ is its wavevector, (see (2.32) *et seq.*),
  $\mathbf{s}$ is the displacement vector from the origin,

and we have omitted the time dependence, by using the instant where the phase vanishes at the origin.

With the wavevector $\mathbf{k}$ in the y–z plane, we have for the wave value in P:

$$\hat{A}_1(x, y, 0) = A_1 e^{ik_y y} = A_1 e^{iky \sin \theta_1}, \tag{19.79}$$

where $k_y = |\mathbf{k}| \sin \theta_1$ is the y component of the wavevector. Note that

$$\nu_{1H} = \frac{k_y}{2\pi} = \frac{\sin \theta_1}{\lambda}$$

is the spatial frequency of the object wave on the hologram plane.

Shown as a broken line in Figure 19.11 is the wave scattered by the object, O; its effect on Plane P may be represented by the general equation

$$\hat{A}_0 = A_0(x, y)e^{i\phi(x,y)}, \tag{19.80}$$

---

[4] By defining the phasor in this form, we imply that the phase increases with the distance from the source. This, in turn, implies that it increases *negatively* with time. We choose this definition since we are primarily concerned with spatial, rather than temporal, variations. With this choice we also avoid the minus sign, which would otherwise appear in many exponents. It should be noted that the choice of sign, having phase increase with time or with displacement, is purely a matter of taste.

where $A_0(x, y)$ represents the amplitude and $\phi(x, y)$ the phase at any point (x, y) on that plane.

The waves represented by $\hat{A}_1$ and $\hat{A}_0$ interfere on P, resulting in a phasor equal to the sum of their phasors. The resulting irradiation at any point (x, y) may then be taken as the square of the absolute value of this sum:

$$E(x, y) = |\hat{A}_1 + \hat{A}_0|^2 = (A_1 e^{ik_y y} + A_0 e^{i\phi})(A_1 e^{-ik_y y} + A_0 e^{-i\phi})$$
$$= A_1^2 + A_0^2(x, y) + A_1 A_0(x, y)[e^{i(\phi - k_y y)} + e^{-i(\phi - k_y y)}]$$
$$= A_1^2 + A_0^2(x, y) + 2A_1 A_0(x, y) \cos(\phi - k_y y). \qquad (19.81)$$

This irradiation is permitted to expose the photographic emulsion for a period t_e, yielding an exposure pattern

$$t_e E(x, y).$$

If the emulsion is then processed, positively, to a gamma value of two, the resulting flux transmittance value will be proportional to the square of the irradiation pattern (19.81) and the amplitude transmittance, according to (19.69), linearly proportional to it:

$$\tilde{\tau}(x, y) = c' E(x, y), \qquad (19.82)$$

with E as given by (19.81).

If the transparency is now illuminated by a reconstruction wave $(\hat{A}_2)$ similar to the original reference wave:

$$\hat{A}_2 = c'' \hat{A}_1. \qquad (19.83)$$

the transmitted wave will have the form:

$$\hat{A}(x, y) = \hat{A}_2 \tilde{\tau} = c'c'' \hat{A}_1 E$$
$$= cA_1\{e^{ik_y y}[A_1^2 + A_0^2(x, y)] + A_1 A_0(x, y)e^{i\phi}$$
$$+ A_1 A_0(x, y)e^{i2k_y y}e^{-i\phi}\}, \qquad (19.84)$$

where c', c'', and $c = c'c''$ are constants of proportionality. Note that the transmitted wave has three components, the middle one being exactly similar to the original object wave, whereas the first and last are multiplied by the exponentials

$$e^{ik_y y} \quad \text{and} \quad e^{i2k_y y}, \qquad (19.85)$$

respectively. This multiplication permits the effective separation of these "spurious" waves from the reconstructed object wave

$$cA_1^2 A_0(x, y)e^{i\phi(x,y)}, \qquad (19.86)$$

so that the latter can be viewed alone. Clearly, then, except for the luminance factor cA_1^2, there will be no physical difference between the wavefield under these reconstruction conditions and the original condition with the object in place. In other words, an observer will see the original object in place, even though he is only viewing the hologram illuminated by plane wave, $\hat{A}_2$.

The last term in (19.84) is simply the original object wave with its phase reversed and multiplied by a factor $e^{i2k_y y}$. The phase reversal means that the object appears located *after* the hologram by the same distance the original object was located before the hologram. Since this *twin image* is located in the observation space, it is real and may be viewed on a screen. The factor $e^{i2k_y y}$ represents a rotation of the direction of wave propagation as discussed presently.

The following considerations show how the exponential factors (19.85) permit the separation of the terms in (19.84).

Noting that a plane wave propagating at an angle θ_1 with the hologram normal has the exponential form (19.79) in the hologram plane, we conclude that a similar factor ($e^{ik'y}$) in the reconstructed wave simulates a similar plane wave at a corresponding angle θ' with the hologram normal where

$$\sin \theta' = \frac{k'}{|\mathbf{k}|}. \tag{19.87}$$

The Fourier transform of this factor is a δ function at infinity in direction θ' and the Fourier transform of a term

$$f e^{ik'y}$$

corresponds to the convolution of $\mathscr{F}[f]$ with a δ function at θ'; that is, the image of the object giving rise to wave f has been displaced through an angle θ'. Accordingly, the real image represented by the last term of (19.84) is located at an angle $2\theta_1$, from the hologram normal, while the first term gives rise to a wave centered on the direction θ_1 and represents a constant term (the original point source) plus the square of the object wave amplitude.

The angular spread within these components will be ($\pm 2\lambda \Delta \nu$) in the first, and ($\pm \lambda \Delta \nu$) in the latter two terms[5] where

$$\Delta \nu = \frac{\sin \varphi}{\lambda} \tag{19.88}$$

is the range of spatial frequencies generated by half the object in the hologram plane. Here φ is half the angular subtense of the object at the

---

[5] Due to the squaring operation, the spatial frequency spectrum in the first term is twice as wide as the spectra in the latter terms. Note that

$$\cos^2 \theta = \tfrac{1}{2}(1 + \cos 2\theta):$$

squaring a trigonometric function doubles its argument.

hologram plane and (19.88) is analogous to (19.79). The three components of (19.84) will thus be spread over the angles:

$$\theta_1 \pm 2\lambda\,\Delta\nu, \qquad \pm\lambda\,\Delta\nu, \qquad 2\theta_1 \pm \lambda\,\Delta\nu \qquad (19.89)$$

respectively, and there will be no overlap provided

$$\theta_1 \geqslant 3\lambda\,\Delta\nu. \qquad (19.90a)$$

The geometry of the three fans is illustrated in Figure 19.12, with dotted, solid, and broken lines representing the three components (19.89), respectively.

If A_0 is much smaller than A_1, so that the second term in the first term of (19.84) is negligible, the angular spread due to this term, too, will be negligible. Condition (19.90a) can then be reduced to

$$\theta_1 \geqslant 2\lambda\,\Delta\nu. \qquad (19.90b)$$

Note that in the original holograms of Gabor the angle θ_1 was zero and therefore the spurious images overlapped the desired image. To minimize the disturbance due to the first term, A_0 had to be kept much smaller than A_1. (The A_1^2-term does not cause a serious disturbance: it is constant in x and y and, therefore, represents only a single point in the far field.) To minimize the disturbance of the last term, the object distance must be made large compared to the object element size, so that the illumination due to the spurious image in the plane of the viewed image is thoroughly blurred and varies there but slowly.

In concluding this analysis, we consider a distant point object that will cause in the hologram plane an irradiation of a form analogous to (19.79):

$$\hat{A}_0 = A_0 \exp{(iky \sin{\theta_0})}, \qquad (19.91)$$

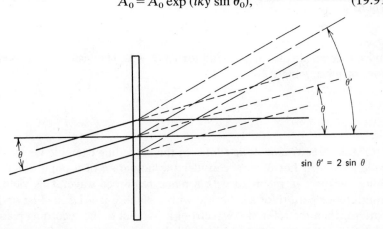

$$\sin{\theta'} = 2\sin{\theta}$$

Figure 19.12 The three terms of the reconstruction wave. The solid, broken and dotted lines represent, respectively, the virtual, real, and zero-order image directions.

where θ_0 is the angle the object wavefront makes with the hologram plane. According to (19.81), with

$$\phi = ky \sin \theta_0,$$

the irradiation is then

$$E(x, y) = A_1^2 + A_0^2 + 2A_0A_1 \cos [ky(\sin \theta_0 - \sin \theta_1)]$$
$$= A_0^2\{1 + R^2 + 2R \cos [k(\sin \theta_0 - \sin \theta_1)y]\}, \quad (19.92)$$

where

$$R = \frac{A_1}{A_0} \quad (19.93)$$

is the ratio of reference to object wave amplitude. The interference will thus yield a sinusoidal fringe pattern with spatial frequency

$$\nu_y = \frac{\sin \theta_0 - \sin \theta_1}{\lambda}$$

$$\approx \frac{\alpha}{\lambda} \quad (19.94)$$

and pitch

$$d = \frac{1}{\nu_y} = \frac{\lambda}{\sin \theta_0 - \sin \theta_1}$$

$$\approx \frac{\lambda}{\alpha}, \quad (19.95)$$

where

$$\alpha = \theta_0 - \theta_1$$

and the approximations are valid for $\theta_0, \theta_1 \ll 1$. The visibility, or modulation, of these fringes is

$$M = \frac{2R}{1 + R^2}. \quad (19.96)$$

19.3.1.2   The Gabor Zone Plate. In the preceding example we treated a general object. Let us now consider the hologram of a point object at a finite distance, a, recorded with a plane reference wave. This yields the interference pattern of a spherical with a plane wave. Let us first find this pattern when the reference wavefront is parallel to the recording plane, P. The phase of the reference wave will then be constant on P and we may take this as our reference phase, so that, on P, the reference phasor may be represented as a real constant:

$$\hat{A}_1 = A_1. \tag{19.97}$$

The spherical object wave will have a phase equal to a constant plus (kd), where d is the distance from the sphere, centered on the object point and its tangent at the axial point, A; see Figure 19.8. Thus its phasor will be similar to that of (19.45):

$$\hat{A}_0 = A_0 e^{i(kd+\phi)} \approx A_0 \exp i\left(\frac{kr^2}{2a} + \phi\right), \tag{19.98}$$

where a is the radius of the tangent sphere, that is, the distance of the point source from P. The resulting interference irradiation distribution will be:

$$E = |\hat{A}_0 + \hat{A}_1|^2 = \left[A_1 + A_0 \cos\left(\frac{kr^2}{2a} + \phi\right)\right]^2 + A_0^2 \sin^2\left(\frac{kr^2}{2a} + \phi\right). \tag{19.99}$$

If

$$A_0 = A_1 = A \tag{19.100}$$

and $\phi = 0$, this simplifies to

$$E = 2A^2\left(1 + \cos\frac{kr^2}{2a}\right). \tag{19.101}$$

The square of this function, corresponding to the resulting illumination is plotted in Figure 19.13 as a solid line; the broken line is the square-wave version (binary-quantization), which the reader may recognize as the transmittance graph across the diameter of a Fresnel zone plate (Section 2.3.1.5). Denoting the radius of the first zero by r_0, we have

$$\frac{kr_0^2}{2a} = \pi$$

and hence

$$r_0^2 = a\lambda. \tag{19.102}$$

Maxima and zeros occur at

$$\frac{r}{r_0} = \sqrt{n} \tag{19.103}$$

for n an even and odd integer, respectively.

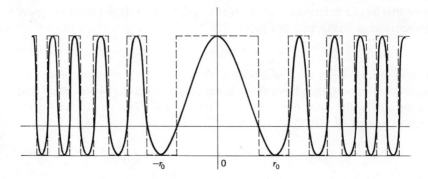

Figure 19.13   Gabor zone plate transmittance distribution: $[1+\cos \alpha x^2]^2$. The broken line shows the corresponding graph for the Fresnel zone plate.

When this irradiation distribution is recorded according to the idealized process postulated in the preceding section, we obtain a record whose amplitude transmittance varies proportionally to the function (19.101). Such a recording with its radial symmetry is sometimes called a *Gabor zone plate*. An illustration thereof is shown in Figure 19.14.

To find the diffraction effects resulting from illuminating a Gabor zone plate with a plane wavefront normal to it, we note that this is simply a special case of our analysis of the preceding section, with $\theta = 0$ and the object a point source at a distance a from the hologram. Thus the resulting wavefront can be resolved into three components: (a) a spherical wavefront diverging from a point at a distance a in front of the hologram; (b) a spherical wavefront converging onto a point at a distance a beyond the hologram; and (c) a uniform plane wavefront representing the mean transmittance.[6]

When the reference wavefront makes an angle θ with the hologram plane, P, we choose our origin at a point $(a \sin \theta)$ above the foot of the

---

[6] This is in contrast with the Fresnel zone plate that has an infinite number of foci at distances

$$\pm \frac{a}{n}, \qquad n = 1, 3, 5, 7 \cdots.$$

The reader may note the analogy with a sinusoidal transmittance diffraction grating that diffracts only into the two first orders, in contrast with the binary diffraction grating that, in general, diffracts light into all integral diffraction orders, just as the Fourier transform of a sinusoid has only zero and fundamental components, whereas that of a square wave has, in general, all components.

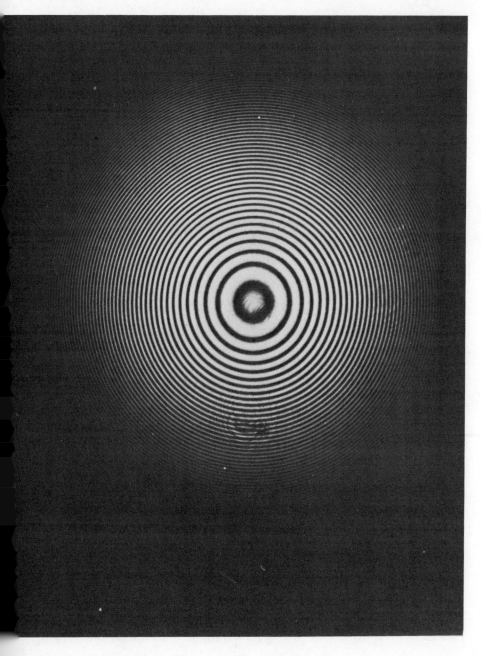

Figure 19.14 Photograph of Gabor zone plate. From W. E. Kock, *Engineering Applications of Lasers and Holography*, Plenum, New York, 1975.

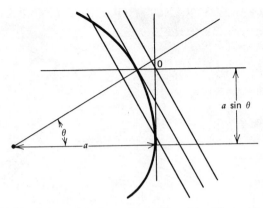

Figure 19.15 Gabor zone plate construction with oblique reference wave.

perpendicular from the object point; see Figure 19.15. We have then again for the object wave:

$$\hat{A}_0 = A_0 \exp i \left(\frac{kr^2}{2a} + \phi\right)$$

$$= A_0 \exp i \left\{\frac{k[x^2 + (y + a \sin \theta)^2]}{2a} + \phi\right\}, \qquad (19.104)$$

and for the reference wave as in (19.79)

$$\hat{A}_1 = A_1 e^{ik_y y} = e^{iky \sin \theta}. \qquad (19.79)$$

The irradiation at any point (x, y) is then equal to the absolute-value-squared of the sum of these. Again assuming (19.100) and noting the identity

$$|e^{iA} + e^{iB}|^2 = [1 + \cos (A - B)], \qquad (19.105)$$

we find for the irradiation pattern here

$$E(x, y) = 2A^2 \left\{1 + \cos \left[\frac{k}{2a}(x^2 + y^2) + \frac{k}{2} a \sin^2 \theta + \phi\right]\right.$$

$$= 2A^2 \left[1 + \cos \frac{kr^2}{2a}\right], \qquad (19.106)$$

where we have set

$$\phi = -\frac{k}{2} a \sin^2 \theta \qquad (19.107)$$

for the last step. This result indicates that the inclination of the wavefront caused a displacement of the Gabor zone plate upward by a distance $(a \sin \theta)$ but left it otherwise unchanged. We note that the above is based on the approximation (19.44) and is therefore valid only for small angles.

In practice, a zone pattern is obtained only in the region where the waves $\hat{A}_0$ and $\hat{A}_1$ overlap, which may be limited to any part of P and need not include the origin.

Indeed, in any small region of the pattern (19.101), at a considerable distance from the origin, the fringe variation is approximately sinusoidal, with a spatial frequency found by differentiating the total number of fringes to that point:

$$\nu_r = \frac{d}{dr}\left(\frac{\phi}{2\pi}\right) = \frac{1}{2\pi}\frac{d}{dr}\left(\frac{kr^2}{2a}\right) = \frac{kr}{2\pi a} = \frac{r}{a\lambda}; \qquad (19.108)$$

this pattern will vary sinusoidally in a direction radially away from the origin with a frequency directly proportional to its distance, r, from this origin.

In conclusion, we derive the interference pattern for the case when both interfering wavefronts are spherical, with point sources at distances a_1 and a_2, respectively, from P, assumed normal to the axis which, in turn, coincides with the line joining the sources. At any point, a distance r from the axial point on P, the difference in phase will be

$$\Delta\phi = \frac{kr^2}{2a_1} - \frac{kr^2}{2a_2} + \phi_0 = \frac{kr^2}{2f} + \phi_0, \qquad (19.109)$$

where

$$f = \left(\frac{1}{a_1} - \frac{1}{a_2}\right)^{-1} \qquad (19.110)$$

is the focal length of the lens, which, if placed at P would image the point a_1 at a_2.

With the two amplitudes equal and ϕ_0 vanishing, this yields, similarly to (19.101),

$$E = 2A^2\left(1 + \cos\frac{kr^2}{2f}\right). \qquad (19.111)$$

Note that (19.111) is a more general form of (19.101) with a replaced by f. This arrangement, too, then yields a Gabor zone plate. The production of actual Gabor zone plates is complicated by nonlinearities of the recording material [37].

Such plates can be used as a mask in recording holograms, eliminating the need for a reference source—for each object point, its twin image as generated by the zone plate, serves as the reference source [38].

19.3.1.3   Reconstruction Variations. Magnification [34a], [39]. In Section 19.3.1.1. we assumed a reconstruction wave identical to the reference wave used in the recording. This is a useful method of reconstruction, yielding aberrationless images. However reconstruction need not be limited to this and the introduction of other reconstruction waves adds flexibility to holography. In this section we discuss imaging effects possible by varying the hologram scale and the reconstruction wave direction, curvature, and wavelength. In general, when the reconstruction wave differs from the original reference wave, image aberrations will be introduced. These do not appear in the formulae given below, which are based on the approximation (19.47). Their detailed discussion is beyond the scope of the present work and the interested reader is referred to Refs. 39 and 40 for details.

Wave Direction. Reference to our discussion of the directionality factor (19.85) shows that changing the direction of the reconstruction wave will change A_2 only in that the factor $e^{ik_y y}$ will be replaced by $\exp ik(lx + my)$, where l and m are the x- and y- direction-cosines, respectively, of the wavevector of the reconstruction wave. Within the limits of the framework assumed in Section 19.3.1.1, the effect of such a rotation will simply be a similar rotation of the three reconstructed waves. The following two examples will illustrate this.

1. If the reconstruction wave, still incident from the left, has the form

$$\hat{A}_2 = A_2 e^{-k_y y},$$

which corresponds to a rotation of the diffracted rays, through 2θ: the virtual image will appear in the direction

$$\theta_v = -\sin^{-1}(2\sin\theta)$$

and the real image on the normal to the hologram.

2. If the reconstruction wavefront is incident from the right in Figure 19.11, antiparallel to the reference wave, a real image will be formed at the original location of the object. This may be useful under conditions where an unaberrated real image is desired.

In both these examples an aberrationless real image is obtained; however, only the second example will work in a thick hologram (Section 19.3.1.6); in the first example, the Bragg conditions, discussed there, are not fulfilled. Also, in the first example an inverted image (mirror image) will be obtained: a right-handed glove would appear as a left-handed one.

Changes in Scale, Wavelength, and Curvature. The formulae we derive here were first developed by R. W. Meier [39] and are often associated with his name. His analysis is in terms of wavefronts. An alternative analysis is based on ray-tracing [41]. Both approaches may be used for aberration analysis, the ray-trace method being superior primarily in systems whose parameters vary rapidly, as in the analysis of effects due to differential emulsion shrinkage [42].

We start by deriving a general expression for the reconstructed wave of a hologram made of an arbitrary point object with a reference wave originating at a point source. We use a Cartesian coordinate system in which the hologram plane coincides with the x–y plane. To allow for enlargement of the hologram after recording and before reconstruction, we prime the coordinates in the hologram plane before enlargement:

$$x = mx', \qquad y = my'. \tag{19.112}$$

To allow for a change in wavelength, we denote by λ' and λ, respectively, the wavelengths used in recording and reconstruction, with k' and k denoting the corresponding wave vector magnitude:

$$\lambda = \mu\lambda', \qquad k = \frac{k'}{\mu}, \tag{19.113}$$

Here m is the magnification factor used in the enlargement and μ is the factor of wavelength change.

The object, reference, and reconstruction waves are, then, all of the form:

$$\hat{A}_k = A_k e^{i\phi k} \tag{19.114}$$

with $k = 0, 1, 2$ for the object, reference, and reconstruction waves, respectively. The phases are, using approximation (19.47),

$$\phi_0 = -k' \frac{2z_0^2 + x_0^2 + y_0^2 - 2x_0 x' - 2y_0 y' + x'^2 + y'^2}{2z_0}$$

$$= -k' \frac{x'^2 + y'^2 - 2x_0 x' - 2y_0 y'}{2z_0} \tag{19.115}$$

$$\phi_1 = -k' \frac{2z_1^2 + x_1^2 + y_1^2 - 2x_1 x' - 2y_1 y' + x'^2 + y'^2}{2z_1}$$

$$= -k' \frac{x'^2 + y'^2 - 2x_1 x' - 2y_1 y'}{2z_1} \tag{19.116}$$

$$\phi_2 = -k \frac{2z_2^2 + x_2^2 + y_2^2 - 2x_2 x - 2y_2 y + x^2 + y^2}{2z_2}$$

$$= -k \frac{x^2 + y^2 - 2x_2 x - 2y_2 y}{2z_2}, \tag{19.117}$$

where the last member in each equation is obtained by letting the phase at the axial point of the hologram vanish. With the assumptions of idealized photoprocessing, the reconstructed wavefront, and its conjugate, will be

$$\hat{A} = \hat{A}_2\hat{A}_0\hat{A}_1^* = cA_0A_1A_2e^{i(\phi_0-\phi_1+\phi_2)} = Ae^{i\phi} \tag{19.118}$$

$$\hat{A}' = \hat{A}_2\hat{A}_0^*\hat{A}_1 = cA_0A_1A_2e^{i(-\phi_0+\phi_1+\phi_2)} = Ae^{i\phi'}. \tag{19.119}$$

On substituting (19.112)–(19.117) into these, we find the phases of $\hat{A}$ and $\hat{A}'$:

$$\phi = \phi_0 - \phi_1 + \phi_2 = -\tfrac{1}{2}k\left\{\frac{\mu\left[\dfrac{x^2+y^2}{m^2}-\dfrac{2x_0x}{m}-\dfrac{2y_0y}{m}\right]}{z_0}\right.$$

$$-\frac{\mu\left[\dfrac{(x^2+y^2)}{m^2}-\dfrac{2x_1x}{m}-\dfrac{2y_1y}{m}\right]}{z_1}$$

$$\left.+\frac{[x^2+y^2-2x_2x-2y_2y]}{z_2}\right\}$$

$$= -k\left[\tfrac{1}{2}(x^2+y^2)\left(\frac{1}{z_2}-\frac{\mu}{z_0m^2}-\frac{\mu}{z_1m^2}\right)\right.$$

$$\left.-x\left(\frac{x_2}{z_2}+\frac{\mu x_0}{mz_0}-\frac{\mu x_1}{mz_1}\right)-y\left(\frac{y_2}{z_2}+\frac{\mu y_0}{mz_0}-\frac{\mu y_1}{mz_1}\right)\right] \tag{19.120}$$

and, similarly,

$$\phi' = -k\left[\tfrac{1}{2}(x^2+y^2)\left(\frac{1}{z_2}-\frac{\mu}{z_0m^2}+\frac{\mu}{z_1m^2}\right)\right.$$

$$\left.-x\left(\frac{x_2}{z_2}-\frac{\mu x_0}{mz_0}+\frac{\mu x_1}{mz_1}\right)-y\left(\frac{y_2}{z_2}-\frac{\mu y_0}{mz_0}+\frac{\mu y_1}{mz_1}\right)\right]. \tag{19.121}$$

If the waves (19.118)–(19.119) are to represent image points, the phases (19.120)–(19.121), too, must be of the form:

$$\phi = -k\frac{(x^2+y^2)-2x_3x-2y_3y}{2z_3}, \tag{19.122}$$

where (x_3, y_3, z_3) is the coordinate of the reconstructed image point. On matching coefficients of (x^2+y^2), x, and y in (19.120) and (19.122), respectively, we find:

$$z_3 = \left(\frac{1}{z_2} \pm \frac{\mu}{m^2 z_0} \mp \frac{\mu}{m^2 z_1}\right)^{-1} = \frac{m^2 z_0 z_1 z_2}{m^2 z_0 z_1 \pm \mu z_1 z_2 \mp \mu z_0 z_2}$$

(19.123)

$$x_3 = \frac{\left(\frac{x_2}{z_2} \pm \frac{\mu x_0}{m z_0} \mp \frac{\mu_1 x_1}{m z_1}\right) m^2 z_0 z_1 z_2}{m^2 z_0 z_1 \pm \mu z_1 z_2 \mp \mu z_0 z_2}$$

$$= \frac{m(m x_2 z_0 z_1 \pm \mu x_0 z_1 z_2 \mp \mu x_1 z_0 z_2)}{m^2 z_0 z_1 \pm \mu z_1 z_2 \mp \mu z_0 z_2}$$

(19.124)

$$y_3 = \frac{m(m y_2 z_0 z_1 \pm \mu y_0 z_1 z_2 \mp \mu y_1 z_0 z_2)}{m^2 z_0 z_1 \pm \mu z_1 z_2 \mp \mu z_0 z_2}.$$

(19.125)

Here the upper and lower signs apply to the images obtained with ϕ and ϕ', respectively.

The lateral magnification, m_L, may be found from (19.124) by differentiation:

$$m_L = \frac{dx_3}{dx_0} = \frac{m}{1 - z_0/z_1 \pm m^2 z_0/\mu z_2}.$$

(19.126a)

Similarly, the angular magnification is

$$m^* = \frac{d(x_3/z_3)}{d(x_0/z_0)} = \pm \frac{\mu}{m}.$$

(19.126b)

Clearly the angular magnification depends only on the enlargement of the hologram copy and the wavelength ratio, whereas the linear magnification depends also on the ratios of the object distance to the reference and reconstruction point source distances.

Equations 19.123–19.125 include the effects of changing the direction of the reconstruction beam, which we already discussed in some detail at the end of Section 19.3.1.1. They do not cover questions of beam intensity and are primarily useful for thin holograms, which are not so sensitive to changes in wavelength and the direction of the reconstruction beam.

Efforts at recording holograms at X-ray wavelengths and reconstructing them with light in the visible part of the spectrum [43] do not appear to have proven useful. However, impressive magnifications have been gotten even with operation restricted to one wavelength [44].

19.3.1.4 Effects of Object Diffusivity. Three-Dimensional Reconstruction.

We define as a nondiffusing object one whose complex transmittance [see (2.140)] or reflectance varies, in the main, but slowly over regions large compared to the wavelength, with rapid variations confined to a small fraction of the total area. The usual photographic transparency is a nondiffusing object. When such an object is irradiated by light having a smooth coherent wavefront, the resulting diffraction pattern will be a relatively sharp rendition of the transmittance for considerable distances from the transparency. The blurring will be significant only when the distance from the transparency is comparable to

$$\frac{s^2}{\lambda},$$

where s is the width of the regions of uniform transmittance and λ is the wavelength of the transmitted light. Consequently, in recording a hologram of such a transparency, under the usual conditions, the object beam will form a blurred projection, or shadowgram, of the object on the hologram. The reference beam will then superimpose the required interference fringes onto the shadowgram to form the hologram. In such a hologram, every area and element receives light from only a small region on the object and, hence, will contribute to the formation of only the small region of the image corresponding to it. When such a hologram is viewed from close by, only a small part of the image can be seen at a time; the viewer must scan the hologram with his eye in order to see all parts of the image. One way to overcome this difficulty is to form a real image on a diffusing screen either directly or by using a lens to reimage the virtual image.

Another frequent consequence of the projection of the object onto the hologram is a large variation of irradiation over the hologram (these variations will be essentially identical to those in the object). As shown in Sections 19.3.2.1 and 19.3.2.2, the ratio of irradiation values due to the object and reference beams, respectively, must be restricted to a narrow region for efficient holography, and this becomes impossible for the large ranges likely to be found in the object transparency.

In addition, nondiffuse objects are prone to spurious interference patterns, essentially Gabor zone plates, due to scattering particles and small defects in the intervening space, as discussed in Section 19.1.4.

All these difficulties are absent in the hologram of a diffuse object.[7] It

[7] We mention, in passing, that controlled phase nonuniformity, synthetically introduced, has been used to obtain the advantages of diffuse holography without some of its disadvantages [45].

is, of course, a simple matter to turn a transparency into a diffuse object by preposing a diffusing screen, such as a sheet of ground glass. When a diffusing screen is irradiated, each object element scatters the radiation over a large angle, so that each such element contributes to the irradiation over the whole hologram. This not only leads to a more uniform irradiation of the latter, it also causes each small patch of the hologram to contain information concerning the whole object and to permit its reconstruction. A small patch of the hologram implies, of course, a small aperture of the image-forming system and, hence, limited resolving power, but the object as a whole is there. Thus diffuse holographic recording ensures against total loss of image elements as caused, for instance, by scratches or other defects which may completely obliterate essential information when coherent holography is used—certainly in dense photographic storage systems.

The potential for three-dimensional recording is another important aspect of holography of diffuse objects. Each hologram patch carries information concerning the appearance of the object from its particular position. By viewing through various patches, one can reconstruct the object as it appeared from various points of view and actually look around images to see images of objects hidden behind other objects. Similarly, when viewed with both eyes simultaneously, each eye sees the image as it appeared originally from that position, and stereoscopic vision results.

On the other hand, diffuse object holography increases the resolution requirements on the recording material. In nondiffuse recording, the maximum angle between reference and object rays equals the angle, θ_1, between the object and reference beams plus the angle through which the object diffracts a significant amount of light (usually a relatively small angle). In the holography of diffuse objects, the angle subtended by the object at the hologram must be added to θ_1, as given by (19.90a). Figure 19.16 illustrates the geometry involved. If overlap is to be avoided the maximum angle, θ_{mx}, between object and reference rays will be at least:

$$\theta_{mx} = \tan^{-1}\frac{(w-h)}{2z_0} + \tan^{-1}\frac{w+h}{2z_0} + \theta_1$$

$$\approx \frac{w}{z_0} + \theta_1 \approx 2\varphi + \theta_1, \tag{19.127}$$

where w and h are the total widths of the object and hologram, respectively, φ is the half-field angle, and

$$\theta_1 = 3\lambda\Delta\nu. \tag{19.90a}$$

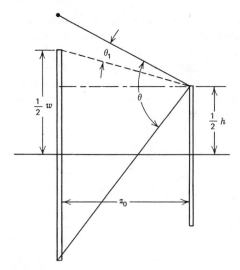

Figure 19.16   Illustrating angle requirements in diffuse object holography.

According to (19.94) an increased angle, θ_{mx}, means an increase in spatial frequency for the resulting interference fringes and hence increased demands on the resolving power of the recording material.

Another drawback of holography of diffuse objects is the introduction of the speckling effect, due to the random phase shifts introduced at the diffusing object, as discussed in Section 19.1.3.

Three-dimensional reconstruction requires some additional discussion. We have already noted in Section 19.3.1.1 that reconstruction generally produces both an imaginary image, duplicating the original object, and a real image appearing in front of the hologram, essentially like its mirror image. When dealing with three-dimensional reconstruction, this implies that the parts of the object that originally faced the observer, face away from him in the real image—an inversion of the depth rendition. Such inverted presentation is called *pseudoscopic* in contrast to the correct *orthoscopic* rendition. The observer relying on stereoscopic or parallax clues for his depth perception (see Section 15.8), will sense various object parts at certain relative distances from his eye, but he may find that a part further away blocks his view of a part closer to him, thus providing contradictory clues to his depth perception. For this reason pseudoscopic viewing may be uncomfortable and disturbing.

Pseudoscopic viewing results when a three-dimensional image is formed at the location of the original object (relative to the hologram), but viewed from the opposite side; or when it is viewed from the same side, but imaged on the opposite side (both are methods for obtaining real images). An orthoscopic real image may be obtained by recording a virtual object. We may, for instance, use a lens system to image the object beyond the hologram plane. The hologram will then be recording it as a virtual object. On reconstruction, the real image will appear in the location of the original virtual object and be orthoscopic (although with depth distortion, unless imaging was close to unity magnification throughout the object depth). In this arrangement, the virtual image will be pseudoscopic.

19.3.1.5. Phase Holograms Exposure variations can also be translated into phase, instead of density, variations. Bleached silver halide holograms (Section 19.3.4.2) and photothermoplastic holograms (Sections 17.4.1.2 and 19.3.4.2) are examples of such techniques. As shown in Section 19.3.3.2, this type of recording is inherently not linear; but for small phase variations it will be only slightly nonlinear, so that reasonably good results can be obtained. Because practical values of light absorption can be very small in pure phase holograms, the diffraction efficiency, and hence brightness obtained here can be very high.

The phase shift introduced into the transmitted wave is generally due to two distinct factors: (a) change in thickness and (b) change in refractive index. Phase shifts due to the former can be thought of as taking place at the surface of the medium, whereas those due to index changes are a process that is cumulative through the thickness of the recording medium. In general, the phase-shift introduced by a hologram element is

$$\phi = \frac{2\pi t(n-1)}{\lambda},\tag{19.128}$$

where t is the thickness and n is the refractive index of the recording medium. The change in phase is, therefore given by

$$\Delta\phi = \frac{\partial\phi}{\partial t}\Delta t + \frac{\partial\phi}{\partial n}\Delta n = \frac{2\pi\,\Delta t(n-1)}{\lambda} + \frac{2\pi t\,\Delta n}{\lambda},\tag{19.129}$$

where the two terms, respectively, relate to the two factors mentioned above.

For an analysis of diffraction in a phase grating, see Sections 14.2.3.3, 14.2.3.4 and 19.3.1.7.

19.3.1.6 Thick Holograms; Bragg Diffraction and Reflection Holograms

Three-Dimensional Interferograms. Our earlier analysis of holographic recording and reconstruction was based on the assumption of a "plane" hologram; that is, that the total action of the hologram may be assumed to take place in a plane. This approximation is valid only if the thickness of the recording is small compared with the spacing between the hologram fringes. By way of illustration, consider a hologram made with the usual He-Ne laser ($\lambda = 0.6328 \ \mu$m) and an angle (θ) of 30° between object and reference beam. The fringe spacing will then be given by [see (19.95)]

$$d = \frac{\lambda}{\sin \theta} = 1.266 \ \mu\text{m}.$$

The usual photographic emulsion, including the high-resolution emulsions used in holography, have a thickness ranging from approximately 5 to 20 μm—certainly not small compared to the above fringe spacing. Consequently, such a hologram can not realistically be treated as a plane diffraction grating. It will consist of exposed layers formed in the emulsion, generally running obliquely through it, rather than fringes in a plane. These layers coincide with regions of constructive interference. When two plane waves interfere, the layers are planes bisecting the angle between the wave vectors and normal to the plane defined by them. See Figure 19.17, where the light solid and dotted lines represent wave peaks and troughs, respectively, advancing in the direction of the wave vectors shown normal to the wavefronts. Points where two wave peaks, or wave troughs, intersect, represent regions of fully constructive interference. They are shown in heavy lines and along them the exposed layers are formed. Points of intersection between a peak and a trough are points of fully destructive interference. They are shown as broken heavy lines and are regions of zero exposure (if the interfering waves are of equal amplitude). These points can be seen to advance horizontally toward the right, generating the heavy horizontal lines shown, as the two waves advance. The generated lines shown are the traces of the planes of uniform exposure all normal to the plane of the figure (parallel to the x–z plane). Note that the planes of uniform exposure bisect the angle between the wave vectors of the two interfering plane waves.

If the interfering wavefronts are spherical, the exposed layers, too, will be curved. In general, these layers will be centered on the locus of points equidistant from the focal points of the interfering waves or from spheres centered on these points and having radii equal to an integral number of wavelengths. For instance, if a plane wave interferes with a spherical one, the region of constructive interference will be paraboloids defined by the center of the sphere as the focus and planes, spaced by λ, as directrices (see Figure 19.18).

Bragg Diffraction. Diffraction at such periodic surfaces is more restricted than that at a plane diffraction grating; it occurs only for certain

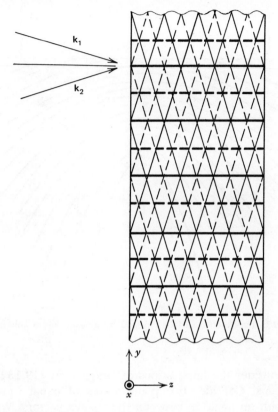

Figure 19.17 Generation of interference layers by plane waves.

angles of incidence. This phenomenon is called Bragg diffraction and has been treated in Section 14.2.3.4. According to it, the angle (θ_i) of incidence of a wave on a three-dimensional grating equals the angle (θ_d) of diffraction:

$$\theta_d = \theta_i. \tag{19.130}$$

In addition, the diffraction grating equation, too, must be satisfied [see (14.116)]:

$$\sin \theta_i + \sin \theta_d = \frac{m\lambda}{d}. \tag{19.131}$$

The simultaneous fulfillment of (19.130) and (19.131) implies

$$\sin \theta_i = \sin \theta_d = \frac{m\lambda}{2d}. \tag{19.132}$$

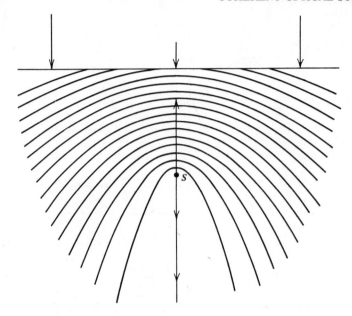

Figure 19.18   Interference surfaces formed by a plane and a spherical wave.

We now consider the implications of Bragg's law (19.132) applied to thick holograms. Consider two plane waves of equal wavelengths, λ, interfering with an angle α between their wave vectors. Without compromising generality, we choose the coordinate system such that the z axis bisects the angle between their wave vectors and the x axis is perpendicular to the plane defined by these vectors. This is illustrated in Figure 19.19, which is a scaled-up version of Figure 19.17, the x axis being perpendicular to the plane of the figure. In triangle ABC, AC represents the wavelength and $\angle ABC = \alpha$, the angle between the interfering wavefronts equals that between their respective wavevectors. Hence

$$\overline{AB} = \frac{\lambda}{\sin \alpha}.$$

Finally in triangle ABD, the spacing, d, between the planes of constructive interference is

$$d = \overline{AD} = \cos \tfrac{1}{2}\alpha \, \frac{\lambda}{\sin \alpha} = \frac{\lambda}{2 \sin \tfrac{1}{2}\alpha}, \tag{19.133}$$

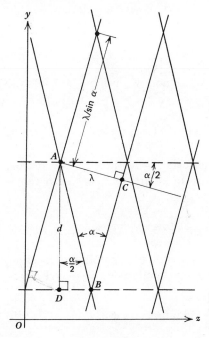

Figure 19.19 Details of the interference of two plane waves and its relationship to the Bragg condition.

with the last step following from the well-known identity

$$\sin \alpha = 2 \sin \tfrac{1}{2}\alpha \cos \tfrac{1}{2}\alpha. \tag{19.134}$$

Turning now to diffraction effects observed with the processed interferogram, we find, according to (19.132) with $m = 1$, and using (19.133), that the angle of incidence must be

$$\sin \theta_i = \frac{\lambda'}{2d} = \frac{\lambda'}{\lambda} \sin \tfrac{1}{2}\alpha \tag{19.135}$$

$$= \sin \tfrac{1}{2}\alpha, \qquad \lambda' = \lambda, \tag{19.135a}$$

where λ' is the wavelength of the incident wave. This is identical with the angle between the original interfering wavefront and the resulting diffracting planes, provided we use the same wavelength. Thus, with $\lambda' = \lambda$ and for efficient diffraction, the light must be incident parallel to the wavevector of one of the original interfering wavefronts.

So far we have considered angles and waves only inside the recording medium. Outside this medium, the sines of the angles of incidence will be larger by a factor of n, the index of refraction of the recording medium relative to that of the surrounding medium. The wavelength, λ_0, outside this medium, too, will be larger by this factor. Thus the angles of incidence of the original waves were

$$\sin \tfrac{1}{2}\alpha_0 = n \sin \tfrac{1}{2}\alpha \qquad (19.136)$$

and the angle required for Bragg diffraction is, using (19.135) and (19.136)

$$\sin \theta_0 = n \sin \theta_i = \left(\frac{\lambda}{\lambda'}\right) n \sin \tfrac{1}{2}\alpha = \left(\frac{\lambda_0}{\lambda_0'}\right) \sin \tfrac{1}{2}\alpha_0, \qquad (19.137)$$

still coinciding with the original interfering wavefronts. The spacing between the diffracting layers is, in terms of externally observed parameters:

$$d = \frac{\lambda}{2 \sin \tfrac{1}{2}\alpha} = \frac{\lambda_0}{2 \sin \tfrac{1}{2}\alpha_0}. \qquad (19.138)$$

In terms of an arbitrary coordinate system, the amplitude resulting from two interfering plane waves of equal amplitude, a, may be represented by the sum

$$\hat{a} = a[\exp i(\mathbf{k}_1 \cdot \mathbf{x} + \phi_1) + \exp i(\mathbf{k}_2 \cdot \mathbf{x} + \phi_2)] \qquad (19.139)$$

and its irradiation by

$$E = \hat{a}\hat{a}^* = 2a^2\{1 + \cos[(\mathbf{k}_1 - \mathbf{k}_2) \cdot \mathbf{x} + (\phi_1 - \phi_2)]\}. \qquad (19.140)$$

This attains its maximum value, four, in the planes defined by

$$(\mathbf{k}_1 - \mathbf{k}_2) \cdot \mathbf{x} + (\phi_1 - \phi_2) = 2m\pi, \qquad m = \cdots -2, -1, 0, 1, 2 \cdots . \qquad (19.141)$$

All of these are perpendicular to the vector $(\mathbf{k}_1 - \mathbf{k}_2)$ and spaced from each other by

$$d = \frac{2\pi m}{|\mathbf{k}_1 - \mathbf{k}_2|} = \frac{m\lambda}{|\mathbf{k}_1^0 - \mathbf{k}_2^0|} = \frac{m\lambda}{2 \sin \tfrac{1}{2}\alpha}, \qquad (19.142)$$

where $\mathbf{k}_{1,2}^0$ designate unit vectors parallel to $\mathbf{k}_{1,2}$, respectively,
     α is the angle between them and (19.142) is in agreement with
     (19.133).

To ascertain the emulsion thickness for which Bragg diffraction becomes significant, we consider normally incident light diffracted at the surface of the grating through an angle

$$\theta_1 = \sin^{-1} \frac{\lambda}{d}.$$

When the thickness, t, is such that all the light diffracted in this manner strikes a region of peak density, we may assume that ordinary diffraction will be very low. This occurs when

$$t \sin \theta_1 = \frac{t\lambda}{d} = d.$$

On substituting here for d from (19.95), we find

$$t = \frac{d^2}{\lambda} = \frac{\lambda}{\alpha^2}. \tag{19.143}$$

The diffraction efficiencies obtained, and the conditions for high-diffraction efficiency, are discussed in the next section.

Reflection Holograms. When the object and reference waves are incident from opposite sides of the hologram and, as usual, the reconstruction beam duplicates the reference beam, the primary (virtual image) wave will leave the hologram on the same side on which the reconstruction is incident. This technique is termed reflection holography. Here, too, Bragg diffraction is the main phenomenon, but the diffracting planes' orientation tends to be parallel to the hologram plane, in contrast to the normal orientation (Figure 19.16) typical for the transmission hologram.

Specifically, if the hologram plane bisects the angle between the object and reference wave vectors, the discussion in connection with Figure 19.16 indicates that the exposure will be in the form of layers parallel to the hologram surface (see Figure 19.20). If the angle of incidence of each of the two waves is θ, the angle between them will be

$$\alpha = \pi - 2\theta$$

and, on substituting this into (19.138), we find for the spacing between the layers:

$$d = \frac{\lambda}{2 \sin \frac{1}{2}\alpha} = \frac{\lambda}{2 \cos \theta}. \tag{19.144}$$

This spacing has its minimum value when the two waves are incident normally to the hologram. We have then

$$d_{mn} = \frac{\lambda}{2}. \tag{19.145}$$

The reader may note that this is a case of standing waves, with $\frac{\lambda}{2}$ between nodes.

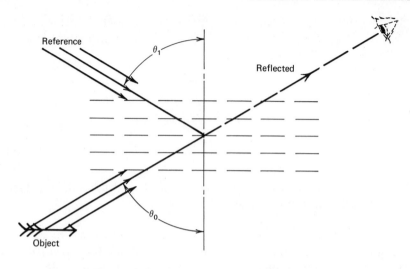

Figure 19.20 Illustrating reflection holography.

In the usual photographic emulsion, whose thickness may be approximately 16 μm, we obtain

$$N = \frac{2nt}{\lambda_0} = 77.6$$

layers, where we have used the most popular wavelength of the He-Ne laser line (0.6328 μm) and

$$n = 1.535$$

for the refractive index of gelatine at that wavelength [46].

When such a hologram is processed and replaced into the original system, with the object removed, a virtual image will appear in the original object location, when the hologram is viewed from the side from which it is illuminated. See Figure 19.20, where the eye of the viewer is shown in dotted lines.

19.3.1.7 *Efficiency of Diffraction* [34b]. The luminance of the image reconstructed from a hologram varies directly with the fraction of the light diffracted into the image. We call this fraction *diffraction efficiency* and present, in the following, brief analyses of its determination for various types of holograms. We assume linearity and therefore need treat the efficiency only for a single, general, object point. We restrict our analysis to sinusoidal interference patterns. This applies, strictly speaking, only to Fourier and Fraunhofer holograms; however, even Fresnel holograms often approximate a sinusoidal pattern. (See Sections 19.3.2.2–19.3.2.4 for the definitions of these hologram types.)

We treat thin and thick holograms separately, and for each of them we consider both the pure absorption and phase holograms. Actual absorption holograms will exhibit some phase modulation as well, and *vice versa*; but usually one of these predominates significantly.

In the thick hologram, more and more energy is diffracted out of the incident reconstruction wavefront as it progresses into the medium. This effect may be analyzed by means of the coupled wave theory [47]. We do not describe this here, and present only some representative results. For a more general treatment of diffraction at sinusoidal hologram patterns, the reader is referred to Refs. 34b and 47. These present a complete analysis, assuming only that the energy transfer is small over a one-wavelength advance and that the wave is incident at, or near, the Bragg angle. Another derivation and a comparison with experimental results can be found in Ref. 48.

A. *Thin Absorption Hologram.* The thin absorption hologram of a single distant point exhibits a sinusoidal transmittance variation. Its diffraction pattern field distribution is obtained as the Fourier transform of the amplitude transmittance $\sqrt{\tau}$. Thus, for irradiation with a uniform plane wave of luminance L_0:

$$L(\theta) = L_0\{\mathscr{F}[\sqrt{\tau_0}(1 + \tilde{M}\cos 2\pi\nu_1 x)]\}^2$$

$$= L_0\tau_0\left[\delta(\nu) + \frac{\tilde{M}}{2}\delta(\nu + \nu_1) + \frac{\tilde{M}}{2}\delta(\nu - \nu_1)\right]^2$$

$$= L_0\tau_0\left\{\delta^2(\nu) + \frac{\tilde{M}^2}{4}[\delta^2(\nu + \nu_1) + \delta^2(\nu - \nu_1)]\right\}, \qquad (19.146)$$

where $\sqrt{\tau_0}$ is the mean amplitude transmittance,
$\tilde{M}$ is the modulation of the amplitude transmittance, and

$$\theta = \sin^{-1}\lambda\nu_1.$$

Note that the diffraction efficiency is proportional to the *square* of the amplitude transmittance modulation.

For maximum diffraction efficiency, the modulation should be unity, and the maximum value of $\sqrt{\tau_0}$ for which this is possible, is one half. This corresponds to a density

$$D_0 = -\log_{10}\tau_0 = 0.6.$$

On substituting these optimum values for τ_0 and M into (19.146) and noting that the incident wave is $L_0 \delta^2(\nu)$, we find that the efficiencies will be

$$\eta_0 = \tfrac{1}{4}$$
$$\eta_1 = \tfrac{1}{16}$$

in the zero and each of the first orders, respectively. It vanishes for all other orders. Since only the first orders yield reconstruction, the maximum reconstruction diffraction efficiency here is clearly 6.25%.

B. *Thin phase holograms.* The diffraction effects for a thin sinusoidal phase grating have been analyzed earlier in connection with light diffraction at ultrasonic waves (Section 14.2.3.3.). There we found for the diffraction efficiency in the first order

$$\eta_1 = J_1^2(\phi_1), \tag{19.147}$$

where ϕ_1 is the amplitude of the sinusoidal phase variations. This attains its maximum value, 0.33857, when $\phi_1 = 1.84118$. According to the first term in (19.129) this corresponds to a change in thickness of

$$\Delta t = 1.841 \cdots \frac{\lambda}{2\pi(n-1)} = \frac{0.293\lambda}{(n-1)}.$$

The reader should note that, in contrast to the absorption hologram, the phase hologram is intrinsically nonlinear, causing both higher-order images and intensity distortion within the first order.

C. *Thick absorption holograms.* As mentioned above, the full analysis of the thick absorption hologram is beyond the scope of this discussion and we present here only the results for two special cases.

1. *Transmission Hologram—Fringe Layers Normal to Medium Layer.* When the absorbing layers are perpendicular to the recording medium surface, coupled wave theory predicts, at Bragg angle incidence, a diffraction efficiency of the form

$$\eta = \tfrac{1}{2}e^{-2\hat{D}} (\cosh M\hat{D} - 1) = \tfrac{1}{2}\tau_0^{(\theta)} (\cosh M\hat{D} - 1), \tag{19.148}$$

where M is the modulation of the absorption coefficient α, that is,

$$M = \frac{\alpha_1}{\alpha_0} \leqslant 1, \tag{19.149}$$

α_0 is the mean value of the absorption coefficient and
α_1 is the amplitude of its sinusoidal variation;

$$\hat{D} = \alpha_0 t \sec \theta \tag{19.150}$$

is the "natural" amplitude density[8] in the direction of incidence, θ_i; and the mean transmittance, measured at an angle, θ, with the surface normal is

$$\tau_0^{(\theta)} = \exp\left(-2\alpha t \sec \theta\right). \tag{19.152}$$

The diffraction efficiency has its maximum value when the modulation is unity and

$$\hat{D}_0 = \log 3, \tag{19.153}$$

with the corresponding density, according to (19.151):

$$D_0 = 2\hat{D}_0 \frac{\cos \theta}{\log 10} = 0.954 \cos \theta. \tag{19.154}$$

The resulting maximum efficiency is

$$\eta_{mx} = \tfrac{1}{27} = 3.7\%.$$

Equation 19.153 can readily be confirmed by setting $M = 1$ in (19.148), differentiating it with respect to $\hat{D}$, equating the result to zero and solving for $\hat{D}$.

Graphs of η as a function of $\hat{D}$, for various values of M, are shown in Figure 19.21.

2. *Reflection Hologram—Fringe Layers Parallel to Medium Layer.* For a reflection hologram with the fringe layers parallel to the surface of the recording medium, coupled wave theory shows that the diffraction efficiency is, at Bragg angle incidence:

$$\eta = \frac{M^2}{[2 + \sqrt{4 - M^2} \coth\left(\tfrac{1}{2}\hat{D}\sqrt{4 - M^2}\right)]^2}, \tag{19.155}$$

where the symbols are as defined following (19.148).

Here the diffraction efficiency grows monotonically with both $\hat{D}$ and M and for $M = 1$ reaches its maximum value:

$$\eta_1 = \left[2 + \sqrt{3} \coth \frac{\sqrt{3}\hat{D}}{2}\right]^{-2}. \tag{19.156}$$

---

[8] We defined here the "natural" density, $\hat{D}$, for convenience. To establish the relationship between it and the common density, D, we note that the amplitude transmittance:

$$\tilde{\tau} = e^{-\alpha t} = e^{-\hat{D}\cos\theta}$$

and hence,

$$\hat{D} = -\sec \theta \log \tilde{\tau} = -\tfrac{1}{2}\sec \theta \log \tau = \tfrac{1}{2}D \sec \theta \log 10$$

$$= 1.1513 \frac{D}{\cos \theta}. \tag{19.151}$$

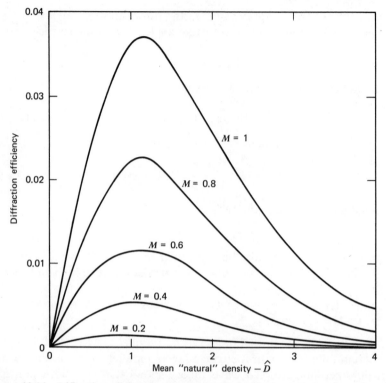

Figure 19.21 Diffraction efficiency versus "natural" density for various values of modulation. Thick sinusoidal transmission absorption grating.

This, in turn, approaches asymptotically its maximum value of

$$\eta_{mx} = (2 + \sqrt{3})^{-2} = 0.0718$$

as $\hat{D}$ approaches infinity.

Curves for diffraction efficiency as a function of $\hat{D}$ are shown in Figure 19.22 for various values of modulation, M.

D. Thick Phase Holograms. The diffraction efficiency for thick phase holograms with iso-phase planes normal to the hologram surface, has been presented in Section 14.2.3.4 in connection with ultrasonic waves. At the Bragg angle, it is

$$\eta = \sin^2 \tfrac{1}{2}\phi_1, \qquad (19.157)$$

where

$$\phi_1 = \frac{2\pi n_1 t}{\lambda \cos \theta} \qquad (19.158)$$

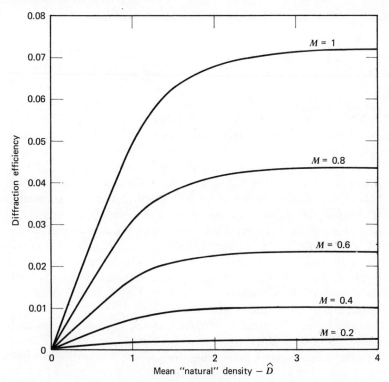

Figure 19.22 Diffraction efficiency versus "natural" density for various values of modulation. Thick sinusoidal reflection absorption grating.

is the phase shift introduced by the amplitude, n_1, of the sinusoidal index variations, as the wavefront, incident at angle θ, passes through the hologram, thickness, t. It is readily seen that here the diffraction efficiency can reach 100% and that this occurs for

$$\phi_1 = m\pi, \qquad m = 1, 2, 3 \cdots . \tag{19.159}$$

In practice, efficiencies close to this value have been obtained (see Table 191).

With the isophase planes parallel to the medium, a reflection hologram is obtained and, when used at the Bragg angle, its efficiency is given by

$$\eta = \tanh^2 \tfrac{1}{2}\phi_1. \tag{19.160}$$

This, too, approaches 100% for a very thick hologram ($\phi_1 \to \infty$).

The results of the coupled wave theory as given above, have been confirmed experimentally. The maximum efficiencies obtainable in theory are compared with those obtained in practice in Table 191 [34b].

19.3.2 Holographic Geometries [34c] and Systems

In this section we describe the various types of hologram geometry. Monochrome holograms are described in Sections 19.3.2.1–19.3.2.6 and summarized in Section 19.3.2.7. Color holography is treated in Section 19.3.2.8.

19.3.2.1 "In-Line" Holography. Gabor's original holograms were made without the offset technique described in Sections 19.2.4 and 19.3.1.1; the collimated coherent beam was incident normal to the object (a photographic transparency) and the recording material, with the source, object, and hologram centers lying along a straight line.

This is the simplest method of holography; there is no need for a special reference beam—the transparent regions of the object provide this satisfactorily if they constitute the major portion of the object. The resolution requirements on the recording material, too, are very much lower in "in-line" holography. The severe requirements on resolution and stability in the usual holographic system emanate primarily from the off-axis method of operation (see Section 19.3.3.3) and hence the requirements are substantially lower here.

However, this advantage is obtained at the expense of the ability to separate the two images generated by the hologram; the wavefronts generating these overlap, precluding their separation geometrically. Since one of them is always out of focus (except in the trivial case of a hologram in the image plane exactly, which is an ordinary photograph), every "in-line" hologram reconstruction is accompanied by a spurious background pattern, which interferes with the detection of the image.

As discussed in Section 19.3.3.2, the ratio of reference-to-object-beam irradiation must be substantially more than unity to ensure linearity in photographic recording. This limits good "in-line" holograms to objects in which the opaque portions constitute but a small fraction of the total object area.

19.3.2.2 Fourier Holography. Already in Section 19.2.1 we noted that with the arrangement of Figure 19.9, with

$$a = b = f, \tag{19.59}$$

we obtain in the diffraction plane (x', y') the Fourier transform of the field distribution in the object plane (x, y). A hologram made with this arrangement (the object in the front focal plane of the lens, the hologram in its rear focal plane, and the reference wave generated by a point source coplanar with the object) is called a *Fourier-transform hologram* or a

Fourier hologram, for short. For instance, the Fourier hologram of a point source is a sinusoidal pattern of the form (19.71) with

$$\nu_0 = \frac{s}{\lambda f}, \tag{19.161}$$

where s is the separation of the object from reference point source.

If we reconstruct the Fourier hologram with a collimated beam as obtained, for instance, from an illuminated pinhole placed at the focal point of a lens, we have essentially the spatial filtering arrangement for image generation, as discussed in Section 19.2.3. We obtain, then, in the image plane the Fourier transform of the filter transmittance, convoluted with a δ function, that is, we obtain an image of the original object, with its location determined by that of the pinhole image. It is interesting to note that there the image position will be unaffected by the lateral position of the hologram. This has proven itself important in optical pattern recognition techniques and is potentially useful in the reconstruction stage of motion picture holography, where invariance of image with hologram motion may be an attractive feature.

If condition (19.59a) is not satisfied, there will be introduced into the object wavefront, $\hat{A}_0$, a factor

$$\exp \frac{ikr'^2(f-a)}{2f^2}, \tag{19.60}$$

where a is the object distance from the lens. The reference wavefront, $\hat{A}_1$, due to the point source, too, will have the same factor introduced into it. Consequently, this factor will drop out of both the image information carrying terms, $\hat{A}_0\hat{A}_1^*$ and $\hat{A}_0^*\hat{A}_1$. Therefore exactly the same hologram will be obtained, as long as the reference wave source is coplanar with the object. A hologram obtained in this manner has been termed *quasi-Fourier hologram*.

If the lens is omitted altogether in the above arrangement, we have the situation in Section 19.2.1, which led to

$$\hat{A}_1 = C e^{ika} e^{ik(X^2+Y^2)/2a} \iint [f(x, y) e^{ik(x^2+y^2)/2a}] e^{ik(xX-yY)/2a} \, dx \, dy. \tag{19.48}$$

This is again identical to the result obtained in the Fourier hologram, except that we obtain the Fourier transform of

$$f'(x, y) = f(x, y) e^{ikr^2/2a}$$

instead of that of $f(x, y)$ alone. Comparison with (19.45) shows that we

obtained simply an image of the original object in contact with a thin lens
of focal length $(-a)$, again with a factor analogous to (19.60):

$$e^{ikr^2/2a}. \tag{19.162}$$

Due to a similar phase factor in the object wave, this latter effect drops
out, just as it did in the quasi-Fourier hologram. Thus the *lensless* Fourier
hologram will appear similar to the usual Fourier-transform hologram,
except for the virtual thin lens appearing in contact with the object. With
phase-insensitive viewing, this will be of no significance.

In summary, the essential characteristic of a Fourier hologram is the
coplanarity of reference point source and object. Incidentally, this re-
stricts strict Fourier holography to two-dimensional objects.

19.3.2.3 Fraunhofer Holography. Already in Section 2.3.1.3 we showed
that a Fourier transform can be obtained as the Fraunhofer (far-field)
diffraction pattern of a transmittance distribution. Hence a hologram
made in the far field of the object (2.136) will be quite similar to a
Fourier hologram, even if the reference wave point source is at infinity
and, hence, not coplanar with the object. The only distinction will be a
phase factor like (19.162) in contact with the true Fourier hologram. This
phase factor can be explained by reference to (2.135), which demonstrates the Fraunhofer diffraction pattern to be equivalent to the Fourier
transform of the aperture transmittance, when the diffraction pattern is
measured relative to a sphere centered at the object [condition (2.137)].
Since here the reference wavefront is a plane, a phase factor equal to (kd)
is introduced, where d is the separation between the reference sphere and
the tangent plane at its vertex, as shown in Figure 19.8 and (19.43). In
effect, this is again a virtual lens (19.45) in contact with the hologram and
will appear as a Gabor zone plate superimposed on the hologram pattern.
The effect of this virtual lens is to destroy the independence of image
location from a lateral hologram displacement. Note that the Fraunhofer
hologram just described, too, is lensless.

Due to the relatively large object distances (2.136) used, the mutual
interference between the twin images will be quite small in such a
Fraunhofer hologram, so that "in-line" holography becomes feasible.
Indeed, one of the first practical applications of holography was the
recording of aerosol particles executed in-line by means of lensless
Fraunhofer holography [49].

Fraunhofer holography does not necessarily require large object distances, only a far-field diffraction pattern at the hologram plane. This can
be obtained also with an object quite close to the hologram, if the plane
wave illuminating it is focused on the hologram, as we showed at the end

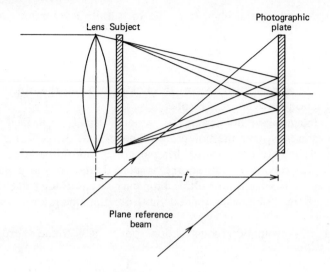

Figure 19.23 Recording Fraunhofer hologram by means of a lens.

of Section 2.3.1.3. In the Fraunhofer holography of large objects, it may therefore be convenient to introduce such a lens; the plane reference wave may then be provided with oblique incidence, bypassing the lens (see Figure 19.23).

19.3.2.4 Fresnel Holography. When the distance of the object from the recording plane is not large relative to its size, and conditions (2.136) or (2.137) are not fulfilled, a *Fresnel hologram* is obtained. This is similar to the Fraunhofer hologram in that each object point contributes a Gabor zone plate to the recording, but the interaction between the various object points is such that the diffraction pattern approximates (2.135) and the object wave distribution $\hat{A}_0$ no longer simulates the Fourier transform of the object transmittance.

This is the usual arrangement used in lensless holography of three-dimensional objects.

19.3.2.5 Image Plane Holography [50]. In the *image plane hologram*, an image of the object to be recorded is formed near the hologram plane. If the object is three-dimensional, part of the image will extend beyond the hologram plane, both ahead of it and behind it. This type of hologram permits reconstruction by light with poor coherence properties and low monochromaticity. This is evident from the results of Section 19.3.3.5, according to which the blur, Δx_3, due to a finite incoherent source size (Δx_2) is

$$\Delta x_3 = \Delta x_2 \frac{z_0}{z_1}, \qquad (19.163)$$

where z_0, z_1 are the distances to the hologram from the object and the reconstruction wave source, respectively.

In the image plane hologram, the "object" usually is, in fact, an image formed by a lens near the hologram plane so that the object distance is very small; it may even vanish for part of a three-dimensional object. Thus, if the reconstruction source is placed at a distance very large compared to its size, the resulting blur may be quite tolerable, provided no part of the "object" extended far from the hologram plane during recording.

For a polychromatic reconstruction source, it is found there that the blur is

$$\Delta x_3 = \frac{z_0 \theta \Delta \lambda}{\lambda}, \qquad (19.164)$$

where θ is the angle the reference (and reconstruction) wavevector makes with the object wavevector and
$\Delta \lambda$ is the spread of wavelengths in the reconstruction flux.

Again we see that a small value of z_0 permits a sizable wavelength spread, provided the object size does not demand too high a value for θ.

Such holograms may even be reconstructed with white light. They are easiest and by far the most efficient to use in displays where special light sources are not practical. They are readily identified by the fact that the image itself is discernible, although blurred, on the hologram with ordinary daylight viewing.

19.3.2.6 Summary of Monochrome Holographic Geometries.
Most of the common holography geometries are summarized in Figure 19.24. The curved lines represent the nodes (or antinodes) in the interference pattern generated by a plane wave incident from above and a spherical wave generated by a scattering object at S. They form parabolae, as shown. Depending on the position of the hologram, the various types of holograms described in this section are obtained. Representative locations and the corresponding hologram types, are indicated in the figure. Fourier holograms are not represented because they require the object and reference beam to be coplanar in a plane parallel to that of the hologram.

A representative arrangement for practical holography is described in Section 19.3.4.5.

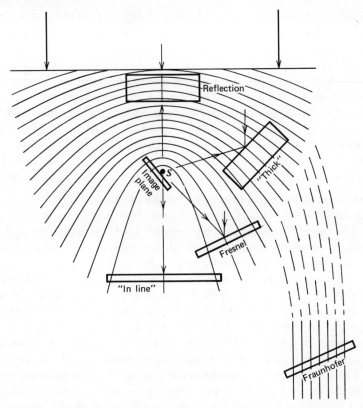

Figure 19.24 Illustrating various types of monochrome holograms. The plane reference wave is incident from above and the point object is located at S. After Collier, Burckhardt, and Lin [34].

19.3.2.7 Color Holography [34d]

General Considerations. To obtain multicolor holograms, flux at more than one wavelength must be used and, to obtain a substantial chromaticity range, at least three wavelengths must be provided. In general, the range of chromaticities obtainable can be found by plotting the chromaticity points of the constituent wavelengths on the chromaticity diagram (e.g., Figure 1.12), the polygon obtained on joining these (in the sequence of ascending or descending wavelength) encloses all the available chromaticities.

For instance, if the wavelengths 0.4762, 0.5208, and 0.6471 μm, obtainable with a krypton ion laser, are used, the resulting triangle includes by far the major portion of all chromaticities. For reasons of practicality, the two argon ion laser lines 0.488 and 0.5145 μm, together with the He-Ne laser line at 0.6328 μm, are usually used, although pure blue and most of the purples are sacrificed thereby.

In color holography, the recording process consists essentially of forming individual holograms in each of the component wavelengths, all superimposed on the same recording. In the reconstruction the same wavelengths are usually provided and, in addition, precautions must be taken to ensure the proper matching up of each spectral component of the incident flux with the corresponding fringe pattern and to guard against mismatching. Such mismatching may be avoided automatically in thick holograms, but requires special techniques in thin holograms.

Thin Color Holograms. Conceptually one of the simplest ways to avoid mismatching of hologram flux constituents is to break up the hologram area into small component holograms, each formed at just one wavelength. In reconstruction, we must then only ensure that each of the constituent spectral components falls only on the associated area elements. This separation may be effected by placing, over the hologram, a filter array, each of whose elements passes only one of the flux components. Both reference and object waves may then consist of a mixture of the spectral components and the fiter will ensure that an array of only monochrome holograms will result. In reconstruction, the same filter, used in proper register, will ensure proper matching of flux-hologram pairs. The filter elements must be small compared to the aperture used in viewing the reconstruction; if the hologram is to be viewed directly, the element size must be small compared to the pupil of the eye, so that several area elements are used at any one time. See Figure 19.25, where the pupil diameter must equal at least D.

It is not necessary to place the filter in actual contact with the hologram. It suffices to have the reference, and reconstruction, waves pass through it, with the filter imaged on the hologram, ensuring that each area element receives reference flux of only the wavelength corresponding to it. This reference flux can not interfere with the flux at other wavelengths reaching the same area element from the object, so that the proper monochrome hologram is formed at each area element.

An alternative technique is based on the fact that the mismatched wavelength-hologram combination will produce a different diffraction angle. Assuming an object point displaced by an angle θ from the reference source, the spatial frequencies of the fringe patterns generated by wavelengths λ_1 and λ_2, respectively, will be

$$\nu_1 = \frac{\sin\theta}{\lambda_1}, \qquad \nu_2 = \frac{\sin\theta}{\lambda_2}. \tag{19.165}$$

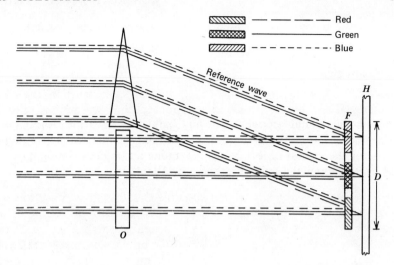

Figure 19.25 Multicolor thin hologram filter array system. O is object; F filter array; and H hologram. Note that the sizes of the filter elements is greatly exaggerated for the purpose of clarity. Only one image element is illustrated.

Interaction of the first fringe pattern with λ_2 and *vice versa*, will produce images at

$$\sin \theta_{12} = \nu_1 \lambda_2 = \left(\frac{\lambda_2}{\lambda_1}\right) \sin \theta$$

$$\sin \theta_{21} = \nu_2 \lambda_1 = \left(\frac{\lambda_1}{\lambda_2}\right) \sin \theta. \qquad (19.166)$$

These spurious images will be separated from the true images by

$$\Delta \theta_{12} = \theta - \sin^{-1}\left[\left(\frac{\lambda_2}{\lambda_1}\right) \sin \theta\right], \qquad \Delta \theta_{21} = \theta - \sin^{-1}\left[\left(\frac{\lambda_1}{\lambda_2}\right) \sin \theta\right],$$
$$(19.167)$$

respectively. Therefore, to avoid overlap, the total image may not extend over more than the smaller of these angles. This restriction can be reduced, and the acceptable object size increased, by having the reference waves incident at different angles (say θ_1 and θ_2). Then, for an object point on the axis,

$$\nu_1 = \frac{\sin \theta_1}{\lambda_1}, \qquad \nu_2 = \frac{\sin \theta_2}{\lambda_2}. \qquad (19.168)$$

If the object has angular extent, its hologram will have a certain spatial bandwidth spread and we must ensure that there is no overlap of the bands of the holograms recorded at the two wavelengths. Assuming

$$\lambda_1 > \lambda_2, \qquad \theta_1 < \theta_2,$$

this means that

$$\nu_{1mx} \leq \nu_{2mn},$$

where the maximum frequency in any band is due to the object point farthest away from the reference source and the minimum frequency occurs for the closest object point. Assuming an object extending over a field of $\pm\varphi$ about the axis, reference to (19.94) indicates that

$$\nu_{1mx} = \frac{\sin \theta_1 + \sin \varphi}{\lambda_1}$$

$$\nu_{2mn} = \frac{\sin \theta_2 - \sin \varphi}{\lambda_2}. \tag{19.169}$$

To find the limit on φ, we equate these and solve for $\sin \varphi$, yielding

$$\sin \varphi = \frac{\lambda_1 \sin \theta_2 - \lambda_2 \sin \theta_1}{\lambda_1 + \lambda_2}. \tag{19.170}$$

For instance with values 0.6328 μm and 0.488 μm for λ_1 and λ_2, and 30° and 50° for θ_1 and θ_2, respectively, we find $\varphi = 12.4°$, showing that an object covering a field close to 25° may be recorded at these two wavelengths without overlap of the reconstructed image with either of the spurious images. If a third wavelength is incorporated, three separate spatial frequency bands must be accommodated, and the permissible object size will be further restricted.

A third possible approach uses individual "coding" of each of the component monochrome holograms; see Section 19.3.6.3. Here the hologram is formed with the reference beam passing through a "phase coder," such as a piece of ordinary ground glass. If the reconstruction beam passes through the identical "phase coder"—in the identical position—the hologram will be properly reconstructed, otherwise the light will be merely scattered. If such a ground glass is used in the mul-tiwavelength reference beam, the coding will be different for each wavelength component, due to the difference in refractive index of the glass at these wavelengths, and each such component will reconstruct an image with only the hologram it had formed originally. Its interaction with the other holograms will cause some light to be scattered over the image, reducing contrast and signal-to-noise ratio.

Thick Color Holograms. In Section 19.3.1.6 we showed that, in thick holograms, efficient diffraction takes place only if the grating obeys Bragg's condition (19.132) and that this condition is automatically met if the reconstruction beam follows the original reference beam in both wavelength and direction relative to the hologram. If the reconstruction beam follows the direction of the original reference beam, but has a different wavelength, the diffraction efficiency drops as the wavelength change increases. This drop will be rapid when the hologram has an appreciable thickness.

If light of several wavelength components is used in recording the hologram, each component will form its own component hologram independently. If the hologram is sufficiently thick, each of the component holograms will satisfy Bragg's condition for only the one component originally responsible for forming it, provided the reconstruction beam is identical to the original reference beam. Thus inter-color cross talk may be eliminated in thick holograms.

If the thickness is sufficient to provide a sharply peaked Bragg diffraction, even white light may be used in the reconstruction. Each component hologram will then select a narrow spectral band around the wavelength responsible for the formation of that component hologram.

Clearly, distortion of the diffracting layers in the hologram medium may have a disruptive effect on the resulting hologram, in general, and in particular in color holography where interaction between various wavelengths must be considered. This is especially severe in reflection holograms, where the spacing between layers tends to be much smaller, so that the same absolute displacement constitutes a larger fractional disturbance. Ordinary shrinkage of silver-halide photographic emulsions, as caused by the fixing process, may shift the diffracted color from red to green. Special precautions must be taken to compensate for the shrinkage, if such color shifting is to be avoided. Triethanolamine may be used to produce a compensatory swelling in the emulsion [34e].

Diffraction Efficiency in Color Holograms. As we showed in Section 19.3.1.7, image luminance depends on the degree of modulation obtained in the hologram. This, in turn, is limited by the range of transmittance (or phase) change permitted by the linearity requirements (see Section 19.3.3.2). If, now, several holograms are superimposed, we must allow for all of them to be in phase at some point, so that their amplitudes add to yield the resulting amplitude. Thus, if there are N components, the amplitude, a_i, of each is limited to

$$a_i \leqslant \frac{a_0}{N}, \qquad (19.171)$$

where a_0 is the amplitude permitted if there were but one hologram.

As noted earlier (19.146), the diffraction efficiency varies with the square of the modulation, so that the diffraction efficiency here is reduced by a factor of N^{-2}. Thus, for instance, the efficiency of a three-color hologram is inherently reduced by a factor of nine below that of an equivalent monochrome hologram.

19.3.3 Holographic System Components: Requirements

19.3.3.1 Coherence and Monochromaticity in Recording. The degree of diffraction of the incident light, and hence the efficiency of the reconstruction, is determined directly by the modulation of the hologram fringe pattern. This modulation, together with the mean hologram transmittance, determines the image luminance obtainable with a given reconstruction light source. The modulation of the fringe pattern is therefore of primary importance and we analyze it here in terms of the degree of coherence between reference and object waves, their direction of polarization and their relative amplitudes.

For incoherently generated flux, the spatial coherence can be found from the van Cittert–Zernike theorem (Section 2.4.2.2) and the temporal coherence from the Fourier transform of the energy spectrum (2.4.2.1). For laser sources, with the usual filtering for the purely axial (TEM$_{00}$) modes, the spatial coherence may be assumed complete. For good temporal coherence, however, a single axial mode must be filtered out. (See Section 19.3.4.1 and 19.3.4.3 for appropriate techniques.)

Treating, first, a source consisting of a single spectral bandwidth Δv_t, we note that its coherence time is given by

$$\Delta t \approx \frac{1}{4\pi \, \Delta v_t}. \qquad (19.172)$$

This is based on our result (3.49) from Fourier theory and is accurate for a spectral band of Gaussian shape. To translate this into terms of wavelength range, $\Delta\lambda$, and coherence length, we substitute into it

$$\Delta v_t = \left|\frac{dv_t}{d\lambda}\right| \Delta\lambda = \left|\frac{d}{d\lambda}\frac{c}{\lambda}\right| \Delta\lambda = c\frac{\Delta\lambda}{\lambda^2}, \qquad (19.173)$$

and

$$\Delta s = c \, \Delta t, \qquad (19.174)$$

where c is the velocity of light. Hence

$$\Delta s \approx \frac{\lambda^2}{4\pi \, \Delta\lambda} \qquad (19.175)$$

is the coherent pathlength. The pathlength differences used in hologram recording or reconstruction must not exceed this value, if significant interference is to be obtained.

On the other hand, if several axial modes are present simultaneously, as, for instance, in the output of a laser, the coherence length will be controlled by the mode structure rather than the shape of the individual lines.

Here the temporal degree of coherence, $\hat{\gamma}(\Delta t)$, is again found as the Fourier transform of the energy spectrum. In the presence of n purely axial modes, this spectrum can be approximated by a sum of n delta functions, spaced

$$\Delta \nu_t = \frac{c}{2L}, \tag{19.176}$$

where L is the length of the laser cavity [see (2.240)].

This Fourier transform is similar to the one we found in our analysis of the diffraction grating of n infinitesimal slits. See (2.171) and Figure 2.18b. Accordingly

$$\hat{\gamma}\left(\frac{s}{c}\right) = \frac{\sin\left(\pi n s/2L\right)}{n \sin\left(\pi s/2L\right)}, \tag{19.177}$$

where we have written ($\Delta t = s/c$) for the time shift and s is the pathlength difference.

For $n = 2$, this reduces to

$$\hat{\gamma}\left(\frac{s}{c}\right) = \cos\frac{\pi s}{2L}. \tag{19.178}$$

As indicated by (19.177), whenever $n(>1)$ axial modes are present, the coherence drops to zero for a pathlength difference

$$s_p = \frac{2mL}{n}; \quad n > 1, \quad m = 1, 2, 3, \ldots, \tag{19.179}$$

and reaches another maximum for any pathlength at which the denominator vanishes, that is, when

$$s_p = 2mL. \tag{19.180}$$

Secondary maxima occur when the sine in the numerator equals unity, that is, when

$$s_s = \frac{(2m+1)L}{n}. \tag{19.181}$$

Note that the location of the secondary maxima depends on the number of active axial modes. If, in holographic recording, we operate at such a secondary maximum, a change in n may cause the disappearance of the interference fringes. Such changes do seem to occur spontaneously in ordinary laser operation where the cavity length changes with temperature, changing the cavity mode wavelengths (shifting the dotted bell-shaped spikes in Figure 7.15) and may, therefore, increase or decrease by unity the number of such modes excited. This has been cited to account for the periodic fading and reappearing of fringes in holographic recording [51].

Turning to polarization effects, we treat the case of light linearly polarized, taking the x axis normal to the plane defined by the reference and object wavevectors. We may write the complex amplitude of the reference wave in the hologram plane [see (19.79)]:

$$\hat{A}_1 = A_1 e^{ik_{1y}y}, \tag{19.182}$$

where $k_{1y} = |\mathbf{k}_1| \sin \theta_1$ is the y component of $\mathbf{k}_1$ and θ_1 is the angle of incidence of the reference wave on the hologram plane. For the object wave we write separately the components polarized parallel and perpendicular to the direction of polarization of the reference wave. These are, respectively,

$$\hat{A}_{0\|} = A_0 \cos \psi e^{ik_{0y}y} \tag{19.183}$$

$$\hat{A}_{0\perp} = A_0 \sin \psi e^{ik_{0y}y}, \tag{19.184}$$

where ψ is the angle between the direction of polarization of the reference and object waves.

Of the three components (19.182)–(19.184), only the first two are capable of interfering and we have for the mean value of their product:

$$\hat{A}_1 \hat{A}_{0\|}^* = A_0 A_1 \cos \psi \hat{\gamma} \exp i(k_{1y} - k_{0y})y$$
$$= A_0 A_1 \gamma \cos \psi \exp i[(k_{1y} - k_{0y})y + \phi]. \tag{19.185}$$

See (2.200) for the significance of

$$\hat{\gamma} = \gamma e^{i\phi},$$

there denoted by $\gamma(r_1, r_2; \tau)$.

The total irradiation at any point y in the hologram plane is then given by

$$E(y) = (\hat{\mathbf{A}}_1 + \hat{\mathbf{A}}_0) \cdot (\hat{\mathbf{A}}_1^* + \hat{\mathbf{A}}_0^*) = A_1^2 + A_0^2 + 2\mathcal{R}(\hat{A}_1 \hat{A}_{0\|}^*). \tag{19.186}$$

See the derivation of (2.183). Hence, using (19.185),

$$E(y) = A_1^2 + A_0^2 + 2A_0 A_1 \gamma \cos \psi \cos \beta(y), \tag{19.187}$$

where we have abbreviated:

$$\beta(y) = (k_{1y} - k_{0y})y + \phi. \tag{19.188}$$

$E(y)$ attains its maximum and minimum values, respectively, when $\cos \beta$ equals plus and minus unity. Hence the modulation of the fringe pattern in the hologram plane is

$$M = \frac{E(y)_{mx} - E(y)_{mn}}{E(y)_{mx} + E(y)_{mn}}$$

$$= \frac{2A_0 A_1 \gamma \cos \psi}{A_0^2 + A_1^2} = \frac{2\sqrt{r}\gamma \cos \psi}{1 + r}, \tag{19.189}$$

where $r = A_0^2/A_1^2$ is the ratio of object to reference wave radiance. Solving (19.189) for r, we find for the required ratio

$$r = \frac{(1 \pm \sqrt{1 - B})^2}{B}, \tag{19.190}$$

where

$$B = \left(\frac{M}{\gamma \cos \psi}\right)^2, \qquad M \leq \cos \psi.$$

Note that the maximum value of the modulation for any given value of r is

$$M = \frac{2\sqrt{r}}{1 + r}. \tag{19.191}$$

Also, that

$$M(r) = M\left(\frac{1}{r}\right) \tag{19.192}$$

and that this attains its upper limit, that is, unity, only if r, γ, and $\cos \psi$ all equal unity.

Thus we see that for high modulation, r should be close to unity. In the following section we shall see that the limited linearity range of recording media demands a value of r less than unity, so that the optimum value of r must be a compromise between the demands of linearity and diffraction efficiency.

Concerning the factor $\cos \psi$, which should be unity, we note that this may be obtained when the reference and object waves are both polarized normal to the "interference plane," that is, the plane defined by the wave vectors of these two waves. Usually, then, it is desirable to have the direction of polarization perpendicular to the surface of the optical bench. If it has a component parallel to the diffraction plane, ψ will differ from zero (for offset holography) and hence $\cos \psi$ will be less than unity [52].

We briefly note that the coherence of object and reference waves is compromised when the object being recorded has a motion having a radial component. This is due to the Doppler shift in the reflected wave. If the reference wave has a similar shift applied to it, we obtain a method for "Doppler mapping," that is, mapping "isovelocity" regions on the object [53].

19.3.3.2   Linearity of Recording. The exposure of the recording material is obtained as the time integral of the irradiation and, assuming the irradiation constant during the exposure time, t, the exposure is

$$H = Et. \tag{19.193}$$

Separating the constant and y dependent terms of E [see (19.187) and (19.189)], we may write

$$H = H_0 + H_1 \cos \beta, \tag{19.194}$$

where

$$H_0 = (A_0^2 + A_1^2)t \tag{19.195}$$

$$H_1 = 2A_0 A_1 \gamma t \cos \psi = H_0 M. \tag{19.196}$$

The exposure (together with the processing, if required) is converted into a change of amplitude transmittance, $\hat{\tau}$. We write this as a phasor since, in general, it is composed of both a magnitude, $\tilde{\tau}$, and a phase, ϕ,

$$\hat{\tau} = \tilde{\tau} e^{i\phi}. \tag{19.197}$$

These represent, respectively, the change in amplitude and phase experienced by the transmitted light. In practice it is usual to have one or the other of these predominate to such an extent, that the other may be neglected. Depending on the significant factor, one speaks then of *absorption* or *phase holography*, respectively.

Let us first treat absorption holography. Here the amplitude transmittance ($\tilde{\tau}$) resulting from the exposure is obtained from the characteristic (transmittance versus exposure) curve. In general, this is not linear. The nonlinearity of the recording material introduces distortion into the recorded interference pattern and this, in turn, lowers the effective diffraction efficiency and produces spurious images [54]. We may, however, limit operation to a region where the curvature is small, so that the characteristic may be approximated by

$$\tilde{\tau} = \tau(H_0) - H_1 \tilde{\tau}'(H_0)\gamma \cos \beta + \tfrac{1}{2} H_1^2 \tilde{\tau}''(H_0)\gamma^2 \cos^2 \beta + \cdots, \tag{19.198}$$

where $\tilde{\tau}'$, $\tilde{\tau}''$ are the first and second derivatives, respectively, of $\tilde{\tau}$ with respect to H.

For the recording to be approximated by a linear process, the terms beyond the first two in (19.198) must be negligible, so that we obtain, on substituting from (19.196),

$$\tilde{\tau} = \tilde{\tau}_0 + H_0 M \tilde{\tau}_0' \gamma \cos \beta. \tag{19.199}$$

The region of linearity must thus extend over the full range of $\tilde{\tau}$, as limited by the values at

$$\cos \beta = \pm 1.$$

According to (19.194) and (19.196) this implies an exposure range

$$H_0(1-M) < H < H_0(1+M). \tag{19.200}$$

The required linear range, R, is thus given by

$$R = \frac{1+M}{1-M}. \tag{19.201}$$

Alternatively, with R given by the recording material characteristic, we find on solving (19.201) for M, that the available R limits the permissible modulation to

$$M \leqslant \frac{R-1}{R+1}. \tag{19.202}$$

Once M is fixed in this manner, it determines the irradiation ratio of the object to reference waves, as given by (19.190).

Turning to phase holography, we find that here linearity demands that the derivative of the phase factor be constant:

$$\frac{d}{dH}(e^{i\phi}) = ie^{i\phi} \frac{d\phi}{dH} = k. \tag{19.203}$$

On solving this, for example, by separating the variables, we find

$$\phi = -i \log H + k.$$

With real H, this implies a complex or imaginary ϕ which, in turn, implies amplitude attenuation. Thus a pure phase factor is inconsistent with linearity. However, for very small values of ϕ, the exponential is close to unity and then, according to (19.203) a constant value of $(d\phi/dH)$ is sufficient for almost linear operation. It is this fact that makes phase holography, with its increased efficiency, practical.

19.3.3.3 Recording Medium: Resolving Power, Size and Mechanical Stability.

Clearly, the recording medium must have a resolving power sufficient to accommodate the spatial frequencies contained in the hologram, and these may be quite high. According to (19.94) this frequency, for any object point, is given by

$$\nu = \frac{\sin\theta_0 - \sin\theta_1}{\lambda} \approx \frac{\Delta\theta}{\lambda}, \qquad (19.204)$$

where $\Delta\theta$ is the angle between the object and reference wave vectors. According to (19.90), this angle must be at least twice the total angular subtense of the object, if an image free of overlap is to be obtained. Thus even if we take an object of half-angle only ($\varphi = 0.1$), we find a minimum spatial frequency of over 630 cycles/mm when light from a He-Ne laser ($\lambda = 0.6328\ \mu$m) is used. (With argon ion laser light of 0.488 μm, it is about 820 cycles/mm.) Therefore we must use a recording medium exhibiting an mtf sufficiently high so as not to attenuate excessively the recorded signal. In effect, this mtf becomes another factor multiplying the modulation of the hologram fringes as given by (19.189).

Equation 19.204 is valid for thin holograms. In thick ones the result is similar, if the interference layers are close to perpendicular to the surface of the medium. In reflection holograms, however, the required spatial frequency band can be very much higher. For instance, when the interfering waves are incident normal to the surface, the layers are oriented parallel to the medium surface and the fringe spacing is half a wavelength as measured in the medium. By way of illustration, assuming again a vacuum wavelength of 0.6328 μm and a refractive index of 1.5, the spatial frequency is

$$\nu = \frac{2n}{\lambda_0} = 4740\ \text{mm}^{-1},$$

a very stringent requirement, which calls for special photographic materials.

It should be noted that in Fourier holography the resolving power of the recording medium affects only the luminance of the image and only indirectly the image detail resolved. This latter is determined primarily by the width, w, of the hologram, or hologram portion utilized in the reconstruction. Specifically, the resolved element size is, according to (9.108)

$$\Delta h = \frac{\lambda z_0}{w}, \qquad (19.205)$$

where z_0 is the object-hologram distance.

Information Capacity. We now estimate for a Fourier transform hologram the number of resolution elements obtainable with a given hologram size and recording medium mtf. Let the limiting spatial frequency
having an acceptable mtf be N. Then $\Delta\theta$ in (19.204) has a maximum
value of

$$\Delta\theta \leqslant \lambda N. \tag{19.206}$$

For an object size, $2h$, $\Delta\theta$ must be allowed a value of, at least,

$$\Delta\theta \geqslant \frac{h}{z_0}, \tag{19.207}$$

even without offset and with an infinitesimal hologram size. Considering
this in conjunction with (19.206), we conclude that the half-object size is
limited to

$$h \leqslant z_0 \lambda N = h_{mx}. \tag{19.208}$$

Hence, recalling result (19.205), we find the number (n) of object
elements (width Δw) resolved across the total object width ($2h$) is

$$n = \frac{2h}{\Delta h} = 2Nw. \tag{19.209}$$

Comparison with (3.59) shows that this is identical to the number of
object elements that would have been resolved across the object if it had
been recorded with ordinary photography on this recording medium.
Note that this effectively allows one object element resolved for each
half-cycle accommodated across the width of the recording at the limiting
frequency, N. We have noted earlier (19.90b) that, if a clean image
without overlap is to be obtained, the reference point must be removed
from the object by the half-object width itself, at least. This limits the
total object size that can be accommodated to h_{mx} and hence to one-half
the amount of (19.209): the number of elements that can be recorded
under these conditions is only half the number that could be recorded in
ordinary photography. This loss of information capacity is caused by the
existence of the twin image and this, in turn, is due to the fact that only
one component of the object wave (the one in phase, or phase opposition,
with the reference wave) is recorded [55]. In the presence of noise,
further losses are due to the condition ($A_0 \ll A_1$) on which (19.90b) is
predicated. When this condition is relaxed, (19.90a) applies and only
one-sixth of the full information capacity can be accommodated. Even
then additional losses are caused by the need to modulate the object wave
onto the reference wave with the degree of modulation limited by the
linearity of the recording material [55].

Fictitious Mask. Up to this point, we ignored the effect of the recording mtf, except for its cutoff frequency. We now analyze the effect of the transfer function of the recording material in more detail. Consider first the Fourier-transform hologram, where object and reference wave source are in the same plane parallel to the hologram. Note that an object point (P) at a distance x from the reference point (R) gives rise to a fringe pattern of spatial frequency approximately equal to [see (19.94)]

$$\nu \approx \frac{\alpha}{\lambda} \approx \frac{x}{z_0 \lambda}, \qquad x \ll z_0, \tag{19.210}$$

where z_0 is the distance from the object plane to the hologram.

Clearly, the mtf of the recording material will attenuate this fringe pattern by a factor $T(x/z_0\lambda)$. This effect is equivalent to an attenuating mask placed over the object, with the luminance attenuation proportional to the emulsion mtf, with the scale given by (19.210) and the origin at the reference source point. This has been called a *"fictitious mask"* [56]. See Figure 19.26a.

In the Fresnel and Fraunhofer holograms, where the reference point source is not in the object plane, the fictitious mask can still be constructed, with the location of the origin now varying with the location of the hologram region under consideration. Refer to Figure 19.26b. The origin is found by drawing a line from the hologram region (A) under consideration to the reference point (R); the point (O) of intersection with the object plane locates the origin. The scale is still given by (19.210).

In this case, the "fictitious mask" limits both the field and the effective aperture, so that the medium mtf may affect resolution in the image also.

Hologram Size. We have already noted that in the holography of diffuse objects, each point on the hologram receives information from every object point. Thus any small section of the hologram should suffice to reconstruct the whole object. Indeed, this is essentially true (if we disregard obscuration effects in three-dimensional objects). Note, however, that the hologram acts as an aperture through which the image is formed. The smaller this aperture, the greater the blur caused by diffraction effects inherent in the limitation on the wavefront extent. Thus the amount of object detail resolvable in the reconstruction depends directly on the size of the hologram used, as given, for instance, by (19.205) [57].

Mechanical Stability. Because of the small pitch of the fringes generally used in holography, mechanical stability, too, becomes a major consideration. Although air turbulence in the beam path may be a factor in fringe

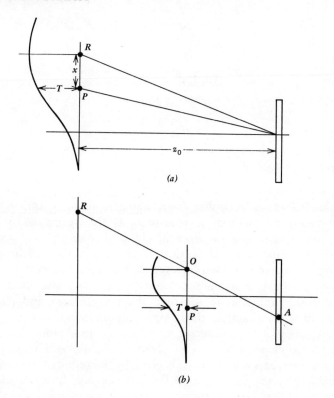

(a)

(b)

Figure 19.26 Illustrating the "fictitious mask" concept: (a) Fourier transform hologram; (b) Fresnel transform hologram.

stability, usually vibrations of the base and components are the prime disturbing factor.

If we assume the motion involved to be sinusoidal, we can readily calculate its effect on the amplitude of the sinusoidal fringe associated with any object point. According to (19.94) its spatial frequency is

$$\nu = \frac{\sin \theta_1 - \sin \theta_0}{\lambda}. \tag{19.94}$$

According to Section 18.4.1.1, sinusoidal motion, amplitude $\frac{1}{2}X$, has a blurring effect, which has associated with it an otf in the form of a zero-order Bessel function:

$$\mathbf{T}_s(\nu) = J_0(\pi X \nu). \tag{18.93}$$

When the image itself is sinusoidal, the otf simply acts to attenuate the amplitude.

Thus (18.93) is an additional factor of the fringe modulation as given by (19.189). Alternatively, it can be treated as an additional "fictitious mask" placed over the object [58]. The first zero of the zero-order Bessel function occurs at 2.405 and, hence, a vibratory motion of amplitude

$$\tfrac{1}{2}X = \frac{2.405}{2\pi\nu} = \frac{0.3827\lambda}{\sin\theta_0 - \sin\theta_1} \tag{19.211}$$

will cause the vanishing of the fringes of an object point at an angle θ_0 from the hologram normal.

19.3.3.4 Sensitivity and Exposure Time.

The *holographic sensitivity* of a holographic recording medium is measured as the reciprocal of the exposure required to produce a given image luminance per unit reconstruction flux. As pointed out in Sections 17.1.3.1 and 17.4.1.1, respectively, recording techniques based on silver halide and photoconductivity photography may have a significant quantum gain and are, therefore, likely to be considerably more sensitive than, say, diazo and photochromic materials. In practice, the fringe modulation, and hence the diffraction efficiency of the hologram increases, up to a point, with the quantity of radiation per unit area received by the hologram from the object. The required exposure time (t) is given by the required quantity of radiation divided by the flux received from the object. In terms of the photographic sensitivity, $k_D(\lambda)$, for density, D, at wavelength, λ,

$$k_D(\lambda) = \frac{1}{Et}.$$

Hence

$$t = (kE)^{-1} = \frac{A}{k\eta\Phi}, \tag{19.212}$$

where E is the irradiation at the hologram,
$\quad A$ is its area,
$\quad \Phi$ is the total laser flux, and
$\quad \eta$ is the efficiency with which the laser flux is transferred to the object and hence to the recording medium.

In diffuse object holography, no more than about 10% of the object flux can be expected to reach the hologram, so that, when the flux required for the reference beam is considered, η can be taken as no more than half that value, that is, 0.05.

By way of illustration, consider a 10-cm^2 ($A = 10^{-2}$ m^2) hologram made with a 1-mW He-Ne laser on E.K. 649F emulsion ($k^{-1} = 0.8$ J/m^2). Equation 19.212 yields for this

$$t \geqslant 160 \text{ sec.}$$

19.3.3.5 Granularity and Form Stability of the Recording Material

Granularity. The granularity of the recording material will cause the appearance of noise in the form of random luminance fluctuations in the hologram image. Even if the granularity introduces only pure phase shifts, interference effects will convert these into amplitude modulation.

By way of illustration, we treat the Fourier hologram, where the image luminance amplitude is the Fourier transform of the transmittance distribution in the hologram plane. Hence, if we use a uniformly exposed grainy recording material as a hologram, we obtain in the image plane a luminance distribution following $|S_g|^2$, where S_g is the granularity spectrum:

$$S_g(\nu_x, \nu_y) = \mathscr{F}[\tilde{\tau}(X, Y)] \tag{19.213}$$

with scale factors, according to (9.85)

$$\nu_x = \frac{x'}{\lambda b}, \qquad \nu_y = \frac{y'}{\lambda b}, \tag{19.214}$$

where $\tilde{\tau}$ is the amplitude transmittance of the grainy hologram,

    X, Y are the coordinates in the hologram plane,

    x', y' are those in the image plane, and

    b is the distance from the hologram to the image plane.

Any actual hologram can be viewed as the superposition of the ideal hologram pattern on a grainy background similar to the one just described. This grainy background will produce, in the image plane, a general luminance distribution veiling the image. This luminance will have a mean value $\overline{L_N}$, say, and a standard deviation therefrom also equal to $\overline{L_N}$, as shown in (19.19).

Potentially even more significant, however, are the random fluctuations that this veil introduces into the image luminance. These are due to the interference of the veil luminance, with its random phase, with the image luminance, with which it is coherent.

We define the magnitude of these fluctuations, the noise, N, by the root-mean-square deviation in the image luminance caused by the interference effects. This may be calculated from the mean square value according to (3.41):

$$N^2 = \text{var}(L) = \overline{L^2} - \overline{L}^2. \tag{19.215}$$

The luminance itself can be calculated from the sum of image and background amplitudes:

$$L = |\sqrt{L_i} + \hat{a}_N|^2 = L_i + \overline{L_N} + \sqrt{L_i}\mathscr{R}(\hat{a}_N),$$

where $\sqrt{L_i}$ is the amplitude of the image luminance,

$\hat{a}_N = a_N e^{i\phi}$ is the phasor amplitude of the granularity luminance, with presumed constant amplitude, a_N, and random phase ϕ relative to the image wave, and

$a_N^2 = \bar{L}_N$ is the mean background luminance.

Thus we may write

$$L = L_i + \bar{L}_N + 2\sqrt{L_i}a_N \cos \phi, \tag{19.216}$$

with ϕ uniformly distributed over the circle. Since $\cos \phi$ vanishes in the mean, we have for the mean value of L:

$$\bar{L} = L_i + \bar{L}_N. \tag{19.217}$$

To find the variance of

$$L_N = \left|\sum a_n e^{i\phi_n}\right|^2, \tag{19.218}$$

we refer to the appendix (19.308) and note that

$$\text{var } L_N = \bar{L}_N^2. \tag{19.219}$$

To find the variance of L, we note that the variance of a sum of random variables equals the sum of their variances and that the variance of L_i vanishes. Hence, from (19.216) and (19.219),

$$N^2 = \text{var } L = \bar{L}_N^2 + 4L_i\bar{L}_N \text{ var } (\cos \phi)$$

$$= L_N^2 + \frac{2L_i\bar{L}_N}{\pi}\int_{-\pi}^{\pi} \cos^2 \phi \, d\phi$$

$$= \bar{L}_N^2 + 2L_i\bar{L}_N. \tag{19.220}$$

The signal, on the other hand, is given by

$$S = \bar{L} = L_i + \bar{L}_N \tag{19.221}$$

and hence the signal-to-noise ratio by

$$R^2 = \frac{S^2}{N^2} = \frac{(L_i + \bar{L}_N)^2}{\bar{L}_N^2 + 2L_i\bar{L}_N}. \tag{19.222}$$

In the situation usually of interest, $L_i \gg L_N$ and then

$$R = \sqrt{\frac{L_i}{2\bar{L}_n}}. \tag{19.223}$$

We note, in conclusion, that the effective detectivity of a photographic medium used for holographic recording may be made substantially higher than its detectivity when used in conventional photography. This advantage is due to the fact that the signal-carrying component of the resultant exposure distribution is directly proportional to the amplitude of the reference wave; see (19.86). The amplitude of the reference wave can thus be used as an amplifier for the signal [59].

Index Matching. Random surface unevenness contributes an important term to the phase portion of the granularity. This, however, can be eliminated by immersing the hologram in a liquid matching the refractive index of the recording medium. For instance, by mounting a silver-halide hologram against a glass plate, with a few drops of xylene to fill the gap between the emulsion and the cover plate, we can eliminate a major part of the phase granularity introduced by the surface distortions. Alternatively, the hologram may be placed inside a trough (a "liquid gate") filled with an index matching liquid [60].

Form Stability of the Recording Material. In addition to meeting the requirements already mentioned, the recording material should maintain its shape so that is has, during reconstruction, the same form as it had during recording. Any distortion or buckling will affect the reconstructed image: shrinking or expansion produce a change in scale [factor m in (19.112)]—and the accompanying displacement, a change in phase—of the imaging contributed by the hologram element considered. The effect of buckling can be analyzed in terms of hologram rotation and bending, and these, in turn, introduce distortion and astigmatism into the imaging [61].

19.3.3.6 Coherence and Monochromaticity in Reconstruction.

After having clarified conditions for holographic recording, we turn to those for reconstruction. Although an original hologram can always be reconstructed by means of a wave identical to the reference wave used in recording, this is a sufficient, but not a necessary condition.

We start with (19.123)–(19.125) and rewrite these for the usual case of an unmagnified hologram ($m = 1$) and for a reconstruction source in the plane of the original reference source ($z_2 = z_1$). This yields a virtual image with

$$z_3 = \frac{z_0 z_1}{z_0 + \mu z_1 - \mu z_0} \tag{19.224}$$

$$x_3 = \frac{x_2 z_0 + \mu x_0 z_1 - \mu x_1 z_0}{z_0 + \mu z_1 - \mu z_0}. \tag{19.225}$$

Since the performance in the y direction is totally analogous to that in the x direction, the analysis in the x direction is representative and sufficient.

Let us now investigate the effect of a finite source size in reconstruction on the assumption that it is quasimonochromatic and of the same wavelength as the recording process ($\mu = 1$). This yields for (19.224) and (19.225):

$$z_3 = z_0 \tag{19.226}$$

$$x_3 - x_0 = \frac{(x_2 - x_1)z_0}{z_1}. \tag{19.227}$$

These imply, respectively, that the image point will be in the object plane and that it is displaced by an amount proportional to the displacement of the reconstruction point source $(x_2 - x_1)$, with the proportionality factor equal to z_0/z_1. A moments thought will show us that, accordingly, *an extended reconstruction source will introduce a blur with a spread function equal to the source luminance distribution with a scale factor z_0/z_1 applied to the coordinates.* Thus, with a distant reference source and a very short object-to-hologram distance, even a sizable reconstruction source may introduce but a negligible blur (see image plane holography, Section 19.3.2.5).

Proceeding to the other extreme, a point source with extended spectrum, we modify (19.224) and (19.225) by introducing $x_1 = x_2$. To simplify the analysis, we also restrict ourselves to a very distant reference source

$$z_1 \gg \frac{z_0(1-\mu)}{\mu}. \tag{19.228}$$

With these restrictions, the equations become

$$z_3 = \frac{z_0}{\mu} \tag{19.229}$$

$$\Delta x = x_3 - x_0 = \frac{x_1(\mu^{-1}-1)z_0}{z_1}, \tag{19.230}$$

where Δx is the shift in the image point due to the change, μ, in the wavelength. Recalling (19.113),

$$\mu = \frac{\lambda_2}{\lambda_1} \tag{19.231}$$

and letting

$$\tan \theta_1 = \frac{x_1}{z_1} \approx \theta_1, \tag{19.232}$$

the reference wave angle, (19.230) becomes

$$\Delta x = \frac{z_0 \theta_1 (\lambda_1 - \lambda_2)}{\lambda_2}. \qquad (19.230a)$$

Conversely, any displacement, Δx, corresponds to a reconstruction wavelength

$$\lambda_2 = \frac{\lambda_1}{1 + \Delta x / z_0 \theta_1}. \qquad (19.233)$$

For a source with continuous spectrum, the luminance at any point Δx from the original image point will clearly be proportional to the energy spectrum of the source at the wavelength given by (19.232). Thus a reconstruction source with an extended energy spectrum $W[\lambda]$, results in a line spread function:

$$L(\Delta x) \sim W\left(\frac{\lambda_1}{1 + \Delta x / z_0 \theta_1}\right), \qquad (19.234)$$

where λ_1 is the wavelength used in the recording process.

Lack of coherence and monochromaticity introduce blurring in another way, as well. Due to diffraction effects, the size of the hologram grating determines the sharpness of the image, even in the absence of aberrations. However, since holographic imaging is based on coherence, hologram size can contribute to sharpness only to the extent that the irradiation is coherent over that size. Thus the van Cittert–Zernike theorem (Section 2.4.2.2) limits the *effective* hologram size for a reference source of finite dimensions. Similarly, a large hologram size will be effective only to the extent that the coherence length of the light, as determined by its monochromaticity, suffices to cover the pathlength differences introduced by the hologram width. Figure 19.27 shows that the pathlength difference

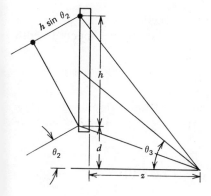

Figure 19.27 Illustrating pathlength difference in a large hologram.

introduced by a hologram, width h, when irradiated by a coherent collimated wave at angle θ_2, is

$$\Delta s = h \sin \theta_2 - \sqrt{z^2 + (h+d)^2} - \sqrt{z^2 + d^2}, \qquad (19.235)$$

where z is the distance of the image point from the hologram plane,
$\qquad \theta_3$ is the azimuth angle of the image point relative to the normal to the hologram, and

$$d + \tfrac{1}{2}h = z \tan \theta_3, \qquad (19.236)$$

is the y coordinate of the image point, relative to the hologram center. Here the calculations are made for rays in the y–z plane, which is taken parallel to the reconstruction wave vector.

If the coherence length of the reconstruction wave is less than this, only a portion of the hologram will contribute coherently to formation of any image point, with a corresponding limitation on the resolution. We may then consider Δs as given by the coherence length and solve (19.235) for h.

19.3.4 Holographic System Components: Practical Considerations

In this section we analyze the requirements to be met by components used in holography and provide a brief survey of those most frequently used there.

19.3.4.1 *Light Sources and Their Filtering.* It was pointed out in Section 19.3.3.1 that the efficiency of the holographic process varies directly with the magnitude of the coherence factor $\gamma(t)$ and how this dictates the monochromaticity of the light source used in holography. According to our result (19.175) there, the relative spectral width must be limited to

$$\frac{\Delta\lambda}{\lambda} \leq \frac{\lambda}{4\pi \, \Delta s}, \qquad (19.237)$$

if coherence is to be maintained over a pathlength difference of Δs. For instance, with a He–Ne laser and $\Delta s = 0.1$ m, the relative spectral width must be no more than

$$\frac{\Delta\lambda}{\lambda} = 0.5 \times 10^{-6}.$$

Comparing this to the spectral purity obtainable from a gaseous discharge running at a reasonable luminance level, say a mercury discharge lamp at 1000 K, we find from (4.21):

$$\frac{\Delta\lambda}{\lambda} = \frac{\Delta\omega}{\omega} = 7.16 \times 10^{-7} \sqrt{\frac{T}{M}} = 1.6 \times 10^{-6}$$

for the Doppler broadening alone. [Here T is the absolute temperature of the gas and M ($=200$) is its molecular weight.] This is too large by a factor of three, at least. But even then, the high demands of spatial coherence, which we have neglected so far, would restrict us to using a very small region of the discharge and would, therefore, result in an extremely low illumination level, implying very long exposure times.

For this reason, laser sources are used almost exclusively in holographic recording. In these, the relative spectral width can be maintained at approximately 10^{-9}; see Section 7.2.21. However, even here care must be taken to eliminate multiple modes.

The transverse modes are quite readily eliminated by spatial filtering. But there will, generally, remain a number of axial modes spaced

$$\Delta \nu_t = \frac{c}{2L}. \tag{19.176}$$

Consequently, coherence, and therefore interference, will then vanish for a pathlength difference equal to the laser cavity length, L, even if only two modes are present [see (19.178)]. If three modes are present, this zero will occur at half this distance, etc. It is therefore highly desirable to eliminate all but one axial mode, if a significant coherence length is needed. This can be accomplished by using a cavity sufficiently short, so that only one mode is contained within the spectral band of the lasing molecule. In a He–Ne laser such a cavity would usually have to be less than approximately 10 cm. On the other hand, this expedient limits severely the power obtainable from the laser. Alternatively, all but one of the axial modes may be eliminated, by coupling two resonant cavities so that, within the spectral band, they have only one mode in common. One of these cavities may be made long, so as to accommodate a large volume of the laser medium and the other one short, to ensure a large mode spacing. Figure 19.28 illustrates two such techniques.

By illuminating various portions of the object by separate beams split off from the main beam and differing from each other, in pathlength, by $2L$, the periodic reappearance of coherence can be used to achieve great depth of field [62].

19.3.4.2 Recording Materials [63]. Let us now compare the desired characteristics of the recording material with those found in practice. A wide variety of recording materials have been used in holography and many of these are superior in one respect or another. However, silver halide emulsions still seem to be the most popular and, apparently, the best in overall performance. In this section we therefore discuss these at length and then briefly survey other media that have been used successfully.

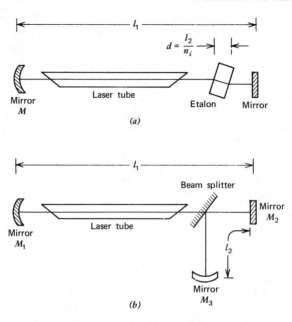

Figure 19.28 Schematics of axial mode filtering systems.

Silver Halide Materials [63a]. A number of silver halide materials are available, meeting the requirements listed in Section 19.3.3. Some of these are listed in Table 192 [64], together with their emulsion thickness and resolution capabilities; where available, approximate values of the following, too, are given: exposure requirements (at 0.514 and 0.6328 μm), maximum slope of the transmittance versus exposure curve, and limits of linearity. See Table 193 for scattering data on some materials [65], [66].

The dependence of diffraction efficiency on exposure level, modulation, and wavelength, are important considerations in planning holography [66].

When very short pulses are used, reciprocity law failure may be significant; see Section 17.2.1.5. Methods of alleviating this have been investigated [67].

Processing techniques have been described that will minimize thickness distortion and even compensate for shrinkage [34e].

Silver halide materials also lend themselves to the preparation of phase holograms [68]. By dissolving the silver grains out of the processed emulsion, the absorption modulation is converted into phase modulation.

In one process, the metallic silver grains are converted into transparent silver salts whose refractive index differs from that of the surrounding gelatine. A process using potassium ferricyanide is described in Ref. 34e. A dry technique, using bromine vapor, too, is available [69]; with it a refractive index modulation of 0.033 and diffraction efficiencies of over 70% have been obtained.

Alternatively, the silver grains may be dissolved out of the emulsion and the adjacent gelatine tanned with the degree of tanning proportional to the original metallic silver concentration. After drying, the emulsion thickness at any point will be increased according to the degree of tanning at that point. The emulsion thickness, however, limits the spatial frequency at which efficient surface deformation can be obtained, so that the former method is recommended for high-resolution holograms.

It has also been found that performance can be improved by bleaching and redarkening, yielding hybrid phase-absorption holograms [70].

Transmission versus exposure curves for two holographic emulsions are shown in Figure 19.29 and the root-efficiency versus fringe modulation curves for the same emulsions are shown in Figure 19.30(a) and (b), for various values of mean exposure [34e]. The spatial frequency there was about 1200 mm^{-1} and the wavelength 0.6328 μm. Since scattering in the emulsion, and hence the mtf, are wavelength-sensitive (see Section 17.2.2.3) the obtainable diffraction efficiency, too, is wavelength-dependent. Representative results are shown in Figure 19.30(c) [71].

Dichromated Gelatine [63b]. Dichromated gelatine film (see Section 17.3.2.1) in holography [72] is usually used partially prehardened. Its major advantages are (a) low absorption and therefore very high diffraction efficiency; (b) very high resolving power; and (c) low granularity. Its sensitivity, however, is rather low.

Photopolymers and Photoresists [63c]. Photopolymer materials (Section 17.3.2.2), too, can be good hologram recording media with full diffraction efficiency and a spatial frequency response good up to 3000 [73]–6000 mm^{-1} [74]. They are fixed by means of a postexposure, which completes the polymerization. In another system, index variations are photoinduced in the polymer [75].

Diazo. Diazo materials (Section 17.3.1) have been considered for holographic recording [76]. They are capable of very high resolution, but suffer from low sensitivity and their limited dynamic range precludes the attainment of a high diffraction efficiency. In terms of granularity they seem to be somewhat superior to silver halide emulsions [76].

Photothermoplastic [63d]. Photothermoplastic films (see Section

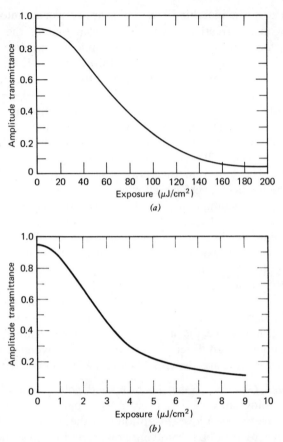

Figure 19.29 Amplitude transmittance versus exposure curves for (*a*) Kodak 649F and (*b*) Agfa-Gevaert 10E70 emulsions. Collier, Burckhardt and Lin [34].

17.4.1.2) represent a useful medium for phase holography [77]. They have the advantage of being reusable many hundreds of times and are developable almost instantaneously *in situ*. They also yield high efficiency holograms.

Ferroelectric Crystals [63*e*]. These crystals (see Section 14.3.7), too, have been employed as reusable phase hologram media [78]. Their high resolution capability and availability in large thicknesses make them promising candidates for high-density storage in the form of thick multiple holograms.

Photochromics [63*f*]. Photochromic materials (Section 17.3.4) are another reusable hologram material. They are particularly interesting because of their potentially large volume storage capabilities. They can be

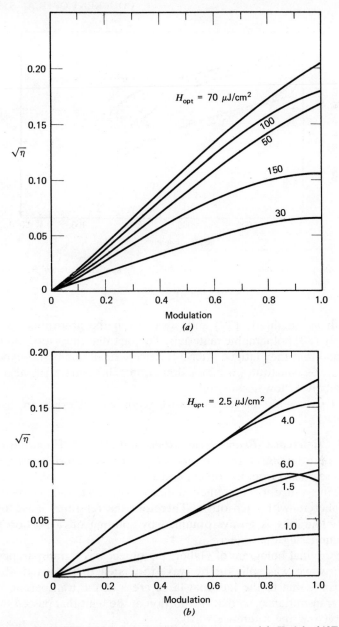

Figure 19.30 Root-efficiency versus modulation curves for (a) Kodak 649F and (b) Agfa-Gevaert 10E70 emulsions for various mean exposure values. Obtained with two plane waves incident at ±22.5°, respectively, to the emulsion normal. Wavelength 0.6328 μm. Each exposure value (in μ J/cm²) is indicated next to the corresponding curve. Collier, Burckhardt and Lin [34].

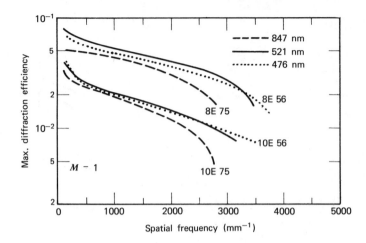

Figure 19.30(*c*) Dependence of diffraction efficiency on spatial frequency. Buschmann and Metz [71].

used both as amplitude [79] and, away from the absorption peak, as phase [79], [80] holographic materials. The fact that they, too, can not be "fixed" and that they fatigue after a relatively small number of exposure-erasure cycles constitutes a major limitation. These materials also suffer from a relatively low sensitivity.

Typical parameters of representative recording materials are listed in Table 194 [74].

19.3.4.3 Reference Beam Generation and Object Illumination. As mentioned at the outset, holography is based on the interference between the wave scattered from the object and a reference wave coherent with it. In the visible range it is extremely difficult to generate independently two waves coherent with each other. Therefore the reference wave used in optical holography is always obtained by splitting off a portion of the wave illuminating the object.

In the original hologram of Gabor, the object was a transparency. The reference wave was split off by wavefront area division and obtained automatically due to the fact that large areas of the transparency had a uniform transmittance, so that the portion of the light that passed through them served as the reference wave.

Reference wave generation techniques, useful also with opaque objects, are illustrated in Figure 19.31. The first two of these are also based on wavefront area division; the object and reference portions of the wave may be recombined by means of a prism (*a*) or a mirror (*b*). Wavefront

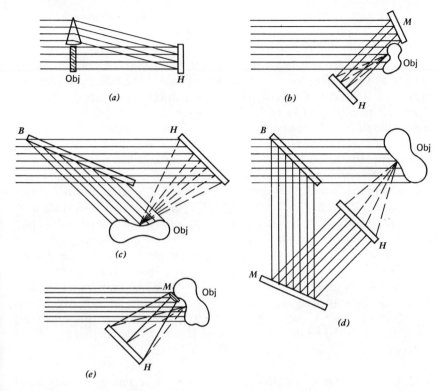

Figure 19.31 Reference wave generation techniques. *M* is mirror. *B* is beam splitter; *H* is hologram. In each figure, the light is incident from the left.

amplitude division is implemented by means of a beamsplitter, as illustrated in (*c*), there.

In reflection holography, the reference beam is incident on the side opposite the one facing the object. A typical arrangement is shown in Figure 19.31*d*. This technique, incidentally, requires the use of a recording material capable of accepting exposure from both sides. Note that photographic films are often provided with an opaque antihalation backing that must be removed before the film can be used for reflection holography. This may be done by means of an appropriate solvent, such as alcohol. Always one must ensure that the difference in total pathlength between object and reference beam does not exceed the coherence length.

Occasionally, the reference beam may be derived from light reflected specularly from the object [81]. This is referred to as a "local reference beam" (LRB) [82] and is especially useful [77] when the object distance is large and unknown, so that it is difficult to ensure the limited optical pathlength difference demanded by the coherence requirements, if an independent reference beam is provided. This difficulty can be circumvented by generating the reference beam close to the hologram plane by means of light derived from part of the object illumination. We may either use part of the object as the reference source [81] or select part of the scattered light by spatial filtering [82]. This technique is illustrated in Figure 19.31*e*. The locally generated reference beam has also been used in image plane holography, where special methods are required [83].

Since the laser beam is usually quite small—no more than a few millimeters in diameter—it must be expanded if a large object is to be illuminated. This may be accomplished by means of an (inverted) telescope arrangement. A combination of two positive lenses spaced so that their focal points coincide (see Figure 9.41) permit filtering the laser beam for axial mode operation by simply placing a pinhole in the common focal plane. If such filtering is not required, a Galilean telescope (Figure 9.43) is more compact. If the laser beam need not be collimated, a far simpler approach is possible: a single positive or negative lens is placed in the laser beam path, causing it to diverge from the lens focal point and covering an object area whose size is determined by the object distance from this focal point. See Figure 19.32.

19.3.4.4 *Mechanical Considerations.* The need to minimize mechanical vibrations, as discussed in Section 19.3.3.3, calls for special mechanical arrangements. The table carrying the hologram recording system must be isolated as well as possible from the vibrations of the environment. The table legs may be placed, for instance, on a stack of alternating layers of sand and rubber. In one system, the heavy plate carrying the recording system is placed on inflated rubber tubes. More sophisticated systems rest on legs that are in the form of pistons in air-filled cylinders, with the air pressure servocontrolled, so as to keep the table top level. Whenever possible, the entire arrangement should be placed in an environment with

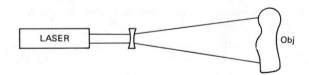

Figure 19.32 Illustrating a simple method for laser beam expansion.

minimum inherent vibrations, that is, remote from heavy traffic and as close as possible to the ground (the basement—not the top floor!). Since air turbulence, too, can cause disturbances, it may be advisable to cover the whole system during recording so as to minimize such turbulence. Note that the usual mechanical shutters may introduce vibrations and that for exposures of 1 sec or more, a handheld opaque sheet may be superior to a high-quality mechanical shutter.

19.3.4.5 *A Representative Holography Arrangement.* The reader may have followed the discussion on the prerequisites of successful holography—in all its details—up to this point, and yet may not be able to make a hologram. To remedy this situation, we present in Figure 19.33 a simple, practical optical arrangement, that should enable the reader to make a reasonably good Fresnel hologram of a three-dimensional opaque object using only readily available optical components and an inexpensive He–Ne laser. All components are readily mounted on a flat table and he must only take the precaution of providing very stable support and place it in a relatively vibration-free environment.

The spatial filter is desirable to provide a pure axial (TEM_{00}) mode. It can be constructed using a low-power microscope objective (say 16 mm f.l.) with a pinhole (say 25 μm in diameter), and the accurate alignment of the pinhole is the one critical adjustment in the arrangement. After leaving the pinhole, the beam will be slightly diverging. If this divergence does not suffice to illuminate the desired object area, it may be increased by the insertion of a negative lens into the laser beam, as shown.

Part of the wavefront is reflected toward the hologram plate or film by means of a mirror, and part by means of the diffuse object, as shown in the figure. On the hologram plane, the radiation scattered from the object interferes with that reflected from the mirror and the resulting interference pattern is recorded photographically. The recording material must, of course, have sufficient resolving power, as discussed in Section 19.3.3.3. At the hologram plane, the ratio of the direct illumination to

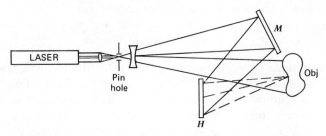

Figure 19.33 Representative simple holography arrangement.

that from the object should be of the order of five [66], the exact value depending on the material used and best determined by trial and error.

19.3.5   Extraordinary Hologram Recording Techniques

***19.3.5.1   Incoherent Illumination Holography* [34f].** It is possible to prepare holograms of incoherently illuminated objects by permitting each object point to generate its own reference wave. To this end, the wavefront from each object point is split into two components, which are, then, permitted to interfere to yield the hologram. Accordingly, each object point generates a hologram independent of the holograms generated by the other object points—all of them recorded simultaneously on the same total hologram. The interference patterns generated by the various object points must differ in such a manner that when the hologram is illuminated by a coherent wave, the diffraction effects will provide an image luminance distribution similar to that of the object.

Incoherent Fourier Holography. To obtain a Fourier holographic recording with incoherently illuminated objects, we first form, by some geometrical optical technique, two coplanar images of the object, one inverted relative to the other; see Figure 19.34*a*. Every pair of corresponding points will then generate a separate (Young's) fringe pattern. The spatial frequency of each pattern will be governed by the distance of the point from the axis of symmetry and its orientation by the direction of its displacement from this axis. See Figure 19.34*b* [84] for a possible arrangement.

Incoherent Fresnel Holography. In Fresnel holography, the two wavefronts are arranged so that the relative displacement between their centers of curvature has a component normal to the hologram plane, which receives, accordingly, two spherical wavefronts differing in radius of curvature and hence generating a Gabor zone pattern (Section 19.3.1.2). Different object points give rise to zone patterns with centers spaced according to the angular separation of the object points. This fact permits reconstruction with a single coherent wave.

Several techniques have been proposed for generating the double spherical wavefronts. These include the following:

1. A lens made of birefringent material splits the incident plane wave into two spherical waves with differing radii of curvature. (Polarizer and analyzer are required to provide coherence and parallelism of polarization, respectively, both of which are essential to interference.)

2. The object wave is passed through a Fresnel zone plate which inherently provides a multiplicity of focal lengths [85].

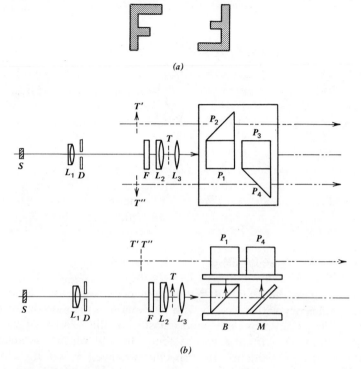

Figure 19.34 (*a*) Primary image for incoherent Fourier holography. (*b*) Optical arrangement: top and side views. Worthington [84].

3. Splitting the wavefront area and providing lenses with differing focal lengths for the two wavefront segments.

4. Using a triangular interferometer arrangement, in conjunction with an afocal (telescope) lens system, which is traversed in opposite directions by two parts of the wave, results in two different magnifications for these two parts (see Figure 19.35) [86].

In another system, each object point is made to generate its own reference wave by passing some of its light through a pinhole that diffracts it over the whole hologram area, in the form of a spherical wave centered on the pinhole. This will interfere with the undiffracted part of the object wave.

Diffraction Efficiency [86]. Because each point on the incoherently illuminated object creates its own interference fringe pattern, and these patterns combine incoherently, the modulation of the resultant hologram is here substantially more limited than in coherent holography.

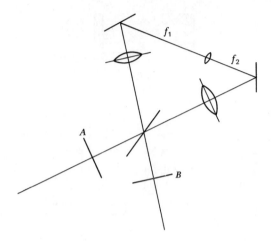

Figure 19.35 Triangular interferometer: afocal lens system arrangement yielding two axial image points from a single axial object point. Cochran [86].

Let us compare the resultant modulation in the two cases. Since the diffraction efficiency varies directly with the square of the modulation [see (19.146)], this would provide a good measure of the relative diffraction efficiency attainable with the two systems. Instead of the modulation itself, we use, for convenience, the "specific variance," v, which is closely related to the modulation:

$$v(f) = \frac{\sigma^2(f)}{\bar{f}^2}. \tag{19.238}$$

In the coherent case, we have for the illumination at any point:

$$f_c = \left(\sum a_j \cos \phi_j\right)^2, \tag{19.239}$$

whereas in the incoherent case it is

$$f_i = \sum (a_j \cos \phi_j)^2 = \tfrac{1}{2} \sum a_j^2 (1 + \cos 2\phi_j)$$
$$= \tfrac{1}{2} c + \tfrac{1}{2} \sum a_j \cos 2\phi_j, \tag{19.240}$$

where a_j, ϕ_j are amplitude and phase of the wave arriving at the hologram point in question from object point j, the summation is taken over all j, and we have abbreviated:

$$c = N a_j^2, \tag{19.241}$$

with N the number of object points. Using the results derived in Appendix 19.1 for v (19.310)–(19.311), we find

$$\bar{f}_c^2 = \frac{c^2}{4}, \tag{19.242}$$

$$\sigma^2(f_c) = \frac{c^2}{2} \tag{19.243}$$

and hence,

$$v(f_c) = 2. \tag{19.244}$$

On the other hand, referring to the results derived there for u (19.291)–(19.292), we find

$$\bar{f}_i^2 = \frac{c^2}{4} \tag{19.245}$$

$$\sigma^2(f_i) = \overline{f_i^2} - \bar{f}_i^2 = \frac{c^2}{4} + \tfrac{1}{2}c\sum a_j \cos \phi_j + \frac{\left(\sum a_j \cos^2 \phi_j\right)^2}{4} - \frac{c^2}{4} = \frac{c}{8}. \tag{19.246}$$

Hence

$$v(f_i) = \frac{1}{2c} = \frac{1}{2N\overline{a_j^2}}. \tag{19.247}$$

Comparison of results (19.244) and (19.247) indicates that the diffraction efficiency is independent of the number of object points in the coherent case but drops with that number when incoherent recording is used.

19.3.5.2 Hologram Replication [34g]. The photographic replication, or copying, of holograms poses special problems because of the high resolution usually required.

Plane Absorption Holograms. Plane, or thin, absorption holograms may, in principle, be copied by contact printing if care is taken to reduce the residual spacing between the two emulsions to a degree compatible with the resolution requirements.

Alternatively, we may holographically record a holographic reconstruction. In that method, the reconstructed image need not be real and it may be possible to use the undiffracted wave as the reference wave. In practice, if fairly coherent illumination is employed, this means that the recording material may be placed close to the hologram to be copied, but not necessarily as close as required in contact printing [87]. (See Ref. 34g for coherence requirements on the source.) It should be noted, however, that both the real and the virtual image generated by the original hologram will each give rise to a real and virtual image. Thus two real and two virtual images will be obtained from the replica. Each pair will be separated, axially, by twice the separation between the original hologram and the replica during the copying process.

Volume Holograms. As shown in Section 19.3.1.6, only one image is obtained in thick holograms. Such holograms may therefore be copied according to the method just described without the disturbing double image effect. To obtain high diffraction efficiency, the copying wave should be similar in shape to the original reference wave.

Plane Phase Holograms. Phase holograms can not be replicated by contact printing since the transmitted wavefront first attains the required amplitude modulation at some distance from the original hologram, where the interference between the diffracted and undiffracted waves becomes significant. Therefore such holograms must be copied via a reconstruction process.

Phase holograms based on surface contour modulation, such as those recorded on photothermoplastic plates, may be replicated by molding or pressing.

19.3.5.3 Synthetic Holograms: Combination of Subholograms. Clearly a hologram image can be synthesized from a number of components by sequentially recording these, taking care that hologram and reference wave remain in place throughout these recordings. The result is then similar to the hologram that would have been obtained had all of the object components been recorded simultaneously.

Components can also be subtracted from a recorded (but not fixed) hologram, by adding the component negatively, that is, with a phase shift of π radians introduced into the object or reference wave. When this is done with proper attention to the exposure level, the subtracted portion will be found to have vanished from the recording [88].

It can be shown that the diffraction efficiency of such a synthesized hologram will be equal to the diffraction efficiency that would have been obtained had the whole object been recorded at once, provided exposure levels and wave amplitude ratios are optimized in each case [89].

19.3.5.4 Synthetic Holograms: Computer Generated [89*]. Essentially, a hologram is but a certain pattern of varying transmittance and/or optical thickness. In the ordinary hologram, this pattern is generated by interference between two or more wavefronts. Clearly, however, the identical pattern could be generated by other means, such as by painting a

scaled-up version of the corresponding pattern on a large canvas and then reducing it photographically. Such synthetic holograms have, indeed, been made, generally from master patterns generated by a computer-controlled plotter. Such plots need not correspond to any real object; they can be made to yield any conceived object—it is only necessary to compute the interference pattern that would result from such an object if it were placed in the given holographic recording system.

Note that patterns generated by a plotter are inherently quantized, at least in one direction, since the plot is inherently one-dimensional. In this sense plotter-generated holograms are similar to television images.

We now describe briefly two methods used for this purpose.

Binary Synthetic Holograms [90]–[91]. Binary holograms, where the master plot consists of only black and white elements, without any grey tones, are the type most readily made. Here we describe one representative method.

The hologram area is divided into an array of $m \times n$ lattice points. Each of these is located at the center of a rectangular cell. Each such cell, in turn, is to impart to the incident plane wavefront an amplitude and a phase equal to those desired for the reconstructed wave at the corresponding lattice point. This is accomplished by making the background opaque and placing into each cell a window whose size is proportional to the desired amplitude and whose location within the cell causes the desired phase shift in a manner we explain presently.

Consider a simple diffraction grating as illustrated in Figure 19.36,

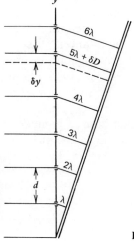

Figure 19.36 Illustrating the retarded phase concept.

where a plane wave, wavelength λ, is incident from the left and the diffracted wavefront is shown by the double line. If the spacing between apertures is d, the diffracted wavefront will be turned through an angle

$$\theta = \frac{\sin^{-1}\lambda}{d} \qquad (19.248)$$

and each aperture contributes to this wavefront with a phase

$$\phi = \frac{2\pi}{\lambda}D = 2\pi n, \qquad (19.249)$$

where D is the distance from the aperture center to the wavefront and n is the serial number of aperture, starting with the reference aperture as zero.

If we now displace one of the apertures a distance δy along the grating (as shown at aperture No. 5 in the figure) consideration of the similar triangles involved shows that the corresponding D value will increase by an amount

$$\delta D = \frac{\lambda}{d}\delta y \qquad (19.250)$$

and the contribution of that aperture will change through a phase difference

$$\delta\phi = \frac{2\pi}{\lambda}\delta D = \frac{2\pi\,\delta y}{d}. \qquad (19.251)$$

Thus the location of the aperture within the spacing period, determines the phase of its contribution.

In practice, the master pattern may be drawn by plotting within each cell a line segment perpendicular to the intended plane of diffraction; setting the length of the line segment according to the required amplitude; and its location within the cell according to the required phase. If the width of the line segment is maintained constant throughout, its area will be proportional to its length. When the master pattern is photographed, the black line segments will produce the desired transparent windows.

The number of cells required is determined by the desired number of resolution elements as follows. Let Δx and δx be the image size and image resolution element size, respectively, in the x direction. The size of the hologram, ΔX, is determined by the resolution requirement [see (19.205), with $z_0 = b$, $w = \Delta X$, and $\Delta h = \delta x$]:

$$\Delta X = \frac{\lambda b}{\delta x}, \qquad (19.252)$$

where b is the distance from the hologram to the image. The spacing between cells is determined by the sampling theorem:

$$\delta X = \frac{1}{2\,\Delta\nu}, \tag{19.253}$$

where $2\,\Delta\nu$ is the total width of the spectrum of the diffraction pattern on the hologram. This spectrum is, however, similar to the image itself, with a scale factor according to (19.57) given by

$$2\,\Delta\nu = \frac{\Delta x}{\lambda b} = \frac{1}{\delta X}, \tag{19.254}$$

with the last step following from (19.253). Hence the number of cells required in the x direction is

$$N_x = \frac{\Delta X}{\delta X} = \frac{\dfrac{\lambda}{\delta x}\,b}{\dfrac{\lambda}{\Delta x}\,b} = \frac{\Delta x}{\delta x} = n_x, \tag{19.255}$$

the number of resolution elements in the image! The same result is obtained in the y direction.

Note that, if the sampling is less dense than given in (19.253), the diffraction will displace the image by less than the total image size, Δx, and the diffracted image will partly overlap the undiffracted one, creating ambiguities in the region of overlap. (This corresponds to the "aliasing" of communication theory; cf. end of Section 18.4.2). We conclude that for unambiguous imaging, the number of cells in the synthetic hologram must equal the number of resolution elements in the image.

As presented here, the phase assigned to any window is the one correct for the center of the cell, whereas the actual window may be displaced from the center. In a modification of this scheme, the "windows" are placed at every location where the phase is an integral multiple of 2π. This ensures the phase at the "window" location to be exact. The "windows" then take the form of thin transparent lines, which trace out the loci of

$$\phi = 2\pi n, \qquad n = 0, \pm 1, \pm 2, \ldots,$$

one for each value of n [91].

Ideally, the "windows" in the cells would be made infinitesimally narrow, to diffract equally in all directions. However, energy considerations demand that they be made as large as possible (in practice approximately $\frac{1}{4}$ of the cell width). This tends to cause a drop-off of efficiency toward the edges of the image (analogous to the envelope, Figure 2.18c). Here the drop-off may, however, be compensated for by simply designing the hologram for an object which is brighter toward the edges, by the same factor.

Kinoforms [22]. We noted earlier that the illumination a hologram receives from a diffuse object tends to be quite uniform, with only the phase changing rapidly over the hologram area. The pure phase "hologram," called *kinoform* is based on this fact. Here the phase of the object wave alone is impressed on the "hologram," and the amplitude is assumed to be uniform throughout, so that the kinoform transmittance, too, is made uniform. A coherent wave passing through such a kinoform will simulate the object wave and an image thereof can be observed by means of this wave.

The kinoform is a hologram in the original sense of the word (a total recording) but differs from the usual hologram in that the phase factor is recorded without resort to interference. In principle, it can be fully efficient without losses due to absorption and the formation of spurious images—although in practice these can not be avoided completely.

To make a kinoform, the desired phase distribution is first calculated from a simulated object whose diffuse character is simulated by introducing a randomly varying phase at each object point. A master pattern is then plotted under computer control, with the desired phase (quantized and taken modulo 2π) translated into a proportional reflectance value. Note that this pattern does contain grey tones in addition to black and white. After photographing the pattern, the resulting transparency is bleached and tanned, so that its optical thickness at any point is proportional to the exposure at that point.

Exposure and bleaching of kinoforms must be carefully controlled so that full black and full white correspond to phase shifts of exactly $\pm\pi$. Deviation from this requirement leads to the generation of spurious secondary images, which may interfere seriously with the main image.

19.3.5.5 Acoustic Holography [92].

Holography is essentially a wave technique and is, therefore, not restricted to electromagnetic radiation. In fact, considerable work has been done in recording holograms of acoustically irradiated objects. Such holograms may then be reconstructed optically after suitable translation into an optically effective form. This technique can make important contributions in imaging objects that are inaccessible to visible radiation. It is a competitor of, and potentially far superior to, roentgenography. We list here some such areas of application.

Applications. The most obvious application of acoustic holography is in medicine, where it may replace roentgenography with its concomitant health hazards and provide superior imaging in such varied tasks as checking fetal development, brain tumor location, breast tumor diagnosis, management of tendon trauma, etc. [93]–[95].

Similarly, in the inspection of metal castings for flaws and other nondestructive testing tasks, acoustic holography is capable, in principle, of providing more information than roentgenology. Other proposed applications are in underwater viewing and in the surveillance of nuclear reactors in which liquid metal coolants may prevent optical viewing. Geophysical investigations, too, have been mentioned as potential areas of application of acoustic holography.

General Description. In acoustical holography, an object embedded in an acoustic medium is irradiated by an acoustic wave, usually generated by means of a piezoelectric transducer. The wave scattered (reflected or transmitted) by the object is made to interfere with a reference wave, permitting the recording of amplitude and phase of the scattered wave. When this is done coherently over a sizable area, the record may be used to generate a hologram, that may then be viewed by the usual optical means.

The attenuation of the acoustic energy in the medium is a major determinant in setting the upper limit on the choice of acoustic frequency, ν_t. As shown in Section 14.2.2.4, this attenuation is exponential. In solids, the exponent is directly proportional to the frequency:

$$I_S(x) = I_0 \exp(-2\alpha\nu_t x), \qquad (19.256)$$

where $I_S(x)$, I_0 are the sound intensities at points x and 0, respectively, and

α is the absorption coefficient.

Some representative values for $\alpha \times 10^6$ are for compression (and shear waves) [96a]:

$$0.061, \qquad 0.123\ (0.019), \qquad 0.321, \qquad 0.635\ (0.437)\ \text{sec/m}$$

for aluminum, fused quartz, flint, and plateglass, respectively. In liquids, the exponent varies with the square of the acoustic frequency:

$$I_L(x) = I_0 \exp(-2\alpha\nu_t^2 x), \qquad (19.257)$$

where [96b]

$$\alpha = 25 \times 10^{-15}\ \text{sec}^2/\text{m} \quad \text{for water}$$
$$= 2.5 \times 10^{-2}\ \text{sec}^2/\text{m} \quad \text{for glycerine, and}$$
$$= 6 \times 10^{-15}\ \text{sec}^2/\text{m} \quad \text{for mercury.}$$

By way of illustration, if there is to be at least 1% of the acoustic intensity remaining after passing the wave through one meter of medium, the maximum permissible frequencies are found to be 9.6 and 121 MHz, respectively, for water and fused quartz.

Usually the major difficulties in acoustic holography are the coherent recording of the acoustic wavefield over a sufficiently large area and the translation of the recording into an optical hologram. The following three subsections are devoted to various solutions to these difficulties.

Another difficulty arises when a major part of the scattering occurs at a point that is not of interest. This unwanted scattered wave may flood the interference plane, swamping the echos from the region under investigation. This may happen, for instance, when the acoustic radiation must enter a body whose interior is to be viewed. Here there is a tendency for the major portion of the energy to be scattered at the (uninteresting) surface. In such cases, an attempt is made to match acoustically the medium in which the transducer is located with that in which the object is embedded. But such matching is often far from perfect. To overcome the resulting swamping by irrelevant scattering, the acoustic wave may be gated: the acoustic energy is generated in short bursts containing, say, a hundred cycles, and the scattered returns are recorded only at the delay corresponding to the arrival of echos from the object.

Reference Source Implementation. In acoustical holography, the reference source, too, would have to be acoustical. It could, however, be an independent source synchronized with the object source.

If the acoustic field is detected by an acousto-electric transducer, the reference source conveniently may be avoided and the phase detected electrically in the transducer output circuit. But even if the transducer is acousto-optic, the phase detection process can be done *after* transduction, either by heterodyne or homodyne techniques. In these, a laser beam is reflected from the vibrating surface of the acoustic medium; the reflected beam will then be frequency modulated by the Doppler effect and can be beaten against the incident beam.

Scanning and Sampling Techniques. Experimentally, the simplest technique for recording an acoustic hologram is based on scanning the interference field by means of a piezoelectric detector. The electrical signal thus obtained may be converted into an equivalent optical pattern by any of the light modulation techniques that may be convenient, for example, by means of a crt display. The resulting luminance pattern may then be photographed to yield a hologram. If the electrical signal is recorded, as on magnetic tape, it may, of course, be reconstructed at some convenient later time.

Instead of scanning the wavefield by moving the detector, the acoustic radiator may be moved while the detector is held stationary. Reciprocity considerations (Section 2.3.1.6) show that the signal thus obtained will be identical to the one that would have been obtained with source and detector positions interchanged and scanning done by means of the detector.

Clearly, a third scanning method is obtained if the object is moved. In one practical scanning method, equivalent to this, source and detector (possibly identical!) are mounted on the same carriage and move together. When this technique is used, the resolution obtained warrants special discussion.

When one component only is moved, the effective hologram aperture size is determined by the area scanned. Specifically, the resolution element has, then, an angular size given by

$$\Delta\beta = \frac{\Lambda}{w}, \tag{19.258}$$

where Λ is the acoustic wavelength and
w is the width of the scan.

This is in complete analogy with the equivalent optical situation [see (9.108)].

When source and detector are mounted in close proximity of each other and moved together, the phase change resulting from the motion is doubled, so that the effective scan length is twice the actual scan length as far as the resulting resolution is concerned. Refer to Figure 19.37. With the source stationary, say at A, when the detector moves from A to B this produces a phase shift

$$\Delta\phi_1 = \frac{2\pi\,\Delta r}{\Lambda} \approx \frac{\pi w^2}{r\Lambda}, \tag{19.259}$$

with the approximation based on the result (19.44). When both move together from A to B, the distances r_0 and $r_0 + \Delta r$ are traversed twice as the wave propagates from source to detector via the object. Thus the

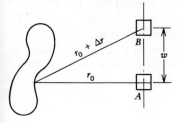

Figure 19.37 Illustrating scan-associated phase change.

phase shift introduced in this case by the scan displacement w is

$$\Delta\phi_2 = \frac{4\pi\,\Delta r}{\Lambda} = 2\,\Delta\phi_1. \tag{19.260}$$

Note that a raster of linear scans covers the scanned field completely in the scan direction, but only samples it in the direction normal to it. For good reconstruction, the raster scans should be spaced no more widely than called for by the sampling theorem (3.53), otherwise the multiple reconstruction images will overlap.

Instead of scanning the acoustic field in the interference plane mechanically, we may place an array of detectors there. If these are spaced sufficiently closely, as demanded by the sampling theorem, perfect reconstruction is possible, in principle.

Image Transducers. It is possible to convert the acoustic field into an optical pattern almost instantaneously, and we discuss some of the proposed techniques here.

1. Conceptually the simplest method of "reading" the acoustic field is the interrogation of the *free liquid-air surface* by means of a light beam. Due to the phenomenon of radiation pressure, this surface will be deformed proportionally to the acoustic energy; here it will follow the node structure of the interference pattern. This surface deformation could then serve directly as a phase hologram.

2. In principle, it should be possible to expose a *silver halide photographic emulsion* to the acoustic field and to obtain blackening of the emulsion in the areas of acoustic excitation, proportional to this excitation. Some work has been done along this line, but efficient holograms do not seem to have been obtained by this technique.

3. *The particle cell* is another system yielding an optical presentation directly. It consists of a thin-walled cell filled with fine, opaque particles in suspension. When exposed to an acoustic field, these particles tend to migrate toward the nodes, resulting in a transmittance variation that follows the structure of the acoustic interference pattern.

4. The *ultrasonic camera* or *Sokolov tube*[9] is an acoustoelectric transducer. It is similar, in principle, to the image orthicon (Section 16.4.3.1) with the photosensitive cathode replaced by a piezoelectric crystal. The acoustic waves excite this plate, which oscillates in response to the waves impinging on it. These oscillations, in turn, cause an alternating voltage to appear on the inner crystal surface. When this surface is scanned by an electron beam, the resulting emission of secondary electrons will be

[9] S. Y. Sokolov (1939).

modulated accordingly, resulting in an output similar to that of an image orthicon. This output can be used to modulate a crt display, which, when photographed, may yield a hologram. Recent developments have made this into a practical device [97].

Thermoplastic films deformed by radiation pressure and liquid crystals [98] too, have been suggested as acousto-optic transducers.

19.3.5.6 Microwave Holography. Holograms have also been recorded with microwaves [99]. One possible arrangement used an array of glow discharge tubes to detect the radiation. This is followed by an array of lamps to generate an optical image. This, in turn, is photographically converted into a transmittance pattern, which then serves as a hologram [100].

19.3.6 Special Purpose Holography

19.3.6.1 Holographic Diffraction Grating. The Fourier hologram of a point source is a sinusoidal grating. Such a grating, if thin, concentrates all energy into the zero and first orders and, if it is thick and Bragg diffraction controls, all the energy is concentrated into a single order. Thus holography is a relatively simple method for the manufacture of efficient diffraction gratings [101]–[103], [75].

Because the Bragg condition limits efficient diffraction to a narrow angular field (at a given wavelength) or to a narrow wavelength range (at a given field angle), thick holographic diffraction gratings can be used to investigate the spectrum of an extended source, a point at a time, without the need of a scanning slit [74].

Thick holograms of straight fringe patterns, with a plane mirror backing, have been substituted for diffraction gratings in the tuning of dye lasers. By using a hologram containing two such patterns superimposed, it is possible to obtain two wavelengths simultaneously and coaxially [104].

19.3.6.2 Composite Holograms [34h]. Several images may be recorded on the same hologram, for individual reconstruction. (The case of simultaneous reconstruction has been discussed in Section 19.3.5.3.) The subholograms may be spatially separate or superimposed on the total hologram.

Spatially Separated Composites. The basic concept of the spatially separated composite hologram is rather simple: several independent holograms are placed next to each other on a single recording. As the viewer moves his eye from place to place behind the recording, the observed image changes.

Strictly speaking any hologram of a three-dimensional object is composite in the sense that the view changes from position to position on the hologram. This immediately points to a number of applications for composite holograms.

1. *Increased Range of Viewing Angles.* In the usual hologram the range of viewing angles is limited by the size of the hologram. The hologram acts like a window through which the object is viewed. The viewer may move his eyes sideways to obtain a view from various angles; but, as long as the "window" is plane, this option is limited to 180° and this only for an infinitely large window. A composite hologram, on the other hand, can provide a complete panoramic view. This is obtained by exposing the hologram in vertical strips, a strip at a time, for example, by moving a vertical slit in front of the hologram plate as the object rotates about a vertical axis; see Figure 19.38. This procedure is usually executed in discrete steps. When binocular viewing is desired, care must be taken to make the rotational steps sufficiently small relative to displacement, so as to minimize the difference between the images viewed simultaneously by the two eyes.

2. *Hyper- and Hypostereoscopic Holograms.* The stereoscopic effect observed in three-dimensional holograms is due to the small differences in the images presented to the two eyes (see Section 15.8.1). These differences may be increased, to yield an enhanced sense of depth, by removing alternate thin strips from the regular hologram. This shrinks the hologram size macroscopically without affecting the spacing in the microscopic fringe pattern. The concomitant lateral displacement of the hologram segments is not accompanied by a lateral image shift, since it is only the reconstruction source location and the fringe frequency that determine the image location.

Similarly, the stereoscopic effect may be reduced by making two

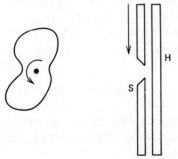

Figure 19.38 Construction of panoramic hologram. *S* is scanning slit and *H* is hologram.

identical holograms, cutting them into strips, and interleaving these strips.

3. *Information Storage* [105]. In some holographic information storage systems, a very large number of holograms is stored in an array on a holographic plate. The hologram to be displayed is then selected by directing the narrow reconstruction beam to the proper location. Since a 1-mm^2 hologram is capable of storing a page containing approximately 10^4 bits of information, a 9×12 cm holographic plate, which can hold 10^4 of these, could store 10^8 bits.

4. *Information Reduction.* When a hologram is to be transmitted over an electrical communication channel, the amount of data to be transmitted may be prohibitive. A significant reduction in resolution is impossible, since this would lose the fringe pattern entirely; reduction of the area, reduces the range of viewing angles. In a third alternative, only a series of narrow strips is transmitted; the received hologram is reconstituted by making a composite in which the transmitted strip is printed twice (or more times) to fill in the known gap between the transmitted strips.

Superimposed Composites and Dense Information Storage [106]. Several holograms may be superimposed on the same medium. If the processing is linear and the medium thick, the interference between these, during reconstruction, may be kept to a negligible level, as mentioned in our discussion of multicolor holography (Section 19.3.2.7). It is necessary only to change the orientation of the reference beam relative to the hologram plane between exposures. Since in these holograms efficient reconstruction is possible only with the reconstruction beam parallel to the original reference beam (assuming an unchanged wavelength), the orientation of the reconstruction beam will determine which of the superimposed holograms will be displayed. Note that the hologram may be rotated about any axis, in its plane or normal to it, to yield separable reconstructions. A reduction in S/N capacity here is due to the necessity to limit the amplitude of each recording so that the sum of the amplitudes will not exceed the operating range of the medium, as already mentioned in connection with multicolor holography.

The possibility of multiple hologram recording in thick media has stimulated much interest in holography for compact information storage. In principle, of course, the information storage capacity of a recording medium is no larger when the information is recorded holographically rather than directly [see, for instance, (19.209)]. However, the exploitation of the Bragg effect becomes very simple in holography and opens the way for convenient high-density information storage.

We note here parenthetically that for dense information storage even in thin media there is a significant advantage in holography. When the information is very tightly packed, a minor defect in the recording medium or process can destroy totally a significant batch of data if direct recording is used. In holographic recording, on the other hand, each data element is spread over a relatively large area element. Here a small defect will affect a large number of data elements—but only slightly.

Another multiplexing technique, based on hologram coding and applicable even to thin holograms, is discussed in conjunction with coded holograms at the end of the next section.

19.3.6.3 *Spatially Modulated Reference Source* [34i]

Ghost Images. In all the forms of holography discussed so far, the reference wave was plane or spherical—effectively derived from a point source or its equivalent. There is, in principle, no need for such a restriction.

Consider two coherent waves, scattered by two objects, incident on a recording medium in the x plane. Let these be described by

$$\hat{A}_1 = A_1 e^{i\phi_1}, \qquad \hat{A}_2 = A_2 e^{i\phi_2}, \qquad (19.261a,b)$$

respectively. The resulting illumination in the x plane will be

$$E(\mathbf{x}) = |A_1 e^{i\phi_1} + A_2 e^{i\phi_2}|^2$$
$$= A_1^2 + A_2^2 - A_1 A_2 [e^{1(\phi_1 - \phi_2)} + e^{-i(\phi_1 - \phi_2)}]. \qquad (19.262)$$

If the medium is processed appropriately, the resulting transmittance, too, will be proportional to this. If, then, it is illuminated by one of the component waves, say $\hat{A}_1$, the transmitted wave will have the form

$$\hat{A}(x) = (A_1^2 + A_2^2) A_1 e^{i\phi_1} + A_1^2 A_2 e^{i(2\phi_1 - \phi_2)} + A_1^2 A_2 e^{i\phi_2}. \quad (19.263)$$

The last term is exactly similar to the wave that originally came from the second object (19.261b) and therefore represents a reconstruction of this object. Clearly, the situation is strictly symmetrical with respect to the two objects, and we could reconstruct the first object by using the light scattered from the second.

One implication of this observation is, that we may dispense with the reference beam when recording a composite object and, in reconstruction, use one component of the object to reconstruct the other. The missing portion of the original object, which will nevertheless appear when the hologram is viewed, is called the *ghost image*. This technique is related to the locally generated reference beam (Section 19.3.4.3).

Reference Source Position Tolerance. Holographic ghost image formation requires that the reconstruction wavefront at the hologram duplicate the reference wavefront during recording. If the reference object is a distant point source, the reference wavefront amplitude will vary but slowly over the hologram and its phase will vary periodically. Hence the exact location of the reconstruction source is not critical, especially if the hologram is thin; only the image location, relative to the reconstruction source, remains constant: any displacement of the reconstruction source will simply result in a similar displacement in the image location. If, however, both the reference source and the object subtend large angles at the hologram plane, their phase variations will not be periodic and hence their interference pattern will vary rapidly in the hologram plane. For such holograms, the position of the reconstruction source, relative to the hologram, must duplicate accurately that of the original reference source. The tolerance on this duplication is similar to the size of the Fraunhofer diffraction lobes for an aperture of the same size and to the speckle patches observed with an aperture of the same size. The size, w, considered here is that of the smaller of the two—object and reference. The tolerance is then of the order of magnitude

$$\frac{\lambda b}{w},$$

where b is the distance from the source to the hologram. For instance, with He–Ne-laser light and with an object (or reference source) subtending $20°$ at the hologram, a misalignment of only 2μm will totally degrade the reconstruction process.

On the other hand, if Fourier transform holography is employed and the object approximates a point, the reconstruction source location is noncritical, even if it is very large. This forms the basis for the technique we describe now.

Character Recognition. Because of the symmetry relationship between object and reference wave, we can look upon the usual hologram of a complex object with a point reference source as the hologram of a point object made with the "object" as a reference source. Whenever the original complex object is presented to the hologram, a point image will appear at the location of the original point source. If the object appears at a different location, the point image, too, will be displaced by the same amount.

This fact has obvious application in character recognition. For instance, if we holographically record a printed letter (or word) together with a point source and then present a printed page to the hologram, a point image will appear for every instance of the letter (or word) appearing on the page and will appear at a location indicating where on the page it appears.

Alternatively, we may record a number of characters sequentially, all appearing at the same location, each with a point source at some other location. Each time one of the original characters is presented, a point image will appear, with the location of the image indicating the particular character presented.

If thin Fourier transform holography is employed, the hologram will be insensitive to a change in reference-source location and all the holograms may be recorded simultaneously: a hologram is recorded of a single point source with an array of the characters to be recognized. During the recognition procedure, the character is presented at a fixed reference position and the corresponding point image will appear with a displacement that is the negative of the displacement of the original letter relative to the point source.

The two approaches just described may be combined into a single system that recognizes simultaneously all characters on a page. The hologram is made of an array of m pages, each page blank except for one character in the upper right-hand corner, say. Here m is the number of characters to be recognized and each page is given another one of these (see Figure 19.39a). The reader will readily ascertain that if we now present a page filled with these characters (Figure 19.39b) as a reconstruction source to the hologram, the resulting image can again be divided

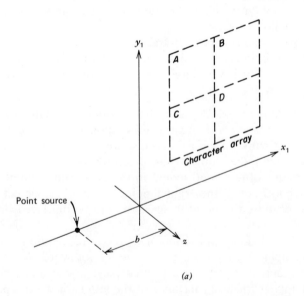

(a)

Figure 19.39 Full-page, multiple-character recognition scheme. (a) Reference source; (b) reconstruction source; (c) reconstructed image.

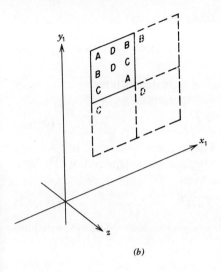

(b)

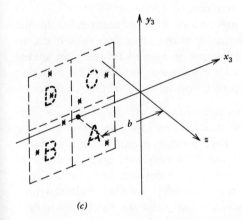

(c)

Figure 19.39 (Continued)

into an array of m pages, each associated with one of the characters. Each of the image pages will contain one point image for each time its letter appears on the sample page and, again, the locations of the point images will correspond to the locations of that character on the sample page (see Figure 19.39c).

Correlation and Matched Filtering. If one of the two wavefronts, say $\hat{a}_2$, considered at the beginning of this section is plane, the corresponding values of a_2 and ϕ_2 will be

$$a_2 = \text{constant}, \qquad \phi_2 = k_y y \qquad (19.264)$$

as in (19.79). Equation 19.262 then takes the form

$$\frac{1}{A_2} E(\mathbf{x}) = A_2 + \frac{A_1^2(\mathbf{x})}{A_2} + A_1(\mathbf{x})e^{i\phi_1(\mathbf{x})}e^{-ik_y y} + A_1(\mathbf{x})e^{-i\phi_1(\mathbf{x})}e^{ik_y y}. \qquad (19.265)$$

The last term here will be recognized as a matched filter to the object function a_1 whose Fourier transform is represented by A_1:

$$a_1 = \mathcal{F}^{-1}[\hat{A}_1], \qquad (19.266)$$

with an obliquity factor $(\exp ik_y y)$. Upon reconstruction with $\hat{A}_1$, the resulting image wavefront will be the absolute-value-squared of (19.263), and the term of interest will have the form:

$$|\hat{A}_1|^2 = |\mathcal{F}[a_1]|^2. \qquad (19.267)$$

The resulting image, g, will be, according to the Wiener-Kinthchine theorem (3.120)

$$g = \mathcal{F}[|A_1|^2] = a_1 \odot a_1. \qquad (19.268)$$

the autocorrelation of the object function.

The conditions (19.264) are met (more or less accurately) in the character recognition schemes discussed in the preceding subsection, so that the holograms described there may be viewed properly as spatial filters matched to the character to be recognized. Similarly, the resulting point images are simply the autocorrelation function of that character.

Digital Computation. Holograms can be used as logical gates and as digital computer memories. In one such scheme [107] the input is in the form of an array of area elements, each producing a phase shift of either 0 or π, corresponding to binary 0 and 1, respectively. Such an array, in conjunction with a reference element at some distance from the array, is used as object when the hologram is originally recorded. Subsequently, similar arrays serve as input (reconstruction sources), to this hologram, which then performs a gating operation.

The operation can be understood in terms of ghost image generation: each element reconstructs the reference source, but with a phase determined by the element phase. If the contributions of each input element are of the same phase (e.g., if they all have the same state as they had in the original object) their reconstructions will interfere constructively; otherwise they will interfere destructively.

Coded Holograms. If, during recording, a scattering screen is interposed in the reference beam, the disturbance introduced thereby will become part of the reference beam and the same screen, or an exact duplicate of it, will be required for reconstruction. Such a screen may serve as a coding device for holograms: the screen becomes a key required to view the coded hologram. An ordinary piece of ground glass may serve this purpose, although we should then have only a single copy of the "key."

As indicated in the opening remarks to this section, the functions of object and reference are essentially symmetrical. Hence a scattering factor applied to the object during recording may be canceled by applying it to the reconstruction beam during the imaging process. See the end of the next section (Section 19.3.6.4) for further discussion of this.

Multiplexing by Coded Holography. Coded holography opens the way to multiplexing even in thin holograms, where the reconstruction beam direction is essentially unrestricted. We may superimpose a number of holograms, each one coded with a different scattering plate, or a scattering plate positioned differently. During reconstruction, the desired hologram is selected by the choice of scattering plate or its position, respectively. For any such choice, only the corresponding hologram will be reconstructed; all others will only contribute to the light scattered over the background.

This scattered light limits the number (N) of holograms that may be multiplexed. The limitations may be estimated as follows. In general, there will be one fringe pattern diffracting light into the image and ($N-1$) such patterns diffracting light diffusely over the image field. Thus in the hologram plane, the signal-to-noise luminance ratio will be

$$\left[\frac{L_i}{\bar{L}_N}\right]_0 = \frac{1}{N-1}, \qquad (19.269)$$

seemingly a rather hopeless situation for $N > 2$. However, if we restrict the image size to a small portion of the field, the concentration factor may increase the signal-to-noise ratio significantly.

For instance, if the object is simply a point source and its location carries the information, we may find the image flux concentrated into an area of the order of magnitude (πr_A^2), where r_A is the radius of the Airy disc as determined by the diameter (w) of the hologram. From (19.205) with $z_0 = b$, this is

$$r_A = \frac{\lambda b}{w}. \qquad (19.270)$$

On the other hand, the other fringe patterns have their light scattered over a patch whose area is of the order of magnitude ($\pi b^2 \theta^2$), where θ is the half-angle subtended by the scattering screen at the hologram and b is the distance from the hologram to the image. The concentration multiplies the ratio of luminances in the signal and noise terms by a factor equal to the ratio of these two areas, so that

$$\frac{L_i}{L_N} = \left(\frac{b\theta}{r_A}\right)^2 \left[\frac{L_i}{L_N}\right]_0 = \frac{(w\theta/\lambda)^2}{N-1}. \qquad (19.271)$$

According to (19.223) the corresponding signal-to-noise ratio is

$$R \approx \frac{w\theta}{\lambda\sqrt{2N}}; \qquad N, R \gg 1. \qquad (19.272)$$

Thus, in principle at least, a large number of such point images could be accommodated.

19.3.6.4 Holograms as "Geometrical Optics" Components. Holograms have also been used in applications where normally geometrical optics components would be employed. Some of these applications are described in the following.

Correction of Lens Aberrations. Consider a lens affected by aberrations. A hologram made of a point source collimated by this lens can act as a corrector plate: when used in conjunction with the lens, it will eliminate the aberrations [108].

High-Resolution Imaging. A large, high-resolution hologram should, in principle, be capable of forming an image several centimeters in diameter with a resolution of a thousand lines per millimeter: tens of thousands of resolution elements across the image diameter. This exceeds by a factor of 10 the results that are normally obtained with highly corrected lenses and has important potential applications in the manufacture of integrated circuits. Here photolithographic masks carrying microscopic details must be imaged onto the semiconductor wafers. This is normally done by contact printing, a process causing the mask to deteriorate rapidly. High-resolution holography would seem to offer a welcome alternative.

Past efforts in this direction have run into difficulties. To avoid the spurious effects that tend to accompany work with coherent illumination (see Section 19.1.4), a diffusing screen must be used in conjunction with the object transparency; this screen, in turn, tends to introduce speckling (see Section 19.1.3). Also, the alignment of reference source, hologram, and image plane tends to be extremely critical. This latter difficulty may be overcome by lensless "para-image–plane" holography, in which object and reference beams are introduced from opposite sides of the photographic emulsion, with the object transparency very close to the emulsion.

The method is illustrated in Figure 19.40. The reference beam is introduced through a prism and enters the emulsion through the substrate via a layer of index-matching liquid. It is totally reflected at the emulsion-air interface and again at the prism surface, leaving the prism antiparallel to the entering beam. The object beam is introduced from the air surface of the emulsion and interferes with the reference beam both before and after its reflection. This inscribes two hologram patterns simultaneously. When the processed recording is illuminated antiparallel to the direction of the exiting reference beam, both hologram patterns contribute to the formation of a real image at the original location of the object transparency. A resolving power of over 600 lines/mm has been obtained with this method.

In a direct system of image plane holography, using a very poor-quality, high-aperture imaging lens, a resolution of 500 lines/mm over a 50 mm diameter has been obtained [109].

Multiple Imaging. In some manufacturing processes, for example, integrated circuit production, an array of many images of a basic master pattern must be formed. This may be accomplished by mechanically translating the recording medium, as in a step-and-repeat camera, or by using a "fly's-eye" lens—an array of miniature lenses, each forming its own image. Holography offers another alternative.

Consider a hologram formed of a two-dimensional array of coherent point sources by means of a reference point source coplanar and coherent

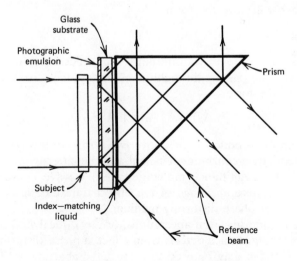

Figure 19.40 Para-image-plane holography.

with them. When this hologram is illuminated by a point source, a lattice of image points will be generated. If a small object pattern is introduced in place of the reference source, each of the points in the image lattice will duplicate the object pattern. This can be understood in terms of the point-elements making up the small object: each of these creates the two-dimensional lattice of points with a corresponding slight displacement. The points generated by each of the reference object elements combine to form the duplicate images.

Holographic lenses have also been suggested for machining by laser, where complex focusing patches may obviate difficult mechanical manipuation [110].

Imaging through a Phase-Distorting Medium. In the preceding section we already pointed out that a hologram taken through a scattering or phase-distorting component will incorporate these distortions and can be used to reverse them. Consider the problem of viewing through a phase-distorting medium. Here a hologram may be made by the interference of waves $\hat{A}_d$ and $\hat{A}_r$, where $\hat{A}_d$ results from a plane wave passing through the distorting component and $\hat{A}_r$ from a reference wave converging onto point, P; see Figure 19.41a. This results in a transmittance distribution containing, among others, a term

$$\hat{A}_d^* \hat{A}_r.$$

When this is now illuminated by a plane wave passing through the object and then through the phase-distorting medium (Figure 19.41b), the resulting wave will have the form

$$\bar{\tau}_0 \hat{A}_d,$$

where $\bar{\tau}_0$ is the complex transmittance (or reflectance) of the object transparency. After passing through the hologram, the wave will have one component of the form

$$(\bar{\tau}_0 \hat{A}_d) \cdot (\hat{A}_d^* \hat{A}_r) = |\hat{A}_d|^2 \bar{\tau}_0 \hat{A}_r \qquad (19.273)$$

resulting from the product of the above two terms. In this result, the first factor is essentially a constant coefficient and the latter two terms represent the object transmittance modulated onto a wave converging onto point P, so that there an image of the object transmittance will appear, unaffected by the phase-distorting medium.

Since a lens suffering from aberrations, or any optical surface suffering from nonabsorbing surface defects is, in effect, a phase-distorting component, this technique may be used to correct the aberrations. The hologram used in these applications may be viewed as a corrector plate [111].

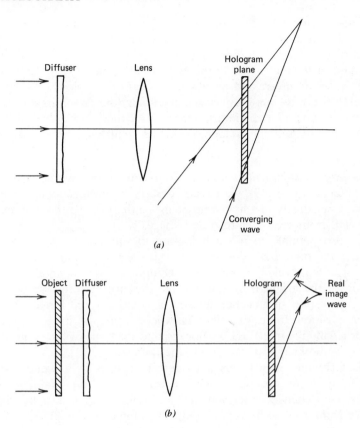

Figure 19.41   Holographic penetration of phase-distorting medium. (*a*) Making hologram; (*b*) viewing.

In another technique, a hologram of an object is formed through a phase-distorting medium with the (undistorted) reference wave incident on the hologram from the same side. In reconstruction, the reconstruction wave is incident from the opposite side, anti-parallel to the original reference wave. If the phase-distorting medium is left in the same relative position, an undistorted real image will appear at the location of the original object.

A word of caution. The above techniques seem to permit visual penetration of atmospheric fog, and other random scattering media, by means of holography. This, however, is not so: the scattering factor to be applied in reconstruction must be identical to the one used in recording.

Thus any factor that changes in time, or is otherwise not accurately duplicable, does not lend itself to these techniques.

In a related technique, both object and reference waves, pass through the phase-distorting layer. If the hologram plane is very close to the phase-distorting layer, and the distortion varies but slowly, both object and reference wave may experience essentially the same phase distortion over the regions of their interference. If so, their respective distortions will cancel out and the hologram will be unaffected by the distorting layer.

Holographic Subtraction. In our discussion of the synthesis of holograms (Section 19.3.5.3), we already described one technique for adding an image negatively to a hologram being synthesized, so that the image is subtracted from the other recorded images [88].

It is also possible to use a hologram to subtract a portion from an object at the time it is being viewed. This technique is based on the fact that a negatively developed hologram is opaque at the location of the bright fringes in the interference pattern. We prepare a hologram of part of a scene, develop it negatively and replace it in the original position. If we now view the whole scene, the fringes due to the part originally recorded will be blocked and that part of the scene will be missing in the image we obtain via the hologram [112].

This technique may be used to detect changes in a scene: the unchanged parts of the scene will be invisible, or only faintly visible, whereas any changed part will appear bright. It may also be used to remove from a scene unwanted and disturbing elements [112].

Scanning. Holograms can also be used to produce a two-dimensional raster scan from a simple rotational motion [113], [114].

19.3.6.5 *Holographic Interferometry* [34j]. Holography may be used for interferometric testing, that is, to detect and measure displacements of the order of magnitude of a wavelength. In the usual interferometry, the surfaces investigated must be specularly reflecting or transmitting; holography enables us to apply the same technique to diffusely reflecting objects.

Holography can be used both for static and dynamic ("real time," "live") interferometry. In addition, it can be used for finding topographic contour lines of a surface and for phase contrast microscopy. Below we devote a subsection to each of these. But first we must discuss the basic concepts involved and then establish the relationship between a displacement and the corresponding interferometric effect when dealing with diffuse holography.

The microscopic contour variations of a diffuse surface introduce large random phase shifts into the beams reflected from neighboring surface elements. However, if the surface is displaced and we compare light reflected from a certain surface element before and after its displacement, the phase difference will be constant, or almost so, for all the elements in a sizable surface patch, if the illumination is coherent over this patch and the microscopic surface structure is not disturbed significantly. Hence, light reflected from the patch in its two positions could interfere, provided the wavefront emanating from one position is "frozen" while the patch moves into the second position. (Of course, the light reflected from one patch will interfere with that from any other patch, but the resulting interference pattern will have large random fluctuations superimposed on it, so that the effect can not be detected.) The "freezing" may be done by holographic recording.

Fringe Luminance and Location. The luminance of an interference fringe varies with $\cos^2 \frac{1}{2}\phi$, where ϕ is the phase difference between the interfering waves. A displacement $\mathbf{s}$ of a reflecting point on the object surface, increases the optical pathlength of the incident light by an amount $\mathbf{s} \cdot \mathbf{u}_i$ and decreases that of the reflected light by an amount $\mathbf{s} \cdot \mathbf{u}_r$, where $\mathbf{u}_i$ and $\mathbf{u}_r$ are unit vectors in the direction of propagation of the incident and reflected waves, respectively. Hence the resultant phase shift is

$$\phi = \frac{2\pi \mathbf{s} \cdot (\mathbf{u}_i - \mathbf{u}_r)}{\lambda} = \frac{2\pi |\mathbf{u}_i - \mathbf{u}_r| \, s \cos \theta}{\lambda}, \qquad (19.274)$$

where θ is the angle between the direction of the displacement and $(\mathbf{u}_i - \mathbf{u}_r)$. When $\mathbf{u}_i$ and $\mathbf{u}_r$ are both parallel to the displacement, the resultant phase shift is maximum and given by

$$\phi_{mx} = \frac{4\pi s}{\lambda}. \qquad (19.275)$$

The location of the fringe in space is more difficult to determine. If it is very close to the moving surface, there will be no difficulty in matching the fringe to the element whose motion was responsible for it. If, however, there is a significant distance between surface and fringe, the association will be hazy, the haziness expressing itself as a blurring of the fringes if the surface is in focus and a blurring of the surface if the fringes are sharp. In principle, this could be overcome by using a small relative aperture, but this enlarges the speckle patches [see (19.22) and subsequent remarks] until they mask the fringes.

For pure rotation through a small angle α, the fringe pattern will appear at a distance

$$h = \frac{x \sin \theta_r \cos^2 \theta_r}{\cos \theta_i + \cos \theta_r}, \qquad (19.276)$$

beyond the surface (see Appendix 19.2). Here x is the distance of the element from the axis of rotation and θ_i, θ_r are the angles of incidence and reflection, respectively of the illuminating beam. Thus, near the axis of rotation ($x = 0$) and also for normal viewing ($\theta_r = 0$), the fringe pattern will be located on the surface ($h = 0$).

Pure translation can be viewed as rotation about an axis at infinity and gives rise to a fringe pattern located at infinity. Each fringe is then associated with the object surface as a whole and not with any individual element of it. On visual observation, the fringe will appear at infinity. It can also be observed at the focal point of a lens placed near the hologram.

If the fringe pattern is sufficiently close to the surface so that the fringe luminance associated with a surface element can be measured accurately, the magnitude and direction of any surface element displacement may be determined as follows. We view a surface element from direction $\mathbf{u}_{r1}$, and then change the direction of viewing. By the time direction $\mathbf{u}_{r2}$ is reached, a number, m, of fringes will have passed across the element being viewed. Comparison with (19.274) shows that this corresponds to an equation:

$$\phi_2 - \phi_1 = 2\pi m = \frac{2\pi \mathbf{s} \cdot [(\mathbf{u}_i - \mathbf{u}_{r2}) - (\mathbf{u}_i - \mathbf{u}_{r1})]}{\lambda} \qquad (19.277)$$

or

$$m\lambda = \mathbf{s} \cdot (\mathbf{u}_{r1} - \mathbf{u}_{r2}). \qquad (19.278)$$

In this equation, the three components of $\mathbf{s}$ are the only unknown quantities. By repeating the procedure for two other directions of viewing, we obtain three equations in three unknown quantities, enabling us to solve for them, provided they are independent (i.e., the three directions are not coplanar).

Static Interferometry. To compare an object surface in two states, it is recorded sequentially in these states on one hologram, each component exposure receiving half of the desired final total exposure. In the unchanged regions of the surface, the two recordings reenforce, while those regions that moved between the exposures, produce interference effects. The effect is as if the surface was present at both locations simultaneously (with the one closer to the viewer semitransparent).

This technique can be used to study straining of members under load and, with the aid of the very short exposures obtainable with pulsed lasers, even under dynamic loading. It can also be used to study air turbulence and shock waves. The first exposure is made through the undisturbed medium and the second one during the disturbance. The refractive index changes associated with the disturbance result in phase shifts and produce interference fringes that permit a detailed analysis of the strain pattern. The major advantages of this technique over standard interferometric techniques are the following:

1. The avoidance of the large interferometric arrangements with their attendant stability problems, and

2. The automatic cancelation of any refractive index nonuniformities present in the associated components, such as the windows of the wind tunnel.

Vibrating surfaces, too, can be studied by the techniques just described. First the stationary surface is recorded and then, after vibrations have set in, a short flash exposure records the surface at any desired phase of the vibrational cycle, the phase being determined by the timing of the flash relative to the vibrational cycle.

When the vibration is recorded over a complete cycle, or any number of such cycles, of the vibration, the resulting hologram will be a summation of the interference effects over the full range of the vibration. This yields an effective luminance, as generated by the interference effect, averaged over a cycle:

$$\bar{L} = \frac{1}{2\pi} \int_{-\pi}^{\pi} L(\alpha) \, d\alpha = \frac{1}{2\pi} \int_{-\pi}^{\pi} L_0 \cos^2 \phi(\alpha) \, d\alpha$$

$$= \tfrac{1}{2} L_0 \left(1 + \frac{1}{2\pi} \int_{-\pi}^{\pi} \cos 2\phi \, d\alpha \right), \tag{19.279}$$

where L_0 is the surface luminance as recorded in the absence of interference effects. Assuming the vibration to be sinusoidal, we have from (19.274):

$$\phi(\alpha) = K s_0 \cos \alpha, \tag{19.280}$$

where s_0 is the amplitude of the vibration and we have written:

$$K = 2\pi \, |\mathbf{u}_i - \mathbf{u}_r| \, \frac{\cos \theta}{\lambda} \tag{19.281}$$

and

$$s = s_0 \cos \alpha. \tag{19.282}$$

On substituting (19.280) into (19.279) and evaluating the integral, we obtain

$$\bar{L} = \tfrac{1}{2}L_0[1 + J_0(2Ks_0)].\qquad(19.283)$$

With s_0 a function of the surface coordinates, one can obtain a reasonably accurate plot of the amplitude distribution of the vibrations. If $2Ks_0$ exceeds the first zero of the zero-order Bessel function, $(j_{01} = 2.405)$, s_0 will be a multivalued function of $J_0(2Ks_0)$ and care must be taken to eliminate the resultant ambiguities. Also note that here the visibility is lower than that obtainable with instantaneous exposures that "freeze" the motion.

If several (say m) displacements are to be compared, the corresponding holograms may be multiplexed and viewed, a pair at a time. This is accomplished by dividing the hologram area into

$$\frac{m(m-1)}{2}$$

patches, with each exposure recorded on $(m-1)$ of these, so that each exposure pair have just one patch in common. Coherent illumination of the common patch will reconstruct the interference pattern of the associated pair [115]. Note that two such holograms will interfere only if they are recorded on the same hologram region [116], since otherwise the coherence between waves received from the same surface area is lost due to surface roughness. In effect, the two hologram regions must be within a single speckle patch.

Instead of exposing the two holograms on the same plate, they may be exposed on two separate plates. This has the advantage that the recording of the object in m different states permits the interferometric comparison of all $m(m-1)/2$ possible combinations using full size holograms. The alignment of the two holograms relative to each other is, of course, critical. In one method, each exposure is made on two photographic plates with their emulsion sides in contact with each other. This permits their relative alignment to be made quite readily with the three-pin arrangement used for aligning single plates (see Figure 19.42). In reconstruction, front and rear plates must be used in accordance with their recording position, so that one front plate always is combined with one rear plate in viewing any combination [117].

Dynamic Interferometry. To obtain dynamically observable interferometry, a holographic recording is made of the object at rest. This is processed *in situ*, or replaced into the system after processing to an accuracy considerably better than a hologram fringe width (19.95). When

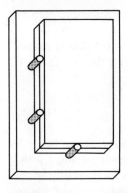

Figure 19.42 Three-pin arrangement for accurate positioning of photographic plates.

the object is now viewed through the hologram, it will be seen very faintly—the bright fringes of the interference pattern coincide with the dark fringes on the (negative) hologram and *vice versa*. If the surface is displaced, however, interference fringes will appear in a manner similar to the one described in the preceding subsection. Thus the motion of the object surface can be followed in detail. This method can be used in wind-tunnel interferometry, as well.

In principle, the motion of a vibrating surface, too, could be followed in this manner. But here the rapidity of the motion is likely to cause difficulties. By observing the surface over a period of time, an average luminance, similar to (19.279), is obtained. Improved visibility and more detailed observations are possible by stroboscopic observation, that is, by using pulsed laser illumination synchronized with the vibration. By controlling the pulse timing relative to the vibration phase, the vibrating surface can be studied at various points during the vibration cycle.

By preparing the hologram appropriately, it is also possible to obtain an image with dark lines marking the loci of points having any desired vibration amplitude [118].

Contour Generation. Double exposure holographic recording can also be used to obtain topographical contours of a surface: instead of moving the surface between the exposures, the wavelength is changed. This changes the distance of any surface point as measured in wavelengths by a factor equal to the wavelength ratio. This, in turn, corresponds to a phase change equal to

$$\Delta\phi = 2\pi \left(\frac{1}{\lambda_1} - \frac{1}{\lambda_2}\right) z = \frac{2\pi z \, \Delta\lambda}{\lambda_1\lambda_2}, \qquad (19.284)$$

where z is the distance from the source to the surface plus that from the surface to the hologram plane and $\Delta\lambda$ is the change in wavelength.

To obtain the distance, Δz, corresponding to one fringe cycle ($\Delta \phi = 2\pi$), we note that this corresponds to a change $2\,\Delta z$ in the value of z, if we assume the directions of incidence and reflection to be almost parallel to the z axis. Substituting these values for $\Delta \phi$ and z in (19.284) and solving for Δz, we find

$$\Delta z = \frac{\lambda_1 \lambda_2}{2\,\Delta \lambda}. \qquad (19.285)$$

Note that the wavelength can be changed without moving the laser: by retuning the laser, if it is continuously tunable and by selecting another spectral line, if it has a multiple discrete spectrum. Alternatively, the object may be immersed in media with different refractive indices. Noting that in a medium of refractive index n_i the wavelength is

$$\lambda_i = \frac{\lambda_0}{n_i}, \qquad (19.286)$$

where λ_0 is the vacuum wavelength, we conclude that

$$\Delta \lambda = \frac{\lambda_0 \, \Delta n}{n_1 n_2}, \qquad (19.287)$$

where Δn is the change in refractive index. Hence on substituting (19.72) and (19.73) into (19.285) we find

$$\Delta z = \frac{\lambda_0}{2\,\Delta n}. \qquad (19.288)$$

With freon under three atmospheres of pressure, Δn is 3×10^{-3} so that contour levels as small as 0.1 mm can be obtained [119].

Phase Contrast Microscopy [34k]. Clear objects, which appear only as a phase change, are difficult to see. To make them visible, schlieren or interferometric methods may be used. In microscopy, Zernike's phase-contrast method (Section 9.5.3.4) is often used. Holographic microscopy offers an alternative. In this technique, the object wave passes through the microscope objective, as usual, while the reference wave is introduced directly into the image plane; see Figure 19.43. We may now prepare a hologram of an empty slide and, after processing, replace it into the image plane. If a phase object is now viewed, it will produce interference and appear with relative luminance

$$\frac{L}{L_0} = c \cos^2 \tfrac{1}{2}\phi, \qquad (19.289)$$

where ϕ is the phase shift introduced by the object. The constant, c,

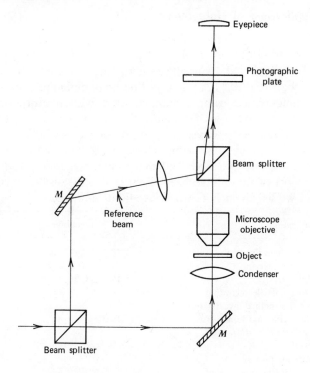

Figure 19.43 Holographic phase contrast microscopy.

depends on the luminance ratio of the original (blank slide) to the present object and can be eliminated completely by a proper adjustment of the latter.

19.3.6.6 Polarization Holography. The state of polarization of the light emanating from any object element, too, may be recorded holographically. The technique used is based on the fact that the interference takes place only between wave components with parallel directions of polarization. By using two reference sources with orthogonal directions of polarization, the hologram associated with each will record only the object luminance component with the respective direction of polarization. In reconstruction, each component may be viewed individually, regardless of the polarization of the reconstruction beam; on the other hand, if the reconstruction sources duplicate the direction of polarization of the original references sources, the reconstructed image will duplicate the polarization effects of the original object wave [120].

19.3.7 Applications of Holography

The realistic display of three-dimensional objects is, perhaps, the most obvious application of holography. It has been used for this purpose in commercial displays and its application to three-dimensional television display has been studied [121]. An on-line holographic television system has been implemented using a lumitron (crt with photothermoplastic screen) [122] and conditions for bandwidth compression have been surveyed [123]. Even video tapes [124] and video disks [125] have been demonstrated in holographic form. However, a large number of additional applications are possible and many of these appear today to be significant. Most of these have been discussed in the earlier parts of this chapter and we list them here for convenient reference.

Application	Section
1. Image enhancement	19.2.2, 19.3.6.3, 19.3.6.4
2. Image generation	19.2.3, 19.2.4, 19.3.5.4
3. Microscopy	19.3.1.3, 19.3.6.5
4. Aerosol particle recording	19.3.2.3
5. Internal structure investigation (in medicine and nondestructive testing)	19.3.5.4
6. Manufacture of diffraction gratings	19.3.6.1
7. Panoramic display	19.3.6.2
8. Dense information storage	19.3.6.2, 19.3.6.3
9. Coded information storage	19.3.6.3
10. Character recognition	19.3.6.3
11. Digital computing	19.3.6.3
12. High-resolution imaging	19.3.6.4
13. Multiple imaging	19.3.6.4
14. Scanning	19.3.6.4
15. Imaging through phase-distorting medium and compensation for lens aberrations	19.3.6.4
16. Detecting changes in object	19.3.6.4
17. Dynamic stress analysis	19.3.6.5
18. Wind tunnel analysis	19.3.6.5
19. Topographic contouring	19.3.6.5
20. Comparison of nearly identical images	19.2.5, 19.3.6.4

APPENDIX 19.1 RANDOM WALK IN TWO DIMENSIONS

When analyzing rough object surfaces under coherent illumination, we occasionally obtain the sum

$$\sum_j a_j e^{i\phi_j}$$

of random terms, each of random magnitude, a_j, and phase, ϕ_j. When this sum is plotted as a vector (phasor) sum, in two dimensions, with the phase defining the vector direction, we obtain a two-dimensional random walk. When this sum contains a large number of steps, the central limit theorem applies. The sum then approaches a Rayleigh distribution, as we now show.

Consider first the distribution of one of the components, say the x component

$$u = \sum_{j=1}^{N} a_j \cos \phi_j \tag{19.290}$$

and its mean value:

$$\bar{u} = \overline{Na_j \cos \phi_j} = N\bar{a}_j \overline{\cos \phi_j} = 0. \tag{19.291}$$

The variance of this sum is

$$\sigma^2(u) = N\sigma^2(a_j \cos \phi_j) = N\overline{(a_j \cos \phi_j - \bar{a}_j \overline{\cos \phi_j})^2}$$

$$= N\overline{a_j^2} \overline{\cos^2 \phi_j} = \tfrac{1}{2}N\overline{a_j^2} = \tfrac{1}{2}c \tag{19.292}$$

and

$$\overline{u^2} = \sigma^2(u) + \bar{u}^2 = \tfrac{1}{2}c, \tag{19.293}$$

where we introduced the abbreviation

$$N\overline{a_j^2} = c \tag{19.294}$$

and made use of the fact that

$$\overline{\cos \phi} = 0, \qquad \overline{\cos^2 \phi} = \tfrac{1}{2} \tag{19.295}$$

for ϕ uniformly distributed over 2π radians.

Applying the central limit theorem (Section 3.4.2) we find the probability density (pd)

$$p_x(u) = \frac{e^{-u^2/c}}{\sqrt{\pi c}}. \tag{19.296}$$

For future reference we also note that

$$\overline{u^4} = \int_{-\infty}^{\infty} u^4 p(u) \, du = (\pi c)^{-1/2} \int_{-\infty}^{\infty} u^4 \exp \frac{-u^2}{c} \, du = \tfrac{3}{4}c^2. \tag{19.297}$$

Symmetry considerations indicate that the two-dimensional distribution must have the form:

$$p(x, y) = p_1(\sqrt{x^2 + y^2}),$$
(19.298)

with

$$p_x(x) = \int_{-\infty}^{\infty} p_1(\sqrt{x^2 + y^2}) \, dy = (\pi c)^{-1/2} \exp \frac{-x^2}{c},$$
(19.299)

and the last equality derived from (19.296). Equation 19.299 is analogous to the relationship between the point and line spread functions (3.16) and (18.156).

The reader will readily confirm that

$$p(x, y) = \frac{1}{\pi c} \exp \frac{-(x^2 + y^2)}{c}$$
(19.300)

satisfies (19.299). This two-dimensional pd may be written as a one-dimensional pd of the variable

$$r = \sqrt{x^2 + y^2}.$$
(19.301)

To obtain this pd, we integrate $p(x, y)$ over a circular element, width dr, concentric with the coordinate system. Within this element, $p(x, y)$ is constant, so that the probability included in this element is simply the product of $p(x, y)$ with the area of the element

$$dA = 2\pi r \, dr.$$
(19.302)

Hence, combining (19.300)–(19.302),

$$p_r(r) \, dr = p(x, y) \, dA = \frac{2r}{c} \exp \frac{-r^2}{c} \, dr.$$

We conclude that the sum of a large number (N) of terms with random magnitude and phase (with the phase uniformly distributed over 2π) will have a resultant magnitude, r, with pd

$$p_r(r) = \frac{2r}{c} e^{-r^2/c},$$
(19.303)

where

$$r = \left| \sum_{j=1}^{N} a_j \exp(i\phi_j) \right|$$
(19.304)

and c is N times the mean squared value of the magnitude, a. p_r is known as the *Rayleigh distribution*.

The mean and second moment of r are, respectively,

$$\bar{r} = \frac{2}{c} \int_0^\infty r^2 \exp \frac{-r^2}{c} \, dr = \tfrac{1}{2}\sqrt{\pi c} \qquad (19.305)$$

$$\overline{r^2} = \frac{2}{c} \int_0^\infty r^3 \exp \frac{-r^2}{c} \, dr = c. \qquad (19.306)$$

The second moment of r^2 is

$$\overline{r^4} = \frac{2}{c} \int_0^\infty r^5 \exp \frac{-r^2}{c} \, dr = 2c^2 \qquad (19.307)$$

and its variance is

$$\text{var}\,(r^2) = \overline{r^4} - \overline{r^2}^2 = 2c^2 - c^2 = c^2 = \overline{r^2}^2. \qquad (19.308)$$

Now consider the function

$$v = u^2 = \left(\sum a_j \cos \phi_j \right)^2. \qquad (19.309)$$

Its mean value will be from (19.293)

$$\bar{v} = \overline{u^2} = \tfrac{1}{2}c \qquad (19.310)$$

and its variance from (19.297) and (19.310):

$$\sigma^2(v) = \overline{v^2} - \bar{v}^2 = \overline{u^4} - \overline{u^2}^2 = \tfrac{3}{4}c^2 - \tfrac{1}{2}c^2 = \tfrac{1}{4}c^2. \qquad (19.311)$$

APPENDIX 19.2   LOCATION OF INTERFERENCE FRINGES

We derive here an expression for the location of interference fringes resulting from a displacement of a surface element that can be represented as a pure rotation. Note that any actual infinitesimal displacement can be represented as such a rotation plus a motion tangent to the element. The axis of the rotation is the intersection of the surface element planes before and after the displacement.

Consider the element located at A (Figure 19.44) and A' before and after the rotation, respectively. Light is incident on A and A' in direction θ_i relative to the surface normal and viewed in direction θ_r after reflection therefrom. The location of the fringe is taken at a point (P) where the path difference between the wave scattered from the two positions is independent of the viewing angle θ_r. We take the x coordinate parallel to the original element plane (OA) with its origin (O) at the axis of rotation. We denote the x and y coordinates of P by x_0 and h, respectively.

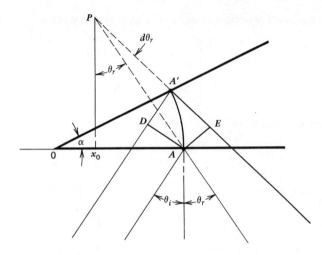

Figure 19.44 Determining location of interference fringe resulting from rotated reflector.

The x coordinate of point A associated with the fringe at P is a function of the viewing angle θ_r:

$$x = x_0 + h \tan \theta_r, \qquad (19.312)$$

and its derivative:

$$\frac{dx}{d\theta_r} = \frac{h}{\cos^2 \theta_r}. \qquad (19.313)$$

The corresponding path difference (s) introduced into the wave from the source to the receiver by the displacement from A to A' can be seen to be

$$s = \overline{A'D} + \overline{A'E} \approx \overline{AA'} \, (\cos \theta_i + \cos \theta_r)$$
$$= \alpha x \, (\cos \theta_i + \cos \theta_r), \qquad (19.314)$$

where α is the infinitesimal angle through which the element was turned and points

D, E are the feet of the perpendiculars dropped from A to the incident and reflected rays passing through A'.[10]

---

[10]Actually AE is perpendicular to the bisector of angle APA'; but, since this angle is infinitesimal, the distinction is negligible.

We now seek the value of h for which the derivative of s with respect to θ_r vanishes; that is,

$$\frac{ds}{d\theta_r} = \alpha\left[\frac{dx}{d\theta_r}(\cos\theta_i + \cos\theta_r) - x\sin\theta_r\right] = 0. \qquad (19.315)$$

Substituting into this from (19.313) and solving for h:

$$h = \frac{x\sin\theta_r\cos^2\theta_r}{\cos\theta_i + \cos\theta_r}. \qquad (19.316)$$

This is identical to (19.276) in the text.

REFERENCES

[1] B. J. Thompson, "Image formation with partially coherent light," in *Progress in Optics*, Vol. 7, E. Wolf, ed., North-Holland, Amsterdam, (1969), pp. 169–230.

[2] R. E. Swing and J. R. Clay, "Ambiguity of the transfer function with partially coherent illumination," *J. Opt. Soc. Am.* **57**, 1180–1189 (1967).

[3] R. E. Kinzly, "Investigations of the influence of the degree of coherence upon images of edge objects," *J. Opt. Soc. Am.* **55**, 1002–1007 (1965).

[4] D. Grimes and B. J. Thompson, "Two-point resolution with partially coherent light," *J. Opt. Soc. Am.* **57**, 1330–1334 (1967).

[5] J. C. Dainty, ed., *Laser Speckle and Related Phenomena*, Springer, Berlin, 1975; (a) T. S. McKechnie, "Speckle reduction," pp. 123–170; (b) J. W. Goodman, "Statistical properties of speckle patterns," pp. 9–75; (c) G. Parry, "Speckle patterns in partially coherent light," pp. 77–121; (d) M. Françon, "Information processing using speckle patterns," pp. 171–201; (e) A. E. Ennos, "Speckle interferometry," pp. 203–253; (f) J. C. Dainty, "Stellar speckle interferometry," pp. 255–280.

[6] P. S. Considine, "Effects of coherence on imaging systems," *J. Opt. Soc. Am.* **56**, 1001–1009 (1966).

[7] S. Lowenthal and D. Joyeux, "Speckle removal by a slowly moving diffuser associated with a motionless diffuser," *J. Opt. Soc. Am.* **61**, 847–851 (1971).

[8] J. C. Dainty, "The statistics of speckle patterns," in *Progress in Optics*, Vol. 14, E. Wolf, ed., North-Holland, Amsterdam 1976, pp. 3–46.

[9] M. Born and E. Wolf, *Principles of Optics*, 4th ed., Pergamon, New York, 1970; (a) Section 10.5.2.

[10] I. Weingärtner, W. Mirandé, and E. Menzel, "Linearität im Mikrodensitometer," *Optik* **34**, 53–60 (1971).

[11] H. Kogelnik and T. Li, "Laser beams and resonators," *Appl. Opt.* **5**, 1550–1567 (1966); also *Proc. IEEE* **54**, 1312–1329 (1966).

[12] A. Yariv, *Introduction to Optical Electronics*, Holt, Rinehart & Winston, New York, 1971; Chapter 3.

[13] L. D. Dickson, "Characteristics of a propagating Gaussian beam," *Appl. Opt.* **9**, 1854–1861 (1970).

[14] J. D. Zook and T. C. Lee, "Geometrical interpretation of Gaussian beam optics," *Appl. Opt.* **11**, 2140–2145 (1972).

[15] A. Vander Lugt, "Coherent optical processing," *Proc. IEEE* **62**, 1300–1319 (1974).

[16] E. L. O'Neill, *Introduction to Statistical Optics*, Addison-Wesley, Reading (1963), Chapter 2, Section 6.6.

[17] J. W. Goodman, *Introduction to Fourier Optics*, McGraw-Hill, New York, 1968, Chapter 7.

[18] A. Papoulis, *Systems and Transforms with Applications in Optics*, McGraw-Hill, New York, 1968, Section 11.4.

[19] E. G. Arfken, *Mathematical Methods for Physicists*, 2nd Ed. Academic, New York 1970, Section 15.4; R. Bracewell, *The Fourier Transform and Its Applications*, McGraw-Hill, New York, 1965, Chapter 6.

[20] J. Tsujiuchi, "Correction of optical images by compensation of aberrations and by spatial frequency filtering," *Progress in Optics*, Vol. 2, E. Wolf, Ed., North-Holland, Amsterdam 1963, pp. 131–180.

[21] A. S. Marathay, "Realization of complex spatial filters with polarized light," *J. Opt. Soc. Am.* **59,** 748–752 (1969).

[22] L. B. Lesem, P. M. Hirsch, and J. A. Jordan, "The kinoform: A new wavefront reconstruction device," *IBM. J. Res. Develop.* **13,** 150–155 (1969).

[23] A. vander Lugt, "Signal detection by complex spatial filtering," *IEEE Trans.* **IT–10,** 139–145 (1964).

[24] E. N. Leith, "Complex spatial filters for image deconvolution," *Proc. IEEE* **65,** 18–28 (1977).

[25] G. W. Stroke, M. Halioua, F. Thon, and D. H. Willasch, "Image improvement and three-dimensional reconstruction using holographic image processing," *Proc. IEEE* **65,** 39–62 (1977).

[26] J. W. Goodman, "Operations achievable with coherent optical information processing systems," *Proc. IEEE* **65,** 29–38 (1977).

[27] J. K. T. Eu, C. Y. C. Liu, and A. W. Lohmann "Spatial filter for differentiation," *Opt. Commun.* **9,** 168–171 (1973).

[28] J. K. T. Eu and A. W. Lohmann, "Isotropic Hilbert spatial filtering," *Opt. Commun.* **9,** 257–262 (1973).

[29] G. W. Stroke, R. G. Zech, "Photographic realization of an image-deconvolution filter for holographic Fourier-transform division," *Jap. J. Appl. Phys.* **7,** 764–766 (1968).

[30] L. M. Feldman, "A selected bibliography on optical spatial filtering," *Opt. Eng.* **11,** 102–112 (1972).

[30*] S. R. Dashiell and A. A. Sawchuk, "Non-linear optical processing: Analysis and synthesis," *Appl. Opt.* **16,** 1009–1025 (1977).

[31] S. H. Lee, S. K. Yao and A. G. Milnes, "Optical image synthesis (complex amplitude addition and subtraction) in real time by a diffraction-grating interferometric method," *J. Opt. Soc. Am.* **60,** 1037–1041 (1970).

[32] S. K. Yao and S. H. Lee, "Synthesis of a spatial filter for the combined operations of subtraction and correlation," *Appl. Opt.* **10,** 1154–1159 (1971).

[33] H. Weinberger and U. Almi, "Interference method for pattern comparison," *Appl. Opt.* **10,** 2482–2487 (1971).

[34] R. J. Collier, C. B. Burckhardt, and L. H. Lin, *Optical Holography*. Academic, New York, 1971; (*a*) Chapter 3; (*b*) Chapter 9; (*c*) Section 8.3; (*d*) Chapter 17; (*e*) Sections 10.8.4–10.8.5; (*f*) Section 20.3; (*g*) Section 20.1; (*h*) Chapter 18; (*i*) Chapter 14; (*j*) Chapter 15; (*k*) Chapter 13.

[35] D. Gabor, "A New Microscopic Principle," *Nature* **161,** 777 (1948) and "Microscopy by reconstructed wavefronts," *Proc. Roy. Soc.* **A197,** 454–487 (1949); (*a*) pp. 484–485.

[36] E. N. Leith and J. Upatnieks, "Reconstructed wavefronts and communication theory," *J. Opt. Soc. Am.* **52,** 1123–1130 (1962).

[37] H. H. M. Chau, "Zone plates produced optically," *Appl. Opt.* **8,** 1209–1211 (1969).

[38] H. H. M. Chau, "Fourier transform hologram by zone plate," *Opt. Commun.* **9,** 350–353 (1973).

[39] R. W. Meier, "Magnification and third-order aberrations in holography", *J. Opt. Soc. Am.* **55,** 987–992 (1965).

[40] J. F. Miles, "Imaging and magnification properties in holography," *Opt. Acta* **19,** 165–186 (1972).

[41] I. Přikryl, "A contribution to hologram imagery," *Opt. Acta* **21,** 517–528 (1974).

[42] J. N. Latta, "Computer-based analysis of holography using ray tracing," *Appl. Opt.* **10,** 2698–2710 (1971).

[43] G. W. Stroke, *An Introduction to Coherent Optics and Holography*, 2nd ed., Academic, New York, 1969, Section VI7.

[44] E. N. Leith and J. Upatnieks, "Microscopy by wavefront reconstruction," *J. Opt. Soc. Am.* **55,** 569–570 (1965).

[45] Y. Tsunoda and Y. Takeda, "Higher density image-storage holograms by a random phase sampling method," *Appl. Opt.* **13,** 2046–2051 (1974).

[46] R. N. Wolfe, J. J. DePalma, and S. B. Saunders, "Measurement of volume reflectance and volume transmittance of turbid media," *J. Opt. Soc. Am.* **55,** 956–962 (1965).

[47] H. Kogelnik, "Coupled wave theory for thick hologram gratings," *Bell Syst. Tech. J.* **48,** 2909–2947 (1969).

[48] A. A. Friesem and J. L. Walker, "Thick absorption recording media in holography," *Appl. Opt.* **9,** 201–214 (1970).

[49] B. J. Thompson, J. H. Ward, and W. R. Zinky, "Application of hologram techniques for particle size analysis," *Appl. Opt.* **6,** 519–526 (1967).

[50] G. B. Brandt, "Image plane holography," *Appl. Opt.* **8,** 1421–1429 (1969).

[51] B. S. K. Chow, "Measurement of gas laser cavity stability," *Opt. Commun.* **11,** 231–234 (1974).

[52] H. W. Rose, T. L. Williamson, and S. A. Collins, "Polarization effects in holography," *Appl. Opt.* **9,** 2394–2396 (1970).

[53] J. W. Goodman, "Temporal filtering properties of holograms," *Appl. Opt.* **6,** 857–859 (1967).

[54] A. A. Friesem and J. S. Zelenka, "Effects of film nonlinearities in holography," *Appl. Opt.* **6,** 1755–1759 (1967).

[55] B. R. Russell, "Resolution limitations in holographic images," *Appl. Opt.* **8,** 971–973 (1969).

[56] R. F. van Ligten, "Influence of photographic film on wavefront reconstruction, I. Plane wavefronts," *J. Opt. Soc. Am.* **56,** 1–9 (1966); II. "Cylindrical wavefronts," 1009–1014.

[57] T. Jannson, "Impulse response and shannon number of holographic systems," *Opt. Commun.* **10,** 232–237 (1974).

[58] L. Levi, "The effect of mechanical vibrations on holograms," *Opt. Commun.* **13,** 252–253 (1975).

[59] J. W. Goodman, R. B. Miles, and R. B. Kimball, "Comparative noise performance of photographic emulsions in holographic and conventional imagery," *J. Opt. Soc. Am.* **58,** 609–614 (1968).

[60] For example, see Ref. 35a.

[61] M. Matsumura, "Evaluation of deformation tolerance of the hologram medium," *J. Opt. Soc. Am.* **64,** 928–933 (1974).

[62] J. Politch and A. Assa, "Application of longitudinal multimode laser coherence properties to increase the holographic depth of field," *Opt. Commun.* **7,** 266–269 (1973).

[63] *Holographic Recording Materials*, H. M. Smith, Ed., Springer, Berlin (1977); (a) K. Biedermann, "Silver halide photographic materials," Ch. 2, pp. 21–74; (b) D. Meyerhofer, "Dichromated gelatin," Ch. 3, pp. 75–99; (c) R. A. Bartolini, "Photoresists," Ch. 7, pp. 209–227; (d) J. C. Urbach, "Thermoplastic hologram recording," Ch. 6, pp. 161–207; (e) D. L. Staebler, "Ferroelectric crystals," Ch. 4, pp. 101–132; (f) R. C. Duncan, "Inorganic photochromic materials," Ch. 5, pp. 133–160.

[64] Based on manufacturers' data.

[65] S. Johansson and K. Biedermann, "Multiple-sine-slit, etc.," *Appl. Opt.* **13**, 2288–2291 (1974).

[66] A. A. Friesem, A. Kozma, and G. F. Adams, "Recording parameters of spatially modulated coherent wavefronts," *Appl. Opt.* **6**, 851–856 (1967).

[67] M. Hercher and B. Ruff, "High-intensity reciprocity failure in Kodak 649-F plates at 6943 Å," *J. Opt. Soc. Am.* **57**, 103–105 (1967).

[68] R. L. van Renesse and F. A. J. Bouts, "Efficiency of bleaching agents for holography," *Optik* **38**, 156–168 (1973).

[69] A. Graube, "Advances in bleaching methods for photographically recorded holograms," *Appl. Opt.* **13**, 2942–2946 (1974).

[70] G. C. Righini, V. Russo, and S. Sottini, "Low noise and good efficiency volume holograms," *Appl. Opt.* **11**, 951–953 (1972).

[71] H. T. Buschmann and H. J. Metz, "Die Wellenlängenabhängigkeit der Übertragungseigenschaften photographischer Materialen für die Holographie", *Opt. Commun.* **2**, 373–376 (1971).

[72] D. Meyerhofer, "Phase holograms in dichromated gelatin," *RCA Rev.* **33**, 110–130 (1972).

[73] B. L. Booth, "Photopolymer material for holography," *Appl. Opt.* **14**, 593–601 (1975).

[74] M. R. B. Forshaw, "Thick holograms: A survey," *Opt. Laser Tech.* **6**, 28–35 (1974).

[75] J. M. Moran and I. P. Kaminow, "Properties of holographic gratings photoinduced in polymethyl methacrylate," *Appl. Opt.* **12**, 1964–1970 (1973).

[76] K. O. Hill and G. W. Jull, "Holographic-recording characteristics of diazo photosensitive films," *Opt. Acta* **21**, 535–545 (1974).

[77] T. C. Lee, "Holographic recording on thermoplastic films," *Appl. Opt.* **13**, 888–895 (1974).

[78] W. J. Burke, D. L. Staebler, W. Phillips, and G. A. Alphonse, "Volume phase holographic storage in ferroelectric crystals," *Opt. Eng.* **17**, 308–316 (1978).

[79] W. J. Tomlinson, "Volume holograms in photochromic materials," *Appl. Opt.* **14**, 2456–2467 (1975).

[80] H. Blume, "Highly efficient dichroic phase holograms in KCl:Na," *Opt. Acta* **21**, 357–363 (1974).

[81] L. Rosen and W. Clark, "Film plane holograms without external source reference beams," *Appl. Phys. Lett.* **10**, 140–142 (1967).

[82] H. J. Caulfield, J. L. Harris, H. W. Hemstreet, and J. G. Cobb, "Local reference beam generation in holography," *Proc. IEEE* **55**, 1758 (1967).

[83] G. B. Brandt, "Image plane holography," *Appl. Opt.* **8**, 1421–1429 (1969).

[84] H. R. Worthington, "Production of holograms with incoherent illumination," *J. Opt. Soc. Am.* **56**, 1397–1398 (1966).

[85] R. Silva and G. L. Rogers, "Decoding of a noncoherent hologram using noncoherent light," *J. Opt. Soc. Am.* **65**, 1448–1450 (1975).

[86] G. Cochran, "New method of making fresnel transforms with incoherent light," *J. Opt. Soc. Am.* **56**, 1513–1517 (1966).

[87] D. B. Brumm, "Copying holograms," *Appl. Opt.* **5**, 1946–1947 (1966).

[88] D. Gabor, G. W. Stroke, R. Restrick, A. Funkhouser, and D. Brumm, "Optical image synthesis by holographic Fourier transformation," *Phys. Lett.* **18**, 116–118 (1965).

[89] M. Lang, G. Goldmann, and P. Graf, "A contribution to the comparison of single exposure and multiple exposure storage holograms," *Appl. Opt.* **10**, 168–173 (1971).

[89*] W-H. Lee, "Computer-generated holograms: Techniques and applications," *Progr. Opt.* **16**, E. Wolf, ed., North-Holland, Amsterdam, 1978, pp. 121–232.

[90] A. W. Lohmann and D. P. Paris, "Binary Fraunhofer holograms, generated by computer," *Appl. Opt.* **6**, 1739–1748 (1967).

[91] P. Chavel and J.-P. Hugonin, "High quality computer holograms: The problem of phase representation," *J. Opt. Soc. Am.* **66**, 989–996 (1976).

[92] P. B. Hildebrand and B. B. Brenden, *An Introduction to Acoustical Holography*, Plenum, New York, 1972.

[93] P. S. Green, ed., *Acoustical Holography*, Vol. **5**, Plenum, New York, 1974.

[94] D. R. Holbrooke, E. E. McCurry, and V. Richards, "Medical uses of acoustical holography," Ref. 93, pp. 415–451.

[95] R. E Anderson, "Potential medical applications for ultrasonic holography," Ref. 93, pp. 505–513.

[96] L. Bergmann, *Der Ultraschall*, 6th Ed. Hirzel, Stuttgart, 1954; (*a*) Tables 88 and 89; (*b*) Table 48.

[97] J. E. Jacobs and D. A. Peterson, "Advances in the Sokoloff tube," Ref. 93, pp. 633–645.

[98] J. L. Fergason, "Liquid crystal detectors," *Acoustical Holography*, Vol. 2, A. T. Metherell and L. Larmore, eds., Plenum, New York, 1970, pp. 53–58.

[99] N. H. Farhat and W. R. Guard, "Holographic imaging at 70 GHz," *Proc. IEEE* **58**, 1955–1956 (1970); see also "Millimeter wave holographic imaging of concealed weapons," *ibid.* **59**, 1383–1384 (1971).

[100] N. S. Kopeika, "Millimetre-wave holography recording with glow discharge detectors," *Int. J. Electron.* **38**, 609–613 (1975).

[101] G. Schmahl and D. Rudolph, "Holographic diffraction gratings," in *Progress in Optics*, Vol. 14, E. Wolf, ed., North-Holland, Amsterdam, 1976, pp. 197–244.

[102] T. Namioka and W. R. Hunter, "A comparison of the efficiency and focused stray light characteristics of a conventionally ruled- and a holographically produced-concave diffraction grating in the vacuum ultraviolet," *Opt. Commun.* **8**, 229–233 (1973).

[103] O. Bryngdahl, "Formation of blazed gratings," *J. Opt. Soc. Am.* **60**, 140–141 (1970).

[104] A. A. Friesem, U. Ganiel, G. Neumann, and D. Peri, "A tunable dye laser with a composite holographic wavelengths selector," *Opt. Commun.* **9**, 149–151 (1973).

[105] B. Hill, "Some Aspects of a large capacity holographic memory," *Appl. Opt.* **11**, 182–191 (1972).

[106] W. J. Burke, D. L. Staebler, W. Phillips, and G. A. Alphonse, "Volume phase holographic storage in ferroelectric crystals," *Opt. Eng.* **17**, 308–316 (1978).

[107] K. Preston, "Digital holographic logic," *Pattern Recognition* **5**, 37–49 (1973).

[108] J. Upatnieks, A. Vander Lugt, and E. Leith, "Correction of lens aberrations by means of holograms," *Appl. Opt.* **5**, 589–593 (1966).

[109] T. Suzuki and J. Tsujiuchi, "A holographic image printing technique with high resolution," *Opt. Commun.* **9**, 360–363 (1973).

[110] J. M. Moran, "Laser machining with a holographic lens," *Appl. Opt.* **10**, 412–415 (1971).

[111] J. Upatnieks, A. Vander Lugt, and E. Leith, "Correction of lens aberrations by means of holograms," *Appl. Opt.* **5**, 589–593 (1966).

[112] K. Bromley, M. A. Monahan, J. F. Bryant, and B. J. Thompson, "Holographic subtraction," *Appl. Opt.* **10**, 174–181 (1971).

[113] D. H. McMahon, A. R. Franklin, and J. B. Thaxter, "Light beam deflection using holographic scanning techniques," *Appl. Opt.* **8**, 399–402 (1969).

[114] O. Bryngdahl and W. H. Lee, "Laser beam scanning using computer-generated holograms," *Appl. Opt.* **15**, 183–194 (1976).

[115] P. Hariharan and Z. S. Hagedus, "Simple multiplexing technique for double exposure hologram interferometry," *Opt. Commun.* **9**, 152–155 (1973).

[116] R. Dandliker, E. Marom, and F. M. Mottier, "Wavefront sampling in holographic interferometry," *Opt. Commun.* **6**, 368–371 (1972).

[117] N. Abramson, "Sandwich hologram interferometry: a new dimension in holographic comparison," *Appl. Opt.* **13**, 2019–2025 (1974); *ibid* **15**, 200–205 (1976).

[118] T. Sato, H. Ogawa, and M. Ueda, "Contour generation of vibrating object, by weighted subtraction of holograms," *Appl. Opt.* **13**, 1280–1282 (1974).

[119] N. Shiotake, T. Tsuruta, and Y. Itoh, "Holographic generation of contour map of diffusely reflecting surface by using immersion method," *Jap. J. Appl. Phys.* **7**, 904–909 (1968).

[120] O. Bryngdahl, "Polarizing holography," *J. Opt. Soc. Am.* **57**, 545–546 (1967); **58**, 702 (1968).

[121] E. N. Leith, J. Upatnieks, B. P. Hildebrand, and K. Haines, "Requirements for a wavefront reconstruction television facsimile system," *J. Soc. Mot. Pict. T. V. Eng.* **74**, 893–896 (1965).

[122] R. J. Doyle and W. E. Glenn, "Remote real-time reconstruction of holograms using the lumatron," *Appl. Opt.* **11**, 1261–1264 (1972).

[123] D. Casasent and R. L. Herold, "Television based Fourier holographic system," *Appl. Opt.* **13**, 2268–2273 (1974).

[124] W. J. Hannan, R. E. Flori, M. Lurie, and R. G. Ryan, "Holotape: A low-cost prerecorded television system, using holographic storage," *J. Soc. Mot. Pict. T. V. Eng.* **82**, 905–915 (1973).

[125] Y. Tsunoda, K. Tatsuno, K. Kataoka, and Y. Takeda, "Holographic video disk: An alternative approach to optical video disks," *Appl. Opt.* **15**, 1398–1403 (1976).

Tables

TABLE 11 SUPPLEMENT. SPECTRAL ENERGY DISTRIBUTION OF C.I.E. STANDARD ILLUMINANT D_{65}

λ (μm)	0.00	0.01	0.02	0.03	0.04	0.05	0.06	0.07	0.08	0.09
0.3	0.03	3.3	20.2	37.1	39.9	44.9	46.6	52.1	50.0	54.6
0.4	82.8	91.5	93.4	86.7	104.9	117.0	117.8	114.9	115.9	108.8
0.5	109.4	107.8	104.8	107.7	104.4	104.0	100.0	96.3	95.8	88.7
0.6	90.0	89.6	87.7	83.3	83.7	80.0	80.2	82.3	78.3	69.7
0.7	71.6	74.3	61.6	69.9	75.1	63.6	46.4	66.8	63.4	64.3
0.8	59.5	52.0	57.4	60.3						

Source. D. B. Judd and G. Wyszeck, *Color in Business, Science and Industry*, 3rd Ed., Wiley, New York (1975), Table 2.1.

TABLE 71 COMPOSITION, THERMAL EXPANSION, AND PHOTOELASTIC CONSTANTS OF SOME GLASSES (Section 10.1.1)

Glass type	No.	S.g.	Thermal expansion			Composition (%)[a]								Photoelastic constant
Name			α_{20} 10^{-6}	α_{90} 10^{-6}	$\Delta L/L^{b}$ 10^{-6}	SiO_2	B_2O_3	Na_2O	K_2O	BaO	PbO	Al_2O_3	ZnO	B^* brewster
1 C-1[c]	523/586	2.53	8.0	8.7	830	73	1	—	15	—	—	—	—	2.5[g]
2 BSC-1[g]	511/635	2.48	7.7	8.3	796	68.2	10	10	9.5	—	—	—	2	
3 BSC-2[h]	517/645	2.53	6.5	7.0	675	68.6	11.4	3.9	10	1.8	—	—	2	2.9
4 LBC-2[h]	573/574	3.21	7.8	8.3	800	48.1	4.5	1	7.5	28.3	—	—	10.1	2.75[g]
5 DBC-1[g]	611/588	3.58	6.4	7.0	669	34.6	11.0	—	—	46.9	—	5	1.1	
6 DBC-3[g]	611/572	3.57	6.1	6.8	641	34.5	10.1	—	—	42	—	5	7.8	2.1
7 CF-1	529/516	2.73	7.3	7.7	745	81	—	11	—	—	3	—	3.5	
8 BF-1[g]	584/460	3.31	8.0	8.4	819	49.8	—	1.2	8.2	13.4	18.7	—	8	
9 DF-2[h]	617/366	3.64	7.3	7.6	742	73.5	—	—	6.5	—	19.5	—	0.5	3.0
10 EDF-3[g,h]	770/293	4.51	7.5	8.0	773	33.7	—	—	4	—	62	—	—	
11 SiO_2 (100%)	458/532	2.20	0.5[f]			100	—	—	—	—	—	—	—	
12 SiO_2 (96%)		2.18			240[d]	96.3	2.9	—	—	—	—	—	—	3.5
13 Plate[c]		2.5			2600[d]	71.5	—	13.4	—	—	—	0.9	—	3.65
14 LX[e]BSC		2.23			960[d]	81	13	3.6	—	—	—	2.2	—	2.5
15 Extra-LX[e]	484/532	2.21	0.03[f]		0[d]									

[a] Components constituting less than 0.5% are not listed.

[b] 0–100°C, except as noted.

[c] C-1 and plate glass contain 10% and 12.8%, respectively, CaO including some MgO.

[d] 0–300°C, except No. 15, which is 0–200°C.

[e] LX stands for low expansion. No. 15 is Corning No. 7971.

[f] Average, 5–50°C.

[g] Composition (and birefringence) data are for glasses with similar, but not identical, refraction and dispersion.

[h] Further refractive indices of these glasses can be found in Table 63.

Composition data based primarily on Ref. 10.5, mechanical data on Ref. 10.13, and optical data of Refs. 10.6 and 10.7.

TABLE 72   TYPES OF OPTICAL

| Glass type | Abbreviations | | | Range | |
	Amer.	Ger.	Fr.	n_D	V
Fluor crown	FC	FK	FC	< 1.49	> 62
Borosilicate crown	BSC	BK	BSC	1.49	62–68
				1.54	62
Crown	C	K	C	< 1.54	55–62
Extra-light flint	ELF	LLF	FeL	< 1.55	45–50
				1.57	45
Light flint	LF	LF	FL	< 1.57	40–45
				1.60	40
Dense flint	DF	F	FD	< 1.60	35–40
				1.65	35
Extra-dense flint	EDF	SF	FDD	< 1.65	20–35
			FeD	1.98	20
Crown flint	CF	KF	CHD	< 1.54	50–55
				1.55	50
Light barium crown	LBC	BaK		1.54	55–62
		(BaLK)	BCL	1.60	55
Dense barium crown	DBC	SK	BCD	1.54–1.62	62
				1.60–1.655	55
Extra-dense barium	EDBC	SSK	BCDD	1.60–1.655	55
crown				1.60–1.67	50
Light barium flint	LBF	BaLF	FBL	1.54–1.60	55
				1.55–1.60	50
Barium flint	BF	BaF	FB	1.55–1.67	50
				1.57	45
				1.585–1.71	42.5
Dense barium flint	DBF	BaSF	FBD	1.585–1.71	42.5
				1.65–1.75	35
				1.75	29
Lanthanum crown	LaC	LaK	—	> 1.62	> 62
				> 1.655	> 55
				> 1.67	> 50
Lanthanum flint	LaF	LaF	—	> 1.67	50
				> 1.75	29–50
				1.98	20
Short flint	—	KzF	—	—	—
Phosphate crown	PC	PK	PC	1.49	> 68
				1.54	> 62
Dense phosphate crown	DPC	PSK	PCD	1.54–1.62	> 62

Source.   Kreidl and Rood [10.7].

GLASSES (Section 10.1.2.2)

		Prototype				
n_D	V	Density, d (g/cm^3)	P_{UV}	P_{IR}	Coeff. of expansion (10^{-7}/°C)	Typical stain class
1.4660	66.3	2.31	1.881	−1.711	93	5
1.4980	67.0	2.44	1.878	−1.720	60	1
1.5240	59.5	2.53	1.814	−1.752	100	1
1.5410	47.3	2.90	1.774	−1.883	83	1
1.5750	41.4	3.21	1.744	−1.950	98	1
1.6210	36.2	3.67	1.717	−2.003	83	1
1.6890	30.9	4.24	1.695	−2.103	81	3
1.5286	51.6	2.73	1.795	−1.840	84	1
1.5725	56.8	3.20	1.797	−1.783	80	2
1.6230	56.9	3.58	1.802	−1.766	72	5
1.6570	50.9	3.69	1.781	−1.834	77	2
1.5620	51.0	3.09	1.781	−1.846	99	1
1.5838	46.0	3.31	1.756	−1.904	98	1
1.6570	36.6	4.01	1.717	−2.015	77	1
1.6910	54.8	4.03	1.803	−1.790	89	5
1.7200	46.0	4.20	1.831	−1.914	73	5
1.6130	44.2	3.18	1.750	−1.871	58	5
1.5182	65.2	2.50	1.884	−1.720	73	2
1.5686	63.2	3.07	1.860	−1.737	67	5

TABLE 73 DATA FOR REFRACTIVE INDEX FORMULAE (Section 10.1.2.2)

	Ammonium dihydrogen phosphate-O	Ammonium dihydrogen phosphate-E	Arsenic trisulfide	Barium fluoride	Cadmium sulfide ordinary	Cadmium sulfide extra ord.	Cadmium telluride[a] irtran 6	Calcium aluminate	Calcium fluoride[a] irtran 3
Symbol	ADP-O	ADP-E	AS S	BA F	CD S-O	CD S-E	IRTR-6	CA AL O	IRTR-3
Ref.	1[b]	1[b]	2[b]	3[b]	4[b]	4[b]	5	6	5
B_0	17.412692	7.96996	6.77304	5.976218	5.235	5.239	2.682384	1.64289	1.4278071
B_1	0.01114254	0.009687319	0.04271328	0.002148552	0.1819	0.2076	0.118029	0.00786	0.0022806966
P_1	6043.4272	2322.2816	0.1201436	0.006096347					
P_2			0.1073729	8232.619					
P_3			0.02407126						
P_4									
P_5			717.7427						
P_1'	0.01327055	0.01285702	0.0225	0.0033396	0.1651	0.1651	0.028	0.028	0.028
P_2'			0.0625	0.01203					
P_3'			0.1225	2151.7					
P_4'			0.2025						
P_5'	400	400	750						
R							3.2768010E-2	-2.31E-4	-9.1939015E-5
A_1							-1.202984E-4	-2.2133E-3	-1.1165792E-3
A_2							2.177336E-8	-1.598E-5	-1.5949659E-6
λ_1	0.21	0.21	0.56	0.26	0.55	0.55	0.9	0.6	0.5
λ_2	1.53	1.53	12	10.35	1.4	1.4	16	4.3	11
$10^6\,\Delta_{mx}$[d]	8	8	20	9	400	400		30	
T(°C)	24.8	24.8	25	25			RT[c]	RT[c]	RT[c]
dn/dT ($10^{-6}/°C$) Vis			20	-15.2					
dn/dT ($10^{-6}/°C$) IR(10 μ)			0				95.1		

	Cesium bromide	Cesium iodide	Fluorite	Germanium	Lithium fluoride	Magnesium fluoride[e] ordinary	Magnesium fluoride[e] extraord.	Magnesium fluoride[a] irtran 1	Magnesium oxide
Symbol	CS BR	CS I	CAF	GE	LIF	MG F-O	MG F-E	IRTR-1	MG O
Ref.	8[b]	9[b]	7[b]	6	11[b]	12	12	5	13[b]
B_0	5.640752	6.397747	5.887152	3.99931	7.0537595	1.36957	1.381	1.3776955	2.956362
B_1	0.0018612								
P_1	41110.49	0.000182436	0.001433973	0.391707	0.00492029			0.001515529	0.0219577
P_2	0.0290764	0.021666540	0.0047478		4091.75				
P_3		0.009353855	4620.306						
P_4		0.01786217							
P_5		87108.45							
P'_1	14390.4	0.00052701	0.00252643	0.028	797.89253			0.028	0.01428322
P'_2	0.024964	0.02149156	0.01007833		0.005316				
P'_3		0.032761	1200.556						
P'_4		0.044944							
P'_5		25921							
R				1.63492E-1				2.1254394E-4	
A_1	-3.338E-6			-6.6E-6				-1.5041172E-3	-1.062387E-2
A_2				5.3E-8				-4.4109708E-6	-2.04968E-5
λ_1	0.36	0.3	0.22	2	0.4	0.4	0.4	0.5	0.36
λ_2	39	52	9.8	13.5	5.9	0.7	0.7	9	5.35
$10^5 \Delta_{mx}$[d]	40	40	9	60	3				5
$T(°C)$	31	24	24		23.6			RT	23.3
dn/dT $(10^{-6}/°C)$ Vis	79	99.4	-10.4	504[f]	-16.4				16[f]
dn/dT $(10^{-6}/°C)$ IR(10 μ)			-5.6						

$$n^{(2)} = A_1\lambda^2 + A_2\lambda^4 + B_0 + B_1/\lambda^2 + \sum P_i/(\lambda^2 - P_i') + R/(\lambda^2 - 0.028)^2$$

[a] Polycrystalline.
[b] The formula yields n^2.
[c] RT = room temperature. O = ordinary, E = extraordinary ray.

[d] Δ_{mx} is the maximum inaccuracy of the analytic approximation.
[e] $n(O) = B_0 + 0.0035821/(\lambda - 0.14925)$, $n(E) = B_0 + 0.0037415/(\lambda - 0.14947)$.
[f] Data for Ge 2.4 μm; for MgO at 5.5 μm.

[1] F. Zernike, JOSA 54, 1215–1220 (1964).
[2] W. S. Rodney et al., JOSA 48, 633–636 (1958).
[3] I. H. Malitson, JOSA 54, 628–632 (1964).
[4] S. J. Czyzak, JOSA 47, 240–243 (1957).
[5] Eastman-Kodak Co., Publ. No. U-72 (1971).
[6] M. Herzberger and C. D. Salzberg, JOSA 52, 420–427 (1962).
[7] I. H. Malitson, Appl. Opt. 2, 1103–1107 (1963).
[8] W. S. Rodney and R. J. Spindler, NBS 51, 123–126 (1953).
[9] W. S. Rodney, JOSA 45, 987–992 (1955).
[10] L. W. Tilton et al. NBS 43, 81–86 (1949).
[11] L. W. Tilton and E. K. Plyler, NBS 47, 25–30 (1951).
[12] A. Duncanson and R. W. H. Stevenson, Proc. Phys. Soc. (Lond.) 72, 1001–1006 (1968).

TABLE 73 (Continued)

	Magnesium oxide[a] irtran 5	Potassium bromide	Poatassium chloride	Potassium dihydrogen phosphate-O[c]	Potassium dihydrogen phosphate-E[c]	Rutile ordinary	Rutile extraord.	Sapphire	Silica glass
Symbol	IRTR-5	K BR	K CL	KDP-O	KDP-E	RUTI-O	RUTI-E	S	SI O
Ref.	5	14[b]	15[b]	1[b]	1[b]	16[b]	16[b]	17[b]	18[b]
B_0	1.7200516	2.361323	2.174967	15.257548	5.361766	5.913	7.197	8.362854	3.001588
B_1		0.007676							
P_1	0.00561194	0.0156569	0.008344206	0.01011279	0.008653247	0.2441	0.3322	0.003865738	0.003257465
P_2			0.00698382	5198.829	1291.174			0.0129684	0.005512145
P_3								1697.044	87.89375
P_4									
P_5	0.028	0.0324	0.0119082	0.01294238	0.01229326	0.0803	0.0843	0.00377588	0.004679148
P'_1			0.025555	400	400	0.0803		0.0122544	0.01351206
P'_2								321.3616	97.934
P'_3									
P'_4									
P'_5									
R	-1.0986148E-5								
A_1	-3.0994558E-3	-3.11497E-4	-5.13495E-4						
A_2	-9.6139613E-6	-5.8613E-8	-1.67587E-7						
λ_1	0.5	0.4	0.185	0.21	0.21	0.43	0.43	0.26	0.21
λ_2	9	22	20.6	1.53	1.53	1.53	1.53	5.6	3.71
$10^5 \Delta_{max}^d$	10	10	10	16	7			6	4
$T(°C)$ RT[c]	22	22	15	24.8	24.8	300	300	24	20
$\dfrac{dn}{dT}(10^{-5}/°C)$ Vis / IR(10μ)	40	40	-250 / -230					13	10

	Silicon	Silver chloride	Sodium chloride	Sphalerite	Strontium titanate	Thallium Bromide iodide	Zinc Selenide[a] irtran 4	Zinc sulfide	Zinc sulfide[a] irtran 2
Symbol	SI	AG CL	NA CL	SPHAL	SR TlO	KRS-5	IRTR-4	ZN S	IRTR-2
Ref.	6	19[b]	15[b]	16[b]	6	10[b]	5	4[b]	5
B_0	3.41696	4.00804	2.330165	5.164	2.28355	5.676927	2.4350823	5.131	2.2569735
B_1			0.005343924						
P_1	0.138497	0.079086	0.01278685	0.1208	0.035906	0.2873017	0.051567572	0.1275	0.032640935
P_2									
P_3									
P_4									
P_5	0.028	0.04584	0.01485	0.0732	0.028	0.1027734	0.028	0.0732	0.028
P_1'			0.02547414						
P_2'									
P_3'									
P_4'									
P_5'									
R	1.3924E-2				1.666E-3		2.4901923E-3		6.0314637E-4
A_1	-2.09E-5	-8.5111E-4	-9.285837E-4		-6.1335E-3	-4.522264E-4	-2.7245212E-4		-5.2705532E-4
A_2	1.48E-7	-1.9762E-7	-2.86086E-7		-1.502E-5	-1.80197E-8	-9.851275E-8		-6.0428638E-7
λ_1	1.3	0.58	0.185	0.36	1.0	0.57	0.5	0.44	0.5
λ_2	11	26	16	1.53	5.3	39.2	20	1.4	13
$10^5\Delta_{mx}$	20	35	10	300	30	60		500	
$T(°C)$	35	24	18			27	RT[c]		RT[c]
$\frac{dn}{dT}(10^{-6}/°C)$ Vis	24	-61	-35.7			-254			
IR(10 μ)	-61		-22			-235			182

[a] Polycrystalline.
[b] The formula yields n^2.
[c] RT = room temperature. O = ordinary, E = extraordinary ray.
[d] Δ_{mx} is the maximum inaccuracy of the analytic approximation.

$$n^{(2)} = A_1\lambda^2 + A_2\lambda^4 + B_0 + B_1/\lambda^2 + \sum P_i/(\lambda^2 - P_i') + R/(\lambda^2 - 0.028)^2$$

[13] R. E. Stephens and I. H. Malitson, NBS 49, 249–252 (1952).
[14] R. E. Stephens et al., JOSA 43, 110–112 (1953).
[15] F. Paschen, Ann. Phys. 26, 120–138 (1908).
[16] J. R. DeVore, JOSA 41, 416–419 (1951).
[17] I. H. Malitson, JOSA 52, 1377–1379 (1962).

[18] I. H. Malitson, JOSA 55, 1205–1209 (1965).
[19] L. W. Tilton et al., JOSA 40, 540–543 (1950).
[20] J. E. Harvey and W. L. Wolfe, JOSA 65, 1267–1268 (1975).

JOSA = J. Opt. Soc. Am.
NBS = J. Res. Nat. Bur. Stds.

TABLE 74 REFRACTIVE INDICES OF CRYSTALLINE MEDIA (Section 10.1.2.2)

(see Table 73 for meaning of symbols heading columns)

	ADP-O	ADP-E	AS S	BA F	CD S-O	CD S-E	IRTR-6	CA AL O	IRTR-3
λ_1	1.6321	1.5728	2.6869	1.5143	2.5610	2.5981	2.8868	1.6637	1.4359
0.1									
0.2									
0.3	1.5640	1.5128		1.5010					
0.4	1.5408	1.4926		1.4848					
0.5	1.5303	1.4837		1.4778					1.4359
0.6	1.5240	1.4788	2.6365	1.4741	2.4836	2.5108		1.6637	1.4334
0.7	1.5195	1.4756	2.5620	1.4719	2.4073	2.4245		1.6577	1.4318
0.8	1.5158	1.4732	2.5209	1.4704	2.3702	2.3825		1.6537	1.4306
0.9	1.5124	1.4713	2.4952	1.4693	2.3488	2.3582	2.8868	1.6508	1.4297
1.0	1.5092	1.4695	2.4777	1.4686	2.3351	2.3426	2.8384	1.6485	1.4289
2.0			2.4261	1.4646			2.7137	1.6357	1.4239
3.0			2.4161	1.4612			2.6949	1.6225	1.4179
4.0			2.4112	1.4567			2.6880	1.6039	1.4097
5.0			2.4073	1.4510			2.6842		1.3990
6.0			2.4033	1.4440			2.6814		1.3856
7.0			2.3990	1.4357			2.6790		1.3693
8.0			2.3940	1.4258			2.6766		1.3498
9.0			2.3883	1.4144			2.6742		1.3269
10.0			2.3815	1.4013			2.6718		1.3002
11.0			2.3737				2.6691		1.2694
12.0			2.3645				2.6663		
13.0							2.6634		
14.0							2.6602		
15.0							2.6569		
16.0							2.6535		
17.0									
18.0									
19.0									
20.0									
25.0									
30.0									
35.0									
40.0									
45.0									
50.0									
λ_2	1.4900	1.4609	2.3645	1.3963	2.3101	2.3140	2.6535	1.5969	1.2694

	CS BR	CS I	CA F	GE	LI F	MG F-O	MG F-E	IRTR-1	MG O
λ_1	1.7539	1.9787	1.4811	4.1083	1.3987	1.3839	1.3959	1.3877	1.7735
0.1									
0.2									
0.3		1.9787	1.4540						
0.4	1.7352	1.8503	1.4419		1.3987	1.3839	1.3959		1.7622
0.5	1.7090	1.8063	1.4365		1.3943	1.3798	1.3917	1.3877	1.7455
0.6	1.6958	1.7852	1.4336		1.3918	1.3775	1.3893	1.3831	1.7367
0.7	1.6882	1.7732	1.4318		1.3902	1.3761	1.3878	1.3809	1.7313
0.8	1.6835	1.7657	1.4305		1.3890			1.3795	1.7276
0.9	1.6802	1.7607	1.4296		1.3880			1.3785	1.7249
1.0	1.6779	1.7572	1.4289		1.3871			1.3778	1.7228
2.0	1.6706	1.7462	1.4239	4.1083	1.3788			1.3720	1.7085
3.0	1.6690	1.7440	1.4179	4.0449	1.3666			1.3640	1.6916
4.0	1.6681	1.7430	1.4096	4.0244	1.3494			1.3526	1.6681
5.0	1.6674	1.7424	1.3990	4.0151	1.3266			1.3374	1.6367
6.0	1.6666	1.7418	1.3856	4.0102				1.3179	
7.0	1.6657	1.7412	1.3693	4.0072				1.2934	
8.0	1.6648	1.7406	1.3498	4.0053				1.2634	
9.0	1.6637	1.7399	1.3268	4.0040				1.2269	
10.0	1.6625	1.7392		4.0032					
11.0	1.6612	1.7383		4.0026					
12.0	1.6598	1.7375		4.0023					
13.0	1.6582	1.7365		4.0021					
14.0	1.6565	1.7355							
15.0	1.6547	1.7344							
16.0	1.6527	1.7332							
17.0	1.6506	1.7319							
18.0	1.6484	1.7306							
19.0	1.6460	1.7291							
20.0	1.6435	1.7276							
25.0	1.6286	1.7188							
30.0	1.6095	1.7077							
35.0	1.5856	1.6943							
40.0		1.6781							
45.0		1.6591							
50.0		1.6366							
∞	1.5625	1.6266	1.3054	4.0021	1.3007	1.3761	1.3878	1.2269	1.6237

TABLE 74   (*Continued*)

	IRTR-5	K BR	K CL	KDP-O	KDP-E	RUTI-O	RUTI-E	SAPPH.	SI C
λ_1	1.7443	1.5912	1.8315	1.6074	1.5510	2.8717	3.2402	1.8373	1.53
0.1									
0.2			1.7191						
0.3			1.5460	1.5456	1.4982			1.8144	1.48
0.4		1.5912	1.5110	1.5245	1.4802			1.7866	1.47
0.5	1.7443	1.5697	1.4970	1.5149	1.4725	2.7114	3.0335	1.7743	1.46
0.6	1.7357	1.5589	1.4898	1.5093	1.4683	2.6049	2.8986	1.7676	1.45
0.7	1.7306	1.5528	1.4857	1.5052	1.4656	2.5512	2.8312	1.7633	1.45
0.8	1.7272	1.5488	1.4830	1.5019	1.4637	2.5197	2.7919	1.7602	1.45
0.9	1.7247	1.5462	1.4812	1.4989	1.4622	2.4995	2.7667	1.7578	1.45
1.0	1.7227	1.5443	1.4799	1.4960	1.4610	2.4856	2.7495	1.7557	1.45
2.0	1.7089	1.5382	1.4754					1.7377	1.43
3.0	1.6920	1.5366	1.4738					1.7122	1.41
4.0	1.6683	1.5355	1.4723					1.6752	
5.0	1.6368	1.5344	1.4706					1.6240	
6.0	1.5962	1.5332	1.4686						
7.0	1.5452	1.5318	1.4662						
8.0	1.4824	1.5302	1.4634						
9.0	1.4060	1.5284	1.4603						
10.0		1.5264	1.4567						
11.0		1.5241	1.4528						
12.0		1.5216	1.4483						
13.0		1.5189	1.4434						
14.0		1.5160	1.4380						
15.0		1.5127	1.4321						
16.0		1.5092	1.4257						
17.0		1.5055	1.4187						
18.0		1.5014	1.4110						
19.0		1.4971	1.4028						
20.0		1.4924	1.3938						
25.0									
30.0									
35.0									
40.0									
45.0									
50.0									
λ_2	1.4060	1.4822	1.3881	1.4793	1.4555	2.4538	2.7100	1.5848	1.

TABLE 74 (*Continued*)

	SI	AG CL	NA CL	SPHAL.	SR TI O	KRS-5	IRTR-4	ZN S	IRTR-2
λ_1	3.5053	2.0688	1.8976	2.7029	2.3161	2.6401	2.7178	2.4880	2.4161
0.1									
0.2			1.7907						
0.3			1.6072						
0.4			1.5677	2.5604					
0.5			1.5518	2.4181			2.7178	2.4191	2.4161
0.6		2.0638	1.5436	2.3633		2.6065	2.6129	2.3613	2.3606
0.7		2.0459	1.5389	2.3353		2.5335	2.5582	2.3317	2.3302
0.8		2.0348	1.5358	2.3189		2.4923	2.5258	2.3143	2.3116
0.9		2.0275	1.5337	2.3082		2.4663	2.5049	2.3031	2.2993
1.0		2.0224	1.5322	2.3009	2.3161	2.4488	2.4905	2.2953	2.2907
2.0	3.4526	2.0061	1.5268		2.2679	2.3977	2.4471		2.2631
3.0	3.4324	2.0023	1.5244		2.2312	2.3885	2.4384		2.2558
4.0	3.4254	1.9998	1.5220		2.1838	2.3849	2.4339		2.2504
5.0	3.4221	1.9974	1.5190		2.1223	2.3827	2.4303		2.2447
6.0	3.4203	1.9948	1.5155			2.3809	2.4266		2.2381
7.0	3.4191	1.9919	1.5114			2.3792	2.4225		2.2304
8.0	3.4184	1.9885	1.5066			2.3775	2.4180		2.2213
9.0	3.4179	1.9846	1.5011			2.3757	2.4130		2.2107
10.0	3.4177	1.9803	1.4949			2.3737	2.4074		2.1985
11.0	3.4177	1.9756	1.4879			2.3716	2.4011		2.1846
12.0		1.9703	1.4801			2.3693	2.3942		2.1688
13.0		1.9644	1.4715			2.3668	2.3865		2.1508
14.0		1.9581	1.4619			2.3641	2.3782		
15.0		1.9511	1.4515			2.3613	2.3690		
16.0		1.9436	1.4401			2.3582	2.3591		
17.0		1.9354				2.3549	2.3483		
18.0		1.9266				2.3515	2.3366		
19.0		1.9171				2.3478	2.3240		
20.0		1.9069				2.3439	2.3105		
25.0		1.8436				2.3211			
30.0						2.2925			
35.0						2.2575			
40.0									
45.0									
50.0									
λ_2	3.4177	1.8282	1.4401	2.2841	2.1007	2.2225	2.3105	2.2800	2.1508

TABLE 75 REFRACTIVE INDICES OF ARSENIC-MODIFIED SELENIUM GLASS (Section 10.1.5)

(λ/μm)	Se : As
1.014	2.5783
2	2.5016
3	2.4882
4	2.4835
5	2.4811
6	2.4798
7	2.4787
8	2.4779
9	2.4772
10	2.4767
11	2.4758
12	2.4749
13	2.4760
14	2.4743

Source. Salzberg and Villa, *J. Opt. Soc. Am.* **47,** 244 (1957).

TABLE 76 REFRACTIVE INDICES OF SCHOTT INFRARED MATERIALS (Section 10.1.5)

λ (μm)	IRG-2 Germanate	IRG-3 La Flint	IRG-N6 CaAl Silicate	IRG-7 Lead Silicate	IRG-9 Fluorphosphate	IRG-11 Ca Aluminate
0.365	1.9780			1.5983	1.5005	
0.4047	1.9492	1.887	1.6259	1.5871	1.4961	
0.4800	1.9177	1.8594	1.6155	1.5743	1.4905	1.6944
0.4861	1.9159	1.8578	1.6148	1.5735	1.4902	1.6935
0.5461	1.9018	1.8455	1.6098	1.5675	1.4875	1.6863
0.5876	1.8948	1.8394	1.6072	1.5644	1.4861	1.6827
0.6438	1.8875	1.8330	1.6044	1.5612	1.4845	1.6788
0.6563	1.8862	1.8318	1.6038	1.5606	1.4842	1.6781
0.7065	1.8815	1.8276	1.6019	1.5585	1.4832	
0.8521	1.8722	1.8194		1.5541	1.4810	1.6704
1.014	1.866	1.8138	1.5951	1.5509	1.4793	1.6668
1.5296	1.8556	1.8034	1.5888	1.5442	1.4755	1.6598
1.9701	1.8494	1.7966	1.5836	1.5389	1.4722	1.6550
2.3254	1.8444	1.7908	1.5790	1.5341	1.4692	1.6509
2.674	1.8392	1.7845	1.5737	1.5286	1.4658	1.6463
3.303	1.8283	1.7709	1.5620	1.5164	1.4583	1.6366
4.258	1.8071	1.7436	1.5379			1.6174
4.586	1.7984	1.7320				1.6095

Source. Infrared-Transmitting Optical Materials, Schott, Mainz (1971).

TABLE 77 REFRACTIVE INDICES OF CALCITE AND QUARTZ
(Section 10.2)

λ (μm)	Calcite		Quartz	
	n_0	n_e	n_0	n_e
0.185			1.65751	
0.198	1.90284[a]	1.57796	1.65087	
0.231	1.80233	1.54541	1.61395	
0.340	1.70078	1.50562	1.56747	1.57737
0.394	1.68374	1.49810	1.55846	1.56805
0.434	1.67552	1.49430	1.55396	1.56339
0.508	1.66527	1.48956	1.54822	1.55746
0.589	1.65835	1.48640	1.54424	1.55335
0.768	1.64974	1.48259	1.53903	1.54794
0.833	1.64772	1.48176	1.53773	1.54661
0.991	1.64380	1.48022	1.53514	1.54392
1.159	1.64051	1.47910	1.53283	1.54152
1.307	1.63789	1.47831	1.53090	1.53951
1.396	1.63637	1.47789	1.52977	1.53832
1.479	1.6349	(1.47752)[c]	1.52865	1.53716
1.541	1.63381	(1.47726)	1.52781	1.53630
1.682	1.63187	(1.47666)	1.52583	1.53422
1.761	1.62974	(1.47633)	1.52468	1.53301
1.946	1.62602	(1.47557)	1.52184	1.53004
2.053	1.62372	(1.47492)	1.52005	1.52823
2.3	1.62099[b]	(1.47476)	1.51561	
2.6			1.50986	
3.0			1.49953	
3.5			1.48451	
4.0			1.46617	
4.2			1.4569	
5.0			1.417	
6.45			1.274	
7.0			1.167	

[a] At 0.2 μm.
[b] At 2.172 μm.
[c] Values in parentheses obtained by interpolation.
Based on Ballard et al. [10.27a].

TABLE 78 REFRACTIVE INDICES OF PLASTIC OPTIC MATERIALS AND THEIR TEMPERATURE DEPENDENCE (Section 10.3)

λ (μm)	Polystyrene (°C)			Polycyclohexyl methacrylate (°C)			Polymethyl methacrylate (°C)
	15	35	55	15	35	55	20
0.4358	1.6176	1.6148	1.6120	1.5184	1.5160	1.5131	1.5019
0.4861	1.6062	1.6034	1.6006	1.5134	1.5010	1.5081	1.4975
0.5461							1.4932
0.5896	1.5923	1.5897	1.5869	1.5071	1.5046	1.5018	1.4913
0.6563	1.5870	1.5843	1.5816	1.5044	1.5021	1.4992	1.4890
0.7679	1.5812	1.5785	1.5758	1.5016	1.4992	1.4964	
V numbers	31.0			56.9			57.8

Source. Raine [10.32].

TABLE 79   PLASTIC GLASSES (Section 10.3)

Name of monomer	Optical properties of polymer	
	Refractive index*	V-number
Allyl methacrylate	1.5196	49.0
Benzhydryl methacrylate	1.5933	31.0
Benzyl methacrylate	1.5680	36.5
n-butyl methacrylate	1.483	49
Tert-butyl methacrylate	1.4638	51
o-chlorobenzhydryl methacrylate	1.6040	30
α-(o-chlorophenyl)-ethyl methacrylate	1.5624	37.5
Cyclohexyl-cyclohexyl methacrylate	1.5250	53
Cyclohexyl methacrylate	1.5064	56.9
p-cyclohexyl-phenyl methacrylate	1.5575	39.0
α-β-diphenyl-ethyl methacrylate	1.5816	30.5
Menthyl methacrylate	1.4890	54.5
Ethylene dimethacrylate	1.5063	53.4
Hexamethylene glycol dimethacrylate	1.5066	56
Methacrylic anhydride	1.5228	48.5
Methyl methacrylate	1.4913	57.8
m-nitro-benzyl methacrylate	1.5845	27.4
2-nitro-2-methyl-propyl methacrylate	1.4868	48
α-phenyl-allyl methacrylate	1.5573	34.8
α-phenyl-n-amyl methacrylate	1.5396	40
α-phenyl-ethyl methacrylate	1.5487	37.5
β-phenyl-ethyl methacrylate	1.5592	36.5
Tetrahydrofurfuryl methacrylate	1.5096	54
Vinyl methacrylate	1.5129	46
Styrene	1.5907	30.8
Vinyl formate	1.4757	55
Phenyl cellosolve methacrylate	1.5624	36.2
p-methoxy-benzyl methacrylate	1.552	32.5
Ethylene chlorohydrin methacrylate	1.517	54
o-chlorostyrene	1.6098	31
Pentachlorophenyl methacrylate	1.608	22.5
Phenyl methacrylate	1.5706	35.0
Vinyl naphtalene	1.6818	20.9
Vinyl thiophene	1.6376	29
Eugenol methacrylate	1.5714	33
m-cresyl methacrylate	1.5683	36.8
o-methyl-p-methoxy styrene	1.5868	30.3
o-methoxy styrene	1.5932	29.7
o-methyl styrene	1.5874	32
Ethyl sulphide dimethacrylate	1.547	44
Allyl cinnamate	1.57	30
Diacetin methacrylate	1.4855	50

TABLE 79 (*Continued*)

Name of monomer	Optical properties of polymer	
	Refractive index*	V-number
Ethylene glycol benzoate methacrylate	1.555	36.8
Ethyl glycolate methacrylate	1.4903	55
p-isopropyl styrene	1.554	35
Bornyl methacrylate	1.5059	54.6
Triethyl carbinyl methacrylate	1.4889	57
Butyl mercaptyl methacrylate	1.5390	41.8
o-chlorobenzyl methacrylate	1.5823	37
α-methallyl methacrylate	1.4917	49
β-methallyl methacrylate	1.5110	47
α-naphthyl methacrylate	1.6411	20.5
Ethyl acrylate	1.4685	58
Cinnamyl methacrylate	1.5951	26.5
Methyl acrylate	1.4793	59
Terpineyl methacrylate	1.514	50
Furfuryl methacrylate	1.5381	39.2
p-methoxy styrene	1.5967	28
β-amino-ethyl methacrylate	1.537	52.5
Methyl α-bromoacrylate	1.5672	46.5
Vinyl benzoate	1.5775	30.7
Phenyl vinyl ketone	1.586	26.0
Vinyl carbazole	1.683	18.8
Lead methacrylate	1.645	28
2-chlorocyclohexyl methacrylate	1.5179	56
1-phenyl-cyclohexyl methacrylate	1.5645	40
Triethoxy-silicol methacrylate	1.436	53
p-bromophenyl methacrylate	1.5964	33
2-3 dibromopropyl methacrylate	1.5739	44
Diethyl-amino-ethyl methacrylate	1.5174	54
1-methyl-cyclohexyl methacrylate	1.5111	54
n-hexyl methacrylate	1.4813	57
2-6-dichlorostyrene	1.6248	31.3
β-bromo-ethyl metacrylate	1.5426	40
μ-polychloroprene	1.5540	36
Methyl α-chloracrylate	1.5172	57
β-naphthyl methacrylate	1.6298	24
Vinyl phenyl sulphide	1.6568	27.5
Methacryl methyl salicylate	1.5707	34
Methyl isopropenyl ketone	1.5200	54.5
Ethylene glycol mono-methacrylate	1.5119	56
N-benzyl methacrylamide	1.5965	34.5
β-phenyl-sulphone ethyl methacrylate	1.5682	39

TABLE 79 (*Continued*)

Name of monomer	Optical properties of polymer	
	Refractive index*	V-number
N-methyl methacrylamide	1.5398	47.5
N-allyl methacrylamide	1.5476	47
Methacryl-phenyl salicylate	1.6006	36
N-β-methoxyethyl methacrylamide	1.5246	53
N-β-phenylethyl methacrylamide	1.5857	37
Cyclohexyl α-ethoxyacrylate	1.4969	58
1-3-dichloropropyl-2-methacrylate	1.5270	56
2-methyl-cyclohexyl methacrylate	1.5028	53
3-methyl-cyclohexyl methacrylate	1.4947	55
3-3-5-trimethyl-cyclohexyl methacrylate	1.485	54
N-vinyl phthalimide	1.6200	24.1
Fluorenyl Mothacrylate	1.6319	23.1
α-naphthyl-carbinyl methacrylate	1.63	25
p-p^3-xylylenyl dimethacrylate	1.5559	37
Cyclohexanediol-1-4 dimethacrylate	1.5067	54.3
Ethylidene dimethacrylate	1.4831	52.9
p-divinyl benzene	1.6150	28.1
Decamethylene glycol dimethacrylate	1.4990	56.3
Vinyl cyclohexene dioxide	1.5303	56.4
Methyl α-methylene butyrolactone	1.5118	53.9
α-methylene butyrolactone	1.5412	56.4
4-dioxolylmethyl methacrylate	1.5084	59.7
Methylene-α-valerolactone	1.5431	47.8
o-methoxy-phenyl metacrylate	1.5705	33.4
Isopropyl methacrylate	1.4728	57.9
Trifluoroisopropyl methacrylate	1.4177	65.3
β-ethoxy-ethyl methacrylate	1.4833	32.0
Name of Polymer		
Condensation resin from di- (*p*-amino-cyclohexyl) methane and sebacic acid	1.5199	52.0
Columbia Resin 39	1.5001	58.8

Source. Raine [10.32]. * D-line at 20°C.

TABLE 80a ABSORPTION COEFFICIENTS OF SOME OPTICAL GLASSES
Units: m^{-1} (Section 10.1.2.3)

λ μm	C-1 523/586	BSC-1 511/635	BSC-2 517/645	LBC-2 573/574	DBC-1 611/588	DBC-3 611/572	CF-1 529/516	BF-1 584/460	DF-2 617/366	EDF-3 720/293
0.3	—[b]	—	480	—	—	—	—	—	—	—
0.32	160	140	120	300	—	—	230	460	—	—
0.34	59	4.4	64	140	390	240	13	53	160	—
0.36	17	10.5	27	74	150	19	3.0	6.2	32	270
0.38	2.8	2.0	5.1	7.8	3.2	15	1.0	2.0	17	76
0.40	0.7	0.5	0	0	0	0.6	0	0	10	36
0.42	0[a]	0	0	0	0	0	0	0	6.7	16
0.44	0	0	0	0	0	0	0	0	4.6	8.3
0.46	0	0	0	0	0	0	0	0	3.0	3.9
0.48	0	0	0	0	0	0	0	0	2.0	1.5
0.50	0	0	0	0	0	0	0	0	1.1	0.7
0.60	0	0	0	0	0	0	0	0	0	0.5
0.70	0	0	0	0	0	0	0	0	0	0
0.80	0	0	1.5	0.8	0.6	0	0	0	0	0
1.0	0	0	5.7	2.8	3.5	0	0	0	0	0
2.0	5.13	12	16	10	43	22	36	12	0	0
3.0	17.4	530	—	510	—	—	470	470	280	350
4.0	—	—	—	—	—	—	—	—	—	440

Cutoff wavelengths and specific gravity

	C-1	BSC-1	BSC-2	LBC-2	DBC-1	DBC-3	CF-1	BF-1	DF-2	EDF-3
$\lambda_{uv}(\mu m)$	0.301	0.300	0.296	0.306	0.328	0.320	0.310	0.316	0.326	0.350
$\lambda_{ir}(\mu m)$	4.00	3.20	3.00	3.20	2.90	2.85	3.35	3.25	3.50	4.10
Spec. grav.	2.53	2.48	2.53	3.21	3.58	3.57	2.73	3.31	3.64	4.51

[a] 0 indicates α below $0.10\,\mathrm{m}^{-1}$.
[b] — indicates α in excess of $530\,\mathrm{m}^{-1}$.
Source. Molby, Ref. 10.13.

TABLE 80*b*   ABSORPTION COEFFICIENTS OF SOME OPTICAL GLASSES
Units: m^{-1} (Section 10.1.2.3)

λ (μm)	Glass type					
	C 511/604	BSC 510/635	LBC 540/597	DBC 623/581	DF 604/381	EDF 699/301
0.28	515	256	—	—	—	—
0.29	221	180	—	—	—	—
0.30	109	101	205	677	—	—
0.31	47.9	59.9	79.2	239	539	—
0.32	18.4	38.5	30.6	118	160	—
0.33	10.0	21.1	13.3	58.3	51.3	—
0.34	5.09	9.51	7.00	27.6	21.6	—
0.35	2.92	4.44	4.21	15.7	10.1	629
0.36	1.83	2.53	2.83	8.54	4.95	178
0.37	1.30	1.74	1.96	5.02	2.67	81.5
0.38	1.09	1.33	1.60	3.13	2.04	28.3
0.39	0.682	0.953	1.19	2.13	1.02	12.5
0.40	0.465	0.617	0.979	1.52	0.610	5.66
0.42	0.534	0.636	0.755	1.04	0.465	1.61
0.44	0.511	0.636	0.751	1.04	0.414	0.702
0.46	0.461	0.520	0.629	0.854	0.297	0.424
0.48	0.368	0.495	0.601	0.647	0.251	0.233
0.50	0.343	0.449	0.481	0.474	0.205	0.116
0.55	0.318	0.357	0.318	0.233	0.204	0.138
0.60	0.343	0.380	0.318	0.327	0.272	0.139
0.70	0.320	0.311	0.364	0.424	0.251	0.116
0.80	0.440	0.405	0.484	0.573	0.391	0.256
1.00	0.562	0.594	0.702	0.829	0.580	0.375
1.20	0.610	0.665	0.705	0.831	0.534	0.447
1.40	1.94	2.58	2.51	2.20	1.01	0.905
1.60	1.37	1.64	1.35	1.23	1.19	0.953
1.80	4.24	4.65	2.86	2.15	3.32	2.37

Adapted from Glass Catalog of Sovirel, Levalois, Perret, France.

TABLE 81 CHANGE OF REFRACTIVE INDEX WITH TEMPERATURE. CHANGES RELATIVE TO 0°C. Units: 10^{-5}.
(Section 10.1.2.5)

λ (nm)	-194°C	-100°C	-80°C	-60°C	-40°C	-20°C	+20°C	40°C	60°C	80°C	100°C	dn/dT[a]
Crown glass (C-1)												
480	-5.8	-9.9	-9.0	-7.5	-5.5	-2.9	+3.6	7.9	12.7	17.9	23.5	0.199
589	+0.3	-5.9	-5.8	-5.1	-3.8	-2.1	+2.7	6.0	9.8	14.0	18.6	0.150
644	+2.2	-4.5	-4.5	-4.1	-3.1	-1.7	+2.3	5.2	8.7	12.4	16.7	0.131
Borosilicate crown glass (BSC-2)												
480	+1.0	-7.6	-7.6	-6.9	-5.3	-2.9	+3.6	8.0	13.0	18.7	25.1	0.200
589	+5.6	-5.0	-5.4	-5.0	-3.8	-2.1	+3.0	6.5	10.8	15.6	21.1	0.160
644	+6.9	-3.6	-4.4	-4.4	-3.7	-2.2	+2.8	6.1	10.2	14.9	20.3	0.152
Light barium crown glass (LBC-2)												
480	+20.2	+3.1	+1.1	-0.25	-0.8	-0.62	+1.4	3.4	6.2	9.6	13.6	0.085
589	+25.4	+6.8	+4.1	+2.1	+0.9	+0.19	+0.4	1.4	3.2	5.7	8.7	0.036
644	+28.4	+8.5	+5.6	+3.1	+1.5	+0.4	+0.14	.9	2.3	4.1	6.4	0.025
Dense barium crown glass (DBC-1)												
480	+9.6	-3.7	-4.6	-4.4	-3.6	-2.1	+3.0	6.5	10.6	15.1	20.1	0.162
589	+14.8	+0.5	-1.3	-2.3	-2.4	-1.7	+2.1	4.6	7.8	11.5	15.7	0.116
644	+16.6	+1.4	-0.4	-1.3	-1.6	-1.2	+0.8	4.1	6.9	10.1	13.7	0.103

Crown flint glass (CF-1)

480	−2.6	−11.2	−11.0	−9.8	−7.5	−4.2	+4.8	10.4	16.9	24.1	32.0	0.261
589	+4.6	−6.5	−7.1	−6.8	−5.4	−3.1	+3.7	8.1	13.4	19.4	26.0	0.205
644	+7.4	−4.8	−5.7	−5.8	−4.8	−2.9	+3.3	7.3	12.1	17.7	23.9	0.184

Barium flint glass (BF-1)

480	−0.4	−9.9	−9.9	−8.9	−6.9	−3.9	+4.5	9.8	15.9	22.7	30.1	0.246
589	+9.3	−2.8	−4.0	−4.5	−4.0	−2.5	+2.8	6.5	10.9	16.1	21.5	0.162
644	+13.6	−1.3	−2.7	−3.3	−3.0	−1.8	+2.4	5.6	9.5	14.1	19.2	0.140

Dense flint glass (DF-2)

480	−26.8	−28.9	−26.0	−21.5	−15.6	−8.3	+9.3	19.7	31.1	43.5	57.0	0.492
589	−9.2	−18.1	−16.8	−14.3	−10.4	−5.6	+7.0	14.8	23.6	33.2	43.7	0.370
644	−4.9	−15.0	−14.6	−12.9	−9.9	−5.6	6.3	13.4	21.3	30.1	39.7	0.334

Extra dense flint glass (EDF-3)

480	−77.3	−60.1	−51.9	−41.8	−29.8	−15.8	+16.5	34.3	53.4	73.6	94.7	0.858
589	−44.8	−40.9	−35.8	−28.9	−20.5	−10.7	+12.2	25.4	39.6	54.7	70.5	0.634
644	−36.0	−35.5	−31.3	−25.4	−18.0	−9.4	+11.0	23.0	35.9	49.6	63.9	0.575

[a] Gradient at 20°C in $10^5 \Delta n/°C$.
Source. Molby [10.13].

TABLE 82 REFRACTIVE INDEX THERMAL COEFFICIENTS FOR SOME CORNING GLASSES[a] (Section 10.1.2.5)

	Glass type and Corning no.		
λ (μm)	Silica 7940	Alumino-silicate 1723	Vycor 7913
0.23	19.6	—	—
0.265	16.6	—	16.3
0.365	14.0	11.1	13.1
0.4	13.4	10.2	12.5
0.5	12.4	9.3	11.9
1.0	11.8	8.2	11.2
1.5	11.5	7.9	10.9
2.0	11.2	8.0	10.9
2.5	11.3	8.1	11.0
3.0	11.2	—	—

[a] Values in units of $(10^{-6}/°C)$.
Adapted from Wray and Neu [10.14].

TABLE 84 REFRACTIVE INDEX AT ROOM TEMPERATURE FOR VARIOUS ANNEALING TEMPERATURES FOR BOROSILICATE CROWN: 517/645 (Section 10.1.3.3)

Temp. (°C)	Index	Stabilization time (hr)
500	1.51800	8000
520	1.51725	800
530	1.51710	250
540	1.51630	90
550	1.51580	50
560	1.51535	5
590	1.51385	0.8
620	1.51280	0.5
630	1.51255	0.6
640	1.51215	0.3
650	1.51185	0.05
660	1.51150	0.05
670	1.51115	0.05

Source. Brandt [10.17].

TABLE 83 VISCOSITY-TEMPERATURE CHARACTERISTICS OF SOME GLASSES (Section 10.1.3.3)

	Silica 100% (°C)	Silica 96% (°C)	Lo-Exp.[a] BSC (°C)	Plate and Crown (°C)	Flint (°C)	Viscosity $\log_{10}$(poise)
Strain point	990	820	520	510	460	14.5
Anneal. point	1050	910	565	550	480	13.0
Soft point	1580	1500	820	730	610	7.6
Flow point	2000[b]	—	(1050)	920	(910)	5.0[a]

Data from Ref. 10.7, except (a) which is from Ref. 10.6 and (b) which is from Ref. 10.16 and corresponds to a viscosity of 10^6 poise.

TABLE 85 SIGNIFICANT IMPURITY CONCENTRATIONS
IN SODA-LIME SILICATE GLASSES (Section 10.1.6)

| Ion | Absorption Peak (nm) | Concentration for 20 dB/km (10^{-9}) | |
		(Absorption at the Peak)	(Absorption at 800 nm)
Cu^{2+}	800	9	9
Fe^{2+}	1100	8	15
Ni^{2+}	650	4	26
V^{3+}	475	18	36
Cr^{3+}	675	8	83
Mn^{3+}	500	18	1800

Source. Maurer [10.23].

TABLE 86 TRADE NAMES OF SOME INFRARED
MATERIALS (Section 10.2)

IRG[a]	2	Germanate glass
	3	Lanthanum dense flint glass
	N6	Calcium-Aluminum Silicate glass
	7	Lead silicate glass
	9	Fluor phosphate glass
	11	Ca Aluminate Glass
Irtran[b]	1	Magnesium fluoride, polycrystalline
	2	Zinc sulfide, polycrystalline
	3	Calcium fluoride, polycrystalline
	4	Zinc selenide, polycrystalline
	5	Magnesium oxide, polycrystalline
	6	Cadmium telluride, polycrystalline
KRS	5	Thallium bromide-iodide
	6	Thallium bromide-chloride

[a] IRG, Schott, Mainz.
[b] Irtran, Eastman Kodak, Rochester.

TABLE 87   OPTICAL MATERIALS FOR THE INFRARED: MECHANICAL

Material	Code or Formula	Approximate Transmission Range (μ)	Melt. Pt. (°C)[c]	Thermal[a] conductivity (10^{-4} cal cm^{-1} sec^{-1}°C^{-1})	
Arsenic selenium glass	Se(As)	1–19	~70*[b]	3.3	(20)[d]
Arsenic trisulfide glass	As$_2$S$_3$	0.6–~11	210*	4	(40)
Barium fluoride	BaF$_2$	0.15–15	1280	280	(13)
Barium titanate	BaTiO$_3$	<0.5–6.9	1600	32	(20)
Borosilicate glass	—	0.3–3	~730	—	
Cadmium sulfide	CdS	0.52–16	900**[b]	380	(14)
Cadmium telluride	CdTe	0.9–15	1041–1050	—	
(polycrystalline, Irtran 6[c])	CdTe	0.9–33	1090	100	(0°C)
Calcium aluminate glasses (IRG 11)	—	0.5–5	~900*	32	(25)
Calcium carbonate	CaCO$_3$	0.3–5.5	Decomposes at 894.4	132	(0)∥c axis[f]
				111	(0)⊥c axis
Calcium fluoride	CaF$_2$	0.13–12	1360	232	(36)
Calcium fluoride (polycrystalline, Irtran 3)	CaF$_2$	0.20–11.5	1360	190	(80)
Cesium bromide	CsBr	0.22–55	636	23	(25)
Cesium iodide	CsI	0.24–70	621	27	(25)
Fused silica	SiO$_2$	0.2–4.5	~1710*	28.2	(41)
Gallium arsenide	GaAs	1–11	1238	—	
Gallium antimonide	GaSb	2.0–~2.5	720	—	
Gallium phosphide	GaP	0.6–4.5	500+	—	
Germanium	Ge	1.8–23	936	1400	(20)
Germanate glasses (IRG 2)	—	0.4–6	~800*	6.4	(20)
Indium antimonide	InSb	7.0–16	523	850	(20)
Indium arsenide	InAs	3.8–~7	942	—	
Lead fluoride	PbF$_2$	0.25–~17	855	—	
Lead selenide	PbSe	5.0–7	1065	100	
Lead sulfide	PbS	3.0–7	1114	16	
Lead telluride	PbTe	4.0–7	917	120	
Lead silicate glasses (IRG 7)	—	0.4–5	~600*	—	
Lithium fluoride	LiF	0.12–9	870	270	(41)
Magnesium fluoride	MgF$_2$	0.11–7.5	~1255	—	
Magnesium fluoride (polycrystalline, Irtran 1)	MgF$_2$	0.45–9.2	1255	350	(56)
Magnesium oxide	MgO	0.25–8.5	2800	600	(20)
Magnesium oxide (polycrystalline, Irtran 5)	MgO	0.39–9.4	~2800	1040	(36)
Potassium bromide	KBr	0.23–40	730	115	(46)
Potassium chloride	KCl	0.21–30	776	156	(42)

928

Coefficient of Expansion $(10^{-6}/°C)^a$	Specific Heat	Hardness (Knoop no. for Given Load)	Modulus of elasticity (GN/m^2)	Solubility
34	—	—	—	Insol. in water
24.6 (33–165)	—	109, 100 g	15.9	Sol. in alkalis
—	—	82, 500 g	53	0.17 g/100 g water at 10°C
6.2 (4–20)‖c axis	0.077 at −98°C	Vickers 200–580	34	—
15.7 (4–20)⊥c axis				
8.3 (25–525)	—	—	—	—
4.2 (27–70)	—	122∓4, 25 g	—	Insol. in water
4.5 (50), 5.9 (600)	—	Vickers 435 at room temp.	—	
5.3 (15–30)	0.045	45	37	Probably insol.
				—
8.3 (25–300)	—	—	—	—
25 (0)‖c axis	0.203 at 0°C	Moh 3	72‖c axis	0.0014 g/100 g
−5.8 (0)⊥c axis		—	88⊥c axis	water at 25°C
24 (20–60)	0.204 at 0°C	158.3, 500 g	76	0.0017 g/100 g water at 26°C
22.3 (25–450)	0.2 at 0°C	200	99	Insol. in water
47.9 (20–50)	0.063 at 20°C	19.5, 200 g	15.9	124.3 g/100 g water at 25°C
50 (25–50)	0.048 at 20°C	—	5.3	44 g/100 g water at 0°C
0.5 (20–900)	0.22	461, 200 g	73	Insol. in water
5.7	—	—		Insol. in water
—	—	—	63	Insol. in water
—				
5.5 (25)	0.074 from 0°C to 100°C	—	103	Insol. in water
9 (26–594)	—	542, 1000 g	90	—
4.9 (20–60)	—	—	43	Insol. in water
5.3	—	—	—	Insol. in water
—	—	—	—	—
—	—	—	—	Insol. in water
—	0.050	—	—	—
—	—	—	—	—
9.8 (25–300)	—	—	48	—
37 (0–100)	0.373 at 10°C	102–113, 600 g	65	0.27 g/100 g water at 18°C
18.8‖c axis	—	—	—	0.0076 g/100 g water at 18°C
13.1⊥c axis	—	—	—	—
11.9 (25–400)	0.23 at 25°C	576	114	Insol. in water
13.8 (20–1000)	0.209 at 0°C	692, 600 g	249	Insol. in water
12.8 (25–450)	0.21 at 0°C	640	332	—
43 (20–60)	0.104 at 0°C	5.9 in ⟨110⟩ direction, 200 g	27	53.48 g/100 g water at 0°C
36 (20–60)	0.162 at 0°C	7.2 in ⟨110⟩ direction, 200 g	30	34.7 g/100 g water at 20°C

TABLE 87

Material	Code or Formula	Approximate Transmission Range (μ)	Melt. Pt. (°C)[c]	Thermal[a] Conductivity (10^{-4} cal cm^{-1} sec^{-1}°C^{-1})
Potassium dihydrogen phosphate	KDP	0.25–1.7	252.6	29 (39)$\|c$ axis 32 (46)$\perp c$ axis
Potassium iodide	KI	<0.38–42	723	—
Quartz (crystalline)	SiO$_2$	0.4–4.5	<1470	255 (50)$\|c$ axis 148 (50)$\perp c$ axis
Sapphire	Al$_2$O$_3$	0.17–6.5	2030	600 (26)$\|c$ axis 550 (23)$\perp c$ axis
Selenium, amorphous	SE	1–20	35*	—
Silicon	Si	1.2–15	1420	3900 (40)
Silver chloride	AgCl	0.4–30	457.7	27.5 (22)
Sodium chloride	NaCl	0.21–26	801	155 (16)
Sodium fluoride	NaF	<0.19–15	980 (tet.) 997 (cub.)	—
Sodium nitrate	NaNO$_3$		306.8	—
Spinel	MgO · 3.5Al$_2$O$_2$	0.9–6.0	2030 to 2060	330 (35)
Strontium titanate	SrTiO$_3$	0.39–6.8	2080	—
Tellurium	Te	3.5–8	449.7	150
Thallium bromide	TlBr	0.44–40	460	14 (43)
Thallium bromide iodide (KRS-5)	Tl(Br-I)	0.5–40	414.5	13 (20)
Thallium bromide chloride (KRS-6)	Tl(Br-Cl)	0.21–34	423.5	17.1 (56)
Thallium chloride	TlCl	0.44–34	430	18 (38)
Titanate silicate glasses	—	0.4–5	800*	—
Titanium dioxide	TiO$_2$	~0.43–6.2	1825	300 (36)$\|c$ axis 210 (44)$\perp c$ axis
Zinc selenide (Irtran 4)	ZnSe	0.48–21.8	1520	310 (54)
Zinc sulfide (Irtran 2)	ZnS	0.57–14.7	1830 to 150 psi	370 (54)
Conventional glasses for comparison	—	0.35–3	600*	20
Metals for comparison	—	—	—	Ag: 10,000

$\|c$, parallel to the c, axis; $\perp c$, perpendicular to the c axis.

[a] At the deg. C temperature given in parentheses.

[b] Soft. point,*; volatile at,**.

[c] Discontinued.

Source. Kreidl and Rood [10.7] with minor additions and modifications. Much of the data from S. S. Ballard, K. A. McCarthy, and W. L. Wolfe, "State-of-the-art report on optical materials for infrared instrumentation," University of Michigan Press, Ann Arbor, 1959.

Coefficient of Expansion ($10^{-6}/°C$)[a]	Specific Heat	Hardness (Knoop no. for Given Load)	Modulus of elasticity (GN/m^2)	Solubility
—	—	—	—	33 g/100 g water at 25°C
42.6 (40)	0.75 at −3°C	—	32	127.5 g/100 g water at 0°C
7.97 (0–80)‖c axis	0.188 from 12°C	741, 500 g	77	Insol. in water
13.37 (0–80)⊥c axis	to 100°C		97	
6.7 (50)‖c axis	0.18 at 25°C	1370, 1000 g	345	Insol. in water
5.0 (50)⊥c axis				
36.8 (40)	—	—	—	—
4.15 (10–50)	0.168 at 25°C	1150	131	Insol. in water
30 (20–60)	0.0848 at 0°C	9.5, 200 g	20	Insol. in water
44 (−50–200)	0.204 at 0°C	15.2, 200 g in ⟨110⟩ direction	40	35.7 g/100 g water at 0°C
36 (20)	0.26 at 0°C	—	—	4.22 g/100 g water at 18°C
12 (50‖c axis	0.247 at 0°C	19.2, 200 g	—	73 g/100 g water at 0°C
11 (50)⊥c axis				
5.9 (40)	—	1140, 100 g	—	Insol. in water
9.4	—	595	—	—
16.75 (40)	4.79×10^{-2} at 300°C	—	—	Insol. in water
51 (20–60)	0.045 at 25°C	11.9, 500 g	30	0.05 g/100 g water at 25°C
58 (20–100)	—	40.2, 200 g	15.9	0.05 g/100 g at room temperature
50 (20–100)	0.0482 at 20°C	29.9, 500 g	21	0.32 g/100 g water at 20°C
53 (20–60)	0.0520 at 0°C	12.8, 500 g	32	0.32 g/100 g water at 20°C
8.3 (25–300)	—	—	—	—
9.19 (40)‖c axis	0.17 at 25°C	879, 500 g	—	Insol. in water; sol. in conc. acid
7.14 (40)⊥c axis				
8.2 (30–450)	0.08 at 25°C	150	71	Insol. in water
7.5 (25–400)	0.12 at 0–100°C	354	97	Insol. in water
—	—	—	—	—
Cu: 14(−90 to −15)	—	—	—	—

TABLE 88 ABSORPTION CHARACTERISTICS OF IRTRAN MATERIALS (Section 10.2)

λ (μm)	Irtran-1			Irtran-2			Irtran-3			Irtran-4			Irtran-5			Irtran-6*		
	n	τ (12 mm)	α (mm^{-1})	n	τ (12 mm)	α (mm^{-1})	n	τ (12 mm)	α (mm^{-1})	n	τ (12 mm)	α (mm^{-1})	n	τ (12 mm)	α (mm^{-1})	n	τ (6 mm)	α (mm^{-1})
0.5	1.3877		0.98	2.4161		1.46	1.4359	0.175	0.140	2.7178	0.085	0.165	1.7443	0.3	0.0877	2.838	0.3	0.116
1	1.3778	0.07	0.217	2.2907	0.14	0.136	1.4289	0.425	0.0660	2.485	0.285	0.0717	1.7227	0.505	0.0449	2.714	0.45	0.0571
2	1.3720	0.5	0.0536	2.2631	0.47	0.0366	1.4239	0.685	0.0264	2.447	0.43	0.0390	1.7089	0.61	0.0296	2.695	0.51	0.0382
3	1.3640	0.64	0.0332	2.2558	0.59	0.0183	1.4179	0.815	0.0120	2.440	0.475	0.0311	1.6920	0.693	0.0194	2.688	0.525	0.0340
4	1.3526	0.795	0.0154	2.2504	0.635	0.0125	1.4097	0.84	0.0097	2.435	0.505	0.0263	1.6684	0.745	0.0140	2.684	0.54	0.0297
5	1.3374	0.71	0.0250	2.2447	0.665	0.0089	1.3990	0.875	0.0065	2.432	0.53	0.0225	1.6368	0.767	0.0123	2.681	0.55	0.0270
6	1.3179	0.66	0.0315	2.2381	0.675	0.0079	1.3856	0.885	0.0058	2.428	0.54	0.0211	1.5962	0.6	0.0336	2.679	0.55	0.0271
7	1.2934	0.235	0.118	2.2304	0.68	0.0075	1.3693	0.85	0.0095	2.423	0.55	0.0198	1.5452	0.17	0.140	2.677	0.55	0.0272
8	1.2634			2.2213	0.675	0.0083	1.3498	0.695	0.0266	2.418	0.56	0.0184	1.4824		0.36	2.674	0.54	0.0302
9	1.2269			2.2107	0.66	0.0104	1.3269	0.395	0.0741	2.413	0.56	0.0186	1.4060		0.92			
10				2.1986	0.645	0.0125	1.3002	0.077	0.211	2.407	0.57	0.0173				2.672	0.538	0.0309
11				2.1846	0.435	0.0451	1.2694		0.553	2.401	0.585	0.0154				2.669	0.545	0.0290
12				2.1688	0.43	0.0465				2.394	0.583	0.0158				2.666	0.535	0.0320
13				2.1508	0.175	0.121				2.386	0.59	0.0151				2.663	0.515	0.0382
14										2.378	0.575	0.0173				2.660	0.539	0.0312
15										2.370	0.53	0.0241				2.667	0.528	0.0346
2.7	1.3667	0.45	0.0625															
5.3	1.3320	0.79	0.0163															

* Discontinued

Values of n and τ from Eastman-Kodak Co., Publ. No. V-72 (1971).

Plastic	Supplier[a]	Density (g/cm³)	Coeff. of expansion (10^{-6}/°C)	Upper limit of stability (°C)	n_D (20°C)	V value
Allyl diglycol carbonate	(1)	1.32	90–100	60–70	1.498	53.6
Polymethyl methacrylate	(2)	1.19	63	70–100	1.492	57.8
Polystyrene	—	1.10	80	70	1.591	30.8
Copolymer styrene-methacrylate	(3)	1.14	66	95	1.533	42.4
Copolymer methyl-styrene-methyl methacrylate	(4)	1.17	—	110–120	1.519	—
Polycarbonate	(5)	1.2	70	120–135	1.586	29.9
Polyester-styrene	—	1.22	80–150	50–120	1.54–1.57	ca. 43
Cellulose ester	—	1.30	90–100	50–60	1.47–1.50	45–50
Copolymer styrene-acrylonitrile	(6)	1.07	70	90	1.569	35.7
Polycyclohexyl methacrylate		1.095	76	105	1.506	56

[a] (1) "CR-39," Pittsburgh Plate Glass, Columbia-Southern Division; (2) "Plexiglas," Rohm and Haas Co.; "Lucite," E. I. DuPont Co.; "Acrylite," American Cyanamid Co.; (3) "Zerlon," Dow Chemical Co.; (4) "Bavick," J. T. Baker Chemical Co.; (5) "Merlon," Mobay Chem. Co.; "Lexan," General Electric Company; (6) "Lustran," Monsanto Chemical; "Tyril," Dow Chemical Co.; "Bakelite C-11," Union Carbide Plastics Co.

Source. Kreidl and Rood [10.7].

TABLE 89b   PHYSICAL PROPERTIES OF PRINCIPAL OPTICAL PLASTICS[a] (Section 10.3)

Properties	ASTM method	Units	Methyl methacrylate (acrylic)	Polystyrene (styrene)	Polycarbonate	Methyl methacrylate styrene copolymer
Refractive Index (n_D)	D 542		1.491	1.590	1.586	1.562
Abbe Value (v)	D 542		57.2	30.9	34.7	35
$dn/dt \times 10^{-5}/°C$			8.5	12.0	14.3	14.0
Haze (%)	D 1003	%	<2	<3	<3.0	<3
Luminous Transmittance (0.125 in. thickness)	D 1003	%	92	88	89	90
Critical Angle (i_c)	D 648-56	degree	42.2	39.0	39.1	39.6
Deflection Temperature		°F				
3.6 F/min, 264 psi			198	180	280	
3.6 F/min, 66 psi			214	230	270	212
Coefficient of Linear Thermal Expansion	D 696-44	in./in./°F $\times 10^{-5}$	3.6	3.5	3.8	3.6
Recommended Max. Cont. Service Temp.		°F	198	180	255	200
Water Absorption (Immersed 24 hr at 73 F)	D 570-63	%	0.3	0.2	0.15	0.15
Specific Gravity (Density)	D 792		1.19	1.06	1.20	1.09
Hardness (0.25-in. sample)	D 785-62		M 97	M 90	M 70	M 75
Impact Strength (Izod Notch)	D 256	ft-lb/in.	0.3–0.5	0.35	12–17	
Dielectric Strength	D 149-64	V/mil	500	500	400	450
Dielectric Constant:	D 150					
60 Hz			3.7	2.6	2.90	3.40
10^6 Hz			2.2	2.45	2.88	2.90
Power Factor:	D 150					
60 Hz			0.05	0.0002	0.0007	0.006
10^6 Hz			0.03	0.0002–0.0004	0.0075	0.013
Volume Resistivity	D 257	ohm-cm	10^{18}	$>10^{16}$	8×10^{16}	10^{15}
Trade Names			Lucite Perspex Plexiglas	Dylene Styron	Lexan Makrolon Merlon	NAS

[a] This information is taken from raw material manufacturers' available published data. Specific material formulation data should be confirmed prior to design and specification.

Source: Price [10.32]

934

TABLE 90   LONG WAVELENGTH PASS FILTERS (Section 11.1.2)

37% Cut (nm)	Description	Slope (nm)	Extent (μm)	Thickness (mm)	Mfr.[a]
150	Quartz	35	4.5	1	
165	Sapphire	65	5.5	1	
170	Water	20	1.4	1	
200	NaCl	20	20	1	
240	$CaCO_3$	35	5.5	2	
250	Glycerine	100	1.4	10	
255	7905 (Vycor)	45	3.5	2	C
280	9700 (Corex)	40	2.6	2	C
290	Wratten 0	70	2.6	0.10	EK
305	7740 (Pyrex)	35	2.6	2	C
310	BSC-2	30	2.6	5	—
315	0160	20	3.0	2	C
320 to 340	Plate glass				
330	DF Type	30	2.6	5	—
345	GG-1	50	2.6	2	S
360	7380	20	2.6	2	C
360	T-1	20	2.6	3.5	BL
360	EDF type	40	2.6	5	—
365	OY-10	20	2.6	2	
370	2876	30	2.6	2	P
380	L-38	30	2.6	2	H
385	GG-13	35	2.6	2	S
385	3850	40	2.6	4	C
390	L-39	30	2.6	2	H
390	Wratten 1A	30	2.6	0.10	EK
400	3-75	57	2.81	—	C
	2C	28	—	—	EK
405	2B	30	—	—	EK
	L-41	16	2.7	—	H
415	3-74	37	2.73	—	C
	3-144	42	2.7	—	C
	GG 420	34	—	3	S
	GG 19	53	—	3	S
420	2A	17	—	—	EK
	L-42	16	2.7	—	H
425	2E	17	—	—	EK
430	GG 435	15	—	3	S
	3-73	26	2.73	—	C
435	Y-44	15	2.75	—	H
455	Y-46	16	2.7	—	H
	3-72	26	2.73	—	C
	GG 9	60	—	1	S
	GG 455	16	—	3	S
465	3	29	—	—	EK
	GG 10	57	—	1	S
470	GG 475	15	—	3	S
	4	20	—	—	EK

Table 90 (*Continued*)

37% Cut (nm)	Description	Slope (nm)	Extent (μm)	Thickness (mm)	Mfr.[a]
475	Y-48	19	2.7	—	H
480	3-71	22	2.81	—	C
490	8	38	—	—	EK
	GG 495	16	—	3	S
495	Y-50	17	2.7	—	H
500	3-70	27	2.73	—	C
	9	58	—	—	EK
515	Y-52	16	2.7	—	H
	OG 515	25	—	3	S
520	3-69	19	2.7	—	C
530	15	26	—	—	EK
	OG 530	15	—	3	S
535	3-68	19	2.7	—	C
	16	29	—	—	EK
540	O-54	14	2.7	—	H
550	21	23	—	—	EK
	OG 550	15	—	3	S
555	3-67	15	2.7	—	C
560	O-56	19	2.7	—	H
565	OG 570	15	—	3	S
	22	23	—	—	EK
570	3-66	16	2.7	—	C
580	2-73	18	2.7	—	C
	23A	20	—	—	EK
	O-58	18	2.8	—	H
585	OG 590	16	—	3	S
	24	20	—	—	EK
595	2-63	15	2.7	—	C
	R-60	18	2.76	—	H
600	26	25	—	—	EK
	2-62	16	2.7	—	C
	25	12	—	—	EK
610	RG 610	15	—	3	S
	2-61	17	2.7	—	C
620	29	8	—	—	EK
	R-62	19	2.76	—	H
	2-60	20	2.7	—	C
625	RG 630	15	—	3	S

37% Cut (nm)	Description	Slope (nm)	Extent (μm)	Thickness (mm)	Mfr.[a]
630	2-59	17	2.7	—	C
640	R-64	20	2.76	—	H
	2-58	20	2.7	—	C
645	RG 645	15	—	3	S
660	R-66	19	2.74	—	H
665	2-64	20	2.7	—	C
	RG 665	15	—	3	S
675	R-68	18	2.74	—	H
695	RG 695	17	—	3	S
700	5850	30	1.05	4	C
710	Wratten 89B	35	2.6	0.10	EK
710	R-4	70	2.6	4	BL
740	Wratten 88A	35	2.6	0.10	EK
750	2930	80	2.6	4	C
750	R-76	200	2.6	2	H
780	Wratten 87	65	2.6	0.10	EK
840	Wratten 87C	65	2.6	0.10	EK
850	2550	160	2.6	2	C
900	OX-5	200	2.6	2	
920	Black carrara	225	2.6	8	P
920	OX-2	250	2.6	3	
950	2540	250	2.6	2	C
960	RG-7	200	2.6	2	S
1000	Se(As) glass	c	19 (50%)	2.5	EK
1000	Selenium	c	19 (50%)	1.5	—
1200	Silicon	c	8.5 (50%)	2	—
1800	Germanium	c	20 (30%)	2	—
2000	KFIR-210	1400	25	1	EK
2400	5021	c	4.5 (30%)	0.75	C
2800	KFIR-220	1500	25	1	EK
4000	Lead telluride	c	7	0.1	—
4300	KFIR-230	2000	25	1	EK
5800	KFIR-240	2700	25	1	EK
7500	KFIR-250	3800	25	1	EK

[a] BL, Bausch & Lomb; C, Corning; EK, Eastman Kodak; H. Hoya; P. Pittsburgh Glass; S. Schott (Jena).

Based in part, on Scharf [11.1a].

TABLE 91 SHORT WAVELENGTH PASS FILTERS (Section 11.1.2)

37% Cut (μm)	Description	Slope (μm)	Extent (μm)	Thickness (mm)	Mfr.[a]
0.390	9863	0.035	0.270	3	C
0.430	B-72	0.350	0.400	2	H
0.460	B-3	0.065	0.330	3.5	BL
0.470	B-2	0.070	0.330	3.5	BL
0.530	5031	0.080	0.400	4	C
0.565	4308	0.120	0.370	4	C
0.580	9788	0.100	0.350	5	C
0.685	B-72	0.350	0.400	2	H
0.720	KG-3	0.230	0.330	2	S
0.740	B-75	0.200	0.300	2	H
0.800	2043	0.350	0.350	2	P
0.800	ON-22	0.500	0.370	2	
0.800	B-76	0.260	0.330	2	H
0.800	Evaporated gold	0.800	0.500	—	—
1.2	Water	0.35	0.3	20	—
1.5	Ammonium dihydrogen phosphate	0.3	0.2	8	—
1.8	Water	1.2	0.3	0.5	—
2.2	Methyl methacrylate	0.2	0.3	1	—
2.65	Common glasses	0.15	0.35	10	—
3.5	7905 (Vycor)	0.2	0.3	3	C
4.2	Silica Glass	1.2	0.2	2	—
4.4	0160	1.7	0.3	1	C
5.2	Mica	1.0	0.4	0.25	—
6.8	Sapphire	1.5	0.2	0.50	—
8.5	Lithium fluoride	1.5	0.2	1	—
8.5	Irtran-1	2.0	1.2	1	EK
12	Calcium fluoride	2.0	0.2	1	—
12	Barium fluoride	2.5	0.2	9	—
15	Irtran-2	2.0	2.0	1	EK
18	Sodium chloride	8	0.2	1	—
25	Silver chloride	10	0.4	5	—
35	Potassium bromide	12	0.2	4	—

[a] BL–Bausch & Lomb. C–Corning. EK–Eastman Kodak. H–Hoya; P–Pittsburgh. S–Schott.

Source. Scharf [11.1a].

TABLE 92   BAND-PASS FILTERS (Section 11.1.3)

Mean λ (mμ)	Description	Half-height bandwidth (nm)	Base bandwidth (nm)	Max. trans. (%)	Rejection region (nm)	Thickness (mm)	Mfr.[a]
320	UG-11	90	130	85	0–680	1	S
320	Evaporated silver	20	—	10	0–	—	—
320	Wratten 12	20	—	1.5	0–500	0.1	EK
320	OX-7	140	175	85	0–650	2	
360	5840	50	70	75	0–700	3.5	C
360	U-2	50	75	70	0–680	2	H
380	OV-1	45	70	75	0–650	2	
380	5850	120	185	90	0–680	4	C
400	5113	60	100	40	0–2000	4	C
410	BG-12	140	180	80	0–680	1	S
420	OB-10	100	150	55	0–800	2	
435	B-43	90	145	50	0–800	2	H
435	Wratten 47B	45	90	50	0–720	0.1	EK
440	5031	140	210	85	0–720	4.5	C
450	Wratten 49	45	70	25	0–720	0.1	EK
455	Wratten 94	30	—	9	0–720	0.1	EK
460	Wratten 48	60	90	30	0–720	0.1	EK
475	BG-18	250	330	80	0–1200	1	S
475	OB-2	150	240	65	0–900	2	
480	BG-1	160	280	70	0–1000	3.5	BL
480	4407	120	210	60	0–1000	7	C
485	Wratten 44	90	135	53	0–700	0.1	EK
485	Wratten 75	35	50	20	0–720	0.1	EK
495	5032	150	200	80	0–700	3.5	C
500	Wratten 65A	55	95	45	0–700	0.1	EK
525	Wratten 55	70	110	70	0–720	0.1	EK
525	VG-14	90	140	45	0–950	1	S
530	Wratten 61	55	90	40	0–720	0.1	EK
535	G-58	60	110	50	0–900	2	H
540	G-11	90	170	60	0–900	2	H
540	Wratten 93	35	—	6	0–720	0.1	EK
550	Wratten 99	35	50	20	0–720	0.1	EK
550	G-9	70	125	40	0–700	5.5	BL
580	Wratten 73	30	—	6	0–720	0.1	EK
610	Wratten 72B	25	—	6	0–720	0.1	EK
750	5970	100	130	17	390–	10	C
800	2600	250	350	80	0–2200	3	C
850	2540+2600	150	250	25	0–	6	C
900	5850	400	600	90	480–1700	4	C
900	9788	1100	1600	60	600–	5	C
900	9782	800	1100	40	560–	5	C
900	5433	700	1000	35	480–	10	C
900	5021	2200	3000	50		0.75	C

[a] BL–Bausch & Lomb; C–Corning; EK–Eastman Kodak; H–Hoya; S–Schott (Jena).

Source. Scharf [11.1a].

TABLE 93 BAND REJECTION FILTERS (Section 11.1.3)

Mean λ (nm)	Supplier	Description	37% width (nm)	Min. trans. (%)	Trans. region (nm)	Thick-ness (mm)
520	Eastman Kodak	Wratten 30	120	0.1	350–2600	0.10
535	Eastman Kodak	Wratten 31	135	0.1	400–2600	0.10
545	Eastman Kodak	Wratten 32	125	0.1	360–2600	0.10
555	Eastman Kodak	Wratten 36	250	0.1	390–2600	0.10
560	Eastman Kodak	Wratten 34	210	0.1	330–2600	0.10
560	Eastman Kodak	Wratten 35	240	0.1	350–2600	0.10
585	Eastman Kodak	Wratten 77A	40	0.1	510–765	8
585	—	Didymium glass	30	0.1	360–720	4
635	Eastman Kodak	Wratten 40	150	0.1	450–2600	0.10
640	Eastman Kodak	Wratten 90	100	0.1	550–2600	0.10
770	—	Didymium glass	100	5	600–2600	4

Source. Scharf [11.1*a*].

TABLE 94 COMPENSATING FILTERS (Section 11.1.4)

Supplier	Description	Mired shift	A.N.S.I. value
Eastman Kodak[a]	Wratten 78	−240	0—55—93
Eastman Kodak[a]	Wratten 78AA	−195	0—43—78
Eastman Kodak[a]	Wratten 78A	−110	0—24—43
Eastman Kodak[a]	Wratten 78C	−25	0—3—7
Eastman Kodak[a]	Wratten 80B	−130	0—16—36
Eastman Kodak[a]	Wratten 80C	−100	0—14—42
Eastman Kodak[a]	Wratten 81	+10	5—1—0
Eastman Kodak[a]	Wratten 81B	+25	14—4—0
Eastman Kodak[a]	Wratten 81EF	+50	38—8—0
Eastman Kodak[a]	Wratten 82	−10	0—2—3
Eastman Kodak[a]	Wratten 82C	−45	0—10—19
Eastman Kodak[a]	Wratten 85	+130	58—20—0
Eastman Kodak[a]	Wratten 85C	+100	46—13—0
Eastman Kodak[a]	Wratten 86	+240	125—34—0
Eastman Kodak[a]	Wratten 86A	+110	65—16—0
Eastman Kodak[a]	Wratten 86C	+25	23—4—0
Jena	FG-4	−20/mm	
Corning	5900	−40/mm	
Chance	OB-8	−50/mm	
Jena	BG-34	−70/mm	
Jena	FG-8	+30/mm	
Corning	3307	+60/mm	
Eastman Kodak	CC 05 R		5—4—0
Eastman Kodak	CC 50 R		38—38—0
Eastman Kodak	CC 05 B		0—2—2
Eastman Kodak	CC 50 B		0—33—32
Eastman Kodak	CC 05 G		4—0—3
Eastman Kodak	CC 50 G		31—0—23
Eastman Kodak	CC 05 Y		5—1—0
Eastman Kodak	CC 50 Y		40—2—0
Eastman Kodak	CC 05 M		1—4—0
Eastman Kodak	CC 50 M		2—36—0
Eastman Kodak	CC 05 C		1—0—3
Eastman Kodak	CC 50 C		0—6—31
Eastman Kodak	Wratten 102[b]		
Eastman Kodak	Wratten 106[c]		

[a] Eastman Kodak supplies these gelatin filters.
[b] Converts barrier-layer photocell to luminosity.
[c] Converts S-4 photocell to luminosity.
Source. Scharf [11.1*a*].

TABLE 95 NEUTRAL DENSITY FILTERS (Section 11.1.5)

Density	Supplier	Description	Neutrality range (nm)
0.10–5.0	Bausch & Lomb	N-1 through N-5	450–600
0.10–5.0	Hoya	N-5 through N-70	400–800
0.10–3.0	Chance	ON-10, 11, 17, 33	450–650
0.10–5.0	Chance	ON-28, 29, 30, 31, 32	400–750
0.1–4.0	Jena	NG-3, 4, 5, 11	400–800
0.5–5.0	Jena	NG-1, 9, 10	450–650
0.1–4.0	Eastman Kodak	Wratten 96	500–700
0.1–5.0	—	Evaporated Inconel	(Material)
0.20–0.60	—	Normal screening	Entire range
0.2–2.0	—	Electroformed screens	Entire range
0.1–3.0	—	Diaphragms	Entire range
0.1–2.0	—	Sector wheels	Entire range
0.4–3.0	—	Nicol prism	350–2000
0.4–3.0	Polaroid Corp.	Type HN-22	450–700
0.4–3.0	Polaroid Corp.	Type KN-36	450–650
0.4–3.0	Polaroid Corp.	Type HR	700–2200
0.4–3.0	Polaroid Corp.	Type HN-38	550–700

Source. Scharf [11.1*a*].

TABLE 96 CORNING COLORED FILTERS—OUTLINE (Section 11.1.2)

Glass No.	CS	Color and properties	Thickness (mm)	Integral Color Parameters[a]		
				τ	x	y
0160	0–54	Clear; Ultraviolet transmitting	2.0	—	—	—
2030	2–64	Red; Sharp cut	3.0			
2403	2–58	Red; Sharp cut	3.0			
2404	2–59	Red; Sharp cut	3.0			
2408	2–60	Red; Sharp cut	3.0			
2412	2–61	Red; Sharp cut	3.0			
2418	2–62	Red; Sharp cut	3.0			
2424	2–63	Red; Sharp cut	3.0			
2434	2–73	Red; Sharp cut	3.0			
2540	7–56	Black; IR transmitting; Visible absorbing	2.5			
2550	7–57	Black; IR transmitting; Visible absorbing	2.0			
2600	7–69	Black; IR transmitting; Visible absorbing	3.0			
3060	3–75	Straw	2.0			
3304	3–76	Dark amber	3.0	0.220	0.606	0.392
3307	3–77	Dark amber	3.0	0.536	0.552	0.431
3384	3–70	Yellow	3.0			
3385	3–71	Yellow	3.0			
3387	3–72	Straw	3.0			
3389	3–73	Straw	3.0			
3391	3–74	Straw	3.0			
3480	3–66	Yellow; Sharp cut	3.0			
3482	3–67	Yellow; Sharp cut	3.0			
3484	3–68	Yellow; Sharp cut	3.0			
3486	3–69	Yellow; Sharp cut	3.0			
3718	3–94	Yellow	3.0	0.905	0.460	0.418
3750	3–79	Yellow; Yellow green fluorescing	5.0			

TABLE 96 (*Continued*)

Glass No.	CS	Color and properties	Thickness (mm)	Integral Color Parameters[a]		
				τ	x	y
3780	3–80	Yellow	2.0	0.744	0.528	0.456
3850	0–51	Clear; UV transmitting	4.0			
3961	1–56	Bluish; IR absorbing; Visible transmitting	2.5			
3962	1–57	Bluish; IR absorbing; Visible transmitting	2.5	0.562	0.412	0.421
3965	1–58	Bluish; IR absorbing; Visible transmitting	2.5	0.696	0.427	0.416
3966	1–59	Bluish; IR absorbing; Visible transmitting	2.5	0.806	0.434	0.412
4010	4–64	Green	4.0	0.097	0.240	0.661
4015	4–65	Yellow green	3.0	0.580	0.449	0.503
4060	4–67	Green	2.0	0.155	0.209	0.429
4084	4–68	Green	4.5	0.215	0.266	0.512
4303	4–72	Blue green	4.0	0.165	0.178	0.347
4305	4–71	Blue green	4.0	0.269	0.226	0.384
4308	4–70	Blue green	4.0	0.377	0.282	0.405
4309	4–69	Blue green	4.0	0.536	0.344	0.414
4445	4–74	Green	2.5	0.258	0.236	0.409
4602	1–75	Bluish; IR absorbing; Visible transmitting	3.0			
4784	4–94	Blue green	5.0	0.439	0.295	0.446
5030	5–57	Blue	5.0	0.035	0.135	0.112
5031	5–56	Blue	4.5	0.155	0.156	0.291
5070	7–62	Amethyst	3.9	0.042	0.490	0.271
5071	7–63	Amethyst	3.9	0.530	0.462	0.388
5073	7–64	Amethyst	3.9	0.317	0.473	0.364
5113	5–58	Blue	4.0	0.0005	0.163	0.013
5120	1–60	Smoky violet; Absorbs yellow	5.2			
5300	4–106	Green	3.9			

5330	1-64	Blue	4.5	0.063	0.182	0.188
5433	5-59	Blue	5.0	0.016	0.145	0.077
5543	5-60	Blue	5.0	0.0075	0.148	0.044
5562	5-61	Blue	5.0	0.029	0.146	0.109
5572	1-61	Blue	5.0	0.236	0.306	0.320
5840	7-60	Black; UV transmitting; Visible absorbing	4.5			
5850	7-59	Purple; UV transmitting; Visible absorbing	4.0	0.0025	0.185	0.023
5860	7-37	Black; UV transmitting; Visible absorbing	5.0			
5874	7-39	Black; UV transmitting; Visible absorbing	5.0			
5900	1-62	Blue	5.5	0.123	0.310	0.329
5970	7-51	Black; UV transmitting; Visible absorbing	5.0			
7380	0-52	Clear; UV transmitting	2.0			
7740	0-53	Clear; UV transmitting	2.0			
7905	9-30	Clear; UV transmitting; Long Range IR transmitting	2.0			
7910	9-54	Clear; UV transmitting	2.0			
8364	7-98	Gray	2.0			
9780	4-76	Blue green	5.0			
9782	4-96	Blue green	5.0			
9788	4-97	Blue green	5.0			
9830	4-77	Green	3.4			
9863	7-54	Black; UV transmitting; Visible absorbing	3.0			

[a] Relative to Std. Illuminant A. Values representative only. A. Werner, private communication.
Source. R. G. Saxton [11.4].

TABLE 97 CORNING COLORED FILTERS—TRANSMITTANCE SPECTRA (Section 11.2)

Transmittance

λ μm	0160	2030	2403	2404	2408	2412	2418	2424	2434	2540	2550	2600
						Corning glass number						
0.22	0.000	0.000	0.000	0.000	0.000	0.000	0.000	0.000	0.000	0.000	0.000	0.000
0.24	0.000	0.000	0.000	0.000	0.000	0.000	0.000	0.000	0.000	0.000	0.000	0.000
0.26	0.000	0.000	0.000	0.000	0.000	0.000	0.000	0.000	0.000	0.000	0.000	0.000
0.28	0.000	0.000	0.000	0.000	0.000	0.000	0.000	0.000	0.000	0.000	0.000	0.000
0.30	0.005	0.000	0.000	0.000	0.000	0.000	0.000	0.000	0.000	0.000	0.000	0.000
0.32	0.642	0.000	0.000	0.000	0.000	0.000	0.000	0.000	0.000	0.000	0.000	0.000
0.34	0.850	0.000	0.000	0.000	0.000	0.000	0.000	0.000	0.000	0.000	0.000	0.000
0.36	0.882	0.000	0.000	0.000	0.000	0.000	0.000	0.000	0.000	0.000	0.000	0.000
0.38	0.890	0.000	0.000	0.000	0.000	0.000	0.000	0.000	0.000	0.000	0.000	0.000
0.40	0.892	0.000	0.000	0.000	0.000	0.000	0.000	0.000	0.000	0.000	0.000	0.000
0.41	0.893	0.000	0.000	0.000	0.000	0.000	0.000	0.000	0.000	0.000	0.000	0.000
0.42	0.896	0.000	0.000	0.000	0.000	0.000	0.000	0.000	0.000	0.000	0.000	0.000
0.43	0.896	0.000	0.000	0.000	0.000	0.000	0.000	0.000	0.000	0.000	0.000	0.000
0.44	0.898	0.000	0.000	0.000	0.000	0.000	0.000	0.000	0.000	0.000	0.000	0.000
0.45	0.899	0.000	0.000	0.000	0.000	0.000	0.000	0.000	0.000	0.000	0.000	0.000
0.46	0.900	0.000	0.000	0.000	0.000	0.000	0.000	0.000	0.000	0.000	0.000	0.000
0.47	0.900	0.000	0.000	0.000	0.000	0.000	0.000	0.000	0.000	0.000	0.000	0.000
0.48	0.900	0.000	0.000	0.000	0.000	0.000	0.000	0.000	0.000	0.000	0.000	0.000
0.49	0.900	0.000	0.000	0.000	0.000	0.000	0.000	0.000	0.000	0.000	0.000	0.000
0.50	0.900	0.000	0.000	0.000	0.000	0.000	0.000	0.000	0.000	0.000	0.000	0.000
0.51	0.900	0.000	0.000	0.000	0.000	0.000	0.000	0.000	0.000	0.000	0.000	0.000
0.52	0.900	0.000	0.000	0.000	0.000	0.000	0.000	0.000	0.000	0.000	0.000	0.000
0.53	0.900	0.000	0.000	0.000	0.000	0.000	0.000	0.000	0.000	0.000	0.000	0.000
0.54	0.900	0.000	0.000	0.000	0.000	0.000	0.000	0.000	0.000	0.000	0.000	0.000
0.55	0.900	0.000	0.000	0.000	0.000	0.000	0.000	0.000	0.000	0.000	0.000	0.000
0.56	0.901	0.000	0.000	0.000	0.000	0.000	0.000	0.000	0.000	0.000	0.000	0.000
0.57	0.904	0.000	0.000	0.000	0.000	0.000	0.000	0.000	0.005	0.000	0.000	0.000
0.58	0.904	0.000	0.000	0.000	0.000	0.000	0.000	0.005	0.200	0.000	0.000	0.000
0.59	0.908	0.000	0.000	0.000	0.000	0.000	0.008	0.170	0.615	0.000	0.000	0.000
0.60	0.910	0.000	0.000	0.000	0.000	0.006	0.250	0.575	0.808	0.000	0.000	0.000

0.61	0.000	0.000	0.000	0.856	0.790	0.660	0.190	0.018	0.000	0.000	0.000	0.910
0.62	0.000	0.000	0.000	0.872	0.848	0.822	0.625	0.265	0.015	0.000	0.000	0.910
0.63	0.000	0.000	0.000	0.881	0.870	0.862	0.828	0.670	0.295	0.018	0.000	0.910
0.64	0.000	0.001	0.000	0.887	0.880	0.874	0.868	0.828	0.660	0.260	0.006	0.910
0.65	0.000	0.003	0.000	0.892	0.887	0.881	0.881	0.866	0.796	0.675	0.028	0.910
0.66	0.000	0.005	0.000	0.895	0.893	0.885	0.885	0.877	0.828	0.838	0.110	0.910
0.67	0.000	0.006	0.000	0.897	0.897	0.887	0.887	0.883	0.842	0.871	0.305	0.910
0.68	0.000	0.009	0.000	0.899	0.900	0.889	0.889	0.886	0.847	0.880	0.550	0.910
0.69	0.000	0.012	0.000	0.900	0.901	0.900	0.900	0.888	0.851	0.885	0.735	0.910
0.70	0.000	0.017	0.000	0.900	0.903	0.900	0.900	0.888	0.852	0.886	0.820	0.910
0.71	0.000	0.023	0.000	0.900	0.903	0.889	0.889	0.888	0.854	0.888	0.853	0.910
0.72	0.040	0.031	0.000	0.899	0.903	0.888	0.888	0.887	0.853	0.889	0.864	0.910
0.73	0.175	0.041	0.000	0.897	0.903	0.887	0.887	0.885	0.851	0.900	0.867	0.910
0.74	0.372	0.055	0.000	0.896	0.903	0.886	0.886	0.884	0.850	0.900	0.867	0.910
0.75	0.547	0.069	0.000	0.895	0.902	0.885	0.885	0.870	0.849	0.900	0.866	0.910
0.80	0.770	0.225	0.005	0.866	0.881	0.857	0.858	0.840	0.827	0.875	0.839	0.910
1.00	0.350	0.780	0.562	0.842	0.857	0.822	0.828	0.845	0.772	0.840	0.801	0.912
1.20	0.000	0.870	0.790	0.849	0.859	0.827	0.837	0.854	0.786	0.845	0.799	0.908
1.40	0.000	0.895	0.850	0.857	0.862	0.840	0.848	0.873	0.809	0.854	0.811	0.909
1.60	0.000	0.904	0.872	0.879	0.880	0.858	0.869	0.870	0.837	0.873	0.839	0.913
1.80	0.000	0.900	0.880	0.872	0.877	0.854	0.864	0.868	0.829	0.870	0.844	0.909
2.00	0.000	0.897	0.880	0.871	0.874	0.851	0.861	0.868	0.827	0.868	0.841	0.904
2.20	0.005	0.875	0.860	0.835	0.837	0.810	0.820	0.825	0.773	0.820	0.833	0.888
2.40	0.049	0.870	0.868	0.812	0.818	0.792	0.803	0.809	0.757	0.803	0.832	0.875
2.60	0.058	0.850	0.858	0.767	0.772	0.723	0.750	0.754	0.695	0.750	0.822	0.868
2.80	0.030	0.480	0.450	0.260	0.050	0.050	0.100	0.100	0.100	0.100	0.600	0.690
3.00	0.022	0.383	0.465	0.400	0.122	0.072	0.070	0.070	0.020	0.070	0.470	0.630
3.20	0.020	0.330	0.390	0.470	0.180	0.142	0.140	0.150	0.074	0.140	0.340	0.500
3.40	0.018	0.245	0.310	0.350	0.150	0.120	0.000	0.140	0.078	0.140	0.260	0.379
3.60	0.021	0.220	0.280	0.015	0.006	0.000	0.000	0.000	0.000	0.000	0.247	0.320
3.80	0.035	0.250	0.285	0.020	0.000	0.000	0.000	0.000	0.000	0.000	0.257	0.310
4.00	0.068	0.275	0.285	0.015	0.000	0.000	0.000	0.000	0.000	0.000	0.274	0.311
4.20	0.065	0.190	0.190	0.017	0.000	0.000	0.000	0.000	0.000	0.000	0.200	0.251
4.40	0.020	0.100	0.050	0.002	0.000	0.000	0.000	0.000	0.000	0.000	0.060	0.110
4.60	0.000	0.000	0.000	0.000	0.000	0.000	0.000	0.000	0.000	0.000	0.000	0.012
4.80	0.000	0.000	0.000	0.000	0.000	0.000	0.000	0.000	0.000	0.000	0.000	0.004
5.00	0.000	0.000	0.000	0.000	0.000	0.000	0.000	0.000	0.000	0.000	0.000	0.000

TABLE 97 (Continued)

Transmittance

λ μm	Corning glass number											
	3060	3304	3307	3384	3385	3387	3389	3391	3480	3482	3484	3486
0.22	0.000	0.000	0.090	0.000	0.000	0.000	0.000	0.000	0.000	0.000	0.000	0.000
0.24	0.000	0.000	0.000	0.000	0.000	0.000	0.000	0.000	0.000	0.000	0.000	0.000
0.26	0.000	0.000	0.000	0.000	0.000	0.000	0.000	0.000	0.000	0.000	0.000	0.000
0.28	0.000	0.000	0.000	0.000	0.000	0.000	0.000	0.000	0.000	0.000	0.000	0.000
0.30	0.000	0.000	0.000	0.000	0.000	0.000	0.005	0.000	0.000	0.000	0.000	0.000
0.32	0.000	0.000	0.038	0.000	0.000	0.000	0.010	0.000	0.000	0.000	0.000	0.000
0.34	0.000	0.000	0.050	0.000	0.000	0.000	0.015	0.000	0.000	0.000	0.000	0.000
0.36	0.000	0.000	0.027	0.000	0.005	0.000	0.020	0.000	0.000	0.000	0.000	0.000
0.38	0.060	0.000	0.016	0.005	0.010	0.010	0.026	0.000	0.000	0.000	0.000	0.000
0.40	0.410	0.000	0.014	0.011	0.016	0.020	0.025	0.075	0.000	0.000	0.000	0.005
0.41	0.517	0.000	0.014	0.011	0.016	0.020	0.105	0.425	0.000	0.000	0.000	0.005
0.42	0.604	0.000	0.016	0.010	0.015	0.019	0.437	0.655	0.000	0.000	0.000	0.005
0.43	0.665	0.000	0.022	0.009	0.013	0.017	0.620	0.747	0.000	0.000	0.000	0.005
0.44	0.710	0.000	0.033	0.005	0.011	0.050	0.714	0.801	0.000	0.000	0.000	0.005
0.45	0.748	0.000	0.049	0.003	0.010	0.325	0.780	0.838	0.000	0.000	0.000	0.005
0.46	0.778	0.000	0.070	0.002	0.008	0.565	0.820	0.860	0.000	0.000	0.000	0.004
0.47	0.800	0.000	0.101	0.001	0.060	0.690	0.848	0.874	0.000	0.000	0.000	0.003
0.48	0.819	0.000	0.143	0.005	0.410	0.763	0.866	0.884	0.000	0.000	0.000	0.003
0.49	0.836	0.003	0.193	0.088	0.640	0.803	0.878	0.890	0.000	0.000	0.000	0.002
0.50	0.850	0.009	0.250	0.350	0.727	0.834	0.886	0.895	0.000	0.000	0.000	0.001
0.51	0.860	0.019	0.315	0.595	0.780	0.854	0.890	0.898	0.000	0.000	0.000	0.045
0.52	0.870	0.037	0.379	0.725	0.817	0.868	0.892	0.900	0.000	0.000	0.003	0.425
0.53	0.875	0.063	0.447	0.789	0.840	0.876	0.894	0.901	0.000	0.000	0.175	0.710
0.54	0.881	0.102	0.504	0.825	0.856	0.883	0.895	0.902	0.000	0.015	0.600	0.792
0.55	0.884	0.146	0.560	0.846	0.866	0.887	0.894	0.903	0.000	0.230	0.774	0.823
0.56	0.886	0.200	0.607	0.860	0.873	0.889	0.893	0.902	0.020	0.675	0.818	0.844
0.57	0.885	0.255	0.648	0.869	0.876	0.890	0.892	0.901	0.325	0.850	0.839	0.859
0.58	0.883	0.310	0.680	0.873	0.878	0.889	0.890	0.900	0.710	0.885	0.854	0.868
0.59	0.882	0.360	0.705	0.876	0.877	0.887	0.886	0.898	0.829	0.894	0.865	0.876
0.60	0.882	0.404		0.877	0.877	0.884		0.896	0.858	0.900	0.873	0.882

0.61	0.882	0.438	0.722	0.877	0.877	0.881	0.884	0.893	0.869	0.903	0.880	0.886
0.62	0.882	0.466	0.735	0.875	0.876	0.875	0.880	0.890	0.876	0.905	0.885	0.888
0.63	0.882	0.488	0.744	0.871	0.874	0.871	0.876	0.886	0.881	0.906	0.888	0.889
0.64	0.883	0.505	0.748	0.865	0.872	0.866	0.872	0.884	0.884	0.907	0.890	0.890
0.65	0.885	0.519	0.750	0.860	0.867	0.860	0.868	0.881	0.885	0.908	0.892	0.890
0.66	0.886	0.531	0.750	0.856	0.863	0.856	0.865	0.876	0.886	0.908	0.893	0.890
0.67	0.888	0.543	0.749	0.850	0.858	0.851	0.860	0.873	0.886	0.908	0.894	0.890
0.68	0.890	0.552	0.745	0.844	0.853	0.846	0.856	0.869	0.886	0.908	0.893	0.889
0.69	0.891	0.561	0.740	0.837	0.847	0.839	0.852	0.865	0.885	0.907	0.892	0.887
0.70	0.892	0.569	0.734	0.831	0.842	0.834	0.847	0.860	0.884	0.907	0.891	0.885
0.71	0.893	0.574	0.727	0.825	0.837	0.827	0.842	0.856	0.882	0.906	0.890	0.883
0.72	0.893	0.575	0.720	0.819	0.831	0.822	0.837	0.852	0.880	0.905	0.888	0.880
0.73	0.892	0.576	0.712	0.813	0.825	0.816	0.831	0.848	0.877	0.905	0.886	0.877
0.74	0.891	0.574	0.702	0.807	0.820	0.810	0.826	0.844	0.874	0.904	0.885	0.875
0.75	0.890	0.570	0.694	0.800	0.814	0.805	0.820	0.840	0.870	0.903	0.882	0.873
0.80	0.871	0.526	0.642	0.770	0.865	0.780	0.775	0.815	0.837	0.878	0.846	0.829
1.00	0.830	0.435	0.516	0.715	0.830	0.725	0.716	0.772	0.801	0.857	0.811	0.781
1.20	0.860	0.429	0.500	0.718	0.858	0.735	0.730	0.782	0.807	0.859	0.819	0.793
1.40	0.901	0.475	0.540	0.750	0.900	0.768	0.768	0.810	0.828	0.870	0.837	0.817
1.60	0.917	0.580	0.635	0.795	0.918	0.812	0.812	0.843	0.852	0.884	0.856	0.841
1.80	0.916	0.627	0.675	0.808	0.915	0.818	0.820	0.849	0.847	0.882	0.852	0.834
2.00	0.908	0.620	0.668	0.800	0.909	0.822	0.817	0.846	0.847	0.884	0.852	0.835
2.20	0.900	0.630	0.675	0.802	0.900	0.823	0.810	0.840	0.805	0.865	0.829	0.798
2.40	0.885	0.651	0.690	0.800	0.885	0.825	0.811	0.842	0.787	0.853	0.817	0.777
2.60	0.860	0.650	0.690	0.785	0.858	0.815	0.800	0.840	0.725	0.818	0.757	0.718
2.80	0.550	0.345	0.390	0.325	0.670	0.360	0.340	0.440	0.050	0.060	0.110	0.060
3.00	0.379	0.320	0.360	0.318	0.348	0.324	0.322	0.423	0.080	0.088	0.190	0.088
3.20	0.315	0.240	0.290	0.268	0.332	0.288	0.290	0.395	0.150	0.158	0.270	0.145
3.40	0.250	0.151	0.190	0.218	0.289	0.280	0.255	0.353	0.120	0.132	0.140	0.090
3.60	0.231	0.130	0.150	0.217	0.266	0.280	0.249	0.351	0.000	0.000	0.000	0.000
3.80	0.258	0.140	0.160	0.228	0.270	0.298	0.267	0.376	0.000	0.000	0.000	0.000
4.00	0.283	0.140	0.175	0.220	0.280	0.290	0.260	0.365	0.000	0.000	0.000	0.000
4.20	0.200	0.090	0.125	0.143	0.210	0.210	0.178	0.288	0.000	0.000	0.000	0.000
4.40	0.100	0.020	0.030	0.025	0.080	0.070	0.040	0.115	0.000	0.000	0.000	0.000
4.60	0.008	0.008	0.010	0.007	0.002	0.003	0.000	0.009	0.000	0.000	0.000	0.000
4.80	0.000	0.005	0.009	0.000	0.000	0.000	0.000	0.000	0.000	0.000	0.000	0.000
5.00	0.000	0.001	0.008	0.000	0.000	0.000	0.000	0.000	0.000	0.000	0.000	0.000

TABLE 97 (Continued)

Transmittance

λ μm	Corning glass number											
	3718	3750	3780	3850	3961	3962	3965	3966	4010	4015	4060	4084
0.22	0.000	0.000	0.000	0.000	0.000	0.000	0.000	0.000	0.000	0.000	0.000	0.000
0.24	0.000	0.000	0.000	0.000	0.000	0.000	0.000	0.000	0.000	0.000	0.000	0.000
0.26	0.000	0.000	0.000	0.000	0.000	0.000	0.000	0.000	0.000	0.000	0.000	0.000
0.28	0.000	0.000	0.000	0.000	0.000	0.000	0.000	0.000	0.000	0.000	0.000	0.000
0.30	0.000	0.000	0.000	0.000	0.000	0.000	0.000	0.000	0.000	0.000	0.000	0.000
0.32	0.004	0.000	0.000	0.000	0.000	0.000	0.000	0.055	0.000	0.000	0.000	0.000
0.34	0.030	0.000	0.000	0.005	0.000	0.018	0.018	0.375	0.000	0.000	0.001	0.018
0.36	0.550	0.215	0.000	0.350	0.020	0.125	0.192	0.630	0.000	0.000	0.021	0.128
0.38	0.665	0.327	0.000	0.675	0.085	0.270	0.430	0.710	0.000	0.000	0.080	0.248
0.40	0.480	0.113	0.000	0.749	0.185	0.395	0.558	0.781	0.000	0.000	0.178	0.216
0.41	0.443	0.088	0.000	0.788	0.218	0.426	0.636	0.788	0.000	0.000	0.228	0.180
0.42	0.465	0.088	0.000	0.812	0.248	0.453	0.651	0.795	0.000	0.000	0.281	0.151
0.43	0.560	0.135	0.000	0.828	0.269	0.474	0.666	0.800	0.000	0.000	0.335	0.136
0.44	0.675	0.255	0.006	0.841	0.290	0.494	0.678	0.806	0.000	0.000	0.388	0.140
0.45	0.748	0.410	0.028	0.850	0.313	0.519	0.693	0.816	0.000	0.005	0.434	0.163
0.46	0.780	0.472	0.058	0.858	0.331	0.538	0.709	0.824	0.006	0.025	0.473	0.200
0.47	0.803	0.570	0.092	0.865	0.346	0.556	0.724	0.831	0.021	0.073	0.506	0.247
0.48	0.800	0.555	0.088	0.870	0.361	0.570	0.737	0.836	0.050	0.145	0.527	0.303
0.49	0.802	0.550	0.095	0.874	0.370	0.582	0.748	0.840	0.100	0.245	0.535	0.370
0.50	0.824	0.597	0.152	0.878	0.376	0.590	0.756	0.842	0.160	0.350	0.528	0.430
0.51	0.862	0.720	0.325	0.881	0.377	0.594	0.762	0.843	0.220	0.455	0.503	0.465
0.52	0.894	0.825	0.595	0.883	0.373	0.593	0.765	0.841	0.252	0.537	0.460	0.467
0.53	0.904	0.853	0.717	0.884	0.364	0.588	0.765	0.838	0.247	0.582	0.400	0.438
0.54	0.905	0.860	0.763	0.884	0.354	0.579	0.761	0.833	0.207	0.594	0.325	0.376
0.55	0.906	0.864	0.783	0.883	0.342	0.569	0.746	0.827	0.153	0.572	0.252	0.303
0.56	0.907	0.867	0.795	0.882	0.331	0.559	0.736	0.821	0.096	0.525	0.183	0.225
0.57	0.907	0.869	0.799	0.880	0.317	0.547	0.724	0.813	0.054	0.457	0.125	0.157
0.58	0.907	0.870	0.804	0.877	0.298	0.529	0.706	0.802	0.026	0.385	0.083	0.107
0.59	0.907	0.870	0.811	0.876	0.276	0.508	0.685	0.790	0.011	0.311	0.053	0.073
0.60	0.908	0.876	0.826	0.876	0.251	0.481	0.662	0.775	0.004	0.245	0.033	0.048

0.61	0.034	0.020	0.190	0.000	0.757	0.636	0.452	0.225	0.875	0.835	0.880	0.908
0.62	0.024	0.012	0.145	0.000	0.739	0.610	0.423	0.299	0.874	0.842	0.881	0.909
0.63	0.018	0.007	0.115	0.000	0.719	0.577	0.392	0.217	0.875	0.848	0.884	0.909
0.64	0.015	0.004	0.098	0.000	0.698	0.546	0.359	0.147	0.876	0.854	0.885	0.910
0.65	0.012	0.001	0.084	0.000	0.675	0.515	0.326	0.125	0.877	0.859	0.887	0.910
0.66	0.010	0.000	0.075	0.000	0.652	0.482	0.297	0.104	0.880	0.864	0.891	0.911
0.67	0.009	0.000	0.075	0.000	0.630	0.450	0.265	0.086	0.883	0.869	0.896	0.913
0.68	0.008	0.000	0.071	0.000	0.603	0.418	0.235	0.063	0.885	0.873	0.900	0.914
0.69	0.007	0.000	0.065	0.000	0.576	0.385	0.206	0.055	0.887	0.877	0.901	0.915
0.70	0.007	0.000	0.067	0.000	0.550	0.352	0.279	0.042	0.888	0.880	0.904	0.915
0.71	0.007	0.000	0.070	0.000	0.524	0.322	0.155	0.032	0.888	0.883	0.905	0.915
0.72	0.007	0.000	0.075	0.000	0.496	0.294	0.133	0.025	0.889	0.885	0.906	0.915
0.73	0.007	0.000	0.080	0.000	0.472	0.266	0.114	0.018	0.888	0.885	0.907	0.915
0.74	0.008	0.000	0.084	0.000	0.448	0.242	0.097	0.014	0.887	0.882	0.907	0.915
0.75	0.008	0.000	0.086	0.000	0.424	0.220	0.084	0.010	0.886	0.882	0.906	0.914
0.80	0.013	0.000	0.109	0.000	0.310	0.120	0.033	0.000	0.863	0.855	0.875	0.902
1.00	0.100	0.018	0.215	0.000	0.158	0.038	0.002	0.000	0.820	0.882	0.860	0.899
1.20	0.303	0.158	0.393	0.007	0.161	0.040	0.002	0.000	0.850	0.898	0.882	0.898
1.40	0.548	0.404	0.549	0.058	0.270	0.100	0.018	0.002	0.894	0.855	0.810	0.880
1.60	0.710	0.612	0.663	0.162	0.390	0.190	0.050	0.007	0.905	0.855	0.805	0.882
1.80	0.791	0.740	0.740	0.299	0.408	0.201	0.057	0.008	0.904	0.907	0.844	0.900
2.00	0.830	0.817	0.783	0.422	0.435	0.228	0.105	0.011	0.895	0.909	0.819	0.897
2.20	0.792	0.840	0.803	0.518	0.475	0.277	0.140	0.021	0.870	0.900	0.720	0.888
2.40	0.808	0.862	0.817	0.597	0.512	0.320	0.005	0.037	0.850	0.890	0.668	0.865
2.60	0.740	0.870	0.813	0.634	0.515	0.339	0.092	0.050	0.800	0.860	0.570	0.840
2.80	0.010	0.520	0.460	0.270	0.085	0.135	0.133	0.022	0.200	0.620	0.075	0.460
3.00	0.033	0.620	0.418	0.260	0.230	0.200	0.150	0.033	0.165	0.465	0.028	0.282
3.20	0.128	0.642	0.345	0.204	0.294	0.247	0.112	0.054	0.120	0.420	0.016	0.252
3.40	0.130	0.631	0.235	0.131	0.200	0.165	0.008	0.048	0.070	0.370	0.007	0.175
3.60	0.006	0.634	0.192	0.121	0.018	0.016	0.015	0.003	0.045	0.351	0.003	0.150
3.80	0.016	0.600	0.200	0.125	0.048	0.035	0.010	0.007	0.067	0.368	0.000	0.168
4.00	0.006	0.557	0.200	0.123	0.022	0.020	0.013	0.006	0.080	0.370	0.000	0.182
4.20	0.001	0.422	0.130	0.070	0.021	0.022	0.001	0.005	0.040	0.300	0.000	0.120
4.40	0.000	0.135	0.015	0.002	0.001	0.000	0.000	0.001	0.005	0.140	0.000	0.020
4.60	0.000	0.012	0.000	0.000	0.000	0.000	0.000	0.000	0.000	0.010	0.000	0.002
4.80	0.000	0.002	0.000	0.000	0.000	0.000	0.000	0.000	0.000	0.000	0.000	0.000
5.00	0.000	0.000	0.000	0.000	0.000	0.000	0.000	0.000	0.000	0.000	0.000	0.000

TABLE 97 (Continued)

Transmittance

λ μm	Corning glass number											
	4303	4305	4308	4309	4445	4602	4784	5030	5031	5070	5071	5073
0.22	0.000	0.000	0.000	0.000	0.000	0.000	0.000	0.000	0.000	0.000	0.000	0.000
0.24	0.000	0.000	0.000	0.000	0.000	0.000	0.000	0.000	0.000	0.000	0.000	0.000
0.26	0.000	0.000	0.000	0.000	0.000	0.000	0.000	0.000	0.000	0.000	0.000	0.000
0.28	0.000	0.000	0.000	0.000	0.000	0.000	0.000	0.000	0.000	0.000	0.000	0.000
0.30	0.000	0.000	0.000	0.000	0.000	0.001	0.000	0.000	0.016	0.012	0.045	0.090
0.32	0.000	0.060	0.000	0.022	0.001	0.106	0.000	0.000	0.145	0.310	0.330	0.540
0.34	0.011	0.380	0.190	0.394	0.018	0.505	0.009	0.038	0.420	0.628	0.340	0.757
0.36	0.188	0.590	0.580	0.740	0.114	0.755	0.200	0.285	0.685	0.745	0.420	0.757
0.38	0.390	0.723	0.740	0.831	0.248	0.827	0.450	0.595	0.820	0.712	0.420	0.810
0.40	0.545	0.750	0.826	0.884	0.400	0.835	0.596	0.770	0.884	0.600	0.665	0.830
0.41	0.588	0.770	0.840	0.887	0.454	0.845	0.627	0.799	0.894	0.430	0.712	0.786
0.42	0.624	0.786	0.850	0.890	0.505	0.846	0.648	0.808	0.895	0.290	0.716	0.705
0.43	0.654	0.798	0.857	0.892	0.550	0.851	0.666	0.797	0.890	0.170	0.694	0.620
0.44	0.680	0.809	0.862	0.894	0.593	0.856	0.680	0.767	0.872	0.097	0.655	0.525
0.45	0.698	0.815	0.867	0.897	0.631	0.857	0.697	0.738	0.865	0.055	0.612	0.436
0.46	0.712	0.814	0.869	0.898	0.659	0.856	0.717	0.702	0.864	0.035	0.568	0.365
0.47	0.715	0.802	0.866	0.897	0.678	0.861	0.735	0.628	0.845	0.023	0.531	0.313
0.48	0.705	0.780	0.856	0.894	0.689	0.866	0.750	0.522	0.805	0.017	0.501	0.275
0.49	0.678	0.740	0.838	0.885	0.687	0.869	0.763	0.406	0.750	0.015	0.482	0.252
0.50	0.636	0.685	0.810	0.872	0.673	0.870	0.768	0.288	0.684	0.013	0.469	0.237
0.51	0.570	0.610	0.770	0.850	0.641	0.869	0.767	0.186	0.601	0.013	0.463	0.231
0.52	0.480	0.525	0.714	0.817	0.586	0.866	0.753	0.105	0.495	0.014	0.461	0.230
0.53	0.387	0.430	0.650	0.781	0.520	0.863	0.725	0.053	0.388	0.016	0.464	0.235
0.54	0.288	0.340	0.576	0.736	0.437	0.865	0.676	0.022	0.295	0.018	0.473	0.245
0.55	0.205	0.255	0.502	0.683	0.355	0.869	0.615	0.007	0.198	0.023	0.486	0.260
0.56	0.132	0.184	0.422	0.627	0.275	0.868	0.525	0.000	0.113	0.030	0.502	0.278
0.57	0.082	0.127	0.345	0.565	0.202	0.863	0.427	0.000	0.057	0.038	0.522	0.300
0.58	0.047	0.087	0.277	0.505	0.144	0.856	0.328	0.000	0.025	0.048	0.540	0.324
0.59	0.026	0.057	0.218	0.447	0.102	0.848	0.235	0.000	0.008	0.058	0.555	0.348
0.60	0.013		0.170	0.393	0.068	0.838	0.157	0.000	0.000		0.571	0.373

0.61	0.006	0.036	0.131	0.341	0.046	0.824	0.102	0.000	0.000	0.070	0.587	0.395
0.62	0.001	0.022	0.100	0.296	0.031	0.806	0.058	0.000	0.000	0.083	0.600	0.415
0.63	0.000	0.013	0.074	0.256	0.020	0.787	0.032	0.000	0.000	0.096	0.612	0.435
0.64	0.000	0.007	0.056	0.221	0.013	0.767	0.017	0.000	0.000	0.108	0.622	0.450
0.65	0.000	0.004	0.042	0.191	0.008	0.745	0.007	0.000	0.000	0.120	0.633	0.466
0.66	0.000	0.002	0.033	0.167	0.006	0.722	0.002	0.000	0.000	0.135	0.644	0.482
0.67	0.000	0.001	0.025	0.146	0.003	0.695	0.000	0.000	0.000	0.148	0.658	0.498
0.68	0.000	0.000	0.020	0.128	0.001	0.665	0.000	0.000	0.000	0.165	0.674	0.515
0.69	0.000	0.000	0.016	0.116	0.000	0.634	0.000	0.000	0.000	0.182	0.686	0.531
0.70	0.000	0.000	0.013	0.104	0.000	0.600	0.000	0.000	0.000	0.200	0.700	0.548
0.71	0.000	0.000	0.010	0.096	0.000	0.565	0.000	0.000	0.000	0.220	0.712	0.566
0.72	0.000	0.000	0.009	0.088	0.000	0.531	0.000	0.004	0.024	0.245	0.725	0.586
0.73	0.000	0.000	0.007	0.083	0.000	0.496	0.000	0.047	0.119	0.268	0.736	0.606
0.74	0.000	0.000	0.006	0.079	0.000	0.463	0.000	0.190	0.330	0.295	0.749	0.675
0.75	0.000	0.000	0.005	0.075	0.000	0.430	0.000	0.440	0.580	0.323	0.759	0.642
0.80	0.000	0.000	0.008	0.080	0.003	0.258	0.000	0.890	0.917	0.505	0.815	0.750
1.00	0.013	0.005	0.045	0.188	0.056	0.019	0.000	0.753	0.868	0.860	0.885	0.872
1.20	0.080	0.060	0.166	0.375	0.269	0.009	0.000	0.455	0.720	0.890	0.897	0.890
1.40	0.210	0.180	0.342	0.542	0.527	0.016	0.003	0.100	0.285	0.892	0.902	0.892
1.60	0.350	0.342	0.500	0.658	0.701	0.035	0.038	0.056	0.162	0.890	0.902	0.892
1.80	0.475	0.482	0.617	0.732	0.792	0.059	0.142	0.052	0.140	0.878	0.890	0.877
2.00	0.560	0.590	0.690	0.772	0.839	0.048	0.275	0.075	0.175	0.860	0.865	0.860
2.20	0.635	0.653	0.720	0.778	0.850	0.038	0.345	0.172	0.295	0.840	0.830	0.840
2.40	0.663	0.704	0.752	0.791	0.870	0.045	0.404	0.330	0.483	0.812	0.795	0.805
2.60	0.260	0.710	0.748	0.770	0.861	0.066	0.340	0.382	0.530	0.804	0.752	0.775
2.80	0.249	0.370	0.370	0.250	0.280	0.018	0.045	0.030	0.030	0.550	0.500	0.500
3.00	0.202	0.250	0.203	0.212	0.395	0.000	0.000	0.010	0.001	0.390	0.308	0.340
3.20	0.135	0.212	0.145	0.168	0.451	0.000	0.000	0.065	0.026	0.222	0.145	0.180
3.40	0.124	0.136	0.084	0.107	0.449	0.000	0.000	0.002	0.006	0.120	0.070	0.090
3.60	0.132	0.120	0.078	0.077	0.470	0.000	0.000	0.000	0.000	0.078	0.032	0.063
3.80	0.132	0.125	0.082	0.083	0.470	0.000	0.000	0.000	0.000	0.078	0.029	0.060
4.00	0.078	0.131	0.094	0.090	0.448	0.000	0.000	0.000	0.000	0.093	0.037	0.075
4.20	0.008	0.073	0.050	0.042	0.320	0.000	0.000	0.000	0.000	0.079	0.020	0.048
4.40	0.000	0.009	0.008	0.002	0.100	0.000	0.000	0.000	0.000	0.020	0.004	0.014
4.60	0.000	0.000	0.000	0.000	0.007	0.000	0.000	0.000	0.000	0.002	0.000	0.000
4.80	0.000	0.000	0.000	0.000	0.000	0.000	0.000	0.000	0.000	0.000	0.000	0.000
5.00	0.000	0.000	0.000	0.000	0.000	0.000	0.000	0.000	0.000	0.000	0.000	0.000

TABLE 97 (Continued)

Transmittance

λ μm	Corning glass number											
	5113	5120	5300	5330	5433	5543	5562	5572	5840	5850	5860	5874
0.22	0.000	0.000	0.000	0.000	0.000	0.000	0.000	0.000	0.000	0.000	0.000	0.000
0.24	0.000	0.000	0.000	0.000	0.000	0.000	0.000	0.000	0.000	0.000	0.000	0.000
0.26	0.000	0.000	0.000	0.000	0.000	0.000	0.000	0.000	0.000	0.000	0.000	0.000
0.28	0.000	0.000	0.000	0.002	0.000	0.000	0.000	0.000	0.001	0.039	0.000	0.000
0.30	0.000	0.000	0.000	0.250	0.000	0.000	0.000	0.045	0.242	0.490	0.008	0.000
0.32	0.000	0.000	0.000	0.622	0.000	0.000	0.000	0.325	0.600	0.790	0.179	0.031
0.34	0.000	0.000	0.000	0.796	0.000	0.000	0.012	0.660	0.682	0.858	0.340	0.228
0.36	0.035	0.018	0.000	0.835	0.100	0.120	0.205	0.805	0.392	0.850	0.085	0.447
0.38	0.200	0.540	0.000	0.865	0.350	0.380	0.495	0.874	0.000	0.788	0.000	0.378
0.40	0.371	0.670	0.000	0.850	0.585	0.600	0.717	0.873	0.000	0.720	0.000	0.032
0.41	0.371	0.790	0.000	0.823	0.636	0.635	0.748	0.865	0.000	0.630	0.000	0.004
0.42	0.337	0.805	0.000	0.783	0.665	0.646	0.761	0.857	0.000	0.522	0.000	0.000
0.43	0.272	0.560	0.000	0.725	0.674	0.635	0.759	0.845	0.000	0.410	0.000	0.000
0.44	0.198	0.386	0.000	0.650	0.665	0.602	0.742	0.832	0.000	0.290	0.000	0.000
0.45	0.118	0.485	0.000	0.555	0.635	0.550	0.713	0.808	0.000	0.175	0.000	0.000
0.46	0.055	0.475	0.008	0.455	0.577	0.465	0.662	0.765	0.000	0.125	0.000	0.000
0.47	0.013	0.370	0.026	0.355	0.467	0.335	0.565	0.693	0.000	0.022	0.000	0.000
0.48	0.000	0.385	0.085	0.270	0.327	0.190	0.435	0.605	0.000	0.005	0.000	0.000
0.49	0.000	0.685	0.125	0.197	0.205	0.090	0.300	0.523	0.000	0.000	0.000	0.000
0.50	0.000	0.660	0.106	0.145	0.120	0.040	0.197	0.430	0.000	0.000	0.000	0.000
0.51	0.000	0.390	0.094	0.110	0.060	0.013	0.110	0.328	0.000	0.000	0.000	0.000
0.52	0.000	0.305	0.064	0.085	0.024	0.002	0.051	0.247	0.000	0.000	0.000	0.000
0.53	0.000	0.175	0.149	0.068	0.008	0.000	0.022	0.216	0.000	0.000	0.000	0.000
0.54	0.000	0.610	0.147	0.055	0.005	0.000	0.012	0.246	0.000	0.000	0.000	0.000
0.55	0.000	0.817	0.087	0.043	0.006	0.000	0.015	0.285	0.000	0.000	0.000	0.000
0.56	0.000	0.230	0.013	0.032	0.007	0.000	0.018	0.258	0.000	0.000	0.000	0.000
0.57	0.000	0.125	0.000	0.022	0.000	0.000	0.011	0.175	0.000	0.000	0.000	0.000
0.58	0.000	0.000	0.000	0.016	0.000	0.000	0.002	0.116	0.000	0.000	0.000	0.000
0.59	0.000	0.006	0.000	0.012	0.000	0.000	0.000	0.113	0.000	0.000	0.000	0.000
0.60	0.000	0.180	0.000		0.000	0.000	0.000		0.000	0.000	0.000	0.000

0.61	0.000	0.000	0.000	0.000	0.123	0.000	0.000	0.000	0.009	0.000	0.545	0.000
0.62	0.000	0.000	0.000	0.000	0.126	0.000	0.000	0.000	0.010	0.000	0.825	0.000
0.63	0.000	0.000	0.000	0.000	0.120	0.000	0.000	0.000	0.013	0.000	0.838	0.000
0.64	0.000	0.000	0.000	0.000	0.111	0.000	0.000	0.000	0.015	0.000	0.878	0.000
0.65	0.000	0.000	0.000	0.000	0.120	0.000	0.000	0.000	0.015	0.000	0.893	0.000
0.66	0.000	0.000	0.000	0.000	0.165	0.000	0.000	0.000	0.014	0.000	0.883	0.000
0.67	0.000	0.000	0.000	0.000	0.265	0.000	0.000	0.000	0.013	0.000	0.820	0.000
0.68	0.000	0.000	0.000	0.000	0.425	0.000	0.000	0.000	0.015	0.000	0.705	0.000
0.69	0.000	0.000	0.029	0.000	0.615	0.001	0.000	0.000	0.022	0.000	0.743	0.000
0.70	0.000	0.000	0.160	0.007	0.756	0.004	0.000	0.000	0.041	0.000	0.860	0.000
0.71	0.017	0.000	0.385	0.020	0.837	0.004	0.001	0.000	0.085	0.000	0.876	0.000
0.72	0.075	0.000	0.615	0.037	0.874	0.004	0.002	0.000	0.165	0.000	0.815	0.000
0.73	0.168	0.000	0.760	0.060	0.889	0.002	0.002	0.000	0.285	0.000	0.435	0.000
0.74	0.257	0.000	0.843	0.086	0.895	0.002	0.002	0.000	0.460	0.000	0.045	0.000
0.75	0.320	0.000	0.878	0.080	0.898	0.001	0.001	0.000	0.645	0.000	0.055	0.000
0.80	0.335	0.000	0.890	0.009	0.895	0.003	0.002	0.005	0.900	0.000	0.030	0.000
1.00	0.218	0.000	0.716	0.000	0.880	0.020	0.015	0.010	0.800	0.003	0.860	0.010
1.20	0.050	0.000	0.169	0.000	0.690	0.047	0.020	0.030	0.570	0.026	0.770	0.071
1.40	0.011	0.004	0.042	0.004	0.640	0.095	0.040	0.060	0.410	0.072	0.550	0.190
1.60	0.012	0.002	0.036	0.002	0.625	0.154	0.090	0.116	0.405	0.200	0.620	0.203
1.80	0.010	0.000	0.048	0.000	0.635	0.223	0.060	0.172	0.425	0.265	0.580	0.100
2.00	0.010	0.000	0.168	0.000	0.750	0.410	0.090	0.343	0.476	0.374	0.580	0.080
2.20	0.018	0.000	0.338	0.000	0.780	0.528	0.247	0.475	0.535	0.533	0.747	0.068
2.40	0.036	0.000	0.492	0.000	0.780	0.600	0.400	0.575	0.552	0.370	0.400	0.030
2.60	0.074	0.000	0.510	0.000	0.745	0.603	0.520	0.580	0.540	0.523	0.530	0.021
2.80	0.088	0.000	0.170	0.000	0.470	0.360	0.532	0.162	0.007	0.265	0.250	0.023
3.00	0.003	0.000	0.128	0.000	0.315	0.280	0.132	0.195	0.012	0.202	0.080	0.040
3.20	0.020	0.000	0.084	0.000	0.200	0.172	0.151	0.131	0.082	0.180	0.072	0.019
3.40	0.053	0.000	0.065	0.000	0.100	0.100	0.105	0.079	0.003	0.131	0.029	0.001
3.60	0.050	0.000	0.070	0.002	0.049	0.053	0.065	0.061	0.000	0.103	0.017	0.000
3.80	0.000	0.000	0.082	0.001	0.037	0.040	0.030	0.068	0.000	0.093	0.021	0.000
4.00	0.000	0.000	0.090	0.005	0.042	0.050	0.032	0.071	0.000	0.112	0.025	0.000
4.20	0.000	0.000	0.055	0.002	0.020	0.017	0.039	0.020	0.000	0.069	0.010	0.000
4.40	0.000	0.000	0.002	0.000	0.002	0.000	0.015	0.000	0.000	0.007	0.001	0.000
4.60	0.000	0.000	0.000	0.000	0.000	0.000	0.000	0.000	0.000	0.000	0.000	0.000
4.80	0.000	0.000	0.000	0.000	0.000	0.000	0.000	0.000	0.000	0.000	0.000	0.000
5.00	0.000	0.000	0.000	0.000	0.000	0.000	0.000	0.000	0.000	0.000	0.000	0.000

TABLE 97 (*Continued*)

Transmittance

λ μm	Corning glass number											
	5900	5970	7380	7740	7905	7910	8364	9780	9782	9788	9830	9863
0.22	0.000	0.000	0.000	0.000	0.000	0.012	0.000	0.000	0.000	0.000	0.000	0.000
0.24	0.000	0.000	0.000	0.000	0.360	0.505	0.000	0.000	0.000	0.000	0.000	0.054
0.26	0.000	0.000	0.000	0.000	0.495	0.780	0.000	0.000	0.000	0.000	0.000	0.482
0.28	0.000	0.000	0.000	0.004	0.590	0.855	0.000	0.000	0.000	0.000	0.000	0.731
0.30	0.000	0.138	0.000	0.321	0.720	0.877	0.000	0.000	0.000	0.000	0.000	0.831
0.32	0.000	0.600	0.000	0.722	0.825	0.900	0.000	0.000	0.000	0.000	0.000	0.862
0.34	0.008	0.799	0.000	0.851	0.880	0.903	0.002	0.015	0.000	0.060	0.001	0.854
0.36	0.150	0.742	0.440	0.889	0.910	0.905	0.083	0.290	0.060	0.470	0.059	0.816
0.38	0.445	0.190	0.795	0.900	0.915	0.906	0.136	0.590	0.445	0.770	0.160	0.620
0.40	0.678	0.029	0.892	0.916	0.920	0.920	0.296	0.725	0.747	0.885	0.130	0.090
0.41	0.688	0.000	0.904	0.915	0.920	0.920	0.273	0.705	0.790	0.895	0.044	0.018
0.42	0.635	0.000	0.910	0.915	0.920	0.921	0.232	0.770	0.818	0.902	0.004	0.003
0.43	0.586	0.000	0.913	0.914	0.920	0.923	0.191	0.778	0.836	0.905	0.000	0.000
0.44	0.522	0.000	0.915	0.913	0.920	0.924	0.157	0.801	0.847	0.906	0.000	0.000
0.45	0.458	0.000	0.916	0.913	0.922	0.925	0.144	0.814	0.855	0.906	0.000	0.000
0.46	0.400	0.000	0.917	0.914	0.922	0.925	0.140	0.823	0.860	0.906	0.000	0.000
0.47	0.350	0.000	0.917	0.915	0.922	0.925	0.141	0.832	0.863	0.906	0.000	0.000
0.48	0.306	0.000	0.918	0.915	0.923	0.925	0.146	0.839	0.863	0.905	0.000	0.000
0.49	0.275	0.000	0.918	0.915	0.930	0.926	0.152	0.843	0.859	0.904	0.014	0.000
0.50	0.246	0.000	0.919	0.915	0.925	0.926	0.166	0.843	0.848	0.900	0.180	0.000
0.51	0.223	0.000	0.919	0.915	0.925	0.927	0.178	0.838	0.825	0.893	0.175	0.000
0.52	0.196	0.000	0.919	0.916	0.925	0.928	0.190	0.824	0.784	0.880	0.018	0.000
0.53	0.172	0.000	0.919	0.916	0.923	0.928	0.196	0.798	0.720	0.862	0.000	0.000
0.54	0.154	0.000	0.919	0.916	0.926	0.929	0.198	0.756	0.627	0.831	0.050	0.000
0.55	0.148	0.000	0.920	0.917	0.923	0.929	0.197	0.697	0.515	0.787	0.265	0.000
0.56	0.151	0.000	0.920	0.917	0.925	0.930	0.199	0.615	0.380	0.728	0.165	0.000
0.57	0.146	0.000	0.919	0.918	0.925	0.930	0.206	0.518	0.255	0.655	0.035	0.000
0.58	0.125	0.000	0.918	0.919	0.925	0.930	0.217	0.414	0.150	0.570	0.004	0.000
0.59	0.102	0.000	0.918	0.920	0.925	0.930	0.222	0.302	0.075	0.475	0.000	0.000
0.60	0.093	0.000	0.920	0.920	0.925	0.930	0.215	0.215	0.032	0.380	0.000	0.000

0.000	0.000	0.290	0.010	0.135	0.196	0.930	0.925	0.920	0.920	0.000	0.087	0.61
0.000	0.000	0.210	0.002	0.080	0.175	0.930	0.925	0.920	0.920	0.000	0.081	0.62
0.000	0.000	0.145	0.000	0.042	0.161	0.930	0.926	0.919	0.920	0.000	0.070	0.63
0.000	0.000	0.094	0.000	0.021	0.156	0.931	0.927	0.919	0.920	0.000	0.061	0.64
0.000	0.000	0.059	0.000	0.008	0.162	0.931	0.927	0.918	0.920	0.000	0.055	0.65
0.000	0.000	0.035	0.000	0.003	0.176	0.932	0.927	0.917	0.920	0.000	0.055	0.66
0.000	0.075	0.020	0.000	0.000	0.200	0.932	0.928	0.916	0.920	0.000	0.059	0.67
0.022	0.380	0.010	0.000	0.000	0.228	0.932	0.927	0.916	0.920	0.007	0.065	0.68
0.106	0.642	0.005	0.000	0.000	0.248	0.932	0.927	0.915	0.921	0.036	0.068	0.69
0.234	0.694	0.001	0.000	0.000	0.251	0.932	0.928	0.915	0.921	0.085	0.068	0.70
0.332	0.666	0.000	0.000	0.000	0.237	0.933	0.926	0.914	0.922	0.145	0.066	0.71
0.383	0.607	0.000	0.000	0.000	0.223	0.933	0.926	0.912	0.922	0.222	0.064	0.72
0.384	0.531	0.000	0.000	0.000	0.210	0.933	0.926	0.910	0.922	0.323	0.060	0.73
0.358	0.445	0.000	0.000	0.000	0.197	0.934	0.927	0.909	0.921	0.385	0.057	0.74
0.322	0.045	0.000	0.000	0.000	0.180	0.934	0.928	0.907	0.921	0.287	0.055	0.75
0.175	0.000	0.000	0.000	0.000	0.110	0.932	0.930	0.890	0.918	0.032	0.050	0.80
0.119	0.000	0.000	0.000	0.000	0.032	0.928	0.930	0.860	0.910	0.032	0.085	1.00
0.016	0.007	0.005	0.000	0.000	0.032	0.928	0.925	0.860	0.910	0.021	0.180	1.20
0.005	0.000	0.080	0.000	0.018	0.062	0.930	0.925	0.870	0.906	0.109	0.295	1.40
0.007	0.000	0.266	0.000	0.131	0.137	0.930	0.931	0.892	0.910	0.088	0.405	1.60
0.011	0.157	0.430	0.011	0.317	0.182	0.930	0.931	0.896	0.903	0.040	0.495	1.80
0.029	0.041	0.512	0.085	0.440	0.171	0.929	0.934	0.897	0.900	0.008	0.590	2.00
0.048	0.000	0.455	0.216	0.440	0.184	0.835	0.934	0.875	0.898	0.002	0.628	2.20
0.060	0.000	0.433	0.278	0.440	0.225	0.890	0.930	0.850	0.890	0.009	0.640	2.40
0.051	0.057	0.252	0.325	0.280	0.258	0.780	0.920	0.820	0.860	0.018	0.630	2.60
0.000	0.020	0.060	0.212	0.060	0.145	0.180	0.908	0.140	0.375	0.030	0.470	2.80
0.000	0.000	0.000	0.040	0.000	0.130	0.695	0.880	0.360	0.425	0.030	0.248	3.00
0.000	0.000	0.000	0.000	0.000	0.125	0.760	0.861	0.490	0.380	0.030	0.121	3.20
0.000	0.000	0.000	0.000	0.000	0.099	0.620	0.670	0.270	0.310	0.020	0.032	3.40
0.000	0.000	0.000	0.000	0.000	0.115	0.080	0.111	0.010	0.270	0.015	0.010	3.60
0.000	0.000	0.000	0.000	0.000	0.142	0.240	0.270	0.040	0.275	0.020	0.010	3.80
0.000	0.000	0.000	0.000	0.000	0.172	0.150	0.170	0.013	0.260	0.030	0.006	4.00
0.000	0.000	0.000	0.000	0.000	0.158	0.230	0.250	0.026	0.180	0.017	0.001	4.20
0.000	0.000	0.000	0.000	0.000	0.078	0.080	0.085	0.004	0.040	0.002	0.000	4.40
0.000	0.000	0.000	0.000	0.000	0.010	0.020	0.050	0.000	0.000	0.000	0.000	4.60
0.000	0.000	0.000	0.000	0.000	0.003	0.000	0.000	0.000	0.000	0.000	0.000	4.80
0.000	0.000	0.000	0.000	0.000	0.000	0.000	0.000	0.000	0.000	0.000	0.000	5.00

TABLE 98 SCHOTT FILTERS—TRANSMITTANCE SPECTRA (Section 11.1.2)

λ nm	UG1	UG3	UG5	UG11	BG1	BG3	BG7	BG12	BG13	BG14	BG18	BG20	BG23	BG24	BG25	BG26	BG28	BG34	BG36
250	—					0.004						5×10^{-5}		0.880					
260	—		0.786			0.079						0.022		0.886					
270	3×10^{-5}		0.833		0.007	0.300						0.072		0.892					
280	0.004		0.854		0.120	0.523						0.161		0.900					
290	0.051		0.874		0.434	0.668						0.272		0.905	0.0004				
300	0.186	0.060	0.886	0.034	0.680	0.745		5×10^{-5}				0.365		0.906	0.025	0.150			0.195
310	0.365	0.292	0.893	0.232	0.806	0.792		0.009				0.430		0.907	0.185	0.411			0.278
320	0.526	0.537	0.893	0.478	0.825	0.814		0.097		0.358	0.005	0.510	0.077	0.908	0.451	0.650	0.004	0.001	0.385
330	0.644	0.694	0.896	0.665	0.848	0.844	0.016	0.295		0.596	0.055	0.571	0.262	0.910	0.647	0.786	0.052	0.030	0.293
340	0.722	0.780	0.886	0.768	0.865	0.844	0.073	0.483	0.030	0.745	0.174	0.722	0.455	0.908	0.765	0.852	0.161	0.130	0.508
350	0.763	0.822	0.886	0.817	0.874	0.854	0.168	0.614	0.116	0.822	0.311	0.389	0.598	0.908	0.829	0.884	0.286	0.303	0.057
360	0.767	0.854	0.874	0.839	0.876	0.863	0.266	0.697	0.266	0.861	0.402	0.404	0.687	0.908	0.858	0.901	0.401	0.474	0.081
370	0.749	0.866	0.866	0.846	0.883	0.863	0.354	0.743	0.426	0.882	0.475	0.872	0.742	0.908	0.893	0.909	0.490	0.611	0.833
380	0.653	0.865	0.829	0.850	0.876	0.863	0.430	0.772	0.546	0.893	0.521	0.878	0.778	0.908	0.896	0.911	0.558	0.711	0.841
390	0.416	0.843	0.722	0.846	0.876	0.863	0.492	0.797	0.628	0.900	0.567	0.884	0.804	0.906	0.896	0.916	0.612	0.781	0.860
400	0.137	0.791	0.526	0.832	0.869	0.848	0.547	0.806	0.683	0.904	0.603	0.877	0.822	0.902	0.891	0.916	0.655	0.821	0.850
410	0.017	0.685	0.322	0.798	0.856	0.825	0.591	0.797	0.720	0.907	0.631	0.888	0.837	0.890	0.878	0.917	0.688	0.842	0.864
420	0.0009	0.598	0.197	0.697	0.837	0.790	0.629	0.798	0.747	0.908	0.658	0.880	0.848	0.875	0.858	0.917	0.710	0.803	0.827
430	5×10^{-5}	0.501	0.136	0.455	0.812	0.739	0.665	0.775	0.768	0.911	0.686	0.733	0.857	0.848	0.833	0.917	0.730	0.783	0.468
440	—	0.414	0.117	0.116	0.788	0.686	0.691	0.751	0.784	0.913	0.713	0.640	0.865	0.807	0.799	0.917	0.746	0.760	0.254
450	—	0.341	0.112	0.004	0.737	0.606	0.714	0.710	0.795	0.913	0.740	0.706	0.871	0.737	0.750	0.917	0.752	0.733	0.263
460	—	0.293	0.101		0.680	0.511	0.734	0.657	0.803	0.913	0.759	0.724	0.874	0.640	0.691	0.917	0.748	0.701	0.339
470	—	0.250	0.078		0.582	0.366	0.744	0.554	0.811	0.913	0.768	0.668	0.875	0.532	0.584	0.916	0.722	0.670	0.250
480	—	0.233	0.056		0.442	0.206	0.750	0.406	0.818	0.910	0.786	0.630	0.872	0.449	0.436	0.913	0.672	0.646	0.331
490	—		0.043		0.300	0.092	0.744	0.261	0.816	0.905	0.795	0.833	0.863	0.395	0.286		0.602	0.623	0.624

λ (nm)																		
500	—	0.220	0.028		0.207	0.040	0.726	0.172	0.813	0.897	0.813	0.828	0.851	0.341	0.185	0.911	0.528	0.603
510	—	0.213	0.017		0.124	0.014	0.690	0.087	0.801	0.884	0.823	0.692	0.828	0.291	0.099	0.907	0.436	0.583
520	—	0.213	0.010		0.064	0.004	0.645	0.039	0.788	0.868	0.813	0.694	0.795	0.244	0.046	0.899	0.343	0.560
530	—	0.217	0.006		0.029	0.0007	0.583	0.016	0.765	0.846	0.804	0.514	0.753	0.204	0.018	0.891	0.254	0.535
540	—	0.226	0.005		0.019	0.0004	0.509	0.009	0.741	0.818	0.795	0.815	0.701	0.193	0.011	0.878	0.203	0.511
550	—	0.241	0.005		0.027	0.0007	0.437	0.013	0.715	0.788	0.759	0.883	0.647	0.186	0.016	0.866	0.191	0.501
560	—	0.261	0.003		0.044	0.002	0.360	0.020	0.677	0.752	0.722	0.866	0.586	0.168	0.027	0.849	0.187	0.503
570	—	0.285	0.002		0.037	0.0009	0.292	0.015	0.640	0.715	0.667	0.226	0.524	0.135	0.021	0.833	0.150	0.545
580	—	0.308	0.0006		0.014	9×10^{-5}	0.229	0.005	0.600	0.675	0.594	0.166	0.463	0.108	0.006	0.814	0.094	0.471
590	—	0.329	0.0008		0.004	8×10^{-6}	0.177	0.0009	0.561	0.635	0.521	0.150	0.406	0.115	0.002	0.794	0.052	0.432
600	—	0.355	0.001		0.004	8×10^{-6}	0.134	0.0007	0.522	0.594	0.411	0.538	0.350	0.133	0.002	0.772	0.040	0.411
610	—	0.377	0.002		0.005	9×10^{-6}	0.100	0.0009	0.480	0.553	0.347	0.785	0.303	0.150	0.002	0.750	0.035	0.400
620	—	0.394	0.003		0.005	3×10^{-5}	0.073	0.0009	0.443	0.517	0.274	0.885	0.260	0.169	0.002	0.730	0.028	0.386
630	—	0.407	0.004		0.005	3×10^{-5}	0.054	0.0006	0.407	0.481	0.174	0.892	0.221	0.194	0.002	0.709	0.021	0.357
640	—	0.422	0.011		0.004	3×10^{-6}	0.042	0.0005	0.374	0.449	0.146	0.909	0.186	0.248	0.0009	0.688	0.015	0.337
650	—	0.448	0.032		0.005	9×10^{-6}	0.028	0.0005	0.342	0.414	0.101	0.909	0.161	0.348	0.0009	0.672	0.012	0.321
660	—	0.468	0.100		0.009	6×10^{-5}	0.023	0.0009	0.310	0.384	0.064	0.909	0.136	0.488	0.003	0.656	0.011	0.311
670	0.0003	0.494	0.238	0.003	0.028	0.0006	0.016	0.003	0.285	0.361	0.046	0.892	0.116	0.628	0.009	0.636	0.012	0.313
680	0.006	0.520	0.419	0.026	0.095	0.007	0.013	0.012	0.262	0.338	0.027	0.852	0.101	0.749	0.040	0.622	0.014	0.316
690	0.050	0.541	0.580	0.081	0.272	0.073	0.010	0.036	0.241	0.321	0.009	0.860	0.087	0.823	0.122	0.609	0.015	0.321
700	0.162	0.560	0.680	0.165	0.523	0.290	0.008	0.066	0.223	0.300	0.009	0.905	0.082	0.862	0.229	0.595	0.015	0.318
710	0.285	0.581	0.730	0.219	0.733	0.565	0.007	0.083	0.205	0.291	0.005	0.909	0.065	0.883	0.309	0.587	0.015	0.314
720	0.364	0.601	0.750	0.249	0.842	0.760	0.006	0.086	0.194	0.279	0.003	0.899	0.065	0.894	0.347	0.578	0.013	0.306
730	0.434	0.620	0.739	0.226	0.895	0.863	0.005	0.086	0.182	0.268	0.002	0.785	0.058	0.895	0.357	0.570	0.011	0.300
740	0.454	0.641	0.722	0.180	0.914	0.908	0.005	0.086	0.173	0.261	0.0009	0.290	0.058	0.896	0.359	0.564	0.010	0.294
750		0.659	0.701	0.132	0.916	0.916	0.005	0.081	0.165	0.257	0.0005	0.312	0.058	0.895	0.355	0.558	0.010	0.289
τ_T	0.00	0.28	0.01		0.06	0.02	0.4	0.05	0.66	0.74	0.69	0.59	0.21	0.05		0.84	0.21	0.50
x	0.245	0.319	0.192		0.153	0.154	0.194	0.149	0.269	0.270	0.259	0.233	0.220	0.150		0.294	0.166	0.267
y	0.038	0.263	0.074		0.070	0.030	0.274	0.057	0.318	0.313	0.335	0.295	0.173	0.059		0.323	0.180	0.288
λ_D (nm)	−561	−546	443		465	455	486	464	488	487	493	486	463	464		487	478	480
P_e	1	0.26	0.81		0.9	0.98	0.48	0.93	0.17	0.17	0.19	0.32	0.53	0.93		0.07	0.67	0.21

Note: several columns also carry a further value (0.580, 0.196, 0.196, 0.139, 0.642, 0.806, 0.761, 0.042, 0.0005, 0.003, 0.235, 0.740, 0.807, 0.786, 0.845, 0.879, 0.872, 0.756, 0.580, 0.696, 0.820, 0.850, 0.781, 0.314, 0.009, 0.010 for rows 500–750) in the rightmost column.

TABLE 98 (Continued)

	BG37	BG38	VG3	VG4	VG5	VG6	VG9	VG10	VG14	GG4	GG9	GG10	GG17	GG19	GG21	GG375
250																
260														0.0003		
270														0.0006		
280														0.004		
290														0.049		
300	0.004										0.005			0.242	0.013	
310	0.077	0.035									0.037			0.388	0.065	
320	0.300	0.194									0.009			0.284	0.142	
330	0.545	0.435	0.008							0.052	0.0002		0.008	0.998	0.236	
340	0.705	0.630	0.075							0.068			0.060	0.018	0.427	
350	0.798	0.740	0.258							0.052			0.369	0.003	0.701	0.027
360	0.845	0.800	0.400							0.043			0.662	0.0008	0.829	0.361
370	0.874	0.837	0.475			0.004		0.0003		0.043			0.732	0.0008	0.856	0.707
380	0.885	0.855	0.525			0.009		0.0003		0.085			0.723	0.005	0.856	0.842
390	0.895	0.865	0.524			0.047		0.0003		0.200			0.614	0.032	0.821	0.883
400	0.897	0.877	0.470	0.0001	0.0002	0.165		0.0004		0.400	0.0004		0.490	0.145	0.773	0.903
410	0.893	0.884	0.361	0.0001	0.0004	0.315	0.0009	0.006		0.541	0.008	0.003	0.416	0.302	0.742	0.909
420	0.883	0.887	0.174	0.0005	0.0009	0.394	0.006	0.026	0.0001	0.601	0.030	0.014	0.420	0.402	0.741	0.916
430	0.870	0.893	0.072	0.003	0.008	0.443	0.020	0.070	0.0009	0.647	0.068	0.038	0.502	0.481	0.775	0.917
440	0.850	0.895	0.108	0.017	0.034	0.485	0.056	0.150	0.005	0.690	0.134	0.090	0.616	0.578	0.817	
450	0.820	0.898	0.150	0.066	0.107	0.529	0.123	0.276	0.020	0.728	0.237	0.182	0.735	0.663	0.861	
460	0.781	0.899	0.153	0.172	0.243	0.579	0.223	0.419	0.056	0.769	0.360	0.306	0.762	0.730	0.869	
470	0.706	0.898	0.080	0.311	0.396	0.622	0.327	0.550	0.115	0.802	0.494	0.433	0.820	0.791	0.886	
480	0.587	0.905	0.029	0.452	0.540	0.660	0.425	0.648	0.187	0.824	0.608	0.548	0.810	0.831	0.886	
490	0.451	0.905	0.034	0.562	0.651	0.701	0.516	0.721	0.268	0.841	0.693	0.645	0.810	0.852	0.883	

λ (nm)															
500	0.348	0.906	0.165	0.650	0.724	0.735	0.583	0.770	0.352	0.855	0.754	0.718	0.826	0.867	0.888
510	0.236	0.906	0.565	0.717	0.774	0.758	0.639	0.805	0.426	0.869	0.802	0.772	0.862	0.880	0.898
520	0.147	0.905	0.655	0.763	0.803	0.765	0.659	0.829	0.463	0.877	0.825	0.819	0.900	0.888	0.910
530	0.084	0.898	0.320	0.783	0.810	0.755	0.654	0.844	0.465	0.885	0.842	0.847	0.914	0.894	0.911
540	0.061	0.894	0.093	0.790	0.806	0.729	0.622	0.851	0.434	0.889	0.854	0.866	0.917	0.894	0.916
550	0.071	0.883	0.216	0.781	0.792	0.690	0.569	0.853	0.376	0.891	0.867	0.870	0.917	0.898	
560	0.096	0.872	0.699	0.763	0.768	0.638	0.505	0.848	0.300	0.891	0.872	0.872	0.917	0.902	
570	0.080	0.852	0.669	0.734	0.737	0.580	0.434	0.840	0.226	0.891	0.867	0.866	0.917	0.898	
580	0.040	0.827	0.448	0.699	0.703	0.521	0.363	0.827	0.163	0.888	0.864	0.854	0.917	0.902	
590	0.015	0.799	0.257	0.657	0.659	0.461	0.295	0.811	0.111	0.883	0.855	0.837	0.917	0.898	
600	0.014	0.763	0.134	0.607	0.622	0.400	0.234	0.791	0.071	0.879	0.854	0.812	0.917	0.898	
610	0.015	0.721	0.047	0.563	0.577	0.344	0.184	0.770	0.046	0.873	0.842	0.786	0.917	0.894	
620	0.015	0.676	0.008	0.517	0.536	0.296	0.144	0.746	0.028	0.869	0.836	0.765	0.917	0.894	
630	0.013	0.628	0.003	0.488	0.497	0.255	0.113	0.729	0.017	0.865	0.829	0.740	0.917	0.894	
640	0.010	0.579	0.007	0.465	0.469	0.224	0.093	0.719	0.012	0.863	0.825	0.728	0.917	0.894	
650	0.010	0.537	0.005	0.445	0.446	0.199	0.077	0.715	0.008	0.861	0.823	0.726	0.917	0.894	
660	0.014	0.485	0.0009	0.424	0.424	0.176	0.064	0.703	0.005	0.859	0.823	0.716	0.917	0.890	
670	0.029	0.428	0.005	0.420	0.414	0.163	0.057	0.715	0.005	0.861	0.825	0.724	0.917	0.894	
680	0.066	0.379	0.272	0.420	0.409	0.155	0.052	0.714	0.004	0.865	0.836	0.750	0.917	0.898	
690	0.128	0.332	0.674	0.405	0.391	0.140	0.045	0.710	0.003	0.867	0.836	0.739	0.917	0.894	
700	0.188	0.286	0.827	0.405	0.388	0.134	0.042	0.720	0.003	0.869	0.836	0.750	0.917	0.898	
710	0.220	0.246	0.858	0.409	0.388	0.130	0.040	0.725	0.003	0.871	0.842	0.772		0.898	
720	0.231	0.209	0.858	0.418	0.388	0.128	0.039	0.743	0.003	0.877	0.854	0.794		0.898	
730	0.231	0.176	0.848	0.443	0.391	0.125	0.038	0.748	0.002	0.881	0.855	0.816		0.898	
740	0.227	0.150	0.827	0.435	0.396	0.124	0.039	0.765	0.002	0.888	0.872	0.835		0.907	
750	0.222	0.126	0.895	0.443	0.396	0.123	0.037	0.733	0.002	0.891	0.874	0.852		0.910	
τ_T	0.12	0.84		0.69	0.71	0.60	0.46	0.80	0.27	0.88	0.83	0.81		0.89	
x	0.157	0.291		0.364	0.352	0.268	0.276	0.358	0.249	0.324	0.373	0.375		0.333	
y	0.105	0.328		0.492	0.477	0.365	0.485	0.434	0.572	0.352	0.440	0.456		0.367	
λ_D (nm)	470	491		563	562	502	537	567	535	568	569	568		569	
P_e	0.83	0.08		0.60	0.53	0.15	0.35	0.42	0.54	0.10	0.48	0.53		0.16	

TABLE 98 (Continued)

	GG385	GG395	GG400*	GG420*	GG435*	GG455*	GG475*	GG495*	OG515*	OG530*	OG550*	OG570*	OG590*	RG6
290														0.160
300														0.267
310														0.351
320														0.412
330														0.455
340	0.002	0.0009												0.484
350	0.004	0.004												0.502
360	0.132	0.065												0.509
370	0.444	0.255												0.519
380	0.666	0.468	9×10^{-5}											0.526
390	0.775	0.629	0.091											0.535
400	0.835	0.735	0.360											0.541
410	0.863	0.789	0.567	0.019										0.548
420	0.879	0.831	0.668	0.332	0.040									0.550
430	0.886	0.848	0.702	0.595	0.440									0.553
440	0.892	0.864	0.742	0.727	0.724	0.008								0.557
450	0.894	0.873	0.765	0.786	0.827	0.330		5×10^{-5}						0.556
460	0.897	0.880	0.792	0.822	0.866	0.689	9×10^{-5}	0.0003						0.551
470	0.900	0.884	0.816	0.838	0.882	0.812	0.041	0.002						0.543
480	0.900	0.890	0.838	0.845	0.882	0.851	0.384	0.014						0.529
490	0.903	0.893	0.852	0.862	0.896	0.868	0.709	0.199						0.505
500			0.868	0.862	0.896	0.877	0.847	0.617	0.0005					0.470
510			0.868	0.868	0.898	0.879	0.911	0.820	0.69	1×10^{-5}				0.427
520			0.868	0.868	0.898	0.879	0.911	0.889	0.483	0.065				0.381
530			0.874	0.874	0.896	0.879	0.913	0.908	0.752	0.489	0.001			0.352
540			0.868	0.874	0.896	0.879		0.913	0.842	0.749	0.117			0.351
550			0.874	0.874	0.898	0.883			0.876	0.845	0.500	0.0006		0.380

λ (nm)												
560	0.882	0.874	0.901	0.883			0.889	0.883	0.757	0.123		0.439
570	0.868	0.874	0.901	0.884			0.906	0.901	0.855	0.514		0.511
580	0.868	0.874	0.896	0.882			0.908	0.907	0.889	0.759	0.006	0.582
590	0.882	0.874	0.890	0.880			0.908	0.907	0.913	0.847	0.264	0.644
600	0.868	0.874	0.887	0.877			0.913	0.913	0.913	0.886	0.689	0.698
610	0.857	0.898	0.890	0.879				0.913		0.900	0.842	0.738
620	0.850	0.864	0.890	0.879						0.908	0.900	0.768
630	0.862	0.890	0.890	0.891						0.913	0.906	0.792
640	0.862	0.890	0.890	0.895							0.913	0.812
650	0.868	0.894	0.890	0.893								0.828
660	0.874	0.891	0.896	0.899								0.839
670	0.882	0.897	0.901	0.901								0.850
680	0.882	0.907	0.907	0.909								0.859
690	0.889	0.912										0.867
700	0.889	0.912	0.911	0.912								0.872
710	0.900	0.912	0.911	0.912								0.876
720	0.900	0.912	0.911	0.912								0.878
730	0.900	0.912	0.911	0.912								0.833
740	0.900	0.912	0.911	0.912								0.885
750	0.900	0.912	0.911	0.912								0.887
760	0.900											
770	0.900											
780	0.900											
790	0.900											
τ_T	0.87	0.88	0.90	0.90	0.89	0.86	0.73	0.66	0.51	0.34	0.17	0.51
x	0.321	0.324	0.325	0.359	0.415	0.440	0.484	0.506	0.554	0.609	0.672	0.349
y	0.345	0.351	0.356	0.417	0.510	0.525	0.508	0.489	0.445	0.391	0.328	0.301
λ_D (nm)	568	569	567	569	570	571	576	579	587	596	612	−501
P_e	0.07	0.09	0.11	0.37	0.80	0.91	0.98	0.99	1.00	1.00	1.00	0.17

TABLE 98 (Continued)

	RG6	RG610*	RG630*	RG645*	RG665*	RG695*	RG715*	RG780*	RG830*	RG1000*	RG9*
600		0.015									
610		0.328	0.002								
620		0.697	0.138								
630		0.838	0.571	0.0004							
640		0.878	0.808	0.055	0.0027						
650		0.902	0.892	0.489	0.205	0.00005					
660		0.905	0.915	0.778	0.633	0.0001	0.00003				
670		0.910	0.915	0.871	0.882	0.0006	0.0001				
680		0.915		0.905	0.879	0.005	0.0006				
690				0.913		0.083	0.006				
700					0.905	0.393	0.079	0.00005	1×10^{-6}		0.0045
710					0.912	0.712	0.340	0.0001	3×10^{-6}		0.064
720						0.853	0.661	0.001	6×10^{-6}		0.264
730						0.900	0.816	0.002	9×10^{-6}		0.528
740						0.913	0.879	0.010	4×10^{-5}		0.722
750							0.907	0.046	8×10^{-5}		0.844
760							0.913	0.145	0.0003		
770								0.350	0.0008		
780								0.585	0.003		
790								0.755	0.010		

λ (nm)					
800	0.895	0.845	0.045	0.00002	0.903
850	0.902	0.890	0.810	0.0027	0.894
900	0.905	0.900	0.900	0.045	0.845
950	0.906	0.914	0.902	0.204	0.779
1000	0.908	0.914	0.902	0.409	0.672
1200	0.913		0.902	0.819	0.015
1400	0.912		0.902	0.889	0.002
1600	0.914		0.902	0.911	0.002
1800	0.916		0.902	0.911	0.002
2000	0.913		0.902	0.911	0.045
2200	0.902		0.902	0.884	0.160
2400	0.883		0.884	0.860	0.328
2500	0.872		0.884	0.844	0.356
τ_T	0.51	0.08	0.04	0.01	0.01
x	0.349	0.703	0.717	0.729	0.733
y	0.301	0.297	0.283	0.271	0.267
λ_D (nm)	−501	627	638	657	674
P_e	0.17	1.0	1.0	1.0	1.0

All filters are 1 mm thick, except those marked with *; these are 3 mm thick.
Source. Schott [11.6].

965

TABLE 99   GLASS COLORING ADDITIVES (Section 11.2.3)

Purple	Manganese oxidized (Mn^{3+})
	Nickel (in K^+ glass)
Blue	Cobalt (with UV transmission, and red transmission)
	Copper oxidized (Cu^+, Cu^{++}) (with red absorption)
	Sulfur (in high B_2O_3 borates)
Daylight	Cobalt + Copper (+ some Manganese)
Blue green (Cyan)	Copper
Green	Copper (in Mg glasses, with TiO_2, with Cr, Fe)
	Chromium (with red transmission, with Cu, Fe)
	Iron (less pure, with uv and ir absorption) (with Cr, Cu)
	Uranium (reduced in phosphate glasses with two sharp transmission peaks) (oxidized, with yellow cast and fluoresence)
	Vanadium (with uv absorption)
	Molybdenum (in high P_2O_5 phosphate glass)
Black	Cobalt + other ions (Mn, Ni, Fe, Cu, Cr) (transmission variable)
	Iron sulfide (reducing melt) (transmission flat)
	Manganese + cobalt, in lead glasses (with high infrared transmission)
Colorless, uv absorbing	Cerium
	Titanium
	Iron (in phosphate glasses)
Colorless, heat absorbing	Iron (in phosphate glasses) (reduced)
Yellow	Uranium (greenish, with fluorescence)
	Cadmium sulfide (steep absorption edge, pure yellow)
	Cerium and titanium oxide, (faint each alone, brilliant together)
	Silver (unstable)
	Silver stain
Amber	Sodium sulfide (reduced with carbon, sloppy absorption curve-amber)
Brown	Manganese—reduced by Sb_2O_3, As_2O_3, FeO, Ce_2O_3) (unstable in sunlight, uv)
	Manganese—Iron
	Titanium—Iron
	Nickel (in sodium glasses)
	Iron Selenide (warm, with flat curve, step-up in red)
	Manganese—Titanium
	Manganese—Cerium (stable in sunlight, uv) (and many combinations)
Orange	Cadmium sulfide + selenium
Red	Cadmium sulfide + more selenium
	Gold (purplish red, light)
	Copper (with green, causing dark appearance)
	Copper stain
	Uranium (in high lead glass)
	Antimony sulfide (not commercial)

Source.   Kreidl [11.8].

TABLE 100 INFRARED CUTOFF
WAVELENGTHS OF SEMICONDUCTOR
MATERIALS (Section 11.2.4)

Material	λ_c (μm)	$n(\lambda_c)$	$n(15\ \mu$m)
Si	1.1	3.5	3.4
Ge	1.8	4.1	4.0
InAs	3.8	3.4	3.2
InSb	7.3	4.0	4.0

Source. Cox, Hass, and Jacobus [11.12].

967

TABLE 101 MATERIALS USED IN MULTILAYER FILM CONSTRUCTION (Section 11.3)

(a) Dielectrics

Materials	Evaporation technique	Refractive index	Region of transparency	Remarks	References
Aluminium oxide (Al_2O_3)	Electron bombardment	1.62 at 0.6 μ {Substrate temperature 300°C 1.59 at 1.6 μ {Substrate temperature 300°C 1.59 at 0.6 μ {Substrate temperature 40°C 1.56 at 1.6 μ {temperature 40°C		Can also be produced by anodic oxidation of Al in ammonium tartrate solution.[1]	2
Antimony trioxide (Sb_2O_3)	Molybdenum boat	2.29 at 366 mμ 2.04 at 546 mμ	300 mμ–>1 μ	Important to avoid overheating otherwise decomposes	3
Antimony sulphide (Sb_2S_3)		3.0 at 589 mμ	500 mμ–10μ	Brief note, Ref. 4, p. 189	5, 6
Bismuth oxide (Bi_2O_3)	Reactive sputtering of bismuth in oxygen	2.45 at 550 mμ			5, 7
Cadmium sulphide (CdS)	Quartz crucible with spiral filament in contact with charge	2.6 at 600 mμ 2.27 at 7 μ	600 mμ–7 μ	Avoid overheating, filament temperature must be $\leqq$1025°C	5, 7
Cadmium telluride (CdTe)	Molybdenum boat	3.05 in near infra-red			Brief Ref. 8
Calcium fluoride (CaF_2)	Molybdenum or tantalum boat	1.23–1.26 at 546 mμ	150 mμ–12 μ		5, 8, 9
Ceric oxide (CeO_2)	Tungsten boat	2.2 at 550 mμ (Ref. 10) 2.18 at 550 mμ. Substrate temperature 50°C 2.42 at 550 mμ. Substrate temperature 350°C (Ref. 11) 2.2 in near infra-red	400 mμ–16 μ	Shows water absorption band at 2.7 μ Tends to form inhomogeneous layers	10, 11, 12
Cerous fluoride	Tungsten boat	1.63 at 550 mμ 1.59 at 2 μ	300 mμ–>5 μ	Hot substrate Crazes on cold substrate (Ref. 8)	8, 10, 13
Chiolite $(5NaF \cdot 3AlF_3)$	Howitzer or tantalum boat			Similar to cryolite	8
Cryolite (Na_3AlF_6)	Howitzer or tantalum boat	1.35 at 550 mμ	<200 mμ–14 μ	Slightly hygroscopic, soft, easily damaged	5, 8, 9, 10

Material	Method	Refractive index	Transmission range	Remarks	References
Germanium (Ge)	Electron bombardment or graphite boat	4.0	1.7 μ–100 μ	Absorption band centred at approx. 25 μ	8, 10
Lanthanum fluoride (LaF$_3$)	Tungsten boat	1.59 at 550 mμ 1.57 at 2.0 μ	220 mμ–>2 μ	Slightly inhomogeneous Substrate heated	10, 13, 14
Lanthanum oxide (La$_2$O$_3$)	Tungsten boat	1.95 at 550 mμ 1.86 at 2.0 μ	350 mμ–>2 μ	Hot substrate (~300°C)	10, 13
Lead chloride (PbCl$_2$)	Platinum boat	2.3 at 550 mμ 2.0 at 10 μ	300 mμ–>14 μ		8, 15
Lead fluoride (PbF$_2$)	Molybdenum boat (Ref. 15) Platinum boat	1.75 at 550 mμ 1.70 at 1μ	240 mμ–>20 μ		8, 10, 16, 17
Lead telluride (PbTe)	Tantalum boat	5.5	3.4 μ–30 μ	Avoid overheating Hot substrate (see text, p. 218)	18
Lithium fluoride (LiF)	Tantalum boat	1.36–1.37 at 546 mμ	110 mμ–7 μ		5, 19
Magnesium fluoride (MgF$_2$)	Tantalum boat	1.38 at 550 mμ 1.35 at 2 μ	210 mμ–10 μ	Films on heated substrates much more robust	5, 8, 9, 10, 20, 21
Neodymium fluoride (NdF$_3$)	Tungsten boat	1.60 at 550 mμ 1.58 at 2 μ	220 mμ–>2 μ	Hot substrate 300°C	10, 13
Neodymium oxide (Nd$_2$O$_3$)	Tungsten boat	2.0 at 550 mμ 1.95 at 2 μ	400 mμ–>2 μ	Hot substrate 300°C Decomposes at high boat temperatures	10, 13
Praseodymium oxide (Pr$_6$O$_{11}$)	Tungsten boat	1.92 at 500 mμ 1.83 at 2 μ	400 mμ–>2 μ	Hot substrate 300°C	13
Silicon (Si)	Electron bombardment with water cooled hearth	3.5	1.1–10 μ		10
Silicon monoxide (SiO)	Tantalum boat or howitzer	2.0 at 550 mμ 1.7 at 6 μ	500 mμ–8 μ	Fast evaporation at low pressure	Brief Ref. in 8; 2, 5, 10, 22, 23
Disilicon trioxide (Si$_2$O$_3$)	Tantalum boat or howitzer	1.52–1.55 at 550 mμ	300 mμ–8 μ		2, 10, 24, 25, 26, 27, 28, 30
Silicon dioxide (SiO$_2$)	Reactive evaporation of SiO in O$_2$ (Refs. 24, 30) Evaporation of Si, SiO$_2$, ZrO$_2$ mixture (Ref. 26) Electron beam (Refs. 2, 29)	1.46 at 500 mμ 1.445 at 1.6 μ	<200 mμ–8 μ (in thin films)		2, 10, 29, 31
Sodium fluoride (NaF)	Mixture in tungsten boat (Ref. 31) (see text, p. 219) Tantalum boat	1.34 visible	<250 mμ–14 μ		Brief note, p. 60 of Ref. 5

TABLE 101 (Continued)
(a) Dielectrics

Materials	Evaporation technique	Refractive index	Region of transparency	Remarks	References
Tellurium (Te)	Tantalum boat	4.9 at 6μ	3.4–20 μ		8, 10, 32, 33
Titanium dioxide (TiO₂)	Electron bombardment (Ref. 29) Reactive evaporation of TiO in O₂ (Refs. 24, 30)	2.2–2.7 at 550 mμ depending on structure (see p. 216)	350 mμ–12 μ	Can also be produced by subsequent oxidation of Ti film	5, 10, 24, 29, 30, 34, 35, 36, 37
Thallous chloride (TlCl)	Tantalum boat	2.6 at 12 μ	Visible region– >20 μ		8, 38
Thorium oxide (ThO₂)	Electron bombardment	1.8 at 550 mμ 1.75 at 2.0 μ	250 mμ–>2 μ	Radioactive	2
Thorium fluoride (ThF₄)	Tantalum boat	1.52 at 400 mμ 1.51 at 750 mμ	200 mμ–>15 μ	Radioactive (see Ref. 39) Note: Thorium oxyfluoride (ThOF₂) actually forms ThF₄ when evaporated, Ref. 39	8, 10, 39, 40, 41
Zinc selenide (ZnSe)	Platinum boat	2.58 at 633 mμ	600 mμ–>15 μ		40
Zinc sulphide (ZnS)	Tantalum boat or howitzer	2.35 at 550 mμ 2.2 at 2.0 μ	380 mμ– approx. 25 μ		5, 8, 10, 12, 20, 41
Zirconium oxide (ZrO₂)	Electron bombardment	2.1 at 550 mμ 2.0 at 2.0 μ			22

Source. McLeod [11.15].

ªReferences.

[2] J. T. Cox, G. Hass, and J. B. Ramsay, "Improved dielectric films for multilayer coatings and mirror protection," *J. Phys.* **25**, 250–254 (1964).

[3] F. A. Jenkins, "Extension du domaine spectral de pouvoir réflecteur élevé des couches multiples dielectrique," *J. Phys. Rad.* **19**, 301–306 (1958).

[4] O. S. Heavens, J. Ring, and S. D. Smith, "Interference filters for the infra-red," *Spectrochim. Acta* **10**, 179–194 (1957).

[5] O. S. Heavens, "Optical properties of thin films," *Rep. Progress in Physics* **23**, 1–65 (1960). Includes brief reviews of the following materials: CaF_2 pp. 42–43, Na_3AlF_6 pp. 43–44, LiF p. 44, ZnS pp. 44–45, CdS 45, CeO_2 pp. 45–46, Sb_2S_3 46, SiO pp. 46–47, TiO_2 p. 47, and Al_2O_3 p. 48.

[6] S. H. Billings, and M. Hyman, Jr., "The infra-red refractive index and dispersion of evaporated stibnite films," *J. Opt. Soc. Am.* **37**, 119–121 (1947).

[7] J. F. Hall, and W. F. C. Ferguson, "Optical properties of cadmium sulphide and zinc sulphide from 0.6 micron to 14 microns," *J. Opt. Soc. Am.* **45**, 714–718 (1955).

[8] A. E. Ennos, "Stresses developed in optical film coatings," *Appl. Opt.* **5**, 51–61 (1966). Results on ZnS, MgF_2, $ThOF_2$, PbF_2, $NaAlF_6$, $5NaF·3AlF_3$, CaF_2, CeF_2, CeF_3, Al_2O_3. Multilayers of ZnS—$ThOF_2$, ZnS—Na_3AlF_6, PbF_2—Na_3AlF_6.

970

[9] O. S. Heavens, and S. D. Smith, "Dielectric thin films," *J. Opt. Soc. Am.* **47**, 469–472 (1957).

[10] E. Ritter, "Gesichtspunkte bei der Stoffauswahl für dünne Schichten in der Optik," *Z. Angew. Math. Phys.* **12**, 275–276 (1961). A very short note giving a table summarizing the optical properties of 17 different materials.

[11] G. Hass, J. B. Ramsay, and R. Thun, "Optical properties and structure of cerium dioxide films," *J. Opt. Soc. Am.* **48**, 324–327 (1958).

[12] J. T. Cox, and G. Hass, "Anti-reflection coatings for germanium and silicon in the infra-red," *J. Opt. Soc. Am.* **48**, 677–680 (1958).

[13] G. Hass, J. B. Ramsay, and R. Thun, "Optical properties of various evaporated rare earth oxides and fluorides," *J. Opt. Soc. Am.* **49**, 116–120 (1959).

[14] A. Bourg, N. Barbaroux, and M. Bourg, "Propriétés optiques et structure de couches minces de fluorure de lanthane," *Opt. Acta* **12**, 151–160 (1965).

[15] S. Penselin, and A. Steudel, "Fabry-Perot Interferometerverspiegelungen aus dielektrischen Vielfachschichten," *Zeit. Phys.* **142**, 21–41 (1955).

[16] British Patent No. 994,638, *Interference Filters*, 1965.

[17] Z. Lés, F. Lés, and L. Gabla, "Semitransparent metallic–dielectric mirrors with low absorption coefficient in the ultraviolet region of the spectrum, 3200–2400 Å," *Acta Phys. Polonica* **23**, 211–214 (1963).

[18] C. S. Evans and J. S. Seeley (Department Applied Physical Sciences, University of Reading), "Properties of Thick Evaporated Layers of Lead Telluride," paper presented at the Colloquium on IV–VI compounds, Paris, July, 1968.

[19] L. G. Schulz, "The structure and growth of evaporation LiF and NaCl films on amorphous substrates," *J. Chem. Phys.* **17**, 1153–1162 (1949).

[20] J. F. Hall and W. F. C. Ferguson, "Dispersion of zinc sulphide and magnesium fluoride films in the visible spectrum," *J. Opt. Soc. Am.* **45**, 74–75 (1955).

[21] J. F. Hall, "Optical properties of magnesium fluoride films in the ultra-violet," *J. Opt. Soc. Am.* **47**, 662–665 (1957).

[22] G. Hass, and C. D. Salzberg, "Optical properties of silicon monoxide in the wavelength region from 0.24 to 14.0 microns," *J. Opt. Soc. Am.* **44**, 181–187 (1954).

[23] M. A. Novice, "Stresses in evaporated silicon monoxide films," *Vacuum* **14**, 385–392 (1964).

[24] British Patent No. 775,002, *Improvements in or Relating to the Manufacture of Thin Light Transmitting Layers*, 1957.

[25] E. Ritter, "Zur Kenntnis der SiO- und Si₂O₃-Phase in dünnen Schichten," *Opt. Acta* **9**, 197–202 (1962).

[26] E. Okamoto, and Y. Hishimuma, "Properties of evaporated thin films of Si₂O₃," *Trans. 3rd Int. Vac. Congress* **2**, part 2, 49–56 (1965). Production of Si₂O₃ films by evaporating mixture of SiO₂, Si and ZrO₂.

[27] A. P. Bradford, G. Hass, M. McFarland, and E. Ritter, "Effect of ultra-violet irradiation on the optical properties of silicon oxide films," *Appl. Opt.* **4**, 971–976 (1965).

[28] A. P. Bradford, and G. Hass, "Increasing the far-ultra-violet reflectance of silicon-oxide protected aluminium mirrors by ultra-violet irradiation," *J. Opt. Soc. Am.* **53**, 1096–1100 (1963).

[29] W. Reichelt, "Fortschritte in der Herstellung von Oxydschichten für optische und elektrische Zwecke," *Trans. 3rd Int. Vac. Congress* **2**, part 2, 25–29 (1965).

[30] U.S. Patent 2,920,002, *Process for the Manufacture of Thin Films*, 1960.

[31] British Patent No. 632,442, *Method of Coating with Quartz by Thermal Evaporation*, 1947.

[32] T. S. Moss, "Optical properties of tellurium in the infra-red," *Proc. Phys. Soc.* **65**, 62–66 (1952).

[33] R. G. Greenler, "Interferometry in the infra-red," *J. Opt. Soc. Am.* **45**, 788–791 (1955).

[34] G. Hass, "Preparation, properties, and optical applications of thin films of titanium dioxide," *Vacuum* **2**, 331–345 (1952).

[35] U.S. Patent 2,784,115, *Method of Producing Titanium Dioxide Coatings*, 1957.

[36] British Patent No. 895,879, *Improvements in and Relating to the Oxidation and/or Transparency of Thin Partly Oxidic Layers*, 1962.

[37] U.S. Patent 3,034,924, *Use of a Rare Earth Metal in Vaporizing Metals and Metal Oxides*, 1962.

[38] British Patent No. 970,071, *Infra-red Filters*, 1964.

[39] W. Heitmann and E. Ritter, "Production and properties of vacuum evaporated films of thorium fluoride," *Appl. Opt.* **7**, 307–309 (1968).

[40] W. Heitmann, "Extrem hochreflektierende dielektrische Spiegelschichten mit Zinkselenid," *Z. Angew. Phys.* **21**, 503–508 (1966).

[41] K. H. Behrndt, and D. W. Doughty, "Fabrication of multilayer dielectric films," *J. Vac. Sci. Tec.* **3**, 264–272 (1966).

971

TABLE 101b TRANSMITTANCE OF METAL FILMS ON GLASS—DEPENDENCE ON REFLECTANCE (Section 11.3)

	Transmittance	
ρ	Aluminum[a]	Silver[b]
0.1	0.72 5	0.68 5
0.2	0.48 2	0.41 5
0.3	0.35 5	0.29 0
0.4	0.31 5	0.18 8
0.5	0.27 5	0.18 8
0.6	0.21 4	0.20 3
0.7	0.12 8	0.18 8
0.8	0.04 9	0.13 0
0.9	—	0.05 8

[a] At 0.6 μm. From Walkenhorst.
[b] At 0.578 μm. From Goos.
Source. Holland [11.21].

TABLE 102 MAXIMUM AND MINIMUM TRANSMITTANCE VALUES OF EVAPORATED POLARIZING GRIDS (Section 11.4.1)

λ (μm)	Aluminum		Gold	
	t_{mx}	t_{mn}	t_{mx}	t_{mn}
0.5	0.058	0.038	0.407	0.292
0.7	0.238	0.022	0.447	0.288
1.0	0.276	0.011	0.520	0.137
1.5	0.316	0.005	0.724	0.064
2.0	0.396	0.003	0.802	0.035
3.0	0.518	—	0.861	0.020
4.0	0.601	—	0.866	0.004
5.0	0.686	—	0.888	0.007
6.0	0.686	—	0.864	0.005

Source. Bird and Parrish [11.31].

TABLE 103   TRANSMITTANCE SPECTRA OF POLAROID SHEET POLARIZERS
(Section 11.4.1)

λ	HN-22 Sheet		HN-32 Sheet		HN-38 Sheet		KN-36 Sheet		HR Sheet	
μm	τ_1	τ_2	τ_1	τ_2	τ_1	τ_2	τ_1	τ_2	τ_1	τ_2
0.375	0.11	0.000005	0.33	0.001	0.54	0.02	0.42	0.002	0.00	0.00
0.40	0.21	0.00001	0.47	0.003	0.67	0.04	0.51	0.001	0.00	0.00
0.45	0.45	0.000003	0.68	0.0005	0.81	0.02	0.65	0.0003	0.00	0.00
0.50	0.55	0.000002	0.75	0.00005	0.86	0.005	0.71	0.00005	0.00	0.00
0.55	0.48	0.000002	0.70	0.00002	0.82	0.0007	0.74	0.00004	0.00	0.00
0.60	0.43	0.000002	0.67	0.00002	0.79	0.0003	0.79	0.00003	0.01	0.00
0.65	0.47	0.000002	0.70	0.00002	0.82	0.0003	0.83	0.00008	0.05	0.00
0.70	0.59	0.000003	0.77	0.00003	0.86	0.0007	0.88	0.02	0.10	0.00
1.0									0.55	0.05
1.5									0.65	0.00
2.0									0.70	0.00
2.5									0.10	0.02

	HNPB (3.5) Sheet				
λ			λ		
μm	τ_1	τ_2	μm	τ_1	τ_2
0.(275)*	(0.250)	(0.0126)	0.340	0.602	0.0002
0.280	0.328	0.0110	0.350	0.568	0.0001
0.290	0.340	0.0040	0.360	0.550	0.0003
0.300	0.372	0.0017	0.370	0.568	0.0007
0.310	0.448	0.0009	0.380	0.604	0.0009
0.320	0.546	0.0006	0.390	0.644	0.0008
0.330	0.611	0.0003	0.400	0.688	0.0005

Source.   R. C. Jones [11.37].

TABLE 104 POSITIONS OF THE CHRISTIANSEN AND
RESTSTRAHLEN PEAKS FOR VARIOUS CRYSTALS
(Section 11.4.3 and 11.4.4)

Crystal	Christiansen	Wavelength (μm) Reststrahlen	Fundamental
LiF	11.2	26	32.6
NaCl	(32)	52.0	61.1
NaBr	37		74.7
NaI	49		85.5
KCl	37	63.4	70.7
KBr	52	81.5	88.3
KI	64	94	102.0
RbCl	45	73.8	84.8
RbBr	65		114.0
RbI	73		129.5
CsCl	50		102.0
CsBr	60		134.0
TlCl	45	91.9	117.0
TlBr	64	117	
TlI	90	151.8	
SiO$_2$	7.3		

Source. Barnes and Bonner [11.46].

TABLE 105 CHRISTIANSEN FILTER BANDWIDTH
(nm) DEPENDENCE ON GRAIN SIZE (Section 11.4.4)

Wavelength (μm)	Grain size (mm)				
	1.49	0.67	0.33	0.22	0.17
0.430	0.8	0.84	—	—	—
0.44	0.88	0.94	1.08	1.06	—
0.45	0.98	1.04	1.14	1.20	—
0.46	1.12	1.16	1.23	1.35	1.65
0.48	1.45	1.42	1.49	1.72	2.02
0.50	1.84	1.72	1.85	2.16	2.55
0.52	2.30	2.08	2.30	2.68	3.22
0.54	2.84	2.48	2.84	3.32	4.02
0.56	3.43	2.91	3.44	4.06	4.92
0.58	—	3.37	4.14	4.92	5.90
t (peak)		0.4	0.36		

Source. Denmark and Cady [11.49].

974

TABLE 106a PROPERTIES OF SOME SELECTIVE SURFACES FOR SOLAR
ENERGY APPLICATION (Section 11.4.6)

Surface	α^a	ε^a	Ref.[d]
"Nickel Black"; containing oxides and sulfides of Ni and Zn, on polished Ni	0.91–0.94	0.11	10
"Nickel Black" on galvanized iron (experimental)	0.89	0.12	10
Same process[b] (commercial)		0.16–0.18	10
"Nickel Black", 2 layers on electroplated Ni on mild steel (α and ε after 6 hr immersion in boiling water)	0.94	0.07	7
"Chrome Black" on steel	0.868	0.088	6
CuO on Ni; made by electrode position of Cu and subsequent oxidation	0.81	0.17	5
Co_3O_4 on silver; by deposition and oxidation	0.90	0.27	5
CuO on Al; by spraying dilute $Cu(NO_3)_2$ solution on hot Al plate and baking	0.93	0.11	4
"Cu Black" on Cu, by treating Cu with solution of NaOH and $NaClO_2$	0.89	0.17	1
Ebanol C on Cu; commercial Cu-blackening treatment giving coatings largely CuO	0.90	0.16	2
CuO on anodized Al; treat Al with hot $Cu(NO_3)_2$—$KMnO_4$ solution and bake	0.85	0.11	9
Al_2O_3—Mo—Al_2O_3Mo—Al_2O_3 interference layers on Mo (ε measured at 500°F)	0.91	0.085	8
PbS crystals on Al	0.89	0.20	11
In_2O_3:Sn, MgF_2-coated[c]	0.9	0.081	3
$TiO_2/Ag/TiO_2^c$	0.54	0.017	3

[a] α = absorptance for solar energy; ε = emittance for long-wave radiation at temperatures typical of flat-plate solar collectors.

[b] Commercial processes.

[c] Ir-reflecting surfaces.

[d] References.

[1] D. J. Close, "Flat plate solar absorbers: The production and testing of a selective surface for copper absorber plates," Report E.D.7, Engr. Sect. Commonwealth Scientific and Industrial Research Organization, Melbourne, Australia (1962).

[2] D. K. Edwards, K. E. Nelson, R. D. Roddick, and J. T. Gier, "Basic studies of the use and control of solar energy," Report No. 60–93 Dept. of Engineering, Univ. of California (October 1960).

[3] J. C. C. Fan and F. J. Bachner, "Transparent heat mirrors for solar energy applications," *Appl. Opt.* **15**, 1012 (1976).

[4] H. C. Hottel and T. A. Unger, "The properties of a copper oxide-aluminum selective black surface absorber of solar energy," *Solar Energy* **3**, 10 (1959).

[d] References (*Continued*)

[5] P. Kokoropoulos, E. Salem, and F. Daniels, "Selective radiation coatings-preparation and high temperature stability," *Solar Energy* **3**, 19 (1959).

[6] G. E. McDonald, "Spectral reflectance properties of black chrome for use as a solar selective coating," *Solar Energy* **17**, 119–122 (1975).

[7] R. N. Schmidt, Honeywell Corp., private communication, 1974.

[8] R. N. Schmidt, K. C. Park, and E. Janssen, "High temperature solar absorber coatings, part II." Tech. Doc. Report No. ML-TDR-64-250 from Honeywell Research Center to Air Force Materials Laboratory (September, 1964).

[9] H. Tabor, "Selective surfaces for solar collectors," *Low Temperature Engineering Applications of Solar energy*, New York, ASHRAE, 1967.

[10] H. Tabor, J. Harris, H. Weinberger, and B. Doron, "Further studies on selective black coatings," *Proc. UN Conf., New Sources of Energy* **4**, 618 (1964).

[11] D. A. Williams, T. A. Lappin, and J. A. Duffie, "Selective radiation properties of particulate coatings," *Trans. ASME, J. Engr. Power* **85A**, 213 (1963).

Source. Based mainly on Duffie and Beckman [11.52].

TABLE 106b. SOLAR-THERMAL CONVERSION EFFICIENCIES WITH TEMPERATURE-DEPENDENT ABSORPTANCE AND EMITTANCE[a] (Section 11.4.6)

	$T_{\rho\lambda}$(°C)	α_s	T_x(°C)[c]	$\varepsilon_T(T_x)$	$\varepsilon_{TH}(T_x)$
MgF_2—Mo—CeO_2	20	0.903	180(1×)	0.081	—
	538	0.900	480(32×)	0.149	—
SiO_2—Si_3N_4—Si—	20	0.735	180(1×)	0.134	—
Cr_2O_3—Ag—Cr_2O_3	500	0.781	480(32×)	0.196	—
Black Cr—Black Ni	20	0.865	230(1×)	0.045	—
	500	0.872	480(32×)	0.122	—
Black Cr—bright Ni	20	0.928	180(1×)	0.131	—
	300	0.925	330(10×)	0.191	—
	500	0.923	530(100×)	0.314	—
	300[b]	0.925	200	0.206	0.20
	300[b]	0.925	300	0.240	0.24

[a] $T_{\rho\lambda}$ and T_x are the reflectance and emittance temperatures, respectively. α_s is the total solar absorptance for $M=2$. ε_T and ε_{TH} are the integrated total emittance and the measured hemispherical total emittance, respectively.

[b] Parameters from normalization of ε_T to ε_{TH} at 300°C.

[c] Concentration ratio in parentheses.

Source. Soule and Smith [11.56].

TABLE 107 COMPOSITION OF THE ATMOSPHERE
(Section 12.1.1.1)

Gas	Molecular weight	Fraction of dry air By volume ($\times 10^{-6}$)	By weight ($\times 10^{-6}$)	Amount (atmo-cm)	Notes
N_2	28.013	780840	755230	624000	
O_2	31.999	209470	231420	167400	
H_2O	18.015	1000–28000	600–17000	800–22000	b d
Ar	39.948	9340	12900	7450	
CO_2	44.010	320	500	260	a
Ne	20.179	18.2	12.7	14.6	
He	4.003	5.24	0.72	4.2	
CH_4	16.043	1.8	1.0	1.4	
Kr	83.80	1.14	3.3	0.91	
CO	28.010	0.06–1	0.06–1	0.05–0.08	a
SO_2	64.06	1	2	1	a
H_2	2.016	0.5	0.04	0.4	
$N_2O[3]$	44.012	0.27	0.5	0.2	
O_3	47.998	0.01–0.1	0.02–0.2	0.25	b c
H_2S	34.08	0.002–0.02	0.002–0.02	0.0015–0.015	
HNO_3	63.016	0–0.005	0–0.01	0–0.004	
Xe	131.30	0.087	0.39	0.07	
NO_2	46.006	0.0005–0.02	0.0008–0.03	0.0004–0.02	a
Rn	222	$0.0^{13}6$	$0.0^{12}5$	5×10^{-14}	
NO	30.006	trace	trace	trace	a
NH_3	17.032	trace	trace	trace	

1 atmo-cm = thickness of layer in cm when reduced to STP
$= 2.687 \times 10^{19}$ molecules cm^{-2}

[a] Greater in industrial areas.
[b] Meteorological or geographical variations.
[c] Increases in ozone layer.
[d] Decreases with height.

Source. Allen [12.1] and McCartney [12.2a].

TABLE 108 ABSOLUTE HUMIDITY OF SATURATED ATMOSPHERE [g/m³] (Section 12.1.1.2)

T (°C)	0	1	2	3	4	5	6	7	8	9
−40	0.1200	0.1075	0.0962	0.0861	0.0769	0.0687	0.0612	0.0545	0.0485	0.0431
−30	0.341	0.308	0.279	0.252	0.227	0.205	0.184 9	0.166	0.149	0.134
−20	0.888	0.810	0.738	0.672	0.611	0.556	0.505	0.458	0.415	0.376
−10	2.145	1.971	1.808	1.658	1.520	1.393	1.275	1.166	1.066	0.973
−0	4.84	4.47	4.13	3.82	3.52	3.25	2.99	2.76	2.54	2.33
0	4.84	5.18	5.55	5.94	6.35	6.79	7.25	7.74	8.26	8.81
10	9.39	10.00	10.64	11.33	12.05	12.81	13.61	14.45	15.34	16.28
20	17.3	18.3	19.4	20.5	21.7	23.0	24.3	25.7	27.2	28.7
30	30.3	32.0	33.7	35.6	37.5	39.5	41.6	43.8	46.1	48.5
40	51.0	53.6	56.3	59.2	62.1	65.2	68.4	79	75.3	78.9

TABLE 109 VAPOR PRESSURE OF ICE ($t < 0°C$) AND WATER ($t > 0°C$) [mm Hg] (Section 12.1.1.2)

T (°C)	0	1	2	3	4	5	6	7	8	9
-40	0.0969	0.0865	0.0771	0.0687	0.0611	0.0543	0.0483	0.0427	0.0379	0.0335
-30	0.287	0.259	0.233	0.209	0.188	0.169	0.151	0.136	0.121	0.108
-20	0.779	0.708	0.642	0.582	0.528	0.478	0.432	0.391	0.353	0.318
-10	1.956	1.790	1.636	1.495	1.365	1.246	1.136	1.035	0.942	0.857
-0	4.58	4.22	3.88	3.57	3.28	3.02	2.77	2.54	2.33	2.14
+0	4.58	4.92	5.29	5.68	6.10	6.54	7.01	7.51	8.04	8.61
10	9.21	9.84	10.52	11.23	11.99	12.79	13.64	14.53	15.48	16.48
20	17.54	18.6	19.8	21.1	22.4	23.8	25.2	26.7	28.4	30.0
30	31.8	33.7	35.7	37.7	39.9	42.2	44.6	47.1	49.7	52.4
40	55.3	58.4	61.5	64.8	68.3	71.9	75.7	79.6	83.7	88.0

Source. Stull [12.3].

TABLE 110 OPTICAL CONSTANTS OF WATER (Section 12.2.1.1)

λ (μm)	$n_I(\lambda)$[a]	ρ(λ) (%)	n(λ)
0.200	110n		1.396
0.225	49n		1.373
0.250	33.5n		1.362
0.275	23.5n		1.354
0.300	16.0n		1.349
0.325	10.8n		1.346
0.350	6.5n		1.343
0.375	3.5n		1.341
0.400	1.86n		1.339
0.425	1.3n		1.338
0.450	1.02n		1.337
0.475	0.935n		1.336
0.500	1.0n		1.335
0.525	1.32n		1.334
0.550	1.96n		1.333
0.575	3.6n		1.333
0.600	10.9n		1.332
0.625	13.9n		1.332
0.650	16.4n		1.331
0.675	22.3n		1.331
0.700	33.5n		1.331
0.725	91.5n		1.330
0.750	156n		1.330
0.775	148n		1.330
0.800	125n		1.329
0.825	182n		1.329
0.850	293n		1.329
0.875	391n		1.328

λ (μm)	$n_I(\lambda)$	n(λ)	ρ(λ) %
3.40	0.0195	1.420	3.37
3.45	0.0132	1.410	
3.50	0.0094	1.400	3.05
3.6	0.00515	1.385	2.8
3.7	0.00360	1.374	2.59
3.8	0.00340	1.364	2.36
3.9	0.00380	1.357	2.25
4.0	0.00460	1.351	(2.2)
4.1	0.00562	1.346	2.18
4.2	0.00688	1.342	2.16
4.3	0.00845	1.338	2.15
4.4	0.0103	1.334	2.14
4.5	0.0134	1.332	2.13
4.6	0.0147	1.330	2.11
4.7	0.0157	1.330	2.10
4.8	0.0150	1.330	2.08
4.9	0.0137	1.328	2.05
5.0	0.0124	1.325	2.02
5.1	0.0111	1.322	1.98
5.2	0.0101	1.317	1.93
5.3	0.0098	1.312	1.86
5.4	0.0103	1.305	1.80
5.5	0.0116	1.298	1.73
5.6	0.0142	1.289	1.66
5.7	0.0203	1.277	1.57
5.8	0.0330	1.262	1.40
5.9	0.0622	1.248	1.50
6.0	0.107	1.265	2.02

λ (μm)	$n_I(\lambda)$	n(λ)	ρ(λ) %
9.8	0.0479	1.229	1.09
10.0	0.0508	1.218	0.99
10.5	0.0662	1.185	(0.81)
11.0	0.0968	1.153	0.72
11.5	0.142	1.126	1.02
12.0	0.199	1.111	1.47
12.5	0.259	1.123	2.15
13.0	0.305	1.146	3.02
13.5	0.343	1.177	3.55
14.0	0.370	1.210	4.10
14.5	0.388	1.241	4.72
15.0	0.402	1.270	5.12
15.5	0.414	1.297	
16.0	0.422	1.325	5.29
16.5	0.428	1.351	
17.0	0.429	1.376	6.15
17.5	0.429	1.401	
18.0	0.426	1.423	6.21
18.5	0.421	1.443	
19.0	0.414	1.461	
19.5	0.404	1.476	
20.0	0.393	1.480	
21.0	0.382	1.487	
22	0.373	1.500	
23	0.367	1.511	
24	0.361	1.521	
25	0.356	1.531	
26	0.350	1.539	

λ	α_a	n_r	n_I
0.900	486n	1.328	
0.925	1.06μ	1.328	
0.950	2.93μ	1.327	
0.975	3.48μ	1.327	
1.0	2.89μ	1.327	1.96
1.2	9.89μ	1.324	1.93
1.4	138μ	1.321	1.90
1.6	85.5μ	1.317	1.87
1.8	115μ	1.312	1.82
2.0	1100μ	1.306	1.74
2.2	289μ	1.296	1.63
2.4	956μ	1.279	1.47
2.6	3.17m	1.242	1.25
2.65	6.7m	1.219	
2.70	0.019	1.188	0.96
2.75	0.059	1.157	(0.8)
2.80	0.115	1.142	1.41
2.85	0.185	1.149	2.0
2.90	0.268	1.201	2.48
2.95	0.298	1.292	2.87
3.00	0.272	1.371	3.4
3.05	0.240	1.426	(4.2)
3.10	0.192	1.467	4.13
3.15	0.135	1.483	(4.15)
3.20	0.0924	1.478	4.0
3.25	0.0610	1.467	(3.83)
3.30	0.0368	1.450	3.65
3.35	0.0261	1.432	(3.5)

λ	α_a	n_r	n_I
6.1	0.131	1.319	2.28
6.2	0.0880	1.363	2.46
6.3	0.0570	1.357	2.34
6.4	0.0449	1.347	2.22
6.5	0.0392	1.339	2.12
6.6	0.0356	1.334	2.05
6.7	0.0337	1.329	2.00
6.8	0.0327	1.324	1.98
6.9	0.0322	1.321	1.97
7.0	0.0320	1.317	1.95
7.1	0.0320	1.314	
7.2	0.0321	1.312	1.87
7.3	0.0322	1.309	
7.4	0.0324	1.307	1.77
7.5	0.0326	1.304	
7.6	0.0328	1.302	1.72
7.7	0.0331	1.299	
7.8	0.0335	1.297	1.68
7.9	0.0339	1.294	
8.0	0.0343	1.291	1.66
8.2	0.0351	1.286	1.64
8.4	0.0361	1.281	1.62
8.6	0.0372	1.275	1.57
8.8	0.0385	1.269	1.51
9.0	0.0399	1.262	1.44
9.2	0.0415	1.255	1.35
9.4	0.0433	1.247	1.27
9.6	0.0454	1.239	1.18

λ	ρ	n_r
27	0.344	1.545
28	0.338	1.549
29	0.333	1.551
30	0.328	1.551
32	0.324	1.546
34	0.329	1.536
36	0.343	1.527
38	0.361	1.522
40	0.385	1.519
42	0.409	1.522
44	0.436	1.530
46	0.462	1.541
48	0.488	1.555
50	0.514	1.587
60	0.587	1.703
70	0.576	1.821
80	0.547	1.886
90	0.536	1.924
100	0.532	1.957
110	0.531	1.966
120	0.526	2.004
130	0.514	2.036
140	0.500	2.056
150	0.495	2.069
160	0.496	2.081
170	0.497	2.094
180	0.499	2.107
190	0.501	2.119
200	0.504	2.130

[a] n, μ, m signify multiplication by 10^{-9}, 10^{-6}, 10^{-3}, respectively. The absorption coefficient, α_a, is related to the imaginary part, n_I, of the refractive index according to: $\alpha_a = 4\pi n_I/\lambda$. Reflectivity values ($\rho$) from Centeno, *J. Opt. Soc. Am.* **31**, 241–247 (1941).

Source. Hale and Querry [12.11].

TABLE 111 VARIATION OF METEOROLOGICAL QUANTITIES
(Section 12.1.1.3)

Latitude	Sea level					Tropopause		
	$\bar{T}$(land)	ΔT	$\bar{T}$(ocean)	$\bar{p}$	$\bar{p}_v$	$\bar{T}$	Height	$\bar{p}$
	°C	°C	°C	mmHg	mmHg	°C	km	mmHg
0°	27	1	27	758	21	−86	17.0	60
10°	26	3	26	759	20	−81	16.6	74
20°	24	6	24	761	18	−74	15.5	97
30°	20	9	20	763	14	−66	13.7	127
40°	13	13	14	761	9	−61	11.8	160
50°	6	17	7	756	5	−58	9.8	198
60°	−2	21	2	751	2	−55	9.0	233
70°	−10	26	0		1	—54	8.1	258
80°	−18	29	−2			−53	7.8	285
90°	−25							

Source. Allen [12.1].

TABLE 112 TYPICAL AEROSOL CONCENTRATIONS
(Section 12.1.1.4)

Atmosphere type	Concentration		Main radii range
	part./m^3	μg/m^3	μm
Clean Country Air	10^8	50	0.1–1.0
Pale blue haze, industr. areas	10^{11}	1000	0.03–0.2
Fog	10^6–5×10^7		3–60
Clouds, higher altitudes	5×10^7–1.5×10^9		2–30
Maritime, 16 km/hr wind		4	
32 km/hr wind		10	
60 km/hr wind		30	
100 km/hr wind		100	

Source. Manson [12.4a].

h	$\log P$	T	$\log \rho$	$\log N$	H	$\log l$
km	in $N\,m^{-2}$	°K	in $g\,cm^{-3}$	in cm^{-3}	km	in cm
0	5.01	288	-2.91	19.41	8.4	-5.2
1	4.95	282	-2.95	19.36	8.3	-5.1
2	4.90	275	-3.00	19.31	8.2	-5.1
3	4.85	269	-3.04	19.28	8.0	-5.0
4	4.79	262	-3.09	19.23	7.8	-5.0
5	4.73	256	-3.13	19.19	7.5	-5.0
6	4.67	249	-3.18	19.14	7.2	-4.9
8	4.55	236	-3.28	19.04	6.8	-4.8
10	4.42	223	-3.38	18.98	6.6	-4.7
15	4.08	217	-3.71	18.61	6.3	-4.4
20	3.75	217	-4.05	18.27	6.4	-4.0
30	3.08	230	-4.74	17.58	6.8	-3.4
40	2.47	253	-5.39	16.92	7.4	-2.7
50	1.91	273	-5.98	16.34	8.1	-2.1
60	1.36	246	-6.50	15.82	7.3	-1.6
70	0.73	216	-7.07	15.26	6.5	-1.1
80	0.00	183	-7.72	14.60	5.5	-0.4
90	-0.81	183	-8.45	13.80	5.5	$+0.4$
100	-1.53	210	-9.30	12.98	6.4	$+1.3$
110	-2.14	260	-10.00	12.29	8.1	$+2.1$
120	-2.57	390	-10.62	11.69	11.8	$+2.7$
150	-3.32	780	-11.67	10.66	24	$+3.7$
200	-4.06	1200	-12.5	9.86	35	$+4.3$
250	-4.55	1400	-13.1	9.3	46	$+4.7$
300	-5.0	1500	-13.6	8.9	54	$+5.1$
400	-5.7	1500	-14.5	8.1	70	$+5.8$
500	-6.4	1600	-15.2	7.4	80	$+6.4$
700	-7.4	1600	-16.5	6.4	110	$+7.3$
1000	-8.4	1600	-17.8	5.2·	150	
2000	-9.1	1800	-18.7	4.3		
3000	-9.3	2000	-19.0	4.0		
5000	-9.4	3000	-19.4	3.6		
10000	-9.6	15000	-20.0	3.0		
20000	-10.0	50000	-20.7	2.0		
30000	-10.6	1×10^5	-21.2	1.0		
50000	-10.8	2×10^5	-21.6	0.6		

h, altitude [km]; p, pressure [N/m^2]; T, temperature [°K]; ρ, mass density [g/cm^3]; N, number density [m^{-3}] (including molecules, atoms, and ions); h, scale height [km]; l, mean free path [m].

Source. Allen [12.1].

TABLE 114 U.S. STANDARD ATMOSPHERE (Section 12.1.2)

Altitude, km	Temperature, K	Pressure, N/m²	Density, g/cm³	Viscosity, poises	Speed of sound, m/sec	Molecular weight
−5	320.68	1.7776+5	1.9311−3	1.9422−4	358.986	28.964
−4	314.17	1.5960	1.7697	1.9123	355.324	28.964
−3	307.66	1.4297	1.6189	1.8820	351.625	28.964
−2	301.15	1.2778	1.4782	1.8515	347.888	28.964
−1	294.65	1.1393	1.3470	1.8206	344.111	28.964
0	288.15	1.0132	1.2250	1.7894	340.294	28.964
1	281.65	8.9876+4	1.1117	1.7579	336.435	28.964
2	275.15	7.9501	1.0066	1.7260	332.532	28.964
3	268.66	7.0121	9.0925−4	1.6938	328.583	28.964
4	262.17	6.1660	8.1935	1.6612	324.589	28.964
5	255.68	5.4048	7.3643	1.6282	320.545	28.964
6	249.19	4.7218	6.6011	1.5949	316.452	28.964
7	242.70	4.1105	5.9002	1.5612	312.306	28.964
8	236.22	3.5652	5.2579	1.5271	308.105	28.964
9	229.73	3.0801	4.6706	1.4926	303.848	28.964
10	223.25	2.6500	4.1351	1.4577	299.532	28.964
11	216.77	2.2700	3.6480	1.4223	295.154	28.964
12	216.65	1.9399	3.1194	1.4216	295.069	28.964
13	216.65	1.6580	2.6660	1.4216	295.069	28.964
14	216.65	1.4170	2.2786	1.4216	295.069	28.964
15	216.65	1.2112	1.9475	1.4216	295.069	28.964
16	216.65	1.0353	1.6647	1.4216	295.069	28.964
17	216.65	8.8497+3	1.4230	1.4216	295.069	28.964
18	216.65	7.5652	1.2165	1.4216	295.069	28.964
19	216.65	6.4675	1.0400	1.4216	295.069	28.964
20	216.65	5.5293	8.8910−5	1.4216	295.069	28.964
25	221.55	2.5492	4.0084	1.4484	298.389	28.964
30	226.51	1.1970	1.8410	1.4753	301.709	28.964
40	250.35	2.8714+2	3.9957−6	1.6009	317.189	28.964
50	270.65	7.9779+1	1.0269	1.7037	329.799	28.964
60	255.77	2.2461	3.0592−7	1.6287	320.606	28.964
70	219.70	5.5205+0	8.7535−8	1.4383	297.139	28.964
80	180.65	1.0366	1.9990	1.216	269.44	28.964
100	210.02	3.0075−2	4.974−10	a	a	28.88
150	892.79	5.0617−4	1.836−12			26.92
200	1235.95	1.3339	3.318−13			25.56
250	1357.28	4.6706−5	9.978−14			24.11
300	1432.11	1.8838	3.585			22.66
400	1487.38	4.0304−6	6.498−15			19.94
500	1499.22	1.0957	1.577			17.94
600	1506.13	3.4502−7	4.640−16			16.84
700	1507.61	1.1918	1.537			16.17

a Equations used to compute viscosity and speed of sound not applicable above 90 km.

Note. A one- or two-digit number (preceded by a plus or minus sign) following the initial entry indicates the power of 10 by which that entry and each succeeding entry of that column should be multiplied.

Source. "U.S. Standard Atmosphere, 1962," NASA, USAF, USWB, Washington, D.C., Dec. 1962.

TABLE 115 ATMOSPHERIC REFRACTION (Δ) VERSUS APPARENT ZENITH ANGLE (Z)[a] (Section 12.2.1.3)

Z°	0°	1°	2°	3°	4°	5°	6°	7°	8°	9°	M[b]	τ_v[c]
0	0.0000	0.0003	0.0006	0.0008	0.0011	0.0014	0.0017	0.0019	0.0022	0.0025	1	0.84
10	0.0028	0.0031	0.0034	0.0037	0.004	0.0043	0.0046	0.0049	0.0052	0.0055	1.95	0.84
20	0.0058	0.0061	0.0065	0.0068	0.0071	0.0075	0.0078	0.0082	0.0086	0.0090	1.064	0.83
30	0.0094	0.0098	0.0102	0.0106	0.0110	0.0114	0.0118	0.0122	0.0126	0.0130	1.154	0.82
40	0.0135	0.0139	0.0143	0.0148	0.0153	0.0159	0.0166	0.0172	0.0179	0.0186	1.304	0.80
50	0.0193	0.0200	0.0207	0.0214	0.0222	0.0231	0.0239	0.0248	0.0257	0.0269	1.553	0.77
60	0.0280	0.0292	0.0303	0.0316	0.0329	0.0344	0.0361	0.0378	0.0397	0.0419	1.995	0.71
70	0.0442	0.0467	0.0492	0.0522	0.0556	0.0594	0.0636	0.0686	0.0742	0.0808	2.904	0.61
80	0.0886	0.0978	0.1092	0.1233	0.1411	0.1644					5.6	0.38

Z°	0′	10′	20′	30′	40′	50′	M[b]	τ_v[c]
85	0.1644	0.1689	0.1739	0.1792	0.1844	0.1900	10.4	0.17
86	0.1958	0.2019	0.2083	0.2153	0.2228	0.2308	12.44	0.12
87	0.2367	0.2486	0.2586	0.2692	0.2803	0.2922	15.36	0.07
88	0.3050	0.3189	0.3339	0.3503	0.3686	0.3886	19.8	
89	0.4103	0.4342	0.4600	0.4883	0.5192	0.5531	27.0	
90	0.5900						38.0	$(10^{-7}-10^{-10})$[d]

[a] Value of Δ in degrees. The last two columns list air mass number and visual transmittance, respectively, both referring to the Z value heading the line.

[b] Values of M due to A. Bemporad, Meteorol. Z. 24, 306–313 (1907); see Refs. 12.1 and 12.2c.

[c] Values of τ based on Ref. 12.31c. Values of τ computed from: $\tau = \exp(-0.172M)$, which, in turn, was derived on the basis of data in the above references. These values refer to a clear day and should be taken as approximations only, especially since the spectral dependence of τ has been neglected in the above formula.

[d] In this range the approximation (c) breaks down. Data given are based on a closer approximation. The values in the last column were calculated neglecting the difference in scale height of the gaseous and aerosol components of the atmosphere—an approximation valid for the larger elevation angles but not near the horizon. There the observed density values are much higher than those implied by the table. The following values have been calculated for the horizon based on the American Standard Atmosphere [12.8a] (scale heights for gases and aerosols are 7.3 km and 1.1 km, respectively):

λ (μm)	0.4	0.45	0.5	0.6	0.7	0.8	0.9	1.06
Density	16.5	13.0	10.8	9.2	8.2	7.0	6.5	5.9

(M. Menat, private communication.)

TABLE 116 SPECTRUM OF CONTINUOUS ATMOSPHERIC ABSORPTION (Section 12.2.2.2)

λ	Molecular Scattering	Ozone	Dust, Clear Conditions		
(μm)	(per atmosphere)[a]	(per 3 mm at S.T.P.)	(per atmosphere)[a]	Total[a]	Transmittance
0.20	7.36	2.4	0.24	20	0.00
0.22	4.76	17	0.21	27	0.00
0.24	3.21	65	0.19	68	0.00
0.26	2.25	88	0.17	89	0.00
0.28	1.63	34	0.157	36	0.00
0.30	1.21	3.2	0.143	4.5	0.011
0.32	0.92	0.24	0.132	1.30	0.273
0.34	0.71	0.02	0.122	0.84	0.43
0.36	0.56	0.00	0.113	0.68	0.51
0.38	0.448	0.000	0.106	0.55	0.58
0.40	0.361	0.000	0.099	0.46	0.63
0.45	0.223	0.001	0.084	0.31	0.73
0.50	0.144	0.012	0.074	0.23	0.79
0.55	0.098	0.031	0.065	0.195	0.82
0.60	0.068	0.044	0.058	0.170	0.84
0.65	0.0495	0.023	0.053	0.126	0.88
0.70	0.0366	0.008	0.048	0.092	0.911
0.80	0.0215	0.001	0.040	0.062	0.939
0.90	0.0133	0.000	0.035	0.048	0.953
1.0	0.0087		0.030	0.039	0.962
1.2	0.0042		0.024	0.028	0.972
1.4	0.0022		0.019	0.021	0.979
1.6	0.0013		0.016	0.017	0.983
1.8	0.0008		0.014	0.015	0.985
2.0	0.0005		0.012	0.013	0.987
3.0	0.0001		0.008	0.008	0.992
5.0	0.0		0.006	0.006	0.994
10.0	0.0		0.005	0.005	0.995

[a] The "natural" density $(-\log_e \tau)$ is given.

Source. Allen [12.1].

TABLE 117 BACKSCATTERING FUNCTION AND EXTINCTION COEFFICIENTS OF MAJOR CLOUD TYPES[a] (Section 12.2.3.1)

λ (μm)	0.488	0.694	1.06	4.0	10.6
Nimbostratus	7.16/128	6.03/130	6.45/132	3.96/147	0.154/136
Altostratus	6.77/108	4.52/109	4.99/112	3.2/130	0.125/83.9
Stratus II	6.04/100	4.76/101	4.62/103	2.87/114	0.131/104
Cumulus Congestus	3.97/69.2	3.01/69.8	3.66/71.3	2.43/81.0	0.0788/67.6
Stratus I	3.13/66.9	2.88/67.9	3.08/69.7	1.47/90.1	0.0742/42.8
Cumulonimbus	2.40/43.5	2.21/43.8	2.19/44.4	0.913/48.2	0.116/50.9
Stratocumulus	2.44/45.3	1.91/46.0	2.08/47.1	0.891/59.6	0.0595/24.8
Fair Weather Cumulus	1.18/21.0	0.868/21.3	1.00/21.9	0.471/27.6	0.023/11.7

[a] The backscattering function, $\beta(\pi)$ [km^{-1} sr^{-1}] and the extinction coefficient, α_T [km^{-1}], are given at five wavelengths, in the form β/α_T.
Source. Carrier, Cato, and von Essen [12.19].

TABLE 118 EXTINCTION AND SCATTERING COEFFICIENTS FOR THREE MODEL ATMOSPHERES CONTAINING PURE WATER AEROSOLS[a] (Section 12.2.3.2)

			Haze C		Haze M		Cloud	
λ (μm)	n_R	n_I	α_T (km^{-1})	Albedo (α_s/α_T)	α_T (km^{-1})	Albedo (α_s/α_T)	α_T (km^{-1})	Albedo (α_s/α_T)
0.45	1.34	0	0.1206	1.0	0.1056	1.0	16.33	1.0
0.70	1.33	0	0.0759	1.0	0.1055	1.0	16.72	1.0
1.61	1.315	0	0.0312	1.0	0.0691	1.0	17.58	1.0
2.25	1.29	0	0.0194	1.0	0.0424	1.0	18.21	1.0
3.07	1.525	0.0682	0.0289	0.620	0.0602	0.721	18.58	0.529
3.90	1.353	0.0059	0.0128	0.930	0.0236	0.948	20.65	0.914
5.30	1.315	0.0143	0.0075	0.800	0.0112	0.826	24.01	0.884
6.05	1.315	0.1370	0.0129	0.260	0.0189	0.297	19.86	0.543
8.15	1.29	0.0472	0.0050	0.385	0.0062	0.410	18.75	0.746
10.0	1.212	0.0601	0.0032	0.202	0.0045	0.178	11.18	0.601
11.5	1.111	0.1831	0.0064	0.051	0.0097	0.044	10.10	0.289
16.6	1.44	0.4000	0.0082	0.104	0.0134	0.075	16.97	0.395

[a] Hazes C and M correspond, respectively to continental and coastal hazes. The assumed particle distributions, N_r in units of (m^3 μm)$^{-1}$, are:

$$\text{Haze } C: N_r = 0, r < 0.03, \quad N = 2.3 \times 10^9 \text{ m}^{-3}, r_c = 5 \ \mu\text{m};$$
$$= 10^9, 0.03 < r < 0.1;$$
$$= 10^5 r^{-4}, r > 0.1.$$

Haze M: $N_r = 5.33 \times 10^{10} r \exp(-8.9444 \sqrt{r})$, $N = 10^8$ m^{-3}; $r_p = 0.05 \ \mu$m.

Cloud: $N_r = 2.373 \times 10^6 r^6 \exp(-1.5r)$, $N = 10^8$ m^{-3}, $r_p = 4 \ \mu$m.

Here r is the particle radius in μm and the subscripts c and p indicate the upper limit and the location of the peak of N_r, respectively. N is the particle density.
Source. Deirmendjian [12.20].

TABLE 119 TRANSMITTANCE THROUGH
RAIN (1.8 km path) (Section 12.2.3.3)

Condition	Rainfall rate (cm/hr)	Transmittance
Light rain	0.25	0.88
Medium rain	1.25	0.74
Heavy rain	2.5	0.65
Cloudburst	10.0	0.38

Source. Hudson [12.6].

TABLE 120 SPECTRA RAYLEIGH SCATTERING CROSS SECTIONS, σ_r, AEROSOL SCATTERING COEFFICIENT AT SEA LEVEL, $\alpha_{SA}(0)$, AND OZONE ABSORPTION COEFFICIENT[a] AT VARIOUS WAVELENGTHS, λ. (Section 12.2.3.4)

λ (μ)	σ_r (m^2)	$\alpha_{SA}(0)$ (km^{-1})	A_0 (cm^{-1})
0.27	8.959×10^{-30}	0.29	2.10×10^2
0.28	7.645	0.27	1.06×10^2
0.30	5.676	0.26	1.01×10^1
0.32	4.309	0.25	8.98×10^{-1}
0.34	3.334	0.24	6.40×10^{-2}
0.36	2.622	0.24	1.80×10^{-3}
0.38	2.091	0.23	0
0.40	1.689	0.200	0
0.45	1.038	0.180	3.50×10^{-3}
0.50	6.735×10^{-31}	0.167	3.45×10^{-2}
0.55	4.563	0.158	9.20×10^{-2}
0.60	3.202	0.150	1.32×10^{-1}
0.65	2.313	0.142	6.20×10^{-2}
0.70	1.713	0.135	2.30×10^{-2}
0.80	9.989×10^{-32}	0.127	1.00×10^{-2}
0.90	6.212	0.120	0
1.06	3.320	0.113	0
1.26	1.600	0.108	0
1.67	5.210×10^{-33}	0.098	0
2.17	1.800	0.085	0
3.50	2.681×10^{-34}	0.070	0
4.00	1.571	0.063	0

[a] Value per atmosphere length of ozone.
Source. Elterman [12.46].

TABLE 121   CONSTANTS FOR USE IN EQUATIONS 12.60 (Section 12.2.4)

Band μm	Band limits (cm^{-1})	Weak band $[A = cw^{1/2}(P+p)^q]$		Transition[a] (cm^{-1})	Strong band $[A = C + D \log w + Q \log (P+p)]$		
		c	q		C	D	Q
Water vapor							
6.3(4.88–8.7)	1150–2050	356	0.30	160	302	218	157
3.2(3–3.57)	2800–3340	40.2	0.30	500	—	—	—
2.7(2.27–3)	3340–4400	316	0.32	200	337	246	150
1.87(1.69–2.08)	4800–5900	152	0.30	275	127	232	144
Carbon dioxide							
15(12.5–18.18)	550–800	3.16	0.44	50	−68	55	47
5.2(5.05–5.35)	1870–1980	0.024	0.40	30	—	—	—
4.8(4.63–5.05)	1980–2160	0.12	0.37	60	—	—	—
4.3(4–4.63)	2160–2500	—	—	50	27.5	34	31.5
2.7(2.63–2.87)	3480–3800	3.15	0.43	50	−137	77	68
2(1.92–2.11)	4750–5200	0.492	0.39	80	−536	138	114

[a] Use weak band expression for values of A less than value given in the transition column and the strong band expression for values of A that are greater. w is expressed in precipitable cm of H_2O and atm cm of CO_2. All logarithms are to the base 10. P and p are the total and partial pressures, respectively, expressed in mm Hg. For atmospheric CO_2, $p = 32 \times 10^{-4} P$; therefore $(P+p) \approx P$. For water vapor under average conditions, $p = 0.01 P$; even under the highest humidities it is rare for p to exceed 0.05 P; thus $(P+p) \approx P$.

Source.   Adapted from Howard *et al.* [12.27].

TABLE 122   COORDINATES, MAGNITUDES, AND COLOR TEMPERATURES OF THE BRIGHTEST HEAVENLY BODIES (Section 12.3.2)

	Body	Constellation	Rt. ascen. (hr)	Declination (deg)	Vis. magn. V	T_c K
	Sun[a]				−26.8	5,900
	Moon				−12.2	5,900
Planets[a]						
	Venus				−4.28	5,900
	Mars				−2.25	5,900
	Jupiter				2.25	5,900
	Mercury				−1.8	5,900
	Saturn				−0.93	5,900
Stars						
	Sirius	α Canis Ma	6.716	−16.647	−1.47	11,200
	Canopus	α Carinae	6.381	−52.668	−0.71	6,200
		α Centauri	14.603	−60.631	−0.28	23,000
	Arcturus	α Bootis	14.223	19.442	−0.06	13,750
	Vega	α Lyrae	18.588	38.737	0.04	11,200
	Rigel	β Orionis	5.202	−8.258	0.08	13,000
	Capella	α Aurigae	5.216	45.950	0.09	4,700
	Procyon	α Canis Mi	7.611	5.355	0.34	5,450
	Betelgeus	α Orionis	5.874	7.400	0.4–1.3	2,810
	Achernar	α Eridani	1.598	−57.490	0.49	15,000
		β Centauri	14.005	−60.133	0.61	23,000
	Altair	α Aquilae	19.806	8.735	0.78	7,500
	Aldebaran	α Tauri	4.551	16.410	0.8	3,130
		α Crucis	12.397	−62.822	0.81	2,810
	Antares	α Scorpii	16.439	−26.323	0.92	2,900
	Spica	α Virginis	13.376	−10.902	0.98	
	Formalhaut	α Piscis A.	22.915	−29.888	1.15	
	Pollux	β Geminorum	7.704	28.148	1.15	3,750
		β Crucis	12.746	−59.415	1.24	
	Deneb	α Cygni	20.662	45.102	1.26	
	Regulus	α Leonis	10.095	12.212	1.36	
		ε Canis Ma	6.944	−28.903	1.50	
	Castor	α Geminorum	7.524	32.000	1.56	

[a] Visual magnitudes given for moon and planets are m_v and correspond to the body at its brightest. Values of coordinates and visual magnitude (V) are from Ref. 12.31f. Those of effective temperature and visual magnitude (m_v) are from Ref. 12.38.

TABLE 123 MOON: RELATIVE INTENSITY VERSUS PHASE (Section 12.3.2)

Phase angle (deg)	I/I_0
0	1.0
5	0.929
10	0.809
20	0.625
30	0.483
40	0.377
50	0.288
60	0.225
70	0.172
80	0.127
90	0.089
100	0.061
110	0.041
120	0.027
130	0.017
140	0.009
150	0.004
160	0.001

Source. Allen [12.1].

TABLE 124 LUNAR ILLUMINANCE ON EARTH (Section 12.3.2)

True altitude of center of Moon	Illuminance on horizontal surface, E, At elongation ϕ_e —lux (or lm m^{-2})			
	$\phi_e = 180°$ (Full Moon)	$\phi_e = 120°$	$\phi_e = 90°$ (1st or 3rd quarter)	$\phi_e = 60°$
−0.8° (moonrise or moonset)	9.74×10^{-4}	2.73×10^{-4}	1.17×10^{-4}	3.12×10^{-5}
0°	1.57×10^{-3}	4.40×10^{-4}	1.88×10^{-4}	5.02×10^{-5}
10°	2.34×10^{-2}	6.55×10^{-3}	2.81×10^{-3}	7.49×10^{-4}
20°	5.87×10^{-2}	1.64×10^{-2}	7.04×10^{-3}	1.88×10^{-3}
30°	0.101	2.83×10^{-2}	1.21×10^{-2}	3.23×10^{-3}
40°	0.143	4.00×10^{-2}	1.72×10^{-2}	4.58×10^{-3}
50°	0.183	5.12×10^{-2}	2.20×10^{-2}	5.86×10^{-3}
60°	0.219	6.13×10^{-2}	2.63×10^{-2}	—
70°	0.243	6.80×10^{-2}	2.92×10^{-2}	—
80°	0.258	7.22×10^{-2}	3.10×10^{-2}	—
90°	0.267	7.48×10^{-2}	—	—

Source. Bond and Henderson, in *RCA Electro-Optics Handbook*, Harrison, N.J. (1974).

TABLE 125 HORIZON-SKY LUMINANCE AND TERRESTRIAL IL-
LUMINANCE (ORDER OF MAGNITUDE) (Section 12.3.3)

Time	Condition	L cd/m^2	E lm/m^2
Day	Clear, sunlight	10^4	10^5
	Clear, shade	10^4	10^4
	Overcast	10^3	10^3
	Heavy overcast	10^2	10^2
Sunset	Overcast	10	10
Sunset $+\frac{1}{4}$ hr	Clear	1	10
Sunset $+\frac{1}{2}$ hr	Clear	10^{-1}	1
Night	Bright moon	10^{-2}	10^{-1}
	Clear, moonless	10^{-3}	10^{-3}
	Overcast, moonless	10^{-4}	10^{-4}

Source. Luminance from Ref. 12.22g.

TABLE 126 DAYLIGHT ILLUMINATION VERSUS SOLAR ELEVATION (ε)[a] (Section 12.3.3)

ε deg	Direct sunlight			Skylight			Total		
	E_H	E_V	E_N	E_H	E_V	E_N	E_H	E_V	E_N
0.8							0.453[b]		
3	0.211	4.03	4.04	2.76	6.32	6.51	2.98	10.3	10.5
5	1.08	12.4	12.4	3.50	8.03	8.26	4.57	20.5	20.7
10	6.35	36.1	36.6	5.29	10.3	10.6	11.6	46.3	47.2
20	22.9	63.1	67.2	8.07	12.3	13.1	31.0	75.3	80.3
30	41.1	71.3	82.2	10.2	13.0	14.2	51.2	84.3	96.4
40	58.6	69.9	91.2	11.7	13.1	14.7	70.3	83.0	106
50	73.7	61.9	96.2	13.0	12.9	15.1	86.8	74.8	111
60	86.1	49.7	99.5	14.1	12.4	15.4	100	62.1	115
70	95.4	34.8	102	15.0	11.0	15.6	110	45.7	117
80	101.0	17.8	102	15.5	8.98	15.7	116	26.7	118
90	103.0	0	103	15.9	6.62	15.9	119	6.62	119

[a] E_H, E_V, E_N are illumination values, in klux, on a surface horizontal, vertical and normal to the sun's rays, respectively.
[b] From Bond and Henderson, in RCA Electro-Optics Handbook, RCA, Harrison, N.J. (1974).
Source. Jones and Condit [12.30].

TABLE 127   TWILIGHT SKY LUMINANCE (NIT) AND ILLUMINANCE ON A HORIZONTAL SURFACE[a] (Section 12.3.3)

A	ε_s	Sea Level						Elevation 2800 m						E (lux)
		$\varepsilon = 0°$	$10°$	$30°$	$50°$	$70°$	$90°$	$\varepsilon = 0°$	$10°$	$30°$	$50°$	$70°$	$90°$	
0°	5°	11,000	11,000	3,800	1,600	860	460	11,000	11,000	1,600	680	380	330	2044
	3°	—	6,500	2,400	1,100	670	380	—	6000	1000	470	260	230	1345
	0°	—	1,600	800	380	230	160	—	1,600	400	180	100	86	430
	−3°	180	160	100	43	32	22	220	160	51	21	14	11	43
	−6°	8.6	7.5	3.2	1.3	0.86	0.65	9.7	7.5	1.9	0.65	0.36	0.24	1.35
	−9°	0.38	0.32	0.077	0.043	0.02	0.016	0.43	0.32	0.06	0.019	0.012	0.008	0.054
	−12°	18 m	16 m	6.5 m	2.5 m	1.7 m	1.3 m	19 m	14 m	3.6 m	1.5 m	1.1 m	0.82 m	
	−15°	3 m	2.3 m	1.2 m	0.67 m	0.61 m	0.43 m	1.9 m	1.1 m	0.51 m	0.36 m	0.27 m	0.22 m	
45°	5°	2700	2800	1700	860	540	460	2,200	1800	1100	540	330	330	
	3°	1800	1900	1400	650	440	380	1600	1500	710	380	250	230	
	0°	650	620	540	300	200	160	670	540	250	140	100	86	
	−3°	100	110	62	37	24	22	120	100	39	20	13	11	
	−6°	4.5	3.8	2.2	0.86	0.67	0.65	4.3	3.8	1.1	0.49	0.33	0.24	
	−9°	0.19	0.16	0.054	0.03	0.017	0.016	0.18	0.16	0.037	0.019	0.011	0.008	
	−12°	8.6 m	7.3 m	4 m	2 m	1.5 m	1.3 m	8.2 m	7.5 m	2.2 m	1.4 m	0.97 m	0.82 m	
	−15°	1.3 m	1.3 m	1.1 m	0.65 m	0.54 m	0.43 m	0.86 m	0.72 m	0.43 m	0.34 m	0.27 m	0.22 m	
90°	5°	1100	1000	860	650	480	460	1400	1100	480	380	320	330	
	3°	790	860	710	540	370	380	970	910	410	300	250	230	
	0°	370	370	340	240	180	160	400	370	180	120	100	86	
	−3°	59	65	38	32	22	22	45	430	24	17	12	11	
	−6°	1.6	0.97	0.97	0.75	0.54	0.65	1.2	0.86	0.54	0.43	0.27	0.24	
	−9°	0.058	0.048	0.037	0.022	0.015	0.016	0.054	0.049	0.022	0.016	0.011	0.008	
	−12°	3.8 m	3.8 m	2.7 m	1.7 m	1.3 m	1.3 m	3.9 m	3.6 m	1.6 m	1.0 m	0.86 m	0.82 m	
	−15°	0.75 m	0.75 m	0.81 m	0.56 m	0.5 m	0.43 m	0.54 m	0.48 m	0.38 m	0.31 m	0.27 m	0.22 m	

135°	5°	1700	1600	1100	690	520	460	1700	1600	770	510	370	330
	3°	1200	1100	750	540	420	380	1100	1100	590	380	260	230
	0°	420	380	320	220	170	160	340	340	220	140	100	86
	−3°	43	32	32	32	22	22	26	23	23	16	12	11
	−6°	0.97	0.75	0.75	0.75	0.54	0.65	0.84	0.75	0.48	0.4	0.27	0.24
	−9°	0.034	0.032	0.027	0.022	0.014	0.016	0.039	0.038	0.022	0.014	0.01	0.008
	−12°	2.7 m	2.7 m	2.2 m	1.6 m	1.1 m	1.3 m	2.8 m	2.5 m	1.5 m	1 m	0.75 m	0.82 m
	−15°	0.6 m	0.59 m	0.54 m	0.48 m	0.46 m	0.43 m	0.54 m	0.54 m	0.39 m	0.3 m	0.24 m	0.22 m
180°	5°	1900	1800	1200	700	540	460	2000	1800	860	540	380	330
	3°	1300	1300	840	580	430	380	1300	1200	650	390	260	230
	0°	450	430	380	250	200	160	410	400	240	150	100	86
	−3°	22	22	43	34	22	22	26	22	23	17	12	11
	−6°	0.97	0.86	0.75	0.75	0.54	0.65	0.84	0.73	0.46	0.38	0.27	0.24
	−9°	0.034	0.032	0.027	0.022	0.016	0.016	0.039	0.032	0.019	0.013	0.01	0.008
	−12°	2.7 m	2.7 m	2.2 m	1.8 m	1.2 m	1.3 m	3 m	3.2 m	1.5 m	1 m	0.75 m	0.82 m
	−15°	0.60 m	0.58 m	0.54 m	0.5 m	0.43 m	0.43 m	0.62 m	0.65 m	0.45 m	0.32 m	0.25 m	0.22 m

[a] A and ε are azimuth and elevation of the subject sky point; ε_s is the solar elevation angle; E is the illuminance on a horizontal surface 2800 m above sea level. A is measured relative to the suns meridian. m stands for a factor of 10^{-3}.

Source. Koomen et al. [12.31].

TABLE 128   ZENITH-SKY LUMINANCE
[L (nit)] AND ILLUMINANCE [E (lux)]
DUE TO IT ON HORIZONTAL SURFACE
AND ITS COLOR TEMPERATURE [$T_c(E)$
IN (K)] AS FUNCTIONS OF SOLAR
ELEVATION (ε). (Section 12.3.3)

ε	$\log_{10} L$	$\log_{10} E$	$c_2/T_c{}^a$
0	2.18	2.91	2.18
$-1°$	1.95	2.66	2.15
$-2°$	1.66	2.35	2.12
$-3°$	1.31	1.96	1.99
$-4°$	0.9	1.51	1.78
$-5°$	0.36	1.01	1.60
$-6°$	-0.18	0.47	1.34
$-7°$	-0.72	-0.05	1.25
$-8°$	-1.26	-0.55	1.13
$-9°$	-1.72	-1.01	0.97
$-10°$	-2.10	-1.42	0.86
$-11°$	-2.42	-1.81	0.90
$-12°$	-2.78	-2.19	1.03
$-13°$	-3.11	-2.52	1.38
$-14°$	-3.38	-2.80	1.82
$-15°$	-3.56	-3.09	2.52
$-16°$	-3.66	-3.19	3.05
$-17°$	-3.69	-3.23	3.33
$-18°$	-3.70	-3.25	3.39

[a] As deduced from measurements at 0.42
and 0.61 μm.
   Source.   Koomen *et al.* [12.31*b*].

Tables 129 and 130, on the following pages, are for the year 1962 and refer to 0^h ephemeris time. To apply them to another year (y) and to standard time (zone-longitude, L_0) at hour, t_h, add a fractional day, D, to the date, where

$$D = \frac{1}{2} - F\left[\frac{y}{4}\right] + \frac{L_0}{360} + \frac{t_h}{24}$$

where $F[\ ]$ indicates the fractional part, and in Jan. and Feb. of a leap year, $(D-1)$ is used in place of D. L_0 is positive west of Greenwich and negative east thereof.

996

TABLE 129 APPARENT DECLINATION OF SUN (DEG) (Section 12.3.4)

DAY	JAN	FEB	MAR	APR	MAY	JUN	JUL	AUG	SEP	OCT	NOV	DEC
1	-23.0	-17.3	-7.8	4.3	14.9	21.9	23.1	18.2	8.5	-2.9	-14.2	-21.7
2	-23.0	-17.0	-7.5	4.7	15.2	22.1	23.1	17.9	8.2	-3.3	-14.5	-21.8
3	-22.9	-16.7	-7.1	5.0	15.5	22.2	23.0	17.7	7.8	-3.7	-14.8	-22.0
4	-22.8	-16.4	-6.7	5.4	15.8	22.3	22.9	17.4	7.4	-4.1	-15.1	-22.1
5	-22.7	-16.1	-6.3	5.8	16.0	22.5	22.8	17.2	7.1	-4.5	-15.5	-22.3
6	-22.6	-15.8	-5.9	6.2	16.3	22.6	22.8	16.9	6.7	-4.8	-15.8	-22.4
7	-22.4	-15.5	-5.5	6.6	16.6	22.7	22.7	16.6	6.3	-5.2	-16.1	-22.5
8	-22.3	-15.2	-5.2	6.9	16.9	22.8	22.5	16.3	6.0	-5.6	-16.4	-22.6
9	-22.2	-14.9	-4.8	7.3	17.2	22.9	22.4	16.1	5.6	-6.0	-16.7	-22.7
10	-22.0	-14.6	-4.4	7.7	17.4	22.9	22.3	15.8	5.2	-6.4	-16.9	-22.8
11	-21.9	-14.2	-4.0	8.1	17.7	23.0	22.2	15.5	4.8	-6.8	-17.2	-22.9
12	-21.7	-13.9	-3.6	8.4	17.9	23.1	22.1	15.2	4.4	-7.1	-17.5	-23.0
13	-21.6	-13.6	-3.2	8.8	18.2	23.2	21.9	14.9	4.1	-7.5	-17.8	-23.1
14	-21.4	-13.2	-2.8	9.1	18.4	23.2	21.8	14.6	3.7	-7.9	-18.0	-23.2
15	-21.2	-12.9	-2.4	9.5	18.7	23.3	21.6	14.3	3.3	-8.2	-18.3	-23.2
16	-21.1	-12.6	-2.0	9.9	18.9	23.3	21.5	14.0	2.9	-8.6	-18.6	-23.3
17	-20.9	-12.2	-1.6	10.2	19.2	23.3	21.3	13.6	2.5	-9.0	-18.8	-23.3
18	-20.7	-11.9	-1.2	10.6	19.4	23.4	21.1	13.3	2.1	-9.4	-19.0	-23.4
19	-20.5	-11.5	-0.8	11.0	19.6	23.4	21.0	13.0	1.8	-9.7	-19.3	-23.4
20	-20.3	-11.2	-0.4	11.3	19.8	23.4	20.8	12.7	1.4	-10.1	-19.5	-23.4
21	-20.0	-10.8	-0.0	11.6	20.0	23.4	20.6	12.3	1.0	-10.4	-19.7	-23.4
22	-19.8	-10.4	0.4	12.0	20.2	23.4	20.4	12.0	0.6	-10.8	-20.0	-23.4
23	-19.6	-10.1	0.7	12.3	20.4	23.4	20.2	11.7	0.2	-11.2	-20.2	-23.4
24	-19.4	-9.7	1.1	12.6	20.6	23.4	20.0	11.3	-0.2	-11.5	-20.4	-23.4
25	-19.1	-9.3	1.5	13.0	20.8	23.4	19.8	11.0	-0.6	-11.8	-20.6	-23.4
26	-18.9	-9.0	1.9	13.3	21.0	23.4	19.6	10.7	-1.0	-12.2	-20.8	-23.4
27	-18.6	-8.6	2.3	13.6	21.2	23.3	19.4	10.3	-1.4	-12.5	-21.0	-23.3
28	-18.4	-8.2	2.7	13.9	21.3	23.3	19.1	10.0	-1.7	-12.9	-21.2	-23.3
29	-18.1	-7.8	3.1	14.2	21.5	23.3	18.9	9.6	-2.1	-13.2	-21.3	-23.3
30	-17.8	-7.5	3.5	14.5	21.7	23.2	18.7	9.3	-2.5	-13.5	-21.5	-23.2
31	-17.6		3.9		21.8		18.4	8.9		-13.9		-23.1

997

TABLE 130 EQUATION OF TIME (MIN) (Section 12.3.4)

DAY	JAN	FEB	MAR	APR	MAY	JUN	JUL	AUG	SEP	OCT	NOV	DEC
1	-3.2	-13.5	-12.6	-4.2	2.8	2.4	-3.6	-6.3	-0.7	10.0	16.3	11.2
2	-3.7	-13.7	-12.4	-3.9	3.0	2.3	-3.7	-6.2	0.1	10.4	16.4	10.9
3	-4.2	-13.8	-12.2	-3.6	3.1	2.1	-3.9	-6.2	0.4	10.7	16.4	10.5
4	-4.6	-13.9	-12.0	-3.3	3.2	1.9	-4.1	-6.1	0.7	11.0	16.4	10.1
5	-5.1	-14.0	-11.8	-3.0	3.3	1.8	-4.3	-6.0	1.0	11.3	16.4	9.7
6	-5.6	-14.1	-11.5	-2.7	3.4	1.6	-4.5	-5.9	1.4	11.6	16.3	9.3
7	-6.0	-14.2	-11.3	-2.4	3.4	1.4	-4.7	-5.8	1.7	11.9	16.3	8.9
8	-6.4	-14.2	-11.1	-2.1	3.5	1.2	-4.8	-5.7	2.0	12.2	16.3	8.4
9	-6.9	-14.3	-10.8	-1.8	3.6	1.0	-5.0	-5.6	2.4	12.5	16.2	8.0
10	-7.3	-14.3	-10.6	-1.6	3.6	0.9	-5.1	-5.4	2.7	12.7	16.1	7.6
11	-7.7	-14.3	-10.3	-1.3	3.7	0.7	-5.3	-5.3	3.1	13.0	16.0	7.1
12	-8.1	-14.3	-10.1	-1.0	3.7	0.5	-5.4	-5.1	3.4	13.3	15.9	6.6
13	-8.5	-14.3	-9.9	-0.8	3.7	0.3	-5.5	-4.9	3.9	13.5	15.8	6.2
14	-8.8	-14.3	-9.5	-0.5	3.7	0.0	-5.7	-4.8	4.1	13.8	15.7	5.7
15	-9.2	-14.3	-9.2	-0.3	3.7	-0.2	-5.9	-4.6	4.5	14.0	15.5	5.2
16	-9.6	-14.2	-9.0	0.0	3.7	-0.4	-6.0	-4.4	4.8	14.2	15.3	4.8
17	-9.9	-14.1	-8.7	0.2	3.7	-0.6	-6.0	-4.2	5.2	14.4	15.2	4.3
18	-10.2	-14.1	-8.4	0.5	3.7	-0.8	-6.1	-4.0	5.6	14.6	15.0	3.7
19	-10.6	-14.0	-8.1	0.7	3.6	-1.0	-6.1	-3.8	5.9	14.8	14.8	3.3
20	-10.9	-13.9	-7.8	0.9	3.6	-1.2	-6.2	-3.5	6.3	15.0	14.5	2.8
21	-11.2	-13.8	-7.5	1.1	3.6	-1.4	-6.3	-3.3	6.6	15.2	14.3	2.3
22	-11.4	-13.7	-7.2	1.3	3.5	-1.7	-6.3	-3.0	7.0	15.4	14.1	1.8
23	-11.7	-13.6	-6.9	1.5	3.4	-1.9	-6.3	-2.8	7.3	15.5	13.8	1.3
24	-12.0	-13.4	-6.6	1.7	3.4	-2.1	-6.4	-2.5	7.7	15.7	13.5	0.8
25	-12.2	-13.3	-6.3	1.9	3.3	-2.3	-6.4	-2.3	8.0	15.8	13.2	0.3
26	-12.4	-13.1	-6.0	2.1	3.2	-2.5	-6.4	-2.0	8.4	15.9	12.9	-0.2
27	-12.7	-12.9	-5.7	2.3	3.1	-2.7	-6.4	-1.7	8.7	16.0	12.6	-0.7
28	-12.9	-12.8	-5.4	2.4	3.0	-2.9	-6.4	-1.4	9.0	16.1	12.3	-1.1
29	-13.1	-12.6	-5.1	2.6	2.9	-3.2	-6.4	-1.2	9.4	16.2	12.0	-1.7
30	-13.2		-4.8	2.7	2.7	-3.4	-6.4	-0.9	9.7	16.2	11.6	-2.2
31	-13.4		-4.5		2.6		-6.3	-0.6		16.3		-2.6

TABLE 131 SOLAR SYSTEM DATA (Section 12.3.5)

Name	Body			Orbit	
	Radius, mean M m	Mass 10^{24} kg	Intensity[a] cd	Axis, semi-major Gm	Period, sidereal Years
Mercury	2.42	0.331	0.6×10^{18}	57.9	0.24085
Venus	6.118	4.87	6.0×10^{18}	108.2	0.61521
Earth	6.371	5.98	2.0×10^{18}	149.6	1.0
Mars	3.385	0.642	10^{18}	227.9	1.8809
Jupiter	70.36	1900	11.0×10^{18}	778.3	11.862
Saturn	58.41	569	2.0×10^{18}	1427	29.458
Uranus	23.55	86.9	0.13×10^{18}	2870	84.015
Neptune	22.3	103	0.04×10^{18}	4497	164.79
Pluto	3	1.02	10^{14}	(5900)	247.7
Sun	696	1.989×10^{6}	3.0×10^{27}	—	—
Moon	1.738	0.0735	0.4×10^{18}	0.384	$29^{d} 12^{h} 44^{m}$ [b]

[a] Intensity given is in the direction of the sun.

[b] For the moon, the period given is the synodic, in days, hours, and minutes.

Source. Based on Refs. 12.31*d* and *e*. See also Ref. 12.1.

TABLE 132 LUMINOUS REFLECTANCE OF NATURAL OBJECTS (%)
(Section 12.3.6)

	Smithsonian[a] tables	Sewing[b] handbook	Krinov[c]
Class A. Water Surfaces			
1. Bay	3–4	—	—
2. Bay and river	6–10	—	—
3. Inland water	5–10	—	5
4. Ocean	3–7	—	—
5. Ocean, deep	3–5	—	—
Class B. Bare Areas and Soils			
1a. Snow, fresh fallen	70–86	—	77
1b. Snow, covered with ice	—	—	75
2. Limestone, clay	—	—	63
3. Calcareous rocks	—	30	—
4. Granite	—	12	—
5. Mountain tops, bare	—	—	24
6a. Sand, dry	—	31	24
6b. Sand, wet	—	18	—
7a. Clay soil, dry	—	15	—
7b. Clay soil, wet	—	7.5	9
8a. Ground, bare, rich soil, dry	10–20	7.2	9
8b. Ground, bare, rich soil, wet	—	5.5	—
8c. Ground, black earth, sand loam	—	—	3
8d. Field, plowed, dry	20–25	—	—
Class C. Vegetative Formations			
1a. Coniferous forest, winter	—	—	3
1b. Coniferous forest, summer	3–10	—	8
1c. Deciduous forest, summer	—	—	10
1d. Deciduous forest, fall	—	—	15
1e. Dark hedges	—	1	—
2. Coniferous forest, summer, from airplane	—	—	3
3a. Meadow, dry grass	3–6	—	8
3b. Grass, lush	15–25	—	10
4. Meadow, low grass, from airplane	—	—	8
5. Field crops, ripe	7	—	15
Class D. Roads and Buildings			
1. Earth roads	—	—	3
2. Black top roads	—	8	9
3a. Concrete road, smooth, dry	—	35	—
3b. Concrete road, smooth, wet	—	15	—
4a. Concrete road, rough, dry	—	35	—
4b. Concrete road, rough, wet	—	25	—
5. Buildings	—	—	9
6. Limestone tiles	—	25	—

[a] R. J. List, ed., Smithsonian Meteorological Tables, Smithsonian Institution, Washington, D.C., 1951.

[b] R. Sewing, "Handbuch der Lichttechnik." Springer, Berlin, 1938.

[c] E. L. Krinov, "Spektral'naia otrazhatel'naia sposobnost' prirodnykh obrazovanii," Laboratoriia Aerometodov, Akad. Nauk SSSR, Moscow, 1947.

Source. Stewart and Hopfield [12.40].

TABLE 133   REFLECTIVITY   OF
WATER (Section 12.3.6)

Solar elevation (deg)	Reflectance
0	1.00
5	0.584
10	0.348
20	0.134
30	0.06
40	0.034
50	0.025
70	0.021
90	0.02

Source. List [12.8].

TABLE 134   EMISSIVITY SPECTRUM OF WATER IN THE INFRARED REGION,
NORMAL (Section 12.3.6)

λ (μm)	ε	λ (μm)	ε	λ (μm)	ε	λ (μm)	ε	λ (μm)	ε
1.00	0.980357	2.90	0.975254	4.60	0.978927	6.50	0.978819	10.2	0.990797
1.05	0.980460	2.95	0.971350	4.70	0.979027	6.60	0.979451	10.4	0.991497
1.10	0.980564	3.00	0.966006	4.80	0.979240	6.70	0.979990	10.6	0.992177
1.20	0.980667	3.02	0.965148	4.90	0.979458	6.80	0.980199	10.8	0.992671
1.30	0.980872	3.07	0.956071	5.00	0.979785	6.90	0.980297	10.9	0.993014
1.40	0.980911	3.10	0.957109	5.10	0.980211	7.00	0.980492	11.0	0.992748
1.50	0.981180	3.16	0.958825	5.20	0.980732	7.20	0.981295	11.1	0.992274
1.60	0.981383	3.20	0.960063	5.30	0.981448	7.40	0.982288	11.2	0.991595
1.70	0.981485	3.30	0.963451	5.40	0.982052	7.60	0.982748	11.3	0.990943
1.80	0.981789	3.40	0.966305	5.50	0.982733	7.80	0.983159	11.4	0.990298
1.90	0.982091	3.50	0.969496	5.60	0.983377	8.00	0.983352	11.5	0.989789
2.00	0.982590	3.60	0.971980	5.70	0.984321	8.20	0.983641	11.6	0.989103
2.20	0.983672	3.70	0.974388	5.80	0.985983	8.40	0.983831	11.7	0.988103
2.40	0.985294	3.80	0.976395	5.85	0.987511	8.60	0.984301	11.8	0.986978
2.50	0.986311	3.90	0.977487·	5.90	0.984996	8.80	0.984851	12.0	0.985289
2.60	0.987473	4.03	0.978134	6.00	0.979843	9.00	0.985562	12.5	0.978449
2.70	0.990431	4.10	0.978239	6.04	0.978344	9.20	0.986514	13.0	0.969827
2.74	0.992534	4.20	0.978449	6.10	0.977177	9.40	0.987342	13.5	0.964539
2.77	0.991010	4.30	0.978548	6.20	0.975399	9.60	0.988236	14.0	0.959012
2.80	0.985759	4.40	0.978644	6.30	0.976561	9.80	0.989148	14.5	0.952793
2.85	0.979957	4.50	0.978734	6.40	0.977653	10.00	0.990084	15.0	0.948781

Source. Bramson [12.40].

TABLE 135 SKY-GROUND LU-
MINANCE RATIO, TYPICAL
(Section 12.4.2)

Sky	Ground	Lum. ratio
Overcast	Snow	1
	Desert	7
	Forest	25
Clear	Snow	0.2
	Desert	1.4
	Forest	5

Source. Middleton [12.22c].

TABLE 136 INTERNATIONAL VISIBILITY CODE
(Section 12.4.3)

Code no.	Meteorologic range km	Description[a]	α_v km^{-1}
0	<0.05	dense fog	>78
1	0.050–0.200	thick fog	20–78
2	0.20–0.50	moderate fog	7.8–20
3	0.5–1.0	light fog	3.9–7.8
4	1–2	thin fog	2.0–3.9
5	2–4	haze	0.98–2.0
6	4–10	light haze	0.39–0.98
7	10–20	clear	0.20–0.39
8	20–50	very clear	0.078–0.20
9	>50	exceptionally clear	<0.078

[a] These terms are not accepted internationally.
Source. Hulburt [12.45].

TABLE 137 CONTRAST ATTENUATION IN THE ATMOSPHERE (Section 12.4.3)

RANGE ρ	0.1	0.2	0.5	1.0	2.0	5.0	10	20	50	100
0.01	0.99603	0.99208	0.98044	0.96164	0.92611	0.83370	0.71482	0.55621	0.33392	0.20042
0.02	0.99193	0.98398	0.96090	0.92474	0.86002	0.71078	0.55132	0.38057	0.19727	0.10043
0.03	0.98770	0.97570	0.94139	0.88926	0.80061	0.61629	0.44538	0.28649	0.13838	0.07434
0.04	0.98334	0.96723	0.92192	0.85515	0.74695	0.54143	0.37121	0.22791	0.10560	0.05574
0.05	0.97885	0.95858	0.90251	0.82234	0.69828	0.48072	0.31641	0.19794	0.08473	0.04424
0.06	0.97423	0.94975	0.88318	0.79079	0.65397	0.43052	0.27431	0.15995	0.07028	0.03642
0.07	0.96946	0.94073	0.86393	0.76045	0.61349	0.38834	0.24096	0.13698	0.05970	0.03077
0.08	0.96456	0.93154	0.84478	0.73128	0.57639	0.35244	0.21392	0.11977	0.05162	0.02640
0.09	0.95951	0.92217	0.82576	0.70322	0.54228	0.32153	0.19156	0.10593	0.04525	0.02315
0.10	0.95431	0.91262	0.80686	0.67624	0.51085	0.29466	0.17278	0.09456	0.04010	0.02046
0.20	0.89392	0.80818	0.62760	0.45731	0.29643	0.14422	0.07772	0.04043	0.01657	0.00836
0.30	0.81742	0.69122	0.47241	0.30025	0.18291	0.08218	0.04285	0.02180	0.00887	0.00446
0.40	0.72560	0.56936	0.34592	0.20913	0.11677	0.05023	0.02576	0.01305	0.00526	0.00264
0.50	0.62224	0.45163	0.24780	0.14142	0.07609	0.03189	0.01620	0.00817	0.00328	0.00164
0.60	0.51397	0.34587	0.17458	0.09564	0.05022	0.02071	0.01046	0.00526	0.00211	0.00106
0.70	0.40879	0.25690	0.12149	0.06467	0.03342	0.01364	0.00687	0.00345	0.00138	0.00069
0.80	0.31382	0.18611	0.08380	0.04373	0.02236	0.00906	0.00455	0.00228	0.00091	0.00046
0.90	0.23358	0.13223	0.05745	0.02958	0.01501	0.00606	0.00304	0.00152	0.00061	0.00030
1.00	0.16949	0.09259	0.03922	0.02000	0.01010	0.00407	0.00204	0.00102	0.00041	0.00020
1.10	0.12057	0.06415	0.02669	0.01352	0.00681	0.00273	0.00137	0.00069	0.00027	0.00014
1.20	0.08451	0.04412	0.01813	0.00915	0.00459	0.00184	0.00092	0.00046	0.00018	0.00008
1.30	0.05859	0.03018	0.01229	0.00619	0.00310	0.00124	0.00062	0.00031	0.00012	0.00006
1.40	0.04031	0.02057	0.00833	0.00418	0.00210	0.00084	0.00042	0.00021	0.00008	0.00004
1.50	0.02758	0.01398	0.00564	0.00283	0.00142	0.00057	0.00028	0.00014	0.00006	0.00003

Range is relative to meteorologic range. ρ is sky-background luminance ratio. See TABLE 135.

TABLE 138   COLOR PARAMETERS FOR DISTANT ACHROMATIC OBJECTS
(Section 12.4.6)

Atmosphere	Assumed contrast C_0	Distance r (km.)	Relative luminance L_r/L_A	Chromaticity coordinates x	y	Dominant wavelength $(m\mu)$	Excitation purity p_e
(a) Pure Air	(Horizon)		1.000	0.3101	0.3163		0
	3.0	10	3.675	0.3210	0.3287	576	0.062
		30	3.139	0.3395	0.3474	577	0.181
		50	2.718	0.3535	0.3587	578	0.230
		100	2.013	0.3691	0.3634	581	0.283
		200	1.378	0.3578	0.3415	586	0.196
		300	1.151	0.3382	0.3260	591	0.100
		400	1.064	0.3252	0.3198	597	0.049
		500	1.028	0.3177	0.3176	599	0.024
	2.0	10	2.783	0.3196	0.3271	576	0.053
		30	2.426	0.3351	0.3428	577	0.138
		50	2.145	0.3460	0.3514	578	0.192
		100	1.675	0.3560	0.3529	581	0.220
		200	1.252	0.3443	0.3344	587	0.140
		300	1.101	0.3295	0.3230	592	0.070
	1.0	10	1.892	0.3171	0.3242	575	0.038
		50	1.573	0.3337	0.3394	578	0.125
		200	1.135	0.3287	0.3261	587	0.076
	−0.5	10	0.554	0.2990	0.3037	475	0.055
		20	0.602	0.2931	0.2975	475	0.084
		30	0.644	0.2898	0.2949	477	0.098
		50	0.714	0.2876	0.2942	478	0.107
		100	0.831	0.2904	0.3006	480	0.090
		200	0.937	0.2997	0.3106	484	0.044
		300	0.975	0.3048	0.3145	487	0.020

−1.0		10	0.108	0.2272	0.2220	475	0.416
		20	0.203	0.2319	0.2303	476	0.384
		30	0.287	0.2365	0.2385	477	0.358
		50	0.427	0.2453	0.2529	478	0.308
		100	0.662	0.2640	0.2797	481	0.207
		200	0.874	0.2882	0.3043	484	0.092
		300	0.950	0.2992	0.3126	487	0.043
(b) Light Haze ($\alpha = 0.0080$. $\lambda^{-2.09}$ km^{-1})	3.0	20	2.738	0.3238	0.3287	579	0.070
		30	2.328	0.3365	0.3403	579	0.135
		40	2.014	0.3409	0.3430	580	0.157
		50	1.777	0.3420	0.3426	581	0.162
		100	1.208	0.3316	0.3300	584	0.095
		150	1.057	0.3191	0.3204	589	0.035
	2.0	20	2.159	0.3216	0.3268	579	0.059
		30	1.885	0.3315	0.3358	579	0.110
		40	1.677	0.3343	0.3374	580	0.124
		50	1.518	0.3345	0.3364	580	0.120
		100	1.139	0.3250	0.3256	584	0.066
		150	1.038	0.3162	0.3191	588	0.023
	1.0	50	1.259	0.3245	0.3281	580	0.070
	−0.5	10	0.621	0.3009	0.3070	477	0.045
		50	0.873	0.3004	0.3083	480	0.045
		100	0.966	0.3060	0.3137	483	0.018
		150	0.991	0.3086	0.3136	484	0.004
	−1.0	10	0.242	0.2674	0.2739	478	0.202
		50	0.745	0.2880	0.2982	480	0.100
		100	0.932	0.3016	0.3109	483	0.038
		150	0.981	0.3071	0.3148	485	0.013

Source. Middleton [12.22f].

TABLE 139 BEAUFORT SCALE OF WIND FORCE (Section 12.4.7.2)

Beaufort Number	General Description	Specifications	Limits of velocity 6 m above level ground				
			m/sec	km/hr	mph	knots	
0	Calm	Smoke rises vertically	Under 0.6	Under 1	Under 1	Under 1	
1	Light air	Wind direction shown by smoke drift but not by vanes	0.6–1.7	1–6	1–3	1–3	
2	Slight breeze	Wind felt on face; leaves rustle; ordinary vane moved by wind	1.8–3.3	7–12	4–7	4–6	
3	Gentle breeze	Leaves and twigs in constant motion; wind extends light flag	3.4–5.2	13–18	8–11	7–10	
4	Moderate breeze	Dust and loose paper; small branches are moved	5.3–7.4	19–26	12–16	10–14	
5	Fresh breeze	Small trees in leaf begin to sway	7.5–9.8	27–35	17–22	15–19	
6	Strong breeze	Large branches in motion; whistling in telegraph wires	9.9–12.4	36–44	23–27	19–24	
7	Moderate gale	Whole trees in motion	12.5–15.2	45–55	28–34	24–30	
8	Fresh gale	Twigs broken off trees; progress generally impeded	15.3–18.2	56–66	35–41	30–35	
9	Strong gale	Slight structural damage occurs; chimney pots re-moved	18.3–21.5	67–77	42–48	36–42	
10	Whole gale	Trees uprooted; considerable structural damage	21.6–25.4	78–90	49–56	42–49	
11	Storm	Very rarely experienced; widespread damage	25.5–29.0	91–104	57–67	49–56	
12	Hurricane		Above 29.0	Above 104	Above 67	Above 56	

Source. Petterssen [12.53].

D/d_0	Long exposure	Short Exposure	
		Far Field	Near Field
0.1	0.00978	0.00988	0.00997
0.5	0.1852	0.208	0.237
1.0	0.445	0.586	0.844
2.0	0.699	1.048	2.36
3.0	0.797	1.202	3.32
3.5	0.826	1.217	3.49
3.8	0.837	1.225	3.50
4.0	0.848	1.234	3.48
5.0	0.878	1.249	3.20
7.0	0.913	1.253	2.52
10	0.939	1.242	2.05
15	0.960	1.222	1.780
20	0.970	1.206	1.654
30	0.980	1.183	1.524
50		1.156	1.407
100		1.124	1.298
200		1.098	1.223
500			1.156
1000			1.120

Source. Fried [12.47].

TABLE 141 EXTINCTION COEFFI-
CIENTS OF PLASTIC (CROFON)
FIBERS (Section 13.2.1.4)

$\lambda(\mu m)$	$\alpha(cm^{-1})$
0.400	0.0137
0.450	0.0084
0.500	0.0057
0.550	0.0044
0.600	0.0037
0.650	0.0030
0.700	0.0033
0.733	0.0099
0.770	0.0032
0.805	0.0060
0.834	0.0053
0.946	0.0099
1.076	0.0165

Source. Brown and Derick [13.41b].

TABLE 142 FIRST 10 ZEROS OF BESSEL FUNCTIONS, J_{0-8} (Section 13.2.3.1)

m	j_{0m}	j_{1m}	j_{2m}	j_{3m}	j_{4m}	j_{5m}	j_{6m}	j_{7m}	j_{8m}
1	2.40483	3.83171	5.13562	6.38016	7.58834	8.77148	9.93611	11.08637	12.22509
2	5.52008	7.01559	8.41724	9.76102	11.06471	12.33860	13.58929	14.82127	16.03777
3	8.65373	10.17347	11.61984	13.01520	14.37254	15.70017	17.00382	18.28758	19.55454
4	11.79153	13.32369	14.79595	16.22347	17.61597	18.98013	20.32079	21.69154	22.94517
5	14.93092	16.47063	17.95982	19.40942	20.82693	22.21780	23.58608	24.93493	26.26681
6	18.07106	19.61586	21.11700	22.58273	24.01902	25.43034	26.82015	28.19119	29.54566
7	21.21164	22.76008	24.27011	25.74817	27.19909	28.62662	30.03372	31.42279	32.79580
8	24.35247	25.90367	27.42057	28.90835	30.37101	31.81172	33.23304	34.63709	36.02562
9	27.49348	29.04683	30.56920	32.06485	33.53714	34.98878	36.42202	37.83872	39.24045
10	30.63461	32.18968	33.71652	35.21867	36.69900	38.15987	39.60324	41.03077	42.44389

Source. Abramowitz and Stegun [13.51].

TABLE 143 MULTIMODE FIBER CHANNEL CAPACITY (Section 13.3.3.2)

Index distribution	Exponent b (Eq. 13.216)	Capacity Mbt · km/sec
Step		18
Eighth order	8	31
Quartic	4	59
Ideal	$2(1-\delta)$	13,800
Ibid, 5% error		870

Source. Miller, Marcatili, and Li [13.34].

TABLE 144 STRESS-OPTIC COEFFICIENTS FOR VARIOUS ISOTROPIC MATERIALS. REPRESENTATIVE VALUES (Section 14.1.3.1)

Material	B (brewster) $(10^{-12}\, m^2/N)$
Allyl diglycol	34.2
Bakelite	53
Cellulose Nitrate	$2 \sim 20$
Epoxy Resins	~ 56
Gelatine	$(1.7 \sim 20)10^3$
Methacrylates:	
Benzyl	45
Polycyclohexyl	5.9
Polymethyl	$-2.7 \sim 4.7$
Polyphenyl	39.8
Polycarbonate	78
Polyethylene	~ 2000
Polymethylene	~ 1700
Polystyrene (glassy)	$8 \sim 10$
(p-Cl)	23
Rubbers:	
Gutta-percha	3080
Natural rubber	2000
Polyurethane	3500

Source. Based on Refs. 14.2 and 14.43*b*
For stress-optic coefficients of optical glasses, see Table 71.

TABLE 145 STRESS-OPTIC COEFFICIENTS FOR SOME GLASSES & CRYSTALLINE MATERIALS (Section 14.1.3.2)

Material	Symmetry	p_{11}	p_{12}	p_{13}	p_{14}	p_{31}	p_{33}	p_{41}	p_{44}	p_{66}	λ (μm)
Glass: Silica	∞	0.121	0.270	p_{12}	0	p_{12}	p_{11}	0	$(p_{11}-p_{12})/2$	$(p_{11}-p_{12})/2$	0.633
#7070*	∞	0.113	0.23	"	0	"	"	0	"	"	0.633
#8363*	∞	0.196	0.185	"	0	"	"	0	"	"	0.633
KCl	m3m	0.215	0.159	"	0	"	"	0	−0.024	p_{44}	0.589
NaCl	m3m	0.137	0.178	"	0	"	"	0	−0.011	"	"
LiF	m3m	0.02	0.130	"	0	"	"	0	−0.045	"	"
MgO	m3m	−0.32	−0.08	"	0	"	"	0	−0.096	"	0.56
CaF$_2$	m3m	0.056	0.228	"	0	"	"	0	0.024	"	0.589
Y$_3$Al$_5$O$_{12}$(YAG)	m3m	−0.029	0.009	"	0	"	"	0	−0.062	"	0.633
Y$_3$Fe$_5$O$_{12}$(YIG)	m3m	0.025	0.073	"	0	"	"	0	0.041	"	"
Y$_3$Ga$_5$O$_{12}$(YGG)	m3m	0.091	0.019	"	0	"	"	0	0.079	"	"
SrTiO$_3$	m3m	0.15	0.095	"	0	"	"	0	0.072	"	"
Diamond	m3m	−0.31	0.09	"	0	"	"	0	−0.12	"	0.589
GaAs	$\bar{4}$3m	−0.165	−0.14	"	0	"	"	0	−0.072	"	1.15
GaP	$\bar{4}$3m	−0.151	−0.082	"	0	"	"	0	−0.074	"	0.633
Ammon. Alum.	m3	0.378	0.465	0.454	0	"	"	0	−0.009	"	0.589
Potass. Alum.	m3	0.275	0.354	0.345	0	"	"	0	−0.005	"	0.589
Calcite	$\bar{3}$2/m	0.095	0.189	0.215	0.006	0.309	0.178	0.010	−0.09	$(p_{11}-p_{12})/2$	0.589
α-Quartz	32	0.138	0.25	0.259	0.029	0.258	0.098	−0.042	−0.0685	"	
LiNbO$_3$	3m	0.036	0.072	0.092	0.070	0.178	0.088	0.155		"	
LiTaO$_3$	3m	0.0804	0.0804	0.094	0.031	0.086	0.150	0.024	0.022	"	
CdS	6mm	0.142	0.066	?	0	0.041	?	0	~−0.054	"	0.589

* Corning code.

Source. Based on Refs. 14.43c and 14.36.

TABLE 146 COLORS OBSERVED AT VARIOUS RETARDATIONS BETWEEN CROSSED POLARIZERS (Section 14.1.4.2)

Retardation (nm)	Color Transmitted	Retardation (nm)	Color Transmitted
0	black	843	yellow green
40	iron grey	866	green yellow
57	lavender grey	910	clear yellow
158	grey blue	948	orange
218	grey	998	brilliant orange red
234	green white	1101	dark violet red
259	off white	1128	bright blue violet
267	yellow white	1151	indigo
275	pale straw yellow	1258	green blue
281	straw yellow	1334	sea green
308	bright yellow	1376	brilliant green
332	brilliant yellow	1426	green yellow
430	brown yellow	1495	flesh colour
505	red orange	1534	crimson
536	red	1621	dull purple
551	deep red	1652	violet grey
565	purple	1682	grey blue
575	violet	1711	dull sea green
583	indigo	1744	blue green
664	sky blue	1811	bright green
728	green blue	1927	bright green grey
747	green	2007	white green
826	bright green	2048	flesh red

Source. Kuske and Robertson [14.2].

TABLE 147 PIEZOELECTRIC TRANSDUCERS FOR ACOUSTOOPTICS (Section 14.2.1.2)

Material	Point group	Density ρ (g/cm³)	Mode[a]	Orientation	Coupling factor	Dielectric constant	Freq. const. (GHz·μm)	Mechanical impedance (10⁶ kg/m²s)
$Ba_2NaNb_5O_{15}$	2mm	5.41	L	Z	0.57	32	3.075	33.3
			S	Y	0.25	227	1.83	19.8
$LiGaO_2$	2mm	4.19	L	Z	0.30	8.5	3.13	26.2
Li_2GeO_3	2mm	3.50	L	Z	0.31	12.1	3.25	22.8
SiO_2	32	2.65	L	X	0.098	4.58	2.87	15.2
			S	Y	0.137	4.58	1.925	10.2
$LiNbO_3$	3m	4.64	L	36°Y[b]	0.49	38.6	3.65	33.9
			S	163°Y[b]	0.62	42.9	2.24	20.8
			S	X	0.68	44.3	2.40	22.3
$LiTaO_3$	3m	7.45	L	47°Y[b]	0.29	42.7	3.70	55.2
			S	X	0.44	42.6	2.11	31.4
$LiIO_3$	6	4.5	L	Z	0.51	6	2.065	18.5
			S	Y	0.60	8	1.26	11.3
AlN	6mm	3.26	L	Z	0.20	8.5	5.2	34.0
ZnO	6mm	5.68	L	Z	0.27	8.8	3.18	36.2
			S	39°Y[c]	0.35	8.6	1.62	18.4
			S	Y	0.31	8.3	1.44	16.4
CdS	6mm	4.82	L	Z	0.15	9.5	2.25	21.7
			S	40°Y[c]	0.21	9.3	1.05	10.1
$Bi_{12}GeO_{20}$	23	9.22	L	(111)	0.19 (0.155)	38.6	1.65	30.4
			S	(110)	0.32 (0.235)	38.6	0.878	16.2

[a] L, longitudinal mode; S, shear mode.

[b] α°Y means that the plate-normal is rotated α° from the Y axis around the X axis. Warner et al., J. Acous. Soc. Am. **42**, 1223 (1967).

[c] The angle between the plate-normal and the c axis is given. Foster et al., IEEE Trans. **SU-15**, 28 (1968). See Refs. 14.6 and 14.7 for the preferred nomenclature.

Source From Uchida and Niizeki [14.8].

TABLE 148 SELECTED ACOUSTOOPTIC MATERIALS (Section 14.2.3)

Material	Optical transmission (μm)	Density ρ(g/cm^3)	Acoustic wave				Optical wave	
			Mode and propagation direction	Velocity V (km/sec)	Attenuation α_0(dB/ μsec GHz^2)	Polarization direction	Refractive index n	λ_o (μm)
Fused Silica	0.2–4.5	2.2	L	5.96	7.2	$\perp$	1.46	0.633
LiNbO$_3$	0.4–4.5	4.64	L[100]	6.57	0.1	35°y	2.2	0.633
TiO$_2$	0.45–6	4.23	L[100]	8.03	—	[010]	2.58	0.633
Sr$_{0.75}$Ba$_{0.25}$Nb$_2$O$_6$	0.4–6	5.4	L[001]	5.5	2.2	$\parallel$	2.3	0.633
Diamond	0.2–5	3.52	L[100]	17.5	2.6	$\parallel$	2.42	0.589
PbMoO$_4$	0.42–5.5	6.95	L[001]	3.63	5.5	$\perp$	2.39	0.633
TeO$_2$	0.35–5	6.0	L[001]	4.2	6.3	$\perp$	2.26	0.633
			S[110]	0.62	17.9	CIR	2.26	0.633
GaP	0.6–10	4.13	L[110]	6.32	3.8	$\perp$	3.31	0.633
As$_{12}$Se$_{55}$Ge$_{33}$	1–14	4.4	L	2.52	1.7	$\perp$	2.7	1.06
As$_2$Se$_3$	0.9–11	4.64	L	2.25	27.5	$\parallel$	2.89	0.633
GaAs	1–11	5.34	L[110]	5.15	15.5	$\parallel$	3.37	1.15
Tl$_3$AsS$_4$	0.6–12	6.2	L[001]	2.15	5	$\parallel$	2.83	0.633
Tl$_3$PSe$_4$	0.85–8	6.31	L[010]	2.0	30	$\parallel$	3.09	1.15
Ge	2–20	5.33	L[111]	5.50	16.5	$\parallel$	4.00	10.6

Source. Chang [14.11].

TABLE 149 FREQUENCY RANGES OF BIRE-
FRINGENT ACOUSTOOPTIC MEDIA
(Section 14.2.3.6)

Crystal	Acoustic wave[a] polarization	ν_{mn} GHz	ν_1[b] GHz	ν_{mx} GHz
α-quartz	L	0.082	1.52	28
	S-f	0.075	1.35	25
	S-s	0.048	0.89	16.5
Al_2O_3	L	0.140	2.9	62
	S-f	0.114	2.4	56
	S-s	0.064	1.3	28
$LiNbO_3$	L	1.060	7.45	27.2
TiO_2	S	2.440	10.7	49.3
Te (10.6 μm)	L	0.290	0.81	2.6

[a] L—longitudinal, S—shear mode, f—fast, s—slow. All
waves propagate in X direction.

[b] ν_1 is the frequency at which the deviation from the simple
Bragg condition becomes dominant. [see (12.143)].

Source. Dixon [14.19].

TABLE 150a FORM OF ELASTIC STIFFNESS (c_{pq}) AND COMPLIANCE (s_{pq}) MATRICES FOR THE 32 CRYSTAL CLASSES[a]
(Appendix 14.1)

	Triclinic 1 $\bar{1}$	Monoclinic 2 m 2/m	Orthorhombic mm2 222 mmm	Tetragonal 4 $\bar{4}$ 4/m	Tetragonal 422, 4mm $\bar{4}2m$, 4/mmm	Trigonal 3 $\bar{3}$	Trigonal 32 3m $\bar{3}m$	Hexagonal 6, 6, 6/m 6/mmm, 622 6mm, $\bar{6}2$	Cubic 23, m3 $\bar{4}3m$, 432 m3m	Isotropic
11	11	11	11	11	11	11	11	11	11	11
12	12	12	12	12	12	12	12	12	12	12
13	13	13[b]	13	13	13	13	13	13	12	12
14	14	.	.	.	.	14	14	.	.	.
15	15	15	.	.	.	15	.	.	.	.
16	16	.	.	16	.	.	.	.	.	.
22	22	22	22	11	11	11	11	11	11	11
23	23	23	23	13	13	13	13	13	12	12
24	24	.	.	.	.	−14	−14	.	.	.
25	25	25	.	.	.	−15	.	.	.	.
26	26	.	.	−16	.	.	.	.	.	.
33	33	33	33	33	33	33	33	33	11	11
34	34	.	.	.	.	.	.	.	.	.
35	35	35	.	.	.	.	.	.	.	.
36	36	.	.	.	.	.	.	.	.	.
44	44	44	44	44	44	44	44	44	44	c
45	45	.	.	.	.	.	.	.	.	.
46	46	46	.	.	.	−15[d]	.	.	.	.
55	55	55	55	44	44	44	44	44	44	c
56	56	.	.	.	.	14[d]	14[d]	.	.	.
66	66	66	66	66	66	c	c	c	44	c
21[e]	21[e]	13	9	11(7)	9(6)	15(7)	12(6)	9(5)	9(3)	9(2)

[a] All these matrices are symmetrical about the main diagonal.
[b] Dots indicate vanishing coefficients.
[c] Stiffness coefficient: $(c_{11}-c_{12})/2$. Compliance coefficient: $2(s_{11}-s_{12})$.
[d] The compliance coefficient is double the indicated value.
[e] The last row gives the number of nonvanishing coefficients in each class. The number of independent coefficients is given in parentheses, whenever it differs from that number.
Source. Based on Nye [14.33].

TABLE 150b FORM OF ELECTROOPTIC (r_{pi}) AND PIEZOELECTRIC (d_{pi}) MATRICES, FOR THE 32 CRYSTAL CLASSES[a] (Section 14.3.1.1) and Appendix 14.1)

Triclinic	Monoclinic		Orthorhombic		Tetragonal					Trigonal			Hexagonal					Cubic
1[d]	2[d]	m[d]	222	mm2[d]	4[d]	$\bar{4}$	422	4mm[d,f]	$\bar{4}2m$	3[d]	32	3m[d]	6[d]	$\bar{6}$	622	6mm[d,f]	$\bar{6}m2$	23 $4\bar{3}m$
11	·[b]	11	·	·	·	·	·	·	·	11	11	·	·	11	·	·	11	·
12	12	·	·	·	·	·	·	·	·	12	·	12	·	12	·	·	·	·
13	·	13	·	13	13	13	·	13	·	13	·	13	13	·	·	13	·	·
21	·	21	·	·	·	·	·	·	·	-11	-11	·	·	-11	·	·	·	·
22	22	·	·	·	·	·	·	·	·	-12	·	-12	·	-12	·	·	-11	·
23	·	23	·	23	13	-13	·	13	·	13	·	13	13	·	·	13	·	·
31	·	31	·	·	·	·	·	·	·	·	·	·	·	·	·	·	·	·
32	32	·	·	·	·	·	·	·	·	·	·	·	·	·	·	·	·	·
33	·	33	·	33	33	·	·	33	·	33	·	33	33	·	·	33	·	·
41	41	·	41	·	41	41	41	·	41	41	41	·	41	·	41	·	·	41
42	·	42	·	42	42	42	·	42	·	42	·	42	42	·	·	42	·	·
43	43	·	·	·	·	·	·	·	·	·	·	·	·	·	·	·	·	·
51	·	51	·	51	42	-42	·	42	·	42	·	42	42	·	·	42	·	·
52	52	·	52	·	-41	41	-41	·	41	-41	-41	·	-41	·	-41	·	·	41
53	·	53	·	·	·	·	·	·	·	·	·	·	·	·	·	·	·	·
61	61	·	·	·	·	·	·	·	·	12[c]	·	12[c]	·	12[c]	·	·	·	·
62	·	62	·	·	·	·	·	·	·	-11[c]	-11[c]	·	·	-11[c]	·	·	-11[c]	·
63	63	·	63	·	·	63	·	·	63	·	·	·	·	·	·	·	·	41
18[e]	8	10	3	5	7(4)	7(4)	2(1)	5(3)	3(2)	13(6)	5(2)	8(4)	7(4)	6(2)	2(1)	5(3)	3(1)	3(1)

[a] The following crystal symmetry classes have a vanishing matrix:

Triclinic: $\bar{1}$
Monoclinic: 2/m
Orthorhombic: mmm
Tetragonal: 4/m, 4/mmm
Trigonal: $\bar{3}$, $\bar{3}m$
Hexagonal: 6/m, 6/mmm
Cubic: m$\bar{3}$, 432, 4/m$\bar{3}m$
Isotropic

[b] Dots indicate vanishing coefficients.

[c] The zeros and identities apply to the piezoelectric matrix as well as to the electrooptic matrix, except that in the piezoelectric matrix the coefficients marked ° must be doubled.

[d] These crystal classes are piezoelectric under hydrostatic compression. They are also pyroelectric.

[e] The last row gives the number of nonvanishing coefficients in each class. The number of independent coefficients is given in parentheses, whenever it differs from that number.

[f] Note that classes 4mm and 6mm have identical matrices.

Source. Based on Nye [14.33].

TABLE 150c FORM OF PHOTOELASTICITY MATRICES FOR THE 32 CRYSTAL CLASSES (Appendix 14.1)

	Tri-clinic	Mono-clinic, Ortho-rhomb.	Tetragonal		Trigonal		Hexagonal		Cubic		Isotropic
			4, 4̄, 4/m	4mm, 4̄2m, 422, 4/mmm	3, 3̄	3m, 32, 3̄m	6, 6̄, 6/m	6̄m2, 6mm, 622, 6/mmm	23, m3	4̄3m, 432, m3m	
11	11	11	11	11	11	11	11	11	11	11	11
12	12	12	12	12	12	12	12	12	12	12	12
13	13	13	13	13	13	13	13	13	13	12	12
14	14	·	·	·	14	14	·	·	·	·	·
15	15	15	·	·	15	·	·	·	·	·	·
16	16	·	16	·	16	·	16	·	·	·	·
21	21	21	12	12	12	12	12	12	13	12	12
22	22	22	11	11	11	11	11	11	11	11	11
23	23	23	13	13	13	13	13	13	12	12	12
24	24	·	·	·	-14	-14	·	·	·	·	·
25	25	25	·	·	-15	·	·	·	·	·	·
26	26	·	-16	·	-16	·	-16	·	·	·	·
31	31	31	31	31	31	31	31	31	12	12	12
32	32	32	31	31	31	31	31	31	13	12	12
33	33	33	33	33	33	33	33	33	11	11	11
34	34	·	·	·	·	·	·	·	·	·	·
35	35	35	·	·	·	·	·	·	·	·	·
36	36	·	·	·	·	·	·	·	·	·	·
41	41	·	·	·	41	41	·	·	·	·	·
42	42	·	·	·	-41	-41	·	·	·	·	·
43	43	·	·	·	·	·	·	·	·	·	·
44	44	44	44	44	44	44	44	44	44	44	b
45	45	·	45	·	45	·	45	·	·	·	·
46	46	46	·	·	46	·	·	·	·	·	·
51	51	51	·	·	-46[c]	·	·	·	·	·	·
52	52	52	·	·	46[c]	·	·	·	·	·	·
53	53	53	·	·	·	·	·	·	·	·	·
54	54	·	-45	·	-45	·	-45	·	·	·	·
55	55	55	44	44	44	44	44	44	44	44	b
56	56	·	·	·	41[c]	41[c]	·	·	·	·	·
61	61	·	61	·	-16[c]	·	-16[c]	·	·	·	·
62	62	·	-61	·	16[c]	·	16[c]	·	·	·	·
63	63	·	·	·	·	·	·	·	·	·	·
64	64	64	·	·	-15	·	·	·	·	·	·
65	65	·	·	·	14	14	·	·	·	·	·
66	66	66	66	66	b	b	b	b	44	44	b
36[a]	20	18(10)	12(7)	30(12)	18(8)	18(8)	12(6)	12(4)	12(3)	12(2)	

[a] The last row gives the number of nonvanishing coefficients in each class. The number of independent coefficients is given in parentheses whenever it differs from that number.

[b] For the elastooptic constant [$p = \partial\Lambda/\partial s$ (dimensionless)], this indicates $(p_{11} - p_{12})/2$. For the piezooptic constant [$\pi = \partial\Lambda/\partial T\,(m^2/N)$], this indicates $(\pi_{11} - \pi_{12})$.

[c] Half the value is to be taken.

Source. Based on Nye [14.33].

TABLE 151. PRINCIPAL AXES TRANSFORMATIONS IN ELECTROOPTIC

Class	Field Direction	$a_1' - a_1$	$a_2' - a_2$
2	[010]	$r_{12}E_y + \dfrac{r_{52}^2 E_y^2}{a_1 - a_3 + (r_{12} - r_{32})E_y}$	$r_{22}E_y$
m	[001]	$r_{31}E_z + \dfrac{r_{53}^2 E_z^2}{a_1 - a_3 + (r_{31} - r_{33})E_z}$	$r_{23}E_z$
m	[100]	$r_{11}E_x + \dfrac{r_{51}^2 E_x^2}{a_1 - a_3 + (r_{11} - r_{31})E_x}$	$r_{21}E_x$
$mm2$	[001]	$r_{13}E_z$	$r_{23}E_z$
$mm2$	[010]	0	$\dfrac{r_{42}^2 E_y^2}{a_2 - a_3}$
$mm2$	[100]	$\dfrac{r_{51}^2 E_x^2}{a_1 - a_3}$	0
222	[001]	$\dfrac{r_{63}^2 E_z^2}{a_1 - a_2}$	$-\dfrac{r_{63}^2 E_z^2}{a_1 - a_2}$
222	[010]	$\dfrac{r_{52}^2 E_y^2}{a_1 - a_3}$	0
222	[100]	0	$\dfrac{r_{41}^2 E_x^2}{a_2 - a_3}$
4	[001]	$r_{13}E_z$	$r_{23}E_z$
4	[010]	$\dfrac{E_y^2(r_{51}^2 - r_{41}^2)}{a_1 - a_3}$	0
4	[100]	0	$\dfrac{E_x^2(r_{51}^2 + r_{41}^2)}{a_1 - a_3}$
4	[110]	$2\dfrac{(r_{41}^2 + r_{51}^2)(E_x^2 + E_y^2)}{a_1 - a_3}$	$-2\dfrac{(r_{41}^2 + r_{51}^2)(E_x^2 + E_y^2)}{a_1 - a_3}$
$\bar 4$	[001]	$E_z\sqrt{r_{13}^2 + r_{63}^2}$	$-E_z\sqrt{r_{13}^2 + r_{63}^2}$
$\bar 4$	[010]	$\dfrac{E_y^2(r_{41}^2 - r_{51}^2)}{a_1 - a_3}$	0
$\bar 4$	[100]	0	$\dfrac{E_x^2(r_{51}^2 + r_{41}^2)}{a_1 - a_3}$

a'_3-a_3	$\tan\theta$		
$r_{32}E_y - \dfrac{r_{52}^2 E_y^2}{a_1-a_3+(r_{12}-r_{32})E_y}$	$\dfrac{2r_{52}E_y}{a_1-a_3+(r_{12}-r_{32})E_y}$ [2]		
$r_{33}E_z - \dfrac{r_{53}^2 E_z^2}{a_1-a_3+(r_{31}-r_{33})E_z}$	$\dfrac{2r_{53}E_z}{a_1-a_3+(r_{31}-r_{33})E_z}$ [2]		
$r_{31}E_x - \dfrac{r_{51}^2 E_x^2}{a_1-a_3+(r_{11}-r_{31})E_x}$	$\dfrac{2r_{51}E_x}{a_1-a_3+(r_{11}-r_{31})E_x}$ [2]		
$r_{33}E_z$	0		
$-\dfrac{r_{42}^2 E_y^2}{a_2-a_3}$	$\dfrac{2r_{42}E_y}{a_2-a_3}$ [1]		
$-\dfrac{r_{51}^2 E_x^2}{a_1-a_3}$	$\dfrac{2r_{51}E_x}{a_1-a_3}$ [2]		
0	$\dfrac{2r_{63}E_z}{a_1-a_2}$ [3]		
$-\dfrac{r_{52}^2 E_y^2}{a_1-a_3}$	$\dfrac{2r_{52}E_y}{a_1-a_3}$ [2]		
$-\dfrac{r_{41}^2 E_x^2}{a_2-a_3}$	$\dfrac{2r_{41}E_x}{a_2-a_3}$ [1]		
$r_{33}E_z$	0		
$-\dfrac{E_y^2(r_{51}^2-r_{41}^2)}{a_1-a_3}$	$\dfrac{2r_{51}E_y}{a_1-a_3}$ [1]	$\dfrac{2r_{41}E_y}{a_3-a_1}$ [2]	
$-\dfrac{E_x^2(r_{51}^2+r_{41}^2)}{a_1-a_3}$	$\dfrac{2r_{41}E_x}{a_1-a_3}$ [1]	$\dfrac{2r_{51}E_x}{a_1-a_3}$ [2]	
0	$\dfrac{2(r_{41}E_x-r_{51}E_y)}{a_1-a_3}$ [1]	$\dfrac{2(r_{51}E_x-r_{41}E_y)}{a_1-a_3}$ [2]	
0	$\pm\dfrac{\pi}{4}$ [3]		
$-\dfrac{E_y^2(r_{41}^2-r_{51}^2)}{a_1-a_3}$	$\dfrac{2r_{51}E_y}{a_3-a_1}$ [1]	$\dfrac{2r_{41}E_y}{a_1-a_3}$ [2]	
$-\dfrac{E_x^2(r_{51}^2+r_{41}^2)}{a_1-a_3}$	$\dfrac{2r_{41}E_x}{a_1-a_3}$ [1]	$\dfrac{2r_{51}E_x}{a_1-a_3}$ [2]	

TABLE 151

Class	Field Direction	$a_1' - a_1$	$a_2' - a_2$
$\bar{4}$	[110]	$-2\dfrac{(r_{41}^2 + r_{51}^2)(E_x^2 + E_y^2)}{a_1 - a_3}$	$2\dfrac{(r_{41}^2 + r_{51}^2)(E_x^2 + E_y^2)}{a_1 - a_3}$
$\bar{4}2m$	[001]	$r_{63}E_z$	$-r_{63}E_z$
$\bar{4}2m$	[010]	$\dfrac{r_{41}^2 E_y^2}{a_1 - a_3}$	0
$\bar{4}2m$	[100]	0	$\dfrac{r_{41}^2 E_x^2}{a_1 - a_3}$
$\bar{4}2m$	[110]	$\dfrac{r_{41}^2(E_x^2 + E_y^2)}{a_1 - a_3}$	$-\dfrac{r_{41}^2(E_x^2 + E_y^2)}{a_1 - a_3}$
422	[001]	0	0
422	[010]	$-\dfrac{r_{41}^2 E_y^2}{a_1 - a_3}$	0
422	[100]	0	$-\dfrac{r_{41}^2 E_x^2}{a_1 - a_3}$
422	[110]	$\dfrac{r_{41}^2(E_x^2 + E_y^2)}{a_1 - a_3}$	$-\dfrac{r_{41}^2(E_x^2 + E_y^2)}{a_1 - a_3}$
4mm	[001]	$r_{13}E_z$	$r_{13}E_z$
4mm	[010]	0	$\dfrac{r_{51}^2 E_y^2}{a_1 - a_3}$
4mm	[100]	$\dfrac{r_{51}^2 E_x^2}{a_1 - a_3}$	0
4mm	[110]	$\dfrac{r_{51}^2(E_x^2 + E_y^2)}{a_1 - a_3}$	$-\dfrac{r_{51}^2(E_x^2 + E_y^2)}{a_1 - a_3}$
3	[0001]	$r_{13}E_z$	$r_{13}E_z$
32	[0001]	0	0
32	[11$\bar{2}$0]	$r_{11}E_x$	$-r_{11}E_x + \dfrac{E_x(a_3 r_{11} + r_{41}^2 E_x)}{a_1 - r_{11}E_x}$
3m	[0001]	$r_{13}E_z$	$r_{13}E_z$
3m	[11$\bar{2}$0]	$-r_{22}E_y$	$r_{22}E_y + \dfrac{r_{51}^2 E_y^2}{a_1 - a_3 + r_{22}E_y}$
$\bar{6}$	[0001]	0	0

$a_3'-a_3$	tan θ	
0	$\dfrac{2(r_{41}E_x - r_{51}E_y)}{a_1-a_3}$ [1]	$\dfrac{2(r_{51}E_x + r_{41}E_y)}{a_1-a_3}$ [2]
0	$\pm\dfrac{\pi}{4}$ [3]	
$-\dfrac{r_{41}^2 E_y^2}{a_1-a_3}$	$\dfrac{2r_{41}E_y}{a_1-a_3}$ [2]	
$-\dfrac{r_{41}^2 E_x^2}{a_1-a_3}$	$\dfrac{2r_{41}E_x}{a_1-a_3}$ [1]	
0	$\dfrac{2r_{41}E_x}{a_1-a_3}$ [1]	$\dfrac{2r_{41}E_y}{a_1-a_3}$ [2]
0	0	
$\dfrac{r_{41}^2 E_y^2}{a_1-a_3}$	$\dfrac{2r_{41}E_y}{a_3-a_1}$ [2]	
$-\dfrac{r_{41}^2 E_x^2}{a_1-a_3}$	$\dfrac{2r_{41}E_x}{a_1-a_3}$ [1]	
0	$\dfrac{2r_{41}E_x}{a_1-a_3}$ [1]	$\dfrac{2r_{41}E_y}{a_3-a_1}$ [2]
$r_{33}E_z$	0	
$-\dfrac{r_{51}^2 E_y^2}{a_1-a_3}$	$\dfrac{2r_{51}E_y}{a_1-a_3}$ [1]	
$-\dfrac{r_{51}^2 E_x^2}{a_1-a_3}$	$\dfrac{2r_{51}E_x}{a_1-a_3}$ [2]	
0	$\dfrac{2r_{51}E_y}{a_1-a_3}$ [1]	$\dfrac{2r_{51}E_x}{a_1-a_3}$ [2]
$r_{33}E_z$	0	
0	0	
$-\dfrac{E_x(a_3 r_{11}+r_{41}^2 E_x)}{a_1 - r_{11}E_x}$	$\dfrac{2r_{41}E_x}{a_1-a_3+r_{41}E_x}$ [1]	
$r_{33}E_z$	0	
$-\dfrac{r_{51}^2 E_y^2}{a_1-a_3+r_{22}E_y}$	$\dfrac{2r_{51}E_y}{a_1-a_3+r_{22}E_y}$ [1]	
0	0	

TABLE 151

Class	Field Direction	$a_1' - a_1$	$a_2' - a_2$
$\bar{6}$	$[10\bar{1}0]$	$-E_y\sqrt{r_{11}^2+r_{22}^2}$	$E_y\sqrt{r_{11}^2+r_{22}^2}$
$\bar{6}$	$[11\bar{2}0]$	$E_x\sqrt{r_{11}^2+r_{22}^2}$	$-E_x\sqrt{r_{11}^2+r_{22}^2}$
$\bar{6}m2$	$[0001]$	0	0
$\bar{6}m2$	$[10\bar{1}0]$	$r_{11}E_y$	$-r_{11}e_y$
$\bar{6}m2$	$[11\bar{2}0]$	$r_{11}E_x$	$-r_{11}E_x$
6	$[0001]$	$r_{13}E_z$	$r_{13}E_z$
6	$[10\bar{1}0]$	$\dfrac{E_y^2(r_{41}^2+r_{51}^2)}{a_1-a_3}$	0
$6mm$	$[11\bar{2}0]$	$\dfrac{r_{51}^2E_x^2}{a_1-a_3}$	0
622	$[0001]$	0	0
622	$[10\bar{1}0]$	$\dfrac{r_{41}^2E_y^2}{a_1-a_3}$	0
622	$[11\bar{2}0]$	0	$\dfrac{r_{41}^2E_x^2}{a_1-a_3}$
$6mm$	$[0001]$	$r_{13}E_z$	$r_{13}E_z$
$6mm$	$[10\bar{1}0]$	0	$\dfrac{r_{51}^2E_y^2}{a_1-a_3}$
$6mm$	$[11\bar{2}0]$	$\dfrac{r_{51}^2E_x^2}{a_1-a_3}$	0
$23,\ \bar{4}3m$	$[001]$	$r_{41}E_z$	$-r_{41}E_z$
$23,\ \bar{4}3m$	$[010]$	$r_{41}E_y$	0
$23,\ \bar{4}3m$	$[100]$	0	$r_{41}E_x$
$23,\ \bar{4}3m$	$[110]$	$r_{41}\sqrt{E_x^2+E_y^2}$	$-\sqrt{E_x^2+E_y^2}$

This table lists in Columns 3–5, the *change* in the x^2, y^2, z^2-coefficients of the ellipsoid in the principal coordinate system (a_1, a_2, a_3, respectively, must be added to the listed expression to yield the coefficient of the ellipsoid). The last column lists the slope that the principal axis makes with the original unprimed axis. [The axis the slope refers to is listed next to it in brackets.] a_1, a_2, a_3 are the coefficients of x^2, y^2, and z^2, respectively, of the original ellipsoid.

Source. After Vlokh and Zheludev [14.35].

	$a_3' - a_3$	$\tan\theta$	
	0	$\pm\dfrac{\pi}{4}$ [3]	
	0	$\pm\dfrac{\pi}{4}$ [3]	
	0	0	
	0	$\pm\dfrac{\pi}{4}$ [3]	
	0	0	
$r_{33}E_z$		0	
$-\dfrac{E_y^2(r_{41}^2+r_{51}^2)}{a_1-a_3}$		$\dfrac{2r_{51}E_y}{a_1-a_3}$ [1]	$\dfrac{2r_{41}E_y}{a_3-a_1}$ [2]
$-\dfrac{r_{51}^2E_x^2}{a_1-a_3}$		$\dfrac{2r_{51}E_x}{a_1-a_3}$ [2]	
	0	0	
$-\dfrac{r_{41}^2E_y^2}{a_1-a_3}$		$\dfrac{2r_{41}E_y}{a_3-a_1}$ [1]	
$-\dfrac{r_{41}^2E_x^2}{a_1-a_3}$		$\dfrac{2r_{41}E_x}{a_1-a_3}$ [1]	
$r_{33}E_z$		0	
$-\dfrac{r_{51}^2E_y^2}{a_1-a_3}$		$\dfrac{2r_{51}E_y}{a_1-a_3}$ [1]	
$-\dfrac{r_{51}^2E_x^2}{a_1-a_3}$		$\dfrac{2r_{51}E_x}{a_1-a_3}$ [2]	
	0	$\pm\dfrac{\pi}{4}$ [3]	
$-r_{41}E_y$		$\pm\dfrac{\pi}{4}$ [2]	
$-r_{41}E_x$		$\pm\dfrac{\pi}{4}$ [1]	
	0	$\pm\dfrac{\pi}{4}$ [1], [2]	

TABLE 152 CHARACTERISTICS OF ELECTROOPTIC MATERIALS (Section 14.3.1.2)

Material[a]	Sym.		T_C (K)	r_{13} pm/V	r_{22} pm/V	r_{33} pm/V	r_{41} pm/V	r_{42} pm/V	r_{63} pm/V	r_c^b pm/V	n_o	n_e	ε_1	ε_3
KH_2PO_4 (KDP)	$\bar{4}2m$	[T]	123				8.6		−10.5		1.51	1.47	42	
		[S]							9.7				44	
KD_2PO_4 (DKDP)	$\bar{4}2m$	[T]	222				8.8		26.4		1.51	1.47		50
		[S]												
$NH_4H_2PO_4$ (ADP)	$\bar{4}2m$	[T]	148				24.5		−8.5		1.53	1.48	58	48
		[S]							5.5				56	15
KH_2AsO_4 (KDA)	$\bar{4}2m$	[T]	97				12.5		10.9		1.57	1.52	58	14
		[S]											54	21
RbH_2AsO_4 (RDA)	$\bar{4}2m$	[T]	110						13		1.56	1.52	53	19
		[S]											41	27
													39	24
$BaTiO_3$	$4mm$	[T]	393					1640		108	2.44	2.37	3000	170
		[S]		8		28		820		23			2000	100
$LiNbO_3$	$3m$	[T]	1470	8.6	7	30.8		28		19	2.286	2.2	78	32
		[S]			3.4					21			43	28
$LiTaO_3$	$3m$	[T]	890	7.9		35.8		20		22	2.176	2.18		47
		[S]		1.4	(1)	2.6				28				43
ZnO	$6mm$	[S]		0.92		1.85					2.0	2.015	8.15	
ZnS	$\bar{4}3m$	[T]					2.1				2.315		16	
		[S]											12.5	
ZnSe	$6mm$	[S]						2			2.705	2.709		
	$\bar{4}3m$	[T]									2.66			
CdS	$6mm$	[T]		1.1		2.4	1.6	3.7			2.743	2.726	10.6	7.8
		[S]											8	7.7

[a] T and S indicate constant stress and strain, respectively. Values are at wavelengths ranging from 0.546 μm–0.633 μm.

[b] $r_c = \gamma_{33} - (n_1/n_3)^3 r_{13}$.

Source. Data based on Kaminow and Turner [14.37].

TABLE 153*a* VERDET'S CONSTANTS FOR SOME LIQUIDS (rd/m · T)[a]
(Section 14.4.2.2)

Material	Formula	Wavelength (μm)					
		0.578	0.6	0.8	1.0	1.5	2
Water	H_2O	4.0	3.7	2.0	1.28		
Benzine	C_6H_6	9.0	8.2	4.5	2.8	1.13	0.64
Carbon disulfide	CS_2	12.5	11.5	6.2	3.9	1.69	0.90
Carbon tetrachloride	CCl_4	4.9	4.7	2.6	1.66	0.73	0.38

TABLE 153*b*   VERDET'S CONSTANTS FOR SOME SOLIDS (rd/m · T)[a]
(Section 14.4.2.2)

Material	Formula	Wavelength (μm)			
		0.436	0.546	0.578	0.589
Quartz	SiO_2	7.6	4.8	4.3	
Zinc Sulfide	ZnS	191	95	82	
Salt	NaCl				9.6
Flint Glass					23

[a] Values at temperatures near 20°C. To convert to the common units (min/G · cm) divide values in this table by 291.

Source. Data from compilations of Condon [14.44*a*] and Cappeller [14.60].

TABLE 154 CHARACTERISTICS OF MAGNETOOPTIC RECORDING MATERIALS (Section 14.4.2.3)

Material	$4\pi M_s$ (G)	Specific sat. rotation F_s (deg/μm)	Write-erase temp.	Resolution (lines/mm)	Sensitivity (nJ/μm²)	Write erase field (Oe)	Absorp. coeff. α (10^7 m^{-1})	Fig. of merit $2F/\alpha$ (deg.)	Operating wavelength (μm)	Operating temp.
MnBi[a]	7,200	90	360°C	2000	0.3–1	600	5	3.6	0.63	20°C
MnBi[a]	5,500	35	180°C	2000	0.1–0.3	600	5	1.4	0.63	20°C
Mn$_{.8}$Ti$_{.2}$Bi[a]	—	21	125°C	—	0.1–0.3	—	5.3	0.8	0.63	20°C
MnGe$_3$[b]	12,400	14	37°C	—	—	—	3.5	0.8	0.55	0°C
MnAlGe	3,600	13	245°C	>40	3	500	4.7	0.54	0.63	100°C
MnGaGe	4,170	8.1	185°C	—	—	500	5.8	0.28	0.63	20°C
CrTe[b]	1,015	5	61°C	—	—	—	2	0.5	0.55	20°C
MnSb[b]	9,600	30	300°C	—	—	—	7.5	0.8	0.55	20°C
MnAs[b]	7,900	4.5	45°C	—	—	—	4.5	0.2	0.60	20°C
Fe$_5$Si$_3$	7,500	4.8	108°C	—	—	—	3.7	0.26	0.55	20°C
EuO[b]	23,700	32	70°K	1000	0.01–0.1	50	0.86	7.5	0.60	77°K
EuO(Fe)[b]	19,200	33	180°K	—	—	—	2.2	~3	0.70	6°K
EuS[b]	14,000	478	16°K	—	—	—	1.65	58	0.69	4.2°K
EuSe[b]	13,200	14	7°K	—	—	—	0.0008	3600	0.75	200°K
CoCr$_2$S$_4$	2,300	1210	225°K	—	—	~100	2.2	110	1.00	20°C
GdCo	3,450	18	120°C	—	0.5	~500	8	0.45	0.45–0.82	15°+3°C
GdIG	7,300	0.7	25°C	50–100	0.1–0.8	~100	0.06	2.3	0.63	(−1°)+28°C
GdIG (Film)	5,800	0.6	−1°C	300	1.25	500	0.06	2	0.60	20°C
Co—P—(Ni-Fe)[b]	—	—	150°C	200	1	20	—	—	—	20°C
NeFe—PdCo—Co[b]	—	—	~140°C	40	0.5	20	—	—	—	20°C
CrO$_2$	1,600	—	134°C	100	2.3	50	—	0.1	0.50	20°C

[a] The first row refers to the low temperature phase of MnBi and the second to its high temperature phase; for MnTiBi, only the high temperature phase is given.
[b] In these media, the easy direction of magnetization lies in the memory plane. Therefore, full utilization of readout figure of merit cannot be realized.
Source. After D. Chen [14.61].

TABLE 155 COMPARISON OF ELECTRO- AND MAGNETOOPTIC EFFECTS
(Section 14.4.4)

Field/light direction	Order	Description of effect	Name of effect	
			Electrooptic	Magnetooptic
∥	1st	Polar. plane rotated, on transm.	—	Faraday
∥	1st	Polar. plane rotated, on reflect.	—	Kerr, m–o
⊥	1st	Birefringence induced	Pockels	Voigt
⊥	2nd	Birefringence induced	Kerr, e–o	Cotton–Mouton
Any	—	Absorption edge shifted toward long λ	Franz–Keldysh	—
∥ & ⊥	All	Spectrum split	Stark	Zeeman

TABLE 156   CHARACTERISTICS OF LIQUID CRYSTALS (Section 14.5.1.5)

Name	Temp. Range (°C)	$\varepsilon_\parallel$	$\varepsilon_\perp$	Ref.
Nematic				
4-methoxybenzylidene-4′-n-butylaniline (MBBA)	21–47	4.6[a]	5.1[a]	69
4-Ethoxybenzylidene-4′-n-butylaniline (EBBA)	35–79	4.9[a]	4.6[a]	
4-Methoxy-4′-n-butylazoxybenzene	19–76			69
p-Azoxyanisole (PAA)	117–137			69
4-n-Hexyl-4′-cyanobiphenyl	14–28			69
n(4′-Ethoxybenzylidene)4-aminobenzonitride (PEBAB)	106–128	21	7	75
High resistivity, nonscattering mixture No. X11643	12–100	5.2	6.8	EK[a]
Dynamic scattering mixture, No. 11643	9–99	5.2	6.8	EK[a]
Cholesteric				
Cholesteryl nonanoate	145–179			69
Smectic A				
P-Phenylbenzal-p-aminoethylbenzoate	121–131			69
Smectic B				
Ethyl-4-ethoxybenzol-4′-aminocinnamate	77–116			69

[a] From "Eastman Liquid Crystal Products," Eastman Kodak Co., Rochester (1973).

n	I	II	III	IV	V	VI	VII	XIV	XV	XVI	XVII
5	—	—	—	—	—	—	56.3–56.0	—	—	—	—
6	—	72.7–71.5	83.0–82.0	—	—	56.0–53.0	68.5–68.0	—	—	—	34.3–31.3
7	—	74.0–72.9	—	—	—	68.0–67.0	64.3–64.1	—	—	—	47.8–47.0
8	—	—	—	50.3–48.1	41.0–37.3	—	67.3–66.7	—	—	—	52.4–51.9
9	77.9–76.5	—	39.5–39.3	—	50.0–45.5	—	—	—	—	—	55.4–55.2
10	77.1–76.1	—	—	—	51.0–49.4	—	70.2–70.1	—	—	—	—
11	—	—	47.0–44.2	49.9–48.7	52.3–51.5	—	70.5	61.9–61.1	—	39.5–39.2	—
12	—	—	44.6–44.3	—	53.0–52.1	—	70.5	63.6–63.1	26.6–26.1	43.5–42.1	—
13	—	—	50.0–48.1	51.1–49.8	53.1–52.5	—	68.7	—	—	43.6–42.6	—
14	—	—	48.7–48.3	47.35	53.9–52.9	—	67.8	—	36.0–35.6	—	—
15	—	—	53.0–52.0	52.4–51.6	54.9–54.1	—	66.5	—	—	41.4–40.3	—
16	—	—	51.7–51.3	49.6–49.4	55.7–55.0	—	65.4	—	44.0–41.7	—	—
17	—	—			56.2–55.6		64.5				
18	—	—			56.4–55.8	57.8–57.4	63.2–63.1				
19	—	—			55.1–49.9		62.5–62.3				
20	—	—				57.0–56.0	60.6–60.4				

[a]The temperature values were obtained by optical microscopy. n = alkyl or acyl chain length, respectively. I, cholesteryl alkanoates; II, S-cholesteryl alkanethioates; III, cholesteryl ω-phenylalkanoates; IV, cholesteryl ω-phenylalkanethioates; V, S-cholesteryl ω-phenylalkanethioates; VI, cholesteryl S-alkyl thiocarbonates; VII, S-cholesteryl alkyl thiocarbonates; XIV, 5α-cholestan-3β-yl alkanoates; XV, 5α-cholestan-3β-yl ω-phenylalkanoates; XVI, 5α-cholestan-3β-yl alkyl carbonates; XVII, 5α-cholestan-3β-yl S-alkyl thiocarbonates.

Source. W. Elser and R. D. Ennulat, in Adv. Liq. Cryst. Vol. 2, G. H. Brown, ed., Academic, New York, 1976, p. 126.

TABLE 158 SCHEMATIC EYE (Section 15.1.2)

	Full Theoretical Eye		Simplified Eye	
	Unaccommodated	Accommodated	Unaccommodated	Accommodated
Refractive index				
Cornea	1.3771	1.3771	1.336	1.336
Aqueous humor	1.3374	1.3374	1.336	1.336
Crystalline lens (total index)	1.42	1.427	1.4208	1.4260
Vitreous humor	1.336	1.336	1.336	1.336
Distance along axis from corneal pole (mm)				
Posterior surface of the cornea	0.55	0.55	—	—
Anterior surface of the lens	3.6	3.2	6.3740	5.7763
Posterior surface of the lens	7.6	7.7	6.3740	5.7763
Radius of curvature (mm)				
Anterior surface of the cornea	7.8	7.8	8	8
Posterior surface of the cornea	6.5	6.5	—	—
Anterior surface of the lens	10.2	6.0	10.2	6
Posterior surface of the lens	6	-5.5	-6	-5.5
Dioptric power				
Anterior surface of the cornea	48.3462	48.3462	42	42
Posterior surface of the cornea	-6.1077	-6.1077	—	—
Anterior surface of the lens	8.0980	14.9333	8.3097	15.0049
Posterior surface of the lens	14	16.5455	14.1265	16.3690
Cornea				
Power	42.3564	42.3564	42	42
Position of principal points : object	-0.0576	-0.0576	0	0
: image	-0.0597	-0.0597	0	0
Focal length: object	-23.6092	-23.6092	-23.8095	-23.8095
	31.5749	31.5749	31.8095	31.8095

Crystalline lens

Power	21.7787	30.6996	22.4362	31.3739
Position of principal				
points : object	6.0218	5.4730	6.3740	5.7763
: image	6.2007	5.6506	6.3740	5.7763
Focal length: object	−61.4087	−43.5641	−59.5466	−42.5832
: image	61.3444	43.5185	59.5466	42.5832

Complete eye

Power	59.9404	67.6767	59.9404	67.6767
Position of principal				
points : object	1.5946	1.8190	1.7858	2.0043
: image	1.9078	2.1915	1.9078	2.1915
Position of focal				
points : object	−15.0887	−12.9571	−14.8974	−12.7718
: image	24.1965	21.9325	24.1965	21.9325
Focal length: object	−16.6832	−14.7761	−16.6832	−14.7761
: image	22.2888	19.7409	22.2888	19.7409
Position of nodal				
points : object	7.2001	6.7838	7.3914	6.9691
: image	7.5133	7.1563	7.5133	7.1563
Accommodation (diopters)	0	6.9633	0	6.9633
Distance along				
axis : image nodal				
point to image focal point				
(at retina), (mm)	16.6832	14.7762	16.6832	14.7762

Source. Wyszecki and Stiles [15.3a].

TABLE 159 PHOTOPIC LUMINOSITY (Section 15.1.3.3)

λ μm	0	1	2	3	4	5	6	7	8	9	Exp
0.36	0.3917	0.4394	0.4930	0.5532	0.6208	0.6965	0.7813	0.8767	0.9840	1.1040	−5
0.37	0.1239	0.1389	0.1556	0.1744	0.1958	0.2202	0.2484	0.2804	0.3153	0.3522	−4
0.38	0.3900	0.4283	0.4691	0.5159	0.5718	0.6400	0.7234	0.8221	0.9351	1.0610	−4
0.39	0.1200	0.1350	0.1515	0.1702	0.1918	0.2170	0.2469	0.2812	0.3185	0.3573	−3
0.40	0.3960	0.4337	0.4730	0.5179	0.5722	0.6400	0.7246	0.8255	0.9412	1.0700	−3
0.41	0.1210	0.1362	0.1531	0.1720	0.1935	0.2180	0.2455	0.2764	0.3118	0.3526	−2
0.42	0.4000	0.4546	0.5159	0.5829	0.6546	0.7300	0.8087	0.8909	0.9768	1.0660	−2
0.43	0.1160	0.1257	0.1358	0.1463	0.1572	0.1684	0.1801	0.1921	0.2045	0.2172	−1
0.44	0.2300	0.2429	0.2561	0.2696	0.2835	0.2980	0.3131	0.3288	0.3452	0.3623	−1
0.45	0.3800	0.3985	0.4177	0.4377	0.4584	0.4800	0.5024	0.5257	0.5498	0.5746	−1
0.46	0.6000	0.6260	0.6528	0.6804	0.7091	0.7390	0.7702	0.8027	0.8367	0.8723	−1
0.47	0.9098	0.9492	0.9905	1.0340	1.0790	1.1260	1.1750	1.2270	1.2800	1.3350	−1
0.48	0.1390	0.1447	0.1505	0.1565	0.1627	0.1693	0.1762	0.1836	0.1913	0.1994	0
0.49	0.2080	0.2171	0.2267	0.2369	0.2475	0.2586	0.2702	0.2823	0.2951	0.3086	0
0.50	0.3230	0.3384	0.3547	0.3717	0.3893	0.4073	0.4256	0.4443	0.4634	0.4829	0
0.51	0.5030	0.5236	0.5445	0.5657	0.5870	0.6082	0.6293	0.6503	0.6709	0.6908	0
0.52	0.7100	0.7282	0.7455	0.7620	0.7778	0.7932	0.8081	0.8225	0.8363	0.8495	0
0.53	0.8620	0.8738	0.8850	0.8955	0.9054	0.9149	0.9237	0.9321	0.9399	0.9472	0
0.54	0.9540	0.9603	0.9660	0.9713	0.9760	0.9803	0.9841	0.9875	0.9903	0.9928	0
0.55	0.9950	0.9967	0.9981	0.9991	0.9997	1.0000	0.9999	0.9993	0.9983	0.9969	0
0.56	0.9950	0.9926	0.9897	0.9864	0.9827	0.9786	0.9741	0.9692	0.9639	0.9581	0
0.57	0.9520	0.9455	0.9385	0.9312	0.9235	0.9154	0.9070	0.8983	0.8892	0.8798	0
0.58	0.8700	0.8599	0.8494	0.8386	0.8276	0.8163	0.8048	0.7931	0.7812	0.7692	0
0.59	0.7570	0.7448	0.7324	0.7200	0.7075	0.6949	0.6822	0.6695	0.6567	0.6438	0

	0	1	2	3	4	5	6	7	8	9	
0.60	0.6310	0.6182	0.6053	0.5925	0.5796	0.5668	0.5540	0.5411	0.5284	0.5156	0
0.61	0.5030	0.4905	0.4780	0.4657	0.4534	0.4412	0.4291	0.4170	0.4050	0.3930	0
0.62	0.3810	0.3689	0.3568	0.3448	0.3328	0.3210	0.3093	0.2979	0.2866	0.2756	0
0.63	0.2650	0.2548	0.2449	0.2353	0.2261	0.2170	0.2082	0.1995	0.1912	0.1830	0
0.64	0.1750	0.1672	0.1596	0.1523	0.1451	0.1382	0.1315	0.1250	0.1188	0.1128	0
0.65	1.0700	1.0150	0.9619	0.9112	0.8626	0.8160	0.7712	0.7283	0.6871	0.6477	−1
0.66	0.6100	0.5740	0.5396	0.5067	0.4755	0.4458	0.4176	0.3908	0.3656	0.3420	−1
0.67	0.3200	0.2996	0.2808	0.2633	0.2471	0.2320	0.2180	0.2050	0.1928	0.1812	−1
0.68	1.7000	1.5900	1.4840	1.3810	1.2830	1.1920	1.1070	1.0270	0.9533	0.8846	−2
0.69	0.8210	0.7624	0.7085	0.6591	0.6138	0.5723	0.5343	0.4996	0.4676	0.4380	−2
0.70	0.4102	0.3838	0.3589	0.3354	0.3134	0.2929	0.2738	0.2560	0.2393	0.2237	−2
0.71	0.2091	0.1954	0.1825	0.1704	0.1590	0.1484	0.1384	0.1291	0.1204	0.1123	−2
0.72	1.0470	0.9766	0.9111	0.8501	0.7932	0.7400	0.6901	0.6433	0.5995	0.5585	−3
0.73	0.5200	0.4839	0.4501	0.4183	0.3887	0.3611	0.3354	0.3114	0.2892	0.2685	−3
0.74	0.2492	0.2313	0.2147	0.1993	0.1850	0.1719	0.1598	0.1486	0.1383	0.1288	−3
0.75	1.2000	1.1190	1.0430	0.9734	0.9085	0.8480	0.7915	0.7386	0.6892	0.6430	−4
0.76	0.6000	0.5598	0.5223	0.4872	0.4545	0.4240	0.3956	0.3692	0.3445	0.3215	−4
0.77	0.3000	0.2799	0.2611	0.2436	0.2272	0.2120	0.1978	0.1845	0.1722	0.1606	−4
0.78	1.4990	1.3990	1.3050	1.2180	1.1360	1.0600	0.9886	0.9217	0.8592	0.8009	−5
0.79	0.7466	0.6960	0.6488	0.6049	0.5639	0.5258	0.4902	0.4570	0.4260	0.3972	−5
0.80	0.3703	0.3452	0.3218	0.3000	0.2797	0.2608	0.2431	0.2267	0.2113	0.1970	−5
0.81	1.8370	1.7120	1.5960	1.4880	1.3870	1.2930	1.2060	1.1240	1.0480	0.9771	−6
0.82	0.9109	0.8493	0.7917	0.7381	0.6881	0.6415	0.5981	0.5576	0.5198	0.4846	−6

TABLE 160   SCOTOPIC LUMINOSITY (Section 15.1.3.3)

λ μm	0	1	2	3	4	5	6	7	8	9	Exp
0.38	0.5893	0.6647	0.7521	0.8537	0.9716	1.1080	1.2680	1.4530	1.6680	1.9180	-3
0.39	0.2209	0.2547	0.2939	0.3394	0.3921	0.4531	0.5236	0.6049	0.6984	0.8059	-2
0.40	0.9292	1.0700	1.2310	1.4130	1.6190	1.8520	2.1130	2.4050	2.7300	3.0890	-2
0.41	0.3483	0.3916	0.4386	0.4897	0.5448	0.6041	0.6677	0.7357	0.8080	0.8849	-1
0.42	0.9661	1.0510	1.1410	1.2350	1.3340	1.4350	1.5410	1.6510	1.7640	1.8790	-1
0.43	0.1998	0.2119	0.2243	0.2369	0.2496	0.2625	0.2755	0.2886	0.3017	0.3149	0
0.44	0.3281	0.3412	0.3543	0.3673	0.3803	0.3931	0.4058	0.4183	0.4307	0.4429	0
0.45	0.4550	0.4669	0.4786	0.4902	0.5015	0.5129	0.5240	0.5349	0.5458	0.5565	0
0.46	0.5672	0.5778	0.5884	0.5991	0.6097	0.6204	0.6312	0.6422	0.6533	0.6644	0
0.47	0.6756	0.6871	0.6986	0.7102	0.7219	0.7337	0.7454	0.7574	0.7693	0.7811	0
0.48	0.7930	0.8048	0.8166	0.8281	0.8397	0.8509	0.8620	0.8730	0.8837	0.8941	0
0.49	0.9043	0.9139	0.9234	0.9324	0.9410	0.9491	0.9568	0.9638	0.9703	0.9763	0
0.50	0.9817	0.9865	0.9904	0.9938	0.9966	0.9984	0.9995	1.0000	0.9995	0.9984	0
0.51	0.9966	0.9936	0.9901	0.9858	0.9808	0.9750	0.9685	0.9612	0.9532	0.9445	0
0.52	0.9352	0.9253	0.9147	0.9036	0.8919	0.8796	0.8668	0.8535	0.8397	0.8257	0
0.53	0.8110	0.7960	0.7807	0.7652	0.7492	0.7332	0.7166	0.7002	0.6834	0.6667	0
0.54	0.6497	0.6327	0.6156	0.5985	0.5814	0.5644	0.5475	0.5306	0.5139	0.4973	0
0.55	0.4808	0.4645	0.4484	0.4325	0.4170	0.4015	0.3864	0.3715	0.3569	0.3427	0
0.56	0.3288	0.3151	0.3018	0.2888	0.2762	0.2639	0.2519	0.2403	0.2291	0.2182	0
0.57	0.2076	0.1975	0.1876	0.1782	0.1690	0.1602	0.1517	0.1436	0.1358	0.1284	0
0.58	1.2120	1.1430	1.0780	1.0150	0.9557	0.8989	0.8449	0.7934	0.7447	0.6986	-1
0.59	0.6548	0.6133	0.5741	0.5372	0.5022	0.4694	0.4383	0.4091	0.3816	0.3558	-1

0.60	0.3315	0.3087	0.2874	0.2674	0.2487	0.2312	0.2147	0.1994	0.1851	0.1718	−1
0.61	1.5930	1.4770	1.3690	1.2690	1.1750	1.0880	1.0070	0.9322	0.8624	0.7974	−2
0.62	0.7374	0.6817	0.6301	0.5822	0.5379	0.4969	0.4590	0.4238	0.3913	0.3613	−2
0.63	0.3335	0.3079	0.2842	0.2623	0.2421	0.2235	0.2062	0.1903	0.1757	0.1621	−2
0.64	1.4970	1.3820	1.2760	1.1780	1.0880	1.0050	0.9281	0.8574	0.7925	0.7325	−3
0.65	0.6772	0.6262	0.5792	0.5358	0.4958	0.4590	0.4249	0.3935	0.3645	0.3377	−3
0.66	0.3129	0.2901	0.2689	0.2493	0.2313	0.2146	0.1991	0.1848	0.1716	0.1593	−3
0.67	1.4800	1.3750	1.2770	1.1870	1.1040	1.0260	0.9543	0.8876	0.8258	0.7686	−4
0.68	0.7155	0.6660	0.6203	0.5777	0.5381	0.5014	0.4673	0.4356	0.4061	0.3787	−4
0.69	0.3533	0.3295	0.3075	0.2870	0.2679	0.2501	0.2336	0.2182	0.2038	0.1905	−4
0.70	1.7800	1.6640	1.5560	1.4540	1.3600	1.2730	1.1910	1.1140	1.0430	0.9763	−5
0.71	0.9143	0.8562	0.8020	0.7513	0.7040	0.6598	0.6184	0.5798	0.5438	0.5099	−5
0.72	0.4783	0.4487	0.4211	0.3951	0.3709	0.3482	0.3270	0.3070	0.2884	0.2710	−5
0.73	0.2546	0.2393	0.2250	0.2115	0.1989	0.1870	0.1759	0.1655	0.1557	0.1466	−5
0.74	1.3790	1.2990	1.2230	1.1510	1.0840	1.0220	0.9625	0.9070	0.8549	0.8057	−6
0.75	0.7596	0.7163	0.6755	0.6371	0.6010	0.5670	0.5351	0.5050	0.4767	0.4500	−6
0.76	0.4249	0.4012	0.3790	0.3580	0.3382	0.3196	0.3021	0.2855	0.2699	0.2552	−6
0.77	0.2413	0.2282	0.2159	0.2042	0.1932	0.1829	0.1731	0.1638	0.1551	0.1468	−6
0.78	0.1390										

TABLE 161 DARK ADAPTATION[a]
(Sections 15.3.3.1 and 15.5.1.4)

$L_0(\text{cd/m}^2) =$	0.3	3	30	300
$t =$ 0	3.71	4.30	4.90	5.60
1 sec.	2.68	3.51	4.70	4.84
2	2.37	3.23	3.97	4.50
5	2.00	2.71	3.42	4.04
10	1.75	2.35	2.96	3.86
20	1.54	1.99	2.56	3.62
40	1.34	1.68	2.19	3.30
60	1.18	1.44	1.99	2.66
2 min	0.98	1.28	1.67	2.38
5	0.82	0.98	1.28	1.74
10	0.80	0.82	0.91	1.12
20	0.70	0.69	0.74	0.90
30	0.59	0.64	0.67	0.73
40	0.52	0.53	0.59	0.68
50	0.49	0.48	0.56	0.60
60	0.44	0.46	0.49	0.53

[a] Absolute threshold as a function of time in dark, for various values of preadaptation luminance (L_0). Threshold values are log [luminance (μ cd/m^2)]. Preadaptation time was 3 min and the test patch was a 5^0 square.

Source. Blanchard [15.2e].

TABLE 162 RETINAL IMAGE POINT SPREAD FUNCTION
[sr^{-1}]^a (Section 15.4.1)

Radius (m rd)	Pupil diameter (mm)						Exp.
	2.0	3.0	3.8	4.9	5.8	6.6	
0.020	3.6	3.4	2.4	2.0	1.5	1.3	6
0.028	3.5	3.3	2.4	2.0	1.5	1.3	6
0.040	3.5	3.1	2.3	1.9	1.5	1.3	6
0.057	3.3	2.9	2.2	1.8	1.4	1.2	6
0.080	3.1	2.6	2.0	1.6	1.3	1.2	6
0.11	2.7	2.3	1.8	1.4	1.2	1.0	6
0.160	2.0	1.7	1.4	1.2	0.95	0.84	6
0.23	1.1	1.0	0.88	0.77	0.67	0.60	6
0.32	4.8	4.9	4.7	4.5	4.2	3.9	5
0.45	2.0	2.2	2.3	2.4	2.4	2.4	5
0.64	0.77	0.87	1.0	1.2	1.2	1.3	5
0.90	2.8	3.2	4.2	4.8	5.6	6.3	4
1.3	1.0	1.2	1.6	1.8	2.2	2.6	4
1.8	4.1	4.8	6.0	6.7	7.8	9.0	3
2.6	1.7	2.0	2.3	2.5	2.8	3.0	3
3.6	0.73	0.85	0.94	1.0	1.1	1.1	3
5.1	3.3	3.7	4.0	4.3	4.4	4.5	2
7.2	1.6	1.6	1.7	1.8	1.8	1.8	2
10	73	73	73	73	73	73	0
14	32	31	31	31	31	31	0
20	13	13	13	13	13	13	0
29	5.4	5.4	5.4	5.5	5.5	5.5	0
41	2.3	2.3	2.3	2.3	2.3	2.3	0
58	0.99	0.99	0.99	0.99	0.99	0.99	0
82	0.45	0.45	0.45	0.45	0.45	0.45	0
116	0.22	0.22	0.22	0.22	0.22	0.22	0

^{a} Each entry must be multiplied by 10^a, where a is the exponent listed in the last column of the corresponding line. Successive radii increase by a factor of $\sqrt{2}$.

Source. Vos, Walraven, and van Meeteren [15.59].

TABLE 163 STANDARD S CURVES (Section 16.2.1.1)

LAMDA	S1	S3	S4	S5	S6	S7	S8	S9	S10	S11	S12	S13
0.30	0.909	0.000	0.048	0.900	0.900	0.430	0.048	0.000	0.000	0.000	0.009	0.565
0.31	0.270	0.000	0.260	0.940	0.810	0.430	0.170	0.075	0.000	0.008	0.010	0.600
0.32	0.520	0.000	0.546	0.970	0.700	0.870	0.568	0.270	0.100	0.059	0.012	0.645
0.33	0.810	0.000	0.685	0.990	0.570	1.140	0.815	0.375	0.250	0.178	0.016	0.690
0.34	1.105	0.000	0.782	1.000	0.460	1.070	0.929	0.440	0.400	0.393	0.023	0.740
0.35	1.350	0.565	0.858	0.990	0.370	0.990	0.974	0.515	0.554	0.621	0.031	0.790
0.36	1.480	0.700	0.915	0.975	0.300	0.940	0.994	0.560	0.678	0.780	0.042	0.835
0.37	1.350	0.860	0.955	0.950	0.240	0.900	0.998	0.620	0.772	0.838	0.058	0.880
0.38	1.100	0.930	0.981	0.925	0.190	0.860	0.992	0.650	0.838	0.870	0.074	0.915
0.39	0.820	0.965	0.995	0.900	0.150	0.840	0.982	0.695	0.872	0.887	0.090	0.940
0.40	0.561	0.985	1.000	0.870	0.115	0.840	0.964	0.725	0.910	0.905	0.106	0.965
0.41	0.360	0.997	0.997	0.840	0.075	0.860	0.945	0.765	0.940	0.929	0.122	0.980
0.42	0.235	1.000	0.986	0.810	0.050	0.880	0.923	0.810	0.965	0.969	0.138	0.990
0.43	0.190	0.998	0.968	0.780	0.025	0.900	0.900	0.875	0.984	0.991	0.154	0.995
0.44	0.177	0.993	0.946	0.750	0.010	0.910	0.875	0.905	0.996	1.000	0.170	1.000
0.45	0.171	0.982	0.920	0.715	0.000	0.930	0.848	0.945	1.000	0.994	0.186	0.992
0.46	0.170	0.970	0.889	0.680	0.000	0.945	0.818	0.970	0.997	0.978	0.204	0.982
0.47	0.174	0.950	0.853	0.645	0.000	0.955	0.785	0.990	0.985	0.958	0.224	0.970
0.48	0.182	0.930	0.810	0.610	0.000	0.960	0.748	1.000	0.967	0.931	0.265	0.952
0.37	0.194	0.900	0.761	0.575	0.000	0.965	0.708	0.990	0.945	0.902	0.510	0.930
0.50	0.209	0.870	0.706	0.535	0.000	0.975	0.665	0.980	0.920	0.864	0.900	0.893
0.51	0.227	0.840	0.645	0.495	0.000	0.980	0.620	0.960	0.892	0.817	0.400	0.852
0.52	0.248	0.810	0.578	0.450	0.000	0.985	0.573	0.920	0.859	0.768	0.058	0.810
0.53	0.271	0.780	0.506	0.405	0.000	0.985	0.527	0.870	0.821	0.705	0.036	0.750
0.54	0.295	0.750	0.431	0.360	0.000	0.990	0.482	0.810	0.779	0.640	0.028	0.680
0.55	0.321	0.720	0.360	0.310	0.000	0.995	0.438	0.740	0.733	0.571	0.022	0.600
0.56	0.349	0.690	0.294	0.260	0.000	0.995	0.396	0.670	0.683	0.496	0.018	0.500
0.57	0.378	0.665	0.235	0.210	0.000	0.995	0.357	0.600	0.630	0.420	0.014	0.410
0.58	0.408	0.640	0.184	0.165	0.000	1.000	0.321	0.510	0.575	0.346	0.011	0.325
0.59	0.439	0.615	0.140	0.130	0.000	1.000	0.287	0.420	0.519	0.268	0.008	0.245
0.60	0.471	0.590	0.104	0.100	0.000	1.000	0.254	0.330	0.463	0.197	0.005	0.170
0.61	0.504	0.565	0.076	0.070	0.000	1.000	0.223	0.220	0.408	0.129	0.000	0.105
0.62	0.539	0.540	0.054	0.050	0.000	1.000	0.194	0.140	0.356	0.079	0.000	0.068
0.63	0.575	0.515	0.038	0.035	0.000	1.000	0.168	0.080	0.309	0.047	0.000	0.043
0.64	0.611	0.492	0.026	0.025	0.000	1.000	0.144	0.035	0.267	0.029	0.000	0.028
0.65	0.647	0.469	0.017	0.015	0.000	1.000	0.123	0.010	0.229	0.018	0.000	0.020
0.66	0.683	0.446	0.011	0.010	0.000	0.995	0.105	0.000	0.195	0.012	0.000	0.012
0.67	0.718	0.424	0.007	0.006	0.000	0.995	0.089	0.000	0.163	0.008	0.000	0.008
0.68	0.753	0.402	0.004	0.003	0.000	0.990	0.075	0.000	0.133	0.006	0.000	0.005
0.69	0.787	0.380	0.002	0.001	0.000	0.985	0.064	0.000	0.106	0.005	0.000	0.002
0.40	0.820	0.360	0.001	0.000	0.000	0.980	0.053	0.000	0.083	0.004	0.000	0.000
0.71	0.851	0.340	0.000	0.000	0.000	0.980	0.044	0.000	0.064	0.003	0.000	0.000
0.72	0.881	0.320	0.000	0.000	0.000	0.970	0.036	0.000	0.048	0.002	0.000	0.000
0.73	0.909	0.301	0.000	0.000	0.000	0.965	0.029	0.000	0.035	0.001	0.000	0.000
0.74	0.933	0.282	0.000	0.000	0.000	0.960	0.023	0.000	0.025	0.000	0.000	0.000
0.75	0.954	0.264	0.000	0.000	0.000	0.945	0.018	0.000	0.017	0.000	0.000	0.000
0.76	0.971	0.246	0.000	0.000	0.000	0.930	0.012	0.000	0.010	0.000	0.000	0.000
0.77	0.984	0.228	0.000	0.000	0.000	0.920	0.008	0.000	0.000	0.000	0.000	0.000
0.78	0.993	0.210	0.000	0.000	0.000	0.900	0.005	0.000	0.000	0.000	0.000	0.000
0.79	0.998	0.192	0.000	0.000	0.000	0.875	0.003	0.000	0.000	0.000	0.000	0.000
0.80	1.000	0.175	0.000	0.000	0.000	0.845	0.002	0.000	0.000	0.000	0.000	0.000
0.81	0.998	0.158	0.000	0.000	0.000	0.805	0.001	0.000	0.000	0.000	0.000	0.000
0.82	0.993	0.140	0.000	0.000	0.000	0.770	0.000	0.000	0.000	0.000	0.000	0.000
0.83	0.983	0.122	0.000	0.000	0.000	0.730	0.000	0.000	0.000	0.000	0.000	0.000
0.84	0.969	0.105	0.000	0.000	0.000	0.695	0.000	0.000	0.000	0.000	0.000	0.000
0.85	0.951	0.088	0.000	0.000	0.000	0.650	0.000	0.000	0.000	0.000	0.000	0.000
0.86	0.929	0.070	0.000	0.000	0.000	0.610	0.000	0.000	0.000	0.000	0.000	0.000
0.87	0.903	0.052	0.000	0.000	0.000	0.565	0.000	0.000	0.000	0.000	0.000	0.000
0.88	0.874	0.035	0.000	0.000	0.000	0.520	0.000	0.000	0.000	0.000	0.000	0.000
0.89	0.841	0.020	0.000	0.000	0.000	0.480	0.000	0.000	0.000	0.000	0.000	0.000

TABLE 163 (*Continued*)

LAMDA	S14
0.90	0.200
0.95	0.360
1.00	0.530
1.05	0.685
1.10	0.865
1.15	0.975
1.20	0.990
1.25	0.935
1.30	0.685
1.35	0.420
1.40	0.190
1.45	0.080
1.50	0.045
1.55	0.020
1.60	0.015

LAMDA	S15	S16	S17	S18	S19	S20	S21	S23	S25
0.30	0.000	0.000	0.000	0.029	0.290	0.000	0.899	0.140	0.555
0.31	0.000	0.000	0.040	0.086	0.540	0.000	0.940	0.250	0.610
0.32	0.000	0.000	0.070	0.151	0.670	0.000	0.980	0.350	0.662
0.33	0.000	0.000	0.100	0.211	0.730	0.000	1.000	0.480	0.705
0.34	0.000	0.000	0.115	0.238	0.770	0.000	0.982	0.590	0.750
0.35	0.000	0.000	0.130	0.254	0.795	0.000	0.950	0.680	0.788
0.36	0.000	0.000	0.140	0.269	0.815	0.000	0.913	0.760	0.820
0.37	0.000	0.000	0.150	0.274	0.830	0.000	0.874	0.825	0.850
0.38	0.000	0.000	0.160	0.280	0.845	0.000	0.835	0.880	0.870
0.39	0.000	0.000	0.170	0.288	0.850	0.760	0.796	0.927	0.885
0.40	0.000	0.000	0.180	0.293	0.860	0.829	0.757	0.960	0.898
0.41	0.000	0.000	0.190	0.298	0.872	0.886	0.713	0.990	0.914
0.42	0.000	0.000	0.200	0.303	0.887	0.930	0.670	1.000	0.940
0.43	0.000	0.000	0.220	0.309	0.905	0.962	0.627	0.995	0.988
0.44	0.000	0.000	0.240	0.313	0.927	0.986	0.584	0.975	1.000
0.45	0.000	0.000	0.265	0.318	0.950	1.000	0.541	0.952	0.999
0.46	0.000	0.000	0.295	0.324	0.975	0.990	0.498	0.927	0.989
0.47	0.000	0.000	0.335	0.330	0.988	0.968	0.455	0.895	0.965
0.48	0.000	0.000	0.380	0.335	0.995	0.932	0.412	0.852	0.933
0.37	0.000	0.000	0.440	0.340	1.000	0.887	0.369	0.820	0.900
0.50	0.000	0.000	0.500	0.346	0.995	0.836	0.327	0.791	0.865
0.51	0.000	0.000	0.570	0.352	0.990	0.780	0.286	0.762	0.830
0.52	0.000	0.000	0.645	0.358	0.975	0.722	0.248	0.733	0.795
0.53	0.000	0.000	0.720	0.365	0.930	0.671	0.210	0.704	0.760
0.54	0.000	0.000	0.800	0.371	0.886	0.620	0.183	0.675	0.698
0.55	0.000	0.000	0.870	0.378	0.820	0.570	0.163	0.645	0.632
0.56	0.000	0.000	0.935	0.385	0.750	0.522	0.145	0.617	0.565
0.57	0.000	0.000	0.980	0.392	0.600	0.480	0.130	0.588	0.465
0.58	0.000	0.000	1.000	0.399	0.450	0.437	0.115	0.559	0.323
0.59	0.000	0.000	0.980	0.407	0.300	0.398	0.105	0.530	0.201
0.60	0.000	0.000	0.945	0.416	0.190	0.360	0.096	0.501	0.160
0.61	0.000	0.000	0.890	0.425	0.130	0.326	0.087	0.472	0.120
0.62	0.000	0.000	0.830	0.435	0.100	0.295	0.078	0.443	0.094
0.63	0.000	0.000	0.760	0.445	0.075	0.266	0.069	0.414	0.070
0.64	0.000	0.000	0.690	0.455	0.062	0.240	0.060	0.385	0.047
0.65	0.000	0.000	0.620	0.466	0.050	0.217	0.051	0.356	0.031
0.66	0.000	0.000	0.530	0.483	0.039	0.193	0.042	0.327	0.022
0.67	0.000	0.000	0.440	0.504	0.030	0.171	0.034	0.298	0.015
0.68	0.000	0.000	0.370	0.539	0.022	0.150	0.028	0.269	0.009
0.69	0.000	0.000	0.300	0.594	0.015	0.134	0.023	0.240	0.005
0.40	0.000	0.000	0.240	0.670	0.008	0.118	0.018	0.211	0.002
0.71	0.000	0.000	0.200	0.800	0.002	0.100	0.013	0.182	0.000
0.72	0.000	0.000	0.160	0.930	0.000	0.087	0.008	0.153	0.000
0.73	0.000	0.000	0.130	1.000	0.000	0.076	0.003	0.126	0.000
0.74	0.000	0.000	0.100	0.960	0.000	0.064	0.001	0.103	0.000
0.75	0.000	0.000	0.075	0.870	0.000	0.053	0.000	0.085	0.000
0.76	0.000	0.000	0.060	0.800	0.000	0.045	0.000	0.068	0.000
0.77	0.000	0.000	0.048	0.744	0.000	0.036	0.000	0.052	0.000
0.78	0.000	0.000	0.037	0.695	0.000	0.029	0.000	0.040	0.000
0.79	0.000	0.000	0.027	0.653	0.000	0.021	0.000	0.029	0.000
0.80	0.000	0.000	0.022	0.612	0.000	0.017	0.000	0.019	0.000
0.81	0.000	0.000	0.017	0.572	0.000	0.000	0.000	0.011	0.000
0.82	0.000	0.000	0.012	0.533	0.000	0.000	0.000	0.005	0.000
0.83	0.000	0.000	0.007	0.496	0.000	0.000	0.000	0.000	0.000
0.84	0.000	0.000	0.003	0.460	0.000	0.000	0.000	0.000	0.000
0.85	0.000	0.000	0.000	0.426	0.000	0.000	0.000	0.000	0.000
0.86	0.000	0.000	0.000	0.394	0.000	0.000	0.000	0.000	0.000
0.87	0.000	0.000	0.000	0.362	0.000	0.000	0.000	0.000	0.000
0.88	0.000	0.000	0.000	0.332	0.000	0.000	0.000	0.000	0.000
0.89	0.000	0.000	0.000	0.303	0.000	0.000	0.000	0.000	0.000

TABLE 163 (*Continued*)

ULTRAVIOLET LAMDA	S6	S7	S8	S17	S19	S23	S25
0.15	0.000	0.000	0.000	0.000	0.000	0.000	0.000
0.16	0.000	0.000	0.000	0.000	0.000	0.000	0.000
0.17	0.000	0.000	0.000	0.000	0.000	0.000	0.000
0.18	0.000	0.000	0.000	0.000	0.000	0.000	0.000
0.19	0.000	0.000	0.000	0.000	0.000	0.000	0.000
0.20	0.000	0.000	0.000	0.000	0.000	0.000	0.000
0.21	0.100	0.000	0.000	0.000	0.000	0.000	0.000
0.22	0.270	0.000	0.000	0.000	0.000	0.000	0.000
0.23	0.460	0.000	0.000	0.000	0.000	0.000	0.000
0.24	0.640	0.000	0.000	0.000	0.000	0.000	0.000
0.25	0.810	0.015	0.000	0.000	0.000	0.000	0.000
0.26	0.930	0.040	0.000	0.000	0.000	0.000	0.000
0.27	0.990	0.080	0.000	0.000	0.000	0.000	0.000
0.33	0.990	0.150	0.000	0.000	0.010	0.000	0.000
0.29	0.950	0.435	0.005	0.000	0.100	0.000	0.000

TABLE 163   (*Continued*)

INFRARED LAMDA	S1	S7	S16	S25
0.90	0.803	0.435	0.000	0.000
0.43	0.761	0.385	0.000	0.000
0.92	0.716	0.335	0.000	0.000
0.93	0.670	0.300	0.000	0.000
0.94	0.622	0.250	0.000	0.000
0.95	0.573	0.200	0.000	0.000
0.96	0.523	0.150	0.000	0.000
0.97	0.473	0.100	0.000	0.000
0.98	0.423	0.060	0.000	0.000
0.99	0.375	0.040	0.000	0.000
1.00	0.329	0.025	0.000	0.000
1.01	0.285	0.015	0.000	0.000
1.02	0.245	0.010	0.000	0.000
1.03	0.208	0.005	0.000	0.000
1.04	0.175	0.000	0.000	0.000
1.05	0.148	0.000	0.000	0.000
1.06	0.124	0.000	0.000	0.000
1.07	0.104	0.000	0.000	0.000
1.08	0.088	0.000	0.000	0.000
1.09	0.074	0.000	0.000	0.000
1.10	0.063	0.000	0.000	0.000
1.11	0.053	0.000	0.000	0.000
1.12	0.044	0.000	0.000	0.000
1.13	0.036	0.000	0.000	0.000
1.14	0.029	0.000	0.000	0.000
1.15	0.023	0.000	0.000	0.000
1.16	0.017	0.000	0.000	0.000
1.17	0.012	0.000	0.000	0.000
1.18	0.008	0.000	0.000	0.000
1.19	0.005	0.000	0.000	0.000
1.20	0.002	0.000	0.000	0.000

Source.   Based on JEDEC [16.16a].

TABLE 164 PHOTOCATHODE CHARACTERISTICS (Section 16.2.1.1)

Device S Number[a]	Photocathode type and envelope	Conversion factor[b] (k) (lum/W)	Typical luminous sensitivity[c] (s_{typ}) (μA/l)	Maximum luminous sensitivity[d] (s_{max}) (μA/l)	λ_{max} (Å)	Typical radiant sensitivity[e] (σ_{typ}) (mA/W)	Typical quantum efficiency[f] (%)	Typical photocathode Dark Emission[g] at 25°C (Am/cm²)
S1	Ag–O–Cs Lime-glass bulb	93.9	25	60	8000	2.35	0.36	$900. \times 10^{-15}$
S3	Ag–O–Rb Lime-glass bulb	286	6.5	20	4200	1.86	0.55	—
S4	Cs–Sb Lime-glass bulb	977	40	110	4000	39.1	12	0.2×10^{-15}
S5[h]	Cs–Sb 9741 glass bulb	1252	40	80	3400	50.1[h]	18[h]	0.3×10^{-15}
S8	Cs–Bi Lime-glass bulb	755	3	20	3650	2.26	0.77	0.13×10^{-15}
S9	Cs–Sb Semitransparent, Lime-glass bulb	683	30	110	4800	20.5	5.3	—
S10	Ag–Bi–O–Cs Semitransparent, Lime-glass bulb	508	40	100	4500	20.3	5.6	70×10^{-15}
S11	Cs–Sb Semitransparent, Lime-glass bulb	804	60	110	4400	48.2	14	3×10^{-15}
S13	Cs–Sb Semitransparent, Fused-silica bulb	795	60	80	4400	47.7	13	4×10^{-15}
S17	Cs–Sb Lime-glass bulb, Reflecting substrate	664	125	160	4900	83	21	1.2×10^{-15}

S number	Description							
S19[i]	Cs–Sb Fused-silica bulb	—	40	70	—	22[i]	11[i]	0.3×10^{-15}
S20	Sb–K–Na–Cs (Multialkali) Semitransparent Lime-glass bulb	428	150	250	4200	64.2	18	0.3×10^{-15}
S21	Cs–Sb Semitransparent, 9741 glass bulb	779	30	60	4400	23.4	6.6	—

[a] The S number is the designation of the spectral-response characteristic of the device and includes the transmission of the device envelope.

[b] k is the conversion factor from A/lm (of 2870°K radiation) to A/W (at the wavelength of peak sensitivity).

[c] s is the luminous sensitivity for the photocathode for 2870°K color-temperature test lamp. In the case of a multiplier photocube, output sensitivity is μs, where μ is the amplification of the multiplier phototube.

[d] Care must be used in converting s_{max} to a σ_{max} figure. Photocathodes having maximum lumen sensitivity frequently have more red sensitivity than normal, and the formula cannot be applied without reevaluation of the spectral response for the particular maximum-sensitivity device.

[e] σ is the radiant sensitivity at the wavelength of maximum response.

[f] 100% quantum efficiency implies one photoelectron per incident quantum, or $e/h\nu = \lambda/12{,}395$, where λ is expressed in angstrom units. Quantum efficiency at λ_{max} is computed by comparing the radiant sensitivity at λ_{max} with the 100%-quantum-efficiency expression above.

[g] Most of these data are obtained from multiplier phototube characteristics. For tubes capable of operating at very high gain factors, the dark emission at the photocathode is taken as the output dark current divided by the gain (or the equivalent minimum anode dark current input multiplied by cathode sensitivity). On tubes where other dc dark-current sources are predominant, the dark noise figure may be used. In this case, if all the noise originates from the photocathode emission, it may be shown that the photocathode dark emission in amperes is approximately $0.4 \times 10^{18} \times$(equivalent noise input in lm times cathode sensitivity in A/lm)^2. The data shown are all given per unit area of the photocathode.

[h] The S5 spectral response is suspected to be in error. The data tabulated conform to the published curve, which is maximum at 3400 Å. Present indications are that the peak value should agree with that of the S4 curve (4000 Å). Typical radiant sensitivity and quantum efficiency would then agree with those for S4 response.

[i] No value for k or λ_{max} is given because the spectral response data are in question. The values quoted for σ and typical quantum efficiency are only typical of measurements made at the specific wavelength 2537 Å and not at the wavelength of peak sensitivity as for the other data.

Source. Engstrom [16.17].

1043

TABLE 165   CHARACTERISTICS OF TYPICAL

Tube	No. of Dynodes	Structure[a]	S^b No.	Sensitivity[c] Cathode $\frac{rad.\ mA}{W}$	lum $\frac{\mu A}{lm}$	Anode rad. kA/W	lum A/lm	Operat. char. typ. Volt. kV	Gain 10^6
MPT's									
1P21	9	c/o	4	40	40	120	120	1	3
1P28	9	c/o	5	50	40	125	100	1	2.5
543C-01	14	V	1	3.2	30	3.2	30	2.5	1
6342A	10	c	11	64	80	25	31	1.25	0.39
931A	9	c/o	4	40	40	80	80	1	2
9634QR	13	V	11	$(82)^e$	90	(1800)	2000	1.45	(22)
9684B	11	B	1	(2.4)	25	(1.9)	20	1.4	(0.8)
9698B	9	B	20	(64)	110	(11.6)	20	1.3	(0.18)
9790B	9	V	20	(64)	130	(25)	50	1.35	(0.38)
C31034A	11	L	*128*	155	1000	62	400	1.5	0.4
C31034D	11	L	*142*	42	150	17	60	1.5	0.4
XP1004	10	L	13	60	70	(84)	100	1.45	1.4
XP1040	14	L	11	60	70	(6000)	(7000)	2.4	100
XP1110	10	L	11	60	60	(30)	30	1.4	(0.5)
XP1143	6	L/o	4	40	40	(0.4)	(0.4)	3.5	0.01
Photo-diodes									
868	0		1	8.4		90×10^{-6}		0.09	8×10^{-6}
929	0		4	44		45×10^{-6}		0.25	10^{-6}

[a] c—circular cage; V—venetian blind; B—box-type; L—linear, focused; o—opaque.
[b] Numbers in italics are mfr's. See Figure 16.6c.
[c] Lumen values refer to radiation of blackbody at 2870°K; watt values to radiation at wavelength of peak sensitivity.
[d] R—RCA; E—EMI; P—Phillips; S—EMR, Schlumberger.
[e] Values in parentheses were estimated on the basis of manufacturers' data.

| Dark current | | Noise-equiv.[c] input | | | | Cath. | |
Cath. fA	Anode (@ sens.)	fW · $Hz^{-1/2}$	plm · $Hz^{-1/2}$	I_{Amx} mA	Remarks	dim. (mm)	Mfr.[d]
(2)	1 nA @ 20 A/1m	(0.63)	(0.63)	0.1	L^3	8×24	R
(10)	5 nA @ 20 A/1m	(1.1)	(1.4)	0.5	uv window	8×24	R
(130)	130 nA @ 30 A/1m	(63)	(6.8)		ir sens.	10 ϕ	S
(16)	4 nA @ 20 A/1m	(1.1)	(0.9)	2	general	43 ϕ	R
(10)	5 nA @ 20 A/1m	(1.4)	(1.4)	0.1	g'nl low cost	8×24	R
(1.4)	30 nA @ 2000 A/1m	0.28	0.22	1	hi. quant. effic.	44 ϕ	E
(6250)	5 μA @ 20 A/1m	550	56	0.5	ir sens.	44 ϕ	E
2.8	0.5 nA @ 20 A/1m	0.28	0.17	1	compact	23 ϕ	E
13	5 nA@50 A/1m	0.95	0.5	1	large cath.	111 ϕ	E
(30)	3 nA @ 100 A/lm	(0.63)	(0.1)	0.01	ir sens.	4×10	R
(30)	10 nA @ 50 A/lm	(2.3)	(0.65)	0.01	ir sens.	4×10	R
10	15 nA @ 100 A/lm	(0.97)	(0.83)	0.1	uv	44 ϕ	P
(20)	2 μA @ 7000 A/lm	(1.3)	(1.1)	0.2	L^3 fast	110 ϕ	P
(40)	20 nA @ 30 A/lm	(1.9)	(1.9)	0.2	comp. & rugged	14 ϕ	P
(10^5)	1 μA @ 0.4 A/lm	(142)	(142)	0.2	1 ns pulse	280 mm^2	P
	100 nA			0.01	gas diode	16×32	R
	12.5 μA			0.005	vac diode	16×21	R

TABLE 166a DARK COUNTING RATE CHARACTERISTICS[a] (Section 16.2.4.7)

Tube type	Photo-cathode type/IEDP (mm)	Temp. (°C)	Rate (pulses/sec)		Dark count[b] uncertainty		Equivalent input[b] flux uncertainty (W)	
			Maximum	Typical	Maximum	Typical	Maximum	Typical
FW118, FW142	S1/2.54	25	6×10^6	3×10^5	$\pm 2.5 \times 10^3$	$\pm 6 \times 10^2$	3.5×10^{-13}	9×10^{-11}
FW118, FW142	S1/2.54	−20	6×10^2	30	± 25	± 6	3.5×10^{-15}	9×10^{-16}
FW118, FW142	S1/2.54	−78	100	1	± 10	± 1	1.5×10^{-15}	1.5×10^{-16}
FW129, FW136	S11/2.54	25	100	10	± 10	± 3	1.2×10^{-16}	3.6×10^{-17}
FW130, FW143	S20/2.54	25	100	10	± 10	± 3	9×10^{-17}	2.7×10^{-17}

[a] For an absolute photoelectron counting efficiency of 80–90%. Manufacturer: ITT.
[b] For a 1-sec observation time.
Source. Eberhardt [16.43].

TABLE 166*b* DARK COUNT DATA FOR MPT's (REFRIGERATED[a])
(Section 16.2.4.7)

Cell		Cathode		Mean No. dark counts/sec	obs/exp s.d. σ_0/σ_e
Type	Serial	Diam.	Type		
9558C	5312	5 cm	S20	18	1.8
9558C	5415	5 cm	S20	33	2.6
9558C	5462	5 cm	S20	15	2.7
9558C	5496	5 cm	S20	22	3.0
9558A	5833	5 cm	S20	31	2.8
9558A	5844	5 cm	S20	14	1.5 (before discharge)
9558A	5844	5 cm	S20	4	1.1 (after discharge)
9502S	5598	1 cm	CsSb(S)	6	1.0
6256SA	11519	1 cm	CsSB(S)	2	1.1
6256SA	11731	1 cm	CsSb(S)	5	1.1
6256SA	11715	1 cm	CsSb(S)	14	1.1

[a] $T < -20°C$, usually $-65°C$, and with 2000 V applied to cells. Manufacturer: EMI.
Source. Rodman and Smith [16.40].

TABLE 167 SEMICONDUCTOR   PHOTODETECTORS,
TYPICAL CHARACTERISTICS (Section 16.3)

Photodetector	Gain	Response time (sec)
Photoconductor	10^5	10^{-3}
p-n Junction	1	10^{-11}
p-i-n Junction	1	10^{-10}
Junction transistor	10^2	10^{-8}
Avalanche photodiode	10^4	10^{-10}
Metal-semiconductor diode	1	10^{-11}
Field-effect transistor	10^2	10^{-7}

Source. Sze [16.54*a*].

TABLE 168 CHARACTERISTICS OF INFRARED DETECTORS (Section 16.3)

Detector material	Operating mode[a]	Useful wavelength range[b] (μ)	Wavelength of peak response (μ)	Resistance (Ohm)[c]	Time constant (μ sec)	D^*(500°K blackbody) cm(Hz)$^{1/2}$ W^{-1} (at frequency indicated)[d]	D^*(At peak response) cm(Hz)$^{1/2}$W^{-1} (at frequency indicated)[d]
Thermal Detectors							
Room-temperature operation							
Thermocouple	thermoelectric	1–40	—	1–10	25,000	3–$12\times10^8(5)$	6–$15\times10^8(5)$
Evaporated thermopile	thermoelectric	1–40	—	100	5,000	$1\times10^8(20)$	$2\times10^8(20)$
Thermistor bolometer	bolometer	0.2–40	—	1–5×10^6	2,000	0.3–$1.2\times10^8(15)$	1–$3\times10^8(15)$
Ferroelectric bolometer	bolometer	1–12	—	—	—	$1.1\times10^8(100)$	—
Golay cell	gas expansion	5–1000	—	—	20,000	1–$5\times10^8(10)$	5–$10\times10^8(10)$
Low temperature operation							
Niobium nitride bolometer (16°K)	superconducting	—	—	0.2	550	$5\times10^9(360)$	$5\times10^9(360)$
Carbon bolometer (2.1°K)	bolometer	40–100	—	$0.12=10^6$	10,000	$4\times10^{10}(13)$	$4\times10^{10}(13)$
Germanium bolometer (2.1°K)	bolometer	5–2000	—	12×10^3	400	$8\times10^{11}(200)$	$8\times10^{11}(200)$
Triglycine sulphate (TGS)	pyroelectric	~500	—	—	~1	2.6×10^8	
Photon Detectors							
Room-temperature operation							
Cadmium sulphide (CdS)	PC	0.4–0.7	0.55	10^{15e}	70,000		1–$5\times10^{12}(90)$
Selenium, amorphous	PC	~0.58	0.46	$10^{13}\ \Omega$ cm	50		50–$100\times10^9(800)$
Antimony trisulphide (Sb$_2$S$_3$)	PC	0.55–0.75	0.65	$10^{12}\ \Omega$ cm	10,000		
Silicon	PV	0.5–1.05	0.84	0.1–1×10^6	100	10^{10}–$10^{11}(90)$	1–$5\times10^{12}(90)$
Lead sulfide (PbS)	PC	0.6–3.0	2.3–2.7	0.5–10×10^6	50–500	1–$7\times10^8(800)$	50–$100\times10^9(800)$
Indium arsenide (InAs)	PV	1–3.7	3.2	20	~1	1–$3\times10^8(900)$	3–$7\times10^9(900)$
Indium arsenide (InAs)	PEM	1.4–3.8	3.4	—	~1	—	$6\times10^9(1000)$
Lead selenide (PbSe)	PC	0.9–4.6	3.8	1–10×10^6	2	0.7–$2\times10^8(800)$	1–$4\times10^9(800)$
Indium antimonide (InSb)	PEM	0.5–7.5	6.2	20	~0.1	$0.8\times10^8(1000)$	$0.3\times10^9(1000)$
Operation at 195°K							
Lead sulfide (PbS)	PC	0.5–3.3	2.6	0.5–5×10^6	80–4000	0.7–$7\times10^9(800)$	20–$70\times10^{10}(800)$
Indium arsenide (InAs)	PV	0.5–3.5	3.2	—	~1	1–$5\times10^9(1800)$	3–$25\times10^{10}(1800)$
Indium arsenide (InAs)	PEM	1.3–3.6	3.2	—	~1	$3\times10^9(1000)$	$20\times10^{10}(1000)$
Lead selenide (PbSe)	PC	0.8–5.1	4.2	10×10^6	30	2–$4\times10^9(800)$	1–$4\times10^{10}(800)$
Indium antimonide (InSb)	PC	0.5–6.5	5.1	20	~1	$1\times10^9(800)$	0.5–$0.9\times10^{10}(800)$

Operation at 77°K

Material	Type[a]	Range (μm)[b]	Peak (μm)	Impedance[c]		D^*[d]	D^*
Lead sulfide (PbS)	PC	0.7–3.8	2.9	1–10×10^6	500–3000	3–8×10^9(800)	8–20×10^{10}(800)
Indium arsenide (InAs)	PV	0.6–3.2	2.9	10×10^6	~2	3–8×10^9(1800)	20–70×10^{10}(1000)
Tellurium (Te)	PC	0.7–4	3.6	1×10^3	60	3×10^9(3000)	5×10^{10}(3000)
Lead telluride (PbTe)	PC	1–5.4	5.0	50–500×10^6	~5	1×10^9(90)	0.8×10^{10}(90)
Lead selenide (PbSe)	PC	0.8–6.6	5.1	5–10×10^6	40	2–6×10^9(800)	1–3×10^{10}(800)
Indium antimonide (InSb)	PV	0.6–5.6	5.1	1–50×10^4	~1	3–20×10^9(900)	3–8×10^{10}(900)
Indium antimonide (InSb)	PC	0.7–5.9	5.3	2–10×10^4	1–10	3–10×10^9(900)	2–6×10^{10}(900)
Gold-doped germanium (p-type)	PC	1–9	5.4	0.1–10×10^6	~1	1–3×10^9(800)	0.3–1×10^{10}(800)
Gold-doped germanium (n-type)	PC	1–5.5	1.5	—	50	0.5–2×10^9(90)	1×10^{10}(90)
Mercury-cadmium-telluride	PV	6–15	10.6	5–50	0.01	109–10^{10}(900)	—

Operation below 50°K

Material	Type[a]	Range (μm)[b]	Peak (μm)	Impedance[c]		D^*[d]	D^*
Zinc-doped germanium-Silicon alloy (Ge-Se: Zn) (48°K)	PC	2–15	10.5	0.1×10^6	~1	5×10^9(100)	1×10^{10}(100)
Mercury-doped germanium (Ge: Hg) (30°K)	PC	3–14	11	2–100×10^4	~1	3–9×10^9(900)	1–1.5×10^{10}(900)
Cadmium-doped germanium (Ge: Cd) (28°K)	PC	6–24	17	20×10^4	~1	4×10^9(400)	1.2×10^{10}(400)
Copper-doped germanium (Ge: Cu) (4.2°K)	PC	6–29	23	0.5–1×10^6	~1	5–10×10^9(1800)	1.5–3×10^{10}(1800)
Zinc-doped germanium (Ge: Zn) (4.2°K)	PC	7–40	37	0.3×10^6	~0.01	3–4×10^9(800)	0.9–1.2×10^{10}(800)

[a] PV = photovoltaic, PC = photoconductive, PEM = photoelectromagnetic.

[b] Wavelengths between which D^* exceeds 0.2 of its peak value.

[c] Value shown is for a square element. For PV detectors the value is the dynamic impedance dV/dI.

[d] All D^* values given for a detector viewing a hemispherical surround at a temperature of 300°K.

[e] For a typical cell [16.58].

Source. Hudson [16.55a] and Kruse et al. [16.1c], based on manufacturers' data.

1049

(a) Indium Arsenide (InAs) [SB, TI]

Type	Photovoltaic.
Element sizes available	Standard elements are circular, ranging from 0.5 to 3 mm in diameter.
Time constant	Less than 2 μsec at all temperatures.
Dynamic impedance	5–10 $\times 10^3$ ohm at 195°K. 3–6 $\times 10^4$ ohm at 77°K.
Responsivity	3 $\times 10^{-1}$ VW^{-1} at 295°K. 5 $\times 10^2$ VW^{-1} at 195°K. 5 $\times 10^3$ VW^{-1} at 77°K.
$\dfrac{D^*\text{(peak)}}{D^*\text{(500°K)}}$	65 at 77°K 40 at 195°K.

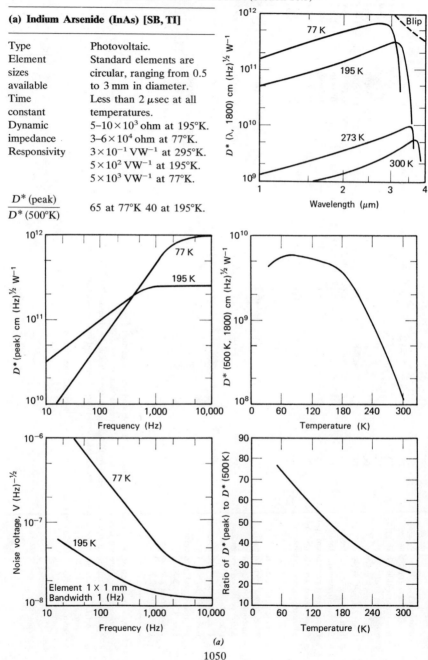

TABLE 169 (*Continued*)

(b) Lead Sulfide (PbS) [II, SB]

Type	Photoconductive.
Element sizes available	From 0.01×0.01 to 25×25 mm. Rectangular elements from 0.01 mm wide. For arrays, minimum element spacing is 0.01 mm.

Temp. (°K)	Time constant (μsec)	Resistance (megohms)	$\mathcal{R}$ (VW⁻¹)	$\dfrac{D^*\text{(peak)}}{D^*\text{(500°K)}}$
295	50–500	0.3–0.6	4×10^3	90
195	800–4000	3–6	3×10^5	25
77	500–3000	6–12	2×10^5	60

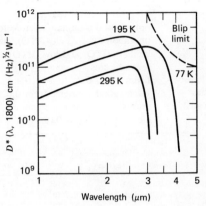

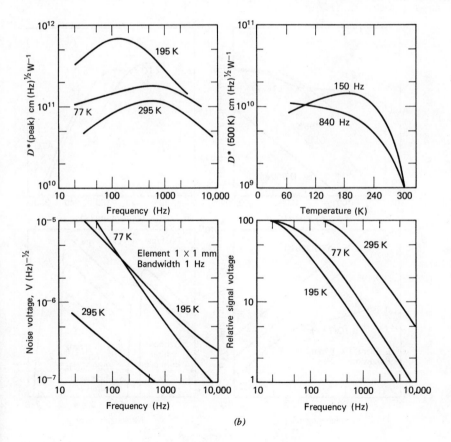

(b)

1051

TABLE 169 *(Continued)*

(c) Lead Selenide (PbSe) [SB]

Type	Photoconductive. Material is optimized during manufacture for operation at ambient (ATO). Intermediate (ITO) and low (LTO) temperatures.
Element sizes available	From 0.25×0.25 to 10×10 mm. Rectangular elements from 0.075 mm wide. For arrays, minimum element spacing is 0.025 mm.

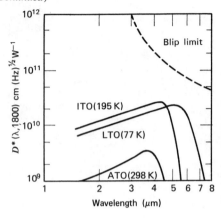

Type	Time constant (μsec)	Resistance (megohms)	$\mathscr{R}$ (VW^{-1})	$\dfrac{D^* \text{ (peak)}}{D^* (500°K)}$
ATO	2	2–5	2×10^3	17
ITO	30	6–12	5×10^4	5.5
LTO	40	0.5–2	2×10^5	3.5–4

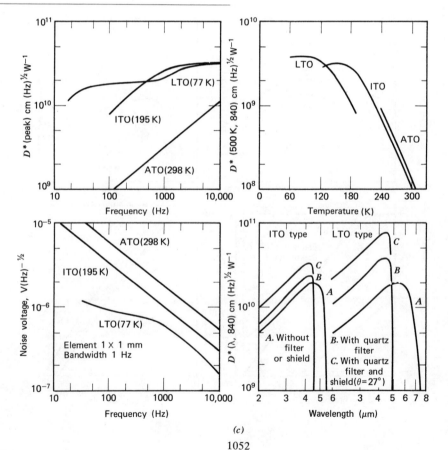

(c)

TABLE 169 (*Continued*)

(d) Indium Antimonide (InSb) [SB, TI]

Type	Photoconductive.
Element sizes available	From 0.5×0.5 to 10×10 mm. Rectangular elements from 0.05 mm wide. For arrays, minimum element spacing is 0.025 mm.
Time constant	Less than 10 μsec, even when shielded.
Resistance	$2-10 \times 10^3$ ohm.
Responsivity	2×10^4 VW^{-1}.

$$\frac{D^* \text{(peak)}}{D^* \text{(500°K)}} \quad 5.7$$

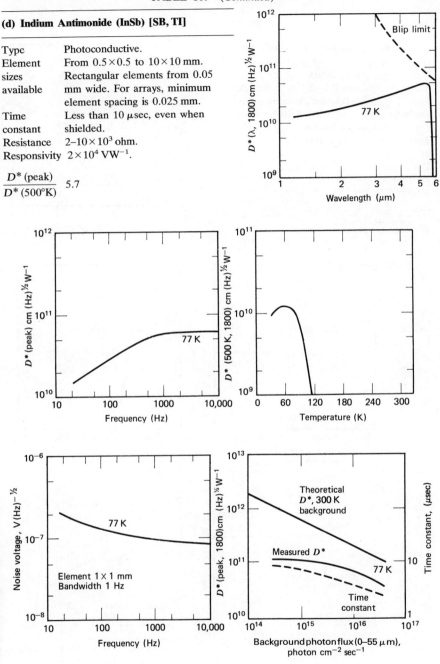

(*d*)

TABLE 169   (Continued)

(e) Indium Antimonide (InSb) [PC, TI]

Type	Photovoltaic.
Element sizes available	From 0.1×0.1 to 10×10 mm. Rectangular elements from 0.1 mm wide. Circular elements with diameters from 0.1 to 10 mm.
Time constant	Less than 1 μsec.
Dynamic impedance	$2-5 \times 10^4$ ohm.
Responsivity	$2 \times 10^4 \mathrm{VW}^{-1}$.
$\dfrac{D^* (\text{peak})}{D^* (500°\text{K})}$	6.2.

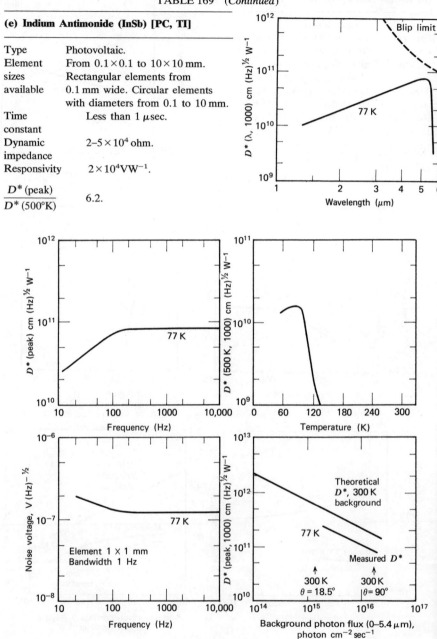

(e)

1082

TABLE 169 (*Continued*)

(f) Mercury-Doped Germanium (Ge:Hg) [SB, TI]

Type	Photoconductive.
Element sizes available	Standard elements are circular ranging from 0.3 to 3 mm in diameter.
Time constant	Less than 0.1 μsec for all temperatures below 25°K.
Dynamic impedance	Varies with bias. shielding, etc. Typically 0.5 megohm for $\theta = 30°$.
Responsivity	1.5×10^5 VW^{-1} ($\theta = 30°$).

$$\frac{D^*\ (\text{peak})}{D^*\ (500°\text{K})} \quad 1.8\ (\text{Irtran-2 window}).$$

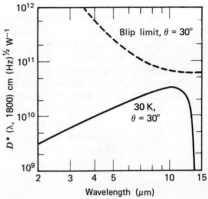

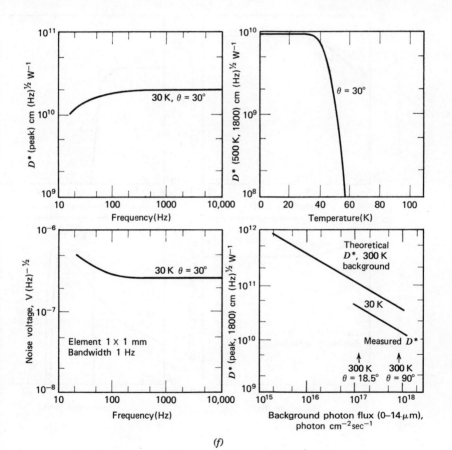

(f)

TABLE 169 *(Continued)*

(g) Copper-Doped Germanium (Ge:Cu) [SB, TI]

Type	Photoconductive.
Element sizes available	Standard elements are circular ranging from 0.3 to 3 mm in diameter.
Time constant	Less than 0.1 μsec for all temperatures below 15°K.
Dynamic impedance	Varies with bias, shielding, etc. Typically 2×10^4 ohm for $\theta = 30°$.
Responsivity	2×10^5 VW^{-1} ($\theta = 30°$).
$\dfrac{D^*\,(\text{peak})}{D^*\,(500°\text{K})}$	2.5 (KRS-5 window) 1.8 (window and cooled filter of Irtran-2).

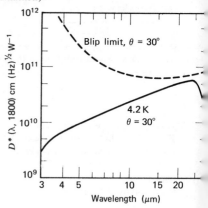

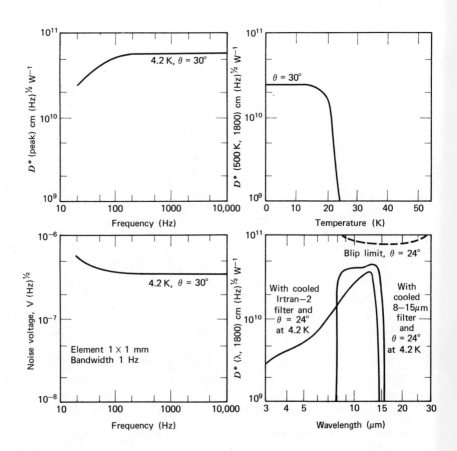

(g)

TABLE 169 (*Continued*)

(h) Thermistor [BE, SC]

Type	Bolometer.
Element sizes available	From 0.1×0.1 to 2.5×2.5 mm. Rectangular elements from 0.1 mm wide. Germanium-immersed elements from 0.1×0.1 to 1×1 mm.
Time constant	1–5 msec.
Resistance	0.1 mm wide. Germanium-immersed elements, resistances of 0.27, 0.5, and 2.7 megohms available.
Responsivity	50 VW^{-1}. 10^3 VW^{-1} (Germanium-immersed element).

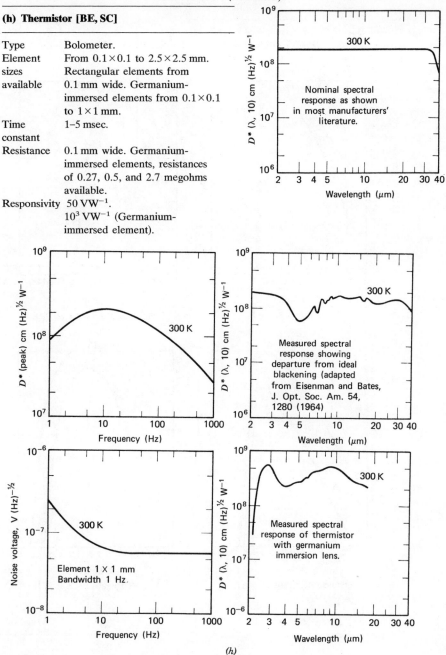

Nominal spectral response as shown in most manufacturers' literature.

Measured spectral response showing departure from ideal blackening (adapted from Eisenman and Bates, J. Opt. Soc. Am. 54, 1280 (1964)

Measured spectral response of thermistor with germanium immersion lens.

Element 1 × 1 mm
Bandwidth 1 Hz.

(*h*)

Sources of data. Indicated in square brackets. BE: Barnes Engineering Co.; II: Infrared Industries; PC: Philco Corp.; SB: Santa Barbara Research Center; SC: Servo Corp. of America; TI: Texas Instruments Inc. From Hudson [16.55].

TABLE 170 S CURVES OF HOMOGENEOUS SEMICONDUCTOR DETECTORS
(Section 16.3)

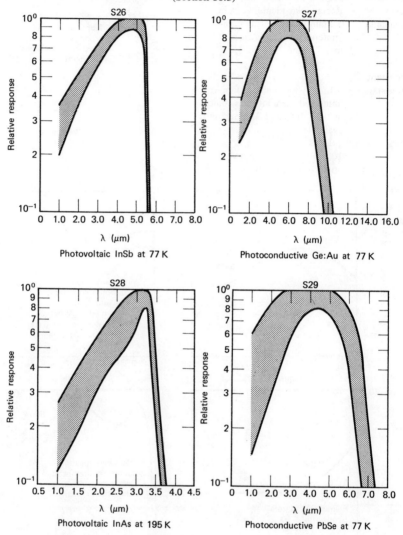

S26

Relative response

λ (μm)

Photovoltaic InSb at 77 K

S27

Relative response

λ (μm)

Photoconductive Ge:Au at 77 K

S28

Relative response

λ (μm)

Photovoltaic InAs at 195 K

S29

Relative response

λ (μm)

Photoconductive PbSe at 77 K

1058

TABLE 170 (*Continued*)

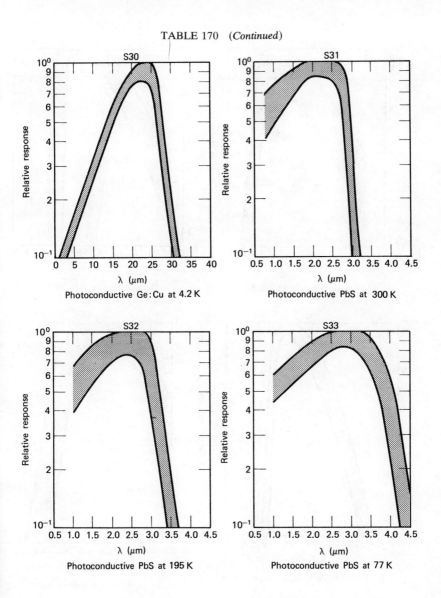

Photoconductive Ge:Cu at 4.2 K

Photoconductive PbS at 300 K

Photoconductive PbS at 195 K

Photoconductive PbS at 77 K

TABLE 170   (*Continued*)

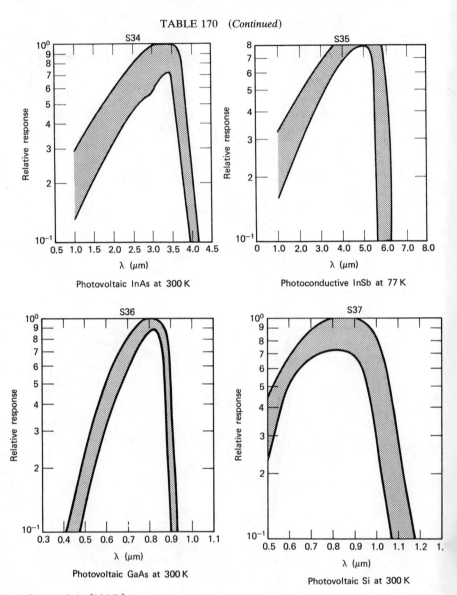

Photovoltaic InAs at 300 K

Photoconductive InSb at 77 K

Photovoltaic GaAs at 300 K

Photovoltaic Si at 300 K

Source: Jedec [16.16b].

TABLE 170   (*Continued*)

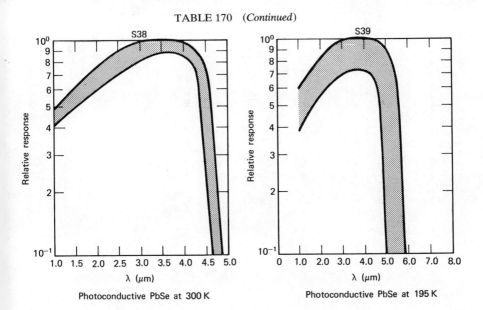

Photoconductive PbSe at 300 K

Photoconductive PbSe at 195 K

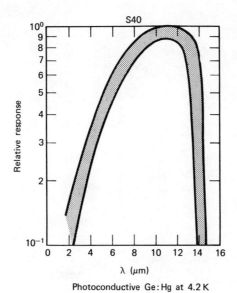

Photoconductive Ge:Hg at 4.2 K

TABLE 171 PERFORMANCE CHARACTERISTICS OF PHOTODIODES (Section 16.3.2.5)

Diode	Wave-length range (μm)	Peak efficiency (%) or responsivity	Sensitive area (cm²)	Capacitance (pF)	Series resistance (Ω)	Response time (seconds)	Dark current	Operating temperature (K)	Comments
Silicon n^+-p	0.4–1	40	2×10^{-5}	0.8 at -23 V	6	130 ps with 50-Ω load	50 pA at -10 V	300	avalanche photo diode
Silicon p-i-n	0.6328	>90	2×10^{-5}	<1	~1	100 ps with 50-Ω load	$<10^{-9}$ A at -40 V	300	optimized for 0.6328 μm
Silicon p-i-n	0.4–1.2	>90 at 0.9 μm >70 at 1.06 μm	5×10^{-2}	3 at -200 V 3 at -200 V	<1 <1	7 ns 7 ns	0.2 μA at -30 V	300	
Metal-i-nSi	0.38–0.8	>70	3×10^{-2}	15 at -100 V		10 ns with 50-Ω load	2×10^{-2} A at -6 V	300	
Au-nSi	0.6328	70	2			<500 ps		300	Schottky barrier, antireflection coating
PtSi-nSi	0.35–0.6	~40	2×10^{-5}	<1		120 ps		300	Schottky barrier avalanche photo-diode
Ag–GaAs	<0.36	50						300	
Ag–ZnS	<0.35	70						300	
Au–ZnS	<0.35	50						300	

Ge n^+-p	0.4–1.55	50 uncoated	2×10^{-5}	0.8 at −16 V	<10	120 ps	2×10^{-8}	300	Germanium avalanche photodiode
Ge p-i-n	1–1.65	60	2.5×10^{-5}	3		25 ns at 500 V		77	illumination entering from side
GaAs point contact	0.6328	40		0.027	30				
InAs p-n	0.5–3.5	>25	3.2×10^{-4}	3 at −5 V	12	$<10^{-6}$		77	
InSb p-n	0.4–5.5	>25	5×10^{-4}	7.1 at −0.2 V	18	5×10^{-6}		77	
InSb p-n	2–5.6		5×10^{-4}				1 MΩ shunt resistance	77	Reverse break down voltages 30 V
$Pb_{1-x}Sn_xTe$ $x = 0.16$	9.5 μm	45 V/W $\eta = 60$	4×10^{-3}			$\sim 10^{-9}$		77	shunt resistance $R_i = 10\ \Omega$
$Pb_{1-x}Sn_xSe$ $x = 0.064$	11.4 μm	3.5 V/W $\eta = 15$	7.8×10^{-3}			$\sim 10^{-9}$		77	shunt resistance $R_i = 2.5\ \Omega$
$Hg_{1-x}Cd_xTe$ $x = 0.17$	15 μm	$\eta \sim 10\text{–}30$	4×10^{-4}		8	$<3 \times 10^{-9}$		77	shunt resistance $R_i > 100\ \Omega$

1063

Source. Melchior et al. [16.80]. For detailed references, see that article.

TABLE 172 PHYSICAL PROPERTIES OF LOW-TEMPERATURE COOLANTS
(Section 16.3.4.4)

Coolant	Boiling temperature[a] (K)	Refrigeration capacity $(W \cdot hr \cdot liter^{-1})$	Spec. grav. or weight-density $(kg \cdot liter^{-1})$
Ice	273.2[b]	—	—
Solid carbon dioxide	194.6[c]	—	—
Liquid oxygen	90.2	67.6	1.14
Liquid argon	87.3	63.5	1.39
Liquid nitrogen	77.3	44.4	0.807
Liquid neon	27.1	28.9	1.21
Liquid hydrogen	20.4	8.79	0.071
Liquid helium	4.2	0.71	0.125

[a] At a pressure of 760 mm Hg.
[b] Melting point.
[c] Sublimation point.
Source. Hudson, Ref. 16.55.

TABLE 173 SYMMETRICAL TWO-TUBE LENS: VOLTAGE RATIO V_2/V_1 AND FOCAL LENGTH f; MIDFOCAL DISTANCES MF AND MIDPRINCIPAL PLANE DISTANCES MP; ALL ABSOLUTE VALUES MEASURED IN TERMS OF TUBE RADII R (Section 16.4.1.5)

V_2/V_1[a]	2	3	4	5	6	8	10	20	30	50	100
M/F_2	26	10.6	6.4	4.7	3.9	3.1	2.6	1.0	0.7	0.4	0.2
MP_2	5	3.3	3.2	3.0	2.8	2.5	2.4	2.6	2.7	2.8	3.0
f_2	30.6	13.9	9.6	7.7	6.7	5.6	5.0	4.1	4.0	3.9	0.5
MF_1	25	11	7	5.3	4.5	3.5	3.0	2.2	1.8	1.6	1.5
MP_1	2.9	2.6	2.0	1.9	1.8	1.6	1.4	1.2	1.1	1.1	1.1
f_1	21.6	8.0	4.8	3.4	2.7	1.9	1.6	1.0	0.7	0.5	0.4

[a] $V_1 < V_2$
Source. Klemperer [16.96d]

TABLE 174   CONSTANTS FOR USE IN MTF
APPROXIMATION: $T(\nu) = (\exp - (\nu/\nu_0)^a$
(Section 16.4.6)

Device	(ν_0, a)
Image orthicon	(8.2, 2.1)
Intensifier/silicon-vidicon	
C21125/C23136	(16, 1.6)
Vidicon 8507A	(20, 1.4)
Image intensifier C21125	(32, 1.1)
Return-beam vidicon C23098	(50, 1.3)
Phosphor screen	(46, 1.1)

Source. Johnson [16.155]

TABLE 175   COMPARISON   OF   VARIOUS   THERMO-
ELECTRIC MATERIALS (Section 16.5.1.2)

Material	Thermo-electric power P (μV/°C)	Thermal conduc-tivity K (W m^{-1} °C^{-1})	Electrical conduc-tivity σ (Ω^{-1} m^{-1})
Silver	+2.9	424	6.1×10^7
Iron	+16	67	1.0×10^7
Nickel	−19	59	1.3×10^7
Antimony	+40	20	2.4×10^6
Bismuth	−60	8.3	8.3×10^5
Tellurium			
(a) (single crystal)	+436	1.8	2.9×10^3
(b) (polycrystalline)	+376	1.5	3.3×10^3
(c) (polycrystalline)	+372	1.0	5.0×10^2
(d) (Baker & Co.)	+119	2	3.2×10^4
Constantan	−38	21.2	2.0×10^6
Chromel-p	+30	20	1.2×10^6
95% Bi–5% Sn	+30	4.5	3.6×10^5
97% Bi–3% Sb	−75	7	5.8×10^5
90% Bi–10% Sb	−78	5.3	6.2×10^5
99.6% Te–0.4% Bi	+191	2.3	3.5×10^4
99.1% Te–0.9% Sb	+139	2.3	4.3×10^4
98.5% Te–1.5% S	+575	1.5–3	2.9×10^2
65% Sb–35% Cd	+106	1.5–4	1.7×10^5
75% Sb–25% Cd	+112	1.5–4	1.4×10^5

Source.   Smith *et al.* [16.9].

1065

TABLE 176   PYROELECTRIC AND RELATED CONSTANTS (Section 16.5.1.3)

Material (Temp., °K)	K ($\mu C/K \cdot cm^2$)	ε''	c' ($J/cm^3 \cdot K$)	Q $K/c'\sqrt{\varepsilon''}$	T_c (K)
TGS (300)	0.02	0.16	1.8	0.028	322
TGS (320)	0.14	0.6	1.8	0.1	322
TGS: alanine	0.03	0.15–0.3	2.5	0.02–0.03	312
TGSe (295)	0.65	1.1	1.9	0.33	
$Sr_{0.67}Ba_{0.33}Nb_2O_6$ (333)	0.11	25	2	0.011	333
$BaTiO_3$ (333)	0.07	300	2(?)	0.002	393
$BaTiO_3$ (381)	10	270	2(?)	0.3	393

Source.   Based on Hadni [16.163], Putley [16.161*b*], and Glass [16.162].

TABLE 177   PHOTOGRAPHIC EMULSIONS: GRAIN SIZES AND NUMBERS (Section 17.1.1.1)

Film Type	Grain cross section area average (μm^2)	Grain density ($10^6/mm^3$)
Lippmann	0.002	
Motion picture, pos.	0.3	118
Fine grain, roll	0.5	52
Portrait	0.6	26
High-speed, roll	0.9	23
X-ray	2.3	6.3

Based on James and Higgins, *Fundamentals of Photographic Theory*, Morgan and Morgan, Dobbs Ferry, N.Y., 1960. With permission.

TABLE 178   DIMENSIONAL CHANGES (FRACTIONAL) IN PHOTOGRAPHIC SUBSTRATES (Section 17.1.1.2)

Material	Thickness (μm)	Coefficients $\times 10^{5a}$ Humidity	Thermal	Processing	Aging $\times 10^4$ (1 yr at room cond's)
Cellulose triacetate	12.5	8	6	-7×10^{-4}	-15
Polyethylene terephthalate	100	2	2	-6 to $+3 \times 10^{-4}$	-3
	62	3	2	-3 to $+3 \times 10^{-4}$	-4
Glass		0	0.9	0	0
Paper		4–14		-8 to $+2 \times 10^{-3}$	

[a] The thermal and humidity coefficients are per degree celsius and per percentage point of relative humidity, respectively.

Source.   Based on data in [17.3*b*].

TABLE 179 STANDARD FILM DIMENSIONS (APPROXIMATE) (Section 17.1.1.4)

Film type	Film width	Frame Width mm	Frame Length mm	Frame Diag. mm	N^a	Perforations Pitch mm	Perforations Dimens.^{b} mm	Perforations Locat.^{c} mm	Typical Objective Efl. mm	Typical Objective Tot. field (deg)
Motion Picture	8 mm	4.9	3.7	6	1	3.8	1.83×1.27	0.9	12.5	27
	16 mm	10	7.5	12	1	7.6	1.83×1.27	0.9	25	27
	35 mm	22	16	27	4	4.75	2.8×2.0	2.0	50	30
	70 mm	52	23	57	5	4.75 (5.94)			100	32
Still, Roll	35 mm	24	36	43	8	4.75			50	47
	70 mm	57	{ 57	80 }		5.94	3.3×2	2.5 I^d	75	56.5
		57	{ 82	100 }		4.75	2.8×2	2.0 II^d		
	No. 120	57				4.75	2.8×2	2.0	100	53
	5 in. (126 mm)									
	9.5 in. (240 mm)									

^{a} Number of perforations per frame.
^{b} Vary considerably. Representative value given.
^{c} Perforation edge from film edge.
^{d} Perforation type.

Cut Film

Film Size Nominal inch	Norm mm	Film Size Nominal cm	Norm mm
2.25×3.25	56×81.4	4.5×6	43.5×58.5
3.25×4.25	81×106	6.5×9	63.5×88.5
4×5	100×125	9×12	88.5×118.5
5×7	125×176	10×15	98.5×148
8×10	202×253		
11×14	279×355		
14×17	355×431		
16×20	406×507		
20×24	507×609		
30×40	761×1015		

Microfiche, Film size: 105×148 mm

Type	1	2	3^e	4	5^e	6	7^e
No. columns	14	12	9	25	16	28	18
No. rows	7	5	7	13	13	15	15
No. frames	98	60	63	325	208	420	270
Frame width	10	11.75	15.5	5.5	8.75	5	7.75
height	12.5	16.5	12.5		7	6.25	6.25
diag	16	20.3	20	8.9	11.2	8	10
Reduction	24	20	24	42	42	48	48

^{e} Designed for standard computer print-out format.
Based on ANSI/NMA Stds. MS2-1978 and MS5 (PH 5.9)-1975

TABLE 180   MULTIPLE EXPOSURE PHOTOGRAPHIC EFFECTS[a]
(Section 17.2.1.7)

Name	First exposure	Interm. procedure	Second exposure	effect
Clayden "Black Lightning"	High intensity light	—	Interm. intensity light	First exposure desensitizes
Villard	X-ray	—	Light	First exposure desensitizes
Sequence Effect	intermediate intensity light		low intensity light	reciprocity law failure
Herschel	Blue light	—	Red light	Second exposure destroys effects of first
Debot	Blue light	Chromic acid	Red light	Emulsion is sensitive to red light
Albert	Light	Chromic acid	Light	First exposure desensitizes
Sabattier	Light	Develop	Light	First exposure desensitizes
Classical Reversal	Light	Develop + Chromic acid	Light	First exposure desensitizes

[a] All of effects, except Sequence and Debot, are reversal effects.

TABLE 181   MTF'S OF HIGH-RESOLUTION EMULSIONS (Section 17.2.2.1)

Mfr	Type	$T(400 \text{ mm}^{-1})$	$T(1500 \text{ mm}^{-1})$
Eastman Kodak	120–02	1.0	1.0
Eastman Kodak	649F	0.75	0.66
Agfa–Gevaert	Scienta 8E75	0.99	0.88
	Scienta 8E75B	0.85	0.59
	Scienta 10E75	0.76	0.62

Source. After Johansson and Biedermann [17.33].

TABLE 182   GRANULARITY CODE
(Section 17.2.3.1)

Description	Selwyn G (μm)	$1000 \, \sigma(D)$[a]
Coarse	≥ 2.65	≥ 44
Moderately coarse	2.11–2.65	35–44
Medium	1.62–2.05	27–34
Fine	1.32–1.56	22–26
Very fine	0.96–1.26	16–21
Extremely fine	0.36–0.9	6–15
Microfine	≤ 0.36	≤ 6

[a] Measured with scanning aperture having diameter of 48 μm, observed by F/2 optical system, on sample having unity density.

Source. After Kowaliski [17.2f] and Eastman Kodak [17.25b].

TABLE 183 INFORMATION CAPACITY OF VARIOUS (KODAK) FILMS (Section 17.2.4.2)

Film	Info. capac.[a] total		Info. capac.[b] "reliable"		Available[b] density levels "reliable"
	k bits / mm²	μm² / bit	k cells / mm²	k bits / mm²	
Royal-X Pan	5	200	1.4	1.4	2
Plus-X Pan	18.6	53.8	2.1	3.3	3
Panatomic-X	28.5	35	4.4	7	3
Fine grain cine pos.			1.1	3.3	8
Microfile #5454			6.4	16	6
High Resolution #649			1600	1600	2

[a] Data from Jones [17.45].
[b] Discrete signal levels, separated 10σ, where σ is the standard deviation of granularity.

Source. Based on Altman and Zweig, *Photogr. Sci. Eng.*, **7**, 173 (1963); from [17.2g].

TABLE 184 EXPOSURE REQUIREMENTS FOR SOME COMMERCIAL PHOTORESISTS (Section 17.3.2.1)

Material	Original thickness (μm)	$\text{Log } E_{mx}$[a] (J/m²)	$\text{Log } E_{mn}$[b] (J/m²)	Required[c] contrast	Resolution[d] (cycles/mm)	Manufacturer[e]
KPR-11	0.41	3.6	2.85	5.6	<1000	K
KTFR	0.37	2.8	1.8	10.0	<1000	K
KMER	0.39	3.0	2.2	6.3	<1000	K
AZ-1350	0.33	3.2	2.64	3.6	>1500	S
AZ-111	0.89	3.2	2.72	3.0	<1000	S
PR-115	1.25	3.05	2.3	5.6		G

[a] E_{mx} is exposure energy required to obtain the best resolution on development while retaining the maximum of initial film thickness.
[b] E_{mn} (1) For positive resists, the exposure energy corresponding to the intersection of the back extrapolation of the linear portion of the characteristic curve with the ordinate corresponding to full developed film thickness. (2) For negative resists, the exposure energy corresponding to the intersection of the extrapolation of the linear portion of the characteristic curve with the abscissa.
[c] Contrast requirement is E_{mx}/E_{mn}.
[d] From [17.56b].
[e] K—Eastman–Kodak Co.; S—Shipley Co. (Newton, MA); G—GAF (General Analine and Film).

Source. Goldrick and Plauger [17.59*].

TABLE 185 PHOTOPOLYMERIZATION IMAGING SYSTEMS (Section 17.3.2.1)

Physical state	"Readout" procedure	Type of image	Applications	References
Solubility	Wash-off	Relief	Letterpress Printing	U.S. Patents 3,081,168; 2,760,868; 2,791,504; 2,760,863.
Solubility	Wash-off	Hydrophilic Hydrophobic Surface	Lithography	J. Photogr. Sci. **18**, 155–156 (1970)
Solubility	Wash-off	Self-supporting Polymers	Templates	Belg. Patent 596,378 (1960)
Solubility	Etching	Resist	Printed Circuits, Integrated Circuits, Chemical Milling	J. Kosar, Light Sensitive Systems, Wiley, p. 141 (1965)
Adhesion	Peel-apart	Pigment	Engineering Reproduction	J. Photogr. Sci. **18**, 156–157 (1970).
Tackiness	Toner Dust-on	Pigment	Image Transfer Color Proofing	TAGA Proc., Vol. 9 (1968)
Viscosity	Diffusion	Dye Imbibition Color Formation	Imaging Color Prints	Brit. Patent 893,063 (1962) Photogr. Sci. Eng. **13**, 84 (1969)
Thermal Transitions	Thermal Transfer	Pigment	Imaging	U.S. Patents 3,060,025 (1962); 3,060,026 (1962); 3,085,088 (1963)
Conductivity	Electroylsis Wash-off	Pigment	Imaging	Photogr. Sci. Eng. **13**, 184 (1969)
Light Scatter	Projection	Diffraction	Holography	Appl. Phys. Lett. **14**, 159 (1969)
Light Scatter	Projection	Scatter	Imaging	Photogr. Sci. Eng. **12**, 177 (1968)

Source. Walker [17.60].

TABLE 186   SENSITIVITY OF REPRESENTATIVE XEROGRAPHIC PHOTOCONDUCTORS
(Section 17.4.1.1)

Material	Speed[a]
Polyvinyl carbazole	0.014
Selenium, amorphous	2
Selenium-tellurium alloy	20
Zinc-cadmium sulfide	1.8
Zinc oxide	0.2
Zinc oxide, dye sensitized	1–2

[a] Equivalent ANSI (ASA) index, relative to Standard Illuminant A.
Source. Based on Ref. 17.69.

TABLE 187 SOME USEFUL TRANSFER AND SPREAD FUNCTIONS[a] (Section 18.1.2.4)

Name	Otf	$\Delta\nu$	Line spread function	Point spread function	Applications
1 Ideal Syst.	1	∞	$\delta(x)$	$\lim\limits_{s\to 0}\dfrac{\pi e^{-\pi^2 r^2/s}}{s}$	
2 Sq. L	$\sin\pi b\nu/\pi b\nu$	$1/2b$	$1/b,\ |x|<\tfrac{1}{2}b$	$1/[\pi b\sqrt{b^2/4-r^2}],\ r<b/2$	Scan. slit
3 Sq. mtf	$1,\ \nu<\nu_c$	ν_c	$\sin 2\pi\nu_c x/\pi x$		Rect. aperture (coher. illum.)
4 Linear mtf	$1-\nu/\nu_c,\ \nu<\nu_c$	$\tfrac{1}{3}\nu_c$	$\nu_c(\sin\pi\nu_c x/\pi\nu_c x)^2$		Rect. aperture (incoher. illum.)
5 Tophat P	$J_1(2\pi R\nu)/\pi R\nu$	$8/3\pi^2 R$	$2\sqrt{1-(x/R)^2}/\pi R$	$1/\pi R^2;\ r<R$	Circ. aperture (incoher. illum.)
6 Airy Disc	$\dfrac{2}{\pi}\left[\cos^{-1}\left(\dfrac{\nu}{\nu_c}\right)-\dfrac{\nu}{\nu_c}\sqrt{1-(\nu/\nu_c)^2}\right]$	$\dfrac{8(15\pi-32)}{45\pi^2}\nu_c$ $=0.2724\nu_c$	$\dfrac{3}{2}\dfrac{H_1(2\pi\nu_c x)}{(2\pi\nu_c x)}$	$J_1^2(\pi\nu_c r)/\pi r^2$	Circ. aperture (incoher. illum.)
	$\dfrac{2}{\pi}\left[\cos^{-1}(\lambda F\nu)-\lambda F\nu\sqrt{1-(\lambda F\nu)^2}\right]$		$\dfrac{3}{2}\dfrac{H_1(2\pi x/\lambda F)}{(2\pi x/\lambda F)}$	$J_1^2(\pi r/\lambda F)/\pi r^2$	
7 Lorentz (Expon. L, sym.)	$a^2/[a^2+(2\pi\nu)^2]$	$\tfrac{1}{8}a$	$\tfrac{1}{2}a\exp(-a|x|)$	$\dfrac{a^2}{2\pi}-K_0(ar)$	Photogr. emulsion natural line
	$b^2/(b^2+\nu^2)$	$\tfrac{1}{4}\pi b$	$\pi b\exp(-2\pi b|x|)$	$2\pi b^2 K_0(2\pi br)$	
8 Expon. L, (one-sided)	$a/(a-i2\pi\nu)$	$\tfrac{1}{4}a$	$ae^{-ax},\ x>0$		Expon. impulse resp.
	$(1+i2\pi\nu/a)/[1+(2\pi\nu/a)^2]$				
9 Expon. T,	$e^{-a\nu}$	$1/2a$	$2a/[a^2+(2\pi x)^2]$	$2\pi a[a^2+(2\pi r)^2]^{-3/2}$	Photogr. emulsion
	$e^{-2\pi b\nu}$	$1/4\pi b$	$b/\pi(b^2+x^2)$	$(b/2\pi)(b^2+r^2)^{-3/2}$	
10 Expon. P	$[1+(2\pi\nu/a)^2]^{-3/2}$	$3a/32$	$(a^2/\pi)K_1(a|x|)|x|$	$(a^2/2\pi)e^{-ar}$	
11 Gaussian	$e^{-2(\pi\nu\sigma)^2}$	$\dfrac{1}{4\sqrt{\pi}\,\sigma}$	$e^{-x^2/2\sigma^2}/\sqrt{2\pi\sigma^2}$	$(2\pi\sigma^2)^{-1}e^{-r^2/2\sigma^2}$	Crt diffusion
	$e^{-s\nu^2}$	$\sqrt{\pi/8s}$	$\sqrt{\pi/s}\,e^{-\pi^2 x^2/s}$	$(\pi/s)e^{-\pi^2 r^2/s}$	

[a] When limits on range are given, the function vanishes outside range. a, b, σ, s are constants; $a=2\pi b$, $s=2\pi^2\sigma^2$, $\nu_c=1/\lambda F$ is the cutoff frequency, F is the effective F/number, λ is wavelength, K_0, K_1 are Bessel functions of the third kind with imaginary argument, and H_1 is the Struve function.

1071

TABLE 188a SPECTRAL MATCHING FACTORS, BLACKBODY-S CURVE COMBINATIONS (Section 18.1.4.1)

S1 Peak at 0.80 Eff.at 2854 K = 0.1856

Temp	0	100	200	300	400	500	600	700	800	900
0	0.000E 00	0.0000E 00	0.328E-23	0.772E-15	0.113E-10	0.358E-08	0.165E-06	0.252E-05	0.192E-04	0.920E-04
1000	0.317E-03	0.863E-03	0.196E-02	0.388E-02	0.689E-02	0.112E-01	0.170E-01	0.244E-01	0.334E-01	0.439E-01
2000	0.558E-01	0.689E-01	0.830E-01	0.979E-01	0.113E 00	0.129E 00	0.145E 00	0.161E 00	0.177E 00	0.192E 00
3000	0.208E 00	0.222E 00	0.237E 00	0.251E 00	0.264E 00	0.276E 00	0.289E 00	0.300E 00	0.311E 00	0.321E 00
4000	0.331E 00	0.340E 00	0.349E 00	0.358E 00	0.365E 00	0.373E 00	0.380E 00	0.387E 00	0.393E 00	0.399E 00
5000	0.404E 00	0.410E 00	0.414E 00	0.419E 00	0.423E 00	0.427E 00	0.431E 00	0.435E 00	0.438E 00	0.441E 00
6000	0.444E 00	0.446E 00	0.448E 00	0.450E 00	0.452E 00	0.454E 00	0.455E 00	0.457E 00	0.458E 00	0.459E 00
7000	0.459E 00	0.460E 00	0.460E 00	0.460E 00	0.460E 00	0.460E 00	0.460E 00	0.459E 00	0.459E 00	0.458E 00
8000	0.457E 00	0.456E 00	0.455E 00	0.454E 00	0.453E 00	0.451E 00	0.450E 00	0.448E 00	0.446E 00	0.444E 00
9000	0.443E 00	0.441E 00	0.438E 00	0.436E 00	0.434E 00	0.432E 00	0.430E 00	0.427E 00	0.425E 00	0.422E 00

S3 Peak at 0.42 Eff.at 2854 K = 0.0580

Temp	0	100	200	300	400	500	600	700	800	900
0	0.000E 00	0.000E 00	0.289E-31	0.515E-20	0.184E-14	0.360E-11	0.531E-09	0.180E-07	0.246E-06	0.184E-05
1000	0.905E-05	0.328E-04	0.952E-04	0.232E-03	0.495E-03	0.948E-03	0.166E-02	0.273E-02	0.422E-02	0.620E-02
2000	0.876E-02	0.119E-01	0.157E-01	0.203E-01	0.255E-01	0.315E-01	0.381E-01	0.455E-01	0.535E-01	0.620E-01
3000	0.712E-01	0.808E-01	0.910E-01	0.101E 00	0.112E 00	0.123E 00	0.134E 00	0.146E 00	0.157E 00	0.169E 00
4000	0.180E 00	0.192E 00	0.203E 00	0.214E 00	0.225E 00	0.236E 00	0.247E 00	0.257E 00	0.267E 00	0.276E 00
5000	0.285E 00	0.294E 00	0.303E 00	0.311E 00	0.318E 00	0.325E 00	0.332E 00	0.339E 00	0.345E 00	0.351E 00
6000	0.356E 00	0.361E 00	0.365E 00	0.369E 00	0.373E 00	0.377E 00	0.380E 00	0.382E 00	0.385E 00	0.387E 00
7000	0.389E 00	0.390E 00	0.392E 00	0.393E 00	0.393E 00	0.394E 00	0.394E 00	0.394E 00	0.394E 00	0.394E 00
8000	0.393E 00	0.393E 00	0.392E 00	0.391E 00	0.390E 00	0.388E 00	0.387E 00	0.385E 00	0.383E 00	0.382E 00
9000	0.380E 00	0.378E 00	0.376E 00	0.373E 00	0.371E 00	0.369E 00	0.366E 00	0.364E 00	0.361E 00	0.359E 00

S4 Peak at 0.40 Eff.at 2854 K = 0.0157

Temp	0	100	200	300	400	500	600	700	800	900
0	0.000E 00	0.000E 00	0.000E 00	0.297E-27	0.447E-20	0.911E-16	0.692E-13	0.803E-11	0.287E-09	0.468E-08
1000	0.438E-07	0.273E-06	0.125E-05	0.455E-05	0.137E-04	0.355E-04	0.813E-04	0.168E-03	0.320E-03	0.576E-03
2000	0.947E-03	0.149E-02	0.227E-02	0.330E-02	0.465E-02	0.635E-02	0.844E-02	0.109E-01	0.139E-01	0.173E-01
3000	0.213E-01	0.257E-01	0.306E-01	0.360E-01	0.419E-01	0.482E-01	0.550E-01	0.621E-01	0.696E-01	0.774E-01

	0	100	200	300	400	500	600	700	800	900
4000	0.855E-01	0.938E-01	0.102E 00	0.111E 00	0.119E 00	0.128E 00	0.137E 00	0.146E 00	0.155E 00	0.164E 00
5000	0.173E 00	0.182E 00	0.190E 00	0.199E 00	0.207E 00	0.215E 00	0.223E 00	0.231E 00	0.238E 00	0.245E 00
6000	0.252E 00	0.259E 00	0.265E 00	0.271E 00	0.277E 00	0.283E 00	0.288E 00	0.293E 00	0.298E 00	0.302E 00
7000	0.306E 00	0.310E 00	0.314E 00	0.317E 00	0.320E 00	0.323E 00	0.326E 00	0.328E 00	0.330E 00	0.332E 00
8000	0.334E 00	0.336E 00	0.337E 00	0.338E 00	0.339E 00	0.340E 00	0.341E 00	0.341E 00	0.342E 00	0.342E 00
9000	0.342E 00	0.342E 00	0.341E 00	0.341E 00	0.341E 00	0.340E 00	0.339E 00	0.338E 00	0.337E 00	0.336E 00

S4 Filtered   Peak at 0.40   Eff.at 2854 K = 0.0155

Temp	0	100	200	300	400	500	600	700	800	900
0	0.000E 00	0.000E 00	0.000E 00	0.297E-27	0.447E-20	0.911E-16	0.692E-13	0.803E-11	0.287E-09	0.488E-08
1000	0.438E-07	0.273E-06	0.125E-05	0.455E-05	0.137E-04	0.355E-04	0.813E-04	0.168E-03	0.320E-03	0.667E-03
2000	0.945E-03	0.149E-02	0.226E-02	0.329E-02	0.463E-02	0.631E-02	0.838E-02	0.108E-01	0.137E-01	0.171E-01
3000	0.210E-01	0.253E-01	0.301E-01	0.353E-01	0.409E-01	0.470E-01	0.534E-01	0.601E-01	0.672E-01	0.745E-01
4000	0.820E-01	0.897E-01	0.976E-01	0.105E 00	0.113E 00	0.121E 00	0.129E 00	0.137E 00	0.145E 00	0.153E 00
5000	0.161E 00	0.168E 00	0.176E 00	0.183E 00	0.190E 00	0.197E 00	0.203E 00	0.209E 00	0.216E 00	0.221E 00
6000	0.227E 00	0.232E 00	0.237E 00	0.242E 00	0.246E 00	0.251E 00	0.255E 00	0.258E 00	0.262E 00	0.265E 00
7000	0.268E 00	0.271E 00	0.273E 00	0.275E 00	0.277E 00	0.279E 00	0.281E 00	0.282E 00	0.284E 00	0.285E 00
8000	0.285E 00	0.286E 00	0.287E 00	0.287E 00	0.287E 00	0.287E 00	0.287E 00	0.287E 00	0.287E 00	0.286E 00
9000	0.286E 00	0.285E 00	0.284E 00	0.284E 00	0.283E 00	0.282E 00	0.281E 00	0.279E 00	0.278E 00	0.277E 00

S5   Peak at 0.34   Eff.at 2854 K = 0.0130

Temp	0	100	200	300	400	500	600	700	800	900
0	0.000E 00	0.000E 00	0.000E 00	0.136E-27	0.280E-20	0.670E-16	0.553E-13	0.676E-11	0.248E-09	0.409E-08
1000	0.384E-07	0.239E-06	0.109E-05	0.395E-05	0.118E-04	0.303E-04	0.692E-04	0.142E-03	0.269E-03	0.476E-03
2000	0.791E-03	0.124E-02	0.188E-02	0.274E-02	0.385E-02	0.526E-02	0.699E-02	0.909E-02	0.115E-01	0.144E-01
3000	0.177E-01	0.215E-01	0.256E-01	0.303E-01	0.353E-01	0.408E-01	0.467E-01	0.529E-01	0.595E-01	0.665E-01
4000	0.738E-01	0.813E-01	0.891E-01	0.971E-01	0.105E 00	0.113E 00	0.122E 00	0.130E 00	0.139E 00	0.148E 00
5000	0.156E 00	0.165E 00	0.173E 00	0.182E 00	0.190E 00	0.199E 00	0.207E 00	0.215E 00	0.223E 00	0.230E 00
6000	0.238E 00	0.245E 00	0.252E 00	0.259E 00	0.266E 00	0.272E 00	0.278E 00	0.284E 00	0.290E 00	0.295E 00
7000	0.300E 00	0.305E 00	0.310E 00	0.314E 00	0.319E 00	0.322E 00	0.326E 00	0.330E 00	0.333E 00	0.336E 00
8000	0.339E 00	0.341E 00	0.344E 00	0.346E 00	0.348E 00	0.350E 00	0.351E 00	0.353E 00	0.354E 00	0.355E 00
9000	0.356E 00	0.357E 00	0.357E 00	0.358E 00	0.358E 00	0.358E 00	0.358E 00	0.358E 00	0.358E 00	0.358E 00

TABLE 188a   (Continued)

S5   Filtered
Peak at 0.34   Eff. at 2854 K = 0.128

Temp	0	100	200	300	400	500	600	700	800	900
0	0.000E 00	0.000E 00	0.000E 00	0.136E-27	0.280E-20	0.670E-16	0.555E-13	0.676E-11	0.248E-09	0.409E-08
1000	0.384E-07	0.239E-06	0.109E-05	0.395E-05	0.118E-04	0.303E-04	0.691E-04	0.142E-03	0.269E-03	0.475E-03
2000	0.789E-03	0.124E-02	0.187E-02	0.272E-02	0.382E-02	0.520E-02	0.690E-02	0.894E-02	0.113E-01	0.141E-01
3000	0.172E-01	0.208E-01	0.247E-01	0.290E-01	0.337E-01	0.387E-01	0.440E-01	0.496E-01	0.555E-01	0.616E-01
4000	0.679E-01	0.743E-01	0.809E-01	0.876E-01	0.943E-01	0.101E 00	0.107E 00	0.114E 00	0.121E 00	0.128E 00
5000	0.134E 00	0.141E 00	0.147E 00	0.153E 00	0.159E 00	0.165E 00	0.171E 00	0.176E 00	0.181E 00	0.186E 00
6000	0.191E 00	0.196E 00	0.200E 00	0.204E 00	0.208E 00	0.212E 00	0.216E 00	0.219E 00	0.222E 00	0.225E 00
7000	0.228E 00	0.230E 00	0.232E 00	0.234E 00	0.236E 00	0.238E 00	0.240E 00	0.241E 00	0.242E 00	0.243E 00
8000	0.244E 00	0.245E 00	0.245E 00	0.246E 00	0.246E 00	0.246E 00	0.246E 00	0.246E 00	0.246E 00	0.246E 00
9000	0.246E 00	0.245E 00	0.245E 00	0.244E 00	0.243E 00	0.242E 00	0.242E 00	0.241E 00	0.240E 00	0.239E 00

S6
Peak at 0.27   Eff. at 2854 K = 0.490E-03

Temp	0	100	200	300	400	500	600	700	800	900
0	0.000E 00	0.000E 00	0.000E 00	0.000E 00	0.906E-32	0.561E-25	0.182E-20	0.294E-17	0.734E-15	0.528E-13
1000	0.159E-11	0.256E-10	0.258E-09	0.180E-08	0.954E-08	0.401E-07	0.140E-06	0.424E-06	0.113E-05	0.270E-05
2000	0.593E-05	0.120E-04	0.229E-04	0.411E-04	0.702E-04	0.114E-03	0.180E-03	0.273E-03	0.402E-03	0.576E-03
3000	0.804E-03	0.109E-02	0.146E-02	0.192E-02	0.247E-02	0.314E-02	0.393E-02	0.486E-02	0.592E-02	0.715E-02
4000	0.853E-02	0.100E-01	0.118E-01	0.137E-01	0.158E-01	0.180E-01	0.205E-01	0.232E-01	0.260E-01	0.290E-01
5000	0.322E-01	0.356E-01	0.391E-01	0.428E-01	0.466E-01	0.506E-01	0.548E-01	0.590E-01	0.634E-01	0.678E-01
6000	0.724E-01	0.770E-01	0.818E-01	0.865E-01	0.914E-01	0.963E-01	0.101E 00	0.106E 00	0.111E 00	0.116E 00
7000	0.121E 00	0.125E 00	0.130E 00	0.135E 00	0.140E 00	0.145E 00	0.150E 00	0.154E 00	0.159E 00	0.164E 00
8000	0.168E 00	0.173E 00	0.177E 00	0.181E 00	0.185E 00	0.189E 00	0.193E 00	0.197E 00	0.201E 00	0.205E 00
9000	0.208E 00	0.212E 00	0.215E 00	0.218E 00	0.222E 00	0.225E 00	0.228E 00	0.230E 00	0.233E 00	0.236E 00

S6   Filtered
Peak at 0.27   Eff. at 2854 K = 0.295E-03

Temp	0	100	200	300	400	500	600	700	800	900
0	0.000E 00	0.000E 00	0.000E 00	0.000E 00	0.906E-32	0.561E-25	0.182E-20	0.294E-17	0.733E-15	0.527E-13
1000	0.158E-11	0.254E-10	0.254E-09	0.176E-08	0.916E-08	0.379E-07	0.130E-06	0.384E-06	0.996E-06	0.232E-05
2000	0.493E-05	0.968E-05	0.177E-04	0.307E-04	0.506E-04	0.796E-04	0.120E-03	0.175E-03	0.248E-03	0.340E-03
3000	0.457E-03	0.598E-03	0.769E-03	0.969E-03	0.120E-02	0.146E-02	0.176E-02	0.209E-02	0.246E-02	0.286E-02

Temp	0	100	200	300	400	500	600	700	800	900
4000	0.329E-02	0.375E-02	0.425E-02	0.476E-02	0.531E-02	0.587E-02	0.646E-02	0.706E-02	0.768E-02	0.831E-02
5000	0.895E-02	0.960E-02	0.102E-01	0.109E-01	0.115E-01	0.122E-01	0.128E-01	0.134E-01	0.141E-01	0.147E-01
6000	0.153E-01	0.159E-01	0.165E-01	0.171E-01	0.176E-01	0.181E-01	0.187E-01	0.192E-01	0.197E-01	0.201E-01
7000	0.206E-01	0.210E-01	0.214E-01	0.218E-01	0.221E-01	0.225E-01	0.228E-01	0.231E-01	0.234E-01	0.237E-01
8000	0.240E-01	0.242E-01	0.244E-01	0.246E-01	0.248E-01	0.250E-01	0.251E-01	0.253E-01	0.254E-01	0.255E-01
9000	0.256E-01	0.257E-01	0.258E-01	0.259E-01	0.259E-01	0.259E-01	0.259E-01	0.260E-01	0.260E-01	0.260E-01

S7

Peak at 0.58 Eff. at 2854 K = 0.1638

Temp	0	100	200	300	400	500	600	700	800	900
0	0.000E 00	0.000E 00	0.452E-27	0.313E-17	0.249E-12	0.206E-09	0.173E-07	0.398E-06	0.405E-05	0.240E-04
1000	0.982E-04	0.305E-03	0.774E-03	0.168E-02	0.324E-02	0.566E-02	0.916E-02	0.139E-01	0.200E-01	0.276E-01
2000	0.366E-01	0.472E-01	0.591E-01	0.724E-01	0.868E-01	0.102E 00	0.118E 00	0.136E 00	0.154E 00	0.172E 00
3000	0.191E 00	0.210E 00	0.229E 00	0.248E 00	0.267E 00	0.286E 00	0.305E 00	0.323E 00	0.341E 00	0.359E 00
4000	0.376E 00	0.393E 00	0.409E 00	0.424E 00	0.440E 00	0.454E 00	0.468E 00	0.482E 00	0.494E 00	0.507E 00
5000	0.518E 00	0.529E 00	0.540E 00	0.550E 00	0.559E 00	0.568E 00	0.576E 00	0.584E 00	0.591E 00	0.598E 00
6000	0.604E 00	0.609E 00	0.615E 00	0.619E 00	0.624E 00	0.627E 00	0.631E 00	0.634E 00	0.636E 00	0.639E 00
7000	0.640E 00	0.642E 00	0.643E 00	0.644E 00	0.645E 00	0.645E 00	0.645E 00	0.644E 00	0.644E 00	0.643E 00
8000	0.642E 00	0.641E 00	0.639E 00	0.637E 00	0.635E 00	0.633E 00	0.631E 00	0.629E 00	0.626E 00	0.623E 00
9000	0.620E 00	0.617E 00	0.614E 00	0.611E 00	0.608E 00	0.604E 00	0.601E 00	0.597E 00	0.593E 00	0.590E 00

S8

Peak at 0.36 Eff. at 2854 K = 0.0217

Temp	0	100	200	300	400	500	600	700	800	900
0	0.000E 00	0.000E 00	0.243E-35	0.206E-23	0.327E-17	0.167E-13	0.488E-11	0.279E-09	0.578E-08	0.609E-07
1000	0.400E-06	0.186E-05	0.673E-05	0.199E-04	0.504E-04	0.112E-03	0.227E-03	0.423E-03	0.734E-03	0.119E-02
2000	0.186E-02	0.276E-02	0.396E-02	0.550E-02	0.741E-02	0.974E-02	0.125E-01	0.157E-01	0.194E-01	0.237E-01
3000	0.284E-01	0.336E-01	0.393E-01	0.455E-01	0.522E-01	0.592E-01	0.667E-01	0.745E-01	0.826E-01	0.910E-01
4000	0.997E-01	0.108E 00	0.117E 00	0.126E 00	0.136E 00	0.145E 00	0.154E 00	0.163E 00	0.173E 00	0.182E 00
5000	0.191E 00	0.200E 00	0.209E 00	0.217E 00	0.226E 00	0.234E 00	0.242E 00	0.249E 00	0.257E 00	0.264E 00
6000	0.271E 00	0.278E 00	0.284E 00	0.290E 00	0.296E 00	0.301E 00	0.307E 00	0.312E 00	0.316E 00	0.321E 00
7000	0.325E 00	0.329E 00	0.332E 00	0.335E 00	0.339E 00	0.341E 00	0.344E 00	0.346E 00	0.348E 00	0.350E 00
8000	0.352E 00	0.353E 00	0.355E 00	0.356E 00	0.357E 00	0.357E 00	0.358E 00	0.358E 00	0.358E 00	0.358E 00
9000	0.358E 00	0.358E 00	0.358E 00	0.357E 00	0.356E 00	0.356E 00	0.355E 00	0.354E 00	0.353E 00	0.352E 00

TABLE 188a (Continued)

S9

Peak at 0.48 Eff. at 2854 K = 0.0239

Temp	0	100	200	300	400	500	600	700	800	900
0	0.000E 00	0.000E 00	0.000E 00	0.245E-28	0.145E-20	0.632E-16	0.751E-13	0.114E-10	0.481E-09	0.865E-08
1000	0.856E-07	0.549E-06	0.255E-05	0.925E-05	0.276E-04	0.705E-04	0.158E-03	0.322E-03	0.602E-03	0.104E-02
2000	0.170E-01	0.264E-02	0.392E-02	0.559E-02	0.771E-02	0.103E-01	0.134E-01	0.171E-01	0.214E-01	0.262E-01
3000	0.316E-01	0.376E-01	0.440E-01	0.510E-01	0.584E-01	0.662E-01	0.744E-01	0.829E-01	0.917E-01	0.100E 00
4000	0.109E 00	0.119E 00	0.128E 00	0.137E 00	0.147E 00	0.156E 00	0.165E 00	0.174E 00	0.183E 00	0.192E 00
5000	0.201E 00	0.209E 00	0.217E 00	0.225E 00	0.233E 00	0.240E 00	0.247E 00	0.254E 00	0.260E 00	0.266E 00
6000	0.272E 00	0.277E 00	0.282E 00	0.287E 00	0.292E 00	0.296E 00	0.300E 00	0.303E 00	0.307E 00	0.310E 00
7000	0.313E 00	0.315E 00	0.318E 00	0.320E 00	0.321E 00	0.323E 00	0.324E 00	0.325E 00	0.326E 00	0.327E 00
8000	0.328E 00	0.328E 00	0.328E 00	0.328E 00	0.328E 00	0.328E 00	0.328E 00	0.327E 00	0.326E 00	0.326E 00
9000	0.325E 00	0.324E 00	0.323E 00	0.321E 00	0.320E 00	0.319E 00	0.317E 00	0.316E 00	0.314E 00	0.312E 00

S10

Peak at 0.45 Eff. at 2854 K = 0.0322

Temp	0	100	200	300	400	500	600	700	800	900
0	0.000E 00	0.000E 00	0.000E 00	0.404E-24	0.138E-17	0.108E-13	0.414E-11	0.284E-09	0.670E-08	0.775E-07
1000	0.546E-06	0.267E-05	0.100E-04	0.305E-04	0.787E-04	0.178E-03	0.363E-03	0.677E-03	0.117E-02	0.191E-02
2000	0.296E-02	0.438E-02	0.624E-02	0.859E-02	0.114E-01	0.149E-01	0.190E-01	0.237E-01	0.290E-01	0.350E-01
3000	0.415E-01	0.487E-01	0.564E-01	0.645E-01	0.732E-01	0.823E-01	0.917E-01	0.101E 00	0.111E 00	0.121E 00
4000	0.131E 00	0.142E 00	0.152E 00	0.163E 00	0.173E 00	0.183E 00	0.193E 00	0.203E 00	0.213E 00	0.223E 00
5000	0.232E 00	0.241E 00	0.250E 00	0.258E 00	0.266E 00	0.274E 00	0.281E 00	0.289E 00	0.295E 00	0.302E 00
6000	0.308E 00	0.313E 00	0.319E 00	0.324E 00	0.329E 00	0.333E 00	0.337E 00	0.341E 00	0.344E 00	0.347E 00
7000	0.350E 00	0.352E 00	0.355E 00	0.357E 00	0.358E 00	0.360E 00	0.361E 00	0.362E 00	0.363E 00	0.364E 00
8000	0.364E 00	0.364E 00	0.364E 00	0.364E 00	0.364E 00	0.363E 00	0.362E 00	0.362E 00	0.361E 00	0.360E 00
9000	0.358E 00	0.357E 00	0.356E 00	0.354E 00	0.353E 00	0.351E 00	0.349E 00	0.347E 00	0.345E 00	0.343E 00

S11

Peak at 0.44 Eff. at 2854 K = 0.0202

Temp	0	100	200	300	400	500	600	700	800	900
0	0.000E 00	0.000E 00	0.000E 00	0.325E-26	0.223E-19	0.277E-15	0.157E-12	0.154E-10	0.509E-09	0.793E-08
1000	0.723E-07	0.442E-06	0.200E-05	0.715E-05	0.212E-04	0.541E-04	0.122E-03	0.249E-03	0.469E-03	0.819E-03
2000	0.134E-02	0.210E-02	0.315E-02	0.453E-02	0.630E-02	0.850E-02	0.111E-01	0.143E-01	0.180E-01	0.222E-01
3000	0.270E-01	0.323E-01	0.381E-01	0.445E-01	0.512E-01	0.585E-01	0.661E-01	0.740E-01	0.823E-01	0.908E-01

	0	100	200	300	400	500	600	700	800	900
4000	0.995E-01	0.108E 00	0.117E 00	0.126E 00	0.135E 00	0.144E 00	0.153E 00	0.162E 00	0.171E 00	0.180E 00
5000	0.189E 00	0.197E 00	0.205E 00	0.213E 00	0.221E 00	0.229E 00	0.236E 00	0.243E 00	0.249E 00	0.256E 00
6000	0.262E 00	0.267E 00	0.273E 00	0.278E 00	0.283E 00	0.287E 00	0.292E 00	0.296E 00	0.299E 00	0.303E 00
7000	0.306E 00	0.309E 00	0.311E 00	0.314E 00	0.316E 00	0.318E 00	0.319E 00	0.321E 00	0.322E 00	0.323E 00
8000	0.324E 00	0.324E 00	0.325E 00	0.325E 00	0.325E 00	0.325E 00	0.325E 00	0.325E 00	0.324E 00	0.324E 00
9000	0.323E 00	0.322E 00	0.321E 00	0.320E 00	0.319E 00	0.318E 00	0.316E 00	0.315E 00	0.314E 00	0.312E 00

S12   Eff. at 2854 K = 0.407E-02   Peak at 0.50

Temp	0	100	200	300	400	500	600	700	800	900
0	0.000E 00	0.000E 00	0.000E 00	0.194E-31	0.372E-23	0.317E-18	0.605E-15	0.136E-12	0.824E-11	0.204E-09
1000	0.270E-08	0.222E-07	0.128E-06	0.559E-06	0.195E-05	0.570E-05	0.144E-04	0.323E-04	0.657E-04	0.122E-03
2000	0.213E-03	0.350E-03	0.546E-03	0.812E-03	0.116E-02	0.160E-02	0.215E-02	0.282E-02	0.360E-02	0.450E-02
3000	0.552E-02	0.666E-02	0.791E-02	0.927E-02	0.107E-01	0.122E-01	0.139E-01	0.156E-01	0.173E-01	0.191E-01
4000	0.210E-01	0.228E-01	0.247E-01	0.266E-01	0.285E-01	0.304E-01	0.322E-01	0.340E-01	0.358E-01	0.375E-01
5000	0.392E-01	0.408E-01	0.424E-01	0.440E-01	0.454E-01	0.468E-01	0.481E-01	0.494E-01	0.506E-01	0.517E-01
6000	0.528E-01	0.538E-01	0.547E-01	0.556E-01	0.564E-01	0.572E-01	0.579E-01	0.585E-01	0.590E-01	0.596E-01
700T	0.600E-01	0.604E-01	0.608E-01	0.611E-01	0.613E-01	0.616E-01	0.617E-01	0.619E-01	0.619E-01	0.620E-01
8000	0.620E-01	0.620E-01	0.620E-01	0.619E-01	0.618E-01	0.616E-01	0.615E-01	0.613E-01	0.611E-01	0.609E-01
9000	0.606E-01	0.603E-01	0.601E-01	0.597E-01	0.594E-01	0.591E-01	0.588E-01	0.584E-01	0.580E-01	0.576E-01

S13   Eff. at 2854 K = 0.0205   Peak at 0.44

Temp	0	100	200	300	400	500	600	700	800	900
0	0.000E 00	0.000E 00	0.000E 00	0.209E-27	0.403E-20	0.945E-16	0.793E-13	0.994E-11	0.377E-09	0.641E-08
1000	0.618E-07	0.392E-06	0.182E-05	0.665E-05	0.200E-04	0.517E-04	0.117E-03	0.242E-03	0.458E-03	0.805E-03
2000	0.133E-02	0.208E-02	0.313E-02	0.452E-02	0.630E-02	0.853E-02	0.112E-01	0.144E-01	0.182E-01	0.226E-01
3000	0.275E-01	0.330E-01	0.390E-01	0.456E-01	0.527E-01	0.603E-01	0.683E-01	0.767E-01	0.856E-01	0.947E-01
4000	0.104E 00	0.113E 00	0.123E 00	0.133E 00	0.143E 00	0.153E 00	0.164E 00	0.174E 00	0.184E 00	0.194E 00
5000	0.204E 00	0.214E 00	0.223E 00	0.233E 00	0.242E 00	0.251E 00	0.260E 00	0.268E 00	0.276E 00	0.284E 00
6000	0.292E 00	0.299E 00	0.306E 00	0.313E 00	0.319E 00	0.325E 00	0.331E 00	0.336E 00	0.342E 00	0.346E 00
7000	0.351E 00	0.355E 00	0.359E 00	0.363E 00	0.366E 00	0.370E 00	0.373E 00	0.375E 00	0.378E 00	0.380E 00
8000	0.382E 00	0.383E 00	0.385E 00	0.386E 00	0.387E 00	0.388E 00	0.389E 00	0.389E 00	0.390E 00	0.390E 00
9000	0.390E 00	0.389E 00	0.389E 00	0.389E 00	0.388E 00	0.387E 00	0.387E 00	0.386E 00	0.385E 00	0.383E 00

TABLE 188a   (Continued)

S14   Peak at 1.50   Eff. at 2854 K = 0.4656

Temp	0	100	200	300	400	500	600	700	800	900
0	0.000E 00	0.000E 00	0.000E 00	0.000E 00	0.000E 00	0.670E-04	0.579E-03	0.257E-02	0.759E-02	0.170E-01
1000	0.317E-01	0.516E-01	0.762E-01	0.104E 00	0.134E 00	0.166E 00	0.199E 00	0.230E 00	0.261E 00	0.290E 00
2000	0.317E 00	0.342E 00	0.365E 00	0.386E 00	0.404E 00	0.421E 00	0.436E 00	0.449E 00	0.460E 00	0.469E 00
3000	0.477E 00	0.484E 00	0.490E 00	0.494E 00	0.498E 00	0.501E 00	0.502E 00	0.503E 00	0.504E 00	0.503E 00
4000	0.503E 00	0.501E 00	0.499E 00	0.497E 00	0.495E 00	0.492E 00	0.489E 00	0.485E 00	0.481E 00	0.478E 00
5000	0.474E 00	0.469E 00	0.465E 00	0.460E 00	0.456E 00	0.451E 00	0.446E 00	0.441E 00	0.436E 00	0.431E 00
6000	0.426E 00	0.421E 00	0.416E 00	0.411E 00	0.406E 00	0.401E 00	0.396E 00	0.391E 00	0.386E 00	0.380E 00
7000	0.375E 00	0.370E 00	0.365E 00	0.360E 00	0.356E 00	0.351E 00	0.346E 00	0.341E 00	0.336E 00	0.332E 00
8000	0.327E 00	0.322E 00	0.318E 00	0.313E 00	0.309E 00	0.304E 00	0.300E 00	0.296E 00	0.292E 00	0.287E 00
9000	0.283E 00	0.279E 00	0.275E 00	0.271E 00	0.267E 00	0.264E 00	0.260E 00	0.256E 00	0.252E 00	0.249E 00

S15   Peak at 0.58   Eff. at 2854 K = 0.0464

Temp	0	100	200	300	400	500	600	700	800	900
0	0.000E 00	0.000E 00	0.800E-34	0.536E-22	0.461E-16	0.162E-12	0.375E-10	0.183E-08	0.342E-07	0.333E-06
1000	0.206E-05	0.911E-05	0.312E-04	0.883E-04	0.213E-03	0.455E-03	0.876E-03	0.155E-02	0.256E-02	0.398E-02
2000	0.590E-02	0.836E-02	0.114E-01	0.151E-01	0.194E-01	0.243E-01	0.299E-01	0.360E-01	0.427E-01	0.498E-01
3000	0.573E-01	0.651E-01	0.732E-01	0.815E-01	0.899E-01	0.983E-01	0.106E 00	0.115E 00	0.123E 00	0.131E 00
4000	0.139E 00	0.147E 00	0.155E 00	0.162E 00	0.169E 00	0.175E 00	0.182E 00	0.188E 00	0.193E 00	0.199E 00
5000	0.204E 00	0.208E 00	0.213E 00	0.217E 00	0.220E 00	0.224E 00	0.227E 00	0.229E 00	0.232E 00	0.234E 00
6000	0.236E 00	0.238E 00	0.239E 00	0.240E 00	0.241E 00	0.242E 00	0.243E 00	0.243E 00	0.243E 00	0.243E 00
7000	0.243E 00	0.243E 00	0.242E 00	0.242E 00	0.241E 00	0.240E 00	0.239E 00	0.238E 00	0.237E 00	0.236E 00
8000	0.234E 00	0.233E 00	0.232E 00	0.230E 00	0.228E 00	0.227E 00	0.225E 00	0.223E 00	0.222E 00	0.220E 00
9000	0.218E 00	0.216E 00	0.214E 00	0.212E 00	0.210E 00	0.208E 00	0.206E 00	0.205E 00	0.203E 00	0.201E 00

S16   Peak at 0.73   Eff. at 2854 K = 0.1067

Temp	0	100	200	300	400	500	600	700	800	900
0	0.000E 00	0.000E 00	0.153E-25	0.198E-16	0.688E-12	0.358E-09	0.228E-07	0.440E-06	0.399E-05	0.219E-04
1000	0.850E-04	0.254E-03	0.628E-03	0.133E-02	0.252E-02	0.434E-02	0.693E-02	0.104E-01	0.147E-01	0.201E-01
2000	0.264E-01	0.336E-01	0.416E-01	0.503E-01	0.597E-01	0.695E-01	0.798E-01	0.902E-01	0.100E 00	0.111E 00
3000	0.122E 00	0.132E 00	0.143E 00	0.153E 00	0.163E 00	0.172E 00	0.182E 00	0.190E 00	0.199E 00	0.207E 00

	0	100	200	300	400	500	600	700	800	900
4000	0.214E 00	0.221E 00	0.228E 00	0.235E 00	0.240E 00	0.246E 00	0.251E 00	0.256E 00	0.260E 00	0.264E 00
5000	0.267E 00	0.271E 00	0.274E 00	0.276E 00	0.278E 00	0.280E 00	0.282E 00	0.284E 00	0.285E 00	0.286E 00
6000	0.286E 00	0.287E 00	0.287E 00	0.287E 00	0.287E 00	0.287E 00	0.287E 00	0.286E 00	0.285E 00	0.284E 00
7000	0.284E 00	0.282E 00	0.281E 00	0.280E 00	0.279E 00	0.277E 00	0.276E 00	0.274E 00	0.272E 00	0.270E 00
8000	0.269E 00	0.267E 00	0.265E 00	0.263E 00	0.261E 00	0.259E 00	0.257E 00	0.254E 00	0.252E 00	0.250E 00
9000	0.248E 00	0.246E 00	0.243E 00	0.241E 00	0.239E 00	0.237E 00	0.234E 00	0.232E 00	0.230E 00	0.227E 00

S17 Peak at 0.49 Eff. at 2854 K = 0.0245

Temp	0	100	200	300	400	500	600	700	800	900
0	0.000E 00	0.000E 00	0.000E 00	0.214E-26	0.253E-19	0.416E-15	0.263E-12	0.264E-10	0.848E-09	0.127E-07
1000	0.111E-06	0.660E-06	0.290E-05	0.101E-04	0.295E-04	0.742E-04	0.165E-03	0.333E-03	0.618E-03	0.106E-02
2000	0.174E-02	0.269E-02	0.399E-02	0.570E-02	0.786E-02	0.105E-01	0.137E-01	0.175E-01	0.219E-01	0.269E-01
3000	0.325E-01	0.387E-01	0.455E-01	0.529E-01	0.607E-01	0.691E-01	0.779E-01	0.870E-01	0.965E-01	0.106E 00
4000	0.116E 00	0.126E 00	0.137E 00	0.147E 00	0.158E 00	0.168E 00	0.179E 00	0.189E 00	0.200E 00	0.210E 00
5000	0.220E 00	0.230E 00	0.239E 00	0.249E 00	0.258E 00	0.267E 00	0.276E 00	0.284E 00	0.292E 00	0.300E 00
6000	0.307E 00	0.314E 00	0.321E 00	0.327E 00	0.333E 00	0.339E 00	0.345E 00	0.350E 00	0.355E 00	0.359E 00
7000	0.363E 00	0.367E 00	0.371E 00	0.374E 00	0.377E 00	0.380E 00	0.383E 00	0.385E 00	0.387E 00	0.389E 00
8000	0.390E 00	0.392E 00	0.393E 00	0.394E 00	0.394E 00	0.395E 00	0.395E 00	0.395E 00	0.395E 00	0.395E 00
9000	0.395E 00	0.395E 00	0.394E 00	0.393E 00	0.392E 00	0.391E 00	0.390E 00	0.389E 00	0.388E 00	0.386E 00

S18 Peak at 0.45 Eff. at 2854 K = 0.0295

Temp	0	100	200	300	400	500	600	700	800	900
0	0.000E 00	0.000E 00	0.000E 00	0.109E-22	0.156E-16	0.702E-13	0.181E-10	0.931E-09	0.174E-07	0.168E-06
1000	0.102E-05	0.446E-05	0.151E-04	0.423E-04	0.102E-03	0.218E-03	0.423E-03	0.758E-03	0.127E-02	0.201E-02
2000	0.304E-02	0.441E-02	0.616E-02	0.836E-02	0.110E-01	0.141E-01	0.178E-01	0.220E-01	0.268E-01	0.320E-01
3000	0.377E-01	0.439E-01	0.505E-01	0.575E-01	0.648E-01	0.724E-01	0.802E-01	0.882E-01	0.964E-01	0.104E 00
4000	0.112E 00	0.121E 00	0.129E 00	0.137E 00	0.145E 00	0.153E 00	0.161E 00	0.168E 00	0.176E 00	0.183E 00
5000	0.189E 00	0.196E 00	0.202E 00	0.208E 00	0.214E 00	0.219E 00	0.224E 00	0.229E 00	0.233E 00	0.237E 00
6000	0.241E 00	0.245E 00	0.248E 00	0.251E 00	0.254E 00	0.256E 00	0.258E 00	0.260E 00	0.262E 00	0.263E 00
7000	0.265E 00	0.266E 00	0.266E 00	0.267E 00	0.268E 00	0.268E 00	0.268E 00	0.268E 00	0.268E 00	0.267E 00
8000	0.267E 00	0.266E 0t	0.266E 00	0.265E 00	0.264E 00	0.263E 00	0.262E 00	0.260E 00	0.259E 00	0.258E 00
9000	0.256E 00	0.255E 00	0.253E 00	0.251E 00	0.250E 00	0.248E 00	0.246E 00	0.244E 00	0.243E 00	0.241E 00

TABLE 188a (Continued)

S19   Peak at 0.33   Eff. at 2854 K = 0.0103

Temp	0	100	200	300	400	500	600	700	800	900
0	0.000E 00	0.000E 00	0.000E 00	0.162E-25	0.999E-19	0.110E-14	0.524E-12	0.415E-10	0.107E-08	0.133E-07
1000	0.990E-07	0.506E-06	0.196E-05	0.614E-05	0.163E-04	0.380E-04	0.796E-04	0.152E-03	0.272E-03	0.457E-03
2000	0.730E-03	0.111E-02	0.163E-02	0.232E-02	0.320E-02	0.430E-02	0.565E-02	0.727E-02	0.919E-02	0.114E-01
3000	0.139-E01	0.169E-01	0.201E-01	0.237E-01	0.277E-01	0.321E-01	0.368E-01	0.419E-01	0.474E-01	0.531E-01
4000	0.592E-01	0.657E-01	0.724E-01	0.794E-01	0.866E-01	0.942E-01	0.101E 00	0.109E 00	0.118E 00	0.126E 00
5000	0.134E 00	0.143E 00	0.151E 00	0.160E 00	0.169E 00	0.178E 00	0.187E 00	0.196E 00	0.205E 00	0.214E 00
6000	0.223E 00	0.232E 00	0.240E 00	0.249E 00	0.258E 00	0.267E 00	0.275E 00	0.284E 00	0.292E 00	0.300E 00
7000	0.308E 00	0.316E 00	0.324E 00	0.332E 00	0.340E 00	0.347E 00	0.354E 00	0.362E 00	0.369E 00	0.375E 00
8000	0.382E 00	0.389E 00	0.395E 00	0.401E 00	0.407E 00	0.413E 00	0.419E 00	0.424E 00	0.429E 00	0.435E 00
9000	0.440E 00	0.445E 00	0.449E 00	0.454E 00	0.458E 00	0.462E 00	0.466E 00	0.470E 00	0.474E 00	0.477E 00

S19 Filtered   Peak at 0.33   Eff. at 2854 K = 0.0100

Temp	0	100	200	300	400	500	600	700	800	900
0	0.000E 00	0.000E 00	0.000E 00	0.162E-25	0.999E-19	0.110E-14	0.524E-12	0.415E-10	0.107E-08	0.133E-07
1000	0.990E-07	0.506E-06	0.196E-05	0.614E-05	0.163E-04	0.380E-04	0.795E-04	0.152E-03	0.272E-03	0.457E-03
2000	0.728E-03	0.110E-02	0.162E-02	0.230E-02	0.317E-02	0.424E-02	0.555E-02	0.711E-02	0.895E-02	0.110E-01
3000	0.134E-01	0.161E-01	0.191E-01	0.223E-01	0.259E-01	0.297E-01	0.337E-01	0.380E-01	0.425E-01	0.472E-01
4000	0.520E-01	0.570E-01	0.621E-01	0.673E-01	0.726E-01	0.779E-01	0.832E-01	0.885E-01	0.938E-01	0.991E-01
5000	0.104E 00	0.109E 00	0.114E 00	0.119E 00	0.124E 00	0.128E 00	0.133E 00	0.137E 00	0.142E 00	0.146E 00
6000	0.150E 00	0.154E 00	0.157E 00	0.161E 00	0.164E 00	0.167E 00	0.170E 00	0.173E 00	0.175E 00	0.178E 00
7000	0.180E 00	0.182E 00	0.184E 00	0.186E 00	0.188E 00	0.189E 00	0.191E 00	0.192E 00	0.193E 00	0.194E 00
8000	0.195E 00	0.195E 00	0.196E 00	0.196E 00	0.197E 00	0.197E 00	0.197E 00	0.197E 00	0.197E 00	0.197E 00
9000	0.197E 00	0.197E 00	0.197E 00	0.196E 00	0.196E 00	0.195E 00	0.195E 00	0.194E 00	0.193E 00	0.192E 00

S20   Peak at 0.42   Eff. at 2854 K = 0.0383

Temp	0	100	200	300	400	500	600	700	800	900
0	0.000E 00	0.000E 00	0.297E-34	0.231E-22	0.280E-16	0.118E-12	0.299E-10	0.152E-08	0.285E-07	0.274E-06
1000	0.166E-05	0.719E-05	0.241E-04	0.669E-04	0.159E-03	0.336E-03	0.642E-03	0.113E-02	0.187E-02	0.292E-02
2000	0.434E-02	0.620E-02	0.855E-02	0.114E-01	0.149E-01	0.189E-01	0.236E-01	0.290E-01	0.349E-01	0.414E-01
3000	0.485E-01	0.562E-01	0.644E-01	0.730E-01	0.821E-01	0.916E-01	0.101E 00	0.111E 00	0.121E 00	0.132E 00

	0	100	200	300	400	500	600	700	800	900
4000	0.142E 00	0.153E 00	0.164E 00	0.174E 00	0.185E 00	0.195E 00	0.206E 00	0.216E 00	0.226E 00	0.236E 00
5000	0.246E 00	0.255E 00	0.264E 00	0.273E 00	0.281E 00	0.289E 00	0.297E 00	0.305E 00	0.312E 00	0.319E 00
6000	0.325E 00	0.331E 00	0.337E 00	0.342E 00	0.347E 00	0.352E 00	0.357E 00	0.361E 00	0.365E 00	0.368E 00
7000	0.371E 00	0.374E 00	0.377E 00	0.379E 00	0.381E 00	0.383E 00	0.385E 00	0.386E 00	0.387E 00	0.388E 00
8000	0.389E 00	0.389E 00	0.390E 00	0.390E 00	0.390E 00	0.390E 00	0.389E 00	0.389E 00	0.388E 00	0.387E 00
9000	0.386E 00	0.385E 00	0.384E 00	0.382E 00	0.381E 00	0.379E 00	0.378E 00	0.376E 00	0.374E 00	0.372E 00

S21

Peak at 0.44   Eff. at 2854 K = 0.0209

Temp	0	100	200	300	400	500	600	700	800	90fi
0	0.000E 00	0.000E 00	0.000E 00	0.633E-27	0.931E-20	0.182E-15	0.133E-12	0.150E-10	0.523E-09	0.834E-08
1000	0.766E-07	0.469E-06	0.211E-05	0.755E-05	0.223E-04	0.568E-04	0.128E-03	0.261E-03	0.489E-03	0.853E-03
2000	0.140E-02	0.218E-02	0.326E-02	0.469E-02	0.652E-02	0.880E-02	0.115E-01	0.148E-01	0.186E-01	0.230E-01
3000	0.280E-01	0.335E-01	0.395E-01	0.461E-01	0.532E-01	0.608E-01	0.688E-01	0.773E-01	0.860E-01	0.951E-01
4000	0.104E 00	0.114E 00	0.123E 00	0.133E 00	0.143E 00	0.153E 00	0.163E 00	0.173E 00	0.183E 00	0.193E 00
5000	0.203E 00	0.213E 00	0.222E 00	0.231E 00	0.241E 00	0.249E 00	0.258E 00	0.266E 00	0.274E 00	0.282E 00
6000	0.290E 00	0.297E 00	0.304E 00	0.310E 00	0.316E 00	0.322E 00	0.328E 00	0.333E 00	0.339E 00	0.343E 00
7000	0.348E 00	0.352E 00	0.356E 00	0.359E 00	0.363E 00	0.366E 00	0.369E 00	0.371E 00	0.374E 00	0.376E 00
8000	0.378E 00	0.379E 00	0.381E 00	0.382E 00	0.383E 00	0.384E 00	0.384E 00	0.385E 00	0.385E 00	0.385E 00
9000	0.385E 00	0.385E 00	0.385E 00	0.384E 00	0.384E 00	0.383E 00	0.382E 00	0.381E 00	0.380E 00	0.379E 00

S23

Peak at 0.23   Eff. at 2854 K = 0.128E-04

Temp	0	100	200	300	400	500	600	700	800	900
0	0.000E 00	0.000E 00	0.000E 00	0.000E 00	0.000E 00	0.000E 00	0.952E-29	0.242E-24	0.467E-21	0.163E-18
1000	0.173E-16	0.782E-15	0.186E-13	0.272E-12	0.272E-11	0.200E-10	0.115E-09	0.543E-09	0.215E-08	0.743E-08
2000	0.226E-07	0.622E-07	0.155E-06	0.360E-06	0.777E-06	0.157E-05	0.302E-05	0.552E-05	0.965E-05	0.162E-04
3000	0.262E-04	0.411E-04	0.626E-04	0.928E-04	0.134E-03	0.189E-03	0.262E-03	0.356E-03	0.476E-03	0.625E-03
4000	0.809E-03	0.103E-02	0.130E-02	0.161E-02	0.199E-02	0.242E-02	0.292E-02	0.349E-02	0.414E-02	0.486E-02
5000	0.567E-02	0.657E-02	0.756E-02	0.865E-02	0.983E-02	0.111E-01	0.124E-01	0.139E-01	0.155E-01	0.172E-01
6000	0.190E-01	0.209E-01	0.229E-01	0.249E-01	0.271E-01	0.294E-01	0.317E-01	0.342E-01	0.367E-01	0.393E-01
7000	0.420E-01	0.447E-01	0.476E-01	0.504E-01	0.534E-01	0.564E-01	0.594E-01	0.625E-01	0.656E-01	0.688E-01
8000	0.719E-01	0.752E-01	0.784E-01	0.816E-01	0.849E-01	0.882E-01	0.914E-01	0.947E-01	0.979E-01	0.101E-00
9000	0.104E 00	0.107E 00	0.110E 00	0.114E 00	0.117E 00	0.120E 00	0.123E 00	0.126E 00	0.129E 00	0.132E 00

TABLE 188a (Continued)

S25

Peak at 0.42 Eff. at 2854 K = 0.0602

Temp	0	100	200	300	400	500	600	700	800	900
0	0.000E 00	0.000E 00	0.745E-30	0.356E-19	0.690E-14	0.955E-11	0.113E-08	0.330E-07	0.405E-06	0.279E-05
1000	0.128E-04	0.445E-04	0.123E-03	0.291E-03	0.605E-03	0.113E-02	0.194E-02	0.313E-02	0.476E-02	0.691E-02
2000	0.964E-02	0.130E-01	0.170E-01	0.217E-01	0.271E-01	0.332E-01	0.400E-01	0.475E-01	0.556E-01	0.643E-01
3000	0.736E-01	0.834E-01	0.937E-01	0.104E 00	0.115E 00	0.126E 00	0.138E 00	0.150E 00	0.162E 00	0.174E 00
4000	0.186E 00	0.199E 00	0.211E 00	0.223E 00	0.235E 00	0.247E 00	0.259E 00	0.271E 00	0.282E 00	0.293E 00
5000	0.304E 00	0.315E 00	0.325E 00	0.335E 00	0.345E 00	0.354E 00	0.363E 00	0.372E 00	0.380E 00	0.388E 00
6000	0.395E 00	0.403E 00	0.409E 00	0.416E 00	0.422E 00	0.428E 00	0.433E 00	0.438E 00	0.443E 00	0.447E 00
7000	0.451E 00	0.455E 00	0.459E 00	0.462E 00	0.465E 00	0.467E 00	0.469E 00	0.471E 00	0.473E 00	0.475E 00
8000	0.476E 00	0.477E 00	0.478E 00	0.478E 00	0.479E 00	0.479E 00	0.479E 00	0.479E 00	0.478E 00	0.478E 00
9000	0.477E 00	0.476E 00	0.475E 00	0.474E 00	0.473E 00	0.471E 00	0.470E 00	0.468E 00	0.466E 00	0.464E 00

Photopic

Peak at 0.555 Eff. at 2854 K = 0.0239

Temp	0	100	200	300	400	500	600	700	800	900
0	0.000E 00	0.000E 00	0.000E 00	0.107E-25	0.419E-19	0.609E-15	0.421E-12	0.467E-10	0.160E-08	0.247E-07
1000	0.217E-06	0.127E-05	0.544E-05	0.183E-04	0.514E-04	0.123E-03	0.264E-03	0.510E-03	0.907E-03	0.150E-02
2000	0.235E-02	0.349E-02	0.497E-02	0.681E-02	0.904E-02	0.116E-01	0.146E-01	0.180E-01	0.218E-01	0.258E-01
3000	0.302E-01	0.348E-01	0.396E-01	0.445E-01	0.496E-01	0.547E-01	0.599E-01	0.650E-01	0.701E-01	0.751E-01
4000	0.800E-01	0.848E-01	0.894E-01	0.938E-01	0.981E-01	0.102E 00	0.105E 00	0.109E 00	0.113E 00	0.116E 00
5000	0.119E 00	0.121E 00	0.124E 00	0.126E 00	0.128E 00	0.130E 00	0.132E 00	0.133E 00	0.135E 00	0.136E 00
6000	0.137E 00	0.137E 00	0.138E 00	0.139E 00	0.139E 00	0.139E 00	0.139E 00	0.139E 00	0.139E 00	0.139E 00
7000	0.138E 00	0.138E 00	0.138E 00	0.137E 00	0.136E 00	0.136E 00	0.135E 00	0.134E 00	0.133E 00	0.132E 00
8000	0.131E 00	0.130E 00	0.129E 00	0.128E 00	0.127E 00	0.126E 00	0.125E 00	0.123E 00	0.122E 00	0.121E 00
9000	0.120E 00	0.118E 00	0.117E 00	0.116E 00	0.115E 00	0.113E 00	0.112E 00	0.111E 00	0.110E 00	0.108E 00

Scotopic

Peak at 0.507 Eff. at 2854 K = 0.0136

Temp	0	100	200	300	400	500	600	700	800	900
0	0.000E 00	0.000E 00	0.000E 00	0.397E-28	0.261E-21	0.733E-17	0.962E-14	0.179E-11	0.929E-10	0.200E-08
1000	0.230E-07	0.168E-06	0.873E-06	0.346E-05	0.111E-04	0.303E-04	0.721E-04	0.153E-03	0.296E-03	0.531E-03
2000	0.890E-03	0.140E-02	0.212E-02	0.307E-02	0.428E-02	0.578E-02	0.758E-02	0.971E-02	0.121E-01	0.149E-01
3000	0.179E-01	0.213E-01	0.249E-01	0.288E-01	0.329E-01	0.371E-01	0.415E-01	0.461E-01	0.507E-01	0.553E-01
4000	0.600E-01	0.647E-01	0.693E-01	0.739E-01	0.784E-01	0.828E-01	0.871E-01	0.913E-01	0.953E-01	0.991E-01
5000	0.102E 00	0.106E 00	0.109E 00	0.112E 00	0.115E 00	0.118E 00	0.121E 00	0.123E 00	0.126E 00	0.128E 00
6000	0.130E 00	0.131E 00	0.133E 00	0.134E 00	0.136E 00	0.137E 00	0.138E 00	0.138E 00	0.139E 00	0.140E 00
7000	0.140E 00	0.141E 00	0.141E 00	0.141E 00	0.141E 00	0.141E 00	0.141E 00	0.141E 00	0.140E 00	0.140E 00
8000	0.139E 00	0.139E 00	0.138E 00	0.138E 00	0.137E 00	0.136E 00	0.136E 00	0.135E 00	0.134E 00	0.133E 00
9000	0.132E 00	0.131E 00	0.130E 00	0.129E 00	0.128E 00	0.127E 00	0.126E 00	0.125E 00	0.124E 00	0.123E 00

TABLE 188*b* SPECTRAL MATCHING FACTORS, P AND S CURVES
(Section 18.1.4.1)

	S1	S3	S4	S5	S6	S7	S8	S9	S10	S11	S12	S13
P1	0.278	775	494	395	0	986	522	840	804	683	145	716
P2	0.314	746	427	346	0	986	480	734	742	594	111	610
P3	0.482	592	163	140	0	996	273	359	472	259	19	256
P4[a]	0.349	766	498	401	7	967	531	661	724	601	116	610
P5	0.410	930	869	724	58	912	836	824	910	892	175	919
P6	0.344	741	438	350	1	978	484	648	700	559	115	566
P7	0.298	824	602	484	10	955	614	750	804	705	133	717
P11	0.207	929	817	637	8	943	767	932	950	917	260	934
P12	0.480	588	135	121	0	998	261	344	466	234	7	222
P13	0.628	482	35	32	0	996	144	81	263	54	0	49
P14	0.386	725	427	344	6	970	474	559	650	511	99	515
P15	0.383	862	701	578	37	940	698	815	854	787	197	816
P16	0.827	918	969	888	156	878	965	700	854	880	94	929
P17	0.301	797	538	434	6	967	567	749	785	667	127	680
P18	0.487	761	569	493	52	954	604	631	713	617	102	644
P19	0.486	583	127	114	0	998	255	325	455	219	11	206
P20	0.393	667	285	237	0	993	372	562	613	428	57	435
P21	0.450	624	203	170	0	992	313	385	511	289	46	279
P22	0.377	732	458	363	4	971	486	592	666	546	118	558
P23	0.271	779	505	404	0	984	526	869	820	704	36	749
P24	0.277	800	550	431	0	976	564	810	808	700	208	719
P25	0.524	557	109	97	0	997	228	266	406	183	5	175
P26	0.476	589	126	114	0	998	259	336	466	223	6	208
P27	0.601	500	44	41	0	997	161	109	294	71	0	63
P28	0.334	722	378	309	0	989	446	687	706	543	92	555

[a] P4 is the silicate-sulfide type.

TABLE 188*b*   (*Continued*)

S14	S15	S16	S17	S18	S19	S20	S21	S25	Phot.	Scot.	
356	723	366	888	666	218	698	721	737	738	769	P1
363	761	374	755	618	209	670	621	714	521	722	P2
404	800	428	343	383	105	499	268	571	171	560	P3
349	605	373	672	641	289	699	612	746	374	433	P4*
264	306	315	893	784	578	895	895	933	329	129	P5
362	667	381	651	614	241	666	573	717	399	505	P6
330	560	354	770	725	352	767	715	804	397	420	P7
305	377	330	954	905	456	879	923	908	600	197	P11
405	862	422	305	370	98	497	233	567	105	606	P12
434	635	478	78	236	57	371	60	473	9	217	P13
362	611	388	561	583	259	653	517	707	283	382	P14
303	488	.338	871	727	425	802	799	843	494	376	P15
192	175	287	855	510	784	902	887	972	45	3	P16
342	641	361	767	688	300	734	683	773	442	539	P17
314	520	356	682	564	393	714	637	775	281	347	P18
406	860	423	276	363	97	492	212	562	90	581	P19
384	815	400	564	495	148	583	447	639	356	707	P20
394	812	410	333	425	138	537	282	603	150	522	P21
362	554	390	613	604	261	657	560	713	390	343	P22
355	719	365	929	670	210	703	759	742	810	861	P23
348	627	363	829	711	274	727	719	768	627	530	P24
413	784	438	244	331	86	460	188	539	87	483	P25
404	888	418	282	367	97	499	215	567	83	615	P26
429	692	463	94	256	63	394	75	490	12	273	P27
369	801	379	699	578	188	645	566	691	462	751	P28

TABLE 188c MISCELLANEOUS MATCHING FACTORS (Section 18.1.4.1)

| Category | Detector | CIE standard illuminant[a] | | | | Phot.[b] Daylight | Blackbody | | Phosphors | | Ref. Wavelength (μm) |
		A	B	C	D_{65}		3000°K	5000°K	P-11	P-16	
Norm. Observ.	Photopic	0.1777	0.2565	0.2898	0.2662	0.3702	0.0303	0.1192	0.2010	0.0032	0.555
	Scotopic	0.1008	0.2196	0.2914	0.2637	0.3354	0.0180	0.1028	0.6033	0.0424	0.507
Photoelectr. Detectors	S10	0.2380	0.4493	0.5886	0.5505	0.6425	0.0414	0.2275	0.9504	0.8593	0.450
	S11	0.1497	0.3526	0.4893	0.4630	0.5163	0.0269	0.1847	0.9159	0.8858	0.440
	S12[c]	0.0302	0.0799	0.1134	0.1025	0.1166	0.0055	0.0389	0.2701	0.0945	0.502
	S20	0.2833	0.4682	0.5939	0.5578	0.6390	0.0483	0.2366	0.8783	0.9057	0.420
	CdSe	0.2563	0.1697	0.1312	0.1392	0.0626	0.0481	0.0871	0.0401	0.0400	0.740
	Si diode	0.8254	0.7379	0.6803	0.6838	0.6707	0.2794	0.4918	0.5371	0.3292	0.940
	Vidicon[d]	0.4157	0.5479	0.6335	0.5798	0.7092	0.0704	0.2594	0.7306	0.3615	0.550
Photographic Emulsions	Blue-Sens.	0.085	0.219	0.2944	0.3340	0.297	0.016	0.142	0.429	1.341	0.400
	Orthochrom.	0.273	0.580	0.7717	0.7200	0.854	0.049	0.241	0.243	0.941	0.400
	Panchrom.	0.259	0.454	0.5706	0.5470	0.648	0.045	0.235	0.688	0.969	0.400
Lum. Efficacy	Phot. (lm/W)[e]	121	174	197	181	252	20.6	81.0	137	2.2	
	Scot. (lm/W)[f]	176	383	509	460	586	31.4	180	1053	74.1	

[a] Matching factors in this category computed assuming the spectra to vanish below 0.35 μm.
[b] Based on Ref. 17.58b.
[c] Representative for CdS.
[d] RCA Type 2 Surface.
[e] Based on 680 lm/W at reference wavelength.
[f] Based on 1746 lm/W at reference wavelength.
Source. Levi [18.3], with additions.

TABLE 189 PERFECT-LENS-MTF, POLYCHROMATIC LIGHT (Section 18.1.4.2)
MTF of Source-Detector Combinations
Blackbody at 3000 K

ν_s c/mm	Normal observer		Photoelectric detectors							Photographic emulsions		
	Pho-topic	Sco-topic	S10	S11	S12	S20	CdSe	Si diode	Vidicon	Blue sensi-tive	Ortho chro-matic	Pan chro-matic
50	0.9637	0.9672	0.9650	0.9676	0.9689	0.9632	0.9513	0.9470	0.9607	0.9692	0.9669	0.9648
100	0.9273	0.9345	0.9301	0.9352	0.9378	0.9264	0.9027	0.8942	0.9214	0.9384	0.9338	0.9297
150	0.8911	0.9018	0.8952	0.9028	0.9067	0.8897	0.8542	0.8415	0.8822	0.9076	0.9007	0.8946
200	0.8549	0.8691	0.8603	0.8705	0.8757	0.8530	0.8059	0.7891	0.8431	0.8769	0.8677	0.8596
250	0.8189	0.8366	0.8256	0.8383	0.8448	0.8165	0.7580	0.7372	0.8042	0.8462	0.8348	0.8248
300	0.7830	0.8042	0.7911	0.8062	0.8139	0.7802	0.7104	0.6858	0.7655	0.8157	0.8020	0.7900
350	0.7473	0.7719	0.7567	0.7742	0.7832	0.7441	0.6633	0.6350	0.7271	0.7853	0.7694	0.7554
400	0.7118	0.7397	0.7225	0.7424	0.7526	0.7082	0.6167	0.5850	0.6889	0.7550	0.7369	0.7211
450	0.6765	0.7077	0.6885	0.7108	0.7222	0.6726	0.5708	0.5358	0.6511	0.7248	0.7046	0.6869
500	0.6416	0.6760	0.6548	0.6794	0.6919	0.6373	0.5255	0.4876	0.6136	0.6949	0.6725	0.6531
550	0.6069	0.6445	0.6214	0.6482	0.6618	0.6024	0.4811	0.4406	0.5766	0.6651	0.6407	0.6195
600	0.5726	0.6132	0.5884	0.6173	0.6320	0.5679	0.4376	0.3949	0.5401	0.6356	0.6091	0.5862
650	0.5387	0.5822	0.5557	0.5866	0.6024	0.5337	0.3951	0.3507	0.5040	0.6063	0.5779	0.5533
700	0.5052	0.5515	0.5234	0.5562	0.5730	0.5001	0.3537	0.3081	0.4686	0.5772	0.5469	0.5208
750	0.4722	0.5211	0.4915	0.5262	0.5440	0.4670	0.3136	0.2674	0.4338	0.5485	0.5163	0.4887
800	0.4397	0.4911	0.4601	0.4966	0.5152	0.4345	0.2749	0.2289	0.3997	0.5200	0.4861	0.4571
850	0.4077	0.4615	0.4292	0.4673	0.4867	0.4025	0.2378	0.1928	0.3664	0.4919	0.4562	0.4260
900	0.3763	0.4323	0.3988	0.4384	0.4587	0.3713	0.2024	0.1599	0.3339	0.4642	0.4269	0.3954
950	0.3455	0.4035	0.3691	0.4100	0.4310	0.3408	0.1690	0.1308	0.3024	0.4368	0.3980	0.3654
1000	0.3154	0.3753	0.3400	0.3821	0.4037	0.3111	0.1380	0.1061	0.2719	0.4099	0.3696	0.3361
1050	0.2861	0.3475	0.3116	0.3547	0.3768	0.2822	0.1097	0.0856	0.2424	0.3834	0.3417	0.3074
1100	0.2576	0.3204	0.2840	0.3279	0.3504	0.2544	0.0846	0.0688	0.2142	0.3573	0.3144	0.2795
1150	0.2299	0.2938	0.2572	0.3017	0.3245	0.2275	0.0626	0.0552	0.1875	0.3318	0.2878	0.2524
1200	0.2032	0.2678	0.2313	0.2761	0.2991	0.2018	0.0440	0.0441	0.1624	0.3068	0.2619	0.2261
1250	0.1775	0.2425	0.2064	0.2512	0.2743	0.1775	0.0290	0.0351	0.1391	0.2824	0.2366	0.2008
1300	0.1528	0.2180	0.1825	0.2271	0.2501	0.1547	0.0181	0.0279	0.1178	0.2587	0.2122	0.1765
1350	0.1295	0.1942	0.1599	0.2038	0.2266	0.1337	0.0114	0.0221	0.0987	0.2356	0.1886	0.1533
1400	0.1074	0.1713	0.1388	0.1813	0.2037	0.1145	0.0082	0.0174	0.0817	0.2132	0.1659	0.1314
1450	0.0869	0.1493	0.1192	0.1599	0.1816	0.0974	0.0062	0.0137	0.0668	0.1916	0.1442	0.1108
1500	0.0680	0.1283	0.1014	0.1395	0.1602	0.0822	0.0047	0.0107	0.0540	0.1708	0.1235	0.0918
1550	0.0513	0.1084	0.0853	0.1203	0.1398	0.0690	0.0037	0.0084	0.0432	0.1509	0.1041	0.0747
1600	0.0369	0.0898	0.0711	0.1024	0.1202	0.0575	0.0029	0.0065	0.0341	0.1321	0.0860	0.0600
1650	0.0252	0.0724	0.0587	0.0861	0.1017	0.0476	0.0023	0.0050	0.0266	0.1143	0.0694	0.0479
1700	0.0162	0.0567	0.0480	0.0715	0.0843	0.0392	0.0018	0.0039	0.0205	0.0979	0.0547	0.0383
1750	0.0098	0.0428	0.0389	0.0587	0.0682	0.0320	0.0015	0.0030	0.0156	0.0830	0.0424	0.0308
1800	0.0055	0.0310	0.0313	0.0476	0.0535	0.0260	0.0012	0.0023	0.0118	0.0700	0.0328	0.0248
1850	0.0029	0.0215	0.0249	0.0383	0.0404	0.0210	0.0009	0.0018	0.0089	0.0590	0.0255	0.0201
1900	0.0015	0.0143	0.0197	0.0305	0.0289	0.0168	0.0007	0.0013	0.0066	0.0500	0.0199	0.0164
1950	0.0007	0.0091	0.0155	0.0241	0.0195	0.0134	0.0006	0.0010	0.0049	0.0425	0.0155	0.0134
2000	0.0003	0.0055	0.0120	0.0188	0.0129	0.0105	0.0005	0.0008	0.0036	0.0362	0.0120	0.0108
2050	0.0002	0.0032	0.0093	0.0146	0.0092	0.0082	0.0004	0.0006	0.0026	0.0308	0.0092	0.0087
2100	0.0001	0.0018	0.0071	0.0112	0.0065	0.0063	0.0003	0.0004	0.0019	0.0260	0.0069	0.0069
2150	0.0000	0.0009	0.0053	0.0085	0.0046	0.0048	0.0002	0.0003	0.0014	0.0218	0.0052	0.0054
2200	0.0000	0.0004	0.0040	0.0064	0.0032	0.0036	0.0002	0.0002	0.0010	0.0180	0.0038	0.0042
2250	0.0000	0.0002	0.0029	0.0047	0.0022	0.0027	0.0001	0.0002	0.0007	0.0147	0.0028	0.0032
2300	0.0000	0.0001	0.0021	0.0034	0.0015	0.0020	0.0001	0.0001	0.0004	0.0113	0.0020	0.0024
2350	0.0000	0.0000	0.0015	0.0025	0.0010	0.0014	0.0001	0.0001	0.0003	0.0092	0.0014	0.0018
2400	0.0000	0.0000	0.0010	0.0017	0.0007	0.0010	0.0000	0.0001	0.0002	0.0071	0.0010	0.0013
2450	0.0000	0.0000	0.0007	0.0012	0.0004	0.0007	0.0000	0.0000	0.0001	0.0052	0.0006	0.0009
2500	0.0000	0.0000	0.0005	0.0008	0.0002	0.0004	0.0000	0.0000	0.0001	0.0037	0.0004	0.0006
2550	0.0000	0.0000	0.0003	0.0005	0.0001	0.0003	0.0000	0.0000	0.0000	0.0025	0.0003	0.0004
2600	0.0000	0.0000	0.0002	0.0003	0.0001	0.0002	0.0000	0.0000	0.0000	0.0016	0.0002	0.0002
2650	0.0000	0.0000	0.0001	0.0001	0.0000	0.0001	0.0000	0.0000	0.0000	0.0009	0.0001	0.0001
2700	0.0000	0.0000	0.0000	0.0001	0.0000	0.0000	0.0000	0.0000	0.0000	0.0005	0.0000	0.0001
2750	0.0000	0.0000	0.0000	0.0000	0.0000	0.0000	0.0000	0.0000	0.0000	0.0002	0.0000	0.0000
2800	0.0000	0.0000	0.0000	0.0000	0.0000	0.0000	0.0000	0.0000	0.0000	0.0000	0.0000	0.0000

TABLE 189   (Continued)

Blackbody at 5000 K

λ's /mm	Normal observer		Photoelectric detectors							Photographic emulsions		
	Pho-topic	Sco-topic	S10	S11	S12	S20	CdSe	Si diode	Vidicon	Blue sensi-tive	Ortho chro-matic	Pan chro-matic
50	0.9643	0.9679	0.9681	0.9698	0.9696	0.9672	0.9537	0.9536	0.9641	0.9721	0.9686	0.9677
100	0.9286	0.9358	0.9362	0.9396	0.9393	0.9345	0.9075	0.9072	0.9283	0.9442	0.9372	0.9353
150	0.8930	0.9037	0.9043	0.9094	0.9090	0.9018	0.8614	0.8610	0.8925	0.9163	0.9058	0.9031
200	0.8575	0.8717	0.8726	0.8793	0.8787	0.8692	0.8156	0.8150	0.8568	0.8885	0.8746	0.8709
250	0.8221	0.8398	0.8409	0.8492	0.8485	0.8366	0.7699	0.7693	0.8213	0.8607	0.8433	0.8388
300	0.7868	0.8080	0.8093	0.8192	0.8184	0.8042	0.7246	0.7240	0.7858	0.8330	0.8122	0.8068
350	0.7517	0.7763	0.7778	0.7894	0.7884	0.7720	0.6798	0.6792	0.7506	0.8054	0.7813	0.7749
400	0.7168	07448	0.7465	0.7597	0.7586	0.7399	0.6354	0.6348	0.7156	0.7779	0.7504	0.7432
450	0.6822	0.7134	0.7154	0.7301	0.7288	0.7080	0.5915	0.5911	0.6809	0.7505	0.7197	0.7117
500	0.6478	0.6823	0.6845	0.7007	0.6993	0.6763	0.5483	0.5482	0.6465	0.7233	0.6892	0.6804
550	0.6137	0.6513	0.6538	0.6715	0.6699	0.6449	0.5057	0.5060	0.6124	0.6962	0.6589	0.6493
600	0.5799	0.6206	0.6234	0.6424	0.6407	0.6138	0.4640	0.4648	0.5786	0.6692	0.6289	0.6185
650	0.5465	0.5901	0.5932	0.6137	0.6118	0.5830	0.4232	0.4245	0.5453	0.6424	0.5991	0.5880
700	0.5135	0.5600	0.5633	0.5851	0.5831	0.5525	0.3833	0.3855	0.5124	0.6159	0.5696	0.5577
750	0.4809	0.5301	0.5338	0.5569	0.5546	0.5224	0.3445	0.3478	0.4800	0.5895	0.5403	0.5279
800	0.4489	0.5006	0.5046	0.5289	0.5264	0.4926	0.3070	0.3115	0.4481	0.5634	0.5114	0.4984
850	0.4173	0.4714	0.4758	0.5012	0.4986	0.4634	0.2708	0.2770	0.4168	0.5376	0.4829	0.4693
900	0.3863	0.4427	0.4475	0.4739	0.4710	0.4346	0.2361	0.2444	0.3861	0.5120	0.4547	0.4406
950	0.3558	0.4143	0.4195	0.4469	0.4439	0.4063	0.2030	0.2144	0.3561	0.4867	0.4269	0.4124
000	0.3261	0.3864	0.3921	0.4204	0.4171	0.3786	0.1719	0.1871	0.3268	0.4617	0.3996	0.3847
050	0.2970	0.3590	0.3652	0.3942	0.3907	0.3514	0.1431	0.1628	0.2984	0.4370	0.3728	0.3576
100	0.2687	0.3322	0.3389	0.3685	0.3647	0.3249	0.1169	0.1412	0.2709	0.4127	0.3464	0.3310
150	0.2412	0.3058	0.3131	0.3433	0.3392	0.2992	0.0933	0.1221	0.2443	0.3887	0.3206	0.3051
200	0.2145	0.2801	0.2880	0.3186	0.3142	0.2741	0.0728	0.1053	0.2189	0.3652	0.2954	0.2798
250	0.1888	0.2550	0.2637	0.2945	0.2897	0.2500	0.0556	0.0905	0.1947	0.3421	0.2707	0.2553
300	0.1642	0.2306	0.2400	0.2709	0.2657	0.2268	0.0422	0.0775	0.1719	0.3194	0.2468	0.2315
350	0.1406	0.2069	0.2173	0.2480	0.2424	0.2047	0.0329	0.0662	0.1505	0.2973	0.2235	0.2086
400	0.1183	0.1840	0.1955	0.2258	0.2197	0.1838	0.0273	0.0564	0.1308	0.2756	0.2010	0.1867
450	0.0974	0.1620	0.1748	0.2043	0.1977	0.1641	0.0231	0.0478	0.1126	0.2545	0.1794	0.1657
500	0.0781	0.1408	0.1552	0.1836	0.1764	0.1458	0.0196	0.0404	0.0962	0.2340	0.1586	0.1460
550	0.0605	0.1207	0.1369	0.1638	0.1559	0.1288	0.0168	0.0340	0.0814	0.2141	0.1389	0.1276
600	0.0451	0.1017	0.1199	0.1449	0.1362	0.1132	0.0144	0.0285	0.0683	0.1950	0.1202	0.1109
650	0.0322	0.0839	0.1043	0.1272	0.1175	0.0989	0.0124	0.0238	0.0568	0.1766	0.1028	0.0960
700	0.0218	0.0675	0.0900	0.1107	0.0997	0.0859	0.0106	0.0198	0.0468	0.1591	0.0868	0.0831
750	0.0139	0.0527	0.0772	0.0956	0.0831	0.0742	0.0091	0.0164	0.0383	0.1425	0.0727	0.0718
800	0.0084	0.0398	0.0657	0.0818	0.0678	0.0636	0.0077	0.0135	0.0310	0.1271	0.0606	0.0619
850	0.0048	0.0289	0.0554	0.0695	0.0538	0.0542	0.0066	0.0111	0.0250	0.1131	0.0505	0.0533
900	0.0026	0.0202	0.0464	0.0585	0.0414	0.0458	0.0055	0.0090	0.0200	0.1003	0.0420	0.0458
950	0.0014	0.0136	0.0386	0.0488	0.0309	0.0384	0.0046	0.0073	0.0158	0.0887	0.0347	0.0392
000	0.0007	0.0087	0.0318	0.0404	0.0229	0.0318	0.0039	0.0058	0.0124	0.0781	0.0284	0.0332
050	0.0004	0.0054	0.0259	0.0331	0.0175	0.0262	0.0032	0.0046	0.0096	0.0683	0.0230	0.0279
00	0.0002	0.0032	0.0209	0.0268	0.0134	0.0213	0.0026	0.0037	0.0074	0.0593	0.0184	0.0232
50	0.0001	0.0017	0.0167	0.0215	0.0101	0.0171	0.0021	0.0028	0.0056	0.0509	0.0146	0.0190
00	0.0000	0.0009	0.0131	0.0170	0.0075	0.0135	0.0017	0.0022	0.0041	0.0432	0.0114	0.0154
50	0.0000	0.0004	0.0102	0.0132	0.0055	0.0105	0.0013	0.0016	0.0030	0.0361	0.0087	0.0122
00	0.0000	0.0002	0.0077	0.0102	0.0040	0.0081	0.0010	0.0012	0.0021	0.0297	0.0066	0.0096
50	0.0000	0.0001	0.0058	0.0076	0.0028	0.0060	0.0008	0.0009	0.0015	0.0239	0.0049	0.0073
00	0.0000	0.0000	0.0042	0.0056	0.0020	0.0044	0.0006	0.0006	0.0010	0.0188	0.0036	0.0054
50	0.0000	0.0000	0.0030	0.0040	0.0013	0.0031	0.0004	0.0004	0.0006	0.0143	0.0025	0.0039
00	0.0000	0.0000	0.0020	0.0027	0.0008	0.0021	0.0003	0.0003	0.0004	0.0105	0.0017	0.0027
50	0.0000	0.0000	0.0013	0.0018	0.0005	0.0014	0.0002	0.0002	0.0002	0.0073	0.0012	0.0018
00	0.0000	0.0000	0.0008	0.0011	0.0003	0.0008	0.0001	0.0001	0.0001	0.0048	0.0007	0.0011
50	0.0000	0.0000	0.0004	0.0006	0.0002	0.0005	0.0001	0.0001	0.0000	0.0029	0.0004	0.0006
00	0.0000	0.0000	0.0002	0.0003	0.0001	0.0002	0.0000	0.0000	0.0000	0.0015	0.0002	0.0003
50	0.0000	0.0000	0.0001	0.0001	0.0000	0.0001	0.0000	0.0000	0.0000	0.0006	0.0001	0.0001
00	0.0000	0.0000	0.0000	0.0000	0.0000	0.0000	0.0000	0.0000	0.0000	0.0002	0.0000	0.0000

TABLE 189   (*Continued*)

P-11 Phosphor

ν_s c/mm	Normal Observer		Photoelectric detectors							Photographic emulsions		
	Pho-topic	Sco-topic	S10	S11	S12	S20	CdSe	Si diode	Vidicon	Blue sensi-tive	Ortho chro-matic	Pan chro-matic
50	0.9675	0.9692	0.9701	0.9702	0.9696	0.9703	0.9700	0.9697	0.9696	0.9701	0.9699	0.9701
100	0.9350	0.9385	0.9403	0.9405	0.9391	0.9405	0.9400	0.9395	0.9393	0.9401	0.9398	0.9403
150	0.9025	0.9078	0.9105	0.9108	0.9087	0.9109	0.9101	0.9093	0.9090	0.9102	0.9098	0.9105
200	0.8701	0.8771	0.8807	0.8811	0.8784	0.8812	0.8802	0.8791	0.8788	0.8804	0.8798	0.8807
250	0.8378	0.8465	0.8510	0.8515	0.8481	0.8516	0.8504	0.8490	0.8486	0.8506	0.8498	0.8510
300	0.8056	0.8160	0.8214	0.8220	0.8180	0.8221	0.8207	0.8190	0.8185	0.8210	0.8200	0.8214
350	0.7735	0.7856	0.7919	0.7926	0.7879	0.7928	0.7911	0.7891	0.7885	0.7914	0.7902	0.7919
400	0.7416	0.7554	0.7625	0.7633	0.7579	0.7635	0.7616	0.7593	0.7587	0.7619	0.7606	0.7625
450	0.7098	0.7253	0.7333	0.7341	0.7281	0.7343	0.7322	0.7297	0.7290	0.7326	0.7311	0.7333
500	0.6783	0.6953	0.7042	0.7051	0.6985	0.7054	0.7030	0.7002	0.6994	0.7034	0.7018	0.7042
550	0.6470	0.6656	0.6753	0.6763	0.6690	0.6765	0.6739	0.6709	0.6700	0.6744	0.6727	0.6753
600	0.6159	0.6360	0.6465	0.6476	0.6398	0.6479	0.6451	0.6418	0.6409	0.6456	0.6437	0.6465
650	0.5851	0.6067	0.6180	0.6192	0.6107	0.6195	0.6165	0.6129	0.6119	0.6170	0.6150	0.6180
700	0.5546	0.5776	0.5897	0.5910	0.5819	0.5913	0.5881	0.5843	0.5832	0.5887	0.5865	0.5897
750	0.5244	0.5488	0.5617	0.5630	0.5534	0.5633	0.5599	0.5559	0.5548	0.5605	0.5582	0.5617
800	0.4945	0.5203	0.5339	0.5353	0.5251	0.5356	0.5320	0.5278	0.5266	0.5327	0.5303	0.5339
850	0.4651	0.4921	0.5064	0.5078	0.4972	0.5082	0.5044	0.5000	0.4987	0.5052	0.5026	0.5064
900	0.4360	0.4643	0.4792	0.4807	0.4696	0.4811	0.4772	0.4725	0.4712	0.4779	0.4753	0.4792
950	0.4074	0.4368	0.4523	0.4539	0.4423	0.4544	0.4502	0.4454	0.4440	0.4510	0.4482	0.4524
1000	0.3792	0.4097	0.4258	0.4275	0.4154	0.4280	0.4237	0.4186	0.4172	0.4245	0.4216	0.4259
1050	0.3516	0.3830	0.3997	0.4014	0.3889	0.4019	0.3975	0.3922	0.3908	0.3984	0.3954	0.3998
1100	0.3245	0.3568	0.3740	0.3758	0.3628	0.3763	0.3717	0.3663	0.3648	0.3726	0.3695	0.3741
1150	0.2979	0.3310	0.3487	0.3505	0.3372	0.3511	0.3464	0.3408	0.3393	0.3473	0.3441	0.3489
1200	0.2720	0.3057	0.3239	0.3258	0.3121	0.3263	0.3215	0.3158	0.3143	0.3225	0.3192	0.3241
1250	0.2467	0.2810	0.2996	0.3015	0.2875	0.3020	0.2971	0.2913	0.2897	0.2982	0.2948	0.2998
1300	0.2222	0.2569	0.2758	0.2777	0.2634	0.2782	0.2733	0.2674	0.2658	0.2744	0.2709	0.2760
1350	0.1984	0.2334	0.2525	0.2544	0.2400	0.2550	0.2500	0.2440	0.2424	0.2512	0.2476	0.2528
1400	0.1754	0.2105	0.2298	0.2318	0.2171	0.2324	0.2273	0.2213	0.2197	0.2285	0.2250	0.2302
1450	0.1533	0.1883	0.2078	0.2098	0.1950	0.2104	0.2053	0.1993	0.1976	0.2065	0.2029	0.2082
1500	0.1321	0.1669	0.1865	0.1884	0.1735	0.1890	0.1840	0.1779	0.1762	0.1852	0.1816	0.1869
1550	0.1119	0.1463	0.1658	0.1678	0.1528	0.1684	0.1634	0.1573	0.1557	0.1647	0.1611	0.1664
1600	0.0929	0.1265	0.1460	0.1479	0.1330	0.1485	0.1436	0.1376	0.1359	0.1449	0.1413	0.1466
1650	0.0752	0.1077	0.1270	0.1288	0.1140	0.1295	0.1246	0.1188	0.1171	0.1260	0.1224	0.1277
1700	0.0588	0.0900	0.1089	0.1107	0.0960	0.1113	0.1066	0.1009	0.0993	0.1081	0.1046	0.1097
1750	0.0441	0.0734	0.0918	0.0935	0.0791	0.0942	0.0897	0.0842	0.0826	0.0912	0.0878	0.0928
1800	0.0315	0.0581	0.0758	0.0774	0.0633	0.0782	0.0740	0.0687	0.0672	0.0755	0.0722	0.0771
1850	0.0211	0.0443	0.0612	0.0626	0.0489	0.0634	0.0595	0.0547	0.0532	0.0612	0.0581	0.0628
1900	0.0133	0.0323	0.0480	0.0493	0.0360	0.0500	0.0466	0.0422	0.0409	0.0485	0.0455	0.0500
1950	0.0079	0.0223	0.0364	0.0375	0.0250	0.0383	0.0353	0.0315	0.0303	0.0375	0.0346	0.0387
2000	0.0044	0.0144	0.0266	0.0274	0.0165	0.0282	0.0258	0.0225	0.0216	0.0283	0.0253	0.0290
2050	0.0023	0.0087	0.0185	0.0192	0.0109	0.0198	0.0180	0.0154	0.0147	0.0207	0.0176	0.0209
2100	0.0011	0.0048	0.0122	0.0127	0.0069	0.0133	0.0119	0.0100	0.0094	0.0147	0.0116	0.0143
2150	0.0005	0.0023	0.0076	0.0079	0.0040	0.0083	0.0074	0.0060	0.0057	0.0100	0.0071	0.0093
2200	0.0002	0.0010	0.0044	0.0046	0.0022	0.0049	0.0044	0.0034	0.0032	0.0064	0.0041	0.0057
2250	0.0001	0.0004	0.0024	0.0024	0.0011	0.0026	0.0024	0.0018	0.0016	0.0039	0.0022	0.0032
2300	0.0000	0.0001	0.0012	0.0012	0.0005	0.0013	0.0012	0.0009	0.0008	0.0022	0.0011	0.0017
2350	0.0000	0.0000	0.0005	0.0005	0.0002	0.0006	0.0005	0.0004	0.0003	0.0011	0.0005	0.0008
2400	0.0000	0.0000	0.0002	0.0002	0.0001	0.0002	0.0002	0.0001	0.0001	0.0005	0.0002	0.0003
2450	0.0000	0.0000	0.0001	0.0001	0.0000	0.0001	0.0001	0.0000	0.0000	0.0002	0.0001	0.0001
2500	0.0000	0.0000	0.0000	0.0000	0.0000	0.0000	0.0000	0.0000	0.0000	0.0000	0.0000	0.0000
2550	0.0000	0.0000	0.0000	0.0000	0.0000	0.0000	0.0000	0.0000	0.0000	0.0000	0.0000	0.0000
2600	0.0000	0.0000	0.0000	0.0000	0.0000	0.0000	0.0000	0.0000	0.0000	0.0000	0.0000	0.0000
2650	0.0000	0.0000	0.0000	0.0000	0.0000	0.0000	0.0000	0.0000	0.0000	0.0000	0.0000	0.0000
2700	0.0000	0.0000	0.0000	0.0000	0.0000	0.0000	0.0000	0.0000	0.0000	0.0000	0.0000	0.0000
2750	0.0000	0.0000	0.0000	0.0000	0.0000	0.0000	0.0000	0.0000	0.0000	0.0000	0.0000	0.0000
2800	0.0000	0.0000	0.0000	0.0000	0.0000	0.0000	0.0000	0.0000	0.0000	0.0000	0.0000	0.0000

TABLE 189 (*Continued*)

P-16 Phosphor

ν_s c/mm	Normal observer		Photoelectric detectors							Photographic emulsions		
	Pho- topic	Sco- topic	S10	S11	S12	S20	CdSe	Si diode	Vidicon	Blue sensi- tive	Ortho chro- matic	Pan chro- matic
50	0.9713	0.9722	0.9749	0.9749	0.9744	0.9749	0.9750	0.9746	0.9744	0.9757	0.9748	0.9751
100	0.9427	0.9443	0.9497	0.9498	0.9488	0.9498	0.9500	0.9493	0.9488	0.9514	0.9495	0.9502
150	0.9141	0.9165	0.9246	0.9248	0.9232	0.9248	0.9251	0.9240	0.9232	0.9272	0.9243	0.9254
200	0.8855	0.8888	0.8996	0.8998	0.8976	0.8997	0.9002	0.8987	0.8977	0.9029	0.8992	0.9005
250	0.8570	0.8611	0.8745	0.8748	0.8721	0.8747	0.8753	0.8734	0.8722	0.8787	0.8740	0.8757
300	0.8285	0.8334	0.8495	0.8499	0.8467	0.8498	0.8505	0.8483	0.8467	0.8546	0.8490	0.8510
350	0.8002	0.8059	0.8246	0.8250	0.8213	0.8249	0.8257	0.8231	0.8214	0.8305	0.8240	0.8263
400	0.7720	0.7784	0.7998	0.8002	0.7960	0.8001	0.8010	0.7980	0.7961	0.8064	0.7990	0.8017
450	0.7438	0.7510	0.7750	0.7755	0.7707	0.7754	0.7763	0.7731	0.7708	0.7825	0.7741	0.7771
500	0.7158	0.7238	0.7503	0.7508	0.7456	0.7507	0.7518	0.7482	0.7457	0.7586	0.7494	0.7527
550	0.6880	0.6967	0.7257	0.7263	0.7206	0.7262	0.7274	0.7234	0.7207	0.7348	0.7247	0.7283
600	0.6603	0.6698	0.7012	0.7019	0.6956	0.7017	0.7030	0.6987	0.6958	0.7111	0.7001	0.7041
650	0.6238	0.6430	0.6769	0.6776	0.6709	0.6774	0.6788	0.6742	0.6710	0.6875	0.6757	0.6799
700	0.6055	0.6164	0.6527	0.6534	0.6462	0.6533	0.6547	0.6498	0.6464	0.6641	0.6514	0.6559
750	0.5785	0.5900	0.6286	0.6294	0.6217	0.6292	0.6308	0.6255	0.6219	0.6407	0.6272	0.6321
800	0.5516	0.5638	0.6047	0.6055	0.5974	0.6053	0.6070	0.6014	0.5976	0.6175	0.6032	0.6084
850	0.5250	0.5379	0.5809	0.5818	0.5733	0.5816	0.5834	0.5775	0.5734	0.5945	0.5794	0.5848
900	0.4987	0.5122	0.5574	0.5583	0.5493	0.5581	0.5599	0.5537	0.5495	0.5716	0.5558	0.5614
950	0.4727	0.4867	0.5340	0.5349	0.5256	0.5348	0.5367	0.5302	0.5257	0.5489	0.5323	0.5382
1000	0.4469	0.4616	0.5108	0.5118	0.5020	0.5116	0.5136	0.5069	0.5022	0.5264	0.5091	0.5152
1050	0.4216	0.4367	0.4878	0.4889	0.4787	0.4887	0.4908	0.4837	0.4789	0.5040	0.4860	0.4925
1100	0.3965	0.4122	0.4651	0.4662	0.4557	0.4660	0.4682	0.4609	0.4559	0.4819	0.4632	0.4699
1150	0.3719	0.3880	0.4426	0.4437	0.4329	0.4435	0.4458	0.4382	0.4331	0.4600	0.4407	0.4476
1200	0.3476	0.3641	0.4204	0.4215	0.4103	0.4213	0.4236	0.4159	0.4105	0.4383	0.4184	0.4255
1250	0.3237	0.3406	0.3984	0.3996	0.3881	0.3994	0.4018	0.3938	0.3883	0.4168	0.3964	0.4037
1300	0.3003	0.3176	0.3768	0.3780	0.3662	0.3777	0.3802	0.3720	0.3664	0.3957	0.3747	0.3821
1350	0.2774	0.2949	0.3554	0.3566	0.3446	0.3564	0.3589	0.3505	0.3447	0.3747	0.3532	0.3609
1400	0.2550	0.2727	0.3343	0.3356	0.3233	0.3354	0.3379	0.3293	0.3235	0.3541	0.3322	0.3400
1450	0.2331	0.2510	0.3136	0.3149	0.3024	0.3146	0.3173	0.3085	0.3025	0.3337	0.3114	0.3193
1500	0.2118	0.2298	0.2932	0.2946	0.2818	0.2943	0.2970	0.2881	0.2820	0.3137	0.2910	0.2991
1550	0.1910	0.2092	0.2733	0.2746	0.2617	0.2743	0.2770	0.2680	0.2619	0.2940	0.2710	0.2791
1600	0.1710	0.1891	0.2537	0.2550	0.2420	0.2547	0.2575	0.2484	0.2421	0.2746	0.2514	0.2596
1650	0.1516	0.1696	0.2345	0.2358	0.2227	0.2355	0.2383	0.2291	0.2228	0.2556	0.2322	0.2405
1700	0.1329	0.1508	0.2157	0.2171	0.2039	0.2168	0.2196	0.2104	0.2040	0.2370	0.2134	0.2217
1750	0.1151	0.1326	0.1974	0.1988	0.1857	0.1985	0.2013	0.1921	0.1857	0.2188	0.1951	0.2035
1800	0.0981	0.1153	0.1796	0.1810	0.1679	0.1807	0.1835	0.1743	0.1679	0.2010	0.1773	0.1857
1850	0.0821	0.0987	0.1623	0.1637	0.1507	0.1634	0.1662	0.1570	0.1507	0.1837	0.1601	0.1683
1900	0.0671	0.0830	0.1456	0.1470	0.1341	0.1467	0.1495	0.1403	0.1341	0.1668	0.1433	0.1516
1950	0.0533	0.0683	0.1295	0.1308	0.1182	0.1305	0.1333	0.1243	0.1181	0.1505	0.1272	0.1353
2000	0.0409	0.0546	0.1139	0.1153	0.1029	0.1150	0.1177	0.1088	0.1027	0.1346	0.1118	0.1197
2050	0.0302	0.0422	0.0991	0.1004	0.0885	0.1001	0.1028	0.0941	0.0882	0.1194	0.0970	0.1047
2100	0.0213	0.0313	0.0850	0.0863	0.0748	0.0860	0.0886	0.0802	0.0744	0.1047	0.0830	0.0904
2150	0.0141	0.0219	0.0717	0.0730	0.0621	0.0726	0.0752	0.0671	0.0616	0.0907	0.0698	0.0769
2200	0.0086	0.0143	0.0593	0.0605	0.0504	0.0602	0.0626	0.0550	0.0497	0.0774	0.0576	0.0642
2250	0.0047	0.0084	0.0479	0.0490	0.0397	0.0487	0.0510	0.0439	0.0390	0.0649	0.0463	0.0524
2300	0.0023	0.0044	0.0376	0.0386	0.0303	0.0383	0.0405	0.0340	0.0295	0.0532	0.0362	0.0416
2350	0.0010	0.0019	0.0285	0.0294	0.0222	0.0291	0.0311	0.0253	0.0213	0.0425	0.0274	0.0320
2400	0.0004	0.0007	0.0208	0.0216	0.0156	0.0213	0.0230	0.0181	0.0147	0.0328	0.0199	0.0237
2450	0.0001	0.0003	0.0144	0.0151	0.0103	0.0148	0.0163	0.0123	0.0095	0.0244	0.0139	0.0168
2500	0.0000	0.0001	0.0094	0.0099	0.0064	0.0097	0.0108	0.0078	0.0056	0.0171	0.0092	0.0112
2550	0.0000	0.0000	0.0058	0.0062	0.0037	0.0060	0.0068	0.0046	0.0031	0.0114	0.0057	0.0070
2600	0.0000	0.0000	0.0031	0.0034	0.0019	0.0033	0.0038	0.0024	0.0014	0.0068	0.0032	0.0040
2650	0.0000	0.0000	0.0016	0.0017	0.0009	0.0017	0.0020	0.0012	0.0006	0.0037	0.0017	0.0020
2700	0.0000	0.0000	0.0006	0.0007	0.0004	0.0007	0.0008	0.0005	0.0002	0.0017	0.0007	0.0009
2750	0.0000	0.0000	0.0002	0.0002	0.0001	0.0002	0.0002	0.0001	0.0000	0.0005	0.0002	0.0002
2800	0.0000	0.0000	0.0000	0.0000	0.0000	0.0000	0.0000	0.0000	0.0000	0.0001	0.0000	0.0000

Source: Levi [18.3]

TABLE 190 SPECIAL NONLINEAR RESPONSE FORMS
(Section 18.3.2)

Name	Mathematical description[a]
a. Thresholding	$f = \begin{cases} f_0, & x < x_0 \\ f_0 + bx, & x > x_0 \end{cases}$
b. Limiting	$f = \begin{cases} f_0 + bx, & x < x_0 \\ f_1 \equiv f_0 + bx_0, & x > x_0 \end{cases}$
c. Quantizing (equi-spaced)	$f = f_0 + n\,\Delta f$, where n is integer nearest $b(x - x_0)/\Delta f$
d. Logarithmic	$f = f_0 + b \log x$
e. Power-law	$f = f_0 + bx^a$

[a] a, b, f_0, x_0, and Δf are real constants.

TABLE 191 MAXIMUM EFFICIENCY FOR VARIOUS HOLOGRAM TYPES
(Section 19.3.1.7)

Hologram Medium: Mode of Diffraction:	Thin transmission		Thick transmission		Thick reflection	
Property Modulated:	Amplitude Trans-mittance	Phase Shift	Absorp-tion Constant	Refrac-tive Index	Absorp-tion Constant	Refrac-tive Index
Maximum theoretical efficiency (%):	6.25	33.9	3.7	100	7.2	100
Maximum efficiency obtained experimentally:	6.0	32.6	3.0	90	3.8	80

Source. Collier *et al.* [19.34b].

TABLE 192 CHARACTERISTICS OF PHOTOGRAPHIC HOLOGRAM RECORDING MATERIALS (Section 19.3.4.2)

Mfr[a]	Name	Form	Emulsion Thickness (μm)	Sensitivity $(J/m^2)^{-1}$ at $\lambda =$ 514 nm	at $\lambda =$ 633 nm	Resol. mm^{-1}
EK	Hi-Resol.	Plate	6	1.1	—	>2000
EK	649-F	Plate	17	1.1	1.1	>2000
		Film	6	1.1	1.1	>2000
AG	8E75	Plate	7		13	>2000
		Film	5		13	3000
AG	10E75	Plate	7		50	3000
		Film	5		50	2800
AG	10E56	Plate	7	50		2800
		Film	5	50		2800
IL	HeNe/1	Plate	9		200	>2000

[a] AG, Agfa Gevaert; EK, Eastman Kodak; IL, Ilford.
Based on manufacturers' data.

TABLE 193
FLUX SCATTERED FROM PHOTOGRAPHIC PLATES[a]
(Section 19.3.4.2)

Mfr	Name	Φ_n/Φ_{in} $(c/mm)^{-2}$
EK	649-F	0.95×10^{-9}
EK	120–02	0.60×10^{-9}
AG	8E75	1.15×10^{-9}
AG	8E75B	0.95×10^{-9}
AG	10E75	1.75×10^{-9}

[a] All plates developed to amplitude transmittance of 0.45 developed in Agfa 80, 20°C, 5 min.
Source. Johansson and Biedermann [19.65].

TABLE 194 THICK RECORDING MEDIA,

	Mode of action	Form of record	Thickness D_{max} [μm]	$E(\eta_{max})$ [J cm^{-2}]
Agfa 8E75 (unbleached)	Photographic	Absorption	7	25×10^{-6}
Kodak 649F (unbleached)	Photographic	Absorption	15	7×10^{-5}
Kodak 649F (bleached)	Photographic	Phase	15	1.8×10^{-3}
DCG	Photo-resist	Phase	20	0.02
α_2-salicylidene aniline	Photochromic	Absorption	10	0.02
Electron-irradiated NaCl	Photochromic	Absorption	15	0.012
Dupont photopolymer	Photopolymer	Phase	10^3	0.02
4128 photopolymer	Photopolymer	Phase	$>10^4$	0.125
UV 57 photopolymer	Photopolymer	Phase	10^3	0.5
γ-irradiated CaF$_2$:La	Photochromic	Absorption	$>10^3$	0.3
Quinone-doped PMMA	Photopolymer	Phase	10^3	4
Fe-doped LiNbO$_3$	Optical damage	Phase	$>10^4$	0.5
Undoped LiNbO$_3$	Optical damage	Phase	$>10^4$	150
As$_{15}$S$_{85}$ glass	Devitrification	Phase	450	9
Undoped PMMA	Photopolymer	Phase	$>10^3$	100
Photothermo-plastic	Photocond. +Thermoplast	Phase	—	10^{-4}

Source. Primarily from Forshaw [19.74].

PERFORMANCE DATA (Section 19.3.4.2)

Resolution limit [lines mm^{-1}]	Spectral range [nm]	Erasability	Stability to light	Time stability	η_{max}(abs) [%]
2000	600–700	No	Excellent	Excellent	6.5
6000	400–650	No	Excellent	Excellent	6.5
6000	400–650	No	Good	Excellent	50
>5000	350–650	No	Excellent	Excellent	80
>3300	458–515	Yes	—	1 day	0.14
1000	476	Yes	—	Poor?	0.03
2000	488	No	Good?	Good?	85
1250	400–700?	No	Good?	Good?	~90
5000	458–515?	No	Good	Good?	~90
>3000	633	Yes	—	Poor?	0.4
6000	400–750	No	Good?	Good?	~90
1500?	400–700	Yes	—	7 days	~90
1500?	400–700	Yes	—	7 days	50
>1000	488	No	Good?	Fair?	18
5000	325	No	Good?	Good?	70
4000+	—	Yes	Excellent	Excellent	30

Author Index

Subject Index

Note: (T) indicates a table. Italic numerals indicate definitions of the term or extensive data.